Data Science and Machine Learning

Praise for the first edition:
"In nine succinct but information-packed chapters, the authors provide a logically structured and robust introduction to the mathematical and statistical methods underpinning the still-evolving field of AI and data science."
– Joacim Rocklöv and Albert A. Gayle, *International Journal of Epidemiology*, Volume 49, Issue 6

"This book organizes the algorithms clearly and cleverly. The way the Python code was written follows the algorithm closely—very useful for readers who wish to understand the rationale and flow of the background knowledge."
– Yin-Ju Lai and Chuhsing Kate Hsiao, *Biometrics*, Volume 77, Issue 4

The purpose of ***Data Science and Machine Learning: Mathematical and Statistical Methods*** is to provide an accessible, yet comprehensive textbook intended for students interested in gaining a better understanding of the mathematics and statistics that underpin the rich variety of ideas and machine learning algorithms in data science.

New in the Second Edition
This expanded edition provides updates across key areas of statistical learning:

- **Monte Carlo Methods**: A new section introducing the *regenerative rejection sampling*—a simpler alternative to MCMC.
- **Unsupervised Learning**: Inclusion of two multidimensional diffusion kernel density estimators, as well as the *bandwidth perturbation matching* method for the optimal data-driven bandwidth selection.
- **Regression**: New automatic bandwidth selection for local linear regression.
- **Feature Selection and Shrinkage**: A new chapter introducing the *klimax method* for model selection in high-dimensions.
- **Reinforcement Learning**: A new chapter on contemporary topics such as policy iteration, temporal difference learning, and policy gradient methods, all complete with Python code.
- **Appendices**: Expanded treatment of linear algebra, functional analysis, and optimization that includes the coordinate-descent method and the novel *Majorization--Minimization method* for constrained optimization.

Key Features:

- Focuses on mathematical understanding.
- Presentation is self-contained, accessible, and comprehensive.
- Extensive list of exercises and worked-out examples.
- Many concrete algorithms with Python code.
- Full color throughout and extensive indexing.
- A single-counter consecutive numbering of all theorems, definitions, equations, etc., for easier text searches.

Zdravko I. Botev, PhD, is the pioneer of several modern statistical methodologies, including the *diffusion kernel density estimator*, the *generalized splitting method* for rare-event simulation, the *bandwidth perturbation matching* method, the *regenerative rejection sampling* method, and the *klimax method* for feature selection. His contributions to computational statistics and data science have been recognized with honours such as the Christopher Heyde Medal from the Australian Academy of Science and the Gavin Brown Prize from the Australian Mathematical Society.

Dirk P. Kroese, PhD, is an Emeritus Professor in Mathematics and Statistics at the University of Queensland. He is known for his significant contributions to the fields of applied probability, mathematical statistics, machine learning, and Monte Carlo methods. He has published over 140 articles and 7 books. He is a pioneer of the well-known Cross-Entropy (CE) method, which is being used around the world to help solve difficult estimation and optimization problems in science, engineering, and finance.

Thomas Taimre, PhD, is a Senior Lecturer of Mathematics and Statistics at The University of Queensland. His research interests range from applied probability and Monte Carlo methods to applied physics and the remarkably universal self-mixing effect in lasers. He has published over 100 articles, holds a patent, and is the coauthor of Handbook of Monte Carlo Methods (Wiley).

Chapman & Hall/CRC Machine Learning & Pattern Recognition

A First Course in Machine Learning
Simon Rogers, Mark Girolami

Statistical Reinforcement Learning: Modern Machine Learning Approaches
Masashi Sugiyama

Sparse Modeling: Theory, Algorithms, and Applications
Irina Rish, Genady Grabarnik

Computational Trust Models and Machine Learning
Xin Liu, Anwitaman Datta, Ee-Peng Lim

Regularization, Optimization, Kernels, and Support Vector Machines
Johan A.K. Suykens, Marco Signoretto, Andreas Argyriou

Machine Learning: An Algorithmic Perspective, Second Edition
Stephen Marsland

Bayesian Programming
Pierre Bessiere, Emmanuel Mazer, Juan Manuel Ahuactzin, Kamel Mekhnacha

Multilinear Subspace Learning: Dimensionality Reduction of Multidimensional Data
Haiping Lu, Konstantinos N. Plataniotis, Anastasios Venetsanopoulos

Data Science and Machine Learning: Mathematical and Statistical Methods
Dirk P. Kroese, Zdravko Botev, Thomas Taimre, Radislav Vaisman

Deep Learning and Linguistic Representation
Shalom Lappin

Artificial Intelligence and Causal Inference
Momiao Xiong

Introduction to Machine Learning with Applications in Information Security, Second Edition
Mark Stamp

Entropy Randomization in Machine Learning
Yuri S. Popkov, Alexey Yu. Popkov, Yuri A. Dubno

Transformers for Machine Learning
Uday Kamath, Kenneth Graham and Wael Emara

The Pragmatic Programmer for Machine Learning
Marco Scutari and Mauro Malvestio

Machine Learning, Animated
Mark Liu

Deep and Shallow: Machine Learning in Music and Audio
Shlomo Dubnov and Ross Greer

A Concise Introduction to Machine Learning, 2nd Edition
A.C. Faul

Data Science and Machine Learning: Mathematical and Statistical Methods, Second Edition
Zdravko I. Botev, Dirk P. Kroese, and Thomas Taimre

For more information on this series please visit: https://www.routledge.com/Chapman--HallCRC-Machine-Learning--Pattern-Recognition/book-series/CRCMACLEAPAT

Data Science and Machine Learning

Mathematical and Statistical Methods

Second Edition

Zdravko I. Botev, Dirk P. Kroese, Thomas Taimre

CRC Press
Taylor & Francis Group
Boca Raton London New York

CRC Press is an imprint of the
Taylor & Francis Group, an **informa** business
A CHAPMAN & HALL BOOK

Designed Cover Image: Jessica Anne Kroese

Second edition published 2026
by CRC Press
2385 NW Executive Center Drive, Suite 320, Boca Raton FL 33431

and by CRC Press
4 Park Square, Milton Park, Abingdon, Oxon, OX14 4RN

CRC Press is an imprint of Taylor & Francis Group, LLC

First edition published by CRC Press 2020

ISBN: 9781032488684 (hbk)
ISBN: 9781032488691 (pbk)
ISBN: 9781003391173 (ebk)

DOI: 10.1201/9781003391173

Typeset in Nimbus Roman No9 L

Publisher's Note: This book has been prepared from camera-ready copy provided by the author.

To George, Sarah, Sofia, and my parents

— ZIB

To Lesley, Elise, and Jessica

— DPK

To my grandparents: Arno, Harry, Juta, and Maila

— TT

CONTENTS

PREFACE

In our present world of automation, cloud computing, algorithms, artificial intelligence, and big data, few topics are as relevant as *data science* and *machine learning*. Their recent popularity lies not only in their applicability to real-life questions, but also in their natural blending of many different disciplines, including mathematics, statistics, computer science, engineering, economics and finance.

To someone starting to learn these topics, the multitude of computational techniques and mathematical ideas may seem overwhelming. Some may be satisfied with only learning how to use off-the-shelf recipes to apply to practical situations. But what if the assumptions of the black-box recipe are violated? Can we still trust the results? How should the algorithm be adapted? To be able to truly understand data science and machine learning it is important to appreciate the mathematical and statistical principles, as well as the derivation of all algorithms.

The purpose of this book is to provide an accessible, yet comprehensive, account of data science and machine learning. It is intended for anyone interested in gaining a better understanding of the mathematics and statistics that underpin the rich variety of ideas and machine learning algorithms in data science. Our viewpoint is that computer languages come and go, but the underlying key ideas and algorithms endure and form the basis for future developments.

Before we turn to a description of the topics in this book and the changes that we made in this *Second Edition*, we would like to say a few words about its philosophy. This book resulted from various courses in data science and machine learning at the Universities of Queensland and New South Wales, Australia. When we taught these courses, we noticed that students were eager to learn not only how to apply algorithms but also to understand how these algorithms actually work. However, many existing textbooks assumed either too much background knowledge (e.g., measure theory and functional analysis) or too little (everything is a black box), and the information overload from often disjointed and contradictory internet sources made it more difficult for students to gradually build up their knowledge and understanding. We therefore wanted to write a book about data science and machine learning that can be read as a linear story, with a substantial "backstory" in the appendices that is articulated in a consistent and unifying mathematical notation. The main narrative starts very simply and builds up gradually to quite an advanced level. The backstory contains all the necessary background, as well as additional information, from linear algebra (Appendix A), functional analysis (Appendix B), multivariate differentiation and optimization (Appendix C), and probability and statistics (Appendix D). Moreover, to make the abstract ideas come alive, we believe it is important that the reader sees actual implementations of the algorithms, directly translated from the theory. We have chosen Python as our programming language, as it is freely available, has clean and straightforward syntax, and has been adopted as the programming language of choice by many researchers and practitioners in data science, machine learning, and artificial intelligence. It has many useful packages for data manipulation (often ported from R) and has been designed to be easy to program. A gentle introduction to Python is given in Appendix E.

To keep the book manageable in size we had to be selective in our choice of topics. The changes and new additions in this *Second Edition* are given at the end of this preface. Important definitions are emphasized in the text through the use of wine-red boldfaced **keywords**, and connections between various concepts are highlighted via page references (indicated by a ☞) in the margin. Key definitions and theorems are displayed in boxes. Whenever feasible we provide proofs of theorems. Finally, we place great importance on *notation*. It is often the case that once a consistent and concise system of notation is in place, seemingly difficult ideas are revealed as intuitive or obvious. We use different fonts to distinguish between different types of objects. Vectors are denoted by letters in boldface italics, $\boldsymbol{x}, \boldsymbol{X}$, and matrices by uppercase letters in boldface roman font, $\mathbf{A}, \mathbf{K}$. We also distinguish between random vectors and their values by using upper and lower case letters, e.g., $\boldsymbol{X}$ (random vector) and $\boldsymbol{x}$ (its value or outcome). Sets are usually denoted by calligraphic letters $\mathcal{G}, \mathcal{H}$. The symbols for probability and expectation are $\mathbb{P}$ and $\mathbb{E}$, respectively. Distributions are indicated by sans serif font, as in Bin and Gamma; exceptions are the ubiquitous notations $\mathcal{N}$ and $\mathcal{U}$ for the normal and uniform distributions.
☞ xxi A summary of the most important symbols and abbreviations is given on Pages xxi–xxvi.

Data science provides the language and techniques necessary for understanding and dealing with data. It involves the design, collection, analysis, and interpretation of numerical data, with the aim of extracting patterns and useful knowledge. Machine learning, which is closely related to data science, deals with the design of algorithms and computer resources to learn from data. The organization of the book follows roughly the typical steps in a data science project: Gathering data to gain information about a research question; cleaning, summarization, and visualization of the data; modeling and analysis of the data; translating decisions about the model into decisions and predictions about the research question. As this is a mathematics and statistics oriented book, most emphasis will be on modeling and analysis.

We start in Chapter 1 with the reading, structuring, summarization, and visualization of data using the data manipulation package **pandas** in Python. Although the material covered in this chapter requires no mathematical knowledge, it forms an obvious starting point for data science: to better understand the nature of the available data. In Chapter 2, we introduce the main ingredients of *statistical learning*. We distinguish between *supervised* and *unsupervised* learning techniques, and discuss how we can assess the predictive performance of (un)supervised learning methods. An important part of statistical learning is the *modeling* of data. We introduce various useful models in data science including linear, multivariate Gaussian, and Bayesian models. Many algorithms in machine learning and data science make use of Monte Carlo techniques, which is the topic of Chapter 3. Monte Carlo can be used for simulation, estimation, and optimization. Chapter 4 is concerned with unsupervised learning, where we discuss techniques such as density estimation, clustering, and principal component analysis. We then turn our attention to supervised learning in Chapter 5, and explain the ideas behind a broad class of regression models. Therein, we also describe how Python's `statsmodels` package can be used to define and analyze linear models. In Chapter 6 we turn our attention to model selection and shrinkage, and introduce the new klimax method for effective estimation of regression coefficients. Chapter 7 builds upon the previous regression chapters by developing the powerful concepts of kernel methods and regularization, which allow the fundamental ideas of Chapter 5 to be expanded in an elegant way, using the theory of reproducing kernel Hilbert spaces. In Chapter 8, we proceed with the classification task, which also belongs to the supervised learning framework, and consider various methods for classification, including Bayes classification, linear and quadratic discriminant analysis, K-nearest neighbors, and support vector machines. In Chapter 9, we explore versatile methods for regression and classification that make use of tree structures. In Chapter 10, we consider the workings of neural networks and deep learning, and show that these learning algorithms have a simple mathematical interpretation. Finally, in Chapter 11 we provide an introduction to reinforcement learning, and illustrate the theory with a wide variety of algorithms. An extensive range of exercises is provided at the end of each chapter.

> Python code and data sets for each chapter can be downloaded from the GitHub site: `https://github.com/DSML-book`

New in the Second Edition

While the overall premise of the first edition remains intact, significant updates and enhancements have been incorporated in this second edition, as detailed below.

- Chapter 2 was updated by providing new proofs for the asymptotic behavior of the model evidence and the posterior distribution in Bayesian learning.
- In Chapter 3 we added the new *regenerative rejection sampling* method, which offers a simple alternative to Markov Chain Monte Carlo simulation.
- Significant changes were made to the kernel density estimation section in Chapter 4. In particular, we describe in detail the theory behind Gaussian kernel density estimation in multiple dimensions, and provide a state-of-the-art bandwidth selection procedure, called *bandwidth perturbation matching*. In addition, we give full details, including Python code, of the *theta kernel density estimator*, which provides superior estimates of densities with bounded support. Also in this case a highly effective automatic bandwidth selection rule is provided.

- Bandwidth perturbation matching can also be applied in a regression setting. We added a novel automatic bandwidth selection method for *Gaussian-weighted local linear regression* to Chapter 5.
- Chapter 6 is completely new and deals with model selection and parameter shrinkage in regression models. We introduce the new *klimax* method. It is shown to outperform well-known methods when the features exhibit strong dependence and their number is larger than the number of observations.
- As modern learning paradigms often involve some form of reinforcement, we have added a new Chapter 11 on *reinforcement learning*, featuring topics such as value and policy iteration, temporal difference learning, and policy gradient methods, all illustrated with algorithms and working code.
- The background material on linear algebra in Appendix A has been restructured and expanded with new treatments on block matrix decompositions and efficient numerical methods for solving linear equations. Functional analysis now appears in Appendix B, and we have added new material on the discrete cosine transform and the efficient computation of convolutions via the fast Fourier transform. The new *Majorization–Minimization* (MM) method for constrained optimization has been added to Appendix C, as well as various other enhancements in optimization theory.
- The *Second Edition* features numerous additional algorithms, Python code, and exercises.
- To make the hard copy easier to handle, we have reduced the book dimensions to 7×10 inch. Another stylistic change is the consecutive numbering of definitions, theorems, equations, figures, etc., using a *single* counter. This facilitates searching in the text.

Acknowledgments

We thank all who through their comments, feedback, and suggestions have contributed to this book, including Qibin Duan, Luke Taylor, Rémi Mouzayek, Harry Goodman, Bryce Stansfield, Ryan Tongs, Dillon Steyl, Bill Rudd, Nan Ye, Christian Hirsch, Chris van der Heide, Sarat Moka, Aapeli Vuorinen, and the anonymous referees. Some of the Python code for Chapters 1 and 5 was adapted from [100]. We thank Benoit Liquet for making this available, and Lauren Jones for translating the R code into Python. The book was test-run during the 2019 *Summer School of the Australian Mathematical Sciences Institute*. More than 80 bright upper-undergraduate (Honours) students used the book for the course *Mathematical Methods for Machine Learning*, taught by Zdravko Botev. We are grateful for the valuable feedback that they provided. Our special thanks go out to Robert Salomone, Liam Berry, Robin Carrick, and Sam Daley, who commented in great detail on earlier versions of the entire book and wrote or improved our Python code. Their enthusiasm, perceptiveness, and kind assistance have been invaluable. We are also deeply indebted to Slava Vaisman, our dear friend and collaborator.

For this *Second Edition* we like to thank Nicky van Foreest and Wan-Ting Chen for finding various errata that we missed in the first edition, and Marcus Hoerger, Nan Ye,

Jun Ju, and Thomas Moerland for their feedback and assistance with the reinforcement learning chapter. David Grubbs deserves a special accolade for his professionalism and attention to detail in his role as Editor for both editions of this book.

Of course, none of this work would have been possible without the loving support, patience, and encouragement from our families, and we thank them with all our hearts.

Zdravko Botev, Dirk Kroese, and Thomas Taimre
Sydney and Brisbane

NOTATION

We could, of course, use any notation we want; do not laugh at notations; invent them, they are powerful. In fact, mathematics is, to a large extent, invention of better notations.

Richard P. Feynman

We have tried to use a notation system that is, in order of importance, simple, descriptive, consistent, and compatible with historical choices. Achieving all of these goals all of the time would be impossible, but we hope that our notation helps to quickly recognize the type or "flavor" of certain mathematical objects (vectors, matrices, random vectors, probability measures, etc.) and clarify intricate ideas.

We make use of various typographical aids, and it will be beneficial for the reader to be aware of some of these.

- Boldface font is used to indicate composite objects, such as column vectors $\boldsymbol{x} = [x_1, \dots, x_n]^\top$ and matrices $\mathbf{X} = [x_{ij}]$. Note also the difference between the upright bold font for matrices and the slanted bold font for vectors.

- Random variables are generally specified with upper case roman letters X, Y, Z and their outcomes with lower case letters x, y, z. Random vectors are thus denoted in upper case slanted bold font: $\boldsymbol{X} = [X_1, \dots, X_n]^\top$.

- Sets of vectors are generally written in calligraphic font, such as $\mathcal{X}$, but the set of real numbers uses the common blackboard bold font $\mathbb{R}$. Expectation and probability also use the latter font.

- Probability distributions use a sans serif font, such as Bin and Gamma. Exceptions to this rule are the "standard" notations $\mathcal{N}$ and $\mathcal{U}$ for the normal and uniform distributions.

- We often omit brackets when it is clear what the argument is of a function or operator. For example, we prefer $\mathbb{E}X^2$ to $\mathbb{E}[X^2]$.

• We employ color to emphasize that certain words refer to a `dataset`, **`function`**, or `package` in Python. All code is written in `typewriter` font. To be compatible with past notation used for regression modeling, we introduced the special blue symbols $\mathbf{X}$ and $\boldsymbol{v}_k$ for the model (design or feature) matrix and its k-th column, respectively.

• Important notation such as $\mathcal{T}$, g, g^* is often defined in a mnemonic way, such as $\mathcal{T}$ for "training", g for "guess", g^* for the "star" (that is, optimal) guess, and ℓ for "loss".

• We will occasionally use a Bayesian notation convention in which the *same* symbol is used to denote different (conditional) probability densities. In particular, instead of writing $f_X(x)$ and $f_{X \mid Y}(x \mid y)$ for the probability density function (pdf) of X and the conditional pdf of X given Y, we simply write $f(x)$ and $f(x \mid y)$. This particular style of notation can be of great descriptive value, despite its apparent ambiguity.

General font/notation rules

x	scalar
$\boldsymbol{x}$	vector
$\boldsymbol{X}$	random vector
$\mathbf{X}$	matrix
$\mathcal{X}$	set
$\widehat{x}$	estimate or approximation
x^*	optimal
$\overline{x}$	average

Common mathematical symbols

$\forall$	for all
$\exists$	there exists
$\in$	belongs to (a set)
$\propto$	is proportional to
$\perp$	is perpendicular to
$\sim$	is distributed as
$\stackrel{\text{iid}}{\sim}$, $\sim_{\text{iid}}$	are independent and identically distributed as
$\stackrel{\text{approx.}}{\sim}$	is approximately distributed as
∇f	gradient of f
$\nabla^2 f$	Hessian of f
Δf	Laplacian of f
$f \in C^p$	f has continuous derivatives of order p
$\approx$	is approximately
$\ll$	is much smaller than

$\oplus$	direct sum
$\odot$	elementwise product, also known as Hadamard product
$\cap$	intersection
$\cup$	union
$\wedge$	$x \wedge y$ is the minimum of x and y
$\vee$	$x \vee y$ is the maximum of x and y
$:=$, $=:$	is defined as
$\xrightarrow{\text{a.s.}}$	converges almost surely to
$\xrightarrow{\text{d}}$	converges in distribution to
$\xrightarrow{\mathbb{P}}$	converges in probability to
$\xrightarrow{L^p}$	converges in L^p-norm to
$\|\cdot\|$	Euclidean or 2-norm
$\langle\cdot,\cdot\rangle$	inner product: $\langle \boldsymbol{x}, \boldsymbol{y}\rangle = (\|\boldsymbol{x}+\boldsymbol{y}\|^2 - \|\boldsymbol{x}-\boldsymbol{y}\|^2)/4$
$\|\cdot\|_{\mathbf{A}}$	Mahalanobis norm: $\|\boldsymbol{x}\|_{\mathbf{A}}^2 = \boldsymbol{x}^\top \mathbf{A}\boldsymbol{x}$ for positive definite matrix $\mathbf{A}$
$\langle\cdot,\cdot\rangle_{\mathbf{A}}$	Mahalanobis inner product: $\langle \boldsymbol{x}, \boldsymbol{y}\rangle_{\mathbf{A}} = (\|\boldsymbol{x}+\boldsymbol{y}\|_{\mathbf{A}}^2 - \|\boldsymbol{x}-\boldsymbol{y}\|_{\mathbf{A}}^2)/4$
$\|\boldsymbol{x}\|_0$	the number of nonzero components of vector $\boldsymbol{x}$
$\|\boldsymbol{x}\|_p$	p-norm of $\boldsymbol{x}$: $(\sum_i \|x_i\|^p)^{1/p}$ when $p \geq 1$
$\|\boldsymbol{x}\|_\infty$	∞-norm of $\boldsymbol{x}$: $\max_i \|x_i\|$
$\lceil x \rceil$	smallest integer larger than x
$\lfloor x \rfloor$	largest integer smaller than x
x_+	the rectified linear unit (ReLU): $x_+ = \max\{x, 0\}$
S_σ^{LAS}	lasso operator: $S_\sigma^{\text{LAS}}(x) = x\big(1 - \sigma/\|x\|\big)_+$
S_σ^{JS}	James–Stein positive-part operator: $S_\sigma^{\text{JS}}(x) = x\left(1 - (\sigma/x)^2\right)_+$

Matrix/vector notation

$\mathbf{I}_p$	p-dimensional identity matrix
$\mathbf{A}^\top, \boldsymbol{x}^\top$	transpose of matrix $\mathbf{A}$ or vector $\boldsymbol{x}$
$\mathbf{A}^{-1}$	inverse of matrix $\mathbf{A}$
$\mathbf{A}^+$	pseudo-inverse of matrix $\mathbf{A}$
$\mathbf{A}^{-\top}$	inverse of matrix $\mathbf{A}^\top$ or transpose of $\mathbf{A}^{-1}$
$\mathbf{A} \succ \mathbf{B}$	matrix $\mathbf{A} - \mathbf{B}$ is positive definite
$\mathbf{A} \succeq \mathbf{B}$	matrix $\mathbf{A} - \mathbf{B}$ is positive semidefinite
$\text{diag}(\boldsymbol{v})$	Diagonal matrix with diagonal entries given by the vector $\boldsymbol{v}$
$\dim(\boldsymbol{x})$	dimension of vector $\boldsymbol{x}$
$\boldsymbol{x}_{\mathcal{A}}$	vector with entries $\{x_i, i \in \mathcal{A}\}$

$\det(\mathbf{A})$	determinant of matrix $\mathbf{A}$
$\lvert\mathbf{A}\rvert$	absolute value of the determinant of matrix $\mathbf{A}$
$\operatorname{tr}(\mathbf{A})$	trace of matrix $\mathbf{A}$
$\boldsymbol{x}^{\boldsymbol{n}}$	multi-index notation for $\prod_i x_i^{n_i}$
$\boldsymbol{x}!$	multi-index notation for the product of factorials $\prod_i (x_i!)$
$\lvert\boldsymbol{x}\rvert$	multi-index notation for $\sum_i \lvert x_i\rvert$

Reserved letters and words

$\mathbb{C}$	set of complex numbers
d	differential symbol
$\mathcal{D}$	Kullback–Leibler divergence
$\mathbb{E}$	expectation
e	the number $2.71828\ldots$
f	probability density (discrete or continuous)
g	prediction function
$\mathbb{1}\{A\}$ or $\mathbb{1}_A$	indicator function of set A
i	the square root of -1
ℓ	risk: expected loss
Loss	loss function
ln	(natural) logarithm
$\mathbb{N}$	set of natural numbers $\{0, 1, \ldots\}$
$\mathcal{O}$	big-O order symbol: $f(x) = \mathcal{O}(g(x))$ if $\lvert f(x)\rvert \leq \text{const} \times g(x)$ as $x \to a$
o	little-o order symbol: $f(x) = o(g(x))$ if $f(x)/g(x) \to 0$ as $x \to a$
$\mathcal{O}_{\mathbb{P}}$	big-O order symbol in probability
$o_{\mathbb{P}}$	little-o order symbol in probability
$\mathbb{P}$	probability measure
π	the number $3.14159\ldots$
$\mathbb{R}$	set of real numbers (one-dimensional Euclidean space)
$\mathbb{R}^n$	n-dimensional Euclidean space
$\mathbb{R}_+$	positive real line: $[0, \infty)$
τ	deterministic training set
$\mathcal{T}$	random training set
$\boldsymbol{v}_k$	the k-th column of the model matrix
$\mathbf{X}$	model (feature) matrix with columns $\{\boldsymbol{v}_k\}$
$\mathbf{X}_{\mathcal{A}}$	model matrix with columns $\{\boldsymbol{v}_k, k \in \mathcal{A}\}$
$\mathbb{Z}$	set of integers $\{\ldots, -1, 0, 1, \ldots\}$

Probability distributions

Ber	Bernoulli
Bessel	Bessel
Beta	beta
Bin	binomial
Exp	exponential
Geom	geometric
Gamma	gamma
GIG	generalized inverse-gamma
F	Fisher–Snedecor F
InvGamma	inverse gamma
$\mathcal{N}$	normal or Gaussian
Pareto	Pareto
Poi	Poisson
t	Student's t
$\mathcal{U}$	uniform
Weib	Weibull distribution

Abbreviations and acronyms

BIC	Bayesian information criterion
BPM	Bandwidth Perturbation Matching
cdf	cumulative distribution function
CMC	crude Monte Carlo
CE	cross-entropy
DCT	discrete cosine transform
DFT	discrete Fourier transform
EM	expectation–maximization
FFT	fast Fourier transform
GP	Gaussian process
IDFT	inverse discrete Fourier transform
iid	independent and identically distributed
JS	(positive-part) James–Stein
JSC	(positive-part) James–Stein information criterion
KDE	Kernel density estimate/estimator
KL	Kullback–Leibler

KLC	Kullback–Leibler information criterion
Klimax	Kullback–Leibler Information Maximization
KKT	Karush–Kuhn–Tucker
Lasso	Least Absolute Shrinkage and Selection Operator
LAS	Lasso information criterion
MAP	maximum *a posteriori*
MCMC	Markov chain Monte Carlo
MCTS	Monte Carlo Tree Search
MDP	Markov Decision Process
MISE	mean integrated squared error
MLE	maximum likelihood estimator/estimate
PCA	principal component analysis
pdf	probability density function (discrete or continuous)
PRESS	predicted residual sum of squares
SVD	singular value decomposition
UCT	Upper Confidence bound for Trees

CHAPTER 1

Importing, Summarizing, and Visualizing Data

This chapter describes where to find useful data sets, how to load them into Python, and how to (re)structure the data. We also discuss various ways in which the data can be summarized via tables and figures. Which type of plots and numerical summaries are appropriate depends on the type of the variable(s) in play. Readers unfamiliar with Python are advised to read Appendix E first.

1.1 Introduction

Data comes in many shapes and forms, but can generally be thought of as being the result of some *random experiment* — an experiment whose outcome cannot be determined in advance, but whose workings are still subject to analysis. Data from a random experiment are often stored in a table or spreadsheet. A statistical convention is to denote variables — often called **features** — as *columns* and the individual items (or units) as *rows*. It is useful to think of three types of columns in such a spreadsheet:

1. The first column is usually an identifier or index column, where each unit/row is given a unique name or ID.

2. Certain columns (features) can correspond to the design of the experiment, specifying, for example, to which experimental group the unit belongs. Often the entries in these columns are *deterministic*; that is, they stay the same if the experiment were to be repeated.

3. Other columns represent the observed measurements of the experiment. Usually, these measurements exhibit *variability*; that is, they would change if the experiment were to be repeated.

There are many data sets available from the Internet and in software packages. A well-known repository of data sets is the Machine Learning Repository maintained by the University of California at Irvine (UCI), found at `https://archive.ics.uci.edu/`.

DOI: 10.1201/9781003391173-1

These data sets are typically stored in a CSV (comma separated values) format, which can be easily read into Python. For example, to access the abalone data set from this website with Python, download the file to your working directory, import the pandas package via

```
import pandas as pd
```

and read in the data as follows:

```
abalone = pd.read_csv('abalone.data',header = None)
```

It is important to add header = None, as this lets Python know that the first line of the CSV does not contain the names of the features, as it assumes so by default. The data set was originally used to predict the age of abalone from physical measurements, such as shell weight and diameter.

Another useful repository of over 1000 data sets from various packages in the R programming language, collected by Vincent Arel-Bundock, can be found at:

https://vincentarelbundock.github.io/Rdatasets/datasets.html.

For example, to read Fisher's famous iris data set from R's datasets package into Python, type:

```
urlprefix = 'https://vincentarelbundock.github.io/Rdatasets/csv/'
dataname = 'datasets/iris.csv'
iris = pd.read_csv(urlprefix + dataname)
```

The iris data set contains four physical measurements (sepal/petal length/width) on 50 specimens (each) of 3 species of iris: setosa, versicolor, and virginica. Note that in this case the headers are included. The output of read_csv is a DataFrame object, ☞ 694 which is pandas's implementation of a spreadsheet; see Section E.12.1. The DataFrame method head gives the first few rows of the DataFrame, including the feature names. The number of rows can be passed as an argument and is 5 by default. For the iris DataFrame, we have:

```
iris.head()
```

```
   rownames  Sepal.Length  ...  Petal.Width  Species
0         1           5.1  ...          0.2   setosa
1         2           4.9  ...          0.2   setosa
2         3           4.7  ...          0.2   setosa
3         4           4.6  ...          0.2   setosa
4         5           5.0  ...          0.2   setosa

[5 rows x 6 columns]
```

The feature names can be obtained via the columns attribute of the DataFrame object, as in iris.columns. Note that the first column is a duplicate index column, whose name (assigned by pandas) is 'rownames'. We can drop this column and reassign the iris object as follows:

```
iris = iris.drop('rownames', axis=1)
```

The data for each feature (corresponding to its specific name) can be accessed by using Python's *slicing* notation []. For example, the object iris['Sepal.Length'] contains the 150 sepal lengths.

The first three rows of the abalone data set from the UCI repository can be found as follows:

```
abalone.head(3)
```

```
   0      1      2      3       4       5       6      7   8
0  M  0.455  0.365  0.095  0.5140  0.2245  0.1010  0.150  15
1  M  0.350  0.265  0.090  0.2255  0.0995  0.0485  0.070   7
2  F  0.530  0.420  0.135  0.6770  0.2565  0.1415  0.210   9
```

Here, the missing headers have been assigned according to the order of the natural numbers. The names should correspond to Sex, Length, Diameter, Height, Whole weight, Shucked weight, Viscera weight, Shell weight, and Rings, as described in the file with the name abalone.names on the UCI website. We can manually add the names of the features to the DataFrame by reassigning the columns attribute, as in:

```
abalone.columns = ['Sex', 'Length', 'Diameter', 'Height',
'Whole weight','Shucked weight', 'Viscera weight', 'Shell weight',
'Rings']
```

1.2 Structuring Features According to Type

We can generally classify features as either quantitative or qualitative. **Quantitative** features possess "numerical quantity", such as height, age, number of births, etc., and can either be *continuous* or *discrete*. Continuous quantitative features take values in a continuous range of possible values, such as height, voltage, or crop yield; such features capture the idea that measurements can always be made more precisely. Discrete quantitative features have a countable number of possibilities, such as a count.

In contrast, **qualitative** features do not have a numerical meaning, but their possible values can be divided into a fixed number of categories, such as {M,F} for gender or {blue, black, brown, green} for eye color. For this reason such features are also called **categorical**. A simple rule of thumb is: if it does not make sense to average the data, it is categorical. For example, it does not make sense to average eye colors. Of course it is still possible to represent categorical data with numbers, such as 1 = blue, 2 = black, 3 = brown, but such numbers carry no quantitative meaning. Categorical features are often called **factors**.

When manipulating, summarizing, and displaying data, it is important to correctly specify the type of the variables (features). We illustrate the structuring of the data using the nutrition_elderly data set from [100], which contains the results of a study involving nutritional measurements of thirteen features (columns) for 226 elderly individuals (rows). The data set can be obtained from:

http://www.biostatisticien.eu/springeR/nutrition_elderly.xls.

Excel files can be read directly into pandas via the read_excel method:

```
xls = 'http://www.biostatisticien.eu/springeR/nutrition_elderly.xls'
nutri = pd.read_excel(xls)
```

This creates a DataFrame object nutri. The first three rows are as follows:

```
pd.set_option('display.max_columns', 8) # to fit display
nutri.head(3)
```

```
   gender  situation  tea ...  cooked_fruit_veg  chocol  fat
0       2          1    0 ...                 4       5    6
1       2          1    1 ...                 5       1    4
2       2          1    0 ...                 2       5    4

[3 rows x 13 columns]
```

You can check the type (or structure) of the variables via the info method of nutri.

```
nutri.info()
```

```
 <class 'pandas.core.frame.DataFrame'>
RangeIndex: 226 entries, 0 to 225
Data columns (total 13 columns):
gender              226 non-null int64
situation           226 non-null int64
tea                 226 non-null int64
coffee              226 non-null int64
height              226 non-null int64
weight              226 non-null int64
age                 226 non-null int64
meat                226 non-null int64
fish                226 non-null int64
raw_fruit           226 non-null int64
cooked_fruit_veg    226 non-null int64
chocol              226 non-null int64
fat                 226 non-null int64
dtypes: int64(13)
memory usage: 23.0 KB
```

All 13 features in nutri are (at the moment) interpreted by Python as *quantitative* variables, indeed as integers, simply because they have been entered as whole numbers. The *meaning* of these numbers becomes clear when we consider the description of the features, given in Table 1.1. Table 1.2 shows how the variable types should be classified.

Table 1.1: Description of the variables in the nutritional study [100].

Feature	Description	Unit or Coding
gender	Gender	1=Male; 2=Female
situation	Family status	1=Single 2=Living with spouse 3=Living with family 4=Living with someone else
tea	Daily consumption of tea	Number of cups
coffee	Daily consumption of coffee	Number of cups
height	Height	cm
weight	Weight (actually: mass)	kg
age	Age at date of interview	Years
meat	Consumption of meat	0=Never 1=Less than once a week 2=Once a week 3=2–3 times a week 4=4–6 times a week 5=Every day
fish	Consumption of fish	As in meat
raw_fruit	Consumption of raw fruits	As in meat
cooked_fruit_veg	Consumption of cooked fruits and vegetables	As in meat
chocol	Consumption of chocolate	As in meat
fat	Type of fat used for cooking	1=Butter 2=Margarine 3=Peanut oil 4=Sunflower oil 5=Olive oil 6=Mix of vegetable oils (e.g., Isio4) 7=Colza oil 8=Duck or goose fat

Table 1.2: The feature types for the data frame nutri.

Qualitative	gender, situation, fat meat, fish, raw_fruit, cooked_fruit_veg, chocol
Discrete quantitative	tea, coffee
Continuous quantitative	height, weight, age

Note that the categories of the qualitative features in the second row of Table 1.2, meat, . . . , chocol have a natural order. Such qualitative features are sometimes called **ordinal**, in contrast to qualitative features without order, which are called **nominal**. We will not make such a distinction in this book.

We can modify the Python value and type for each categorical feature, using the **replace** and **astype** methods. For categorical features, such as gender, we can replace the value 1 with 'Male' and 2 with 'Female', and change the type to 'category' as follows:

```
DICT = {1:'Male', 2:'Female'} # dictionary specifies replacement
nutri['gender'] = nutri['gender'].replace(DICT).astype('category')
```

The structure of the other categorical-type features can be changed in a similar way. Continuous features such as `height` should have type `float`:

```
nutri['height'] = nutri['height'].astype(float)
```

We can repeat this for the other variables (see Exercise 2) and save this modified data frame as a CSV file, by using the **pandas** method **to_csv**.

```
nutri.to_csv('nutri.csv',index=False)
```

1.3 Summary Tables

It is often useful to summarize a large spreadsheet of data in a more condensed form. A table of counts or a table of frequencies makes it easier to gain insight into the underlying distribution of a variable, especially if the data are qualitative. Such tables can be obtained with the methods **describe** and **value_counts**.

As a first example, we load the `nutri` `DataFrame`, which we restructured and saved (see previous section) as `'nutri.csv'`, and then construct a summary for the feature (column) `'fat'`.

```
nutri = pd.read_csv('nutri.csv')
nutri['fat'].describe()
```

```
count           226
unique            8
top       sunflower
freq             68
Name: fat, dtype: object
```

We see that there are 8 different types of fat used and that sunflower has the highest count, with 68 out of 226 individuals using this type of cooking fat. The counts for the different fat types can be obtained via the method **value_counts**.

```
nutri['fat'].value_counts()
```

```
sunflower    68
peanut       48
olive        40
margarine    27
Isio4        23
butter       15
duck          4
colza         1
Name: fat, dtype: int64
```

Column labels are also attributes of a `DataFrame`, and `nutri.fat`, for example, is exactly the same object as `nutri['fat']`.

It is also possible to use **crosstab** to *cross tabulate* between two or more variables, giving a **contingency table**:

```
pd.crosstab(nutri.gender, nutri.situation)
```

```
situation  Couple  Family  Single
gender
Female         56       7      78
Male           63       2      20
```

We see, for example, that the proportion of single men is substantially smaller than the proportion of single women in the data set of elderly people. To add row and column totals to a table, use `margins=True`.

```
pd.crosstab(nutri.gender, nutri.situation, margins=True)
```

```
situation  Couple  Family  Single  All
gender
Female         56       7      78  141
Male           63       2      20   85
All           119       9      98  226
```

1.4 Summary Statistics

In the following, $\boldsymbol{x} = [x_1, \ldots, x_n]^\top$ is a column vector of n numbers. For our nutri data, the vector $\boldsymbol{x}$ could, for example, correspond to the heights of the $n = 226$ individuals.

The **sample mean** of $\boldsymbol{x}$, denoted by $\overline{x}$, is simply the average of the data values:

$$\overline{x} = \frac{1}{n}\sum_{i=1}^{n} x_i .$$

Using the **mean** method in Python for the nutri data, we have, for instance:

```
nutri['height'].mean()
```

```
163.96017699115043
```

The p-**sample quantile** $(0 < p < 1)$ of $\boldsymbol{x}$ is a value x such that at least a fraction p of the data is less than or equal to x and at least a fraction $1 - p$ of the data is greater than or equal to x. The **sample median** is the sample 0.5-quantile. The p-sample quantile is also called the $100 \times p$ **percentile**. The 25, 50, and 75 sample percentiles are called the first, second, and third **quartiles** of the data. For the nutri data they are obtained as follows:

```
nutri['height'].quantile(q=[0.25,0.5,0.75])
```

```
0.25    157.0
0.50    163.0
0.75    170.0
```

The sample mean and median give information about the *location* of the data, while the distance between sample quantiles (say the 0.1 and 0.9 quantiles) gives some indication of the *dispersion* (spread) of the data. Other measures for dispersion are the **sample range**, $\max_i x_i - \min_i x_i$, the **sample variance**

$$s^2 = \frac{1}{n-1}\sum_{i=1}^{n}(x_i - \overline{x})^2, \tag{1.3}$$

☞ 664 and the **sample standard deviation** $s = \sqrt{s^2}$. For the nutri data, the range (in cm) is:

```
nutri['height'].max() - nutri['height'].min()
```

```
48.0
```

The variance (in cm^2) is:

```
round(nutri['height'].var(), 2)   # round to two decimal places
```

```
81.06
```

And the standard deviation can be found via:

```
round(nutri['height'].std(), 2)
```

```
9.0
```

We already encountered the **describe** method in the previous section for summarizing qualitative features, via the most frequent count and the number of unique elements. When applied to a *quantitative* feature, it returns instead the minimum, maximum, mean, and the three quartiles. For example, the `'height'` feature in the nutri data has the following summary statistics.

```
nutri['height'].describe()
```

```
count    226.000000
mean     163.960177
std        9.003368
min      140.000000
25\%      157.000000
50\%      163.000000
75\%      170.000000
max      188.000000
Name: height, dtype: float64
```

1.5 Visualizing Data

In this section we describe various methods for visualizing data. The main point we would like to make is that the way in which variables are visualized should always be adapted to the variable types; for example, qualitative data should be plotted differently from quantitative data.

For the rest of this section, it is assumed that matplotlib.pyplot, pandas, and numpy, have been imported in the Python code as follows:

```
import matplotlib.pyplot as plt
import pandas as pd
import numpy as np
```

1.5.1 Plotting Qualitative Variables

Suppose we wish to display graphically how many elderly people are living by themselves, as a couple, with family, or otherwise. Recall that the data are given in the `situation` column of our nutri data. Assuming that we already *restructured the data*, as in Section 1.2, we can make a **barplot** of the number of people in each category via the `plt.bar` function of the standard matplotlib plotting library. The inputs are the x-axis positions, heights, and widths of each bar respectively. ☞ 3

```
width = 0.35 # the width of the bars
x = [0, 0.8, 1.6] # the bar positions on x-axis
situation_counts=nutri['situation'].value_counts()
plt.bar(x, situation_counts, width, edgecolor = 'black')
plt.xticks(x, situation_counts.index)
plt.show()
```

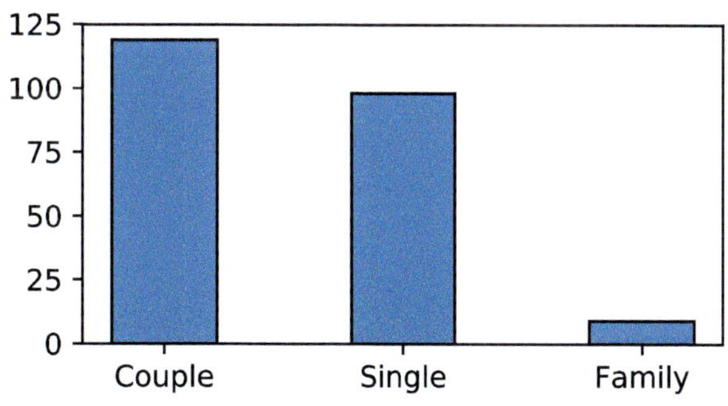

Figure 1.4: Barplot for the qualitative variable `'situation'`.

1.5.2 Plotting Quantitative Variables

We now present a few useful methods for visualizing quantitative data, again using the nutri data set. We will first focus on continuous features (e.g., `'age'`) and then add some specific graphs related to discrete features (e.g., `'tea'`). The aim is to describe the variability present in a single feature. This typically involves a central tendency, where observations tend to gather around, with fewer observations further away. The main aspects of the distribution are the *location* (or center) of the variability, the *spread* of the variability (how far the values extend from the center), and the *shape* of the variability; e.g., whether or not values are spread symmetrically on either side of the center.

1.5.2.1 Boxplot

A **boxplot** can be viewed as a graphical representation of the five-number summary of the data consisting of the minimum, maximum, and the first, second, and third quartiles. Figure 1.5 gives a boxplot for the `'age'` feature of the `nutri` data.

```
plt.boxplot(nutri['age'],widths=width,vert=False)
plt.xlabel('age')
plt.show()
```

The `widths` parameter determines the width of the boxplot, which is by default plotted vertically. Setting `vert=False` plots the boxplot horizontally, as in Figure 1.5.

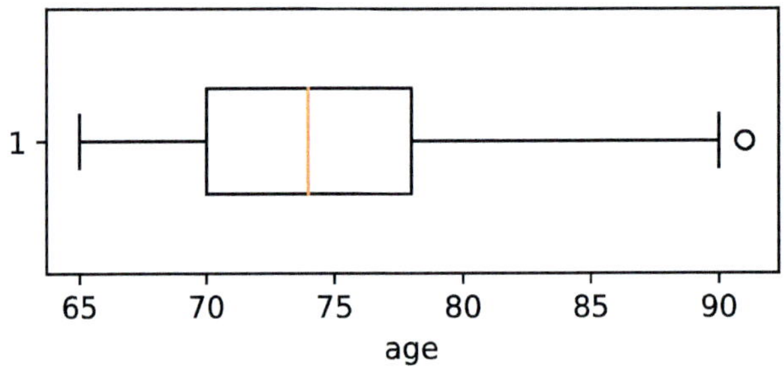

Figure 1.5: Boxplot for `'age'`.

The box is drawn from the first quartile (Q_1) to the third quartile (Q_3). The vertical line inside the box signifies the location of the median. So-called "whiskers" extend to either side of the box. The size of the box is called the **interquartile range**: IQR = $Q_3 - Q_1$. The left whisker extends to the largest of (a) the minimum of the data and (b) $Q_1 - 1.5$ IQR. Similarly, the right whisker extends to the smallest of (a) the maximum of the data and (b) $Q_3 + 1.5$ IQR. Any data point outside the whiskers is indicated by a small hollow dot, indicating a suspicious or deviant point (outlier). Note that a boxplot may also be used for discrete quantitative features.

1.5.2.2 Histogram

A **histogram** is a common graphical representation of the distribution of a quantitative feature. We start by breaking the range of the values into a number of *bins* or *classes*. We tally the counts of the values falling in each bin and then make the plot by drawing rectangles whose bases are the bin intervals and whose heights are the counts. In Python we can use the function `plt.hist`. For example, Figure 1.6 shows a histogram of the 226 ages in `nutri`, constructed via the following Python code.

```
weights = np.ones_like(nutri.age)/nutri.age.count()
plt.hist(nutri.age,bins=9,weights=weights,facecolor='cyan',
         edgecolor='black', linewidth=1)
plt.xlabel('age')
plt.ylabel('Proportion of Total')
plt.show()
```

Here 9 bins were used. Rather than using raw counts (the default), the vertical axis here gives the percentage in each class, defined by $\frac{\text{count}}{\text{total}}$. This is achieved by choosing the "weights" parameter to be equal to the vector with entries 1/266, with length 226. Various plotting parameters have also been changed.

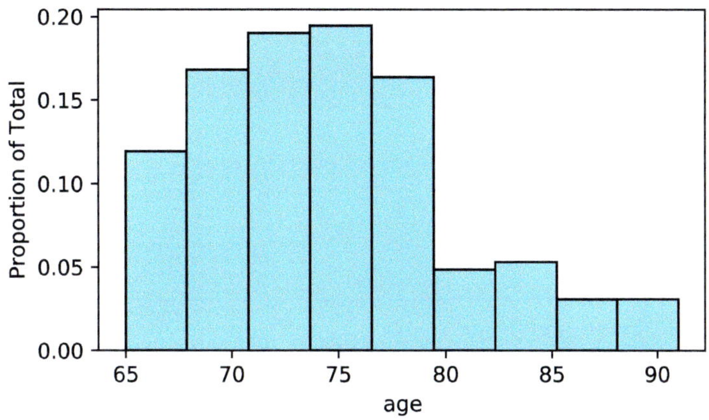

Figure 1.6: Histogram of `'age'`.

Histograms can also be used for discrete features, although it may be necessary to explicitly specify the bins and placement of the ticks on the axes.

1.5.2.3 Empirical Cumulative Distribution Function

The **empirical cumulative distribution function**, denoted by F_n, is a step function which jumps an amount k/n at observation values, where k is the number of tied observations at that value. For observations $x_1, \ldots, x_n$, $F_n(x)$ is the fraction of observations less than or equal to x, i.e.,

$$F_n(x) = \frac{\text{number of } x_i \leq x}{n} = \frac{1}{n} \sum_{i=1}^{n} \mathbb{1}\{x_i \leq x\}, \tag{1.7}$$

where $\mathbb{1}$ denotes the **indicator** function; that is, $\mathbb{1}\{x_i \leq x\}$ is equal to 1 when $x_i \leq x$ and 0 otherwise. To produce a plot of the empirical cumulative distribution function we can use the `plt.step` function. The result for the age data is shown in Figure 1.8. The empirical cumulative distribution function for a discrete quantitative variable is obtained in the same way.

```
x = np.sort(nutri.age)
y = np.linspace(0,1,len(nutri.age))
plt.xlabel('age')
plt.ylabel('Fn(x)')
plt.step(x,y)
plt.xlim(x.min(),x.max())
plt.show()
```

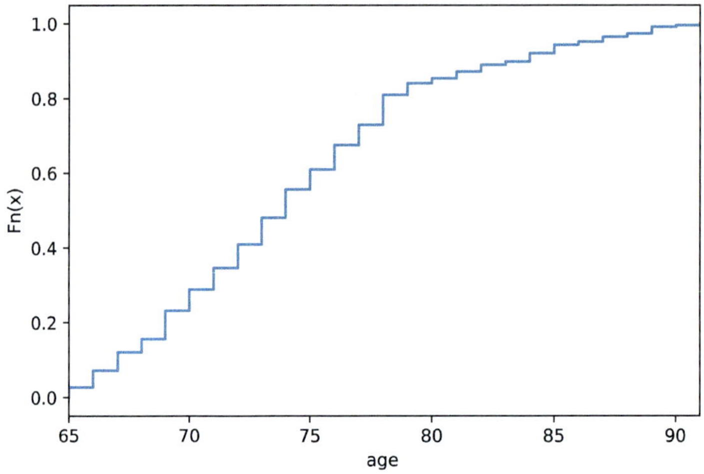

Figure 1.8: Plot of the empirical distribution function for the continuous quantitative feature `'age'`.

1.5.3 Data Visualization in a Bivariate Setting

In this section, we present a few useful visual aids to explore relationships between two features. The graphical representation will depend on the type of the two features.

1.5.3.1 Two-way Plots for Two Categorical Variables

Comparing barplots for two categorical variables involves introducing subplots to the figure. Figure 1.9 visualizes the contingency table of Section 1.3, which cross-tabulates the family status (situation) with the gender of the elderly people. It simply shows two barplots next to each other in the same figure.

The figure was made using the **seaborn** package, which was specifically designed to simplify statistical visualization tasks.

```
import seaborn as sns
sns.countplot(x='situation', hue = 'gender', data=nutri,
     hue_order = ['Male', 'Female'], palette = ['SkyBlue','Pink'],
      saturation = 1, edgecolor='black')
plt.legend(loc='upper center')
plt.xlabel('')
plt.ylabel('Counts')
plt.show()
```

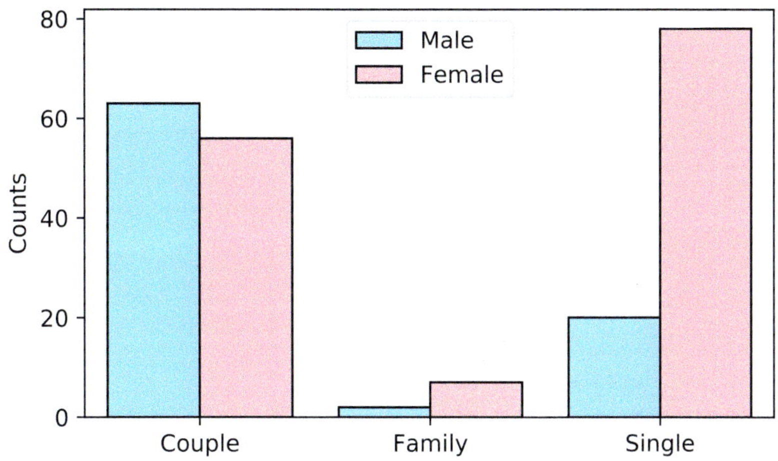

Figure 1.9: Barplot for two categorical variables.

1.5.3.2 Plots for Two Quantitative Variables

We can visualize patterns between two quantitative features using a **scatterplot**. This can be done with `plt.scatter`. The following code yields a scatterplot of `'weight'` against `'height'` for the `nutri` data.

```
plt.scatter(nutri.height, nutri.weight, s=12, marker='o')
plt.xlabel('height')
plt.ylabel('weight')
plt.show()
```

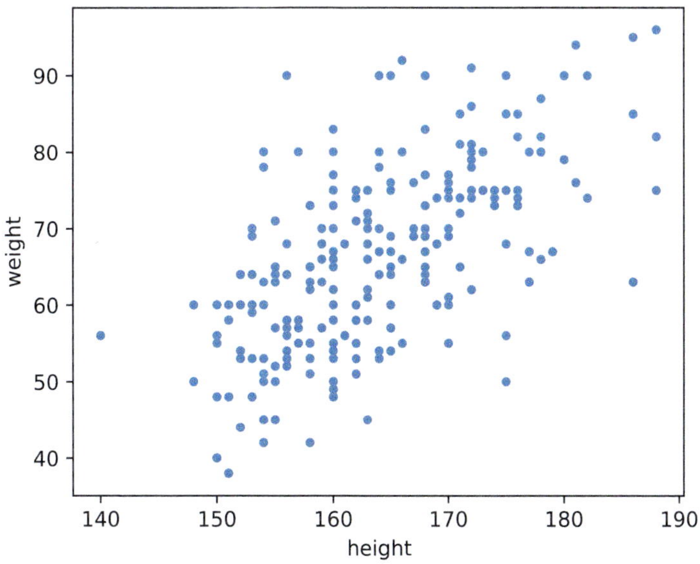

Figure 1.10: Scatterplot of `'weight'` against `'height'`.

The next Python code illustrates that it is possible to produce highly sophisticated scatter plots, such as in Figure 1.11. The figure shows the birth weights (mass) of babies whose mothers smoked (blue triangles) or not (red circles). In addition, straight lines were fitted to the two groups, suggesting that birth weight decreases with age when the mother smokes, but increases when the mother does not smoke! The question is whether these trends are statistically significant or due to chance. We will revisit this data set later ☞ 238 on in the book.

```
urlprefix = 'https://vincentarelbundock.github.io/Rdatasets/csv/'
dataname = 'MASS/birthwt.csv'
bwt = pd.read_csv(urlprefix + dataname)
bwt = bwt.drop('rownames', axis=1)   # drop rownames column
styles = {0: ['o','red'], 1: ['^','blue']}
for k in styles:
    grp = bwt[bwt.smoke==k]
    m,b = np.polyfit(grp.age, grp.bwt, 1) # fit a straight line
    plt.scatter(grp.age, grp.bwt, c=styles[k][1], s=15, linewidth=0,
        marker = styles[k][0])
    plt.plot(grp.age, m*grp.age + b, '-', color=styles[k][1])

plt.xlabel('age')
plt.ylabel('birth weight (g)')
plt.legend(['non-smokers','smokers'],prop={'size':8},
             loc=(0.5,0.8))
plt.show()
```

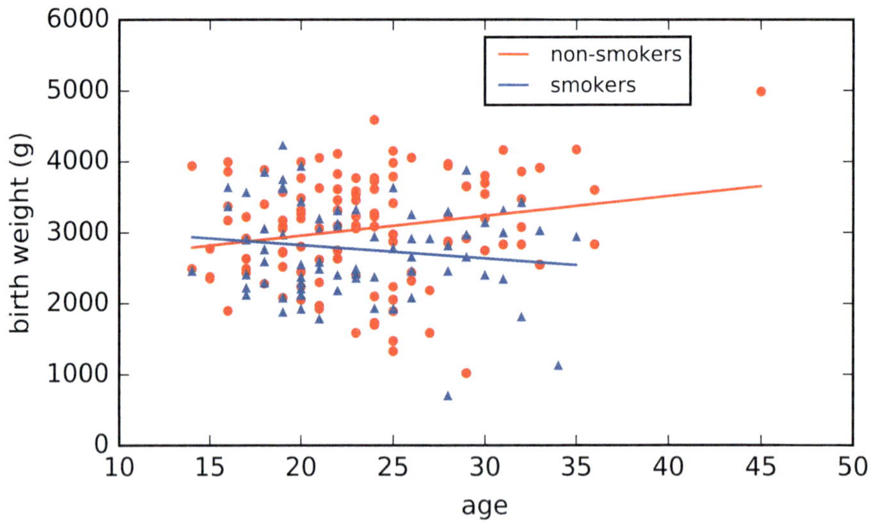

Figure 1.11: Birth weight against age for smoking and non-smoking mothers.

1.5.3.3 Plots for One Qualitative and One Quantitative Variable

In this setting, it is interesting to draw boxplots of the quantitative feature for each level of the categorical feature. Assuming the variables are structured correctly, the function **plt.boxplot** can be used to produce Figure 1.12, using the following code:

```
males = nutri[nutri.gender == 'Male']
females = nutri[nutri.gender == 'Female']
plt.boxplot([males.coffee,females.coffee],notch=True,widths=(0.5,0.5))
plt.xlabel('gender')
plt.ylabel('coffee')
plt.xticks([1,2],['Male','Female'])
plt.show()
```

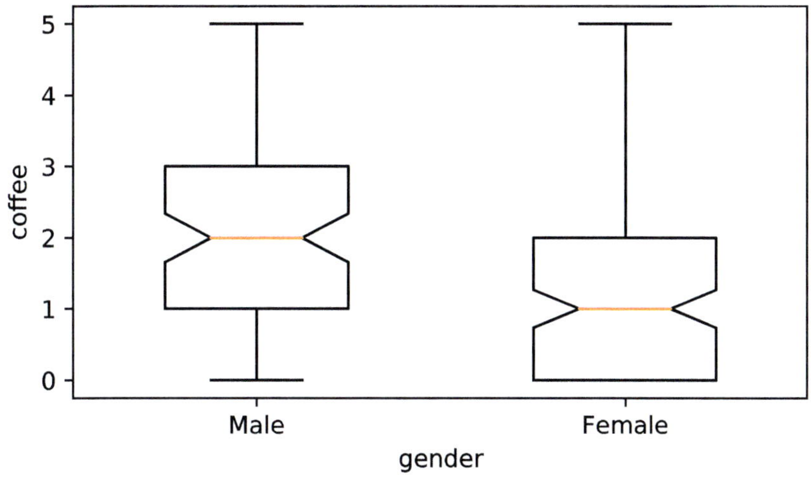

Figure 1.12: Boxplots of a quantitative feature `'coffee'` as a function of the levels of a categorical feature `'gender'`. Note that we used a different, "notched", style boxplot this time.

Further Reading

The focus in this book is on the mathematical and statistical analysis of data, and for the rest of the book we assume that the data is available in a suitable form for analysis. However, a large part of practical data science involves the *cleaning* of data; that is, putting it into a form that is amenable to analysis with standard software packages. Standard Python modules such as `numpy` and `pandas` can be used to reformat rows, rename columns, remove faulty outliers, merge rows, and so on. McKinney, the creator of `pandas`, gives many practical case studies in [121]. Effective data visualization techniques are beautifully illustrated in [92].

Exercises

Before you attempt these exercises, make sure you have up-to-date versions of the relevant Python packages, specifically `matplotlib`, `pandas`, and `seaborn`. An easy way to ensure this is to update packages via the Anaconda Navigator, as explained in Appendix E.

1. Visit the UCI Repository `https://archive.ics.uci.edu/`. Read the description of the data and download the Mushroom data set `agaricus-lepiota.data`. Using

pandas, read the data into a DataFrame called mushroom, via read_csv.

(a) How many features are in this data set?

(b) What are the initial names and types of the features?

(c) Rename the first feature (index 0) to 'edibility' and the sixth feature (index 5) to 'odor'. Hint: The column names in pandas are immutable; so individual columns cannot be modified directly. However it is possible to assign the entire column names list via mushroom.columns = newcols.

(d) The 6th column lists the various odors of the mushrooms: encoded as 'a', 'c',.... Replace these with the names 'almond', 'creosote', etc. (categories corresponding to each letter can be found on the website). Also replace the 'edibility' categories 'e' and 'p' with 'edible' and 'poisonous'.

(e) Make a contingency table cross-tabulating 'edibility' and 'odor'.

(f) Which mushroom odors should be avoided, when gathering mushrooms for consumption?

(g) What proportion of odorless mushroom samples were safe to eat?

2. Change the type and value of variables in the nutri data set according to Table 1.1 and save the data as a CSV file. The modified data should have eight categorical features, three floats, and two integer features.

3. It frequently happens that a table with data needs to be restructured before the data can be analyzed using standard statistical software. As an example, consider the test scores in Table 1.13 of 5 students before and after specialized tuition.

Table 1.13: Student scores.

Student	Before	After
1	75	85
2	30	50
3	100	100
4	50	52
5	60	65

This is not in the standard format described in Section 1.1. In particular, the student scores are divided over two columns, whereas the standard format requires that they are collected in one column, e.g., labelled 'Score'. Reformat (by hand) the table in standard format, using three features:

- 'Score', taking continuous values,
- 'Time', taking values 'Before' and 'After',
- 'Student', taking values from 1 to 5.

Useful methods for reshaping tables in pandas are melt, stack, and unstack.

4. Create a similar barplot as in Figure 1.9, but now plot the corresponding *proportions* of males and females in each of the three situation categories. That is, the heights of the bars should sum up to 1 for both barplots with the same `'gender'` value. Hint: `seaborn` does not have this functionality built in, instead you need to first create a contingency table and use `matplotlib.pyplot` to produce the figure.

5. The `iris` data set, mentioned in Section 1.1, contains various features, including ☞ 2
`'Petal.Length'` and `'Sepal.Length'`, of three species of iris: setosa, versicolor, and virginica.

(a) Load the data set into a `pandas` `DataFrame` object.

(b) Using `matplotlib.pyplot`, produce boxplots of `'Petal.Length'` for each the three species, in one figure.

(c) Make a histogram with 20 bins for `'Petal.Length'`.

(d) Produce a similar scatterplot for `'Sepal.Length'` against `'Petal.Length'` to that of the left plot in Figure 1.14. Note that the points should be colored according to the `'Species'` feature as per the legend in the right plot of the figure.

(e) Using the `kdeplot` method of the `seaborn` package, reproduce the right plot of Figure 1.14, where kernel density plots for `'Petal.Length'` are given. ☞ 144

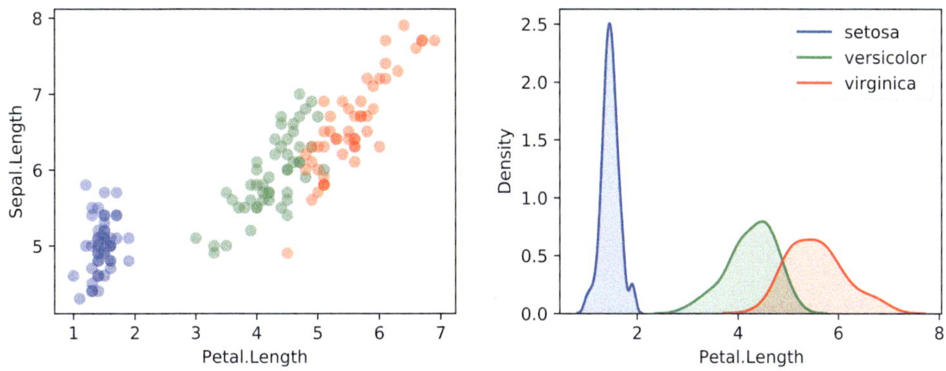

Figure 1.14: Left: scatterplot of `'Sepal.Length'` against `'Petal.Length'`. Right: kernel density estimates of `'Petal.Length'` for the three species of iris.

6. Import the data set `EuStockMarkets` from the same website as the `iris` data set above. The data set contains the daily closing prices of four European stock indices during the 1990s, for 260 working days per year.

(a) Create a vector of times (working days) for the stock prices, between 1991.496 and 1998.646 with increments of 1/260.

(b) Reproduce Figure 1.15. Hint: Use a dictionary to map column names (stock indices) to colors.

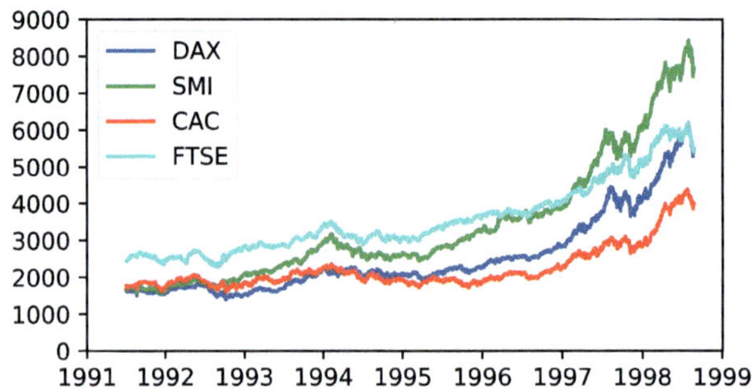

Figure 1.15: Closing stock indices for various European stock markets.

7. Consider the `KASANDR` data set from the UCI Machine Learning Repository, which can be downloaded from https://archive.ics.uci.edu/ml/machine-learning-databases/00385/de.tar.bz2.

This archive file has a size of 900Mb, so it may take a while to download. Uncompressing the file (e.g., via 7-Zip) yields a directory `de` containing two large CSV files: `test_de.csv` and `train_de.csv`, with sizes 372Mb and 3Gb, respectively. Such large data files can still be processed efficiently in `pandas`, provided there is enough memory. The files contain records of user information from Kelkoo web logs in Germany as well as meta-data on users, offers, and merchants. The data sets have 7 attributes and 1919561 and 15844717 rows, respectively. The data sets are anonymized via hex strings.

(a) Load `train_de.csv` into a `pandas` `DataFrame` object `de`, using

```
read_csv('train_de.csv',delimiter='\t').
```

If not enough memory is available, load `test_de.csv` instead. Note that entries are separated here by tabs, not commas. Time how long it takes for the file to load, using the `time` package. (It took 38 seconds for `train_de.csv` to load on one of our computers.)

(b) How many unique users and merchants are in this data set?

8. Visualizing data involving more than two features requires careful design, which is often more of an art than a science.

(a) Go to Vincent Arel-Bundocks's website (URL given in Section 1.1) and read the `Orange` data set into a `pandas` `DataFrame object` called `orange`. Remove its first (unnamed) column.

(b) The data set contains the circumferences of 5 orange trees at various stages in their development. Find the names of the features.

(c) In Python, import `seaborn` and visualize the growth curves (circumference against age) of the trees, using the `regplot` and `FacetGrid` methods.

CHAPTER 2

STATISTICAL LEARNING

The purpose of this chapter is to introduce the reader to some common concepts and themes in statistical learning. We discuss the difference between supervised and unsupervised learning, and how we can assess the predictive performance of supervised learning. We also examine the central role that the linear and Gaussian properties play in the modeling of data. We conclude with a section on Bayesian learning. The required probability and statistics background is given in Appendix D.

2.1 Introduction

Although structuring and visualizing data are important aspects of data science, the main challenge lies in the mathematical analysis of the data. When the goal is to interpret the model and quantify the uncertainty in the data, this analysis is usually referred to as **statistical learning**. In contrast, when the emphasis is on making predictions using large-scale data, then it is common to speak about **machine learning** or **data mining**.

There are two major goals for modeling data: 1) to accurately predict some future quantity of interest, given some observed data, and 2) to discover unusual or interesting patterns in the data. To achieve these goals, one must rely on knowledge from three important pillars of the mathematical sciences.

Function approximation. Building a mathematical model for data usually means understanding how one data variable depends on another data variable. The most natural way to represent the relationship between variables is via a mathematical function or map. We usually assume that this mathematical function is not completely known, but can be approximated well given enough computing power and data. Thus, data scientists have to understand how best to approximate and represent functions using the least amount of computer processing and memory.

Optimization. Given a class of mathematical models, we wish to find the best possible model in that class. This requires some kind of efficient search or optimization procedure. The optimization step can be viewed as a process of fitting or calibrating a function to observed data. This step usually requires knowledge of optimization algorithms and efficient computer coding or programming.

Probability and Statistics. In general, the data used to fit the model is viewed as a realization of a random process or numerical vector, whose probability law determines the accuracy with which we can predict future observations. Thus, in order

DOI: 10.1201/9781003391173-2

to quantify the uncertainty inherent in making predictions about the future, and the sources of error in the model, data scientists need a firm grasp of probability theory and statistical inference.

2.2 Supervised and Unsupervised Learning

Given an input or **feature** vector $\boldsymbol{x}$, one of the main goals of machine learning is to predict an output or **response** variable y. For example, $\boldsymbol{x}$ could be a digitized signature and y a binary variable that indicates whether the signature is genuine or false. Another example is where $\boldsymbol{x}$ represents the weight and smoking habits of an expecting mother and y the birth weight of the baby. The data science attempt at this prediction is encoded in a mathematical function g, called the **prediction function**, which takes as an input $\boldsymbol{x}$ and outputs a guess $g(\boldsymbol{x})$ for y (denoted by $\widehat{y}$, for example). In a sense, g encompasses all the information about the relationship between the variables $\boldsymbol{x}$ and y, excluding the effects of chance and randomness in nature.

In **regression** problems, the response variable y can take any real value. In contrast, when y can only lie in a finite set, say $y \in \{0, \dots, c-1\}$, then predicting y is conceptually the same as classifying the input $\boldsymbol{x}$ into one of c categories, and so prediction becomes a **classification** problem.

We can measure the accuracy of a prediction $\widehat{y}$ with respect to a given response y by using some **loss function** $\text{Loss}(y, \widehat{y})$. In a regression setting the usual choice is the squared-error loss $(y - \widehat{y})^2$. In the case of classification, the zero–one (also written 0–1) loss function $\text{Loss}(y, \widehat{y}) = \mathbb{1}\{y \neq \widehat{y}\}$ is often used, which incurs a loss of 1 whenever the predicted class $\widehat{y}$ is not equal to the class y. Later on in this book, we will encounter various other useful loss functions, such as the cross-entropy and hinge loss functions (see, e.g., Chapter 8).

> The word *error* is often used as a measure of distance between a "true" object y and some approximation $\widehat{y}$ thereof. If y is real-valued, the absolute error $|y - \widehat{y}|$ and the squared error $(y - \widehat{y})^2$ are both well-established error concepts, as are the norm $\|\mathbf{y} - \widehat{\mathbf{y}}\|$ and squared norm $\|\mathbf{y} - \widehat{\mathbf{y}}\|^2$ for vectors. The squared error $(y - \widehat{y})^2$ is just one example of a loss function.

It is unlikely that any mathematical function g will be able to make accurate predictions for all possible pairs $(\boldsymbol{x}, y)$ one may encounter in Nature. One reason for this is that, even with the same input $\boldsymbol{x}$, the output y may be different, depending on chance circumstances or randomness. For this reason, we adopt a probabilistic approach and assume that each pair $(\boldsymbol{x}, y)$ is the outcome of a random pair $(\boldsymbol{X}, Y)$ that has some joint probability density $f(\boldsymbol{x}, y)$. We then assess the predictive performance via the expected loss, usually called the **risk**, for g:

$$\ell(g) = \mathbb{E}\, \text{Loss}(Y, g(\boldsymbol{X})). \tag{2.1}$$

For example, in the classification case with zero–one loss function the risk is equal to the probability of incorrect classification: $\ell(g) = \mathbb{P}[Y \neq g(\boldsymbol{X})]$. In this context, the prediction function g is called a **classifier**. Given the distribution of (X, Y) and any loss function, we can in principle find the best possible $g^* := \text{argmin}_g \mathbb{E}\,\text{Loss}(Y, g(X))$ that yields the smallest risk $\ell^* := \ell(g^*)$. We will see in Chapter 8 that in the classification case with $y \in \{0, \dots, c-1\}$ and $\ell(g) = \mathbb{P}[Y \neq g(X)]$, we have ☞ 353

$$g^*(\boldsymbol{x}) = \underset{y \in \{0,\dots,c-1\}}{\text{argmax}} \; f(y \,|\, \boldsymbol{x}),$$

where $f(y \,|\, \boldsymbol{x}) = \mathbb{P}[Y = y \,|\, \boldsymbol{X} = \boldsymbol{x}]$ is the conditional probability of $Y = y$ given $\boldsymbol{X} = \boldsymbol{x}$. As already mentioned, for regression the most widely-used loss function is the squared-error loss. In this setting, the optimal prediction function g^* is often called the **regression function**. The following theorem specifies its exact form.

Theorem 2.2 (Optimal Prediction Function for Squared-Error Loss): For the squared-error loss $\text{Loss}(y, \widehat{y}) = (y - \widehat{y})^2$, the optimal prediction function g^* is equal to the conditional expectation of Y given $\boldsymbol{X} = \boldsymbol{x}$:

$$g^*(\boldsymbol{x}) = \mathbb{E}[Y \,|\, \boldsymbol{X} = \boldsymbol{x}].$$

Proof. Let $g^*(\boldsymbol{x}) = \mathbb{E}[Y \,|\, \boldsymbol{X} = \boldsymbol{x}]$. For any function g, the squared-error risk satisfies

$$\begin{aligned}
\mathbb{E}(Y - g(\boldsymbol{X}))^2 &= \mathbb{E}[(Y - g^*(\boldsymbol{X}) + g^*(\boldsymbol{X}) - g(\boldsymbol{X}))^2] \\
&= \mathbb{E}(Y - g^*(\boldsymbol{X}))^2 + 2\mathbb{E}[(Y - g^*(\boldsymbol{X}))(g^*(\boldsymbol{X}) - g(\boldsymbol{X}))] + \mathbb{E}(g^*(\boldsymbol{X}) - g(\boldsymbol{X}))^2 \\
&\geq \mathbb{E}(Y - g^*(\boldsymbol{X}))^2 + 2\mathbb{E}[(Y - g^*(\boldsymbol{X}))(g^*(\boldsymbol{X}) - g(\boldsymbol{X}))] \\
&= \mathbb{E}(Y - g^*(\boldsymbol{X}))^2 + 2\mathbb{E}\left\{(g^*(\boldsymbol{X}) - g(\boldsymbol{X}))\mathbb{E}[Y - g^*(\boldsymbol{X}) \,|\, \boldsymbol{X}]\right\}.
\end{aligned}$$

In the last equation we used the tower property. By the definition of the conditional expectation, we have $\mathbb{E}[Y - g^*(\boldsymbol{X}) \,|\, \boldsymbol{X}] = 0$. It follows that $\mathbb{E}(Y - g(\boldsymbol{X}))^2 \geq \mathbb{E}(Y - g^*(\boldsymbol{X}))^2$, showing that g^* yields the smallest squared-error risk. □ ☞ 635

One consequence of Theorem 2.2 is that, conditional on $\boldsymbol{X} = \boldsymbol{x}$, the (random) response Y can be written as

$$Y = g^*(\boldsymbol{x}) + \varepsilon(\boldsymbol{x}), \tag{2.3}$$

where $\varepsilon(\boldsymbol{x})$ can be viewed as the random deviation of the response from its conditional mean at $\boldsymbol{x}$. This random deviation satisfies $\mathbb{E}\,\varepsilon(\boldsymbol{x}) = 0$. Further, the conditional variance of the response Y at $\boldsymbol{x}$ can be written as $\mathbb{V}\text{ar}\,\varepsilon(\boldsymbol{x}) = v^2(\boldsymbol{x})$ for some unknown positive function v. Note that, in general, the probability distribution of $\varepsilon(\boldsymbol{x})$ is unspecified.

Since, the optimal prediction function g^* depends on the typically unknown joint distribution of $(\boldsymbol{X}, Y)$, it is not available in practice. Instead, all that we have available is a finite number of (usually) independent and identically distributed (abbreviated as iid) random variables from the joint density $f(\boldsymbol{x}, y)$. We denote this iid sample by $\mathcal{T} = \{(\boldsymbol{X}_1, Y_1), \dots, (\boldsymbol{X}_n, Y_n)\}$ and call it the **training set** with n examples; $\mathcal{T}$ is a mnemonic for training. It will be important to distinguish between a random training set $\mathcal{T}$ and its (deterministic) outcome $\{(\boldsymbol{x}_1, y_1), \dots, (\boldsymbol{x}_n, y_n)\}$. We will use the notation τ

for the latter. We will also add the subscript n in τ_n when we wish to emphasize the size of the training set.

Our goal is thus to "learn" the unknown g^* using the n examples in the training set $\mathcal{T}$. Let us denote by $g_{\mathcal{T}}$ the best (by some criterion) approximation for g^* that we can construct from $\mathcal{T}$. Note that $g_{\mathcal{T}}$ is a random function. A particular outcome is denoted by g_{τ}. It is often useful to think of a teacher–learner metaphor, whereby the function $g_{\mathcal{T}}$ is a **learner** who learns the unknown functional relationship $g^* : \boldsymbol{x} \mapsto y$ from the training data $\mathcal{T}$. We can imagine a "teacher" who provides n examples of the true relationship between the output Y_i and the input $\boldsymbol{X}_i$ for $i = 1, \ldots, n$, and thus "trains" the learner $g_{\mathcal{T}}$ to predict the output of a new input $\boldsymbol{X}$, for which the correct output Y is not provided by the teacher (is unknown).

The above setting is called **supervised learning**, because one tries to learn the functional relationship between the feature vector $\boldsymbol{x}$ and response y in the presence of a teacher who provides n examples. It is common to speak of "explaining" or predicting y on the basis of $\boldsymbol{x}$, where $\boldsymbol{x}$ is a vector of **explanatory variables**.

An example of supervised learning is email spam detection. The goal is to train the learner $g_{\mathcal{T}}$ to accurately predict whether any future email, as represented by the feature vector $\boldsymbol{x}$, is spam or not. The training data consists of the feature vectors of a number of different email examples as well as the corresponding labels (spam or not spam). For instance, a feature vector could consist of the number of times sales-pitch words like "free", "sale", or "miss out" occur within a given email.

As seen from the above discussion, most questions of interest in supervised learning can be answered if we know the conditional pdf $f(y \mid \boldsymbol{x})$, because we can then in principle work out the function value $g^*(\boldsymbol{x})$.

In contrast, **unsupervised learning** makes no distinction between response and explanatory variables, and the objective is simply to learn the structure of the unknown distribution of the data. In other words, we need to learn $f(\boldsymbol{x})$. In this case the guess $g(\boldsymbol{x})$ is an approximation of $f(\boldsymbol{x})$ and the risk is of the form

$$\ell(g) = \mathbb{E}\,\mathrm{Loss}(f(\boldsymbol{X}), g(\boldsymbol{X})).$$

An example of unsupervised learning is when we wish to analyze the purchasing behaviors of the customers of a grocery shop that has a total of, say, a hundred items on sale. A feature vector here could be a binary vector $\boldsymbol{x} \in \{0, 1\}^{100}$ representing the items bought by a customer on a visit to the shop (a 1 in the k-th position if a customer bought item $k \in \{1, \ldots, 100\}$ and a 0 otherwise). Based on a training set $\tau = \{\boldsymbol{x}_1, \ldots, \boldsymbol{x}_n\}$, we wish to find any interesting or unusual purchasing patterns. In general, it is difficult to know if an unsupervised learner is doing a good job, because there is no teacher to provide examples of accurate predictions.

The main methodologies for unsupervised learning include *clustering*, *principal component analysis*, and *kernel density estimation*, which will be discussed in more
☞ 133 detail in Chapter 4.

In the next three sections we will focus on supervised learning. The main supervised learning methodologies are *regression* and *classification*, which are covered in Chapters 5,
☞ 197 6, and 8. More advanced supervised learning techniques, including *reproducing kernel Hilbert spaces*, *tree methods*, and *deep learning*, will be discussed in Chapters 7, 9, and 10. A different learning paradigm is given in Chapter 11, which introduces *reinforcement*
☞ 461 *learning*.

2.3 Training and Test Loss

Given an arbitrary prediction function g, it is typically not possible to compute its risk $\ell(g)$ in (2.1). However, using the training sample $\mathcal{T}$, we can approximate $\ell(g)$ via the empirical (sample average) risk

$$\ell_{\mathcal{T}}(g) = \frac{1}{n}\sum_{i=1}^{n} \text{Loss}(Y_i, g(\boldsymbol{X}_i)), \tag{2.4}$$

which we call the **training loss**. The training loss is thus an unbiased estimator (a random variable) of the risk (the expected loss) for a prediction function g, based on the training data $\mathcal{T}$.

To approximate the optimal prediction function g^* (the minimizer of the risk $\ell(g)$) we first select a suitable collection of approximating functions $\mathcal{G}$ and then take our *learner* to be the function in $\mathcal{G}$ that minimizes the training loss; that is, the random function:

$$g_{\mathcal{T}}^{\mathcal{G}} = \underset{g\in\mathcal{G}}{\text{argmin}}\, \ell_{\mathcal{T}}(g). \tag{2.5}$$

For example, the simplest and most useful $\mathcal{G}$ is the set of *linear* functions of $\boldsymbol{x}$; that is, the set of all functions $g : \boldsymbol{x} \mapsto \boldsymbol{\beta}^\top \boldsymbol{x}$ for some real-valued vector $\boldsymbol{\beta}$.

We suppress the superscript $\mathcal{G}$ when it is clear which function class is used. Note that minimizing the training loss over all possible functions g (rather than over all $g \in \mathcal{G}$) does not lead to a meaningful optimization problem, as any function g for which $g(\boldsymbol{X}_i) = Y_i$ for all i gives minimal training loss. In particular, for a squared-error loss, the training loss will be 0. Unfortunately, such functions have a poor ability to predict new (that is, independent from $\mathcal{T}$) pairs of data. This poor generalization performance is called **overfitting**.

> By choosing g a function that predicts the training data exactly (and is, for example, 0 otherwise), the squared-error training loss is zero. Minimizing the training loss is not the ultimate goal!

The prediction accuracy of new pairs of data is measured by the **generalization risk** of the learner. For a *fixed* training set τ it is defined as

$$\ell(g_{\tau}^{\mathcal{G}}) = \mathbb{E}\,\text{Loss}(Y, g_{\tau}^{\mathcal{G}}(\boldsymbol{X})), \tag{2.6}$$

where $(\boldsymbol{X}, Y)$ is distributed according to $f(\boldsymbol{x}, y)$. In the discrete case the generalization risk is therefore:

$$\ell(g_{\tau}^{\mathcal{G}}) = \sum_{\boldsymbol{x},y} \text{Loss}(y, g_{\tau}^{\mathcal{G}}(\boldsymbol{x}))\, f(\boldsymbol{x}, y).$$

Replace the sum with an integral for the continuous case.

The situation is illustrated in Figure 2.7, where the distribution of $(\boldsymbol{X}, Y)$ is indicated by the red dots. The training set (points in the shaded regions) determines a fixed prediction function shown as a straight line. Three possible outcomes of $(\boldsymbol{X}, Y)$ are shown (black dots). The amount of loss for each point is shown as the length of the dashed lines. The generalization risk is the average loss over all possible pairs $(\boldsymbol{x}, y)$, weighted by the corresponding $f(\boldsymbol{x}, y)$.

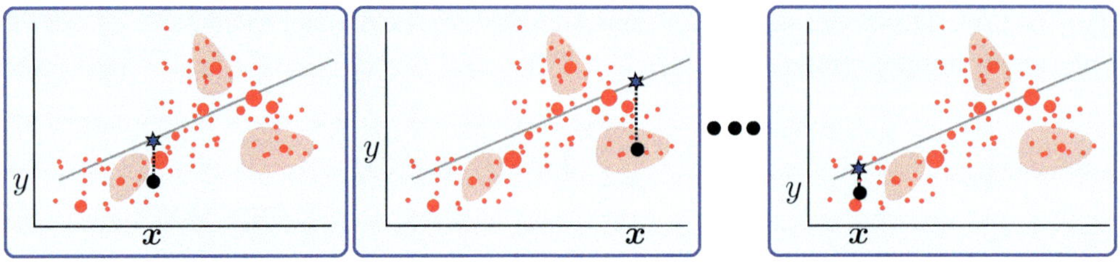

Figure 2.7: The generalization risk for a fixed training set is the weighted-average loss over all possible pairs $(\boldsymbol{x}, y)$.

For a (random) training set $\mathcal{T}$, the generalization risk is thus a random variable that depends on $\mathcal{T}$ (and $\mathcal{G}$). If we average the generalization risk over all possible instances of $\mathcal{T}$, we obtain the **expected generalization risk**:

$$\mathbb{E}\,\ell(g_{\mathcal{T}}^{\mathcal{G}}) = \mathbb{E}\,\mathrm{Loss}(Y, g_{\mathcal{T}}^{\mathcal{G}}(\boldsymbol{X})), \tag{2.8}$$

where $(\boldsymbol{X}, Y)$ in the expectation above is independent of $\mathcal{T}$. In the discrete case, we have:

$$\begin{aligned}\mathbb{E}\,\ell(g_{\mathcal{T}}^{\mathcal{G}}) &= \sum_{\tau} \ell(g_{\tau}^{\mathcal{G}})\,\mathbb{P}(\mathcal{T} = \tau) = \sum_{\boldsymbol{x}_1, y_1, \ldots, \boldsymbol{x}_n, y_n} \ell(g_{\tau}^{\mathcal{G}}) f(\boldsymbol{x}_1, y_1) \cdots f(\boldsymbol{x}_n, y_n)\\ &= \sum_{\boldsymbol{x}, y, \boldsymbol{x}_1, y_1, \ldots, \boldsymbol{x}_n, y_n} \mathrm{Loss}(y, g_{\tau}^{\mathcal{G}}(\boldsymbol{x})) f(\boldsymbol{x}, y) f(\boldsymbol{x}_1, y_1) \cdots f(\boldsymbol{x}_n, y_n).\end{aligned}$$

Figure 2.9 gives an illustration.

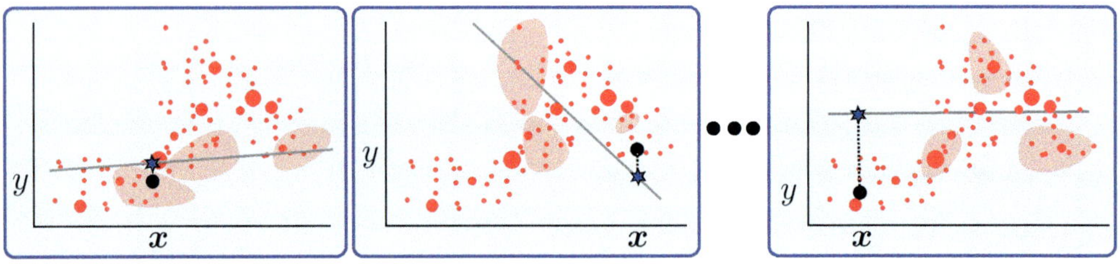

Figure 2.9: The expected generalization risk is the weighted-average loss over all possible pairs $(\boldsymbol{x}, y)$ and over all training sets.

For any outcome τ of the training data, we can estimate the generalization risk without bias by taking the sample average

$$\ell_{\mathcal{T}'}(g_{\tau}^{\mathcal{G}}) := \frac{1}{n'} \sum_{i=1}^{n'} \mathrm{Loss}(Y_i', g_{\tau}^{\mathcal{G}}(\boldsymbol{X}_i')), \tag{2.10}$$

where $\{(\boldsymbol{X}_1', Y_1'), \ldots, (\boldsymbol{X}_{n'}', Y_{n'}')\} =: \mathcal{T}'$ is a so-called **test sample**. The test sample is independent of $\mathcal{T}$, and has the same distribution $\mathcal{T}$; that is, it is an iid sample from $f(\boldsymbol{x}, y)$ for some sample size n'. We call the estimator (2.10) the **test loss**.

Table 2.11: Summary of definitions for supervised learning.

$\boldsymbol{x}$	Fixed explanatory (feature) vector.
$\boldsymbol{X}$	Random explanatory (feature) vector.
y	Fixed (real-valued) response.
Y	Random response.
$f(\boldsymbol{x}, y)$	Joint pdf of $\boldsymbol{X}$ and Y, evaluated at $(\boldsymbol{x}, y)$.
$f(y \mid \boldsymbol{x})$	Conditional pdf of Y given $\boldsymbol{X} = \boldsymbol{x}$, evaluated at y.
τ or τ_n	Fixed training data $\{(\boldsymbol{x}_i, y_i), i = 1, \dots, n\}$.
$\mathcal{T}$ or $\mathcal{T}_n$	Random training data $\{(\boldsymbol{X}_i, Y_i), i = 1, \dots, n\}$.
$\mathbf{X}$	Matrix of explanatory variables, with n rows $\boldsymbol{x}_i^\top, i = 1, \dots, n$ and $\dim(\boldsymbol{x})$ feature columns; one of the features may be the constant 1.
$\boldsymbol{y}$	Vector of response variables $(y_1, \dots, y_n)^\top$.
g	Prediction (guess) function.
$\text{Loss}(y, \widehat{y})$	Loss incurred when predicting response y with $\widehat{y}$.
$\ell(g)$	Risk for prediction function g; that is, $\mathbb{E}\,\text{Loss}(Y, g(\boldsymbol{X}))$.
g^*	Optimal prediction function; that is, $\text{argmin}_g\, \ell(g)$.
$g^{\mathcal{G}}$	Optimal prediction function in function class $\mathcal{G}$; that is, $\text{argmin}_{g \in \mathcal{G}}\, \ell(g)$.
$\ell_\tau(g)$	Training loss for prediction function g; that is, the sample average estimate of $\ell(g)$ based on a fixed training sample τ.
$\ell_{\mathcal{T}}(g)$	The same as $\ell_\tau(g)$, but now for a random training sample $\mathcal{T}$.
$g_\tau^{\mathcal{G}}$ or g_τ	The *learner*: $\text{argmin}_{g \in \mathcal{G}}\, \ell_\tau(g)$. That is, the optimal prediction function based on a fixed training set τ and function class $\mathcal{G}$. We suppress the superscript $\mathcal{G}$ if the class is implicit.
$g_{\mathcal{T}}^{\mathcal{G}}$ or $g_{\mathcal{T}}$	The learner, where we have replaced τ with a random training set $\mathcal{T}$.

For a random training set $\mathcal{T}$ we can define $\ell_{\mathcal{T}'}(g_{\mathcal{T}}^{\mathcal{G}})$ similarly. It is then crucial to assume that $\mathcal{T}$ is independent of $\mathcal{T}'$. Table 2.11 summarizes the main definitions and notation for supervised learning.

To compare the predictive performance of various learners in the function class $\mathcal{G}$, as measured by the test loss, we can use the *same* fixed training set τ and test set τ' for all learners. When there is an abundance of data, the "overall" data set is usually (randomly) divided into a training and test set, as depicted in Figure 2.12. We then use the training data to construct various learners $g_\tau^{\mathcal{G}_1}, g_\tau^{\mathcal{G}_2}, \dots$, and use the test data to select the best (with the smallest test loss) among these learners. In this context the test set is called the **validation set**. Once the best learner has been chosen, a third "test" set can be used to assess the predictive performance of the best learner. The training, validation, and test sets can again be obtained from the overall data set via a random allocation. When the overall data set is of modest size, it is customary to perform the validation phase (model selection) on the training set only, using cross-validation. This is the topic of Section 2.5.2. ☞ 39

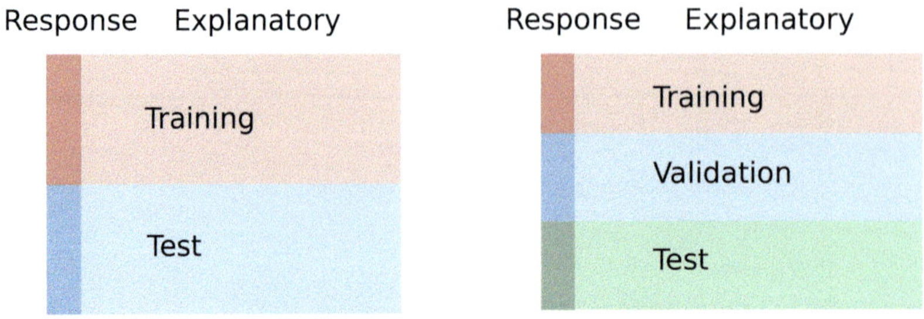

Figure 2.12: Statistical learning algorithms often require the data to be divided into training and test data. If the latter is used for model selection, a third set is needed for testing the performance of the selected model.

We next consider a concrete example that illustrates the concepts introduced so far.

■ **Example 2.13 (Polynomial Regression)** In what follows, it will appear that we have arbitrarily replaced the symbols $\mathbf{x}, g, \mathcal{G}$ with $u, h, \mathcal{H}$, respectively. The reason for this switch of notation will become clear at the end of the example.

The data (depicted as dots) in Figure 2.14 are $n = 100$ points $(u_i, y_i), i = 1, \ldots, n$ drawn from iid random points $(U_i, Y_i), i = 1, \ldots, n$, where the $\{U_i\}$ are uniformly distributed on the interval $(0, 1)$ and, given $U_i = u_i$, the random variable Y_i has a normal distribution with expectation $10 - 140u_i + 400u_i^2 - 250u_i^3$ and variance $\ell^* = 25$. This is an example of a **polynomial regression model**. Using a squared-error loss, the optimal prediction function $h^*(u) = \mathbb{E}[Y \mid U = u]$ is thus

$$h^*(u) = 10 - 140u + 400u^2 - 250u^3,$$

which is depicted by the dashed curve in Figure 2.14.

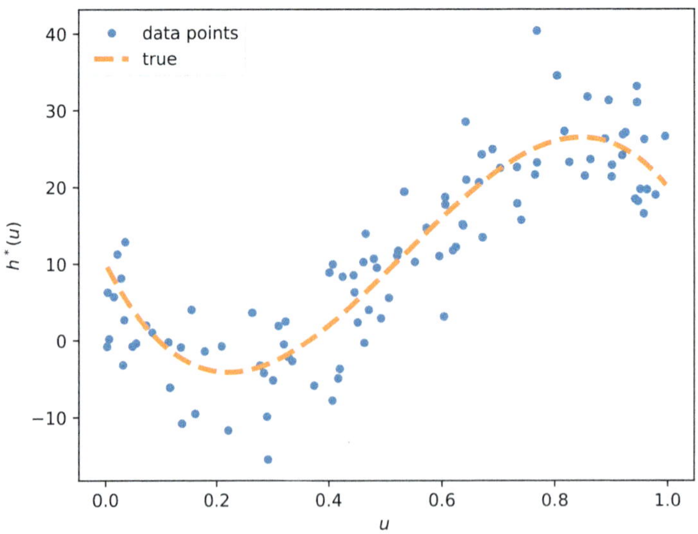

Figure 2.14: Training data and the optimal polynomial prediction function h^*.

To obtain a good estimate of $h^*(u)$ based on the training set $\tau = \{(u_i, y_i), i = 1, \dots, n\}$, we minimize the outcome of the training loss (2.4):

$$\ell_\tau(h) = \frac{1}{n}\sum_{i=1}^{n}(y_i - h(u_i))^2, \tag{2.15}$$

over a suitable set $\mathcal{H}$ of candidate functions. Let us take the set $\mathcal{H}_p$ of polynomial functions in u of order $p - 1$:

$$h(u) := \beta_1 + \beta_2 u + \beta_3 u^2 + \cdots + \beta_p u^{p-1} \tag{2.16}$$

for $p = 1, 2, \dots$ and parameter vector $\boldsymbol{\beta} = [\beta_1, \beta_2, \dots, \beta_p]^\top$. This function class contains the best possible $h^*(u) = \mathbb{E}[Y \mid U = u]$ for $p \geq 4$. Note that optimization over $\mathcal{H}_p$ is a parametric optimization problem, in that we need to find the best $\boldsymbol{\beta}$. Optimization of (2.15) over $\mathcal{H}_p$ is not straightforward, unless we notice that (2.16) is a *linear* function in $\boldsymbol{\beta}$. In particular, if we map each feature u to a feature vector $\boldsymbol{x} = [1, u, u^2, \dots, u^{p-1}]^\top$, then the right-hand side of (2.16) can be written as the function

$$g(\boldsymbol{x}) = \boldsymbol{x}^\top \boldsymbol{\beta},$$

which is linear in $\boldsymbol{x}$ (as well as $\boldsymbol{\beta}$). The optimal $h^*(u)$ in $\mathcal{H}_p$ for $p \geq 4$ then corresponds to the function $g^*(\boldsymbol{x}) = \boldsymbol{x}^\top \boldsymbol{\beta}^*$ in the set $\mathcal{G}_p$ of linear functions from $\mathbb{R}^p$ to $\mathbb{R}$, where $\boldsymbol{\beta}^* = [10, -140, 400, -250, 0, \dots, 0]^\top$. Thus, instead of working with the set $\mathcal{H}_p$ of polynomial functions we may prefer to work with the set $\mathcal{G}_p$ of linear functions. This brings us to a very important idea in statistical learning:

> Expand the feature space to obtain a *linear* prediction function.

Let us now reformulate the learning problem in terms of the new explanatory (feature) variables $\boldsymbol{x}_i = [1, u_i, u_i^2, \dots, u_i^{p-1}]^\top, i = 1, \dots, n$. It will be convenient to arrange these feature vectors into a matrix $\mathbf{X}$ with rows $\boldsymbol{x}_1^\top, \dots, \boldsymbol{x}_n^\top$:

$$\mathbf{X} = \begin{bmatrix} 1 & u_1 & u_1^2 & \cdots & u_1^{p-1} \\ 1 & u_2 & u_2^2 & \cdots & u_2^{p-1} \\ \vdots & \vdots & \vdots & \ddots & \vdots \\ 1 & u_n & u_n^2 & \cdots & u_n^{p-1} \end{bmatrix}. \tag{2.17}$$

Collecting the responses $\{y_i\}$ into a column vector $\boldsymbol{y}$, the training loss (2.4) can now be written compactly as

$$\frac{1}{n}\|\boldsymbol{y} - \mathbf{X}\boldsymbol{\beta}\|^2. \tag{2.18}$$

To find the optimal learner (2.5) in the class $\mathcal{G}_p$ we need to pick a suitable, possibly the unique, minimizer $\widehat{\boldsymbol{\beta}}$ of (2.18):

$$\widehat{\boldsymbol{\beta}} \in \underset{\boldsymbol{\beta}}{\operatorname{argmin}} \|\boldsymbol{y} - \mathbf{X}\boldsymbol{\beta}\|^2. \tag{2.19}$$

Using the "belongs to" symbol $\in$, rather than the "equals to" symbol in (2.19), accounts for the possibility that the minimizer of the squared norm $\|\boldsymbol{y} - \mathbf{X}\boldsymbol{\beta}\|^2$ is not unique. When the set $\operatorname{argmin}_{\boldsymbol{\beta}} \|\boldsymbol{y} - \mathbf{X}\boldsymbol{\beta}\|^2$ is a singleton, that is, the minimizer is unique, then $\widehat{\boldsymbol{\beta}} = \operatorname{argmin}_{\boldsymbol{\beta}} \|\boldsymbol{y} - \mathbf{X}\boldsymbol{\beta}\|^2$ is called the **ordinary least-squares** estimate of $\boldsymbol{\beta}$.

As is illustrated in Figure 2.20, to find $\widehat{\boldsymbol{\beta}}$, we choose $\mathbf{X}\widehat{\boldsymbol{\beta}}$ to be equal to the orthogonal projection of $\boldsymbol{y}$ onto the linear space spanned by the columns of the matrix $\mathbf{X}$; that is, $\mathbf{X}\widehat{\boldsymbol{\beta}} = \mathbf{P}\boldsymbol{y}$, where $\mathbf{P}$ is the **projection matrix**.

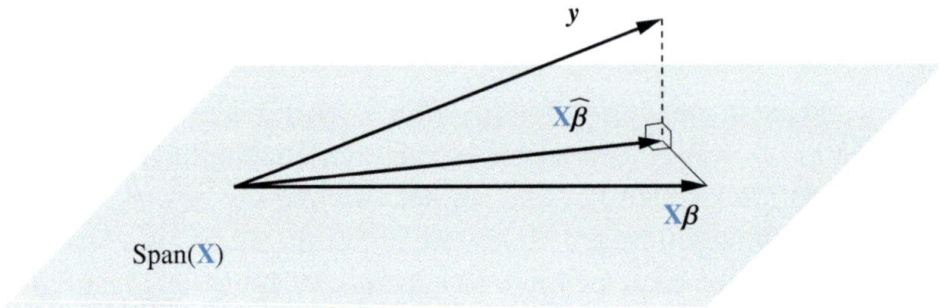

Figure 2.20: $\mathbf{X}\widehat{\boldsymbol{\beta}}$ is the orthogonal projection of $\boldsymbol{y}$ onto the linear space spanned by the columns of the matrix $\mathbf{X}$.

☞ 521 According to Theorem A.17, the projection matrix is given by

$$\mathbf{P} = \mathbf{X}\,\mathbf{X}^{+}, \tag{2.21}$$

☞ 516 where the $p \times n$ matrix $\mathbf{X}^{+}$ is the **pseudo-inverse** of $\mathbf{X}$. If $\mathbf{X}$ happens to be of *full column*
☞ 514 *rank* (so that none of the columns can be expressed as a linear combination of the other columns), then $\mathbf{X}^{+} = (\mathbf{X}^{\top}\mathbf{X})^{-1}\mathbf{X}^{\top}$. In any case, from $\mathbf{X}\widehat{\boldsymbol{\beta}} = \mathbf{P}\boldsymbol{y}$ and $\mathbf{P}\mathbf{X} = \mathbf{X}$, we can see that all minimizers of (2.18) satisfy the **normal equations**:

$$\mathbf{X}^{\top}\mathbf{X}\boldsymbol{\beta} = \mathbf{X}^{\top}\mathbf{P}\boldsymbol{y} = (\mathbf{P}\mathbf{X})^{\top}\boldsymbol{y} = \mathbf{X}^{\top}\boldsymbol{y}. \tag{2.22}$$

In Exercise 8 it is shown that set of all possible minimizers of $\|\boldsymbol{y} - \mathbf{X}\boldsymbol{\beta}\|^2$, or equivalently the set of all solutions to (2.22), can be written explicitly as:

$$\mathbf{X}^{+}\boldsymbol{y} + (\mathbf{I}_p - \mathbf{X}^{+}\mathbf{X})\,\boldsymbol{u}, \quad \boldsymbol{u} \in \mathbb{R}^p,$$

where $\mathbf{I}_p$ is the p-dimensional identity matrix. Since $(\mathbf{I}_p - \mathbf{X}^{+}\mathbf{X})\,\boldsymbol{u}$ and $\mathbf{X}^{+}\boldsymbol{y}$ are orthogonal to each other, we have by Pythagoras:

$$\|\mathbf{X}^{+}\boldsymbol{y} + (\mathbf{I}_p - \mathbf{X}^{+}\mathbf{X})\,\boldsymbol{u}\|^2 = \|\mathbf{X}^{+}\boldsymbol{y}\|^2 + \|(\mathbf{I}_p - \mathbf{X}^{+}\mathbf{X})\,\boldsymbol{u}\|^2 \geq \|\mathbf{X}^{+}\boldsymbol{y}\|^2.$$

In other words, the unique **minimum-norm least-squares** minimizer of (2.18) is:

$$\widehat{\boldsymbol{\beta}} = \mathbf{X}^{+}\boldsymbol{y}. \tag{2.23}$$

Figure 2.24 shows the trained learners for various values of p:

$$h_{\tau}^{\mathcal{H}_p}(u) = g_{\tau}^{\mathcal{G}_p}(\boldsymbol{x}) = \boldsymbol{x}^{\top}\widehat{\boldsymbol{\beta}}$$

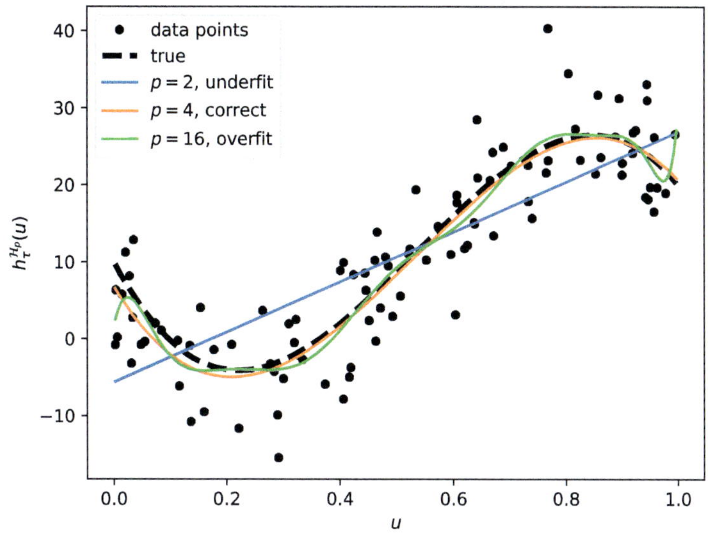

Figure 2.24: Training data with fitted curves for $p = 2, 4$, and 16. The true cubic polynomial curve for $p = 4$ is also plotted (dashed line).

We see that for $p = 16$ the fitted curve lies closer to the data points, but is further away from the dashed true polynomial curve, indicating that we overfit. The choice $p = 4$ (the true cubic polynomial) is better than $p = 16$, or indeed $p = 2$ (straight line).

Each function class $\mathcal{G}_p$ gives a different learner $g_\tau^{\mathcal{G}_p}$, $p = 1, 2, \ldots$. To assess which is better, we should not simply take the one that gives the smallest training loss. We can, at least theoretically, always get a *zero* training loss by taking $p = n$, because for any set of n points there exists a polynomial of degree $n - 1$ that interpolates all points!

Instead, we assess the predictive performance of the learners using the test loss (2.10), computed from a test data set. If we collect all n' test feature vectors in a matrix $\mathbf{X}'$ and the corresponding test responses in a vector $\boldsymbol{y}'$, then, similar to (2.18), the test loss can be written compactly as

$$\ell_{\tau'}(g_\tau^{\mathcal{G}_p}) = \frac{1}{n'} \|\boldsymbol{y}' - \mathbf{X}'\widehat{\boldsymbol{\beta}}\|^2,$$

where $\widehat{\boldsymbol{\beta}}$ is given by (2.23), using the training data.

Figure 2.25 shows a plot of the test loss against the number of parameters in the vector $\boldsymbol{\beta}$; that is, p. The graph has a characteristic "bath-tub" shape and is at its lowest for $p = 4$, correctly identifying the polynomial order 3 for the true model. Note that the test loss, as an estimate for the generalization risk (2.10), becomes numerically unreliable after $p = 16$. The reader may check that the graph for the training loss exhibits a similar numerical instability for large p, and in fact fails to numerically decrease to 0 for large p, contrary to what it should do in theory. The numerical problems arise from the fact that for large p the columns of the (Vandermonde) matrix $\mathbf{X}$ are of vastly different magnitudes and so floating point errors quickly become very large.

Finally, observe that the lower bound for the test loss is here around 21, which corresponds to an estimate of the minimal (squared-error) risk $\ell^* = 25$.

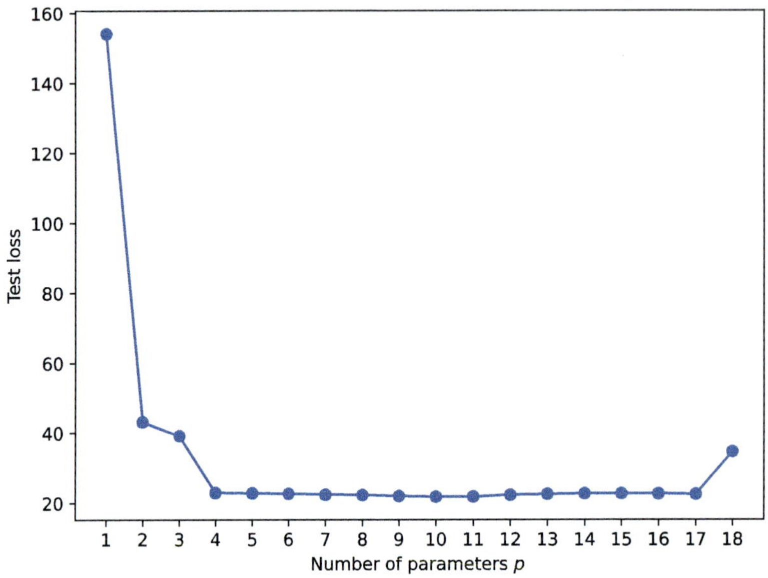

Figure 2.25: Test loss as function of the number of parameters p of the model.

This script shows how the training data were generated and plotted in Python:

polyreg1.py

```
import numpy as np
from numpy.random import rand , randn
from numpy.linalg import norm , solve
import matplotlib.pyplot as plt
def generate_data(beta , sig, n):
   u = np.random.rand(n, 1)
   y = (u ** np.arange(0, 4)) @ beta + sig * np.random.randn(n, 1)
   return u, y

np.random.seed(12)
beta = np.array([[10, -140, 400, -250]]).T
n = 100
sig = 5
u, y = generate_data(beta , sig, n)
xx = np.arange(np.min(u), np.max(u)+5e-3, 5e-3)
yy = np.polyval(np.flip(beta), xx)
plt.plot(u, y, '.', markersize=8)
plt.plot(xx, yy, '--',linewidth=3)
plt.xlabel(r'$u$')
plt.ylabel(r'$h^*(u)$')
plt.legend(['data points','true'])
plt.show()
```

The following code, which imports the code above, fits polynomial models with $p = 1, \ldots, K = 18$ parameters to the training data and plots a selection of fitted curves, as shown in Figure 2.24.

polyreg2.py

```
from polyreg1 import *
max_p = 18
p_range = np.arange(1, max_p + 1, 1)
X = np.ones((n, 1))
betahat, trainloss = {}, {}

for p in p_range:  # p is the number of parameters
    if p > 1:
        X = np.hstack((X, u**(p-1)))  # add column to matrix

    betahat[p] = solve(X.T @ X, X.T @ y)
    trainloss[p] = (norm(y - X @ betahat[p])**2/n)

p = [2, 4, 16]  # select three curves

#replot the points and true line and store in the list "plots"
plots = [plt.plot(u, y, 'k.', markersize=8)[0],
         plt.plot(xx, yy, 'k--',linewidth=3)[0]]
# add the three curves
for i in p:
    yy = np.polyval(np.flip(betahat[i]), xx)
    plots.append(plt.plot(xx, yy)[0])

plt.xlabel(r'$u$')
plt.ylabel(r'$h^{\mathcal{H}_p}_{\tau}(u)$')
plt.legend(plots,('data points', 'true','$p=2$, underfit',
                  '$p=4$, correct','$p=16$, overfit'))
plt.savefig('polyfitpy.pdf',format='pdf')
plt.show()
```

The last code snippet which imports the previous code, generates the test data and plots the graph of the test loss, as shown in Figure 2.25.

polyreg3.py

```
from polyreg2 import *
u_test, y_test = generate_data(beta, sig, n) # generate test data

MSE = []
X_test = np.ones((n, 1))

for p in p_range:
    if p > 1:
        X_test = np.hstack((X_test, u_test**(p-1)))

    y_hat = X_test @ betahat[p]  # predictions
    MSE.append(np.sum((y_test - y_hat)**2/n))

plt.plot(p_range, MSE, 'b', p_range, MSE, 'bo')
plt.xticks(ticks=p_range)
plt.xlabel('Number of parameters $p$')
plt.ylabel('Test loss')
```

■

Example 2.26 (Minimum-Norm Least-Squares via Coordinate Descent) If there is no unique minimizer of (2.18), it is common practice to opt for the unique minimum-norm least-squares estimator (2.23). Some software packages do not have a routine that computes the pseudo-inverse $\mathbf{X}^+$, and even if available, the routine may be too slow — most algorithms for computing the pseudo-inverse matrix are iterative and require storage of the orthogonal bases for the column and row space of $\mathbf{X}$.

☞ 606 Instead, the coordinate descent Algorithm C.68 can be easily adapted to compute $\mathbf{X}^+\boldsymbol{y}$ without the explicit construction of $\mathbf{X}^+$. First, note that if we apply Algorithm C.68 to the minimization of the quadratic function

$$f(\boldsymbol{\beta}) := \|\boldsymbol{y} - \mathbf{X}\boldsymbol{\beta}\|^2 - \|\boldsymbol{y}\|^2 = \boldsymbol{\beta}^\top [\mathbf{X}^\top \mathbf{X}]\boldsymbol{\beta} - 2\boldsymbol{\beta}^\top [\mathbf{X}^\top \boldsymbol{y}],$$

we will not necessarily obtain the minimum-norm minimizer (2.23), but some element of the set $\operatorname{argmin}_{\boldsymbol{\beta}} f(\boldsymbol{\beta})$. Exercise 8 shows that the minimum-norm minimizer $\mathbf{X}^+\boldsymbol{y}$ lies in the row space of $\mathbf{X}$, that is, $\mathbf{X}^+\boldsymbol{y} = \mathbf{X}^\top\boldsymbol{\alpha}$ for some vector $\boldsymbol{\alpha} \in \mathbb{R}^n$. Hence, we can always compute a minimizer of $f(\mathbf{X}^\top\boldsymbol{\alpha})$, that is,

$$\widehat{\boldsymbol{\alpha}} \in \operatorname*{argmin}_{\boldsymbol{\alpha} \in \mathbb{R}^n} \boldsymbol{\alpha}^\top [\mathbf{X}\mathbf{X}^\top \mathbf{X}\mathbf{X}^\top]\boldsymbol{\alpha} - 2\boldsymbol{\alpha}^\top [\mathbf{X}\mathbf{X}^\top \boldsymbol{y}],$$

using the coordinate descent Algorithm C.68, and then deliver the desired minimum-norm least-squares estimate $\mathbf{X}^\top\widehat{\boldsymbol{\alpha}} = \mathbf{X}^+\boldsymbol{y}$. This recipe is efficient when n is not much larger than p. If n is much larger than p, then it is more efficient to compute a vector $\boldsymbol{u} \in \mathbb{R}^n$ such that $\mathbf{X}^+\boldsymbol{y} = \mathbf{X}^\top\mathbf{X}\boldsymbol{u}$; see Exercises 8 and 9. ■

2.4 Tradeoffs in Statistical Learning

The art of machine learning in the supervised case is to make the generalization risk (2.6) or expected generalization risk (2.8) as small as possible, while using as few computational resources as possible. In pursuing this goal, a suitable class $\mathcal{G}$ of prediction functions has to be chosen. This choice is driven by various factors, such as

- the complexity of the class (e.g., is it rich enough to adequately approximate, or even contain, the optimal prediction function g^*?),
- the ease of training the learner via the optimization program (2.5),
- how accurately the training loss (2.4) estimates the risk (2.1) within class $\mathcal{G}$,
- the feature types (categorical, continuous, etc.).

As a result, the choice of a suitable function class $\mathcal{G}$ usually involves a tradeoff between conflicting factors. For example, a learner from a simple class $\mathcal{G}$ can be trained very quickly, but may not approximate g^* very well, whereas a learner from a rich class $\mathcal{G}$ that contains g^* may require a lot of computing resources to train.

To better understand the relation between model complexity, computational simplicity, and estimation accuracy, it is useful to decompose the generalization risk into several parts, so that the tradeoffs between these parts can be studied. We will consider two such decompositions: the approximation–estimation tradeoff and the bias–variance tradeoff.

We can decompose the generalization risk (2.6) into the following three components:

$$\ell(g_\tau^{\mathcal{G}}) = \underbrace{\ell^*}_{\text{irreducible risk}} + \underbrace{\ell(g^{\mathcal{G}}) - \ell^*}_{\text{approximation error}} + \underbrace{\ell(g_\tau^{\mathcal{G}}) - \ell(g^{\mathcal{G}})}_{\text{statistical error}}, \tag{2.27}$$

where $\ell^* := \ell(g^*)$ is the **irreducible risk** and $g^{\mathcal{G}} := \operatorname{argmin}_{g \in \mathcal{G}} \ell(g)$ is the best learner within class $\mathcal{G}$. No learner can predict a new response with a smaller risk than ℓ^*.

The second component is the **approximation error**; this measures the difference between the irreducible risk and the best possible risk that can be obtained by selecting the best prediction function in the selected class of functions $\mathcal{G}$. Determining a suitable class $\mathcal{G}$ and minimizing $\ell(g)$ over this class is purely a problem of numerical and functional analysis, as the training data τ are not present. For a fixed $\mathcal{G}$ that does not contain the optimal g^*, the approximation error cannot be made arbitrarily small and may be the dominant component in the generalization risk. The only way to reduce the approximation error is by expanding the class $\mathcal{G}$ to include a larger set of possible functions.

The third component is the **statistical (estimation) error**. It depends on the training set τ and, in particular, on how well the learner $g_\tau^{\mathcal{G}}$ estimates the best possible prediction function, $g^{\mathcal{G}}$, within class $\mathcal{G}$. For any sensible estimator this error should decay to zero (in probability or expectation) as the training size tends to infinity. ☞ 645

The **approximation–estimation tradeoff** pits two competing demands against each other. The first is that the class $\mathcal{G}$ has to be simple enough so that the statistical error is not too large. The second is that the class $\mathcal{G}$ has to be rich enough to ensure a small approximation error. Thus, there is a tradeoff between the approximation and estimation errors.

For the special case of the squared-error loss, the generalization risk is equal to $\ell(g_\tau^{\mathcal{G}}) = \mathbb{E}(Y - g_\tau^{\mathcal{G}}(\boldsymbol{X}))^2$; that is, the expected squared error[1] between the predicted value $g_\tau^{\mathcal{G}}(\boldsymbol{X})$ and the response Y. Recall that in this case the optimal prediction function is given by $g^*(\boldsymbol{x}) = \mathbb{E}[Y \mid \boldsymbol{X} = \boldsymbol{x}]$. The decomposition (2.27) can now be interpreted as follows:

1. The first component, $\ell^* = \mathbb{E}(Y - g^*(\boldsymbol{X}))^2$, is the *irreducible error*, as no prediction function will yield a smaller expected squared error.

2. The second component, the approximation error $\ell(g^{\mathcal{G}}) - \ell(g^*)$, is equal to $\mathbb{E}(g^{\mathcal{G}}(\boldsymbol{X}) - g^*(\boldsymbol{X}))^2$. We leave the proof (which is similar to that of Theorem 2.2) as an exercise; see Exercise 2. Thus, the approximation error (defined as a risk difference) can here be interpreted as the expected squared error between the optimal predicted value and the optimal predicted value within the class $\mathcal{G}$.

3. For the third component, the statistical error, $\ell(g_\tau^{\mathcal{G}}) - \ell(g^{\mathcal{G}})$ there is no direct interpretation as an expected squared error *unless* $\mathcal{G}$ is the class of *linear* functions; that is, $g(\boldsymbol{x}) = \boldsymbol{x}^\top \boldsymbol{\beta}$ for some vector $\boldsymbol{\beta}$. In this case we can write (see Exercise 3) the statistical error as $\ell(g_\tau^{\mathcal{G}}) - \ell(g^{\mathcal{G}}) = \mathbb{E}(g_\tau^{\mathcal{G}}(\boldsymbol{X}) - g^{\mathcal{G}}(\boldsymbol{X}))^2$.

[1] Colloquially called *mean squared error*.

Thus, when using a squared-error loss, the generalization risk for a linear class $\mathcal{G}$ can be decomposed as:

$$\ell(g_\tau^{\mathcal{G}}) = \mathbb{E}(g_\tau^{\mathcal{G}}(\boldsymbol{X}) - Y)^2 = \ell^* + \underbrace{\mathbb{E}(g^{\mathcal{G}}(\boldsymbol{X}) - g^*(\boldsymbol{X}))^2}_{\text{approximation error}} + \underbrace{\mathbb{E}(g_\tau^{\mathcal{G}}(\boldsymbol{X}) - g^{\mathcal{G}}(\boldsymbol{X}))^2}_{\text{statistical error}}. \tag{2.28}$$

Note that in this decomposition the statistical error is the only term that depends on the training set.

■ **Example 2.29 (Polynomial Regression (cont.))** We continue Example 2.13. Here $\mathcal{G} = \mathcal{G}_p$ is the class of linear functions of $\boldsymbol{x} = [1, u, u^2, \ldots, u^{p-1}]^\top$, and $g^*(\boldsymbol{x}) = \boldsymbol{x}^\top \boldsymbol{\beta}^*$. Conditional on $\boldsymbol{X} = \boldsymbol{x}$ we have that $Y = g^*(\boldsymbol{x}) + \varepsilon(\boldsymbol{x})$, with $\varepsilon(\boldsymbol{x}) \sim \mathcal{N}(0, \ell^*)$, where $\ell^* = \mathbb{E}(Y - g^*(\boldsymbol{X}))^2 = 25$ is the irreducible error. We wish to understand how the approximation and statistical errors behave as we change the complexity parameter p.

First, we consider the approximation error. Any function $g \in \mathcal{G}_p$ can be written as

$$g(\boldsymbol{x}) = h(u) = \beta_1 + \beta_2 u + \cdots + \beta_p u^{p-1} = [1, u, \ldots, u^{p-1}]\, \boldsymbol{\beta},$$

and so $g(\boldsymbol{X})$ is distributed as $[1, U, \ldots, U^{p-1}]\boldsymbol{\beta}$, where $U \sim \mathcal{U}(0, 1)$. Similarly, $g^*(\boldsymbol{X})$ is distributed as $[1, U, U^2, U^3]\boldsymbol{\beta}^*$. It follows that an expression for the approximation error is:

$$\int_0^1 \left([1, u, \ldots, u^{p-1}]\, \boldsymbol{\beta} - [1, u, u^2, u^3]\, \boldsymbol{\beta}^*\right)^2 \mathrm{d}u.$$

To minimize this error, we set the gradient with respect to $\boldsymbol{\beta}$ to zero and obtain the p
☞ 581 linear equations

$$\int_0^1 \left([1, u, \ldots, u^{p-1}]\, \boldsymbol{\beta} - [1, u, u^2, u^3]\, \boldsymbol{\beta}^*\right) \mathrm{d}u = 0,$$
$$\int_0^1 \left([1, u, \ldots, u^{p-1}]\, \boldsymbol{\beta} - [1, u, u^2, u^3]\, \boldsymbol{\beta}^*\right) u\, \mathrm{d}u = 0,$$
$$\vdots$$
$$\int_0^1 \left([1, u, \ldots, u^{p-1}]\, \boldsymbol{\beta} - [1, u, u^2, u^3]\, \boldsymbol{\beta}^*\right) u^{p-1} \mathrm{d}u = 0.$$

Let

$$\mathbf{H}_p = \int_0^1 [1, u, \ldots, u^{p-1}]^\top [1, u, \ldots, u^{p-1}]\, \mathrm{d}u$$

be the $p \times p$ **Hilbert matrix**, which has (i, j)-th entry given by $\int_0^1 u^{i+j-2}\, \mathrm{d}u = 1/(i+j-1)$. Then, the above system of linear equations can be written as $\mathbf{H}_p \boldsymbol{\beta} = \widetilde{\mathbf{H}} \boldsymbol{\beta}^*$, where $\widetilde{\mathbf{H}}$ is the $p \times 4$ upper left sub-block of $\mathbf{H}_{\widetilde{p}}$ and $\widetilde{p} = \max\{p, 4\}$. The solution, which we denote by $\boldsymbol{\beta}_p$, is:

$$\boldsymbol{\beta}_p = \begin{cases} \frac{65}{6}, & p = 1, \\ [-\frac{20}{3}, 35]^\top, & p = 2, \\ [-\frac{5}{2}, 10, 25]^\top, & p = 3, \\ [10, -140, 400, -250, 0, \ldots, 0]^\top, & p \geq 4. \end{cases} \tag{2.30}$$

Hence, the approximation error $\mathbb{E}\left(g^{\mathcal{G}_p}(\boldsymbol{X}) - g^*(\boldsymbol{X})\right)^2$ is given by

$$\int_0^1 \left([1, u, \ldots, u^{p-1}]\, \boldsymbol{\beta}_p - [1, u, u^2, u^3]\, \boldsymbol{\beta}^*\right)^2 \mathrm{d}u = \begin{cases} \frac{32225}{252} \approx 127.9, & p = 1, \\ \frac{1625}{63} \approx 25.8, & p = 2, \\ \frac{625}{28} \approx 22.3, & p = 3, \\ 0, & p \geq 4. \end{cases} \tag{2.31}$$

Notice how the approximation error becomes smaller as p increases. In this particular example the approximation error is in fact zero for $p \geq 4$. In general, as the class of approximating functions $\mathcal{G}$ becomes more complex, the approximation error goes down.

Next, we illustrate the typical behavior of the statistical error. Since $g_\tau(\boldsymbol{x}) = \boldsymbol{x}^\top \widehat{\boldsymbol{\beta}}$, the statistical error can be written as

$$\int_0^1 \left([1, \ldots, u^{p-1}](\widehat{\boldsymbol{\beta}} - \boldsymbol{\beta}_p)\right)^2 \mathrm{d}u = (\widehat{\boldsymbol{\beta}} - \boldsymbol{\beta}_p)^\top \mathbf{H}_p (\widehat{\boldsymbol{\beta}} - \boldsymbol{\beta}_p). \tag{2.32}$$

Figure 2.33 illustrates the decomposition (2.28) of the generalization risk for the *same* training set that was used to compute the test loss in Figure 2.25. Recall that test loss gives an estimate of the generalization risk, using independent test data. Comparing the two figures, we see that in this case the two match closely. The global minimum of the statistical error is approximately 0.28, with minimizer $p = 4$. Since the approximation error is monotonically decreasing to zero, $p = 4$ is also the global minimizer of the generalization risk.

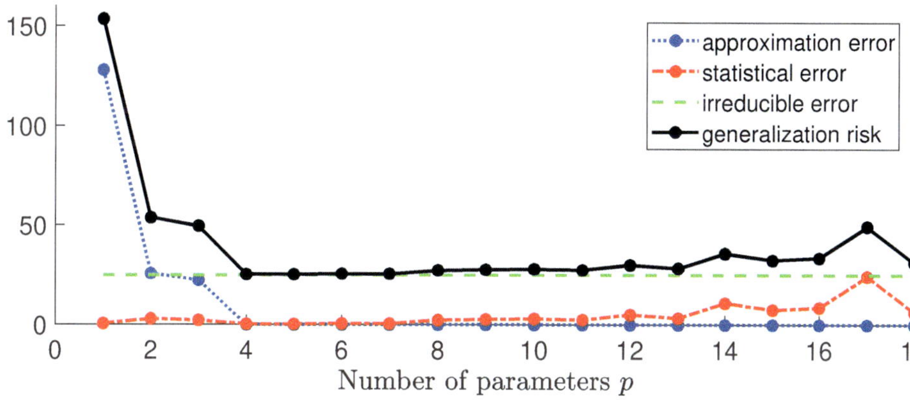

Figure 2.33: The generalization risk for a particular training set is the sum of the irreducible error, the approximation error, and the statistical error. The approximation error decreases to zero as p increases, whereas the statistical error has a tendency to increase after $p = 4$.

Note that the statistical error depends on the estimate $\widehat{\boldsymbol{\beta}}$, which in its turn depends on the training set τ. We can obtain a better understanding of the statistical error by considering its *expected* behavior; that is, averaged over many training sets. This is explored in Exercise 15. ■

Using again a squared-error loss, a second decomposition (for general $\mathcal{G}$) starts from

$$\ell(g_\tau^{\mathcal{G}}) = \ell^* + \ell(g_\tau^{\mathcal{G}}) - \ell(g^*),$$

where the statistical error and approximation error are combined. Using similar reasoning as in the proof of Theorem 2.2, we have

$$\ell(g_\tau^{\mathcal{G}}) = \mathbb{E}(g_\tau^{\mathcal{G}}(\boldsymbol{X}) - Y)^2 = \ell^* + \mathbb{E}\left(g_\tau^{\mathcal{G}}(\boldsymbol{X}) - g^*(\boldsymbol{X})\right)^2 = \ell^* + \mathbb{E}D^2(\boldsymbol{X}, \tau),$$

where $D(\boldsymbol{x}, \tau) := g_\tau^{\mathcal{G}}(\boldsymbol{x}) - g^*(\boldsymbol{x})$. Now consider the random variable $D(\boldsymbol{x}, \mathcal{T})$ for a random training set $\mathcal{T}$. The expectation of its square is:

$$\begin{aligned}\mathbb{E}\left(g_{\mathcal{T}}^{\mathcal{G}}(\boldsymbol{x}) - g^*(\boldsymbol{x})\right)^2 &= \mathbb{E}D^2(\boldsymbol{x}, \mathcal{T}) = (\mathbb{E}D(\boldsymbol{x}, \mathcal{T}))^2 + \mathbb{V}\mathrm{ar}\, D(\boldsymbol{x}, \mathcal{T}) \\ &= \underbrace{(\mathbb{E}g_{\mathcal{T}}^{\mathcal{G}}(\boldsymbol{x}) - g^*(\boldsymbol{x}))^2}_{\text{pointwise squared bias}} + \underbrace{\mathbb{V}\mathrm{ar}\, g_{\mathcal{T}}^{\mathcal{G}}(\boldsymbol{x})}_{\text{pointwise variance}}\,.\end{aligned} \tag{2.34}$$

If we view the learner $g_{\mathcal{T}}^{\mathcal{G}}(\boldsymbol{x})$ as a function of a random training set, then the **pointwise squared bias** term is a measure for how close $g_{\mathcal{T}}^{\mathcal{G}}(\boldsymbol{x})$ is on average to the true $g^*(\boldsymbol{x})$, whereas the **pointwise variance** term measures the deviation of $g_{\mathcal{T}}^{\mathcal{G}}(\boldsymbol{x})$ from its expected value $\mathbb{E}g_{\mathcal{T}}^{\mathcal{G}}(\boldsymbol{x})$. The squared bias can be reduced by making the class of functions $\mathcal{G}$ more complex. However, decreasing the bias by increasing the complexity often leads to an increase in the variance term. We are thus seeking learners that provide an optimal balance between the bias and variance, as expressed via a minimal generalization risk. This is called the **bias–variance tradeoff**.

Note that the *expected* generalization risk (2.8) can be written as $\ell^* + \mathbb{E}D^2(\boldsymbol{X}, \mathcal{T})$, where $\boldsymbol{X}$ and $\mathcal{T}$ are independent. It therefore decomposes as

$$\mathbb{E}\,\ell(g_{\mathcal{T}}^{\mathcal{G}}) = \ell^* + \underbrace{\mathbb{E}\left(\mathbb{E}[g_{\mathcal{T}}^{\mathcal{G}}(\boldsymbol{X}) \mid \boldsymbol{X}] - g^*(\boldsymbol{X})\right)^2}_{\text{expected squared bias}} + \underbrace{\mathbb{E}[\mathbb{V}\mathrm{ar}[g_{\mathcal{T}}^{\mathcal{G}}(\boldsymbol{X}) \mid \boldsymbol{X}]]}_{\text{expected variance}}\,. \tag{2.35}$$

2.5 Estimating Risk

The most straightforward way to quantify the generalization risk (2.6) is to estimate it via the test loss (2.10). However, the generalization risk depends inherently on the training set, and so different training sets may yield significantly different estimates. Moreover, when there is a limited amount of data available, reserving a substantial proportion of the data for testing rather than training may be uneconomical. In this section we consider different methods for estimating risk measures which aim to circumvent these difficulties.

2.5.1 In-Sample Risk

We mentioned that, due to the phenomenon of overfitting, the training loss of the learner, $\ell_\tau(g_\tau)$ (for simplicity, here we omit $\mathcal{G}$ from $g_\tau^{\mathcal{G}}$), is not a good estimate of the generalization risk $\ell(g_\tau)$ of the learner. One reason for this is that we use the same data for both training the model and assessing its risk. How should we then estimate the generalization risk or expected generalization risk?

To simplify the analysis, suppose that we wish to estimate the average accuracy of the predictions of the learner g_τ at the n feature vectors $\boldsymbol{x}_1, \ldots, \boldsymbol{x}_n$ (these are part of the training set τ). In other words, we wish to estimate the **in-sample risk** of the learner g_τ:

$$\ell_{\text{in}}(g_\tau) = \frac{1}{n} \sum_{i=1}^{n} \mathbb{E}\,\text{Loss}(Y_i', g_\tau(\boldsymbol{x}_i)), \tag{2.36}$$

where each response Y_i' is drawn from $f(y \,|\, \boldsymbol{x}_i)$, independently. Even in this simplified setting, the training loss of the learner will be a poor estimate of the in-sample risk. Instead, the proper way to assess the prediction accuracy of the learner at the feature vectors $\boldsymbol{x}_1, \ldots, \boldsymbol{x}_n$, is to draw new response values $Y_i' \sim f(y \,|\, \boldsymbol{x}_i)$, $i = 1, \ldots, n$, that are independent from the responses $y_1, \ldots, y_n$ in the training data, and then estimate the in-sample risk of g_τ via

$$\frac{1}{n} \sum_{i=1}^{n} \text{Loss}(Y_i', g_\tau(\boldsymbol{x}_i)).$$

For a fixed training set τ, we can compare the training loss of the learner with the in-sample risk. Their difference,

$$\text{op}_\tau = \ell_{\text{in}}(g_\tau) - \ell_\tau(g_\tau),$$

is called the **optimism** (of the training loss), because it measures how much the training loss underestimates (is optimistic about) the unknown in-sample risk. Mathematically, it is simpler to work with the **expected optimism**:

$$\mathbb{E}[\text{op}_{\mathcal{T}} \,|\, \boldsymbol{X}_1 = \boldsymbol{x}_1, \ldots, \boldsymbol{X}_n = \boldsymbol{x}_n] =: \mathbb{E}_{\mathbf{X}}\, \text{op}_{\mathcal{T}},$$

where the expectation is taken over a random training set $\mathcal{T}$, conditional on $\boldsymbol{X}_i = \boldsymbol{x}_i$, $i = 1, \ldots, n$. For ease of notation, we have abbreviated the expected optimism to $\mathbb{E}_{\mathbf{X}}\, \text{op}_{\mathcal{T}}$, where $\mathbb{E}_{\mathbf{X}}$ denotes the expectation operator conditional on $\boldsymbol{X}_i = \boldsymbol{x}_i, i = 1, \ldots, n$. As in Example 2.13, the feature vectors are stored as the rows of an $n \times p$ matrix $\mathbf{X}$. It turns out that the expected optimism for various loss functions can be expressed in terms of the (conditional) covariance between the observed and predicted response.

Theorem 2.37 (Expected Optimism): For the squared-error loss and 0–1 loss with 0–1 response, the expected optimism is

$$\mathbb{E}_{\mathbf{X}}\, \text{op}_{\mathcal{T}} = \frac{2}{n} \sum_{i=1}^{n} \mathbb{C}\text{ov}_{\mathbf{X}}(g_{\mathcal{T}}(\boldsymbol{x}_i), Y_i). \tag{2.38}$$

Proof. In what follows, all expectations are taken conditional on $\boldsymbol{X}_1 = \boldsymbol{x}_1, \ldots, \boldsymbol{X}_n = \boldsymbol{x}_n$. Let Y_i be the response for $\boldsymbol{x}_i$ and let $\widehat{Y}_i = g_{\mathcal{T}}(\boldsymbol{x}_i)$ be the predicted value. Note that the latter depends on $Y_1, \ldots, Y_n$. Also, let Y_i' be an independent copy of Y_i for the *same* $\boldsymbol{x}_i$, as in (2.36). In particular, Y_i' has the same distribution as Y_i and is statistically independent of all $\{Y_j\}$, including Y_i, and therefore is also independent of $\widehat{Y}_i$. We have

$$\begin{aligned}
\mathbb{E}_{\mathbf{X}}\, \mathrm{op}_{\mathcal{T}} &= \frac{1}{n}\sum_{i=1}^{n} \mathbb{E}_{\mathbf{X}}\left[(Y_i' - \widehat{Y}_i)^2 - (Y_i - \widehat{Y}_i)^2\right] \\
&= \frac{2}{n}\sum_{i=1}^{n} \mathbb{E}_{\mathbf{X}}\left[(Y_i - Y_i')\widehat{Y}_i\right] \\
&= \frac{2}{n}\sum_{i=1}^{n} \left(\mathbb{E}_{\mathbf{X}}[Y_i\widehat{Y}_i] - \mathbb{E}_{\mathbf{X}}Y_i\, \mathbb{E}_{\mathbf{X}}\widehat{Y}_i\right) \\
&= \frac{2}{n}\sum_{i=1}^{n} \mathbb{C}\mathrm{ov}_{\mathbf{X}}(\widehat{Y}_i, Y_i).
\end{aligned}$$

The proof for the 0–1 loss with 0–1 response is left as Exercise 4. □

In summary, the expected optimism indicates how much, on average, the training loss deviates from the expected in-sample risk. Since the covariance of independent random variables is zero, the expected optimism is zero if the learner $g_{\mathcal{T}}$ is statistically independent from the responses $Y_1, \ldots, Y_n$.

Example 2.39 (Polynomial Regression (cont.)) We continue Example 2.29, where the components of the response vector $\boldsymbol{Y} = [Y_1, \ldots, Y_n]^\top$ are independent and normally distributed with variance $\ell^* = 25$ (the irreducible error) and expectations:

$$\mathbb{E}_{\mathbf{X}} Y_i = g^*(\boldsymbol{x}_i) = \boldsymbol{x}_i^\top \boldsymbol{\beta}^*, \quad i = 1, \ldots, n.$$

Using the formula (2.23) for the least-squares estimator $\widehat{\boldsymbol{\beta}}$, the expected optimism (2.38) is

$$\begin{aligned}
\frac{2}{n}\sum_{i=1}^{n} \mathbb{C}\mathrm{ov}_{\mathbf{X}}\left(\boldsymbol{x}_i^\top \widehat{\boldsymbol{\beta}}, Y_i\right) &= \frac{2}{n}\mathrm{tr}\left(\mathbb{C}\mathrm{ov}_{\mathbf{X}}\left(\mathbf{X}\widehat{\boldsymbol{\beta}}, \boldsymbol{Y}\right)\right) \\
&= \frac{2}{n}\mathrm{tr}\left(\mathbb{C}\mathrm{ov}_{\mathbf{X}}\left(\mathbf{X}\mathbf{X}^+\boldsymbol{Y}, \boldsymbol{Y}\right)\right) \\
&= \frac{2\mathrm{tr}\left(\mathbf{X}\mathbf{X}^+ \mathbb{C}\mathrm{ov}_{\mathbf{X}}\left(\boldsymbol{Y}, \boldsymbol{Y}\right)\right)}{n} \\
&= \frac{2\ell^* \mathrm{tr}\left(\mathbf{X}\mathbf{X}^+\right)}{n} \\
&= \frac{2\ell^* p}{n}.
\end{aligned}$$

☞ 515 In the last equation we used the cyclic property of the trace (Theorem A.7):

$$\mathrm{tr}(\mathbf{X}\mathbf{X}^+) = \mathrm{tr}(\mathbf{X}^+\mathbf{X}) = \mathrm{tr}(\mathbf{I}_p),$$

assuming that $\mathrm{rank}(\mathbf{X}) = p$. Therefore, an estimate for the in-sample risk (2.36) is:

$$\widehat{\ell}_{\mathrm{in}}(g_{\tau}) = \ell_{\tau}(g_{\tau}) + 2\ell^* p/n, \tag{2.40}$$

where we have assumed that the irreducible risk ℓ^* is known. Figure 2.41 shows that this estimate is very close to the test loss from Figure 2.25. Hence, instead of computing the test loss to assess the best model complexity p, we could simply have minimized the training loss plus the correction term $2\ell^* p/n$. In practice, ℓ^* also has to be estimated somehow.

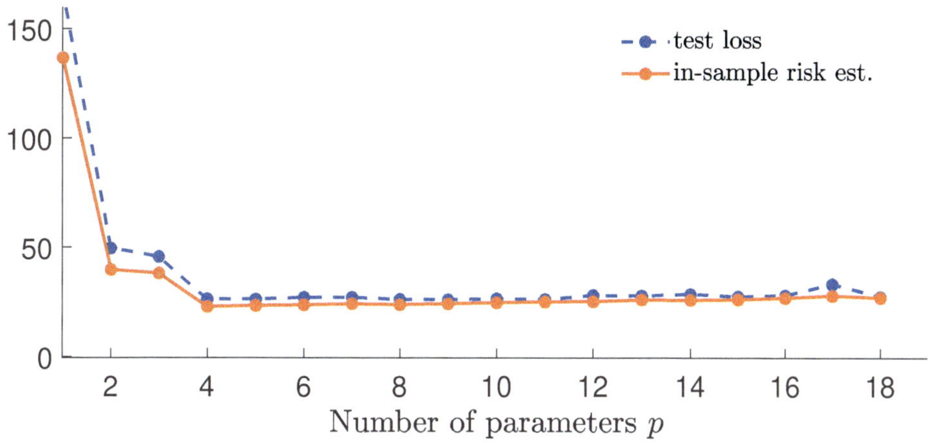

Figure 2.41: In-sample risk estimate $\widehat{\ell_{\text{in}}}(g_{\tau})$ as a function of the number of parameters p of the model. The test loss is superimposed as a blue dashed curve.

■

2.5.2 Cross-Validation

In general, for complex function classes $\mathcal{G}$, it is very difficult to derive simple formulas of the approximation and statistical errors, let alone for the generalization risk or expected generalization risk. As we saw, when there is an abundance of data, the easiest way to assess the generalization risk for a given training set τ is to obtain a test set τ' and evaluate the test loss (2.10). When a sufficiently large test set is not available but computational resources are cheap, one can instead gain direct knowledge of the expected generalization risk via a computationally intensive method called **cross-validation**. ☞ 24

The idea is to make multiple identical copies of the data set, and to partition each copy into different training and test sets, as illustrated in Figure 2.42. Here, there are four copies of the data set (consisting of response and explanatory variables). Each copy is divided into a test set (colored blue) and training set (colored pink). For each of these sets, we estimate the model parameters using only training data and then predict the responses for the test set. The average loss between the predicted and observed responses is then a measure for the predictive power of the model.

In particular, suppose we partition a data set $\mathcal{T}$ of size n into K **folds** $C_1, \ldots, C_K$ of sizes $n_1, \ldots, n_K$ (hence, $n_1 + \cdots + n_K = n$). Typically $n_k \approx n/K$, $k = 1, \ldots, K$.

Let ℓ_{C_k} be the test loss when using C_k as test data and all remaining data, denoted $\mathcal{T}_{-k}$, as training data. Each ℓ_{C_k} is an unbiased estimator of the generalization risk for training set $\mathcal{T}_{-k}$; that is, for $\ell(g_{\mathcal{T}_{-k}})$.

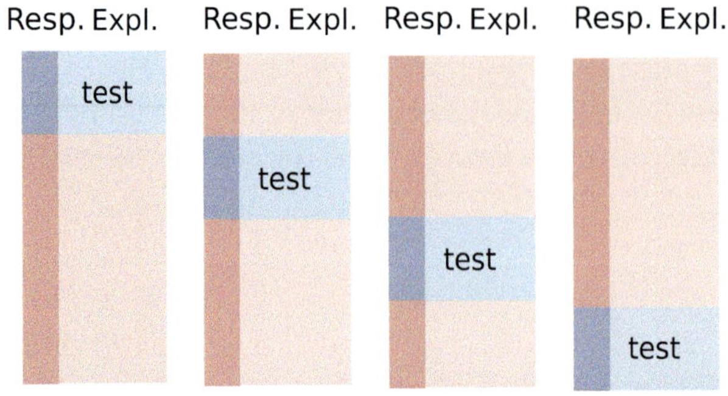

Figure 2.42: An illustration of four-fold cross-validation, representing four copies of the same data set. The data in each copy is partitioned into a training set (pink) and a test set (blue). The darker columns represent the response variable and the lighter ones the explanatory variables.

The **K-fold cross-validation** loss is the weighted average of these risk estimators:

$$\begin{aligned}\mathrm{CV}_K &= \sum_{k=1}^{K} \frac{n_k}{n} \ell_{C_k}(g_{\mathcal{T}_{-k}}) \\ &= \frac{1}{n}\sum_{k=1}^{K}\sum_{i \in C_k} \mathrm{Loss}(y_i, g_{\mathcal{T}_{-k}}(\boldsymbol{x}_i)) \\ &= \frac{1}{n}\sum_{i=1}^{n} \mathrm{Loss}(y_i, g_{\mathcal{T}_{-\kappa(i)}}(\boldsymbol{x}_i)),\end{aligned}$$

where the function $\kappa : \{1, \ldots, n\} \to \{1, \ldots, K\}$ assigns each of the n observations to one of the K folds. That is, $\kappa(i)$ denotes the fold to which the i-th observation belongs. In the special case where $K = n$, each observation forms its own fold, so $\kappa(i) = i$. As the average is taken over varying training sets $\{\mathcal{T}_{-k}\}$, it estimates the expected generalization risk $\mathbb{E}\,\ell(g_{\mathcal{T}})$, rather than the generalization risk $\ell(g_\tau)$ for the particular training set τ.

■ **Example 2.43 (Polynomial Regression (cont.))** For the polynomial regression example, we can calculate a K-fold cross-validation loss with a nonrandom partitioning of the training set using the following code, which imports the previous code for the polynomial regression example. We omit the full plotting code.

polyregCV.py

```
from polyreg3 import *

K_vals = [5, 10, 100]  # number of folds
cv = np.zeros((len(K_vals), max_p))   # cv loss
X = np.ones((n, 1))

for p in p_range:
  if p > 1:
    X = np.hstack((X, u**(p-1)))
```

```
    j = 0
    for K in K_vals:
      loss = []
      for k in range(1, K+1):
         # integer indices of test samples
         test_ind = ((n/K)*(k-1) + np.arange(1,n/K+1)-1).astype('int')
         train_ind = np.setdiff1d(np.arange(n), test_ind)

         X_train, y_train = X[train_ind, :], y[train_ind, :]
         X_test, y_test = X[test_ind, :], y[test_ind]

         # fit model and evaluate test loss
         betahat = solve(X_train.T @ X_train, X_train.T @ y_train)
         loss.append(norm(y_test - X_test @ betahat) ** 2)

      cv[j, p-1] = sum(loss)/n
      j += 1

# basic plotting
plt.plot(p_range, cv[0, :], 'k-.')
plt.plot(p_range, cv[1, :], 'r')
plt.plot(p_range, cv[2, :], 'b--')
plt.show()
```

Figure 2.44 shows the cross-validation loss for $K \in \{5, 10, 100\}$. The case $K = 100$ corresponds to the **leave-one-out cross-validation**, which can be computed more efficiently using the formula in Theorem 5.14. ☞ 204

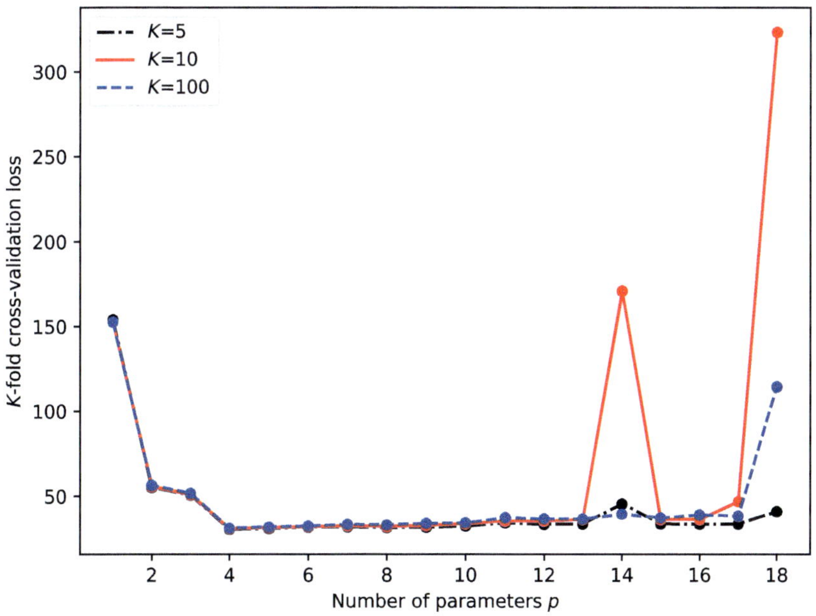

Figure 2.44: K-fold cross-validation for the polynomial regression example.

2.6 Modeling Data

The first step in any data analysis is to *model* the data in one form or another. For example, in an *unsupervised* learning setting with data represented by a vector $\boldsymbol{x} = [x_1, \ldots, x_p]^\top$, a very general model is to assume that $\boldsymbol{x}$ is the outcome of a random vector $\boldsymbol{X} = [X_1, \ldots, X_p]^\top$ with some unknown pdf f. The model can then be refined by assuming a specific form of f.

When given a sequence of such data vectors $\boldsymbol{x}_1, \ldots, \boldsymbol{x}_n$, one of the simplest models is to assume that the corresponding random vectors $\boldsymbol{X}_1, \ldots, \boldsymbol{X}_n$ are independent and ☞ 633 identically distributed (iid). We write

$$\boldsymbol{X}_1, \ldots, \boldsymbol{X}_n \overset{\text{iid}}{\sim} f \quad \text{or} \quad \boldsymbol{X}_1, \ldots, \boldsymbol{X}_n \overset{\text{iid}}{\sim} \mathsf{Dist},$$

to indicate that the random vectors form an iid sample from a sampling pdf f or sampling distribution Dist. This model formalizes the notion that the knowledge about one variable does not provide extra information about another variable. The main theoretical use of independent data models is that the joint density of the random vectors $\boldsymbol{X}_1, \ldots, \boldsymbol{X}_n$ is ☞ 633 simply the *product* of the marginal ones; see Theorem D.21. Specifically,

$$f_{\boldsymbol{X}_1, \ldots, \boldsymbol{X}_n}(\boldsymbol{x}_1, \ldots, \boldsymbol{x}_n) = f(\boldsymbol{x}_1) \cdots f(\boldsymbol{x}_n).$$

In most models of this kind, our approximation or model for the sampling distribution is specified up to a small number of parameters. That is, $g(\boldsymbol{x})$ is of the form $g(\boldsymbol{x} \mid \boldsymbol{\beta})$ which is known up to some parameter vector $\boldsymbol{\beta}$. Examples for the one-dimensional case ($p = 1$) include the $\mathcal{N}(\mu, \sigma^2)$, $\mathsf{Bin}(n, p)$, and $\mathsf{Exp}(\lambda)$ distributions. See Tables D.11 and ☞ 629 D.12 for other common sampling distributions.

Typically, the parameters are unknown and must be estimated from the data. In a nonparametric setting the whole sampling distribution would be unknown. To visualize the underlying sampling distribution from outcomes $\boldsymbol{x}_1, \ldots, \boldsymbol{x}_n$ one can use graphical representations such as histograms, density plots, and empirical cumulative distribution ☞ 11 functions, as discussed in Chapter 1.

If the order in which the data were collected (or their labeling) is not informative or relevant, then the joint pdf of $\boldsymbol{X}_1, \ldots, \boldsymbol{X}_n$ satisfies the symmetry:

$$f_{\boldsymbol{X}_1, \ldots, \boldsymbol{X}_n}(\boldsymbol{x}_1, \ldots, \boldsymbol{x}_n) = f_{\boldsymbol{X}_{\pi_1}, \ldots, \boldsymbol{X}_{\pi_n}}(\boldsymbol{x}_{\pi_1}, \ldots, \boldsymbol{x}_{\pi_n}) \tag{2.45}$$

for any permutation $\pi_1, \ldots, \pi_n$ of the integers $1, \ldots, n$. We say that the infinite sequence $\boldsymbol{X}_1, \boldsymbol{X}_2, \ldots$ is **exchangeable** if this permutational invariance (2.45) holds for any finite subset of the sequence. As we shall see in Section 2.9 on Bayesian learning, it is common to assume that the random vectors $\boldsymbol{X}_1, \ldots, \boldsymbol{X}_n$ are a subset of an exchangeable sequence and thus satisfy (2.45). Note that while iid random variables are exchangeable, the converse is not necessarily true. Thus, the assumption of an exchangeable sequence of random vectors is weaker than the assumption of iid random vectors.

Figure 2.46 illustrates the modeling tradeoffs. The keywords within the triangle represent various modeling paradigms. A few keywords have been highlighted, symbolizing their importance in modeling. The specific meaning of the keywords does not concern us here, but the point is there are many models to choose from, depending on what assumptions are made about the data.

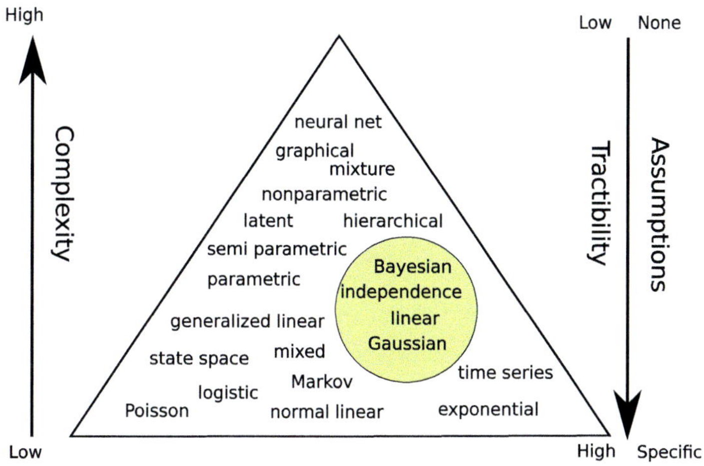

Figure 2.46: Illustration of the modeling dilemma. Complex models are more generally applicable, but may be difficult to analyze. Simple models may be highly tractable, but may not describe the data accurately. The triangular shape signifies that there are a great many specific models but not so many generic ones.

On the one hand, models that make few assumptions are more widely applicable, but at the same time may not be very mathematically tractable or provide insight into the nature of the data. On the other hand, very specific models may be easy to handle and interpret, but may not match the data very well. This tradeoff between the tractability and applicability of the model is very similar to the approximation–estimation tradeoff described in Section 2.4.

In the typical *unsupervised* setting we have a training set $\tau = \{\boldsymbol{x}_1, \ldots, \boldsymbol{x}_n\}$ that is viewed as the outcome of n iid random variables $\boldsymbol{X}_1, \ldots, \boldsymbol{X}_n$ from some unknown pdf f. The objective is then to learn or estimate f from the finite training data. To put the learning in a similar framework as for supervised learning discussed in the preceding Sections 2.3–2.5, we begin by specifying a class of probability density functions $\mathcal{G}_p := \{g(\cdot \mid \boldsymbol{\theta}), \boldsymbol{\theta} \in \boldsymbol{\Theta}\}$, where $\boldsymbol{\theta}$ is a parameter in some subset $\boldsymbol{\Theta}$ of $\mathbb{R}^p$. We now seek the best g in $\mathcal{G}_p$ to minimize some risk. Note that $\mathcal{G}_p$ may not necessarily contain the true f even for very large p.

> We stress that our notation $g(\boldsymbol{x})$ has a different meaning in the supervised and unsupervised case. In the supervised case, g is a prediction function for a response y; in the unsupervised setting, g is an approximation of a density f.

For each $\boldsymbol{x}$ we measure the discrepancy between the true model $f(\boldsymbol{x})$ and the hypothesized model $g(\boldsymbol{x} \mid \boldsymbol{\theta})$ using the loss function

$$\text{Loss}(f(\boldsymbol{x}), g(\boldsymbol{x} \mid \boldsymbol{\theta})) = \ln \frac{f(\boldsymbol{x})}{g(\boldsymbol{x} \mid \boldsymbol{\theta})} = \ln f(\boldsymbol{x}) - \ln g(\boldsymbol{x} \mid \boldsymbol{\theta}).$$

The expected value of this loss (that is, the risk) is thus

$$\ell(g) = \mathbb{E} \ln \frac{f(\boldsymbol{X})}{g(\boldsymbol{X} \mid \boldsymbol{\theta})} = \int f(\boldsymbol{x}) \ln \frac{f(\boldsymbol{x})}{g(\boldsymbol{x} \mid \boldsymbol{\theta})} \, \mathrm{d}\boldsymbol{x}. \tag{2.47}$$

The integral in (2.47) provides a fundamental way to measure the distance between two densities and is called the **Kullback–Leibler (KL) divergence**[2] between f and $g(\cdot \mid \boldsymbol{\theta})$. Note that the KL divergence is not symmetric in f and $g(\cdot \mid \boldsymbol{\theta})$. Moreover, it is always greater than or equal to 0 (see Exercise 20) and equal to 0 when $f = g(\cdot \mid \boldsymbol{\theta})$.

Using similar notation as for the supervised learning setting in Table 2.11, define $g^{\mathcal{G}_p}$ as the global minimizer of the risk in the class $\mathcal{G}_p$; that is, $g^{\mathcal{G}_p} = \operatorname{argmin}_{g \in \mathcal{G}_p} \ell(g)$. If we define

$$\begin{aligned}\boldsymbol{\theta}^* &= \underset{\boldsymbol{\theta}}{\operatorname{argmin}}\, \mathbb{E}\, \mathrm{Loss}(f(\boldsymbol{X}), g(\boldsymbol{X} \mid \boldsymbol{\theta})) = \underset{\boldsymbol{\theta}}{\operatorname{argmin}} \int \big(\ln f(\boldsymbol{x}) - \ln g(\boldsymbol{x} \mid \boldsymbol{\theta})\big) f(\boldsymbol{x})\, \mathrm{d}\boldsymbol{x} \\ &= \underset{\boldsymbol{\theta}}{\operatorname{argmax}} \int f(\boldsymbol{x}) \ln g(\boldsymbol{x} \mid \boldsymbol{\theta})\, \mathrm{d}\boldsymbol{x} = \underset{\boldsymbol{\theta}}{\operatorname{argmax}}\, \mathbb{E} \ln g(\boldsymbol{X} \mid \boldsymbol{\theta}),\end{aligned}$$

then $g^{\mathcal{G}_p} = g(\cdot \mid \boldsymbol{\theta}^*)$ and learning $g^{\mathcal{G}_p}$ is equivalent to learning (or estimating) $\boldsymbol{\theta}^*$. To learn $\boldsymbol{\theta}^*$ from a training set $\tau = \{\boldsymbol{x}_1, \ldots, \boldsymbol{x}_n\}$ we then minimize the training loss,

$$\frac{1}{n} \sum_{i=1}^{n} \mathrm{Loss}(f(\boldsymbol{x}_i), g(\boldsymbol{x}_i \mid \boldsymbol{\theta})) = -\frac{1}{n} \sum_{i=1}^{n} \ln g(\boldsymbol{x}_i \mid \boldsymbol{\theta}) + \frac{1}{n} \sum_{i=1}^{n} \ln f(\boldsymbol{x}_i),$$

giving:

$$\widehat{\boldsymbol{\theta}}_n := \underset{\boldsymbol{\theta}}{\operatorname{argmax}}\, \frac{1}{n} \sum_{i=1}^{n} \ln g(\boldsymbol{x}_i \mid \boldsymbol{\theta}). \tag{2.48}$$

As the logarithm is an increasing function, this is equivalent to

$$\widehat{\boldsymbol{\theta}}_n := \underset{\boldsymbol{\theta}}{\operatorname{argmax}} \prod_{i=1}^{n} g(\boldsymbol{x}_i \mid \boldsymbol{\theta}),$$

where $\prod_{i=1}^{n} g(\boldsymbol{x}_i \mid \boldsymbol{\theta})$ is the **likelihood** of the data; that is, the joint density of the $\{\boldsymbol{X}_i\}$ evaluated at the points $\{\boldsymbol{x}_i\}$. We therefore have recovered the classical **maximum likelihood estimate** of $\boldsymbol{\theta}^*$. ☞ 665

When the risk $\ell(g(\cdot \mid \boldsymbol{\theta}))$ is convex in $\boldsymbol{\theta}$ over a convex set $\boldsymbol{\Theta}$, we can find the maximum likelihood estimator by setting the gradient of the training loss to zero; that is, we solve

$$-\frac{1}{n} \sum_{i=1}^{n} \boldsymbol{S}(\boldsymbol{x}_i \mid \boldsymbol{\theta}) = \boldsymbol{0},$$

where $\boldsymbol{S}(\boldsymbol{x} \mid \boldsymbol{\theta}) := \frac{\partial \ln g(\boldsymbol{x} \mid \boldsymbol{\theta})}{\partial \boldsymbol{\theta}}$ is the gradient of $\ln g(\boldsymbol{x} \mid \boldsymbol{\theta})$ with respect to $\boldsymbol{\theta}$ and is often called the **score**.

■ **Example 2.49 (Exponential Model)** Suppose we have the data $\tau_n = \{x_1, \ldots, x_n\}$, which is modeled as a realization of n positive iid random variables: $X_1, \ldots, X_n \sim_{\text{iid}} f(x)$. We select the class of approximating functions $\mathcal{G}$ to be the parametric class $\{g : g(x \mid \theta) =$

[2]Sometimes called cross-entropy distance.

$\theta \exp(-x\,\theta), x > 0, \theta > 0\}$. In other words, we look for the best $g^{\mathcal{G}}$ within the family of exponential distributions with unknown parameter $\theta > 0$. The likelihood of the data is

$$\prod_{i=1}^{n} g(x_i \mid \theta) = \prod_{i=1}^{n} \theta \exp(-\theta x_i) = \exp(-\theta\, n\, \overline{x}_n + n \ln \theta)$$

and the score is $S(x \mid \theta) = -x + \theta^{-1}$. Thus, maximizing the likelihood with respect to θ is the same as maximizing $-\theta\, n\, \overline{x}_n + n \ln \theta$ or solving $-\sum_{i=1}^{n} S(x_i \mid \theta)/n = \overline{x}_n - \theta^{-1} = 0$. In other words, the solution to (2.48) is the maximum likelihood estimate $\widehat{\theta}_n = 1/\overline{x}_n$. ■

In a *supervised* setting, where the data is represented by a vector $\boldsymbol{x}$ of explanatory variables and a response y, the general model is that $(\boldsymbol{x}, y)$ is an outcome of $(\boldsymbol{X}, Y) \sim f$ for some unknown f. And for a training sequence $(\boldsymbol{x}_1, y_1), \ldots, (\boldsymbol{x}_n, y_n)$ the default model assumption is that $(\boldsymbol{X}_1, Y_1), \ldots, (\boldsymbol{X}_n, Y_n) \sim_{\text{iid}} f$. As explained in Section 2.2, the analysis primarily involves the conditional pdf $f(y \mid \boldsymbol{x})$ and in particular (when using the squared-error loss) the conditional expectation $g^*(\boldsymbol{x}) = \mathbb{E}[Y \mid \boldsymbol{X} = \boldsymbol{x}]$. The resulting representation (2.3) allows us to then write the response at $\boldsymbol{X} = \boldsymbol{x}$ as a function of the feature $\boldsymbol{x}$ plus an error term: $Y = g^*(\boldsymbol{x}) + \varepsilon(\boldsymbol{x})$.

This leads to the simplest and most important model for supervised learning, where we choose a *linear* class $\mathcal{G}$ of prediction or guess functions and assume that it is rich enough to contain the true g^*. If we further assume that, conditional on $\boldsymbol{X} = \boldsymbol{x}$, the error term ε does not depend on $\boldsymbol{x}$, that is, $\mathbb{E}\,\varepsilon = 0$ and $\mathbb{V}\mathrm{ar}\,\varepsilon = \sigma^2$, then we obtain the following model.

Definition 2.50 (Linear Model): In a **linear model** the response Y depends on a p-dimensional explanatory variable $\boldsymbol{x} = [x_1, \ldots, x_p]^\top$ via the linear relationship

$$Y = \boldsymbol{x}^\top \boldsymbol{\beta} + \varepsilon, \tag{2.51}$$

where $\mathbb{E}\,\varepsilon = 0$ and $\mathbb{V}\mathrm{ar}\,\varepsilon = \sigma^2$.

Note that (2.51) is a model for a single pair $(\boldsymbol{x}, Y)$. The model for the training set $\{(\boldsymbol{x}_i, Y_i)\}$ is simply that each Y_i satisfies (2.51) (with $\boldsymbol{x} = \boldsymbol{x}_i$) and that the $\{Y_i\}$ are independent. Gathering all responses in the vector $\boldsymbol{Y} = [Y_1, \ldots, Y_n]^\top$, we can write

$$\boldsymbol{Y} = \mathbf{X}\boldsymbol{\beta} + \boldsymbol{\varepsilon}, \tag{2.52}$$

where $\boldsymbol{\varepsilon} = [\varepsilon_1, \ldots, \varepsilon_n]^\top$ is a vector of iid copies of ε and $\mathbf{X}$ is the so-called **model matrix**, with rows $\boldsymbol{x}_1^\top, \ldots, \boldsymbol{x}_n^\top$. Linear models are fundamental building blocks of statistical learning algorithms. For this reason, a large part of Chapter 5 is devoted to linear regression models. ☞ 197

■ **Example 2.53 (Polynomial Regression (cont.))** For our running Example 2.13, we see that the data is described by a linear model of the form (2.52), with model matrix $\mathbf{X}$ given in (2.17). ■ ☞ 26

Before we discuss a few other models in the following sections, we would like to emphasize a number of points about modeling.

- *Any* model for data is likely to be *wrong*. For example, real data (as opposed to computer-generated data) are often assumed to come from a normal distribution, which is never exactly true. However, an important advantage of using a normal distribution is that it has many nice mathematical properties, as we will see in Section 2.7.
- Most data models depend on a number of unknown parameters, which need to be estimated from the observed data.
- Any model for real-life data needs to be *checked* for suitability. An important criterion is that data simulated from the model should resemble the observed data, at least for a certain choice of model parameters.

Here are some guidelines for choosing a model. Think of the data as a spreadsheet or data frame, as in Chapter 1, where rows represent the data units and the columns the data features (variables, groups).

- First establish the *type* of the features (quantitative, qualitative, discrete, continuous, etc.).
- Assess whether the data can be assumed to be independent across rows or columns.
- Decide on the level of generality of the model. For example, should we use a simple model with a few unknown parameters or a more generic model that has a large number of parameters? Simple specific models are easier to fit to the data (low estimation error) than more general models, but the fit itself may not be accurate (high approximation error). The tradeoffs discussed in Section 2.4 play an important role here.
- Decide on using a classical (frequentist) or Bayesian model. Section 2.9 gives a short introduction to Bayesian learning. ☞ 49

2.7 Multivariate Normal Models

A standard model for numerical observations $x_1, \dots, x_n$ (forming, e.g., a column in a spreadsheet or data frame) is that they are the outcomes of iid normal random variables

$$X_1, \dots, X_n \overset{\text{iid}}{\sim} \mathcal{N}(\mu, \sigma^2).$$

It is helpful to view a normally distributed random variable as a simple transformation of a standard normal random variable. To wit, if Z has a standard normal distribution, then $X = \mu + \sigma Z$ has a $\mathcal{N}(\mu, \sigma^2)$ distribution. The generalization to n dimensions is discussed in Appendix D.7. We summarize the main points: Let $Z_1, \dots, Z_n \sim_{\text{iid}} \mathcal{N}(0, 1)$. ☞ 638
The pdf of $\mathbf{Z} = [Z_1, \dots, Z_n]^\top$ (that is, the joint pdf of $Z_1, \dots, Z_n$) is given by

$$f_{\mathbf{Z}}(\mathbf{z}) = \prod_{i=1}^{n} \frac{1}{\sqrt{2\pi}} \mathrm{e}^{-\frac{1}{2}z_i^2} = (2\pi)^{-\frac{n}{2}} \mathrm{e}^{-\frac{1}{2}\mathbf{z}^\top \mathbf{z}}, \quad \mathbf{z} \in \mathbb{R}^n. \tag{2.54}$$

We write $\mathbf{Z} \sim \mathcal{N}(\mathbf{0}, \mathbf{I}_n)$ and say that $\mathbf{Z}$ has a standard normal distribution in $\mathbb{R}^n$. Let

$$\boldsymbol{X} = \boldsymbol{\mu} + \mathbf{B}\,\mathbf{Z} \tag{2.55}$$

for some $m \times n$ matrix $\mathbf{B}$ and m-dimensional vector $\boldsymbol{\mu}$. Then $\boldsymbol{X}$ has expectation vector $\boldsymbol{\mu}$ and covariance matrix $\boldsymbol{\Sigma} = \mathbf{B}\mathbf{B}^\top$; see (D.33) and (D.34). This leads to the following definition. ☞ 636

Definition 2.56 (Multivariate Normal Distribution): An m-dimensional random vector $\boldsymbol{X}$ that can be written in the form (2.55) for some m-dimensional vector $\boldsymbol{\mu}$ and $m \times n$ matrix $\mathbf{B}$, with $\mathbf{Z} \sim \mathcal{N}(\mathbf{0}, \mathbf{I}_n)$, is said to have a **multivariate normal** or **multivariate Gaussian** distribution with mean vector $\boldsymbol{\mu}$ and covariance matrix $\boldsymbol{\Sigma} = \mathbf{B}\mathbf{B}^\top$. We write $\boldsymbol{X} \sim \mathcal{N}(\boldsymbol{\mu}, \boldsymbol{\Sigma})$.

The m-dimensional density of a multivariate normal distribution has a very similar form to the density of the one-dimensional normal distribution and is given in the next theorem. We leave the proof as an exercise; see Exercise 5. ☞ 62

Theorem 2.57 (Density of a Multivariate Random Vector): Let $\boldsymbol{X} \sim \mathcal{N}(\boldsymbol{\mu}, \boldsymbol{\Sigma})$, where the $m \times m$ covariance matrix $\boldsymbol{\Sigma}$ is invertible. Then $\boldsymbol{X}$ has pdf

$$f_{\boldsymbol{X}}(\boldsymbol{x}) = \frac{1}{\sqrt{(2\pi)^m\,|\boldsymbol{\Sigma}|}}\,\mathrm{e}^{-\frac{1}{2}\,(\boldsymbol{x}-\boldsymbol{\mu})^\top\boldsymbol{\Sigma}^{-1}(\boldsymbol{x}-\boldsymbol{\mu})}, \quad \boldsymbol{x} \in \mathbb{R}^m. \tag{2.58}$$

Figure 2.59 shows the pdfs of two bivariate (that is, two-dimensional) normal distributions. In both cases the mean vector is $\boldsymbol{\mu} = [0, 0]^\top$ and the variances (the diagonal elements of $\boldsymbol{\Sigma}$) are 1. The correlation coefficients (or, equivalently here, the covariances) are respectively $\varrho = 0$ and $\varrho = 0.8$.

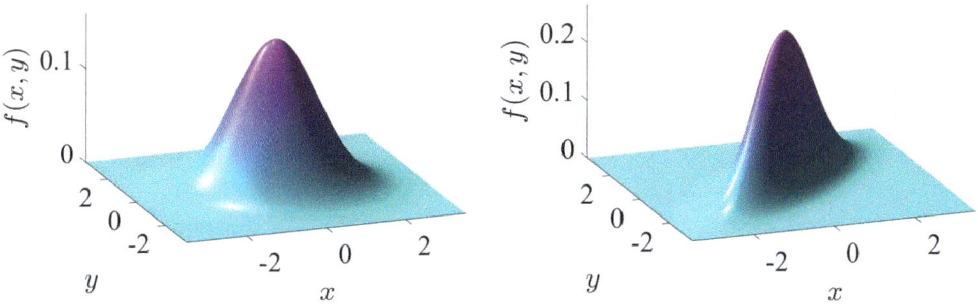

Figure 2.59: Pdfs of bivariate normal distributions with means zero, variances 1, and correlation coefficients 0 (left) and 0.8 (right).

The main reason why the multivariate normal distribution plays an important role in data science and machine learning is that it satisfies the following properties, the details and proofs of which can be found in Appendix D.7: ☞ 638

1. Affine combinations are normal.
2. Marginal distributions are normal.
3. Conditional distributions are normal.

2.8 Normal Linear Models

Normal linear models combine the simplicity of the linear model with the tractability of the Gaussian distribution. They are the principal model for traditional statistics, and include the classic linear regression and analysis of variance models.

Definition 2.60 (Normal Linear Model): Let $\varepsilon \sim \mathcal{N}(0, \sigma^2)$. In a **normal linear model** the response Y depends on a p-dimensional explanatory variable $\boldsymbol{x} = [x_1, \dots, x_p]^\top$, via the linear relationship

$$Y = \boldsymbol{x}^\top \boldsymbol{\beta} + \varepsilon. \tag{2.61}$$

Thus, a normal linear model is a linear model (in the sense of Definition 2.50) with normal error terms. Similar to (2.52), the corresponding normal linear model for the whole training set $\{(\boldsymbol{x}_i, Y_i)\}$ has the form

$$\boldsymbol{Y} = \mathbf{X}\boldsymbol{\beta} + \boldsymbol{\varepsilon}, \tag{2.62}$$

where $\mathbf{X}$ is the model matrix comprised of rows $\boldsymbol{x}_1^\top, \dots, \boldsymbol{x}_n^\top$ and $\boldsymbol{\varepsilon} \sim \mathcal{N}(\mathbf{0}, \sigma^2 \mathbf{I}_n)$. As a consequence, we can write

$$\boldsymbol{Y} = \mathbf{X}\boldsymbol{\beta} + \sigma \boldsymbol{Z},$$

☞ 47 where $\boldsymbol{Z} \sim \mathcal{N}(\mathbf{0}, \mathbf{I}_n)$, so that $\boldsymbol{Y} \sim \mathcal{N}(\mathbf{X}\boldsymbol{\beta}, \sigma^2 \mathbf{I}_n)$. It follows from (2.58) that the joint density of $\boldsymbol{Y}$ is given by

$$g(\boldsymbol{y} \mid \boldsymbol{\beta}, \sigma^2, \mathbf{X}) = (2\pi\sigma^2)^{-\frac{n}{2}} \, \mathrm{e}^{-\frac{1}{2\sigma^2} \|\boldsymbol{y} - \mathbf{X}\boldsymbol{\beta}\|^2}. \tag{2.63}$$

Estimation of the parameter $\boldsymbol{\beta}$ can be performed via the least-squares method, as discussed in Example 2.13. An estimate can also be obtained via the maximum likelihood method. This simply means finding the parameters σ^2 and $\boldsymbol{\beta}$ that maximize the likelihood of the outcome $\boldsymbol{y}$, given by the right-hand side of (2.63). It is clear that for every value of σ^2 the likelihood is maximal when $\|\boldsymbol{y} - \mathbf{X}\boldsymbol{\beta}\|^2$ is minimal. As a consequence, the maximum likelihood estimate for $\boldsymbol{\beta}$ is the same as the least-squares estimate (2.23). We leave it ☞ 69 as an exercise (see Exercise 23) to show that the maximum likelihood estimate of σ^2 is equal to

$$\widehat{\sigma^2} = \frac{\|\boldsymbol{y} - \mathbf{X}\widehat{\boldsymbol{\beta}}\|^2}{n}, \tag{2.64}$$

where $\widehat{\boldsymbol{\beta}}$ is the maximum likelihood estimate (least squares estimate in this case) of $\boldsymbol{\beta}$.

2.9 Bayesian Learning

In Bayesian unsupervised learning, we seek to approximate the unknown joint density $f(\boldsymbol{x}_1, \ldots, \boldsymbol{x}_n)$ of the training data $\mathcal{T}_n = \{\boldsymbol{X}_1, \ldots, \boldsymbol{X}_n\}$ via a joint pdf of the form

$$\int \left(\prod_{i=1}^{n} g(\boldsymbol{x}_i \,|\, \boldsymbol{\theta}) \right) w(\boldsymbol{\theta}) \, \mathrm{d}\boldsymbol{\theta}, \tag{2.65}$$

where $g(\cdot \,|\, \boldsymbol{\theta})$ belongs to a family of parametric densities $\mathcal{G}_p := \{g(\cdot \,|\, \boldsymbol{\theta}),\ \boldsymbol{\theta} \in \boldsymbol{\Theta}\}$ (viewed as a family of pdfs conditional on a parameter $\boldsymbol{\theta}$ in some set $\boldsymbol{\Theta} \subset \mathbb{R}^p$) and $w(\boldsymbol{\theta})$ is a pdf that belongs to a (possibly different) family of densities $\mathcal{W}_p$. Note how the joint pdf (2.65) satisfies the permutational invariance (2.45) and can thus be useful as a model for training data which is part of an exchangeable sequence of random variables.

> Following standard practice in a Bayesian context, instead of writing $f_X(x)$ and $f_{X \,|\, Y}(x \,|\, y)$ for the pdf of X and the conditional pdf of X given Y, one simply writes $f(x)$ and $f(x \,|\, y)$. If Y is a different random variable, its pdf (at y) is thus denoted by $f(y)$.

We will use the same symbol g for different (conditional) approximating probability densities and f for the different (conditional) true and unknown probability densities. Using Bayesian notation, we can write $g(\tau \,|\, \boldsymbol{\theta}) = \prod_{i=1}^{n} g(\boldsymbol{x}_i \,|\, \boldsymbol{\theta})$ and thus the approximating joint pdf (2.65) can then be written as $\int g(\tau \,|\, \boldsymbol{\theta}) \, w(\boldsymbol{\theta}) \, \mathrm{d}\boldsymbol{\theta}$ and the true unknown joint pdf as $f(\tau) = f(\boldsymbol{x}_1, \ldots, \boldsymbol{x}_n)$.

Once $\mathcal{G}_p$ and $\mathcal{W}_p$ are specified, selecting an approximating function $g(\boldsymbol{x})$ of the form

$$g(\boldsymbol{x}) = \int g(\boldsymbol{x} \,|\, \boldsymbol{\theta}) \, w(\boldsymbol{\theta}) \, \mathrm{d}\boldsymbol{\theta}$$

is equivalent to selecting a suitable w from $\mathcal{W}_p$. Similar to (2.47), we can use the Kullback–Leibler risk to measure the discrepancy between the proposed approximation (2.65) and the true $f(\tau)$:

$$\ell(g) = \mathbb{E} \ln \frac{f(\mathcal{T})}{\int g(\mathcal{T} \,|\, \boldsymbol{\theta}) \, w(\boldsymbol{\theta}) \, \mathrm{d}\boldsymbol{\theta}} = \int f(\tau) \ln \frac{f(\tau)}{\int g(\tau \,|\, \boldsymbol{\theta}) \, w(\boldsymbol{\theta}) \, \mathrm{d}\boldsymbol{\theta}} \, \mathrm{d}\tau. \tag{2.66}$$

The main difference with (2.47) is that since the training data is not necessarily iid (it may be exchangeable, for example), the expectation must be with respect to the joint ☞ 42 density of $\mathcal{T}$, not with respect to the marginal $f(\boldsymbol{x})$ (as in the iid case).

Minimizing the training loss is equivalent to maximizing the likelihood of the training data τ; that is, solving the optimization problem

$$\max_{w \in \mathcal{W}_p} \int g(\tau \,|\, \boldsymbol{\theta}) \, w(\boldsymbol{\theta}) \, \mathrm{d}\boldsymbol{\theta},$$

where the maximization is over an appropriate class $\mathcal{W}_p$ of density functions that is believed to result in the smallest Kullback–Leibler risk.

Suppose that we have a rough guess, denoted $w_0(\boldsymbol{\theta})$, for the best $w \in \mathcal{W}_p$ that minimizes the Kullback–Leibler risk. We can always increase the resulting likelihood $L_0 := \int g(\tau \,|\, \boldsymbol{\theta})\, w_0(\boldsymbol{\theta})\, \mathrm{d}\boldsymbol{\theta}$ by instead using the density $w_1(\boldsymbol{\theta}) := w_0(\boldsymbol{\theta})\, g(\tau \,|\, \boldsymbol{\theta})/L_0$, giving a likelihood $L_1 := \int g(\tau \,|\, \boldsymbol{\theta})\, w_1(\boldsymbol{\theta})\, \mathrm{d}\boldsymbol{\theta}$. To see this, write L_0 and L_1 as expectations with respect to w_0. In particular, we can write

$$L_0 = \mathbb{E}_{w_0}\, g(\tau \,|\, \boldsymbol{\theta}) \quad \text{and} \quad L_1 = \mathbb{E}_{w_1}\, g(\tau \,|\, \boldsymbol{\theta}) = \mathbb{E}_{w_0} g^2(\tau \,|\, \boldsymbol{\theta})/L_0.$$

It follows that

$$L_1 - L_0 = \frac{1}{L_0}\mathbb{E}_{w_0}\left[g^2(\tau \,|\, \boldsymbol{\theta}) - L_0^2\right] = \frac{1}{L_0}\mathbb{V}\mathrm{ar}_{w_0}[g(\tau \,|\, \boldsymbol{\theta})] \geq 0. \tag{2.67}$$

We may thus expect to obtain better predictions using w_1 instead of w_0, because w_1 has taken into account the observed data τ and increased the likelihood of the model. In fact, if we iterate this process (see Exercise 25) and create a sequence of densities $w_1, w_2, \ldots$ such that $w_t(\boldsymbol{\theta}) \propto w_{t-1}(\boldsymbol{\theta})\, g(\tau \,|\, \boldsymbol{\theta})$, then $w_t(\boldsymbol{\theta})$ concentrates more and more of its probability mass at the maximum likelihood estimator $\widehat{\boldsymbol{\theta}}$ (see (2.48)) and in the limit equals a (degenerate) point-mass pdf at $\widehat{\boldsymbol{\theta}}$. In other words, in the limit we recover the maximum likelihood method: $g_\tau(\boldsymbol{x}) = g(\boldsymbol{x} \,|\, \widehat{\boldsymbol{\theta}})$. Thus, unless the class of densities $\mathcal{W}_p$ is restricted to be non-degenerate, maximizing the likelihood as much as possible leads to a degenerate choice for $w(\boldsymbol{\theta})$.

In many situations, the maximum likelihood estimate $g(\tau \,|\, \widehat{\boldsymbol{\theta}})$ is either not an appropriate approximation to $f(\tau)$ (see Example 2.79), or simply fails to exist (see Exercise 20
☞ 192 in Chapter 4). In such cases, given an initial non-degenerate guess $w_0(\boldsymbol{\theta}) = g(\boldsymbol{\theta})$, one can obtain a more appropriate and non-degenerate approximation to $f(\tau)$ by taking $w(\boldsymbol{\theta}) = w_1(\boldsymbol{\theta}) \propto g(\tau \,|\, \boldsymbol{\theta})\, g(\boldsymbol{\theta})$ in (2.65), giving the following Bayesian learner of $f(\boldsymbol{x})$:

$$g_\tau(\boldsymbol{x}) := \int g(\boldsymbol{x} \,|\, \boldsymbol{\theta}) \frac{g(\tau \,|\, \boldsymbol{\theta})\, g(\boldsymbol{\theta})}{\int g(\tau \,|\, \boldsymbol{\vartheta})\, g(\boldsymbol{\vartheta})\, \mathrm{d}\boldsymbol{\vartheta}}\, \mathrm{d}\boldsymbol{\theta}, \tag{2.68}$$

☞ 632 where $\int g(\tau \,|\, \boldsymbol{\vartheta})\, g(\boldsymbol{\vartheta})\, \mathrm{d}\boldsymbol{\vartheta} = g(\tau)$. Using Bayes' formula for probability densities,

$$g(\boldsymbol{\theta} \,|\, \tau) = \frac{g(\tau \,|\, \boldsymbol{\theta})\, g(\boldsymbol{\theta})}{g(\tau)}, \tag{2.69}$$

we can write $w_1(\boldsymbol{\theta}) = g(\boldsymbol{\theta} \,|\, \tau)$. With this notation, we have the following definitions.

Definition 2.70 (Prior, Likelihood, and Posterior): Let τ and $\mathcal{G}_p := \{g(\cdot \,|\, \boldsymbol{\theta}), \boldsymbol{\theta} \in \boldsymbol{\Theta}\}$ be the training set and family of approximating functions.

- A pdf $g(\boldsymbol{\theta})$ that reflects our *a priori* beliefs about $\boldsymbol{\theta}$ is called the **prior** pdf.
- The conditional pdf $g(\tau \,|\, \boldsymbol{\theta})$ is called the **likelihood**.
- Inference about $\boldsymbol{\theta}$ is given by the **posterior** pdf $g(\boldsymbol{\theta} \,|\, \tau)$, which is proportional to the product of the prior and the likelihood:

$$g(\boldsymbol{\theta} \,|\, \tau) \propto g(\tau \,|\, \boldsymbol{\theta})\, g(\boldsymbol{\theta}).$$

Remark 2.71 (Early Stopping) Bayes iteration is an example of an "early stopping" heuristic for maximum likelihood optimization, where we exit after only one step. As observed above, if we keep iterating, we obtain the maximum likelihood estimate (MLE). In a sense the Bayes rule provides a regularization of the MLE. Regularization is discussed in more detail in Chapter 7; see also Example 2.79. The early stopping rule is also of benefit in regularization; see Exercise 18 in Chapter 7.

On the one hand, the initial guess $g(\boldsymbol{\theta})$ conveys the *a priori* (prior to training the Bayesian learner) information about the optimal density in $\mathcal{W}_p$ that minimizes the Kullback–Leibler risk. Using this prior $g(\boldsymbol{\theta})$, the Bayesian approximation to $f(\boldsymbol{x})$ is the **prior predictive density**:

$$g(\boldsymbol{x}) = \int g(\boldsymbol{x} \mid \boldsymbol{\theta})\, g(\boldsymbol{\theta})\, \mathrm{d}\boldsymbol{\theta}.$$

On the other hand, the posterior pdf conveys improved knowledge about this optimal density in $\mathcal{W}_p$ after training with τ. Using the posterior $g(\boldsymbol{\theta} \mid \tau)$, the Bayesian learner of $f(\boldsymbol{x})$ is the **posterior predictive density**:

$$g_{\mathcal{T}}(\boldsymbol{x}) = g(\boldsymbol{x} \mid \tau) = \int g(\boldsymbol{x} \mid \boldsymbol{\theta})\, g(\boldsymbol{\theta} \mid \tau)\, \mathrm{d}\boldsymbol{\theta},$$

where we have assumed that $g(\boldsymbol{x} \mid \boldsymbol{\theta}, \tau) = g(\boldsymbol{x} \mid \boldsymbol{\theta})$; that is, the likelihood depends on τ only through the parameter $\boldsymbol{\theta}$.

The choice of the prior is typically governed by two considerations:

1. the prior should be simple enough to facilitate the computation or simulation of the posterior pdf;

2. the prior should be general enough to model ignorance of the parameter of interest.

Priors that do not convey much knowledge of the parameter are said to be *uninformative*. The uniform or *flat* prior in Example 2.79 (to follow) is frequently used.

For the purpose of analytical and numerical computations, we can view $\boldsymbol{\theta}$ as a random vector with prior density $g(\boldsymbol{\theta})$, which after training is updated to the posterior density $g(\boldsymbol{\theta} \mid \tau)$.

The above thinking allows us to write $g(\boldsymbol{x} \mid \tau) \propto \int g(\boldsymbol{x} \mid \boldsymbol{\theta})\, g(\tau \mid \boldsymbol{\theta})\, g(\boldsymbol{\theta})\, \mathrm{d}\boldsymbol{\theta}$, for example, thus ignoring any constants that do not depend on the argument of the densities.

Example 2.72 (Normal Model) Suppose that the training data $\mathcal{T} = \{X_1, \ldots, X_n\}$ is modeled using a likelihood $g(x \mid \boldsymbol{\theta})$, given by the pdf of

$$X \mid \boldsymbol{\theta} \sim \mathcal{N}(\mu, \sigma^2),$$

where $\boldsymbol{\theta} := [\mu, \sigma^2]^\top$. Next, we need to specify the prior distribution of $\boldsymbol{\theta}$ to complete the model. We can specify prior distributions for μ and σ^2 separately and then take their product to obtain the prior for vector $\boldsymbol{\theta}$ (assuming independence). A possible prior distribution for μ is

$$\mu \sim \mathcal{N}(\nu, \phi^2). \tag{2.73}$$

It is typical to refer to any parameters of the prior density as **hyperparameters** of the Bayesian model. Instead of giving directly a prior for σ^2 (or σ), it turns out to be convenient to give the following prior distribution to $1/\sigma^2$:

$$\frac{1}{\sigma^2} \sim \mathsf{Gamma}(\alpha, \beta). \tag{2.74}$$

The smaller α and β are, the less informative is the prior. Under this prior, σ^2 is said to have an **inverse gamma**[3] distribution. If $1/Z \sim \mathsf{Gamma}(\alpha, \beta)$, then the pdf of Z is proportional to $\exp(-\beta/z)/z^{\alpha+1}$ (Exercise 24). The Bayesian posterior is then given by:

☞ 69

$$\begin{aligned} g(\mu, \sigma^2 \mid \tau) &\propto g(\mu) \times g(\sigma^2) \times g(\tau \mid \mu, \sigma^2) \\ &\propto \exp\left\{-\frac{(\mu-\nu)^2}{2\phi^2}\right\} \times \frac{\exp\{-\beta/\sigma^2\}}{(\sigma^2)^{\alpha+1}} \times \frac{\exp\{-\sum_i (x_i-\mu)^2/(2\sigma^2)\}}{(\sigma^2)^{n/2}} \\ &\propto (\sigma^2)^{-n/2-\alpha-1} \exp\left\{-\frac{(\mu-\nu)^2}{2\phi^2} - \frac{\beta}{\sigma^2} - \frac{(\mu-\overline{x}_n)^2 + S_n^2}{2\sigma^2/n}\right\}, \end{aligned}$$

where $S_n^2 := \frac{1}{n}\sum_i x_i^2 - \overline{x}_n^2 = \frac{1}{n}\sum_i (x_i - \overline{x}_n)^2$ is the (scaled) sample variance. All inference about (μ, σ^2) is then represented by the posterior pdf. To facilitate computations it is helpful to find out if the posterior belongs to a recognizable family of distributions. For example, the conditional pdf of μ given σ^2 and τ is

$$g(\mu \mid \sigma^2, \tau) \propto \exp\left\{-\frac{(\mu-\nu)^2}{2\phi^2} - \frac{(\mu-\overline{x}_n)^2}{2\sigma^2/n}\right\},$$

which after simplification can be recognized as the pdf of

$$(\mu \mid \sigma^2, \tau) \sim \mathcal{N}\big(\omega_n \overline{x}_n + (1-\omega_n)\nu,\ \omega_n\, \sigma^2/n\big), \tag{2.75}$$

where we have defined the weight parameter: $\omega_n := \frac{n}{\sigma^2} \big/ \big(\frac{1}{\phi^2} + \frac{n}{\sigma^2}\big)$. We can then see that the posterior mean $\mathbb{E}[\mu \mid \sigma^2, \tau] = \omega_n \overline{x}_n + (1-\omega_n)\nu$ is a weighted linear combination of the prior mean ν and the sample average $\overline{x}_n$. Further, as $n \to \infty$, the weight $\omega_n \to 1$ and thus the posterior mean approaches the maximum likelihood estimate $\overline{x}_n$. ■

It is sometimes possible to use a prior $g(\boldsymbol{\theta})$ that is not a *bona fide* probability density, in the sense that $\int g(\boldsymbol{\theta})\, \mathrm{d}\boldsymbol{\theta} = \infty$, as long as the resulting posterior $g(\boldsymbol{\theta} \mid \tau) \propto g(\tau \mid \boldsymbol{\theta}) g(\boldsymbol{\theta})$ is a proper pdf. Such a prior is called an **improper prior**.

■ **Example 2.76 (Normal Model (cont.))** An example of an improper prior is obtained from (2.73) when we let $\phi \to \infty$ (the larger ϕ is, the more uninformative is the prior).

[3]Reciprocal gamma distribution would have been a better name.

Then, $g(\mu) \propto 1$ is a flat prior, but $\int g(\mu)\,\mathrm{d}\mu = \infty$, making it an improper prior. Nevertheless, the posterior is a proper density, and in particular the conditional posterior of $(\mu \mid \sigma^2, \tau)$ simplifies to

$$(\mu \mid \sigma^2, \tau) \sim \mathcal{N}(\overline{x}_n, \sigma^2/n),$$

because the weight parameter ω_n goes to 1 as $\phi \to \infty$. The improper prior $g(\mu) \propto 1$ also allows us to simplify the posterior marginal for σ^2:

$$g(\sigma^2 \mid \tau) = \int g(\mu, \sigma^2 \mid \tau)\,\mathrm{d}\mu \propto (\sigma^2)^{-(n-1)/2-\alpha-1} \exp\left\{-\frac{\beta + nS_n^2/2}{\sigma^2}\right\},$$

which we recognize as the density corresponding to

$$\frac{1}{\sigma^2}\,\Big|\,\tau \sim \mathsf{Gamma}\left(\alpha + \frac{n-1}{2},\ \beta + \frac{n}{2}S_n^2\right).$$

In addition to $g(\mu) \propto 1$, we can also use an improper prior for σ^2. If we take the limit $\alpha \to 0$ and $\beta \to 0$ in (2.74), then we also obtain the improper prior $g(\sigma^2) \propto 1/\sigma^2$ (or equivalently $g(1/\sigma^2) \propto 1/\sigma^2$). In this case, the posterior marginal density for σ^2 implies that:

$$\frac{nS_n^2}{\sigma^2}\,\Big|\,\tau \sim \chi^2_{n-1}$$

and the posterior marginal density for μ implies that:

$$\frac{\mu - \overline{x}_n}{S_n/\sqrt{n-1}}\,\Big|\,\tau \sim \mathsf{t}_{n-1}. \tag{2.77}$$

In general, deriving a simple formula for the posterior density of $\boldsymbol{\theta}$ is either impossible or too tedious. Instead, the Monte Carlo methods in Chapter 3 can be used to simulate (approximately) from the posterior for the purposes of inference and prediction. ■

One way in which a distributional result such as (2.77) can be useful is in the construction of a 95% **credible interval** $\mathcal{I}$ for the parameter μ; that is, an interval $\mathcal{I}$ such that the probability $\mathbb{P}[\mu \in \mathcal{I} \mid \tau]$ is equal to 0.95. For example, the symmetric 95% credible interval is

$$\mathcal{I} = \left[\overline{x}_n - \frac{S_n}{\sqrt{n-1}}\,t_{n-1;0.975},\ \overline{x}_n + \frac{S_n}{\sqrt{n-1}}\,t_{n-1;0.975}\right],$$

where $t_{n-1;0.975}$ is the 0.975-quantile of the t_{n-1} distribution. Note that the credible interval is not a random object and that the parameter μ is interpreted as a random variable with a distribution. This is unlike the case of classical confidence intervals, where the parameter is nonrandom, but the interval is (the outcome of) a random object. ☞ 666

As a generalization of the 95% Bayesian credible interval we can define a $1-\alpha$ **credible region**, which is any set $\mathcal{R}$ satisfying

$$\mathbb{P}[\boldsymbol{\theta} \in \mathcal{R} \mid \tau_n] = \int_{\boldsymbol{\theta} \in \mathcal{R}} g(\boldsymbol{\theta} \mid \tau_n)\,\mathrm{d}\boldsymbol{\theta} \geq 1 - \alpha. \tag{2.78}$$

Example 2.79 (Bayesian Regularization of Maximum Likelihood) Consider modeling the number of deaths during birth in a maternity ward. Suppose that the hospital data consists of $\tau = \{x_1, \ldots, x_n\}$, with $x_i = 1$ if the i-th baby has died during birth and $x_i = 0$ otherwise, for $i = 1, \ldots, n$. A possible Bayesian model for the data is $\theta \sim \mathcal{U}(0, 1)$ (uniform prior) with $(X_1, \ldots, X_n \mid \theta) \sim_{\text{iid}} \mathsf{Ber}(\theta)$, where θ represents the probability of death. The likelihood is therefore

$$g(\tau \mid \theta) = \prod_{i=1}^{n} \theta^{x_i} (1 - \theta)^{1-x_i} = \theta^s (1 - \theta)^{n-s},$$

where $s = x_1 + \cdots + x_n$ is the total number of deaths. Since $g(\theta) = 1$, the posterior pdf is

$$g(\theta \mid \tau) \propto \theta^s (1 - \theta)^{n-s}, \quad \theta \in [0, 1],$$

which is the pdf of the $\mathsf{Beta}(s + 1, n - s + 1)$ distribution. The normalization constant is $(n + 1)\binom{n}{s}$. The posterior pdf is shown in Figure 2.80 for $(s, n) = (0, 100)$. It is not

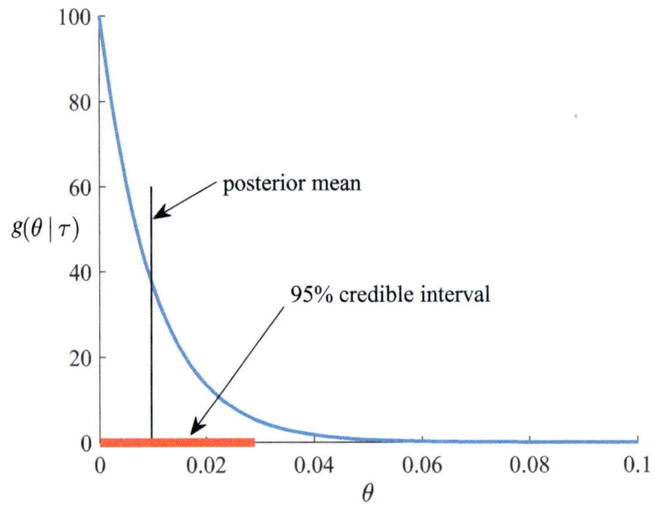

Figure 2.80: Posterior pdf for θ, with $n = 100$ and $s = 0$.

difficult to see that the **maximum a posteriori** (MAP) estimate (the mode or maximizer of the posterior density) is

$$\underset{\theta}{\operatorname{argmax}}\, g(\theta \mid \tau) = \frac{s}{n},$$

which agrees with the maximum likelihood estimate for the true unknown probability of death. Figure 2.80 also shows that the left one-sided 95% credible interval for θ is $[0, 0.0292]$, where 0.0292 is the 0.95 quantile (rounded) of the $\mathsf{Beta}(1, 101)$ distribution.

Observe that when $(s, n) = (0, 100)$ the maximum likelihood estimate $\widehat{\theta} = 0$ infers that deaths at birth are not possible. We know that this inference is wrong — the probability of death can never be zero, it is simply (and fortunately) too small to be inferred accurately from a sample size of $n = 100$. In contrast to the maximum likelihood estimate, the posterior mean $\mathbb{E}[\theta \mid \tau] = (s + 1)/(n + 2)$ is not zero for $(s, n) = (0, 100)$ and provides the more reasonable point estimate of 0.0098 for the probability of death.

In addition, while computing a Bayesian credible interval poses no conceptual difficulties, it is not simple to derive a confidence interval for the maximum likelihood estimate of $\widehat{\theta}$, because the likelihood as a function of θ is not differentiable at $\theta = 0$. As a result of this lack of smoothness, the usual confidence intervals based on the normal approximation cannot be used. ■

We now return to the unsupervised learning setting of Section 2.6, but consider this from a Bayesian perspective. Recall from (2.66) that the Kullback–Leibler risk for an approximating function g is

$$\int f(\tau')[\ln f(\tau') - \ln g(\tau')]\, d\tau',$$

where τ' denotes the test data of size $n' = n$. Since $\int f(\tau') \ln f(\tau')\, d\tau'$ plays no role in minimizing the risk, we consider instead the **cross-entropy risk**, defined as ☞ 134

$$\ell(g) = -\int f(\tau') \ln g(\tau')\, d\tau'.$$

Note that the smallest possible cross-entropy risk is $\ell_n^* = -\int f(\tau') \ln f(\tau')\, d\tau' \leq \ell(g)$.

Assuming that the sets $\mathcal{T}_n$ and $\mathcal{T}_n'$ are comprised of $n + n' = 2n$ iid random variables with density f, we can show (Exercise 28) that the expected generalization risk satisfies

$$\mathbb{E}\,\ell(g_{\mathcal{T}_n}) = \mathbb{E} \ln g(\mathcal{T}_n) - \mathbb{E} \ln g(\mathcal{T}_{2n}) \leq -n\ell_1^* - \mathbb{E} \ln g(\mathcal{T}_{2n}), \tag{2.81}$$

where $g(\tau_n)$ and $g(\tau_{2n})$ are the prior predictive densities of τ_n and τ_{2n}, respectively. In this context, the prior predictive density $g(\mathcal{T}_n)$ is usually called the **model evidence** or **marginal likelihood** for the class $\mathcal{G}_p$. Observe that the expectation of the negative of the marginal log-likelihood, $-\mathbb{E} \ln g(\mathcal{T}_n)$, is the same as the cross-entropy risk between the true unknown joint density $f(\tau_n)$ and the prior predictive density $g(\tau_n)$. This implies, for example, that equality in (2.81) is achieved if and only if $g \equiv f$.

The primary use of the model evidence is in the selection of a model from the set of models $\{\mathcal{G}_p, p = 1, 2, \ldots\}$, indexed by the dimension p of $\boldsymbol{\theta}$. Rather than selecting the model that minimizes the cross-entropy risk, in Bayesian learning we instead select the model that maximizes the model evidence. This is because, provided that all models are assigned equal prior probability, selecting the model with the highest model evidence is equivalent to selecting the model with the maximum Bayesian posterior probability; see Example 2.91. Chapter 6 is entirely dedicated to applying this idea in the context of the Gaussian linear model.

One of the difficulties in using the model evidence is that its integral representation $\int g(\tau_n \mid \boldsymbol{\theta})\, g(\boldsymbol{\theta})\, d\boldsymbol{\theta} = g(\tau_n)$ is rarely available as a simple closed-form formula and typically requires sophisticated approximation methods. One option is to use Markov Chain Monte Carlo estimation methods. Alternatively, when n is much larger than p, we may use the MAP parameter $\overline{\boldsymbol{\theta}}_n := \operatorname{argmax}_{\boldsymbol{\theta}}\, g(\boldsymbol{\theta} \mid \tau_n)$ in an approximation of the model evidence, as in Lemma 2.82 below. ☞ 88

To pave the way for Lemma 2.82, recall the definition of $\mathbf{A}$-strong convexity in (C.37), and the definition of *uniform Lipschitz continuity*. ☞ 592 ☞ 658

Lemma 2.82 (Laplace's Approximation of the Model Evidence): Let $\mathbf{Z}_n \sim \mathcal{N}(\overline{\boldsymbol{\theta}}_n, \boldsymbol{\Sigma}_n/n)$, where the covariance matrices satisfy $\ln|\boldsymbol{\Sigma}_n| = \mathcal{O}(1)$ as $n \to \infty$. Then, the logarithm of the model evidence is lower bounded:

$$\ln g(\tau_n) \geq \mathbb{E} \ln[g(\tau_n \mid \mathbf{Z}_n) g(\mathbf{Z}_n)] + \frac{p}{2} \ln(2\pi \mathrm{e}/n) + \frac{1}{2} \ln|\boldsymbol{\Sigma}_n|. \tag{2.83}$$

Further, if the gradients $\{\nabla \ln g(\boldsymbol{\theta} \mid \tau_n)/n\}$ are uniformly Lipschitz continuous and each $\ln g(\boldsymbol{\theta} \mid \tau_n)/n$ is $\mathbf{A}$-strongly concave in $\boldsymbol{\theta}$, then as $n \to \infty$:

$$\ln g(\tau_n) = \ln[g(\tau_n \mid \overline{\boldsymbol{\theta}}_n) g(\overline{\boldsymbol{\theta}}_n)] - \frac{p}{2} \ln n + \mathcal{O}(1). \tag{2.84}$$

☞ 657 *Proof.* The lower bound (2.83) follows from applying (D.95) in Theorem D.94, where $\ell_n(\boldsymbol{\theta}) := -\ln[g(\tau_n \mid \boldsymbol{\theta}) g(\boldsymbol{\theta})]/n = -\ln g(\boldsymbol{\theta} \mid \tau_n)/n - \ln g(\tau_n)/n$ with global minimizer $\overline{\boldsymbol{\theta}}_n$. Similarly, the expansion (2.84) follows from (D.98) in Lemma D.97 using the same $\ell_n(\boldsymbol{\theta})$ and $\overline{\boldsymbol{\theta}}_n$ as above (and noting that ℓ_n is $\mathbf{A}$-strongly convex), where the gradients $\{\nabla \ell_n\}$ are uniformly Lipschitz continuous. □

We first comment on the assumptions behind Laplaces's approximation (2.84) in Lemma 2.82. It is possible to relax the $\mathbf{A}$-strong concavity assumption to a *local* $\mathbf{A}$-strong concavity that holds only in a small neighborhood of $\overline{\boldsymbol{\theta}}_n$. More importantly, it is always possible to select the prior $g(\boldsymbol{\theta})$ such that the log-posterior $\ln g(\boldsymbol{\theta} \mid \tau_n)$ is $\mathbf{A}$-strongly concave. For example, if the log-likelihood $\ln g(\tau_n \mid \boldsymbol{\theta})$ is concave and the prior $g(\boldsymbol{\theta})$ is the pdf of the $\mathcal{N}(\mathbf{0}, \mathbf{A}^{-1})$ distribution, then the log-posterior is guaranteed to be $\mathbf{A}$-strongly concave.

When the model evidence is difficult to compute, Lemma 2.82 suggests the we can instead maximize the lower bound on the right-hand side of (2.83), where the expectation with respect to $\mathbf{Z}_n$ can be replaced with a Monte Carlo approximation.

Alternatively, (2.84) in Lemma 2.82 suggests that we can use the following large-sample approximation:

$$-2 \ln g(\tau_n) = \underbrace{-2 \ln g(\tau_n \mid \overline{\boldsymbol{\theta}}_n) + p \ln n}_{= \text{BIC}} - 2 \ln g(\overline{\boldsymbol{\theta}}_n) + \mathcal{O}(1). \tag{2.85}$$

The asymptotic approximation on the right-hand side of (2.85) is called the **Bayesian information criterion** (BIC). Again, we prefer the class $\mathcal{G}_p$ with the smallest BIC. The BIC is an accurate approximation to $-2 \ln g(\tau_n)$ when n is significantly larger than p and the prior is such that $\ln g(\overline{\boldsymbol{\theta}}_n) = \mathcal{O}(1)$.

The BIC provides an alternative to the *Akaike information criterion* (AIC) for model
☞ 139 selection. In general, we use the AIC when the goal is to make accurate predictions (AIC approximately minimizes the cross-entropy risk) and we use the BIC when the goal is to select a parsimonious (and possibly the one true) model from the set $\{\mathcal{G}_p, p = 1, 2, \ldots\}$.

Finally, we mention the important result that for large n the Bayesian posterior density $g(\boldsymbol{\theta} \mid \tau_n)$ can frequently be approximated by the normal density $\varphi_{\boldsymbol{\Sigma}_n/n}(\cdot - \overline{\boldsymbol{\theta}})$ of the random variable $\mathbf{Z}_n \sim \mathcal{N}(\overline{\boldsymbol{\theta}}_n, \boldsymbol{\Sigma}_n/n)$, where $-n\boldsymbol{\Sigma}_n^{-1}$ is the Hessian of the log-posterior at its mode $\overline{\boldsymbol{\theta}}_n$.

Theorem 2.86 (Asymptotic Normality of Bayesian Posterior): Suppose that $-\ln g(\boldsymbol{\theta} \mid \tau_n)/n$ is $\mathbf{A}$-strongly convex in $\boldsymbol{\theta}$ and its second-order derivatives are uniformly continuous (in both n and $\boldsymbol{\theta}$) with Hessian matrix $\boldsymbol{\Sigma}_n^{-1}$ at $\overline{\boldsymbol{\theta}}_n$. Then,

$$\mathbb{E} \ln \frac{\varphi_{\boldsymbol{\Sigma}_n/n}(\mathbf{Z}_n - \overline{\boldsymbol{\theta}}_n)}{g(\mathbf{Z}_n \mid \tau_n)} \longrightarrow 0, \quad n \uparrow \infty.$$

In other words, the Kullback–Leibler divergence between the pdf of the $\mathcal{N}(\overline{\boldsymbol{\theta}}_n, \boldsymbol{\Sigma}_n/n)$ distribution and the posterior $g(\cdot \mid \tau_n)$ converges to zero as $n \uparrow \infty$.

Proof. Using the differential entropy of $\mathbf{Z}_n$; see (2.102): ☞ 67

$$\mathcal{H}(\mathbf{Z}_n) = -\int_{\mathbb{R}^p} \varphi_{\boldsymbol{\Sigma}_n/n}(\boldsymbol{z}) \ln \varphi_{\boldsymbol{\Sigma}_n/n}(\boldsymbol{z})\, \mathrm{d}\boldsymbol{z} = \tfrac{p}{2} \ln(2\pi \mathrm{e}/n) + \tfrac{1}{2} \ln |\boldsymbol{\Sigma}_n|,$$

the Kullback–Leibler divergence can be written as

$$\begin{aligned}\mathbb{E} \ln \frac{\varphi_{\boldsymbol{\Sigma}_n/n}(\mathbf{Z}_n - \overline{\boldsymbol{\theta}}_n)}{g(\mathbf{Z}_n \mid \tau_n)} &= \mathbb{E} \ln \varphi_{\boldsymbol{\Sigma}_n/n}(\mathbf{Z}_n - \overline{\boldsymbol{\theta}}_n) - \mathbb{E} \ln g(\mathbf{Z}_n \mid \tau_n) \\ &= -\frac{p}{2} \ln(2\pi \mathrm{e}/n) - \frac{1}{2} \ln |\boldsymbol{\Sigma}_n| - \mathbb{E} \ln[g(\tau_n \mid \mathbf{Z}_n) g(\mathbf{Z}_n)] + \ln g(\tau_n).\end{aligned}$$

Applying (D.100) in Theorem D.99 to the $\mathbf{A}$-convex function $\ell_n(\boldsymbol{\theta}) := -\ln[g(\tau_n \mid \boldsymbol{\theta}) g(\boldsymbol{\theta})]/n$ with global minimizer $\overline{\boldsymbol{\theta}}_n$ yields: ☞ 657

$$\ln g(\tau_n) = \ln[g(\tau_n \mid \overline{\boldsymbol{\theta}}_n) g(\overline{\boldsymbol{\theta}}_n)] + \frac{p}{2} \ln(2\pi/n) + \frac{1}{2} \ln |\boldsymbol{\Sigma}_n| + o(1).$$

Hence, substituting $\ln g(\tau_n)$ simplifies the Kullback–Leibler divergence to:

$$\begin{aligned}\mathbb{E} \ln \frac{\varphi_{\boldsymbol{\Sigma}_n/n}(\mathbf{Z}_n - \overline{\boldsymbol{\theta}}_n)}{g(\mathbf{Z}_n \mid \tau_n)} &= \mathbb{E} \ln \frac{g(\tau_n \mid \overline{\boldsymbol{\theta}}_n) g(\overline{\boldsymbol{\theta}}_n)}{g(\tau_n \mid \mathbf{Z}_n) g(\mathbf{Z}_n)} - \frac{p}{2} + o(1) \\ &= \underbrace{n \left(\mathbb{E}\, \ell_n(\mathbf{Z}_n) - \ell_n(\overline{\boldsymbol{\theta}}_n) \right) - \frac{p}{2}}_{= o(1)} + o(1) = o(1),\end{aligned}$$

where in the last line we applied (D.101), completing the proof. □

The asymptotic normality result of Theorem 2.86 is illustrated numerically in Exercise 30. Additionally, Exercise 30 suggests that the uniform continuity assumption in Theorem 2.86 can be relaxed.

A common application of the asymptotic normality result of Theorem 2.86 is in the approximate computation of a credible region $\mathcal{R}$ for the parameter $\boldsymbol{\theta}$, like the one in (2.78). In particular, with the help of Theorem 2.86, Exercise 31 shows that

$$\sup_{\mathcal{R}} \left| \mathbb{P}[\boldsymbol{\theta} \in \mathcal{R} \mid \tau_n] - \mathbb{P}[\mathbf{Z}_n \in \mathcal{R}] \right| \to 0, \quad n \uparrow \infty. \tag{2.87}$$

In other words, the probability that the parameter lies in $\mathcal{R}$ under the posterior distribution becomes indistinguishable (as $n \uparrow \infty$) from the probability that the Gaussian random vector $\mathbf{Z}_n$ lies in $\mathcal{R}$.

Although the above Bayesian theory has been presented in an unsupervised learning setting, it can be readily extended to the supervised case. We only need to relabel the training set $\mathcal{T}_n$. In particular, when (as is typical for regression models) the training responses $Y_1, \dots, Y_n$ are considered as random variables but the corresponding feature vectors $\boldsymbol{x}_1, \dots, \boldsymbol{x}_n$ are viewed as being fixed, then $\mathcal{T}_n$ is the collection of random responses $\{Y_1, \dots, Y_n\}$. Alternatively, we can simply identify $\mathcal{T}_n$ with the response vector $\boldsymbol{Y} = [Y_1, \dots, Y_n]^\top$. We will adopt this notation in the next example.

Example 2.88 (Polynomial Regression (cont.)) Consider Example 2.29 once again, but now in a Bayesian framework, where the prior knowledge on the model parameters $(\sigma^2, \boldsymbol{\beta})$ is specified by $g(\sigma^2) \propto 1/\sigma^2$ and $\boldsymbol{\beta} \mid \sigma^2 \sim \mathcal{N}(\mathbf{0}, \sigma^2 \mathbf{\Gamma}^2)$ with $\mathbf{\Gamma}$ being a diagonal matrix of positive hyperparameters. Let $\mathbf{\Sigma} := (\mathbf{X}^\top \mathbf{X} + \mathbf{\Gamma}^{-2})^{-1} = \mathbf{\Gamma}(\mathbf{\Gamma}\mathbf{X}^\top\mathbf{X}\mathbf{\Gamma} + \mathbf{I}_p)^{-1}\mathbf{\Gamma}$. Then, the posterior can be written as:

$$g(\boldsymbol{\beta}, \sigma^2 \mid \boldsymbol{y}) = \frac{\exp\left(-\frac{\|\boldsymbol{y}-\mathbf{X}\boldsymbol{\beta}\|^2}{2\sigma^2}\right)}{(2\pi\sigma^2)^{n/2}} \times \frac{\exp\left(-\frac{\|\mathbf{\Gamma}^{-1}\boldsymbol{\beta}\|^2}{2\sigma^2}\right)}{(2\pi\sigma^2)^{p/2}\,|\mathbf{\Gamma}|} \times \frac{1}{\sigma^2} \Big/ g(\boldsymbol{y})$$

$$= \frac{(\sigma^2)^{-(n+p)/2-1}}{(2\pi)^{(n+p)/2}\,|\mathbf{\Gamma}|} \exp\left(-\frac{\|\mathbf{\Sigma}^{-1/2}(\boldsymbol{\beta} - \overline{\boldsymbol{\beta}})\|^2}{2\sigma^2} - \frac{(n+p+2)\,\overline{\sigma}^2}{2\sigma^2}\right) \Big/ g(\boldsymbol{y}),$$

where $\overline{\boldsymbol{\beta}} := \mathbf{\Sigma}\mathbf{X}^\top \boldsymbol{y}$ and $\overline{\sigma}^2 := \boldsymbol{y}^\top(\mathbf{I} - \mathbf{X}\mathbf{\Sigma}\mathbf{X}^\top)\boldsymbol{y}/(n+p+2)$ are the MAP parameters, which may be viewed as estimates of the "true" $\boldsymbol{\beta}^*$ and variance ℓ^*. In addition, the model evidence $g(\boldsymbol{y})$ for $\mathcal{G}_p$ is:

$$g(\boldsymbol{y}) = \iint g(\boldsymbol{\beta}, \sigma^2, \boldsymbol{y})\, \mathrm{d}\boldsymbol{\beta}\, \mathrm{d}\sigma^2$$

$$= \frac{|\mathbf{\Sigma}|^{1/2}}{(2\pi)^{n/2}|\mathbf{\Gamma}|} \int_0^\infty \frac{\exp\left(-\frac{(n+p+2)\,\overline{\sigma}^2}{2\sigma^2}\right)}{(\sigma^2)^{n/2+1}}\, \mathrm{d}\sigma^2$$

$$= \frac{|\mathbf{\Sigma}|^{1/2}\Gamma(n/2)}{|\mathbf{\Gamma}|(\pi(n+p+2)\,\overline{\sigma}^2)^{n/2}},$$

where $\Gamma(n) := \int_0^\infty \mathrm{e}^{-x} x^{n-1} \mathrm{d}x$ denotes the gamma function, not to be confused with the boldface notation for the diagonal matrix $\mathbf{\Gamma}$. Therefore, twice the negative of the marginal log-likelihood is

$$-2 \ln g(\boldsymbol{y}) = n \ln\left[\pi(n+p+2)\,\overline{\sigma}^2\right] - 2 \ln \Gamma(n/2) + 2 \ln |\mathbf{\Gamma}| - \ln |\mathbf{\Sigma}|.$$

On the other hand, the minus of the log-likelihood of $\boldsymbol{Y}$ can be written as

$$-\ln g(\boldsymbol{y} \mid \boldsymbol{\beta}, \sigma^2) = \frac{\|\boldsymbol{y} - \mathbf{X}\boldsymbol{\beta}\|^2}{2\sigma^2} + \frac{n}{2}\ln(2\pi\sigma^2)$$

$$= \frac{\|\mathbf{\Sigma}^{-1/2}(\boldsymbol{\beta} - \overline{\boldsymbol{\beta}})\|^2}{2\sigma^2} + \frac{(n+p+2)\,\overline{\sigma}^2}{2\sigma^2} + \frac{n}{2}\ln(2\pi\sigma^2).$$

Therefore, the BIC approximation (2.85) is

$$-2 \ln g(\boldsymbol{y} \mid \overline{\boldsymbol{\beta}}, \overline{\sigma}^2) + (p+1)\ln n = n \ln(2\pi\mathrm{e}\,\overline{\sigma}^2) + (p+1)\ln n + (p+2), \tag{2.89}$$

where the extra $\ln n$ term in $(p + 1)\ln n$ is due to the inclusion of σ^2 in $\boldsymbol{\theta} = (\sigma^2, \boldsymbol{\beta})$. Figure 2.90 shows $-2\ln g(\mathbf{y})$ and the corresponding BIC approximation (2.89), where we used a hyperparameter $\boldsymbol{\Gamma} = 10^2 \times \mathbf{I}_p$ for the prior density of $\boldsymbol{\beta}$. We can see that both approximations exhibit a pronounced minimum at $p = 4$, thus identifying the true polynomial regression model. Compare the overall qualitative shape of the cross-entropy risk estimate with the shape of the square-error risk estimate in Figure 2.44.

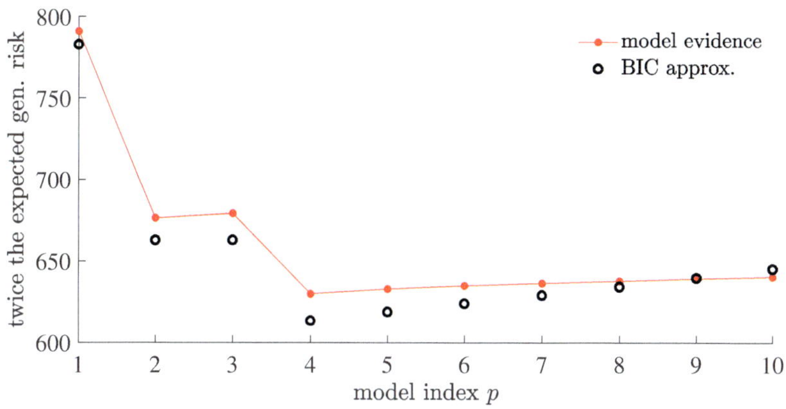

Figure 2.90: The BIC and marginal likelihood used for model selection.

It is possible to give the model complexity parameter p a Bayesian treatment, in which we define a prior density on the set of all models under consideration. For example, let $g(p)$, $p = 1, \ldots, m$ be a prior density on m candidate models. Treating the model complexity index p as an additional parameter to $\boldsymbol{\theta} \in \mathbb{R}^p$, and applying Bayes' formula, the posterior for $(\boldsymbol{\theta}, p)$ can be written as:

$$\begin{aligned} g(\boldsymbol{\theta}, p \mid \tau) &= g(\boldsymbol{\theta} \mid p, \tau) \times g(p \mid \tau) \\ &= \underbrace{\frac{g(\tau \mid \boldsymbol{\theta}, p)\, g(\boldsymbol{\theta} \mid p)}{g(\tau \mid p)}}_{\text{posterior of } \boldsymbol{\theta} \text{ given model } p} \times \underbrace{\frac{g(\tau \mid p)\, g(p)}{g(\tau)}}_{\text{posterior of model } p}. \end{aligned}$$

The model evidence for a fixed p is now interpreted as the prior predictive density of τ, conditional on the model p:

$$g(\tau \mid p) = \int g(\tau \mid \boldsymbol{\theta}, p)\, g(\boldsymbol{\theta} \mid p)\, \mathrm{d}\boldsymbol{\theta},$$

and the quantity

$$g(\tau) = \sum_{p=1}^{m} g(\tau \mid p)\, g(p)$$

is interpreted as the marginal likelihood of all the m candidate models. Finally, a simple method for model selection is to pick the index $\widehat{p}$ with the largest posterior probability:

$$\widehat{p} = \underset{p}{\operatorname{argmax}}\, g(p \mid \tau) = \underset{p}{\operatorname{argmax}}\, g(\tau \mid p)\, g(p).$$

■ **Example 2.91 (Polynomial Regression (cont.))** Let us revisit Example 2.88 by giving the parameter $p = 1, \ldots, m$, with $m = 10$, a Bayesian treatment. Recall that we used the notation $\tau = \boldsymbol{y}$ in that example. We assume that the prior $g(p) = 1/m$ is flat and uninformative so that the posterior is given by

$$g(p \mid \boldsymbol{y}) \propto g(\boldsymbol{y} \mid p) = \frac{|\boldsymbol{\Sigma}|^{1/2}\,\Gamma(n/2)}{|\boldsymbol{\Gamma}|(\pi(n+p+2)\,\overline{\sigma}^2)^{n/2}},$$

where all quantities in $g(\boldsymbol{y} \mid p)$ are computed using the first p columns of $\mathbf{X}$. Figure 2.92 shows the resulting posterior density $g(p \mid \boldsymbol{y})$. The figure also shows the posterior density $\widehat{g}(\boldsymbol{y} \mid p) / \sum_{p=1}^{10} \widehat{g}(\boldsymbol{y} \mid p)$, where

$$\widehat{g}(\boldsymbol{y} \mid p) := \exp\left(-\frac{n\ln(2\pi\mathrm{e}\,\overline{\sigma}^2) + (p+1)\ln n + (p+2)}{2}\right)$$

is derived from the BIC approximation (2.89). In both cases, there is a clear maximum at $p = 4$, suggesting that a third-degree polynomial is the most appropriate model for the data.

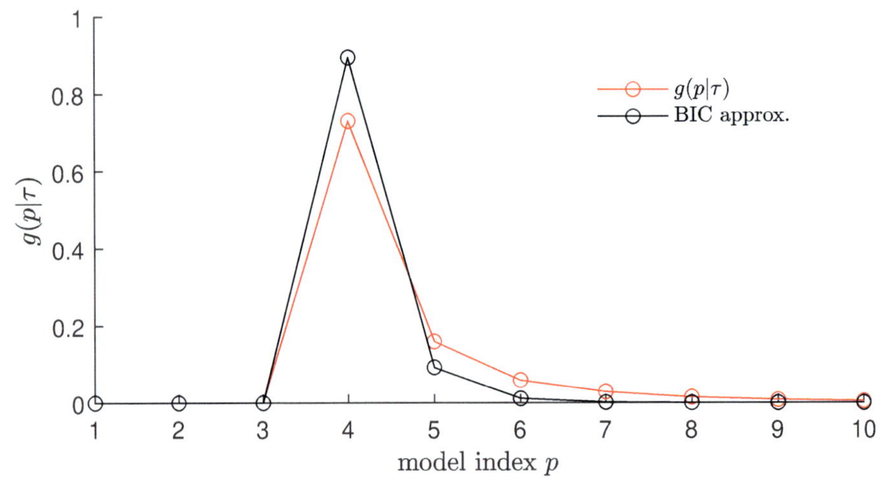

Figure 2.92: Posterior probabilities for each polynomial model of degree $p - 1$.

■

Suppose that we wish to compare two models, say model $p = 1$ and model $p = 2$. Instead of computing the posterior $g(p \mid \tau)$ explicitly, we can compare the posterior odds ratio:

$$\frac{g(p=1 \mid \tau)}{g(p=2 \mid \tau)} = \frac{g(p=1)}{g(p=2)} \times \underbrace{\frac{g(\tau \mid p=1)}{g(\tau \mid p=2)}}_{\text{Bayes factor } B_{1\mid 2}}.$$

This gives rise to the **Bayes factor** $B_{i \mid j}$, whose value signifies the strength of the evidence in favor of model i over model j. In particular $B_{i \mid j} > 1$ means that the evidence in favor for model i is larger.

Example 2.93 (Savage–Dickey Ratio) Suppose that we have two models. Model $p = 2$ has a likelihood $g(\tau \mid \mu, \nu, p = 2)$, depending on two parameters. Model $p = 1$ has the same functional form for the likelihood but now ν is fixed to some (known) ν_0; that is, $g(\tau \mid \mu, p = 1) = g(\tau \mid \mu, \nu = \nu_0, p = 2)$. We also assume that the prior information on μ for model 1 is the same as that for model 2, conditioned on $\nu = \nu_0$. That is, we assume $g(\mu \mid p = 1) = g(\mu \mid \nu = \nu_0, p = 2)$. As model 2 contains model 1 as a special case, the latter is said to be *nested* inside model 2. We can formally write (see also Exercise 34):

$$\begin{aligned} g(\tau \mid p = 1) &= \int g(\tau \mid \mu, p = 1)\, g(\mu \mid p = 1)\, \mathrm{d}\mu \\ &= \int g(\tau \mid \mu, \nu = \nu_0, p = 2)\, g(\mu \mid \nu = \nu_0, p = 2)\, \mathrm{d}\mu \\ &= g(\tau \mid \nu = \nu_0, p = 2) = \frac{g(\tau, \nu = \nu_0 \mid p = 2)}{g(\nu = \nu_0 \mid p = 2)}. \end{aligned}$$

Hence, the Bayes factor simplifies to

$$B_{1\,|\,2} = \frac{g(\tau \mid p = 1)}{g(\tau \mid p = 2)} = \frac{g(\tau, \nu = \nu_0 \mid p = 2)}{g(\nu = \nu_0 \mid p = 2)} \Big/ g(\tau \mid p = 2) = \frac{g(\nu = \nu_0 \mid \tau, p = 2)}{g(\nu = \nu_0 \mid p = 2)}.$$

In other words, $B_{1\,|\,2}$ is the ratio of the posterior density to the prior density of ν, evaluated at $\nu = \nu_0$ and both under the unrestricted model $p = 2$. This ratio of posterior to prior densities is called the **Savage–Dickey density ratio**.

Whether to use a classical (frequentist) or Bayesian model is largely a question of convenience. Classical inference is useful because it comes with a huge repository of ready-to-use results, and requires no (subjective) prior information on the parameters. Bayesian models are useful because the whole theory is based on the elegant Bayes' formula, and uncertainty in the inference (e.g., confidence intervals) can be quantified much more naturally (e.g., credible intervals). A usual practice is to "Bayesify" a classical model, simply by adding some prior information on the parameters.

Further Reading

A popular textbook on statistical learning is [81]. Accessible treatments of mathematical statistics can be found, for example, in [41], [101], and [178]. More advanced treatments are given in [15], [39], and [108]. A good overview of modern-day statistical inference is given in [54]. Classical references on pattern classification and machine learning are [17] and [53]. For advanced learning theory including information theory and Rademacher complexity, we refer to [45] and [153]. An applied reference for Bayesian inference is [69]. For a survey of numerical techniques relevant to computational statistics, see [128].

Exercises

1. Suppose that the loss function is the piecewise linear function

$$\text{Loss}(y, \widehat{y}) = \alpha\, (\widehat{y} - y)_+ + \beta\, (y - \widehat{y})_+, \quad \alpha, \beta > 0,$$

where c_+ is equal to c if $c > 0$, and zero otherwise. Show that the minimizer of the risk $\ell(g) = \mathbb{E}\,\text{Loss}(Y, g(\boldsymbol{X}))$ satisfies

$$\mathbb{P}[Y < g^*(\boldsymbol{x}) \mid \boldsymbol{X} = \boldsymbol{x}] = \frac{\beta}{\alpha + \beta}.$$

In other words, $g^*(\boldsymbol{x})$ is the $\beta/(\alpha+\beta)$ quantile of Y, conditional on $\boldsymbol{X} = \boldsymbol{x}$.

2. Show that, for the squared-error loss, the risk $\ell(g)$ of any prediction function g satisfies:

$$\ell(g) = \ell^* + \mathbb{E}(g(\boldsymbol{X}) - g^*(\boldsymbol{X}))^2, \tag{2.94}$$

where $\ell^* = \ell(g^*)$ is the irreducible risk. Hence, deduce that the approximation error $\ell(g^{\mathcal{G}}) - \ell(g^*)$ in (2.27) is equal to $\mathbb{E}(g^{\mathcal{G}}(\boldsymbol{X}) - g^*(\boldsymbol{X}))^2$. Hint: Expand $\ell(g^{\mathcal{G}}) = \mathbb{E}(Y - g^*(\boldsymbol{X}) + g^*(\boldsymbol{X}) - g^{\mathcal{G}}(\boldsymbol{X}))^2$.

3. Suppose $\mathcal{G}$ is the class of *linear* functions. A linear function evaluated at a feature $\boldsymbol{x}$ can be described as $g(\boldsymbol{x}) = \boldsymbol{\beta}^\top \boldsymbol{x}$ for some parameter vector $\boldsymbol{\beta}$ of appropriate dimension. Denote $g^{\mathcal{G}}(\boldsymbol{x}) = \boldsymbol{x}^\top \boldsymbol{\beta}^{\mathcal{G}}$ and $g^{\mathcal{G}}_{\tau}(\boldsymbol{x}) = \boldsymbol{x}^\top \widehat{\boldsymbol{\beta}}$. Show that

$$\mathbb{E}\left(g^{\mathcal{G}}_{\tau}(\boldsymbol{X}) - g^*(\boldsymbol{X})\right)^2 = \mathbb{E}\left(\boldsymbol{X}^\top \widehat{\boldsymbol{\beta}} - \boldsymbol{X}^\top \boldsymbol{\beta}^{\mathcal{G}}\right)^2 + \mathbb{E}\left(\boldsymbol{X}^\top \boldsymbol{\beta}^{\mathcal{G}} - g^*(\boldsymbol{X})\right)^2.$$

Hence, deduce that the statistical error in (2.27) is $\ell(g^{\mathcal{G}}_{\tau}) - \ell(g^{\mathcal{G}}) = \mathbb{E}\,(g^{\mathcal{G}}_{\tau}(\boldsymbol{X}) - g^{\mathcal{G}}(\boldsymbol{X}))^2$.

4. Show that formula (2.38) holds for the 0–1 loss with 0–1 response.

5. Let $\boldsymbol{X}$ be an n-dimensional normal random vector with mean vector $\boldsymbol{\mu}$ and invertible covariance matrix $\boldsymbol{\Sigma}$. Show that $\boldsymbol{X}$ has joint probability density

$$f_{\boldsymbol{X}}(\boldsymbol{x}) = \frac{1}{\sqrt{(2\pi)^n\,|\boldsymbol{\Sigma}|}}\, \mathrm{e}^{-\frac{1}{2}(\boldsymbol{x}-\boldsymbol{\mu})^\top \boldsymbol{\Sigma}^{-1}(\boldsymbol{x}-\boldsymbol{\mu})}, \quad \boldsymbol{x} \in \mathbb{R}^n.$$

6. This exercise gives a limiting characterization of the pseudo-inverse and pseudo-determinant of a matrix.

(a) Using the compact SVD of $\mathbf{X} \in \mathbb{R}^{n \times p}$ with $\text{rank}(\mathbf{X}) = r$, show that the pseudo-inverse $\mathbf{X}^+$ can be represented as the limits:

☞ 532

$$\mathbf{X}^+ = \lim_{\lambda \downarrow 0} (\lambda \mathbf{I}_p + \mathbf{X}^\top \mathbf{X})^{-1} \mathbf{X}^\top = \lim_{\lambda \downarrow 0} \mathbf{X}^\top (\lambda \mathbf{I}_n + \mathbf{X}\mathbf{X}^\top)^{-1}. \tag{2.95}$$

(b) Suppose that $\mathbf{A} \in \mathbb{R}^{n \times n}$ is a square matrix with $\text{rank}(\mathbf{A}) = r < n$. The **pseudo-determinant** of $\mathbf{A}$ is defined as:

$$\det(\mathbf{A})^+ := \lim_{\varepsilon \downarrow 0} \frac{\det(\mathbf{A} + \varepsilon \mathbf{I}_n)}{\varepsilon^{n-r}}.$$

☞ 528

Use Theorem A.31 and Example A.32 to show that $\det(\mathbf{A})^+$ equals the product of all the nonzero eigenvalues of $\mathbf{A}$.

7. Use Definition A.10 of the pseudo-inverse $\mathbf{A}^+$ to prove the following identities. ☞ 516

(a)

$$\mathbf{A}^+ = (\mathbf{A}^\top\mathbf{A})^+\mathbf{A}^\top = \mathbf{A}^\top(\mathbf{A}\mathbf{A}^\top)^+.$$

(b)

$$(\mathbf{A}\mathbf{B})^+ = \mathbf{B}^+\mathbf{A}^+,$$

where $\mathbf{A} \in \mathbb{R}^{n\times p}$ and $\mathbf{B} \in \mathbb{R}^{p\times m}$ have linearly independent columns and rows, respectively.

8. Let $\mathbf{X} \in \mathbb{R}^{n\times p}$ be a model matrix and let $\widehat{\boldsymbol{\beta}} := \mathbf{X}^+\boldsymbol{y}$ be the (minimum-norm) least-squares estimate.

(a) Prove the Pythagorean identity

$$\|\boldsymbol{y} - \mathbf{X}\boldsymbol{\beta}\|^2 = \|\boldsymbol{y} - \mathbf{X}\widehat{\boldsymbol{\beta}}\|^2 + \|\mathbf{X}(\widehat{\boldsymbol{\beta}} - \boldsymbol{\beta})\|^2,$$

where $\boldsymbol{\beta} \in \mathbb{R}^p$ is arbitrary. Hence, conclude that for all $\boldsymbol{\beta} \in \mathbb{R}^p$

$$\|\boldsymbol{y} - \mathbf{X}\widehat{\boldsymbol{\beta}}\|^2 \leq \|\boldsymbol{y} - \mathbf{X}\boldsymbol{\beta}\|^2.$$

(b) Show that if $\mathbf{X}$ has linearly independent columns (full column rank), then $\widehat{\boldsymbol{\beta}}$ is the unique global minimizer of $\|\boldsymbol{y} - \mathbf{X}\boldsymbol{\beta}\|^2$. That is, show that

$$\|\boldsymbol{y} - \mathbf{X}\boldsymbol{\beta}\|^2 > \|\boldsymbol{y} - \mathbf{X}\widehat{\boldsymbol{\beta}}\|^2$$

for any $\boldsymbol{\beta} \neq \widehat{\boldsymbol{\beta}}$.

(c) Let

$$\mathcal{V} := \{\boldsymbol{v} \in \mathbb{R}^p : \boldsymbol{v} = \mathbf{X}^\top\boldsymbol{\alpha}, \boldsymbol{\alpha} \in \mathbb{R}^n\}$$

be the linear subspace of $\mathbb{R}^p$ spanned by the rows of $\mathbf{X}$, and define the **null space** of $\mathbf{X}$ as:

$$\mathcal{V}^\perp := \{\boldsymbol{v} \in \mathbb{R}^p : \mathbf{X}\boldsymbol{v} = \mathbf{0}\}.$$

Show that $\mathcal{V}^\perp$ is the orthogonal complement of $\mathcal{V}$, that is, $\boldsymbol{a}^\top\boldsymbol{b} = 0$ for any $\boldsymbol{a} \in \mathcal{V}$ and $\boldsymbol{b} \in \mathcal{V}^\perp$.

(d) Show that $\widehat{\boldsymbol{\beta}} \in \mathcal{V}$ and that

$$\mathcal{V}^\perp = \{\boldsymbol{v} \in \mathbb{R}^p : \boldsymbol{v} = (\mathbf{I}_p - \mathbf{X}^+\mathbf{X})\boldsymbol{u},\ \boldsymbol{u} \in \mathbb{R}^p\}.$$

(e) Show that the set of all vectors of the form $\widehat{\boldsymbol{\beta}}+\boldsymbol{u}$, $\boldsymbol{u} \in \mathcal{V}^\perp$ is the set of all minimizers of $\|\boldsymbol{y} - \mathbf{X}\boldsymbol{\beta}\|^2$, that is,

$$\underset{\boldsymbol{\beta}}{\operatorname{argmin}} \|\boldsymbol{y} - \mathbf{X}\boldsymbol{\beta}\|^2 = \{\widehat{\boldsymbol{\beta}} + \boldsymbol{u}, \boldsymbol{u} \in \mathcal{V}^\perp\}.$$

In other words, a minimizer of $\|\boldsymbol{y} - \mathbf{X}\boldsymbol{\beta}\|^2$ is the minimum-norm minimizer if and only if it lies in the column space $\mathcal{V}$ of $\mathbf{X}$.

(f) Show that of all minimizers of the form $\widehat{\boldsymbol{\beta}} + \boldsymbol{u}$, where $\boldsymbol{u} \in \mathcal{V}^{\perp}$, the minimizer $\widehat{\boldsymbol{\beta}}$ is the one with the smallest norm, that is,

$$\|\widehat{\boldsymbol{\beta}} + \boldsymbol{u}\|^2 \geq \|\widehat{\boldsymbol{\beta}}\|^2, \quad \text{for all } \boldsymbol{u} \in \mathcal{V}^{\perp}.$$

☞ 594 9. Use the Lagrangian method to find the unique solution of the normal equations (2.22) that has minimal Euclidean norm. In other words, find the solution to the constrained optimization:

$$\min_{\boldsymbol{\beta}} \frac{1}{2}\|\boldsymbol{\beta}\|^2$$
$$\text{subject to:} \quad \mathbf{X}^\top \boldsymbol{y} - \mathbf{X}^\top \mathbf{X}\boldsymbol{\beta} = \mathbf{0}.$$

☞ 606 Hence, explain how one can use the coordinate descent Algorithm C.68 to compute the minimum-norm least-squares estimate (2.18).

10. Consider minimizing the quadratic function $f(\boldsymbol{\beta}) := (\|\boldsymbol{y} - \mathbf{X}\boldsymbol{\beta}\|^2 - \|\boldsymbol{y}\|^2)/2$ using the method of gradient descent in Example C.61; that is, starting from some $\boldsymbol{b}_0$ we iterate ☞ 603 until convergence:

$$\boldsymbol{b}_{t+1} = \boldsymbol{b}_t - \alpha_t\, \boldsymbol{g}_t, \quad t = 0, 1, \ldots,$$

where $\boldsymbol{g}_t := \nabla f(\boldsymbol{b}_t) = \mathbf{X}^\top(\mathbf{X}\boldsymbol{b}_t - \boldsymbol{y})$.

(a) Prove that the optimal step size α_t is:

$$\underset{\alpha}{\operatorname{argmin}}\, f(\boldsymbol{b}_t - \alpha\, \boldsymbol{g}_t) = \frac{\|\boldsymbol{g}_t\|^2}{\|\mathbf{X}\boldsymbol{g}_t\|^2}.$$

Hence, write pseudo-code that implements the gradient descent algorithm.

(b) Assuming that $\boldsymbol{b}_0 = \mathbf{0}$ and that the gradient descent iteration converges, prove that the algorithm must converge to the minimum-norm least-squares estimate (2.18).

11. Suppose that in the polynomial regression Example 2.13 we select the linear class of functions $\mathcal{G}_p$ with $p \geq 4$. Then, $g^* \in \mathcal{G}_p$ and the approximation error is zero, because $g^{\mathcal{G}_p}(\boldsymbol{x}) = g^*(\boldsymbol{x}) = \boldsymbol{x}^\top\boldsymbol{\beta}$, where $\boldsymbol{\beta} = [10, -140, 400, -250, 0, \ldots, 0]^\top \in \mathbb{R}^p$. Use ☞ 635 the tower property to show that the learner $g_{\tau}(\boldsymbol{x}) = \boldsymbol{x}^\top\widehat{\boldsymbol{\beta}}$ with $\widehat{\boldsymbol{\beta}} = \mathbf{X}^+\boldsymbol{y}$, assuming $\text{rank}(\mathbf{X}) \geq 4$, is **unbiased**:

$$\mathbb{E}\, g_{\mathcal{T}}(\boldsymbol{x}) = g^*(\boldsymbol{x}).$$

12. (Exercise 11 continued.) Observe that the learner $g_{\mathcal{T}}$ can be written as a linear combination of the response variable: $g_{\mathcal{T}}(\boldsymbol{x}) = \boldsymbol{x}^\top\mathbf{X}^+\boldsymbol{Y}$. Prove that for any learner of the form $\boldsymbol{x}^\top\mathbf{A}\boldsymbol{y}$, where $\mathbf{A} \in \mathbb{R}^{p\times n}$ is some matrix and that satisfies $\mathbb{E}_{\mathbf{X}}[\boldsymbol{x}^\top\mathbf{A}\boldsymbol{Y}] = g^*(\boldsymbol{x})$, we have

$$\mathbb{V}\text{ar}_{\mathbf{X}}[\boldsymbol{x}^\top\mathbf{X}^+\boldsymbol{Y}] \leq \mathbb{V}\text{ar}_{\mathbf{X}}[\boldsymbol{x}^\top\mathbf{A}\boldsymbol{Y}],$$

where the equality is achieved for $\mathbf{A} = \mathbf{X}^+$. This is called the **Gauss–Markov inequality**. Hence, using the Gauss–Markov inequality deduce that for the unconditional variance:

$$\mathbb{V}\text{ar}\, g_{\mathcal{T}}(\boldsymbol{x}) \leq \mathbb{V}\text{ar}[\boldsymbol{x}^\top\mathbf{A}\boldsymbol{Y}].$$

Deduce that $\mathbf{A} = \mathbf{X}^+$ also minimizes the expected generalization risk.

13. Consider again the polynomial regression analysis in Example 2.13. Use the fact that $\mathbb{E}_{\mathbf{X}} \widehat{\boldsymbol{\beta}} = \mathbf{X}^{+}\boldsymbol{h}^*(\boldsymbol{u})$, where $\boldsymbol{h}^*(\boldsymbol{u}) = \mathbb{E}[\boldsymbol{Y} \,|\, \boldsymbol{U} = \boldsymbol{u}] = [h^*(u_1), \ldots, h^*(u_n)]^\top$, to show that the expected in-sample risk is:

$$\mathbb{E}_{\mathbf{X}}\, \ell_{\text{in}}(g_{\mathcal{T}}) = \ell^* + \frac{\|\boldsymbol{h}^*(\boldsymbol{u})\|^2 - \|\mathbf{X}\mathbf{X}^{+}\boldsymbol{h}^*(\boldsymbol{u})\|^2}{n} + \frac{\ell^* p}{n}.$$

Also, use Theorem D.27 to show that the expected statistical error is: ☞ 634

$$\mathbb{E}_{\mathbf{X}}\, (\widehat{\boldsymbol{\beta}} - \boldsymbol{\beta})^\top \mathbf{H}_p (\widehat{\boldsymbol{\beta}} - \boldsymbol{\beta}) = \ell^* \text{tr}(\mathbf{X}^{+}(\mathbf{X}^{+})^\top \mathbf{H}_p) + (\mathbf{X}^{+}\boldsymbol{h}^*(\boldsymbol{u}) - \boldsymbol{\beta})^\top \mathbf{H}_p (\mathbf{X}^{+}\boldsymbol{h}^*(\boldsymbol{u}) - \boldsymbol{\beta}).$$

14. Consider the setting of the polynomial regression in Example 2.29. Use Theorem D.92 to prove that ☞ 656

$$\sqrt{n}\,(\widehat{\boldsymbol{\beta}}_n - \boldsymbol{\beta}_p) \xrightarrow{\text{d}} \mathcal{N}\left(\mathbf{0},\ \ell^* \mathbf{H}_p^{-1} + \mathbf{H}_p^{-1}\mathbf{M}_p\mathbf{H}_p^{-1}\right), \tag{2.96}$$

where $\mathbf{M}_p := \mathbb{E}[\boldsymbol{X}\boldsymbol{X}^\top (g^*(\boldsymbol{X}) - g^{\mathcal{G}_p}(\boldsymbol{X}))^2]$ is the matrix with (i, j)-th entry:

$$\int_0^1 u^{i+j-2}(h^{\mathcal{H}_p}(u) - h^*(u))^2 \,\text{d}u,$$

and $\mathbf{H}_p^{-1}$ is the $p \times p$ **inverse Hilbert matrix** with (i, j)-th entry:

$$(-1)^{i+j}(i + j - 1)\binom{p + i - 1}{p - j}\binom{p + j - 1}{p - i}\binom{i + j - 2}{i - 1}^2.$$

Observe that $\mathbf{M}_p = \mathbf{0}$ for $p \geq 4$, so that the matrix $\mathbf{M}_p$ term is due to choosing a restrictive class $\mathcal{G}_p$ that does not contain the true prediction function.

15. In Example 2.29 we saw that the statistical error can be expressed (see (2.32)) as

$$\int_0^1 \left([1, \ldots, u^{p-1}](\widehat{\boldsymbol{\beta}} - \boldsymbol{\beta}_p)\right)^2 \text{d}u = (\widehat{\boldsymbol{\beta}} - \boldsymbol{\beta}_p)^\top \mathbf{H}_p (\widehat{\boldsymbol{\beta}} - \boldsymbol{\beta}_p).$$

By Exercise 14 the random vector $\mathbf{Z}_n := \sqrt{n}(\widehat{\boldsymbol{\beta}}_n - \boldsymbol{\beta}_p)$ has asymptotically a multivariate normal distribution with mean vector $\mathbf{0}$ and covariance matrix

$$\mathbf{V} := \ell^* \mathbf{H}_p^{-1} + \mathbf{H}_p^{-1}\mathbf{M}_p\mathbf{H}_p^{-1}.$$

Use Theorem D.27 to show that the *expected* statistical error is asymptotically ☞ 634

$$\mathbb{E}\,(\widehat{\boldsymbol{\beta}} - \boldsymbol{\beta}_p)^\top \mathbf{H}_p (\widehat{\boldsymbol{\beta}} - \boldsymbol{\beta}_p) \simeq \frac{\ell^* p}{n} + \frac{\text{tr}(\mathbf{M}_p \mathbf{H}_p^{-1})}{n} + o(1/n), \quad n \to \infty. \tag{2.97}$$

Plot this large-sample approximation of the expected statistical error and compare it with the outcome of the statistical error.

We note a subtle technical detail: In general, convergence in distribution does not imply convergence in L^p-norm (see Example D.72), and so here we have implicitly assumed that ☞ 648

$$\|\mathbf{Z}_n\| \xrightarrow{\text{d}} \textsf{Dist.} \Rightarrow \|\mathbf{Z}_n\| \xrightarrow{L^2} \text{constant} := \lim_{n \uparrow \infty} \mathbb{E}\|\mathbf{Z}_n\|.$$

16. Consider again Example 2.29. The result in (2.96) suggests that $\mathbb{E}\widehat{\boldsymbol{\beta}}_n \to \boldsymbol{\beta}_p$ as $n \to \infty$, where $\boldsymbol{\beta}_p$ is the solution in the class $\mathcal{G}_p$ given in (2.30). Thus, the large-sample approximation of the pointwise bias of the learner at $\boldsymbol{x} = [1, \ldots, u^{p-1}]^\top$:

$$g_{\mathcal{T}_n}^{\mathcal{G}_p}(\boldsymbol{x}) = \langle \boldsymbol{x}, \widehat{\boldsymbol{\beta}}_n \rangle$$

is the following:

$$\mathbb{E}\, g_{\mathcal{T}_n}^{\mathcal{G}_p}(\boldsymbol{x}) - g^*(\boldsymbol{x}) = [1, \ldots, u^{p-1}]\, \boldsymbol{\beta}_p - [1, u, u^2, u^3]\, \boldsymbol{\beta}^* + o(1), \quad n \to \infty.$$

Use Python to reproduce Figure 2.98, which shows the (large-sample) pointwise squared bias of the learner for $p \in \{1, 2, 3\}$. Note how the bias is larger near the endpoints $u = 0$ and $u = 1$. Explain why the areas under the curves correspond to the approximation errors.

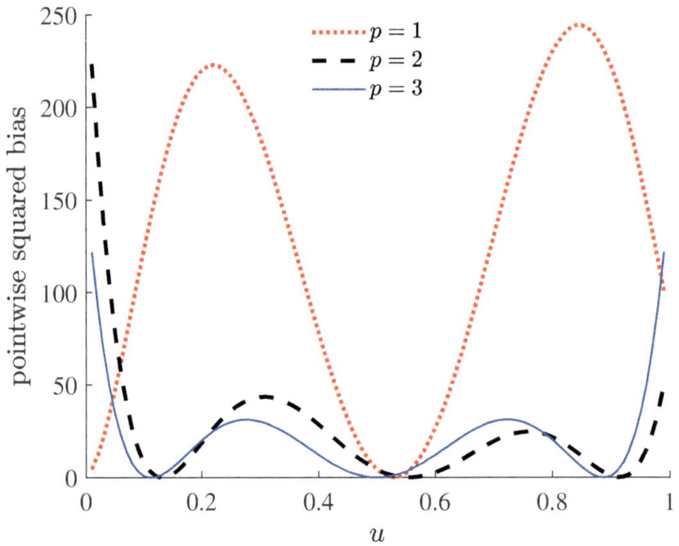

Figure 2.98: The large-sample pointwise squared bias of the learner for $p = 1, 2, 3$. The bias is zero for $p \geq 4$.

17. For our running Example 2.29 we can use (2.96) to derive a large-sample approximation of the pointwise variance of the learner $g_{\mathcal{T}_n}(\boldsymbol{x}) = \boldsymbol{x}^\top \widehat{\boldsymbol{\beta}}_n$. In particular, show that for large n

$$\mathbb{V}\mathrm{ar}\, g_{\mathcal{T}_n}(\boldsymbol{x}) = \frac{\ell^* \boldsymbol{x}^\top \mathbf{H}_p^{-1} \boldsymbol{x}}{n} + \frac{\boldsymbol{x}^\top \mathbf{H}_p^{-1} \mathbf{M}_p \mathbf{H}_p^{-1} \boldsymbol{x}}{n} + o(1/n), \quad n \to \infty. \tag{2.99}$$

Figure 2.100 shows this (large-sample) variance of the learner for different values of the predictor u and model index p. Observe that the variance ultimately increases in p and that it is smaller at $u = 1/2$ than closer to the endpoints $u = 0$ or $u = 1$. Since the bias is also larger near the endpoints, we deduce that the pointwise mean squared error (2.34) is larger near the endpoints of the interval $[0, 1]$ than near its middle. In other words, the error is much smaller in the center of the data cloud than near its periphery.

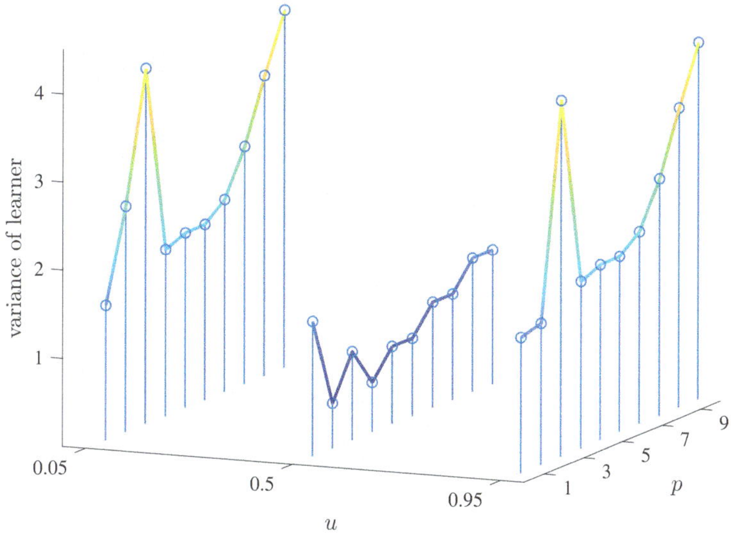

Figure 2.100: The pointwise variance of the learner for various pairs of p and u.

18. Denote the pdf of the $\mathcal{N}(\mathbf{0}, \boldsymbol{\Sigma})$ distribution by $\varphi_{\boldsymbol{\Sigma}}(\cdot)$, and let

$$\mathcal{D}(\boldsymbol{\mu}_0, \boldsymbol{\Sigma}_0 \,|\, \boldsymbol{\mu}_1, \boldsymbol{\Sigma}_1) = \int_{\mathbb{R}^d} \varphi_{\boldsymbol{\Sigma}_0}(\boldsymbol{x} - \boldsymbol{\mu}_0) \ln \frac{\varphi_{\boldsymbol{\Sigma}_0}(\boldsymbol{x} - \boldsymbol{\mu}_0)}{\varphi_{\boldsymbol{\Sigma}_1}(\boldsymbol{x} - \boldsymbol{\mu}_1)} \, \mathrm{d}\boldsymbol{x}$$

be the Kullback–Leibler divergence between the densities of the $\mathcal{N}(\boldsymbol{\mu}_0, \boldsymbol{\Sigma}_0)$ and $\mathcal{N}(\boldsymbol{\mu}_1, \boldsymbol{\Sigma}_1)$ distributions on $\mathbb{R}^d$.

(a) Show that $2\mathcal{D}(\boldsymbol{\mu}_0, \boldsymbol{\Sigma}_0 \,|\, \boldsymbol{\mu}_1, \boldsymbol{\Sigma}_1)$ is equal to:

$$\mathrm{tr}(\boldsymbol{\Sigma}_1^{-1}\boldsymbol{\Sigma}_0) - \ln |\boldsymbol{\Sigma}_1^{-1}\boldsymbol{\Sigma}_0| + (\boldsymbol{\mu}_1 - \boldsymbol{\mu}_0)^\top \boldsymbol{\Sigma}_1^{-1}(\boldsymbol{\mu}_1 - \boldsymbol{\mu}_0) - d. \tag{2.101}$$

(b) The **differential entropy** of a random vector $\boldsymbol{X} \sim f$ is defined as

$$\mathcal{H}(\boldsymbol{X}) = -\int f(\boldsymbol{x}) \ln f(\boldsymbol{x}) \, \mathrm{d}\boldsymbol{x}.$$

Prove that the differential entropy of a d-dimensional Gaussian random vector $\boldsymbol{X} \sim \mathcal{N}(\boldsymbol{\mu}, \boldsymbol{\Sigma})$ is given by

$$\mathcal{H}(\boldsymbol{X}) = \frac{d}{2} \ln(2\pi \mathrm{e}) + \frac{1}{2} \ln |\boldsymbol{\Sigma}|. \tag{2.102}$$

19. Let $h : \boldsymbol{x} \mapsto \mathbb{R}$ be a convex function and let $\boldsymbol{X}$ be a random variable. Use the subgradient definition of convexity to prove **Jensen's inequality**: ☞ 590

$$\mathbb{E}\, h(\boldsymbol{X}) \geq h(\mathbb{E}\boldsymbol{X}). \tag{2.103}$$

20. Using Jensen's inequality, show that the Kullback–Leibler divergence between probability densities f and g is always positive; that is,

$$\mathbb{E} \ln \frac{f(\boldsymbol{X})}{g(\boldsymbol{X})} \geq 0,$$

where $\boldsymbol{X} \sim f$.

21. The purpose of this exercise is to prove the following **Vapnik–Chernovenkis bound**: for any *finite* class $\mathcal{G}$ (containing only a finite number $|\mathcal{G}|$ of possible functions) and a general *bounded* loss function, $l \leq \text{Loss} \leq u$, the expected statistical error is bounded from above according to:

$$\mathbb{E}\,\ell(g^{\mathcal{G}}_{\mathcal{T}_n}) - \ell(g^{\mathcal{G}}) \leq \frac{(u-l)\sqrt{2\ln(2|\mathcal{G}|)}}{\sqrt{n}}. \tag{2.104}$$

Note how this bound conveniently does not depend on the distribution of the training set $\mathcal{T}_n$ (which is typically unknown), but only on the complexity (i.e., cardinality) of the class $\mathcal{G}$. We can break up the proof of (2.104) into the following four parts:

(a) For a general function class $\mathcal{G}$, training set $\mathcal{T}$, risk function ℓ, and training loss $\ell_{\mathcal{T}}$, we have, by definition, $\ell(g^{\mathcal{G}}) \leq \ell(g)$ and $\ell_{\mathcal{T}}(g^{\mathcal{G}}_{\mathcal{T}}) \leq \ell_{\mathcal{T}}(g)$ for all $g \in \mathcal{G}$. Show that

$$\ell(g^{\mathcal{G}}_{\mathcal{T}}) - \ell(g^{\mathcal{G}}) \leq \sup_{g \in \mathcal{G}} |\ell_{\mathcal{T}}(g) - \ell(g)| + \ell_{\mathcal{T}}(g^{\mathcal{G}}) - \ell(g^{\mathcal{G}}),$$

where we used the notation sup (supremum) for the least upper bound. Since $\mathbb{E}\ell_{\mathcal{T}}(g) = \mathbb{E}\ell(g)$, we obtain, after taking expectations on both sides of the inequality above:

$$\mathbb{E}\,\ell(g^{\mathcal{G}}_{\mathcal{T}}) - \ell(g^{\mathcal{G}}) \leq \mathbb{E} \sup_{g \in \mathcal{G}} |\ell_{\mathcal{T}}(g) - \ell(g)|.$$

(b) If X is a zero-mean random variable taking values in the interval $[l, u]$, then the following **Hoeffding's inequality** states that the moment generating function satisfies

$$\mathbb{E}\,\mathrm{e}^{tX} \leq \exp\left(\frac{t^2(u-l)^2}{8}\right), \quad t \in \mathbb{R}. \tag{2.105}$$

Prove this result by using the fact that the line segment joining points $(l, \exp(tl))$ and $(u, \exp(tu))$ bounds the convex function $x \mapsto \exp(tx)$ for $x \in [l, u]$; that is:

$$\mathrm{e}^{tx} \leq \mathrm{e}^{tl}\frac{u-x}{u-l} + \mathrm{e}^{tu}\frac{x-l}{u-l}, \quad x \in [l, u].$$

(c) Let $Z_1, \ldots, Z_n$ be (possibly dependent and non-iid) zero-mean random variables with moment generating functions that satisfy $\mathbb{E}\exp(tZ_k) \leq \exp(t^2\eta^2/2)$ for all k and some parameter η. Use Jensen's inequality (2.103) to prove that for any $t > 0$, ☞ 631

$$\mathbb{E}\max_k Z_k = \frac{1}{t}\mathbb{E}\ln\max_k \mathrm{e}^{tZ_k} \leq \frac{1}{t}\ln n + \frac{t\eta^2}{2}.$$

From this derive that

$$\mathbb{E} \max_k Z_k \leq \eta\sqrt{2 \ln n}.$$

Finally, show that this last inequality implies that

$$\mathbb{E} \max_k |Z_k| \leq \eta\sqrt{2 \ln(2n)}.$$

(d) Returning to the objective of this exercise, denote the elements of $\mathcal{G}$ by $g_1, \ldots, g_{|\mathcal{G}|}$, and let $Z_k = \ell_{\mathcal{T}_n}(g_k) - \ell(g_k)$. By part (a) it is sufficient to bound $\mathbb{E} \max_k |Z_k|$. Show that the $\{Z_k\}$ satisfy the conditions of (c) with $\eta = (u - l)/\sqrt{n}$. For this you will need to apply part (b) to the random variable $\mathrm{Loss}(g(\boldsymbol{X}), Y) - \ell(g)$, where $(\boldsymbol{X}, Y)$ is a generic data point. Now complete the proof of (2.104).

22. Consider the problem in Exercise 21a above. Show that

$$|\ell_{\mathcal{T}}(g_{\mathcal{T}}^{\mathcal{G}}) - \ell(g^{\mathcal{G}})| \leq 2 \sup_{g \in \mathcal{G}} |\ell_{\mathcal{T}}(g) - \ell(g)| + \ell_{\mathcal{T}}(g^{\mathcal{G}}) - \ell(g^{\mathcal{G}}).$$

From this, conclude:

$$\mathbb{E}\, |\ell_{\mathcal{T}}(g_{\mathcal{T}}^{\mathcal{G}}) - \ell(g^{\mathcal{G}})| \leq 2\mathbb{E} \sup_{g \in \mathcal{G}} |\ell_{\mathcal{T}}(g) - \ell(g)|.$$

The last bound allows us to assess how close the training loss $\ell_{\mathcal{T}}(g_{\mathcal{T}}^{\mathcal{G}})$ is to the optimal risk $\ell(g^{\mathcal{G}})$ within class $\mathcal{G}$.

23. Show that for the normal linear model $\boldsymbol{Y} \sim \mathcal{N}(\mathbf{X}\boldsymbol{\beta}, \sigma^2 \mathbf{I}_n)$, the maximum likelihood estimator of σ^2 is identical to the method of moments estimator (2.64).

24. Let $X \sim \mathsf{Gamma}(\alpha, \lambda)$. Show that the pdf of $Z = 1/X$ is equal to

$$\frac{\lambda^{\alpha} (z)^{-\alpha-1} \mathrm{e}^{-\lambda (z)^{-1}}}{\Gamma(\alpha)}, \qquad z > 0.$$

25. Consider the sequence $w_0, w_1, \ldots$, where $w_0 = g(\boldsymbol{\theta})$ is a non-degenerate initial guess and $w_t(\boldsymbol{\theta}) \propto w_{t-1}(\boldsymbol{\theta}) g(\tau \,|\, \boldsymbol{\theta})$, $t > 1$. We assume that $g(\tau \,|\, \boldsymbol{\theta})$ is not the constant function (with respect to $\boldsymbol{\theta}$) and that the maximum likelihood value

$$g(\tau \,|\, \widehat{\boldsymbol{\theta}}) = \max_{\boldsymbol{\theta}} g(\tau \,|\, \boldsymbol{\theta}) < \infty$$

exists (is bounded). Let

$$l_t := \int g(\tau \,|\, \boldsymbol{\theta}) w_t(\boldsymbol{\theta})\, \mathrm{d}\boldsymbol{\theta}.$$

Show that $\{l_t\}$ is a strictly increasing and bounded sequence. Hence, conclude that its limit is $g(\tau \,|\, \widehat{\boldsymbol{\theta}})$.

26. Consider the Bayesian model for $\tau = \{x_1, \ldots, x_n\}$ with likelihood $g(\tau \,|\, \mu)$ such that $(X_1, \ldots, X_n \,|\, \mu) \sim_{\text{iid}} \mathcal{N}(\mu, 1)$ and prior pdf $g(\mu)$ such that $\mu \sim \mathcal{N}(\nu, 1)$ for some hyperparameter ν. Define a sequence of densities $w_t(\mu), t \geq 2$ via $w_t(\mu) \propto w_{t-1}(\mu)\, g(\tau \,|\, \mu)$,

starting with $w_1(\mu) = g(\mu)$. Let a_t and b_t denote the mean and precision[4] of μ under the posterior $g_t(\mu \mid \tau) \propto g(\tau \mid \mu) w_t(\mu)$. Show that $g_t(\mu \mid \tau)$ is a normal density with precision $b_t = b_{t-1} + n$, $b_0 = 1$ and mean $a_t = (1 - \gamma_t) a_{t-1} + \gamma_t \overline{x}_n$, $a_0 = \nu$, where $\gamma_t := n/(b_{t-1} + n)$. Hence, deduce that $g_t(\mu \mid \tau)$ converges to a degenerate density with a point-mass at $\overline{x}_n$.

27. Consider again Example 2.76, where we have a normal model with improper prior $g(\boldsymbol{\theta}) = g(\mu, \sigma^2) \propto 1/\sigma^2$. Show that the prior predictive pdf is an improper density $g(x) \propto 1$, but that the posterior predictive density is

$$g(x \mid \tau) \propto \left(1 + \frac{(x - \overline{x}_n)^2}{(n+1) S_n^2}\right)^{-n/2}.$$

Deduce that $\frac{X - \overline{x}_n}{S_n \sqrt{(n+1)/(n-1)}} \sim \mathsf{t}_{n-1}$.

28. Assuming that $\boldsymbol{X}_1, \boldsymbol{X}_2, \ldots \overset{\text{iid}}{\sim} f$, show that (2.81) holds.

29. Let $g_{\mathcal{T}_n}(\tau_n') = g(\tau_n' \mid \mathcal{T}_n) = \int g(\tau_n' \mid \boldsymbol{\theta})\, g(\boldsymbol{\theta} \mid \mathcal{T}_n)\, \mathrm{d}\boldsymbol{\theta}$ be the posterior predictive density of τ_n' after observing $\mathcal{T}_n$. Show that the expected cross-entropy generalization risk of $g_{\mathcal{T}_n}$ can then be decomposed into a bias and variance component:

$$\mathbb{E}\,\ell(g_{\mathcal{T}_n}) = \ell_n^* + \underbrace{\int f(\tau_n') \ln \frac{f(\tau_n')}{\mathbb{E}\, g(\tau_n' \mid \mathcal{T}_n)}\, \mathrm{d}\tau_n'}_{\text{"bias" component}} + \underbrace{\mathbb{E} \int f(\tau_n') \ln \frac{\mathbb{E}\, g(\tau_n' \mid \mathcal{T}_n)}{g(\tau_n' \mid \mathcal{T}_n)}\, \mathrm{d}\tau_n'}_{\text{"variance" component}}.$$

30. In the Bayesian model in Example 2.79 we find that the posterior pdf $g(\theta \mid \tau)$ is equal to the pdf of the $\mathsf{Beta}(s+1, n-s+1)$ distribution.

(a) Show that the second derivative of the log-posterior at the posterior mode $\overline{\theta} := s/n$ is given by

$$\frac{\mathrm{d}^2}{\mathrm{d}\theta^2} \ln g(\theta \mid \tau)\Big|_{\theta = \overline{\theta}} = -\frac{n}{\overline{\theta}(1 - \overline{\theta})}.$$

(b) Let $X_n \sim \mathsf{Beta}(s+1, n-s+1)$, where $\overline{\theta} = s/n$. Verify that

$$\lim_{n \to \infty} \frac{n}{\overline{\theta}(1 - \overline{\theta})} \mathbb{V}\mathrm{ar} X_n = 1.$$

(c) Theorem 2.86 suggests that for large n, the posterior pdf is approximately normal with mean $\overline{\theta}$ and variance $\overline{\theta}(1 - \overline{\theta})/n$. Verify this numerically by plotting the posterior pdf and normal approximation for the cases $n = 10, \overline{\theta} = 0.5$ and $n = 100, \overline{\theta} = 1/4$.

As a matter of fact, the conditions of Theorem 2.86 are not satisfied here, because the second-order derivative, $[\theta(1 - \theta)]^{-1}$, of $\ln g(\theta \mid \tau)/n$ is not uniformly continuous on $(0, 1)$. The current exercise thus suggests that the claim of Theorem 2.86 is valid under less stringent assumptions.

[4]The precision is the reciprocal of the variance.

31. The purpose of this exercise is to establish (2.87) with the help of Theorem 2.86. To this end we derive the following intermediate results.

(a) By locating the global minimum of the function

$$\xi(x) := x \ln \frac{x}{y} + (1-x) \ln \frac{1-x}{1-y} - 2(x-y)^2,$$

prove the inequality

$$x \ln \frac{x}{y} + (1-x) \ln \frac{1-x}{1-y} \geq 2(x-y)^2, \quad 1 > x \geq y > 0.$$

(b) Let $\boldsymbol{X}$ and $\boldsymbol{Y}$ be random vectors drawn from densities f and g. Show that for any set A (to which we can assign a probability), we have:

$$\mathbb{E} \ln \frac{f(\boldsymbol{X})}{g(\boldsymbol{X})} \geq \mathbb{P}[\boldsymbol{X} \in A] \ln \frac{\mathbb{P}[\boldsymbol{X} \in A]}{\mathbb{P}[\boldsymbol{Y} \in A]} + \mathbb{P}[\boldsymbol{X} \in A^c] \ln \frac{\mathbb{P}[\boldsymbol{X} \in A^c]}{\mathbb{P}[\boldsymbol{Y} \in A^c]}.$$

Hint: Consider splitting the expectation as:

$$\mathbb{E} \ln \frac{f(\boldsymbol{X})}{g(\boldsymbol{X})} = \mathbb{E}\left[\ln \frac{f(\boldsymbol{X})}{g(\boldsymbol{X})} \,\middle|\, \boldsymbol{X} \in A \right] \mathbb{P}[\boldsymbol{X} \in A] + \mathbb{E}\left[\ln \frac{f(\boldsymbol{X})}{g(\boldsymbol{X})} \,\middle|\, \boldsymbol{X} \in A^c \right] \mathbb{P}[\boldsymbol{X} \in A^c],$$

and then applying Jensen's inequality (2.103) from Exercise 19 to the conditional expectations.

(c) Use parts (a) and (b) to show that

$$\sqrt{\frac{1}{2} \mathbb{E} \ln \frac{f(\boldsymbol{X})}{g(\boldsymbol{X})}} \geq \sup_A \big| \mathbb{P}[\boldsymbol{X} \in A] - \mathbb{P}[\boldsymbol{Y} \in A] \big|.$$

Hence, use Theorem 2.86 to deduce the inequality (2.87).

32. Suppose that $\tau = \{x_1, \ldots, x_n\}$ are observations of iid continuous and strictly positive random variables, and that there are two possible models for their pdf. The first model $p = 1$ is

$$g(x \mid \theta, p = 1) = \theta \exp(-\theta x)$$

and the second $p = 2$ is

$$g(x \mid \theta, p = 2) = \left(\frac{2\theta}{\pi} \right)^{1/2} \exp\left(-\frac{\theta x^2}{2} \right).$$

For both models, assume that the prior for θ is a gamma density

$$g(\theta) = \frac{b^t}{\Gamma(t)} \theta^{t-1} \exp(-b\theta),$$

with the same hyperparameters b and t. Find a formula for the Bayes factor, $g(\tau \mid p = 1)/g(\tau \mid p = 2)$, for comparing these models.

33. Suppose that we have a total of m possible models with prior probabilities $g(p)$, $p = 1, \ldots, m$. Show that the posterior probability of model $g(p \mid \tau)$ can be expressed in terms of all the $p(p-1)$ Bayes factors:

$$g(p = i \mid \tau) = \left(1 + \sum_{j \neq i} \frac{g(p = j)}{g(p = i)} B_{j \mid i}\right)^{-1}.$$

34. Given the data $\tau = \{x_1, \ldots, x_n\}$, suppose that we use the likelihood $(X \mid \boldsymbol{\theta}) \sim \mathcal{N}(\mu, \sigma^2)$ with parameter $\boldsymbol{\theta} = (\mu, \sigma^2)^\top$ and wish to compare the following two nested models.

(a) Model $p = 1$, where $\sigma^2 = \sigma_0^2$ is known and this is incorporated via the prior

$$g(\boldsymbol{\theta} \mid p = 1) = g(\mu \mid \sigma^2, p = 1)\, g(\sigma^2 \mid p = 1) = \frac{1}{\sqrt{2\pi}\sigma} \mathrm{e}^{-\frac{(\mu - x_0)^2}{2\sigma^2}} \times \delta(\sigma^2 - \sigma_0^2).$$

(b) Model $p = 2$, where both mean and variance are unknown with prior

$$g(\boldsymbol{\theta} \mid p = 2) = g(\mu \mid \sigma^2)\, g(\sigma^2) = \frac{1}{\sqrt{2\pi}\sigma} \mathrm{e}^{-\frac{(\mu - x_0)^2}{2\sigma^2}} \times \frac{b^t (\sigma^2)^{-t-1} \mathrm{e}^{-b/\sigma^2}}{\Gamma(t)}.$$

Show that the prior $g(\boldsymbol{\theta} \mid p = 1)$ can be viewed as the limit of the prior $g(\boldsymbol{\theta} \mid p = 2)$ when $t \to \infty$ and $b = t\sigma_0^2$. Hence, conclude that

$$g(\tau \mid p = 1) = \lim_{\substack{t \to \infty \\ b = t\sigma_0^2}} g(\tau \mid p = 2)$$

and use this result to calculate $B_{1 \mid 2}$. Check that the formula for $B_{1 \mid 2}$ agrees with the Savage–Dickey density ratio:

$$\frac{g(\tau \mid p = 1)}{g(\tau \mid p = 2)} = \frac{g(\sigma^2 = \sigma_0^2 \mid \tau)}{g(\sigma^2 = \sigma_0^2)},$$

where $g(\sigma^2 \mid \tau)$ and $g(\sigma^2)$ are the posterior and prior, respectively, under model $p = 2$.

CHAPTER 3

MONTE CARLO METHODS

Many algorithms in machine learning and data science make use of Monte Carlo techniques. This chapter gives an introduction to the three main uses of Monte Carlo simulation: to (1) simulate random objects and processes in order to observe their behavior, (2) estimate numerical quantities by repeated sampling, and (3) solve complicated optimization problems through randomized algorithms.

3.1 Introduction

Briefly put, **Monte Carlo simulation** is the generation of random data by means of a computer. These data could arise from simple models, such as those described in Chapter 2, or from very complicated models describing real-life systems, such as the positions of vehicles on a complex road network, or the evolution of security prices in the stock market. In many cases, Monte Carlo simulation simply involves random sampling from certain probability distributions. The idea is to repeat the random experiment that is described by the model many times to obtain a large quantity of data that can be used to answer questions about the model. The three main uses of Monte Carlo simulation are:

Sampling. Here the objective is to gather information about a random object by observing many realizations of it. For instance, this could be a random process that mimics the behavior of some real-life system such as a production line or telecommunications network. Another usage is found in Bayesian statistics, where Markov chains are often used to sample from a posterior distribution. ☞ 50

Estimation. In this case the emphasis is on estimating certain numerical quantities related to a simulation model. An example is the evaluation of multi-dimensional integrals via Monte Carlo techniques. This is achieved by writing the integral as the expectation of a random variable, which is then approximated by the sample mean. Appealing to the Law of Large Numbers guarantees that this approximation will ☞ 653 eventually converge when the sample size becomes large.

Optimization. Monte Carlo simulation is a powerful tool for the optimization of complicated objective functions. In many applications these functions are deterministic

DOI: 10.1201/9781003391173-3

and randomness is introduced artificially in order to more efficiently search the domain of the objective function. Monte Carlo techniques are also used to optimize noisy functions, where the function itself is random; for example, when the objective function is the output of a Monte Carlo simulation.

The Monte Carlo method dramatically changed the way in which statistics is used in today's analysis of data. The ever-increasing complexity of data requires radically different statistical models and analysis techniques from those that were used 20 to 100 years ago. By using Monte Carlo techniques, the data analyst is no longer restricted to using basic (and often inappropriate) models to describe data. Now, any probabilistic model that can be simulated on a computer can serve as the basis for statistical analysis. This Monte Carlo revolution has had an impact on both Bayesian and frequentist statistics. In particular, in frequentist statistics, Monte Carlo methods are often referred to as resampling techniques. An important example is the well-known bootstrap method [55], where statistical quantities such as confidence intervals and P-values for statistical tests can simply be determined by simulation without the need of a sophisticated analysis of the underlying probability distributions; see, for example, [41] for basic applications. The impact on Bayesian statistics has been even more profound, through the use of Markov chain Monte Carlo (MCMC) techniques [125, 71]. MCMC samplers construct a Markov process which converges in distribution to a desired (often high-dimensional) density. This convergence in distribution justifies using a finite run of the Markov process as an approximate random realization from the target density. The MCMC approach has rapidly gained popularity as a versatile heuristic approximation, partly due to its simple computer implementation and inbuilt mechanism to tradeoff between computational cost and accuracy; namely, the longer one runs the Markov process, the better the approximation. Nowadays, MCMC methods are indispensable for analyzing posterior distributions for inference and model selection; see also [73, 140].

The following three sections elaborate on these three uses of Monte Carlo simulation in turn.

3.2 Monte Carlo Sampling

In this section we describe a variety of Monte Carlo sampling methods, from the building block of simulating uniform random numbers to MCMC samplers.

3.2.1 Generating Random Numbers

At the heart of any Monte Carlo method is a **random number generator**: a procedure that produces a stream of uniform random numbers on the interval (0,1). Since such numbers are usually produced via deterministic algorithms, they are not truly random. However, for most applications all that is required is that such pseudo-random numbers are statistically indistinguishable from genuine random numbers $U_1, U_2, \ldots$ that are uniformly distributed on the interval (0,1) and are independent of each other; we write $U_1, U_2, \ldots \sim_{\text{iid}} \mathcal{U}(0, 1)$. For example, in Python the `rand` method of the `numpy.random` module is widely used for this purpose.

Most random number generators at present are based on linear recurrence relations. One of the most important random number generators is the **multiple-recursive generator** (MRG) of *order* k, which generates a sequence of integers $X_k, X_{k+1}, \ldots$ via the linear recurrence

$$X_t = (a_1 X_{t-1} + \cdots + a_k X_{t-k}) \bmod m, \quad t = k, k+1, \ldots \tag{3.1}$$

for some **modulus** m and **multipliers** $\{a_i, i = 1, \ldots, k\}$. Here "mod" refers to the modulo operation: $n \bmod m$ is the remainder when n is divided by m. The recurrence is initialized by specifying k "seeds", $X_0, \ldots, X_{k-1}$. To yield fast algorithms, all but a few of the multipliers should be 0. When m is a large integer, one can obtain a stream of pseudo-random numbers $U_k, U_{k+1}, \ldots$ between 0 and 1 from the sequence $X_k, X_{k+1}, \ldots$, simply by setting

$$U_t = \frac{X_t}{m}.$$

It is also possible to set a small modulus, in particular $m = 2$. The output function for such **modulo 2 generators** is then typically of the form

$$U_t = \sum_{i=1}^{w} X_{tw+i-1} 2^{-i}$$

for some $w \leq k$, e.g., $w = 32$ or 64. Examples of modulo 2 generators are the **feedback shift register** generators, the most popular of which are the **Mersenne twisters**; see, for example, [110] and [120]. MRGs with excellent statistical properties can be implemented efficiently by combining several simpler MRGs and carefully choosing their respective moduli and multipliers. One of the most successful is L'Ecuyer's `MRG32k3a` generator; see [106]. From now on, we assume that the reader has a sound random number generator available.

3.2.2 Simulating Random Variables

Simulating a random variable X from an arbitrary (that is, not necessarily uniform) distribution invariably involves the following two steps:

1. Simulate uniform random numbers $U_1, \ldots, U_k$ on $(0, 1)$ for some $k = 1, 2, \ldots$.

2. Return $X = g(U_1, \ldots, U_k)$, where g is some real-valued function.

The construction of suitable functions g is as much of an art as a science. Many simulation methods may be found, for example, in [98] and the accompanying website:

https://people.smp.uq.edu.au/DirkKroese/montecarlohandbook/.

Two of the most useful general procedures for generating random variables are the *inverse-transform* and the *rejection sampling* methods. Before we discuss these, we show one possible way to simulate standard normal random variables. In Python we can generate standard normal random variables via the `randn` method of the `numpy.random` module.

■ **Example 3.2 (Simulating Standard Normal Random Variables)** If X and Y are independent standard normally distributed random variables (that is, $X, Y \sim_{\text{iid}} \mathcal{N}(0, 1)$), then their joint pdf is

$$f(x, y) = \frac{1}{2\pi} e^{-\frac{1}{2}(x^2+y^2)}, \quad (x, y) \in \mathbb{R}^2,$$

☞ 637 which is a radially symmetric function. In Example D.38 we see that, in polar coordinates, the angle Θ that the random vector $[X, Y]^\top$ makes with the positive x-axis is $\mathcal{U}(0, 2\pi)$ distributed (as would be expected from the radial symmetry) and the radius R has pdf $f_R(r) = r\, e^{-r^2/2}, r > 0$. Moreover, R and Θ are independent. We will see shortly, in ☞ 79 Example 3.10, that R has the same distribution as $\sqrt{-2 \ln U}$ with $U \sim \mathcal{U}(0, 1)$. So, to simulate $X, Y \sim_{\text{iid}} \mathcal{N}(0, 1)$, the idea is to first simulate R and Θ independently and then return $X = R\cos(\Theta)$ and $Y = R\sin(\Theta)$ as a pair of independent standard normal random variables. This leads to the Box–Muller approach for generating standard normal random variables.

Algorithm 3.3: Normal Random Variable Simulation (Box–Muller)

Output: Independent standard normal random variables X and Y.

Simulate two independent random variables, U_1 and U_2, from $\mathcal{U}(0, 1)$.

$X \leftarrow (-2 \ln U_1)^{1/2} \cos(2\pi U_2)$

$Y \leftarrow (-2 \ln U_1)^{1/2} \sin(2\pi U_2)$

return X, Y

■

Once a standard normal number generator is available, simulation from any n-dimensional normal distribution $\mathcal{N}(\boldsymbol{\mu}, \boldsymbol{\Sigma})$ is relatively straightforward. The first step is to find an $n \times n$ matrix $\mathbf{B}$ that decomposes $\boldsymbol{\Sigma}$ into the matrix product $\mathbf{B}\mathbf{B}^\top$. In fact there exist many such decompositions. One of the more important ones is the *Cholesky factorization*, ☞ 537 which is a special case of the *LU decomposition*; see Section A.8.1 for more information on such decompositions. In Python, the function `cholesky` of `numpy.linalg` can be used to produce such a matrix $\mathbf{B}$.

Once the Cholesky factorization is determined, it is easy to simulate $\boldsymbol{X} \sim \mathcal{N}(\boldsymbol{\mu}, \boldsymbol{\Sigma})$ as, by definition, it is the affine transformation $\boldsymbol{\mu} + \mathbf{B}\boldsymbol{Z}$ of an n-dimensional standard normal random vector.

Algorithm 3.4: Normal Random Vector Simulation

Input: $\boldsymbol{\mu}, \boldsymbol{\Sigma}$

Output: $\boldsymbol{X} \sim \mathcal{N}(\boldsymbol{\mu}, \boldsymbol{\Sigma})$

Determine the Cholesky factorization $\boldsymbol{\Sigma} = \mathbf{B}\mathbf{B}^\top$.

Simulate $\boldsymbol{Z} = [Z_1, \ldots, Z_n]^\top$ by drawing $Z_1, \ldots, Z_n \sim_{\text{iid}} \mathcal{N}(0, 1)$.

$\boldsymbol{X} \leftarrow \boldsymbol{\mu} + \mathbf{B}\boldsymbol{Z}$

return $\boldsymbol{X}$

■ **Example 3.5 (Simulating from a Bivariate Normal Distribution)** The Python program below draws $N = 1000$ iid samples from the two bivariate ($n = 2$) normal pdfs in Figure 2.59. The resulting point clouds are given in Figure 3.6. ☞ 47

bvnormal.py

```
import numpy as np
from numpy.random import randn
import matplotlib.pyplot as plt

N = 1000
r = 0.0   #change to 0.8 for other plot
Sigma = np.array([[1, r], [r, 1]])
B = np.linalg.cholesky(Sigma)
x = B @ randn(2,N)
plt.scatter([x[0,:]],[x[1,:]], alpha = 0.4, s = 4)
```

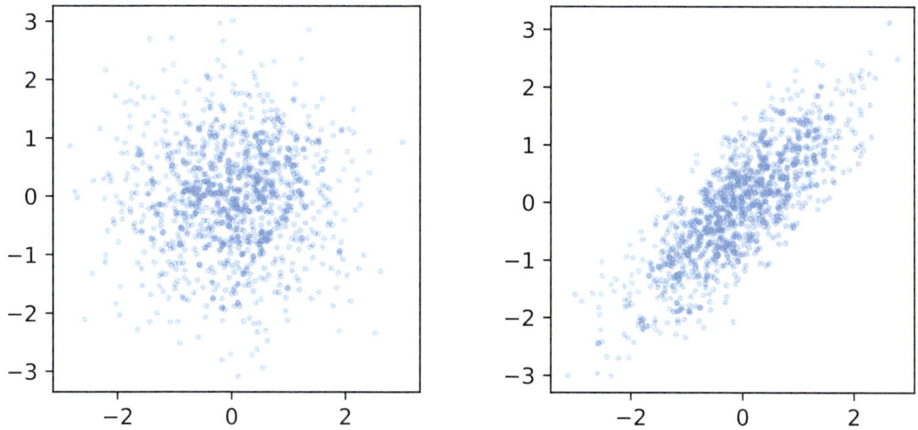

Figure 3.6: 1000 realizations of bivariate normal distributions with means zero, variances 1, and correlation coefficients 0 (left) and 0.8 (right).

■

In some cases, the covariance matrix $\mathbf{\Sigma}$ has special structure which can be exploited to create even faster generation algorithms, as illustrated in the following example.

■ **Example 3.7 (Simulating Normal Vectors in $O(n^2)$ Time)** Suppose that the random vector $\boldsymbol{X} = [X_1, \ldots, X_n]^\top$ represents the values at times $t_0 + k\delta$, $k = 0, \ldots, n-1$ of a zero-mean *Gaussian process* $(X(t), t \geq 0)$ that is **weakly stationary**, meaning that ☞ 341 $\mathbb{C}\text{ov}(X(s), X(t))$ depends only on $t - s$. Then clearly the covariance matrix of $\boldsymbol{X}$, say $\mathbf{A}_n$, is a symmetric *Toeplitz matrix*. Suppose for simplicity that $\mathbb{V}\text{ar}\, X(t) = 1$. Then the ☞ 549 covariance matrix is in fact a correlation matrix, and will have the following structure:

$$\mathbf{A}_n := \begin{bmatrix} 1 & a_1 & \cdots & a_{n-2} & a_{n-1} \\ a_1 & 1 & \ddots & & a_{n-2} \\ \vdots & \ddots & \ddots & \ddots & \vdots \\ a_{n-2} & & \ddots & \ddots & a_1 \\ a_{n-1} & a_{n-2} & \cdots & a_1 & 1 \end{bmatrix}.$$

☞ 552 Using the *Levinson–Durbin* algorithm we can compute a lower diagonal matrix $\mathbf{L}_n$ and a diagonal matrix $\mathbf{D}_n$ in $O(n^2)$ time such that $\mathbf{L}_n \mathbf{A}_n \mathbf{L}_n^\top = \mathbf{D}_n$; see Theorem A.90. If we simulate $\mathbf{Z}_n \sim \mathcal{N}(\mathbf{0}, \mathbf{I}_n)$, then the solution $\boldsymbol{X}$ of the linear system:

$$\mathbf{L}_n \boldsymbol{X} = \mathbf{D}_n^{1/2} \mathbf{Z}_n$$

has the desired distribution $\mathcal{N}(\mathbf{0}, \mathbf{A}_n)$. The linear system is solved in $O(n^2)$ time via forward substitution, as compared to the usual $O(n^3)$ time. ■

3.2.2.1 Inverse-Transform Method

The inverse-transform method is a powerful method for simulating real-valued random variables from an arbitrary distribution. It is based on the following observation:

Lemma 3.8 (Inverse-Transform): Let $U \sim \mathcal{U}(0, 1)$ and let $X \sim F$ be a real-valued random variable with cumulative distribution function (cdf) F whose inverse function is F^{-1}. Then $F^{-1}(U)$ has the same distribution as X.

Proof. For every x we have $\mathbb{P}[F^{-1}(U) \leq x] = \mathbb{P}[U \leq F(x)] = F(x)$. In other words, the cdf of $F^{-1}(U)$ is the same as the cdf of X. □

Note that every cdf has a unique inverse function defined by $F^{-1}(u) = \inf\{x : F(x) \geq u\}$. If, for each u, the equation $F(x) = u$ has a unique solution x, this definition coincides with the usual interpretation of the inverse function. Lemma 3.8 leads to the following algorithm for simulating a random variable X with cdf F:

Algorithm 3.9: Inverse-Transform Method

Input: Cumulative distribution function F.
Output: Random variable X distributed according to F.

1 Generate U from $\mathcal{U}(0, 1)$.
2 $X \leftarrow F^{-1}(U)$
3 **return** X

The inverse-transform method works both for continuous and discrete distributions. After importing `numpy` as `np`, simulating numbers $0, \ldots, k-1$ according to probabilities $p_0, \ldots, p_{k-1}$ can be done via `np.argmax(np.cumsum(p) > np.random.rand())`, where `p` is the vector of the probabilities.

Example 3.10 (Example 3.2 (cont.)) One remaining issue in Example 3.2 is how to simulate the radius R when we only know its density $f_R(r) = r\,\mathrm{e}^{-r^2/2}, r > 0$. We can use the inverse-transform method for this, but first we need to determine its cdf. The cdf of R is, by integration of the pdf, $F_R(r) = 1 - \mathrm{e}^{-\frac{1}{2}r^2}, r > 0$ and its inverse is found by solving $u = F_R(r)$ in terms of r, giving

$$F_R^{-1}(u) = \sqrt{-2\ln(1-u)}, \quad u \in (0,1).$$

Thus R has the same distribution as $\sqrt{-2\ln(1-U)}$, with $U \sim \mathcal{U}(0,1)$. Since $1-U$ also has a $\mathcal{U}(0,1)$ distribution, R has also the same distribution as $\sqrt{-2\ln U}$. ■

3.2.2.2 Rejection Sampling

Rejection sampling is used to sample from a "difficult" probability density function $f(\boldsymbol{x})$, with $\boldsymbol{x} \in \mathbb{R}^n$, which is proportional to a known function $f_\propto(\boldsymbol{x})$; that is,

$$f(\boldsymbol{x}) = \frac{f_\propto(\boldsymbol{x})}{\int f_\propto(\boldsymbol{x})\,\mathrm{d}\boldsymbol{x}}.$$

The algorithm proceeds to simulate from an "simple" **proposal** pdf $g(\boldsymbol{x})$ satisfying $f_\propto(\boldsymbol{x}) \leq c\,g(\boldsymbol{x})$ for some constant $c \geq \int f_\propto(\boldsymbol{x})\,\mathrm{d}\boldsymbol{x}$, and then accepting or rejecting the drawn sample with a certain *probability of acceptance*. When $n = 1$ simulation from g can be accomplished via the inverse-transform method.

Lemma 3.11 (Rejection Sampling): Let $\boldsymbol{X} \sim g$ and $U \sim \mathcal{U}(0,1)$, independently. Then, conditional on the acceptance event $A := \{c\,g(\boldsymbol{X})U \leq f_\propto(\boldsymbol{X})\}$, the pdf of $\boldsymbol{X}$ is f. Moreover, the probability of acceptance is $\mathbb{P}[A] = \int f_\propto(\boldsymbol{x})\,\mathrm{d}\boldsymbol{x}/c$.

Proof. For any $\boldsymbol{x}$, conditional on $\boldsymbol{X} = \boldsymbol{x}$, the random variable $Y := c\,g(\boldsymbol{X})U$ is uniformly distributed on $[0, c\,g(\boldsymbol{x})]$. This means that the joint distribution of $(\boldsymbol{X}, Y)$ has density:

$$h(\boldsymbol{x}, y) = g(\boldsymbol{x})\frac{\mathbb{1}_{[0,c\,g(\boldsymbol{x})]}(y)}{c\,g(\boldsymbol{x})} = \frac{1}{c}\mathbb{1}_{[0,c\,g(\boldsymbol{x})]}(y).$$

Conditioning on the acceptance event $A = \{Y \leq f_\propto(\boldsymbol{X})\}$, the conditional joint density becomes:

$$h(\boldsymbol{x}, y \,|\, A) = \frac{h(\boldsymbol{x}, y)\mathbb{1}_{[0,f_\propto(\boldsymbol{x})]}(y)}{\mathbb{P}[A]} = \frac{\frac{1}{c}\mathbb{1}_{[0,f_\propto(\boldsymbol{x})]}(y)}{\mathbb{P}[A]},$$

where the probability of acceptance is:

$$\mathbb{P}[A] = \int\int_0^{f_\propto(\boldsymbol{x})} \frac{1}{c}\,\mathrm{d}y\,\mathrm{d}\boldsymbol{x} = \frac{1}{c}\int f_\propto(\boldsymbol{x})\,\mathrm{d}\boldsymbol{x}.$$

To find the marginal density of $\boldsymbol{X}$ conditional on A, integrate out y:

$$\int_{\mathbb{R}} h(\boldsymbol{x}, y \,|\, A)\,\mathrm{d}y = \frac{f_\propto(\boldsymbol{x})}{c\,\mathbb{P}[A]} = \frac{f_\propto(\boldsymbol{x})}{\int f_\propto(\boldsymbol{z})\,\mathrm{d}\boldsymbol{z}} = f(\boldsymbol{x}).$$

□

The efficiency of the rejection sampling is usually expressed in terms of the probability of acceptance. Algorithm 3.12 gives the pseudo-code.

Algorithm 3.12: Rejection Sampling

Input: Pdf g and constant c such that $f_\propto(\boldsymbol{x})/g(\boldsymbol{x}) \leq c$ for all $\boldsymbol{x}$.
Output: Random vector $\boldsymbol{X}$ distributed according to pdf f.

1 **repeat**
2   Simulate $\boldsymbol{X} \sim g$ and $U \sim \mathcal{U}(0, 1)$, independently.
3     $W \leftarrow f_\propto(\boldsymbol{X})/g(\boldsymbol{X})$
4 **until** $W \geq c\,U$
5 **return** $\boldsymbol{X}$

■ **Example 3.13 (Simulating Gamma Random Variables)** We can use rejection sampling to simulate random variables from $\mathsf{Gamma}(\alpha, \lambda)$. Consider, for example, the ☞ 629 Gamma distribution with $\alpha = 2$ and $\lambda = 1$. Its pdf,

$$f(x) = \frac{\lambda^\alpha x^{\alpha-1} \mathrm{e}^{-\lambda x}}{\Gamma(\alpha)}, \quad x \geq 0,$$

where Γ is the gamma function $\Gamma(\alpha) := \int_0^\infty \mathrm{e}^{-x} x^{\alpha-1}\,\mathrm{d}x$, $\alpha > 0$, is depicted by the blue solid curve in Figure 3.14. This pdf happens to lie completely under the graph of $c\,g(x)$, where $c = 1.6$ and $g(x) = 0.4\exp(-0.4x), x \geq 0$ is the pdf of the exponential distribution $\mathsf{Exp}(0.4)$. Hence, we can simulate from this particular Gamma distribution ☞ 629 by accepting or sample from the $\mathsf{Exp}(0.4)$ distribution subject to the condition in line 4 of Algorithm 3.12. Simulating from the $\mathsf{Exp}(0.4)$ distribution can be done via the inverse-transform method: simulate $U \sim \mathcal{U}(0, 1)$ and return $X = -\ln(U)/0.4$.

The following Python code implements Algorithm 3.12 for this example.

accrejgamma.py

```python
from math import exp, gamma, log
from numpy.random import rand

alpha = 2
lam = 1
c = 1.6

def f(x): return lam**alpha * x**(alpha-1) * exp(-lam*x)/gamma(alpha)
def g(x): return 0.4*exp(-0.4*x)

while True:
    X = - log(rand())/0.4
    U = rand()
    W = f(X)/g(X)
    if W >= c*U:
        break

print(X)
```

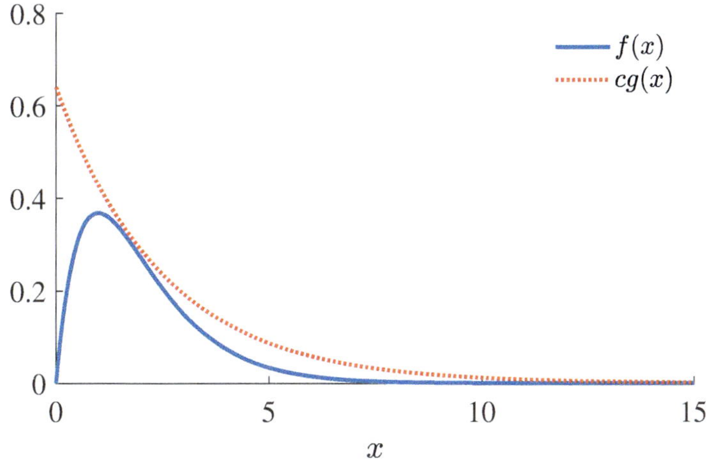

Figure 3.14: The pdf g of the Exp(0.4) distribution multiplied by $c = 1.6$ dominates the pdf f of the Gamma(2, 1) distribution.

3.2.2.3 Regenerative Rejection Sampling

Expanding upon the concept of rejection sampling, we introduce a new technique called **regenerative rejection sampling**. This method is closely related to the importance sampling strategy in Section 3.3.3 and offers a simple alternative to Markov Chain Monte Carlo simulation, discussed in Section 3.2.5.

☞ 103
☞ 88

The pseudo-code for regenerative rejection sampling is presented in Algorithm 3.15, and is followed by an explanation of why it works, based on the theory of regenerative stochastic processes. In the algorithm, it is assumed that the proposal pdf $g(\boldsymbol{x}) > 0$ for all $\boldsymbol{x}$ such that $f(\boldsymbol{x}) > 0$.

Algorithm 3.15: Regenerative Rejection Sampling

Input: Proposal pdf g and time t.
Output: Random vector $\boldsymbol{X}$ approximately distributed according to target pdf f.

```
N ← 0
repeat
    N ← N + 1
    Simulate X_N ~ g.
    W_N ← f_∝(X_N)/g(X_N)
until W_1 + ··· + W_N > t
return X ← X_N
```

To understand the working of Algorithm 3.15, we need some knowledge of the theory of regenerative processes.

Definition 3.16 (Regenerative Process): A stochastic process $\{\boldsymbol{Y}_t, t \geq 0\}$ is said to be **regenerative** if there exist random times $T_0 < T_1 < T_2 < \cdots$, where the differences $T_i - T_{i-1}, i = 1, 2, \ldots$ are iid, and for each i the conditional distribution of the process $\{\boldsymbol{Y}_{T_i+t}, t \geq 0\}$ given $\{\boldsymbol{Y}_s, s \leq T_i\}$ is the same as the distribution of $\{\boldsymbol{Y}_{T_0+t}, t \geq 0\}$.

Thus, starting from an initial time T_0, the process "resets" itself probabilistically at each subsequent **regeneration time** T_i, in the sense that its future behavior depends only on its state at T_i and not on its past, and is identical to the process's behavior starting from T_0. Often T_0 is set to 0. The interval between two successive regeneration times is called a **cycle**. The distribution of the length C of a cycle length is said to be of **lattice** type if C takes values in a lattice $\{a + bn, n \in \mathbb{Z}\}$ for some values of a and $b \neq 0$. A distribution not of this type is called **nonlattice**. One of the main results in the theory of regenerative processes is that, under very mild conditions, a regenerative process has a limiting distribution; see [98, Theorem A.9.1]:

Theorem 3.17 (Regeneration Theorem): A regenerative process $\{\boldsymbol{Y}_t, t \geq 0\}$ with right-continuous paths and nonlattice cycle lengths of finite expectation converges in distribution as $t \to \infty$ to a variable $\boldsymbol{Y}$, such that for all functions H:

$$\widetilde{\mathbb{E}} H(\boldsymbol{Y}) = \frac{\mathbb{E} \int_{T_0}^{T_1} H(\boldsymbol{Y}_s)\, \mathrm{d}s}{\mathbb{E}[T_1 - T_0]}, \tag{3.18}$$

provided that the expectation exists. Here $\widetilde{\mathbb{E}}$ denotes the expectation operator under which $\boldsymbol{Y}$ is distributed according to its limiting distribution.

An important corollary of Theorem 3.17 is the theoretical justification of the regenerative rejection sampling method.

Lemma 3.19 (Convergence of Regenerative Rejection Sampling): The output $\boldsymbol{X}$ of Algorithm 3.15 converges to its limiting distribution with pdf f as $t \to \infty$, provided that $\mathbb{E}W_1 < \infty$ and the distribution of W_1 is nonlattice.

Proof. Recall that $\boldsymbol{X}_1, \boldsymbol{X}_2, \ldots$ denotes the iid sequence of random vectors drawn from the proposal pdf g, and that $W_i = f_\propto(\boldsymbol{X}_i)/g(\boldsymbol{X}_i)$, $i = 1, 2, \ldots$. Consider the regenerative process $\{\boldsymbol{Y}_t, t \geq 0\}$ with regeneration times $T_0 = 0$ and $T_i = T_{i-1} + W_i$ and taking the value $\boldsymbol{X}_i$ on the cycle $[T_{i-1}, T_i)$, $i = 1, 2, \ldots$. Thus, the $\{W_i\}$ are the cycle lengths of the regenerative process. If $\mathbb{E}W_1 < \infty$ and the distribution of W_1 is nonlattice, then Theorem 3.17 implies that $\boldsymbol{Y}_t \xrightarrow{\mathrm{d}} \boldsymbol{Y}$ as $t \to \infty$, where $\boldsymbol{Y}$ has pdf f. Namely, by (3.18) we have for any function H:

$$\widetilde{\mathbb{E}}H(\boldsymbol{Y}) = \frac{\mathbb{E}_g \int_{T_0}^{T_1} H(\boldsymbol{Y}_s)\mathrm{d}s}{\mathbb{E}_g W_1} = \frac{\mathbb{E}_g H(\boldsymbol{X}_1) \int_{T_0}^{T_1} 1\,\mathrm{d}s}{\mathbb{E}_g W_1} = \frac{\mathbb{E}_g H(\boldsymbol{X}_1) W_1}{\mathbb{E}_g W_1}$$
$$= \frac{\int H(\boldsymbol{x}) \frac{f_\propto(\boldsymbol{x})}{g(\boldsymbol{x})} g(\boldsymbol{x})\,\mathrm{d}\boldsymbol{x}}{\int \frac{f_\propto(\boldsymbol{x})}{g(\boldsymbol{x})} g(\boldsymbol{x})\,\mathrm{d}\boldsymbol{x}} = \int H(\boldsymbol{x}) f(\boldsymbol{x})\,\mathrm{d}\boldsymbol{x}.$$

□

An advantage of the regenerative rejection sampling Algorithm 3.15 over the usual rejection sampling Algorithm 3.12 is that the regenerative rejection sampling does not require any prior knowledge of the constant c in the inequality $f_\propto(\boldsymbol{x}) \leq c g(\boldsymbol{x})$, as is required for ordinary rejection sampling. This flexibility is particularly beneficial in scenarios where c is either difficult to compute, or potentially even infinite ($c = \infty$). A disadvantage of regenerative rejection sampling is that, much like Markov Chain Monte Carlo, the method provides only approximate sampling. Consequently, achieving high-quality simulations may necessitate a large value for t, potentially resulting in a large number of simulations from the proposal pdf g.

Example 3.20 (Simulating Gamma Random Variables (cont.)) Consider Example 3.13 again, but now using the regenerative rejection sampling Algorithm 3.15. Recall that we wish to simulate from the target pdf $f(x) = x\mathrm{e}^{-x}$, for $x \geq 0$. Suppose that the proposal pdf is $g(x) = \mathrm{e}^{-x}$, $x \geq 0$. In this case we are able to analytically derive the pdf of the random variable $\boldsymbol{X}$ returned by Algorithm 3.15 for any given time t. In particular, let $X_1, X_2, \ldots$ be an iid sequence from g and repeat the construction of the regenerative process $\{Y_t, t \geq 0\}$ above. That is, Y_t takes the value X_i on each cycle $[T_{i-1}, T_i)$ with corresponding cycle length $W_i = f(X_i)/g(X_i) = X_i$, $i = 1, 2, \ldots$. Defining

$$N_t := \max\{n \in \mathbb{N} : T_n \leq t\},$$

we can write $Y_t = X_{N_t}, t \geq 0$. From Exercise 24 we find that the pdf of X_{N_t} here is:

$$f_{X_{N_t}}(x) = \begin{cases} x\,\mathrm{e}^{-x}, & x < t, \\ (1+t)\,\mathrm{e}^{-x}, & x \geq t. \end{cases}$$

Thus, the pdf has a discontinuity at $x = t$ and is identical to the target up to time t. It follows that the integrated absolute difference between $f_{X_{N_t}}$ and $f \equiv f_{X_\infty}$ is:

$$\int_0^\infty \left| f_{X_{N_t}}(x) - f_{X_\infty}(x) \right| \mathrm{d}x = \int_t^\infty \mathrm{e}^{-x}(x - (1+t))\,\mathrm{d}x = 2\,\mathrm{e}^{-t}(t+1),$$

which goes to 0 at an exponential rate in t. This also happens to be equal to twice the *total variation* distance: $\sup_A \left|\mathbb{P}[X_{N_t} \in A] - \mathbb{P}[X_\infty \in A]\right|$; see Exercise 25. We can thus determine how large t needs to be to attain a certain error bound.

The following Python program implements Algorithm 3.15 with a Gamma(α, λ) target and Exp(λ) proposal distribution, for the specific case $\alpha = 2$ and $\lambda = 1$. Figure 3.21 shows kernel density estimates of the pdfs $f_{X_{N_t}}$ for different values of t, obtained from 10^5 iid samples. One can clearly see the large discontinuity at $x = 1$, for the case $t = 1$, and the much smaller discontinuity at $x = 3$, for the case $t = 3$. When $t = 10$ the estimate is visually indistinguishable from the target pdf (indicated by the dashed curve).

☞ 153

rrs.py

```
from math import exp, log
from numpy.random import rand

alpha = 2; lam = 1

def f_prop(x): return  x**(alpha-1) * exp(-lam*x)

def g(x): return lam*exp(-lam*x)

N = 0; W = 0
t = 10
while True:
    N += 1
    X = - log(rand())/lam
    W += f_prop(X)/g(X)
    if W > t:
        break
print(X)
```

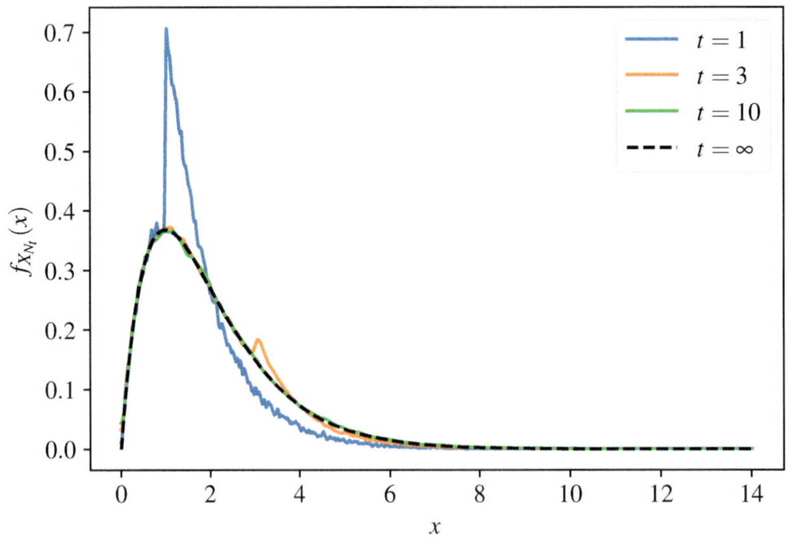

Figure 3.21: Kernel density plots of 10^5 data points simulated via regenerative rejection sampling for different values of t.

3.2.3 Simulating Random Vectors and Processes

Techniques for generating random vectors and processes are as diverse as the class of random processes themselves; see, for example, [98]. We highlight a few general scenarios.

When $X_1, \ldots, X_n$ are *independent* random variables with pdfs $f_i, i = 1, \ldots, n$, so that their joint pdf is $f(\boldsymbol{x}) = f_1(x_1) \cdots f_n(x_n)$, the random vector $\boldsymbol{X} = [X_1, \ldots, X_n]^\top$ can be simply simulated by drawing each component $X_i \sim f_i$ individually — for example, via the inverse-transform method or rejection sampling. ☞ 633

For *dependent* components $X_1, \ldots, X_n$, we can, as a consequence of the *product rule* of probability, represent the joint pdf $f(\boldsymbol{x})$ as ☞ 635

$$f(\boldsymbol{x}) = f(x_1, \ldots, x_n) = f_1(x_1)\, f_2(x_2 \,|\, x_1) \cdots f_n(x_n \,|\, x_1, \ldots, x_{n-1}), \tag{3.22}$$

where $f_1(x_1)$ is the marginal pdf of X_1 and $f_k(x_k \,|\, x_1, \ldots, x_{k-1})$ is the conditional pdf of X_k given $X_1 = x_1, X_2 = x_2, \ldots, X_{k-1} = x_{k-1}$. Provided the conditional pdfs are known, one can generate $\boldsymbol{X}$ by first generating X_1, then, given $X_1 = x_1$, generate X_2 from $f_2(x_2 \,|\, x_1)$, and so on, until generating X_n from $f_n(x_n \,|\, x_1, \ldots, x_{n-1})$.

The latter method is particularly applicable for generating Markov chains. Recall from Section D.10 that a **Markov chain** is a stochastic process $\{X_t, t = 0, 1, 2, \ldots\}$ that satisfies the **Markov property**; meaning that for all t and s the conditional distribution of X_{t+s} given $X_u, u \leq t$, is the same as that of X_{t+s} given only X_t. As a result, each conditional density $f_t(x_t \,|\, x_1, \ldots, x_{t-1})$ can be written as a one-step **transition density** $q_t(x_t \,|\, x_{t-1})$; that is, the probability density to go from state x_{t-1} to state x_t in one step. In many cases of interest the chain is **time-homogeneous**, meaning that the transition density q_t does not depend on t. Such Markov chains can be generated *sequentially*, as given in Algorithm 3.23. ☞ 660

Algorithm 3.23: Simulate a Markov Chain

Input: Number of steps N, initial pdf f_0, transition density q.

Draw X_0 from the initial pdf f_0.
for $t = 1$ **to** N **do**
  Draw X_t from the distribution corresponding to the density $q(\cdot \,|\, X_{t-1})$
return $X_0, \ldots, X_N$

■ **Example 3.24 (Markov Chain Simulation)** For time-homogeneous Markov chains with a discrete state space, we can visualize the one-step transitions by means of a **transition graph**, where arrows indicate possible transitions between states and the labels describe the corresponding probabilities. Figure 3.25 shows (on the left) the transition graph of the Markov chain $\{X_t, t = 0, 1, 2, \ldots\}$ with state space $\{1, 2, 3, 4\}$ and one-step transition matrix

$$\mathbf{P} = \begin{bmatrix} 0 & 0.2 & 0.5 & 0.3 \\ 0.5 & 0 & 0.5 & 0 \\ 0.3 & 0.7 & 0 & 0 \\ 0.1 & 0 & 0 & 0.9 \end{bmatrix}.$$

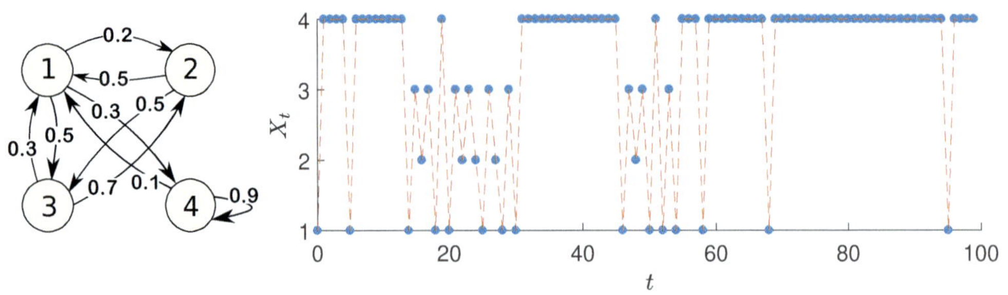

Figure 3.25: The transition graph (left) and a typical path (right) of the Markov chain.

In the same figure (on the right) a typical outcome (path) of the Markov chain is shown. The path was simulated using the Python program below. In this implementation the Markov chain always starts in state 1. We will revisit Markov chains, and in particular
☞ 88 Markov chains with continuous state spaces, in Section 3.2.5.

MCsim.py

```
import numpy as np
import matplotlib.pyplot as plt

n = 101
P = np.array([[0, 0.2, 0.5, 0.3],
              [0.5, 0, 0.5, 0],
              [0.3, 0.7, 0, 0],
              [0.1, 0, 0, 0.9]])
x = np.array(np.ones(n, dtype=int))
x[0] = 0
for t in range(0,n-1):
    x[t+1] = np.argmax(np.cumsum(P[x[t],:])> np.random.rand())
x = x + 1  # add 1 to all elements of the vector x
plt.plot(np.array(range(0,n)),x, 'o')
plt.plot(np.array(range(0,n)),x, '--')
plt.show()
```

■

3.2.4 Resampling

The idea behind **resampling** is very simple: an iid sample $\tau := \{x_1, \ldots, x_n\}$ from some unknown cdf F represents our best knowledge of F if we make no further *a priori* assumptions about it. If it is not possible to simulate more samples from F, the best way to "repeat" the experiment is to *resample* from the original data by drawing from the
☞ 11 empirical cdf F_n; see (1.7). That is, we draw each x_i with equal probability and repeat this N times, according to Algorithm 3.26 below. As we draw here "with replacement", multiple instances of the original data points may occur in the resampled data.

Algorithm 3.26: Sampling from an Empirical Cdf.

Input: Original iid sample $x_1, \ldots, x_n$ and sample size N.
Output: Iid sample $X_1^*, \ldots, X_N^*$ from the empirical cdf.

1 **for** $t = 1$ **to** N **do**
2   Draw $U \sim \mathcal{U}(0, 1)$
3   Set $I \leftarrow \lceil nU \rceil$
4   Set $X_t^* \leftarrow x_I$
5 **return** $X_1^*, \ldots, X_N^*$

In Step 3, $\lceil nU \rceil$ returns the *ceiling* of nU; that is, it is the smallest integer larger than or equal to nU. Consequently, I is drawn uniformly at random from the set of indices $\{1, \ldots, n\}$.

By sampling from the empirical cdf we can thus (approximately) repeat the experiment that gave us the original data as many times as we like. This is useful if we want to assess the properties of certain statistics obtained from the data. For example, suppose that the original data τ gave the statistic $t(\tau)$. By resampling we can gain information about the *distribution* of the corresponding random variable $t(\mathcal{T})$.

Example 3.27 (Quotient of Uniforms) Let $U_1, \ldots, U_n, V_1, \ldots, V_n$ be iid $\mathcal{U}(0, 1)$ random variables and define $X_i = U_i/V_i$, $i = 1, \ldots, n$. Suppose we wish to investigate the distribution of the sample median $\widetilde{X}$ and sample mean $\overline{X}$ of the (random) data $\mathcal{T} := \{X_1, \ldots, X_n\}$. Since we know the model for $\mathcal{T}$ exactly, we can generate a large number, N say, of independent copies of it, and for each of these copies evaluate the sample medians $\widetilde{X}_1, \ldots, \widetilde{X}_N$ and sample means $\overline{X}_1, \ldots, \overline{X}_N$. For $n = 100$ and $N = 1000$ the empirical cdfs might look like the left and right curves in Figure 3.28, respectively. Contrary to what you might have expected, the distributions of the sample median and sample mean do not match at all. The sample median is quite concentrated around 1, whereas the distribution of the sample mean is much more spread out.

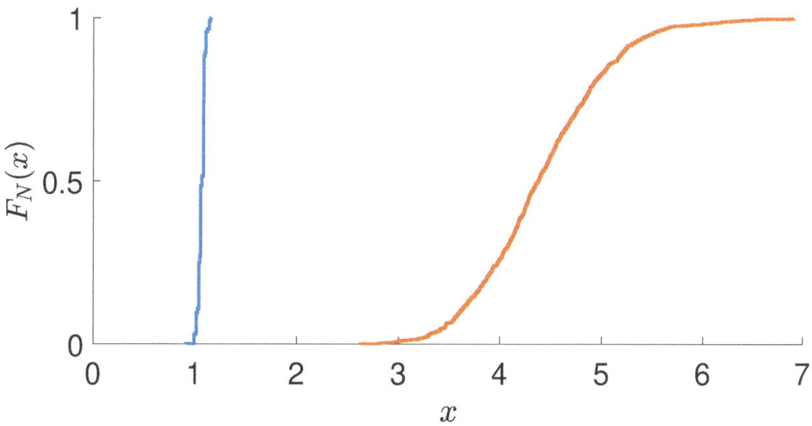

Figure 3.28: Empirical cdfs of the medians of the resampled data (left curve) and sample means (right curve) of the resampled data.

Instead of sampling completely new data, we could also *reuse* the original data by resampling them via Algorithm 3.26. This gives independent copies $\widetilde{X}_1^*, \ldots, \widetilde{X}_N^*$ and $\overline{X}_1^*, \ldots, \overline{X}_N^*$, for which we can again plot the empirical cdf. The results will be similar to the previous case. In fact, in Figure 3.28 the cdf of the *resampled* sample medians and sample means are plotted. The corresponding Python code is given below. The essential point of this example is that resampling of data can greatly add to the understanding of the probabilistic properties of certain measurements on the data, *even if the underlying model is not known*. See Exercise 12 for a further investigation of this example.

☞ 128

quotunif.py

```
import numpy as np
from numpy.random import rand, choice
import matplotlib.pyplot as plt
from statsmodels.distributions.empirical_distribution import ECDF

n = 100
N = 1000
x = rand(n)/rand(n)  # data
med = np.zeros(N)
ave = np.zeros(N)
for i in range(0,N):
    s = choice(x, n, replace=True) # resampled data
    med[i] = np.median(s)
    ave[i] = np.mean(s)

med_cdf = ECDF(med)
ave_cdf = ECDF(ave)
plt.plot(med_cdf.x, med_cdf.y)
plt.plot(ave_cdf.x, ave_cdf.y)
plt.show()
```

■

3.2.5 Markov Chain Monte Carlo

Markov chain Monte Carlo (MCMC) is a Monte Carlo sampling technique for (approximately) generating samples from an arbitrary distribution — often referred to as the **target** distribution. The basic idea is to run a Markov chain long enough such that its limiting distribution is close to the target distribution. Often such a Markov chain is constructed to be reversible, so that the detailed balance equations (D.105) can be used. Depending on the starting position of the Markov chain, the initial random variables in the Markov chain may have a distribution that is significantly different from the target (limiting) distribution. The random variables that are generated during this **burn-in period** are often discarded. The remaining random variables form an *approximate* and *dependent* sample from the target distribution.

☞ 662

In the next two sections we discuss two popular MCMC samplers: the Metropolis–Hastings sampler and the Gibbs sampler.

3.2.5.1 Metropolis–Hastings Sampler

The Metropolis–Hastings sampler [125] is similar to the rejection sampling in that it ☞ 79 simulates a trial state, which is then accepted or rejected according to some random mechanism. Specifically, suppose we wish to sample from a target pdf $f(\boldsymbol{x})$, where $\boldsymbol{x}$ takes values in some d-dimensional set. The aim is to construct a Markov chain $\{\boldsymbol{X}_t, t = 0, 1, \ldots\}$ in such a way that its limiting pdf is f. Suppose the Markov chain is in state $\boldsymbol{x}$ at time t. A transition of the Markov chain from state $\boldsymbol{x}$ is carried out in two phases. First a **proposal** state $\boldsymbol{Y}$ is drawn from a transition density $q(\boldsymbol{y} \mid \boldsymbol{x})$. This state is accepted as the new state, with **acceptance probability**

$$\alpha(\boldsymbol{x}, \boldsymbol{y}) = \min\left\{\frac{f(\boldsymbol{y})\, q(\boldsymbol{x} \mid \boldsymbol{y})}{f(\boldsymbol{x})\, q(\boldsymbol{y} \mid \boldsymbol{x})}, 1\right\}, \tag{3.29}$$

or rejected otherwise. In the latter case the chain remains in state $\boldsymbol{x}$. The algorithm just described can be summarized as follows:

Algorithm 3.30: Metropolis–Hastings Sampler

Input: Initial state $\boldsymbol{X}_0$, sample size N, target pdf $f(\boldsymbol{x})$, proposal density $q(\boldsymbol{y} \mid \boldsymbol{x})$.

Output: $\boldsymbol{X}_1, \ldots, \boldsymbol{X}_N$ (dependent), approximately distributed according to $f(\boldsymbol{x})$.

```
for t = 0 to N − 1 do
    Draw Y ~ q(y | X_t)                  // draw a proposal
    α ← α(X_t, Y)                        // acceptance probability as in (3.29)
    Draw U ~ U(0, 1)
    if U ≤ α then X_{t+1} ← Y
    else X_{t+1} ← X_t
return X_1, ..., X_N
```

The fact that the limiting distribution of the Metropolis–Hastings Markov chain is equal to the target distribution (under general conditions) is a consequence of the following result.

Theorem 3.31 (Local Balance for the Metropolis–Hastings Sampler): The transition density of the Metropolis–Hastings Markov chain satisfies the detailed ☞ 662 balance equations.

Proof. We prove the theorem for the discrete case only. Because a transition of the Metropolis–Hastings Markov chain consists of two steps, the one-step transition probability to go from $\boldsymbol{x}$ to $\boldsymbol{y}$ is not $q(\boldsymbol{y} \mid \boldsymbol{x})$ but

$$\widetilde{q}(\boldsymbol{y} \mid \boldsymbol{x}) = \begin{cases} q(\boldsymbol{y} \mid \boldsymbol{x})\, \alpha(\boldsymbol{x}, \boldsymbol{y}), & \text{if } \boldsymbol{y} \neq \boldsymbol{x}, \\ 1 - \sum_{\boldsymbol{z} \neq \boldsymbol{x}} q(\boldsymbol{z} \mid \boldsymbol{x})\, \alpha(\boldsymbol{x}, \boldsymbol{z}), & \text{if } \boldsymbol{y} = \boldsymbol{x}. \end{cases} \tag{3.32}$$

We thus need to show that

$$f(\boldsymbol{x})\, \widetilde{q}(\boldsymbol{y} \mid \boldsymbol{x}) = f(\boldsymbol{y})\, \widetilde{q}(\boldsymbol{x} \mid \boldsymbol{y}) \quad \text{for all } \boldsymbol{x}, \boldsymbol{y}. \tag{3.33}$$

With the acceptance probability as in (3.29), we need to check (3.33) for three cases:

(a) $\boldsymbol{x} = \boldsymbol{y}$,

(b) $\boldsymbol{x} \neq \boldsymbol{y}$ and $f(\boldsymbol{y})q(\boldsymbol{x}\,|\,\boldsymbol{y}) \leq f(\boldsymbol{x})q(\boldsymbol{y}\,|\,\boldsymbol{x})$, and

(c) $\boldsymbol{x} \neq \boldsymbol{y}$ and $f(\boldsymbol{y})q(\boldsymbol{x}\,|\,\boldsymbol{y}) > f(\boldsymbol{x})q(\boldsymbol{y}\,|\,\boldsymbol{x})$.

Case (a) holds trivially. For case (b), $\alpha(\boldsymbol{x},\boldsymbol{y}) = f(\boldsymbol{y})q(\boldsymbol{x}\,|\,\boldsymbol{y})/(f(\boldsymbol{x})q(\boldsymbol{y}\,|\,\boldsymbol{x}))$ and $\alpha(\boldsymbol{y},\boldsymbol{x}) = 1$. Consequently,

$$\widetilde{q}(\boldsymbol{y}\,|\,\boldsymbol{x}) = f(\boldsymbol{y})q(\boldsymbol{x}\,|\,\boldsymbol{y})/f(\boldsymbol{x}) \quad \text{and} \quad \widetilde{q}(\boldsymbol{x}\,|\,\boldsymbol{y}) = q(\boldsymbol{x}\,|\,\boldsymbol{y}),$$

so that (3.33) holds. Similarly, for (c) we have $\alpha(\boldsymbol{x},\boldsymbol{y}) = 1$ and $\alpha(\boldsymbol{y},\boldsymbol{x}) = f(\boldsymbol{x})q(\boldsymbol{y}\,|\,\boldsymbol{x})/(f(\boldsymbol{y})q(\boldsymbol{x}\,|\,\boldsymbol{y}))$. It follows that,

$$\widetilde{q}(\boldsymbol{y}\,|\,\boldsymbol{x}) = q(\boldsymbol{y}\,|\,\boldsymbol{x}) \quad \text{and} \quad \widetilde{q}(\boldsymbol{x}\,|\,\boldsymbol{y}) = f(\boldsymbol{x})q(\boldsymbol{y}\,|\,\boldsymbol{x})/f(\boldsymbol{y}),$$

so that (3.33) holds again. □

☞ 661 Thus if the Metropolis–Hastings Markov chain is ergodic, then its limiting pdf is $f(\boldsymbol{x})$. A fortunate property of the algorithm, which is important in many applications, is that in order to evaluate the acceptance probability $\alpha(\boldsymbol{x},\boldsymbol{y})$ in (3.29), one only needs to know the target pdf $f(\boldsymbol{x})$ *up to a constant*; that is $f(\boldsymbol{x}) = c\, f_{\propto}(\boldsymbol{x})$ for some known function $f_{\propto}(\boldsymbol{x})$ but unknown constant c.

The efficiency of the algorithm depends of course on the choice of the proposal transition density $q(\boldsymbol{y}\,|\,\boldsymbol{x})$. Ideally, we would like $q(\boldsymbol{y}\,|\,\boldsymbol{x})$ to be "close" to the target $f(\boldsymbol{y})$, irrespective of $\boldsymbol{x}$. We discuss two common approaches.

1. Choose the proposal transition density $q(\boldsymbol{y}\,|\,\boldsymbol{x})$ independent of $\boldsymbol{x}$; that is, $q(\boldsymbol{y}\,|\,\boldsymbol{x}) = g(\boldsymbol{y})$ for some pdf $g(\boldsymbol{y})$. An MCMC sampler of this type is called an **independence sampler**. The acceptance probability is thus

$$\alpha(\boldsymbol{x},\boldsymbol{y}) = \min\left\{\frac{f(\boldsymbol{y})\,g(\boldsymbol{x})}{f(\boldsymbol{x})\,g(\boldsymbol{y})},\ 1\right\}.$$

2. If the proposal transition density is symmetric (that is, $q(\boldsymbol{y}\,|\,\boldsymbol{x}) = q(\boldsymbol{x}\,|\,\boldsymbol{y})$), then the acceptance probability has the simple form

$$\alpha(\boldsymbol{x},\boldsymbol{y}) = \min\left\{\frac{f(\boldsymbol{y})}{f(\boldsymbol{x})},\ 1\right\}, \tag{3.34}$$

and the MCMC algorithm is called a **random walk sampler**. A typical example is when, for a given current state $\boldsymbol{x}$, the proposal state $\boldsymbol{Y}$ is of the form $\boldsymbol{Y} = \boldsymbol{x}+\boldsymbol{Z}$, where $\boldsymbol{Z}$ is generated from some spherically symmetric distribution, such as $\mathcal{N}(\boldsymbol{0},\mathbf{I})$.

We now give an example illustrating the second approach.

■ **Example 3.35 (Random Walk Sampler)** Consider the two-dimensional pdf

$$f(x_1,x_2) = c\,\mathrm{e}^{-\frac{1}{4}\sqrt{x_1^2+x_2^2}}\left(\sin\left(2\sqrt{x_1^2+x_2^2}\right)+1\right), \quad -2\pi < x_1 < 2\pi,\ -2\pi < x_2 < 2\pi, \tag{3.36}$$

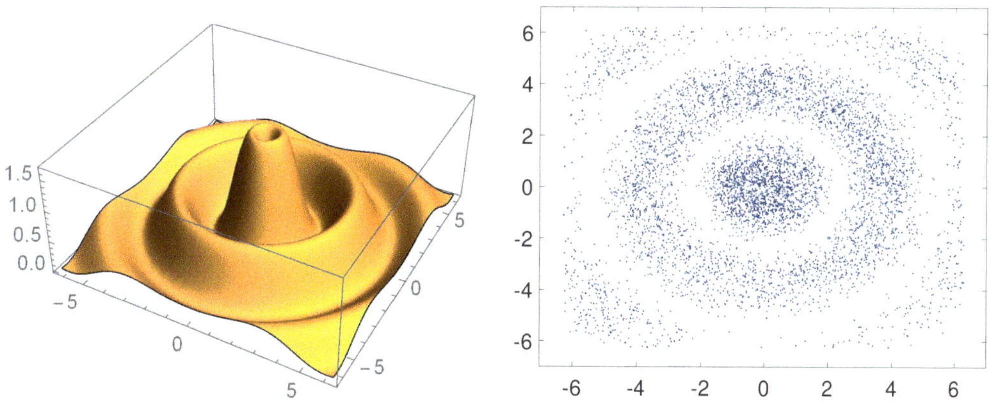

Figure 3.37: Left panel: the two-dimensional target pdf. Right panel: points from the random walk sampler are approximately distributed according to the target pdf.

where c is an unknown normalization constant. The graph of this pdf (unnormalized) is depicted in the left panel of Figure 3.37.

The following Python program implements a random walk sampler to (approximately) draw $N = 10^4$ dependent samples from the pdf f. At each step, given a current state $\boldsymbol{x}$, a proposal $\boldsymbol{Y}$ is drawn from the $\mathcal{N}(\boldsymbol{x}, \mathbf{I})$ distribution. That is, $\boldsymbol{Y} = \boldsymbol{x} + \boldsymbol{Z}$, with $\boldsymbol{Z}$ bivariate standard normal. We see in the right panel of Figure 3.37 that the sampler works correctly. The starting point for the Markov chain is chosen as $(0, 0)$. Note that the normalization constant c is never required to be specified in the program.

rwsamp.py

```
import numpy as np
import matplotlib.pyplot as plt
from numpy import pi, exp, sqrt, sin
from numpy.random import rand, randn

N = 10000
a = lambda x: -2*pi < x
b = lambda x: x < 2*pi
f = lambda x1, x2: exp(-sqrt(x1**2+x2**2)/4)*(
        sin(2*sqrt(x1**2+x2**2))+1)*a(x1)*b(x1)*a(x2)*b(x2)

xx = np.zeros((N,2))
x = np.zeros((1,2))
for i in range(1,N):
    y = x + randn(1,2)
    alpha = np.amin((f(y[0][0],y[0][1])/f(x[0][0],x[0][1]),1))
    r = rand() < alpha
    x = r*y + (1-r)*x
    xx[i,:] = x

plt.scatter(xx[:,0], xx[:,1], alpha =0.4,s =2)
plt.axis('equal')
plt.show()
```

3.2.5.2 Gibbs Sampler

The **Gibbs sampler** [71] uses a somewhat different methodology from the Metropolis–Hastings algorithm and is particularly useful for generating n-dimensional random vectors. The key idea of the Gibbs sampler is to update the components of the random vector one at a time, by sampling them from *conditional* pdfs. Thus, Gibbs sampling can be advantageous if it is easier to sample from the conditional distributions than from the joint distribution.

Specifically, suppose that we wish to sample a random vector $\boldsymbol{X} = [X_1, \ldots, X_n]^\top$ according to a target pdf $f(\boldsymbol{x})$. Let $f(x_i \,|\, x_1, \ldots, x_{i-1}, x_{i+1}, \ldots, x_n)$ represent the conditional pdf[1] of the i-th component, X_i, given the other components: $x_1, \ldots, x_{i-1}, x_{i+1}, \ldots, x_n$. The Gibbs sampling algorithm is as follows:

Algorithm 3.38: Gibbs Sampler

Input: Initial point $\boldsymbol{X}_0$, sample size N, and target pdf f.
Output: $\boldsymbol{X}_1, \ldots, \boldsymbol{X}_N$ approximately distributed according to f.

1 **for** $t = 0$ **to** $N - 1$ **do**
2   Draw Y_1 from the conditional pdf $f(y_1 \,|\, X_{t,2}, \ldots, X_{t,n})$.
3   **for** $i = 2$ **to** n **do**
4   Draw Y_i from the conditional pdf $f(y_i \,|\, Y_1, \ldots, Y_{i-1}, X_{t,i+1}, \ldots, X_{t,n})$.
5   $\boldsymbol{X}_{t+1} \leftarrow \boldsymbol{Y}$
6 **return** $\boldsymbol{X}_1, \ldots, \boldsymbol{X}_N$

There exist many variants of the Gibbs sampler, depending on the steps required to update $\boldsymbol{X}_t$ to $\boldsymbol{X}_{t+1}$ — called the **cycle** of the Gibbs algorithm. In the algorithm above, the cycle consists of Steps 2–5, in which the components are updated in a fixed order $1 \to 2 \to \cdots \to n$. For this reason Algorithm 3.38 is also called the **systematic Gibbs sampler**.

In the **random-order Gibbs sampler**, the order in which the components are updated ☞ 127 in each cycle is a random permutation of $\{1, \ldots, n\}$ (see Exercise 9). Other modifications are to update the components in blocks (i.e., several at the same time), or to update only a random selection of components. The variant where in each cycle only a single random component is updated is called the **random Gibbs sampler**. In the **reversible Gibbs sampler** a cycle consists of the coordinate-wise updating:

$$1 \to 2 \to \cdots \to n - 1 \to n \to n - 1 \to \cdots \to 2 \to 1.$$

In all cases, except for the systematic Gibbs sampler, the resulting Markov chain $\{\boldsymbol{X}_t, t =$ ☞ 661 $1, 2, \ldots\}$ is *reversible* and hence its limiting distribution is precisely $f(\boldsymbol{x})$.

Unfortunately, the systematic Gibbs Markov chain is not reversible and so the detailed balance equations are not satisfied. However, a similar result holds, due to Hammersley and Clifford, under the so-called **positivity condition**: if at a point $\boldsymbol{x} = (x_1, \ldots, x_n)$ all marginal densities $f(x_i) > 0, i = 1, \ldots, n$, then the joint density $f(\boldsymbol{x}) > 0$.

[1]In this section we employ a Bayesian notation style, using the same letter f for different (conditional) densities.

Theorem 3.39 (Hammersley–Clifford Balance for the Gibbs Sampler): Let $q_{1\to n}(\boldsymbol{y}\,|\,\boldsymbol{x})$ denote the transition density of the systematic Gibbs sampler, and let $q_{n\to 1}(\boldsymbol{x}\,|\,\boldsymbol{y})$ be the transition density of the reverse move, in the order $n \to n-1 \to \cdots \to 1$. Then, if the positivity condition holds,

$$f(\boldsymbol{x})\, q_{1\to n}(\boldsymbol{y}\,|\,\boldsymbol{x}) = f(\boldsymbol{y})\, q_{n\to 1}(\boldsymbol{x}\,|\,\boldsymbol{y}). \tag{3.40}$$

Proof. For the forward move we have:

$$q_{1\to n}(\boldsymbol{y}\,|\,\boldsymbol{x}) = f(y_1\,|\,x_2,\ldots,x_n) f(y_2\,|\,y_1,x_3,\ldots,x_n)\cdots f(y_n\,|\,y_1,\ldots,y_{n-1}),$$

and for the reverse move:

$$q_{n\to 1}(\boldsymbol{x}\,|\,\boldsymbol{y}) = f(x_n\,|\,y_1,\ldots,y_{n-1}) f(x_{n-1}\,|\,y_1,\ldots,y_{n-2},x_n)\cdots f(x_1\,|\,x_2,\ldots,x_n).$$

Consequently,

$$\begin{aligned}
\frac{q_{1\to n}(\boldsymbol{y}\,|\,\boldsymbol{x})}{q_{n\to 1}(\boldsymbol{x}\,|\,\boldsymbol{y})} &= \prod_{i=1}^{n} \frac{f(y_i\,|\,y_1,\ldots,y_{i-1},x_{i+1},\ldots,x_n)}{f(x_i\,|\,y_1,\ldots,y_{i-1},x_{i+1},\ldots,x_n)} \\
&= \prod_{i=1}^{n} \frac{f(y_1,\ldots,y_i,x_{i+1},\ldots,x_n)}{f(y_1,\ldots,y_{i-1},x_i,\ldots,x_n)} \\
&= \frac{f(\boldsymbol{y}) \prod_{i=1}^{n-1} f(y_1,\ldots,y_i,x_{i+1},\ldots,x_n)}{f(\boldsymbol{x}) \prod_{j=2}^{n} f(y_1,\ldots,y_{j-1},x_j,\ldots,x_n)} \\
&= \frac{f(\boldsymbol{y}) \prod_{i=1}^{n-1} f(y_1,\ldots,y_i,x_{i+1},\ldots,x_n)}{f(\boldsymbol{x}) \prod_{j=1}^{n-1} f(y_1,\ldots,y_j,x_{j+1},\ldots,x_n)} = \frac{f(\boldsymbol{y})}{f(\boldsymbol{x})}.
\end{aligned}$$

The result follows by rearranging the last identity. The positivity condition ensures that we do not divide by 0 along the line. □

Intuitively, the long-run proportion of transitions $\boldsymbol{x} \to \boldsymbol{y}$ for the "forward move" chain is equal to the long-run proportion of transitions $\boldsymbol{y} \to \boldsymbol{x}$ for the "reverse move" chain. To verify that the Markov chain $\boldsymbol{X}_0, \boldsymbol{X}_1, \ldots$ for the systematic Gibbs sampler indeed has limiting pdf $f(\boldsymbol{x})$, we need to check that the global balance equations (D.104) hold. By integrating (in the continuous case) both sides in (3.40) with respect to $\boldsymbol{x}$, we see that indeed ☞ 661

$$\int f(\boldsymbol{x})\, q_{1\to n}(\boldsymbol{y}\,|\,\boldsymbol{x})\,\mathrm{d}\boldsymbol{x} = f(\boldsymbol{y}).$$

■ **Example 3.41 (Gibbs Sampler for the Bayesian Normal Model)** Gibbs samplers are often applied in Bayesian statistics, to sample from the posterior pdf. Consider for instance the Bayesian normal model ☞ 52

$$\begin{aligned}
f(\mu,\sigma^2) &= 1/\sigma^2 \\
(\boldsymbol{x}\,|\,\mu,\sigma^2) &\sim \mathcal{N}(\mu\mathbf{1},\sigma^2\mathbf{I}).
\end{aligned}$$

Here the prior for (μ, σ^2) is *improper*. That is, it is not a pdf in itself, but by obstinately applying Bayes' formula it does yield a proper posterior pdf. In some sense this prior conveys the least amount of information about μ and σ^2. Following the same procedure as in Example 2.76, we find the posterior pdf:

$$f(\mu, \sigma^2 \,|\, \boldsymbol{x}) \propto \left(\sigma^2\right)^{-n/2-1} \exp\left\{-\frac{1}{2}\frac{\sum_i (x_i - \mu)^2}{\sigma^2}\right\}. \tag{3.42}$$

Note that μ and σ^2 here are the "variables" and $\boldsymbol{x}$ is a fixed data vector. To simulate samples μ and σ^2 from (3.42) using the Gibbs sampler, we need the distributions of both $(\mu \,|\, \sigma^2, \boldsymbol{x})$ and $(\sigma^2 \,|\, \mu, \boldsymbol{x})$. To find $f(\mu \,|\, \sigma^2, \boldsymbol{x})$, view the right-hand side of (3.42) as a function of μ only, regarding σ^2 as a constant. This gives

$$\begin{aligned} f(\mu \,|\, \sigma^2, \boldsymbol{x}) &\propto \exp\left\{-\frac{n\mu^2 - 2\mu \sum_i x_i}{2\sigma^2}\right\} = \exp\left\{-\frac{\mu^2 - 2\mu\overline{x}}{2(\sigma^2/n)}\right\} \\ &\propto \exp\left\{-\frac{1}{2}\frac{(\mu - \overline{x})^2}{\sigma^2/n}\right\}. \qquad (3.43) \end{aligned}$$

This shows that $(\mu \,|\, \sigma^2, \boldsymbol{x})$ has a normal distribution with mean $\overline{x}$ and variance σ^2/n.

Similarly, to find $f(\sigma^2 \,|\, \mu, \boldsymbol{x})$, view the right-hand side of (3.42) as a function of σ^2, regarding μ as a constant. This gives

$$f(\sigma^2 \,|\, \mu, \boldsymbol{x}) \propto (\sigma^2)^{-n/2-1} \exp\left\{-\frac{1}{2}\sum_{i=1}^n (x_i - \mu)^2/\sigma^2\right\}, \tag{3.44}$$

☞ 629 showing that $(\sigma^2 \,|\, \mu, \boldsymbol{x})$ has an inverse-gamma distribution with parameters $n/2$ and $\sum_{i=1}^n (x_i - \mu)^2/2$. The Gibbs sampler thus involves the repeated simulation of

$$\begin{aligned} (\mu \,|\, \sigma^2, \boldsymbol{x}) &\sim \mathcal{N}\left(\overline{x},\ \sigma^2/n\right), \\ (\sigma^2 \,|\, \mu, \boldsymbol{x}) &\sim \mathsf{InvGamma}\left(n/2,\ \sum_{i=1}^n (x_i - \mu)^2/2\right). \end{aligned}$$

Simulating $X \sim \mathsf{InvGamma}(\alpha, \lambda)$ is achieved by first generating $Z \sim \mathsf{Gamma}(\alpha, \lambda)$ and then returning $X = 1/Z$.

> ⚠ In our parameterization of the $\mathsf{Gamma}(\alpha, \lambda)$ distribution, λ is the *rate* parameter. Many software packages instead use the *scale* parameter $c = 1/\lambda$. Be aware of this when simulating Gamma random variables.

The Python script below defines a small data set of size $n = 10$ (which was randomly simulated from a standard normal distribution), and implements the systematic Gibbs sampler to simulate from the posterior distribution, using $N = 10^5$ samples.

gibbsamp.py

```
import numpy as np
import matplotlib.pyplot as plt

x = np.array([[-0.9472, 0.5401, -0.2166, 1.1890, 1.3170,
               -0.4056, -0.4449, 1.3284, 0.8338, 0.6044]])
n = x.size
sample_mean = np.mean(x); sample_var = np.var(x)
sig2 = np.var(x); mu=sample_mean

N = 10**5
gibbs_sample = np.array(np.zeros((N, 2)))
for k in range(N):
    mu = sample_mean + np.sqrt(sig2/n)*np.random.randn()
    V = np.sum((x-mu)**2)/2
    sig2 = 1/np.random.gamma(n/2, 1/V)
    gibbs_sample[k,:]= np.array([mu, sig2])
plt.scatter(gibbs_sample[:,0], gibbs_sample[:,1],alpha =0.1,s =1)
plt.plot(np.mean(x), np.var(x),'wo')
```

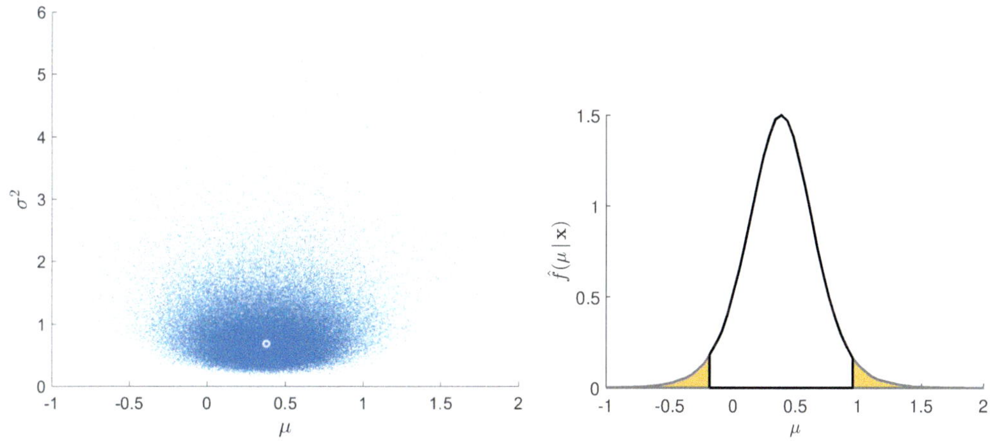

Figure 3.45: Left: approximate draws from the posterior pdf $f(\mu, \sigma^2 \,|\, \boldsymbol{x})$ obtained via the Gibbs sampler. Right: estimate of the posterior pdf $f(\mu \,|\, \boldsymbol{x})$.

The left panel of Figure 3.45 shows the (μ, σ^2) points generated by the Gibbs sampler. Also shown, via the white circle, is the point $(\overline{x}, s^2)$, where $\overline{x} = 0.3798$ is the sample mean and $s^2 = 0.6810$ the sample variance. This posterior point cloud visualizes the considerable uncertainty in the estimates. By projecting the (μ, σ^2) points onto the μ-axis — that is, by ignoring the σ^2 values — one obtains (approximate) samples from the posterior pdf of μ; that is, $f(\mu \,|\, \boldsymbol{x})$. The right panel of Figure 3.45 shows a kernel density estimate (see Section 4.4) of this pdf. The corresponding 0.025 and 0.975 sample quantiles were found to be -0.2054 and 0.9662, respectively, giving the 95% credible interval $(-0.2054, 0.9662)$ for μ, which contains the true expectation 0. Similarly, an estimated 95% credible interval for σ^2 is $(0.3218, 2.2485)$, which contains the true variance 1. ☞ 158

■

3.3 Monte Carlo Estimation

In this section we describe how Monte Carlo simulation can be used to estimate complicated integrals, probabilities, and expectations. A number of variance reduction techniques are introduced as well, including the versatile cross-entropy method.

3.3.1 Crude Monte Carlo

The most common setting for Monte Carlo estimation is the following: Suppose we wish to compute the expectation $\mu = \mathbb{E}Y$ of some (say continuous) random variable Y with pdf f, but the integral $\mathbb{E}Y = \int y f(y)\,\mathrm{d}y$ is difficult to evaluate. For example, if Y is a complicated function of other random variables, it would be difficult to obtain an exact expression for $f(y)$. The idea of **crude Monte Carlo** — sometimes abbreviated as CMC — is to approximate μ by simulating many independent copies $Y_1, \ldots, Y_N$ of Y and then take their *sample mean* $\overline{Y}$ as an estimator of μ. All that is needed is an algorithm to simulate such copies.

☞ 653 By the Law of Large Numbers, $\overline{Y}$ converges to μ as $N \to \infty$, provided the expectation ☞ 654 of Y exists. Moreover, by the Central Limit Theorem, $\overline{Y}$ approximately has a $\mathcal{N}(\mu, \sigma^2/N)$ distribution for large N, provided that the variance $\sigma^2 = \mathbb{V}\mathrm{ar}Y < \infty$. This enables the construction of an approximate $(1-\alpha)$ **confidence interval** for μ:

$$\left(\overline{Y} - z_{1-\alpha/2}\frac{S}{\sqrt{N}},\quad \overline{Y} + z_{1-\alpha/2}\frac{S}{\sqrt{N}}\right), \tag{3.46}$$

where S is the sample standard deviation of the $\{Y_i\}$ and z_γ denotes the γ-quantile of the ☞ 666 $\mathcal{N}(0,1)$ distribution; see also Section D.13. Instead of specifying the confidence interval, one often reports only the sample mean and the **estimated standard error**: $S/\sqrt{N}$, or the **estimated relative error**: $S/(\overline{Y}\sqrt{N})$. The basic estimation procedure for independent data is summarized in Algorithm 3.47 below.

Algorithm 3.47: Crude Monte Carlo for Independent Data

Input: Simulation algorithm for $Y \sim f$, sample size N, confidence level $1-\alpha$.
Output: Point estimate and approximate $(1-\alpha)$ confidence interval for $\mu = \mathbb{E}Y$.

Simulate $Y_1, \ldots, Y_N \overset{\text{iid}}{\sim} f$.
$\overline{Y} \leftarrow \frac{1}{N}\sum_{i=1}^N Y_i$
$S^2 \leftarrow \frac{1}{N-1}\sum_{i=1}^N (Y_i - \overline{Y})^2$
return $\overline{Y}$ and the interval (3.46).

It is often the case that the output Y is a function of some underlying random vector or stochastic process; that is, $Y = H(\boldsymbol{X})$, where H is a real-valued function and $\boldsymbol{X}$ is a random vector or process. The beauty of Monte Carlo for estimation is that (3.46) holds regardless of the dimension of $\boldsymbol{X}$.

■ **Example 3.48 (Monte Carlo Integration)** In **Monte Carlo integration**, simulation is used to evaluate complicated integrals. Consider, for example, the integral

$$\mu = \int_{-\infty}^{\infty} \int_{-\infty}^{\infty} \int_{-\infty}^{\infty} \sqrt{|x_1 + x_2 + x_3|}\, \mathrm{e}^{-(x_1^2+x_2^2+x_3^2)/2}\, \mathrm{d}x_1\, \mathrm{d}x_2\, \mathrm{d}x_3.$$

Defining $Y = |X_1 + X_2 + X_3|^{1/2}(2\pi)^{3/2}$, with $X_1, X_2, X_3 \stackrel{\text{iid}}{\sim} \mathcal{N}(0, 1)$, we can write $\mu = \mathbb{E}Y$. Using the following Python program, with a sample size of $N = 10^6$, we obtained an estimate $\overline{Y} = 17.031$ with an approximate 95% confidence interval $(17.017, 17.046)$.

mcint.py

```
import numpy as np
from numpy import pi

c = (2*pi)**(3/2)
H = lambda x: c*np.sqrt(np.abs(np.sum(x,axis=1)))
N = 10**6
z = 1.96
x = np.random.randn(N,3)
y = H(x)
mY = np.mean(y)
sY = np.std(y)
RE = sY/mY/np.sqrt(N)
print('Estimate = {:3.3f}, CI = ({:3.3f},{:3.3f})'.format(
        mY, mY*(1-z*RE), mY*(1+z*RE)))
```

```
Estimate = 17.031, CI = (17.017,17.046)
```

■

■ **Example 3.49 (Example 2.13 (cont.))** We return to the bias–variance tradeoff in Example 2.13. Figure 2.25 gives estimates of the (squared-error) generalization risk (2.6) as a function of the number of parameters in the model. But how accurate are these estimates? Because we know in this case the exact model for the data, we can use Monte Carlo simulation to estimate the generalization risk (for a fixed training set) and the expected generalization risk (averaged over all training sets) precisely. All we need to do is repeat the data generation, fitting, and validation steps many times and then take averages of the results. The following Python code repeats 100 times:

☞ 26
☞ 30
☞ 23

1. Simulate the training set of size $n = 100$.
2. Fit models up to size $k = 8$.
3. Estimate the test loss using a test set with the same sample size $n = 100$.

Figure 3.50 shows that there is some variation in the test losses, due to the randomness in both the training and test sets. To obtain an accurate estimate of the expected generalization risk (2.8), take the average of the test losses. We see that for $k \leq 8$ the estimate in Figure 2.25 is close to the true expected generalization risk.

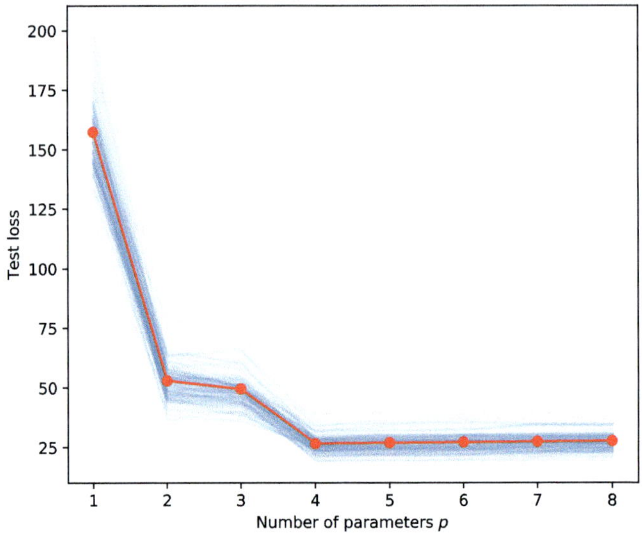

Figure 3.50: Independent estimates of the test loss show some variability.

CMCtestloss.py

```
import numpy as np, matplotlib.pyplot as plt
from numpy.random import rand, randn
from numpy.linalg import solve

def generate_data(beta, sig, n):
    u = rand(n, 1)
    y = (u ** np.arange(0, 4)) @ beta + sig * randn(n, 1)
    return u, y

beta = np.array([[10, -140, 400, -250]]).T
n = 100
sig = 5
betahat = {}
plt.figure(figsize=[6,5])
totMSE = np.zeros(8)
max_p = 8
p_range = np.arange(1, max_p + 1, 1)

for N in range(0,100):
    u, y = generate_data(beta, sig, n)  #training data
    X = np.ones((n, 1))
    for p in p_range:
       if p > 1:
         X = np.hstack((X, u**(p-1)))
       betahat[p] = solve(X.T @ X, X.T @ y)

    u_test, y_test = generate_data(beta, sig, n)  #test data
    MSE = []
    X_test = np.ones((n, 1))
    for p in p_range:
       if p > 1:
           X_test = np.hstack((X_test, u_test**(p-1)))
```

```
        y_hat = X_test @ betahat[p] # predictions
        MSE.append(np.sum((y_test - y_hat)**2/n))

    totMSE = totMSE + np.array(MSE)
    plt.plot(p_range, MSE,'C0',alpha=0.1)

plt.plot(p_range,totMSE/N,'r-o')
plt.xticks(ticks=p_range)
plt.xlabel('Number of parameters $p$')
plt.ylabel('Test loss')
plt.tight_layout()
plt.savefig('MSErepeatpy.pdf',format='pdf')
plt.show()
```

■

3.3.2 Bootstrap Method

The bootstrap method [55] combines CMC estimation with the resampling procedure of Section 3.2.4. The idea is as follows: Suppose we wish to estimate a number μ via some estimator $Y = H(\mathcal{T})$, where $\mathcal{T} := \{X_1, \dots, X_n\}$ is an iid sample from some unknown cdf F. It is assumed that Y does not depend on the order of the $\{X_i\}$. To assess the quality (for example, accuracy) of the estimator Y, one could draw independent replications $\mathcal{T}_1, \dots, \mathcal{T}_N$ of $\mathcal{T}$ and find sample estimates for quantities such as the variance $\mathbb{V}\mathrm{ar}Y$, the bias $\mathbb{E}Y - \mu$, and the mean squared error $\mathbb{E}(Y - \mu)^2$. However, it may be too time-consuming or simply not feasible to obtain such replications. An alternative is to *resample* the original data. To reiterate, given an outcome $\tau = \{x_1, \dots, x_n\}$ of $\mathcal{T}$, we simulate an iid sample $\mathcal{T}^* := \{X_1^*, \dots, X_n^*\}$ from the empirical cdf F_n, via Algorithm 3.26 (hence the resampling size is $N = n$ here).

☞ 86
☞ 87

The rationale is that the empirical cdf F_n is close to the actual cdf F and gets closer as n gets larger. Hence, any quantities depending on F, such as $\mathbb{E}_F g(Y)$, where g is a function, can be approximated by $\mathbb{E}_{F_n} g(Y)$. The latter is usually still difficult to evaluate, but it can be simply estimated via CMC as

$$\frac{1}{K}\sum_{i=1}^{K} g(Y_i^*),$$

where $Y_1^*, \dots, Y_K^*$ are independent random variables, each distributed as $Y^* = H(\mathcal{T}^*)$. This seemingly self-referent procedure is called **bootstrapping** — alluding to Baron von Münchausen, who pulled himself out of a swamp by his own bootstraps. As an example, the bootstrap estimate of the expectation of Y is

$$\widehat{\mathbb{E}Y} = \overline{Y}^* = \frac{1}{K}\sum_{i=1}^{K} Y_i^*,$$

which is simply the sample mean of $\{Y_i^*\}$. Similarly, the bootstrap estimate for $\mathbb{V}\mathrm{ar}Y$ is the sample variance

$$\widehat{\mathbb{V}\mathrm{ar}Y} = \frac{1}{K-1}\sum_{i=1}^{K}(Y_i^* - \overline{Y}^*)^2. \tag{3.51}$$

Bootstrap estimators for the bias and MSE are $\overline{Y}^* - Y$ and $\frac{1}{K}\sum_{i=1}^{K}(Y_i^* - Y)^2$, respectively. Note that for these estimators the unknown quantity μ is replaced with its original estimator Y. Confidence intervals can be constructed in the same fashion. We mention two variants: the **normal method** and the **percentile method**. In the normal method, a $1 - \alpha$ confidence interval for μ is given by

$$(Y \pm z_{1-\alpha/2} S^*),$$

where S^* is the bootstrap estimate of the standard deviation of Y; that is, the square root of (3.51). In the percentile method, the upper and lower bounds of the $1 - \alpha$ confidence interval for μ are given by the $1 - \alpha/2$ and $\alpha/2$ quantiles of Y, which in turn are estimated via the corresponding sample quantiles of the bootstrap sample $\{Y_i^*\}$.

The following example illustrates the usefulness of the bootstrap method for *ratio estimation* and also introduces the *renewal reward process* model for data.

■ **Example 3.52 (Bootstrapping the Ratio Estimator)** A common scenario in stochastic simulation is that the output of the simulation consists of independent pairs of data $(C_1, R_1), (C_2, R_2), \ldots$, where each C is interpreted as the length of a period of time — a so-called *cycle* — and R is the *reward* obtained during that cycle. Such a collection of random variables $\{(C_i, R_i)\}$ is called a **renewal reward process**. Typically, the reward R_i depends on the cycle length C_i. Let A_t be the *average reward* earned by time t; that is, $A_t = \sum_{i=1}^{N_t} R_i/t$, where:

$$N_t = \max\{n : C_1 + \cdots + C_n \leqslant t\}$$

☞ 130 counts the number of complete cycles at time t. It can be shown, see Exercise 20, that if the expectations of the cycle length and reward are finite, then A_t converges to the constant $\mathbb{E}R/\mathbb{E}C$. This ratio can thus be interpreted as the **long-run average reward**. Estimation of the ratio $\mathbb{E}R/\mathbb{E}C$ from data $(C_1, R_1), \ldots, (C_n, R_n)$ is easy: take the **ratio estimator**

$$A = \frac{\overline{R}}{\overline{C}}.$$

However, this estimator A is not unbiased and it is not obvious how to derive confidence intervals. Fortunately, the bootstrap method can come to the rescue: simply resample the pairs $\{(C_i, R_i)\}$, obtain ratio estimators $A_1^*, \ldots, A_K^*$, and from these compute quantities of interest such as confidence intervals.

☞ 85 As a concrete example, let us return to the Markov chain in Example 3.24. Recall that the chain starts at state 1 at time 0. After a certain amount of time T_1, the process returns to state 1. The time steps $0, \ldots, T_1 - 1$ form a natural "cycle" for this process, as from time T_1 onwards the process behaves probabilistically *exactly the same* as when it started, *independently* of $X_0, \ldots, X_{T_1-1}$. Thus, if we define $T_0 = 0$, and let T_i be the i-th time that the chain returns to state 1, then we can break up the time interval into independent cycles of lengths $C_i = T_i - T_{i-1}$, $i = 1, 2, \ldots$. Now suppose that during the i-th cycle a reward

$$R_i = \sum_{t=T_{i-1}}^{T_i - 1} \varrho^{t-T_{i-1}}\, r(X_t)$$

is received, where $r(i)$ is some fixed reward for visiting state $i \in \{1, 2, 3, 4\}$ and $\varrho \in (0, 1)$ is a discounting factor. Clearly, $\{(C_i, R_i)\}$ is a renewal reward process. Figure 3.53 shows the outcomes of 1000 pairs (C, R), using $r(1) = 4, r(2) = 3, r(3) = 10, r(4) = 1$, and $\varrho = 0.9$.

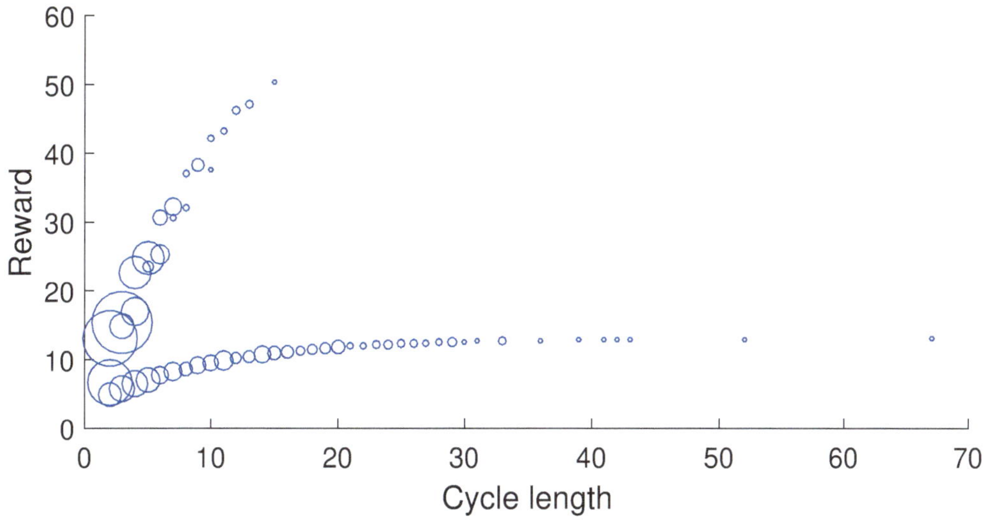

Figure 3.53: Each circle represents a (cycle length, reward) pair. The varying circle sizes indicate the number of occurrences for a given pair. For example, (2,15.43) is the most likely pair. It corresponds to the cycle path $1 \to 3 \to 2 \to 1$.

The long-run average reward is estimated as 2.50 for our data. But how accurate is this estimate? Figure 3.54 shows a density plot of the bootstrapped ratio estimates, where we independently resampled the data pairs 1000 times.

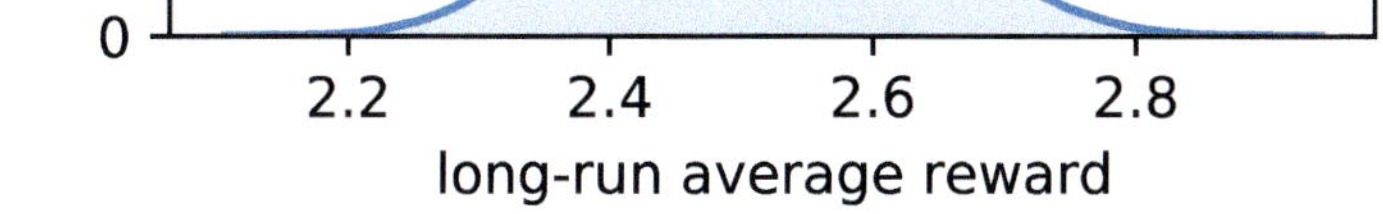

Figure 3.54: Density plot of the bootstrapped ratio estimates for the Markov chain renewal reward process.

Figure 3.54 indicates that the true long-run average reward lies between 2.2 and 2.8 with high confidence. More precisely, the 99% bootstrap confidence interval (percentile method) is here (2.27, 2.77). The following Python script spells out the procedure.

ratioest.py

```
import numpy as np
import matplotlib.pyplot as plt
import seaborn as sns
from numba import jit

np.random.seed(123)
n = 1000
P = np.array([[0, 0.2, 0.5, 0.3],
              [0.5, 0, 0.5, 0],
              [0.3, 0.7, 0, 0],
              [0.1, 0, 0, 0.9]])
r = np.array([4, 3, 10, 1])
Corg = np.zeros((n, 1))
Rorg = np.zeros((n, 1))
rho = 0.9

@jit(nopython=True)
def generate_cyclereward(n, Corg, Rorg):
    for i in range(n):
        t = 1
        xreg = 1  # regenerative state (out of 1,2,3,4)
        reward = r[0]
        x = np.argmax(np.cumsum(P[xreg-1, :]) > np.random.rand()) + 1
        while x != xreg:
            t += 1
            reward += rho**(t-1) * r[x-1]
            x = np.argmax(np.cumsum(P[x-1, :]) > np.random.rand()) + 1
        Corg[i, 0] = t
        Rorg[i, 0] = reward
    return Corg, Rorg

Corg, Rorg = generate_cyclereward(n, Corg, Rorg)
Aorg = np.mean(Rorg) / np.mean(Corg)

K = 5000
A = np.zeros((K, 1))
C = np.zeros((n, 1))
R = np.zeros((n, 1))
for i in range(K):
    ind = np.ceil(n * np.random.rand(1, n)).astype(int)[0] - 1
    C = Corg[ind]
    R = Rorg[ind]
    A[i, 0] = np.mean(R) / np.mean(C)

plt.xlabel('long-run average reward')
plt.ylabel('density')
sns.kdeplot(A.flatten(), fill=True)
plt.show()
```

■

3.3.3 Variance Reduction

The estimation of performance measures in Monte Carlo simulation can be made more efficient by utilizing known information about the simulation model. Variance reduction techniques include antithetic variables, control variables, importance sampling, conditional Monte Carlo, and stratified sampling; see, for example, [98, Chapter 9] and [28, 109]. We shall only deal with control variables and importance sampling here.

Suppose Y is the output of a simulation experiment. A random variable $\widetilde{Y}$, obtained from the same simulation run, is called a **control variable** for Y if Y and $\widetilde{Y}$ are correlated (negatively or positively) and the expectation of $\widetilde{Y}$ is known. The use of control variables for variance reduction is based on the following theorem. We leave its proof to Exercise 21. ☞ 131

Theorem 3.55 (Control Variable Estimation): Let $Y_1, \ldots, Y_N$ be the output of N independent simulation runs and let $\widetilde{Y}_1, \ldots, \widetilde{Y}_N$ be the corresponding control variables, with $\mathbb{E}\widetilde{Y}_k = \widetilde{\mu}$ known. Let $\varrho_{Y,\widetilde{Y}}$ be the correlation coefficient between each Y_k and $\widetilde{Y}_k$. For each $\alpha \in \mathbb{R}$ the estimator

$$\widehat{\mu}^{(c)} = \frac{1}{N} \sum_{k=1}^{N} \left[Y_k - \alpha \left(\widetilde{Y}_k - \widetilde{\mu} \right) \right] \tag{3.56}$$

is an unbiased estimator for $\mu = \mathbb{E}Y$. The minimal variance of $\widehat{\mu}^{(c)}$ is

$$\mathbb{V}\mathrm{ar}\, \widehat{\mu}^{(c)} = \frac{1}{N} (1 - \varrho^2_{Y,\widetilde{Y}})\, \mathbb{V}\mathrm{ar}\, Y, \tag{3.57}$$

which is obtained for $\alpha = \varrho_{Y,\widetilde{Y}} \sqrt{\mathbb{V}\mathrm{ar} Y / \mathbb{V}\mathrm{ar} \widetilde{Y}}$.

From (3.57) we see that, by using the optimal α in (3.56), the variance of the control variate estimator is a factor $1 - \varrho^2_{Y,\widetilde{Y}}$ smaller than the variance of the crude Monte Carlo estimator. Thus, if $\widetilde{Y}$ is highly correlated with Y, a significant variance reduction can be achieved. The optimal α is usually unknown, but it can be easily estimated from the sample covariance matrix of $\{(Y_k, \widetilde{Y}_k)\}$. ☞ 665

In the next example, we estimate the multiple integral in Example 3.48 using control variables.

■ **Example 3.58 (Monte Carlo Integration (cont.))** The random variable $Y = |X_1 + X_2 + X_3|^{1/2}(2\pi)^{3/2}$ is positively correlated with the random variable $\widetilde{Y} = X_1^2 + X_2^2 + X_3^2$, for the same choice of $X_1, X_2, X_3 \overset{\text{iid}}{\sim} \mathcal{N}(0, 1)$. As $\mathbb{E}\widetilde{Y} = \mathbb{V}\mathrm{ar}(X_1 + X_2 + X_3) = 3$, we can use it as a control variable to estimate the expectation of Y. The following Python program is based on Theorem 3.55. It imports the crude Monte Carlo sampling code from Example 3.48. ☞ 96

mcintCV.py

```
from mcint import *

Yc = np.sum(x**2, axis=1) # control variable data
yc = 3  # true expectation of control variable
C = np.cov(y,Yc) # sample covariance matrix
cor = C[0][1]/np.sqrt(C[0][0]*C[1][1])
alpha = C[0][1]/C[1][1]

est = np.mean(y-alpha*(Yc-yc))
RECV = np.sqrt((1-cor**2)*C[0][0]/N)/est  #relative error

print('Estimate = {:3.3f}, CI = ({:3.3f},{:3.3f}), Corr = {:3.3f}'.
      format(est, est*(1-z*RECV), est*(1+z*RECV),cor))
```

```
Estimate = 17.045, CI = (17.032,17.057), Corr = 0.480
```

A typical estimate of the correlation coefficient $\varrho_{Y,\widetilde{Y}}$ is 0.48, which gives a reduction of the variance with a factor $1 - 0.48^2 \approx 0.77$ — a simulation speed-up of 23% compared with crude Monte Carlo. Although the gain is small in this case, due to the modest correlation between Y and $\widetilde{Y}$, little extra work was required to achieve this variance reduction. ■

One of the most important variance reduction techniques is **importance sampling**. This technique is especially useful for the estimation of very small probabilities. The standard setting is the estimation of a quantity

$$\mu = \mathbb{E}_f H(\boldsymbol{X}) = \int H(\boldsymbol{x})\, f(\boldsymbol{x})\, \mathrm{d}\boldsymbol{x}, \tag{3.59}$$

where H is a real-valued function and f the probability density of a random vector $\boldsymbol{X}$, called the **nominal pdf**. The subscript f is added to the expectation operator to indicate that it is taken with respect to the density f.

Let g be another probability density such that $g(\boldsymbol{x}) = 0$ implies that $H(\boldsymbol{x})\, f(\boldsymbol{x}) = 0$. Using the density g we can represent μ as

$$\mu = \int H(\boldsymbol{x})\, \frac{f(\boldsymbol{x})}{g(\boldsymbol{x})}\, g(\boldsymbol{x})\, \mathrm{d}\boldsymbol{x} = \mathbb{E}_g \left[H(\boldsymbol{X})\, \frac{f(\boldsymbol{X})}{g(\boldsymbol{X})} \right]. \tag{3.60}$$

Consequently, if $\boldsymbol{X}_1, \ldots, \boldsymbol{X}_N \sim_{\text{iid}} g$, then

$$\widehat{\mu} = \frac{1}{N} \sum_{k=1}^{N} H(\boldsymbol{X}_k)\, \frac{f(\boldsymbol{X}_k)}{g(\boldsymbol{X}_k)} \tag{3.61}$$

is an unbiased estimator of μ. This estimator is called the **importance sampling estimator** and g is called the importance sampling density. The ratio of densities, $f(\boldsymbol{x})/g(\boldsymbol{x})$, is called the **likelihood ratio**. The importance sampling pseudo-code is given in Algorithm 3.62.

Algorithm 3.62: Importance Sampling Estimation

Input: Function H, importance sampling density g such that $g(\boldsymbol{x}) = 0$ for all $\boldsymbol{x}$ for which $H(\boldsymbol{x})f(\boldsymbol{x}) = 0$, sample size N, confidence level $1 - \alpha$.

Output: Point estimate and approximate $(1 - \alpha)$ confidence interval for $\mu = \mathbb{E}H(\boldsymbol{X})$, where $\boldsymbol{X} \sim f$.

1 Simulate $\boldsymbol{X}_1, \ldots, \boldsymbol{X}_N \overset{\text{iid}}{\sim} g$ and let $Y_i = H(\boldsymbol{X}_i)f(\boldsymbol{X}_i)/g(\boldsymbol{X}_i)$, $i = 1, \ldots, N$.

2 Estimate μ via $\widehat{\mu} = \overline{Y}$ and determine an approximate $(1 - \alpha)$ confidence interval as

$$\mathcal{I} := \left(\widehat{\mu} - z_{1-\alpha/2}\frac{S}{\sqrt{N}},\ \widehat{\mu} + z_{1-\alpha/2}\frac{S}{\sqrt{N}}\right),$$

where z_γ denotes the γ-quantile of the $\mathcal{N}(0, 1)$ distribution and S is the sample standard deviation of $Y_1, \ldots, Y_N$.

3 **return** $\widehat{\mu}$ and the interval $\mathcal{I}$.

■ **Example 3.63 (Importance Sampling)** Let us examine the workings of importance sampling by estimating the area, μ say, under the graph of the function

(3.64) $$M(x_1, x_2) = \mathrm{e}^{-\frac{1}{4}\sqrt{x_1^2+x_2^2}}\left(\sin\left(2\sqrt{x_1^2 + x_2^2}\right) + 1\right), \qquad (x_1, x_2) \in \mathbb{R}^2.$$

We saw a similar function in Example 3.35 (but note the different domain). A natural ☞ 90 approach to estimate the area is to truncate the domain to the square $[-b, b]^2$, for large enough b, and to estimate the integral (with $H(\boldsymbol{x}) := (2b)^2 M(\boldsymbol{x})$):

$$\mu_b = \int_{-b}^{b}\int_{-b}^{b} (2b)^2 M(\boldsymbol{x}) f(\boldsymbol{x})\, \mathrm{d}\boldsymbol{x} = \mathbb{E}_f H(\boldsymbol{X})$$

via crude Monte Carlo, where $f(\boldsymbol{x}) = 1/(2b)^2, \boldsymbol{x} \in [-b, b]^2$, is the pdf of the uniform distribution on $[-b, b]^2$. Here is the Python code which does just that.

impsamp1.py

```
import numpy as np
from numpy import exp, sqrt, sin, pi, log, cos
from numpy.random import rand

b = 1000
H = lambda x1, x2: (2*b)**2 * exp(-sqrt(x1**2+x2**2)/4)*(sin(2*sqrt(
        x1**2+x2**2))+1)*(x1**2 + x2**2 < b**2)
f = 1/((2*b)**2)
N = 10**6
X1 = -b + 2*b*rand(N,1)
X2 = -b + 2*b*rand(N,1)
Z = H(X1,X2)
estCMC = np.mean(Z).item()  # to obtain scalar
RECMC = np.std(Z)/estCMC/sqrt(N).item()
print('CI = ({:3.3f},{:3.3f}), RE = {: 3.3f}'.format(estCMC*(1-1.96*
    RECMC), estCMC*(1+1.96*RECMC),RECMC))
```

```
CI = (82.663,135.036), RE =  0.123
```

For a truncation level of $b = 1000$ and a sample size of $N = 10^6$, a typical estimate is 108.8, with an estimated relative error of 0.123. We have two sources of error here. The first is the error in approximating μ by μ_b. However, as the function H decays exponentially fast, $b = 1000$ is more than enough to ensure this error is negligible. The second type of error is the statistical error, due to the estimation process itself. This can be quantified by the estimated relative error, and can be reduced by increasing the sample size.

Let us now consider an importance sampling approach in which the importance sampling pdf g is radially symmetric and decays exponentially in the radius, similar to the function H. In particular, we simulate (X_1, X_2) in a way akin to Example 3.2, by first generating a radius $R \sim \mathsf{Exp}(\lambda)$ and an angle $\Theta \sim \mathcal{U}(0, 2\pi)$, and then returning $X_1 = R\cos(\Theta)$ and $X_2 = R\sin(\Theta)$. By the Transformation Rule (Theorem D.36) we then have

☞ 76
☞ 636

$$g(\boldsymbol{x}) = f_{R,\Theta}(r,\theta)\frac{1}{r} = \lambda\,\mathrm{e}^{-\lambda r}\frac{1}{2\pi}\frac{1}{r} = \frac{\lambda \mathrm{e}^{-\lambda\sqrt{x_1^2+x_2^2}}}{2\pi\sqrt{x_1^2+x_2^2}}, \quad \boldsymbol{x} \in \mathbb{R}^2 \setminus \{\boldsymbol{0}\}.$$

The following code, which imports the one given above, implements the importance sampling steps, using the parameter $\lambda = 0.1$.

impsamp2.py

```
from impsamp1 import *

lam = 0.1;
g = lambda x1, x2: lam*exp(-sqrt(x1**2 + x2**2)*lam)/sqrt(x1**2 + x2**2)
    /(2*pi);
U = rand(N,1); V = rand(N,1)
R = -log(U)/lam
X1 = R*cos(2*pi*V)
X2 = R*sin(2*pi*V)
Z = H(X1,X2)*f/g(X1,X2)
estIS = np.mean(Z).item()  # obtain scalar
REIS = np.std(Z)/estIS/sqrt(N).item()
print('CI = ({:3.3f},{:3.3f}), RE = {: 3.3f}'.format(estIS*(1-1.96*REIS)
    , estIS*(1+1.96*REIS),REIS))
```

```
CI = (100.723,101.077), RE =  0.001
```

A typical estimate is 100.90 with an estimated relative error of $1 \cdot 10^{-4}$, which gives a substantial variance reduction. In terms of approximate 95% confidence intervals, we have (82.7,135.0) in the CMC case versus (100.7,101.1) in the importance sampling case. Of course, we could have reduced the truncation level b to improve the performance of CMC, but then the approximation error might become more significant. For the importance sampling case, the relative error is hardly affected by the threshold level, but does depend on the choice of λ. We chose λ such that the decay rate is slower than the decay rate of the function H, which is 0.25. ■

As illustrated in the above example, a main difficulty in importance sampling is how to choose the importance sampling distribution. A poor choice of g may seriously affect the accuracy of both the estimate and the confidence interval. The theoretically optimal choice g^* for the importance sampling density minimizes the variance of $\widehat{\mu}$ and is therefore the solution to the functional minimization program

$$\min_g \mathbb{V}\mathrm{ar}_g \left(H(\boldsymbol{X}) \frac{f(\boldsymbol{X})}{g(\boldsymbol{X})} \right). \tag{3.65}$$

It is not difficult to show that the **optimal importance sampling pdf** is

$$g^*(\boldsymbol{x}) = \frac{|H(\boldsymbol{x})| \, f(\boldsymbol{x})}{\mu}. \tag{3.66}$$

Moreover, if either $H(\boldsymbol{x}) \geq 0$ or $H(\boldsymbol{x}) \leq 0$ for all $\boldsymbol{x}$, then the estimator $\widehat{\mu}$ is *constant* under g^*, because then $\mathbb{V}\mathrm{ar}_{g^*}\widehat{\mu} = \mathbb{V}\mathrm{ar}_{g^*}(H(\boldsymbol{X})f(\boldsymbol{X})/g(\boldsymbol{X})) = \mathbb{V}\mathrm{ar}_{g^*}\mu = 0$; see also Exercise 22. An obvious difficulty is that the evaluation of the optimal importance ☞ 131 sampling density g^* is usually not possible, since $g^*(\boldsymbol{x})$ in (3.66) depends on the unknown quantity μ. Nevertheless, one can typically choose a good importance sampling density g "close" to the minimum variance density g^*.

One of the main considerations for choosing a good importance sampling pdf is that the estimator (3.61) should have finite variance. This is equivalent to the requirement that

$$\mathbb{E}_g \left[H^2(\boldsymbol{X}) \frac{f^2(\boldsymbol{X})}{g^2(\boldsymbol{X})} \right] = \mathbb{E}_f \left[H^2(\boldsymbol{X}) \frac{f(\boldsymbol{X})}{g(\boldsymbol{X})} \right] < \infty. \tag{3.67}$$

This suggests that g should not have lighter tails than f and that, preferably, the likelihood ratio, f/g, should be bounded.

3.4 Monte Carlo for Optimization

In this section we describe several Monte Carlo methods for optimization. Such randomized algorithms can be useful for solving optimization problems with many local optima and complicated constraints, possibly involving a mix of continuous and discrete variables. Randomized algorithms are also used to solve *noisy* optimization problems, in which the objective function is unknown and has to be obtained via Monte Carlo simulation.

3.4.1 Simulated Annealing

Simulated annealing is a Monte Carlo technique for minimization that emulates the physical state of atoms in a metal when the metal is heated up and then slowly cooled down. When the cooling is performed very slowly, the atoms settle down to a minimum-energy state. Denoting the state as $\boldsymbol{x}$ and the energy of a state as $S(\boldsymbol{x})$, the probability distribution of the (random) states is described by the **Boltzmann pdf**

$$f(\boldsymbol{x}) \propto \mathrm{e}^{-\frac{S(\boldsymbol{x})}{kT}}, \quad \boldsymbol{x} \in \mathcal{X},$$

where k is Boltzmann's constant and T is the temperature.

Going beyond the physical interpretation, suppose that $S(\boldsymbol{x})$ is an arbitrary function to be minimized, with $\boldsymbol{x}$ taking values in some discrete or continuous set $\mathcal{X}$. The **Gibbs pdf** corresponding to $S(\boldsymbol{x})$ is defined as

$$f_T(\boldsymbol{x}) = \frac{\mathrm{e}^{-\frac{S(\boldsymbol{x})}{T}}}{z_T}, \qquad \boldsymbol{x} \in \mathcal{X},$$

provided that the normalization constant $z_T := \sum_{\boldsymbol{x}} \exp(-S(\boldsymbol{x})/T)$ is finite. Note that this is simply the Boltzmann pdf with the Boltzmann constant k removed. As $T \to 0$, the pdf becomes more and more peaked around the set of global minimizers of S.

The idea of simulated annealing is to create a sequence of points $\boldsymbol{X}_1, \boldsymbol{X}_2, \ldots$ that are approximately distributed according to pdfs $f_{T_1}(\boldsymbol{x}), f_{T_2}(\boldsymbol{x}), \ldots$, where $T_1, T_2, \ldots$ is a sequence of "temperatures" that decreases (is "cooled") to 0 — known as the **annealing schedule**. If each $\boldsymbol{X}_t$ were sampled *exactly* from f_{T_t}, then $\boldsymbol{X}_t$ would converge to a global minimizer of $S(\boldsymbol{x})$ as $T_t \to 0$. However, in practice sampling is *approximate* and convergence to a global minimizer is not assured. A generic simulated annealing algorithm is as follows:

Algorithm 3.68: Simulated Annealing

Input: Annealing schedule $T_0, T_1, \ldots,$, function S, initial value $\boldsymbol{x}_0$.
Output: Approximations to the global minimizer $\boldsymbol{x}^*$ and minimum value $S(\boldsymbol{x}^*)$.

Set $\boldsymbol{X}_0 \leftarrow \boldsymbol{x}_0$ and $t \leftarrow 1$.
while not stopping **do**
    Approximately simulate $\boldsymbol{X}_t$ from $f_{T_t}(\boldsymbol{x})$.
    $t \leftarrow t + 1$
return $\boldsymbol{X}_t, S(\boldsymbol{X}_t)$

A popular annealing schedule is **geometric cooling**, where $T_t = \beta T_{t-1}, t = 1, 2, \ldots,$ for a given initial temperature T_0 and a **cooling factor** $\beta \in (0, 1)$. Appropriate values for T_0 and β are problem-dependent and this has traditionally required tuning on the part of the user. A possible stopping criterion is to stop after a fixed number of iterations, or when the temperature is "small enough".

Approximate sampling from a Gibbs distribution is most often carried out via MCMC. For each iteration t, the Markov chain should theoretically run for a large number of steps to accurately sample from the Gibbs pdf f_{T_t}. However, in practice, one often only runs a *single* step of the Markov chain, before updating the temperature, as in Algorithm 3.69 below.

To sample from a Gibbs distribution f_T, this algorithm uses a random walk Metropolis–Hastings sampler. From (3.34), the acceptance probability of a proposal $\boldsymbol{y}$ is thus ☞ 90

$$\alpha(\boldsymbol{x}, \boldsymbol{y}) = \min\left\{\frac{\mathrm{e}^{-\frac{1}{T}S(\boldsymbol{y})}}{\mathrm{e}^{-\frac{1}{T}S(\boldsymbol{x})}},\ 1\right\} = \min\left\{\mathrm{e}^{-\frac{1}{T}(S(\boldsymbol{y})-S(\boldsymbol{x}))},\ 1\right\}.$$

Hence, if $S(\boldsymbol{y}) < S(\boldsymbol{x})$, then the proposal is aways accepted. Otherwise, the proposal is accepted with probability $\exp(-\frac{1}{T}(S(\boldsymbol{y}) - S(\boldsymbol{x})))$.

Algorithm 3.69: Simulated Annealing with a Random Walk Sampler

Input: Objective function S, starting state $\boldsymbol{X}_0$, initial temperature T_0, number of iterations N, symmetric proposal density $q(\boldsymbol{y}\,|\,\boldsymbol{x})$, constant β.

Output: Approximate minimizer and minimum value of S.

for $t = 0$ **to** $N - 1$ **do**
  Simulate a new state $\boldsymbol{Y}$ from the symmetric proposal $q(\boldsymbol{y}\,|\,\boldsymbol{X}_t)$.
  if $S(\boldsymbol{Y}) < S(\boldsymbol{X}_t)$ **then**
    $\boldsymbol{X}_{t+1} \leftarrow \boldsymbol{Y}$
  else
    Draw $U \sim \mathcal{U}(0,1)$.
    if $U \leqslant \mathrm{e}^{-(S(\boldsymbol{Y})-S(\boldsymbol{X}_t))/T_t}$ **then**
      $\boldsymbol{X}_{t+1} \leftarrow \boldsymbol{Y}$
    else
      $\boldsymbol{X}_{t+1} \leftarrow \boldsymbol{X}_t$
  $T_{t+1} \leftarrow \beta T_t$
return $\boldsymbol{X}_N$ and $S(\boldsymbol{X}_N)$

■ **Example 3.70 (Simulated Annealing for Minimization)** Let us minimize the "wiggly" function depicted in the bottom panel of Figure 3.71 and given by:

$$S(x) = \begin{cases} -\mathrm{e}^{-x^2/100}\sin(13x - x^4)^5\sin(1 - 3x^2)^2, & \text{if } -2 \leqslant x \leqslant 2, \\ \infty, & \text{otherwise.} \end{cases}$$

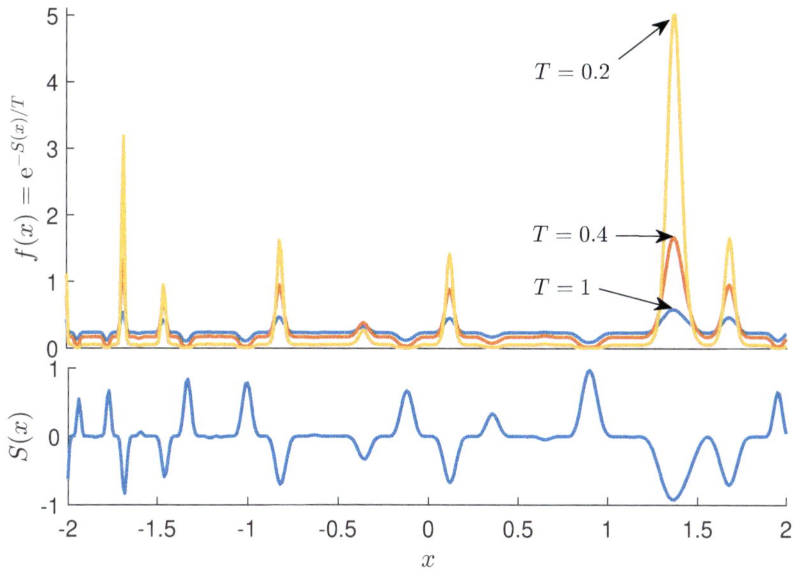

Figure 3.71: Lower panel: the "wiggly" function $S(x)$. Upper panel: three (normalized) Gibbs pdfs for temperatures $T = 1, 0.4, 0.2$. As the temperature decreases, the Gibbs pdf converges to the pdf that has all its mass concentrated at the minimizer of S.

The function has many local minima and maxima, with a global minimizer around 1.4. The figure also illustrates the relationship between S and the (unnormalized) Gibbs pdf f_T. A Python module for this function is given next.

wiggly.py

```
import numpy as np

def wiggly(x):
    y = -np.exp(x**2/100) * np.sin(13*x - x**4)**5 * np.sin(1 - 3*x**2)
        **2
    if x < -2 or x > 2:
        y = float('inf')
    return y
```

The following Python code implements a slight variant of Algorithm 3.69 where, instead of stopping after a fixed number of iterations, the algorithm stops when the temperature is lower than some threshold (here 10^{-3}).

Instead of stopping after a fixed number N of iterations or when the temperature is low enough, it is useful to stop when consecutive function values are closer than some distance ε to each other, or when the best found function value has not changed over a fixed number d of iterations.

For a "current" state x, the proposal state Y is here drawn from the $\mathcal{N}(x, 0.5^2)$ distribution. We use geometric cooling with decay parameter $\beta = 0.999$ and initial temperature $T_0 = 1$. We set the initial state to $x_0 = 0$. Figure 3.72 depicts a realization of the sequence of states x_t for $t = 0, 1, \ldots$. After initially fluctuating wildly, the sequence settles down to a value around 1.37, with $S(1.37) = -0.92$, corresponding to the global optimizer and minimum, respectively.

simann.py

```
import numpy as np
import matplotlib.pyplot as plt
from wiggly import wiggly

S = wiggly
beta = 0.999
sig = 0.5
T = 1
x = 0
xx = []
Sx = S(x)

while T > 10**(-3):
    T = beta * T
    y = x + sig * np.random.randn()
    Sy = S(y)

    alpha = min(np.exp(-(Sy - Sx) / T), 1)
```

```
    if np.random.uniform() < alpha:
        x = y
        Sx = Sy

    xx.append(x)

print(f'minimizer = {x:.3f}, minimum = {Sx:.3f}')

plt.xlabel('Number of iterations')
plt.ylabel('State')
plt.plot(xx)
plt.show()
```

```
minimizer = 1.365, minimum = -0.958
```

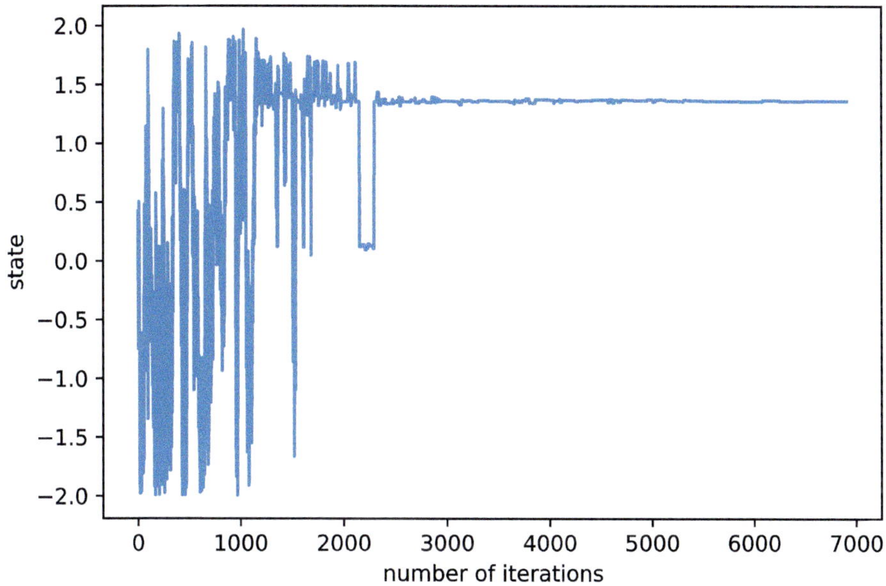

Figure 3.72: Typical states generated by the simulated annealing algorithm.

■

3.4.2 Cross-Entropy Method

The **cross-entropy** (CE) method [145] is a simple Monte Carlo algorithm that can be used for both optimization and estimation.

The basic idea of the CE method for minimizing a function S on a set $\mathcal{X}$ is to define a parametric family of probability densities $\{f(\cdot \mid \boldsymbol{v}), \boldsymbol{v} \in \mathcal{V}\}$ on $\mathcal{X}$ and to iteratively update the parameter $\boldsymbol{v}$ so that $f(\cdot \mid \boldsymbol{v})$ places more mass on states $\boldsymbol{x}$ that have smaller S values than on the previous iteration. In particular, the CE algorithm has two basic phases:

- *Sampling*: Samples $\boldsymbol{X}_1, \dots, \boldsymbol{X}_N$ are drawn independently according to $f(\cdot \mid \boldsymbol{v})$. The objective function S is evaluated at these points.

- *Updating*: A new parameter $\boldsymbol{v}'$ is selected on the basis of those $\boldsymbol{X}_i$ for which $S(\boldsymbol{X}_i) \leq \gamma$ for some level γ. These $\{\boldsymbol{X}_i\}$ form the **elite sample** set, $\mathcal{E}$.

At each iteration the level parameter γ is chosen as the worst of the $N^{\text{elite}} := \lceil \varrho N \rceil$ best performing samples, where $\varrho \in (0, 1)$ is the **rarity parameter** — typically, $\varrho = 0.1$ or $\varrho = 0.01$. The parameter $\boldsymbol{v}$ is updated as a smoothed average $\alpha \boldsymbol{v}' + (1 - \alpha)\boldsymbol{v}$, where $\alpha \in (0, 1)$ is the **smoothing parameter** and

$$\boldsymbol{v}' := \underset{\boldsymbol{v} \in \mathcal{V}}{\operatorname{argmax}} \sum_{\boldsymbol{X} \in \mathcal{E}} \ln f(\boldsymbol{X} \mid \boldsymbol{v}). \tag{3.73}$$

The updating rule (3.73) is the result of minimizing the Kullback–Leibler divergence between the conditional density of $\boldsymbol{X} \sim f(\boldsymbol{x} \mid \boldsymbol{v})$ given $S(\boldsymbol{X}) \leq \gamma$, and $f(\boldsymbol{x}; \boldsymbol{v})$; see [145]. ☞ 665 Note that (3.73) yields the *maximum likelihood estimator* (MLE) of $\boldsymbol{v}$ based on the elite samples. Hence, for many specific families of distributions, explicit solutions can be found. An important example is where $\boldsymbol{X} \sim \mathcal{N}(\boldsymbol{\mu}, \operatorname{diag}(\boldsymbol{\sigma}^2))$; that is, $\boldsymbol{X}$ has independent Gaussian components. In this case, the mean vector $\boldsymbol{\mu}$ and the vector of variances $\boldsymbol{\sigma}^2$ are simply updated via the sample mean and sample variance of the elite samples. This is known as **normal updating**. A generic CE procedure for minimization is given in Algorithm 3.77.

Algorithm 3.77: Cross-Entropy Method for Minimization

Input: Function S, initial sampling parameter $\boldsymbol{v}_0$, sample size N, rarity parameter ϱ, smoothing parameter α.
Output: Approximate minimum of S and optimal sampling parameter $\boldsymbol{v}$.

```
Initialize v_0, set N^elite ← ⌈ϱN⌉ and t ← 0.
while a stopping criterion is not met do
    t ← t + 1
    Simulate an iid sample X_1, ..., X_N from the density f(· | v_{t-1}).
    Evaluate the performances S(X_1), ..., S(X_N) and sort them from smallest
      to largest: S_(1), ..., S_(N).
    Let γ_t be the sample ϱ-quantile of the performances:
      (3.74)    γ_t ← S_(N^elite).
    Determine the set of elite samples E_t = {X_i : S(X_i) ≤ γ_t}.
    Let v'_t be the MLE of the elite samples:
      (3.75)    v'_t ← argmax_v Σ_{X ∈ E_t} ln f(X | v).
    Update the sampling parameter as
      (3.76)    v_t ← α v'_t + (1 − α) v_{t-1}.
return γ_t, v_t
```

The CE algorithm produces a sequence of pairs $(\gamma_1, \boldsymbol{v}_1), (\gamma_2, \boldsymbol{v}_2), \ldots,$ such that γ_t converges (approximately) to the minimal function value, and $f(\cdot \mid \boldsymbol{v}_t)$ to a degenerate pdf that (approximately) concentrates all its mass at a minimizer of S, as $t \to \infty$. A possible stopping condition is to stop when the sampling distribution $f(\cdot \mid \boldsymbol{v}_t)$ is sufficiently close to a degenerate distribution. For normal updating this means that the standard deviation is sufficiently small.

> The output of the CE algorithm could also include the overall best function value and corresponding solution.

The following Python module defines a generic CE function for continuous optimization. It can be applied to functions with scalar and vector-valued inputs. In the **CEM** method below, `f` is the function that is to be minimized. At each iteration of the CE algorithm, `n_samples` are generated, in the multi-dimensional case, from a multivariate distribution with mean vector `mu` and a diagonal covariance matrix given by the squared values of the vector `sigma` of standard deviations. In line 16, `fx` is a one-dimensional array that stores the function values of the samples. The samples are then sorted in ascending order according to their function values, and the first `n_el` samples form the elite samples. The updating of the parameter vectors `mu` and `sigma` is done in Lines 20 and 21.

ce_min.py

```python
import numpy as np

def CEM(f, mu, sigma, n_samples=100, n_el=10, eps=1e-5, n_iter=100,
    alpha=1.0, verbose=False, **kwargs):
    iter = 0

    # Ensure mu and sigma are arrays
    mu = np.atleast_1d(mu)
    sigma = np.atleast_1d(sigma)

    while np.max(sigma) > eps and iter < n_iter:
        if mu.size == 1:
            x = np.random.normal(mu, sigma, n_samples)
        else:
            x = np.random.multivariate_normal(mu, np.diag(sigma**2),
                n_samples)

        fx = np.array([f(x[i], **kwargs) for i in range(n_samples)])
        elite = x[np.argsort(fx)[:n_el]]

        # Update parameters based on the elite samples
        mu = alpha * np.mean(elite, axis=0) + (1 - alpha) * mu
        sigma = alpha * np.std(elite, axis=0) + (1 - alpha) * sigma
        iter += 1


        if verbose:
            print('f(mu)= {}, mu: {}, sigma: {}'.format(f(mu), mu, sigma
                ))

    return mu
```

Example 3.78 (Cross-Entropy Method for Minimization) We minimize the same ☞ 109 function as in Example 3.70, but instead use the **CEM** function above. In this case, the samples are generated from a $\{\mathcal{N}(\mu, \sigma^2)\}$ for the sampling step (Step 4 of Algorithm 3.77), starting with $\mu = 0$ and $\sigma = 3$. The choice of the initial parameter is quite arbitrary, as long as σ is large enough to sample a wide range of points. For the other parameters, we take the default settings in **CEM**; that is, we take $N = 100$ samples at each iteration, keep the $N^{\text{elite}} = 10$ smallest ones as the elite samples, and apply no smoothing ($\alpha = 1$). Note that in this one-dimensional case, the parameters μ and σ are simply updated via the sample mean and sample standard deviation of the elite samples. Figure 3.79 shows how the pdfs of the $\mathcal{N}(\mu_t, \sigma_t^2)$ sampling distributions degenerate to the point mass at the global minimizer 1.366.

CEmethod.py

```python
import numpy as np
from ce_min import CEM
from wiggly import wiggly

np.set_printoptions(precision=3)
mu = 1.0
sigma = 3.0
minimizer = CEM(wiggly, mu=mu, sigma=sigma, verbose=True)
```

```
S(mu)= [0.071], mu: [0.414], sigma: [0.922]
S(mu)= [0.063], mu: [0.81], sigma: [0.831]
S(mu)= [-0.033], mu: [1.212], sigma: [0.69]
S(mu)= [-0.588], mu: [1.447], sigma: [0.117]
S(mu)= [-0.958], mu: [1.366], sigma: [0.007]
S(mu)= [-0.958], mu: [1.366], sigma: [0.]
S(mu)= [-0.958], mu: [1.366], sigma: [3.535e-05]
S(mu)= [-0.958], mu: [1.366], sigma: [2.023e-06]
```

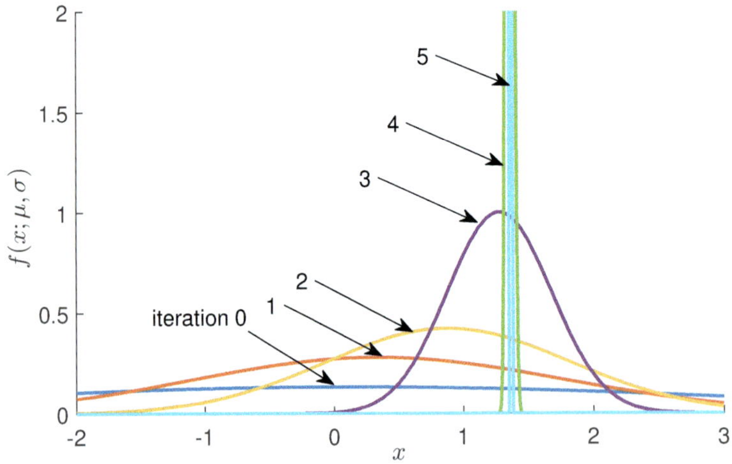

Figure 3.79: The normal pdfs of the first six sampling distributions, truncated to the interval $[-2, 3]$. The initial sampling distribution is $\mathcal{N}(0, 3^2)$.

3.4.3 Splitting for Optimization

Minimizing a function $S(\boldsymbol{x})$, $\boldsymbol{x} \in \mathcal{X}$ is closely related to drawing a random sample from a **level set** of the form $\{\boldsymbol{x} \in \mathcal{X} : S(\boldsymbol{x}) \leq \gamma\}$. Suppose S has minimum value γ^* attained at $\boldsymbol{x}^*$. As long as $\gamma \geq \gamma^*$, this level set contains the minimizer. Moreover, if γ is close to γ^*, the volume of this level set will be small. So, a randomly selected point from this set is expected to be close to $\boldsymbol{x}^*$. Thus, by gradually decreasing the level parameter γ, the level sets will gradually shrink towards the set $\{\boldsymbol{x}^*\}$. Indeed, the CE method was developed with exactly this connection in mind; see, e.g., [144]. Note that the CE method employs a *parametric* sampling distribution to obtain samples from the level sets (the elite samples). In [52] a *non-parametric* sampling mechanism is introduced that uses an evolving collection of particles. The resulting optimization algorithm, called **splitting for continuous optimization** (SCO), provides a fast and accurate way to optimize complicated continuous functions. The details of SCO are given in Algorithm 3.80.

Algorithm 3.80: Splitting for Continuous Optimization (SCO)

Input: Objective function S, sample size N, rarity parameter ϱ, scale factor w, bounded region $\mathcal{B} \subset \mathcal{X}$ that is known to contain a global minimizer, and maximum number of attempts `MaxTry`.

Output: Final iteration number t and sequence $(\boldsymbol{X}_{\text{best},1}, b_1), \ldots, (\boldsymbol{X}_{\text{best},t}, b_t)$ of best solutions and function values at each iteration.

Simulate $\mathcal{Y}_0 = \{\boldsymbol{Y}_1, \ldots, \boldsymbol{Y}_N\}$ uniformly on $\mathcal{B}$. Set $t \leftarrow 0$ and $N^{\text{elite}} \leftarrow \lceil N\varrho \rceil$.
while stopping condition is not satisfied **do**
  Determine the N^{elite} smallest values, $S_{(1)} \leq \cdots \leq S_{(N^{\text{elite}})}$, of $\{S(\boldsymbol{X}), \boldsymbol{X} \in \mathcal{Y}_t\}$, and store the corresponding vectors, $\boldsymbol{X}_{(1)}, \ldots, \boldsymbol{X}_{(N^{\text{elite}})}$, in $\mathcal{X}_{t+1}$. Set $b_{t+1} \leftarrow S_{(1)}$ and $\boldsymbol{X}_{\text{best},t+1} \leftarrow \boldsymbol{X}_{(1)}$.
  Draw $B_i \sim \mathsf{Bernoulli}(\frac{1}{2})$, $i = 1, \ldots, N^{\text{elite}}$, with $\sum_{i=1}^{N^{\text{elite}}} B_i = N \bmod N^{\text{elite}}$.
  for $i = 1$ **to** N^{elite} **do**
    $R_i \leftarrow \left\lfloor \frac{N}{N^{\text{elite}}} \right\rfloor + B_i$ `// random splitting factor`
    $\boldsymbol{Y} \leftarrow \boldsymbol{X}_{(i)}$; $\boldsymbol{Y}' \leftarrow \boldsymbol{Y}$
    for $j = 1$ **to** R_i **do**
      Draw $I \in \{1, \ldots, N^{\text{elite}}\} \setminus \{i\}$ uniformly and let $\boldsymbol{\sigma}_i \leftarrow w|\boldsymbol{X}^{(i)} - \boldsymbol{X}^{(I)}|$.
      Simulate a uniform permutation $\boldsymbol{\pi} = (\pi_1, \ldots, \pi_n)$ of $(1, \ldots, n)$.
      for $k = 1$ **to** n **do**
        for `Try` $= 1$ **to** `MaxTry` **do**
          $Y'(\pi_k) \leftarrow Y(\pi_k) + \sigma_i(\pi_k)Z, \quad Z \sim \mathcal{N}(0, 1)$
          if $S(\boldsymbol{Y}') < S(\boldsymbol{Y})$ **then** $\boldsymbol{Y} \leftarrow \boldsymbol{Y}'$ and **break**.
      Add $\boldsymbol{Y}$ to $\mathcal{Y}_{t+1}$
  $t \leftarrow t + 1$
return $\{(\boldsymbol{X}_{\text{best},k}, b_k), k = 1, \ldots, t\}$

At iteration $t = 0$, the algorithm starts with a population of particles

$$\mathcal{Y}_0 = \{\boldsymbol{Y}_1, \dots, \boldsymbol{Y}_N\}$$

that are uniformly generated on some bounded region $\mathcal{B}$, which is large enough to contain a global minimizer. The function values of all particles in $\mathcal{Y}_0$ are sorted, and the best $N^{\text{elite}} = \lceil N\varrho \rceil$ form the elite particle set $\mathcal{X}_1$, exactly as in the CE method. Next, the elite particles are "split" into $\lfloor N/N^{\text{elite}} \rfloor$ children particles, adding one extra child to some of the elite particles to ensure that the total number of children is again N. The purpose of Line 4 is to randomize which elite particles receive an extra child. Lines 8–15 describe how the children of the i-th elite particle are generated. First, in Line 9, we select one of the *other* elite particles uniformly at random. The same line defines an n-dimensional vector $\boldsymbol{\sigma}_i$ whose components are the absolute differences between the vectors $\boldsymbol{X}_{(i)}$ and $\boldsymbol{X}_{(I)}$, multiplied by a constant w. That is,

$$\boldsymbol{\sigma}_i = w\,|\boldsymbol{X}_{(i)} - \boldsymbol{X}_{(I)}| := w \begin{bmatrix} |X_{(i),1} - X_{(I),1}| \\ |X_{(i),2} - X_{(I),2}| \\ \vdots \\ |X_{(i),n} - X_{(I),n}| \end{bmatrix}.$$

☞ 127 Next, a uniform random permutation $\boldsymbol{\pi}$ of $(1, \dots, n)$ is simulated (see Exercise 9). Lines 11–14 describe how, starting from a candidate child point $\boldsymbol{Y}$, each coordinate of $\boldsymbol{Y}$ is re-sampled, in the order determined by $\boldsymbol{\pi}$, by adding a standard normal random variable to that component, multiplied by the corresponding component of $\boldsymbol{\sigma}_i$ (Line 13). If the resulting $\boldsymbol{Y}'$ has a function value that is less than that of $\boldsymbol{Y}$, then the new candidate is accepted. Otherwise, the *same* coordinate is tried again. If no improvement is found in `MaxTry` attempts, the original component is retained. This process is performed for all elite samples, to produce the first-generation population $\mathcal{Y}_1$. The procedure is then repeated for iterations $t = 1, 2, \dots$, until some stopping criterion is met, e.g., when the best found function value does not change for a number of consecutive iterations, or when the total number of function evaluations exceeds some threshold. The best found function value and corresponding argument (particle) are returned at the conclusion of the algorithm.

The input variable `MaxTry` governs how much computational time is dedicated to updating a component. In most cases we have encountered, the choices $w = 0.5$ and `MaxTry` $= 5$ work well. Empirically, relatively high value for ϱ work well, such as $\varrho = 0.4, 0.8$, or even $\varrho = 1$. The latter case means that at each stage t *all* samples from $\mathcal{Y}_{t-1}$ carry over to the elite set $\mathcal{X}_t$.

■ **Example 3.81 (Test Problem 112)** Hock and Schittkowski [84] provide a rich source of test problems for multiextremal optimization. A challenging one is Problem 112, where the goal is to find $\boldsymbol{x}$ so as to minimize the function

$$S(\boldsymbol{x}) = \sum_{j=1}^{10} x_j \left(c_j + \ln \frac{x_j}{x_1 + \cdots + x_{10}} \right),$$

subject to the following set of constraints:

$$\begin{aligned} x_1 + 2x_2 + 2x_3 + x_6 + x_{10} - 2 &= 0, \\ x_4 + 2x_5 + x_6 + x_7 - 1 &= 0, \\ x_3 + x_7 + x_8 + 2x_9 + x_{10} - 1 &= 0, \\ x_j &\geq 0.000001, \quad j = 1, \ldots, 10, \end{aligned}$$

where the constants $\{c_i\}$ are given in Table 3.82.

Table 3.82: Constants for Test Problem 112.

$c_1 = -6.089$	$c_2 = -17.164$	$c_3 = -34.054$	$c_4 = -5.914$	$c_5 = -24.721$
$c_6 = -14.986$	$c_7 = -24.100$	$c_8 = -10.708$	$c_9 = -26.662$	$c_{10} = -22.179$

The best known minimal value in [84] was -47.707579. In [127] a better solution was found, -47.760765, using a genetic algorithm. The corresponding solution vector was completely different from the one in [84]. A further improvement,-47.76109081, was found in [97], using the CE method, giving a similar solution vector to that in [127]:

0.04067247	0.14765159	0.78323637	0.00141368	0.48526222
0.00069291	0.02736897	0.01794290	0.03729653	0.09685870

To obtain a solution with SCO, we first converted this 10-dimensional problem into a 7-dimensional one by defining the objective function

$$S_7(\boldsymbol{y}) = S(\boldsymbol{x}),$$

where

$$x_2 = y_1, \quad x_3 = y_2, \quad x_5 = y_3, \quad x_6 = y_4, \quad x_7 = y_5, \quad x_9 = y_6, \quad x_{10} = y_7,$$

and

$$\begin{aligned} x_1 &= 2 - (2y_1 + 2y_2 + y_4 + x_7), \\ x_4 &= 1 - (2y_3 + y_4 + y_5), \\ x_8 &= 1 - (y_2 + y_5 + 2y_6 + y_7), \end{aligned}$$

subject to $x_1, \ldots, x_{10} \geq 0.000001$, where the $\{x_i\}$ are taken as functions of the $\{y_i\}$. We then adopted a penalty approach (see Section C.4) by adding a penalty function to the original objective function: ☞ 608

$$\widetilde{S}_7(\boldsymbol{y}) = S(\boldsymbol{x}) + 1000 \sum_{i=1}^{10} \max\{-(x_i - 0.000001), 0\},$$

where, again, the $\{x_i\}$ are defined in terms of the $\{y_i\}$ as above.

Optimizing this last function with SCO, we found, in less time than the other algorithms, a slightly smaller function value: -47.761090859365858, with solution vector

0.040668102417464	0.147730393049955	0.783153291185250
0.001414221643059	0.485246633088859	0.000693172682617
0.027399339496606	0.017947274343948	0.037314369272343
0.096871356429511		

in line with the earlier solutions. ■

3.4.4 Noisy Optimization

In **noisy optimization**, the objective function is unknown, but estimates of function values are available, e.g., via simulation. For example, to find an optimal prediction function g in supervised learning, the exact risk $\ell(g) = \mathbb{E}\,\text{Loss}(Y, g(\boldsymbol{x}))$ is usually unknown and only estimates of the risk are available. Optimizing the risk is thus typically a noisy optimization problem. Noisy optimization features prominently in simulation studies where the behavior of some system (e.g., vehicles on a road network) is simulated under certain parameters (e.g., the lengths of the traffic light intervals) and the aim is to choose those parameters optimally (e.g., to maximize the traffic throughput). For each parameter setting the exact value for the objective function is unknown but estimates can be obtained via the simulation.

☞ 20

In general, suppose the goal is to minimize a function S, where S is unknown, but an estimate of $S(\boldsymbol{x})$ can be obtained for any choice of $\boldsymbol{x} \in \mathcal{X}$. Because the gradient ∇S is unknown, one cannot directly apply classical optimization methods. The **stochastic approximation** method mimics the classical gradient descent method by replacing a deterministic gradient with an estimate $\widehat{\nabla S}(\boldsymbol{x})$. A simple estimator for the i-th component of $\nabla S(\boldsymbol{x})$ (that is, $\partial S(\boldsymbol{u})/\partial x_i$), is the **central difference estimator**

$$\left(\widehat{S}(\boldsymbol{x} + \boldsymbol{e}_i\,\delta/2) - \widehat{S}(\boldsymbol{x} - \boldsymbol{e}_i\,\delta/2)\right)/\,\delta, \tag{3.83}$$

where $\boldsymbol{e}_i$ denotes the i-th unit vector, and $\widehat{S}(\boldsymbol{x} + \boldsymbol{e}_i\,\delta/2)$ and $\widehat{S}(\boldsymbol{x} - \boldsymbol{e}_i\,\delta/2)$ can be any estimators of $S(\boldsymbol{x} + \boldsymbol{e}_i\,\delta/2)$ and $S(\boldsymbol{x} - \boldsymbol{e}_i\,\delta/2)$, respectively. The difference parameter $\delta > 0$ should be small enough to reduce the bias of the estimator, but large enough to keep the variance of the estimator small.

> To reduce the variance in the estimator (3.83) it is important to have $\widehat{S}(\boldsymbol{x} + \boldsymbol{e}_i\,\delta/2)$ and $\widehat{S}(\boldsymbol{x} - \boldsymbol{e}_i\,\delta/2)$ positively correlated. This can for example be achieved by using **common random numbers** in the simulation.

☞ 603

In direct analogy to gradient descent methods, the stochastic approximation method produces a sequence of iterates, starting with some $\boldsymbol{x}_1 \in \mathcal{X}$, via

$$\boldsymbol{x}_{t+1} = \boldsymbol{x}_t - \beta_t\,\widehat{\nabla S}(\boldsymbol{x}_t), \tag{3.84}$$

where $\beta_1, \beta_2, \ldots$ is a sequence of strictly positive step sizes. A generic stochastic approximation algorithm for minimizing a function S is thus as follows:

Algorithm 3.85: Stochastic Approximation

Input: A mechanism to estimate any gradient $\nabla S(\boldsymbol{x})$ and step sizes $\beta_1, \beta_2, \ldots$.
Output: Approximate optimizer of S.

```
Initialize x_1 ∈ X. Set t ← 1.
while a stopping criterion is not met do
    Obtain an estimated gradient ∇S^(x_t) of S at x_t.
    Determine a step size β_t.
    Set x_{t+1} ← x_t − β_t ∇S^(x_t).
    t ← t + 1
return x_t
```

When $\widehat{\nabla S}(\boldsymbol{x}_t)$ is an *unbiased* estimator of $\nabla S(\boldsymbol{x}_t)$ in (3.84) the stochastic approximation Algorithm 3.85 is referred to as the **Robbins–Monro** algorithm. When finite differences are used to estimate $\widehat{\nabla S}(\boldsymbol{x}_t)$, as in (3.83), the resulting algorithm is known as the **Kiefer–Wolfowitz** algorithm. In Section 10.4.1 we will see how stochastic gradient descent is employed in deep learning to minimize the training loss, based on a "minibatch" of training data. ☞ 440

It can be shown [99, 27] that, under certain regularity conditions on S, the sequence $\boldsymbol{x}_1, \boldsymbol{x}_2, \ldots$ converges to the true minimizer $\boldsymbol{x}^*$ when the step sizes decrease slowly enough to 0; in particular, when

$$\sum_{t=1}^{\infty} \beta_t = \infty \quad \text{and} \quad \sum_{t=1}^{\infty} \beta_t^2 < \infty. \tag{3.86}$$

In practice, one rarely uses step sizes that satisfy (3.86), as the convergence of the sequence will be too slow to be of practical use.

An alternative approach to stochastic approximation is the **stochastic counterpart** method, also called **sample average approximation**. It can be applied in situations where the noisy objective function is of the form

$$S(\boldsymbol{x}) = \mathbb{E}\widetilde{S}(\boldsymbol{x}, \boldsymbol{\xi}), \quad \boldsymbol{x} \in \mathscr{X}, \tag{3.87}$$

where $\boldsymbol{\xi}$ is a random vector that can be simulated and $\widetilde{S}(\boldsymbol{x}, \boldsymbol{\xi})$ can be evaluated exactly. The idea is to replace the optimization of (3.87) with that of the sample average

$$\widehat{S}(\boldsymbol{x}) = \frac{1}{N} \sum_{i=1}^{N} \widetilde{S}(\boldsymbol{x}, \boldsymbol{\xi}_i), \quad \boldsymbol{x} \in \mathscr{X}, \tag{3.88}$$

where $\boldsymbol{\xi}_1, \ldots, \boldsymbol{\xi}_N$ are iid copies of $\boldsymbol{\xi}$. Note that $\widehat{S}$ is a deterministic function of $\boldsymbol{x}$ and so can be optimized using any optimization algorithm. A solution to this sample average version is taken to be an estimator of a solution $\boldsymbol{x}^*$ to the original problem (3.87).

■ **Example 3.89 (Determining Good Importance Sampling Parameters)** The selection of good importance sampling parameters can be viewed as a stochastic optimization problem. Consider, for instance, the importance sampling estimator in Example 3.63. Recall that the nominal distribution is the uniform distribution on the square $[-b, b]^2$, ☞ 105 with pdf

$$f_b(\boldsymbol{x}) = \frac{1}{(2b)^2}, \quad \boldsymbol{x} \in [-b, b]^2,$$

where b is large enough to ensure that μ_b is close to μ; in that example, we chose $b = 1000$. The importance sampling pdf is

$$g_\lambda(\boldsymbol{x}) = f_{R,\Theta}(r, \theta)\frac{1}{r} = \lambda \mathrm{e}^{-\lambda r} \frac{1}{2\pi} \frac{1}{r} = \frac{\lambda \mathrm{e}^{-\lambda\sqrt{x_1^2 + x_2^2}}}{2\pi\sqrt{x_1^2 + x_2^2}}, \quad \boldsymbol{x} = (x_1, x_2) \in \mathbb{R}^2 \setminus \{\boldsymbol{0}\},$$

which depends on a free parameter λ. In the example we chose $\lambda = 0.1$. Is this the best choice? Maybe $\lambda = 0.05$ or 0.2 would have resulted in a more accurate estimate. The important thing to realize is that the "effectiveness" of λ can be measured in terms of the ☞ 104 variance of the estimator $\widehat{\mu}$ in (3.61), which is given by

$$\frac{1}{N}\mathbb{V}\mathrm{ar}_{g_\lambda}\left(H(\boldsymbol{X})\frac{f(\boldsymbol{X})}{g_\lambda(\boldsymbol{X})}\right) = \frac{1}{N}\mathbb{E}_{g_\lambda}\left[H^2(\boldsymbol{X})\frac{f^2(\boldsymbol{X})}{g_\lambda^2(\boldsymbol{X})}\right] - \frac{\mu^2}{N} = \frac{1}{N}\mathbb{E}_f\left[H^2(\boldsymbol{X})\frac{f(\boldsymbol{X})}{g_\lambda(\boldsymbol{X})}\right] - \frac{\mu^2}{N}.$$

Hence, the optimal parameter λ^* minimizes $S(\lambda) = \mathbb{E}_f[H^2(\boldsymbol{X})f(\boldsymbol{X})/g_\lambda(\boldsymbol{X})]$, which is unknown, but can be estimated from simulation. To solve this stochastic minimization problem, we first use stochastic approximation. Thus, at each step of the algorithm, the gradient of $S(\lambda)$ is estimated from realizations of $\widehat{S}(\lambda) = H^2(\boldsymbol{X})f(\boldsymbol{X})/g_\lambda(\boldsymbol{X})$, where $\boldsymbol{X} \sim f_b$. As in the original problem (that is, the estimation of μ), the parameter b should be large enough to avoid any bias in the estimator of λ^*, but also small enough to ensure a small variance. The following Python code implements a particular instance of Algorithm 3.85. For sampling from f_b here, we used $b = 100$ instead of $b = 1000$, as this will improve the crude Monte Carlo estimation of λ^*, without noticeably affecting the bias. The gradient of $S(\lambda)$ is estimated in Lines 11–17, using the central difference estimator (3.83). Notice how for the $S(\lambda - \delta/2)$ and $S(\lambda + \delta/2)$ the *same* random vector $\boldsymbol{X} = [X_1, X_2]^\top$ is used. This significantly reduces the variance of the gradient estimator; ☞ 131 see also Exercise 23. The step size β_t should be such that $\beta_t \widehat{\nabla S}(\boldsymbol{x}_t) \approx \lambda_t$. Given the large gradient here, we choose $\beta_0 = 10^{-7}$ and decrease it each step by a factor of 0.99. Figure 3.90 shows how the sequence $\lambda_0, \lambda_1, \ldots$ decreases towards approximately 0.125, which we take as an estimator for the optimal importance sampling parameter λ^*.

stochapprox.py

```
import numpy as np
from numpy import pi
import matplotlib.pyplot as plt

b = 100    # choose b large enough, but not too large
delta = 0.01
H = lambda x1, x2: (2*b)**2*np.exp(-np.sqrt(x1**2 + x2**2)/4)*(np.sin(2*
    np.sqrt(x1**2+x2**2)+1))*(x1**2+x2**2<b**2)
f = 1/(2*b)**2
g = lambda x1, x2, lam: lam*np.exp(-np.sqrt(x1**2+x2**2)*lam)/np.sqrt(x1
    **2+x2**2)/(2*pi)
beta = 10**-7   # step size very small, as the gradient is large
lam = 0.25
lams = np.array([lam])
N=10**4
for i in range(200):
    x1 = -b + 2*b*np.random.rand(N,1)
    x2 = -b + 2*b*np.random.rand(N,1)
    lamL = lam - delta/2
    lamR = lam + delta/2
    estL = np.mean(H(x1,x2)**2*f/g(x1, x2, lamL))
    estR = np.mean(H(x1,x2)**2*f/g(x1, x2, lamR))  # use SAME x1,x2
    gr = (estR-estL)/delta  # gradient
```

```
        lam = lam - gr*beta  # gradient descend
        lams = np.hstack((lams, lam))
        beta = beta*0.99

lamsize = range(0, (lams.size))
plt.plot(lamsize, lams)
plt.show()
```

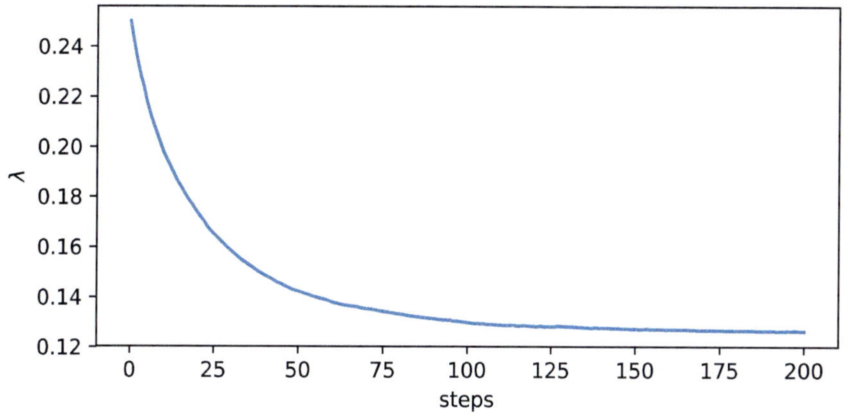

Figure 3.90: The stochastic optimization algorithm produces a sequence $\lambda_t, t = 0, 1, 2, \ldots$ that tends to an approximate estimate of the optimal importance sampling parameter $\lambda^* \approx 0.125$.

Next, we estimate λ^* using a stochastic counterpart approach. As the objective function $S(\lambda)$ is of the form (3.87) (with λ taking the role of $\boldsymbol{x}$ and $\boldsymbol{X}$ the role of $\boldsymbol{\xi}$), we obtain the sample average

$$\widehat{S}(\lambda) = \frac{1}{N} \sum_{i=1}^{N} H^2(\boldsymbol{X}_i) \frac{f(\boldsymbol{X}_i)}{g_\lambda(\boldsymbol{X}_i)}, \tag{3.91}$$

where $\boldsymbol{X}_1, \ldots, \boldsymbol{X}_N \sim_{\text{iid}} f_b$. Once the $\boldsymbol{X}_1, \ldots, \boldsymbol{X}_N \sim_{\text{iid}} f_b$ have been simulated, $\widehat{S}(\lambda)$ is a deterministic function of λ, which can be optimized by any means. We take the most basic approach and simply evaluate the function for $\lambda = 0.01, 0.02, \ldots, 0.3$ and select the minimizing λ on this grid. The code is given below and Figure 3.92 shows $\widehat{S}(\lambda)$ as a function of λ. The minimum value found was $0.60 \cdot 10^4$ for minimizer $\widehat{\lambda}^* = 0.12$, which is in accordance with the value obtained via stochastic approximation. The sensitivity of this estimate can be assessed from the graph: for a wide range of values (say from 0.04 to 0.15) $\widehat{S}$ stays rather flat. So any of these values could be used in an importance sampling procedure to estimate μ. However, very small values (less than 0.02) and large values (greater than 0.25) should be avoided. Our original choice of $\lambda = 0.1$ was therefore justified and we could not have done much better.

stochcounterpart.py

```
from stochapprox import *

lams = np.linspace(0.01, 0.31, 1000)
res = np.array([])
for i in range(lams.size):
    lam = lams[i]
    np.random.seed(1)
    g = lambda x1, x2: lam*np.exp(-np.sqrt(x1**2+x2**2)*lam)/np.sqrt(x1
        **2+x2**2)/(2*pi)
    X = -b + 2*b*np.random.rand(N,1)
    Y = -b + 2*b*np.random.rand(N,1)
    Z = H(X,Y)**2*f/g(X,Y)
    estCMC = np.mean(Z)
    res = np.hstack((res, estCMC))

plt.plot(lams, res)
plt.xlabel(r'$\lambda$')
plt.ylabel(r'$\hat{S}(\lambda)$')
plt.ticklabel_format(style='sci', axis='y', scilimits=(0,0))
```

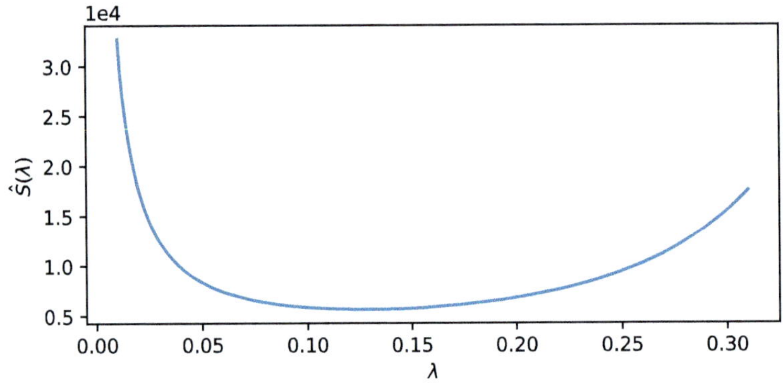

Figure 3.92: The stochastic counterpart method replaces the unknown $S(\lambda)$ (that is, the scaled variance of the importance sampling estimator) with its sample average, $\widehat{S}(\lambda)$. The minimum value of $\widehat{S}$ is attained around $\lambda = 0.12$.

■

A third method for stochastic optimization is the cross-entropy method. In particular, ☞ 112 Algorithm 3.77 can easily be modified to minimize *noisy* functions $S(\boldsymbol{x}) = \mathbb{E}\widetilde{S}(\boldsymbol{x}, \boldsymbol{\xi})$, as defined in (3.87). The only change required in the algorithm is that every function value $S(\boldsymbol{x})$ be replaced by its estimate $\widehat{S}(\boldsymbol{x})$. Depending on the level of noise in the function, the sample size N might have to be increased considerably.

■ **Example 3.93 (Cross-Entropy Method for Noisy Optimization)** To explore the use of the CE method for noisy optimization, take the following noisy discrete optimization problem. Suppose there is a "black box" that contains an unknown binary sequence of n bits. If one feeds the black box any input vector, it will first scramble the input by independently flipping the bits (changing 0 to 1 and 1 to 0) with a probability θ and then return the number of bits that do not match the true (unknown) binary sequence. This is illustrated in Figure 3.94 for $n = 10$. The input to the black box is a binary vector. Inside the black box the digits of the input vector are scrambled by flipping bits with probability θ. The output is the number of bits of the scrambled vector that do not match the true (unknown) binary vector.

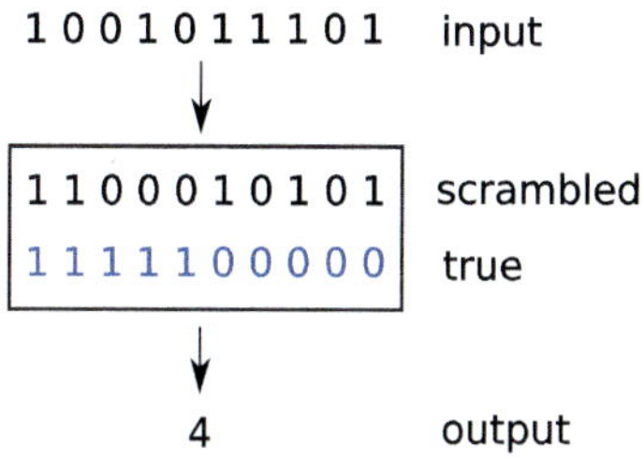

Figure 3.94: A noisy optimization function as a black box.

Denoting by $S(\boldsymbol{x})$ the true number of matching digits for a binary input vector $\boldsymbol{x}$, the black box thus returns a noisy estimate $\widehat{S}(\boldsymbol{x})$. The objective is to estimate the binary sequence inside the black box, by feeding it with many input vectors and observing their output. Or, to put it in a different way, to minimize $S(\boldsymbol{x})$ using $\widehat{S}(\boldsymbol{x})$ as a proxy. Since there are 2^n possible input vectors, it is infeasible to try all possible vectors $\boldsymbol{x}$ even for moderate n.

The following Python program implements the noisy function $\widehat{S}(\boldsymbol{x})$ for $n = 100$. Each input bit is flipped with a rather high probability $\theta = 0.4$, so that the output is a poor indicator of how many bits actually match the true vector. This true vector has 1s at positions $1, \ldots, 50$ and 0s at $51, \ldots, 100$.

Snoisy.py

```python
import numpy as np

def Snoisy(X):   #takes a matrix
    n = X.shape[1]
    N = X.shape[0]
    # true binary vector
    xorg = np.hstack((np.ones((1,n//2)), np.zeros((1,n//2))))
    theta = 0.4 # probability to flip the input
    # storing the number of bits unequal to the true vector
    s = np.zeros(N)
    for i in range(0,N):
       # determine which bits to flip
       flip = (np.random.uniform(size=(n)) < theta).astype(int)
       ind = flip>0
       X[i][ind] = 1-X[i][ind]
       s[i] = (X[i] != xorg).sum()
    return s
```

The CE code below to optimize $S(\boldsymbol{x})$ is quite similar to the continuous optimization code in Example 3.78. However, instead of sampling iid random variables $X_1, \ldots, X_N$ from a normal distribution, we now sample iid binary vectors $\boldsymbol{X}_1, \ldots, \boldsymbol{X}_N$ from a $\mathsf{Ber}(\boldsymbol{p})$ ☞ 114 distribution. More precisely, given a row vector of probabilities $\boldsymbol{p} = [p_1, \dots, p_n]$, we independently simulate the components $X_1, \dots, X_n$ of each binary vector $\boldsymbol{X}$ according to $X_i \sim \mathsf{Ber}(p_i)$, $i = 1, \dots, n$. After each iteration, the vector $\boldsymbol{p}$ is updated as the (vector) mean of the elite samples. The sample size is $N = 1000$ and the number of elite samples is 100. The components of the initial sampling vector $\boldsymbol{p}$ are all equal to $1/2$; that is, the $\boldsymbol{X}$ are initially uniformly sampled from the set of all binary vectors of length $n = 100$. At each subsequent iteration the parameter vector is updated via the mean of the elite samples and evolves towards a degenerate vector $\boldsymbol{p}^*$ with only 1s and 0s. Sampling from such a $\mathsf{Ber}(\boldsymbol{p}^*)$ distribution gives an outcome $\boldsymbol{x}^* = \boldsymbol{p}^*$, which can be taken as an estimate for the minimizer of S; that is, the true binary vector hidden in the black box. The algorithm stops when $\boldsymbol{p}$ has degenerated sufficiently.

Figure 3.95 shows the evolution of the vector of probabilities $\boldsymbol{p}$. This figure may be seen as the discrete analogue of Figure 3.79. We see that, despite the high noise, the CE method is able to find the true state of the black box, and hence the minimum value of S.

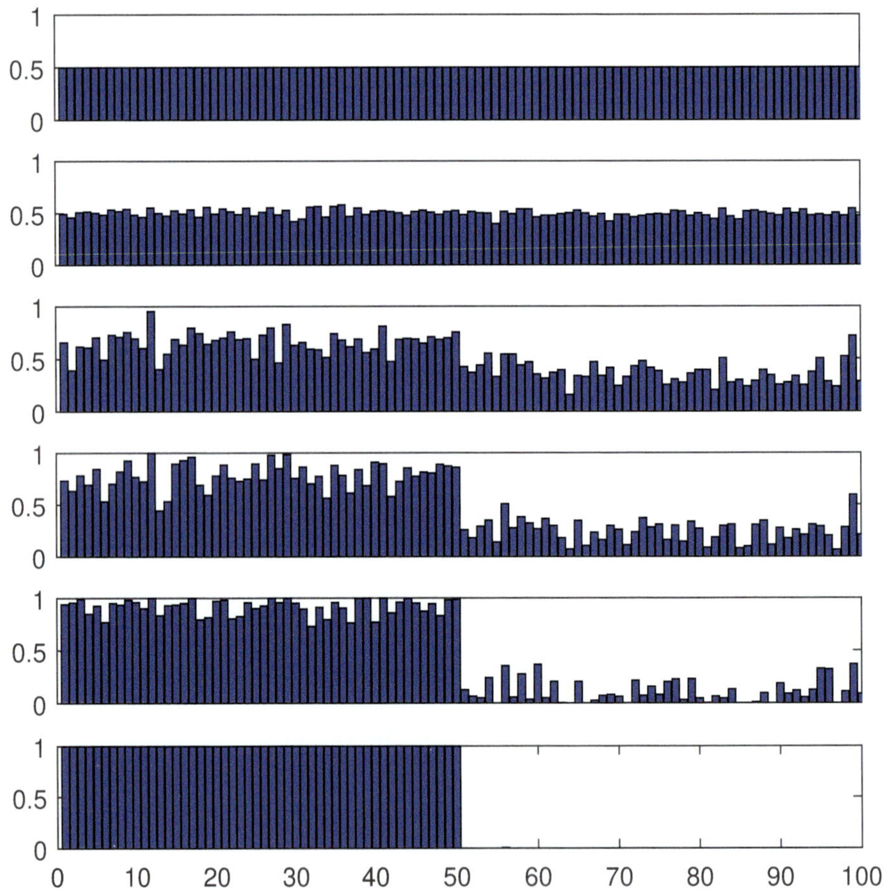

Figure 3.95: Evolution of the vector of probabilities $\boldsymbol{p} = [p_1, \dots, p_n]$ towards the degenerate solution.

CEnoisy.py

```
from Snoisy import Snoisy
import numpy as np
n = 100
rho = 0.1
N = 1000; Nel = int(N*rho); eps = 0.01
p = 0.5*np.ones(n)
i = 0
pstart = p
ps = np.zeros((1000,n))
ps[0] = pstart
pdist = np.zeros((1,1000))
while np.max(np.minimum(p,1-p)) > eps:
    i += 1
    X = (np.random.uniform(size=(N,n)) < p).astype(int)
    X_tmp = np.array(X, copy=True)
    SX = Snoisy(X_tmp)
    ids = np.argsort(SX,axis=0)
    Elite = X[ids[0:Nel],:]
    p = np.mean(Elite,axis=0)
    ps[i] = p
print(p)
```

Further Reading

The encyclopedic articles [96] and [168] explore the importance of Monte Carlo methods in today's quantitative investigations. The *Handbook of Monte Carlo Methods* [98] provides a comprehensive overview of Monte Carlo simulation that explores the latest topics, techniques, and real-world applications. Popular books on simulation and the Monte Carlo method include [64], [103], and [146]. A classic reference on random variable generation is [50], but for more recent studies of fast generators for normal random variables see [25, 26, 42]. Easy introductions to stochastic simulation are given in [72], [139], and [142]. More advanced theory can be found in [7].

Markov chain Monte Carlo is covered extensively in [73, 140]. The theoretical rate of convergence of the simplest Metropolis–Hastings sampler, the independence sampler, is well understood [124]. In particular, the independence sampler converges geometrically fast provided that $f(\boldsymbol{x})/g(\boldsymbol{x})$ is bounded by a finite constant for all $\boldsymbol{x}$, that is, the proposal pdf g dominates the target f uniformly. Interestingly, Example 3.20 shows that this uniform dominance condition is not necessary for the geometrically fast convergence of the rejection sampling Algorithm 3.15.

The research monograph on the cross-entropy method is [145] and a tutorial is provided in [48]. A range of optimization applications of the CE method is given in [24]. Theoretical results on adaptive tuning schemes for simulated annealing may be found, for example, in [155]. There are several established ways for gradient estimation. These include the finite difference method, infinitesimal perturbation analysis, the score function method, and the method of weak derivatives; see, for example, [74, Chapter 7].

Exercises

☞ 76 1. We can modify the Box–Muller method in Example 3.2 to draw X and Y uniformly on the unit disc, $\{(x, y) \in \mathbb{R}^2 : x^2 + y^2 \leq 1\}$, in the following way: Independently draw a radius R and an angle $\Theta \sim \mathcal{U}(0, 2\pi)$, and return $X = R\cos(\Theta), Y = R\sin(\Theta)$. The question is how to draw R.

(a) Show that the cdf of R is given by $F_R(r) = r^2$ for $0 \leq r \leq 1$ (with $F_R(r) = 0$ and $F_R(r) = 1$ for $r < 0$ and $r > 1$, respectively).

(b) Explain how to simulate R using the inverse-transform method.

(c) Simulate 100 independent draws of $[X, Y]^\top$ according to the method described above.

2. A simple rejection sampling approach to simulate a vector $\boldsymbol{X}$ in the unit d-ball $\{\boldsymbol{x} \in \mathbb{R}^d : \|\boldsymbol{x}\| \leq 1\}$ is to first generate $\boldsymbol{X}$ uniformly in the hyper cube $[-1, 1]^d$ and then to accept the point only if $\|\boldsymbol{X}\| \leq 1$. Determine an analytic expression for the probability of acceptance as a function of d and plot this for $d = 1, \ldots, 50$.

3. Let the random variable X have pdf

$$f(x) = \begin{cases} \frac{1}{2}x\,, & 0 \leq x < 1, \\ \frac{1}{2}\,, & 1 \leq x \leq \frac{5}{2}\,. \end{cases}$$

Simulate a random variable from $f(x)$, using

(a) the inverse-transform method;

(b) rejection sampling, with the proposal density

$$g(x) = \frac{8}{25}\,x\,, \quad 0 \leq x \leq \frac{5}{2}\,.$$

4. Construct simulation algorithms for the following distributions:

(a) The Weib(α, λ) distribution, with cdf $F(x) = 1 - \mathrm{e}^{-(\lambda x)^\alpha}, x \geq 0$, where $\lambda > 0$ and $\alpha > 0$.

(b) The Pareto(α, λ) distribution, with pdf $f(x) = \alpha\lambda(1 + \lambda x)^{-(\alpha+1)}, x \geq 0$, where $\lambda > 0$ and $\alpha > 0$.

5. We wish to sample from the pdf

$$f(x) = x\,\mathrm{e}^{-x}, \quad x \geq 0,$$

using rejection sampling with the proposal pdf $g(x) = \mathrm{e}^{-x/2}/2, x \geq 0$.

(a) Find the smallest C for which $Cg(x) \geq f(x)$ for all x.

(b) What is the efficiency of this rejection sampling method?

6. Let $[X, Y]^\top$ be uniformly distributed on the triangle with corners $(0,0)$, $(1,2)$, and $(-1,1)$. Give the distribution of $[U, V]^\top$ defined by the linear transformation

$$\begin{bmatrix} U \\ V \end{bmatrix} = \begin{bmatrix} 1 & 2 \\ 3 & 4 \end{bmatrix} \begin{bmatrix} X \\ Y \end{bmatrix}.$$

7. Explain how to generate a random variable from the *extreme value distribution*, which has cdf

$$F(x) = 1 - \mathrm{e}^{-\exp(\frac{x-\mu}{\sigma})}, \quad -\infty < x < \infty, \quad (\sigma > 0),$$

via the inverse-transform method.

8. Write a program that generates and displays 100 random vectors that are uniformly distributed within the ellipse

$$5\,x^2 + 21\,x\,y + 25\,y^2 = 9.$$

Hint: Consider generating uniformly distributed samples within the circle of radius 3 and use the fact that linear transformations preserve uniformity to transform the circle to the given ellipse.

9. Suppose that $X_i \sim \mathsf{Exp}(\lambda_i)$, independently, for all $i = 1, \ldots, n$. Let $\mathbf{\Pi} = [\Pi_1, \ldots, \Pi_n]^\top$ be the random permutation induced by the ordering $X_{\Pi_1} < X_{\Pi_2} < \cdots < X_{\Pi_n}$, and define $Z_1 := X_{\Pi_1}$ and $Z_j := X_{\Pi_j} - X_{\Pi_{j-1}}$ for $j = 2, \ldots, n$.

(a) Determine an $n \times n$ matrix $\mathbf{A}$ such that $\boldsymbol{Z} = \mathbf{A}\boldsymbol{X}$ and show that $\det(\mathbf{A}) = 1$.

(b) Denote the joint pdf of $\boldsymbol{X}$ and $\mathbf{\Pi}$ as

$$f_{\boldsymbol{X},\mathbf{\Pi}}(\boldsymbol{x}, \boldsymbol{\pi}) = \prod_{i=1}^{n} \lambda_{\pi_i} \exp\left(-\lambda_{\pi_i} x_{\pi_i}\right) \times \mathbb{1}\{x_{\pi_1} < \cdots < x_{\pi_n}\}, \quad \boldsymbol{x} \geq \mathbf{0}, \ \boldsymbol{\pi} \in \mathcal{P}_n,$$

where $\mathcal{P}_n$ is the set of all $n!$ permutations of $\{1, \ldots, n\}$. Use the multivariate transformation formula (D.35) to show that ☞ 636

$$f_{\boldsymbol{Z},\mathbf{\Pi}}(\boldsymbol{z}, \boldsymbol{\pi}) = \exp\left(-\sum_{i=1}^{n} z_i \sum_{k \geq i} \lambda_{\pi_k}\right) \prod_{i=1}^{n} \lambda_i, \qquad \boldsymbol{z} \geq \mathbf{0}, \ \boldsymbol{\pi} \in \mathcal{P}_n.$$

Hence, conclude that the probability mass function of the random permutation $\mathbf{\Pi}$ is:

$$\mathbb{P}[\mathbf{\Pi} = \boldsymbol{\pi}] = \prod_{i=1}^{n} \frac{\lambda_{\pi_i}}{\sum_{k \geq i} \lambda_{\pi_k}}, \qquad \boldsymbol{\pi} \in \mathcal{P}_n.$$

(c) Write pseudo-code to simulate a *uniform* random permutation $\mathbf{\Pi} \in \mathcal{P}_n$; that is, such that $\mathbb{P}[\mathbf{\Pi} = \boldsymbol{\pi}] = \frac{1}{n!}$, and explain how this uniform random permutation can be used to reshuffle a training set τ_n.

10. Consider the Markov chain with transition graph given in Figure 3.96, starting in state 1.

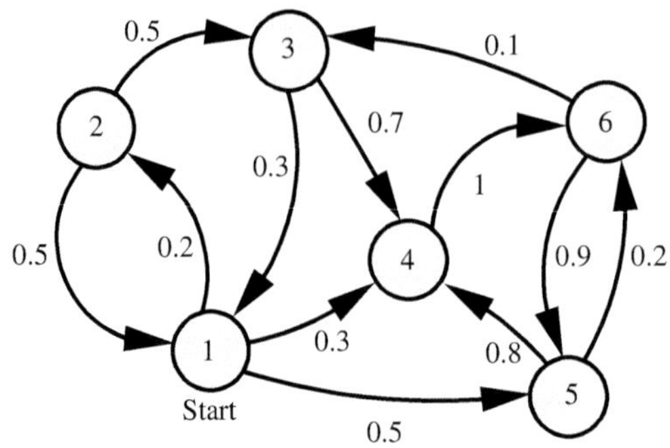

Figure 3.96: The transition graph for the Markov chain $\{X_t, t = 0, 1, 2, \ldots\}$.

(a) Construct a computer program to simulate the Markov chain, and show a realization for $N = 100$ steps.

(b) Compute the limiting probabilities that the Markov chain is in state 1,2,. . . ,6, by ☞ 661 solving the global balance equations (D.104).

(c) Verify that the exact limiting probabilities correspond to the average fraction of times that the Markov process visits states 1,2,. . . ,6, for a large number of steps N.

☞ 662 11. As a generalization of Example D.106, consider a random walk on an arbitrary undirected connected graph with a finite vertex set $\mathcal{V}$. For any vertex $v \in \mathcal{V}$, let $d(v)$ be the number of neighbors of v — called the *degree* of v. The random walk can jump to each one of the neighbors with probability $1/d(v)$ and can be described by a Markov chain. Show that, if the chain is *aperiodic*, the limiting probability that the chain is in state v is equal to $d(v)/\sum_{v' \in \mathcal{V}} d(v')$.

☞ 87 12. Let $U, V \sim_{\text{iid}} \mathcal{U}(0, 1)$. The reason why in Example 3.27 the sample mean and sample median behave very differently is that $\mathbb{E}[U/V] = \infty$, while the median of U/V is finite. Show this, and compute the median. Hint: Start by determining the cdf of $Z = U/V$ by writing it as an expectation of an indicator function.

13. Consider the problem of generating samples from $Y \sim \mathsf{Gamma}(2, 10)$.

(a) Direct simulation: Let $U_1, U_2 \sim_{\text{iid}} \mathcal{U}(0, 1)$. Show that $-\ln(U_1)/10 - \ln(U_2)/10 \sim \mathsf{Gamma}(2, 10)$. Hint: Derive the distribution of $-\ln(U_1)/10$ and use Example D.15.

☞ 631

(b) Simulation via MCMC: Implement an independence sampler to simulate from the $\mathsf{Gamma}(2, 10)$ target pdf

$$f(x) = 100\, x\, \mathrm{e}^{-10x}, \quad x \geq 0,$$

using proposal transition density $q(y \mid x) = g(y)$, where $g(y)$ is the pdf of an $\mathsf{Exp}(5)$ random variable. Generate $N = 500$ samples, and compare the true cdf with the empirical cdf of the data.

14. Let $\mathbf{X} = [X, Y]^\top$ be a random column vector with a bivariate normal distribution with expectation vector $\boldsymbol{\mu} = [1, 2]^\top$ and covariance matrix

$$\boldsymbol{\Sigma} = \begin{bmatrix} 1 & a \\ a & 4 \end{bmatrix}.$$

(a) What are the conditional distributions of $(Y \mid X = x)$ and $(X \mid Y = y)$? Hint: Use Theorem D.53. ☞ 641

(b) Implement a Gibbs sampler to draw 10^3 samples from the bivariate distribution $\mathcal{N}(\boldsymbol{\mu}, \Sigma)$ for $a = 0$, 1, and 1.75, and plot the resulting samples.

15. Here the objective is to sample from the 2-dimensional pdf

$$f(x, y) = c\,\mathrm{e}^{-(xy+x+y)}, \quad x \geq 0, \quad y \geq 0,$$

for some normalization constant c, using a Gibbs sampler. Let $(X, Y) \sim f$.

(a) Find the conditional pdf of X given $Y = y$, and the conditional pdf of Y given $X = x$.

(b) Write working Python code that implements the Gibbs sampler and outputs 1000 points that are approximately distributed according to f.

(c) Describe how the normalization constant c could be estimated via Monte Carlo simulation, using random variables $X_1, \ldots, X_N, Y_1, \ldots, Y_N \overset{\text{iid}}{\sim} \mathsf{Exp}(1)$.

16. We wish to estimate $\mu = \int_{-2}^{2} \mathrm{e}^{-x^2/2}\,\mathrm{d}x = \int H(x) f(x)\,\mathrm{d}x$ via Monte Carlo simulation using two different approaches: (1) defining $H(x) = 4\,\mathrm{e}^{-x^2/2}$ and f the pdf of the $\mathcal{U}[-2, 2]$ distribution and (2) defining $H(x) = \sqrt{2\pi}\,\mathbb{1}\{-2 \leq x \leq 2\}$ and f the pdf of the $\mathcal{N}(0, 1)$ distribution.

(a) For both cases estimate μ via the estimator $\widehat{\mu}$

$$\widehat{\mu} = N^{-1} \sum_{i=1}^{N} H(X_i). \tag{3.97}$$

Use a sample size of $N = 1000$.

(b) For both cases estimate the relative error κ of $\widehat{\mu}$ using $N = 100$.

(c) Give a 95% confidence interval for μ for both cases using $N = 100$.

(d) From part (b), assess how large N should be such that the relative width of the confidence interval is less than 0.01, and carry out the simulation with this N. Compare the result with the true value of μ.

17. Consider estimation of the tail probability $\mu = \mathbb{P}[X \geq \gamma]$ of some random variable X, where γ is large. The crude Monte Carlo estimator of μ is

$$\widehat{\mu} = \frac{1}{N} \sum_{i=1}^{N} Z_i, \tag{3.98}$$

where $X_1, \ldots, X_N$ are iid copies of X and $Z_i = \mathbb{1}\{X_i \geq \gamma\}$, $i = 1, \ldots, N$.

(a) Show that $\widehat{\mu}$ is unbiased; that is, $\mathbb{E}\,\widehat{\mu} = \mu$.

(b) Express the relative error of $\widehat{\mu}$, i.e.,

$$\mathrm{RE} = \frac{\sqrt{\mathbb{V}\mathrm{ar}\,\widehat{\mu}}}{\mathbb{E}\,\widehat{\mu}},$$

in terms of N and μ.

(c) Explain how to estimate the relative error of $\widehat{\mu}$ from outcomes $x_1, \ldots, x_N$ of $X_1, \ldots, X_N$, and how to construct a 95% confidence interval for μ.

(d) An unbiased estimator Z of μ is said to be **logarithmically efficient** if

$$\lim_{\gamma \to \infty} \frac{\ln \mathbb{E} Z^2}{\ln \mu^2} = 1. \tag{3.99}$$

Show that the CMC estimator (3.98) with $N = 1$ is not logarithmically efficient.

18. One of the test cases in [97] involves the minimization of the *Hougen function*. Implement a cross-entropy and a simulated annealing algorithm to carry out this optimization task.

19. In the **binary knapsack problem**, the goal is to solve the optimization problem:

$$\max_{\boldsymbol{x} \in \{0,1\}^n} \boldsymbol{p}^\top \boldsymbol{x},$$

subject to the constraints

$$\mathbf{A}\boldsymbol{x} \leq \boldsymbol{c},$$

where $\boldsymbol{p}$ and $\boldsymbol{w}$ are $n \times 1$ vectors of non-negative numbers, $\mathbf{A} = (a_{ij})$ is an $m \times n$ matrix, and $\boldsymbol{c}$ is an $m \times 1$ vector. The interpretation is that $x_j = 1$ or 0 depending on whether item j with value p_j is packed into the knapsack or not , $j = 1, \ldots, n$; The variable a_{ij} represents the i-th attribute (e.g., volume, weight) of the j-th item. Associated with each attribute is a maximal capacity, e.g., c_1 could be the maximum volume of the knapsack, c_2 the maximum weight, etc.

Write a CE program to solve the `Sento1.dat` knapsack problem at `http://people.brunel.ac.uk/~mastjjb/jeb/orlib/files/mknap2.txt`, as described in [24].

20. Let $(C_1, R_1), (C_2, R_2), \ldots$ be a renewal reward process, with $\mathbb{E}R_1 < \infty$ and $\mathbb{E}C_1 < \infty$. Let $A_t = \sum_{i=1}^{N_t} R_i/t$ be the average reward at time $t = 1, 2, \ldots$, where $N_t = \max\{n : T_n \leq t\}$ and we have defined $T_n = \sum_{i=1}^{n} C_i$ as the time of the n-th renewal.

(a) Show that $T_n/n \xrightarrow{\text{a.s.}} \mathbb{E}C_1$ as $n \to \infty$.

(b) Show that $N_t \xrightarrow{\text{a.s.}} \infty$ as $t \to \infty$.

(c) Show that $N_t/t \xrightarrow{\text{a.s.}} 1/\mathbb{E}C_1$ as $t \to \infty$. Hint: Use the fact that $T_{N_t} \leq t \leq T_{N_t+1}$ for all $t = 1, 2, \ldots$.

(d) Show that

$$A_t \xrightarrow{\text{a.s.}} \frac{\mathbb{E}R_1}{\mathbb{E}C_1} \quad \text{as} \quad t \to \infty.$$

21. Prove Theorem 3.55. ☞ 103

22. Prove that if $H(\boldsymbol{x}) \geq 0$ the importance sampling pdf g^* in (3.66) gives the zero-variance importance sampling estimator $\widehat{\mu} = \mu$.

23. Let X and Y be random variables (not necessarily independent) and suppose we wish to estimate the expected difference $\mu = \mathbb{E}[X - Y] = \mathbb{E}X - \mathbb{E}Y$.

(a) Show that if X and Y are *positively correlated*, the variance of $X - Y$ is smaller than if X and Y are *independent*.

(b) Suppose now that X and Y have cdfs F and G, respectively, and are simulated via the inverse-transform method: $X = F^{-1}(U), Y = G^{-1}(V)$, with $U, V \sim \mathcal{U}(0, 1)$, not necessarily independent. Intuitively, one might expect that if U and V are positively correlated, the variance of $X - Y$ would be smaller than if U and V are independent. Show that this is not always the case by providing a counter-example.

(c) Continuing (b), assume now that F and G are continuous. Show that the variance of $X - Y$ by taking *common random numbers* $U = V$ is no larger than when U and V are independent. Hint: Use the following lemma of Hoeffding [63]: If (X, Y) have joint cdf H with marginal cdfs of X and Y being F and G, respectively, then

$$\mathbb{C}\mathrm{ov}(X, Y) = \int_{-\infty}^{\infty} \int_{-\infty}^{\infty} (H(x, y) - F(x)\, G(y))\, \mathrm{d}x\, \mathrm{d}y,$$

provided $\mathbb{C}\mathrm{ov}(X, Y)$ exists.

24. Let $W_1, W_2, \ldots \stackrel{\text{iid}}{\sim} \mathsf{Exp}(\lambda)$. Define $T_n := W_1 + \cdots + W_n$ for $n \geq 1$ and $T_0 = 0$, as well as $N_t := \max\{k \in \mathbb{N} : T_k \leq t\}$. Think of the $\{T_n, n = 1, 2, \ldots\}$ as the arrival times of customers to a post office and N_t as the total number of customer that have arrived at or before time t. Readers familiar with applied probability will recognize the $\{T_n\}$ as the points a *Poisson process* with rate λ. The process $\{N_t, t \geq 0\}$ is the associated *Poisson counting process*. This process has stationary and independent increments, meaning that for every $t_1 < t_2 < t_3 < t_4$, the increments $N_{t_2} - N_{t_1}$ and $N_{t_4} - N_{t_3}$ are independent, and for every $t, u \in \mathbb{R}_+$, the increment $N_{t+u} - N_t$ has the same distribution as N_u; see, for example, [95].

(a) Show that $T_n \sim \mathsf{Gamma}(n, \lambda)$ and $N_t \sim \mathsf{Poi}(\lambda t)$.

(b) Define the *remaining waiting time* process $\{R_t, t \geq 0\}$ via $R_t = T_{N_t+1} - t$; that is, the time to wait, from t onward, until the next customer arrives. Using the stationarity and independent increments property of $\{N_t, t \geq 0\}$, prove that $R_t \sim \mathsf{Exp}(\lambda)$.

(c) Define the *elapsed waiting time* process $\{E_t, t \geq 0\}$ via $E_t := t - T_{N_t}$; that is, the time since the last customer before time t arrived. Again using the stationarity and independent increments property of $\{N_t, t \geq 0\}$, show that

$$\mathbb{P}[E_t \leq u] = \begin{cases} 1 - \exp(-\lambda u), & u \in [0, t), \\ 1, & u \geq t. \end{cases}$$

(d) Show that the cdf of the random variable $Y_t := T_{N_t+1} - T_{N_t} = E_t + R_t$ is:

$$\mathbb{P}[Y_t \leq y] = \begin{cases} 1 - (1 + \lambda y) \exp(-\lambda y), & y < t \\ 1 - (1 + \lambda t) \exp(-\lambda y), & y \geq t. \end{cases}$$

You may use that E_t and R_t are independent, due to the memoryless property of the exponential distribution. What is the limiting distribution of Y_t as $t \uparrow \infty$?

25. Let $\boldsymbol{X}$ and $\boldsymbol{Y}$ be random vectors in $\mathbb{R}^n$ drawn from densities f and g, respectively. Show that

$$2 \sup_A \left| \mathbb{P}[\boldsymbol{X} \in A] - \mathbb{P}[\boldsymbol{Y} \in A] \right| = \int_{\mathbb{R}^n} \left| f(\boldsymbol{x}) - g(\boldsymbol{x}) \right| \mathrm{d}\boldsymbol{x}.$$

Hint: Use the triangle inequality, $| \int_A f(\boldsymbol{x}) \, \mathrm{d}\boldsymbol{x} - \int_A g(\boldsymbol{x}) \, \mathrm{d}\boldsymbol{x} | \leq \int_A |f(\boldsymbol{x}) - g(\boldsymbol{x})| \, \mathrm{d}\boldsymbol{x}$, with $A = \{\boldsymbol{x} : f(\boldsymbol{x}) \geq g(\boldsymbol{x})\}$.

CHAPTER 4

Unsupervised Learning

When there is no distinction between response and explanatory variables, unsupervised methods are required to learn the structure of the data. In this chapter we look at various unsupervised learning techniques, such as density estimation, clustering, and principal component analysis. Important tools in unsupervised learning include the cross-entropy training loss, mixture models, the Expectation–Maximization algorithm, and the Singular Value Decomposition.

4.1 Introduction

In contrast to supervised learning, where an "output" (response) variable y is explained by an "input" (explanatory) vector $\boldsymbol{x}$, in unsupervised learning there is no response variable and the overall goal is to extract useful information and patterns from the data, e.g., in the form $\tau = \{\boldsymbol{x}_1, \ldots, \boldsymbol{x}_n\}$ or as a matrix $\mathbf{X}^\top = [\boldsymbol{x}_1, \ldots, \boldsymbol{x}_n]$. In essence, the objective of unsupervised learning is to learn about the underlying probability distribution of the data.

We start in Section 4.2 by setting up a framework for unsupervised learning that is similar to the framework used for supervised learning in Section 2.3. That is, we formulate ☞ 23 unsupervised learning in terms of risk and loss minimization; but now involving the cross-entropy risk, rather than the squared-error risk. In a natural way this leads to fundamental learning concepts such as likelihood, Fisher information, and the Akaike information criterion. Section 4.3 introduces the Expectation–Maximization (EM) algorithm as a useful method for maximizing likelihood functions when their solution cannot be found easily in closed form.

If the data forms an iid sample from some unknown distribution, the "empirical distribution" of the data provides valuable information about the unknown distribution. In Section 4.4 we formalize the concept of the empirical distribution (a generalization of the empirical cdf) and explain how we can produce an estimate of the underlying ☞ 11 probability density function of the data using kernel density estimators.

Most unsupervised learning techniques focus on identifying certain traits of the underlying distribution, such as its local maximizers. A related idea is to partition the data into clusters of points that are in some sense "similar" to each other. In Section 4.6

DOI: 10.1201/9781003391173-4

☞ 161 we formulate the clustering problem in terms of a mixture model. In particular, the data are assumed to come from a mixture of (usually Gaussian) distributions, and the objective is to recover the parameters of the mixture distributions from the data. The principal tool for parameter estimation in mixture models is the EM algorithm.

Section 4.7 discusses a more heuristic approach to clustering, where the data are grouped according to certain "cluster centers", whose positions are found by solving an optimization problem. Section 4.8 describes how clusters can be constructed in a hierarchical manner.

Finally, in Section 4.9 we discuss the unsupervised learning technique called Principal Component Analysis (PCA), which is an important tool for reducing the dimensionality of the data.

We will revisit various unsupervised learning techniques in subsequent chapters on *supervised* learning. For example, cross-entropy training loss minimization will be ☞ 242 important in logistic regression (Section 5.8) and classification (Chapter 8), and PCA ☞ 353 can be used for variable selection and dimensionality reduction, to make models easier to train and increase their predictive power; see e.g., Sections 7.8 and 8.29.

4.2 Risk and Loss in Unsupervised Learning

In unsupervised learning, the training data $\mathcal{T} := \{\boldsymbol{X}_1, \ldots, \boldsymbol{X}_n\}$ only consists of (what are usually assumed to be) independent copies of a feature vector $\boldsymbol{X}$; there is no response data. Suppose our objective is to learn the unknown pdf f of $\boldsymbol{X}$ based on an outcome $\tau = \{\boldsymbol{x}_1, \ldots, \boldsymbol{x}_n\}$ of the training data $\mathcal{T}$. Conveniently, we can follow the same line ☞ 23 of reasoning as for *supervised* learning, discussed in Sections 2.3–2.5. Table 4.6 gives a summary of definitions for the case of unsupervised learning. Compare this with ☞ 25 Table 2.11 for the supervised case.

Similar to supervised learning, we wish to find a function g, which is now a probability density (continuous or discrete), that best approximates the pdf f in terms of minimizing a risk

$$\ell(g) := \mathbb{E}\,\mathrm{Loss}(f(\boldsymbol{X}), g(\boldsymbol{X})), \tag{4.1}$$

where Loss is a loss function. In (2.47), we already encountered the Kullback–Leibler risk

$$\ell(g) := \mathbb{E} \ln \frac{f(\boldsymbol{X})}{g(\boldsymbol{X})} = \mathbb{E} \ln f(\boldsymbol{X}) - \mathbb{E} \ln g(\boldsymbol{X}). \tag{4.2}$$

If $\mathcal{G}$ is a class of functions that contains f, then minimizing the Kullback–Leibler risk over $\mathcal{G}$ will yield the (correct) minimizer f. Of course, the problem is that minimization of (4.2) depends on f, which is generally not known. However, since the term $\mathbb{E} \ln f(\boldsymbol{X})$ does not depend on g, it plays no role in the minimization of the Kullback–Leibler risk. By removing this term, we obtain the **cross-entropy risk** (for discrete $\boldsymbol{X}$ replace the integral with a sum):

$$\ell(g) := -\mathbb{E} \ln g(\boldsymbol{X}) = -\int f(\boldsymbol{x}) \ln g(\boldsymbol{x})\, \mathrm{d}\boldsymbol{x}. \tag{4.3}$$

Table 4.6: Summary of definitions for unsupervised learning.

$\boldsymbol{x}$	Fixed feature vector.
$\boldsymbol{X}$	Random feature vector.
$f(\boldsymbol{x})$	Pdf of $\boldsymbol{X}$ evaluated at the point $\boldsymbol{x}$.
τ or τ_n	Fixed training data $\{\boldsymbol{x}_i, i = 1, \ldots, n\}$.
$\mathcal{T}$ or $\mathcal{T}_n$	Random training data $\{\boldsymbol{X}_i, i = 1, \ldots, n\}$.
g	Approximation of the pdf f.
$\text{Loss}(f(\boldsymbol{x}), g(\boldsymbol{x}))$	Loss incurred when approximating $f(\boldsymbol{x})$ with $g(\boldsymbol{x})$.
$\ell(g)$	Risk $\mathbb{E}\,\text{Loss}(f(\boldsymbol{X}), g(\boldsymbol{X}))$ for approximation function g.
$g^{\mathcal{G}}$	Optimal approximation function in function class $\mathcal{G}$; that is, $\text{argmin}_{g \in \mathcal{G}}\, \ell(g)$.
$\ell_\tau(g)$	Training loss for approximation function (guess) g; that is, the sample average estimate of $\ell(g)$ based on a fixed training sample τ.
$\ell_{\mathcal{T}}(g)$	The same as $\ell_\tau(g)$, but now for a random training sample $\mathcal{T}$.
$g_\tau^{\mathcal{G}}$ or g_τ	The *learner*: $\text{argmin}_{g \in \mathcal{G}}\, \ell_\tau(g)$. That is, the optimal approximation function based on a fixed training set τ and function class $\mathcal{G}$. We suppress the superscript $\mathcal{G}$ if the function class is implicit.
$g_{\mathcal{T}}^{\mathcal{G}}$ or $g_{\mathcal{T}}$	The learner for a random training set $\mathcal{T}$.

Thus, minimizing the cross-entropy risk (4.3) over all $g \in \mathcal{G}$, again gives the minimizer f, provided that $f \in \mathcal{G}$. Unfortunately, solving (4.3) is also infeasible in general, as it still depends on f. Instead, we seek to minimize the **cross-entropy training loss**:

$$\ell_\tau(g) := \frac{1}{n}\sum_{i=1}^{n} \text{Loss}(f(\boldsymbol{x}_i), g(\boldsymbol{x}_i)) = -\frac{1}{n}\sum_{i=1}^{n} \ln g(\boldsymbol{x}_i) \tag{4.4}$$

over the class of functions $\mathcal{G}$, where $\tau = \{\boldsymbol{x}_1, \ldots, \boldsymbol{x}_n\}$ is an iid sample from f. This optimization is doable without knowing f and is equivalent to solving the maximization problem

$$\max_{g \in \mathcal{G}} \sum_{i=1}^{n} \ln g(\boldsymbol{x}_i). \tag{4.5}$$

A key step in setting up the learning procedure is to select a suitable function class $\mathcal{G}$ over which to optimize. The standard approach is to parameterize g with a parameter $\boldsymbol{\theta}$ and let $\mathcal{G}$ be the class of functions $\{g(\cdot \mid \boldsymbol{\theta}), \boldsymbol{\theta} \in \Theta\}$ for some p-dimensional parameter set Θ. For the remainder of Section 4.2, we will be using this function class, as well as the cross-entropy risk.

The function $\boldsymbol{\theta} \mapsto g(\boldsymbol{x} \mid \boldsymbol{\theta})$ is called the **likelihood function**. It gives the likelihood of the observed feature vector $\boldsymbol{x}$ under $g(\cdot \mid \boldsymbol{\theta})$, as a function of the parameter $\boldsymbol{\theta}$. The natural logarithm of the likelihood function is called the **log-likelihood** function and its gradient with respect to $\boldsymbol{\theta}$ is called the **score function**, denoted $\boldsymbol{S}(\boldsymbol{x} \mid \boldsymbol{\theta})$; that is,

$$\boldsymbol{S}(\boldsymbol{x} \mid \boldsymbol{\theta}) := \frac{\partial \ln g(\boldsymbol{x} \mid \boldsymbol{\theta})}{\partial \boldsymbol{\theta}} = \frac{\frac{\partial g(\boldsymbol{x} \mid \boldsymbol{\theta})}{\partial \boldsymbol{\theta}}}{g(\boldsymbol{x} \mid \boldsymbol{\theta})}. \tag{4.7}$$

The random score $\boldsymbol{S}(\boldsymbol{X} \mid \boldsymbol{\theta})$, with $\boldsymbol{X} \sim g(\cdot \mid \boldsymbol{\theta})$, is of particular interest. In many cases, its expectation is *equal to the zero vector*; namely,

$$\begin{aligned}\mathbb{E}_{\boldsymbol{\theta}} \boldsymbol{S}(\boldsymbol{X} \mid \boldsymbol{\theta}) &= \int \frac{\frac{\partial g(\boldsymbol{x} \mid \boldsymbol{\theta})}{\partial \boldsymbol{\theta}}}{g(\boldsymbol{x} \mid \boldsymbol{\theta})} g(\boldsymbol{x} \mid \boldsymbol{\theta})\, \mathrm{d}\boldsymbol{x} \\ &= \int \frac{\partial g(\boldsymbol{x} \mid \boldsymbol{\theta})}{\partial \boldsymbol{\theta}}\, \mathrm{d}\boldsymbol{x} = \frac{\partial \int g(\boldsymbol{x} \mid \boldsymbol{\theta})\, \mathrm{d}\boldsymbol{x}}{\partial \boldsymbol{\theta}} = \frac{\partial 1}{\partial \boldsymbol{\theta}} = \mathbf{0},\end{aligned} \tag{4.8}$$

provided that the interchange of differentiation and integration is justified. This is true for a large number of distributions, including the normal, exponential, and binomial distributions. Notable exceptions are distributions whose support depends on the distributional parameter; for example the $\mathcal{U}(0, \theta)$ distribution.

> It is important to see whether expectations are taken with respect to $\boldsymbol{X} \sim g(\cdot \mid \boldsymbol{\theta})$ or $\boldsymbol{X} \sim f$. We use the expectation symbols $\mathbb{E}_{\boldsymbol{\theta}}$ and $\mathbb{E}$ to distinguish the two cases.

From now on we simply assume that the interchange of differentiation and integration is permitted; see, e.g., [105] for sufficient conditions. The covariance matrix of the random score $\boldsymbol{S}(\boldsymbol{X} \mid \boldsymbol{\theta})$ is called the **Fisher information matrix**, which we denote by $\mathbf{F}$ or $\mathbf{F}(\boldsymbol{\theta})$ to show its dependence on $\boldsymbol{\theta}$. Since the expected score is $\mathbf{0}$, we have

$$\mathbf{F}(\boldsymbol{\theta}) = \mathbb{E}_{\boldsymbol{\theta}}[\boldsymbol{S}(\boldsymbol{X} \mid \boldsymbol{\theta})\, \boldsymbol{S}(\boldsymbol{X} \mid \boldsymbol{\theta})^\top]. \tag{4.9}$$

☞ 582 A related matrix is the expected Hessian matrix of $-\ln g(\boldsymbol{X} \mid \boldsymbol{\theta})$:

$$\mathbf{H}(\boldsymbol{\theta}) := \mathbb{E}\left[-\frac{\partial \boldsymbol{S}(\boldsymbol{X} \mid \boldsymbol{\theta})}{\partial \boldsymbol{\theta}}\right] = -\mathbb{E}\begin{bmatrix} \frac{\partial^2 \ln g(\boldsymbol{X} \mid \boldsymbol{\theta})}{\partial^2 \theta_1} & \frac{\partial^2 \ln g(\boldsymbol{X} \mid \boldsymbol{\theta})}{\partial \theta_1 \partial \theta_2} & \cdots & \frac{\partial^2 \ln g(\boldsymbol{X} \mid \boldsymbol{\theta})}{\partial \theta_1 \partial \theta_p} \\ \frac{\partial^2 \ln g(\boldsymbol{X} \mid \boldsymbol{\theta})}{\partial \theta_2 \partial \theta_1} & \frac{\partial^2 \ln g(\boldsymbol{X} \mid \boldsymbol{\theta})}{\partial^2 \theta_2} & \cdots & \frac{\partial^2 \ln g(\boldsymbol{X} \mid \boldsymbol{\theta})}{\partial \theta_2 \partial \theta_p} \\ \vdots & \vdots & \ddots & \vdots \\ \frac{\partial^2 \ln g(\boldsymbol{X} \mid \boldsymbol{\theta})}{\partial \theta_p \partial \theta_1} & \frac{\partial^2 \ln g(\boldsymbol{X} \mid \boldsymbol{\theta})}{\partial \theta_p \partial \theta_2} & \cdots & \frac{\partial^2 \ln g(\boldsymbol{X} \mid \boldsymbol{\theta})}{\partial^2 \theta_p} \end{bmatrix}. \tag{4.10}$$

Note that the expectation here is with respect to $\boldsymbol{X} \sim f$. It turns out that if $f = g(\cdot \mid \boldsymbol{\theta})$, the two matrices are the *same*; that is,

$$\mathbf{F}(\boldsymbol{\theta}) = \mathbf{H}(\boldsymbol{\theta}), \tag{4.11}$$

provided that we may swap the order of differentiation and integration (expectation). This result is called the **information matrix equality**. We leave the proof as Exercise 1.

The matrices $\mathbf{F}(\boldsymbol{\theta})$ and $\mathbf{H}(\boldsymbol{\theta})$ play important roles in approximating the cross-entropy risk for large n. To set the scene, let $g^{\mathcal{G}} = g(\cdot \mid \boldsymbol{\theta}^*)$ be the minimizer of the cross-entropy risk

$$r(\boldsymbol{\theta}) := -\mathbb{E} \ln g(\boldsymbol{X} \mid \boldsymbol{\theta}).$$

We assume that r, as a function of $\boldsymbol{\theta}$, is well-behaved; in particular, that in the neighborhood of $\boldsymbol{\theta}^*$ it is strictly convex and twice continuously differentiable (this holds true, for example, if g is a Gaussian density). It follows that $\boldsymbol{\theta}^*$ is a root of $\mathbb{E}\, \boldsymbol{S}(\boldsymbol{X} \mid \boldsymbol{\theta})$, because

$$\mathbf{0} = \frac{\partial r(\boldsymbol{\theta}^*)}{\partial \boldsymbol{\theta}} = -\frac{\partial\, \mathbb{E} \ln g(\boldsymbol{X} \mid \boldsymbol{\theta}^*)}{\partial \boldsymbol{\theta}} = -\mathbb{E} \frac{\partial \ln g(\boldsymbol{X} \mid \boldsymbol{\theta}^*)}{\partial \boldsymbol{\theta}} = -\mathbb{E}\, \boldsymbol{S}(\boldsymbol{X} \mid \boldsymbol{\theta}^*),$$

again provided that the order of differentiation and integration (expectation) can be swapped. In the same way, $\mathbf{H}(\boldsymbol{\theta})$ is then the Hessian matrix of r. Let $g(\cdot \mid \widehat{\boldsymbol{\theta}}_n)$ be the minimizer of the training loss

$$r_{\mathcal{T}_n}(\boldsymbol{\theta}) := -\frac{1}{n} \sum_{i=1}^{n} \ln g(\boldsymbol{X}_i \mid \boldsymbol{\theta}),$$

where $\mathcal{T}_n = \{\boldsymbol{X}_1, \ldots, \boldsymbol{X}_n\}$ is a random training set. Let r^* be the smallest possible cross-entropy risk, taken over all functions; clearly, $r^* = -\mathbb{E} \ln f(\boldsymbol{X})$, where $\boldsymbol{X} \sim f$. Similar to the supervised learning case, we can decompose the generalization risk, $\ell(g(\cdot \mid \widehat{\boldsymbol{\theta}}_n)) = r(\widehat{\boldsymbol{\theta}}_n)$, into

$$r(\widehat{\boldsymbol{\theta}}_n) = r^* + \underbrace{r(\boldsymbol{\theta}^*) - r^*}_{\text{approx. error}} + \underbrace{r(\widehat{\boldsymbol{\theta}}_n) - r(\boldsymbol{\theta}^*)}_{\text{statistical error}} = r(\boldsymbol{\theta}^*) - \mathbb{E} \ln \frac{g(\boldsymbol{X} \mid \boldsymbol{\theta}^*)}{g(\boldsymbol{X} \mid \widehat{\boldsymbol{\theta}}_n)}.$$

The following theorem specifies the asymptotic behavior of the components of the generalization risk. In the proof we assume that the Hessian (4.10) is continuous and $\mathbb{E}\|\widehat{\boldsymbol{\theta}}_n - \boldsymbol{\theta}^*\|^2 = \mathcal{O}(1/n)$ as $n \to \infty$.

Theorem 4.12 (Approximating the Cross-Entropy Risk): It holds asymptotically ($n \to \infty$) that

$$\mathbb{E} r(\widehat{\boldsymbol{\theta}}_n) - r(\boldsymbol{\theta}^*) = \operatorname{tr}\big(\mathbf{F}(\boldsymbol{\theta}^*)\, \mathbf{H}^{-1}(\boldsymbol{\theta}^*)\big) / (2n) + o(1/n), \tag{4.13}$$

where

$$r(\boldsymbol{\theta}^*) = \mathbb{E} r_{\mathcal{T}_n}(\widehat{\boldsymbol{\theta}}_n) + \operatorname{tr}\big(\mathbf{F}(\boldsymbol{\theta}^*)\, \mathbf{H}^{-1}(\boldsymbol{\theta}^*)\big) / (2n) + o(1/n). \tag{4.14}$$

Proof. A Taylor expansion of $r(\widehat{\boldsymbol{\theta}}_n)$ around $\boldsymbol{\theta}^*$ gives the statistical error ☞ 586

$$r(\widehat{\boldsymbol{\theta}}_n) - r(\boldsymbol{\theta}^*) = (\widehat{\boldsymbol{\theta}}_n - \boldsymbol{\theta}^*)^\top \underbrace{\frac{\partial r(\boldsymbol{\theta}^*)}{\partial \boldsymbol{\theta}}}_{=\mathbf{0}} + \frac{1}{2} (\widehat{\boldsymbol{\theta}}_n - \boldsymbol{\theta}^*)^\top \mathbf{H}(\overline{\boldsymbol{\theta}}_n)(\widehat{\boldsymbol{\theta}}_n - \boldsymbol{\theta}^*),$$

where $\overline{\boldsymbol{\theta}}_n$ lies on the line segment between $\boldsymbol{\theta}^*$ and $\widehat{\boldsymbol{\theta}}_n$. By assumption, $\widehat{\boldsymbol{\theta}}_n \xrightarrow{L^2} \boldsymbol{\theta}^*$ and hence $\overline{\boldsymbol{\theta}}_n \xrightarrow{L^2} \boldsymbol{\theta}^*$, so that the continuity of $\mathbf{H}$ implies that $\mathbf{H}(\overline{\boldsymbol{\theta}}_n) \xrightarrow{L^2} \mathbf{H}(\boldsymbol{\theta}^*)$ as $n \uparrow \infty$. The

matrix $\mathbf{H}(\boldsymbol{\theta}^*)$ is positive definite because $r(\boldsymbol{\theta})$ is strictly convex at $\boldsymbol{\theta}^*$ by assumption, and therefore invertible. It is important to realize that $\widehat{\boldsymbol{\theta}}_n$ is in fact an M-estimator of $\boldsymbol{\theta}^*$. In
☞ 656 particular, in the notation of Theorem D.92, we have $\boldsymbol{\psi} = \mathbf{S}$, $\mathbf{A} = \mathbf{H}(\boldsymbol{\theta}^*)$, and $\mathbf{B} = \mathbf{F}(\boldsymbol{\theta}^*)$. Consequently, by that same theorem,

$$\sqrt{n}\,(\widehat{\boldsymbol{\theta}}_n - \boldsymbol{\theta}^*) \xrightarrow{\text{d}} \mathcal{N}\big(\mathbf{0}, \mathbf{H}^{-1}(\boldsymbol{\theta}^*)\,\mathbf{F}(\boldsymbol{\theta}^*)\,\mathbf{H}^{-\top}(\boldsymbol{\theta}^*)\big).$$

The latter asymptotic normality combined with Slutsky's Theorem D.79 implies that

$$n(\widehat{\boldsymbol{\theta}}_n - \boldsymbol{\theta}^*)^\top \mathbf{H}(\overline{\boldsymbol{\theta}}_n)(\widehat{\boldsymbol{\theta}}_n - \boldsymbol{\theta}^*) \xrightarrow{\text{d}} \operatorname{tr}(\mathbf{F}(\boldsymbol{\theta}^*)\,\mathbf{H}(\boldsymbol{\theta}^*)\,\mathbf{Z}\mathbf{Z}^\top), \tag{4.15}$$

☞ 648 where $\mathbf{Z} \sim \mathcal{N}(\mathbf{0}, \mathbf{I}_p)$. Since the left-hand size of (4.15) is an L^2 random variable and $\mathbb{E}\operatorname{tr}(\mathbf{F}(\boldsymbol{\theta}^*)\mathbf{H}(\boldsymbol{\theta}^*)\mathbf{Z}\mathbf{Z}^\top) = \operatorname{tr}(\mathbf{F}(\boldsymbol{\theta}^*)\,\mathbf{H}(\boldsymbol{\theta}^*))$, we can take expectations on both sides of (4.15) to obtain:

$$2n\,\mathbb{E}[r(\widehat{\boldsymbol{\theta}}_n) - r(\boldsymbol{\theta}^*)] = \mathbb{E}[n(\widehat{\boldsymbol{\theta}}_n - \boldsymbol{\theta}^*)^\top \mathbf{H}(\overline{\boldsymbol{\theta}}_n)(\widehat{\boldsymbol{\theta}}_n - \boldsymbol{\theta}^*)] \longrightarrow \operatorname{tr}(\mathbf{F}(\boldsymbol{\theta}^*)\,\mathbf{H}(\boldsymbol{\theta}^*)).$$

This establishes (4.13). Next, we consider a Taylor expansion of $r_{\mathcal{T}_n}(\boldsymbol{\theta}^*)$ around $\widehat{\boldsymbol{\theta}}_n$:

$$r_{\mathcal{T}_n}(\boldsymbol{\theta}^*) = r_{\mathcal{T}_n}(\widehat{\boldsymbol{\theta}}_n) + (\boldsymbol{\theta}^* - \widehat{\boldsymbol{\theta}}_n)^\top \underbrace{\frac{\partial r_{\mathcal{T}_n}(\widehat{\boldsymbol{\theta}}_n)}{\partial \boldsymbol{\theta}}}_{=\mathbf{0}} + \frac{1}{2}(\boldsymbol{\theta}^* - \widehat{\boldsymbol{\theta}}_n)^\top \mathbf{H}_{\mathcal{T}_n}(\overline{\boldsymbol{\theta}}_n)(\boldsymbol{\theta}^* - \widehat{\boldsymbol{\theta}}_n), \tag{4.16}$$

where $\mathbf{H}_{\mathcal{T}_n}(\overline{\boldsymbol{\theta}}_n) := -\frac{1}{n}\sum_{i=1}^n \frac{\partial \mathbf{S}(X_i \,|\, \overline{\boldsymbol{\theta}}_n)}{\partial \boldsymbol{\theta}}$ is the Hessian of $r_{\mathcal{T}_n}(\boldsymbol{\theta})$ at some $\overline{\boldsymbol{\theta}}_n$ between $\widehat{\boldsymbol{\theta}}_n$ and $\boldsymbol{\theta}^*$. Taking expectations on both sides of (4.16), we obtain

$$r(\boldsymbol{\theta}^*) = \mathbb{E} r_{\mathcal{T}_n}(\widehat{\boldsymbol{\theta}}_n) + \frac{1}{2}\mathbb{E}\,(\boldsymbol{\theta}^* - \widehat{\boldsymbol{\theta}}_n)^\top \mathbf{H}_{\mathcal{T}_n}(\overline{\boldsymbol{\theta}}_n)(\boldsymbol{\theta}^* - \widehat{\boldsymbol{\theta}}_n).$$

If $\mathbf{H}_{\mathcal{T}_n}(\overline{\boldsymbol{\theta}}_n) \xrightarrow{L^2} \mathbf{H}(\boldsymbol{\theta}^*)$ as $n \uparrow \infty$, the same arguments for the derivation of (4.15) yield:

$$n\,\mathbb{E}\,(\boldsymbol{\theta}^* - \widehat{\boldsymbol{\theta}}_n)^\top \mathbf{H}_{\mathcal{T}_n}(\overline{\boldsymbol{\theta}}_n)(\boldsymbol{\theta}^* - \widehat{\boldsymbol{\theta}}_n) \longrightarrow \operatorname{tr}\big(\mathbf{F}(\boldsymbol{\theta}^*)\,\mathbf{H}^{-1}(\boldsymbol{\theta}^*)\big), \quad n \to \infty.$$

Therefore, asymptotically as $n \to \infty$, we have (4.14). □

Theorem 4.12 has a number of interesting consequences:

☞ 36 1. Similar to Section 2.5.1, the training loss $\ell_{\mathcal{T}_n}(g_{\mathcal{T}_n}) = r_{\mathcal{T}_n}(\widehat{\boldsymbol{\theta}}_n)$ tends to underestimate the risk $\ell(g^{\mathcal{G}}) = r(\boldsymbol{\theta}^*)$, because the training set $\mathcal{T}_n$ is used to both train $g \in \mathcal{G}$ (that is, estimate $\boldsymbol{\theta}^*$) and to estimate the risk. The relation (4.14) tells us that on average the training loss underestimates the true risk by $\operatorname{tr}\big(\mathbf{F}(\boldsymbol{\theta}^*)\,\mathbf{H}^{-1}(\boldsymbol{\theta}^*)\big)/(2n)$.

2. Adding equations (4.13) and (4.14), yields the following asymptotic approximation to the expected generalization risk:

$$\mathbb{E}\, r(\widehat{\boldsymbol{\theta}}_n) = \mathbb{E}\, r_{\mathcal{T}_n}(\widehat{\boldsymbol{\theta}}_n) + \frac{1}{n}\operatorname{tr}\big(\mathbf{F}(\boldsymbol{\theta}^*)\,\mathbf{H}^{-1}(\boldsymbol{\theta}^*)\big) + o(1/n). \tag{4.17}$$

The first term on the right-hand side of (4.17) can be estimated (without bias) via the training loss $r_{\mathcal{T}_n}(\widehat{\boldsymbol{\theta}}_n)$. As for the second term, we have already mentioned that when the true model $f \in \mathcal{G}$, then $\mathbf{F}(\boldsymbol{\theta}^*) = \mathbf{H}(\boldsymbol{\theta}^*)$. Therefore, when $\mathcal{G}$ is deemed to be a sufficiently rich class of models parameterized by a p-dimensional vector $\boldsymbol{\theta}$, we may approximate the second term as $\mathrm{tr}(\mathbf{F}(\boldsymbol{\theta}^*)\mathbf{H}^{-1}(\boldsymbol{\theta}^*))/n \approx \mathrm{tr}(\mathbf{I}_p)/n = p/n$. This suggests the following heuristic approximation to the (expected) generalization risk:

$$\mathbb{E}\, r(\widehat{\boldsymbol{\theta}}_n) \approx r_{\mathcal{T}_n}(\widehat{\boldsymbol{\theta}}_n) + \frac{p}{n}. \tag{4.18}$$

3. Multiplying both sides of (4.17) by $2n$ and substituting $\mathrm{tr}\left(\mathbf{F}(\boldsymbol{\theta}^*)\mathbf{H}^{-1}(\boldsymbol{\theta}^*)\right) \approx p$, we obtain the approximation:

$$2n\, r(\widehat{\boldsymbol{\theta}}_n) \approx -2\sum_{i=1}^{n} \ln g(\boldsymbol{X}_i \,|\, \widehat{\boldsymbol{\theta}}_n) + 2p. \tag{4.19}$$

The right-hand side of (4.19) is called the **Akaike information criterion** (AIC). Just like (4.18), the AIC approximation can be used to compare the difference in generalization risk of two or more learners. We prefer the learner with the smallest (estimated) generalization risk.

Suppose that, for a training set $\mathcal{T}$, the training loss $r_{\mathcal{T}}(\boldsymbol{\theta})$ has a unique minimum point $\widehat{\boldsymbol{\theta}}$ which lies in the interior of Θ. If $r_{\mathcal{T}}(\boldsymbol{\theta})$ is a differentiable function with respect to $\boldsymbol{\theta}$, then we can find the optimal parameter $\widehat{\boldsymbol{\theta}}$ by solving

$$\frac{\partial r_{\mathcal{T}}(\boldsymbol{\theta})}{\partial \boldsymbol{\theta}} = \underbrace{\frac{1}{n}\sum_{i=1}^{n} \boldsymbol{S}(\boldsymbol{X}_i \,|\, \boldsymbol{\theta})}_{\boldsymbol{S}_{\mathcal{T}}(\boldsymbol{\theta})} = \boldsymbol{0}.$$

In other words, the maximum likelihood estimate $\widehat{\boldsymbol{\theta}}$ for $\boldsymbol{\theta}$ is obtained by solving the root of the average score function, that is, by solving

$$\boldsymbol{S}_{\mathcal{T}}(\boldsymbol{\theta}) = \boldsymbol{0}. \tag{4.20}$$

It is often not possible to find $\widehat{\boldsymbol{\theta}}$ in an explicit form. In that case one needs to solve the equation (4.20) numerically. There exist many standard techniques for root-finding, e.g., via **Newton's method** (see Section C.3.1), whereby, starting from an initial guess $\boldsymbol{\theta}_0$, ☞ 600
subsequent iterates are obtained via the iterative scheme

$$\boldsymbol{\theta}_{t+1} = \boldsymbol{\theta}_t + \mathbf{H}_{\mathcal{T}}^{-1}(\boldsymbol{\theta}_t)\, \boldsymbol{S}_{\mathcal{T}}(\boldsymbol{\theta}_t),$$

where

$$\mathbf{H}_{\mathcal{T}}(\boldsymbol{\theta}) := \frac{-\,\partial \boldsymbol{S}_{\mathcal{T}}(\boldsymbol{\theta})}{\partial \boldsymbol{\theta}} = \frac{1}{n}\sum_{i=1}^{n} -\frac{\partial \boldsymbol{S}(\boldsymbol{X}_i \,|\, \boldsymbol{\theta})}{\partial \boldsymbol{\theta}}$$

is the average Hessian matrix of $\{-\ln g(\boldsymbol{X}_i \,|\, \boldsymbol{\theta})\}_{i=1}^{n}$. Under $f = g(\cdot \,|\, \boldsymbol{\theta})$, the expectation of $\mathbf{H}_{\mathcal{T}}(\boldsymbol{\theta})$ is equal to the information matrix $\mathbf{F}(\boldsymbol{\theta})$, which does not depend on the data. This suggests an alternative iterative scheme, called **Fisher's scoring method**:

$$\boldsymbol{\theta}_{t+1} = \boldsymbol{\theta}_t + \mathbf{F}^{-1}(\boldsymbol{\theta}_t)\, \boldsymbol{S}_{\mathcal{T}}(\boldsymbol{\theta}_t), \tag{4.21}$$

which is not only easier to implement (if the information matrix can be readily evaluated), but also is more numerically stable.

Example 4.22 (Maximum Likelihood for the Gamma Distribution) We wish to approximate the density of the $\mathsf{Gamma}(\alpha^*, \lambda^*)$ distribution for some true but unknown parameters α^* and λ^*, on the basis of a training set $\tau = \{x_1, \ldots, x_n\}$ of iid samples from this distribution. Choosing our approximating function $g(\cdot \mid \alpha, \lambda)$ in the same class of gamma densities,

$$g(x \mid \alpha, \lambda) = \frac{\lambda^\alpha x^{\alpha-1} \mathrm{e}^{-\lambda x}}{\Gamma(\alpha)}, \quad x \geq 0, \tag{4.23}$$

with $\alpha > 0$ and $\lambda > 0$, we seek to solve (4.20). Taking the logarithm in (4.23), the log-likelihood function is given by

$$l(\alpha, \lambda \mid x) := \alpha \ln \lambda - \ln \Gamma(\alpha) + (\alpha - 1) \ln x - \lambda x.$$

It follows that

$$\boldsymbol{S}(x \mid \alpha, \lambda) = \begin{bmatrix} \frac{\partial}{\partial \alpha} l(\alpha, \lambda \mid x) \\ \frac{\partial}{\partial \lambda} l(\alpha, \lambda \mid x) \end{bmatrix} = \begin{bmatrix} \ln \lambda - \psi(\alpha) + \ln x \\ \frac{\alpha}{\lambda} - x \end{bmatrix},$$

where ψ is the derivative of $\ln \Gamma$: the so-called **digamma function**. Hence,

$$\mathbf{H}(\alpha, \lambda) = -\mathbb{E} \begin{bmatrix} \frac{\partial^2}{\partial \alpha^2} l(\alpha, \lambda \mid X) & \frac{\partial^2}{\partial \alpha \partial \lambda} l(\alpha, \lambda \mid X) \\ \frac{\partial^2}{\partial \alpha \partial \lambda} l(\alpha, \lambda \mid X) & \frac{\partial^2}{\partial \lambda^2} l(\alpha, \lambda \mid X) \end{bmatrix} = -\mathbb{E} \begin{bmatrix} -\psi'(\alpha) & \frac{1}{\lambda} \\ \frac{1}{\lambda} & -\frac{\alpha}{\lambda^2} \end{bmatrix} = \begin{bmatrix} \psi'(\alpha) & -\frac{1}{\lambda} \\ -\frac{1}{\lambda} & \frac{\alpha}{\lambda^2} \end{bmatrix}.$$

Fisher's scoring method (4.21) can now be used to solve (4.20), with

$$\boldsymbol{S}_\tau(\alpha, \lambda) = \begin{bmatrix} \ln \lambda - \psi(\alpha) + n^{-1} \sum_{i=1}^n \ln x_i \\ \frac{\alpha}{\lambda} - n^{-1} \sum_{i=1}^n x_i \end{bmatrix}$$

and $\mathbf{F}(\alpha, \lambda) = \mathbf{H}(\alpha, \lambda)$. ■

4.3 Expectation–Maximization (EM) Algorithm

The **Expectation–Maximization** algorithm (EM) is a general algorithm for maximization of complicated (log-)likelihood functions, through the introduction of auxiliary variables.

> ⚠ To simplify the notation in this section, we use a Bayesian notation system, where the same symbol is used for different (conditional) probability densities.

As in the previous section, given independent observations $\tau = \{\boldsymbol{x}_1, \ldots, \boldsymbol{x}_n\}$ from some unknown pdf f, the objective is to find the best approximation to f in a function class $\mathcal{G} = \{g(\cdot \mid \boldsymbol{\theta}), \boldsymbol{\theta} \in \Theta\}$ by solving the maximum likelihood problem:

$$\boldsymbol{\theta}^* = \underset{\boldsymbol{\theta} \in \Theta}{\operatorname{argmax}}\, g(\tau \mid \boldsymbol{\theta}), \tag{4.24}$$

where $g(\tau \mid \boldsymbol{\theta}) := g(\boldsymbol{x}_1 \mid \boldsymbol{\theta}) \cdots g(\boldsymbol{x}_n \mid \boldsymbol{\theta})$. The key element of the EM algorithm is the augmentation of the data τ with a suitable vector of **latent variables**, $\boldsymbol{z}$, such that

$$g(\tau \mid \boldsymbol{\theta}) = \int g(\tau, \boldsymbol{z} \mid \boldsymbol{\theta})\, \mathrm{d}\boldsymbol{z}.$$

The function $\boldsymbol{\theta} \mapsto g(\tau, \boldsymbol{z} \mid \boldsymbol{\theta})$ is usually referred to as the **complete-data likelihood** function. The choice of the latent variables is guided by the desire to make the maximization of $g(\tau, \boldsymbol{z} \mid \boldsymbol{\theta})$ much easier than that of $g(\tau \mid \boldsymbol{\theta})$.

Suppose p denotes an arbitrary density of the latent variables $\boldsymbol{z}$. Then, we can write:

$$\begin{aligned}
\ln g(\tau \mid \boldsymbol{\theta}) &= \int p(\boldsymbol{z}) \ln g(\tau \mid \boldsymbol{\theta})\, \mathrm{d}\boldsymbol{z} \\
&= \int p(\boldsymbol{z}) \ln \left(\frac{g(\tau, \boldsymbol{z} \mid \boldsymbol{\theta}) / p(\boldsymbol{z})}{g(\boldsymbol{z} \mid \tau, \boldsymbol{\theta}) / p(\boldsymbol{z})} \right) \mathrm{d}\boldsymbol{z} \\
&= \int p(\boldsymbol{z}) \ln \left(\frac{g(\tau, \boldsymbol{z} \mid \boldsymbol{\theta})}{p(\boldsymbol{z})} \right) \mathrm{d}\boldsymbol{z} - \int p(\boldsymbol{z}) \ln \left(\frac{g(\boldsymbol{z} \mid \tau, \boldsymbol{\theta})}{p(\boldsymbol{z})} \right) \mathrm{d}\boldsymbol{z} \\
&= \int p(\boldsymbol{z}) \ln \left(\frac{g(\tau, \boldsymbol{z} \mid \boldsymbol{\theta})}{p(\boldsymbol{z})} \right) \mathrm{d}\boldsymbol{z} + \mathcal{D}(p, g(\cdot \mid \tau, \boldsymbol{\theta})),
\end{aligned} \tag{4.25}$$

where $\mathcal{D}(p, g(\cdot \mid \tau, \boldsymbol{\theta}))$ is the Kullback–Leibler divergence from the density p to $g(\cdot \mid \tau, \boldsymbol{\theta})$. ☞ 43
Since $\mathcal{D} \geq 0$, it follows that

$$\ln g(\tau \mid \boldsymbol{\theta}) \geq \int p(\boldsymbol{z}) \ln \left(\frac{g(\tau, \boldsymbol{z} \mid \boldsymbol{\theta})}{p(\boldsymbol{z})} \right) \mathrm{d}\boldsymbol{z} =: \mathcal{L}(p, \boldsymbol{\theta})$$

for all $\boldsymbol{\theta}$ and any density p of the latent variables. In other words, $\mathcal{L}(p, \boldsymbol{\theta})$ is a lower bound on the log-likelihood that involves the complete-data likelihood. The EM algorithm then aims to increase this lower bound as much as possible by starting with an initial guess $\boldsymbol{\theta}^{(0)}$ and then, for $t = 1, 2, \ldots$, solving the following two steps:

1. $p^{(t)} = \mathrm{argmax}_p\, \mathcal{L}(p, \boldsymbol{\theta}^{(t-1)})$,

2. $\boldsymbol{\theta}^{(t)} = \mathrm{argmax}_{\boldsymbol{\theta} \in \Theta}\, \mathcal{L}(p^{(t)}, \boldsymbol{\theta})$.

The first optimization problem can be solved explicitly. Namely, by (4.25), we have that

$$p^{(t)} = \underset{p}{\mathrm{argmin}}\, \mathcal{D}(p, g(\cdot \mid \tau, \boldsymbol{\theta}^{(t-1)})) = g(\cdot \mid \tau, \boldsymbol{\theta}^{(t-1)}).$$

That is, the optimal density is the conditional density of the latent variables given the data τ and the parameter $\boldsymbol{\theta}^{(t-1)}$. The second optimization problem can be simplified by writing $\mathcal{L}(p^{(t)}, \boldsymbol{\theta}) = Q^{(t)}(\boldsymbol{\theta}) - \mathbb{E}_{p^{(t)}} \ln p^{(t)}(\boldsymbol{Z})$, where

$$Q^{(t)}(\boldsymbol{\theta}) := \mathbb{E}_{p^{(t)}} \ln g(\tau, \boldsymbol{Z} \mid \boldsymbol{\theta})$$

is the expected complete-data log-likelihood under $\boldsymbol{Z} \sim p^{(t)}$. Consequently, the maximization of $\mathcal{L}(p^{(t)}, \boldsymbol{\theta})$ with respect to $\boldsymbol{\theta}$ is equivalent to finding

$$\boldsymbol{\theta}^{(t)} = \underset{\boldsymbol{\theta} \in \Theta}{\mathrm{argmax}}\, Q^{(t)}(\boldsymbol{\theta}).$$

This leads to the following generic EM algorithm.

Algorithm 4.27: Generic EM Algorithm

Input: Data τ, initial guess $\boldsymbol{\theta}^{(0)}$.
Output: Approximation of the maximum likelihood estimate.

$t \leftarrow 1$
while a stopping criterion is not met **do**
  Expectation Step: Find $p^{(t)}(\boldsymbol{z}) := g(\boldsymbol{z} \mid \tau, \boldsymbol{\theta}^{(t-1)})$ and compute the expectation

$$Q^{(t)}(\boldsymbol{\theta}) := \mathbb{E}_{p^{(t)}} \ln g(\tau, \mathbf{Z} \mid \boldsymbol{\theta}). \tag{4.26}$$

  Maximization Step: Let $\boldsymbol{\theta}^{(t)} \leftarrow \operatorname{argmax}_{\boldsymbol{\theta} \in \Theta} Q^{(t)}(\boldsymbol{\theta})$.
  $t \leftarrow t + 1$
return $\boldsymbol{\theta}^{(t)}$

A possible stopping criterion is to stop when

$$\left| \frac{\ln g(\tau \mid \boldsymbol{\theta}^{(t)}) - \ln g(\tau \mid \boldsymbol{\theta}^{(t-1)})}{\ln g(\tau \mid \boldsymbol{\theta}^{(t)})} \right| \leq \varepsilon$$

for some small tolerance $\varepsilon > 0$.

Remark 4.28 (Properties of the EM Algorithm) The identity (4.25) can be used to show that the likelihood $g(\tau \mid \boldsymbol{\theta}^{(t)})$ does not decrease with every iteration of the algorithm. This property is one of the strengths of the algorithm. For example, it can be used to debug computer implementations of the EM algorithm: if the likelihood is observed to decrease at any iteration, then one has detected a bug in the program.

The convergence of the sequence $\{\boldsymbol{\theta}^{(t)}\}$ to a global maximum (if it exists) is highly dependent on the initial value $\boldsymbol{\theta}^{(0)}$ and, in many cases, an appropriate choice of $\boldsymbol{\theta}^{(0)}$ may not be clear. Typically, practitioners run the algorithm from different random starting points over Θ, to ascertain empirically that a suitable optimum is achieved. ■

Example 4.29 (Censored Data) Suppose the lifetime (in years) of a certain type of machine is modeled via a $\mathcal{N}(\mu, \sigma^2)$ distribution. To estimate μ and σ^2, the lifetimes of n (independent) machines are recorded up to c years. Denote these *censored* lifetimes by $x_1, \ldots, x_n$. The $\{x_i\}$ are thus realizations of iid random variables $\{X_i\}$, distributed as $\min\{Y, c\}$, where $Y \sim \mathcal{N}(\mu, \sigma^2)$.

☞ 632

By the law of total probability (see (D.19)), the marginal pdf of each X can be written as:

$$g(x \mid \mu, \sigma^2) = \underbrace{\Phi((c-\mu)/\sigma)}_{\mathbb{P}[Y<c]} \frac{\varphi_{\sigma^2}(x-\mu)}{\Phi((c-\mu)/\sigma)} \mathbb{1}\{x < c\} + \underbrace{\overline{\Phi}((c-\mu)/\sigma)}_{\mathbb{P}[Y \geq c]} \mathbb{1}\{x = c\},$$

where $\varphi_{\sigma^2}(\cdot)$ is the pdf of the $\mathcal{N}(0, \sigma^2)$ distribution, Φ is the cdf of the standard normal distribution, and $\overline{\Phi} := 1 - \Phi$. It follows that the likelihood of the data $\tau = \{x_1, \ldots, x_n\}$

as a function of the parameter $\boldsymbol{\theta} := [\mu, \sigma^2]^\top$ is:

$$g(\tau \mid \boldsymbol{\theta}) = \prod_{i:x_i<c} \frac{\exp\left(-\frac{(x_i-\mu)^2}{2\sigma^2}\right)}{\sqrt{2\pi\sigma^2}} \times \prod_{i:x_i=c} \overline{\Phi}((c-\mu)/\sigma).$$

Let n_c be the total number of x_i such that $x_i = c$. Using n_c latent variables $\boldsymbol{z} = [z_1, \ldots, z_{n_c}]^\top$, we can write the joint pdf:

$$g(\tau, \boldsymbol{z} \mid \boldsymbol{\theta}) = \frac{1}{(2\pi\sigma^2)^{n/2}} \exp\left(-\frac{\sum_{i:x_i<c}(x_i-\mu)^2}{2\sigma^2} - \frac{\sum_{i=1}^{n_c}(z_i-\mu)^2}{2\sigma^2}\right) \mathbb{1}\left\{\min_i z_i \geq c\right\},$$

so that $\int g(\tau, \boldsymbol{z} \mid \boldsymbol{\theta})\, \mathrm{d}\boldsymbol{z} = g(\tau \mid \boldsymbol{\theta})$. We can thus apply the EM algorithm to maximize the likelihood, as follows:

For the E(xpectation)-step, we have for a fixed $\boldsymbol{\theta}$:

$$g(\boldsymbol{z} \mid \tau, \boldsymbol{\theta}) = \prod_{i=1}^{n_c} g(z_i \mid \tau, \boldsymbol{\theta}),$$

where $g(z \mid \tau, \boldsymbol{\theta}) = \mathbb{1}\{z \geq c\}\, \varphi_{\sigma^2}(z-\mu)/\overline{\Phi}((c-\mu)/\sigma)$ is simply the pdf of the $\mathcal{N}(\mu, \sigma^2)$ distribution, truncated to $[c, \infty)$.

For the M(aximization)-step, we compute the expectation of the complete log-likelihood with respect to a fixed $g(\boldsymbol{z} \mid \tau, \boldsymbol{\theta})$ and use the fact that $Z_1, \ldots, Z_{n_c}$ are iid:

$$\mathbb{E} \ln g(\tau, \boldsymbol{Z} \mid \boldsymbol{\theta}) = -\frac{\sum_{i:x_i<c}(x_i-\mu)^2}{2\sigma^2} - \frac{n_c \mathbb{E}(Z-\mu)^2}{2\sigma^2} - \frac{n}{2}\ln\sigma^2 - \frac{n}{2}\ln(2\pi),$$

where Z has a $\mathcal{N}(\mu, \sigma^2)$ distribution, truncated to $[c, \infty)$. To maximize the last expression with respect to μ we set the derivative with respect to μ to zero, and obtain:

$$\mu = \frac{n_c \mathbb{E}Z + \sum_{i:x_i<c} x_i}{n}.$$

Similarly, setting the derivative with respect to σ^2 to zero gives:

$$\sigma^2 = \frac{n_c \mathbb{E}(Z-\mu)^2 + \sum_{i:x_i<c}(x_i-\mu)^2}{n}.$$

In summary, the EM iterates for $t = 1, 2, \ldots$ are as follows:

E-step. Given the current estimate $\boldsymbol{\theta}_t := [\mu_t, \sigma_t^2]^\top$, compute the expectations $\nu_t := \mathbb{E}Z$ and $\zeta_t^2 := \mathbb{E}(Z-\mu_t)^2$, where $Z \sim \mathcal{N}(\mu_t, \sigma_t^2)$, conditional on $Z \geq c$; that is,

$$\nu_t := \mu_t + \sigma_t^2 \frac{\varphi_{\sigma_t^2}(c-\mu_t)}{\overline{\Phi}((c-\mu_t)/\sigma_t)}$$

$$\zeta_t^2 := \sigma_t^2 \left(1 + (c-\mu_t)\frac{\varphi_{\sigma_t^2}(c-\mu_t)}{\overline{\Phi}((c-\mu_t)/\sigma_t)}\right).$$

M-step. Update the estimate to $\boldsymbol{\theta}_{t+1} := [\mu_{t+1}, \sigma^2_{t+1}]^\top$ via the formulas:

$$\mu_{t+1} = \frac{n_c \nu_t + \sum_{i:x_i<c} x_i}{n}$$

$$\sigma^2_{t+1} = \frac{n_c \zeta_t^2 + \sum_{i:x_i<c} (x_i - \mu_{t+1})^2}{n}.$$

■

4.4 Gaussian Kernel Density Estimation

☞ 11 In Section 1.5.2.3 we saw how the empirical cdf $\widehat{F}_n$, obtained from an iid training set $\tau_n = \{x_1, \ldots, x_n\}$ from an unknown distribution on $\mathbb{R}$, gives an estimate of the unknown cdf F of this sampling distribution. The function $\widehat{F}_n$ is a genuine cdf, as it is right-continuous, increasing, and lies between 0 and 1. The corresponding discrete probability distribution is called the **empirical distribution** of the data. A random variable X distributed according to this empirical distribution takes the values $x_1, \ldots, x_n$ with equal probability $1/n$. The empirical distribution concept naturally generalizes to higher dimensions: a random vector $\boldsymbol{X}$ that is distributed according to the empirical distribution of $\boldsymbol{x}_1, \ldots, \boldsymbol{x}_n$ has discrete pdf $\mathbb{P}[\boldsymbol{X} = \boldsymbol{x}_i] = 1/n, i = 1, \ldots, n$. Sampling from such a distribution — in other words *resampling* the original data — was discussed in
☞ 86 Section 3.2.4. The preeminent usage of such sampling is the bootstrap method, discussed
☞ 99 in Section 3.3.2.

In a way, the empirical distribution is the natural answer to the unsupervised learning question: what is the underlying probability distribution of the data? However, the empirical distribution is, by definition, a discrete distribution, whereas the true sampling distribution might be continuous. For continuous data it makes sense to also consider estimation of the pdf of the data. A common approach is to estimate the density via a *kernel density estimate* (KDE), the most prevalent learner to carry this out is given next.

Definition 4.30 (Gaussian KDE): Let $\boldsymbol{x}_1, \ldots, \boldsymbol{x}_n \in \mathbb{R}^d$ be the outcomes of an iid sample from a continuous pdf f. A **Gaussian kernel density estimate** of f is a mixture of normal pdfs, of the form

$$g_{\tau_n}(\boldsymbol{x} \mid \sigma) = \frac{1}{n} \sum_{i=1}^{n} \frac{1}{(2\pi)^{d/2}\sigma^d} \, \mathrm{e}^{-\frac{\|\boldsymbol{x}-\boldsymbol{x}_i\|^2}{2\sigma^2}}, \quad \boldsymbol{x} \in \mathbb{R}^d, \tag{4.31}$$

where $\sigma > 0$ is called the **bandwidth**.

We see that g_{τ_n} in (4.31) is the average of a collection of n normal pdfs, where each normal distribution is centered at the data point $\boldsymbol{x}_i$ and has covariance matrix $\sigma^2 \mathbf{I}_d$. Equivalently, we can view the Gaussian KDE as the pdf of the sum of independent random vectors $\boldsymbol{X} \sim F_n$ and $\boldsymbol{Y} \sim \mathcal{N}(\mathbf{0}, \sigma^2 \mathbf{I}_d)$. Consequently, the pdf is a convolution, and fast Fourier methods can be employed to compute it efficiently, as explained next.

4.4.1 Fast Computation of the Gaussian KDE

In order to use the fast Fourier transform to compute the Gaussian KDE, the data is first *binned*. For one-dimensional data, this means that an interval $[a, b]$ that contains the data is discretized into a regular grid $u_0, u_1, \ldots, u_m$, defining the edges of m bins of size $\mathrm{d}u = (b - a)/m$. Let $p_1, \ldots, p_m$ denote the proportion of observations in each bin and define the discrete random variable $\widetilde{X}$ to take value u_i with probability p_i, $i = 1, \ldots, m$. Similarly, divide the interval $[-6\sigma, 6\sigma]$ into a regular grid $y_0, \ldots, y_r$ with r bins of the same size $\mathrm{d}u$. For $i = 1, \ldots, r$, let v_i be the value of the $\mathcal{N}(0, \sigma^2)$ pdf evaluated at y_i. Let $\widetilde{Y}$ be a random variable that takes value y_i with probability:

$$q_i = \frac{v_i}{\sum_{k=1}^{r} v_k}, \qquad i = 1, \ldots, r,$$

independent of $\widetilde{X}$. The discrete pdf of $\widetilde{X} + \widetilde{Y}$ is given by the **convolution**

$$z_i = [\boldsymbol{p} * \boldsymbol{q}]_i = \sum_{k=1}^{m} p_k \, q_{i-k+1}, \qquad i = 1, \ldots, m + r - 1,$$

where $q_j := 0$ for $j \leq 0$. This full convolution then needs to be restricted to the grid $u_1, \ldots, u_m$ and rescaled by the grid size to yield an approximation of the Gaussian KDE (4.31). This approximation can be very accurate and fast to compute (via the FFT) if m is chosen to be a large enough power of 2. Here is a basic Python implementation that uses the `Convolution` module from Example B.36.

kdebin.py

```
import numpy as np
from Convolution import conv_same


def kdeBin(data, sig):
    n = len(data)
    range_data = np.ptp(data)  # range of the data
    m = 2**16
    a = np.min(data) - 0.3 * range_data
    b = np.max(data) + 0.3 * range_data
    u = np.linspace(a, b, m + 1) # first grid
    du = (b - a) / m  # bin width
    count, _ = np.histogram(data, bins=u)
    p = count/n
    grid = np.arange(-6*sig, 6*sig + du, du) # second grid
    v = np.exp(-0.5 * (grid/sig)**2)
    q = v/(np.sum(v))
    pdf = conv_same(p,q)/du
    mesh = u[1:]
    return pdf, mesh
```

■ **Example 4.32 (Binned KDE)** The Python code below illustrates how one can dramatically speed up the computation of the Gaussian KDE, without sacrificing much accuracy.

kdebinvsgauss

```python
import numpy as np
import matplotlib.pyplot as plt
from kdebin import kdeBin
import time


def gaus_kde(mesh, data, sig):
    data = np.asarray(data)
    const = 1 / (sig * np.sqrt(2 * np.pi))
    return np.mean(const * np.exp(-0.5 * ((mesh[:, None] - data) / sig)
        ** 2), axis=1)

data = np.random.randn(10**4)
sig = 0.1
start_fbin = time.perf_counter()
fbin, mesh = kdeBin(data, sig)
end_fbin = time.perf_counter()
print("fbin computation time:", end_fbin - start_fbin, "seconds")
start_fgaus = time.perf_counter()
fgaus = gaus_kde(mesh,data, sig)
end_fgaus = time.perf_counter()
print("fgaus computation time:", end_fgaus - start_fgaus, "seconds")
plt.plot(mesh, fbin - fgaus) # show difference
plt.xlabel(r'$x$')
plt.ylabel(r'Difference of KDEs')
plt.show()
```

```
fbin computation time: 0.009879900000669295 seconds
fgaus computation time: 18.533656299998256 seconds
```

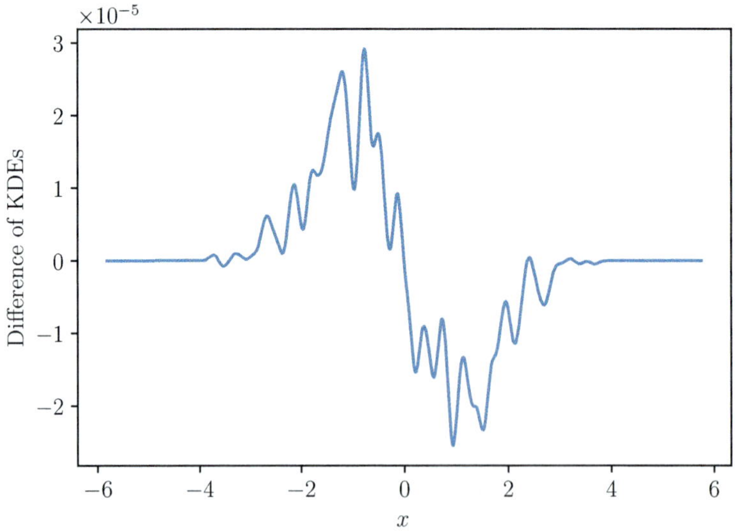

Figure 4.33: Difference between the binned and Gaussian KDEs.

4.4.2 Optimal Gaussian Kernel Bandwidth Selection

A major question is how to choose the bandwidth σ so as to best approximate the unknown pdf f. Choosing very small σ will result in a "spiky" estimate, whereas a large σ will produce an over-smoothed estimate that may not identify important peaks that are present in the unknown pdf. Figure 4.34 illustrates this phenomenon. In this case the data are comprised of 20 points uniformly drawn from the unit square. The true pdf is thus 1 on $[0, 1]^2$ and 0 elsewhere.

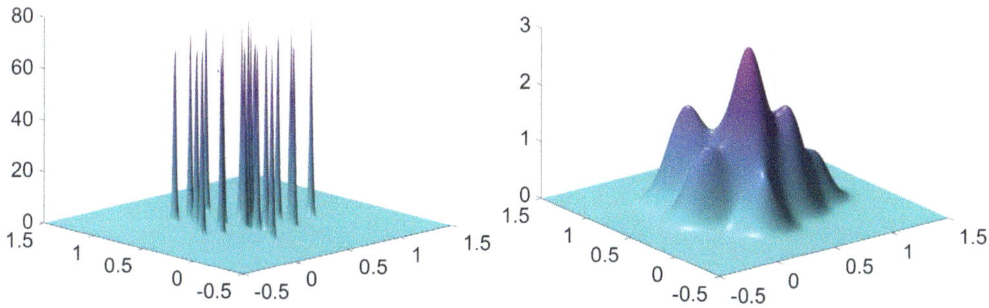

Figure 4.34: Two 2-dimensional Gaussian KDEs, with $\sigma = 0.01$ (left) and $\sigma = 0.1$ (right).

If we denote $\varphi_{\sigma^2}(z) := \frac{1}{(2\pi\sigma^2)^{d/2}} \mathrm{e}^{-\frac{\|z\|^2}{2\sigma^2}}$, $z \in \mathbb{R}^d$ to be the pdf of the $\mathcal{N}(\mathbf{0}, \sigma^2 \mathbf{I}_d)$ distribution, then we can write the Gaussian KDE in (4.31) as

$$g_{\mathcal{T}_n}(\boldsymbol{x} \mid \sigma) = \frac{1}{n} \sum_{i=1}^{n} \varphi_{\sigma^2}(\boldsymbol{x} - \boldsymbol{x}_i). \tag{4.35}$$

Note that the Gaussian pdf is a kernel function of the form (7.23), which is the reason for the KDE appellation. To explain the important issue of bandwidth selection, let $\mathcal{T}_n = \{\boldsymbol{x}_1, \ldots, \boldsymbol{x}_n\}$ be the observed data from the unknown pdf f. First, we define the loss function as ☞ 327

$$\mathrm{Loss}(f(\boldsymbol{x}), g(\boldsymbol{x})) := \frac{(f(\boldsymbol{x}) - g(\boldsymbol{x}))^2}{f(\boldsymbol{x})}.$$

This gives the risk $\ell(g) := \mathbb{E}_f \mathrm{Loss}(f(\boldsymbol{X}), g(\boldsymbol{X})) = \|f - g\|^2$, where $\|\cdot\|$ is the norm on the $L^2(\mathbb{R}^d, \mathrm{d}\boldsymbol{x})$ space of square integrable functions; see Definition B.4. Second, we bypass the selection of a class of approximation functions by choosing the learner to be specified by (4.31) for a fixed σ. Then, a natural choice for σ is the minimizer of the *expected* generalization risk, (that is, averaged over all training sets): ☞ 564

$$\mathbb{E}\ell(g_{\mathcal{T}_n}(\cdot \mid \sigma)) = \mathbb{E}\|f - g_{\mathcal{T}_n}(\cdot \mid \sigma)\|^2.$$

This is called the **mean integrated squared error** (MISE). It can be decomposed into an **integrated squared bias** and **integrated variance** component:

$$\|f - \mathbb{E}g_{\mathcal{T}_n}(\cdot \mid \sigma)\|^2 + \int \mathbb{V}\mathrm{ar}(g_{\mathcal{T}_n}(\boldsymbol{x} \mid \sigma))\, \mathrm{d}\boldsymbol{x}.$$

Under the assumption that f is fourth-order continuously differentiable, we can derive a large-sample expansion of the MISE. Specifically, assuming that $\sigma = \sigma_n \to 0$ as $n \uparrow \infty$, the Taylor expansions in Example C.25 show that the integrated squared bias of $g_{\mathcal{T}_n}$ is given by ☞ 587

$$\|f - \mathbb{E}g_{\mathcal{T}_n}(\cdot \mid \sigma)\|^2 = \frac{\sigma^4}{4}\|\Delta f\|^2 - \frac{\sigma^6}{8}\|\Delta^{3/2} f\|^2 + O(\sigma^8).$$

Additionally, if we assume that $n\,\sigma^d \to 0$ as $n \uparrow \infty$, then the integrated pointwise variance is given by

$$\int \mathbb{V}\mathrm{ar}(g_{\mathcal{T}_n}(\boldsymbol{x} \mid \sigma))\,\mathrm{d}\boldsymbol{x} = \frac{1}{(4\pi)^{d/2} n\,\sigma^d} - \frac{\|f\|^2}{n} + O(\sigma^2/n).$$

It follows that the bandwidth σ that minimizes the MISE $\mathbb{E}\|f - g_{\mathcal{T}_n}(\cdot \mid \sigma)\|^2$ has the large-sample asymptotic expansion (see Exercise 12):

$$\sigma_n := \widetilde{\sigma}_n + \frac{3\|\Delta^{3/2} f\|^2}{4(d+4)\|\Delta f\|^2}\,\widetilde{\sigma}_n^3 + o(n^{-3/(d+4)}), \tag{4.36}$$

where $\widetilde{\sigma}_n$ is the minimizer of the leading asymptotic terms of the MISE:

$$\widetilde{\sigma}_n := \underset{\sigma>0}{\operatorname{argmin}} \left\{ \frac{\sigma^4}{4}\|\Delta f\|^2 + \frac{1}{(4\pi)^{d/2} n\,\sigma^d} \right\} = \left(\frac{d}{n(4\pi)^{d/2}\,\|\Delta f\|^2} \right)^{1/(d+4)}.$$

Substituting the bandwidth (4.36) into $\mathbb{E}\|f - g_{\mathcal{T}_n}(\cdot \mid \sigma)\|^2$ yields the minimum:

$$\begin{aligned} \mathbb{E}\|f - g_{\mathcal{T}_n}(\cdot \mid \sigma_n)\|^2 &= \frac{d+4}{4(4\pi)^{d/2} n\,\sigma_n^d} + O(\sigma_n^6) \\ &= \frac{d+4}{4d}\left[\frac{\|\Delta f\|^{d/2} d}{(4\pi)^{d/2} n} \right]^{4/(d+4)} + O(n^{-6/(d+4)}). \end{aligned} \tag{4.37}$$

At first glance, it appears that to compute the leading term $\widetilde{\sigma}_n$ in (4.36), one needs to first estimate the functional $\|\Delta f\|^2$, and correspondingly a lot of research has been dedicated to the estimation of such functionals [158, 159, 177]. However, as we show in Theorem 4.38 below, there is no need to separately estimate $\|\Delta f\|^2$ in order to successfully estimate the bandwidth value σ_n in (4.36)!

To lay the groundwork for Theorem 4.38, define the perturbation

$$\widetilde{\varphi}_c(\boldsymbol{x}) := \varphi_1(\boldsymbol{x}) - \varphi_c(\boldsymbol{x})$$

for some constant $c > 0$ and the radially symmetric function for all $\boldsymbol{x}, \boldsymbol{x}' \in \mathbb{R}^d$:

$$\begin{aligned} \psi_c(\boldsymbol{x} - \boldsymbol{x}') &:= \int \widetilde{\varphi}_c(\boldsymbol{u} - \boldsymbol{x})\,\widetilde{\varphi}_c(\boldsymbol{u} - \boldsymbol{x}')\,\mathrm{d}\boldsymbol{u} \\ &= \varphi_2(\boldsymbol{x} - \boldsymbol{x}') - 2\varphi_{1+c}(\boldsymbol{x} - \boldsymbol{x}') + \varphi_{2c}(\boldsymbol{x} - \boldsymbol{x}'). \end{aligned}$$

In addition, let $c_d > 0$ be the constant that satisfies the equation:

$$\psi_c(\boldsymbol{0}) = (4\pi)^{-d/2}(1-c)^2.$$

Theorem 4.38 (Bandwidth Perturbation Matching Estimator): If X, X' are iid draws from f, then the bandwidth σ_n in (4.36) satisfies:

$$\mathbb{E}\,\psi_{c_d}\left(\frac{X - X'}{\sigma_n}\right) = \frac{d}{4n}\psi_{c_d}(\mathbf{0}) + O(n^{-(d+6)/(d+4)}). \tag{4.39}$$

Proof. First, define the sets $\mathcal{S}_k := \{\boldsymbol{m} \in \mathbb{N}^d : m_1 + \cdots + m_d = k,\ m_i \text{ is even } \forall i\}$. Then, in the multi-index notation of Taylor's Theorem C.20, where $\boldsymbol{x}^{\boldsymbol{m}} := \prod_i x_i^{m_i}$, the function ψ_c satisfies the moment conditions: ☞ 586

$$\int_{\mathbb{R}^d} \boldsymbol{x}^{\boldsymbol{m}} \psi_c(\boldsymbol{x})\, \mathrm{d}\boldsymbol{x} = \begin{cases} 6(1-c)^2, & \max_i m_i = 4,\ \boldsymbol{m} \in \mathcal{S}_4 \\ 2(1-c)^2, & \max_i m_i < 4,\ \boldsymbol{m} \in \mathcal{S}_4 \\ 0, & |\boldsymbol{m}| \le 5,\ \boldsymbol{m} \in \mathcal{S}_4^c. \end{cases}$$

By a change of variable, we have

$$\mathbb{E}\,\psi_{c_d}\left(\frac{X - X'}{\sigma}\right) = \sigma^d \iint f(\boldsymbol{x}) f(\boldsymbol{x} + \sigma \boldsymbol{x}')\, \psi_{c_d}(\boldsymbol{x}')\, \mathrm{d}\boldsymbol{x}\, \mathrm{d}\boldsymbol{x}'.$$

Hence, an application of Taylor's Theorem C.20 to $f(\boldsymbol{x} + \sigma\boldsymbol{x}')$ as $\sigma \to 0$ yields:

$$\begin{aligned}
\frac{\mathbb{E}\,\psi_{c_d}\left(\frac{X-X'}{\sigma}\right)}{\sigma^d} &= \sum_{|\boldsymbol{m}|\le 5} \frac{\sigma^{|\boldsymbol{m}|}}{\boldsymbol{m}!} \int f(\boldsymbol{x}) \frac{\partial^{|\boldsymbol{m}|} f}{(\partial \boldsymbol{x})^{\boldsymbol{m}}}(\boldsymbol{x})\, \mathrm{d}\boldsymbol{x} \int (\boldsymbol{x}')^{\boldsymbol{m}} \psi_{c_d}(\boldsymbol{x}')\, \mathrm{d}\boldsymbol{x}' + O(\sigma^6) \\
&= \frac{\sigma^4 (1 - c_d)^2}{4} \sum_{\boldsymbol{m} \in \mathcal{S}_4} \int f(\boldsymbol{x}) \frac{\partial^4 f}{(\partial \boldsymbol{x})^{\boldsymbol{m}}}\, \mathrm{d}\boldsymbol{x} + O(\sigma^6) \\
&= \frac{\sigma^4 (1 - c_d)^2}{4} \|\Delta f\|^2 + O(\sigma^6),
\end{aligned}$$

where in the last equation we used the integration by parts identity:

$$\int f(\boldsymbol{x}) \frac{\partial^{|\boldsymbol{m}'+\boldsymbol{m}|} f}{(\partial \boldsymbol{x})^{\boldsymbol{m}'+\boldsymbol{m}}}(\boldsymbol{x})\, \mathrm{d}\boldsymbol{x} = \int \frac{\partial^{|\boldsymbol{m}'|} f}{(\partial \boldsymbol{x})^{\boldsymbol{m}'}}(\boldsymbol{x}) \frac{\partial^{|\boldsymbol{m}|} f}{(\partial \boldsymbol{x})^{\boldsymbol{m}}}(\boldsymbol{x})\, \mathrm{d}\boldsymbol{x},$$

which assumes that $|\boldsymbol{m}' + \boldsymbol{m}|$ is an even integer and the derivatives of f of all orders decay to zero as $\|\boldsymbol{x}\| \to \infty$. Since the bandwidth σ_n in (4.36) is the solution to the equation

$$\frac{\sigma^{d+4}(1 - c_d)^2}{4} \|\Delta f\|^2 = \frac{d}{4n}\psi_{c_d}(\mathbf{0}) + O(\sigma^2/n),$$

it follows that σ_n is also the solution to (4.39), completing the proof. □

Table 4.40 shows the value of the constant $c_d > 0$ in Theorem 4.38 for $d = 1, \ldots, 12$ and Figure 4.41 shows the corresponding function ψ_{c_2} for $d = 2$.

Given the training set $\mathcal{T}_n = \{X_1, \ldots, X_n\}$, Theorem 4.38 suggests that a natural estimator of (4.36) is the solution of the stochastic counterpart of (4.39), namely:

$$\frac{2}{n(n-1)} \sum_{i>j} \psi_{c_d}\left(\frac{X_i - X_j}{\sigma}\right) - \frac{d}{4n}\psi_{c_d}(\mathbf{0}) = 0. \tag{4.42}$$

Table 4.40: The first twelve constants c_d satisfying the nonlinear equation $\psi_c(\mathbf{0}) = (4\pi)^{-d/2}(1-c)^2$. Interestingly, $c_d < 1$ only for the first three physical dimensions.

d	c_d	d	c_d
1	$2.049\,923 \times 10^{-1}$	7	1.655 638
2	$6.180\,340 \times 10^{-1}$	8	1.733 003
3	$9.637\,103 \times 10^{-1}$	9	1.791 497
4	1.222 694	10	1.836 151
5	1.412 687	11	1.870 538
6	1.552 256	12	1.897 226

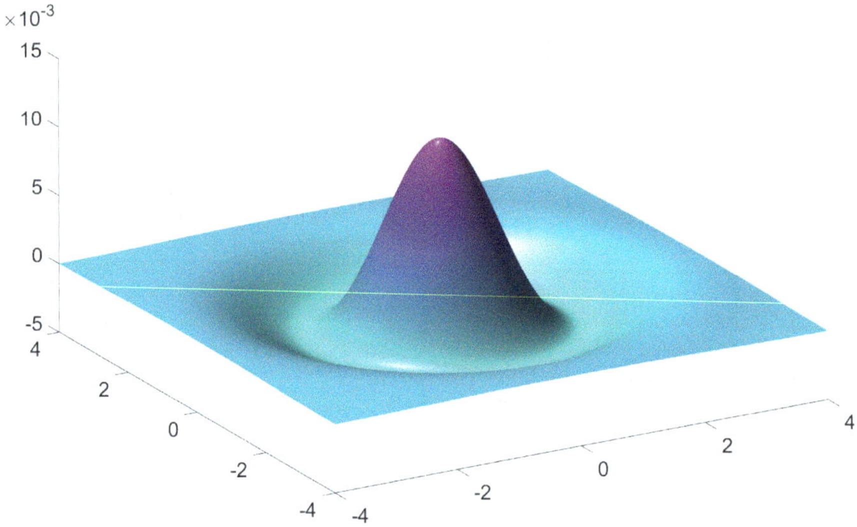

Figure 4.41: The function ψ_{c_2} in two dimensions.

Exercise 9 shows that the equation (4.42) is guaranteed to have at least one root. When n is very small, it is possible that there is more than one root, in which case we recommend picking the largest one. We call this root of (4.42) the **bandwidth perturbation matching** (BPM) estimator. The name is motivated by the fact that the BPM function:

$$\delta_{\mathcal{T}_n}(\boldsymbol{x} \mid \sigma_n) := \frac{g_{\mathcal{T}_n}(x \mid \sigma_n) - g_{\mathcal{T}_n}(x \mid \sqrt{c_d}\,\sigma_n)}{1 - c_d}$$

matches the asymptotic bias and variance of the learner $g_{\mathcal{T}_n}(\cdot \mid \sigma_n)$; see Exercise 11. As a result, its mean integrated squared error $\mathbb{E}\|\delta_{\mathcal{T}_n}(\cdot \mid \sigma_n)\|^2$ is asymptotically the same as $\mathbb{E}\|f - g_{\mathcal{T}_n}(\cdot \mid \sigma_n)\|^2$, but with the crucial distinction that the unknown f is not required for its computation. This allows us to use $\mathbb{E}\|\delta_{\mathcal{T}_n}(\cdot \mid \sigma_n)\|^2$ as a surrogate for the unknown $\mathbb{E}\|f - g_{\mathcal{T}_n}(\cdot \mid \sigma_n)\|^2$ in (4.37).

Also, it is noteworthy that

$$\widehat{\Delta f}(\boldsymbol{x}) := \frac{\delta_{\mathcal{T}_n}(\boldsymbol{x}\,|\,\sigma)}{\sigma^2/2} = \frac{g_{\mathcal{T}_n}(\boldsymbol{x}\,|\,\sigma) - g_{\mathcal{T}_n}(\boldsymbol{x}\,|\,\sqrt{c_d}\,\sigma)}{\sigma^2(1-c_d)/2}, \tag{4.43}$$

is a **Richardson extrapolation estimator** of $\Delta f(\boldsymbol{x})$, whose squared norm can be expressed via the function ψ_{c_d}; see Exercise 11 and Example C.24. ☞ 587

■ **Example 4.44 (Bandwidth Selection)** A common approach in Gaussian kernel density estimation is to choose the bandwidth via Silverman's rule of tumb, which in the one-dimensional case is:

$$\sigma_{\text{RoT}} = s\left(\frac{4}{3n}\right)^{1/5},$$

where s is the sample standard deviation of the data; see Exercise 8. It is well known, however, that this rule of thumb can perform poorly when the data exhibit multiple, well-separated modes, as it tends to oversmooth and obscure important structure in the distribution.

In this example we compare our new BPM approach with R's `dpik` method from the `Kernsmooth` package for bandwidth selection, which is based on the Sheather–Jones automatic bandwidth selection procedure as described in [177]. The latter method alleviates some of the problems of Silverman's rule of thumb, but still ultimately depends on such rules of thumb when estimating functionals like $\|\Delta f\|^2$.

The data is simulated from the distribution:

$$\frac{1}{5}\sum_{k=0}^{4} \mathcal{N}(80k,\ (k+1)^2),$$

which is shorthand for the distribution with pdf

$$f(x) = \frac{1}{5}\sum_{k=0}^{4} \varphi_{(k+1)^2}(x - 80k).$$

The pdf has 5 separated modes with unequal widths, as shown in Figure 4.45.

The figure also displays two Gaussian KDEs: one using a bandwidth automatically selected by R's dpik function, which yields a value of 9.3, and the other using our new BPM method, which produces a much smaller bandwidth of 0.65. We see that `dpik` significantly oversmooths, whereas BPM gives an accurate fit to the true pdf. BPM reduced the integrated square error of `dpik` by a factor of 24.3, achieving the value 6.5×10^{-4} against the value 1.60×10^{-2} for `dpik`. Incidently, Silverman's rule of thumb gives a bandwidth of 30.0 in this case, and so fares even worse than `dpik`. Table 4.97 in Exercise 15 shows the results of similar experiments for a wide range of test distributions, and indicates that BPM is generally very effective in finding the optimal bandwdith automatically.

The following code can be used to create Figure 4.45. It uses the function `kdeBPM` from the module **bpm**, which implements the bandwidth perturbation method. The module can be found on the book's Github site.

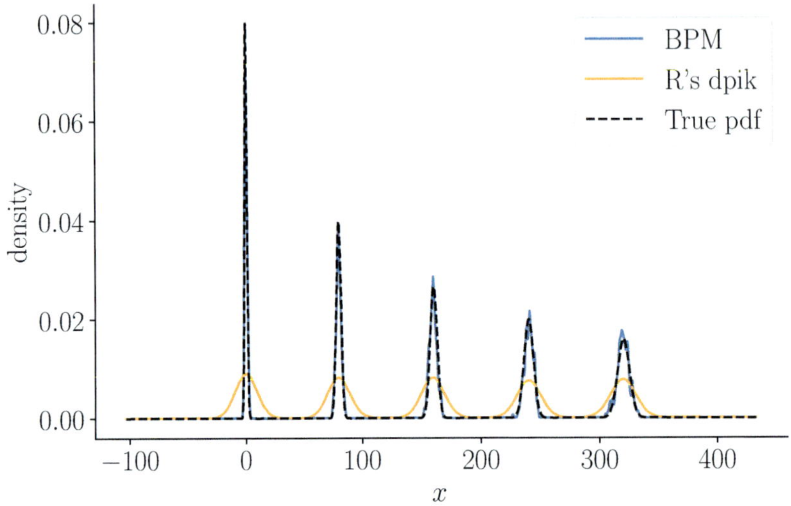

Figure 4.45: Comparison of kernel density estimates with automatic bandwidth selection for a test distribution with five modes.

fivemodes.py

```
import numpy as np
import matplotlib.pyplot as plt
from scipy.stats import norm
from bpm import kdeBPM


Distributions = [{          # dictionay of mixture distributions
    "name": "five_modes",
    "components": [(1/5, 80 * k, k + 1) for k in range(5)]
}]


def simulate_mixture(mixture, n):
    weights, means, stds = zip(*mixture["components"])
    choices = np.random.choice(len(weights), size=n, p=weights)
    return np.array([np.random.normal(means[i], stds[i]) for i in
        choices])


def true_pdf(x, components):
    return sum(w * norm.pdf(x, m, s) for w, m, s in components)


# Simulate data from mixture distribution
name = "five_modes"
mixture = next(d for d in Distributions if d["name"] == name)
n_samples = 1000
samples = simulate_mixture(mixture, n_samples)
```

```python
# Estimate densities and true pdf
density_r, mesh, sig_r = kdeBPM(samples, 9.3)   # using rpik's bandwidth
density_b, _, sig_b = kdeBPM(samples)
pdf_vals = true_pdf(mesh, mixture["components"])

# Compute ISE
delta = mesh[1] - mesh[0]
ise_r = np.sum((pdf_vals - density_r)**2) * delta
ise_b = np.sum((pdf_vals - density_b)**2) * delta

# Plot
plt.plot(mesh, density_b, label='BPM')
plt.plot(mesh, density_r, color="orange", label="R's dpik")
plt.plot(mesh, pdf_vals, color="black", linestyle="dashed", label='True
    PDF')
plt.xlabel(r"$x$")
plt.ylabel("Density")
plt.legend()
plt.show()

print("Bandwidths:", sig_b, sig_r)
print("ISE - BPM:", ise_b, "| R's dpik:", ise_r)
```

```
Bandwidths: 0.569091   9.3
ISE - BPM: 0.0006581 | R's dpik: 0.016005
```

4.5 Theta Kernel Density Estimation

The Gaussian KDE in (4.35) is defined in terms of the function $\kappa(\boldsymbol{x},\boldsymbol{x}') := \varphi_{\sigma^2}(\boldsymbol{x}-\boldsymbol{x}')$, which is a (reproducing) **kernel** — in this case the **Gaussian kernel**; see Example 7.22. It is natural to ask whether other kernels κ can provide viable or improved density estimators by taking the average of kernel values $\{\kappa(\boldsymbol{x},\boldsymbol{x}_i)\}$, as in (4.35). One such alternative kernel that is particularly useful for density estimation is the (multivariate) **theta kernel**, defined as ☞ 327

$$\kappa_{\sigma^2}(\boldsymbol{x},\boldsymbol{x}') := \sum_{\boldsymbol{k}\in\mathbb{Z}^d} \mathrm{e}^{-\|\boldsymbol{k}\|^2\pi^2\sigma^2/2}\cos(\pi\boldsymbol{k}^\top\boldsymbol{x})\prod_{j=1}^{d}\cos(\pi k_j x_j'), \quad \boldsymbol{x},\boldsymbol{x}' \in [0,1]^d.$$

Like the Gaussian kernel it depends on a "bandwidth" parameter $\sigma > 0$.

The theta kernel density estimator allows us to estimate densities on bounded subsets of $\mathbb{R}^d$ in a way that achieves consistent estimation at the boundaries of the subset; see Example 4.52 below. Note that there is no loss of generality in assuming that the training data τ_n lies in $[0,1]^d$, because any bounded region of $\mathbb{R}^d$ can be mapped to the unit hypercube via a suitable one-to-one transformation. Similar to (4.35), the KDE is of the form

$$g_{\tau_n}(\boldsymbol{x}\,|\,\sigma) = \frac{1}{n}\sum_{i=1}^{n}\kappa_{\sigma^2}(\boldsymbol{x},\boldsymbol{x}_i),$$

which, written out, gives

$$g_{\tau_n}(\boldsymbol{x} \mid \sigma) = \sum_{\boldsymbol{k} \in \mathbb{Z}^d} \gamma_{\boldsymbol{k}} \, \mathrm{e}^{-\|\boldsymbol{k}\|^2 \pi^2 \sigma^2/2} \cos(\pi \boldsymbol{k}^\top \boldsymbol{x}),$$

where we define

$$\gamma_{\boldsymbol{k}} := \frac{1}{n} \sum_{i=1}^{n} \prod_{j=1}^{d} \cos(\pi k_j x_{i,j})$$

and, analogously

$$\Gamma_{\boldsymbol{k}} := \frac{1}{n} \sum_{i=1}^{n} \prod_{j=1}^{d} \cos(\pi k_j X_{i,j}).$$

We can think of $\gamma_{\boldsymbol{k}}$ and $\Gamma_{\boldsymbol{k}}$ as respectively the sample average estimate and estimator of

$$c_{\boldsymbol{k}} := \mathbb{E} \prod_{j=1}^{d} \cos(\pi k_j X_j) = \int_{[0,1]^d} f(\boldsymbol{x}) \prod_{j=1}^{d} \cos(\pi k_j x_j) \, \mathrm{d}\boldsymbol{x}, \tag{4.46}$$

where f is the unknown true density that we wish to estimate. The properties of the theta kernel discussed in Example B.28 ensure that $g_{\tau_n}(\cdot \mid \sigma)$ is a probability density on $[0,1]^d$.

The Gaussian kernel $\varphi_t(\boldsymbol{x} - \boldsymbol{x}')$ and theta kernel $\kappa_t(\boldsymbol{x}, \boldsymbol{x}')$ are related in the sense that they are both solutions to the **heat equation**

$$\frac{\partial}{\partial t} h(t, \boldsymbol{x}) = \frac{1}{2} \Delta h(t, \boldsymbol{x}),$$

with the initial condition $h(0, \boldsymbol{x} - \boldsymbol{x}') = \delta(\boldsymbol{x} - \boldsymbol{x}')$, were δ is the Dirac delta function at $\mathbf{0}$; see [21]. The key difference between them is that the Gaussian kernel is defined on $\mathbb{R}^d$ whereas the theta kernel is restricted to $[0,1]^d$.

Physically, $\varphi_t(\boldsymbol{x} - \boldsymbol{x}')$ and $\kappa_t(\boldsymbol{x}, \boldsymbol{x}')$ represent the densities of the heat distribution at position $\boldsymbol{x}$ and time t, given that a unit heat source was deposited at position $\boldsymbol{x}'$ at time 0. For the Gaussian kernel, the heat diffuses throughout the entire $\mathbb{R}^d$ space, whereas for the theta kernel the heat is confined to $[0,1]^d$.

As a result, the kernel density g_τ can be interpreted as the average heat contribution from initial point heat sources at positions $\boldsymbol{x}_1, \ldots, \boldsymbol{x}_n$ diffusing through either $\mathbb{R}^d$ or $[0,1]^d$. In the case of the Gaussian KDE the heat dissipates (becomes 0) as $t \to \infty$, while for the theta KDE the heat distribution eventually becomes uniform over $[0,1]^d$.

One consequence of the above interpretation is that, away from the boundaries of $[0,1]^d$, the difference between the theta kernel and the Gaussian kernel becomes negligible as t becomes smaller, in the sense that

$$\lim_{\sigma \downarrow 0} \frac{\kappa_{\sigma^2}(\boldsymbol{x}, \boldsymbol{x}_i)}{\varphi_{\sigma^2}(\boldsymbol{x} - \boldsymbol{x}_i)} = 1, \quad \boldsymbol{x} \in (0,1)^d .$$

As the heat equation can be efficiently solved on a finite domain using FFT-related methods, the theta KDE can also be computed efficiently in a very similar manner to the Gaussian kernel in Section 4.4.1. This is the topic of the next section.

4.5.1 Fast Computation of the Theta KDE

We now show how the theta KDE can be computed efficiently using FFT methods. For simplicity, we will focus on the one-dimensional case with theta kernel:

$$\begin{aligned}\kappa_{\sigma^2}(x, x_i) &= \sum_{k=-\infty}^{\infty} \mathrm{e}^{-k^2\pi^2\sigma^2/2} \cos(k\pi x)\cos(k\pi x_i),\\ &= \sum_{k=0}^{\infty} \alpha_k \mathrm{e}^{-k^2\pi^2\sigma^2/2} \cos(k\pi x)\cos(k\pi x_i), \quad x, x_i \in [0,1],\end{aligned}$$

where $\alpha_0 := 1$ and $\alpha_k := 2,\ k \geq 1$.

Similar to Section 4.4.1, we bin the data into m bins via a regular grid:

$$[u_0, \ldots, u_m] = [0, 1/m, \ldots, 1],$$

and count the number of data points in each bin. Let p_i be the proportion of observations in the i-th bin, $i = 0, \ldots, m-1$. The number of bins m should be large enough, say $m = 2^{16}$, such that on the interval $[0,1]$ we can accurately approximate the theta KDE using the truncated sum:

$$g_\tau(x \mid \sigma) \approx \sum_{k=0}^{m-1} \alpha_k \gamma_k \, \mathrm{e}^{-k^2\pi^2\sigma^2/2} \cos(k\pi x), \tag{4.47}$$

where the product coefficients $\{\alpha_k\gamma_k\}_{k=0}^{m-1}$ are given by

$$\alpha_k\gamma_k = \alpha_k \frac{1}{n}\sum_{i=1}^{n} \cos(k\pi x_i), \quad k = 0, 1, \ldots, m-1.$$

To compute these coefficients efficiently, note that $\alpha_k\gamma_k = \alpha_k \mathbb{E}\cos(k\pi\widehat{X})$, where $\widehat{X}$ is distributed according to the empirical distribution of the data τ_n. If we replace $\widehat{X}$ with a random variable $\widetilde{X}$ that takes the midpoint value of the i-th bin, $(2i+1)(2m)$, with probability p_i, then an accurate approximation of $\alpha_k\gamma_k$ is

$$\alpha_k\gamma_k \approx \alpha_k \mathbb{E}\cos(k\pi\widetilde{X}) = \alpha_k \sum_{i=0}^{m-1} p_i \cos\left(\frac{\pi(2i+1)k}{2m}\right), \quad k = 0, 1, \ldots, m-1. \tag{4.48}$$

The right-hand side of the last approximation is exactly the discrete cosine transform (DCT) of $\{p_i\}_{i=0}^{m-1}$, as given in Section B.2.3, and can be computed efficiently via the FFT, as in Algorithm B.37. ☞ 579

In addition to using the DCT to compute $\{\alpha_k\gamma_k\}$, the right-hand side of (4.47) evaluated at the midpoints of the bins can be computed via the inverse cosine transform (IDCT) of the *smoothed* coefficients $\{\alpha_k\gamma_k \mathrm{e}^{-k^2\pi^2\sigma^2/2}\}$ times m. Hence, using the theta kernel we can evaluate $g_\tau(\cdot \mid \sigma)$ on a uniform grid of size m in $O(m \ln m)$ operations. The procedure is summarized in the following algorithm.

Algorithm 4.49: Theta Kernel Estimation in One Dimension

Input: Data $x_1, \dots, x_n \in [0, 1]$, number of bins m, bandwidth σ.
Output: Theta KDE at the midpoints of a grid.
Create a uniform grid, $0, 1/m, 2/m, \dots, 1$, which partitions $[0, 1]$ into m bins.
Let $p_0, \dots, p_{m-1}$ be the proportion of data points in each bin.
Compute the DCT $\{\alpha_k \gamma_k\}_{k=0}^{m-1}$ of $\{p_i\}_{i=0}^{m-1}$ via Algorithm B.37.
Let $a_k = \alpha_k \gamma_k \, \mathrm{e}^{-k^2\pi^2\sigma^2/2}$ for $k = 0, \dots, m-1$.
Compute the IDCT $\{b_i\}_{i=0}^{m-1}$ of $\{a_k\}_{k=0}^{m-1}$ via Algorithm B.38.
return $\{m b_k\}$

4.5.2 Optimal Theta Kernel Bandwidth Selection

Recall that the choice of the bandwidth parameter σ is crucial for the statistical performance of the Gaussian KDE, as illustrated in Section 4.4.2. Similarly, the judicious choice of the bandwidth parameter ensures that the theta KDE neither oversmooths nor undersmooths. As a matter of fact, the selection of the theta kernel bandwidth is more straightforward than the selection of the Gaussian kernel bandwidth, because it does not require the derivation of large-sample (asymptotic) approximations to the integrated squared bias and variance that comprise the MISE. To understand why such asymptotic approximations are unnecessary, we first provide a simple formula for the MISE in Theorem 4.50 below.

In the theorem, the random "signs" $S_1, \dots, S_d$ are iid and drawn from the **Rademacher distribution**:

$$\mathbb{P}[S = 1] = \mathbb{P}[S = -1] = 1/2.$$

We also use the notation $\boldsymbol{S} \odot \boldsymbol{k} = [S_1 k_1, S_2 k_2, \dots, S_d k_d]^\top$.

Theorem 4.50 (MISE of the Theta KDE): The mean integrated squared error $\mathbb{E}\ell(g_{\mathcal{T}_n}(\cdot \mid \sigma))$ has series expansion:

$$\sum_{\boldsymbol{k} \in \mathbb{Z}^d} \left(\mathrm{e}^{-\frac{1}{2}\|\boldsymbol{k}\|^2\pi^2\sigma^2} - 1\right)^2 c_{\boldsymbol{k}}^2 + \frac{1}{n} \sum_{\boldsymbol{k} \in \mathbb{Z}^d} \mathrm{e}^{-\|\boldsymbol{k}\|^2\pi^2\sigma^2} \left(\mathbb{E}\, c_{\boldsymbol{k}+\boldsymbol{S}\odot\boldsymbol{k}} - c_{\boldsymbol{k}}^2\right),$$

where the coefficients $\{c_{\boldsymbol{k}}\}$ are given in (4.46).

The proof of the theorem is left as Exercise 17, which additionally shows that an unbiased estimator of the squared coefficient $c_{\boldsymbol{k}}^2$ is

$$\widehat{c_{\boldsymbol{k}}^2} := \frac{n}{n-1}\Gamma_{\boldsymbol{k}}^2 - \frac{1}{n-1}\mathbb{E}[\Gamma_{\boldsymbol{k}+\boldsymbol{S}\odot\boldsymbol{k}} \mid \mathcal{T}_n], \tag{4.51}$$

where the conditional expectation averages over $\boldsymbol{S}$, but not over $\mathcal{T}_n$. The combination of (4.51) with the fact that $\mathbb{E}[\Gamma_{\boldsymbol{k}+\boldsymbol{S}\odot\boldsymbol{k}} \mid \mathcal{T}_n]$ is an unbiased estimator of $\mathbb{E}\, c_{\boldsymbol{k}+\boldsymbol{S}\odot\boldsymbol{k}}$, yields the unbiased estimator of the MISE in Theorem 4.50:

$$\ell_{\mathcal{T}_n}(\sigma) := \sum_{\boldsymbol{k} \in \mathbb{Z}^d} \left(\mathrm{e}^{-\frac{1}{2}\|\boldsymbol{k}\|^2\pi^2\sigma^2} - 1\right)^2 \widehat{c_{\boldsymbol{k}}^2} + \frac{1}{n} \sum_{\boldsymbol{k} \in \mathbb{Z}^d} \mathrm{e}^{-\|\boldsymbol{k}\|^2\pi^2\sigma^2} \left(\mathbb{E}[\Gamma_{\boldsymbol{k}+\boldsymbol{S}\odot\boldsymbol{k}} \mid \mathcal{T}_n] - \widehat{c_{\boldsymbol{k}}^2}\right).$$

This immediately suggests that an estimator of the MISE-optimal bandwidth parameter is:

$$\widehat{\sigma}_n := \underset{\sigma>0}{\operatorname{argmin}}\, \ell_{\mathcal{T}_n}(\sigma).$$

Coming back to the one-dimensional setting of Section 4.5.1, all the formulas above can be simplified, as follows. First, since $c_0 = \Gamma_0 = 1$, we may write

$$\ell_{\mathcal{T}_n}(\sigma) := 2\sum_{k=1}^{\infty}\left(\mathrm{e}^{-\frac{1}{2}k^2\pi^2\sigma^2} - 1\right)^2 \widehat{c_k^2} + \frac{2}{n}\sum_{k=1}^{\infty}\mathrm{e}^{-k^2\pi^2\sigma^2}\left(\mathbb{E}[\Gamma_{k+Sk} \mid \mathcal{T}_n] - \widehat{c_k^2}\right).$$

Second, for $k = 1, 2, \ldots$, we have:

$$\mathbb{E}[\Gamma_{k+S\,k} \mid \mathcal{T}_n] = \frac{\Gamma_{k-k} + \Gamma_{k+k}}{2} = \frac{1 + \Gamma_{2k}}{2},$$

and therefore an estimate for $\alpha_k^2 c_k^2$ is

$$\alpha_k^2\,\widehat{c_k^2} = \frac{n}{n-1}\left(\alpha_k \gamma_k\right)^2 - \frac{2 + \alpha_{2k}\gamma_{2k}}{4(n-1)}, \quad k = 1, 2, \ldots,$$

where we recall that the $\{\alpha_k \gamma_k\}$ can be computed efficiently via the DCT in (4.48) and that $\alpha_k = \alpha_{2k} = 2$ for $k = 1, 2, \ldots$. Finally, the estimate for the MISE is:

$$\ell_{\tau_n}(\sigma) = \frac{1}{2}\sum_{k=1}^{\infty}\left(\mathrm{e}^{-\frac{1}{2}k^2\pi^2\sigma^2} - 1\right)^2\left(\alpha_k^2\,\widehat{c_k^2}\right) + \frac{1}{2n}\sum_{k=1}^{\infty}\mathrm{e}^{-k^2\pi^2\sigma^2}\left(2 + \alpha_{2k}\gamma_{2k} - \alpha_k^2\,\widehat{c_k^2}\right).$$

The Python code below implements the one-dimensional version of the MISE-optimal bandwidth for the theta KDE. The code does not assume that the data lies in [0, 1], but simply adjusts the estimated bandwidth by multiplication with the range of the data:

$$R := \max_i x_i - \min_i x_i.$$

thetakde.py

```
import numpy as np
from scipy import optimize
from DCT import dct, idct
import matplotlib.pyplot as plt


def mise(h, ag2, ag2k, N):
    B = (np.arange(1, len(ag2) + 1) ** 2) * np.pi ** 2 * h ** 2
    val = 0.5 * np.sum(ag2 * (np.exp(-B / 2) - 1) ** 2)
    val += 1 / (2 * N) * np.sum(np.exp(-B) * (2 + ag2k - ag2))
    return val


def kde(data, m=2**16, MIN=None, MAX=None, sig=None, resamp=False):
    data = np.asarray(data)
    m = int(2**np.ceil(np.log2(m)))
    n = data.size
```

```python
    MIN = data.min() if MIN is None else MIN
    MAX = data.max() if MAX is None else MAX

    # Compute the full range and the grid spacing.
    R = MAX - MIN
    dx = R / m
    mesh = np.linspace(MIN, MAX, m) + dx / 2

    # Create the histogram.
    bins = np.linspace(MIN - 1e-7, MAX + 1e-7, m + 1)
    hist, _ = np.histogram(data, bins=bins)
    p = hist / n
    ag = dct(p)  # DCT to compute {alpha_k*gamma_k}.

    # Optimize the bandwidth if not provided.
    if sig is None:
        ag2k = ag[2:m:2]
        ag2 = n / (n - 1) * ag[1:m // 2] ** 2 - (2 + ag2k) / (n - 1)

        def mise1(logh):
            return mise(np.exp(logh), ag2, ag2k, n)

        res = optimize.minimize_scalar(mise1, bounds=(-40, 10), method='
            bounded')
        h_opt = np.exp(res.x)  # Optimal bandwidth in unit interval
        sig = h_opt * R        # Convert to data space bandwidth
    else:
        h_opt = sig / R        # Convert provided bandwidth to unit
            interval

    k = np.arange(m, dtype=np.float64)
    ak = ag * np.exp(-(k ** 2) * np.pi ** 2 * h_opt ** 2 / 2)
    density = idct(ak) / dx  # Perform the IDCT

    # When there is duplicate data due to resampling
    if resamp:
        udata = np.unique(data)
        density, _, usig = kde(udata, m, np.min(udata), np.max(udata))
        density, _, sig = kde(data, m, MIN, MAX, usig)

    return density, mesh, sig
```

Theoretically the outcome, τ_n, of an iid sample drawn from a continuous density f, will almost surely never exhibit ties. However, if there are duplicate values in τ_n, the Python code above removes them and estimates the bandwidth using only the unique values.

Example 4.52 (Comparison of Gaussian and theta KDEs) Figure 4.34 illustrates a common problem with the Gaussian KDE: for distributions on a bounded domain, such as the uniform distribution on $[0, 1]^2$, the Gaussian KDE assigns positive probability mass *outside* this domain. An important advantage of the theta KDE is that it largely avoids this boundary effect. We illustrate this boundary effect with an example.

The following Python program draws an iid sample from the Exp(1) distribution and constructs a Gaussian kernel density estimate. We see in Figure 4.53 that with an appropriate choice of the bandwidth a good fit at the boundary $x = 0$ can be achieved, but at the cost of undersmoothing the rest of the density. The theta KDE can handle the boundary without undersmoothing, and chooses the bandwidth automatically to achieve a superior fit.

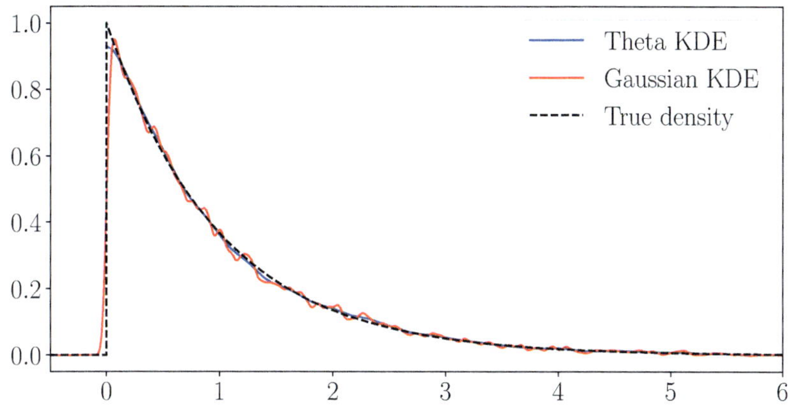

Figure 4.53: Density estimates for Exp(1)-distributed data.

thetakdeexp.py

```
import matplotlib.pyplot as plt
import numpy as np
from thetakde import kde
from kdebin import kdeBin

f = lambda x: np.exp(-x)*(x >= 0) # True PDF
n = 10**4 # Sample Size
x = -np.log(np.random.uniform(size=n))# Generate Data via IT method

density, mesh1, sig = kde(x)
gkde, mesh = kdeBin(x,sig)

plt.plot(mesh1, density, 'b', label='Theta KDE')
plt.plot(mesh, gkde, 'r', label ='Gaussian KDE')
plt.plot(mesh, f(mesh), 'k--', label='True density')
plt.xlim(-0.5, 6)
plt.legend()
```

■

■ **Example 4.54 (Convergence Rate)** Equation (4.37) shows that, theoretically and for a general smooth pdf f, the MISE of a kernel density estimator decreases with the sample size n at a rate of $O(n^{-4/5})$, assuming that an optimal bandwidth is used.

This theoretical insight is confirmed in the following numerical experiment, whereby we compute the theta KDE using data drawn from a standard normal pdf for a range of sample sizes n. For each sample size, we generate 100 independent data sets, apply the theta KDE to estimate the underlying density, and compute the integrated square errors between the estimated and true Gaussian density. The average of these values (estimating the corresponding MISE) are then plotted on a log-log scale to visualize the rate of convergence.This empirical rate is then compared to the theoretical rate of $n^{-4/5}$. Figure 4.55 shows the results of this experiment.

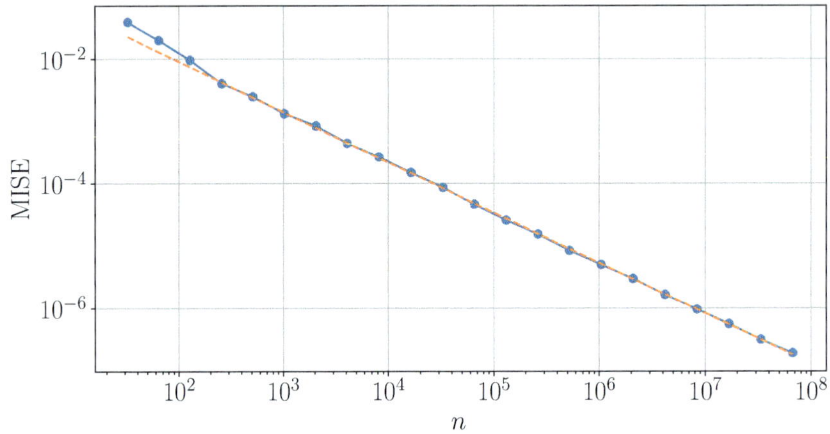

Figure 4.55: Convergence rate of the MISE for the theta KDE, based on 100 repetitions. The dashed line with slope $-4/5$ indicates the theoretical rate.

■

4.6 Clustering via Mixture Models

Clustering is concerned with the grouping of unlabeled feature vectors into clusters, such that samples within a cluster are more similar to each other than samples belonging to different clusters. Usually, it is assumed that the number of clusters is known in advance, but otherwise no prior information is given about the data. Applications of clustering can be found in the areas of communication, data compression and storage, database searching, pattern matching, and object recognition.

A common approach to clustering analysis is to assume that the data comes from a mixture of (usually Gaussian) distributions, and thus the objective is to estimate the parameters of the mixture model by maximizing the likelihood function for the data. Direct optimization of the likelihood function in this case is not a simple task, due to necessary constraints on the parameters (more about this later) and the complicated nature of the likelihood function, which in general has a great number of local maxima and saddle-points. A popular method to estimate the parameters of the mixture model ☞ 140 is the EM algorithm, which was discussed in a more general setting in Section 4.3. In this section we explain the basics of mixture modeling and explain the workings of the EM method in this context. In addition, we show how direct optimization methods can be used to maximize the likelihood.

4.6.1 Mixture Models

Let $\mathcal{T} := \{X_1, \ldots, X_n\}$ be iid random vectors taking values in some set $\mathcal{X} \subseteq \mathbb{R}^d$, each X_i being distributed according to the **mixture density**

$$g(x \mid \theta) = w_1 \phi_1(x) + \cdots + w_K \phi_K(x), \quad x \in \mathcal{X}, \tag{4.56}$$

where $\phi_1, \ldots, \phi_K$ are probability densities (discrete or continuous) on $\mathcal{X}$, and the positive **weights** $w_1, \ldots, w_K$ sum up to 1. This mixture pdf can be interpreted in the following way. Let Z be a discrete random variable taking values $1, 2, \ldots, K$ with probabilities $w_1, \ldots, w_K$, and let X be a random vector whose conditional pdf, given $Z = z$, is ϕ_z. By the product rule (D.29), the joint pdf of Z and X is given by ☞ 635

$$\phi_{Z,X}(z, x) = \phi_Z(z)\, \phi_{X \mid Z}(x \mid z) = w_z\, \phi_z(x)$$

and the marginal pdf of X is found by summing the joint pdf over the values of z, which gives (4.56). A random vector $X \sim g$ can thus be simulated in two steps:

1. First, draw Z according to the probabilities $\mathbb{P}[Z = z] = w_z$, $z = 1, \ldots, K$.

2. Then draw X according to the pdf ϕ_Z.

As $\mathcal{T}$ only contain the $\{X_i\}$ variables, the $\{Z_i\}$ are viewed as *latent* variables. We can interpret Z_i as the hidden label of the cluster to which X_i belongs.

Typically, each ϕ_k in (4.56) is assumed to be known up to some parameter vector η_k. It is customary[1] in clustering analysis to work with *Gaussian* mixtures; that is, each density ϕ_k is Gaussian with some unknown expectation vector μ_k and covariance matrix Σ_k. We gather all unknown parameters, including the weights $\{w_k\}$, into a parameter vector θ. As usual, $\tau = \{x_1, \ldots, x_n\}$ denotes the outcome of $\mathcal{T}$. As the components of $\mathcal{T}$ are iid, their (joint) pdf is given by

$$g(\tau \mid \theta) := \prod_{i=1}^{n} g(x_i \mid \theta) = \prod_{i=1}^{n} \sum_{k=1}^{K} w_k\, \phi_k(x_i \mid \mu_k, \Sigma_k). \tag{4.57}$$

Following the same reasoning as for (4.5), we can estimate θ from an outcome τ by maximizing the log-likelihood function

$$l(\theta \mid \tau) := \sum_{i=1}^{n} \ln g(x_i \mid \theta) = \sum_{i=1}^{n} \ln \left(\sum_{k=1}^{K} w_k\, \phi_k(x_i \mid \mu_k, \Sigma_k) \right). \tag{4.58}$$

However, finding the maximizer of $l(\theta \mid \tau)$ is not easy in general, since the function is typically multiextremal.

Example 4.59 (Clustering via Mixture Models) The data depicted in Figure 4.60 consists of 300 data points that were independently generated from three bivariate normal distributions, whose parameters are given in that same figure. For each of these three distributions, exactly 100 points were generated. Ideally, we would like to cluster the data into three clusters that correspond to the three cases.

[1]Other common mixture distributions include Student t and Beta distributions.

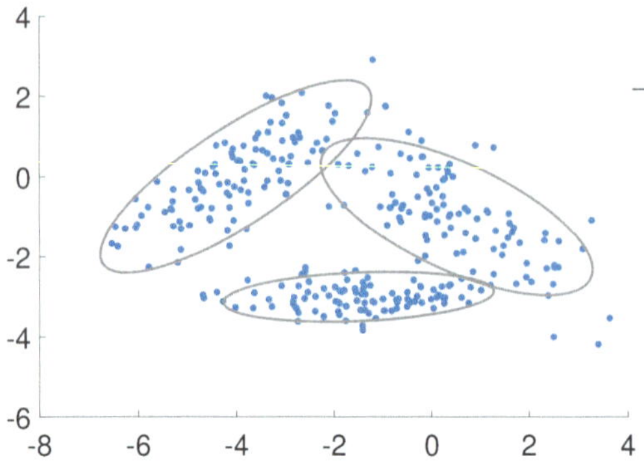

cluster	mean vector	covariance matrix
1	$\begin{bmatrix} -4 \\ 0 \end{bmatrix}$	$\begin{bmatrix} 2 & 1.4 \\ 1.4 & 1.5 \end{bmatrix}$
2	$\begin{bmatrix} 0.5 \\ -1 \end{bmatrix}$	$\begin{bmatrix} 2 & -0.95 \\ -0.95 & 1 \end{bmatrix}$
3	$\begin{bmatrix} -1.5 \\ -3 \end{bmatrix}$	$\begin{bmatrix} 2 & 0.1 \\ 0.1 & 0.1 \end{bmatrix}$

Figure 4.60: Cluster the 300 data points (left) into three clusters, without making any assumptions about the probability distribution of the data. In fact, the data were generated from three bivariate normal distributions, whose parameters are listed on the right.

To cluster the data into three groups, a possible model for the data is to assume that the points are iid draws from an (unknown) mixture of three 2-dimensional Gaussian distributions. This is a sensible approach, although in reality the data were not simulated in this way. It is instructive to understand the difference between the two models. In the mixture model, each cluster label Z takes the value $\{1, 2, 3\}$ with equal probability, and hence, drawing the labels independently, the total number of points in each cluster would be $\mathsf{Bin}(300, 1/3)$ distributed. However, in the actual simulation, the number of points in each cluster is exactly 100. Nevertheless, the mixture model would be an accurate (although not exact) model for these data. Figure 4.61 displays the "target" Gaussian mixture density for the data in Figure 4.60; that is, the mixture with equal weights and with the exact parameters as specified in Figure 4.60.

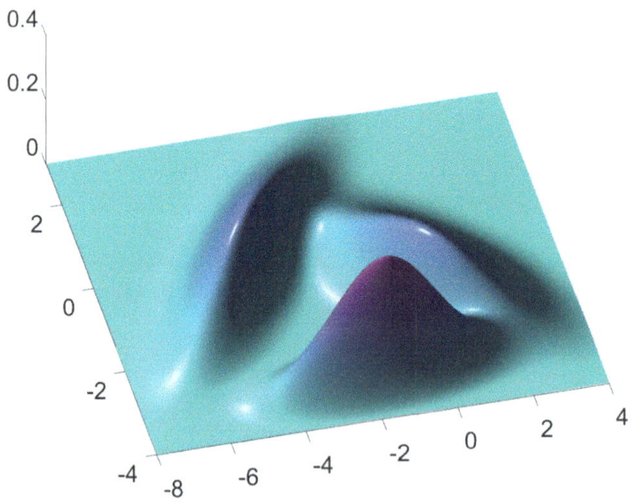

Figure 4.61: The target mixture density for the data in Figure 4.60.

In the next section we carry out the clustering via the EM algorithm. ■

4.6.2 EM Algorithm for Mixture Models

As we saw in Section 4.3, instead of maximizing the log-likelihood function (4.58) directly from the data $\tau = \{\boldsymbol{x}_1, \ldots, \boldsymbol{x}_n\}$, the EM algorithm first *augments* the data with the vector of latent variables — in this case the hidden cluster labels $\boldsymbol{z} = \{z_1, \ldots, z_n\}$. The idea is that τ is only the *observed* part of the complete random data $(\mathcal{T}, \boldsymbol{Z})$, which were generated via the two-step procedure described above. That is, for each data point $\boldsymbol{X}$, first draw the cluster label $Z \in \{1, \ldots, K\}$ according to probabilities $\{w_1, \ldots, w_K\}$ and then, given $Z = z$, draw $\boldsymbol{X}$ from ϕ_z. The joint pdf of $\mathcal{T}$ and $\boldsymbol{Z}$ is

$$g(\tau, \boldsymbol{z} \mid \boldsymbol{\theta}) = \prod_{i=1}^{n} w_{z_i}\, \phi_{z_i}(\boldsymbol{x}_i),$$

which is of a much simpler form than (4.57). It follows that the **complete-data log-likelihood** function

$$\widetilde{l}(\boldsymbol{\theta} \mid \tau, \boldsymbol{z}) = \sum_{i=1}^{n} \ln[w_{z_i}\, \phi_{z_i}(\boldsymbol{x}_i)] \tag{4.62}$$

is often easier to maximize than the original log-likelihood (4.58), for any given $(\tau, \boldsymbol{z})$. But, of course the latent variables $\boldsymbol{z}$ are not observed and so $\widetilde{l}(\boldsymbol{\theta} \mid \tau, \boldsymbol{z})$ cannot be evaluated. In the E-step of the EM algorithm, the complete-data log-likelihood is replaced with the expectation $\mathbb{E}_p\, \widetilde{l}(\boldsymbol{\theta} \mid \tau, \boldsymbol{Z})$, where the subscript p in the expectation indicates that $\boldsymbol{Z}$ is distributed according to the conditional pdf of $\boldsymbol{Z}$ given $\mathcal{T} = \tau$; that is, with pdf

$$p(\boldsymbol{z}) = g(\boldsymbol{z} \mid \tau, \boldsymbol{\theta}) \propto g(\tau, \boldsymbol{z} \mid \boldsymbol{\theta}). \tag{4.63}$$

Note that $p(\boldsymbol{z})$ is of the form $p_1(z_1) \cdots p_n(z_n)$ so that, given $\mathcal{T} = \tau$, the components of $\boldsymbol{Z}$ are independent of each other. The EM algorithm for mixture models can now be formulated as follows:

Algorithm 4.64: EM Algorithm for Mixture Models

Input: Data τ, initial guess $\boldsymbol{\theta}^{(0)}$.

Output: Approximation of the maximum likelihood estimate.

```
t ← 1
while a stopping criterion is not met do
    Expectation Step: Find p^(t)(z) := g(z | τ, θ^(t-1)) and
        Q^(t)(θ) := E_{p^(t)} l~(θ | τ, Z).
    Maximization Step: Let θ^(t) ← argmax_θ Q^(t)(θ).
    t ← t + 1
return θ^(t)
```

A simple stopping rule is to stop when $\left|l(\boldsymbol{\theta}^{(t)} \mid \tau) - l(\boldsymbol{\theta}^{(t-1)} \mid \tau)\right| / \left|l(\boldsymbol{\theta}^{(t)} \mid \tau)\right| < \varepsilon$ for some small tolerance $\varepsilon > 0$. As was mentioned in Section 4.3, the sequence of log-likelihood values *does not decrease* with each iteration. Under certain continuity conditions, the sequence $\{\boldsymbol{\theta}^{(t)}\}$ is guaranteed to converge to a local maximizer of the log-likelihood l. Convergence to a global maximizer (if it exists) depends on the appropriate choice for the starting value. Typically, the algorithm is run from different random starting points.

For the case of Gaussian mixtures, each $\phi_k = \varphi(\cdot \mid \boldsymbol{\mu}_k, \boldsymbol{\Sigma}_k), k = 1, \ldots, K$ is the density of a d-dimensional Gaussian distribution. Let $\boldsymbol{\theta}^{(t-1)}$ be the current guess for the optimal parameter vector, consisting of the weights $\{w_k^{(t-1)}\}$, mean vectors $\{\boldsymbol{\mu}_k^{(t-1)}\}$, and covariance matrices $\{\boldsymbol{\Sigma}_k^{(t-1)}\}$. We first determine $p^{(t)}$ — the pdf of $\boldsymbol{Z}$ conditional on $\mathcal{T} = \tau$ — for the given guess $\boldsymbol{\theta}^{(t-1)}$. As mentioned before, the components of $\boldsymbol{Z}$ given $\mathcal{T} = \tau$ are independent, so it suffices to specify the discrete pdf, $p_i^{(t)}$ say, of each Z_i given the observed point $\boldsymbol{X}_i = \boldsymbol{x}_i$. The latter can be found from Bayes' formula:

$$p_i^{(t)}(k) \propto w_k^{(t-1)} \, \phi_k(\boldsymbol{x}_i \mid \boldsymbol{\mu}_k^{(t-1)}, \boldsymbol{\Sigma}_k^{(t-1)}), \quad k = 1, \ldots, K. \tag{4.65}$$

Next, in view of (4.62), the function $Q^{(t)}(\boldsymbol{\theta})$ can be written as

$$\begin{aligned} Q^{(t)}(\boldsymbol{\theta}) &= \mathbb{E}_{p^{(t)}} \sum_{i=1}^{n} \left(\ln w_{Z_i} + \ln \phi_{Z_i}(\boldsymbol{x}_i \mid \boldsymbol{\mu}_{Z_i}, \boldsymbol{\Sigma}_{Z_i})\right) \\ &= \sum_{i=1}^{n} \mathbb{E}_{p_i^{(t)}} \left[\ln w_{Z_i} + \ln \phi_{Z_i}(\boldsymbol{x}_i \mid \boldsymbol{\mu}_{Z_i}, \boldsymbol{\Sigma}_{Z_i})\right], \end{aligned}$$

where the $\{Z_i\}$ are independent and Z_i is distributed according to $p_i^{(t)}$ in (4.65). This completes the *E-step*. In the *M-step* we maximize $Q^{(t)}$ with respect to the parameter $\boldsymbol{\theta}$; that is, with respect to the $\{w_k\}$, $\{\boldsymbol{\mu}_k\}$, and $\{\boldsymbol{\Sigma}_k\}$. In particular, we maximize

$$\sum_{i=1}^{n} \sum_{k=1}^{K} p_i^{(t)}(k) \left[\ln w_k + \ln \phi_k(\boldsymbol{x}_i \mid \boldsymbol{\mu}_k, \boldsymbol{\Sigma}_k)\right],$$

under the condition $\sum_k w_k = 1$. Using Lagrange multipliers, combined with the fact that $\sum_{k=1}^{K} p_i^{(t)}(k) = 1$, gives the solution for the $\{w_k\}$:

$$w_k = \frac{1}{n} \sum_{i=1}^{n} p_i^{(t)}(k), \quad k = 1, \ldots, K. \tag{4.66}$$

Solving for $\boldsymbol{\mu}_k$ and $\boldsymbol{\Sigma}_k$ now follows from maximizing $\sum_{i=1}^{n} p_i^{(t)}(k) \ln \phi_k(\boldsymbol{x}_i \mid \boldsymbol{\mu}_k, \boldsymbol{\Sigma}_k)$, leading to

$$\boldsymbol{\mu}_k = \frac{\sum_{i=1}^{n} p_i^{(t)}(k) \, \boldsymbol{x}_i}{\sum_{i=1}^{n} p_i^{(t)}(k)}, \quad k = 1, \ldots, K \tag{4.67}$$

and

$$\boldsymbol{\Sigma}_k = \frac{\sum_{i=1}^{n} p_i^{(t)}(k) \, (\boldsymbol{x}_i - \boldsymbol{\mu}_k)(\boldsymbol{x}_i - \boldsymbol{\mu}_k)^\top}{\sum_{i=1}^{n} p_i^{(t)}(k)}, \quad k = 1, \ldots, K, \tag{4.68}$$

which are very similar to the well-known formulas for the MLEs of the parameters of a Gaussian distribution. After assigning the solution parameters to $\boldsymbol{\theta}^{(t)}$ and increasing the iteration counter t by 1, the steps (4.65), (4.66), (4.67), and (4.68) are repeated until convergence is reached. Convergence of the EM algorithm is very sensitive to the choice of initial parameters. It is therefore recommended to try various different starting conditions. For a further discussion of the theoretical and practical aspects of the EM algorithm we refer to [122].

■ **Example 4.69 (Clustering via EM)** We return to the data in Example 4.59, depicted in Figure 4.60, and adopt the model that the data comes from a mixture of three bivariate Gaussian distributions.

The Python code below implements the EM procedure described in Algorithm 4.64. The initial mean vectors $\{\boldsymbol{\mu}_k\}$ of the bivariate Gaussian distributions are chosen (from visual inspection) to lie roughly in the middle of each cluster, in this case these are $[-2, -3]^\top$, $[-4, 1]^\top$, and $[0, -1]^\top$. The corresponding covariance matrices are initially chosen as identity matrices, which is appropriate given the observed spread of the data in Figure 4.60. Finally, the initial weights are $1/3, 1/3, 1/3$. For simplicity, the algorithm stops after 100 iterations, which in this case is more than enough to guarantee convergence. The code and data are available from the book's website in the GitHub folder **Chapter4**.

EMclust.py

```
import numpy as np
from scipy.stats import multivariate_normal

Xmat = np.genfromtxt('clusterdata.csv', delimiter=',')
K = 3
n, D = Xmat.shape

W = np.array([[1/3,1/3,1/3]])
M  = np.array([[-2.0,-4,0],[-3,1,-1]], dtype=np.float32)
# Note that if above *all* entries were written as integers,
# M would be defined to be of integer type, which will give
# the wrong answer

C = np.zeros((3,2,2))

C[:,0,0] = 1
C[:,1,1] = 1

p = np.zeros((3,300))

for i in range(0,100):

#E-step
    for k in range(0,K):
        mvn = multivariate_normal( M[:,k].T, C[k,:,:] )
        p[k,:] = W[0,k]*mvn.pdf(Xmat)

# M Step
    p = (p/sum(p,0))   #normalize
    W = np.mean(p,1).reshape(1,3)
```

```
for k in range(0,K):
    M[:,k] = (Xmat.T @ p[k,:].T)/sum(p[k,:])
    xm = Xmat.T - M[:,k].reshape(2,1)
    C[k,:,:] = xm @ (xm*p[k,:]).T/sum(p[k,:])
```

The estimated parameters of the mixture distribution are given on the right-hand side of Figure 4.70. After relabeling of the clusters, we can observe a close match with the parameters in Figure 4.60.

The ellipses on the left-hand side of Figure 4.70 show a close match between the 95% probability ellipses[2] of the original Gaussian distributions (in gray) and the estimated ones. A natural way to cluster each point $\boldsymbol{x}_i$ is to assign it to the cluster k for which the conditional probability $p_i(k)$ is maximal (with ties resolved arbitrarily). This gives the clustering of the points into red, green, and blue clusters in the figure.

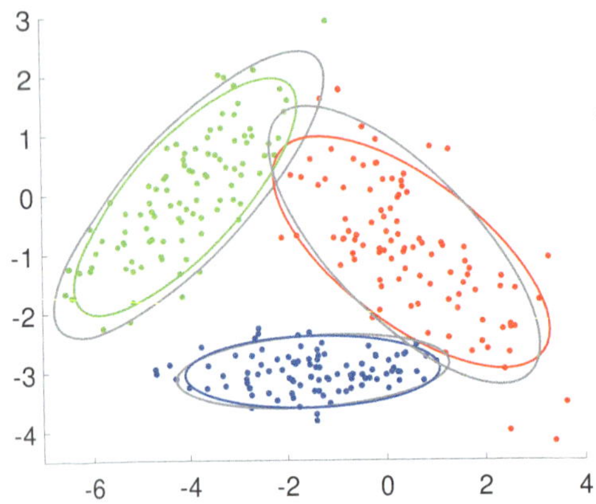

weight	mean vector	covariance matrix
0.33	$\begin{bmatrix} -1.51 \\ -3.01 \end{bmatrix}$	$\begin{bmatrix} 1.75 & 0.03 \\ 0.03 & 0.095 \end{bmatrix}$
0.32	$\begin{bmatrix} -4.08 \\ -0.033 \end{bmatrix}$	$\begin{bmatrix} 1.37 & 0.92 \\ 0.92 & 1.03 \end{bmatrix}$
0.35	$\begin{bmatrix} 0.36 \\ -0.88 \end{bmatrix}$	$\begin{bmatrix} 1.93 & -1.20 \\ -1.20 & 1.44 \end{bmatrix}$

Figure 4.70: The results of the EM clustering algorithm applied to the data depicted in Figure 4.60.

■

As an alternative to the EM algorithm, one can of course use continuous multiextremal optimization algorithms to directly optimize the log-likelihood function $l(\boldsymbol{\theta} \mid \tau) = \ln g(\tau \mid \boldsymbol{\theta})$ in (4.58) over the set Θ of all possible $\boldsymbol{\theta}$. This is done for example in [22], demonstrating superior results to EM when there are few data points. Closer investigation of the likelihood function reveals that there is a hidden problem with any maximum likelihood approach for clustering if Θ is chosen as large as possible — i.e., any mixture distribution is possible. To demonstrate this problem, consider Figure 4.71, depicting the probability density function, $g(\cdot \mid \boldsymbol{\theta})$ of a mixture of two Gaussian distributions, where

$$\boldsymbol{\theta} = [w, \mu_1, \sigma_1^2, \mu_2, \sigma_2^2]^\top$$

is the vector of parameters for the mixture distribution. The log-likelihood function is given by $l(\boldsymbol{\theta} \mid \tau) = \sum_{i=1}^4 \ln g(x_i \mid \boldsymbol{\theta})$, where $x_1, \ldots, x_4$ are the data (indicated by dots in the figure).

[2]For each mixture component, the contour of the corresponding bivariate normal pdf is shown that encloses 95% of the probability mass.

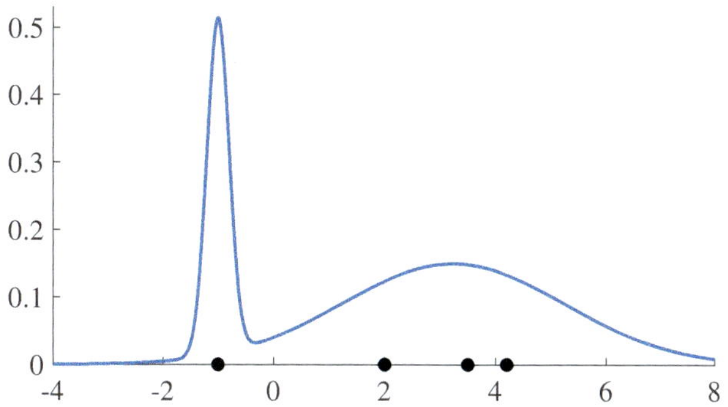

Figure 4.71: Mixture of two Gaussian distributions.

It is clear that by fixing the mixing constant w at 0.25 (say) and centering the first cluster at x_1, one can obtain an arbitrarily large likelihood value by taking the variance of the first cluster to be arbitrarily small. Similarly, for higher dimensional data, by choosing "point" or "line" clusters, or in general "degenerate" clusters, one can make the value of the likelihood infinite. This is a manifestation of the familiar overfitting problem for the training loss that we already encountered in Chapter 2. Thus, the unconstrained maximization of the log-likelihood function is an ill-posed problem, irrespective of the choice of the optimization algorithm!

Two possible solutions to this "overfitting" problem are:

1. Restrict the parameter set Θ in such a way that degenerate clusters (sometimes called spurious clusters) are not allowed.
2. Run the given algorithm and if the solution is degenerate, discard it and run the algorithm afresh. Keep restarting the algorithm until a non-degenerate solution is obtained.

The first approach is usually applied to multiextremal optimization algorithms and the second is used for the EM algorithm.

4.7 Clustering via Vector Quantization

In the previous section we introduced clustering via mixture models, as a form of parametric density estimation (as opposed to the nonparametric density estimation in Section 4.4). The clusters were modeled in a natural way via the latent variables and the EM algorithm provided a convenient way to assign the cluster members. In this section we consider a more heuristic approach to clustering by ignoring the distributional properties of the data. The resulting algorithms tend to scale better with the number of samples n and the dimensionality d.

In mathematical terms, we consider the following clustering (also called data segmentation) problem. Given a collection $\tau = \{\boldsymbol{x}_1, \ldots, \boldsymbol{x}_n\}$ of data points in some d-dimensional space $\mathcal{X}$, divide this data set into K clusters (groups) such that some loss function is minimized. A convenient way to determine these clusters is to first divide up the entire space $\mathcal{X}$, using some distance function $\text{dist}(\cdot, \cdot)$ on this space. A standard choice is the Euclidean (or L_2) distance:

$$\text{dist}(\boldsymbol{x}, \boldsymbol{x}') = \|\boldsymbol{x} - \boldsymbol{x}'\| = \sqrt{\sum_{i=1}^{d} (x_i - x_i')^2}.$$

Other commonly used distance measures on $\mathbb{R}^d$ include the **Manhattan distance**:

$$\sum_{i=1}^{d} |x_i - x_i'|$$

and the **maximum distance**:

$$\max_{i=1,\ldots,d} |x_i - x_i'|.$$

On the set of strings of length d, an often-used distance measure is the **Hamming distance**:

$$\sum_{i=1}^{d} \mathbb{1}\{x_i \neq x_i'\},$$

that is, the number of mismatched characters. For example, the Hamming distance between 010101 and 011010 is 4.

We can partition the space $\mathcal{X}$ into regions as follows: First, we choose K points $\boldsymbol{c}_1, \ldots, \boldsymbol{c}_K$ called **cluster centers**. For each $k = 1, \ldots, K$, let

$$\mathcal{R}_k = \{\boldsymbol{x} \in \mathcal{X} : \text{dist}(\boldsymbol{x}, \boldsymbol{c}_k) \leq \text{dist}(\boldsymbol{x}, \boldsymbol{c}_i) \text{ for all } i \neq k\}$$

be the set of points in $\mathcal{X}$ that lie closer to $\boldsymbol{c}_k$ than any other center. The regions or *cells* $\{\mathcal{R}_k\}$ divide the space $\mathcal{X}$ into what is called a **Voronoi tessellation**. Figure 4.72 shows a Voronoi tessellation of the plane into ten regions, using the Euclidean distance. Note that here the boundaries between the Voronoi cells are straight line segments. In particular, if cell $\mathcal{R}_i$ and $\mathcal{R}_j$ share a border, then a point on this border must satisfy $\|\boldsymbol{x} - \boldsymbol{c}_i\| = \|\boldsymbol{x} - \boldsymbol{c}_j\|$; that is, it must lie on the line that passes through the point $(\boldsymbol{c}_j + \boldsymbol{c}_i)/2$ (that is, the midway point of the line segment between $\boldsymbol{c}_i$ and $\boldsymbol{c}_j$) and be perpendicular to $\boldsymbol{c}_j - \boldsymbol{c}_i$.

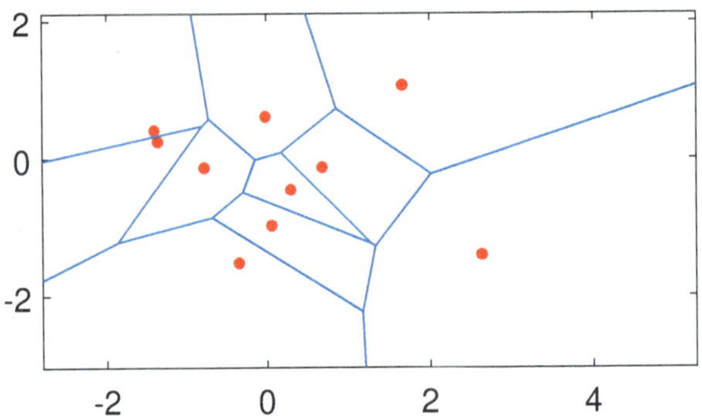

Figure 4.72: A Voronoi tessellation of the plane into ten cells, determined by the (red) centers.

Once the centers (and thus the cells $\{\mathcal{R}_k\}$) are chosen, the points in τ can be clustered according to their nearest center. Points on the boundary have to be treated separately.

This is a moot point for continuous data, as generally no data points will lie exactly on the boundary.

The main remaining issue is how to choose the centers so as to cluster the data in some optimal way. In terms of our (unsupervised) learning framework, we wish to approximate a vector $\boldsymbol{x}$ via one of $\boldsymbol{c}_1, \ldots, \boldsymbol{c}_K$, using a piecewise constant vector-valued function

$$g(\boldsymbol{x} \mid \mathbf{C}) := \sum_{k=1}^{K} \boldsymbol{c}_k \, \mathbb{1}\{\boldsymbol{x} \in \mathcal{R}_k\},$$

where $\mathbf{C}$ is the $d \times K$ matrix $[\boldsymbol{c}_1, \ldots, \boldsymbol{c}_K]$. Thus, $g(\boldsymbol{x} \mid \mathbf{C}) = \boldsymbol{c}_k$ when $\boldsymbol{x}$ falls in region $\mathcal{R}_k$ (we ignore ties). Within this class $\mathcal{G}$ of functions, parameterized by $\mathbf{C}$, our aim is to minimize the training loss. In particular, for the squared-error loss, $\text{Loss}(\boldsymbol{x}, \boldsymbol{x}') = \|\boldsymbol{x} - \boldsymbol{x}'\|^2$, the training loss is

$$\ell_{\tau_n}(g(\cdot \mid \mathbf{C})) = \frac{1}{n} \sum_{i=1}^{n} \|\boldsymbol{x}_i - g(\boldsymbol{x}_i \mid \mathbf{C})\|^2 = \frac{1}{n} \sum_{k=1}^{K} \sum_{\boldsymbol{x} \in \mathcal{R}_k \cap \tau_n} \|\boldsymbol{x} - \boldsymbol{c}_k\|^2. \tag{4.73}$$

Thus, the training loss minimizes the average squared distance between the centers. This framework also combines both the encoding and decoding steps in **vector quantization** [180]. Namely, we wish to "quantize" or "encode" the vectors in τ in such a way that each vector is represented by one of K source vectors $\boldsymbol{c}_1, \ldots, \boldsymbol{c}_K$, such that the loss (4.73) of this representation is minimized.

Most well-known clustering and vector quantization methods update the vector of centers, starting from some initial choice and using iterative (typically gradient-based) procedures. It is important to realize that in this case (4.73) is seen as a function of the centers, where each point $\boldsymbol{x}$ is assigned to the nearest center, thus determining the clusters. It is well known that this type of problem — optimization with respect to the centers — is highly multiextremal and, depending on the initial clusters, gradient-based procedures tend to converge to a *local minimum* rather than a global minimum.

4.7.1 K-Means

One of the simplest methods for clustering is the K-means method. It is an iterative method where, starting from an initial guess for the centers, new centers are formed by taking sample means of the current points in each cluster. The new centers are thus the **centroids** of the points in each cell. Although there exist many different varieties of the K-means algorithm, they are all essentially of the form given in Algorithm 4.74.

Thus, at each iteration, for a given choice of centers, each point in τ is assigned to its nearest center. After all points have been assigned, the centers are recomputed as the centroids of all the points in the current cluster (Line 8). A typical stopping criterion is to stop when the centers no longer change very much. As the algorithm is quite sensitive to the choice of the initial centers, it is prudent to try multiple starting values, e.g., chosen randomly from the bounding box of the data points.

We can see the K-means method as a deterministic (or "hard") version of the probabilistic (or "soft") EM algorithm as follows: Suppose in the EM algorithm we have Gaussian mixtures with a fixed covariance matrix $\boldsymbol{\Sigma}_k = \sigma^2 \mathbf{I}_d$, $k = 1, \ldots, K$, where σ^2 should be

Algorithm 4.74: K**-Means**

Input: Collection of points $\tau = \{\boldsymbol{x}_1, \ldots, \boldsymbol{x}_n\}$, number of clusters K, initial centers $\boldsymbol{c}_1, \ldots, \boldsymbol{c}_K$.

Output: Cluster centers and cells (regions).

while a stopping criterion is not met **do**
- $\mathcal{R}_1, \ldots, \mathcal{R}_K \leftarrow \emptyset$ (empty sets).
- **for** $i = 1$ **to** n **do**
  - $\boldsymbol{d} \leftarrow [\mathrm{dist}(\boldsymbol{x}_i, \boldsymbol{c}_1), \ldots, \mathrm{dist}(\boldsymbol{x}_i, \boldsymbol{c}_K)]$ // distances to centers
  - $k \leftarrow \mathrm{argmin}_j\, d_j$
  - $\mathcal{R}_k \leftarrow \mathcal{R}_k \cup \{\boldsymbol{x}_i\}$ // assign x_i to cluster k
- **for** $k = 1$ **to** K **do**
  - $\boldsymbol{c}_k \leftarrow \dfrac{\sum_{\boldsymbol{x}\in\mathcal{R}_k} \boldsymbol{x}}{|\mathcal{R}_k|}$ // compute the new center as a centroid of points

return $\{\boldsymbol{c}_k\}, \{\mathcal{R}_k\}$

thought of as being infinitesimally small. Consider iteration t of the EM algorithm. Having obtained the expectation vectors $\boldsymbol{\mu}_k^{(t-1)}$ and weights $w_k^{(t-1)}$, $k = 1, \ldots, K$, each point $\boldsymbol{x}_i$ is assigned a cluster label Z_i according to the probabilities $p_i^{(t)}(k)$, $k = 1, \ldots, K$ given in (4.65). But for $\sigma^2 \to 0$ the probability distribution $\{p_i^{(t)}(k)\}$ becomes degenerate, putting all its probability mass on $\mathrm{argmin}_k \|\boldsymbol{x}_i - \boldsymbol{\mu}_k\|^2$. This corresponds to the K-means rule of assigning $\boldsymbol{x}_i$ to its nearest cluster center. Moreover, in the M-step (4.67) each cluster center $\boldsymbol{\mu}_k^{(t)}$ is now updated according to the average of the $\{\boldsymbol{x}_i\}$ that have been assigned to cluster k. We thus obtain the same deterministic updating rule as in K-means.

Example 4.75 (K-means Clustering) We cluster the data from Figure 4.60 via K-means, using the Python implementation below. Note that the data points are stored as a 300 × 2 matrix `Xmat`. We take the same starting centers as in the EM example: $\boldsymbol{c}_1 = [-2, -3]^\top, \boldsymbol{c}_2 = [-4, 1]^\top$, and $\boldsymbol{c}_3 = [0, -1]^\top$. Note also that *squared* Euclidean distances are used in the computations, as these are slightly faster to compute than Euclidean distances (as no square root computations are required) while yielding exactly the same cluster center evaluations.

Kmeans.py

```
import numpy as np
Xmat = np.genfromtxt('clusterdata.csv', delimiter=',')
K = 3
n, D = Xmat.shape
c  = np.array([[-2.0,-4,0],[-3,1,-1]])  # initialize centers
cold = np.zeros(c.shape)
dist2 = np.zeros((K,n))
while np.abs(c - cold).sum() > 0.001:
    cold = c.copy()
    for i in range(0,K):    # compute the squared distances
```

```
        dist2[i,:] = np.sum((Xmat - c[:,i].T)**2, 1)

    label = np.argmin(dist2,0)  # assign the points to nearest centroid
    minvals = np.amin(dist2,0)
    for i in range(0,K):   # recompute the centroids
      c[:,i] = np.mean(Xmat[np.where(label == i),:],1).reshape(1,2)

print('Loss = {:3.3f}'.format(minvals.mean()))
```

```
Loss = 2.288
```

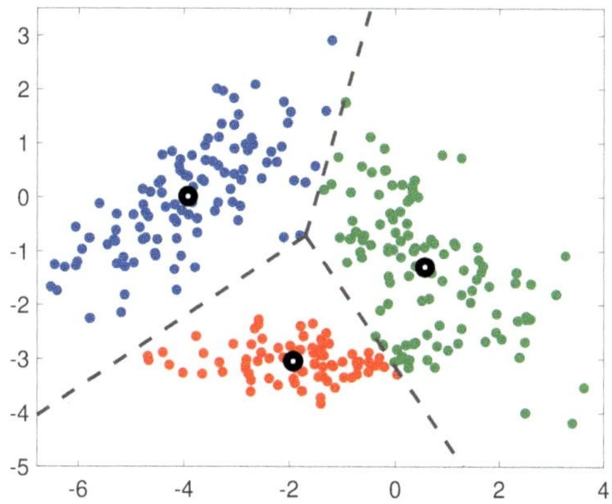

Figure 4.76: Results of the K-means algorithm applied to the data in Figure 4.60. The thick black circles are the centroids and the dotted lines define the cell boundaries.

The algorithm found the following cluster centers: $\boldsymbol{c}_1 = [-1.9286, -3.0416]^\top$, $\boldsymbol{c}_2 = [-3.9237, 0.0131]^\top$, and $\boldsymbol{c}_3 = [0.5611, -1.2980]^\top$, giving the clustering depicted in Figure 4.76. The corresponding loss (4.73) was found to be 2.288. ■

4.7.2 Clustering via Continuous Multiextremal Optimization

As already mentioned, the exact minimization of the loss function (4.73) is difficult to accomplish via standard local search methods such as gradient descent, as the function is highly multimodal. However, nothing is preventing us from using global optimization methods such as the CE or SCO methods discussed in Sections 3.4.2 and 3.4.3. ☞ 111

■ **Example 4.77 (Clustering via CE)** We take the same data set as in Example 4.75 and cluster the points via minimization of the loss (4.73) using the CE method. The Python code below is very similar to the code in Example 3.78, except that now we are dealing with a six-dimensional optimization problem. The loss function is implemented in the function `Scluster`, which essentially reuses the squared distance computation of the K-means code in Example 4.75. The CE program typically converges to a loss of 2.287, corresponding to the (global) minimizers $\boldsymbol{c}_1 = [-1.9286, -3.0416]^\top, \boldsymbol{c}_2 = [-3.8681, 0.0456]^\top$, and $\boldsymbol{c}_3 = [0.5880, -1.3526]^\top$, which slightly differs from the local minimizers for the K-means algorithm. ☞ 114

clustCE.py

```
import numpy as np
np.set_printoptions(precision=4)

Xmat = np.genfromtxt('clusterdata.csv', delimiter=',')
K = 3
n, D = Xmat.shape

def Scluster(c):
    n, D = Xmat.shape
    dist2 = np.zeros((K,n))
    cc = c.reshape(D,K)
    for i in range(0,K):
        dist2[i,:] = np.sum((Xmat - cc[:,i].T)**2, 1)
    minvals = np.amin(dist2,0)
    return minvals.mean()

numvar = K*D
mu = np.zeros(numvar)  #initialize centers
sigma = np.ones(numvar)*2
rho = 0.1
N = 500; Nel = int(N*rho); eps = 0.001

func = Scluster
best_trj = np.array(numvar)
best_perf = np.Inf
trj = np.zeros(shape=(N,numvar))

while(np.max(sigma)>eps):
    for i in range(0,numvar):
        trj[:,i] = (np.random.randn(N,1)*sigma[i]+ mu[i]).reshape(N,)
    S = np.zeros(N)
    for i in range(0,N):
        S[i] = func(trj[i])

    sortedids = np.argsort(S) # from smallest to largest
    S_sorted = S[sortedids]
    best_trj = np.array(n)
    best_perf = np.Inf
    eliteids = sortedids[range(0,Nel)]
    eliteTrj = trj[eliteids,:]
    mu = np.mean(eliteTrj,axis=0)
    sigma = np.std(eliteTrj,axis=0)

    if(best_perf>S_sorted[0]):
        best_perf = S_sorted[0]
        best_trj = trj[sortedids[0]]

print(best_perf)
print(best_trj.reshape(2,3))
```

```
2.2874901831572947
[[-3.9238 -1.8477  0.5895]
 [ 0.0134 -3.0292 -1.2442]]
```

■

4.8 Hierarchical Clustering

It is sometimes useful to determine data clusters in a hierarchical manner; an example is the construction of evolutionary relationships between animal species. Establishing a hierarchy of clusters can be done in a bottom-up or a top-down manner. In the bottom-up approach, also called **agglomerative clustering**, the data points are merged in larger and larger clusters until all the points have been merged into a single cluster. In the top-down or **divisive clustering** approach, the data set is divided up into smaller and smaller clusters. The left panel of Figure 4.78 depicts a hierarchy of clusters.

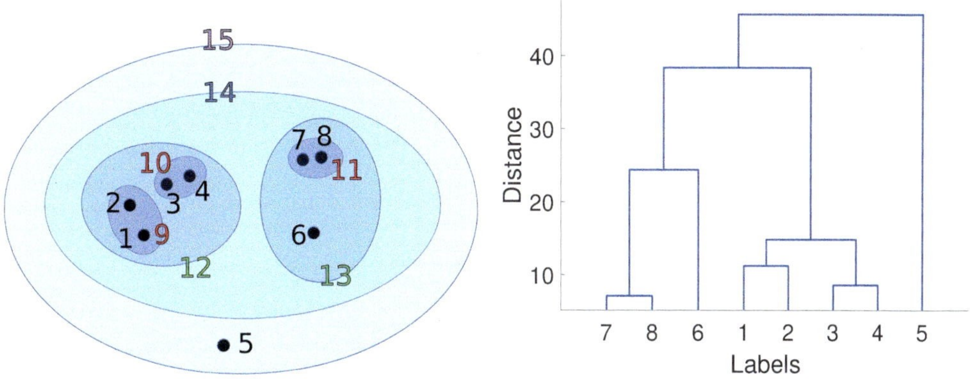

Figure 4.78: Left: a cluster hierarchy of 15 clusters. Right: the corresponding dendrogram.

In Figure 4.78, each cluster is given a cluster identifier. At the lowest level are clusters comprised of the original data points (identifiers $1, \ldots, 8$). The union of clusters 1 and 2 form a cluster with identifier 9, and the union of 3 and 4 form a cluster with identifier 10. In turn the union of clusters 9 and 10 constitutes cluster 12, and so on.

The right panel of Figure 4.78 shows a convenient way to visualize cluster hierarchies using a **dendrogram** (from the Greek *dendro* for tree). A dendrogram not only summarizes how clusters are merged or split, but also shows the distance between clusters, here on the vertical axis. The horizontal axis shows which cluster each data point (label) belongs to.

Many different types of hierarchical clustering can be performed, depending on how the distance is defined between two data points and between two clusters. Denote the data set by $\mathcal{X} = \{\boldsymbol{x}_i, i = 1, \ldots, n\}$. As in Section 4.7, let $\text{dist}(\boldsymbol{x}_i, \boldsymbol{x}_j)$ be the distance between data points $\boldsymbol{x}_i$ and $\boldsymbol{x}_j$. The default choice is the Euclidean distance $\text{dist}(\boldsymbol{x}_i, \boldsymbol{x}_j) = \|\boldsymbol{x}_i - \boldsymbol{x}_j\|$.

Let $\mathcal{I}$ and $\mathcal{J}$ be two disjoint subsets of $\{1, \ldots, n\}$. These sets correspond to two disjoint subsets (that is, clusters) of $\mathcal{X}$: $\{\boldsymbol{x}_i, i = \mathcal{I}\}$ and $\{\boldsymbol{x}_j, j = \mathcal{J}\}$. We denote the distance between these two clusters by $d(\mathcal{I}, \mathcal{J})$. By specifying the function d, we indicate how the clusters are linked. For this reason it is also referred to as the **linkage** criterion. We give a number of examples:

- **Single linkage.** The closest distance between the clusters.

$$d_{\min}(\mathcal{I}, \mathcal{J}) := \min_{i \in \mathcal{I}, j \in \mathcal{J}} \text{dist}(\boldsymbol{x}_i, \boldsymbol{x}_j).$$

- **Complete linkage.** The furthest distance between the clusters.

$$d_{\max}(\mathcal{I}, \mathcal{J}) := \max_{i \in \mathcal{I}, j \in \mathcal{J}} \operatorname{dist}(\boldsymbol{x}_i, \boldsymbol{x}_j).$$

- **Group average.** The mean distance between the clusters. Note that this depends on the cluster sizes.

$$d_{\text{avg}}(\mathcal{I}, \mathcal{J}) := \frac{1}{|\mathcal{I}|\,|\mathcal{J}|} \sum_{i \in \mathcal{I}} \sum_{j \in \mathcal{J}} \operatorname{dist}(\boldsymbol{x}_i, \boldsymbol{x}_j).$$

For these linkage criteria, $\mathcal{X}$ is usually assumed to be $\mathbb{R}^d$ with the Euclidean distance.

Another notable measure for the distance between clusters is *Ward's minimum variance linkage* criterion. Here, the distance between clusters is expressed as the additional amount of "variance" (expressed in terms of the sum of squares) that would be introduced if the two clusters were merged. More precisely, for any set $\mathcal{K}$ of indices (labels) let $\overline{\boldsymbol{x}}_{\mathcal{K}} = \sum_{k \in \mathcal{K}} \boldsymbol{x}_k / |\mathcal{K}|$ denote its corresponding cluster mean. Then

$$d_{\text{Ward}}(\mathcal{I}, \mathcal{J}) := \sum_{k \in \mathcal{I} \cup \mathcal{J}} \|\boldsymbol{x}_k - \overline{\boldsymbol{x}}_{\mathcal{I} \cup \mathcal{J}}\|^2 - \left(\sum_{i \in \mathcal{I}} \|\boldsymbol{x}_i - \overline{\boldsymbol{x}}_{\mathcal{I}}\|^2 + \sum_{j \in \mathcal{J}} \|\boldsymbol{x}_j - \overline{\boldsymbol{x}}_{\mathcal{J}}\|^2 \right). \tag{4.79}$$

It can be shown (see Exercise 18) that the Ward linkage depends only on the cluster means and the cluster sizes for $\mathcal{I}$ and $\mathcal{J}$:

$$d_{\text{Ward}}(\mathcal{I}, \mathcal{J}) = \frac{|\mathcal{I}|\,|\mathcal{J}|}{|\mathcal{I}| + |\mathcal{J}|} \|\overline{\boldsymbol{x}}_{\mathcal{I}} - \overline{\boldsymbol{x}}_{\mathcal{J}}\|^2.$$

> In software implementations, the Ward linkage function is often rescaled by multiplying it by a factor of 2. In this way, the distance between one-point clusters $\{\boldsymbol{x}_i\}$ and $\{\boldsymbol{x}_j\}$ is the *squared* Euclidean distance $\|\boldsymbol{x}_i - \boldsymbol{x}_j\|^2$.

Having chosen a distance on $\mathcal{X}$ and a linkage criterion, a general agglomerative clustering algorithm proceeds in the following "greedy" manner.

Algorithm 4.80: Greedy Agglomerative Clustering

Input: Distance function dist, linkage function d, number of clusters K.
Output: The label sets for the tree.

Initialize the set of cluster identifiers: $\mathcal{I} = \{1, \ldots, n\}$.
Initialize the corresponding label sets: $\mathcal{L}_i = \{i\}$, $i \in \mathcal{I}$.
Initialize a distance matrix $\mathbf{D} = [d_{ij}]$ with $d_{ij} = d(\{i\}, \{j\})$.
for $k = n + 1$ **to** $2n - K$ **do**
  Find i and $j > i$ in $\mathcal{I}$ such that d_{ij} is minimal.
  Create a new label set $\mathcal{L}_k := \mathcal{L}_i \cup \mathcal{L}_j$.
  Add the new identifier k to $\mathcal{I}$ and remove the old identifiers i and j from $\mathcal{I}$.
  Update the distance matrix $\mathbf{D}$ with respect to the identifiers i, j, and k.
return $\mathcal{L}_i, i = 1, \ldots, 2n - K$

Initially, the distance matrix $\mathbf{D}$ contains the (linkage) distances between the one-point clusters containing one of the data points $\boldsymbol{x}_1, \ldots, \boldsymbol{x}_n$, and hence with identifiers $1, \ldots, n$. Finding the shortest distance amounts to a table lookup in $\mathbf{D}$. When the closest clusters are found, they are merged into a new cluster, and a new identifier k (the smallest positive integer that has not yet been used as an identifier) is assigned to this cluster. The old identifiers i and j are removed from the cluster identifier set $\mathcal{I}$. The matrix $\mathbf{D}$ is then updated by adding a k-th column and row that contain the distances between k and any $m \in \mathcal{I}$. This updating step could be computationally quite costly if the cluster sizes are large and the linkage distance between the clusters depends on all points within the clusters. Fortunately, for many linkage functions, the matrix $\mathbf{D}$ can be updated in an efficient manner.

Suppose that at some stage in the algorithm, clusters $\mathcal{I}$ and $\mathcal{J}$, with identifiers i and j, respectively, are merged into a cluster $\mathcal{K} = \mathcal{I} \cup \mathcal{J}$ with identifier k. Let $\mathcal{M}$, with identifier m, be a previously assigned cluster. An update rule of the linkage distance d_{km} between $\mathcal{K}$ and $\mathcal{M}$ is called a **Lance–Williams** update if it can be written in the form

$$d_{km} = \alpha\, d_{im} + \beta\, d_{jm} + \gamma\, d_{ij} + \delta\, |d_{im} - d_{jm}|,$$

where $\alpha, \ldots, \delta$ depend only on simple characteristics of the clusters involved, such as the number of elements within the clusters. Table 4.81 shows the update constants for a number of common linkage functions. For example, for single linkage, d_{im} is the minimal distance between $\mathcal{I}$ and $\mathcal{M}$, and d_{jm} is the minimal distance between $\mathcal{J}$ and $\mathcal{M}$. The smallest of these is the minimal distance between $\mathcal{K}$ and $\mathcal{M}$. That is,

$$d_{km} = \min\{d_{im}, d_{jm}\} = d_{im}/2 + d_{jm}/2 - |d_{im} - d_{jm}|/2.$$

Table 4.81: Constants for the Lance–Williams update rule for various linkage functions, with n_i, n_j, n_m denoting the number of elements in the corresponding clusters.

Linkage	α	β	γ	δ
Single	$1/2$	$1/2$	0	$-1/2$
Complete	$1/2$	$1/2$	0	$1/2$
Group avg.	$\dfrac{n_i}{n_i + n_j}$	$\dfrac{n_j}{n_i + n_j}$	0	0
Ward	$\dfrac{n_i + n_m}{n_i + n_j + n_m}$	$\dfrac{n_j + n_m}{n_i + n_j + n_m}$	$\dfrac{-n_m}{n_i + n_j + n_m}$	0

In practice, Algorithm 4.80 is run until a single cluster is obtained. Instead of returning the label sets of all $2n - 1$ clusters, a **linkage matrix** is returned that contains the same information. At the end of each iteration (Line 8) the linkage matrix stores the merged labels i and j, as well as the (minimal) distance d_{ij}. Other information such as the number of elements in the merged cluster can also be stored. Dendrograms and cluster labels can be directly constructed from the linkage matrix. In the following example, the linkage matrix is returned by the method `agg_cluster`.

■ **Example 4.82 (Agglomerative Hierarchical Clustering)** The Python program below gives a basic implementation of Algorithm 4.80 using the Ward linkage function. The methods **fcluster** and **dendrogram** from the **scipy** module can be used to identify the labels in a cluster and to draw the corresponding dendrogram.

AggCluster.py

```
import numpy as np
from scipy.spatial.distance import cdist

def update_distances(D,i,j, sizes): # distances for merged cluster
    n = D.shape[0]
    d = np.inf * np.ones(n+1)
    for k in range(n): # Update distances
        d[k] = ((sizes[i]+sizes[k])*D[i,k] +
        (sizes[j]+sizes[k])*D[j,k] -
        sizes[k]*D[i,j])/(sizes[i] + sizes[j] + sizes[k])

    infs =  np.inf * np.ones(n) # array of infinity
    D[i,:],D[:,i],D[j,:],D[:,j] =  infs,infs,infs,infs # deactivate
    new_D = np.inf * np.ones((n+1,n+1))
    new_D[0:n,0:n] = D # copy old matrix into new_D
    new_D[-1,:], new_D[:,-1] = d,d # add new row and column
    return new_D

def agg_cluster(X):
    n = X.shape[0]
    sizes = np.ones(n)
    D = cdist(X, X,metric = 'sqeuclidean') # initialize dist. matrix.
    np.fill_diagonal(D, np.inf * np.ones(D.shape[0]))
    Z = np.zeros((n-1,4))  #linkage matrix encodes hierarchy tree
    for t in range(n-1):
        i,j = np.unravel_index(D.argmin(), D.shape) # minimizer pair
        sizes = np.append(sizes, sizes[i] + sizes[j])
        Z[t,:]=np.array([i, j, np.sqrt(D[i,j]), sizes[-1]])
        D = update_distances(D, i,j, sizes)  # update distance matr.
    return Z

import scipy.cluster.hierarchy as h

X = np.genfromtxt('clusterdata.csv',delimiter=',') # read the data
Z = agg_cluster(X)  # form the linkage matrix

h.dendrogram(Z) # SciPy can produce a dendrogram from Z
# fcluster function assigns cluster ids to all points based on Z
cl = h.fcluster(Z, criterion = 'maxclust', t=3)

import matplotlib.pyplot as plt
plt.figure(2), plt.clf()
cols = ['red','green','blue']
colors = [cols[i-1] for i in cl]
plt.scatter(X[:,0], X[:,1],c=colors)
plt.show()
```

Note that the distance matrix is initialized with the squared Euclidean distance, so that the Ward linkage is rescaled by a factor of 2. Also, note that the linkage matrix stores the square root of the minimal cluster distances rather than the distances themselves. We leave it as an exercise to check that by using these modifications the results agree with the `linkage` method from `scipy`; see Exercise 19. ■

In contrast to the bottom-up (agglomerative) approach to hierarchical clustering, the divisive approach starts with one cluster, which is divided into two clusters that are as "dissimilar" as possible, which can then be further divided, and so on. We can use the same linkage criteria as for agglomerative clustering to divide a parent cluster into two child clusters by *maximizing* the distance between the child clusters. Although it is a natural to try to group together data by separating dissimilar ones as far as possible, the implementation of this idea tends to scale poorly with n. The problem is related to the well-known **max-cut problem**: given an $n \times n$ matrix of positive costs $c_{ij}, i, j \in \{1, \ldots, n\}$, partition the index set $\mathcal{I} = \{1, \ldots, n\}$ into two subsets $\mathcal{J}$ and $\mathcal{K}$ such that the total cost across the sets, that is, $\sum_{j \in \mathcal{J}} \sum_{k \in \mathcal{K}} d_{jk}$, is maximal. If instead we maximize according to the *average* distance, we obtain the group average linkage criterion.

■ **Example 4.83 (Divisive Clustering via CE)** The Python code below is used to divide a small data set (of size 300) into two parts according to maximal group average linkage. It uses a short cross-entropy algorithm similar to the one presented in Example 3.93. Given a vector of probabilities $\{p_i, i = 1, \ldots, n\}$, the algorithm generates an $n \times n$ matrix ☞ 122 of Bernoulli random variables with success probability p_i for column i. For each row, the 0s and 1s divide the index set into two clusters, and the corresponding average linkage distance is computed. The matrix is then sorted row-wise according to these distances. Finally, the probabilities $\{p_i\}$ are updated according to the mean values of the best 10% rows. The process is repeated until the $\{p_i\}$ degenerate to a binary vector. This then presents the (approximate) solution on Figure 4.84.

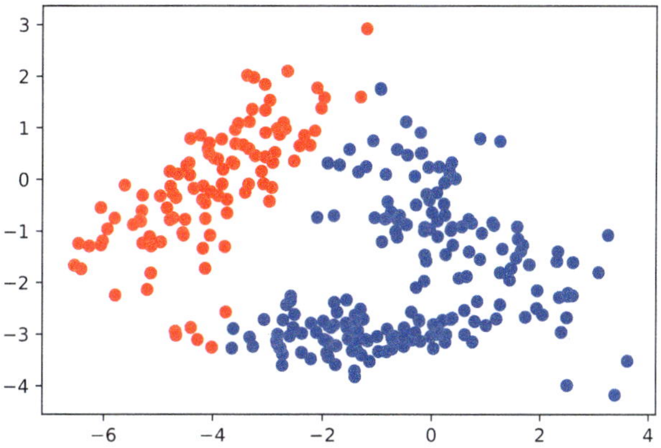

Figure 4.84: Division of the data in Figure 4.60 into two clusters, via the cross-entropy method.

clustCE2.py

```
import numpy as np
from numpy import genfromtxt
from scipy.spatial.distance import squareform
from scipy.spatial.distance import pdist
import matplotlib.pyplot as plt


def S(x,D):
   V1 = np.where(x==0)[0] # {V1,V2} is the partition
   V2 = np.where(x==1)[0]
   tmp = D[V1]
   tmp = tmp[:,V2]
   return np.mean(tmp) # the size of the cut


def maxcut(D,N,eps,rho,alpha):
   n = D.shape[1]
   Ne = int(rho*N)
   p = 1/2*np.ones(n)
   p[0] = 1.0
   while (np.max(np.minimum(p,np.subtract(1,p))) > eps):
       x = np.array(np.random.uniform(0,1,(N,n))<=p, dtype=np.int64)
       sx = np.zeros(N)
       for i in range(N):
           sx[i] = S(x[i],D)

       sortSX = np.flip(np.argsort(sx))
       #print("gamma = ",sx[sortSX[Ne-1]], " best=",sx[sortSX[0]])
       elIds = sortSX[0:Ne]
       elites = x[elIds]
       pnew = np.mean(elites, axis=0)
       p = alpha*pnew + (1.0-alpha)*p

   return np.round(p)

Xmat = genfromtxt('clusterdata.csv', delimiter=',')
n = Xmat.shape[0]
D = squareform(pdist(Xmat))
N = 1000
eps = 10**-2
rho = 0.1
alpha = 0.9

# CE
pout = maxcut(D,N,eps,rho, alpha);

cutval = S(pout,D)
print("cutvalue ",cutval)

#plot
V1 = np.where(pout==0)[0]
xblue = Xmat[V1]
V2 = np.where(pout==1)[0]
```

```
xred = Xmat[V2]
plt.scatter(xblue[:,0],xblue[:,1], c="blue")
plt.scatter(xred[:,0],xred[:,1], c="red")
```

```
cutvalue  4.625207676517948
```

4.9 Principal Component Analysis (PCA)

The main idea of **principal component analysis** (PCA) is to reduce the dimensionality of a data set consisting of many variables. PCA is a *feature reduction* (or *feature extraction*) mechanism, that helps us to handle high-dimensional data with more features than is convenient to interpret.

4.9.1 Motivation: Principal Axes of an Ellipsoid

Consider a d-dimensional normal distribution with mean vector $\mathbf{0}$ and covariance matrix $\boldsymbol{\Sigma}$. The corresponding pdf (see (2.58)) is ☞ 47

$$f(\boldsymbol{x}) = \frac{1}{\sqrt{(2\pi)^n \, |\boldsymbol{\Sigma}|}} \, \mathrm{e}^{-\frac{1}{2}\boldsymbol{x}^\top \boldsymbol{\Sigma}^{-1}\boldsymbol{x}}, \quad \boldsymbol{x} \in \mathbb{R}^d.$$

If we were to draw many iid samples from this pdf, the points would roughly have an *ellipsoid* pattern, as illustrated in Figure 3.6, and correspond to the contours of f: sets of points $\boldsymbol{x}$ such that $\boldsymbol{x}^\top \boldsymbol{\Sigma}^{-1}\boldsymbol{x} = c$, for some $c \geq 0$. In particular, consider the ellipsoid ☞ 77

$$\boldsymbol{x}^\top \boldsymbol{\Sigma}^{-1}\boldsymbol{x} = 1, \quad \boldsymbol{x} \in \mathbb{R}^d. \tag{4.85}$$

Let $\boldsymbol{\Sigma} = \mathbf{B}\mathbf{B}^\top$, where $\mathbf{B}$ is for example the (lower) Cholesky matrix. Then, as explained in Example A.35, the ellipsoid (4.85) can also be viewed as the linear transformation of d-dimensional unit sphere via matrix $\mathbf{B}$. Moreover, the **principal axes** of the ellipsoid can be found via a **singular value decomposition** (SVD) of $\mathbf{B}$ (or $\boldsymbol{\Sigma}$); see Section A.7.3 and Example A.45. In particular, suppose that an SVD of $\mathbf{B}$ is ☞ 543 ☞ 529 ☞ 534

$$\mathbf{B} = \mathbf{U}\mathbf{D}\mathbf{V}^\top \qquad \text{(note that an SVD of } \boldsymbol{\Sigma} \text{ is then } \mathbf{U}\mathbf{D}^2\mathbf{U}^\top).$$

The columns of the matrix $\mathbf{UD}$ correspond to the principal axes of the ellipsoid, and the relative magnitudes of the axes are given by the elements of the diagonal matrix $\mathbf{D}$. If some of these magnitudes are small compared to the others, a reduction in the dimension of the space may be achieved by *projecting* each point $\boldsymbol{x} \in \mathbb{R}^d$ onto the subspace spanned by the main (say $k \ll d$) columns of $\mathbf{U}$ — the so-called **principal components**. Suppose without loss of generality that the first k principal components are given by the first k columns of $\mathbf{U}$, and let $\mathbf{U}_k$ be the corresponding $d \times k$ matrix.

With respect to the standard basis $\{\boldsymbol{e}_i\}$, the vector $\boldsymbol{x} = x_1\boldsymbol{e}_1 + \cdots + x_d\boldsymbol{e}_d$ is represented by the d-dimensional vector $[x_1, \ldots, x_d]^\top$. With respect to the orthonormal basis $\{\boldsymbol{u}_i\}$ formed by the columns of matrix $\mathbf{U}$, the representation of $\boldsymbol{x}$ is $\mathbf{U}^\top\boldsymbol{x}$. Similarly, the projection of any point $\boldsymbol{x}$ onto the subspace spanned by the first k principal vectors is represented by the k-dimensional vector $\mathbf{U}_k^\top \boldsymbol{x}$, with respect to the orthonormal basis formed by the columns of $\mathbf{U}_k$. So, the idea is that if a point $\boldsymbol{x}$ lies close to its projection $\mathbf{U}_k \mathbf{U}_k^\top \boldsymbol{x}$, we may represent it via k numbers instead of d, using the combined features given by the k principal components. See Section A.5 for a review of projections and orthonormal bases. ☞ 521

Example 4.86 (Principal Components) Consider the matrix

$$\boldsymbol{\Sigma} = \begin{bmatrix} 14 & 8 & 3 \\ 8 & 5 & 2 \\ 3 & 2 & 1 \end{bmatrix},$$

which can be written as $\boldsymbol{\Sigma} = \mathbf{B}\mathbf{B}^\top$, with

$$\mathbf{B} = \begin{bmatrix} 1 & 2 & 3 \\ 0 & 1 & 2 \\ 0 & 0 & 1 \end{bmatrix}.$$

Figure 4.87 depicts the ellipsoid $\boldsymbol{x}^\top \boldsymbol{\Sigma}^{-1} \boldsymbol{x} = 1$, which can be obtained by linearly transforming the points on the unit sphere by means of the matrix $\mathbf{B}$.

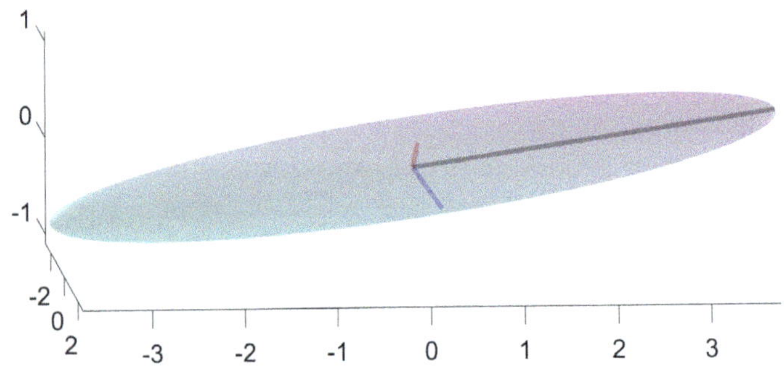

Figure 4.87: A "surfboard" ellipsoid where one principal axis is significantly larger than the other two.

The principal axes and sizes of the ellipsoid are found through a singular value decomposition $\mathbf{B} = \mathbf{U}\mathbf{D}\mathbf{V}^\top$, where $\mathbf{U}$ and $\mathbf{D}$ are

$$\mathbf{U} = \begin{bmatrix} 0.8460 & 0.4828 & 0.2261 \\ 0.4973 & -0.5618 & -0.6611 \\ 0.1922 & -0.6718 & 0.7154 \end{bmatrix} \quad \text{and} \quad \mathbf{D} = \begin{bmatrix} 4.4027 & 0 & 0 \\ 0 & 0.7187 & 0 \\ 0 & 0 & 0.3160 \end{bmatrix}.$$

The columns of $\mathbf{U}$ show the directions of the principal axes of the ellipsoid, and the diagonal elements of $\mathbf{D}$ indicate the relative magnitudes of the principal axes. We see that the first principal component is given by the first column of $\mathbf{U}$, and the second principal component by the second column of $\mathbf{U}$.

The projection of the point $\boldsymbol{x} = [1.052, 0.6648, 0.2271]^\top$ onto the one-dimensional space spanned by the first principal component $\boldsymbol{u}_1 = [0.8460, 0.4972, 0.1922]^\top$ is $\boldsymbol{z} = \boldsymbol{u}_1 \boldsymbol{u}_1^\top \boldsymbol{x} = [1.0696, 0.6287, 0.2429]^\top$. With respect to the basis vector $\boldsymbol{u}_1$, $\boldsymbol{z}$ is represented by the number $\boldsymbol{u}_1^\top \boldsymbol{z} = 1.2643$. That is, $\boldsymbol{z} = 1.2643 \boldsymbol{u}_1$. ■

4.9.2 PCA and Singular Value Decomposition (SVD)

In the setting above, we did not consider any data set drawn from a multivariate pdf f. The whole analysis rested on linear algebra. In **principal component analysis** (PCA) we start with data $\boldsymbol{x}_1, \ldots, \boldsymbol{x}_n$, where each $\boldsymbol{x}$ is d-dimensional. PCA does not require assumptions how the data were obtained, but to make the link with the previous section, we can think of the data as iid draws from a multivariate normal pdf.

Let us collect the data in a matrix $\mathbf{X}$ in the usual way; that is, ☞ 45

$$\mathbf{X} = \begin{bmatrix} x_{11} & x_{12} & \ldots & x_{1d} \\ x_{21} & x_{22} & \ldots & x_{2d} \\ \vdots & \vdots & \vdots & \vdots \\ x_{n1} & x_{n2} & \ldots & x_{nd} \end{bmatrix} = \begin{bmatrix} \boldsymbol{x}_1^\top \\ \boldsymbol{x}_2^\top \\ \vdots \\ \boldsymbol{x}_n^\top \end{bmatrix}.$$

The matrix $\mathbf{X}$ will be the PCA's input. Under this setting, the data consists of points in d-dimensional space, and our goal is to present the data using n feature vectors of dimension $k < d$.

In accordance with the previous section, we assume that underlying distribution of the data has expectation vector $\mathbf{0}$. In practice, this means that before PCA is applied, the data needs to be *centered* by subtracting the *column* mean in every column:

$$x'_{ij} = x_{ij} - \overline{x}_j,$$

where $\overline{x}_j = \frac{1}{n} \sum_{i=1}^n x_{ij}$.

We assume from now on that the data comes from a general d-dimensional distribution with mean vector $\mathbf{0}$ and some covariance matrix $\boldsymbol{\Sigma}$. The covariance matrix $\boldsymbol{\Sigma}$ is by definition equal to the expectation of the random matrix $\boldsymbol{X}\boldsymbol{X}^\top$, and can be estimated from the data $\boldsymbol{x}_1, \ldots, \boldsymbol{x}_n$ via the sample average

$$\widehat{\boldsymbol{\Sigma}} = \frac{1}{n} \sum_{i=1}^n \boldsymbol{x}_i \boldsymbol{x}_i^\top = \frac{1}{n} \mathbf{X}^\top \mathbf{X}.$$

As $\widehat{\boldsymbol{\Sigma}}$ is a covariance matrix, we may conduct the same analysis for $\widehat{\boldsymbol{\Sigma}}$ as we did for $\boldsymbol{\Sigma}$ in the previous section. Specifically, suppose $\widehat{\boldsymbol{\Sigma}} = \mathbf{U}\mathbf{D}^2\mathbf{U}^\top$ is an SVD of $\widehat{\boldsymbol{\Sigma}}$ and let $\mathbf{U}_k$ be the matrix whose columns are the k principal components; that is, the k columns of $\mathbf{U}$ corresponding to the largest diagonal elements in $\mathbf{D}^2$. Note that we have used $\mathbf{D}^2$ instead of $\mathbf{D}$ to be compatible with the previous section. The transformation $\boldsymbol{z}_i = \mathbf{U}_k \mathbf{U}_k^\top \boldsymbol{x}_i$ maps each vector $\boldsymbol{x}_i \in \mathbb{R}^d$ (thus, with d features) to a vector $\boldsymbol{z}_i \in \mathbb{R}^d$ lying in the subspace spanned by the columns of $\mathbf{U}_k$. With respect to this basis, the point $\boldsymbol{z}_i$ has representation $\boldsymbol{z}_i = \mathbf{U}_k^\top(\mathbf{U}_k \mathbf{U}_k^\top \boldsymbol{x}_i) = \mathbf{U}_k^\top \boldsymbol{x}_i \in \mathbb{R}^k$ (thus with k features). The corresponding covariance matrix of the $\boldsymbol{z}_i, i = 1, \ldots, n$ is diagonal. The diagonal elements $\{d_{\ell\ell}\}$ of $\mathbf{D}$ can be interpreted as standard deviations of the data in the directions of the principal components. The quantity $v = \sum_{\ell=1} d_{\ell\ell}^2$ (that is, the trace of $\mathbf{D}^2$) is thus a measure for the amount of variance in the data. The proportion $d_{\ell\ell}^2/v$ indicates how much of the variance in the data is explained by the ℓ-th principal component.

Another way to look at PCA is by considering the question: How can we best project the data onto a k-dimensional subspace in such a way that the total squared distance

☞ 521 between the projected points and the original points is minimal? From Section A.5, we know that any orthogonal projection to a k-dimensional subspace $\mathcal{V}_k$ can be represented by a matrix $\mathbf{U}_k\mathbf{U}_k^\top$, where $\mathbf{U}_k = [\boldsymbol{u}_1, \ldots, \boldsymbol{u}_k]$ and the $\{\boldsymbol{u}_\ell, \ell = 1, \ldots, k\}$ are orthogonal vectors of length 1 that span $\mathcal{V}_k$. The above question can thus be formulated as the minimization program:

$$\min_{\boldsymbol{u}_1,\ldots,\boldsymbol{u}_k} \sum_{i=1}^n \|\boldsymbol{x}_i - \mathbf{U}_k\mathbf{U}_k^\top \boldsymbol{x}_i\|^2. \tag{4.88}$$

Now observe that

$$\begin{aligned}
\frac{1}{n}\sum_{i=1}^n \|\boldsymbol{x}_i - \mathbf{U}_k\mathbf{U}_k^\top \boldsymbol{x}_i\|^2 &= \frac{1}{n}\sum_{i=1}^n (\boldsymbol{x}_i^\top - \boldsymbol{x}_i^\top\mathbf{U}_k\mathbf{U}_k^\top)(\boldsymbol{x}_i - \mathbf{U}_k\mathbf{U}_k^\top \boldsymbol{x}_i) \\
&= \underbrace{\frac{1}{n}\sum_{i=1}^n \|\boldsymbol{x}_i\|^2}_{c} - \frac{1}{n}\sum_{i=1}^n \boldsymbol{x}_i^\top\mathbf{U}_k\mathbf{U}_k^\top\boldsymbol{x}_i = c - \frac{1}{n}\sum_{i=1}^n\sum_{\ell=1}^k \mathrm{tr}(\boldsymbol{x}_i^\top\boldsymbol{u}_\ell\boldsymbol{u}_\ell^\top\boldsymbol{x}_i) \\
&= c - \frac{1}{n}\sum_{\ell=1}^k\sum_{i=1}^n \boldsymbol{u}_\ell^\top\boldsymbol{x}_i\boldsymbol{x}_i^\top\boldsymbol{u}_\ell = c - \sum_{\ell=1}^k \boldsymbol{u}_\ell^\top\widehat{\boldsymbol{\Sigma}}\boldsymbol{u}_\ell,
\end{aligned}$$

☞ 515 where we have used the cyclic property of a trace (Theorem A.7) and the fact that $\mathbf{U}_k\mathbf{U}_k^\top$ can be written as $\sum_{\ell=1}^k \boldsymbol{u}_\ell\boldsymbol{u}_\ell^\top$. It follows that the minimization problem(4.88) is equivalent to the maximization problem

$$\max_{\boldsymbol{u}_1,\ldots,\boldsymbol{u}_k} \sum_{\ell=1}^k \boldsymbol{u}_\ell^\top\widehat{\boldsymbol{\Sigma}}\boldsymbol{u}_\ell. \tag{4.89}$$

This maximum can be at most $\sum_{\ell=1}^k d_{\ell\ell}^2$ and is attained precisely when $\boldsymbol{u}_1, \ldots, \boldsymbol{u}_k$ are the first k principal components of $\widehat{\boldsymbol{\Sigma}}$.

■ **Example 4.90 (Singular Value Decomposition)** The following data set consists of independent samples from the three-dimensional Gaussian distribution with mean vector $\mathbf{0}$ and covariance matrix $\boldsymbol{\Sigma}$ given in Example 4.86:

$$\mathbf{X} = \begin{bmatrix} 3.1209 & 1.7438 & 0.5479 \\ -2.6628 & -1.5310 & -0.2763 \\ 3.7284 & 3.0648 & 1.8451 \\ 0.4203 & 0.3553 & 0.4268 \\ -0.7155 & -0.6871 & -0.1414 \\ 5.8728 & 4.0180 & 1.4541 \\ 4.8163 & 2.4799 & 0.5637 \\ 2.6948 & 1.2384 & 0.1533 \\ -1.1376 & -0.4677 & -0.2219 \\ -1.2452 & -0.9942 & -0.4449 \end{bmatrix}.$$

After replacing $\mathbf{X}$ with its centered version, an SVD $\mathbf{U}\mathbf{D}^2\mathbf{U}^\top$ of $\widehat{\mathbf{\Sigma}} = \mathbf{X}^\top\mathbf{X}/n$ yields the principal component matrix $\mathbf{U}$ and diagonal matrix $\mathbf{D}$:

$$\mathbf{U} = \begin{bmatrix} -0.8277 & 0.4613 & 0.3195 \\ -0.5300 & -0.4556 & -0.7152 \\ -0.1843 & -0.7613 & 0.6216 \end{bmatrix} \quad \text{and} \quad \mathbf{D} = \begin{bmatrix} 3.3424 & 0 & 0 \\ 0 & 0.4778 & 0 \\ 0 & 0 & 0.1038 \end{bmatrix}.$$

We also observe that, apart from the sign of the first column, the principal component matrix $\mathbf{U}$ is similar to that in Example 4.86. Likewise for the matrix $\mathbf{D}$. We see that 97.90% of the total variance is explained by the first principal component. Figure 4.91 shows the projection of the centered data onto the subspace spanned by this principal component, which is then followed by the Python code.

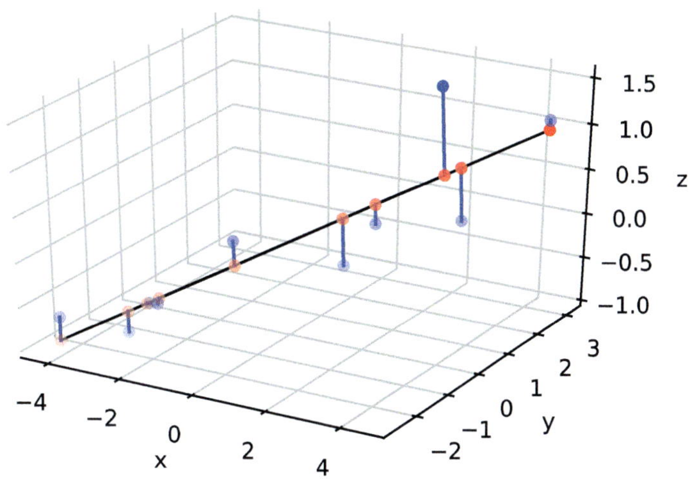

Figure 4.91: Data from the "surfboard" pdf is projected onto the subspace spanned by the largest principal component.

PCAdat.py

```
import numpy as np
import matplotlib.pyplot as plt


# Load the data from the CSV file
X = np.genfromtxt('pcadat.csv', delimiter=',')
n = X.shape[0]

# Gram matrix
X = X - X.mean(axis=0)
G = X.T @ X
U, _, _ = np.linalg.svd(G/n)

# Projected points
Y = X @ np.outer(U[:,0], U[:,0])

# Plotting
fig = plt.figure()
```

```
ax = fig.add_subplot(111, projection='3d')
ax.xaxis.set_pane_color((0, 0, 0, 0))
ax.plot(Y[:,0], Y[:,1], Y[:,2], c='k', linewidth=1)
ax.scatter(X[:,0], X[:,1], X[:,2], c='b')
ax.scatter(Y[:,0], Y[:,1], Y[:,2], c='r')

for i in range(n):
    ax.plot([X[i,0], Y[i,0]], [X[i,1], Y[i,1]], [X[i,2], Y[i,2]], 'b')

ax.set_xlabel('x')
ax.set_ylabel('y')
ax.set_zlabel('z')

plt.show()
```

■

Next is an application of PCA to Fisher's famous iris data set, already mentioned ☞ 2 in Section 1.1, and Exercise 1.5.

■ Example 4.92 (PCA for the Iris Data Set) The iris data set has measurements on four features of the iris plant: sepal length and width, and petal length and width, for a total of 150 specimens. The full data set also contains the species name, but for the purpose of this example we ignore it.

☞ 17 Figure 1.14 shows that there is a significant correlation between the different features. Can we perhaps describe the data using fewer features by taking certain linear combinations of the original features? To investigate this, let us perform a PCA, first centering the data. The following Python code implements the PCA. It is assumed that a CSV file irisX.csv has been made that contains the iris data set (without the species information).

PCAiris.py

```
import numpy as np
np.set_printoptions(precision=4)

X = np.genfromtxt('IrisX.csv',delimiter=',')
n = X.shape[0]
X = X - np.mean(X, axis=0)

[U,D2,UT]= np.linalg.svd((X.T @ X)/n)
print('U = \n', U); print('\n diag(D^2) = ', D2)

z =  U[:,0].T @ X.T
```

```
U =
 [[-0.3614 -0.6566  0.582   0.3155]
 [ 0.0845 -0.7302 -0.5979 -0.3197]
 [-0.8567  0.1734 -0.0762 -0.4798]
 [-0.3583  0.0755 -0.5458  0.7537]]

 diag(D^2) =  [4.2001 0.2411 0.0777 0.0237]
```

The output above shows the principal component matrix (which we called $\mathbf{U}$) as well as the diagonal of matrix $\mathbf{D}^2$. We see that a large proportion of the variance, $4.2001/(4.2001+0.2411+0.0777+0.0237) = 92.46\%$, is explained by the first principal component. Thus, it makes sense to transform each data point $\boldsymbol{x} \in \mathbb{R}^4$ to $\boldsymbol{u}_1^\top \boldsymbol{x} \in \mathbb{R}$. Figure 4.93 shows the kernel density estimate of the transformed data. Interestingly, we see two modes, indicating at least two clusters in the data.

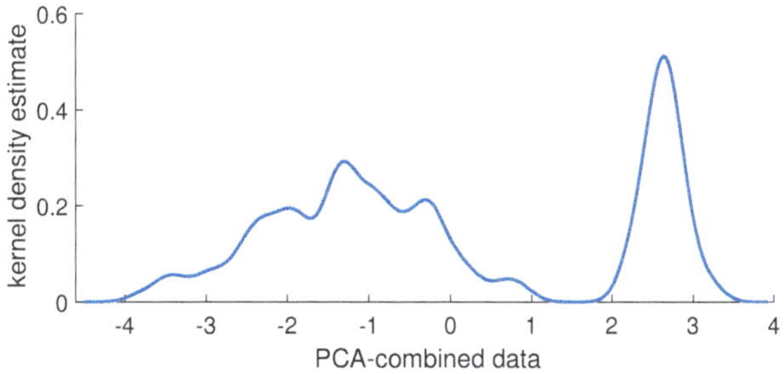

Figure 4.93: Kernel density estimate of the PCA-combined `iris` data.

■

Further Reading

Various information-theoretic measures to quantify uncertainty, including the Shannon entropy and Kullback–Leibler divergence, may be found in [45]. The Fisher information, the prominent information measure in statistics, is discussed in detail in [108]. Akaike's information criterion appeared in [3]. The EM algorithm was introduced in [49] and [122] gives an in-depth treatment. Convergence proofs for the EM algorithm may be found in [31, 184]. A classical reference on kernel density estimation is [158], and [19, 21] are the main references for the theta kernel density estimator. For more on density estimation using the heat equation, see [44]. Kernel density estimation can also be beneficial in smoothing discrete data, see [2, 23]. We also note that when the data is generated by a computer simulation model, then more efficient and bespoke density estimation algorithms exist, see [102, 107] and the references therein.

Theory and applications on finite mixture models may be found in [123]. For more details on clustering applications and algorithms as well as references on data compression, vector quantization, and pattern recognition, we refer to [1, 53, 151, 180]. A useful modification of the K-means algorithm is the *fuzzy K-means* algorithm; see [14]. A popular way to choose the starting positions in K-means is given by the K-means++ heuristic, introduced in [5].

Exercises

1. This exercise is to show that the Fisher information matrix $\mathbf{F}(\boldsymbol{\theta})$ in (4.9) is equal to the matrix $\mathbf{H}(\boldsymbol{\theta})$ in (4.10), in the special case where $f = g(\cdot \mid \boldsymbol{\theta})$, and under the assumption that integration and differentiation orders can be interchanged.

(a) Let $\boldsymbol{h}$ be a vector-valued function and k a real-valued function. Prove the following **quotient rule for differentiation**:

$$\frac{\partial[\boldsymbol{h}(\boldsymbol{\theta})/k(\boldsymbol{\theta})]}{\partial\boldsymbol{\theta}} = \frac{1}{k(\boldsymbol{\theta})}\frac{\partial\boldsymbol{h}(\boldsymbol{\theta})}{\partial\boldsymbol{\theta}} - \frac{1}{k^2(\boldsymbol{\theta})}\frac{\partial k(\boldsymbol{\theta})}{\partial\boldsymbol{\theta}}\boldsymbol{h}(\boldsymbol{\theta})^\top. \tag{4.94}$$

(b) Now take $\boldsymbol{h}(\boldsymbol{\theta}) = \frac{\partial g(\boldsymbol{X}\,|\,\boldsymbol{\theta})}{\partial\boldsymbol{\theta}}$ and $k(\boldsymbol{\theta}) = g(\boldsymbol{X}\,|\,\boldsymbol{\theta})$ in (4.94) and take expectations with respect to $\mathbb{E}_{\boldsymbol{\theta}}$ on both sides to show that

$$-\mathbf{H}(\boldsymbol{\theta}) = \underbrace{\mathbb{E}_{\boldsymbol{\theta}}\left[\frac{1}{g(\boldsymbol{X}\,|\,\boldsymbol{\theta})}\frac{\partial\frac{\partial g(\boldsymbol{X}\,|\,\boldsymbol{\theta})}{\partial\boldsymbol{\theta}}}{\partial\boldsymbol{\theta}}\right]}_{\mathbf{A}} - \mathbf{F}(\boldsymbol{\theta}).$$

(c) Finally show that $\mathbf{A}$ is the zero matrix.

2. Plot the mixture of $\mathcal{N}(0,1)$, $\mathcal{U}(0,1)$, and $\mathsf{Exp}(1)$ distributions, with weights $w_1 = w_2 = w_3 = 1/3$.

3. Denote the pdfs in Exercise 2 by f_1, f_2, f_3, respectively. Suppose that X is simulated via the two-step procedure: First, draw Z from $\{1,2,3\}$, then draw X from f_Z. How likely is it that the outcome $x = 0.5$ of X has come from the uniform pdf f_2?

4. Simulate an iid training set of size 100 from the $\mathsf{Gamma}(2.3, 0.5)$ distribution, and implement the Fisher scoring method in Example 4.22 to find the maximum likelihood estimate. Plot the true and approximate pdfs.

5. Let $\mathcal{T} = \{\boldsymbol{X}_1,\ldots,\boldsymbol{X}_n\}$ be iid data from a pdf $g(\boldsymbol{x}\,|\,\boldsymbol{\theta})$ with Fisher matrix $\mathbf{F}(\boldsymbol{\theta})$. Explain why, under the conditions where (4.8) holds,

$$\boldsymbol{S}_{\mathcal{T}}(\boldsymbol{\theta}) := \frac{1}{n}\sum_{i=1}^{n}\boldsymbol{S}(\boldsymbol{X}_i\,|\,\boldsymbol{\theta})$$

for large n has approximately a multivariate normal distribution with expectation vector $\mathbf{0}$ and covariance matrix $\mathbf{F}(\boldsymbol{\theta})/n$.

6. Figure 4.95 shows a Gaussian KDE with bandwidth $\sigma = 0.2$ on the points $-0.5, 0, 0.2, 0.9$, and 1.5. Reproduce the plot in Python. Using the same bandwidth, plot also the KDE for the same data, but now replacing the Gaussian probability density with the uniform density $\mathbb{1}\{-\sigma \leqslant z \leqslant \sigma\}/(2\sigma)$ on $z \in [-\sigma, \sigma]$.

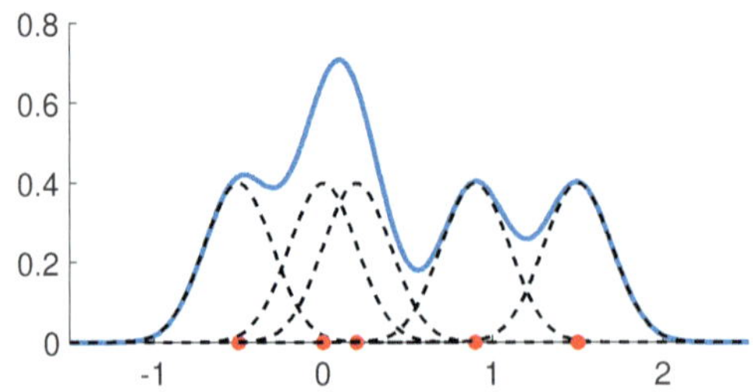

Figure 4.95: The Gaussian KDE (solid line) is the equally weighted mixture of normal pdfs centered around the data and with standard deviation $\sigma = 0.2$ (dashed).

7. The generalization risk for the Gaussian kernel density estimator (4.35) is

$$\ell(g_{\tau_n}(\cdot\,|\,\sigma)) = \mathbb{E}_f \text{Loss}(f(X), g_{\tau_n}(X\,|\,\sigma)) = \|f - g_{\tau_n}(\cdot\,|\,\sigma)\|^2.$$

(a) Show that the minimizer of $\ell(g_{\tau_n}(\cdot\,|\,\sigma))$ is the same as

$$\underset{\sigma>0}{\text{argmin}} \left\{ \frac{1}{n^2} \sum_{i=1}^{n} \sum_{j=1}^{n} \varphi_{2\sigma^2}(\boldsymbol{x}_i - \boldsymbol{x}_j) - 2\mathbb{E}_f g_{\tau_n}(\boldsymbol{X}\,|\,\sigma) \right\}.$$

(b) Show that the n-fold *cross-validation* estimate of the term $\mathbb{E}_f g_{\tau_n}(\boldsymbol{X}\,|\,\sigma)$ is: ☞ 39

$$\frac{2}{n(n-1)} \sum_{i=1}^{n} \sum_{j=i+1}^{n} \varphi_{\sigma^2}(\boldsymbol{x}_i - \boldsymbol{x}_j).$$

8. Silverman's **Gaussian rule of thumb** for selecting the optimal bandwidth parameter in (4.35) assumes that f is the density of the $\mathcal{N}(\overline{\boldsymbol{x}}, s^2\mathbf{I}_d)$ distribution, where $\overline{\boldsymbol{x}}$ and s^2 are the sample mean and variance of the data, respectively. Show that under the Gaussian rule of thumb, formula (4.36) simplifies to

$$s\left(\frac{4}{(d+2)n}\right)^{1/(d+4)}.$$

9. Prove that there exists at least one root of the equation (4.42) when $n > d/4$.

10. For fixed x', the Gaussian kernel function

$$\varphi_t(x - x') := \frac{1}{\sqrt{2\pi t}} \mathrm{e}^{-\frac{1}{2}\frac{(x-x')^2}{t}}$$

is the solution to Fourier's *heat equation*

$$\frac{\partial}{\partial t}\varphi_t(x - x') = \frac{1}{2}\frac{\partial^2}{\partial x^2}\varphi_t(x - x'), \quad x \in \mathbb{R},\ t > 0,$$

with initial condition $\varphi_0(x - x') = \delta(x - x')$ (the Dirac function at x'). Show this. As a consequence, the Gaussian KDE is the solution to the same heat equation, but now with initial condition $\frac{1}{n}\sum_{i=1}^{n}\delta(x - x_i)$. This is the motivation for the theta KDE [21], which is a solution to the same heat equation but now on a *bounded* interval.

11. Define the **Richardson extrapolation density estimator**: ☞ 587

$$\widetilde{g}_{\mathcal{T},c}(\cdot\,|\,\sigma) := \frac{g_{\mathcal{T}}(\cdot\,|\,\sqrt{c}\,\sigma) - c\, g_{\mathcal{T}}(\cdot\,|\,\sigma)}{1 - c},$$

where the constant $c > 0, c \neq 1$.

(a) Show that (4.43) and

$$\frac{\widetilde{g}_{\mathcal{T},c}(\boldsymbol{x}\,|\sqrt{c}\,\sigma) - \widetilde{g}_{\mathcal{T},c}(\boldsymbol{x}\,|\,\sigma)}{c(1 - c^2)\sigma^4/8}$$

are Richardson extrapolation estimators of $\Delta f(\boldsymbol{x})$ and $\Delta^2 f(\boldsymbol{x}) = \sum_{i,j} \frac{\partial^4 f}{\partial x_i^2 \partial x_j^2}(\boldsymbol{x})$, respectively.

(b) Show that

$$\sigma^d(1-c)^2\mathbb{E}\left\|\widetilde{g}_{\mathcal{T}_n,c}(\cdot\,|\,\sigma)-g_{\mathcal{T}_n}(\cdot\,|\,\sigma)\right\|^2=\frac{\psi_c(\mathbf{0})}{n}+\left(1-\frac{1}{n}\right)\mathbb{E}\,\psi_c\left(\frac{\boldsymbol{X}-\boldsymbol{X}'}{\sigma}\right),$$

where $\boldsymbol{X}\ \boldsymbol{X}'$ are iid draws from f.

Hint: The ψ_c in (4.39) can be written as the inner product:

$$\psi_c(\boldsymbol{x}-\boldsymbol{y})=\left\langle\widetilde{\varphi}_c(\cdot-\boldsymbol{x}),\widetilde{\varphi}_c(\cdot-\boldsymbol{y})\right\rangle,$$

where $\widetilde{\varphi}_c(\boldsymbol{x})=\varphi_1(\boldsymbol{x})-\varphi_c(\boldsymbol{x})$ and $\langle\cdot,\cdot\rangle$ is the inner product on the $L^2(\mathbb{R}^d,\mathrm{d}\boldsymbol{x})$ space of square integrable functions; see Definition B.4. ☞ 564

(c) Show that the large-sample integrated variance of the estimator $\widetilde{g}_{\mathcal{T}_n,c_d}(\cdot\,|\,\sigma)$ is the same as the integrated variance of $g_{\mathcal{T}_n}(\cdot\,|\,\sigma)$ as $\sigma=\sigma_n\to 0$ and $n\sigma_n^d\to\infty$ for $n\to\infty$.

(d) Prove that σ_n in (4.36) is not only the (asymptotic) solution to (4.39), but also the solution to the equivalent equation:

$$\mathbb{E}\left\|\widetilde{g}_{\mathcal{T}_n,c_d}(\cdot\,|\,\sigma)-g_{\mathcal{T}_n}(\cdot\,|\,\sigma)\right\|^2=\frac{d+4}{4(4\pi)^{d/2}n\,\sigma^d}+o(n^{-4/(d+4)}). \tag{4.96}$$

Note that the left-hand side of (4.96) is the same as (4.37), except that the true unknown f is replaced with the Richardson extrapolation density estimator $\widetilde{g}_{\mathcal{T}_n,c_d}(\cdot\,|\,\sigma)$. As a matter of fact,

$$\mathbb{E}\left\|\widetilde{g}_{\mathcal{T}_n,c_d}(\cdot\,|\,\sigma)-g_{\mathcal{T}_n}(\cdot\,|\,\sigma)\right\|^2=\mathbb{E}\|\delta_{\mathcal{T}_n}(\cdot\,|\,\sigma)\|^2.$$

12. Derive the asymptotic expansion of the minimizer (4.36) of the mean integrated squared error $\mathbb{E}\|f-g_{\mathcal{T}_n}(\cdot\,|\,\sigma)\|^2$.

13. Let $\widetilde{g}$ be the Richardson extrapolation estimator in Exercise 11.

(a) Show that the expected generalization risk $\mathbb{E}\left\|f-g_{\mathcal{T}_n}(\cdot\,|\,\sigma)\right\|^2$ equals

$$\mathbb{E}\left\|\widetilde{g}_{\mathcal{T}_n,c_d}(\cdot\,|\,\sigma)-g_{\mathcal{T}_n}(\cdot\,|\,\sigma)\right\|^2+\frac{\sigma^6 c_d}{8}\|\Delta^{3/2}f\|^2-\frac{\|f\|^2}{n}+O(\sigma^8+\sigma^2/n).$$

(b) Let $\breve{\sigma}_n$ be the root of the equation (see Theorem 4.38):

$$\mathbb{E}\,\psi_{c_d}\left(\frac{\boldsymbol{X}-\boldsymbol{X}'}{\sigma}\right)=\frac{d}{4n}\psi_{c_d}(\mathbf{0}).$$

Show that the relative error of $\breve{\sigma}_n$ in approximating the bandwidth σ_n in (4.36) satisfies:

$$\frac{\breve{\sigma}_n}{\sigma_n}-1=\frac{(c-1/2)\|\Delta^{3/2}f\|^2}{2(d+4)\|\Delta f\|^2}\,\widetilde{\sigma}_n^2+o(\widetilde{\sigma}_n^2).$$

14. Suppose that f is the pdf of the $\mathcal{N}(\mathbf{0},s^2\mathbf{I}_d)$ distribution and we wish to estimate f via the density estimator (4.35). Define $\xi:=\sigma^2/s^2$.

(a) Show that the MISE satisfies the identity:

$$(4\pi s^2)^{d/2}\mathbb{E}\,\|f - g_{\mathcal{T}_n}(\cdot\,|\,\sigma)\|^2 = 1 + \frac{1/n}{\xi^{d/2}} + \frac{1-1/n}{(1+\xi)^{d/2}} - \frac{2}{(1+\xi/2)^{d/2}}.$$

(b) Use the negative Binomial series:

$$(1+\xi)^{-d/2} = \sum_{k=0}^{\infty} \alpha_k\,\xi^k,$$

where $\alpha_0 := 1$ and

$$\alpha_k := (-1)^k \frac{d(d+2)(d+4)\cdots(d+2(k-1))}{2^k k!}, \qquad k = 1, 2, \ldots.$$

to show that an asymptotic expansion of the square of the minimizer of the MISE $\mathbb{E}\|f - g_{\mathcal{T}_n}(\cdot\,|\,\sigma)\|^2$ is given by the formula:

$$\sigma_n^2 = s^2\left[\frac{4}{(d+2)n}\right]^{2/(d+4)} + \frac{3s^2}{4}\left[\frac{4}{(d+2)n}\right]^{4/(d+4)} + o(n^{-4/(d+4)}).$$

(c) Show that (4.39) is equivalent to solving:

$$\frac{1}{(1+\xi)^{d/2}} - \frac{2}{\left(1+\frac{1+c_d}{2}\xi\right)^{d/2}} + \frac{1}{(1+c_d\xi)^{d/2}} = \frac{d(1-c_d)^2}{4n\xi^{d/2}} + o(n^{-4/(d+4)}).$$

(d) Show that the solution to (4.39) has the asymptotic expansion:

$$s^2\left[\frac{4}{n(d+2)}\right]^{2/(d+4)} + s^2\frac{c_d+1}{2}\left[\frac{4}{n(d+2)}\right]^{4/(d+4)} + o(n^{-4/(d+4)}).$$

15. Using the code in Example 4.44 as a guide, write a Python program to reproduce Table 4.97. This table present a comparison of the mean integrated squared errors (MISE) across 15 data sets using three different kernel density estimation methods. The first method employs a Gaussian kernel with bandwidth selected via the `dpik` function from R's `KernSmooth` package, as described in [177]. The second method applies the theta kernel density estimator, with bandwidth selection performed automatically as detailed in Section 4.5. The third method uses a Gaussian kernel, but with automatic bandwidth selection as outlined in Section 4.4. The first column of Table 4.97 identifies the distribution, as given in Table 4.98. The second column reports the estimated mean integrated squared error (MISE) using Method 1. The third and fourth columns show the ratios of the MISEs obtained from Methods 2 and 3, respectively, to the MISE reported in the second column. In other words, each value in the third and fourth columns represents the MISE of the corresponding method divided by the MISE from Method 1. The last three columns report the bandwidths found by Methods 1, 2, and 3. We see that Methods 2 and 3 yield very similar results and, in most cases, significantly improve on Method 1. For each distribution, each MISE was estimated using 20 independent draws from the distribution. The sample size in each case is 1000.

Table 4.97: Comparison of Gaussian and theta kernel density estimators. Sample size: 1000. Number of repetitions: 20.

Dist.	imseR	qT	qB	bwR	bwT	bwB
1	3.58×10^{-2}	0.189	0.192	0.176	0.024	0.055
2	1.88×10^{-2}	0.449	0.564	0.088	0.023	0.040
3	1.14×10^{-2}	0.863	0.883	0.058	0.022	0.047
4	2.96×10^{-3}	1.020	1.150	0.227	0.049	0.205
5	2.12×10^{-2}	0.424	0.435	0.147	0.021	0.039
6	5.81×10^{-3}	0.889	0.927	0.223	0.036	0.113
7	1.12×10^{-2}	1.370	1.630	0.028	0.020	0.026
8	2.78×10^{-2}	0.064	0.073	0.560	0.009	0.152
9	2.18×10^{-3}	1.180	1.380	0.199	0.040	0.149
10	1.33×10^{-1}	0.084	0.085	0.149	0.014	0.038
11	1.21×10^{-2}	0.528	0.558	0.196	0.028	0.063
12	1.75×10^{-2}	0.045	0.049	3.650	0.002	0.429
13	1.61×10^{-2}	0.040	0.040	9.310	0.001	0.573
14	7.22×10^{-3}	0.065	0.065	47.600	0.001	0.764
15	2.25×10^{-2}	0.480	0.511	0.160	0.022	0.039

Table 4.98: Test pdfs for kernel density estimation [21, 118]. Here $\mathcal{N}(\mu, \sigma^2)$ stands for the Gaussian density with mean μ and variance σ^2.

Name	Pdf
1: claw	$\frac{1}{2}\mathcal{N}(0,1) + \sum_{k=0}^{4} \frac{1}{10}\mathcal{N}\left(\frac{k}{2} - 1, \left(\frac{1}{10}\right)^2\right)$
2: strongly skewed	$\sum_{k=0}^{7} \frac{1}{8}\mathcal{N}\left(3\left(\left(\frac{2}{3}\right)^k - 1\right), \left(\frac{2}{3}\right)^{2k}\right)$
3: kurtotic unimodal	$\frac{2}{3}\mathcal{N}(0,1) + \frac{1}{3}\mathcal{N}\left(0, \left(\frac{1}{10}\right)^2\right)$
4: double claw	$\frac{49}{100}\mathcal{N}\left(-1, \left(\frac{2}{3}\right)^2\right) + \frac{49}{100}\mathcal{N}\left(1, \left(\frac{2}{3}\right)^2\right) + \frac{1}{350}\sum_{k=0}^{6}\mathcal{N}\left(\frac{k-3}{2}, \left(\frac{1}{100}\right)^2\right)$
5: discrete comb	$\frac{2}{7}\sum_{k=0}^{2}\mathcal{N}\left(\frac{12k-15}{7}, \left(\frac{2}{7}\right)^2\right) + \frac{1}{21}\sum_{k=8}^{10}\mathcal{N}\left(\frac{2k}{7}, \left(\frac{1}{21}\right)^2\right)$
6: asymmetric double claw	$\frac{46}{100}\sum_{k=0}^{1}\mathcal{N}\left(2k-1, \left(\frac{2}{3}\right)^2\right) + \sum_{k=1}^{3}\frac{1}{300}\mathcal{N}\left(-\frac{k}{2}, \left(\frac{1}{100}\right)^2\right) + \sum_{k=1}^{3}\frac{7}{300}\mathcal{N}\left(\frac{k}{2}, \left(\frac{7}{100}\right)^2\right)$
7: outlier	$\frac{1}{10}\mathcal{N}(0,1) + \frac{9}{10}\mathcal{N}\left(0, \left(\frac{1}{10}\right)^2\right)$
8: separated bimodal	$\frac{1}{2}\mathcal{N}(-12, \frac{1}{4}) + \frac{1}{2}\mathcal{N}(12, \frac{1}{4})$
9: skewed bimodal	$\frac{3}{4}\mathcal{N}(0,1) + \frac{1}{4}\mathcal{N}\left(\frac{3}{2}, \left(\frac{1}{3}\right)^2\right)$
10: bimodal	$\frac{1}{2}\mathcal{N}\left(0, \left(\frac{1}{10}\right)^2\right) + \frac{1}{2}\mathcal{N}\left(5, 1\right)$
11: asymmetric claw	$\frac{1}{2}\mathcal{N}(0,1) + \sum_{k=-2}^{2}\frac{2^{1-k}}{31}\mathcal{N}\left(k + \frac{1}{2}, \left(\frac{2^{-k}}{10}\right)^2\right)$
12: trimodal	$\frac{1}{3}\sum_{k=0}^{2}\mathcal{N}(80k, (k+1)^4)$
13: 5-modes	$\frac{1}{5}\sum_{k=0}^{4}\mathcal{N}(80k, (k+1)^2)$

Continued on next page

Table 4.98 – continued from previous page

Name	Pdf
14: 10-modes	$\frac{1}{10}\sum_{k=0}^{9}\mathcal{N}(100k,\ (k+1)^2)$
15: smooth comb	$\sum_{k=0}^{5}\frac{2^{5-k}}{63}\mathcal{N}\Big(\frac{65-96/2^k}{21},\ \frac{(32/63)^2}{2^{2k}}\Big)$

16. The theta kernel derives its name from **Jacobi theta function** ϑ_3, defined as

$$\vartheta_3(z,q) = 1 + 2\sum_{k=1}^{\infty} q^{n^2}\cos(2nz), \qquad z, q \in \mathbb{C}.$$

(a) Show that the theta kernel

$$\kappa(x,x') = \sum_{k=-\infty}^{\infty} \mathrm{e}^{-\pi^2 k^2 \sigma^2/2}\cos(\pi k x)\cos(\pi k x')$$

can also be written as

$$\begin{aligned}\kappa(x,x') &= \sum_{k=-\infty}^{\infty} \mathrm{e}^{-2\pi^2 k^2 \sigma^2}\cos(2\pi k(x-x')) \\ &= 1 + 2\sum_{k=1}^{\infty} \mathrm{e}^{-2\pi^2 k^2 \sigma^2}\cos(2\pi k(x-x')) \\ &= \vartheta_3(z,q),\end{aligned}$$

with $z = \pi(x - x')$ and $q = \exp(-2\pi^2\sigma^2)$.
Hint: Use the cosine-product formula (B.27) or, equivalently, the trigonometric identity

$$\cos(a-b) = \cos a \cos b - \sin a \sin b$$

and the facts that the cosine function is even (symmetric around 0) and the sine function is odd.

(b) If X is a continuous random variable with pdf f_X on $\mathbb{R}$, then (under some mild conditions on f_X, see [98]) the random variable $Y = (X \bmod p)$ has pdf

$$f_Y(y) = \sum_{k=-\infty}^{\infty} f_X(y+k), \qquad y \in [0,p).$$

We can use this to simulate random variables from the pdf $\kappa(x,\cdot)$ for each $x \in (0,1)$: First draw $X \sim \mathcal{N}(x,\sigma^2)$ and then "wrap" X onto the interval $[0,1)$ via $Y = (X \bmod 1)$. Prove that Y indeed has pdf $\kappa(x,\cdot)$. Hint: Use Example B.28.

17. Prove Theorem 4.50 by using the following steps.

(a) For each $\boldsymbol{k} \in \mathbb{Z}^d$ define

$$h_{\boldsymbol{k}}(\boldsymbol{x}) := \prod_{j=1}^{d} \cos(\pi k_j x_j), \quad \boldsymbol{x} \in \mathbb{R}^d.$$

☞ 572 Use the identity (B.27) and orthogonality of the cosine functions,

$$\int_0^1 \cos(\pi m u) \cos(\pi n u)\, du = \begin{cases} 1, & n = m = 0, \\ 1/2, & |n| = |m| \neq 0, \\ 0, & \text{otherwise}, \end{cases}$$

to prove that $\{2^{\|\boldsymbol{k}\|_0/2} h_{\boldsymbol{k}}, \boldsymbol{k} \in \mathbb{Z}^d\}$ is a redundant orthonormal set for $L^2([0,1]^d)$; ☞ 565 see Example B.8. Hence, deduce that the integrated squared bias of the theta kernel estimator is:

$$\|f - \mathbb{E} g_{\mathcal{T}_n}(\cdot \mid \sigma)\|^2 = \sum_{\boldsymbol{k} \in \mathbb{Z}^d} \left(\mathrm{e}^{-\|\boldsymbol{k}\|^2 \pi^2 \sigma^2/2} - 1\right)^2 c_{\boldsymbol{k}}^2.$$

(b) Using the identity (B.27), show that

$$\mathbb{E} h_{\boldsymbol{k}}^2(\boldsymbol{X}) = \mathbb{E}\, c_{(\boldsymbol{S}+\boldsymbol{S}')\odot \boldsymbol{k}} = \mathbb{E}\, c_{\boldsymbol{k}+\boldsymbol{S}\odot\boldsymbol{k}},$$

where $S_1, S_1', \ldots, S_d, S_d'$ are iid with Rademacher distribution and $\boldsymbol{X} \sim f$. Hence, conclude that the integrated variance of the theta kernel density estimator is:

$$\mathbb{E}\|g_{\mathcal{T}_n}(\cdot \mid \sigma) - \mathbb{E} g_{\mathcal{T}_n}(\cdot \mid \sigma)\|^2 = \frac{1}{n} \sum_{\boldsymbol{k} \in \mathbb{Z}^d} \mathrm{e}^{-\|\boldsymbol{k}\|^2 \pi^2 \sigma^2} \left(\mathbb{E}\, c_{\boldsymbol{k}+\boldsymbol{S}\odot\boldsymbol{k}} - c_{\boldsymbol{k}}^2\right).$$

(c) Show that an unbiased estimator of $c_{\boldsymbol{k}}^2$ is (4.51).

18. Show that the Ward linkage given in (4.79) is equal to

$$d_{\text{Ward}}(\mathcal{I}, \mathcal{J}) = \frac{|\mathcal{I}|\,|\mathcal{J}|}{|\mathcal{I}| + |\mathcal{J}|} \|\overline{\boldsymbol{x}}_{\mathcal{I}} - \overline{\boldsymbol{x}}_{\mathcal{J}}\|^2.$$

19. Carry out the agglomerative hierarchical clustering of Example 4.82 via the `linkage` method from `scipy.cluster.hierarchy`. Show that the linkage matrices are the same. Give a scatterplot of the data, color coded into $K = 3$ clusters.

20. Suppose that we have the data $\tau_n = \{x_1, \ldots, x_n\}$ in $\mathbb{R}$ and decide to train the two-component Gaussian mixture model

$$g(x \mid \boldsymbol{\theta}) = w_1 \frac{1}{\sqrt{2\pi\sigma_1^2}} \exp\left(-\frac{(x-\mu_1)^2}{2\sigma_1^2}\right) + w_2 \frac{1}{\sqrt{2\pi\sigma_2^2}} \exp\left(-\frac{(x-\mu_2)^2}{2\sigma_2^2}\right),$$

where the parameter vector $\boldsymbol{\theta} = [\mu_1, \mu_2, \sigma_1, \sigma_2, w_1, w_2]^\top$ belongs to the set

$$\Theta = \{\boldsymbol{\theta} : w_1 + w_2 = 1, w_1 \in [0,1], \mu_i \in \mathbb{R}, \sigma_i > 0, \ \forall i\}.$$

Suppose that the training is via the maximum likelihood in (2.48). Show that

$$\sup_{\theta \in \Theta} \frac{1}{n} \sum_{i=1}^{n} \ln g(x_i \mid \theta) = \infty.$$

In other words, find a sequence of values for $\theta \in \Theta$ such that the likelihood grows without bound. How can we restrict the set Θ to ensure that the likelihood remains bounded?

21. A d-dimensional normal random vector $X \sim \mathcal{N}(\mu, \Sigma)$ can be defined via an affine transformation, $X = \mu + \Sigma^{1/2} Z$, of a standard normal random vector $Z \sim \mathcal{N}(\mathbf{0}, \mathbf{I}_d)$, where $\Sigma^{1/2}(\Sigma^{1/2})^\top = \Sigma$. In a similar way, we can define a d-dimensional Student random vector $X \sim \mathsf{t}_\alpha(\mu, \Sigma)$ via a transformation

$$X = \mu + \frac{1}{\sqrt{S}} \Sigma^{1/2} Z, \tag{4.99}$$

where, $Z \sim \mathcal{N}(\mathbf{0}, \mathbf{I}_d)$ and $S \sim \mathsf{Gamma}(\frac{\alpha}{2}, \frac{\alpha}{2})$ are independent, $\alpha > 0$, and $\Sigma^{1/2}(\Sigma^{1/2})^\top = \Sigma$. Note that we obtain the multivariate normal distribution as a limiting case for $\alpha \to \infty$.

(a) Show that the density of the $\mathsf{t}_\alpha(\mathbf{0}, \mathbf{I}_d)$ distribution is given by

$$t_\alpha(x) := \frac{\Gamma((\alpha + d)/2)}{(\pi\alpha)^{d/2}\Gamma(\alpha/2)} \left(1 + \frac{1}{\alpha}\|x\|^2\right)^{-\frac{\alpha+d}{2}}.$$

By the transformation rule (D.37), it follows that the density of $X \sim \mathsf{t}_\alpha(\mu, \Sigma)$ is given by $t_{\alpha,\Sigma}(x - \mu)$, where ☞ 637

$$t_{\alpha,\Sigma}(x) := \frac{1}{|\Sigma^{1/2}|} t_\alpha(\Sigma^{-1/2} x).$$

Hint: Conditional on $S = s$, X has a $\mathcal{N}(\mathbf{0}, \mathbf{I}_d/s)$ distribution.

(b) We wish to fit a $\mathsf{t}_\nu(\mu, \Sigma)$ distribution to given data $\tau = \{x_1, \dots, x_n\}$ in $\mathbb{R}^d$ via the EM method. We use the representation (4.99) and augment the data with the vector $S = [S_1, \dots, S_n]^\top$ of hidden variables. Show that the complete-data likelihood is given by

$$g(\tau, s \mid \theta) = \prod_i \frac{(\alpha/2)^{\alpha/2} s_i^{(\alpha+d)/2-1} \exp(-\frac{s_i}{2}\alpha - \frac{s_i}{2}\|\Sigma^{-1/2}(x_i - \mu)\|^2)}{\Gamma(\alpha/2)(2\pi)^{d/2}|\Sigma^{1/2}|}. \tag{4.100}$$

(c) Show that, as a consequence, conditional on the data τ and parameter θ, the hidden data are mutually independent, and

$$(S_i \mid \tau, \theta) \sim \mathsf{Gamma}\left(\frac{\alpha + d}{2}, \frac{\alpha + \|\Sigma^{-1/2}(x_i - \mu)\|^2}{2}\right), \quad i = 1, \dots, n.$$

(d) At iteration t of the EM algorithm, let $g^{(t)}(s) = g(s \mid \tau, \theta^{(t-1)})$ be the density of the missing data, given the observed data τ and the current parameter guess $\theta^{(t-1)}$.

Verify that the expected complete-data log-likelihood is given by:

$$\mathbb{E}_{g^{(t)}} \ln g(\tau, \boldsymbol{S} \mid \boldsymbol{\theta}) = \frac{n\alpha}{2} \ln \frac{\alpha}{2} - \frac{nd}{2} \ln(2\pi) - n \ln \Gamma\left(\frac{\alpha}{2}\right) - \frac{n}{2} \ln |\boldsymbol{\Sigma}| + \frac{\alpha + d - 2}{2} \sum_{i=1}^{n} \mathbb{E}_{g^{(t)}} \ln S_i - \sum_{i=1}^{n} \frac{\alpha + \|\boldsymbol{\Sigma}^{-1/2}(\boldsymbol{x}_i - \boldsymbol{\mu})\|^2}{2} \mathbb{E}_{g^{(t)}} S_i.$$

Show that

$$\mathbb{E}_{g^{(t)}} S_i = \frac{\alpha^{(t-1)} + d}{\alpha^{(t-1)} + \|(\boldsymbol{\Sigma}^{(t-1)})^{-1/2}(\boldsymbol{x}_i - \boldsymbol{\mu}^{(t-1)})\|^2} =: w_i^{(t-1)}$$

$$\mathbb{E}_{g^{(t)}} \ln S_i = \psi\left(\frac{\alpha^{(t-1)} + d}{2}\right) - \ln\left(\frac{\alpha^{(t-1)} + d}{2}\right) + \ln w_i^{(t-1)},$$

where $\psi := (\ln \Gamma)'$ is *digamma* function.

(e) Finally, show that in the M-step of the EM algorithm $\boldsymbol{\theta}^{(t)}$ is updated from $\boldsymbol{\theta}^{(t-1)}$ as follows:

$$\boldsymbol{\mu}^{(t)} = \frac{\sum_{i=1}^{n} w_i^{(t-1)} \boldsymbol{x}_i}{\sum_{i=1}^{n} w_i^{(t-1)}}$$

$$\boldsymbol{\Sigma}^{(t)} = \frac{1}{n} \sum_{i=1}^{n} w_i^{(t-1)} (\boldsymbol{x}_i - \boldsymbol{\mu}^{(t)})(\boldsymbol{x}_i - \boldsymbol{\mu}^{(t)})^\top,$$

and $\alpha^{(t)}$ is defined implicitly through the solution of the nonlinear equation:

$$\ln\left(\frac{\alpha}{2}\right) - \psi\left(\frac{\alpha}{2}\right) + \psi\left(\frac{\alpha^{(t)} + d}{2}\right) - \ln\left(\frac{\alpha^{(t)} + d}{2}\right) + 1 + \frac{\sum_{i=1}^{n}\left(\ln(w_i^{(t-1)}) - w_i^{(t-1)}\right)}{n} = 0.$$

22. A generalization of both the gamma and inverse-gamma distribution is the **generalized inverse-gamma distribution**, which has density

(4.101) $$f(s) = \frac{(a/b)^{p/2}}{2K_p(\sqrt{ab})} s^{p-1} \mathrm{e}^{-\frac{1}{2}(as + b/s)}, \quad a, b, s > 0, \ p \in \mathbb{R},$$

where K_p is the **modified Bessel function of the second kind**, which can be defined as the integral

(4.102) $$K_p(x) = \int_0^\infty \mathrm{e}^{-x \cosh(t)} \cosh(pt) \, \mathrm{d}t, \quad x > 0, \ p \in \mathbb{R}.$$

We write $S \sim \mathsf{GIG}(a, b, p)$ to denote that S has a pdf of the form (4.101). The function K_p has many interesting properties. Special cases include

$$K_{1/2}(x) = \sqrt{\frac{x\pi}{2}} \mathrm{e}^{-x} \frac{1}{x}$$

$$K_{3/2}(x) = \sqrt{\frac{x\pi}{2}} \mathrm{e}^{-x} \left(\frac{1}{x} + \frac{1}{x^2}\right)$$

$$K_{5/2}(x) = \sqrt{\frac{x\pi}{2}} \mathrm{e}^{-x} \left(\frac{1}{x} + \frac{3}{x^2} + \frac{3}{x^3}\right).$$

More generally, K_p satisfies the recursion

(4.103) $$K_{p+1}(x) = K_{p-1}(x) + \frac{2p}{x} K_p(x).$$

(a) Using the change of variables $\mathrm{e}^z = s\sqrt{a/b}$, show that

$$\int_0^\infty s^{p-1}\mathrm{e}^{-\frac{1}{2}(as+b/s)}\,\mathrm{d}s = 2K_p(\sqrt{ab})(b/a)^{p/2}.$$

(b) Let $S \sim \mathsf{GIG}(a,b,p)$. Show that

(4.104) $$\mathbb{E}S = \frac{\sqrt{b}\,K_{p+1}(\sqrt{ab})}{\sqrt{a}\,K_p(\sqrt{ab})}$$

and

(4.105) $$\mathbb{E}S^{-1} = \frac{\sqrt{a}\,K_{p+1}(\sqrt{ab})}{\sqrt{b}\,K_p(\sqrt{ab})} - \frac{2p}{b}.$$

23. In Exercise 21 we viewed the multivariate Student t_α distribution as a **scale-mixture** of the $\mathcal{N}(\mathbf{0}, \mathbf{I}_d)$ distribution. In this exercise, we consider a similar transformation, but now $\boldsymbol{\Sigma}^{1/2}\boldsymbol{Z} \sim \mathcal{N}(\mathbf{0}, \boldsymbol{\Sigma})$ is not divided but is *multiplied* by $\sqrt{S}$, with $S \sim \mathsf{Gamma}(\alpha/2, \alpha/2)$:

(4.106) $$\boldsymbol{X} = \boldsymbol{\mu} + \sqrt{S}\,\boldsymbol{\Sigma}^{1/2}\,\boldsymbol{Z},$$

where S and $\boldsymbol{Z}$ are independent and $\alpha > 0$.

(a) Show, using Exercise 22, that for $\boldsymbol{\Sigma}^{1/2} = \mathbf{I}_d$ and $\boldsymbol{\mu} = \mathbf{0}$, the random vector $\boldsymbol{X}$ has a d-dimensional **Bessel distribution**, with density:

$$\kappa_\alpha(\boldsymbol{x}) := \frac{2^{1-(\alpha+d)/2}\alpha^{(\alpha+d)/4}\,\|\boldsymbol{x}\|^{(\alpha-d)/2}}{\pi^{d/2}\Gamma(\alpha/2)} K_{(\alpha-d)/2}\left(\|\boldsymbol{x}\|\sqrt{\alpha}\right), \quad \boldsymbol{x} \in \mathbb{R}^d,$$

where K_p is the modified Bessel function of the second kind given in (4.102). We write $\boldsymbol{X} \sim \mathsf{Bessel}_\alpha(\mathbf{0}, \mathbf{I}_d)$. A random vector $\boldsymbol{X}$ is said to have a $\mathsf{Bessel}_\alpha(\boldsymbol{\mu}, \boldsymbol{\Sigma})$ distribution if it can be written in the form (4.106). By the transformation rule (D.37), its density is given by $\frac{1}{\sqrt{|\boldsymbol{\Sigma}|}}\kappa_\alpha(\boldsymbol{\Sigma}^{-1/2}(\boldsymbol{x} - \boldsymbol{\mu}))$. Special instances of the Bessel pdf include:

$$\begin{aligned}
\kappa_2(x) &= \frac{\exp(-\sqrt{2}\,|x|)}{\sqrt{2}} \\
\kappa_4(x) &= \frac{1 + 2\,|x|}{2}\exp(-2\,|x|) \\
\kappa_4(x_1, x_2, x_3) &= \frac{1}{\pi}\exp\left(-2\sqrt{x_1^2 + x_2^2 + x_3^2}\right) \\
\kappa_{d+1}(\boldsymbol{x}) &= \frac{((d+1)/2)^{d/2}\sqrt{\pi}}{(2\pi)^{d/2}\Gamma((d+1)/2)}\exp\left(-\sqrt{d+1}\,\|\boldsymbol{x}\|\right), \quad \boldsymbol{x} \in \mathbb{R}^d.
\end{aligned}$$

Note that k_2 is the (scaled) pdf of the double-exponential or **Laplace** distribution.

(b) Given the data $\tau = \{\boldsymbol{x}_1, \ldots, \boldsymbol{x}_n\}$ in $\mathbb{R}^d$, we wish to fit a Bessel pdf to the data by employing the EM algorithm, augmenting the data with the vector $\boldsymbol{S} = [S_1, \ldots, S_n]^\top$ of missing data. We assume that α is known and $\alpha > d$. Show that conditional on τ (and given $\boldsymbol{\theta}$), the missing data vector $\boldsymbol{S}$ has independent components, with $S_i \sim \mathsf{GIG}(\alpha, b_i, (\alpha - d)/2)$, with $b_i := \|\Sigma^{-1/2}(\boldsymbol{x}_i - \boldsymbol{\mu})\|^2$, $i = 1, \ldots, n$.

(c) At iteration t of the EM algorithm, let $g^{(t)}(\boldsymbol{s}) = g(\boldsymbol{s} \mid \tau, \boldsymbol{\theta}^{(t-1)})$ be the density of the missing data, given the observed data τ and the current parameter guess $\boldsymbol{\theta}^{(t-1)}$. Show that the expected complete-data log-likelihood is given by:

$$Q^{(t)}(\boldsymbol{\theta}) := \mathbb{E}_{g^{(t)}} \ln g(\tau, \boldsymbol{S} \mid \boldsymbol{\theta}) = -\frac{1}{2}\sum_{i=1}^{n} b_i(\boldsymbol{\theta})\, w_i^{(t-1)} + \text{constant}, \tag{4.107}$$

where $b_i(\boldsymbol{\theta}) = \|\Sigma^{-1/2}(\boldsymbol{x}_i - \boldsymbol{\mu})\|^2$ and

$$w_i^{(t-1)} := \frac{\sqrt{\alpha}\, K_{(\alpha-d+2)/2}\left(\sqrt{\alpha\, b_i(\boldsymbol{\theta}^{(t-1)})}\right)}{\sqrt{b_i(\boldsymbol{\theta}^{(t-1)})}\, K_{(\alpha-d)/2}\left(\sqrt{\alpha\, b_i(\boldsymbol{\theta}^{(t-1)})}\right)} - \frac{\alpha - d}{b_i(\boldsymbol{\theta}^{(t-1)})}, \quad i = 1, \ldots, n.$$

(d) From (4.107) derive the M-step of the EM algorithm. That is, show how $\boldsymbol{\theta}^{(t)}$ is updated from $\boldsymbol{\theta}^{(t-1)}$.

24. Consider the ellipsoid $E = \{\boldsymbol{x} \in \mathbb{R}^d : \boldsymbol{x}\Sigma^{-1}\boldsymbol{x} = 1\}$ in (4.85). Let $\mathbf{U}\mathbf{D}^2\mathbf{U}^\top$ be an SVD of $\boldsymbol{\Sigma}$. Show that the linear transformation $\boldsymbol{x} \mapsto \mathbf{U}^\top\mathbf{D}^{-1}\boldsymbol{x}$ maps the points on E onto the unit sphere $\{z \in \mathbb{R}^d : \|z\| = 1\}$.

25. Figure 4.91 shows how the centered "surfboard" data are projected onto the first column of the principal component matrix $\mathbf{U}$. Suppose we project the data instead onto the plane spanned by the first *two* columns of $\mathbf{U}$. What are a and b in the representation $ax_1 + bx_2 = x_3$ of this plane?

26. Figure 4.93 suggests that we can assign each feature vector $\boldsymbol{x}$ in the `iris` data set to one of two clusters, based on the value of $\boldsymbol{u}_1^\top\boldsymbol{x}$, where $\boldsymbol{u}_1$ is the first principal component. Plot the sepal lengths against petal lengths and color the points for which $\boldsymbol{u}_1^\top\boldsymbol{x} < 1.5$ differently to points for which $\boldsymbol{u}_1^\top\boldsymbol{x} \geq 1.5$. To which species of iris do these clusters correspond?

## CHAPTER 5

REGRESSION

Many supervised learning techniques can be gathered under the name "regression". The purpose of this chapter is to explain the mathematical ideas behind regression models and their practical aspects. We analyze the fundamental linear model in detail, and also discuss nonlinear and generalized linear models.

5.1 Introduction

Francis Galton observed in an article in 1889 that the heights of adult offspring are, on the whole, more "average" than the heights of their parents. Galton interpreted this as a degenerative phenomenon, using the term "regression" to indicate this "return to mediocrity". Nowadays, **regression** refers to a broad class of supervised learning techniques where the aim is to predict a quantitative response (output) variable y via a function $g(\boldsymbol{x})$ of an explanatory (input) vector $\boldsymbol{x} = [x_1, \ldots, x_p]^\top$, consisting of p features, each of which can be continuous or discrete. For instance, regression could be used to predict the birth weight of a baby (the response variable) from the weight of the mother, her socio-economic status, and her smoking habits (the explanatory variables).

Let us recapitulate the framework of supervised learning established in Chapter 2. The aim is to find a prediction function g that best guesses[1] what the random output Y ☞ 19 will be for a random input vector $\boldsymbol{X}$. The joint pdf $f(\boldsymbol{x}, y)$ of $\boldsymbol{X}$ and Y is unknown, but a training set $\tau = \{(\boldsymbol{x}_1, y_1), \ldots, (\boldsymbol{x}_n, y_n)\}$ is available, which is thought of as the outcome of a random training set $\mathcal{T} = \{(\boldsymbol{X}_1, Y_1), \ldots, (\boldsymbol{X}_n, Y_n)\}$ of iid copies of $(\boldsymbol{X}, Y)$. Once we have selected a loss function $\text{Loss}(y, \widehat{y})$, such as the **squared-error loss**

$$\text{Loss}(y, \widehat{y}) = (y - \widehat{y})^2, \tag{5.1}$$

then the "best" prediction function g is defined as the one that minimizes the **risk** $\ell(g) = \mathbb{E}\,\text{Loss}(Y, g(\boldsymbol{X}))$. We saw in Section 2.2 that for the squared-error loss this optimal prediction function is the conditional expectation

$$g^*(\boldsymbol{x}) = \mathbb{E}[Y \mid \boldsymbol{X} = \boldsymbol{x}].$$

[1]Recall the mnemonic use of "g" for "guess"

DOI: 10.1201/9781003391173-5

As the squared-error loss is the most widely-used loss function for regression, we will adopt this loss function in most of this chapter.

The optimal prediction function g^* has to be learned from the training set τ by minimizing the training loss

$$\ell_\tau(g) = \frac{1}{n}\sum_{i=1}^{n}(y_i - g(\boldsymbol{x}_i))^2 \tag{5.2}$$

over a suitable class of functions $\mathcal{G}$. Note that in the above definition, the training set τ is assumed to be fixed. For a random training set $\mathcal{T}$, we will write the training loss as $\ell_{\mathcal{T}}(g)$. The function $g_\tau^{\mathcal{G}}$ that minimizes the training loss is the function we use for prediction — the so-called **learner**. When the function class $\mathcal{G}$ is clear from the context, we drop the superscript in the notation.

☞ 21 As we already saw in (2.3), conditional on $\boldsymbol{X} = \boldsymbol{x}$, the response Y can be written as

$$Y = g^*(\boldsymbol{x}) + \varepsilon(\boldsymbol{x}),$$

where $\mathbb{E}\,\varepsilon(\boldsymbol{x}) = 0$. This motivates a standard modeling assumption in supervised learning, in which the responses $Y_1, \ldots, Y_n$, conditional on the explanatory variables:

$$\boldsymbol{X}_1 = \boldsymbol{x}_1, \ldots, \boldsymbol{X}_n = \boldsymbol{x}_n,$$

are assumed to be of the form

$$Y_i = g(\boldsymbol{x}_i) + \varepsilon_i, \quad i = 1, \ldots, n,$$

where the $\{\varepsilon_i\}$ are independent with $\mathbb{E}\,\varepsilon_i = 0$ and $\mathbb{V}\mathrm{ar}\,\varepsilon_i = \sigma^2$ for some function $g \in \mathcal{G}$ and variance σ^2. The above model is usually further specified by assuming that g is completely known up to an unknown parameter vector; that is,

$$Y_i = g(\boldsymbol{x}_i \,|\, \boldsymbol{\beta}) + \varepsilon_i, \quad i = 1, \ldots, n. \tag{5.3}$$

While the model (5.3) is described *conditional* on the explanatory variables, it will be convenient to make one further model simplification, and view (5.3) as if the $\{\boldsymbol{x}_i\}$ were *fixed*, while the $\{Y_i\}$ are random.

⚠ For the remainder of this chapter, we assume that the training feature vectors $\{\boldsymbol{x}_i\}$ are fixed and only the responses are random; that is, $\mathcal{T} = \{(\boldsymbol{x}_1, Y_1), \ldots, (\boldsymbol{x}_n, Y_n)\}$.

The advantage of the model (5.3) is that the problem of estimating the *function* g from the training data is reduced to the (much simpler) problem of estimating the *parameter vector* $\boldsymbol{\beta}$. An obvious disadvantage is that functions of the form $g(\cdot \,|\, \boldsymbol{\beta})$ may not accurately approximate the true unknown g^*. The remainder of this chapter deals with the analysis of models of the form (5.3). In the important case where the function $g(\cdot \,|\, \boldsymbol{\beta})$ is *linear*, the analysis proceeds through the class of linear models. If, in addition, the error terms $\{\varepsilon_i\}$ are assumed to be *Gaussian*, this analysis can be carried out using the rich theory of normal linear models.

5.2 Linear Regression

The most basic regression model involves a linear relationship between the response and a single explanatory variable. In particular, we have measurements $(x_1, y_1), \ldots, (x_n, y_n)$ that lie approximately on a straight line, as in Figure 5.4.

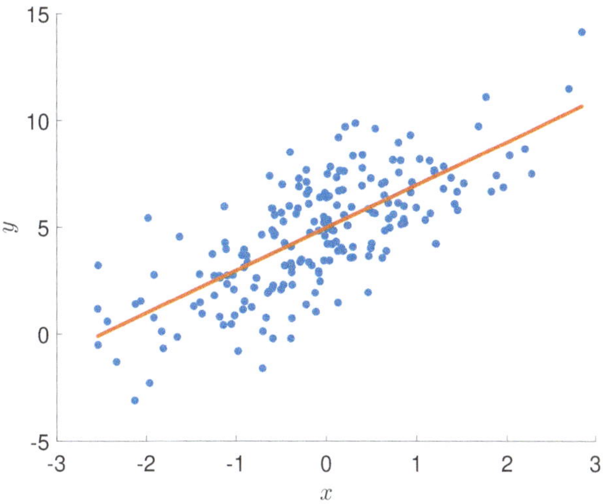

Figure 5.4: Data from a simple linear regression model.

Following the general scheme captured in (5.3), a simple model for these data is that the $\{x_i\}$ are fixed and variables $\{Y_i\}$ are random such that

$$Y_i = \beta_0 + \beta_1 x_i + \varepsilon_i, \quad i = 1, \ldots, n, \tag{5.5}$$

for certain *unknown* parameters β_0 and β_1. The $\{\varepsilon_i\}$ are assumed to be independent with expectation 0 and unknown variance σ^2. The unknown line

$$y = \underbrace{\beta_0 + \beta_1 x}_{g(x \mid \boldsymbol{\beta})}$$

is called the **regression line**. Thus, we view the responses as random variables that would lie exactly on the regression line, were it not for some "disturbance" or "error" term represented by the $\{\varepsilon_i\}$. The extent of the disturbance is modeled by the parameter σ^2. The model in (5.5) is called **simple linear regression**. This model can easily be extended to incorporate more than one explanatory variable, as follows:

Definition 5.6 (Multiple Linear Regression Model): In a **multiple linear regression model** the response Y depends on a d-dimensional explanatory vector $\boldsymbol{x} = [x_1, \ldots, x_d]^\top$, via the linear relationship

$$Y = \beta_0 + \beta_1 x_1 + \cdots + \beta_d x_d + \varepsilon, \tag{5.7}$$

where $\mathbb{E}\,\varepsilon = 0$ and $\mathbb{V}\mathrm{ar}\,\varepsilon = \sigma^2$.

Thus, the data lie approximately on a d-dimensional affine hyperplane

$$y = \underbrace{\beta_0 + \beta_1 x_1 + \cdots + \beta_d x_d}_{g(\boldsymbol{x} \mid \boldsymbol{\beta})},$$

where we define $\boldsymbol{\beta} = [\beta_0, \beta_1, \ldots, \beta_d]^\top$. The function $g(\boldsymbol{x} \mid \boldsymbol{\beta})$ is linear in $\boldsymbol{\beta}$, but not linear in the feature vector $\boldsymbol{x}$, due to the constant β_0. However, *augmenting* the feature space with the constant 1, the mapping $[1, \boldsymbol{x}^\top]^\top \mapsto g(\boldsymbol{x} \mid \boldsymbol{\beta}) := [1, \boldsymbol{x}^\top]\, \boldsymbol{\beta}$ becomes linear
☞ 45 in the feature space and so (5.7) becomes a *linear model* (see Section 2.50). Most software packages for regression include 1 as a feature by default.

Note that in (5.7) we only specified the model for a single pair $(\boldsymbol{x}, Y)$. The model for the training set $\mathcal{T} = \{(\boldsymbol{x}_1, Y_1), \ldots, (\boldsymbol{x}_n, Y_n)\}$ is simply that each Y_i satisfies (5.7) (with $\boldsymbol{x} = \boldsymbol{x}_i$) and that the $\{Y_i\}$ are independent. Setting $\boldsymbol{Y} = [Y_1, \ldots, Y_n]^\top$, we can write the multiple linear regression model for the training data compactly as

$$\boldsymbol{Y} = \mathbf{X}\boldsymbol{\beta} + \boldsymbol{\varepsilon}, \tag{5.8}$$

where $\boldsymbol{\varepsilon} = [\varepsilon_1, \ldots, \varepsilon_n]^\top$ is a vector of iid copies of ε and $\mathbf{X}$ is the **model matrix** given by

$$\mathbf{X} = \begin{bmatrix} 1 & x_{11} & x_{12} & \cdots & x_{1d} \\ 1 & x_{21} & x_{22} & \cdots & x_{2d} \\ \vdots & \vdots & \vdots & \vdots & \vdots \\ 1 & x_{n1} & x_{n2} & \cdots & x_{nd} \end{bmatrix} = \begin{bmatrix} 1 & \boldsymbol{x}_1^\top \\ 1 & \boldsymbol{x}_2^\top \\ \vdots & \vdots \\ 1 & \boldsymbol{x}_n^\top \end{bmatrix}.$$

Example 5.9 (Multiple Linear Regression Model) Figure 5.10 depicts a realization of the multiple linear regression model

$$Y_i = x_{i1} + x_{i2} + \varepsilon_i, \quad i = 1, \ldots, 100,$$

where $\varepsilon_1, \ldots, \varepsilon_{100} \sim_{\text{iid}} \mathcal{N}(0, 1/16)$. The fixed feature vectors (vectors of explanatory variables) $\boldsymbol{x}_i = [x_{i1}, x_{i2}]^\top$, $i = 1, \ldots, 100$ lie in the unit square.

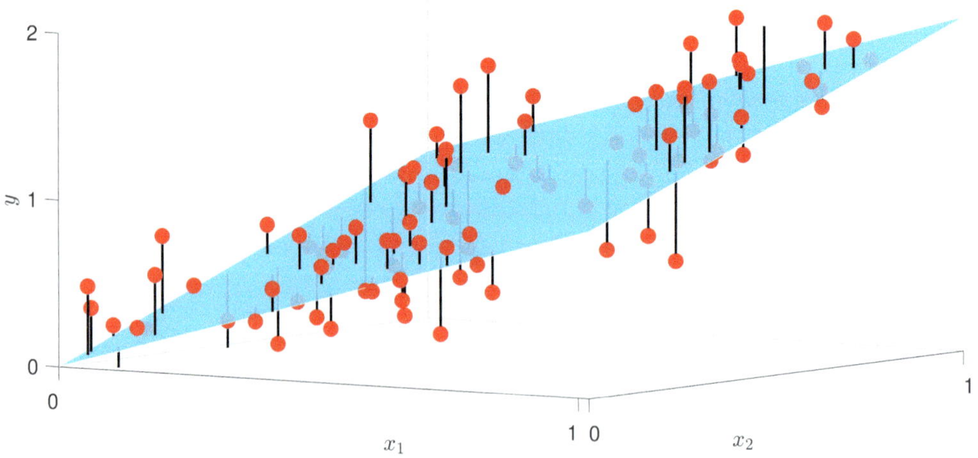

Figure 5.10: Data from a multiple linear regression model.

■

5.3 Analysis via Linear Models

Analysis of data from a linear regression model is greatly simplified through the linear model representation (5.8). In this section we present the main ideas for parameter estimation and model selection for a general linear model of the form

$$\boldsymbol{Y} = \mathbf{X}\boldsymbol{\beta} + \boldsymbol{\varepsilon}, \tag{5.11}$$

where $\mathbf{X}$ is an $n \times p$ matrix, $\boldsymbol{\beta} = [\beta_1, \dots, \beta_p]^\top$ a vector of p parameters, and $\boldsymbol{\varepsilon} = [\varepsilon_1, \dots, \varepsilon_n]^\top$ an n-dimensional vector of independent error terms, with $\mathbb{E}\,\varepsilon_i = 0$ and $\mathbb{V}\mathrm{ar}\,\varepsilon_i = \sigma^2$, $i = 1, \dots, n$. Note that the model matrix $\mathbf{X}$ is assumed to be fixed, and $\boldsymbol{Y}$ and $\boldsymbol{\varepsilon}$ are random. A specific outcome of $\boldsymbol{Y}$ is denoted by $\boldsymbol{y}$ (in accordance with the notation in Section 2.8). ☞ 48

Note that the multiple linear regression model in (5.8) was defined using a different parameterization; in particular, there we used $\boldsymbol{\beta} = [\beta_0, \beta_1, \dots, \beta_d]^\top$. So, when applying the results in the present section to such models, be aware that $p = d + 1$. Also, in this section a feature vector $\boldsymbol{x}$ includes the constant 1, so that $\mathbf{X}^\top = [\boldsymbol{x}_1, \dots, \boldsymbol{x}_n]$.

In the rest of this chapter we will assume that $n > p = d + 1$. The important case of $p \geq n$ is dealt with in Chapter 6. ☞ 261

5.3.1 Parameter Estimation

The linear model $\boldsymbol{Y} = \mathbf{X}\boldsymbol{\beta} + \boldsymbol{\varepsilon}$ contains two unknown parameters, $\boldsymbol{\beta}$ and σ^2, which have to be estimated from the training data τ. To estimate $\boldsymbol{\beta}$, we can repeat exactly the same reasoning used in our recurring polynomial regression Example 2.13 as follows: For a ☞ 26 linear prediction function $g(\boldsymbol{x}) = \boldsymbol{x}^\top\boldsymbol{\beta}$, the (squared-error) training loss can be written as

$$\ell_\tau(g) = \frac{1}{n}\,\|\boldsymbol{y} - \mathbf{X}\boldsymbol{\beta}\|^2,$$

and the optimal learner g_τ minimizes this quantity, leading to the least-squares estimate $\widehat{\boldsymbol{\beta}}$, which satisfies the normal equations

$$\mathbf{X}^\top\mathbf{X}\,\boldsymbol{\beta} = \mathbf{X}^\top\boldsymbol{y}. \tag{5.12}$$

The corresponding training loss can be taken as an estimate of σ^2; that is,

$$\widehat{\sigma^2} = \frac{1}{n}\,\|\,\boldsymbol{y} - \mathbf{X}\widehat{\boldsymbol{\beta}}\,\|^2. \tag{5.13}$$

To justify the latter, note that σ^2 is the second moment of the model errors $\varepsilon_i, i = 1, \dots, n$, in (5.11) and could be estimated via the method of moments (see Section D.12.1) using ☞ 664 the sample average $n^{-1}\sum_i \varepsilon_i^2 = \|\boldsymbol{\varepsilon}\|^2/n = \|\boldsymbol{Y} - \mathbf{X}\boldsymbol{\beta}\|^2/n$, if $\boldsymbol{\beta}$ were known. By replacing $\boldsymbol{\beta}$ with its estimator, we arrive at (5.13). Note that no distributional properties of the $\{\varepsilon_i\}$ were used other than $\mathbb{E}\,\varepsilon_i = 0$ and $\mathbb{V}\mathrm{ar}\,\varepsilon_i = \sigma^2, i = 1, \dots, n$. The vector $\boldsymbol{e} := \boldsymbol{y} - \mathbf{X}\widehat{\boldsymbol{\beta}}$ is called the vector of **residuals** and approximates the (unknown) vector of model errors $\boldsymbol{\varepsilon}$. The quantity $\|\boldsymbol{e}\|^2 = \sum_{i=1}^n e_i^2$ is called the **residual sum of squares** (RSS). Dividing the RSS by $n - p$ gives an unbiased estimate of σ^2, which we call the estimated **residual squared error** (RSE); see Exercise 14.

In terms of the notation given in the summary Table 2.11 for supervised learning, we thus have:

☞ 25

1. The (observed) training data is $\tau = \{\mathbf{X}, \boldsymbol{y}\}$.

2. The function class $\mathcal{G}$ is the class of linear functions of $\boldsymbol{x}$; that is $\mathcal{G} = \{g(\cdot \mid \boldsymbol{\beta}) : \boldsymbol{x} \mapsto \boldsymbol{x}^\top \boldsymbol{\beta},\ \boldsymbol{\beta} \in \mathbb{R}^p\}$.

3. The (squared-error) training loss is $\ell_\tau(g(\cdot \mid \boldsymbol{\beta})) = \|\boldsymbol{y} - \mathbf{X}\boldsymbol{\beta}\|^2/n$.

4. The learner g_τ is given by $g_\tau(\boldsymbol{x}) = \boldsymbol{x}^\top \widehat{\boldsymbol{\beta}}$, where $\widehat{\boldsymbol{\beta}} = \operatorname{argmin}_{\boldsymbol{\beta} \in \mathbb{R}^p} \|\boldsymbol{y} - \mathbf{X}\boldsymbol{\beta}\|^2$.

5. The minimal training loss is $\ell_\tau(g_\tau) = \|\boldsymbol{y} - \mathbf{X}\widehat{\boldsymbol{\beta}}\|^2/n = \widehat{\sigma^2}$.

5.3.2 Model Selection and Prediction

Even if we restrict the learner to be a linear function, there is still the issue of which explanatory variables (features) to include. While including too few features may result in large *approximation error* (underfitting), including too many may result in large *statistical error* (overfitting). As discussed in Section 2.4, we need to select the features which provide the best tradeoff between the approximation and statistical errors, so that the (expected) generalization risk of the learner is minimized. Depending on how the (expected) generalization risk is estimated, there are a number of strategies for feature selection:

☞ 32

1. Use *test data* $\tau' = (\mathbf{X}', \boldsymbol{y}')$ that are obtained independently from the training data τ, to estimate the generalization risk $\mathbb{E}\,\|Y - g_\tau(\boldsymbol{X})\|^2$ via the test loss (2.10). Then choose the collection of features that minimizes the test loss. When there is an abundance of data, part of the data can be reserved as test data, while the remaining data is used as training data.

☞ 24

2. When there is a limited amount of data, we can use *cross-validation* to estimate the expected generalization risk $\mathbb{E}\,\|Y - g_{\mathcal{T}}(\boldsymbol{X})\|^2$ (where $\mathcal{T}$ is a random training set), as explained in Section 2.5.2. This is then minimized over the set of possible choices for the explanatory variables.

☞ 39

3. When one has to choose among many potential explanatory variables, the *klimax method* and *regularized least-squares* become important. Such methods offer another approach to model selection, via the regularization (or homotopy) paths. This will be the topic of Chapter 6.

4. Rather than using computer-intensive techniques, such as the ones above, one can use *theoretical* estimates of the expected generalization risk, such as the in-sample risk, AIC, and BIC, as in Section 2.5, and minimize this to determine a good set of explanatory variables.

☞ 36

5. All of the above approaches do not assume any distributional properties of the error terms $\{\varepsilon_i\}$ in the linear model, other than that they are independent with expectation 0 and variance σ^2. If, however, they are assumed to have a *normal* (Gaussian) distribution, (that is, $\{\varepsilon_i\} \sim_{\text{iid}} \mathcal{N}(0, \sigma^2)$), then the inclusion and exclusion of variables can be decided by means of *hypotheses tests*. This is the classical approach to model selection, and will be discussed in Section 5.4. As a consequence of the central limit theorem, one can use the same approach when the error terms are not necessarily normal, provided their variance is finite and the sample size n is large.

6. Finally, when using a Bayesian approach, comparison of two models can be achieved by computing their so-called *Bayes factor* (see Section 2.9). ☞ 49

All of the above strategies can be thought of as specifications of a simple rule formulated by William of Occam, which can be interpreted as:

> *When presented with competing models, choose the simplest one that explains the data.*

This age-old principle, known as **Occam's razor**, is mirrored in a famous quote of Einstein:

> *Everything should be made as simple as possible, but not simpler.*

In linear regression, the number of parameters or predictors is usually a reasonable measure of the simplicity of the model.

5.3.3 Cross-Validation

We start by considering the n-fold cross-validation, also called **leave-one-out cross-validation**, for the linear model (5.11). We partition the data into n data sets, leaving out precisely one observation per data set, which we then predict based on the $n-1$ remaining observations; see Section 2.5.2 for the general case. Let $\widehat{y}_{-i}$ denote the prediction for ☞ 39 the i-th observation using all the data except y_i. The error in the prediction, $y_i - \widehat{y}_{-i}$, is called a **predicted residual** — in contrast to an ordinary residual, $e_i = y_i - \widehat{y}_i$, which is the difference between an observation and its fitted value $\widehat{y}_i = g_\tau(\boldsymbol{x}_i)$ obtained using the whole sample. In this way, we obtain the collection of predicted residuals $\{y_i - \widehat{y}_{-i}\}_{i=1}^n$ and summarize them through the **predicted residual sum of squares** (**PRESS**):

$$\text{PRESS} = \sum_{i=1}^{n} (y_i - \widehat{y}_{-i})^2.$$

Dividing the PRESS by n gives an estimate of the expected generalization risk.

In general, computing the PRESS is computationally intensive as it involves training and predicting n separate times. For linear models, however, the predicted residuals ☞ 201 can be calculated quickly using only the ordinary residuals and the projection matrix $\mathbf{P} = \mathbf{X}\mathbf{X}^+$ onto the linear space spanned by the columns of the model matrix $\mathbf{X}$ (see (2.21)). The i-th diagonal element $\mathbf{P}_{ii}$ of the projection matrix is called the i-th **leverage**, ☞ 28 and it can be shown that $0 \leq \mathbf{P}_{ii} \leq 1$ (see Exercise 12).

Theorem 5.14 (PRESS for Linear Models): Consider the linear model (5.11), where model matrix $\mathbf{X} \in \mathbb{R}^{n\times p}$ is of full rank with $n > p$. Given an outcome $\boldsymbol{y} = [y_1, \dots, y_n]^\top$ of $\boldsymbol{Y}$, the fitted values can be obtained as $\widehat{\boldsymbol{y}} = \mathbf{P}\boldsymbol{y}$, where $\mathbf{P} = \mathbf{X}\mathbf{X}^+ = \mathbf{X}(\mathbf{X}^\top\mathbf{X})^{-1}\mathbf{X}^\top$ is the projection matrix. If the leverage value $p_i := \mathbf{P}_{ii} \neq 1$ for all $i = 1, \dots, n$, then the predicted residual sum of squares can be written as

$$\text{PRESS} = \sum_{i=1}^{n} \left(\frac{e_i}{1 - p_i}\right)^2,$$

where $e_i = y_i - \widehat{y}_i = y_i - (\mathbf{X}\widehat{\boldsymbol{\beta}})_i$ is the i-th residual.

Proof. It suffices to show that the i-th predicted residual can be written as $y_i - \widehat{y}_{-i} = e_i/(1 - p_i)$. Let $\mathbf{X}_{-i}$ denote the model matrix $\mathbf{X}$ with the i-th row, $\boldsymbol{x}_i^\top$, removed, and define $\boldsymbol{y}_{-i}$ similarly. Then, the least-squares estimate for $\boldsymbol{\beta}$ using all but the i-th observation is $\widehat{\boldsymbol{\beta}}_{-i} = (\mathbf{X}_{-i}^\top\mathbf{X}_{-i})^{-1}\mathbf{X}_{-i}^\top\boldsymbol{y}_{-i}$. Writing $\mathbf{X}^\top\mathbf{X} = \mathbf{X}_{-i}^\top\mathbf{X}_{-i} + \boldsymbol{x}_i\boldsymbol{x}_i^\top$, we have by
☞ 541 the Sherman–Morrison formula

$$(\mathbf{X}_{-i}^\top\mathbf{X}_{-i})^{-1} = (\mathbf{X}^\top\mathbf{X})^{-1} + \frac{(\mathbf{X}^\top\mathbf{X})^{-1}\boldsymbol{x}_i\boldsymbol{x}_i^\top(\mathbf{X}^\top\mathbf{X})^{-1}}{1 - \boldsymbol{x}_i^\top(\mathbf{X}^\top\mathbf{X})^{-1}\boldsymbol{x}_i},$$

where $\boldsymbol{x}_i^\top(\mathbf{X}^\top\mathbf{X})^{-1}\boldsymbol{x}_i = p_i < 1$. Also, $\mathbf{X}_{-i}^\top\boldsymbol{y}_{-i} = \mathbf{X}^\top\boldsymbol{y} - \boldsymbol{x}_i y_i$. Combining all these identities, we have

$$\begin{aligned}
\widehat{\boldsymbol{\beta}}_{-i} &= (\mathbf{X}_{-i}^\top\mathbf{X}_{-i})^{-1}\mathbf{X}_{-i}^\top\boldsymbol{y}_{-i} \\
&= \left((\mathbf{X}^\top\mathbf{X})^{-1} + \frac{(\mathbf{X}^\top\mathbf{X})^{-1}\boldsymbol{x}_i\boldsymbol{x}_i^\top(\mathbf{X}^\top\mathbf{X})^{-1}}{1 - p_i}\right)(\mathbf{X}^\top\boldsymbol{y} - \boldsymbol{x}_i y_i) \\
&= \widehat{\boldsymbol{\beta}} + \frac{(\mathbf{X}^\top\mathbf{X})^{-1}\boldsymbol{x}_i\boldsymbol{x}_i^\top\widehat{\boldsymbol{\beta}}}{1 - p_i} - (\mathbf{X}^\top\mathbf{X})^{-1}\boldsymbol{x}_i y_i - \frac{(\mathbf{X}^\top\mathbf{X})^{-1}\boldsymbol{x}_i p_i y_i}{1 - p_i} \\
&= \widehat{\boldsymbol{\beta}} + \frac{(\mathbf{X}^\top\mathbf{X})^{-1}\boldsymbol{x}_i\boldsymbol{x}_i^\top\widehat{\boldsymbol{\beta}}}{1 - p_i} - \frac{(\mathbf{X}^\top\mathbf{X})^{-1}\boldsymbol{x}_i y_i}{1 - p_i} \\
&= \widehat{\boldsymbol{\beta}} - \frac{(\mathbf{X}^\top\mathbf{X})^{-1}\boldsymbol{x}_i(y_i - \boldsymbol{x}_i^\top\widehat{\boldsymbol{\beta}})}{1 - p_i} = \widehat{\boldsymbol{\beta}} - \frac{(\mathbf{X}^\top\mathbf{X})^{-1}\boldsymbol{x}_i e_i}{1 - p_i}.
\end{aligned}$$

It follows that the predicted value for the i-th observation is given by

$$\widehat{y}_{-i} = \boldsymbol{x}_i^\top\widehat{\boldsymbol{\beta}}_{-i} = \boldsymbol{x}_i^\top\widehat{\boldsymbol{\beta}} - \frac{\boldsymbol{x}_i^\top(\mathbf{X}^\top\mathbf{X})^{-1}\boldsymbol{x}_i e_i}{1 - p_i} = \widehat{y}_i - \frac{p_i e_i}{1 - p_i}.$$

Hence, $y_i - \widehat{y}_{-i} = e_i + p_i e_i/(1 - p_i) = e_i/(1 - p_i)$. □

Since the PRESS formula in Theorem 5.14 is equal to n times the n-fold cross-validation, it is natural to consider if a similar formula exists for K-fold cross-validation.
☞ 39 Using the notation from Section 2.5.2, let $C_k \equiv \{\mathbf{X}_k, \boldsymbol{y}_k\}$ be the test set in the K-th fold and $\widehat{\boldsymbol{\beta}}_{\mathcal{T}_{-k}}$ be the estimate of $\boldsymbol{\beta}$ using all the remaining data (that is, $\mathcal{T}_{-k} \equiv \mathcal{T}\backslash C_k$) as

training set. Note that with this notation, we can write

$$\mathbf{X} = \begin{bmatrix} \mathbf{X}_1 \\ \vdots \\ \mathbf{X}_K \end{bmatrix}, \qquad \boldsymbol{y} = \begin{bmatrix} \boldsymbol{y}_1 \\ \vdots \\ \boldsymbol{y}_K \end{bmatrix},$$

where both matrix $\mathbf{X}_k$ and vector $\boldsymbol{y}_k$ have n_k rows. Then, using the Woodbury identity and similar computations to the one in the proof of Theorem 5.14, we can show that ☞ 541

$$\widehat{\boldsymbol{\beta}}_{\mathcal{T}_{-k}} = \widehat{\boldsymbol{\beta}} - (\mathbf{X}^\top\mathbf{X})^{-1}\mathbf{X}_k^\top(\mathbf{I}_{n_k} - \mathbf{X}_k(\mathbf{X}^\top\mathbf{X})^{-1}\mathbf{X}_k^\top)^{-1}(\boldsymbol{y}_k - \mathbf{X}_k\widehat{\boldsymbol{\beta}}).$$

Additionally, the predictive residuals $\boldsymbol{y}_k - \mathbf{X}_k\widehat{\boldsymbol{\beta}}_{\mathcal{T}_{-k}}$ can be expressed in terms of the ordinary residuals:

$$\boldsymbol{y}_k - \mathbf{X}_k\widehat{\boldsymbol{\beta}}_{\mathcal{T}_{-k}} = (\mathbf{I}_{n_k} - \mathbf{X}_k(\mathbf{X}^\top\mathbf{X})^{-1}\mathbf{X}_k^\top)^{-1}(\boldsymbol{y}_k - \mathbf{X}_k\widehat{\boldsymbol{\beta}}).$$

Therefore, the formula for the K-fold cross-validation simplifies as follows:

$$\mathrm{CV}_K = \sum_{k=1}^{K} \frac{n_k}{n} \frac{\|\boldsymbol{y}_k - \mathbf{X}_k\widehat{\boldsymbol{\beta}}_{\mathcal{T}_{-k}}\|^2}{n_k} = \frac{1}{n}\sum_{k=1}^{K} \|(\mathbf{I}_{n_k} - \mathbf{P}_{n_k})^{-1}(\boldsymbol{y}_k - \mathbf{X}_k\widehat{\boldsymbol{\beta}})\|^2,$$

where $\mathbf{P}_{n_k}$ is the $n_k \times n_k$ diagonal sub-block of $\mathbf{P} = \mathbf{X}(\mathbf{X}^\top\mathbf{X})^{-1}\mathbf{X}^\top$. For example, for $K = 2$ we have

$$\mathbf{P} = \begin{bmatrix} \mathbf{P}_{n_1} & * \\ * & \mathbf{P}_{n_2} \end{bmatrix},$$

and for $K = n$ each $\mathbf{P}_{n_k}$ is simply the k-th diagonal element of $\mathbf{P}$, in agreement with the PRESS formula in Theorem 5.14.

■ **Example 5.15 (Polynomial Regression (cont.))** We return to Example 2.13, where ☞ 26 we estimated the generalization risk for various polynomial prediction functions using independent validation data. Instead, let us estimate the expected generalization risk via cross-validation (thus using only the training set) and apply Theorem 5.14 to compute the PRESS. ☞ 204

polyregpress.py

```
import numpy as np
import matplotlib.pyplot as plt

def generate_data(beta , sig, n):
    u = np.random.rand(n, 1)
    y = u ** np.arange(0, 4) @ beta.reshape(4,1) + (
                                   sig * np.random.randn(n, 1))
    return u, y

np.random.seed(12)
beta = np.array([[10.0, -140, 400, -250]]).T;
sig=5; n = 10**2;
u,y = generate_data(beta,sig,n)
```

```
X = np.ones((n, 1))
K = 12 #maximum number of parameters
press = np.zeros(K+1)
for k in range(1,K):
    if k > 1:
        X = np.hstack((X, u**(k-1))) # add column to matrix
    P = X @ np.linalg.pinv(X) # projection matrix
    e = y - P @ y

    press[k] = np.sum((e/(1-np.diag(P).reshape(n,1)))**2)

plt.plot(press[1:K]/n)
```

The PRESS values divided by $n = 100$ for the constant, linear, quadratic, cubic, and quartic order polynomial regression models are, respectively, 152.487, 56.249, 51.606, 30.999, and 31.634. Hence, the cubic polynomial regression model has the lowest PRESS, indicating that it has the best predictive performance. ■

5.3.4 In-Sample Risk and Akaike Information Criterion

☞ 36 In Section 2.5.1 we introduced the **in-sample risk** as a measure for the accuracy of the prediction function. To recapitulate, given a fixed data set τ with associated response vector $\boldsymbol{y}$ and $n \times p$ matrix of explanatory variables $\mathbf{X}$, the in-sample risk of a prediction function g is defined as

$$\ell_{\text{in}}(g) := \mathbb{E}_{\mathbf{X}} \operatorname{Loss}(Y, g(\boldsymbol{X})), \tag{5.16}$$

where $\mathbb{E}_{\mathbf{X}}$ signifies that the expectation is taken under a different probability model, in which $\boldsymbol{X}$ takes the values $\boldsymbol{x}_1, \ldots, \boldsymbol{x}_n$ with equal probability, and given $\boldsymbol{X} = \boldsymbol{x}_i$ the random variable Y is drawn from the conditional pdf $f(y \mid \boldsymbol{x}_i)$. The difference between the in-sample risk and the training loss is called the **optimism**. For the squared-error ☞ 37 loss, Theorem 2.37 expresses the expected optimism of a learner $g_{\mathcal{T}}$ as two times the average covariance between the predicted values and the responses.

If the conditional variance of the error $Y - g^*(\boldsymbol{X})$ given $\boldsymbol{X} = \boldsymbol{x}$ does not depend on $\boldsymbol{x}$, then the expected in-sample risk of a learner g_τ, averaged over all training sets, has a simple expression:

Theorem 5.17 (Expected In-Sample Risk for Linear Models): Let $\mathbf{X} \in \mathbb{R}^{n \times p}$ be a full-rank model matrix with $n > p$. If $\mathbb{V}\mathrm{ar}[Y - g^*(\boldsymbol{X}) \mid \boldsymbol{X} = \boldsymbol{x}] =: v^2$ does not depend on $\boldsymbol{x}$ and ℓ^* denotes the irreducible risk, then the expected in-sample risk (with respect to the squared-error loss) for a random learner $g_{\mathcal{T}}$ is given by

$$\mathbb{E}_{\mathbf{X}}\, \ell_{\text{in}}(g_{\mathcal{T}}) = \mathbb{E}_{\mathbf{X}}\, \ell_{\mathcal{T}}(g_{\mathcal{T}}) + 2\ell^* p/n. \tag{5.18}$$

Proof. The expected optimism is, by definition, $\mathbb{E}_{\mathbf{X}}[\ell_{\text{in}}(g_{\mathcal{T}}) - \ell_{\mathcal{T}}(g_{\mathcal{T}})]$ which, for the squared-error loss, is equal to $2\ell^* p/n$, using exactly the same reasoning as in Example 2.39. Note that here $\ell^* = v^2$. □

Equation (5.18) is the basis of the following model comparison heuristic: Estimate the irreducible risk $\ell^* = v^2$ via $\widehat{v^2}$, using a model with relatively high complexity. Then choose the linear model with the lowest value of

$$\|\mathbf{y} - \mathbf{X}\widehat{\boldsymbol{\beta}}\|^2 + 2\,\widehat{v^2}p.$$

We can also use the Akaike information criterion (AIC) as a heuristic for model comparison. We discussed the AIC in the unsupervised learning setting in Section 4.2, but the arguments used there can also be applied to the supervised case, under the in-sample model for the data. In particular, let $\boldsymbol{Z} = (\boldsymbol{X}, Y)$. We wish to predict the joint density ☞ 134

$$f(\boldsymbol{z}) = f(\boldsymbol{x}, y) := \frac{1}{n}\sum_{i=1}^{n} \mathbb{1}_{\{\boldsymbol{x}=\boldsymbol{x}_i\}}\, f(y\,|\,\boldsymbol{x}_i),$$

using a prediction function $g(\boldsymbol{z}\,|\,\boldsymbol{\theta})$ from a family $\mathcal{G} := \{g(\boldsymbol{z}\,|\,\boldsymbol{\theta}), \boldsymbol{\theta} \in \mathbb{R}^q\}$, where

$$g(\boldsymbol{z}\,|\,\boldsymbol{\theta}) = g(\boldsymbol{x}, y\,|\,\boldsymbol{\theta}) := \frac{1}{n}\sum_{i=1}^{n} \mathbb{1}_{\{\boldsymbol{x}=\boldsymbol{x}_i\}}\, g_i(y\,|\,\boldsymbol{\theta}).$$

Following Section 4.2, the in-sample cross-entropy risk in this case is $r(\boldsymbol{\theta}) := -\mathbb{E}_{\mathbf{X}} \ln g(\boldsymbol{Z}\,|\,\boldsymbol{\theta})$, and to approximate the optimal parameter $\boldsymbol{\theta}^*$ we minimize the corresponding training loss

$$r_{\tau_n}(\boldsymbol{\theta}) := -\frac{1}{n}\sum_{j=1}^{n} \ln g(\boldsymbol{z}_j\,|\,\boldsymbol{\theta}).$$

The optimal parameter $\widehat{\boldsymbol{\theta}}_n$ for the training loss is thus found by minimizing

$$-\frac{1}{n}\sum_{j=1}^{n} \left(-\ln n + \ln g_j(y_j\,|\,\boldsymbol{\theta})\right).$$

That is, it is the maximum likelihood estimate of $\boldsymbol{\theta}$:

$$\widehat{\boldsymbol{\theta}}_n = \underset{\boldsymbol{\theta}}{\operatorname{argmax}} \sum_{i=1}^{n} \ln g_i(y_i\,|\,\boldsymbol{\theta}).$$

Under the assumption that $f = g(\cdot\,|\,\boldsymbol{\theta}^*)$ for some parameter $\boldsymbol{\theta}^*$, we have from Theorem 4.12 that the estimated in-sample generalization risk can be approximated as ☞ 137

$$\mathbb{E}_{\mathbf{X}}\, r(\widehat{\boldsymbol{\theta}}_n) \approx r_{\mathcal{T}_n}(\widehat{\boldsymbol{\theta}}_n) + \frac{q}{n} = \ln n - \frac{1}{n}\sum_{j=1}^{n} \ln g_j(y_j\,|\,\widehat{\boldsymbol{\theta}}_n) + \frac{q}{n},$$

where q is the number of parameters (typically larger than p for a linear model with a $n \times p$ design matrix).

This leads to the heuristic of selecting the learner $g(\cdot\,|\,\widehat{\boldsymbol{\theta}}_n)$ with the smallest value of the AIC:

$$-2\sum_{i=1}^{n} \ln g_i(y_i\,|\,\widehat{\boldsymbol{\theta}}_n) + 2q. \tag{5.19}$$

■ **Example 5.20 (Normal Linear Model)** For the normal linear model $Y \sim \mathcal{N}(\boldsymbol{x}^\top\boldsymbol{\beta}, \sigma^2)$ (see (2.61)), with a p-dimensional vector $\boldsymbol{\beta}$, we have ($\boldsymbol{\theta} = (\boldsymbol{\beta}, \sigma^2)$)

☞ 48

$$g_i(y_i \,|\, \boldsymbol{\beta}, \sigma^2) = \frac{1}{\sqrt{2\pi\sigma^2}} \exp\Big(-\big(y_i - \boldsymbol{x}_i^\top\boldsymbol{\beta}\big)^2/(2\sigma^2)\Big), \quad i = 1, \dots, n,$$

so that the AIC is

$$n\ln(2\pi) + n\ln\widehat{\sigma}^2 + \|\boldsymbol{y} - \mathbf{X}\widehat{\boldsymbol{\beta}}\|^2/\widehat{\sigma}^2 + 2q, \tag{5.21}$$

where $(\widehat{\boldsymbol{\beta}}, \widehat{\sigma}^2)$ is the maximum likelihood estimate and $q = p + 1$ is the number of parameters (including σ^2). For model comparison we may remove the $n\ln(2\pi)$ term if all the models are normal linear models. ■

⚠ Certain software packages report the AIC without the $n\ln\widehat{\sigma}^2$ term in (5.21). This may lead to sub-optimal model selection if normal models are compared with non-normal ones.

5.3.5 Categorical Features

Suppose that, as described in Chapter 1, the data is given in the form of a spreadsheet or data frame with n rows and $p+1$ columns, where the first element of row i is the response variable y_i, and the remaining p elements form the vector of explanatory variables $\boldsymbol{x}_i^\top$. When all the explanatory variables (features, predictors) are *quantitative*, then the model matrix $\mathbf{X}$ can be directly read off from the data frame as the $n \times p$ matrix with rows $\boldsymbol{x}_i^\top, i = 1, \dots, n$.

However, when some explanatory variables are *qualitative* (categorical), such a one-to-one correspondence between data frame and model matrix no longer holds. The solution is to include *indicator* or *dummy* variables.

Linear models with continuous responses and categorical explanatory variables often arise in **factorial experiments**. These are controlled statistical experiments in which the aim is to assess how a response variable is affected by one or more **factors** tested at several **levels**. A typical example is an agricultural experiment where one wishes to investigate how the yield of a food crop depends on factors such as location, pesticide, and fertilizer.

■ **Example 5.22 (Crop Yield)** The data in Table 5.23 lists the yield of a food crop for four different crop treatments (e.g., strengths of fertilizer) on four different blocks (plots).

Table 5.23: Crop yield for different treatments and blocks.

	Treatment			
Block	1	2	3	4
1	9.2988	9.4978	9.7604	10.1025
2	8.2111	8.3387	8.5018	8.1942
3	9.0688	9.1284	9.3484	9.5086
4	8.2552	7.8999	8.4859	8.9485

The corresponding data frame, given in Table 5.24, has 16 rows and 3 columns: one column for the crop yield (the response variable), one column for the Treatment, with levels 1, 2, 3, 4, and one column for the Block, also with levels 1, 2, 3, 4. The values 1, 2, 3, and 4 have no quantitative meaning (it does not make sense to take their average, for example) — they merely identify the category of the treatment or block.

Table 5.24: Crop yield data organized as a data frame in standard format.

Yield	Treatment	Block
9.2988	1	1
8.2111	1	2
9.0688	1	3
8.2552	1	4
9.4978	2	1
8.3387	2	2
⋮	⋮	⋮
9.5086	4	3
8.9485	4	4

■

In general, suppose there are r factor (categorical) variables $u_1, \ldots, u_r$, where the j-th factor has p_j mutually exclusive levels, denoted by $1, \ldots, p_j$. In order to include these categorical variables in a linear model, a common approach is to introduce an **indicator feature** $x_{jk} = \mathbb{1}\{u_j = k\}$ for each factor j at level k. Thus, $x_{jk} = 1$ if the value of factor j is k and 0 otherwise. Since $\sum_k \mathbb{1}\{u_j = k\} = 1$, it suffices to consider only $p_j - 1$ of these indicator features for each factor j (this prevents the model matrix from being rank deficient). For a single response Y, the feature vector $\boldsymbol{x}^\top$ is thus a row vector of binary variables that indicates which levels were observed for each factor. The model assumption is that Y depends in a linear way on the indicator features, apart from an error term. That is,

$$Y = \beta_0 + \sum_{j=1}^{r} \sum_{k=2}^{p_j} \beta_{jk} \underbrace{\mathbb{1}\{u_j = k\}}_{x_{jk}} + \varepsilon,$$

where we have omitted one indicator feature (corresponding to level 1) for each factor j. For independent responses $Y_1, \ldots, Y_n$, where each Y_i corresponds to the factor values $u_{i1}, \ldots, u_{ir}$, let $x_{ijk} = \mathbb{1}\{u_{ij} = k\}$. Then, the linear model for the data becomes

$$Y_i = \beta_0 + \sum_{j=1}^{r} \sum_{k=2}^{p_j} \beta_{jk} x_{ijk} + \varepsilon_i, \tag{5.25}$$

where the $\{\varepsilon_i\}$ are independent with expectation 0 and some variance σ^2. By gathering the β_0 and $\{\beta_{jk}\}$ into a vector $\boldsymbol{\beta}$, and the $\{x_{ijk}\}$ into a matrix $\mathbf{X}$, we have again a linear model of the form (5.11). The model matrix $\mathbf{X}$ has n rows and $1 + \sum_{j=1}^{r}(p_j - 1)$ columns. Using the above convention that the β_{j1} parameters are subsumed in the parameter β_0

(corresponding to the "constant" feature), we can interpret β_0 as a baseline response when using the explanatory vector $\boldsymbol{x}^\top$ for which $x_{j1} = 1$ for all factors $j = 1, \ldots, r$. The other parameters $\{\beta_{jk}\}$ can be viewed as **incremental effects** relative to this baseline effect. For example, β_{12} describes by how much the response is expected to change if level 2 is used instead of level 1 for factor 1.

Example 5.26 (Single-Factor ANOVA Model) The simplest ANOVA model is the single-factor one, where the response is modeled as:

$$Y_i = \beta_0 + \beta_{12}\mathbb{1}\{u_i = 2\} + \cdots + \beta_{1p}\mathbb{1}\{u_i = p\} + \varepsilon_i,$$

and for $p = 4$ can be arranged as:

$$\boldsymbol{Y} = \begin{bmatrix} \mathbf{1}_{n_1} & \mathbf{0} & \mathbf{0} & \mathbf{0} \\ \mathbf{1}_{n_2} & \mathbf{1}_{n_2} & \mathbf{0} & \mathbf{0} \\ \mathbf{1}_{n_3} & \mathbf{0} & \mathbf{1}_{n_3} & \mathbf{0} \\ \mathbf{1}_{n_p} & \mathbf{0} & \mathbf{0} & \mathbf{1}_{n_p} \end{bmatrix} \begin{bmatrix} \beta_0 \\ \beta_{12} \\ \beta_{13} \\ \beta_{14} \end{bmatrix} + \boldsymbol{\varepsilon},$$

where $\boldsymbol{Y}$ of length $n = n_1 + \cdots + n_p$ is organized as

$$\underbrace{Y_1, \ldots, Y_{n_1}}_{\text{level 1}}, \underbrace{Y_{n_1+1}, \ldots, Y_{n_1+n_2}}_{\text{level 2}}, \ldots, \underbrace{Y_{n-n_p+1}, \ldots, Y_{n_p}}_{\text{level } p}.$$

Example 5.27 (Crop Yield (cont.)) In Example 5.22, the linear model (5.25) has eight parameters: $\beta_0, \beta_{12}, \beta_{13}, \beta_{14}, \beta_{22}, \beta_{23}, \beta_{24}$, and σ^2. The model matrix $\mathbf{X}$ depends on how the crop yields are organized in a vector $\boldsymbol{y}$ and on the ordering of the factors. Let us order $\boldsymbol{y}$ column-wise from Table 5.23, as in $\boldsymbol{y} = [9.2988, 8.2111, 9.0688, 8.2552, 9.4978, \ldots, 8.9485]^\top$, and let Treatment be Factor 1 and Block be Factor 2. Then we can write (5.25) as

$$\boldsymbol{Y} = \underbrace{\begin{bmatrix} \mathbf{1} & \mathbf{0} & \mathbf{0} & \mathbf{0} & \mathbf{C} \\ \mathbf{1} & \mathbf{1} & \mathbf{0} & \mathbf{0} & \mathbf{C} \\ \mathbf{1} & \mathbf{0} & \mathbf{1} & \mathbf{0} & \mathbf{C} \\ \mathbf{1} & \mathbf{0} & \mathbf{0} & \mathbf{1} & \mathbf{C} \end{bmatrix}}_{\mathbf{X}} \underbrace{\begin{bmatrix} \beta_0 \\ \beta_{12} \\ \beta_{13} \\ \beta_{14} \\ \beta_{22} \\ \beta_{23} \\ \beta_{24} \end{bmatrix}}_{\boldsymbol{\beta}} + \boldsymbol{\varepsilon}, \quad \text{where} \quad \mathbf{C} = \begin{bmatrix} 0 & 0 & 0 \\ 1 & 0 & 0 \\ 0 & 1 & 0 \\ 0 & 0 & 1 \end{bmatrix},$$

and with $\mathbf{1} = [1, 1, 1, 1]^\top$ and $\mathbf{0} = [0, 0, 0, 0]^\top$. Estimation of $\boldsymbol{\beta}$ and σ^2, model selection, and prediction can now be carried out in the usual manner for linear models.

In the context of factorial experiments, the model matrix is often called the **design matrix**, as it specifies the design of the experiment; e.g., how many replications are taken for each combination of factor levels. The model (5.25) can be extended by adding products of indicator variables as new features. Such features are called **interaction** terms.

5.3.6 Nested Models

Let $\mathbf{X}$ be a $n \times p$ model matrix of the form $\mathbf{X} = [\mathbf{X}_1, \mathbf{X}_2]$, where $\mathbf{X}_1$ and $\mathbf{X}_2$ are model matrices of dimension $n \times k$ and $n \times (p-k)$, respectively. The linear models $\boldsymbol{Y} = \mathbf{X}_1\boldsymbol{\beta}_1 + \boldsymbol{\varepsilon}$ and $\boldsymbol{Y} = \mathbf{X}_2\boldsymbol{\beta}_2 + \boldsymbol{\varepsilon}$ are said to be *nested within* the linear model $\boldsymbol{Y} = \mathbf{X}\boldsymbol{\beta} + \boldsymbol{\varepsilon}$. This simply means that certain features in $\mathbf{X}$ are ignored in each of the first two models. Note that $\boldsymbol{\beta}$, $\boldsymbol{\beta}_1$, and $\boldsymbol{\beta}_2$ are parameter vectors of dimension p, k, and $p - k$, respectively. In what follows, we assume that $n \geq p$ and that all model matrices are full-rank.

Suppose we wish to assess whether to use the full model matrix $\mathbf{X}$ or the reduced model matrix $\mathbf{X}_1$. Let $\widehat{\boldsymbol{\beta}}$ be the estimate of $\boldsymbol{\beta}$ under the full model (that is, obtained via (5.12)), and let $\widehat{\boldsymbol{\beta}_1}$ denote the estimate of $\boldsymbol{\beta}_1$ for the reduced model. Let $\boldsymbol{Y}^{(2)} = \mathbf{X}\widehat{\boldsymbol{\beta}}$ be the projection of $\boldsymbol{Y}$ onto the space $\mathrm{Span}(\mathbf{X})$ spanned by the columns of $\mathbf{X}$; and let $\boldsymbol{Y}^{(1)} = \mathbf{X}_1\widehat{\boldsymbol{\beta}_1}$ be the projection of $\boldsymbol{Y}$ onto the space $\mathrm{Span}(\mathbf{X}_1)$ spanned by the columns of $\mathbf{X}_1$ only; see Figure 5.28. In order to decide whether the features in $\mathbf{X}_2$ are needed, we may compare the estimated error terms of the two models, as calculated by (5.13); that is, by the residual sum of squares divided by the number of observations n. If the outcome of this comparison is that there is little difference between the model error for the full and reduced model, then it is appropriate to adopt the reduced model, as it has fewer parameters than the full model, while explaining the data just as well. The comparison is thus between the squared norms $\|\boldsymbol{Y} - \boldsymbol{Y}^{(2)}\|^2$ and $\|\boldsymbol{Y} - \boldsymbol{Y}^{(1)}\|^2$. Because of the nested nature of the linear models, $\mathrm{Span}(\mathbf{X}_1)$ is a subspace of $\mathrm{Span}(\mathbf{X})$ and, consequently, the orthogonal projection of $\boldsymbol{Y}^{(2)}$ onto $\mathrm{Span}(\mathbf{X}_1)$ is the same as the orthogonal projection of $\boldsymbol{Y}$ onto $\mathrm{Span}(\mathbf{X}_1)$; that is, $\boldsymbol{Y}^{(1)}$. By Pythagoras' theorem, we thus have the decomposition $\|\boldsymbol{Y}^{(2)} - \boldsymbol{Y}^{(1)}\|^2 + \|\boldsymbol{Y} - \boldsymbol{Y}^{(2)}\|^2 = \|\boldsymbol{Y} - \boldsymbol{Y}^{(1)}\|^2$. This is also illustrated in Figure 5.28.

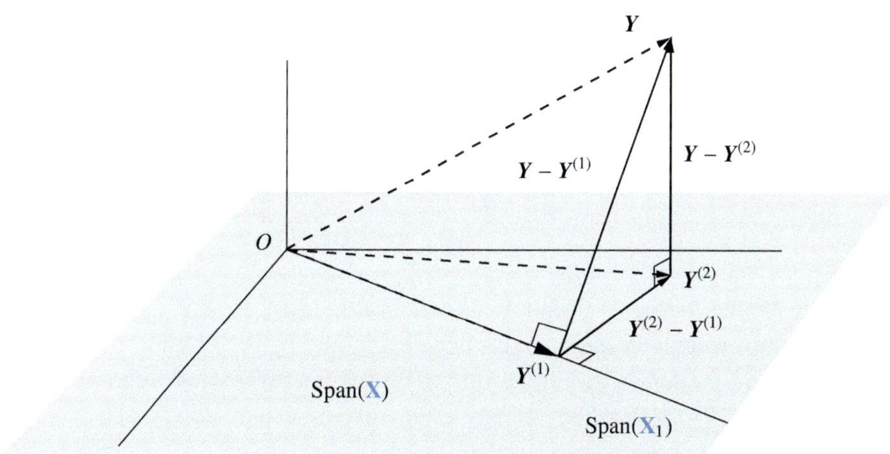

Figure 5.28: The residual sum of squares for the full model corresponds to $\|\boldsymbol{Y} - \boldsymbol{Y}^{(2)}\|^2$ and for the reduced model it is $\|\boldsymbol{Y} - \boldsymbol{Y}^{(1)}\|^2$. By Pythagoras's theorem, the difference is $\|\boldsymbol{Y}^{(2)} - \boldsymbol{Y}^{(1)}\|^2$.

The above decomposition can be generalized to more than two model matrices. Suppose that the model matrix can be decomposed into d submatrices: $\mathbf{X} = [\mathbf{X}_1, \mathbf{X}_2, \ldots, \mathbf{X}_d]$, where the matrix $\mathbf{X}_i$ has p_i columns and n rows, $i = 1, \ldots, d$. Thus, the number of columns[2] in the full model matrix is $p = p_1 + \cdots + p_d$. This creates an increasing sequence of "nested" model matrices: $\mathbf{X}_1, [\mathbf{X}_1, \mathbf{X}_2], \ldots, [\mathbf{X}_1, \mathbf{X}_2, \ldots, \mathbf{X}_d]$, from (say) the baseline normal model matrix $\mathbf{X}_1 = \mathbf{1}$ to the full model matrix $\mathbf{X}$. Think of each model matrix corresponding to specific variables in the model.

We follow a similar projection procedure as in Figure 5.28: First project $\boldsymbol{Y}$ onto $\text{Span}(\mathbf{X})$ to yield the vector $\boldsymbol{Y}^{(d)}$, then project $\boldsymbol{Y}^{(d)}$ onto $\text{Span}([\mathbf{X}_1, \ldots, \mathbf{X}_{d-1}])$ to obtain $\boldsymbol{Y}^{(d-1)}$, and so on, until $\boldsymbol{Y}^{(2)}$ is projected onto $\text{Span}(\mathbf{X}_1)$ to yield $\boldsymbol{Y}^{(1)} = \overline{Y}\mathbf{1}$ (in the case that $\mathbf{X}_1 = \mathbf{1}$).

By applying Pythagoras' theorem, we have the decomposition:

$$\underbrace{\|\boldsymbol{Y} - \boldsymbol{Y}^{(1)}\|^2}_{\text{df}=n-p_1} = \underbrace{\|\boldsymbol{Y} - \boldsymbol{Y}^{(d)}\|^2}_{\text{df}=n-p} + \underbrace{\|\boldsymbol{Y}^{(d)} - \boldsymbol{Y}^{(d-1)}\|^2}_{\text{df}=p_d} + \cdots + \underbrace{\|\boldsymbol{Y}^{(2)} - \boldsymbol{Y}^{(1)}\|^2}_{\text{df}=p_2}. \tag{5.29}$$

Software packages typically report the sums of squares as well as the corresponding **degrees of freedom** (df): $n - p, p_d, \ldots, p_2$.

5.3.7 Coefficient of Determination

To assess how a linear model $\boldsymbol{Y} = \mathbf{X}\boldsymbol{\beta} + \boldsymbol{\varepsilon}$ compares to the default model $\boldsymbol{Y} = \beta_0\mathbf{1} + \boldsymbol{\varepsilon}$, we can compare the variance of the original data, estimated via $\sum_i (Y_i - \overline{Y})^2/n = \|\boldsymbol{Y} - \overline{Y}\mathbf{1}\|^2/n$, with the variance of the fitted data; estimated via $\sum_i (\widehat{Y}_i - \overline{Y})^2/n = \|\widehat{\boldsymbol{Y}} - \overline{Y}\mathbf{1}\|^2/n$, where $\widehat{\boldsymbol{Y}} = \mathbf{X}\widehat{\boldsymbol{\beta}}$. The sum $\sum_i (Y_i - \overline{Y})^2 = \|\boldsymbol{Y} - \overline{Y}\mathbf{1}\|^2$ is sometimes called the **total sum of squares** (TSS), and the quantity

$$R^2 = \frac{\|\widehat{\boldsymbol{Y}} - \overline{Y}\mathbf{1}\|^2}{\|\boldsymbol{Y} - \overline{Y}\mathbf{1}\|^2} \tag{5.30}$$

is called the **coefficient of determination** of the linear model. In the notation of Figure 5.28, $\widehat{\boldsymbol{Y}} = \boldsymbol{Y}^{(2)}$ and $\overline{Y}\mathbf{1} = \boldsymbol{Y}^{(1)}$, so that

$$R^2 = \frac{\|\boldsymbol{Y}^{(2)} - \boldsymbol{Y}^{(1)}\|^2}{\|\boldsymbol{Y} - \boldsymbol{Y}^{(1)}\|^2} = \frac{\|\boldsymbol{Y} - \boldsymbol{Y}^{(1)}\|^2 - \|\boldsymbol{Y} - \boldsymbol{Y}^{(2)}\|^2}{\|\boldsymbol{Y} - \boldsymbol{Y}^{(1)}\|^2} = \frac{\text{TSS} - \text{RSS}}{\text{TSS}}.$$

Note that R^2 lies between 0 and 1. An R^2 value close to 1 indicates that a large proportion of the variance in the data has been explained by the model.

Many software packages also give the **adjusted coefficient of determination**, or simply the adjusted R^2, defined by

$$R^2_{\text{adjusted}} = 1 - (1 - R^2)\frac{n-1}{n-p}.$$

The regular R^2 is always *non-decreasing* in the number of parameters (see Exercise 17), but this may not indicate better predictive power. The adjusted R^2 compensates for this increase by decreasing the regular R^2 as the number of variables increases. This heuristic adjustment can make it easier to compare the quality of two competing models.

[2]As always, we assume the columns are linearly independent.

5.4 Inference for Normal Linear Models

So far we have not assumed any specific probability distribution for the random vector of errors $\boldsymbol{\varepsilon} = [\varepsilon_1, \ldots, \varepsilon_n]^\top$ in a linear model $\boldsymbol{Y} = \mathbf{X}\boldsymbol{\beta} + \boldsymbol{\varepsilon}$. When the error terms $\{\varepsilon_i\}$ are assumed to be normally distributed (that is, $\{\varepsilon_i\} \sim_{\text{iid}} \mathcal{N}(0, \sigma^2)$), whole new avenues open up for inference on linear models. In Section 2.8 we already saw that for such *normal linear models*, estimation of $\boldsymbol{\beta}$ and σ^2 can be carried out via maximum likelihood methods, yielding the same estimators from (5.12) and (5.13). ☞ 48

The following theorem lists the properties of these estimators. In particular, it shows that $\widehat{\boldsymbol{\beta}}$ and $\widehat{\sigma^2} n/(n-p)$ are independent and unbiased estimators of $\boldsymbol{\beta}$ and σ^2, respectively.

Theorem 5.31 (Properties of the Estimators for a Normal Linear Model): Consider the linear model $\boldsymbol{Y} = \mathbf{X}\boldsymbol{\beta} + \boldsymbol{\varepsilon}$, with $\boldsymbol{\varepsilon} \sim \mathcal{N}(\mathbf{0}, \sigma^2 \mathbf{I}_n)$, where $\boldsymbol{\beta}$ is a p-dimensional vector of parameters and σ^2 a dispersion parameter. Provided that $\text{rank}(\mathbf{X}) = p < n$, then the following results hold.

1. The maximum likelihood estimators $\widehat{\boldsymbol{\beta}}$ and $\widehat{\sigma^2}$ are independent.
2. $\widehat{\boldsymbol{\beta}} \sim \mathcal{N}(\boldsymbol{\beta}, \, \sigma^2(\mathbf{X}^\top\mathbf{X})^+)$.
3. $n\,\widehat{\sigma^2}/\sigma^2 \sim \chi^2_{n-p}$.

Proof. Using the pseudo-inverse (Definition A.10), we can write the random vector $\widehat{\boldsymbol{\beta}}$ as $\mathbf{X}^+\boldsymbol{Y}$, which is a linear transformation of a normal random vector. Consequently, $\widehat{\boldsymbol{\beta}}$ has a multivariate normal distribution; see Theorem D.46. The mean vector and covariance matrix follow from the same theorem: ☞ 516 ☞ 640

$$\mathbb{E}\,\widehat{\boldsymbol{\beta}} = \mathbf{X}^+\,\mathbb{E}\boldsymbol{Y} = \mathbf{X}^+\mathbf{X}\,\boldsymbol{\beta} = \boldsymbol{\beta}$$

and

$$\mathbb{C}\text{ov}(\widehat{\boldsymbol{\beta}}) = \mathbf{X}^+\sigma^2\mathbf{I}_n(\mathbf{X}^+)^\top = \sigma^2(\mathbf{X}^\top\mathbf{X})^+.$$

To show that $\widehat{\boldsymbol{\beta}}$ and $\widehat{\sigma^2}$ are independent, define $\boldsymbol{Y}^{(2)} = \mathbf{X}\widehat{\boldsymbol{\beta}}$. Note that $\boldsymbol{Y}/\sigma$ has a $\mathcal{N}(\boldsymbol{\mu}, \mathbf{I}_n)$ distribution, with expectation vector $\boldsymbol{\mu} = \mathbf{X}\boldsymbol{\beta}/\sigma$. A direct application of Theorem D.57 now shows that $(\boldsymbol{Y} - \boldsymbol{Y}^{(2)})/\sigma$ is independent of $\boldsymbol{Y}^{(2)}/\sigma$. Since $\widehat{\boldsymbol{\beta}} = \mathbf{X}^+\mathbf{X}\widehat{\boldsymbol{\beta}} = \mathbf{X}^+\boldsymbol{Y}^{(2)}$ and $\widehat{\sigma^2} = \|\boldsymbol{Y} - \boldsymbol{Y}^{(2)}\|^2/n$, it follows that $\widehat{\sigma^2}$ is independent of $\widehat{\boldsymbol{\beta}}$. Finally, by the same theorem, the random variable $\|\boldsymbol{Y} - \boldsymbol{Y}^{(2)}\|^2/\sigma^2$ has a χ^2_{n-p} distribution, as $\boldsymbol{Y}^{(2)}$ has the same expectation vector as $\boldsymbol{Y}$. □ ☞ 643

As a corollary, each estimator $\widehat{\beta}_i$ of β_i has a normal distribution with expectation β_i and variance $\sigma^2\boldsymbol{u}_i^\top\mathbf{X}^+(\mathbf{X}^+)^\top\boldsymbol{u}_i = \sigma^2\|\boldsymbol{u}_i^\top\mathbf{X}^+\|^2$, where $\boldsymbol{u}_i = [0, \ldots, 0, 1, 0, \ldots, 0]^\top$ is the i-th unit vector; in other words, the variance is $\sigma^2[(\mathbf{X}^\top\mathbf{X})^+]_{ii}$.

It is of interest to test whether certain regression parameters β_i are 0 or not, since if $\beta_i = 0$, the i-th explanatory variable has no direct effect on the expected response and so could be removed from the model. A standard procedure is to conduct a hypothesis test

☞ 667 (see Section D.14 for a review of hypothesis testing) to test the null hypothesis $H_0 : \beta_i = 0$ against the alternative $H_1 : \beta_i \neq 0$, using the test statistic

$$T = \frac{\widehat{\beta}_i / \|\boldsymbol{u}_i^\top \mathbf{X}^+\|}{\sqrt{\text{RSE}}}, \tag{5.32}$$

where RSE is the residual squared error; that is RSE = RSS/$(n-p)$. This test statistic has a t_{n-p} distribution under H_0. To see this, write $T = Z/\sqrt{V/(n-p)}$, with

$$Z = \frac{\widehat{\beta}_i}{\sigma \|\boldsymbol{u}_i^\top \mathbf{X}^+\|} \quad \text{and} \quad V = n\,\widehat{\sigma^2}/\sigma^2.$$

Then, by Theorem 5.31, $Z \sim \mathcal{N}(0,1)$ under H_0, $V \sim \chi^2_{n-p}$, and Z and V are independent.
☞ 644 The result now follows directly from Corollary D.59.

5.4.1 Comparing Two Normal Linear Models

Suppose we have the following normal linear model for data $\boldsymbol{Y} = [Y_1, \ldots, Y_n]^\top$:

$$\boldsymbol{Y} = \underbrace{\mathbf{X}_1\boldsymbol{\beta}_1 + \mathbf{X}_2\boldsymbol{\beta}_2}_{\mathbf{X}\boldsymbol{\beta}} + \boldsymbol{\varepsilon}, \quad \boldsymbol{\varepsilon} \sim \mathcal{N}(\mathbf{0}, \sigma^2\mathbf{I}_n), \tag{5.33}$$

where $\boldsymbol{\beta}_1$ and $\boldsymbol{\beta}_2$ are unknown vectors of dimension k and $p-k$, respectively; and $\mathbf{X}_1$ and $\mathbf{X}_2$ are full-rank model matrices of dimensions $n \times k$ and $n \times (p-k)$, respectively. Above we implicitly defined $\mathbf{X} = [\mathbf{X}_1, \mathbf{X}_2]$ and $\boldsymbol{\beta}^\top = [\boldsymbol{\beta}_1^\top, \boldsymbol{\beta}_2^\top]$. Suppose we wish to test the hypothesis $H_0 : \boldsymbol{\beta}_2 = \mathbf{0}$ against $H_1 : \boldsymbol{\beta}_2 \neq \mathbf{0}$. Following Section 5.3.6, the idea is to compare the residual sum of squares for both models, expressed as $\|\boldsymbol{Y} - \boldsymbol{Y}^{(2)}\|^2$ and $\|\boldsymbol{Y} - \boldsymbol{Y}^{(1)}\|^2$. Using Pythagoras' theorem we saw that $\|\boldsymbol{Y} - \boldsymbol{Y}^{(2)}\|^2 - \|\boldsymbol{Y} - \boldsymbol{Y}^{(1)}\|^2 = \|\boldsymbol{Y}^{(2)} - \boldsymbol{Y}^{(1)}\|^2$, and so it makes sense to base the decision whether to retain or reject H_0 on the basis of the quotient of $\|\boldsymbol{Y}^{(2)} - \boldsymbol{Y}^{(1)}\|^2$ and $\|\boldsymbol{Y} - \boldsymbol{Y}^{(2)}\|^2$. This leads to the following test statistics.

Theorem 5.34 (Test Statistic for Comparing Two Normal Linear Models): For the model (5.33), let $\boldsymbol{Y}^{(2)}$ and $\boldsymbol{Y}^{(1)}$ be the projections of $\boldsymbol{Y}$ onto the space spanned by the p columns of $\mathbf{X}$ and the k columns of $\mathbf{X}_1$, respectively. Then under $H_0 : \boldsymbol{\beta}_2 = \mathbf{0}$ the test statistic

$$T = \frac{\|\boldsymbol{Y}^{(2)} - \boldsymbol{Y}^{(1)}\|^2/(p-k)}{\|\boldsymbol{Y} - \boldsymbol{Y}^{(2)}\|^2/(n-p)} \tag{5.35}$$

has an $\mathsf{F}(p-k, n-p)$ distribution.

Proof. Define $\boldsymbol{X} := \boldsymbol{Y}/\sigma$ with expectation $\boldsymbol{\mu} := \mathbf{X}\boldsymbol{\beta}/\sigma$, and let $\boldsymbol{X}_k := \boldsymbol{Y}^{(1)}/\sigma$ and $\boldsymbol{X}_p := \boldsymbol{Y}^{(2)}/\sigma$, with expectations $\boldsymbol{\mu}_j$, $j = k, p$. Note that $\boldsymbol{\mu}_p = \boldsymbol{\mu}$ and, under H_0, $\boldsymbol{\mu}_k = \boldsymbol{\mu}_p$. We
☞ 643 can directly apply Theorem D.57 to find that $\|\boldsymbol{Y} - \boldsymbol{Y}^{(2)}\|^2/\sigma^2 = \|\boldsymbol{X} - \boldsymbol{X}_p\|^2 \sim \chi^2_{n-p}$ and, under H_0, $\|\boldsymbol{Y}^{(2)} - \boldsymbol{Y}^{(1)}\|^2/\sigma^2 = \|\boldsymbol{X}_p - \boldsymbol{X}_k\|^2 \sim \chi^2_{p-k}$. Moreover, these random variables are independent of each other. The proof is completed by applying Theorem D.58. □

Note that H_0 is rejected for large values of T. The testing procedure thus proceeds as follows:

1. Compute the outcome, t say, of the test statistic T in (5.35).
2. Evaluate the P-value $\mathbb{P}[T \geq t]$, with $T \sim \mathsf{F}(p-k, n-p)$.
3. Reject H_0 if this P-value is too small, say less than 0.05.

For nested models $[\mathbf{X}_1, \mathbf{X}_2, \ldots, \mathbf{X}_i]$, $i = 1, 2, \ldots, d$, as in Section 5.3.6, the F test statistic in Theorem 5.34 can now be used to test whether certain $\mathbf{X}_i$ are needed or not. ☞ 214
In particular, software packages will report the outcomes of

(5.36)
$$F_i = \frac{\|\boldsymbol{Y}^{(i)} - \boldsymbol{Y}^{(i-1)}\|^2 / p_i}{\|\boldsymbol{Y} - \boldsymbol{Y}^{(d)}\|^2/(n-p)},$$

in the order $i = 2, 3, \ldots, d$. Under the null hypothesis that $\boldsymbol{Y}^{(i)}$ and $\boldsymbol{Y}^{(i-1)}$ have the same expectation (that is, adding $\mathbf{X}_i$ to $\mathbf{X}_{i-1}$ has no additional effect on reducing the approximation error), the test statistic F_i has an $\mathsf{F}(p_i, n-p)$ distribution, and the corresponding P-values quantify the strength of the decision to include an additional variable in the model or not. This procedure is called **analysis of variance** (ANOVA).

⚠ Note that the output of an ANOVA table depends on the order in which the variables are considered.

■ **Example 5.37 (Crop Yield (cont.))** We continue Examples 5.22 and 5.27. Decompose the linear model as

$$\boldsymbol{Y} = \underbrace{\begin{bmatrix}\mathbf{1}\\\mathbf{1}\\\mathbf{1}\\\mathbf{1}\end{bmatrix}}_{\mathbf{X}_1} \underbrace{\beta_0}_{\boldsymbol{\beta}_1} + \underbrace{\begin{bmatrix}\mathbf{0}&\mathbf{0}&\mathbf{0}\\\mathbf{1}&\mathbf{0}&\mathbf{0}\\\mathbf{0}&\mathbf{1}&\mathbf{0}\\\mathbf{0}&\mathbf{0}&\mathbf{1}\end{bmatrix}}_{\mathbf{X}_2} \underbrace{\begin{bmatrix}\beta_{12}\\\beta_{13}\\\beta_{14}\end{bmatrix}}_{\boldsymbol{\beta}_2} + \underbrace{\begin{bmatrix}\mathbf{C}\\\mathbf{C}\\\mathbf{C}\\\mathbf{C}\end{bmatrix}}_{\mathbf{X}_3} \underbrace{\begin{bmatrix}\beta_{22}\\\beta_{23}\\\beta_{24}\end{bmatrix}}_{\boldsymbol{\beta}_3} + \boldsymbol{\varepsilon}.$$

Is the crop yield dependent on treatment levels as well as blocks? We first test whether we can remove Block as a factor in the model against it playing a significant role in explaining the crop yields. Specifically, we test $\boldsymbol{\beta}_3 = \mathbf{0}$ versus $\boldsymbol{\beta}_3 \neq \mathbf{0}$ using Theorem 5.34. Now the vector $\boldsymbol{Y}^{(2)}$ is the projection of $\boldsymbol{Y}$ onto the $(p = 7)$-dimensional space spanned by the columns of $\mathbf{X} = [\mathbf{X}_1, \mathbf{X}_2, \mathbf{X}_3]$; and $\boldsymbol{Y}^{(1)}$ is the projection of $\boldsymbol{Y}$ onto the $(k = 4)$-dimensional space spanned by the columns of $\mathbf{X}_{12} := [\mathbf{X}_1, \mathbf{X}_2]$. The test statistic, T_{12} say, under H_0 has an $\mathsf{F}(3, 9)$ distribution.

The Python code below calculates the outcome of the test statistic T_{12} and the corresponding P-value. We find $t_{12} = 34.9998$, which gives a P-value 2.73×10^{-5}. This shows that the block effects are extremely important for explaining the data.

Using the extended model (including the block effects), we can test whether $\boldsymbol{\beta}_2 = \mathbf{0}$ or not; that is, whether the treatments have a significant effect on the crop yield in the presence of the Block factor. This is done in the last six lines of the code below. The outcome of the test statistic is 4.4878, with a P-value of 0.0346. By including the block effects, we effectively reduce the uncertainty in the model and are able to more accurately assess the effects of the treatments, to conclude that the treatment seems to have an effect on the crop yield. A closer look at the data shows that within each block (row) the crop yield roughly increases with the treatment level.

crop.py

```
import numpy as np
from scipy.stats import f
from numpy.linalg import lstsq, norm

yy = np.array([9.2988, 9.4978, 9.7604, 10.1025,
      8.2111, 8.3387, 8.5018,  8.1942,
      9.0688, 9.1284, 9.3484,  9.5086,
      8.2552, 7.8999, 8.4859,  8.9485]).reshape(4,4).T

nrow, ncol = yy.shape[0], yy.shape[1]
n = nrow * ncol
y = yy.reshape(16,)
X_1 = np.ones((n,1))

KM = np.kron(np.eye(ncol),np.ones((nrow,1)))
KM[:,0]
X_2 = KM[:,1:ncol]
IM = np.eye(nrow)
C = IM[:,1:nrow]

X_3 = np.vstack((C, C))
X_3 = np.vstack((X_3, C))
X_3 = np.vstack((X_3, C))

X = np.hstack((X_1,X_2))
X = np.hstack((X,X_3))

p = X.shape[1] #number of parameters in full model
betahat = lstsq(X, y,rcond=None)[0]  #estimate under the full model

ym = X @ betahat

X_12 = np.hstack((X_1, X_2)) #omitting the block effect
k = X_12.shape[1] #number of parameters in reduced model
betahat_12 = lstsq(X_12, y,rcond=None)[0]
y_12 = X_12 @ betahat_12
T_12=(n-p)/(p-k)*(norm(y-y_12)**2 -  norm(y-ym)**2)/norm(y-ym)**2
pval_12 = 1 - f.cdf(T_12,p-k,n-p)

X_13 = np.hstack((X_1, X_3)) #omitting the treatment effect
k = X_13.shape[1] #number of parameters in reduced model
betahat_13 = lstsq(X_13, y,rcond=None)[0]
y_13 = X_13 @ betahat_13
T_13=(n-p)/(p-k)*(norm(y-y_13)**2 - norm(y-ym)**2)/norm(y-ym)**2
pval_13 = 1 - f.cdf(T_13,p-k,n-p)
```

5.4.2 Confidence and Prediction Intervals

As in all supervised learning settings, linear regression is most useful when we wish to predict how a new response variable will behave on the basis of a new explanatory vector $\boldsymbol{x}$. For example, it may be difficult to measure the response variable, but by knowing the estimated regression line and the value for $\boldsymbol{x}$, we will have a reasonably good idea what Y or the expected value of Y is going to be.

Thus, consider a new $\boldsymbol{x}$ and let $Y \sim \mathcal{N}(\boldsymbol{x}^\top \boldsymbol{\beta}, \sigma^2)$, with $\boldsymbol{\beta}$ and σ^2 unknown. First we are going to look at the *expected* value of Y, that is $\mathbb{E}Y = \boldsymbol{x}^\top \boldsymbol{\beta}$. Since $\boldsymbol{\beta}$ is unknown, we do not know $\mathbb{E}Y$ either. However, we can estimate it via the estimator $\widehat{Y} = \boldsymbol{x}^\top \widehat{\boldsymbol{\beta}}$, where $\widehat{\boldsymbol{\beta}} \sim \mathcal{N}(\boldsymbol{\beta}, \sigma^2(\mathbf{X}^\top \mathbf{X})^+)$, by Theorem 5.31. Being linear in the components of $\boldsymbol{\beta}$, $\widehat{Y}$ therefore has a normal distribution with expectation $\boldsymbol{x}^\top \boldsymbol{\beta}$ and variance $\sigma^2 \|\boldsymbol{x}^\top \mathbf{X}^+\|^2$. Let $Z \sim \mathcal{N}(0, 1)$ be the standardized version of $\widehat{Y}$ and $V = \|\boldsymbol{Y} - \mathbf{X}\widehat{\boldsymbol{\beta}}\|^2/\sigma^2 \sim \chi^2_{n-p}$. Then the random variable

$$T := \frac{(\boldsymbol{x}^\top \widehat{\boldsymbol{\beta}} - \boldsymbol{x}^\top \boldsymbol{\beta}) \,/\, \|\boldsymbol{x}^\top \mathbf{X}^+\|}{\|\boldsymbol{Y} - \mathbf{X}\widehat{\boldsymbol{\beta}}\| \,/\sqrt{(n-p)}} = \frac{Z}{\sqrt{V/(n-p)}} \tag{5.38}$$

has, by Corollary D.59, a t_{n-p} distribution. After rearranging the identity ☞ 644

$$\mathbb{P}[|T| \leq t_{n-p;1-\alpha/2}] = 1 - \alpha,$$

where $t_{n-p;1-\alpha/2}$ is the $(1 - \alpha/2)$ quantile of the t_{n-p} distribution, we arrive at the stochastic **confidence interval**

$$\boldsymbol{x}^\top \widehat{\boldsymbol{\beta}} \pm t_{n-p;1-\alpha/2} \sqrt{\text{RSE}}\, \|\boldsymbol{x}^\top \mathbf{X}^+\|, \tag{5.39}$$

where we have identified $\|\boldsymbol{Y} - \mathbf{X}\widehat{\boldsymbol{\beta}}\|^2/(n-p)$ with RSE. This confidence interval quantifies the uncertainty in the learner (regression surface).

A **prediction interval** for a new response Y is different from a confidence interval for $\mathbb{E}Y$. Here the idea is to construct an interval such that Y lies in this interval with a certain guaranteed probability. Note that now we have *two* sources of variation:

1. $Y \sim \mathcal{N}(\boldsymbol{x}^\top \boldsymbol{\beta}, \sigma^2)$ itself is a random variable.

2. Estimating $\boldsymbol{x}^\top \boldsymbol{\beta}$ via $\widehat{Y}$ brings another source of variation.

We can construct a $(1 - \alpha)$ prediction interval, by finding two random bounds such that the random variable Y lies between these bounds with probability $1 - \alpha$. We can reason as follows: Firstly, note that $Y \sim \mathcal{N}(\boldsymbol{x}^\top \boldsymbol{\beta}, \sigma^2)$ and $\widehat{Y} \sim \mathcal{N}(\boldsymbol{x}^\top \boldsymbol{\beta}, \sigma^2 \|\boldsymbol{x}^\top \mathbf{X}^+\|^2)$ are independent. It follows that $Y - \widehat{Y}$ has a normal distribution with expectation 0 and variance

$$\sigma^2(1 + \|\boldsymbol{x}^\top \mathbf{X}^+\|^2). \tag{5.40}$$

Secondly, letting $Z \sim \mathcal{N}(0, 1)$ be the standardized version of $Y - \widehat{Y}$, and repeating the steps used for the construction of the confidence interval (5.39), we arrive at the prediction interval

$$\boldsymbol{x}^\top \widehat{\boldsymbol{\beta}} \pm t_{n-p;1-\alpha/2} \sqrt{\text{RSE}} \sqrt{1 + \|\boldsymbol{x}^\top \mathbf{X}^+\|^2}. \tag{5.41}$$

This prediction interval captures the uncertainty from an as-yet-unobserved response as well as the uncertainty in the parameters of the regression model itself.

■ **Example 5.42 (Confidence Limits in Simple Linear Regression)** The following program draws $n = 100$ samples from a simple linear regression model with parameters $\boldsymbol{\beta} = [6, 13]^\top$ and $\sigma = 2$, where the x-coordinates are evenly spaced on the interval $[0, 1]$. The parameters are estimated in the third block of the code. Estimates for $\boldsymbol{\beta}$ and σ are $[6.03, 13.09]^\top$ and $\widehat{\sigma} = 1.60$, respectively. The program then proceeds by calculating the 95% numeric confidence and prediction intervals for various values of the explanatory variable. Figure 5.43 shows the results.

confpred.py

```
import numpy as np
import matplotlib.pyplot as plt
from scipy.stats import t
from  numpy.linalg import inv, lstsq, norm
np.random.seed(123)

n = 100
x = np.linspace(0.01,1,100).reshape(n,1)
# parameters
beta = np.array([6,13])
sigma = 2
Xmat = np.hstack((np.ones((n,1)), x)) #design matrix
y = Xmat @ beta + sigma*np.random.randn(n)

# solve the normal equations
betahat = lstsq(Xmat, y,rcond=None)[0]
# estimate for sigma
sqMSE = norm(y - Xmat @ betahat)/np.sqrt(n-2)

tquant = t.ppf(0.975,n-2) # 0.975 quantile
ucl = np.zeros(n) #upper conf. limits
lcl = np.zeros(n) #lower conf. limits
upl = np.zeros(n)
lpl = np.zeros(n)
rl = np.zeros(n)  # (true) regression line
u = 0

for i in range(n):
    u = u + 1/n;
    xvec = np.array([1,u])
    sqc = np.sqrt(xvec.T @ inv(Xmat.T @ Xmat) @ xvec)
    sqp = np.sqrt(1 + xvec.T @ inv(Xmat.T @ Xmat) @ xvec)
    rl[i] = xvec.T @ beta;
    ucl[i] = xvec.T @ betahat + tquant*sqMSE*sqc;
    lcl[i] = xvec.T @ betahat - tquant*sqMSE*sqc;
    upl[i] = xvec.T @ betahat + tquant*sqMSE*sqp;
    lpl[i] = xvec.T @ betahat - tquant*sqMSE*sqp;

plt.plot(x,y, '.')
plt.plot(x,rl,'b')
plt.plot(x,ucl,'k:')
plt.plot(x,lcl,'k:')
plt.plot(x,upl,'r--')
plt.plot(x,lpl,'r--')
```

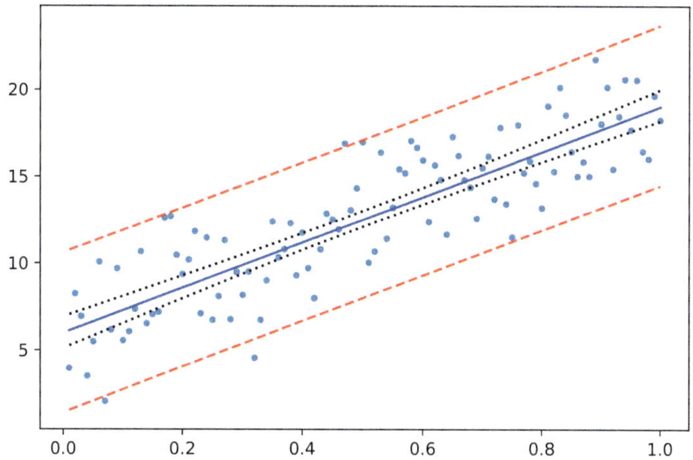

Figure 5.43: The true regression line (blue, solid) and the upper and lower 95% prediction curves (red, dashed) and confidence curves (dotted).

■

5.5 Nonlinear Regression Models

So far we have been mostly dealing with linear regression models, in which the prediction function is of the form $g(\boldsymbol{x} \mid \boldsymbol{\beta}) = \boldsymbol{x}^\top \boldsymbol{\beta}$. In this section we discuss some strategies for handling general prediction functions $g(\boldsymbol{x} \mid \boldsymbol{\beta})$, where the functional form is known up to an unknown parameter vector $\boldsymbol{\beta}$. So the regression model becomes

$$Y_i = g(\boldsymbol{x}_i \mid \boldsymbol{\beta}) + \varepsilon_i, \quad i = 1, \ldots, n, \tag{5.44}$$

where $\varepsilon_1, \ldots, \varepsilon_n$ are independent with expectation 0 and unknown variance σ^2. The model can be further specified by assuming that the error terms have a normal distribution.

Table 5.45 gives some common examples of nonlinear prediction functions for data taking values in $\mathbb{R}$.

Table 5.45: Common nonlinear prediction functions for one-dimensional data.

Name	$g(x \mid \boldsymbol{\beta})$	$\boldsymbol{\beta}$
Exponential	$a\,\mathrm{e}^{bx}$	a, b
Power law	$a\,x^b$	a, b
Logistic	$(1 + \mathrm{e}^{a+bx})^{-1}$	a, b
Weibull	$1 - \exp(-x^b/a)$	a, b
Polynomial	$\sum_{k=0}^{p-1} \beta_k x^k$	$p, \{\beta_k\}_{k=0}^{p-1}$

The logistic and polynomial prediction functions in Table 5.45 can be readily generalized to higher dimensions. For example, for $\boldsymbol{x} \in \mathbb{R}^2$ a general second-order polynomial prediction function is of the form

$$g(\boldsymbol{x} \mid \boldsymbol{\beta}) = \beta_0 + \beta_1 x_1 + \beta_2 x_2 + \beta_{11} x_1^2 + \beta_{22} x_2^2 + \beta_{12} x_1 x_2. \tag{5.46}$$

This function can be viewed as a second-order approximation to a general smooth prediction function $g(x_1, x_2)$; see also Exercise 6. Polynomial regression models are also called *response surface* models. The generalization of the above logistic prediction to $\mathbb{R}^d$ is

$$g(\boldsymbol{x} \mid \boldsymbol{\beta}) = (1 + \mathrm{e}^{-\boldsymbol{x}^\top \boldsymbol{\beta}})^{-1}. \tag{5.47}$$

This function will make its appearance in Section 5.8 and later on in Chapters 8 and 10.

The first strategy for performing regression with nonlinear prediction functions is to extend the feature space to obtain a simpler (ideally linear) prediction function in the extended feature space. We already saw an application of this strategy in Example 2.13 ☞ 26 for the polynomial regression model, where the original feature u was extended to the feature vector $\boldsymbol{x} = [1, u, u^2, \ldots, u^{p-1}]^\top$, yielding a linear prediction function. In a similar way, the right-hand side of the polynomial prediction function in (5.46) can be viewed as a linear function of the extended feature vector $\boldsymbol{\phi}(\boldsymbol{x}) = [1, x_1, x_2, x_1^2, x_2^2, x_1 x_2]^\top$. The function $\boldsymbol{\phi}$ is called a **feature map**.

The second strategy is to transform the response variable y and possibly also the explanatory variable $\boldsymbol{x}$ such that the transformed variables $\widetilde{y}, \widetilde{\boldsymbol{x}}$ are related in a simpler (ideally linear) way. For example, for the exponential prediction function $y = a\,\mathrm{e}^{-bx}$, we have $\ln y = \ln a - bx$, which is a linear relation between $\ln y$ and $[1, x]^\top$.

■ **Example 5.48 (Chlorine)** Table 5.49 lists the free chlorine concentration (in mg per liter) in a swimming pool, recorded every 8 hours for 4 days. A simple chemistry-based model for the chlorine concentration y as a function of time t is $y = a\,\mathrm{e}^{-bt}$, where a is the initial concentration and $b > 0$ is the reaction rate.

Table 5.49: Chlorine concentration (in mg/L) as a function of time (hours).

Hours	Concentration	Hours	Concentration
0	1.0056	56	0.3293
8	0.8497	64	0.2617
16	0.6682	72	0.2460
24	0.6056	80	0.1839
32	0.4735	88	0.1867
40	0.4745	96	0.1688
48	0.3563		

The exponential relationship $y = a\,\mathrm{e}^{-bt}$ suggests that a log transformation of y will result in a *linear* relationship between $\ln y$ and the feature vector $[1, t]^\top$. Thus, if for some given data $(t_1, y_1), \ldots, (t_n, y_n)$, we plot $(t_1, \ln y_1), \ldots, (t_n, \ln y_n)$, these points should approximately lie on a straight line, and hence the simple linear regression model applies. The left panel of Figure 5.50 illustrates that the transformed data indeed lie approximately on a straight line. The estimated regression line is also drawn here. The intercept and slope are $\beta_0 = -0.0555$ and $\beta_1 = -0.0190$ here. The original (non-transformed) data is shown in the right panel of Figure 5.50, along with the fitted curve $y = \widehat{a}\,\mathrm{e}^{-\widehat{b}t}$, where $\widehat{a} = \exp(\widehat{\beta}_0) = 0.9461$ and $\widehat{b} = -\widehat{\beta}_1 = 0.0190$.

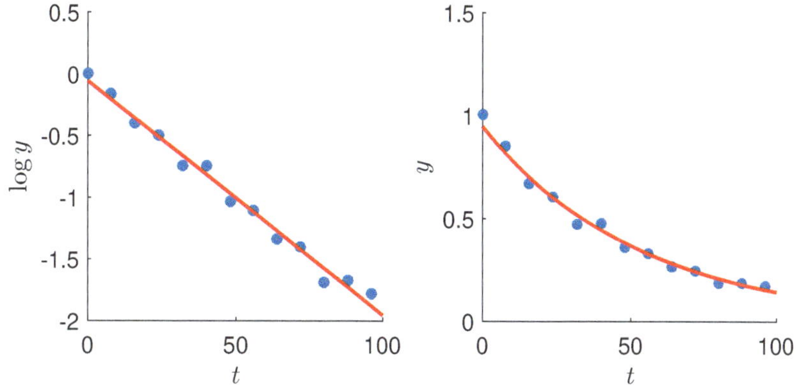

Figure 5.50: The chlorine concentration seems to have an exponential decay.

Recall that for a general regression problem the learner $g_\tau(\boldsymbol{x})$ for a given training set τ is obtained by minimizing the training (squared-error) loss

$$\ell_\tau(g(\cdot\,|\,\boldsymbol{\beta})) = \frac{1}{n}\sum_{i=1}^{n}(y_i - g(\boldsymbol{x}_i\,|\,\boldsymbol{\beta}))^2. \tag{5.51}$$

The third strategy for regression with nonlinear prediction functions is to directly minimize (5.51) by any means possible, as illustrated in the next example.

Example 5.52 (Hougen Function) In [10] the reaction rate y of a certain chemical reaction is posited to depend on three input variables: quantities of hydrogen x_1, n-pentane x_2, and isopentane x_3. The functional relationship is given by the **Hougen function**:

$$y = \frac{\beta_1 x_2 - x_3/\beta_5}{1 + \beta_2 x_1 + \beta_3 x_2 + \beta_4 x_3},$$

where $\beta_1, \ldots, \beta_5$ are the unknown parameters. The objective is to estimate the model parameters $\{\beta_i\}$ from the data, as given in Table 5.53.

Table 5.53: Data for the Hougen function.

x_1	x_2	x_3	y	x_1	x_2	x_3	y
470	300	10	8.55	470	190	65	4.35
285	80	10	3.79	100	300	54	13.00
470	300	120	4.82	100	300	120	8.50
470	80	120	0.02	100	80	120	0.05
470	80	10	2.75	285	300	10	11.32
100	190	10	14.39	285	190	120	3.13
100	80	65	2.54				

The estimation is carried out via the least-squares method. The objective function to minimize is thus

$$\ell_\tau(g(\cdot \mid \boldsymbol{\beta})) = \frac{1}{13}\sum_{i=1}^{13}\left(y_i - \frac{\beta_1 x_{i2} - x_{i3}/\beta_5}{1+\beta_2 x_{i1} + \beta_3 x_{i2} + \beta_4 x_{i3}}\right)^2, \tag{5.54}$$

where the $\{y_i\}$ and $\{x_{ij}\}$ are given in Table 5.53.

☞ 605 This is a highly nonlinear optimization problem, for which standard nonlinear least-squares methods do not work well. Instead, one can use global optimization methods ☞ 111 such as CE and SCO (see Sections 3.4.2 and 3.4.3). Using the CE method, we found the minimal value 0.02299 for the objective function, which is attained at

$$\widehat{\boldsymbol{\beta}} = [1.2526,\ 0.0628,\ 0.0400,\ 0.1124,\ 1.1914]^\top.$$

■

5.6 Gaussian-Weighted Local Linear Regression

When the optimal $g^*(x) = \mathbb{E}[Y \mid X = x]$ is a one-dimensional smooth function, then a very flexible nonlinear model for the data $\tau = \{(x_1, y_1), \ldots, (x_n, y_n)\}$ is Gaussian-weighted **local linear regression**. In this case, the learner g_τ at any point x is of the form $g_\tau(x \mid \lambda) = \widehat{\beta}_0(x) + \widehat{\beta}_1(x)\, x$, where

$$(\widehat{\beta}_0(x), \widehat{\beta}_1(x)) := \underset{\beta_0,\beta_1}{\operatorname{argmin}} \sum_{i=1}^{n} \varphi_{\lambda^2}(x_i - x)\left(y_i - \beta_0 - \beta_1 x_i\right)^2 \tag{5.55}$$

and $\varphi_{\lambda^2}(x) = \exp(-x^2/(2\lambda^2))/\sqrt{2\pi\lambda^2}$ is the Gaussian pdf of the $\mathcal{N}(0, \lambda^2)$ distribution with **bandwidth** parameter $\lambda > 0$. Note that, in the absence of the Gaussian weights $\{\varphi_{\lambda^2}(x_i - x)\}$, the estimates (5.55) do not depend on x and coincide with the least-squares estimates for the simple linear regression (5.5). In contrast, in the presence of the Gaussian weights, the estimates (5.55) provide a *localized* least-squares fit around x, the extent of which is determined by the value of the bandwidth, similar to the bandwidth ☞ 144 in kernel density estimation in Section 4.4.

Given the training data τ, the learner is uniquely determined by the training set τ and the value of the bandwidth. Before we discuss the effect that the bandwidth has on the learner, we first provide a convenient formula for the computation of the learner $g_\tau(\cdot \mid \lambda)$ on a grid of values.

5.6.1 Computation of the Learner $g_\tau(\cdot \mid \lambda)$

Let $\boldsymbol{v} := [x_1, \ldots, x_n]^\top$ be the **model vector** containing all the n input variables. We write $\boldsymbol{V}$ for the random vector $[X_1, \ldots, X_n]^\top$. If we define the diagonal matrix of weights:

$$\mathbf{D}(\boldsymbol{v}) := \operatorname{diag}(\exp(-x_1^2/2), \ldots, \exp(-x_n^2/2)),$$

then (5.55) can be written in matrix form as

$$(\widehat{\beta}_0(x), \widehat{\beta}_1(x)) = \underset{\beta_0,\beta_1}{\operatorname{argmin}} \left\| \mathbf{D}\left(\frac{\boldsymbol{v} - x\mathbf{1}}{\lambda\sqrt{2}}\right)(\boldsymbol{y} - \beta_0\mathbf{1} - \beta_1\boldsymbol{v}) \right\|^2.$$

Moreover, if $\mathbf{X}(\boldsymbol{u}) := [\mathbf{1}, \boldsymbol{u}]$ for any $\boldsymbol{u} \in \mathbb{R}^n$ and

$$\boldsymbol{l}(\boldsymbol{u}) := \mathbf{D}(\boldsymbol{u})\mathbf{X}(\boldsymbol{u})\left(\mathbf{X}(\boldsymbol{u})^\top \mathbf{D}(\boldsymbol{u})\mathbf{X}(\boldsymbol{u})\right)^{-1} \begin{bmatrix} 1 \\ 0 \end{bmatrix}, \tag{5.56}$$

then Exercise 19 establishes the following formula for the value of the learner at x:

$$g_\mathcal{T}(x \mid \lambda) = \boldsymbol{l}\left(\frac{\boldsymbol{v} - x\mathbf{1}}{\lambda}\right)^\top \boldsymbol{y}.$$

This immediately yields the following Python code for the computation of the learner over a grid of values.

gwllr.py

```
def gwllr(x, y, grid, lam):
    n, m  = len(x), len(grid)
    yhat = np.full(m, np.nan)

    for i in range(m):
        u = (x - grid[i]) / lam  # Center and scale predictor x
        D = np.diag(np.exp(-0.5 * u**2))
        X = np.column_stack((np.ones(n), u))
        l = D @ X @ np.linalg.solve(X.T @ D @ X, [1, 0])
        yhat[i] = l.T @ y  # Local linear estimator

    return yhat
```

■ **Example 5.57 (Chlorine Data Set Revisited)** To illustrate the effect of the bandwidth on the performance of the learner $g_\mathcal{T}(\cdot \mid \lambda)$, we apply **gwllr.py** to the chlorine data set in Table 5.49.

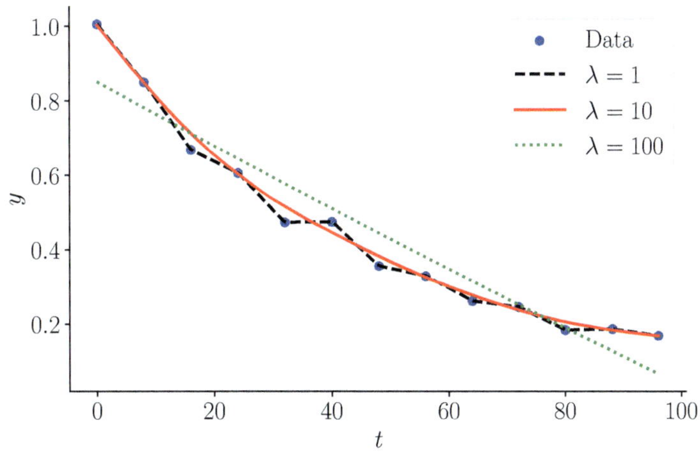

Figure 5.58: The effect of the size of the bandwidth on the Gaussian-weighted local linear regression learner.

Figure 5.58 shows that the learner overfits the data for small bandwidths such as $\lambda = 1$, where the learner is effectively interpolating the data. At the other extreme, for large bandwidths such as $\lambda = 100$, the learner underfits, because it eventually coincides with the simple linear regression line of best fit. A compromise between the extremes of over- and underfitting can be achieved with a judicious choice of λ. Indeed, Figure 5.58 suggests that the value of $\lambda = 10$ provides a reasonable model for the chlorine data that is comparable to the one displayed on Figure 5.50. ■

5.6.2 Selecting λ via Bandwidth Perturbation Matching

As Example 5.57 illustrates, the choice of λ is crucial to the performance of the learner $g_{\mathcal{T}}(\cdot \mid \lambda)$. In order to derive a reliable data-driven bandwidth selection method, we first need to find a suitable metric of statistical accuracy. Instead of the usual squared-error loss $(y - g_{\mathcal{T}}(x \mid \lambda))^2$, consider the loss function

$$\text{Loss}(g^*(x), g_{\mathcal{T}}(x \mid \lambda)) = (g^*(x) - g_{\mathcal{T}}(x \mid \lambda))^2,$$

which measures the squared error between the true regression function g^* and the learner $g_{\mathcal{T}}(\cdot \mid \lambda)$, evaluated at x. This loss yields the risk

$$\ell(g_{\mathcal{T}}(\cdot \mid \lambda)) = \mathbb{E}_f (g^*(X) - g_{\mathcal{T}}(X \mid \lambda))^2 = \int f(x)(g^*(x) - g_{\mathcal{T}}(x \mid \lambda))^2 \, \mathrm{d}x,$$

where f is the pdf of the input variable X. Note that both f and g^* are generally unknown.

Having defined suitable loss and risk functions, we next need to understand how $\lambda = \lambda_n$ affects the conditional expectation of the risk given the predictors $\mathbf{V}$; that is, $\mathbb{E}_{\mathbf{V}}\, \ell(g_{\mathcal{T}_n}(\cdot \mid \lambda_n))$. To this end, we define the Gaussian-weighted averages:

$$S_{r,n}(x) := \frac{1}{n} \sum_{i=1}^{n} \varphi_{\lambda_n^2}(X_i - x) \left(\frac{X_i - x}{\lambda_n} \right)^r, \quad r = 0, 1, 2, \ldots,$$

with expectation $\mathbb{E} S_{r,n}(x) = \mathbb{E}[Z^r f(x + \lambda_n Z)]$, where $Z \sim \mathcal{N}(0, 1)$. Then, recalling the notation for little-o order in probability $o_{\mathbb{P}}$ and convergence in probability $\xrightarrow{\mathbb{P}}$, we make the following assumptions.

☞ 652

Assumptions 5.59 (Local Linear Regression): For $Y = g^*(x) + \varepsilon(x)$, where $g^*(x) = \mathbb{E}[Y \mid X = x]$, assume the following:

1. **Bandwidth Asymptotics.** $\lambda_n \downarrow 0$ as $n \to \infty$ in such a way that $n\lambda_n \to \infty$.

2. **Gaussian-weighted Averages.** $S_{r,n}(x) \xrightarrow{\mathbb{P}} \mathbb{E}[Z^r f(x + \lambda_n Z)]$, but such that:

$$S_{r,n}(x) = \begin{cases} f(x)\, \mathbb{E} Z^r + o_{\mathbb{P}}(1), & \text{odd } r, \\ \lambda_n f'(x)\, \mathbb{E} Z^{r+1} + o_{\mathbb{P}}(\lambda_n), & \text{even } r. \end{cases}$$

3. **Finite Support for f.** The density f of X has support on a bounded interval $[a, b]$ and $\inf_{x \in [a,b]} f(x) > \epsilon$ for some $\epsilon > 0$.

4. **Boundedness.** The second derivatives of g^*, the density f, and the variance $\sigma^2 : x \mapsto \mathbb{E}[\varepsilon^2(X) \mid X = x]$, are all continuous and bounded.

Under Assumptions 5.59, we have the following statistical properties of the learner, whose proof is delegated to Exercise 20.

Lemma 5.60 (Local Linear Regression): Let $\Delta g^*(x)$ be the second derivative of g^* at x, and let $\widetilde{\lambda}_n$ be a bandwidth such that $\widetilde{\lambda}_n = \mathcal{O}(\lambda_n)$ and $\lambda_n = \mathcal{O}(\widetilde{\lambda}_n)$. Then, conditional on V, the learners $g_{\mathcal{T}_n}(x \mid \lambda_n)$ and $g_{\mathcal{T}_n}(x \mid \widetilde{\lambda}_n)$ satisfy the following asymptotic bias and covariance formulas:

$$\text{Bias:} \qquad \mathbb{E}_V\, g_{\mathcal{T}_n}(x \mid \lambda_n) - g^*(x) = \frac{\lambda_n^2}{2}\Delta g^*(x) + o_{\mathbb{P}}(\lambda_n^2)$$

$$\text{Covariance:} \qquad \mathbb{C}\mathrm{ov}_V\left(g_{\mathcal{T}_n}(x \mid \lambda_n),\, g_{\mathcal{T}_n}(x \mid \widetilde{\lambda}_n)\right) = \frac{\sigma^2(x)/f(x)}{n\sqrt{2\pi(\lambda_n^2 + \widetilde{\lambda}_n^2)}} + o_{\mathbb{P}}\left(\frac{1}{n\lambda_n}\right).$$

Lemma 5.60 suggests that the bias of $g_{\mathcal{T}_n}(x \mid \lambda_n)$ is large when either the curvature $|\Delta g^*(x)|$ of $g^*(x)$ is large, or the bandwidth λ_n is large. On the other hand, the variance of $g_{\mathcal{T}_n}(x \mid \lambda_n)$ is large when either the ratio $\sigma^2(x)/f(x)$ is large, or the bandwidth λ_n is small.

Importantly, Lemma 5.60 and the bias–variance decomposition (2.34) allow us to derive the leading-order asymptotic term of the conditional risk: ☞ 36

$$\begin{aligned}
\mathbb{E}_V\, \ell(g_{\mathcal{T}_n}(\cdot \mid \lambda_n)) &= \int f(x)\, \mathbb{E}_V \left(g^*(x) - g_{\mathcal{T}_n}(x \mid \lambda_n)\right)^2 \mathrm{d}x \\
&= \int f(x)\Big(\left(g^*(x) - \mathbb{E}_V\, g_{\mathcal{T}_n}(x \mid \lambda_n)\right)^2 + \mathbb{V}\mathrm{ar}_V\, g_{\mathcal{T}_n}(x \mid \lambda_n)\Big)\mathrm{d}x.
\end{aligned}$$

More precisely, Lemma 5.60 with $\lambda_n = \widetilde{\lambda}_n$ yields:

$$\mathbb{E}_V\, \ell(g_{\mathcal{T}_n}(\cdot \mid \lambda_n)) = \frac{\lambda_n^4}{4}\mathbb{E}(\Delta g^*(X))^2 + \frac{\|\sigma\|^2}{\sqrt{4\pi}\, n\lambda_n} + o_{\mathbb{P}}(\lambda_n^4) + o_{\mathbb{P}}\left(\frac{1}{n\lambda_n}\right),$$

where $\|\cdot\|$ is the norm on the $L^2([a,b],\mathrm{d}x)$ space of square integrable functions; see Definition B.4. It follows that the minimizer of the leading-order asymptotic terms of ☞ 564 $\mathbb{E}_V\, \ell(g_{\mathcal{T}_n}(\cdot \mid \lambda_n))$ is:

$$\text{(5.61)} \qquad \widetilde{\lambda}_n := \underset{\lambda > 0}{\operatorname{argmin}} \left\{\frac{\lambda^4}{4}\mathbb{E}(\Delta g^*(X))^2 + \frac{\|\sigma\|^2}{\sqrt{4\pi}\, n\lambda}\right\} = \left(\frac{\|\sigma\|^2}{n\sqrt{4\pi}\,\mathbb{E}(\Delta g^*(X))^2}\right)^{1/5}.$$

Substituting the bandwidth (5.61) into $\mathbb{E}_V\, \ell(g_{\mathcal{T}_n}(\cdot \mid \lambda_n))$ yields the minimum

$$\text{(5.62)} \qquad \begin{aligned}
\mathbb{E}_V\, \ell(g_{\mathcal{T}_n}(\cdot \mid \widetilde{\lambda}_n)) &= \frac{5\|\sigma\|^2}{4\sqrt{4\pi} n \widetilde{\lambda}_n} + o_{\mathbb{P}}(\widetilde{\lambda}_n^4) \\
&= \frac{5}{4}\left[\frac{\|\sigma\|^2 \sqrt{\mathbb{E}(\Delta g^*(X))^2}}{\sqrt{4\pi} n}\right]^{4/5} + o_{\mathbb{P}}(n^{-4/5}).
\end{aligned}$$

We are now ready to state the main result, which is then used to derive an estimator of the asymptotically optimal bandwidth (5.61). To set the scene, let $c = 0.2049923\ldots$ satisfy the equation

$$1 + \frac{1}{\sqrt{c}} - \frac{2\sqrt{2}}{\sqrt{1+c}} = (1-c)^2,$$

and define the **bandwidth perturbation matching** (BPM) function:

$$\delta_{\mathcal{T}}(x \mid \lambda) := \frac{g_{\mathcal{T}}(x \mid \lambda) - g_{\mathcal{T}}(x \mid \sqrt{c}\lambda)}{1-c}.$$

As we prove in Theorem 5.63 below, the BPM function $\delta_{\mathcal{T}_n}(x \mid \lambda_n)$ has the *same* asymptotic bias and variance as $g_{\mathcal{T}_n}(x \mid \lambda_n)$, and so its conditional risk

$$\int f(x)\,\mathbb{E}_{\mathbf{Y}}\,[\delta^2_{\mathcal{T}_n}(x \mid \lambda_n)]\,\mathrm{d}x$$

asymptotically *matches* $\mathbb{E}_{\mathbf{Y}}\,\ell(g_{\mathcal{T}_n}(\cdot \mid \lambda_n))$ in (5.62), but with the important distinction that the unknown g^* is not required for its computation. This distinction allows us to use the conditional risk of $\delta_{\mathcal{T}_n}(\cdot \mid \lambda_n)$ as a surrogate for the unknown $\mathbb{E}_{\mathbf{Y}}\,\ell(g_{\mathcal{T}_n}(\cdot \mid \lambda_n))$ in (5.62), as described in the following result.

Theorem 5.63 (Bandwidth Perturbation Matching Estimator): The BPM function $\delta_{\mathcal{T}_n}(\cdot \mid \lambda_n)$ has the same asymptotic bias and variance as $g_{\mathcal{T}_n}(\cdot \mid \lambda_n)$. Moreover, the asymptotically optimal bandwidth in (5.61) satisfies the equation:

$$\int f(x)\,\mathbb{E}_{\mathbf{Y}}\,[\delta^2_{\mathcal{T}_n}(x \mid \lambda)]\,\mathrm{d}x - \frac{5\|\sigma\|^2}{4\sqrt{4\pi}\,n\lambda} = o_{\mathbb{P}}(n^{-4/5}). \tag{5.64}$$

Proof. For simplicity of notation let $\lambda := \lambda_n$. First, an application of the bias formula in Lemma 5.60 to $g^*(x) - \mathbb{E}_{\mathbf{Y}}\,g_{\mathcal{T}_n}(x \mid \lambda)$ and $g^*(x) - \mathbb{E}_{\mathbf{Y}}\,g_{\mathcal{T}_n}(x \mid \sqrt{c}\lambda)$ yields:

$$\begin{aligned}
\mathbb{E}_{\mathbf{Y}}\,\delta_{\mathcal{T}_n}(x \mid \lambda) &= \frac{\mathbb{E}_{\mathbf{Y}}\,[g_{\mathcal{T}_n}(x \mid \lambda) - g^*(x)] - \mathbb{E}_{\mathbf{Y}}\,[g_{\mathcal{T}_n}(x \mid \sqrt{c}\lambda) - g^*(x)]}{1-c} \\
&= \frac{1}{2}\Delta g^*(x)\frac{\lambda^2 - (\sqrt{c}\lambda)^2}{1-c} + o_{\mathbb{P}}(\lambda^2) = \frac{\lambda^2}{2}\Delta g^*(x) + o_{\mathbb{P}}(\lambda^2).
\end{aligned}$$

Second, repeated application of the covariance formula in Lemma 5.60 yields:

$$\begin{aligned}
&\mathbb{V}\mathrm{ar}_{\mathbf{Y}}\,\delta_{\mathcal{T}_n}(x \mid \lambda) = \\
&= \frac{\mathbb{V}\mathrm{ar}_{\mathbf{Y}}\,(g_{\mathcal{T}_n}(x \mid \lambda)) + \mathbb{V}\mathrm{ar}_{\mathbf{Y}}\,(g_{\mathcal{T}_n}(x \mid \sqrt{c}\lambda)) - 2\mathbb{C}\mathrm{ov}_{\mathbf{Y}}\,(g_{\mathcal{T}_n}(x \mid \lambda), g_{\mathcal{T}_n}(x \mid \sqrt{c}\lambda))}{(1-c)^2} \\
&= \frac{\sigma^2(x)/f(x)}{\sqrt{2\pi}\,n}\,\frac{\frac{1}{\sqrt{\lambda^2+\lambda^2}} + \frac{1}{\sqrt{c\lambda^2+c\lambda^2}} - \frac{2}{\sqrt{\lambda^2+c\lambda^2}}}{(1-c)^2} + o_{\mathbb{P}}\left(\frac{1}{n\lambda}\right) \\
&= \frac{\sigma^2(x)/f(x)}{\sqrt{4\pi}\,n\lambda}\,\frac{1 + \frac{1}{\sqrt{c}} - \frac{2\sqrt{2}}{\sqrt{1+c}}}{(1-c)^2} + o_{\mathbb{P}}\left(\frac{1}{n\lambda}\right) = \frac{\sigma^2(x)/f(x)}{\sqrt{4\pi}\,n\lambda} + o_{\mathbb{P}}\left(\frac{1}{n\lambda}\right).
\end{aligned}$$

In other words, the BPM function has the same bias and variance as the learner $g_{\mathcal{T}_n}(x \mid \lambda_n)$. Moreover, using the identity $\mathbb{E}_V\, \delta^2_{\mathcal{T}_n} = (\mathbb{E}_V\, \delta_{\mathcal{T}_n})^2 + \mathbb{V}\mathrm{ar}_V\, \delta_{\mathcal{T}_n}$, we deduce that

$$\int f(x)\, \mathbb{E}_V\, [\delta^2_{\mathcal{T}_n}(x \mid \lambda)]\, \mathrm{d}x = \frac{\lambda^4}{4}\mathbb{E}\big(\Delta g^*(X)\big)^2 + \frac{\|\sigma\|^2}{\sqrt{4\pi}\, n\lambda} + o_{\mathbb{P}}(\lambda^4) + o_{\mathbb{P}}\Big(\frac{1}{n\lambda}\Big).$$

When evaluated at $\lambda = \widetilde{\lambda}_n$ given in (5.61), the last equation becomes

$$\int f(x)\, \mathbb{E}_V\, [\delta^2_{\mathcal{T}_n}(x \mid \widetilde{\lambda}_n)]\, \mathrm{d}x = \frac{5\|\sigma\|^2}{4\sqrt{4\pi}n\widetilde{\lambda}_n} + o_{\mathbb{P}}(\widetilde{\lambda}_n^4),$$

which matches (5.64), because $\widetilde{\lambda}_n^4 = O(n^{-4/5})$. □

At face value, to compute the asymptotically optimal bandwidth (5.61) we need to estimate the values of the two unknown quantities $\mathbb{E}(\Delta g^*(X))^2$ and $\|\sigma\|^2$. However, Theorem 5.63 asserts that the bandwidth (5.61) is also a solution to equation (5.64), which does not depend on $\mathbb{E}(\Delta g^*(X))^2$. The last observation suggest that the computation of (5.61) requires only knowledge of $\|\sigma\|^2$. The latter can be readily estimated when the error variance is constant: $\sigma^2(x) = \sigma^2$ for all $x \in [a, b]$.

With constant error variance, $\|\sigma\|^2 = \int_a^b \sigma^2 \mathrm{d}x = (b-a)\sigma^2$, and we need to estimate both σ^2 and $b - a$. The length $b - a$ can be estimated by the range

$$R_n := \max_i X_i - \min_i X_i$$

of the input data, and the error variance σ^2 can be estimated via the leave-one-out or n-fold cross-validation estimator (see Exercise 21):

$$\widehat{\sigma}^2_{\lambda_n} := \frac{1}{n}\sum_{i=1}^{n}\left(\frac{Y_i - g_{\mathcal{T}_n}(X_i \mid \lambda_n)}{1 - \mathbf{L}_{i,i}}\right)^2 = \sigma^2 + o_{\mathbb{P}}(n^{-1/2}), \tag{5.65}$$

where $\mathbf{L}_{i,i} := l_i\big((V - X_i\mathbf{1})/\lambda_n\big)$ is the i-th diagonal element of the $n \times n$ matrix:

$$\mathbf{L} := \begin{bmatrix} \boldsymbol{l}\big((V - X_1\mathbf{1})/\lambda_n\big)^\top \\ \vdots \\ \boldsymbol{l}\big((V - X_n\mathbf{1})/\lambda_n\big)^\top \end{bmatrix}.$$

Equation (5.64) in Theorem 5.63, together with the error variance estimator (5.65), indicate that a reasonable estimator for the asymptotically optimal bandwidth (5.61) is the root of the function:

$$h(\lambda) := \frac{\frac{1}{n}\sum_{i=1}^n \delta^2_{\mathcal{T}_n}(X_i \mid \lambda)}{5R_n\, \widehat{\sigma}^2_\lambda / \big(4\sqrt{4\pi}n\lambda\big)} - 1. \tag{5.66}$$

We call the smallest root $\widehat{\lambda}_n$ of (5.66) the **BPM estimator** of $\widetilde{\lambda}_n$. Note that (5.66) typically has another large root that corresponds to the line of best fit (Example 5.57 illustrated how the learner converges to the line of best fit for large λ). Of course, this large root of (5.66) is not of interest, because it violates the bandwidth asymptotics of $\lambda_n \downarrow 0$ in Assumptions 5.59.

■ **Example 5.67 (Chlorine Data Set Revisited (cont.))** For the chlorine data set, the function h is plotted in Figure 5.68. We find here that the BPM estimate of the optimal bandwidth is $\widehat{\lambda}_n = 13.04847$, which gives a similar fit as the curve in Figure 5.58 with $\lambda = 10$.

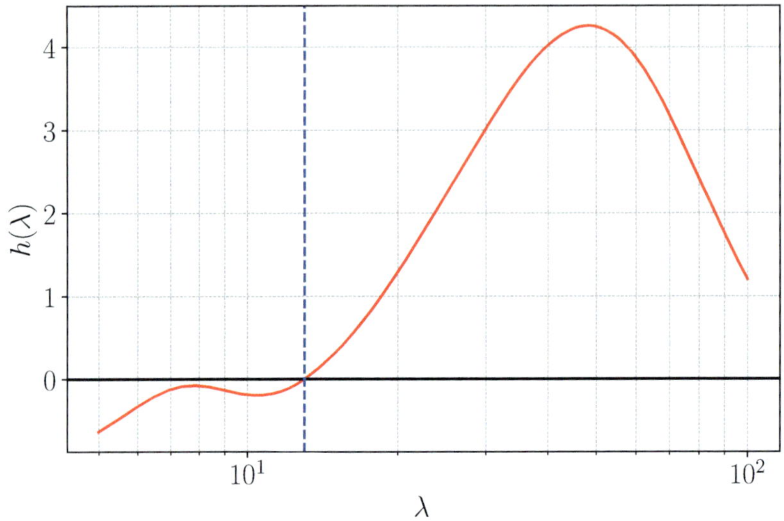

Figure 5.68: The BPM estimate $\widehat{\lambda}_n$ is the smallest root of the function h in (5.66).

The Python code below is a basic implementation that can be improved upon by, among other things, computing $\boldsymbol{l}(\boldsymbol{u})$ via the more efficient formula (5.93) in Exercise 19, rather than via (5.56).

localreg.py

```
import numpy as np
import matplotlib.pyplot as plt
from scipy.optimize import brentq
from gwllr import gwllr


def kernMat(lam, x):
    n = len(x)
    L = np.zeros((n, n))
    for i in range(n):
        u = (x - x[i]) / lam
        D = np.exp(-0.25 * u**2)
        X = np.column_stack((D, D * u))
        l = np.linalg.solve(X.T @ X, np.array([1, 0]))
        L[i, :] = (D * (X @ l)).T
    return L

def bpmEst(lam, x, y):
    c = 0.2049923
    n = len(x)
    r = np.max(x) - np.min(x)
```

```
    L = kernMat(lam, x)
    Lc = kernMat(np.sqrt(c) * lam, x)
    delta = (L @ y - Lc @ y) / (1 - c)
    sig2hat = np.sum(((y - L @ y) / (1 - np.diag(L))) ** 2) / n
    num = np.sum(delta ** 2) / n
    den = 5 * sig2hat * r / (4 * np.sqrt(4 * np.pi) * n * lam)
    val = num / den - 1
    return val


def h(lam): return bpmEst(lam, t, y)

data = np.loadtxt("chlorine.csv", delimiter=",")
t, y = data[:, 0], data[:, 1]
lam_grid = np.linspace(5, 100, 200)
h_values = np.array([h(lam) for lam in lam_grid])

for i in range(1, len(lam_grid)):
    if h(lam_grid[i - 1]) * h(lam_grid[i]) < 0:
        lam_star = brentq(h, lam_grid[i - 1], lam_grid[i])
        break
print(f"Optimal lambda from fzero: {lam_star:.5f}")

grid = np.linspace(min(t), max(t), 200)
yhat = gwllr(t, y, grid, lam=lam_star)

plt.scatter(t, y, color="blue", label="Data")
plt.plot(grid, yhat, color="red", linewidth=2)
plt.xlabel("t")
plt.ylabel("y")
plt.show()

plt.semilogx(lam_grid, h_values, 'r')
plt.xlabel("lambda")
plt.ylabel("h(lambda)")
plt.axhline(0, color='k')
plt.axvline(lam_star, color='b', linestyle='--')
plt.grid(which='both', linestyle='--', linewidth=0.5)
plt.show()
```

```
Optimal lambda from fzero: 13.04847
```

■

Finally, we mention that a common alternative to the BPM estimator is the minimizer of the leave-one-out cross-validation loss (5.65), that is,

$$\widehat{\lambda}_{\text{CV}} := \underset{\lambda>0}{\operatorname{argmin}} \sum_{i=1}^{n} \left(\frac{Y_i - g_{\mathcal{T}_n}(X_i \mid \lambda)}{1 - l_i\big((V - X_i\mathbf{1})/\lambda\big)} \right)^2 .$$

Whilst $\widehat{\lambda}_{\text{CV}}$ is typically very close in value to the BPM estimator $\widehat{\lambda}_n$ for large n, it can be highly variable and unstable for small n. Occasionally, with small n, the cross-validation loss (5.65) may not even exhibit a local minimum. For these reasons, we recommend using the BPM estimator instead.

5.7 Linear Models in Python

In this section we describe how to define and analyze linear models using Python and the data science module **statsmodels**. We encourage the reader to regularly refer back to the theory in the preceding sections of this chapter, so as to avoid using Python merely as a black box without understanding the underlying principles. To run the code start by importing the following code snippet:

```
import matplotlib.pyplot as plt
import pandas as pd
import statsmodels.api as sm
from statsmodels.formula.api import ols
```

5.7.1 Modeling

Although specifying a normal[3] linear model in Python is relatively easy, it requires some subtlety. The main thing to realize is that Python treats quantitative and qualitative (that is, categorical) explanatory variables differently. In **statsmodels**, ordinary least-squares linear models are specified via the function **ols** (short for ordinary least-squares). The main argument of this function is a formula of the form

$$\texttt{y} \sim \texttt{x1} + \texttt{x2} + \cdots + \texttt{xd}, \tag{5.69}$$

where `y` is the name of the response variable and `x1`, ..., `xd` are the names of the explanatory variables. If all variables are *quantitative*, this describes the linear model

$$Y_i = \beta_0 + \beta_1 x_{i1} + \beta_2 x_{i2} + \cdots + \beta_d x_{id} + \varepsilon_i, \quad i = 1, \ldots, n, \tag{5.70}$$

where x_{ij} is the j-th explanatory variable for the i-th observation and the errors ε_i are independent normal random variables such that $\mathbb{E}\varepsilon_i = 0$ and $\mathbb{V}\mathrm{ar}\,\varepsilon_i = \sigma^2$. Or, in matrix form: $\boldsymbol{Y} = \mathbf{X}\boldsymbol{\beta} + \boldsymbol{\varepsilon}$, with

$$\boldsymbol{Y} = \begin{bmatrix} Y_1 \\ \vdots \\ Y_n \end{bmatrix}, \quad \mathbf{X} = \begin{bmatrix} 1 & x_{11} & \cdots & x_{1d} \\ 1 & x_{21} & \cdots & x_{2d} \\ \vdots & \vdots & \ddots & \vdots \\ 1 & x_{n1} & \cdots & x_{nd} \end{bmatrix}, \quad \boldsymbol{\beta} = \begin{bmatrix} \beta_0 \\ \vdots \\ \beta_d \end{bmatrix}, \text{ and } \boldsymbol{\varepsilon} = \begin{bmatrix} \varepsilon_1 \\ \vdots \\ \varepsilon_n \end{bmatrix}.$$

Thus, the first column is always taken as an "intercept" parameter, unless otherwise specified. To remove the intercept term, add `-1` to the `ols` formula, as in `ols('y~x-1')`.

For any linear model, the model matrix can be retrieved via the construction:

```
model_matrix = pd.DataFrame(model.exog,columns=model.exog_names)
```

Let us look at some examples of linear models. In the first model the variables `x1` and `x2` are both considered (by Python) to be quantitative.

[3]For the rest of this section, we assume all linear models to be normal.

```
myData = pd.DataFrame({'y' : [10,9,4,2,4,9],
     'x1' : [7.4,1.2,3.1,4.8,2.8,6.5],
     'x2' : [1,1,2,2,3,3]})
mod = ols("y~x1+x2", data=myData)
mod_matrix = pd.DataFrame(mod.exog,columns=mod.exog_names)
print(mod_matrix)
```

```
   Intercept   x1   x2
0        1.0  7.4  1.0
1        1.0  1.2  1.0
2        1.0  3.1  2.0
3        1.0  4.8  2.0
4        1.0  2.8  3.0
5        1.0  6.5  3.0
```

Suppose the second variable is actually qualitative; e.g., it represents a color, and the levels 1, 2, and 3 stand for red, blue, and green. We can account for such a categorical variable by using the **astype** method to redefine the data type (see Section 1.2). ☞ 3

```
myData['x2'] = myData['x2'].astype('category')
```

Alternatively, a categorical variable can be specified in the model formula by wrapping it with C(). Observe how this changes the model matrix.

```
mod2 = ols("y~x1+C(x2)", data=myData)
mod2_matrix = pd.DataFrame(mod2.exog,columns=mod2.exog_names)
print(mod2_matrix)
```

```
   Intercept  C(x2)[T.2]  C(x2)[T.3]   x1
0        1.0         0.0         0.0  7.4
1        1.0         0.0         0.0  1.2
2        1.0         1.0         0.0  3.1
3        1.0         1.0         0.0  4.8
4        1.0         0.0         1.0  2.8
5        1.0         0.0         1.0  6.5
```

Thus, if a **statsmodels** formula of the form (5.69) contains factor (qualitative) variables, the model is no longer of the form (5.70), but contains indicator variables for each level of the factor variable, except the first level.

For the case above, the corresponding linear model is

$$Y_i = \beta_0 + \beta_1 x_{i1} + \alpha_2 \, \mathbb{1}\{x_{i2} = 2\} + \alpha_3 \, \mathbb{1}\{x_{i2} = 3\} + \varepsilon_i, \quad i = 1, \ldots, 6, \tag{5.71}$$

where we have used parameters α_2 and α_3 to correspond to the indicator features of the qualitative variable. The parameter α_2 describes how much the response is expected to change if the factor x_2 switches from level 1 to 2. A similar interpretation holds for α_3. Such parameters can thus be viewed as incremental effects.

It is also possible to model **interaction** between two variables. For two continuous variables, this simply adds the products of the original features to the model matrix. Adding interaction terms in Python is achieved by replacing "+" in the formula with "*", as the following example illustrates.

```
mod3 = ols("y~x1*C(x2)", data=myData)
mod3_matrix = pd.DataFrame(mod3.exog,columns=mod3.exog_names)
print(mod3_matrix)
```

```
   Intercept  C(x2)[T.2]  C(x2)[T.3]   x1  x1:C(x2)[T.2]  x1:C(x2)[T.3]
0        1.0         0.0         0.0  7.4            0.0            0.0
1        1.0         0.0         0.0  1.2            0.0            0.0
2        1.0         1.0         0.0  3.1            3.1            0.0
3        1.0         1.0         0.0  4.8            4.8            0.0
4        1.0         0.0         1.0  2.8            0.0            2.8
5        1.0         0.0         1.0  6.5            0.0            6.5
```

5.7.2 Analysis

Let us consider some easy linear regression models by using the student survey data set survey.csv from the book's GitHub site, which contains measurements such as height, weight, sex, etc., from a survey conducted among $n = 100$ university students. Suppose we wish to investigate the relation between the shoe size (explanatory variable) and the height (response variable) of a person. First, we load the data and draw a scatterplot of the points (height versus shoe size); see Figure 5.72 (without the fitted line).

```
survey = pd.read_csv('survey.csv')
plt.scatter(survey.shoe, survey.height)
plt.xlabel("Shoe size")
plt.ylabel("Height")
```

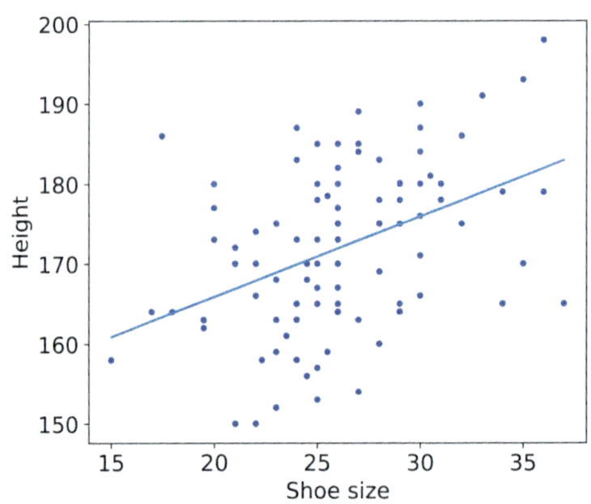

Figure 5.72: Scatterplot of height (cm) against shoe size (cm), with the fitted line.

We observe a slight increase in the height as the shoe size increases, although this relationship is not very distinct. We analyze the data through the simple linear regression model:
☞ 199

$$Y_i = \beta_0 + \beta_1 x_i + \varepsilon_i, \quad i = 1, \ldots, n.$$

In statsmodels this is performed via the ols method as follows:

```
model = ols("height~shoe", data=survey) # define the model
fit = model.fit()  #fit the model defined above
b0, b1 = fit.params
print(fit.params)
```

```
Intercept     145.777570
shoe            1.004803
dtype: float64
```

The above output gives the least-squares estimates of β_0 and β_1. For this example, we have $\widehat{\beta}_0 = 145.778$ and $\widehat{\beta}_1 = 1.005$. Figure 5.72, which includes the regression line, was obtained as follows:

```
plt.plot(survey.shoe, b0 + b1*survey.shoe)
plt.scatter(survey.shoe, survey.height)
plt.xlabel("Shoe size")
plt.ylabel("Height")
```

Although **ols** performs a complete analysis of the linear model, not all its calculations need to be presented. A summary of the results can be obtained with the method **summary**.

```
print(fit.summary())
```

```
Dep. Variable:              height   R-squared:                       0.178
Model:                         OLS   Adj. R-squared:                  0.170
Method:              Least Squares   F-statistic:                     21.28
No. Observations:              100   Prob (F-statistic):           1.20e-05
Df Residuals:                   98   Log-Likelihood:                -363.88
Df Model:                        1   AIC:                             731.8
Covariance Type:         nonrobust   BIC:                             737.0
====================================================================
              coef    std err          t      P>|t|     [0.025     0.975]
--------------------------------------------------------------------
Intercept  145.7776     5.763     25.296     0.000    134.341    157.214
shoe         1.0048     0.218      4.613     0.000      0.573      1.437
====================================================================
Omnibus:                     1.958   Durbin-Watson:                   1.772
Prob(Omnibus):               0.376   Jarque-Bera (JB):                1.459
Skew:                       -0.072   Prob(JB):                        0.482
Kurtosis:                    2.426   Cond. No.                         164.
```

The main output items are the following:

- `coef`: Estimates of the parameters of the regression line.
- `std error`: Standard deviations of the estimators of the regression line. These are the square roots of the variances of the $\{\widehat{\beta}_i\}$ obtained in (5.40). ☞ 217
- `t`: Realization of Student's test statistics associated with the hypotheses $H_0 : \beta_i = 0$ and $H_1 : \beta_i \neq 0, i = 0, 1$. In particular, the outcome of T in (5.32). ☞ 214
- `P>|t|`: P-value of Student's test (two-sided test).

- `[0.025 0.975]`: 95% confidence intervals for the parameters.

☞ 212

- `R-Squared:` Coefficient of determination R^2 (percentage of variation explained by the regression), as defined in (5.30).

- `Adj. R-Squared:` adjusted R^2 (explained in Section 5.3.7).

☞ 214

- `F-statistic:` Realization of the F test statistic (5.35) associated with testing the full model against the default model. The associated degrees of freedom (`Df Model` = 1 and `Df Residuals` = $n - 2$) are given. The P-value is given in the field `Prob (F-statistic)`.

☞ 208

- `AIC:` The AIC number in (5.21); that is, minus two times the log-likelihood plus two times the number of model parameters (which is 3 here).

You can access all the numerical values as they are attributes of the `fit` object. First check which names are available, as in `dir(fit)`. Then access the values via the dot construction. For example, the following extracts the P-value for the slope.

```
fit.pvalues.iloc[1]
```

```
1.1994e-05
```

The results show strong evidence for a linear relationship between shoe size and height (or, more accurately, strong evidence that the slope of the regression line is not zero), as the P-value for the corresponding test is very small ($1.2 \cdot 10^{-5}$). The estimate of the slope indicates that the difference between the average height of students whose shoe size is different by one cm is 1.0048 cm.

Only 17.84% of the variability of student height is explained by the shoe size. We therefore need to add other explanatory variables to the model (multiple linear regression) to increase the model's predictive power.

5.7.3 Analysis of Variance (ANOVA)

We continue the student survey example of the previous section, but now add an extra variable, and also consider an analysis of variance of the model. Instead of "explaining" the student height via their shoe size, we include `weight` as an explanatory variable. The corresponding `ols` formula for this model is

$$\texttt{height} \sim \texttt{shoe} + \texttt{weight},$$

meaning that each random height, denoted by `Height`, satisfies

$$\texttt{Height} = \beta_0 + \beta_1 \texttt{shoe} + \beta_2 \texttt{weight} + \varepsilon,$$

where ε is a normally distributed error term with mean 0 and variance σ^2. Thus, the model has 4 parameters. Before analyzing the model we present a scatterplot of all pairs of variables, using `scatter_matrix`.

```
model = ols("height~shoe+weight", data=survey)
fit = model.fit()
axes = pd.plotting.scatter_matrix(survey[['height','shoe','weight']])
plt.show()
```

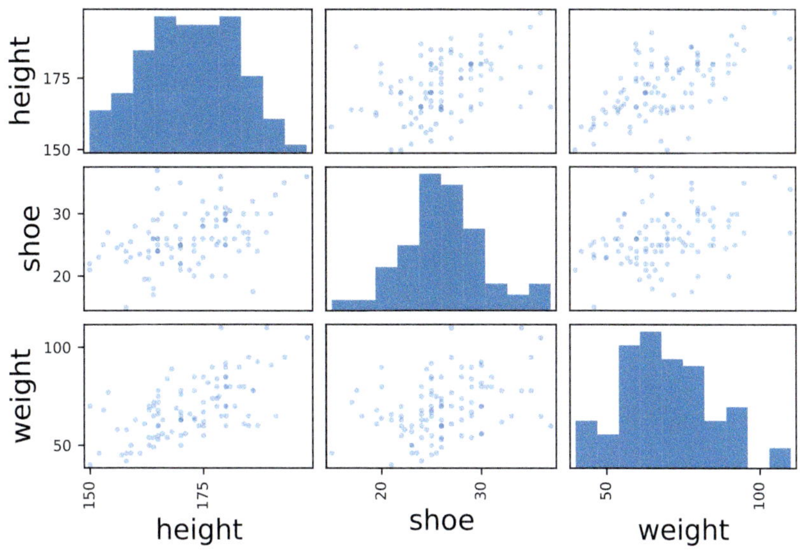

Figure 5.73: Scatterplot of all pairs of variables: height (cm), shoe (cm), and weight (kg).

As for the simple linear regression model in the previous section, we can analyze the model using the **summary** method (below we have omitted some output):

```
fit.summary()
```

```
Dep. Variable:                height   R-squared:                   0.430
Model:                           OLS   Adj. R-squared:              0.418
Method:                Least Squares   F-statistic:                 36.61
No. Observations:                100   Prob (F-statistic):       1.43e-12
Df Residuals:                     97   Log-Likelihood:          -345.58
Df Model:                          2   AIC:                         697.2
                                       BIC:                         705.0
======================================================================
                coef    std err          t      P>|t|     [0.025    0.975]
----------------------------------------------------------------------
Intercept   132.2677      5.247     25.207      0.000    121.853   142.682
shoe          0.5304      0.196      2.703      0.008      0.141     0.920
weight        0.3744      0.057      6.546      0.000      0.261     0.488
```

The `F-statistic` is used to test whether the full model (here with two explanatory variables) is better at "explaining" the height than the default model. The corresponding null hypothesis is $H_0 : \beta_1 = \beta_2 = 0$. The assertion of interest is H_1: at least one of the coefficients β_j $(j = 1, 2)$ is significantly different from zero. Given the result of this test (P-value = $1.429 \cdot 10^{-12}$), we can conclude that at least one of the explanatory variables is associated with height. The individual Student tests indicate that:

- shoe size is linearly associated with student height, after adjusting for weight, with P-value 0.0081. At the same weight, an increase of one cm in shoe size corresponds to an increase of 0.53 cm in average student height;

- weight is linearly associated with student height, after adjusting for shoe size (the P-value is actually $2.82 \cdot 10^{-09}$; the reported value of 0.000 should be read as "less than 0.001"). At the same shoe size, an increase of one kg in weight corresponds to an increase of 0.3744 cm in average student height.

Further understanding is extracted from the model by conducting an analysis of variance. The standard `statsmodels` function is `anova_lm`. The input to this function is the fit object (obtained from `model.fit()`) and the output is a `DataFrame` object.

```
table = sm.stats.anova_lm(fit)
print(table)
```

```
            df       sum_sq      mean_sq          F        PR(>F)
shoe       1.0  1840.467359  1840.467359  30.371310  2.938651e-07
weight     1.0  2596.275747  2596.275747  42.843626  2.816065e-09
Residual  97.0  5878.091294    60.598879        NaN           NaN
```

The meaning of the columns is as follows:

- `df` : The degrees of freedom of the variables, according to the sum of squares
☞ 212
decomposition (5.29). As both `shoe` and `weight` are quantitative variables, their degrees of freedom are both 1 (each corresponding to a single column in the overall model matrix). The degrees of freedom for the residuals is $n - p = 100 - 3 = 97$.

- `sum_sq`: The sum of squares according to (5.29). The total sum of squares is the sum of all the entries in this column. The residual error in the model that cannot be explained by the variables is RSS ≈ 5878.

- `mean_sq`: The sum of squares divided by their degrees of freedom. Note that the residual square error RSE $= \text{RSS}/(n - p) = 60.6$ is an unbiased estimate of the
☞ 213
model variance σ^2; see Section 5.4.

☞ 215

- `F`: These are the outcomes of the test statistic (5.36).

- `PR(>F)`: These are the P-values corresponding to the test statistic in the preceding column and are computed using an F distribution whose degrees of freedom are given in the `df` column.

The ANOVA table indicates that the `shoe` variable explains a reasonable amount of the variation in the model, as evidenced by a sum of squares contribution of 1840 out of $1840 + 2596 + 5878 = 10314$ and a very small P-value. After `shoe` is included in the model, it turns out that the `weight` variable explains even more of the remaining variability, with an even smaller P-value. The remaining sum of squares (5878) is 57% of the total sum of squares, yielding a 43% reduction, in accordance with the R^2 value reported in the summary for the `ols` method. As mentioned in Section 5.4.1, the *order* in which the ANOVA is conducted is important. To illustrate this, consider the output of the following commands.

```
model = ols("height~weight+shoe", data=survey)
fit = model.fit()
table = sm.stats.anova_lm(fit)
print(table)
```

```
            df        sum_sq      mean_sq          F        PR(>F)
weight     1.0   3993.860167  3993.860167  65.906502  1.503553e-12
shoe       1.0    442.882938   442.882938   7.308434  8.104688e-03
Residual  97.0   5878.091294    60.598879        NaN           NaN
```

We see that weight as a single model variable explains much more of the variability than shoe did. If we now also include shoe, we only obtain a small (but according to the P-value still significant) reduction in the model variability.

5.7.4 Confidence and Prediction Intervals

In statsmodels a method for computing confidence or prediction intervals from a dictionary of explanatory variables is get_prediction. It simply executes formula (5.39) or (5.41). A simpler version is predict, which only returns the predicted value. ☞ 217

Continuing the student survey example, suppose we wish to predict the height of a person with shoe size 30 cm and weight 75 kg. Confidence and prediction intervals can be obtained as given in the code below. The new explanatory variable is entered as a dictionary. Notice that the 95% prediction interval (for the corresponding random response) is much wider than the 95% confidence interval (for the expectation of the random response).

```
x = {'shoe': [30.0], 'weight': [75.0]}  # new input (dict)
pred = fit.get_prediction(x)
pred.summary_frame(alpha=0.05).unstack()
```

```
mean           0    176.261722    # predicted value
mean_se        0      1.054015
mean_ci_lower  0    174.169795    # lower bound for CI
mean_ci_upper  0    178.353650    # upper bound for CI
obs_ci_lower   0    160.670610    # lower bound for PI
obs_ci_upper   0    191.852835    # upper bound for PI
dtype: float64
```

5.7.5 Model Validation

We can perform an analysis of residuals to examine whether the underlying assumptions of the (normal) linear regression model are verified. Various plots of the residuals can be used to inspect whether the assumptions on the errors $\{\varepsilon_i\}$ are satisfied. Figure 5.74 gives two such plots. The first is a scatterplot of the residuals $\{e_i\}$ against the fitted values $\widehat{y}_i$. When the model assumptions are valid, the residuals, as approximations of the model error, should behave approximately as iid normal random variables for each of the fitted values, with a constant variance. In this case we see no strong aberrant structure in this plot. The residuals are fairly evenly spread and symmetrical about the $y = 0$ line (not shown). The second plot is a quantile–quantile (or qq) plot. This is a useful way to check for normality of the error terms, by plotting the sample quantiles of the residuals against the theoretical quantiles of the standard normal distribution. Under the model assumptions, the points should lie approximately on a straight line. For the current case there does not seem to be an extreme departure from normality. Drawing a histogram or density plot of the residuals will also help to verify the normality assumption. The following code was used.

```
plt.plot(fit.fittedvalues,fit.resid,'.')
plt.xlabel("fitted values")
plt.ylabel("residuals")
sm.qqplot(fit.resid)
```

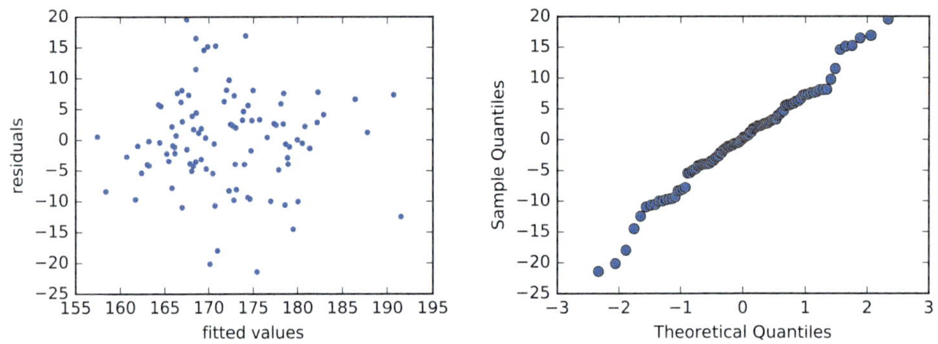

Figure 5.74: Left: residuals against fitted values. Right: a qq plot of the residuals. Neither shows clear evidence against the model assumptions of constant variance and normality.

5.7.6 Variable Selection

Among the large number of possible explanatory variables, we wish to select those which best explain the observed responses. By eliminating redundant explanatory variables, we reduce the statistical error without increasing the approximation error, and thus reduce the (expected) generalization risk of the learner.

In this section, we briefly present two methods for variable selection. They are illustrated on a few variables from the data set birthwt discussed in Section 1.5.3.2.
☞ 13 The data set contains information on the birth weights (masses) of babies, as well as various characteristics of the mother, such as whether she smokes, her age, etc. We wish to explain the child's weight at birth using various characteristics of the mother, her family history, and her behavior during pregnancy. The response variable is weight at birth (quantitative variable bwt, expressed in grams); the explanatory variables are given below.

The data can be obtained as explained in Section 1.5.3.2, or from statsmodels in the following way:

```
bwt = sm.datasets.get_rdataset("birthwt","MASS").data
```

Here is some information about the explanatory variables that we will investigate.

```
age:   mother's age in years
lwt:   mother's weight in lbs
race:  mother's race (1 = white, 2 = black, 3 = other)
smoke: smoking status during pregnancy (0 = no, 1 = yes)
ptl:   no. of previous premature labors
ht:    history of hypertension (0 = no, 1 = yes)
ui:    presence of uterine irritability (0 = no, 1 = yes)
ftv:   no. of physician visits during first trimester
bwt:   birth weight in grams
```

We can see the structure of the variables via `bwt.info()`. Check yourself that all variables are defined as *quantitative* (`int64`). However, the variables `race`, `smoke`, `ht`, and `ui` should really be interpreted as *qualitative* (factors). To fix this, we could redefine them with the method **astype**, similar to what we did in Chapter 1. Alternatively, we could use the `C()` construction in a `statsmodels` formula to let the program know that certain variables are factors. We will use the latter approach.

> For *binary* features it does not matter whether the variables are interpreted as factorial or numerical as the numerical and summary results are identical.

We consider the explanatory variables `lwt`, `age`, `ui`, `smoke`, `ht`, and two recoded binary variables `ftv1` and `ptl1`. We define `ftv1` = 1 if there was at least one visit to a physician, and `ftv1` = 0 otherwise. Similarly, we define `ptl1` = 1 if there is at least one preterm birth in the family history, and `ptl1` = 0 otherwise.

```
ftv1 = (bwt['ftv']>=1).astype(int)
ptl1 = (bwt['ptl']>=1).astype(int)
```

5.7.6.1 Forward Selection and Backward Elimination

The **forward selection** method is an iterative method for variable selection. In the first iteration we consider which feature `f1` is the most significant in terms of its P-value in the models `bwt`~`f1`, with `f1` $\in$ {`lwt`, `age`, ...}. This feature is then selected into the model. In the second iteration, the feature `f2` that has the smallest P-value in the models `bwt`~`f1` + `f2` is selected, where `f2` $\neq$ `f1`, and so on. Usually only features are selected that have a P-value of at most 0.05. The following Python program automates this procedure. Instead of selecting on the P-value one could select on the AIC or BIC value.

forwardselection.py

```
import statsmodels.api as sm
from statsmodels.formula.api import ols
```

```
bwt = sm.datasets.get_rdataset("birthwt","MASS").data
ftv1 = (bwt['ftv']>=1).astype(int)
ptl1 = (bwt['ptl']>=1).astype(int)

remaining_features = {'lwt', 'age', 'C(ui)', 'smoke',
                      'C(ht)', 'ftv1', 'ptl1'}
selected_features = []
while remaining_features:
  PF = []  #list of (P value, feature)
  for f in remaining_features:
    temp = selected_features + [f]  #temporary list of features
    formula = 'bwt~' + '+'.join(temp)
    fit = ols(formula,data=bwt).fit()
    pval= fit.pvalues.iloc[-1]
    if pval < 0.05:
       PF.append((pval,f))
  if PF:  #if not empty
     PF.sort(reverse=True)
     (best_pval, best_f) = PF.pop()
     remaining_features.remove(best_f)
     print('feature {} with P-value = {:.2E}'.
            format(best_f, best_pval))
     selected_features.append(best_f)
  else:
     break
```

```
feature C(ui) with P-value = 7.52E-05
feature C(ht) with P-value = 1.08E-02
feature lwt with P-value = 6.01E-03
feature smoke with P-value = 7.27E-03
```

In **backward elimination** we start with the complete model (all features included) and at each step, we remove the variable with the highest P-value, as long as it is not significant (greater than 0.05). We leave it as an exercise to verify that the order in which the features are removed is: `age`, `ftv1`, and `ptl1`. In this case, forward selection and backward elimination result in the same model, but this need not be the case in general.

This way of model selection has the advantage of being easy to use and of treating the question of variable selection in a systematic manner. The main drawback is that variables are included or deleted based on purely statistical criteria, without taking into account the aim of the study. This usually leads to a model which may be satisfactory from a statistical point of view, but in which the variables are not necessarily the most relevant when it comes to understanding and interpreting the data in the study.

Of course, we can choose to investigate any combination of features, not just the ones suggested by the above variable selection methods. For example, let us see if the mother's weight, her age, her race, and whether she smokes explain the baby's birthweight.

```
formula = 'bwt~lwt+age+C(race)+ smoke'
bwt_model = ols(formula, data=bwt).fit()
print(bwt_model.summary())
```

```
                            OLS Regression Results
==============================================================================
```

```
Dep. Variable: bwt                R-squared: 0.148
Model: OLS                        Adj. R-squared: 0.125
Method: Least Squares             F-statistic: 6.373
No. Observations: 189             Prob (F-statistic):   1.76e-05
Df Residuals: 183                 Log-Likelihood: -1498.4
Df Model: 5                       AIC: 3009.
                                  BIC: 3028.
======================================================================
                 coef    std err        t     P>|t|   [0.025      0.975]
----------------------------------------------------------------------
Intercept   2839.4334    321.435    8.834    0.000  2205.239   3473.628
C(race)[T.2] -510.5015   157.077   -3.250    0.001  -820.416   -200.587
C(race)[T.3] -398.6439   119.579   -3.334    0.001  -634.575   -162.713
smoke       -401.7205    109.241   -3.677    0.000  -617.254   -186.187
lwt            3.9999      1.738    2.301    0.022     0.571      7.429
age           -1.9478      9.820   -0.198    0.843   -21.323     17.427
======================================================================
Omnibus: 3.916                      Durbin-Watson: 0.458
Prob(Omnibus): 0.141                Jarque-Bera (JB): 3.718
Skew: -0.343                        Prob(JB): 0.156
Kurtosis: 3.038                     Cond. No. 899.
```

Given the result of Fisher's global test given by `Prob (F-Statistic)` in the summary (P-value = 1.76×10^{-5}), we can conclude that at least one of the explanatory variables is associated with child weight at birth, after adjusting for the other variables. The individual Student tests indicate that:

- the mother's weight is linearly associated with child weight, after adjusting for age, race, and smoking status (P-value = 0.022). At the same age, race, and smoking status, an increase of one pound in the mother's weight corresponds to an increase of 4 g in the average child weight at birth;
- the age of the mother is not significantly linearly associated with child weight at birth, when mother weight, race, and smoking status are already taken into account (P-value = 0.843);
- weight at birth is significantly lower for a child born to a mother who smokes, compared to children born to non-smoking mothers of the same age, race, and weight, with a P-value of 0.00031 (to see this, inspect `bwt_model.pvalues`). At the same age, race, and mother weight, the child's weight at birth is 401.720 g less for a smoking mother than for a non-smoking mother;
- regarding the interpretation of the variable `race`, we note that the first level of this categorical variable corresponds to white mothers. The estimate of −510.501 g for `C(race)[T.2]` represents the difference in the child's birth weight between black mothers and white mothers (reference group), and this result is significantly different from zero (P-value = 0.001) in a model adjusted for the mother's weight, age, and smoking status.

5.7.6.2 Interaction

We can also include interaction terms in the model. Let us see whether there is any interaction effect between `smoke` and `age` via the model

$$\texttt{Bwt} = \beta_0 + \beta_1 \texttt{age} + \beta_2 \texttt{smoke} + \beta_3 \texttt{age} \times \texttt{smoke} + \varepsilon.$$

In Python this can be done as follows (below we have removed some output):

```
formula = 'bwt~age*smoke'
bwt_model = ols(formula, data=bwt).fit()
print(bwt_model.summary())
```

```
                            OLS Regression Results
================================================================
Dep. Variable: bwt          R-squared: 0.069
Model: OLS                  Adj. R-squared: 0.054
Method: Least Squares       F-statistic: 4.577
No. Observations: 189       Prob (F-statistic):  0.00407
Df Residuals: 183           Log-Likelihood: -1506.8
Df Model: 5                 AIC: 3009.
                            BIC: 3028.
=================================================================
                  coef    std err      t       P>|t|    [0.025    0.975]
-----------------------------------------------------------------
Intercept        2406.1   292.190    8.235    0.000   1829.6    2982.5
smoke             798.2   484.342    1.648    0.101   -157.4    1753.7
age                27.7    12.149    2.283    0.024      3.8      51.7
age:smoke         -46.6    20.447   -2.278    0.024    -86.9      -6.2
```

We observe that the estimate for β_3 (−46.6) is significantly different from zero (P-value = 0.024). We therefore conclude that the effect of the mother's age on the child's weight depends on the smoking status of the mother. The results on association between mother age and child weight must therefore be presented separately for the smoking and the non-smoking group. For non-smoking mothers (`smoke = 0`), the mean child weight at birth increases on average by 27.7 grams for each year of the mother's age. This is statistically significant, as can be seen from the 95% confidence intervals for the parameters (which does not contain zero):

```
bwt_model.conf_int()
```

```
                      0            1
Intercept   1829.605754  2982.510194
age            3.762780    51.699977
smoke       -157.368023  1753.717779
age:smoke    -86.911405    -6.232425
```

Similarly, for smoking mothers, there seems to be a decrease in birthweight, $\widehat{\beta}_1 + \widehat{\beta}_3 = 27.7 - 46.6 = -18.9$, but this is not statistically significant; see Exercise 8.

5.8 Generalized Linear Models

The normal linear model in Section 2.8 deals with continuous response variables — such as height and crop yield — and continuous or discrete explanatory variables. Given the feature vectors $\{\boldsymbol{x}_i\}$, the responses $\{Y_i\}$ are independent of each other, and each has a normal distribution with mean $\boldsymbol{x}_i^\top \boldsymbol{\beta}$, where $\boldsymbol{x}_i^\top$ is the i-th row of the model matrix $\mathbf{X}$. Generalized linear models allow for arbitrary response distributions, including *discrete* ones.

Definition 5.75 (Generalized Linear Model): In a **generalized linear model** (GLM) the expected response for a given feature vector $\boldsymbol{x} = [x_1, \dots, x_p]^\top$ is of the form

$$\mathbb{E}[Y \mid \boldsymbol{X} = \boldsymbol{x}] = h(\boldsymbol{x}^\top \boldsymbol{\beta}) \tag{5.76}$$

for some function h, which is called the **activation function**. The distribution of Y (for a given $\boldsymbol{x}$) may depend on additional *dispersion* parameters that model the randomness in the data that is not explained by $\boldsymbol{x}$.

The *inverse* of function h is called the **link function**. As for the linear model, (5.76) is a model for a single pair $(\boldsymbol{x}, Y)$. Using the model simplification introduced at the end of Section 5.1, the corresponding model for a whole training set $\mathcal{T} = \{(\boldsymbol{x}_i, Y_i)\}$ is that the $\{\boldsymbol{x}_i\}$ are fixed and that the $\{Y_i\}$ are independent; each Y_i satisfying (5.76) with $\boldsymbol{x} = \boldsymbol{x}_i$. Writing $\boldsymbol{Y} = [Y_1, \dots, Y_n]^\top$ and defining $\boldsymbol{h}$ as the multivalued function with components h, we have

$$\mathbb{E}_{\mathbf{X}} \boldsymbol{Y} = \boldsymbol{h}(\mathbf{X}\boldsymbol{\beta}),$$

where $\mathbf{X}$ is the (model) matrix with rows $\boldsymbol{x}_1^\top, \dots, \boldsymbol{x}_n^\top$. A common assumption is that $Y_1, \dots, Y_n$ come from the same family of distributions, e.g., normal, Bernoulli, or Poisson. The central focus is the parameter vector $\boldsymbol{\beta}$, which summarizes how the matrix of explanatory variables $\mathbf{X}$ affects the response vector $\boldsymbol{Y}$. The class of generalized linear models can encompass a wide variety of models. Obviously the normal linear model (2.61) is a generalized linear model, with $\mathbb{E}[Y \mid \boldsymbol{X} = \boldsymbol{x}] = \boldsymbol{x}^\top \boldsymbol{\beta}$, so that h is the identity function. In this case, $Y_i \sim \mathcal{N}(\boldsymbol{x}_i^\top \boldsymbol{\beta}, \sigma^2)$, $i = 1, \dots, n$, where σ^2 is a dispersion parameter.

■ **Example 5.77 (Logistic Regression)** In a **logistic regression** or **logit** model, we assume that the response variables $Y_1, \dots, Y_n$ are independent and distributed according to

$$Y_i \sim \mathsf{Ber}(h(\boldsymbol{x}_i^\top \boldsymbol{\beta})),$$

where h here is defined as the cdf of the **logistic distribution**:

$$h(x) = \frac{1}{1 + \mathrm{e}^{-x}}. \tag{5.78}$$

Large values of $\boldsymbol{x}_i^\top \boldsymbol{\beta}$ thus lead to a high probability that $Y_i = 1$, and small (negative) values of $\boldsymbol{x}_i^\top \boldsymbol{\beta}$ cause Y_i to be 0 with high probability. Estimation of the parameter vector $\boldsymbol{\beta}$ from the observed data is not as straightforward as for the ordinary linear model, but can be accomplished via the minimization of a suitable training loss, as explained below.

As the $\{Y_i\}$ are independent, the pdf of $\boldsymbol{Y} = [Y_1, \dots, Y_n]^\top$ is

$$g(\boldsymbol{y} \mid \boldsymbol{\beta}, \mathbf{X}) = \prod_{i=1}^{n} [h(\boldsymbol{x}_i^\top \boldsymbol{\beta})]^{y_i} [1 - h(\boldsymbol{x}_i^\top \boldsymbol{\beta})]^{1-y_i}.$$

Maximizing the log-likelihood $\ln g(\boldsymbol{y} \mid \boldsymbol{\beta}, \mathbf{X})$ with respect to $\boldsymbol{\beta}$ gives the maximum likelihood estimator of $\boldsymbol{\beta}$. In a supervised learning framework, this is equivalent to minimizing:

$$
\begin{aligned}
-\frac{1}{n}\ln g(\mathbf{y}\,|\,\boldsymbol{\beta},\mathbf{X}) &= -\frac{1}{n}\sum_{i=1}^{n}\ln g(y_i\,|\,\boldsymbol{\beta},\boldsymbol{x}_i) \\
&= -\frac{1}{n}\sum_{i=1}^{n}\left[y_i \ln h(\boldsymbol{x}_i^\top\boldsymbol{\beta}) + (1-y_i)\ln(1-h(\boldsymbol{x}_i^\top\boldsymbol{\beta}))\right].
\end{aligned}
\tag{5.79}
$$

☞ 135 By comparing (5.79) with (4.4), we see that we can interpret (5.79) as the *cross-entropy training loss* between a true conditional pdf $f(y\,|\,\boldsymbol{x})$ and an approximation pdf $g(y\,|\,\boldsymbol{\beta},\boldsymbol{x})$ using the loss function

$$
\text{Loss}(f(y\,|\,\boldsymbol{x}), g(y\,|\,\boldsymbol{\beta},\boldsymbol{x})) := -\ln g(y\,|\,\boldsymbol{\beta},\boldsymbol{x}) = -y\ln h(\boldsymbol{x}^\top\boldsymbol{\beta}) - (1-y)\ln(1-h(\boldsymbol{x}^\top\boldsymbol{\beta})).
$$

Minimizing (5.79) in terms of $\boldsymbol{\beta}$ actually constitutes a *convex* optimization problem. Since $\ln h(\boldsymbol{x}^\top\boldsymbol{\beta}) = -\ln(1+\mathrm{e}^{-\boldsymbol{x}^\top\boldsymbol{\beta}})$ and $\ln(1-h(\boldsymbol{x}^\top\boldsymbol{\beta})) = -\boldsymbol{x}^\top\boldsymbol{\beta} - \ln(1+\mathrm{e}^{-\boldsymbol{x}^\top\boldsymbol{\beta}})$, the cross-entropy training loss (5.79) can be rewritten as

$$
r_\tau(\boldsymbol{\beta}) := \frac{1}{n}\sum_{i=1}^{n}\left[(1-y_i)\boldsymbol{x}_i^\top\boldsymbol{\beta} + \ln\left(1+\mathrm{e}^{-\boldsymbol{x}_i^\top\boldsymbol{\beta}}\right)\right].
$$

We leave it as Exercise 9 to show that the gradient $\nabla r_\tau(\boldsymbol{\beta})$ and Hessian $\mathbf{H}(\boldsymbol{\beta})$ of $r_\tau(\boldsymbol{\beta})$ are given by

$$
\nabla r_\tau(\boldsymbol{\beta}) = \frac{1}{n}\sum_{i=1}^{n}(\mu_i - y_i)\,\boldsymbol{x}_i \tag{5.80}
$$

and

$$
\mathbf{H}(\boldsymbol{\beta}) = \frac{1}{n}\sum_{i=1}^{n}\mu_i(1-\mu_i)\,\boldsymbol{x}_i\boldsymbol{x}_i^\top, \tag{5.81}
$$

respectively, where $\mu_i := h(\boldsymbol{x}_i^\top\boldsymbol{\beta})$.

☞ 590 Notice that $\mathbf{H}(\boldsymbol{\beta})$ is a positive semidefinite matrix for all values of $\boldsymbol{\beta}$, implying the convexity of $r_\tau(\boldsymbol{\beta})$. Consequently, we can find an optimal $\boldsymbol{\beta}$ efficiently; e.g., via Newton's ☞ 600 method. Specifically, given an initial value $\boldsymbol{\beta}_0$, for $t = 1, 2, \ldots$, iteratively compute

$$
\boldsymbol{\beta}_t = \boldsymbol{\beta}_{t-1} - \mathbf{H}^{-1}(\boldsymbol{\beta}_{t-1})\,\nabla r_\tau(\boldsymbol{\beta}_{t-1}), \tag{5.82}
$$

until the sequence $\boldsymbol{\beta}_0, \boldsymbol{\beta}_1, \boldsymbol{\beta}_2, \ldots$ is deemed to have converged, using some pre-determined convergence criterion.

Figure 5.83 shows the outcomes of 100 independent Bernoulli random variables, where each success probability, $(1+\exp(-(\beta_0+\beta_1 x)))^{-1}$, depends on x and $\beta_0 = -3$, $\beta_1 = 10$. The true logistic curve is also shown (dashed line). The minimum training loss curve (red line) is obtained via the Newton scheme (5.82), giving estimates $\widehat{\beta}_0 = -2.66$ and $\widehat{\beta}_1 = 10.08$. The Python code is given below.

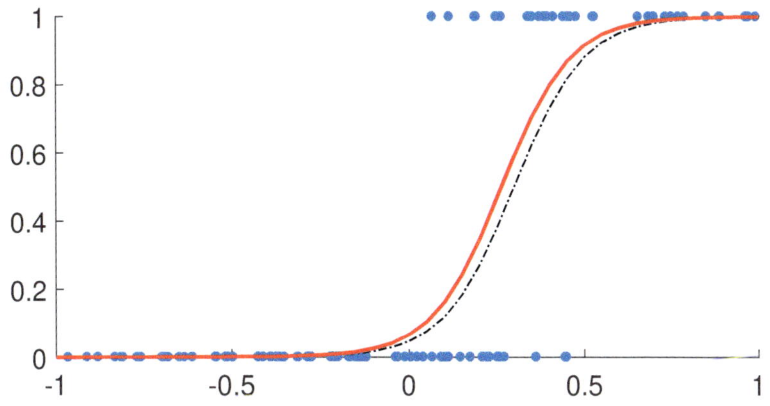

Figure 5.83: Logistic regression data (blue dots), fitted curve (red), and true curve (black dashed).

logreg1d.py

```
import numpy as np
import matplotlib.pyplot as plt
from  numpy.linalg import solve

n = 100                                      # sample size
x = (2*np.random.rand(n)-1).reshape(n,1)  # explanatory variables
beta = np.array([-3, 10])
Xmat = np.hstack((np.ones((n,1)), x))
p = 1/(1 + np.exp(-Xmat @ beta))
y = np.random.binomial(1,p,n)                # response variables

# initial guess
betat = solve(Xmat.T @ Xmat, Xmat.T @ y)

grad = np.array([2,1])                             # gradient

while (np.sum(np.abs(grad)) > 1e-5) :       # stopping criteria
    mu = 1/(1+np.exp(-Xmat @ betat))
    # gradient
    delta = (mu - y).reshape(n,1)
    grad = np.sum(np.multiply( np.hstack((delta,delta)),Xmat), axis=0).T
    # Hessian
    H = Xmat.T @ np.diag(np.multiply(mu,(1-mu))) @ Xmat
    betat = betat - solve(H,grad)
    print(betat)

plt.plot(x,y, '.') # plot data

xx = np.linspace(-1,1,40).reshape(40,1)
XXmat = np.hstack( (np.ones((len(xx),1)), xx))
yy = 1/(1 + np.exp(-XXmat @ beta))
plt.plot(xx,yy,'r-')                         #true logistic curve
yy = 1/(1 + np.exp(-XXmat @ betat));
plt.plot(xx,yy,'k--')
```

Further Reading

An excellent overview of regression is provided in [51] and an accessible mathematical treatment of *linear* regression models can be found in [152]. For extensions to *nonlinear* regression we refer the reader to [10]. Classic texts covering Gaussian-weighted local polynomial regression include [56] and [177]. A practical introduction to multilevel/hierarchical models is given in [70]. For further discussion on regression with discrete responses (classification) we refer to Chapter 8 and the further reading therein. On the important question of how to handle missing data, the classic reference is [112] (see also [122]) and a modern applied reference is [172].

☞ 353

Exercises

1. Following his mentor Francis Galton, the mathematician/statistician Karl Pearson conducted comprehensive studies comparing hereditary traits between members of the same family. Figure 5.84 depicts the measurements of the heights of 1078 fathers and their adult sons (one son per father). The data is available from the book's GitHub site as `pearson.csv`.

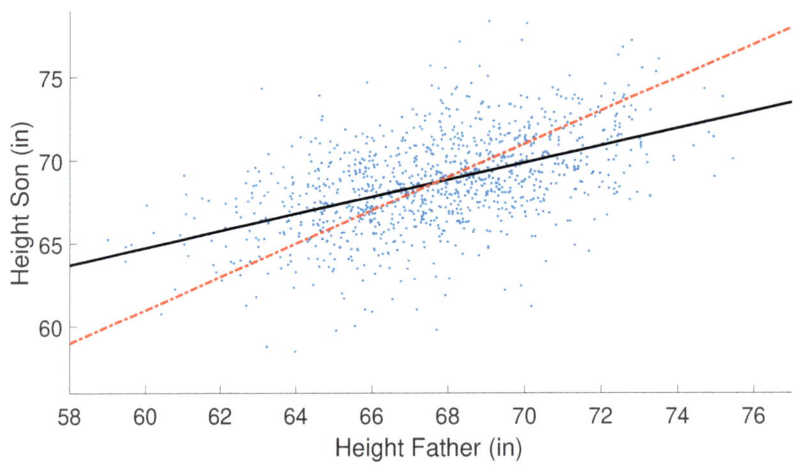

Figure 5.84: A scatterplot of heights from Pearson's data.

   (a) Show that sons are on average 1 inch taller than the fathers.

   (b) We could try to "explain" the height of the son by taking the height of his father and adding 1 inch. The prediction line $y = x + 1$ (red dashed) is given Figure 5.84. The black solid line is the fitted regression line. This line has a slope less than 1, and demonstrates Galton's "regression" to the average. Find the intercept and slope of the fitted regression line.

2. For the simple linear regression model, show that the values for $\widehat{\beta}_1$ and $\widehat{\beta}_0$ that solve

the equations (5.12) are:

$$\widehat{\beta}_1 = \frac{\sum_{i=1}^n (x_i - \overline{x})(y_i - \overline{y})}{\sum_{i=1}^n (x_i - \overline{x})^2} \tag{5.85}$$

$$\widehat{\beta}_0 = \overline{y} - \widehat{\beta}_1 \overline{x}, \tag{5.86}$$

provided that not all x_i are the same.

3. Define the $n \times n$ **centering matrix** $\mathbf{C} := \mathbf{I}_n - \frac{1}{n}\mathbf{1}\mathbf{1}^\top$.

(a) Use Theorem A.17 to show that $\mathbf{C}$ is the orthogonal projection onto the orthogonal complement of $\mathcal{V} = \{\alpha \mathbf{1}, \alpha \in \mathbb{R}\}$. ☞ 521

(b) Use the properties of $\mathbf{C}$ to prove that

$$\sum_{i=1}^n (x_i - \overline{x}_n)^2 = \sum_{i=1}^n x_i^2 - n(\overline{x}_n)^2.$$

(c) Define $\widehat{\beta}_0 := \operatorname{argmin}_{\beta_0} \|\boldsymbol{y} - \boldsymbol{w}\beta_0 - \mathbf{X}\boldsymbol{\beta}\|^2$ and the $n \times n$ projection matrix $\boldsymbol{\Pi}_0^\perp := \mathbf{I}_n - \boldsymbol{w}\boldsymbol{w}^+$, which coincides with $\mathbf{C}$ when the vector $\boldsymbol{w} = \mathbf{1}$. Show that the β_0 can be eliminated from the minimization via substitution with $\widehat{\beta}_0$:

$$\min_{\beta_0} \|\boldsymbol{y} - \boldsymbol{w}\beta_0 - \mathbf{X}\boldsymbol{\beta}\|^2 = \|\boldsymbol{y} - \boldsymbol{w}\widehat{\beta}_0 - \mathbf{X}\boldsymbol{\beta}\|^2 = \|\boldsymbol{\Pi}_0^\perp \boldsymbol{y} - \boldsymbol{\Pi}_0^\perp \mathbf{X}\boldsymbol{\beta}\|^2.$$

4. Edwin Hubble discovered that the universe is expanding. If v is a galaxy's recession velocity (relative to any other galaxy) and d is its distance (from that same galaxy), Hubble's law states that

$$v = Hd,$$

where H is known as Hubble's constant. The following are distance (in millions of light-years) and velocity (thousands of miles per second) measurements made on five galactic clusters.

distance	68	137	315	405	700
velocity	2.4	4.7	12.0	14.4	26.0

State the regression model and estimate H.

5. Let $\mathbf{C} := \mathbf{I}_n - \mathbf{1}\mathbf{1}^\top/n$ be a centering matrix and let

$$\boldsymbol{Y} = \mathbf{1}\beta_0 + \mathbf{C}\mathbf{X}\boldsymbol{\beta} + \boldsymbol{\varepsilon}$$

be a Gaussian linear model with $\boldsymbol{\varepsilon} \sim \mathcal{N}(\mathbf{0}, \sigma^2 \mathbf{I}_n)$ and model matrix $\mathbf{X} \in \mathbb{R}^{n \times p}$ that excludes the constant feature $\mathbf{1}$. Show that the least-squares estimators $\widehat{\beta}_0$ and $\widehat{\boldsymbol{\beta}}$ are independent.

6. The multiple linear regression model (5.7) can be viewed as a first-order approximation of the general model

$$Y = g(\boldsymbol{x}) + \varepsilon, \tag{5.87}$$

where $\mathbb{E}\,\varepsilon = 0$, $\mathbb{V}\mathrm{ar}\,\varepsilon = \sigma^2$, and $g(\boldsymbol{x})$ is some (un)known function of a d-dimensional vector $\boldsymbol{x}$ of explanatory variables. To see this, replace $g(\boldsymbol{x})$ with its first-order Taylor approximation around some point $\boldsymbol{x}_0$ and write this as $\beta_0 + \boldsymbol{x}^\top \boldsymbol{\beta}$. Express β_0 and $\boldsymbol{\beta}$ in terms of g and $\boldsymbol{x}_0$.

7. Table 5.88 shows data from an agricultural experiment where crop yield was measured for two levels of pesticide and three levels of fertilizer. There are three responses for each combination.

Table 5.88: Crop yields for pesticide and fertilizer combinations.

	Fertilizer		
Pesticide	Low	Medium	High
No	3.23, 3.20, 3.16	2.99, 2.85, 2.77	5.72, 5.77, 5.62
Yes	6.78, 6.73, 6.79	9.07, 9.09, 8.86	8.12, 8.04, 8.31

(a) Organize the data in standard form, where each row corresponds to a single measurement and the columns correspond to the response variable and the two factor variables.

(b) Let Y_{ijk} be the response for the k-th replication at level i for factor 1 and level j for factor 2. To assess which factors best explain the response variable, we use the ANOVA model

$$Y_{ijk} = \mu + \alpha_i + \beta_j + \gamma_{ij} + \varepsilon_{ijk}, \tag{5.89}$$

where $\sum_i \alpha_i = \sum_j \beta_j = \sum_i \gamma_{ij} = \sum_j \gamma_{ij} = 0$. Define $\boldsymbol{\beta} = [\mu, \alpha_1, \alpha_2, \beta_1, \beta_2, \beta_3, \gamma_{11}, \gamma_{12}, \gamma_{13}, \gamma_{21}, \gamma_{22}, \gamma_{23}]^\top$. Give the corresponding 18×12 model matrix.

(c) Note that the parameters are linearly dependent in this case. For example, $\alpha_2 = -\alpha_1$ and $\gamma_{13} = -(\gamma_{11} + \gamma_{12})$. To retain only 6 linearly independent variables consider the 6-dimensional parameter vector $\widetilde{\boldsymbol{\beta}} = [\mu, \alpha_1, \beta_1, \beta_2, \gamma_{11}, \gamma_{12}]^\top$. Find the matrix $\mathbf{M}$ such that $\mathbf{M}\widetilde{\boldsymbol{\beta}} = \boldsymbol{\beta}$.

(d) Give the model matrix corresponding to $\widetilde{\boldsymbol{\beta}}$.

8. Show that for the birthweight data in Section 5.7.6.2 there is no significant decrease in birthweight for smoking mothers. Hint: Create a new variable `nonsmoke` = 1−`smoke`, which reverses the encoding for the smoking and non-smoking mothers. Then, the parameter $\beta_1 + \beta_3$ in the original model is the same as the parameter β_1 in the model

$$\texttt{Bwt} = \beta_0 + \beta_1 \texttt{age} + \beta_2 \texttt{nonsmoke} + \beta_3 \texttt{age} \times \texttt{nonsmoke} + \varepsilon.$$

Now find a 95% for β_3 and see if it contains zero.

9. Prove (5.80) and (5.81).

10. In the **Tobit regression** model with normally distributed errors, the response is modeled as:

$$Y_i = \begin{cases} Z_i, & \text{if } u_i < Z_i \\ u_i, & \text{if } Z_i \leq u_i \end{cases}, \qquad \mathbf{Z} \sim \mathcal{N}(\mathbf{X}\boldsymbol{\beta}, \sigma^2 \mathbf{I}_n),$$

where the model matrix $\mathbf{X}$ and the thresholds $u_1, \ldots, u_n$ are given. Typically, $u_i = 0, i = 1, \ldots, n$. Suppose we wish to estimate $\boldsymbol{\theta} := (\boldsymbol{\beta}, \sigma^2)$ via the Expectation–Maximization method, similar to the censored data Example 4.29. Let $\boldsymbol{y} = [y_1, \ldots, y_n]^\top$ be the vector of observed data. ☞ 142

(a) Show that the likelihood of $\boldsymbol{y}$ is:

$$g(\boldsymbol{y} \mid \boldsymbol{\theta}) = \prod_{i: y_i > u_i} \varphi_{\sigma^2}(y_i - \boldsymbol{x}_i^\top \boldsymbol{\beta}) \times \prod_{i: y_i = u_i} \Phi((u_i - \boldsymbol{x}_i^\top \boldsymbol{\beta})/\sigma),$$

where Φ is the cdf of the $\mathcal{N}(0, 1)$ distribution and φ_{σ^2} the pdf of the $\mathcal{N}(0, \sigma^2)$ distribution.

(b) Let $\overline{\boldsymbol{y}}$ and $\underline{\boldsymbol{y}}$ be vectors that collect all $y_i > u_i$ and $y_i = u_i$, respectively. Denote the corresponding matrix of predictors by $\overline{\mathbf{X}}$ and $\underline{\mathbf{X}}$, respectively. For each observation $y_i = u_i$ introduce a latent variable z_i and collect these into a vector $\boldsymbol{z}$. For the same indices i collect the corresponding u_i into a vector $\boldsymbol{c}$. Show that the complete-data likelihood is given by

$$g(\boldsymbol{y}, \boldsymbol{z} \mid \boldsymbol{\theta}) = \frac{1}{(2\pi\sigma^2)^{n/2}} \exp\left(-\frac{\|\overline{\boldsymbol{y}} - \overline{\mathbf{X}}\boldsymbol{\beta}\|^2}{2\sigma^2} - \frac{\|\boldsymbol{z} - \underline{\mathbf{X}}\boldsymbol{\beta}\|^2}{2\sigma^2} \right) \mathbb{1}\{\boldsymbol{z} \leq \boldsymbol{c}\}.$$

(c) For the E-step, show that, for a fixed $\boldsymbol{\theta}$,

$$g(\boldsymbol{z} \mid \boldsymbol{y}, \boldsymbol{\theta}) = \prod_i g(z_i \mid \boldsymbol{y}, \boldsymbol{\theta}),$$

where each $g(z_i \mid \boldsymbol{y}, \boldsymbol{\theta})$ is the pdf of the $\mathcal{N}((\underline{\mathbf{X}}\boldsymbol{\beta})_i, \sigma^2)$ distribution, truncated to the interval $(-\infty, c_i]$.

(d) For the M-step, compute the expectation of the complete log-likelihood

$$-\frac{n}{2} \ln \sigma^2 - \frac{n}{2} \ln(2\pi) - \frac{\|\overline{\boldsymbol{y}} - \overline{\mathbf{X}}\boldsymbol{\beta}\|^2}{2\sigma^2} - \frac{\mathbb{E}\|\mathbf{Z} - \underline{\mathbf{X}}\boldsymbol{\beta}\|^2}{2\sigma^2}.$$

Then, derive the formulas for $\boldsymbol{\beta}$ and σ^2 that maximize the expectation of the complete log-likelihood.

11. Download data set `WomenWage.csv` from the book's website. This data set is a tidied-up version of the women's wages data set from [129]. The first column of the data (`hours`) is the response variable Y. It shows the hours spent in the labor force by married women in the 1970s. We want to understand what factors determine the participation rate of women in the labor force. The predictor variables are:

Table 5.90: Features for the women's wage data set.

Feature	Description
`kidslt6`	Number of children younger than 6 years.
`kidsge6`	Number of children older than 6 years.
`age`	Age of the married woman.
`educ`	Number of years of formal education.
`exper`	Number of years of "work experience".
`nwifeinc`	Non-wife income, that is, the income of the husband.
`expersq`	The square of `exper`, to capture any nonlinear relationships.

We observe that some of the responses are $Y = 0$, that is, some women did not participate in the labor force. For this reason, we model the data using the Tobit regression model, in which the response Y is given as:

$$Y_i = \begin{cases} Z_i, & \text{if } Z_i > 0 \\ 0, & \text{if } Z_i \leq 0 \end{cases}, \qquad \mathbf{Z} \sim \mathcal{N}(\mathbf{X}\boldsymbol{\beta}, \sigma^2 \mathbf{I}_n).$$

With $\boldsymbol{\theta} = (\boldsymbol{\beta}, \sigma^2)$, the likelihood of the data $\boldsymbol{y} = [y_1, \ldots, y_n]^\top$ is:

$$g(\boldsymbol{y} \,|\, \boldsymbol{\theta}) = \textstyle\prod_{i:y_i>0} \varphi_{\sigma^2}(y_i - \boldsymbol{x}_i^\top \boldsymbol{\beta}) \times \prod_{i:y_i=0} \Phi((u_i - \boldsymbol{x}_i^\top \boldsymbol{\beta})/\sigma),$$

where Φ is the standard normal cdf. In Exercise 10, we derived the EM algorithm for maximizing the log-likelihood.

(a) Write down the EM algorithm in pseudo-code as it applies to this Tobit regression.

(b) Implement the EM algorithm pseudo-code in Python. Comment on which factor you think is important in determining the labor participation rate of women living in the USA in the 1970s.

12. Let $\mathbf{P}$ be a projection matrix.

(a) Show that the diagonal elements of $\mathbf{P}$ all lie in the interval $[0, 1]$. In particular, for $\mathbf{P} = \mathbf{X}\mathbf{X}^+$ in Theorem 5.14, show that the leverage value $p_i := \mathbf{P}_{ii}$ satisfies $0 \leq p_i \leq 1$ for all i.

(b) Show that $\boldsymbol{u}_i^\top \mathbf{P} \boldsymbol{u}_j = p_{ij} \in [-1/2, 1/2]$ for $i \neq j$.

(c) If $\mathbf{P} = \mathbf{X}\mathbf{X}^+$ and $\mathbf{X}$ includes the constant feature $\mathbf{1}$, show that

$$\boldsymbol{u}_i^\top \mathbf{P} \boldsymbol{u}_i = p_{ii} \geq 1/n.$$

13. Consider the linear model $\boldsymbol{Y} = \mathbf{X}\boldsymbol{\beta} + \boldsymbol{\varepsilon}$ in (5.11), with $\mathbf{X}$ being the $n \times p$ model matrix and $\boldsymbol{\varepsilon}$ having expectation vector $\mathbf{0}$ and covariance matrix $\sigma^2 \mathbf{I}_n$. Suppose that $\widehat{\boldsymbol{\beta}}_{-i}$ is the least-squares estimate obtained by omitting the i-th observation, Y_i; that is,

$$\widehat{\boldsymbol{\beta}}_{-i} = \underset{\boldsymbol{\beta}}{\operatorname{argmin}} \sum_{j \neq i} (Y_j - \boldsymbol{x}_j^\top \boldsymbol{\beta})^2,$$

where $\mathbf{x}_j^\top$ is the j-th row of $\mathbf{X}$. Let $\widehat{Y}_{-i} = \mathbf{x}_i^\top \widehat{\boldsymbol{\beta}}_{-i}$ be the corresponding fitted value at $\mathbf{x}_i$. Also, define $\boldsymbol{B}_i$ as the least-squares estimator of $\boldsymbol{\beta}$ based on the response data

$$\boldsymbol{Y}^{(i)} := [Y_1, \ldots, Y_{i-1}, \widehat{Y}_{-i}, Y_{i+1}, \ldots, Y_n]^\top.$$

(a) Prove that $\widehat{\boldsymbol{\beta}}_{-i} = \boldsymbol{B}_i$; that is, the linear model obtained from fitting all responses except the i-th is the same as the one obtained from fitting the data $\boldsymbol{Y}^{(i)}$.

(b) Use the previous result to verify that

$$Y_i - \widehat{Y}_{-i} = (Y_i - \widehat{Y}_i)/(1 - \mathbf{P}_{ii}),$$

where $\mathbf{P} = \mathbf{X}\mathbf{X}^+$ is the projection matrix onto the columns of $\mathbf{X}$. Hence, deduce the PRESS formula in Theorem 5.14. ☞ 204

14. Take the linear model $\boldsymbol{Y} = \mathbf{X}\boldsymbol{\beta} + \boldsymbol{\varepsilon}$, where $\mathbf{X}$ is an $n \times p$ model matrix with $\mathbb{E}\boldsymbol{\varepsilon} = \mathbf{0}$ and $\mathbb{C}\mathrm{ov}(\boldsymbol{\varepsilon}) = \sigma^2 \mathbf{I}_n$. Let $\mathbf{P} = \mathbf{X}\mathbf{X}^+$ be the projection matrix onto the columns of $\mathbf{X}$.

(a) Using the properties of the pseudo-inverse (see Definition A.10), show that $\mathbf{P}\mathbf{P}^\top = \mathbf{P}$. ☞ 516

(b) Let $\boldsymbol{E} = \boldsymbol{Y} - \widehat{\boldsymbol{Y}}$ be the (random) vector of residuals, where $\widehat{\boldsymbol{Y}} = \mathbf{P}\boldsymbol{Y}$. Show that the residual vector has expectation $\mathbf{0}$ and covariance matrix $\sigma^2(\mathbf{I}_n - \mathbf{P})$. Hence, deduce that under a normal linear model $\boldsymbol{E} \sim \mathcal{N}(\mathbf{0}, \sigma^2(\mathbf{I}_n - \mathbf{P}))$.

(c) Show that σ^2 can be unbiasedly estimated via

$$S^2 := \frac{1}{n-p}\|\boldsymbol{Y} - \widehat{\boldsymbol{Y}}\|^2 = \frac{1}{n-p}\|\boldsymbol{Y} - \mathbf{X}\widehat{\boldsymbol{\beta}}\|^2. \tag{5.91}$$

Hint: Use the cyclic property of the trace as in Example 2.39.

15. Consider a normal linear model $\boldsymbol{Y} = \mathbf{X}\boldsymbol{\beta} + \boldsymbol{\varepsilon}$, where $\mathbf{X}$ is an $n \times p$ model matrix and $\boldsymbol{\varepsilon} \sim \mathcal{N}(\mathbf{0}, \sigma^2 \mathbf{I}_n)$. Exercise 14 shows that for any such model the i-th standardized residual $E_i/(\sigma\sqrt{1 - \mathbf{P}_{ii}})$ has a standard normal distribution. This motivates the use of the leverage $\mathbf{P}_{ii}$ to assess whether the i-th observation is an outlier depending on the size of the i-th residual relative to $\sqrt{1 - \mathbf{P}_{ii}}$. A more robust approach is to include an estimate for σ using all data except the i-th observation. This gives rise to the **studentized residual** T_i, defined as

$$T_i := \frac{E_i}{S_{-i}\sqrt{1 - \mathbf{P}_{ii}}},$$

where S_{-i} is an estimate of σ obtained by fitting all the observations except the i-th and $E_i = Y_i - \widehat{Y}_i$ is the i-th (random) residual. Exercise 14 shows that we can take, for example,

$$S_{-i}^2 = \frac{1}{n-1-p}\|\boldsymbol{Y}_{-i} - \mathbf{X}_{-i}\widehat{\boldsymbol{\beta}}_{-i}\|^2, \tag{5.92}$$

where $\mathbf{X}_{-i}$ is the model matrix $\mathbf{X}$ with the i-th row removed, is an unbiased estimator of σ^2. We wish to compute S_{-i}^2 efficiently, using S^2 in (5.91), as the latter will typically be available once we have fitted the linear model. To this end, define $\boldsymbol{u}_i$ as the i-th unit vector $[0, \ldots, 0, 1, 0, \ldots, 0]^\top$, and let

$$\boldsymbol{Y}^{(i)} := \boldsymbol{Y} - (Y_i - \widehat{Y}_{-i})\boldsymbol{u}_i = \boldsymbol{Y} - \frac{E_i}{1 - \mathbf{P}_{ii}}\boldsymbol{u}_i,$$

where we have used the fact that $Y_i - \widehat{Y}_{-i} = E_i/(1 - \mathbf{P}_{ii})$, as derived in the proof of Theorem 5.14. Now apply Exercise 13 to prove that

$$S_{-i}^2 = \frac{(n-p)\,S^2 - E_i^2/(1 - \mathbf{P}_{ii})}{n - p - 1}.$$

16. Using the notation from Exercises 13–15, **Cook's distance** for observation i is defined as

$$D_i := \frac{\|\widehat{\boldsymbol{Y}} - \widehat{\boldsymbol{Y}}^{(i)}\|^2}{p\,S^2}.$$

It measures the change in the fitted values when the i-th observation is removed, relative to the residual variance of the model (estimated via S^2).

By using similar arguments as those in Exercise 15, show that

$$D_i = \frac{\mathbf{P}_{ii}\,E_i^2}{(1 - \mathbf{P}_{ii})^2 p\,S^2}.$$

It follows that there is no need to "omit and refit" the linear model in order to compute Cook's distance for the i-th response.

17. Prove that if we add an additional feature to the general linear model, then R^2, the coefficient of determination, is necessarily non-decreasing in value and hence cannot be used to compare models with different numbers of predictors.

18. Let $\boldsymbol{X} := [X_1, \ldots, X_n]^\top$ and $\boldsymbol{\mu} := [\mu_1, \ldots, \mu_n]^\top$. In the fundamental Theorem D.55, we use the fact that if $X_i \sim \mathcal{N}(\mu_i, 1)$, $i = 1, \ldots, n$ are independent, then $\|\boldsymbol{X}\|^2$ has (per definition) a noncentral χ_n^2 distribution. Show that $\|\boldsymbol{X}\|^2$ has moment generating function

$$\frac{\mathrm{e}^{t\|\boldsymbol{\mu}\|^2/(1-2t)}}{(1-2t)^{n/2}}, \quad t < 1/2,$$

and so the distribution of $\|\boldsymbol{X}\|^2$ depends on $\boldsymbol{\mu}$ only through the norm $\|\boldsymbol{\mu}\|$.

19. Recall that the Gaussian-weighted local linear learner evaluated at x is $g_\tau(x \mid \lambda) = \widehat{\beta}_0(x) + \widehat{\beta}_1(x)x$, where the parameters $(\widehat{\beta}_0(x), \widehat{\beta}_1(x))$ are determined via the least-squares optimization (5.55).

(a) Prove that the learner can be written as $g_\tau(x \mid \lambda) = \langle \boldsymbol{y}, \boldsymbol{l}((\boldsymbol{v} - x\mathbf{1})/\lambda)\rangle$, where $\boldsymbol{l}(\boldsymbol{u})$ is given in (5.56).

(b) Show that $\boldsymbol{l}(\boldsymbol{u})$ defined in (5.56) can be written as

$$\boldsymbol{l}(\boldsymbol{u}) = \frac{1}{\|\mathbf{1}\|_{\mathbf{D}}^2 - \langle \mathbf{1}, \boldsymbol{u}\rangle_{\mathbf{D}}^2 / \|\boldsymbol{u}\|_{\mathbf{D}}^2} \mathbf{D}(\boldsymbol{u})\mathbf{1} - \frac{\langle \mathbf{1}, \boldsymbol{u}\rangle_{\mathbf{D}} / \|\boldsymbol{u}\|_{\mathbf{D}}^2}{\|\mathbf{1}\|_{\mathbf{D}}^2 - \langle \mathbf{1}, \boldsymbol{u}\rangle_{\mathbf{D}}^2 / \|\boldsymbol{u}\|_{\mathbf{D}}^2} \mathbf{D}(\boldsymbol{u})\, \boldsymbol{u}, \tag{5.93}$$

where $\langle \boldsymbol{a}, \boldsymbol{b}\rangle_{\mathbf{D}} := \boldsymbol{a}^\top \mathbf{D}(\boldsymbol{u})\boldsymbol{b}$ and $\|\boldsymbol{a}\|_{\mathbf{D}} := \sqrt{\langle \boldsymbol{a}, \boldsymbol{a}\rangle_{\mathbf{D}}}$ are the Mahalanobis inner product and norm, respectively.

20. The purpose of this exercise is to prove the statistical properties of the Gaussian-weighted local linear regression in Lemma 5.60 under Assumptions 5.59.

(a) Let $\boldsymbol{U} := (\boldsymbol{V} - x\mathbf{1})/\lambda_n$. Use the identity (5.93) from Exercise 19 to show that

$$l_i(\boldsymbol{U}) = (1 + o_{\mathbb{P}}(1)) \frac{\varphi_{\lambda_n^2}(X_i - x)}{n\, f(x)} \left(1 - \frac{f'(x)}{f(x)}(X_i - x)\right), \quad i = 1, \ldots, n.$$

(b) Use the result from part (a) to prove that

$$\sum_i l_i(\boldsymbol{U}) U_i^2 = 1 + o_{\mathbb{P}}(1).$$

Hence, using the fact that $\mathbf{X}(\boldsymbol{u})^\top \boldsymbol{l}(\boldsymbol{u}) = [1, 0]^\top$ for any $\boldsymbol{u}$, deduce the conditional bias formula in Lemma 5.60.

(c) Let $\lambda'_n = O(\lambda_n)$ and $\lambda_n = O(\lambda'_n)$. Use the result of part (a) to prove that

$$\sum_{i=1}^n l_i\left(\frac{\boldsymbol{V} - x\mathbf{1}}{\lambda_n}\right) l_i\left(\frac{\boldsymbol{V} - x\mathbf{1}}{\lambda'_n}\right) \sigma^2(X_i) = \frac{\varphi_{\lambda_n^2 + \lambda_n'^2}(0)}{n} \frac{\sigma^2(x)}{f(x)} + o_{\mathbb{P}}\big(1/(n\lambda_n)\big).$$

Hence, deduce the conditional covariance formula in Lemma 5.60.

Hint: Use the fact that

$$\int_{\mathbb{R}} \varphi_{\lambda_n^2}(x - u) \sigma^2(u) f(u) \left(\frac{u - x}{\lambda_n}\right)^r \mathrm{d}u = \begin{cases} \sigma^2(x) f(x) + o_{\mathbb{P}}(1), & r \in \{0, 2\} \\ \lambda_n (\sigma^2(x) f(x))' + o_{\mathbb{P}}(\lambda_n), & r = 1. \end{cases}$$

21. Consider the Gaussian-weighted local linear learner $g_\tau(x \mid \lambda) = \widehat{\beta}_0(x) + \widehat{\beta}_1(x)$, where the parameters $(\widehat{\beta}_0(x), \widehat{\beta}_1(x))$ are determined via the least-squares optimization (5.55). Let $g_{\tau_{\neg i}}(x \mid \lambda)$ be the *leave-one-out* learner trained on the data set $\tau_{\neg i} := \tau \setminus \{(x_i, y_i)\}$ that excludes the i-th pair (x_i, y_i), but is otherwise the same as τ.

(a) If $\mathbb{E}[(Y - g^*(X))^2 \mid X = x] = \sigma^2$ is constant for all x, show that

$$\mathbb{E}_{\boldsymbol{V}} \frac{1}{n} \sum_{i=1}^n (Y_i - g_{\mathcal{T}_{\neg i}}(X_i \mid \lambda_n))^2 = \sigma^2 + \frac{1}{n} \sum_{i=1}^n \mathbb{E}_{\boldsymbol{V}} \left(g^*(X_i) - g_{\mathcal{T}_{\neg i}}(X_i \mid \lambda_n)\right)^2.$$

(b) Let $\widehat{\tau}_i$ be the same as τ, except that y_i is replaced with $g_{\tau_{\neg i}}(x_i \mid \lambda)$. Show that

$$g_{\tau_{\neg i}}(x_i \mid \lambda) = g_{\widehat{\tau}_i}(x_i \mid \lambda),$$

where $g_{\widehat{\tau}_i}(x \mid \lambda) = \widehat{\beta}_{0,i}(x) + \widehat{\beta}_{1,i}(x)\, x$ is the learner trained on $\widehat{\tau}_i$, that is, $(\widehat{\beta}_{0,i}(x), \widehat{\beta}_{1,i}(x))$ is the minimizer of the Gaussian-weighted sum of squares:

$$\varphi_{\lambda^2}(x - x_i)\left(g_{\tau_{\neg i}}(x_i \mid \lambda) - \beta_0 - x_i\beta_1\right)^2 + \sum_{j \neq i} \varphi_{\lambda^2}(x - x_j)\left(y_j - \beta_0 - x_j\beta_1\right)^2.$$

(c) Use the result from part (b) to prove that

$$y_i - g_{\mathcal{T}_{-i}}(x_i \mid \lambda) = \frac{y_i - g_{\mathcal{T}}(x_i \mid \lambda)}{1 - l_i\big((\boldsymbol{v} - x_i\mathbf{1})/\lambda\big)},$$

where $\boldsymbol{l}(\boldsymbol{u})$ is defined in (5.56). Hint: Recall that $g_{\mathcal{T}}(x_i \mid \lambda) = \boldsymbol{l}\big((\boldsymbol{v} - x_i\mathbf{1})/\lambda\big)^\top \boldsymbol{y}$.

22. Show that

$$\widehat{\Delta g^*}(x) := \frac{\delta_{\mathcal{T}_n}(x \mid \lambda_n)}{\lambda_n^2/2},$$

where the BPM function $\delta_{\mathcal{T}_n}(x \mid \lambda_n) := \big(g_{\mathcal{T}_n}(x \mid \lambda_n) - g_{\mathcal{T}_n}(x \mid \sqrt{c}\,\lambda_n)\big)/(1-c)$, is a
☞ 587 **Richardson extrapolation estimator** of $\Delta g^*(x)$ (see Example C.24) such that:

$$\mathbb{E}_{\mathcal{V}}\, \widehat{\Delta g^*}(x) = \Delta g^*(x) + o_{\mathbb{P}}(1).$$

23. Carry out a logistic regression analysis on a (partial) *wine* data set classification problem. The data can be loaded using the following code.

```
from sklearn import datasets
import numpy as np
data = datasets.load_wine()
X = data.data[:, [9,10]]
y = np.array(data.target==1,dtype=np.uint)
X = np.append(np.ones(len(X)).reshape(-1,1),X,axis=1)
```

The model matrix has three features, including the constant feature. Instead of using Newton's method (5.82) to estimate $\boldsymbol{\beta}$, implement a simple gradient descent procedure

$$\boldsymbol{\beta}_t = \boldsymbol{\beta}_{t-1} - \alpha \nabla r_{\tau}(\boldsymbol{\beta}_{t-1}),$$

with learning rate $\alpha = 0.0001$, and run it for 10^6 steps. Your procedure should deliver three coefficients; one for the intercept and the rest for the explanatory variables. Solve the same problem using the `Logit` method of `statsmodels.api` and compare the results.

24. Consider again Example 5.77, where we train the learner via the Newton iteration
☞ 244 (5.82). If $\mathbf{X}^\top := [\boldsymbol{x}_1, \ldots, \boldsymbol{x}_n]$ defines the matrix of predictors and $\boldsymbol{\mu}_t := \boldsymbol{h}(\mathbf{X}\boldsymbol{\beta}_t)$, then the gradient (5.80) and Hessian (5.81) for Newton's method can be written as:

$$\nabla r_{\tau}(\boldsymbol{\beta}_t) = \frac{1}{n}\mathbf{X}^\top(\boldsymbol{\mu}_t - \boldsymbol{y}) \quad \text{and} \quad \mathbf{H}(\boldsymbol{\beta}_t) = \frac{1}{n}\mathbf{X}^\top \mathbf{D}_t \mathbf{X},$$

where $\mathbf{D}_t := \operatorname{diag}(\boldsymbol{\mu}_t \odot (\mathbf{1} - \boldsymbol{\mu}_t))$ is a diagonal matrix. Show that the Newton iteration (5.82) can be written as the **iterative reweighted least-squares** method:

$$\boldsymbol{\beta}_t = \underset{\boldsymbol{\beta}}{\operatorname{argmin}}\ \|\widetilde{\boldsymbol{y}}_{t-1} - \mathbf{X}\boldsymbol{\beta}\|_{\mathbf{D}_{t-1}},$$

where $\widetilde{\boldsymbol{y}}_{t-1} := \mathbf{X}\boldsymbol{\beta}_{t-1} + \mathbf{D}_{t-1}^{-1}(\boldsymbol{y} - \boldsymbol{\mu}_{t-1})$ is the so-called **adjusted response** and $\|\boldsymbol{z}\|_{\mathbf{A}}^2 = \boldsymbol{z}^\top \mathbf{A}\boldsymbol{z}$ is the squared Mahalanobis norm. Hint: Use the fact that $(\mathbf{M}^\top\mathbf{M})^{-1}\mathbf{M}^\top\boldsymbol{z}$ is the minimizer of $\|\mathbf{M}\boldsymbol{\beta} - \boldsymbol{z}\|^2$ and $\|\boldsymbol{z}\|_{\mathbf{A}} = \|\mathbf{A}^{1/2}\boldsymbol{z}\|$.

25. In **multi-output linear regression**, the response variable is a real-valued vector of dimension, say, m. Similar to (5.11), the model can be written in matrix notation:

$$\mathbf{Y} = \mathbf{X}\mathbf{B} + \begin{bmatrix} \boldsymbol{\varepsilon}_1^\top \\ \vdots \\ \boldsymbol{\varepsilon}_n^\top \end{bmatrix},$$

where:

- $\mathbf{Y}$ is an $n \times m$ matrix of n independent responses (stored as row vectors of length m);
- $\mathbf{X}$ is the usual $n \times p$ model matrix;
- $\mathbf{B}$ is an $p \times m$ matrix of model parameters;
- $\boldsymbol{\varepsilon}_1, \ldots, \boldsymbol{\varepsilon}_n \in \mathbb{R}^m$ are independent error terms with $\mathbb{E}\,\boldsymbol{\varepsilon} = \mathbf{0}$ and $\mathbb{E}\,\boldsymbol{\varepsilon}\boldsymbol{\varepsilon}^\top = \boldsymbol{\Sigma}$.

We wish to learn the matrix parameters $\mathbf{B}$ and $\boldsymbol{\Sigma}$ from the training set $\{\mathbf{Y}, \mathbf{X}\}$. To this end, consider minimizing the training loss:

$$\frac{1}{n}\mathrm{tr}\left((\mathbf{Y} - \mathbf{X}\mathbf{B})\,\boldsymbol{\Sigma}^{-1}\,(\mathbf{Y} - \mathbf{X}\mathbf{B})^\top\right),$$

where $\mathrm{tr}(\cdot)$ is the trace of a matrix. ☞ 515

(a) Show that the minimizer of the training loss, denoted $\widehat{\mathbf{B}}$, satisfies the normal equations:

$$\mathbf{X}^\top\mathbf{X}\,\widehat{\mathbf{B}} = \mathbf{X}^\top\mathbf{Y}.$$

(b) Noting that

$$(\mathbf{Y} - \mathbf{X}\mathbf{B})^\top(\mathbf{Y} - \mathbf{X}\mathbf{B}) = \sum_{i=1}^{n} \boldsymbol{\varepsilon}_i\boldsymbol{\varepsilon}_i^\top,$$

explain why

$$\widehat{\boldsymbol{\Sigma}} := \frac{(\mathbf{Y} - \mathbf{X}\widehat{\mathbf{B}})^\top(\mathbf{Y} - \mathbf{X}\widehat{\mathbf{B}})}{n}$$

is a method-of-moments estimator of $\boldsymbol{\Sigma}$, just like the one given in (5.13).

26. Consider the linear regression model $Y = \boldsymbol{x}^\top\boldsymbol{\beta} + \varepsilon$, where the error term ε follows the **generalized normal distribution** with density: ☞ 45

$$\frac{q}{\sigma\Gamma(1/q)2^{1+1/q}}\exp\left(-\frac{|\varepsilon/\sigma|^q}{2}\right), \quad \varepsilon \in \mathbb{R}$$

for some parameters $q > 0$ and $\sigma > 0$.

(a) Plot and identify the density of ε for $q = 1$, $q = 2$, and $q \uparrow \infty$.
Hint: Use the asymptotic expansion as $z \downarrow 0$:

$$\ln\Gamma(z) = -\ln(z) - \gamma z - O(z^2),$$

where $\gamma = 0.57721\ldots$ is the **Euler–Mascheroni** constant.

(b) Given the training set $(\boldsymbol{x}_1, y_1), \ldots, (\boldsymbol{x}_n, y_n)$, find the maximum likelihood estimates of $\boldsymbol{\beta}$ and σ.

27. Define the **quantile loss** function

$$\mathrm{Loss}_q(y, \widehat{y}) := (y - \widehat{y})\left(q - \mathbb{1}_{\{y \le \widehat{y}\}}\right),$$

where $q \in (0, 1)$. In this exercise we study the computation of the **quantile regression** estimate:

$$\widehat{\boldsymbol{\beta}}_q := \underset{\boldsymbol{\beta}}{\mathrm{argmin}} \sum_{i=1}^{n} \mathrm{Loss}_q(y_i, \boldsymbol{x}_i^\top \boldsymbol{\beta}),$$

where $\{(\boldsymbol{x}_1, y_1), \ldots, (\boldsymbol{x}_n, y_n)\}$ is a training set with features in $\mathbb{R}^p$.

(a) Show that for any random variable X:

$$c^* := \inf\{c : \mathbb{P}[X < c] \ge q\} \in \underset{c}{\mathrm{argmin}}\, \mathbb{E}\, \mathrm{Loss}_q(X, c).$$

☞ 61 Compare this result with that in Exercise 1.1.

(b) Using the fact that

$$2\mathrm{Loss}_q(y, \widehat{y}) = (2q - 1)(y - \widehat{y}) + |\widehat{y} - y|,$$

derive a linear program which helps compute $\widehat{\boldsymbol{\beta}}_q$ for a given quantile value q.

(c) Given the so-called **lasso shrinkage operator**, defined as:

$$S_\lambda^{\mathrm{LAS}}(\alpha) := \underset{x}{\mathrm{argmin}} \left\{ \frac{(x - \alpha)^2}{2} + \lambda |x| \right\} = \begin{cases} 0, & |\alpha| \le \lambda, \\ \alpha\left(1 - \frac{\lambda}{|\alpha|}\right), & |\alpha| > \lambda \ge 0, \end{cases} \tag{5.94}$$

☞ 618 show the ADMM minimization method for computing $\widehat{\boldsymbol{\beta}}_q$ is equivalent to the iteration of the following steps:

$$\begin{aligned}
&\boldsymbol{\beta} \leftarrow \mathbf{X}^+(\boldsymbol{y} - \boldsymbol{z} - \boldsymbol{v}/\varrho) \\
&\boldsymbol{r} \leftarrow \boldsymbol{y} - \mathbf{X}\boldsymbol{\beta}, \quad z_k \leftarrow S_{\varrho^{-1}}^{\mathrm{LAS}}(r_k - (v_k - 1 + 2q)/\varrho), \quad k = 1, \ldots, n \\
&\boldsymbol{v} \leftarrow \boldsymbol{v} + \varrho(\boldsymbol{z} - \boldsymbol{r}).
\end{aligned}$$

(d) Suppose that we have been provided the QR decomposition of $\mathbf{X} \in \mathbb{R}^{n \times p}$ with
☞ 522 $n > p$. Explain why one iteration of the ADMM algorithm can then be computed at a cost of $O(n\,p)$. What will be the overall cost of computing the sequence of quantile estimates:

$$\widehat{\boldsymbol{\beta}}_0, \widehat{\boldsymbol{\beta}}_{1/m}, \ldots, \widehat{\boldsymbol{\beta}}_{1-1/m}, \widehat{\boldsymbol{\beta}}_1$$

one after the other for some large integer m? The following pseudo-code may help with your analysis.

Algorithm 5.95: ADMM for Computing $\widehat{\beta}_{k/m}$ for $k = 0, \ldots, m$.

Input: Integer m and $\mathbf{X}, \boldsymbol{y}$, error tol. ε and tuning parameter ϱ, say $\varrho = 1$
Output: Quantile estimates $\widehat{\boldsymbol{\beta}}_0, \widehat{\boldsymbol{\beta}}_{1/m}, \ldots, \widehat{\boldsymbol{\beta}}_{1-1/m}, \widehat{\boldsymbol{\beta}}_1$

QR $\leftarrow \mathbf{X}$ // computing QR via Algorithm A.22 at $O(np^2)$ cost
$z \leftarrow \mathbf{0}_n$, $\boldsymbol{v} \leftarrow \mathbf{0}_n$ // initialize for $q = 0$
for $q = 0, 1/m, \ldots, 1$ **do** // loop over quantiles
  repeat // iterate ADMM steps using a warm start for z, v
    $\boldsymbol{r} \leftarrow \boldsymbol{y} - z - \boldsymbol{v}/\varrho$ // reuse the z, v from previous quantile loop
    $\boldsymbol{r} \leftarrow \boldsymbol{y} - \mathbf{Q}(\mathbf{Q}^\top \boldsymbol{r})$ // costs $O(np)$ per iteration
    $z \leftarrow \boldsymbol{r} - (\boldsymbol{v} - 1 + 2q)/\varrho$
    $z_k \leftarrow S^{\text{LAS}}_{\varrho^{-1}}(z_k)$ for $k = 1, \ldots, n$
    $\boldsymbol{v} \leftarrow \boldsymbol{v} + \varrho(z - \boldsymbol{r})$
  until $\|z - \boldsymbol{r}\| < \varepsilon$
  $\widehat{\boldsymbol{\beta}}_q \leftarrow \mathbf{Q}^\top(\boldsymbol{y} - z - \boldsymbol{v}/\varrho)$
  $\widehat{\boldsymbol{\beta}}_q \leftarrow \mathbf{R}^{-1}\widehat{\boldsymbol{\beta}}_q$ // backward substitution at $O(p^2)$ cost
return $\widehat{\boldsymbol{\beta}}_0, \widehat{\boldsymbol{\beta}}_{1/m}, \ldots, \widehat{\boldsymbol{\beta}}_{1-1/m}, \widehat{\boldsymbol{\beta}}_1$

28. While the least-squares estimate calls for the solution of the linear system of equations (5.12), some extensions call for the solution of a *linear program*. One such extension is the **least-absolute-deviation lasso** regression estimate, defined as: ☞ 594

$$\widehat{\boldsymbol{\beta}}_{\text{LAD}}(\lambda) := \underset{\boldsymbol{\beta}}{\operatorname{argmin}} \left\{ \frac{\|\boldsymbol{y} - \mathbf{X}\boldsymbol{\beta}\|_1}{n} + \lambda \sum_i \omega_i \, |\beta_i| \right\},$$

where $\omega_i \geq 0$ is a user-specified weight given to coefficient β_i and $\lambda \geq 0$ is a parameter giving weight to the term $\sum_i \omega_i |\beta_i|$ (the $\lambda = 0$ case is studied in Exercise 26).

(a) Derive a linear program which helps compute $\widehat{\boldsymbol{\beta}}_{\text{LAD}}$ for a given λ.

(b) Use Exercise 27 to show that if $c^* = \inf\{c \in \mathbb{R} : \mathbb{P}[X < c] \geq 1/2\}$, for some random variable $X \in \mathbb{R}$, then $c^* \in \operatorname{argmin}_c \mathbb{E}|X - c|$. Use this result to derive the coordinate-descent update (C.66) for computing $\widehat{\boldsymbol{\beta}}_{\text{LAD}}$ for a given λ. ☞ 606

(c) Let $\lambda_i := n\lambda\omega_i$ for all i and $\boldsymbol{\Lambda} := \operatorname{diag}(\lambda_1, \ldots, \lambda_p)$. Show the ADMM minimization method for computing $\widehat{\boldsymbol{\beta}}_{\text{LAD}}$ is equivalent to the iteration of the following steps ($z^\top = [z_1^\top, z_2^\top]^\top \in \mathbb{R}^{n+p}$ and $\boldsymbol{v}^\top = [\boldsymbol{v}_1^\top, \boldsymbol{v}_2^\top]^\top \in \mathbb{R}^{n+p}$, where $z_1, \boldsymbol{v}_1 \in \mathbb{R}^n$): ☞ 618

$$\boldsymbol{\beta} \leftarrow (\mathbf{X}^\top\mathbf{X} + \boldsymbol{\Lambda}^2)^{-1}(\mathbf{X}^\top(\boldsymbol{y} - z_1) - \boldsymbol{\lambda} \odot z_2 - (\mathbf{X}^\top\boldsymbol{v}_1 + \boldsymbol{\lambda} \odot \boldsymbol{v}_2)/\varrho)$$

$$\widetilde{\boldsymbol{r}} \leftarrow \begin{bmatrix} \boldsymbol{y} - \mathbf{X}\boldsymbol{\beta} \\ -\boldsymbol{\lambda} \odot \boldsymbol{\beta} \end{bmatrix}, \quad z_k \leftarrow S^{\text{LAS}}_{\varrho^{-1}}(\widetilde{r}_k - v_k/\varrho), \quad k = 1, \ldots, p + n$$

$$\boldsymbol{v} \leftarrow \boldsymbol{v} + \varrho(z - \widetilde{\boldsymbol{r}}),$$

where S^{LAS} is the lasso shrinkage operator in (5.94).

(d) Implement the three optimization methods discussed above (linear programming, coordinate descent and ADMM) in Python and compare their performance on some simulated examples of your choice.

29. (Exercise 28 continued) Yet another example of an estimate that calls for the solution
☞ 594 of a *linear program* is the **supremum-norm lasso** regression estimate:

$$\widehat{\boldsymbol{\beta}}_{\text{SUP}}(\lambda) := \underset{\boldsymbol{\beta}}{\operatorname{argmin}} \left\{ \|\boldsymbol{y} - \mathbf{X}\boldsymbol{\beta}\|_\infty + \lambda \|\boldsymbol{\beta}\|_1 \right\},$$

where $\lambda \geq 0$ is a weight parameter (the $\lambda = 0$ case is studied in Exercise 26).

(a) Derive a linear program which helps compute $\widehat{\boldsymbol{\beta}}_{\text{SUP}}$ for a given λ.

(b) An alternative definition of the supremum-norm lasso is via the following constrained optimization:

$$\breve{\boldsymbol{\beta}}_{\text{SUP}}(\eta) := \underset{\boldsymbol{\beta}:\|\boldsymbol{\beta}\|_1 \leq \eta}{\operatorname{argmin}} \|\boldsymbol{y} - \mathbf{X}\boldsymbol{\beta}\|_\infty,$$

where there is a one-to-one correspondence between the parameters η and λ. Show that this form of the supremum-norm lasso can also be computed via a linear program.

☞ 58 30. Consider Example 2.88 with $\boldsymbol{\Gamma} = \operatorname{diag}(\sqrt{\gamma_1}, \ldots, \sqrt{\gamma_p})$ for some nonnegative vector $\boldsymbol{\gamma} \in \mathbb{R}^p$, so that twice the negative logarithm of the *model evidence* can be written as

$$-2 \ln g(\boldsymbol{y}) = l(\boldsymbol{\gamma}) := n \ln[\boldsymbol{y}^\top(\mathbf{I} - \mathbf{X}\boldsymbol{\Sigma}\mathbf{X}^\top)\boldsymbol{y}] + 2 \ln |\boldsymbol{\Gamma}| - \ln |\boldsymbol{\Sigma}| + c,$$

where c is a constant that depends only on n.

☞ 541 (a) Use the *Woodbury identities* (A.60) and (A.61) to show that

$$\begin{aligned} \mathbf{I} - \mathbf{X}\boldsymbol{\Sigma}\mathbf{X}^\top &= (\mathbf{I} + \mathbf{X}\boldsymbol{\Gamma}^2\mathbf{X}^\top)^{-1} \\ 2 \ln |\boldsymbol{\Gamma}| - \ln |\boldsymbol{\Sigma}| &= \ln |\mathbf{I} + \mathbf{X}\boldsymbol{\Gamma}^2\mathbf{X}^\top|. \end{aligned}$$

Deduce that $l(\boldsymbol{\gamma}) = n \ln[\boldsymbol{y}^\top \mathbf{G} \boldsymbol{y}] - \ln |\mathbf{G}| + c$, where $\mathbf{G} := (\mathbf{I} + \mathbf{X}\boldsymbol{\Gamma}^2\mathbf{X}^\top)^{-1}$.

(b) Let $[\boldsymbol{v}_1, \ldots, \boldsymbol{v}_p] := \mathbf{X}$ denote the p columns/predictors of $\mathbf{X}$. Show that

$$\mathbf{G}^{-1} = \mathbf{I} + \sum_{k=1}^{p} \gamma_k \boldsymbol{v}_k \boldsymbol{v}_k^\top.$$

Explain why setting $\gamma_k = 0$ has the effect of excluding the k-th predictor from the regression model. How can this observation be used for model selection?

☞ 586 (c) Use the identities (C.19) to deduce the formulas for the gradient and Hessian of l:

$$\begin{aligned} \frac{\partial l}{\partial \gamma_i} &= \boldsymbol{v}_i^\top \mathbf{G} \boldsymbol{v}_i - n \frac{(\boldsymbol{v}_i^\top \mathbf{G} \boldsymbol{y})^2}{\boldsymbol{y}^\top \mathbf{G} \boldsymbol{y}} \\ \frac{\partial^2 l}{\partial \gamma_i \, \partial \gamma_j} &= (n-1)(\boldsymbol{v}_i^\top \mathbf{G} \boldsymbol{v}_j)^2 - n \left[\boldsymbol{v}_i^\top \mathbf{G} \boldsymbol{v}_j - \frac{(\boldsymbol{v}_i^\top \mathbf{G} \boldsymbol{y})(\boldsymbol{v}_j^\top \mathbf{G} \boldsymbol{y})}{\boldsymbol{y}^\top \mathbf{G} \boldsymbol{y}} \right]^2. \end{aligned} \tag{5.96}$$

(d) One method to determine which predictors in $\mathbf{X}$ are important is to compute

$$\widehat{\boldsymbol{\gamma}} := \underset{\boldsymbol{\gamma} \geq \mathbf{0}}{\operatorname{argmin}}\, l(\boldsymbol{\gamma})$$

using, for example, the interior-point minimization Algorithm C.102 with gradient and Hessian computed from (5.96). Write Python code to compute $\widehat{\boldsymbol{\gamma}}$ and use it to select the best polynomial model in Example 2.88. ☞ 623

31. (Exercise 30 continued.) Consider again Example 2.88 with $\boldsymbol{\Gamma} = \operatorname{diag}(\sqrt{\gamma_1}, \dots, \sqrt{\gamma_p})$ for some nonnegative model-selection parameter $\boldsymbol{\gamma} \in \mathbb{R}^p$. A Bayesian choice for $\boldsymbol{\gamma}$ is the maximizer of the marginal likelihood $g(\boldsymbol{y} \mid \boldsymbol{\gamma})$; that is, ☞ 58

$$\widehat{\boldsymbol{\gamma}} = \underset{\boldsymbol{\gamma} \geq \mathbf{0}}{\operatorname{argmax}} \iint g(\boldsymbol{\beta}, \sigma^2, \boldsymbol{y} \mid \boldsymbol{\gamma})\, \mathrm{d}\boldsymbol{\beta}\, \mathrm{d}\sigma^2,$$

where

$$\ln g(\boldsymbol{\beta}, \sigma^2, \boldsymbol{y} \mid \boldsymbol{\gamma}) = -\frac{\|\boldsymbol{y} - \mathbf{X}\boldsymbol{\beta}\|^2 + \|\boldsymbol{\Gamma}^{-1}\boldsymbol{\beta}\|^2}{2\sigma^2} - \ln|\boldsymbol{\Gamma}| - \frac{n+p}{2}\ln(2\pi\sigma^2) - \ln\sigma^2.$$

To maximize $g(\boldsymbol{y} \mid \boldsymbol{\gamma})$, one can use the *EM algorithm* with $\boldsymbol{\beta}$ and σ^2 acting as *latent variables* in the *complete-data log-likelihood* $\ln g(\boldsymbol{\beta}, \sigma^2, \boldsymbol{y} \mid \boldsymbol{\gamma})$. Define ☞ 140

$$\begin{aligned} \boldsymbol{\Sigma} &:= (\boldsymbol{\Gamma}^{-2} + \mathbf{X}^\top\mathbf{X})^{-1} \\ \overline{\boldsymbol{\beta}} &:= \boldsymbol{\Sigma}\mathbf{X}^\top\boldsymbol{y} \\ \widehat{\sigma}^2 &:= \left(\|\boldsymbol{y}\|^2 - \boldsymbol{y}^\top\mathbf{X}\overline{\boldsymbol{\beta}}\right)/n. \end{aligned} \tag{5.97}$$

(a) Show that the conditional density of the latent variables $\boldsymbol{\beta}$ and σ^2 is such that

$$\begin{aligned} \left(\sigma^{-2} \,\middle|\, \boldsymbol{\gamma}, \boldsymbol{y}\right) &\sim \mathsf{Gamma}\left(\frac{n}{2}, \frac{n}{2}\widehat{\sigma}^2\right) \\ \left(\boldsymbol{\beta} \,\middle|\, \boldsymbol{\gamma}, \sigma^2, \boldsymbol{y}\right) &\sim \mathcal{N}\left(\overline{\boldsymbol{\beta}}, \sigma^2\boldsymbol{\Sigma}\right). \end{aligned}$$

(b) Use Theorem D.27 to show that the expected complete-data log-likelihood is ☞ 634

$$-\frac{\|\boldsymbol{\Gamma}^{-1}\overline{\boldsymbol{\beta}}\|^2}{2\widehat{\sigma}^2} - \frac{\operatorname{tr}(\boldsymbol{\Gamma}^{-2}\boldsymbol{\Sigma}) + 2\ln|\boldsymbol{\Gamma}|}{2} + c_1,$$

where c_1 is a constant that does not depend on $\boldsymbol{\gamma}$.

(c) Use Theorem A.9 to simplify the expected complete-data log-likelihood and to show that it is maximized at $\gamma_i = \Sigma_{ii} + (\overline{\beta}_i/\widehat{\sigma})^2$ for $i = 1, \dots, p$. Hence, deduce the following E and M steps in the EM algorithm: ☞ 516

E-step. Given $\boldsymbol{\gamma}$, update $(\boldsymbol{\Sigma}, \overline{\boldsymbol{\beta}}, \widehat{\sigma}^2)$ via the formulas (5.97).

M-step. Given $(\boldsymbol{\Sigma}, \overline{\boldsymbol{\beta}}, \widehat{\sigma}^2)$, update $\boldsymbol{\gamma}$ via $\gamma_i = \Sigma_{ii} + (\overline{\beta}_i/\widehat{\sigma})^2$, $i = 1, \dots, p$.

(d) Write Python code to compute $\widehat{\gamma}$ via the EM algorithm, and use it to select the best polynomial model in Example 2.88. A possible stopping criterion is to terminate the EM iterations when

$$\ln g(\mathbf{y} \mid \gamma_{t+1}) - \ln g(\mathbf{y} \mid \gamma_t) < \varepsilon$$

for some small $\varepsilon > 0$, where the marginal log-likelihood is

$$\ln g(\mathbf{y} \mid \boldsymbol{\gamma}) = -\frac{n}{2} \ln(n\pi\widehat{\sigma}^2) - \ln |\mathbf{\Gamma}| + \frac{1}{2} \ln |\mathbf{\Sigma}| + \ln \Gamma(n/2).$$

32. Consider the F test statistic in (5.36) used for (block) feature selection for the nested models

$$[\mathbf{X}_1, \mathbf{X}_2, \dots, \mathbf{X}_i], \quad i = 1, \dots, d.$$

If $[\mathbf{Q}_1, \dots, \mathbf{Q}_i]$ are the mutually orthonormal matrices from Algorithm A.27 that form a ☞ 525 basis for Span $\{\mathbf{X}_1, \dots, \mathbf{X}_i\}$, prove that

$$\|\mathbf{Y}^{(i)} - \mathbf{Y}^{(i-1)}\|^2 = \|\mathbf{Q}_i^\top \mathbf{Y}\|^2, \quad i = 2, \dots, d.$$

☞ 239 Hence, implement in Python the forward selection method in Section 5.7.6.1, that uses the magnitude of $\|\mathbf{Q}_i^\top \mathbf{Y}\|^2/p_i$ as a criterion for determining the first $k < d$ most relevant (block) features. The following pseudo-code may help with the implementation.

Algorithm 5.98: Forward Stepwise Block Feature Selection

Input: $\mathbf{X} \in \mathbb{R}^{n \times (p_1 + \cdots + p_d)}$; number of blocks selected $k < d$, and $\mathbf{y} \in \mathbb{R}^n$
Output: index set $\mathcal{A}$ of k block features in $\mathbf{X}$; feature $\mathbf{1}$ always included

$\mathbf{Q} \leftarrow \mathbf{1}/\sqrt{n}$ // constant feature is always present
$\mathcal{A} \leftarrow \emptyset$ // start with no features selected
repeat // stop once k blocks are selected
  $\alpha \leftarrow -\infty$ // comparison value for computing $\text{argmax}_{i \in \overline{\mathcal{A}}} \|\mathbf{Q}_i^\top \mathbf{y}\|/p_i$
  for $i \in \overline{\mathcal{A}}$ **do** // loop over all blocks currently not selected
    $\mathbf{UR} \leftarrow \mathbf{X}_i - \mathbf{Q}\mathbf{Q}^\top \mathbf{X}_i$ // QR decomposition
    if $\|\mathbf{U}^\top \mathbf{y}\|^2/p_i > \alpha$ **then** // if better maximum found, store it
      $\mathbf{V} \leftarrow \mathbf{U}$
      $\alpha \leftarrow \|\mathbf{U}^\top \mathbf{y}\|^2/p_i$
      $j \leftarrow i$ // index of selected block of feature
  $\mathbf{Q} \leftarrow [\mathbf{Q} \;\; \mathbf{V}]$ // orthonormal basis for selected blocks
  $\mathcal{A} \leftarrow \mathcal{A} \cup \{j\}$ // add to index set of selected blocks
until $|\mathcal{A}| = k$
return $\mathcal{A}$

CHAPTER 6

Feature Selection and Shrinkage

We introduce the klimax method for model selection and shrinkage. The method enables the accurate estimation of regression coefficients of interest, even when the number of features is much larger than the number of observations and the features exhibit strong interdependence. We analyze the statistical properties of the klimax method and provide numerical experiments on both computer-simulated and real data.

6.1 Introduction to the Klimax Method

In Chapter 5 we explored linear regression models where the number of features was fixed and the sample size could grow to infinity. While the methods discussed in the current chapter are applicable to this scenario, they can also address the more difficult and interesting case where the number of features vastly exceeds the sample size and is even allowed to grow indefinitely.

To start, consider the normal linear regression model: ☞ 48

$$Y = \mathbf{1}\beta_0^* + \mathbf{X}\boldsymbol{\beta}^* + \boldsymbol{\varepsilon}, \quad \boldsymbol{\varepsilon} \sim \mathcal{N}(\mathbf{0}, \varsigma^2 \mathbf{I}_n), \tag{6.1}$$

where $\mathbf{X} \in \mathbb{R}^{n\times p}$ is the model matrix, $\boldsymbol{Y} \in \mathbb{R}^n$ is the random response vector, and β_0^*, $\boldsymbol{\beta}^*$, and ς^2 are the "true" unknown regression parameters. As in Chapter 5, the model matrix $\mathbf{X}$ is assumed to be given (fixed) and the outcome of $\boldsymbol{Y}$ is denoted by $\boldsymbol{y}$. Note that here the vector $\mathbf{1}$, corresponding to the constant feature, is not part of the model matrix $\mathbf{X} =: [\boldsymbol{v}_1, \boldsymbol{v}_2, \dots, \boldsymbol{v}_p]$. We call the p columns of $\mathbf{X}$ the **model vectors**, as opposed to the feature vectors, which are the columns of $\mathbf{X}^\top$.

Unlike our treatment of the linear model in Chapter 5, here we allow for the possibility that $p \geq n$ and that some or even most of the features in $\mathbf{X}$ are irrelevant in explaining the expected response $\mathbf{1}\beta_0^* + \mathbf{X}\boldsymbol{\beta}^*$. In other words, if $\mathcal{S} := \{i \geq 1 : \beta_i^* \neq 0\}$ is the set of nonzero regression coefficients, so that $\boldsymbol{Y} = \mathbf{1}\beta_0^* + \sum_{i\in\mathcal{S}} \boldsymbol{v}_i\beta_i^* + \boldsymbol{\varepsilon}$, it is possible that the cardinality of $\mathcal{S}$ is much smaller than p.

The main goal of this chapter is to describe the methodology for the simultaneous selection of the set $\mathcal{S}$ of relevant features — a process frequently called **model selection** or **feature selection** — and the estimation of the corresponding regression coefficients $\boldsymbol{\beta}_{\mathcal{S}}^*$, where $\boldsymbol{\beta}_{\mathcal{S}}^*$ denotes the vector with entries $\{\beta_i^*, i \in \mathcal{S}\}$. To this end, we consider the

DOI: 10.1201/9781003391173-6

☞ 49 Bayesian model for the response data $\boldsymbol{y}$:

$$\begin{aligned}(\boldsymbol{Y} \mid \boldsymbol{\beta}) &\sim \mathcal{N}(\mathbf{1}\beta_0 + \mathbf{X}\boldsymbol{\beta},\ \sigma^2\mathbf{I}_n), &&\text{(Likelihood)}\\ \boldsymbol{\beta} &\sim \mathcal{N}(\mathbf{0}, \sigma^2\boldsymbol{\Gamma}^2), &&\text{(Prior)}\end{aligned}$$

where $\boldsymbol{\Gamma} := \mathrm{diag}(\sqrt{\gamma_1}, \ldots, \sqrt{\gamma_p})$ is a diagonal matrix of positive **shrinkage parameters** $[\gamma_1, \ldots, \gamma_p]^\top =: \boldsymbol{\gamma}$. Here both the intercept β_0 and σ^2 are treated as hyperparameters, and we refer to $\sigma > 0$ as the **regularization parameter**. Later on, we will observe that larger values of σ tend to favor simpler models, whereas smaller values of σ encourage more complex models.

> When $p \gg n$ the optimal choice of the squared regularization parameter σ^2 can be significantly different from the true variance parameter ς^2.

In Exercise 1 we show that for this Bayesian model we have:

$$\boldsymbol{Y} \sim \mathcal{N}\big(\mathbf{1}\beta_0,\ \sigma^2\mathbf{G}^{-1}\big), \qquad \text{(Model Evidence)} \tag{6.2}$$

$$(\boldsymbol{\beta} \mid \boldsymbol{Y}) \sim \mathcal{N}\big(\boldsymbol{\Gamma}^2\mathbf{X}^\top\mathbf{G}(\boldsymbol{Y} - \mathbf{1}\beta_0),\ \sigma^2\boldsymbol{\Sigma}\big), \qquad \text{(Posterior)} \tag{6.3}$$

where $\mathbf{G} = \mathbf{G}(\boldsymbol{\gamma}) := (\mathbf{I}_n + \mathbf{X}\boldsymbol{\Gamma}^2\mathbf{X}^\top)^{-1}$ and $\boldsymbol{\Sigma} = \boldsymbol{\Sigma}(\boldsymbol{\gamma}) := \boldsymbol{\Gamma}(\mathbf{I}_p + \boldsymbol{\Gamma}\mathbf{X}^\top\mathbf{X}\boldsymbol{\Gamma})^{-1}\boldsymbol{\Gamma}$ are covariance matrices that depend on the shrinkage parameter $\boldsymbol{\gamma} \geq \mathbf{0}$. Note that we will frequently omit an explicit reference to $\boldsymbol{\gamma}$.

From (6.3) a natural point estimate of $\boldsymbol{\beta}^*$ is the posterior mode $\boldsymbol{\Gamma}^2\mathbf{X}^\top\mathbf{G}(\boldsymbol{y} - \mathbf{1}\beta_0)$. A practical implementation of this MAP estimate requires knowledge of the shrinkage parameter $\boldsymbol{\gamma} \geq \mathbf{0}$ and β_0. This leads us to the following.

Definition 6.4 (Klimax Method): Given a regularization parameter σ, let $\widehat{\boldsymbol{\gamma}}$ and $\widehat{\beta}_0$ be estimates of the shrinkage and intercept parameters. Then, the **Kullback–Leibler information maximization** (klimax) method estimates $\boldsymbol{\beta}^*$ via:

$$\overline{\boldsymbol{\beta}} = \widehat{\boldsymbol{\Gamma}}^2\mathbf{X}^\top\mathbf{G}(\widehat{\boldsymbol{\gamma}})(\boldsymbol{y} - \mathbf{1}\widehat{\beta}_0). \tag{6.5}$$

The estimate (6.5) of $\boldsymbol{\beta}^*$ is fundamental to all methods for model selection and estimation presented in this chapter. Importantly, if our estimate $\widehat{\gamma}_k = 0$, then the corresponding $\overline{\beta}_k$ is zero as well, which implies the selection of a model that excludes the k-th feature. We now discuss a number of possible estimates of $(\boldsymbol{\gamma}, \beta_0)$.

☞ 49 As in Section 2.9, a natural estimate of $(\boldsymbol{\gamma}, \beta_0)$ is a (possibly local) maximizer of the the model evidence (also known as prior predictive pdf or marginal likelihood) $g(\boldsymbol{y})$ of the response $\boldsymbol{y}$; that is, the pdf of $\boldsymbol{Y}$ in (6.2). Equivalently, this is a maximizer of the marginal log-likelihood

$$\ln g(\boldsymbol{y}) = -\frac{\|\boldsymbol{y} - \mathbf{1}\beta_0\|_{\mathbf{G}}^2}{2\sigma^2} + \frac{1}{2}\ln|\mathbf{G}| - \frac{n}{2}\ln(2\pi\sigma^2), \tag{6.6}$$

where $\langle \boldsymbol{v}_i, \boldsymbol{v}_j\rangle_{\mathbf{G}} := \boldsymbol{v}_i^\top\mathbf{G}\boldsymbol{v}_j$ and $\|\boldsymbol{y}\|_{\mathbf{G}} := \sqrt{\langle \boldsymbol{y}, \boldsymbol{y}\rangle_{\mathbf{G}}}$ are the **Mahalanobis inner product** and **Mahalanobis norm**, respectively.

For a given value of the regularization parameter σ, maximizing $\ln g(\mathbf{y})$ over the shrinkage parameter $\boldsymbol{\gamma} \geq \mathbf{0}$ and intercept parameter β_0 is equivalent to minimizing the **Kullback–Leibler information criterion**:

$$\text{KLC}(\boldsymbol{\gamma}, \beta_0) := \|\mathbf{y} - \mathbf{1}\beta_0\|^2_{\mathbf{G}} - \sigma^2 \ln |\mathbf{G}|. \tag{6.7}$$

Note that $\text{KLC}(\boldsymbol{\gamma}, \beta_0)/(2\sigma^2) + n\ln(2\pi\sigma^2)/2$ is the negative of the marginal log-likelihood in (6.6), and is thus an estimate of the cross-entropy risk, $-\mathbb{E}\ln g(\boldsymbol{Y})$, between the prior predictive distribution (6.2) and the true distribution $\mathcal{N}(\mathbf{1}\beta_0^* + \mathbf{X}\boldsymbol{\beta}^*, \varsigma^2\mathbf{I}_n)$ of $\boldsymbol{Y}$.

While the estimate $(\widehat{\boldsymbol{\gamma}}, \widehat{\beta}_0) = \operatorname{argmin}_{\boldsymbol{\gamma}\geq\mathbf{0},\beta_0} \text{KLC}(\boldsymbol{\gamma}, \beta_0)$ is the most important one in this chapter, it is not the only viable approach to estimating $(\boldsymbol{\gamma}, \beta_0)$. Depending on how the shrinkage and intercept parameters $\boldsymbol{\gamma}$ and β_0 are estimated, we distinguish three main cases for the estimate (6.5):

Klimax estimate. If $(\boldsymbol{\gamma}, \beta_0)$ is estimated via minimization of (6.7), then we refer to (6.5) as the **klimax estimate**.

James–Stein estimate. If $(\boldsymbol{\gamma}, \beta_0)$ is estimated via minimization of (6.7), in which the logarithm $\ln|\mathbf{G}| = \ln|\mathbf{I}_n + \mathbf{X}\boldsymbol{\Gamma}^2\mathbf{X}^\top|$ is replaced with its separable approximation $\sum_{i=1}^p \ln(1 + \gamma_i\|\boldsymbol{v}_i\|^2)$, then we refer to (6.5) as the (positive-part) **James–Stein estimate**.

Lasso estimate. If the columns of $\mathbf{X}$ are centered and $(\boldsymbol{\gamma}, \beta_0)$ is estimated via minimization of (6.7), in which $\ln|\mathbf{G}| = \ln|\mathbf{I}_n + \mathbf{X}\boldsymbol{\Gamma}^2\mathbf{X}^\top|$ is replaced by its linear approximation $\sum_{i=1}^p \gamma_i\|\boldsymbol{v}_i\|^2$, then we refer to (6.5) as the **lasso estimate**.

Each of these three major cases will be motivated and described in detail in Section 6.2. But first, we consider a few simple applications of the klimax estimate.

6.1.1 Examples of the Klimax Method

We start with two examples that highlight the workings of the klimax method — the first example uses real-world data and the second one uses computer-simulated data.

■ **Example 6.8 (Diabetes Data Set)** A commonly used data set to illustrate feature selection is the diabetes data set [169], which may be downloaded from the book's website as `Diabetes.csv`. The data set of $n = 442$ diabetes patients has $p = 10$ (standardized) features, as follows.

	Feature	Description
1	`age`	The age of the patient.
2	`sex`	The gender of the patient.
3	`bmi`	Body Mass Index.
4	`map`	Mean Arterial Pressure.
5	`tc`	Total Cholesterol.
6	`ldl`	Low-Density Lipoprotein ("bad" cholesterol).
7	`hdl`	High-Density Lipoprotein ("good" cholesterol).
8	`tch`	Total Cholesterol, similar to `tc`.
9	`ltg`	Log-transformed glucose level.
10	`glu`	Glucose level.

The response variable is a quantification of the progression of the patient's condition one year after baseline.

Figure 6.9 shows the klimax estimate $\overline{\boldsymbol{\beta}}$ of $\boldsymbol{\beta}^*$ as a function of the regularization parameter σ for each of the 10 regression parameters. Note that for large σ most of the estimated coefficients are zero, thereby eliminating the corresponding features from the model. In fact, for $\sigma > 10^3$, all the coefficients (except the intercept) are estimated as exactly zero, corresponding to the selection of the default constant model $\mathbf{1}\widehat{\beta}_0$ for the response. In contrast, when $\sigma \downarrow 0$, Exercise 15 shows that the klimax estimate $\overline{\boldsymbol{\beta}}$ converges to the minimum-norm least-squares estimate $\mathbf{X}^+\boldsymbol{y}$, corresponding to the inclusion of all the features into the regression model.

The trajectories thus obtained are called **regularization paths** (also known as *homotopy paths* or *coefficient profiles*) for the features. Inspection of such paths helps assess which of the features are relevant to explaining the variability in the observed responses. The dotted red line in the figure highlights the six non-zero parameter values associated with the point estimate of 79.4 for the optimal value of σ (we explain how the optimal σ is selected in Section 6.4). These non-zero parameter values correspond to the numbered features 3, 9, 4, 7, 2, and 5, in the order in which they appear in Figure 6.9 from right to left. We see that the features `glu` and `tch`, which are (likely) dependent on `ltg` and `tc`, are not selected, and neither are `age` and `ldl`.

In Section 6.2 we explain how the regularization path is computed efficiently.

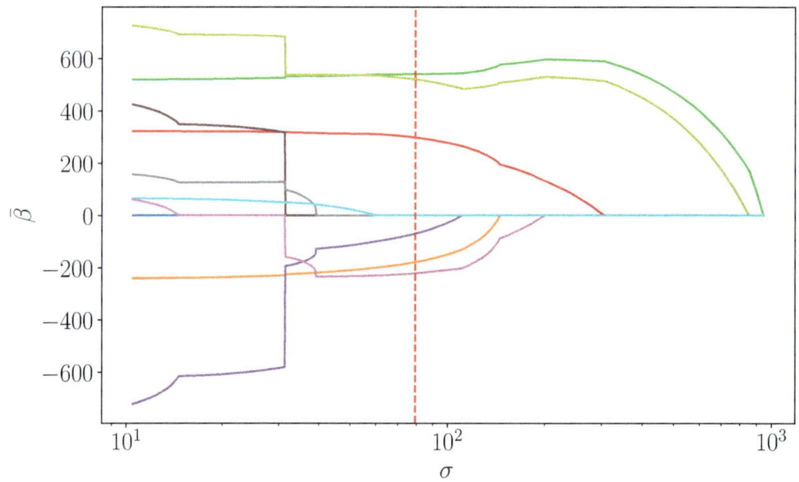

Figure 6.9: Regularization paths of the klimax estimate for the diabetes data set.

■

■ **Example 6.10 (Simulated Data Set)** With the diabetes data set it is not feasible to mathematically verify how well the "true" model and parameters are recovered by the klimax method. Moreover, the diabetes data set follows the traditional format where the number of features p is significantly smaller than the sample size n. In order to illustrate the workings of the klimax method in settings with a known true model, and allowing for $p > n$, we consider the following flexible family of artificial data sets.

Let $s = |\mathcal{S}| = \|\boldsymbol{\beta}^*\|_0$ be the number of nonzero coefficients in $\boldsymbol{\beta}^*$, where we assume that $s < n$. The data set is generated from the linear model (6.1), but without the intercept term, that is, $\beta_0^* = 0$. The model matrix $\mathbf{X}$ is constructed from a sequence of random vectors $\mathbf{Z}_1, \mathbf{Z}_2, \ldots \sim_{\text{iid}} \mathcal{N}(\mathbf{0}, \mathbf{I}_n)$, as follows: The relevant model vectors are defined as $\boldsymbol{v}_i := \mathbf{Z}_i$ for $i = 1, \ldots, s$, so that $\mathbf{X}_\mathcal{S} = [\boldsymbol{v}_1, \ldots, \boldsymbol{v}_s]$. Let $\mathbf{\Pi}_\mathcal{S}^\perp$ denote the projection matrix onto the orthogonal complement of Span $\{\boldsymbol{v}_1, \ldots, \boldsymbol{v}_s\}$. The irrelevant model vectors are ☞ 521 defined as

$$\boldsymbol{v}_i := \mathbf{X}_\mathcal{S}\boldsymbol{\eta} + \mathbf{\Pi}_\mathcal{S}^\perp \mathbf{Z}_i, \quad i \in \mathcal{S}^c,$$

where $\boldsymbol{\eta} \in \mathbb{R}^s$ controls the linear dependence of all features in $\mathcal{S}^c$ on the features in $\mathcal{S}$. By construction, the relevant and irrelevant model vectors are linearly independent when $\boldsymbol{\eta} = \mathbf{0}$, that is, $\langle \boldsymbol{v}_i, \boldsymbol{v}_j \rangle = 0$ for $i \in \mathcal{S}, j \in \mathcal{S}^c$. When $\boldsymbol{\eta} \neq \mathbf{0}$, Exercise 2 shows that the relevant and irrelevant model vectors are statistically dependent:

$$\frac{1}{n}\mathbb{E}\langle \boldsymbol{v}_i, \boldsymbol{v}_j \rangle = \begin{cases} \mathbb{1}_{\{i=j\}}, & i, j \in \mathcal{S} \\ \eta_i, & i \in \mathcal{S},\ j \in \mathcal{S}^c \\ \|\boldsymbol{\eta}\|^2 + (1 - s/n)\mathbb{1}_{\{i=j\}}, & i, j \in \mathcal{S}^c. \end{cases} \tag{6.11}$$

Finally, we assume that $\beta_i^* = 0$ for $i \in \mathcal{S}^c = \{s+1, \ldots, p\}$ (irrelevant features) and $\beta_i^* = Z_i + 2s$ for $i \in \mathcal{S} = \{1, \ldots, s\}$ (relevant features), where $Z_1, Z_2, \ldots \sim_{\text{iid}} \mathcal{N}(0, 1)$ are independent of everything else. In this setup, we can adjust the parameters s and $\boldsymbol{\eta}$ to study the performance of any klimax method.

As a specific example, consider the case where $s = 5$ and $\boldsymbol{\eta} = [0.2, 0.4, 0.6, 0.8, 1]^\top$ for a model matrix with $n = 10^2$ rows and $p = 10^3$ columns. The response vector is simulated from (6.1) using $\varsigma = 1$. The data may be found in `intro_example_2.csv`. The top panel of Figure 6.12 shows the klimax regularization paths.

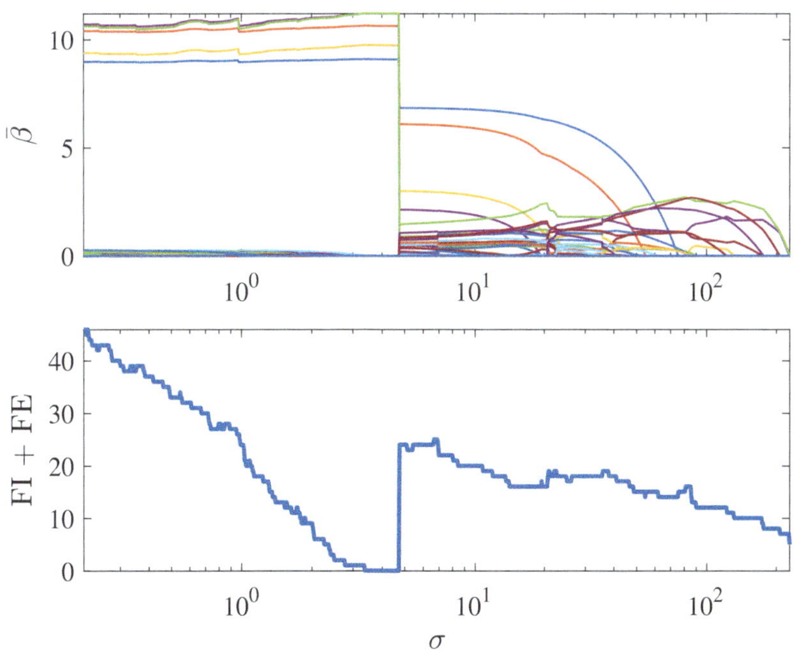

Figure 6.12: Top: Regularization paths of the klimax estimate using the artificial data set. Bottom: total number of false inclusions (FI) and false exclusions (FE) as a function of the regularization parameter σ.

Varying σ from large to small, gradually the correct features emerge, while the irrelevant features drop out, and around $\sigma = 4.7$ all the irrelevant features are quickly dropped. Crucial to any feature selection method is the prevention of **false exclusions** and **false inclusions**. False exclusions refer to relevant features that are omitted from the model, while false inclusions are irrelevant features that are selected into the model. The bottom panel of Figure 6.12 shows the sum of the total number of false inclusions and false exclusions as a function of σ. We can see that for $\sigma \in [3.4, 4.7]$ the correct model features are identified, as the total number of false inclusions and false exclusions is exactly 0. This number increases again as σ decreases. Note that for the choice $\sigma = \varsigma = 1$ a significant number of false inclusions occur, although the corresponding parameter values are small. ■

6.1.2 Special Cases of the Klimax Estimator

We now consider two examples of the estimator (6.5) that are amenable to simple analytical formulas for $\widehat{\boldsymbol{\gamma}}$. A useful identity for the derivation of these analytical formulas is the **push-through identity**

$$\mathbf{A}^\top(\mathbf{I}_n + \mathbf{A}\mathbf{A}^\top)^{-1} = (\mathbf{I}_p + \mathbf{A}^\top\mathbf{A})^{-1}\mathbf{A}^\top, \tag{6.13}$$

which follows from $\mathbf{A}^\top(\mathbf{I}_n + \mathbf{A}\mathbf{A}^\top) = (\mathbf{I}_p + \mathbf{A}^\top\mathbf{A})\mathbf{A}^\top$ for any $n \times p$ matrix $\mathbf{A}$. While the examples we consider below make oversimplifying assumptions and are thus not commonly used in practice, they still provide valuable insight into the workings of the klimax method.

6.1.2.1 James–Stein Estimator

Recall that, for a fixed model matrix $\mathbf{X}$, the response $\boldsymbol{Y} \sim \mathcal{N}(\mathbf{1}\beta_0^* + \mathbf{X}\boldsymbol{\beta}^*, \varsigma^2\mathbf{I}_n)$ for some unknown parameters $\beta_0^*, \boldsymbol{\beta}^*$, and ς. Here we further assume that $p < n$ and all model vectors are orthogonal to each other with squared length n; that is, $\mathbf{X}^\top\mathbf{X} = n\,\mathbf{I}_p$. We also assume that $\mathbf{X}^\top\mathbf{1} = \mathbf{0}$. For given parameters β_0 and $\boldsymbol{\gamma}$, we can use (6.13) to write the estimate (6.5) as

$$\overline{\boldsymbol{\beta}} = \boldsymbol{\Gamma}(\mathbf{I}_p + \boldsymbol{\Gamma}\mathbf{X}^\top\mathbf{X}\boldsymbol{\Gamma})^{-1}\boldsymbol{\Gamma}\mathbf{X}^\top(\boldsymbol{y} - \mathbf{1}\beta_0) = \boldsymbol{\Gamma}^2(\mathbf{I}_p + n\boldsymbol{\Gamma}^2)^{-1}\mathbf{X}^\top\boldsymbol{y}.$$

In other words, the elements of $\overline{\boldsymbol{\beta}}$ are given by

$$\overline{\beta}_k = \omega_k\widehat{\beta}_k, \quad k = 1, \dots, p,$$

where $\omega_k := \gamma_k/(\gamma_k + 1/n)$ is a weight in $[0, 1)$ and

$$\widehat{\boldsymbol{\beta}} := \mathbf{X}^\top\boldsymbol{Y}/n \sim \mathcal{N}(\boldsymbol{\beta}^*, \varsigma^2\mathbf{I}_p/n)$$

☞ 27 is the ordinary least-squares estimator of $\boldsymbol{\beta}^*$. Further, after simplification of (6.7), see Exercise 3, the KLC becomes:

$$\text{KLC}(\boldsymbol{\gamma}, \beta_0) = \|\boldsymbol{y} - \mathbf{1}\beta_0\|^2 - n\sum_k \omega_k\widehat{\beta}_k^2 - \sigma^2\sum_k \ln(1 - \omega_k). \tag{6.14}$$

Let us assume for simplicity that $\varsigma^2 = \mathbb{V}\mathrm{ar}(Y)$ is known and we set $\sigma^2 = \varsigma^2$. If we further take $\gamma_k = \gamma$ for all k and some $\gamma > 0$, then the β_0 and $\omega = \gamma/(\gamma + 1/n)$ that maximize the KLC are given by

$$\widehat{\beta}_0 = \frac{\sum_i y_i}{n} \quad \text{and} \quad \widehat{\omega} = \left[1 - \frac{\varsigma^2 p/n}{\|\widehat{\boldsymbol{\beta}}\|^2}\right]_+ .$$

Defining the (positive-part) **James–Stein operator**:

$$S_\lambda^{\text{JS}}(x) := x[1 - (\lambda/x)^2]_+, \tag{6.15}$$

we can then write the estimate (6.5) and the corresponding estimator as

$$\overline{\beta}_k = S_{\lambda_k}^{\text{JS}}(\widehat{\beta}_k), \quad \text{where} \quad \lambda_k := \varsigma \frac{\sqrt{p}\,|\widehat{\beta}_k|}{\sqrt{n}\,\|\widehat{\boldsymbol{\beta}}\|}, \quad k = 1, \ldots, p. \tag{6.16}$$

We call this the (positive-part) **James–Stein estimate** (or estimator, depending on the context).

Notice that (6.16) implies that $|\overline{\beta}_k| \leq |\widehat{\beta}_k|$ for all k. In other words, the James–Stein operator *reduces the magnitude of the unbiased ordinary least-squares estimator*. This phenomenon is known as **shrinkage**. While shrinkage introduces bias in the estimator, this bias is frequently offset by a significant reduction of the estimator variance, resulting in an overall reduction of the mean squared error. As a matter of fact, Exercise 4 shows that when $p \geq 5$:

$$\mathbb{E}\|\overline{\boldsymbol{\beta}} - \boldsymbol{\beta}^*\|^2 < \mathbb{E}\|\widehat{\boldsymbol{\beta}} - \boldsymbol{\beta}^*\|^2.$$

That is, the James–Stein estimator (6.16) enjoys smaller mean squared error than the unbiased least-squares estimator $\widehat{\boldsymbol{\beta}} = \mathbf{X}^\top \boldsymbol{y}/n$.

6.1.2.2 Ridge Regression Estimator

A special case of the klimax method is when the shrinkage parameter is of the form $\boldsymbol{\gamma} = \mathbf{1}/\lambda$ for some common parameter $\lambda > 0$. For simplicity, assume that $\beta_0^* = 0$, so that the intercept feature can be ignored. Then, a particular case of (6.5) is the **ridge estimate** with shrinkage parameter λ:

$$\widehat{\boldsymbol{\beta}}_\lambda := \mathrm{diag}(\mathbf{1}/\lambda)\mathbf{X}^\top \mathbf{G}(\mathbf{1}/\lambda)\boldsymbol{y}.$$

Using the push-through identity (6.13), we can equivalently represent the ridge estimate with the simpler formula:

$$\widehat{\boldsymbol{\beta}}_\lambda = (\lambda \mathbf{I}_p + \mathbf{X}^\top\mathbf{X})^{-1}\mathbf{X}^\top \boldsymbol{y}. \tag{6.17}$$

For a given estimate of σ, a reasonable choice for $\boldsymbol{\gamma} = \mathbf{1}/\lambda$ is the minimizer of KLC$(\mathbf{1}/\lambda, 0)$ in (6.7). This idea is explored in Exercise 8 under the assumption that p is fixed and n grows to infinity.

Alternatively, the following result shows that if λ is chosen judiciously, then the ridge learner, $\mathbf{X}\widehat{\boldsymbol{\beta}}_\lambda$, enjoys a smaller expected in-sample risk than the minimum-norm least-squares learner $\mathbf{X}\widehat{\boldsymbol{\beta}} = \mathbf{X}\mathbf{X}^+\boldsymbol{Y}$. ☞ 28

Lemma 6.18 (Ridge Estimator): Let $\boldsymbol{Y}'$ be a test response data for the given $\mathbf{X}$. Define $\boldsymbol{\Omega} := (\mathbf{X}^\top\mathbf{X}/n)^+$ and let the optimal shrinkage parameter be

$$\lambda^* := \frac{\varsigma^2 \mathrm{tr}(\boldsymbol{\Omega})}{(\boldsymbol{\beta}^*)^\top\boldsymbol{\Omega}\boldsymbol{\beta}^* + 3\varsigma^2\mathrm{tr}(\boldsymbol{\Omega}^2)/n}.$$

Then, the ridge learner has smaller expected in-sample risk than $\mathbf{X}\widehat{\boldsymbol{\beta}}$:

$$\mathbb{E}\|\boldsymbol{Y}' - \mathbf{X}\widehat{\boldsymbol{\beta}}\|^2 \geq \mathbb{E}\|\boldsymbol{Y}' - \mathbf{X}\widehat{\boldsymbol{\beta}}_{\lambda^*}\|^2 + \frac{\varsigma^2}{n}\lambda^*\mathrm{tr}(\boldsymbol{\Omega}).$$

The proof is delegated to Exercise 5. Lemma 6.18 suggests that applying an optimal amount of shrinkage will improve the accuracy of statistical predictions. However, the optimal shrinkage value λ^* is not available in practice, because $\boldsymbol{\beta}^*$ and ς^2 are unknown. When $n \gg p$, we can replace the true $\boldsymbol{\beta}^*$ and ς^2 with $\widehat{\boldsymbol{\beta}}$ and $\|\boldsymbol{Y} - \mathbf{X}\widehat{\boldsymbol{\beta}}\|^2/(n-p)$, respectively, to obtain a practical *plug-in estimator* of λ^*.

Typically, the ridge estimator is only guaranteed to improve upon the least-squares estimator $\widehat{\boldsymbol{\beta}}$ of $\boldsymbol{\beta}^*$ asymptotically, that is, for a sufficiently large sample size n; see Exercise 8. This is in contrast to the James–Stein estimator (6.16) of $\boldsymbol{\beta}^*$, where an improvement over $\widehat{\boldsymbol{\beta}}$ is guaranteed for any finite sample size n. In this respect, the James–Stein estimator is preferable to the ridge estimator.

Finally, two methods for the efficient computation of the ridge estimate (6.17) are explored in Exercise 9.

6.1.3 Eliminating the Intercept

In the klimax method the main focus is on determining the set $\mathcal{S}$ and the corresponding nonzero coefficients $\boldsymbol{\beta}^*_{\mathcal{S}}$, both of which are determined from the shrinkage parameter $\boldsymbol{\gamma}$. In contrast, the intercept β^*_0, corresponding to the constant feature 1, is not of interest by itself and is not subjected to any shrinkage. In other words, the intercept is a **nuisance parameter** — its value is only needed to help determine the parameter of interest, which in this case is the shrinkage parameter.

We do not treat the regularization parameter σ as a nuisance parameter, and thus postpone discussion of the estimation of the optimal σ until Section 6.4.

As a nuisance parameter, the intercept can always be expressed in terms of the shrinkage parameter via ($\mathbf{G} = \mathbf{G}(\boldsymbol{\gamma})$):

$$\underset{\beta_0}{\mathrm{argmin}}\,\|\boldsymbol{y} - \mathbf{1}\beta_0\|^2_{\mathbf{G}} = (\mathbf{G}^{1/2}\mathbf{1})^+\mathbf{G}^{1/2}\boldsymbol{y} = \langle\mathbf{1}, \boldsymbol{y}\rangle_{\mathbf{G}}/\|\mathbf{1}\|^2_{\mathbf{G}}. \tag{6.19}$$

As a result, one way to eliminate the intercept in (6.7) is to replace β_0 with (6.19) and thus compute

$$\widehat{\boldsymbol{\gamma}} = \underset{\boldsymbol{\gamma}\geq\mathbf{0}}{\mathrm{argmin}}\,\textsc{klc}\big(\boldsymbol{\gamma}, \langle\mathbf{1}, \boldsymbol{y}\rangle_{\mathbf{G}}/\|\mathbf{1}\|^2_{\mathbf{G}}\big).$$

Unfortunately, the resulting optimization over $\boldsymbol{\gamma}$ is cumbersome and unwieldy. Instead, we recommend the following two simpler methods, in which a preprocessing of the training set eliminates the nuisance intercept parameter from any subsequent estimation of the shrinkage parameter.

6.1.3.1 Centering method

Let $\mathbf{\Pi}_0^{\perp} := \mathbf{I}_n - \mathbf{1}\mathbf{1}^{+}$ be the projection matrix onto the space orthogonal to Span $\{\mathbf{1}\}$, where $\mathbf{1}$ is the model vector corresponding to the zero-th coefficient β_0. From Exercise 5.3 we know that $\mathbf{\Pi}_0^{\perp}$ is also an $n \times n$ *centering matrix*. ☞ 247

In the **centering method** the shrinkage parameter γ is estimated from the centered training set $\{\widetilde{\mathbf{X}}, \widetilde{\boldsymbol{y}}\}$, where $\widetilde{\mathbf{X}} := \mathbf{\Pi}_0^{\perp}\mathbf{X}$ and $\widetilde{\boldsymbol{y}} := \mathbf{\Pi}_0^{\perp}\boldsymbol{y}$, rather than from the original $\{\mathbf{X}, \boldsymbol{y}\}$. Thus, rather than minimizing (6.7) jointly over γ and β_0, we instead minimize solely over γ:

$$\widetilde{\boldsymbol{y}}^{\top}\left(\mathbf{I}_n + \widetilde{\mathbf{X}}\mathbf{\Gamma}^2\widetilde{\mathbf{X}}^{\top}\right)^{-1}\widetilde{\boldsymbol{y}} + \sigma^2 \ln |\mathbf{I}_n + \widetilde{\mathbf{X}}\mathbf{\Gamma}^2\widetilde{\mathbf{X}}^{\top}|.$$

Once $\widehat{\gamma}$ is computed without any involvement of the nuisance parameter β_0, the final estimate (be it klimax, lasso, or James–Stein) for $(\boldsymbol{\beta}^*, \beta_0^*)$ is:

$$\overline{\boldsymbol{\beta}} = \widehat{\mathbf{\Gamma}}^2\widetilde{\mathbf{X}}^{\top}\left(\mathbf{I}_n + \widetilde{\mathbf{X}}\widehat{\mathbf{\Gamma}}^2\widetilde{\mathbf{X}}^{\top}\right)^{-1}\widetilde{\boldsymbol{y}}, \quad \widehat{\beta}_0 = \langle \mathbf{1}, \boldsymbol{y} - \mathbf{X}\overline{\boldsymbol{\beta}}\rangle / n. \tag{6.20}$$

A rationale for using the centering method is provided by Exercise 23, which shows that, when $\mathbf{X}$ is replaced by its centered version $\mathbf{\Pi}_0^{\perp}\mathbf{X}$, the formula (6.19) always yields $\overline{y}$, in effect centering the response as well. Additionally, we recall from Exercise 5.3 that the elimination of the intercept from the ordinary least-squares training loss yields:

$$\min_{\beta_0} \|\boldsymbol{y} - \mathbf{1}\beta_0 - \mathbf{X}\boldsymbol{\beta}\|^2 = \|\widetilde{\boldsymbol{y}} - \widetilde{\mathbf{X}}\boldsymbol{\beta}\|^2.$$

6.1.3.2 Restricted Likelihood Method

Let $\overline{\mathbf{Q}} \in \mathbb{R}^{n\times n}$ be an orthonormal matrix with its first row proportional to the constant feature, that is:

$$\overline{\mathbf{Q}} = \begin{bmatrix} \frac{1}{\sqrt{n}}\mathbf{1}^{\top} \\ \mathbf{Q} \end{bmatrix},$$

where $\mathbf{Q} \in \mathbb{R}^{(n-1)\times n}$ collects all the remaining rows of $\overline{\mathbf{Q}}$. As a consequence, $\mathbf{Q}\mathbf{Q}^{\top} = \mathbf{I}_{n-1}$ and $\mathbf{Q}\mathbf{1} = \mathbf{0}$. If $\widetilde{\mathbf{X}} := \mathbf{Q}\mathbf{X}$, then (6.2) and Theorem D.46 imply that ☞ 640

$$\widetilde{\boldsymbol{Y}} := \mathbf{Q}\boldsymbol{Y} \sim \mathcal{N}\left(\mathbf{0}, \sigma^2\left(\mathbf{I}_{n-1} + \widetilde{\mathbf{X}}\mathbf{\Gamma}^2\widetilde{\mathbf{X}}^{\top}\right)\right).$$

In the **restricted likelihood method** the shrinkage parameter γ is estimated from the transformed training set $\{\widetilde{\mathbf{X}}, \widetilde{\boldsymbol{y}}\}$, rather than from the original training set $\{\mathbf{X}, \boldsymbol{y}\}$. Thus, instead of minimizing (6.7), we minimize:

$$\widetilde{\boldsymbol{y}}^{\top}\left(\mathbf{I}_{n-1} + \widetilde{\mathbf{X}}\mathbf{\Gamma}^2\widetilde{\mathbf{X}}^{\top}\right)^{-1}\widetilde{\boldsymbol{y}} + \sigma^2 \ln |\mathbf{I}_{n-1} + \widetilde{\mathbf{X}}\mathbf{\Gamma}^2\widetilde{\mathbf{X}}^{\top}|.$$

Observe that the elimination of the nuisance parameter β_0 has reduced the number of rows of $\mathbf{X}$ and $\boldsymbol{y}$ from n to $n-1$. Similar to the centering method, the final estimate (be it klimax, lasso, or James–Stein) for $(\boldsymbol{\beta}^*, \beta_0^*)$ is:

$$\overline{\boldsymbol{\beta}} = \widehat{\mathbf{\Gamma}}^2\widetilde{\mathbf{X}}^{\top}\left(\mathbf{I}_{n-1} + \widetilde{\mathbf{X}}\widehat{\mathbf{\Gamma}}^2\widetilde{\mathbf{X}}^{\top}\right)^{-1}\widetilde{\boldsymbol{y}}, \quad \widehat{\beta}_0 = \langle \mathbf{1}, \boldsymbol{y} - \mathbf{X}\overline{\boldsymbol{\beta}}\rangle / n.$$

When $\overline{\mathbf{Q}}$ is a linear transformation related to the FFT, then the matrix $\mathbf{Q}$ need not be constructed explicitly and the transformed training set $\{\widetilde{\mathbf{X}}, \widetilde{\boldsymbol{y}}\}$ can be computed very

fast. We thus recommend that the matrix $\overline{\mathbf{Q}}$ encode the discrete cosine transform (DCT)
☞ 579 Algorithm B.37. In other words, the transformed training set is computed in $O(pn \ln n)$ time by taking the DCT of $\mathbf{X}$ and $\boldsymbol{y}$, removing the first row of the output, and dividing by $\sqrt{2n}$.

6.2 Coordinate-Descent for the Klimax Method

☞ 606 In this section we describe how the coordinate-descent method can be used to efficiently compute regularization paths corresponding to the klimax, James–Stein, and lasso estimates. Computation of these estimates for a given σ requires the minimization over $\boldsymbol{\gamma} \geq \mathbf{0}$ of:

$$\|\boldsymbol{y}\|_{\mathbf{G}}^2 + \sigma^2 \times \begin{cases} -\ln|\mathbf{G}|, & \text{(klimax objective)} \\ \sum_i \ln\left(1 + \|\boldsymbol{v}_i\|^2 \gamma_i\right), & \text{(James–Stein objective)} \\ \sum_i \|\boldsymbol{v}_i\|^2 \gamma_i & \text{(lasso objective).} \end{cases}$$

Note that in all cases, we have assumed that the intercept term has been eliminated using any of the methods in Section 6.1.3. Thus, the first objective for the klimax estimate above corresponds to $\text{KLC}(\boldsymbol{\gamma}, 0)$ in (6.7).

Before we proceed, we summarize some of the notation we use in Table 6.21.

Table 6.21: Summary of definitions for feature selection and shrinkage.

$Y, \boldsymbol{y}$	Random response vector and its outcome.
$\beta_0^*, \boldsymbol{\beta}^*$	True values of regression coefficients in the normal model.
$\mathcal{S}$	True set of features, excluding intercept: $\mathcal{S} = \{i \geq 1 : \beta_i^* \neq 0\}$.
ς^2	Variance of response $\varsigma^2 = \mathbb{V}\mathrm{ar}(Y)$.
σ	Regularization parameter.
$\boldsymbol{\gamma}$	Shrinkage parameter.
$\boldsymbol{v}_1, \ldots, \boldsymbol{v}_p$	Model vectors; i.e., columns of model matrix $\mathbf{X}$.
$\mathcal{A}$ and a	Set of activated features $\mathcal{A} = \{i : \gamma_i > 0\}$ with $a = \lvert\mathcal{A}\rvert$.
$\mathbf{X}_{\mathcal{A}}$	Model matrix with columns $\{\boldsymbol{v}_i, i \in \mathcal{A}\}$.
$\boldsymbol{\beta}_{\mathcal{A}}$	Vector with entries $\{\beta_i, i \in \mathcal{A}\}$.
$\boldsymbol{\Gamma}$ and $\boldsymbol{\Gamma}_{\mathcal{A}}$	Diagonal matrices $\boldsymbol{\Gamma}^2 = \mathrm{diag}(\boldsymbol{\gamma})$ and $\boldsymbol{\Gamma}_{\mathcal{A}}^2 = \mathrm{diag}(\boldsymbol{\gamma}_{\mathcal{A}})$.
$\mathbf{G}$ and $\mathbf{G}(\boldsymbol{\gamma})$	Covariance matrix $\left(\mathbf{I}_n + \mathbf{X}\,\mathrm{diag}(\boldsymbol{\gamma})\mathbf{X}^\top\right)^{-1}$.
$\mathbf{G}_{\neg i}$	Covariance matrix $\left(\mathbf{I}_n + \sum_{j \neq i} \gamma_j \boldsymbol{v}_j \boldsymbol{v}_j^\top\right)^{-1}$.
$\lVert \cdot \rVert_{\mathbf{G}}$	Mahalanobis norm $\lVert\boldsymbol{y}\rVert_{\mathbf{G}}^2 = \boldsymbol{y}^\top \mathbf{G} \boldsymbol{y}$.
$\langle \boldsymbol{v}, \boldsymbol{y} \rangle_{\mathbf{G}}$	Mahalanobis inner product $\langle \boldsymbol{v}, \boldsymbol{y} \rangle_{\mathbf{G}} = \boldsymbol{v}^\top \mathbf{G} \boldsymbol{y}$.
S_λ^{JS}	Positive-part James–Stein operator $S_\lambda^{\text{JS}}(x) = x[1 - (\lambda/x)^2]_+$.
S_λ^{LAS}	Lasso operator $S_\lambda^{\text{LAS}}(x) = x[1 - \lvert\lambda/x\rvert]_+$.
$\boldsymbol{\Pi}_{\mathcal{A}}$	Projection matrix onto Span $\{\mathbf{X}_{\mathcal{A}}\}$.
$\boldsymbol{\Pi}_{\mathcal{A}}^\perp$	Projection matrix $\mathbf{I}_n - \boldsymbol{\Pi}_{\mathcal{A}}$.

⚠ Without loss of generality, henceforth we will assume that the intercept β_0 has been eliminated as a nuisance parameter via a preprocessing of the training data.

6.2.1 Klimax Estimate

One simple method for minimizing the Kullback–Leibler information criterion in (6.7) is the method of coordinate descent. In order to formulate a coordinate-descent algorithm for computing the estimate $\widehat{\gamma}$, we first derive the first-order necessary conditions for the minimizer of (6.7).

From the identities (C.19), we see that the gradient of $\text{KLC}(\gamma) := \text{KLC}(\gamma, 0)$ is ☞ 586

$$\frac{\partial(\text{KLC})}{\partial\gamma_i} = \sigma^2 v_i^\top \mathbf{G} v_i - (v_i^\top \mathbf{G}\mathbf{y})^2 = \sigma^2 \|v_i\|_{\mathbf{G}}^2 - \langle v_i, \mathbf{y}\rangle_{\mathbf{G}}^2, \quad i = 1, \dots, p.$$

If $\mathcal{A} := \{i : \gamma_i > 0\}$ is the set of **activated features**, then Example C.45 implies that ☞ 596

$$\frac{\partial(\text{KLC})}{\partial\gamma_i} = 0, \quad \forall i \in \mathcal{A},$$
$$\frac{\partial(\text{KLC})}{\partial\gamma_i} \geq 0, \quad \forall i \in \mathcal{A}^c.$$

In other words, the necessary **klimax-KKT** conditions for a minimizer of $\text{KLC}(\gamma, 0)$ in (6.7) are:

$$\begin{aligned} |\langle v_i, \mathbf{y}\rangle_{\mathbf{G}}| &= \sigma\|v_i\|_{\mathbf{G}}, \quad \forall i \in \mathcal{A}, \\ |\langle v_i, \mathbf{y}\rangle_{\mathbf{G}}| &\leq \sigma\|v_i\|_{\mathbf{G}}, \quad \forall i \in \mathcal{A}^c. \end{aligned} \tag{6.22}$$

Example C.45 also provides a sufficient condition for a minimizer. Namely, if the point γ satisfies (6.22) and the Hessian matrix at $\gamma_{\mathcal{A}}$ is positive definite:

$$\mathbf{H} := \frac{\partial^2(\text{KLC})}{\partial\gamma_{\mathcal{A}}\partial\gamma_{\mathcal{A}}^\top} \succ \mathbf{0},$$

then γ is a local minimizer. From (C.19) the second-order derivatives in $\mathbf{H}$ are given by:

$$\frac{\partial^2(\text{KLC})}{\partial\gamma_i\partial\gamma_j} = 2\langle \mathbf{y}, v_i\rangle_{\mathbf{G}}\langle v_i, v_j\rangle_{\mathbf{G}}\langle v_j, \mathbf{y}\rangle_{\mathbf{G}} - \sigma^2\langle v_i, v_j\rangle_{\mathbf{G}}^2, \quad \forall i, j \in \mathcal{A}. \tag{6.23}$$

Even though a point satisfying the klimax-KKT conditions (6.22) is not necessarily the global minimizer of $\text{KLC}(\gamma)$, we will still denote it as $\widehat{\gamma}$. The corresponding coefficient estimate will again be referred to as the **klimax estimate** of $\boldsymbol{\beta}^*$. The klimax-KKT conditions can be conveniently expressed in terms of the James–Stein operator in (6.15).

Lemma 6.24 (Klimax KKT Conditions): Let $\lambda_i := \sigma \big/ \sqrt{v_i^\top \mathbf{G}_{\neg i} v_i}$ for all i, then the klimax-KKT conditions (6.22) can be written via the James–Stein operator:

$$\gamma_i = \frac{1}{\sigma^2} \frac{v_i^\top \mathbf{G}_{\neg i} \mathbf{y}}{v_i^\top \mathbf{G}_{\neg i} v_i} S_{\lambda_i}^{\text{JS}} \left(\frac{v_i^\top \mathbf{G}_{\neg i} \mathbf{y}}{v_i^\top \mathbf{G}_{\neg i} v_i} \right), \quad i = 1, \ldots, p. \tag{6.25}$$

Also, the klimax-KKT conditions (6.22) are equivalent to the set of equations:

$$\beta_i = S_{\lambda_i}^{\text{JS}} \left(\frac{\langle v_i, \mathbf{y} \rangle_{\mathbf{G}}}{\|v_i\|_{\mathbf{G}}^2} \right), \quad \gamma_i = \frac{\beta_i}{\sigma^2} \frac{\langle v_i, \mathbf{y} \rangle_{\mathbf{G}}}{\|v_i\|_{\mathbf{G}}^2}, \quad i = 1, \ldots, p. \tag{6.26}$$

☞ 541 *Proof.* From the Sherman–Morrison formula, we have

$$\mathbf{G} = \mathbf{G}_{\neg i} - \frac{\gamma_i \mathbf{G}_{\neg i} v_i v_i^\top \mathbf{G}_{\neg i}}{1 + \gamma_i v_i^\top \mathbf{G}_{\neg i} v_i}. \tag{6.27}$$

Therefore, defining $\mu_i := v_i^\top \mathbf{G}_{\neg i} \mathbf{y}$ and $\nu_i := v_i^\top \mathbf{G}_{\neg i} v_i$, we have $\|v_i\|_{\mathbf{G}}^2 = \nu_i/(1 + \gamma_i \nu_i)$ and $\langle v_i, \mathbf{y} \rangle_{\mathbf{G}} = \mu_i/(1 + \gamma_i \nu_i)$. Substituting with the latter identities into (6.22) yields:

$$\begin{aligned} \mu_i^2 &= \sigma^2 \nu_i (1 + \gamma_i \nu_i), \quad \forall i \in \mathcal{A}, \\ \mu_i^2 &\leq \sigma^2 \nu_i, \quad \forall i \in \mathcal{A}^c, \end{aligned}$$

so that $\gamma_i = \mu_i^2/(\sigma^2 \nu_i^2) - 1/\nu_i$ for $i \in \mathcal{A}$. On the other hand, (6.25) can be written as

$$\gamma_i = \frac{1}{\sigma^2} \frac{\mu_i}{\nu_i} S_{\lambda_i}^{\text{JS}} \left(\frac{\mu_i}{\nu_i} \right) = \frac{\mu_i^2}{\sigma^2 \nu_i^2} \left[1 - \frac{\sigma^2 \nu_i}{\mu_i^2} \right]_+ = \begin{cases} \frac{\mu_i^2}{\sigma^2 \nu_i^2} - \frac{1}{\nu_i}, & i \in \mathcal{A}, \\ 0, & i \in \mathcal{A}^c, \end{cases}$$

which establishes that (6.22)⇔(6.25). Next, we show that (6.25)⇔(6.26). Since $\mu_i/\nu_i = \langle v_i, \mathbf{y} \rangle_{\mathbf{G}} / \|v_i\|_{\mathbf{G}}^2$, we can see that (6.26)⇒(6.25). It remains to show that (6.25)⇒(6.26). Since $\beta_i = \gamma_i \langle v_i, \mathbf{y} \rangle_{\mathbf{G}} = 0$ for $i \in \mathcal{A}^c$, we need only prove the implication for $i \in \mathcal{A}$. Assuming that $i \in \mathcal{A}$, we have $\mu_i^2/\nu_i = \sigma^2(1 + \gamma_i \nu_i)$, and therefore:

$$\begin{aligned} S_{\lambda_i}^{\text{JS}} \left(\frac{\langle v_i, \mathbf{y} \rangle_{\mathbf{G}}}{\|v_i\|_{\mathbf{G}}^2} \right) = S_{\lambda_i}^{\text{JS}} \left(\frac{\mu_i}{\nu_i} \right) &= \frac{\mu_i}{\nu_i} \left[1 - \frac{\sigma^2/\nu_i}{\mu_i^2/\nu_i^2} \right]_+ \\ &= \frac{\mu_i}{\nu_i} \left[1 - \frac{1}{1 + \gamma_i \nu_i} \right]_+ \\ &= \frac{\gamma_i \mu_i}{1 + \gamma_i \nu_i} = \gamma_i \langle v_i, \mathbf{y} \rangle_{\mathbf{G}} = \beta_i, \end{aligned}$$

where we recall that $\boldsymbol{\beta} = \boldsymbol{\Gamma}^2 \mathbf{X}^\top \mathbf{G} \mathbf{y}$. This completes the proof. □

One corollary of Lemma 6.24 is that (6.25) is a coordinate-descent cycle for minimizing $\text{KLC}(\boldsymbol{\gamma})$. As a result, the equivalent updating (6.26) immediately yields the following simple algorithm for computing a solution to (6.22) for a given value of σ.

Algorithm 6.28: Computing Klimax Estimate (6.5) for a Given σ

Input: Initial guess $\boldsymbol{\gamma}$, training set $\{\mathbf{X}, \boldsymbol{y}\}$, tolerance ε, and σ.
Output: Solution to (6.22) and klimax estimate (6.5) for the given σ.

```
repeat                                          // iterate over CD cycles
    γ_old ← γ
    LL^T ← I_n + X diag(γ) X^T                  // compute Cholesky factor L via Algorithm A.69
    for i = 1, ..., p do                        // coordinate-descent cycle
        Ly ← L^{-1} y
        Lv ← L^{-1} v_i                         // backward substitution at O(n^2) cost
        vGy ← ⟨Lv, Ly⟩
        vGv ← ||Lv||^2
        λ_i ← σ sqrt((1 − γ_i vGv)/vGv)
        β_i ← S^JS_{λ_i}(vGy/vGv)
        γ_new ← (β_i/σ^2)(vGy/vGv)              // update coefficients via (6.26)
        if γ_new ≠ γ_i then                     // if change in γ_i, then recompute Cholesky factor
            γ_i ← γ_new
            LL^T ← I_n + X diag(γ) X^T          // Cholesky factor at O(n^3) cost
until ||γ − γ_old|| < ε
return γ, β
```

The following Python codes implements the klimax algorithm, as well as the James–Stein operator.

Klimax.py

```python
import numpy as np
from numpy.linalg import solve, cholesky, norm


def S_js(x, lam):
    return np.maximum(0, 1 - (lam / x)**2) * x


def klimax(gam, X, y, sig, max_iter=10**3, eps=1e-6):
    n, p = X.shape
    b = gam.copy()
    for _ in range(max_iter):
        gold = gam.copy()
        L = cholesky(np.eye(n) + X @ (gam * X.T))

        for i in range(p):
            Ly = solve(L, y)
            Lv = solve(L, X[:, i])
            vGy = Lv.T @ Ly
            vGv = Lv.T @ Lv
            lam = sig * np.sqrt((1 - gam[i] * vGv) / vGv)
            b[i] = S_js(vGy / vGv, lam)
            gam_new = b[i] / sig**2 * vGy / vGv
```

```
            if gam_new != gam[i]:
                gam[i] = gam_new
                L = cholesky(np.eye(n) + X @ (gam * X.T))

        if norm(gold - gam) < eps:
            break

    return gam,  b
```

While the brevity and simplicity of Algorithm 6.28 is the main reason for its presentation here, it is not a computationally efficient (and hence not the recommended) algorithm for computing the klimax estimate for a given σ. To begin with, Line 14 of Algorithm 6.28 computes the lower triangular Cholesky factor $\mathbf{L}$ of $\mathbf{G}^{-1}(\boldsymbol{\gamma})$ after a co-
☞ 545 ordinate of $\boldsymbol{\gamma}$ is updated using the costly Algorithm A.69. Since a coordinate update results in a rank-1 perturbation of $\mathbf{G}^{-1}(\boldsymbol{\gamma})$, it is computationally more efficient to update an existing Cholesky factor $\mathbf{L}$ via Algorithm A.76. This lowers the computing cost per coordinate-descent cycle from $O(pn^3)$ to $O(pn^2)$.

Even better, if $\mathcal{A}$, with $|\mathcal{A}| = a$, is the set of activated features in (6.22), so that $\boldsymbol{\gamma}_{\mathcal{A}} > \mathbf{0}$ and $\boldsymbol{\gamma}_{\mathcal{A}^c} = \mathbf{0}$, we can lower the cost per coordinate-descent cycle to $O(pa^2)$ using preprocessing computations of $O(pna)$, as follows: Suppose that at a given coordinate-descent iteration we store in memory the QR decomposition of the activated model vectors, $\mathbf{X}_{\mathcal{A}} = \mathbf{Q}_{\mathcal{A}}\mathbf{R}_{\mathcal{A}}$, where $\mathbf{R} := \mathbf{Q}_{\mathcal{A}}^\top\mathbf{X}$ is an $a \times p$ matrix with columns $\boldsymbol{r}_1, \ldots, \boldsymbol{r}_p$. The QR decomposition is computed via Algorithm A.22 in $O(na^2)$ time and $\mathbf{R}$ is
☞ 524 computed in $O(pna)$ time. Then, we can evaluate the Mahalanobis norm $\|\boldsymbol{v}_i\|_{\mathbf{G}}^2$ via the identity proved in Exercise 11:

$$\|\boldsymbol{v}_i\|_{\mathbf{G}}^2 = (\|\boldsymbol{v}_i\|^2 - \|\boldsymbol{r}_i\|^2)\mathbb{1}_{\{i\in\mathcal{A}^c\}} + \|\boldsymbol{r}_i\|_{\mathbf{G}_{\mathcal{A}}}^2, \qquad i = 1, \ldots, p,$$

where

$$\mathbf{G}_{\mathcal{A}} := \left(\mathbf{I}_a + \mathbf{R}_{\mathcal{A}}\boldsymbol{\Gamma}_{\mathcal{A}}^2\mathbf{R}_{\mathcal{A}}^\top\right)^{-1} \in \mathbb{R}^{a\times a}, \tag{6.29}$$

and $\|\cdot\|_{\mathbf{G}_{\mathcal{A}}} : \mathbb{R}^a \to \mathbb{R}$ is a Mahalanobis norm over the subspace $\mathbb{R}^a$ of $\mathbb{R}^n$. While $(\|\boldsymbol{v}_i\|^2 - \|\boldsymbol{r}_i\|^2)\mathbb{1}_{\{i\in\mathcal{A}^c\}}$ is simple to compute and independent of the value of $\boldsymbol{\gamma}_{\mathcal{A}}$, the Mahalanobis norm $\|\boldsymbol{r}_i\|_{\mathbf{G}_{\mathcal{A}}}$ can either be evaluated from scratch in $O(a^3)$ time, or its value updated in $O(a^2)$ time, because a coordinate change in $\boldsymbol{\gamma}_{\mathcal{A}}$ results in a rank-1 perturbation of $\mathbf{G}_{\mathcal{A}}^{-1}$. One can similarly evaluate the Mahalanobis inner product $\langle \boldsymbol{v}_i, \boldsymbol{y}\rangle_{\mathbf{G}}$. The full details of the fast klimax algorithm are given in Exercises 11 and 12. The
☞ 307 corresponding pseudo-code can be found in Algorithm 6.86.

■ **Remark 6.30 (The Trivial Klimax-KKT Solution)** There is a trivial solution to the klimax-KKT conditions (6.22). Direct substitution verifies that $\widehat{\boldsymbol{\gamma}} = \mathbf{0}$ is a solution to (6.22) for any value of the regularization parameter such that

$$\sigma \geq \sigma_* := \max_i \frac{|\langle \boldsymbol{v}_i, \boldsymbol{y}\rangle|}{\|\boldsymbol{v}_i\|}. \tag{6.31}$$

This is the empty set solution, in which the regularization is so strong that the activated set $\mathcal{A} = \emptyset$; see also Exercise 14. ■

One consequence of all shrinkage parameters being zero for $\sigma \geq \sigma_*$ in (6.31) is that we need only compute and analyze (6.5) for $\sigma < \sigma_*$. It is for this reason that we will typically compute the regularization path of (6.5) on a logarithmically-spaced grid of m values for σ:

$$\sigma_j = \exp(\ln \sigma_0 + j/m(\ln \sigma_* - \ln \sigma_0)), \quad j = 1, \ldots, m, \tag{6.32}$$

where σ_0 equals σ_* times a small enough number, say 10^{-5}.

For example, we can use Algorithm 6.28 to compute the regularization path of the klimax estimator (6.5) over the logarithmically-spaced grid of σ-values (6.32).

Algorithm 6.33: Regularization Path on a Grid of σ-values

Input: Training set $\{\mathbf{X}, \boldsymbol{y}\}$, integer m, error tolerance ε.

Output: Klimax regularization path $(\sigma_1, \boldsymbol{\beta}_1), \ldots, (\sigma_m, \boldsymbol{\beta}_m)$.

1 $\sigma_m \leftarrow \max_i |\langle \boldsymbol{v}_i, \boldsymbol{y} \rangle| / \|\boldsymbol{v}_i\|$ `//` σ_* `from` (6.31) `corresponding to` $\mathcal{A} = \emptyset$

2 $\sigma_0 \leftarrow \sigma_m / 10^5$

3 $\sigma_j \leftarrow \exp(\ln \sigma_0 + j/m(\ln \sigma_m - \ln \sigma_0))$ for $j = 1, \ldots, m$ `//` σ`-grid in` (6.32)

4 $\boldsymbol{\gamma}_m \leftarrow \mathbf{0}$, $\boldsymbol{\beta}_m \leftarrow \mathbf{0}$ `// trivial klimax-KKT solution at` $\sigma_m = \sigma_*$ `in` (6.31)

5 **for** $j = m-1, m-2 \ldots, 2, 1$ **do** `// loop over grid points in reverse order`

6     $(\boldsymbol{\gamma}_j, \boldsymbol{\beta}_j) \leftarrow \text{klimaxEst}(\boldsymbol{\gamma}_{j+1}, \mathbf{X}, \boldsymbol{y}, \varepsilon, \sigma_j)$ `// implements Algorithm 6.28`

7 **return** $(\sigma_1, \boldsymbol{\beta}_1), \boldsymbol{\beta}_2, \ldots, (\sigma_m, \boldsymbol{\beta}_m)$

Notice how in Line 6 of Algorithm 6.33 we exploit the idea of a **warm start** for the coordinate-descent iteration. Namely, we first compute $\boldsymbol{\gamma}_m = \boldsymbol{\beta}_m = \mathbf{0}$ (the sparsest possible estimate) and then try to compute the estimate $\boldsymbol{\beta}_{m-1}$, corresponding to σ_{m-1}, using $\boldsymbol{\beta}_m$ (or rather its equivalent $\boldsymbol{\gamma}_m$) as the initial guess for the coordinate-descent iteration. Once $\boldsymbol{\beta}_{m-1}$ is computed, we then proceed to compute $\boldsymbol{\beta}_{m-2}$ using $\boldsymbol{\beta}_{m-1}$ as the initial guess and so on, until we compute all the estimates for each σ_j with $j = m, m-1, m-2, \ldots, 2, 1$. Below is a Python module for the regularization path computation and grid construction. The module also includes a method for eliminating the intercept using the discrete cosine transform. The code for the discrete cosine transform can be found in Section B.2.3. ☞ 578

RegPath.py

```python
import numpy as np
from Klimax import klimax
from DCT import dct


def regularPath(X, y, grid, E):
    n_grid = len(grid)
    n, p = X.shape
    path = np.full((n_grid, p), np.nan)
    b = np.zeros((p, 1))
    gam = np.zeros((p, 1))

    for k in range(n_grid-1, -1, -1):
        print(k)
        sig = grid[k]
```

```
        gam, b = klimax(gam, X, y, sig)
        path[k, :] = b.flatten()

    return path


def makegrid(X, y, n_grid, c_grid):
    vv = np.sum(X * X, axis=0).reshape(-1, 1)
    Xy = X.T @ y
    a = np.abs(Xy) / np.sqrt(vv)
    sigm = np.max(a)
    E = {np.argmax(a)}
    grid = np.exp(np.linspace(np.log(sigm) - c_grid, np.log(sigm),
        n_grid))
    return grid, E


def eliminate_intercept(X, y):
    n, p = X.shape
    y = dct(y.flatten()).reshape(-1, 1)
    for k in range(p):
        X[:, k] = dct(X[:, k].flatten())
    return X[1:, :] / np.sqrt(2 * n), y[1:] / np.sqrt(2 * n)
```

■ **Example 6.34 (Diabetes Data Set (cont.))** The following Python code produces the regularization paths in Figure 6.9 for the diabetes data in Example 6.8. Note that we eliminate the intercept by transforming the data according to the restricted likelihood method in Section 6.1.3.2.

diabetes.py

```
import numpy as np
import matplotlib.pyplot as plt
from RegPath import regularPath, makegrid, eliminate_intercept

# Load the data
yX = np.loadtxt('Diabetes.csv', delimiter=',', skiprows=1)
y = yX[:, 0].reshape(-1, 1)
X = yX[:, 1:]
n, p = X.shape

X, y = eliminate_intercept(X, y)
n,_ = X.shape  # reduce row dimension by 1

# Create a grid of sigma values
n_grid = 100
c_grid = 4.5
grid, E = makegrid(X,y, n_grid, c_grid)

# Compute the regularization paths
path = regularPath(X, y, grid, E)

# Plot results
```

```
plt.figure(1,figsize=(10,6))
plt.semilogx(grid,path)
plt.xlabel(r'$\sigma$')
plt.ylabel(r'$\bar{\beta}$')
```

■

Penalized Least-Squares Form of Klimax Estimate

We finish this section by pointing out that the klimax estimate $\overline{\boldsymbol{\beta}}$, with $\widehat{\boldsymbol{\gamma}}$ minimizing $\text{KLC}(\boldsymbol{\gamma})$ globally, is also the global minimizer of the **penalized least-squares** objective (see Exercise 16):

$$\frac{\|\boldsymbol{y} - \mathbf{X}\boldsymbol{\beta}\|^2}{2} + \sigma^2 \rho_{\mathbf{X}}(\mathbf{D}\boldsymbol{\beta}/\sigma), \tag{6.35}$$

where $\mathbf{D} := \text{diag}([\|\boldsymbol{v}_1\|, \ldots, \|\boldsymbol{v}_p\|])$ and the **klimax penalty** $\rho_{\mathbf{X}}$ is implicitly defined via:

$$\rho_{\mathbf{X}}(\boldsymbol{\beta}) := \frac{1}{2}\min_{\boldsymbol{\gamma}\geq\mathbf{0}}\left\{\ln|\mathbf{I}_n + \mathbf{X}\boldsymbol{\Gamma}^2\mathbf{X}^\top| + \sum_{i:\beta_i\neq 0}\frac{\beta_i^2}{\|\boldsymbol{v}_i\|^2\gamma_i}\right\}.$$

Interestingly, since the multivariate penalty $\rho_{\mathbf{X}}$ depends on the model matrix $\mathbf{X}$, it is in general not **separable**; that is, it cannot be expressed as a sum $\sum_{i=1}^p \rho_i(\beta_i)$ for some univariate penalty functions $\{\rho_i\}$. In contrast, as we shall see in the next two sections, the corresponding penalties for the James–Stein and lasso estimates are separable.

6.2.2 (Positive-Part) James–Stein Estimate

A special case of the klimax method can be derived by estimating $(\boldsymbol{\gamma}, \beta_0)$ as the minimizer of (6.7), where $\ln|\mathbf{I}_n + \mathbf{X}\boldsymbol{\Gamma}^2\mathbf{X}^\top|$ is replaced with the upper bound:

$$\ln|\mathbf{I}_n + \mathbf{X}\boldsymbol{\Gamma}^2\mathbf{X}^\top| = \ln|\mathbf{I}_p + \boldsymbol{\Gamma}\mathbf{X}^\top\mathbf{X}\boldsymbol{\Gamma}| \leq \sum_{i=1}^p \ln(1 + \|\boldsymbol{v}_i\|^2\gamma_i).$$

This upper bound follows from the identity (A.61) and the bound on the log-determinant in Example A.40. Notice that $\sum_{i=1}^p \ln(1 + \|\boldsymbol{v}_i\|^2\gamma_i)$ is the **separable approximation** to ☞ 531 $\ln|\mathbf{I}_n + \mathbf{X}\boldsymbol{\Gamma}^2\mathbf{X}^\top|$ with equality achieved when $\mathbf{X}^\top\mathbf{X}$ is a diagonal matrix.

Then, for a given regularization parameter σ, the Kullback–Leibler information criterion $\text{KLC}(\boldsymbol{\gamma}) = \text{KLC}(\boldsymbol{\gamma}, 0)$ in (6.7) is upper bounded by the **James–Stein information criterion**:

$$\text{JSC}(\boldsymbol{\gamma}) := \|\boldsymbol{y}\|_{\mathbf{G}}^2 + \sigma^2\sum_{i=1}^p \ln(1 + \|\boldsymbol{v}_i\|^2\gamma_i) \geq \text{KLC}(\boldsymbol{\gamma}). \tag{6.36}$$

Any stationary point of (6.36) satisfies the **James–Stein KKT** conditions:

$$\begin{aligned} |\langle \boldsymbol{v}_i, \boldsymbol{y}\rangle_{\mathbf{G}}| &= \frac{\sigma\|\boldsymbol{v}_i\|}{\sqrt{1 + \gamma_i\|\boldsymbol{v}_i\|^2}}, \quad \forall i \in \mathcal{A}, \\ |\langle \boldsymbol{v}_i, \boldsymbol{y}\rangle_{\mathbf{G}}| &\leq \sigma\|\boldsymbol{v}_i\|, \quad \forall i \in \mathcal{A}^c. \end{aligned} \tag{6.37}$$

In summary, the (positive-part) **James–Stein estimate** is defined as $\overline{\boldsymbol{\beta}} = \widehat{\boldsymbol{\Gamma}}^2\mathbf{X}^\top\mathbf{G}(\widehat{\boldsymbol{\gamma}})\boldsymbol{y}$, where $\widehat{\boldsymbol{\gamma}}$ is a solution of the James–Stein KKT conditions (6.37).

There is no simple formula for a coordinate-descent method minimizing (6.36). However, we next show that (6.36) can be rewritten in a penalized least-squares form, which is amenable to a remarkably simple coordinate-descent algorithm.

Penalized Least-Squares Form of James–Stein Estimate

Whilst the James–Stein KKT conditions (6.37) are useful for deriving the theoretical properties of the James–Stein estimator, they do not provide a simple method for its computation. Instead, the penalized least-squares form of (6.37) will yield a simple coordinate-descent algorithm for the computation of the James–Stein estimate. To this end, we note that for all estimates of the form $\boldsymbol{\beta} = \boldsymbol{\Gamma}^2\mathbf{X}^\top\mathbf{G}(\boldsymbol{\gamma})\boldsymbol{y}$, which includes the James–Stein estimate, there is an alternative way to compute the residual $\boldsymbol{r} := \boldsymbol{y} - \mathbf{X}\boldsymbol{\beta}$, as follows:

Lemma 6.38 (Residual): If $\boldsymbol{\beta} = \boldsymbol{\Gamma}^2\mathbf{X}^\top\mathbf{G}(\boldsymbol{\gamma})\boldsymbol{y}$ for $\boldsymbol{\gamma} \geq \mathbf{0}$, then $\boldsymbol{y} - \mathbf{X}\boldsymbol{\beta} = \mathbf{G}(\boldsymbol{\gamma})\boldsymbol{y}$.

Proof. We have that $\mathbf{X}\boldsymbol{\beta} = \mathbf{X}\boldsymbol{\Gamma}^2\mathbf{X}^\top\mathbf{G}\boldsymbol{y} = (\mathbf{G}^{-1} - \mathbf{I}_n)\mathbf{G}\boldsymbol{y} = \boldsymbol{y} - \mathbf{G}\boldsymbol{y}$, implying the result. □

Let $\boldsymbol{r}_{\neg i} := \boldsymbol{y} - \mathbf{X}\boldsymbol{\beta} + \beta_i\boldsymbol{v}_i = \boldsymbol{r} + \beta_i\boldsymbol{v}_i$ be the residual that excludes the i-th feature. Then, an application of Lemma 6.38 yields:

$$\langle \boldsymbol{v}_i, \boldsymbol{y}\rangle_{\mathbf{G}} = \langle \boldsymbol{v}_i, \boldsymbol{r}\rangle = \langle \boldsymbol{v}_i, \boldsymbol{r}_{\neg i}\rangle - \|\boldsymbol{v}_i\|^2\beta_i.$$

Since $\beta_i = \gamma_i\langle \boldsymbol{v}_i, \boldsymbol{y}\rangle_{\mathbf{G}} = \gamma_i\langle \boldsymbol{v}_i, \boldsymbol{r}\rangle$, we obtain $\langle \boldsymbol{v}_i, \boldsymbol{r}_{\neg i}\rangle = (1 + \gamma_i\|\boldsymbol{v}_i\|^2)\langle \boldsymbol{v}_i, \boldsymbol{r}\rangle$. Hence, the James–Stein KKT conditions (6.37) can be written in terms of the residuals $\boldsymbol{r}$ and $\boldsymbol{r}_{\neg i}$:

$$\text{(6.39)} \qquad \begin{aligned} |\langle \boldsymbol{v}_i, \boldsymbol{r}\rangle| &= \frac{\sigma\|\boldsymbol{v}_i\|}{\sqrt{\langle \boldsymbol{v}_i, \boldsymbol{r}_{\neg i}\rangle / \langle \boldsymbol{v}_i, \boldsymbol{r}\rangle}}, \quad \forall i \in \mathcal{A}, \\ |\langle \boldsymbol{v}_i, \boldsymbol{r}\rangle| &\leq \sigma\|\boldsymbol{v}_i\|, \quad \forall i \in \mathcal{A}^c. \end{aligned}$$

The equivalence of (6.37) and (6.39) then helps us to express the James–Stein KKT conditions (6.37) via the James–Stein operator in (6.15).

Lemma 6.40 (James–Stein KKT Conditions): Let $\lambda_i := \sigma/\|\boldsymbol{v}_i\|$ for all i, then the James–Stein KKT conditions (6.37) can be written as:

$$\text{(6.41)} \qquad \beta_i = S^{\text{JS}}_{\lambda_i}\left(\frac{\langle \boldsymbol{v}_i, \boldsymbol{r}_{\neg i}\rangle}{\|\boldsymbol{v}_i\|^2}\right) = S^{\text{JS}}_{\lambda_i}\left(\beta_i + \frac{\langle \boldsymbol{v}_i, \boldsymbol{r}\rangle}{\|\boldsymbol{v}_i\|^2}\right), \quad i = 1, \ldots, p.$$

Proof. Assuming (6.37)⇔(6.39), we have: $\langle \boldsymbol{v}_i, \boldsymbol{r}\rangle\langle \boldsymbol{v}_i, \boldsymbol{r}_{\neg i}\rangle = \sigma^2\|\boldsymbol{v}_i\|^2$, $\forall i \in \mathcal{A}$. Hence,

$$(\langle \boldsymbol{v}_i, \boldsymbol{r}_{\neg i}\rangle - \|\boldsymbol{v}_i\|^2\beta_i)\langle \boldsymbol{v}_i, \boldsymbol{r}_{\neg i}\rangle = \sigma^2\|\boldsymbol{v}_i\|^2, \quad \forall i \in \mathcal{A}.$$

Solving the last equation for β_i yields:

$$\beta_i = \frac{\langle \boldsymbol{v}_i, \boldsymbol{r}_{\neg i}\rangle}{\|\boldsymbol{v}_i\|^2} - \frac{\sigma^2}{\langle \boldsymbol{v}_i, \boldsymbol{r}_{\neg i}\rangle} = S^{\text{JS}}_{\lambda_i}\left(\frac{\langle \boldsymbol{v}_i, \boldsymbol{r}_{\neg i}\rangle}{\|\boldsymbol{v}_i\|^2}\right), \quad \forall i \in \mathcal{A}.$$

Since $\{S^{\text{JS}}_{\lambda_i}(\langle \boldsymbol{v}_i, \boldsymbol{r}_{\neg i}\rangle / \|\boldsymbol{v}_i\|^2) = 0,\ \forall i \in \mathcal{A}^c\} \Leftrightarrow \{\langle \boldsymbol{v}_i, \boldsymbol{r}\rangle^2 \le \sigma^2 \|\boldsymbol{v}_i\|^2,\ \forall i \in \mathcal{A}^c\}$, equation (6.41) also works for all $i \in \mathcal{A}^c$. This shows that (6.37)$\Leftrightarrow$(6.39)$\Leftrightarrow$(6.41). □

Remarkably, the alternative KKT conditions (6.41) agree with the coordinate-descent cycle that minimizes the *penalized least-squares* objective (6.35), where $\rho_{\mathbf{X}}(\boldsymbol{\beta})$ is replaced with the *separable* **James–Stein penalty** $\rho^{\text{JS}}(\boldsymbol{\beta}) := \sum_{i=1}^{p} \rho(\beta_i)$ and

$$\text{(6.42)} \qquad \rho(\beta) := \frac{\sqrt{\beta^4 + 4\beta^2} - \beta^2}{4} + \frac{1}{2}\ln\left(1 + \frac{\sqrt{\beta^4 + 4\beta^2} + \beta^2}{2}\right).$$

The univariate penalty function (6.42) behaves like the absolute value function near the origin: $\rho(\beta) = |\beta| + O(\beta^2)$ as $\beta \to 0$; see Exercise 18. As a result, the formula (6.41) allows us to compute the James–Stein estimate via the following coordinate-descent algorithm.

Algorithm 6.43: James–Stein Penalized Least-Squares Coordinate Descent

input: Initial $\boldsymbol{\beta}$ and training data $\mathbf{X}, \boldsymbol{y}$, tolerance ε, and regularization σ
output: Solution to equations (6.41), or equivalently, (6.39)

$\boldsymbol{r} \leftarrow \boldsymbol{y} - \mathbf{X}\boldsymbol{\beta}$ `// residual can be supplied as input, or computed from scratch`
repeat `// cycle coordinate descent until convergence`
  $\boldsymbol{\beta}_{\text{old}} \leftarrow \boldsymbol{\beta}$
  for $i = 1, \ldots, p$ **do** `// coordinate-descent loop`
    $\lambda_i \leftarrow \sigma / \|\boldsymbol{v}_i\|$
    $\beta_{\text{new}} \leftarrow S^{\text{JS}}_{\lambda_i}(\beta_i + \boldsymbol{v}_i^\top \boldsymbol{r} / \|\boldsymbol{v}_i\|^2)$ `// update via (6.41)`
    if $\beta_{\text{new}} \ne \beta_i$ **then** `// check if update needed`
      $\boldsymbol{r} \leftarrow \boldsymbol{r} + (\beta_i - \beta_{\text{new}})\boldsymbol{v}_i$ `// update residual`
      $\beta_i \leftarrow \beta_{\text{new}}$
until $\|\boldsymbol{\beta} - \boldsymbol{\beta}_{\text{old}}\| < \varepsilon$
return $\boldsymbol{\beta}$

In general, there are no theoretical guarantees that the coordinate-descent Algorithm 6.43 will always converge to a point satisfying the James–Stein KKT conditions (6.39). In practice, the algorithm invariably converges to a local minimizer of the penalized least-squares: $\|\boldsymbol{y} - \mathbf{X}\boldsymbol{\beta}\|^2/2 + \sigma^2 \rho^{\text{JS}}(\boldsymbol{\beta})$.

■ **Example 6.44 (Simulated Data (cont.))** When the relevant and irrelevant features are linearly independent, that is, $\langle \boldsymbol{v}_i, \boldsymbol{v}_j\rangle = 0$ for $i \in \mathcal{S}, j \in \mathcal{S}^c$, the practical performance of the James–Stein Algorithm 6.43 is similar to that of the klimax Algorithm 6.28. However, this is not the case with the simulated data set in Example 6.10, where $\boldsymbol{\eta} \ne \mathbf{0}$. The top panel of Figure 6.45 shows the regularization paths of the James–Stein estimate. Unlike the paths in Figure 6.12, as σ becomes smaller, more features enter the model, but hardly any features leave the model. As a result, the total number of false inclusions and exclusions never drops to zero; see the bottom panel of Figure 6.45.

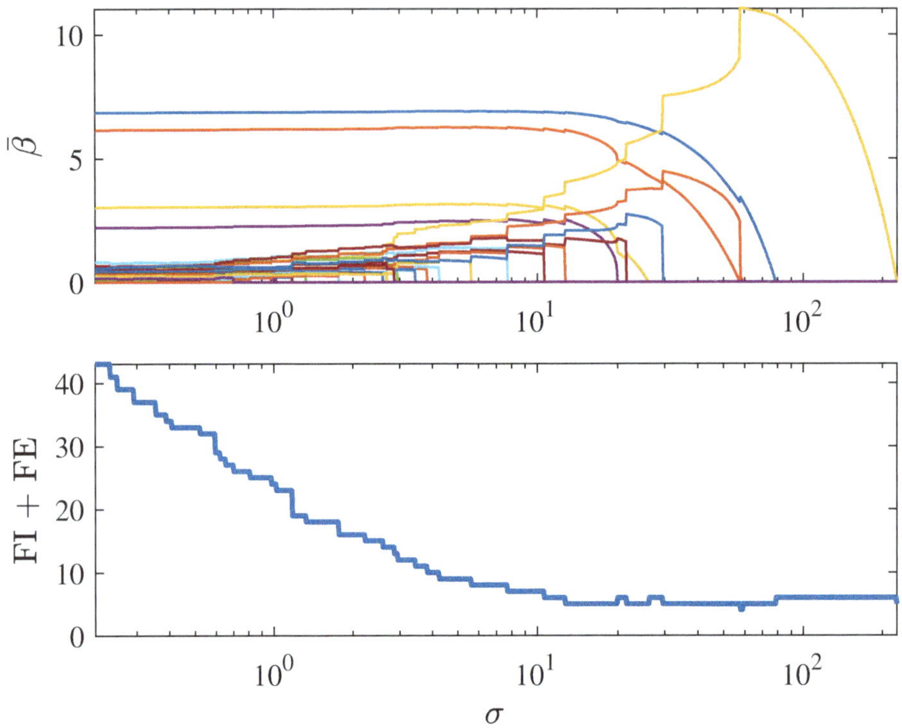

Figure 6.45: Top: Regularization paths of the James–Stein estimate using the simulated data in Example 6.10. Bottom: total number of false inclusions and false exclusions as a function of the regularization parameter σ.

■

6.2.3 Lasso Estimate

In this section we derive the widely used lasso estimate as a special case of (6.5), in which the shrinkage and model selection parameters $\boldsymbol{\gamma}$ are the minimizer of a linearization to (6.7). In the lasso context, it is common to eliminate the intercept coefficient β_0 via the preprocessing centering method discussed in Section 6.1.3.1.

> Henceforth, in the computation of the lasso estimate we assume that both $\mathbf{X}$ and the response $\boldsymbol{y}$ have been centered, that is, $\mathbf{X} \leftarrow \boldsymbol{\Pi}_0^{\perp}\mathbf{X}$ and $\boldsymbol{y} \leftarrow \boldsymbol{\Pi}_0^{\perp}\boldsymbol{y}$.

First, note that (6.7) is not convex as a function of $\boldsymbol{\gamma}$, because it is the sum of the convex term $\|\boldsymbol{y}\|_{\mathbf{G}}^2$ and the concave term $\ln|\mathbf{I}_n + \mathbf{X}\boldsymbol{\Gamma}^2\mathbf{X}^\top|$. However, it is possible to linearize the concave term in order to obtain a convex objective with a unique (global) minimum. Specifically, using the inequality $\ln(1+x) \leq x$, we obtain a convex objective by replacing the concave term with its first-order Taylor approximation around the point

$\boldsymbol{\gamma} = \mathbf{0}$:

$$\ln|\mathbf{I}_n + \mathbf{X}\boldsymbol{\Gamma}^2\mathbf{X}^\top| \leq \sum_{i=1}^{p} \ln(1 + \|\boldsymbol{v}_i\|^2\gamma_i) \leq \sum_{i=1}^{p} \|\boldsymbol{v}_i\|^2\gamma_i = \mathrm{tr}(\mathbf{X}\boldsymbol{\Gamma}^2\mathbf{X}^\top).$$

Of course, this linear approximation is also an upper bound due to the concavity of the log-determinant. In summary, rather than minimizing $\textsc{klc}(\boldsymbol{\gamma}, 0)$ in (6.7), we instead minimize the simpler and convex **lasso information criterion**:

$$\textsc{las}(\boldsymbol{\gamma}) := \|\boldsymbol{y}\|_{\mathbf{G}}^2 + \sigma^2 \sum_{i=1}^{p} \gamma_i \|\boldsymbol{v}_i\|^2.$$

Since the $\textsc{las}$ is convex and the constraint $\boldsymbol{\gamma} \geq \mathbf{0}$ is linear, the necessary and sufficient conditions for a minimizer are the following **lasso-klimax KKT conditions**:

$$\text{(6.46)} \qquad \begin{aligned} |\langle \boldsymbol{v}_i, \boldsymbol{y}\rangle_{\mathbf{G}}| &= \sigma\|\boldsymbol{v}_i\|, \quad \forall i \in \mathcal{A}, \\ |\langle \boldsymbol{v}_i, \boldsymbol{y}\rangle_{\mathbf{G}}| &\leq \sigma\|\boldsymbol{v}_i\|, \quad \forall i \in \mathcal{A}^c. \end{aligned}$$

The **lasso estimate** for the true $\boldsymbol{\beta}^*$ is (6.5), where $\widehat{\boldsymbol{\gamma}}$ is the solution to the lasso-klimax KKT conditions (6.46). If, in addition, an estimate of the intercept coefficient β_0 is required, then its value is given in (6.20). To analyze and solve (6.46), define the **lasso operator**:

$$S_\lambda^{\text{LAS}}(x) := x(1 - \lambda/|x|)_+.$$

Figure 6.47 compares the lasso and James–Stein operators. Notice that the James–Stein operator converges to the identity as $|x| \to \infty$, but that the lasso operator always remains at a distance of $\lambda = 2$ from the identity. This subtle difference between the operators has far-reaching effects on the statistical accuracy of estimators that use either the lasso or the James–Stein operator.

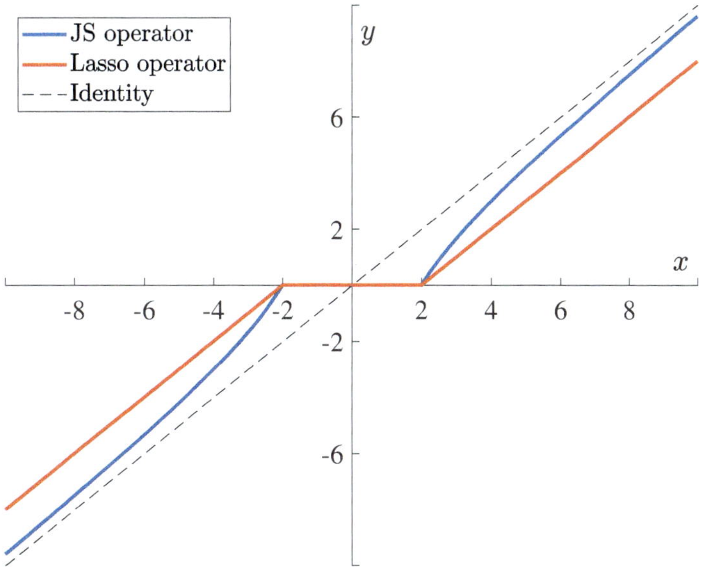

Figure 6.47: James–Stein and lasso operators for $\lambda = 2$.

The lasso operator allows us to reformulate the conditions (6.46), as follows.

Lemma 6.48 (Lasso-Klimax KKT Conditions): Let $\lambda_i := \sigma\|v_i\|/v_i^\top \mathbf{G}_{\neg i} v_i$ for all i, then (6.46) can be expressed via the lasso operator:

$$\gamma_i = \frac{\left|S^{\mathrm{LAS}}_{\lambda_i}\left(\frac{\langle v_i, \mathbf{y}\rangle_{\mathbf{G}}}{\|v_i\|^2_{\mathbf{G}}}\right)\right|}{\sigma\|v_i\|}, \quad i = 1, \dots, p. \tag{6.49}$$

Also, the lasso-klimax KKT conditions (6.46) are equivalent to the equations:

$$\beta_i = S^{\mathrm{LAS}}_{\lambda_i}\left(\frac{\langle v_i, \mathbf{y}\rangle_{\mathbf{G}}}{\|v_i\|^2_{\mathbf{G}}}\right), \quad \gamma_i = \frac{|\beta_i|}{\sigma\|v_i\|}, \quad i = 1, \dots, p. \tag{6.50}$$

Proof. The proof is similar to that of Lemma 6.24, where we showed that $\mu_i = v_i^\top \mathbf{G}_{\neg i}\mathbf{y}$ and $\nu_i = v_i^\top \mathbf{G}_{\neg i} v_i$ satisfy $\|v_i\|^2_{\mathbf{G}} = \nu_i/(1+\gamma_i\nu_i)$ and $\langle v_i, \mathbf{y}\rangle_{\mathbf{G}} = \mu_i/(1+\gamma_i\nu_i)$. As a result, (6.46) can be written as:

$$|\mu_i| = \sigma\|v_i\|(1+\gamma_i\nu_i), \quad \forall i \in \mathcal{A},$$
$$|\mu_i| \le \sigma\|v_i\|, \quad \forall i \in \mathcal{A}^c.$$

In other words, $\gamma_i = \frac{|\mu_i|}{\sigma\|v_i\|\nu_i} - \frac{1}{\nu_i}$ for $i \in \mathcal{A}$. On the other hand, (6.49) can be written as

$$\begin{aligned}\gamma_i = \frac{\left|S^{\mathrm{LAS}}_{\lambda_i}(\mu_i/\nu_i)\right|}{\sigma\|v_i\|} &= \frac{|\mu_i|}{\sigma\|v_i\|\nu_i}\left[1 - \frac{\sigma\|v_i\|/\nu_i}{|\mu_i|/\nu_i}\right]_+ \\ &= \begin{cases}\frac{|\mu_i|}{\sigma\|v_i\|\nu_i} - \frac{1}{\nu_i}, & i \in \mathcal{A}, \\ 0, & i \in \mathcal{A}^c,\end{cases}\end{aligned}$$

which implies that (6.46)⇔(6.49). We next show that (6.46)⇔(6.50).

As in the proof of Lemma 6.24, we only need to show that (6.46) implies (6.50) for all $i \in \mathcal{A}$. Since $|\mu_i| = \sigma\|v_i\|(1+\gamma_i\nu_i)$ for $i \in \mathcal{A}$, we can write:

$$\begin{aligned}S^{\mathrm{LAS}}_{\lambda_i}\left(\frac{\langle v_i, \mathbf{y}\rangle_{\mathbf{G}}}{\|v_i\|^2_{\mathbf{G}}}\right) = S^{\mathrm{LAS}}_{\lambda_i}\left(\frac{\mu_i}{\nu_i}\right) &= \frac{\mu_i}{\nu_i}\left[1 - \frac{\sigma\|v_i\|/\nu_i}{|\mu_i|/\nu_i}\right]_+ \\ &= \frac{\mu_i}{\nu_i}\left[1 - \frac{1}{1+\gamma_i\nu_i}\right]_+ \\ &= \gamma_i\frac{\mu_i}{1+\gamma_i\nu_i} = \gamma_i\langle v_i, \mathbf{y}\rangle_{\mathbf{G}} = \beta_i,\end{aligned}$$

completing the proof. □

A useful corollary of Lemma 6.48 is that (6.49) is a coordinate-descent cycle for minimizing the lasso information criterion LAS. As in the case for the computation of the klimax estimate in Algorithm 6.28, the equivalence (6.49)⇔(6.50) leads to a simple algorithm for computing the lasso estimate.

> The only modification needed to transform Algorithm 6.28 into an algorithm for computing the lasso estimate is to replace lines 9 and 11 in Algorithm 6.28 with $\lambda_i \leftarrow \sigma\|v_i\|(1-\gamma_i\|v_i\|^2_{\mathbf{G}})/\|v_i\|^2_{\mathbf{G}}$ and the update (6.50), respectively.

A more computationally efficient algorithm, using Gram–Schmidt orthogonalization and rank-1 Cholesky updates, is discussed in Exercise 13.

Example 6.51 (Simulated Data Set (cont.)) The practical performance of the lasso estimate on the simulated data set in Example 6.10 is qualitatively different from that of the klimax estimate. The top panel of Figure 6.52 shows that, as σ ranges from large to small values, many irrelevant features enter the model at an early stage, but fail to exit at a later stage. This results in an ever increasing number of false inclusions, as seen in the bottom panel of Figure 6.52. Another important observation is that the regularization paths of the lasso estimate are much smoother and less jagged than those of the klimax and James–Stein estimates. The reason for this less chaotic behavior of the paths is the fact that the lasso criterion LAS is a convex function with a unique global minimum.

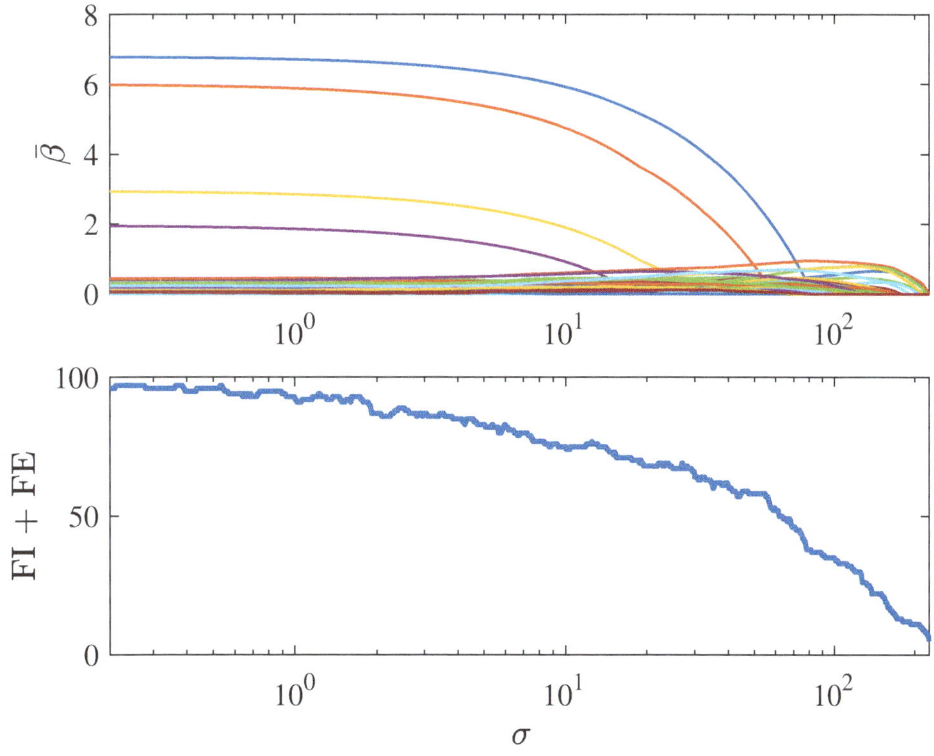

Figure 6.52: Top: Regularization paths of the lasso estimate using the simulated data set in Example 6.10. Bottom: total number of false inclusions and false exclusions as a function of the regularization parameter σ.

Penalized Least-Squares Form of Lasso Estimate

An alternative method to compute the lasso estimate without any reference to the variable $\boldsymbol{\gamma}$ is the *penalized least-squares* form of the lasso. The lasso estimate $\overline{\boldsymbol{\beta}}$ is the minimizer of the penalized least-squares objective (6.35), where $\rho_{\mathbf{X}}(\boldsymbol{\beta})$ is replaced with the *separable* **lasso penalty** $\rho^{\text{LAS}}(\boldsymbol{\beta}) = \|\boldsymbol{\beta}\|_1$, that is,

$$\overline{\boldsymbol{\beta}} = \underset{\boldsymbol{\beta}}{\operatorname{argmin}} \left\{ \frac{\|\boldsymbol{y} - \mathbf{X}\boldsymbol{\beta}\|^2}{2} + \sigma \sum_{i=1}^{p} \|\boldsymbol{v}_i\| \, |\beta_i| \right\}.$$

The absolute value norm $\|\cdot\|_1$ in the definition of ρ^{LAS} is the source for the *least absolute shrinkage and selection operator* or lasso acronym.

The necessary and sufficient conditions for computing the lasso minimizer are the following (see Exercise 19) **lasso-KKT conditions**:

$$\text{(6.53)} \qquad \begin{aligned} \boldsymbol{v}_i^\top(\boldsymbol{y} - \mathbf{X}\boldsymbol{\beta}) &= \sigma\|\boldsymbol{v}_i\|\operatorname{sign}(\beta_i), && \forall i \in \{i : \beta_i \neq 0\}, \\ |\boldsymbol{v}_i^\top(\boldsymbol{y} - \mathbf{X}\boldsymbol{\beta})| &\leq \sigma\|\boldsymbol{v}_i\|, && \forall i \in \{i : \beta_i = 0\}. \end{aligned}$$

In other words, rather than computing $\widehat{\boldsymbol{\gamma}}$ via the lasso-klimax KKT conditions (6.46) and then delivering (6.5), we can instead simply solve the penalized least-squares form (6.53). This equivalence is due to the following result.

Theorem 6.54 (Lasso-Klimax Conditions): Let $\boldsymbol{\beta} = \boldsymbol{\Gamma}^2\mathbf{X}^\top\mathbf{G}(\boldsymbol{\gamma})\boldsymbol{y}$ for all $\boldsymbol{\gamma} \geq \mathbf{0}$. Then, $\boldsymbol{\gamma}$ satisfies the *lasso-klimax KKT conditions* (6.46) if and only if $\boldsymbol{\beta}$ satisfies the *lasso-KKT conditions* (6.53). Moreover, we have $\gamma_i = |\beta_i| / (\sigma\|\boldsymbol{v}_i\|)$ for all i.

Proof. Define $\mathcal{B} := \{i : \beta_i \neq 0\}$ and recall from Lemma 6.38 that the residual $\boldsymbol{y} - \mathbf{X}\boldsymbol{\beta} = \mathbf{G}\boldsymbol{y}$. First, assume that (6.46) holds. Then, $|\beta_i| = \gamma_i|\langle \boldsymbol{v}_i, \boldsymbol{y}\rangle_{\mathbf{G}}| = \sigma\|\boldsymbol{v}_i\|\gamma_i > 0$ for all $i \in \mathcal{A}$ and $\boldsymbol{\beta}_{\mathcal{A}^c} = \mathbf{0} = \boldsymbol{\gamma}_{\mathcal{A}^c}$. In other words, $\mathcal{A} \equiv \mathcal{B}$ and the identity $\boldsymbol{\beta}_{\mathcal{B}} = \boldsymbol{\Gamma}_{\mathcal{B}}\mathbf{X}_{\mathcal{B}}^\top\mathbf{G}\boldsymbol{y} = \boldsymbol{\Gamma}_{\mathcal{B}}\mathbf{X}_{\mathcal{B}}^\top(\boldsymbol{y} - \mathbf{X}\boldsymbol{\beta})$ implies that $\operatorname{sign}(\beta_i) = \operatorname{sign}(\boldsymbol{v}_i^\top(\boldsymbol{y} - \mathbf{X}\boldsymbol{\beta}))$ for all $i \in \mathcal{B}$. The previous results combined with $\boldsymbol{y} - \mathbf{X}\boldsymbol{\beta} = \mathbf{G}\boldsymbol{y}$ prove that (6.46) $\Rightarrow$ (6.53).

Next, assume that (6.53) holds. Since $\beta_i = \gamma_i\langle \boldsymbol{v}_i, \boldsymbol{y}\rangle_{\mathbf{G}} = \gamma_i\sigma\|\boldsymbol{v}_i\|\operatorname{sign}(\beta_i) \neq 0$ for all $i \in \mathcal{B}$, we must have that $\boldsymbol{\gamma}_{\mathcal{B}} > \mathbf{0}$. On the other hand, $0 = \beta_i = \gamma_i\langle \boldsymbol{v}_i, \boldsymbol{y}\rangle_{\mathbf{G}}$ for some $i \in \mathcal{B}^c$ implies that, either $\gamma_i = 0$, or $\boldsymbol{v}_i^\top\mathbf{G}\boldsymbol{y} = 0$. We now argue that $\langle \boldsymbol{v}_i, \boldsymbol{y}\rangle_{\mathbf{G}} = 0 \Rightarrow \gamma_i = 0$, implying that $\boldsymbol{\gamma}_{\mathcal{B}^c} = 0$. This is because the inequality in the KKT conditions $|\langle \boldsymbol{v}_i, \boldsymbol{y}\rangle_{\mathbf{G}}| \leq \sigma\|\boldsymbol{v}_i\|$ is strict, that is $0 < \sigma\|\boldsymbol{v}_i\|$, and therefore the partial derivative $\partial(\text{LAS})/\partial\gamma_i > 0$ is strictly positive. The stationarity of a KKT point then implies that we must have $\gamma_i = 0$.

□

Theorem 6.54 suggests a penalized least-squares coordinate-descent algorithm that is derived from the lasso-KKT conditions (6.53); see Exercise 20.

> The only modification needed to transform the James–Stein Algorithm 6.43 into a lasso coordinate-descent algorithm is to replace the James–Stein shrinkage operator S^{JS} in Line 6 of Algorithm 6.43 with the lasso shrinkage operator S^{LAS}.

Both the lasso-klimax KKT conditions (6.46) and the lasso-KKT conditions (6.53) can be used as necessary and sufficient conditions for the lasso estimate, but the coordinate-descent algorithms derived from these conditions are distinct in both theory and implementation; see Exercises 13 and 20 for a detailed analysis.

We finish by pointing out that the **elastic net estimate** [186], defined as the minimizer of the penalized least-squares objective (for some parameter $\alpha \in (0, 1]$ and positive weights $\{\omega_i\}$):

$$\frac{\|\boldsymbol{y} - \mathbf{X}\boldsymbol{\beta}\|^2}{2} + \sigma \sum_{i=1}^{p} \omega_i \|\boldsymbol{v}_i\| \left(\alpha|\beta_i| + \frac{1-\alpha}{2}\beta_i^2\right), \tag{6.55}$$

can be recast as a special case of the lasso estimate, and is thus another instance of the klimax method; see Exercise 34.

6.3 Statistical Properties of Estimators

In this section we compare theoretically the statistical accuracy of the klimax, James–Stein, and lasso estimators. To this end we first introduce a number of theoretical assumptions.

For simplicity of notation let us assume that the intercept term has been accounted for using the restricted likelihood method in Section 6.1.3.2. The true model becomes:

$$\boldsymbol{Y} = \mathbf{X}\boldsymbol{\beta}^* + \boldsymbol{\varepsilon}, \qquad \boldsymbol{\varepsilon} \sim \mathcal{N}(\mathbf{0}, \varsigma^2 \mathbf{I}_n),$$

where $\mathbf{X} \in \mathbb{R}^{n\times p}$ is a model matrix. Recall that $\mathcal{S} = \{i \geq 1 : \beta_i^* \neq 0\}$ is the set of relevant features. To avoid trivial cases, we preclude $\mathcal{S}$ from being the empty set. In the subsequent theoretical analysis we distinguish between estimators that use the true $\mathcal{S}$ and those that use an estimator of $\mathcal{S}$.

A klimax, James–Stein, or lasso estimator that satisfies its eponymous KKT conditions with the activated set $\mathcal{A} \equiv \mathcal{S}$ is said to be an **oracle estimator**.

Definition 6.56 (Oracle Estimator): Consider the estimator $\overline{\boldsymbol{\beta}} = \boldsymbol{\Gamma}^2\mathbf{X}^\top\mathbf{G}(\boldsymbol{\gamma})\boldsymbol{Y}$ for some $\boldsymbol{\gamma}$. Assuming the KKT conditions listed below hold with $\mathcal{A} \equiv \mathcal{S}$ and $\boldsymbol{y}$ replaced by $\boldsymbol{Y}$, we designate $\overline{\boldsymbol{\beta}}$ as the following estimator of $\boldsymbol{\beta}^*$:

1. **Klimax oracle estimator**, if (6.22) holds;
2. **James–Stein oracle estimator**, if (6.37) holds;
3. **Lasso oracle estimator**, if (6.46) holds;
4. **Least-squares oracle estimator**, if (6.46) holds, as $\sigma \downarrow 0$.

In Exercise 15 we show that the trio of klimax, James–Stein, and lasso estimates all converge to the minimum-norm least-squares estimate as the regularization parameter $\sigma \downarrow 0$. This observation suggests a simpler equivalent definition of the *least-squares* *oracle estimator.* Namely, the **least-squares oracle estimator** of $\boldsymbol{\beta}^*$ is the vector $\mathring{\boldsymbol{\beta}}$ satisfying:

$$\mathring{\boldsymbol{\beta}}_{\mathcal{S}^c} = \mathbf{0} \quad \text{and} \quad \mathring{\boldsymbol{\beta}}_{\mathcal{S}} = \mathbf{X}_{\mathcal{S}}^{+}\boldsymbol{Y}.$$

In other words, the least-squares oracle estimator is simply the ordinary least-squares estimator computed from the model matrix of the true model. The least-squares oracle estimator allows us to define the regression **signal strength** as the quantity:

$$\alpha_n := \frac{\sqrt{n}}{\sigma} \min_{i \in \mathcal{S}} |\mathring{\beta}_i|,$$

where in Corollary 6.59 below we argue that the regularization parameter σ should be proportional to $\varsigma\sqrt{\ln p}$.

According to the definition, all else being equal, the "signal" is stronger when either n is large, or the magnitude of the smallest relevant least-squares estimate is large. Conversely, the "signal" is weaker when either the number of features p is large, or the variance of the response ς^2 is large.

Observe that the computation of the klimax oracle estimator ostensibly requires prior knowledge of the feature set $\mathcal{S}$. We will show that under the technical conditions below, with high probability the klimax oracle estimator corresponds to a local minimizer of the KLC, thus making its computation feasible, irrespective of any prior knowledge of the set $\mathcal{S}$.

☞ 652 We now list the technical conditions, making use of the notation $o_{\mathbb{P}}$ and $O_{\mathbb{P}}$.

Assumptions 6.57 (Klimax): For the model $\boldsymbol{Y} = \mathbf{X}\boldsymbol{\beta}^* + \boldsymbol{\varepsilon}$, assume the following:

1. **Model Boundedness.** $|\mathcal{S}|$ does not depend on n, where $n > |\mathcal{S}|$.

2. **Feature Growth.** As $n \uparrow \infty$ the p satisfies $\ln(p)/n = o(1)$.

3. **Minimal Signal Strength.** As $n \uparrow \infty$ it holds that $1/\alpha_n = o_{\mathbb{P}}(1)$.

4. **Sparse Eigenvalue Lower Bound.** For all n the model matrix satisfies:

$$\min_{\boldsymbol{v} \in \mathbb{R}^{\mathcal{S}}} \frac{\|\mathbf{X}_{\mathcal{S}}\boldsymbol{v}\|^2}{n\|\boldsymbol{v}\|^2} \geq \underline{\lambda} \quad \text{for some } \underline{\lambda} > 0.$$

5. **Klimax Irrepresentability.** As $n \uparrow \infty$ it holds that:

$$\max_{i \in \mathcal{S}^c} \frac{\|\mathbf{X}_{\mathcal{S}}^{+}\boldsymbol{v}_i\|_1}{\|\boldsymbol{\Pi}_{\mathcal{S}}^{\perp}\boldsymbol{v}_i\|/\sqrt{n}} = o_{\mathbb{P}}(\alpha_n).$$

6. **Klimax Regularization.** The regularization parameter σ satisfies:

$$\sigma \geq \max_{i \in \mathcal{S}^c} \frac{|\boldsymbol{Y}^\top \boldsymbol{\Pi}_{\mathcal{S}}^{\perp}\boldsymbol{v}_i|}{\|\boldsymbol{\Pi}_{\mathcal{S}}^{\perp}\boldsymbol{v}_i\|}(1 + o_{\mathbb{P}}(1)).$$

Before proceeding, we make a few comments about the intuition behind each of Assumptions 6.57.

1. It is possible to weaken the *model boundedness* assumption and allow the model size to grow as a function of n and p. However, in this case the model growth will need to be offset with a corresponding growth in the signal strength α_n.
2. The *feature growth* assumption allows the number of features p to be larger than n and grow to infinity, but p cannot grow too fast. For example, a growth of $p = \exp(n)$ is precluded.
3. Since $\mathring{\boldsymbol{\beta}}_{\mathcal{S}} = \boldsymbol{\beta}^*_{\mathcal{S}} + \mathbf{X}^+_{\mathcal{S}}\boldsymbol{\varepsilon}$, the *minimal signal strength* can equivalently be restated as the requirement that $\sigma/(\sqrt{n}\min_{i\in\mathcal{S}}|\beta^*_i|) = o_{\mathbb{P}}(1)$; see Exercise 32.
4. The strictly positive $\underline{\lambda}$ can be interpreted as the smallest eigenvalue of the matrix $\mathbf{X}^\top_{\mathcal{S}}\mathbf{X}_{\mathcal{S}}/n$. The *sparse eigenvalue lower bound* assumption is necessary to ensure that the true model features in $\mathbf{X}_{\mathcal{S}}$ are linearly independent for all n (or at least for all sufficiently large n) and that a unique least-squares oracle estimator $\mathring{\boldsymbol{\beta}}$ exists.
5. Intuitively, the *klimax irrepresentability* requires that all irrelevant features (those in $\mathcal{S}^c$) are not strongly represented in the $\mathrm{Span}(\mathbf{X}_{\mathcal{S}})$ of the relevant features, where the strength of the representation is measured by the 1-norm of $\mathbf{X}^+_{\mathcal{S}}\boldsymbol{v}_i$ for $i \in \mathcal{S}^c$.

   Exercise 29 shows that the *klimax irrepresentability* is slightly less stringent than the *Gram–Schmidt bound*: $\max_{i\in\mathcal{S}^c}\|\boldsymbol{v}_i\|/\|\boldsymbol{\Pi}^\perp_{\mathcal{S}}\boldsymbol{v}_i\| = o_{\mathbb{P}}(\alpha_n)$. Since the strength of the linear dependence between model vectors in $\mathcal{S}$ and $\mathcal{S}^c$ can be measured by $\max_{i\in\mathcal{S}^c}\|\boldsymbol{v}_i\|/\|\boldsymbol{\Pi}^\perp_{\mathcal{S}}\boldsymbol{v}_i\|$, the *Gram–Schmidt bound* ensures that the strength of this linear dependence is dominated by the strength of the signal α_n.

Within a Bayesian context, the hyperparameter σ^2 is typically interpreted as an estimate of the response variance $\varsigma^2 = \mathbb{V}\mathrm{ar}\, Y$. However, contrary to this interpretation, the *klimax regularization* in Assumptions 6.57 postulates values of σ^2 typically larger than ς^2. Indeed, Corollary 6.59 below shows that $1/\sigma^2 = \mathcal{O}_{\mathbb{P}}(1/\ln p)$; that is, σ^2 must increase at a rate of no less than $\ln p$. We first need a useful lemma from probability regarding the maximum of Gaussian random variables.

Lemma 6.58 (Maximum of Gaussians): If $V_1, V_2, \ldots \sim \mathcal{N}(0,1)$, not necessarily independent, then

$$\mathbb{P}\Big[\max_{i\le p}|V_i| \ge \sqrt{2\ln p}\,\Big] \le \frac{1}{\sqrt{2\ln p}}.$$

Proof. Combining Mill's ratio in Theorem D.43 with the union bound (D.1) yields: ☞ 639

$$\begin{aligned}
\mathbb{P}\Big[\max_{i\le p}|V_i| \ge v\Big] &\le \sum_{i=1}^{p}\mathbb{P}[|V_i| \ge v] \quad \text{(using union bound)}\\
&\le 2p\,\overline{\Phi}(v) \le 2p\,\mathrm{e}^{-v^2/2}/(v\sqrt{2\pi}) \quad \text{(using Mill's ratio)}\\
&\le \exp(-v^2/2 + \ln p - \ln v), \quad v > 0.
\end{aligned}$$

Using the above inequality with $\upsilon := \sqrt{2 \ln p}$ immediately yields the desired result. □

Corollary 6.59 (Growth of the Regularization Parameter) An as application of Lemma 6.58, define

$$V_i := \frac{\boldsymbol{\varepsilon}^\top \boldsymbol{\Pi}_S^\perp \boldsymbol{v}_i}{\varsigma \|\boldsymbol{\Pi}_S^\perp \boldsymbol{v}_i\|}, \qquad \forall i \in \mathcal{S}^c.$$

Then, $V_i \sim \mathcal{N}(0, 1)$ for each $i \in \mathcal{S}^c$ and the *klimax regularization* condition can be written as

$$\sigma \geq \varsigma \max_{i \in \mathcal{S}^c} |V_i|.$$

Moreover, by Lemma 6.58 we conclude that

$$\sigma \geq \varsigma \max_{i \in \mathcal{S}^c} |V_i| = \mathcal{O}_{\mathbb{P}}(\sqrt{\ln |\mathcal{S}^c|}).$$

Since $p = |\mathcal{S}^c| + |\mathcal{S}| = \mathcal{O}(|\mathcal{S}^c|)$, the *klimax regularization* condition in Assumptions 6.57 implies that the regularization parameter σ satisfies $1/\sigma^2 = \mathcal{O}_{\mathbb{P}}(1/\ln p)$. ■

Assumptions 6.57 and Corollary 6.59 help us prove (see Exercise 26) the following.

Theorem 6.60 (Klimax Estimator Consistency): With probability approaching unity as $n \uparrow \infty$, the shrinkage parameter $\widehat{\boldsymbol{\gamma}}$ of the klimax oracle estimator is a strict local minimizer of KLC such that $\widehat{\boldsymbol{\gamma}}_{\mathcal{S}} > \mathbf{0}$. Also, $\|\mathring{\boldsymbol{\beta}} - \overline{\boldsymbol{\beta}}\| = \mathcal{O}_{\mathbb{P}}\big(\alpha_n^{-1} \sqrt{\ln(p)/n}\big)$.

By definition the klimax oracle estimator always avoids false inclusions, because $\widehat{\boldsymbol{\gamma}}_{\mathcal{S}^c} = \mathbf{0}$. However, it is not clear if the oracle estimator can prevent false exclusions; that is, prevent any $\widehat{\gamma}_i = 0$ for $i \in \mathcal{S}$. One corollary of Theorem 6.60 is that with increasing probability (as $n \uparrow \infty$), the klimax oracle estimator $\overline{\boldsymbol{\beta}}$ avoids false exclusions and converges to $\mathring{\boldsymbol{\beta}}$ (and hence $\boldsymbol{\beta}^*$) in probability.

A second corollary of Theorem 6.60 is that as n grows larger, we can be increasingly more confident that we have correctly identified the true feature set $\mathcal{S}$ by examining the strictly positive components of a local minimizer $\widehat{\boldsymbol{\gamma}}$ of KLC.

Finally, the rapid convergence of $\|\mathring{\boldsymbol{\beta}} - \overline{\boldsymbol{\beta}}\|^2 = \mathcal{O}_{\mathbb{P}}(\alpha_n^{-2} \ln(p)/n)$ allows us to construct a consistent estimator of the regression parameter ς^2. In particular, Exercise 31 provides a proof of the following result.

Lemma 6.61 (Consistent Estimator of ς^2): Let $\overline{\boldsymbol{\beta}}$ be the klimax oracle estimator of $\boldsymbol{\beta}^*$. Then, under Assumptions 6.57 a consistent estimator of ς^2 is:

$$\frac{\|\boldsymbol{Y} - \mathbf{X}\overline{\boldsymbol{\beta}}\|^2}{n - \|\overline{\boldsymbol{\beta}}\|_0} = \frac{\|\boldsymbol{Y} - \mathbf{X}_{\mathcal{S}}\overline{\boldsymbol{\beta}}_{\mathcal{S}}\|^2}{n - |\mathcal{S}|} = \varsigma^2 + \mathcal{O}_{\mathbb{P}}\big(1/\sqrt{n} + \alpha_n^{-2} \ln(p)/n\big).$$

A similar result as Theorem 6.60 holds for the James–Stein oracle estimator under Assumptions 6.57, but without the need for the *klimax irrepresentability*; see Exercise 27 for details of the proof.

Theorem 6.62 (James–Stein Estimator Consistency): Using Assumptions 6.57, except for the *klimax irrepresentability*, with probability approaching unity as $n \uparrow \infty$, the shrinkage parameter $\widehat{\gamma}$ of the James–Stein oracle estimator $\overline{\beta}$ is a strict local minimizer of JSC such that $\widehat{\gamma}_S > \mathbf{0}$. Moreover, $\|\mathring{\beta}-\overline{\beta}\| = O_{\mathbb{P}}(\alpha_n^{-1}\sqrt{\ln(p)/n})$.

For the analysis of the lasso estimator we use the same 1-4 Assumptions 6.57, but replace assumptions 5 and 6 with the ones below and add an extra seventh assumption.

Assumptions 6.63 (Lasso): For the model $Y = \mathbf{X}\beta^* + \varepsilon$, additionally assume.

5. Lasso Irrepresentability. For some $\delta > 0$ the constraint holds:

$$\max_{i \in S^c} \frac{\|\mathbf{X}_S^+ v_i\|_1}{\|v_i\|/\sqrt{n}} \leq 1 - \delta.$$

6. Lasso Regularization. The regularization parameter σ satisfies:

$$\sigma \geq \max_i \frac{|Y^\top \Pi_S^\perp v_i|}{\|v_i\|\,\delta}.$$

7. Feature Normalization. Model vectors satisfy: $\|v_i\|^2 = n$ for all $i \in S$.

We comment on the intuition behind the lasso-specific Assumptions 6.63.

Lasso Irrepresentability. The *lasso irrepresentability*, where the specific upper bound of $1-\delta$ is required, is a more stringent condition than the *klimax irrepresentability*, where we merely require growth slower than that of the signal strength.

Lasso Regularization. The *lasso regularization* for the lasso estimator requires knowledge of the (typically unknown) constant δ and is thus mainly of theoretical interest. In contrast, in Section 6.4 we show that the *klimax regularization* for the klimax estimator yields a practical method for the near-optimal selection of σ.

Feature Normalization. For the purposes of our proofs and theoretical arguments, the *feature normalization* is equivalent to the *sparse eigenvalue upper bound* assumption, that is, for all n the model matrix satisfies:

$$\max_{v \in \mathbb{R}^{|S|}} \frac{\|\mathbf{X}_S v\|^2}{n\|v\|^2} \leq \overline{\lambda} < \infty. \tag{6.64}$$

This is because the feature normalization implies that the sum of the eigenvalues of $\mathbf{X}_S^\top \mathbf{X}_S/n$ is $\mathrm{tr}(\mathbf{X}_S^\top \mathbf{X}_S)/n = |S|$, and so $\overline{\lambda} \leq |S| < \infty$. Alternatively, we may replace the *feature normalization* with the requirement that model vectors satisfy $\max_{i \in S} \|v_i\|^2 = O(n)$.

> Many authors do not explicitly identify the *feature normalization* as an assumption, but deem it a preprocessing step that is mainly applied for convenience. In reality, rescaling of the model vectors affects the *signal strength* assumed for the original data, and hence affects the statistical performance of the lasso estimator.

Under the additional Assumptions 6.63, we prove (see Exercise 28) the following.

Theorem 6.65 (Lasso Estimator Consistency): Using conditions 1, 2, and 3 in Assumptions 6.57, as well as all Assumptions 6.63, with probability approaching unity as $n \uparrow \infty$, the shrinkage parameter $\widehat{\boldsymbol{\gamma}}$ of the lasso oracle estimator $\overline{\boldsymbol{\beta}}$ is the global minimizer of LAS such that $\widehat{\boldsymbol{\gamma}}_{\mathcal{S}} > \mathbf{0}$. Moreover, $\|\mathring{\boldsymbol{\beta}} - \overline{\boldsymbol{\beta}}\| = \mathcal{O}_{\mathbb{P}}\big(\sqrt{\ln(p)/n}\big)$.

Notice how the presence of a strong signal (that is, large α_n) does not help reduce the norm $\|\mathring{\boldsymbol{\beta}} - \overline{\boldsymbol{\beta}}\| = \mathcal{O}_{\mathbb{P}}\big(\sqrt{\ln(p)/n}\big)$ for the lasso estimator. In contrast, the same norm for the klimax and James–Stein estimators is of order $\mathcal{O}_{\mathbb{P}}\big(\alpha_n^{-1}\sqrt{\ln(p)/n}\big)$, showing that these estimators are able to take advantage of the presence of a strong signal. On the positive side, the lasso information criterion is convex and thus has an easily computable global minimum.

Example 6.66 (Simulated Data (cont.)) In the setup of Example 6.10, assume that $\boldsymbol{\eta}$ satisfies the condition

$$\boldsymbol{\eta}^\top(\mathbf{I}_s - \mathbf{X}_{\mathcal{S}}^\top\mathbf{X}_{\mathcal{S}}/n)\boldsymbol{\eta} \geq \min_{i\in\mathcal{S}^c} \|\boldsymbol{\Pi}_{\mathcal{S}}^\perp \mathbf{Z}_i\|^2/n. \tag{6.67}$$

For instance, Exercise 30 suggests that (6.67) holds with high probability for any $\boldsymbol{\eta}$ with

$$\|\boldsymbol{\eta}\|^2 \geq c/(1-\overline{\lambda}),$$

where the constant $c > 1 > \overline{\lambda}$. It also holds numerically for the simulated data set with

$$\boldsymbol{\eta} = [0.2, 0.4, 0.6, 0.8, 1]^\top$$

in Example 6.10. Given (6.67), from $\|\boldsymbol{\eta}\|_1 \geq \|\boldsymbol{\eta}\|_2$ we deduce that

$$\max_{i\in\mathcal{S}^c} \frac{\|\mathbf{X}_{\mathcal{S}}^+ \boldsymbol{v}_i\|_1^2}{\|\boldsymbol{v}_i\|^2/n} = \frac{n\|\boldsymbol{\eta}\|_1^2}{\min_{i\in\mathcal{S}^c}\|\boldsymbol{v}_i\|^2}$$
$$\geq \frac{n\|\boldsymbol{\eta}\|_2^2}{\|\mathbf{X}_{\mathcal{S}}\boldsymbol{\eta}\|_2^2 + \min_{i\in\mathcal{S}^c}\|\boldsymbol{\Pi}_{\mathcal{S}}^\perp\mathbf{Z}_i\|_2^2} \geq 1,$$

indicating the violation of the lasso irrepresentability in Assumptions 6.63. In other words, with non-vanishing probability the lasso estimator makes false inclusions, that is, selects irrelevant features from $\mathcal{S}^c$. This is the main reason why we observe a high number of false inclusions on Figure 6.52. That Figure 6.52 is not mere bad luck can be illustrated by repeating the simulation experiment a number of independent times.

Table 6.68 summarizes the average performance of the klimax, James–Stein, and lasso estimators over ten independent replications of the simulated data set in Example 6.10. The only difference to the data generation recipe described in Example 6.10 is that here we vary the parameters n and p. For each combination of n and p we report the average of the false inclusions (FI), the false exclusions (FE), and the norms $\|\mathbf{X}(\overline{\boldsymbol{\beta}} - \boldsymbol{\beta}^*)\|$ and $\|\overline{\boldsymbol{\beta}} - \boldsymbol{\beta}^*\|$, over ten replications. For each replication we use the regularization parameter σ that minimizes the total number of false inclusions and exclusions (FE+FI).

Table 6.68: Comparison of klimax, James–Stein, and lasso estimators.

Estimator	n	p	FI	FE	$\|\mathbf{X}(\overline{\boldsymbol{\beta}} - \boldsymbol{\beta}^*)\|$	$\|\overline{\boldsymbol{\beta}} - \boldsymbol{\beta}^*\|$
klimax	50	100	0.1	0	2.32	0.42
James–Stein	50	100	2.5	1.4	6.66	16.38
lasso	50	100	33.3	1.8	6.38	15.54
klimax	100	200	0	0	1.89	0.20
James–Stein	100	200	2.6	1.1	9.29	15.44
lasso	100	200	54.4	2.2	8.53	16.35
klimax	200	400	0	0	1.75	0.14
James–Stein	200	400	1.5	0.5	6.91	7.33
lasso	200	400	123.7	1.4	10.75	14.62
klimax	400	800	0	0	2.03	0.11
James–Stein	400	800	0	0	2.01	0.11
lasso	400	800	206.1	1.4	13.64	14.57
klimax	800	1600	0	0	2.17	0.077
James–Stein	800	1600	0	0	2.16	0.076
lasso	800	1600	378.2	1.0	16.33	13.75

We can see from Table 6.68 that as $n \uparrow \infty$, both the klimax and James–Stein estimator are more likely to identify the true model, whereas this is not the case for the lasso estimator, which suffers from a high number of false inclusions. In other words, this simulation experiment illustrates the model selection consistency of the klimax and James–Stein estimators, as asserted by Theorems 6.60 and 6.62, respectively. In contrast, the lasso estimator is not consistent due to the violation of the lasso irrepresentability condition. ■

We finish this theoretical section by pointing out that it is possible to construct simulated data sets, for which each of the three estimators performs better than the other two. In general, when the regularization paths of the three estimators look qualitatively similar on the same data set, it is because there is little to no linear and statistical dependence between the relevant and irrelevant model vectors. However, when the regularization paths of the three estimators are significantly different, this is usually due to the presence of strong linear and statistical dependence amongst the model vectors.

6.4 Practical Selection of Regularization Parameter

A conceptually simple, albeit computationally costly, method of selecting the regularization parameter σ is to let it be the minimizer of the K-fold cross-validation loss. In other words, if C_k represents the test set of the k-th fold and $\overline{\boldsymbol{\beta}}_{\mathcal{T}_{\neg k}}$ is the estimate (6.5) (for a given σ) computed from the training set $\mathcal{T}_{\neg k} \equiv \mathcal{T} \backslash C_k$, then the cross-validatory estimate of σ is:
☞ 39

$$\widehat{\sigma}_{\text{CV}} := \underset{\sigma>0}{\operatorname{argmin}} \sum_{k=1}^{K} \sum_{i \in C_k} \left(y_i - \left\langle \boldsymbol{x}_i, \overline{\boldsymbol{\beta}}_{\mathcal{T}_{\neg k}} \right\rangle\right)^2. \tag{6.69}$$

This approach is widely used for the lasso estimate and can also be used for the James–Stein estimate. Unfortunately, as Example 6.74 below illustrates, the cross-validatory approach is frequently prohibitively expensive and for this reason we do not recommend its use for the klimax estimate. Instead, we recommend two alternative methods, as discussed in the next two sections.

6.4.1 Monte Carlo Estimator of σ

☞ 32
Similar to the approximation-estimation tradeoff in Section 2.4, in estimating the regularization parameter σ we have to pit two competing demands against each other.

On the one hand, using a gratuitously small regularization parameter σ results in a great number of false inclusions (inclusions of irrelevant features). For instance, when $\sigma \downarrow 0$, Exercise 15 shows that the klimax estimator converges to the minimum-norm least-squares estimator $\mathbf{X}^{+}\boldsymbol{Y}$, which although unbiased, will suffer from large variance, because the rank of $\mathbf{X}$ will be large and most likely equal to n.

On the other hand, using an unnecessarily large regularization parameter σ results in a great number of false exclusions (exclusions of relevant features). For instance, any σ larger than σ_* in (6.31) will exclude all features in $\mathcal{S}$, resulting in the trivial $\overline{\boldsymbol{\beta}} = \mathbf{0}$.

One way to balance these two competing demands on σ is to select the smallest value of σ that will (at least for large n) prevent false inclusions. In other words, σ must be as small as possible whilst satisfying the set of inequalities in the klimax-KKT conditions (6.22); that is,

$$\sigma \geq \max_{i \in \mathcal{S}^c} \frac{|\langle \boldsymbol{v}_i, \boldsymbol{Y} \rangle_{\mathbf{G}}|}{\|\boldsymbol{v}_i\|_{\mathbf{G}}}. \tag{6.70}$$

Since the right-hand-side of the inequality (6.70) also depends on σ via the shrinkage parameter, finding the smallest σ satisfying (6.70) is not trivial. Nevertheless, it can be shown (see the proof of Theorem 6.60 in Exercise 15) that any regularization value

$$\sigma \geq \max_{i \in \mathcal{S}^c} \frac{\left|\left\langle \mathbf{\Pi}_{\mathcal{S}}^{\perp} \boldsymbol{v}_i, \boldsymbol{Y} \right\rangle\right|}{\|\mathbf{\Pi}_{\mathcal{S}}^{\perp} \boldsymbol{v}_i\|}(1 + o_{\mathbb{P}}(1))$$

will be guaranteed to satisfy (6.70), at least asymptotically. In other words, the *klimax regularization* condition in Assumptions 6.57 is a sufficient condition for (6.70) to hold with probability tending to unity as $n \longrightarrow \infty$.

In summary, given knowledge of the set $\mathcal{S}$, an estimator of the smallest acceptable value for the regularization parameter is

$$\widehat{\sigma}(\mathcal{S}) = \max_{i\in\mathcal{S}^c} \frac{|\langle \mathbf{\Pi}_{\mathcal{S}}^{\perp} \boldsymbol{v}_i, \boldsymbol{Y}\rangle|}{\|\mathbf{\Pi}_{\mathcal{S}}^{\perp}\boldsymbol{v}_i\|},$$

where we can replace $\boldsymbol{Y}$ with

$$\boldsymbol{\varepsilon} = \boldsymbol{Y} - \mathbf{X}_{\mathcal{S}}\boldsymbol{\beta}^*_{\mathcal{S}} \sim \mathcal{N}(\mathbf{0}, \varsigma^2 \mathbf{I}_n),$$

because $\mathbf{\Pi}_{\mathcal{S}}^{\perp}\mathbf{X}_{\mathcal{S}} = \mathbf{0}$. As in Corollary 6.59, if we define

$$V_i := \frac{\langle \mathbf{\Pi}_{\mathcal{S}}^{\perp} \boldsymbol{v}_i, \boldsymbol{\varepsilon}/\varsigma\rangle}{\|\mathbf{\Pi}_{\mathcal{S}}^{\perp}\boldsymbol{v}_i\|}, \quad \forall i \in \mathcal{S}^c,$$

then

$$\widehat{\sigma}(\mathcal{S}) = \varsigma \max_{i\in\mathcal{S}^c} |V_i| = \varsigma \|\boldsymbol{V}_{\mathcal{S}^c}\|_\infty$$

and $\boldsymbol{V}_{\mathcal{S}^c}$ is a zero-mean multivariate Gaussian vector with covariance:

$$\mathbb{C}\text{ov}(V_i, V_j) = \frac{\langle \mathbf{\Pi}_{\mathcal{S}}^{\perp} \boldsymbol{v}_i, \mathbf{\Pi}_{\mathcal{S}}^{\perp} \boldsymbol{v}_j\rangle}{\|\mathbf{\Pi}_{\mathcal{S}}^{\perp}\boldsymbol{v}_i\|\,\|\mathbf{\Pi}_{\mathcal{S}}^{\perp}\boldsymbol{v}_j\|}, \quad \forall i, j \in \mathcal{S}^c.$$

We can easily simulate copies of $\boldsymbol{V}_{\mathcal{S}^c}$ and its norm $\|\boldsymbol{V}_{\mathcal{S}^c}\|_\infty$, provided that the set $\mathcal{S}$ is known. This suggests the following methodology for the simultaneous estimation of the set $\mathcal{S}$ and the corresponding estimator $\widehat{\sigma}(\mathcal{S})$: Let $\mathcal{A}(\sigma)$ be the set of activated features corresponding to the nonzero entries of the klimax estimator computed at σ, and let

$$\widehat{\varsigma}^2(\mathcal{A}(\sigma)) := \frac{\|\boldsymbol{Y} - \mathbf{X}_{\mathcal{A}(\sigma)}\overline{\boldsymbol{\beta}}_{\mathcal{A}(\sigma)}\|^2}{n - |\mathcal{A}(\sigma)|} \tag{6.71}$$

be the corresponding estimator of ς^2, motivated by the consistency result in Lemma 6.61. Then, the **klimax regularization estimator** of the regularization parameter is the solution $\widehat{\sigma}$ to the equation:

$$\sigma = \widehat{\varsigma}(\mathcal{A}(\sigma))\|\boldsymbol{V}_{\mathcal{A}^c(\sigma)}\|_\infty, \tag{6.72}$$

where, conditional on $\mathbf{Z} \sim \mathcal{N}(\mathbf{0}, \mathbf{I}_n)$, we define

$$V_i := \frac{\langle \mathbf{\Pi}_{\mathcal{A}(\sigma)}^{\perp} \boldsymbol{v}_i, \mathbf{Z}\rangle}{\|\mathbf{\Pi}_{\mathcal{A}(\sigma)}^{\perp}\boldsymbol{v}_i\|}, \quad \forall i \in \mathcal{A}^c(\sigma).$$

Using N iid draws $\mathbf{Z}_1, \ldots, \mathbf{Z}_N \sim_{\text{iid}} \mathcal{N}(\mathbf{0}, \mathbf{I}_n)$, we can simulate N independent copies of the random function $\sigma \mapsto \|\boldsymbol{V}_{\mathcal{A}^c(\sigma)}\|_\infty$ in (6.72), and therefore obtain N replications $\widehat{\sigma}_1, \ldots, \widehat{\sigma}_N$ of the solution to (6.72).

The empirical distribution of the random sample $\widehat{\sigma}_1, \ldots, \widehat{\sigma}_N$ is thus a Monte Carlo approximation of the large-sample distribution of the smallest σ satisfying (6.70). ☞ 73

Algorithm 6.73: Monte Carlo Simulation of Regularization Parameter σ

input: Simulated $\mathbf{Z}_1, \ldots, \mathbf{Z}_N \overset{\text{iid}}{\sim} \mathcal{N}(\mathbf{0}, \mathbf{I}_n)$, data $\{\mathbf{X}, \boldsymbol{y}\}$, and regularization path $(\sigma_1, \boldsymbol{\beta}_1), \ldots, (\sigma_m, \boldsymbol{\beta}_m)$ computed in Algorithm 6.33.

output: Approximate solution to equation (6.72) for N replications of $V_{\mathcal{A}^c(\sigma)}$.

$\mathbf{Z} \leftarrow [\mathbf{Z}_1, \ldots, \mathbf{Z}_N]$ // collect all Gaussian vectors in a matrix

for $k = 1, \ldots, m$ **do** // regularization path over $\sigma_1, \ldots, \sigma_m$

  $\boldsymbol{b} \leftarrow \boldsymbol{\beta}_k$, $\mathcal{A} \leftarrow \{j : b_j \neq 0\}$ // find the activated set at σ_k

  $\widehat{\varsigma}_k^2 \leftarrow \|\boldsymbol{y} - \mathbf{X}_{\mathcal{A}} \boldsymbol{b}_{\mathcal{A}}\|^2 / (n - |\mathcal{A}|)$ // estimate of ς^2 at σ_k via (6.71)

  $\mathbf{QR} \leftarrow \mathbf{X}_{\mathcal{A}}$ // QR decomposition of $\mathbf{X}_{\mathcal{A}}$ via Algorithm A.22

  $\mathbf{R} \leftarrow \mathbf{Q}^\top \mathbf{X}$

  $\mathbf{V} \leftarrow \mathbf{Z} - \mathbf{Q}(\mathbf{Q}^\top \mathbf{Z})$ // compute projection $\mathbf{\Pi}_{\mathcal{A}}^{\perp} \mathbf{Z}$

  $\boldsymbol{l} \leftarrow (\mathbf{X} \odot \mathbf{X})^\top \mathbf{1} - (\mathbf{R} \odot \mathbf{R})^\top \mathbf{1}$ // compute $l_i := \|\mathbf{\Pi}_{\mathcal{A}}^{\perp} \boldsymbol{v}_i\|^2$ for all i

  $\mathbf{V} \leftarrow \mathrm{diag}(\boldsymbol{l})^{-1/2} \mathbf{X}^\top \mathbf{V}$ // compute $\boldsymbol{v}_i^\top \mathbf{\Pi}_{\mathcal{A}}^{\perp} \mathbf{Z}_j / \|\mathbf{\Pi}_{\mathcal{A}}^{\perp} \boldsymbol{v}_i\|$ for all i, j

  $M_{k,j} \leftarrow \max_{i \in \mathcal{A}^c} |\mathbf{V}_{i,j}|$ for $j = 1, \ldots, N$

for $j = 1, \ldots, N$ **do** // for each replication find a root on path

  $f_k \leftarrow \sigma_k - \widehat{\varsigma}_k M_{k,j}$ for each $k = 1, \ldots, m$

  $k^* \leftarrow \mathrm{argmax}\{k \geq 2 : f_k f_{k-1} < 0\}$

  $\widehat{\sigma}_j \leftarrow \sigma_{k^*}$ // largest (approximate) root of (6.72)

return $\widehat{\sigma}_1, \ldots, \widehat{\sigma}_N$

Algorithm 6.73 is implemented in the Python code below. Note that both the Python code and the algorithm assume that $n > |\mathcal{A}|$.

sigDist.py

```python
import numpy as np
from numpy.linalg import norm, qr

def sigDist(Z, grid, X, y,  path):
    n_samples, _ = X.shape
    _, num_rep = Z.shape
    n_grid = len(grid)
    sumX2 = np.sum(X**2, axis=0)
    M = np.full((n_grid, num_rep), np.nan)
    varY = np.full(n_grid, np.nan)

    for k in range(n_grid):
        b = path[k, :].reshape(-1, 1)
        A = np.nonzero(b)[0]
        if A.size > 0:
            X_A = X[:, A]
            Q, _ = np.linalg.qr(X_A, mode='reduced')
            varY[k] = np.linalg.norm(y - X_A @ b[A])**2 / (n_samples - A
                .size)
            R = Q.T @ X
            V = Z - Q @ (Q.T @ Z)
            sum_r_squared = np.sum(R * R, axis=0)
            ell = np.maximum(sumX2 - sum_r_squared, 0).reshape(-1, 1)
        else:
```

```
            V = Z
            ell = sumX2.reshape(-1, 1)
        ell[A] = np.inf
        V = (X.T @ V)/np.sqrt(ell)

        M[k, :] = np.max(np.abs(V), axis=0)

    root_idx = np.zeros(num_rep, dtype=int)
    for k in range(num_rep):  # Loop over each simulated draw
        f = grid.T - np.sqrt(varY) * M[:, k]  # Compute f for current
            draw
        # Find where f changes sign (f[i] and f[i+1] have opposite signs
            )
        sign_changes = (f[:-1] * f[1:] < 0).nonzero()[0]
        if sign_changes.size:
            root_idx[k] = sign_changes[-1]

    return np.sort(root_idx)
```

Example 6.74 (Diabetes Data Set (cont.)) We can now explain how the optimal σ is selected for the diabetes data set in Example 6.8. First, the klimax regularization paths are computed using **diabetes.py** in Example 6.34. Then, we simulate 10^3 replications of the klimax regularization estimator via the **sigDist.py** module above. The following Python code carries out these steps and produces Figure 6.75. The top panel replicates the klimax regularization paths of Figure 6.9. The bottom panel shows a histogram (on the same logarithmic scale as the top panel) of the random sample:

$$\widehat{\sigma}_1, \ldots, \widehat{\sigma}_{10^3}.$$

The optimal choice for σ in Figure 6.9 corresponds to the median of the random sample. In addition, Figure 6.75 shows the 5- and 95-percentiles of $\widehat{\sigma}_1, \ldots, \widehat{\sigma}_{10^3}$ that indicate a range of "reasonable" models for the diabetes data.

diabetes_CI.py

```
from diabetes import *
from sigDist import *

# Find the optimal sigma
Z = np.random.randn(n, 1000)
Igrid = sigDist(Z, grid, X, y, path) # random sigmas are grid[Igrid]
medSig = np.median(grid[Igrid])

# Plot the results
plt.figure(1,figsize=(10,6))
h1 = plt.subplot(2,1,1)
plt.semilogx(grid,path)
xlim = h1.get_xlim()
plt.xlabel(r'$\sigma$')
plt.ylabel(r'$\bar{\beta}$')
plt.axvline(x=medSig, color='red', linestyle='--')
plt.axvline(x=np.quantile(grid[Igrid],0.95), linestyle='--')
```

```
plt.axvline(x=np.quantile(grid[Igrid],0.05), linestyle='--')

h2 = plt.subplot(2,1,2)
plt.hist(grid[Igrid], bins=50, edgecolor='black',density=True)
plt.xscale('log')
h2.set_xlim(xlim)
plt.axvline(x=medSig, color='red', linestyle='--')
plt.axvline(x=np.quantile(grid[Igrid],0.95), linestyle='--')
plt.axvline(x=np.quantile(grid[Igrid],0.05), linestyle='--')
```

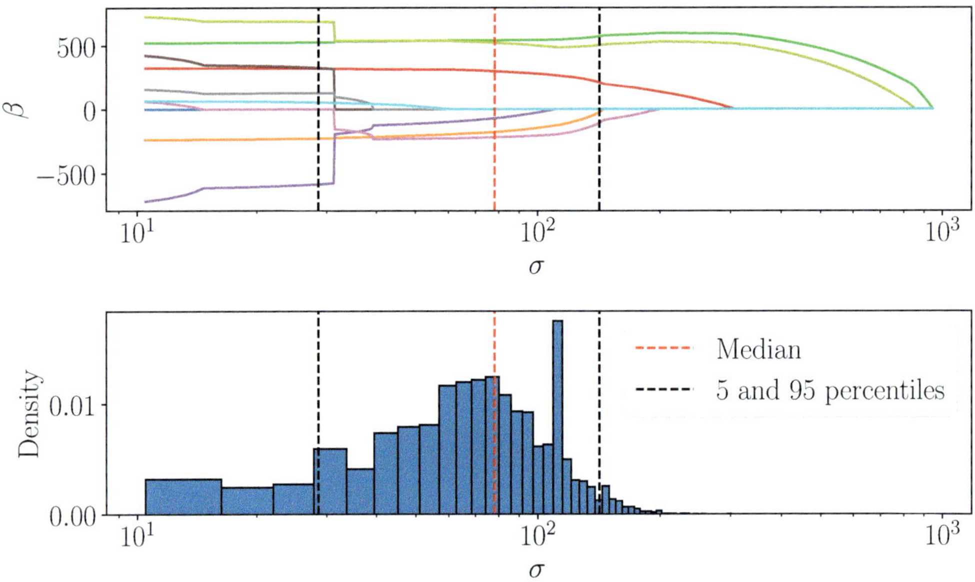

Figure 6.75: Top: klimax regularization paths for the diabetes data set. Bottom: histogram of the random sample $\widehat{\sigma}_1, \ldots, \widehat{\sigma}_{10^3}$. The corresponding 0.05, 0.5 (median), and 0.95 quantiles are indicated by vertical dashed lines.

As mentioned at the beginning of this section, one simple choice for σ is the minimizer $\widehat{\sigma}_{\text{cv}}$ in (6.69) of the cross-validation loss. By randomly reshuffling the training data one can also obtain confidence intervals for σ via cross validation. The reader my verify that, for example, with $K = 2$ folds and 20 trials (i.e., 20 random reshuffles of the data), the median and confidence bounds agree well with those of the Monte Carlo approach above. However, even with such few trials (as opposed to the 10^3 in the implementation above) cross-validation is computationally expensive and introduces too much variability. ■

6.4.2 Plug-In Estimator of σ For Large p

When p is large and, in particular, when $p = p_n \uparrow \infty$ as $n \uparrow \infty$, it is possible to further simplify the estimation of the smallest σ that satisfies the klimax regularization condition (6.70). This simpler estimator is motivated by the following guarantee.

Lemma 6.76 (Klimax Regularization Guarantee): Under Assumptions 6.57, but excluding the *klimax regularization* condition, and provided that

$$\sigma \geq (1+\varepsilon)\varsigma\sqrt{2\ln|\mathcal{S}^c|}$$

for some arbitrary small $\varepsilon > 0$, the *klimax regularization* condition holds with probability approaching unity as $p = p_n \uparrow \infty$.

Proof. As in Corollary 6.59, we define $V_i := \langle \mathbf{\Pi}_{\mathcal{S}}^{\perp} \boldsymbol{v}_i, \boldsymbol{Y}/\varsigma\rangle / \|\mathbf{\Pi}_{\mathcal{S}}^{\perp}\boldsymbol{v}_i\| \sim \mathcal{N}(0,1)$ for all $i \in \mathcal{S}^c$. Satisfying the *klimax regularization* condition then corresponds to the event $A_n := \{\sigma \geq \varsigma\|\boldsymbol{V}_{\mathcal{S}^c}\|_\infty(1+|W_n|)\}$ for some $W_n \xrightarrow{\mathbb{P}} 0$. If $B_n := \{|W_n| \leq \varepsilon\}$, then

$$\begin{aligned}
\mathbb{P}[A_n] &\geq \mathbb{P}[A_n \cap B_n] \\
&\geq \mathbb{P}\big[\sigma \geq \varsigma\|\boldsymbol{V}_{\mathcal{S}^c}\|_\infty(1+\varepsilon),\ B_n\big] \\
&\geq \mathbb{P}\big[\|\boldsymbol{V}_{\mathcal{S}^c}\|_\infty \leq \sqrt{2\ln|\mathcal{S}^c|},\ B_n\big], \quad \text{using } \sigma \geq (1+\varepsilon)\varsigma\sqrt{2\ln|\mathcal{S}^c|} \\
&\geq \mathbb{P}[B_n] - \mathbb{P}\big[\|\boldsymbol{V}_{\mathcal{S}^c}\|_\infty \geq \sqrt{2\ln|\mathcal{S}^c|}\big] \\
&\geq \mathbb{P}[B_n] - (2\ln(p-|\mathcal{S}|))^{-1/2} \to 1, \quad p = p_n \uparrow \infty,
\end{aligned}$$

where in the last line we used Lemma 6.58 and the fact that $\lim_n \mathbb{P}[B_n] = 1$. □

The guarantee of Lemma 6.76 suggests that, given knowledge of the set $\mathcal{S}$ and ς, a reasonable approximation of the asymptotically optimal regularization parameter is $\sigma = \varsigma\sqrt{2\ln|\mathcal{S}^c|} = \varsigma\sqrt{2\ln(p-|\mathcal{S}|)}$. This brings us back to the warning at the beginning of Section 6.1 that for p much larger than n, the optimal regularization parameter is different from the model standard deviation ς. The cost of having to estimate $\mathcal{S}$ is encapsulated in the correction factor $\sqrt{2\ln(p-|\mathcal{S}|)}$ required to match σ with ς. In contrast, when $\mathcal{S}$ is known (so that without loss of generality we may choose $p = |\mathcal{S}|$), Exercise 8 shows that the optimal σ converges to ς as $n \uparrow \infty$, obviating the need for a correction factor.

Of course, a practical application of the approximation $\sigma \approx \varsigma\sqrt{2\ln|\mathcal{S}^c|}$ requires the reliable estimation of both the set $\mathcal{S}$ and the parameter ς. This leads to the following simple methodology for the simultaneous estimation of σ, ς, and $\mathcal{S}$.

Let $\mathcal{A}(\sigma)$ be the set of activated features corresponding to the nonzero entries of the klimax estimator computed at σ, and let $\widehat{\varsigma}^2(\mathcal{A}(\sigma))$ in (6.71) be the corresponding estimator of ς^2. Then, the **plug-in klimax regularization estimator** of the parameter $\sigma = \varsigma\sqrt{2\ln|\mathcal{S}^c|}$ is the solution to the equation:

$$\sigma = \widehat{\varsigma}(\mathcal{A}(\sigma))\sqrt{2\ln(p-|\mathcal{A}(\sigma)|)}, \tag{6.77}$$

obtained by plugging in our best guess $\mathcal{A}(\sigma)$ for the true $\mathcal{S}$. Observe that equation (6.77) is simply equation (6.72), but with the stochastic term $\|\boldsymbol{V}_{\mathcal{A}(\sigma)}\|_\infty$ replaced by its probabilistic bound $\sqrt{2\ln|\mathcal{A}^c(\sigma)|}$ from Corollary 6.59.

One convenient graphical method to identify the root of the function $f(\sigma) := \widehat{\varsigma}(\mathcal{A}(\sigma))\sqrt{2\ln|\mathcal{A}^c(\sigma)|}/\sigma - 1$ is to plot $f(\sigma)$ versus the log of σ.

Example 6.78 (Simulated Data Set (cont.)) Recall that for the data in Example 6.10, we know the model exactly, namely $\mathcal{S} = \{1, 2, 3, 4, 5\}$. This allows us to verify if the solution to (6.77) is helpful in estimating $\mathcal{S}$ from the data.

The top panel of Figure 6.79 displays a section of the regularization paths from Figure 6.12. Furthermore, the green shaded area indicates the range for σ with no false inclusions or exclusions. Additionally, the red dotted line stands for the solution $\widehat{\sigma} \approx 3.30$ to (6.77) and the black dotted line stands for $\varsigma\sqrt{2\ln|\mathcal{S}^c|} = \sqrt{2\ln(10^3 - 5)} \approx 3.72$. These two lines are close, but the plug-in klimax estimate $\widehat{\sigma}$ slightly underestimates the asymptotically optimal $\varsigma\sqrt{2\ln|\mathcal{S}^c|}$. The middle panel shows the function $f(\sigma)$ whose root yields the estimate $\widehat{\sigma} \approx 3.30$.

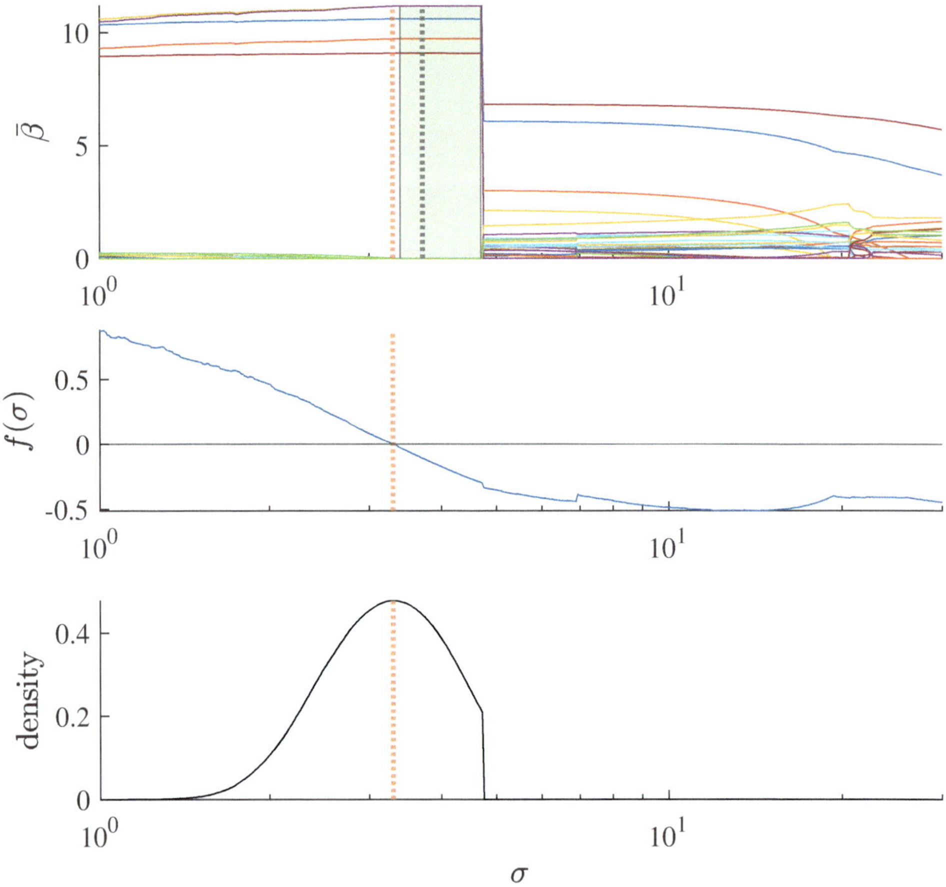

Figure 6.79: Top: A close-up of a section of the regularization paths of the klimax estimate using the simulated data set in Example 6.10. The black dotted vertical line represents the asymptotically optimal value $\varsigma\sqrt{2\ln|\mathcal{S}^c|} \approx 3.72$. Middle: The function $f(\sigma) = \widehat{\varsigma}(\mathcal{A}(\sigma))\sqrt{2\ln|\mathcal{A}^c(\sigma)|}/\sigma - 1$ and its root $\widehat{\sigma}$ indicated by a vertical line at 3.30. Bottom: The function $\sigma_k \mapsto \exp(-\textsc{klc}(\widehat{\boldsymbol{\gamma}}_k)/(2\widehat{\sigma}^2))$, normalized to be a probability density over the grid $\sigma_1, \ldots, \sigma_m$.

Finally, we comment on the bottom panel of Figure 6.79. Let $\widehat{\boldsymbol{\gamma}}_1, \ldots, \widehat{\boldsymbol{\gamma}}_m$ be the shrinkage parameters of the klimax estimate, computed over the grid $\sigma_1, \ldots, \sigma_m$; see Line 6 of Algorithm 6.33. The bottom panel of Figure 6.79 shows a density function over $\sigma_1, \ldots, \sigma_m$, whose probability value at σ_k is given by:

$$\mathbb{P}[\mathcal{A}(\sigma_k) \mid \mathbf{X}, \boldsymbol{y}] = \frac{\exp\left(-\textsc{klc}(\widehat{\boldsymbol{\gamma}}_k)/(2\widehat{\sigma}^2)\right)}{\sum_{j=1}^m \exp\left(-\textsc{klc}(\widehat{\boldsymbol{\gamma}}_j)/(2\widehat{\sigma}^2)\right)}, \quad k = 1, \ldots, m.$$

This density has the same interpretation as the posterior density $g(p \mid \tau)$ displayed on Figure 2.92. Namely, $\mathbb{P}[\mathcal{A}(\sigma_k) \mid \mathbf{X}, \boldsymbol{y}]$ is the Bayesian posterior probability for the estimated model $\mathcal{A}(\sigma_k)$ at grid point σ_k. As usual, high-density regions of this posterior density represent models that are compatible with the observed data. ☞ 60

Naive computation of $\textsc{klc}(\widehat{\boldsymbol{\gamma}}_1), \ldots, \textsc{klc}(\widehat{\boldsymbol{\gamma}}_m)$ can be expensive, as it requires calculating the Mahalanobis norm $\|\boldsymbol{y}\|_{\mathbf{G}}$ and log-determinant $\ln|\mathbf{G}|$ for each of the m grid points $\sigma_1, \ldots, \sigma_m$. Instead, for any grid point with corresponding activated set $\mathcal{A}$, we compute the Kullback–Leibler information criterion via the identity

$$\|\boldsymbol{y}\|_{\mathbf{G}}^2 - \widehat{\sigma}^2 \ln|\mathbf{G}| = \|\boldsymbol{y}\|^2 - \|\mathbf{Q}_{\mathcal{A}}^\top \boldsymbol{y}\|^2 + \|\mathbf{L}_{\mathcal{A}}^{-1} \mathbf{Q}_{\mathcal{A}}^\top \boldsymbol{y}\|^2 + 2\widehat{\sigma}^2 \ln|\mathbf{L}_{\mathcal{A}}|,$$

where $\mathbf{L}_{\mathcal{A}}$ is the Cholesky factor of $\mathbf{G}_{\mathcal{A}}^{-1} = \mathbf{L}_{\mathcal{A}} \mathbf{L}_{\mathcal{A}}^\top$ in (6.29), and $\mathbf{Q}_{\mathcal{A}}$ is the Q-factor in the QR decomposition of the activated model vectors $\mathbf{X}_{\mathcal{A}}$; see Exercise 11 for more details. ■

In summary, if we have to recommend a single reliable estimator of $\boldsymbol{\beta}^*$ for a broad range of data sets, then our choice will be the klimax estimator, with its regularization parameter σ selected either as the median of the output of Algorithm 6.73, or, in the case of p being extremely large, through the solution of equation (6.77).

If both n and p are very large and the model matrix $\mathbf{X}$ has no special structure, then we recommend using Algorithm 6.86 in Exercise 12 as the workhorse for any efficient software implementation of the klimax estimate. Finally, when the model matrix $\mathbf{X}$ is amenable to fast matrix-vector multiplication (see Section A.8.5), then Algorithm 6.97 in Exercise 36 can be a competitive alternative to Algorithm 6.86. ☞ 554

Further Reading

A special case of the klimax method, not discussed here, is the *broken adaptive ridge* regression [46]. Yet another estimate within the klimax framework is the *bridge estimate*; see Exercise 17 and [68] for more theory and references.

The *adaptive lasso* [36] is an iterative lasso estimation procedure that also belongs to the klimax framework. Namely, at each iteration t, given the corresponding lasso shrinkage estimate $\widehat{\boldsymbol{\gamma}}_t$, the term $\ln|\mathbf{I} + \mathbf{X}\boldsymbol{\Gamma}^2\mathbf{X}^\top|$ in $\textsc{klc}(\boldsymbol{\gamma})$ is linearized around $\boldsymbol{\gamma} = \widehat{\boldsymbol{\gamma}}_t$ (rather than around $\boldsymbol{\gamma} = \mathbf{0}$, as in Section 6.2.3), yielding the "adapted" lasso information criterion $\textsc{las}_t(\boldsymbol{\gamma}) := \|\boldsymbol{y}\|_{\mathbf{G}} + \sigma^2 \sum_i [\boldsymbol{v}_i^\top \mathbf{G}(\widehat{\boldsymbol{\gamma}}_t)\boldsymbol{v}_i]\gamma_i$, where the usual squared norm $\|\boldsymbol{v}_i\|^2$ is replaced with the squared Mahalanobis norm $\boldsymbol{v}_i^\top \mathbf{G}(\widehat{\boldsymbol{\gamma}}_t)\boldsymbol{v}_i$. The lasso shrinkage estimate is then updated to $\widehat{\boldsymbol{\gamma}}_{t+1} := \operatorname{argmin}_{\boldsymbol{\gamma}} \textsc{las}_t(\boldsymbol{\gamma})$, and $\widehat{\boldsymbol{\gamma}}_{t+1}$ is itself used to update the lasso information criterion to $\textsc{las}_{t+1}(\boldsymbol{\gamma})$. The process is repeated until convergence of the sequence $\widehat{\boldsymbol{\gamma}}_1, \widehat{\boldsymbol{\gamma}}_2, \ldots$.

A good reference for the theory in Section 6.3 is the monograph [57]. In addition, a key source for the technical Assumptions 6.57 is [60], which proves model-selection consistency under the *complete feature normalization* and *restricted eigenvalue bound* (6.96) in Exercise 29 below. In the analysis of [60], the constant r_ξ appearing in (6.96) is tied explicitly to the particular separable penalty function $\sum_{i=1}^{p} \rho(\beta_i)$ used in the penalized least-squares estimation.

By comparison, Exercise 29 shows that the *klimax irrepresentability* condition in Assumptions 6.57 is less stringent than the *restricted eigenvalue bound* in [60]. Crucially, the *klimax irrepresentability* does not involve any penalty-specific constants, ensuring that it applies to a broader class of model matrices.

Finally, we mention that the lasso estimation can be given a fully Bayesian treatment the involves Monte Carlo and MCMC simulation, see [20, 134].

Exercises

1. This is to prove the Bayesian distributional results (6.2) and (6.3).

   (a) Show that the matrix $\mathbf{G} = (\mathbf{I}_n + \mathbf{X}\boldsymbol{\Gamma}^2\mathbf{X}^\top)^{-1}$ can be written as $\mathbf{G} = \mathbf{I}_n - \mathbf{X}\boldsymbol{\Sigma}\mathbf{X}^\top$, with
☞ 541 $\boldsymbol{\Sigma} = \boldsymbol{\Gamma}(\mathbf{I}_p + \boldsymbol{\Gamma}\mathbf{X}^\top\mathbf{X}\boldsymbol{\Gamma})^{-1}\boldsymbol{\Gamma}$, by applying the Woodbury identity (A.60).

   (b) From the assumption that the error term $\boldsymbol{\varepsilon} \sim \mathcal{N}(\mathbf{0}, \sigma^2\mathbf{I}_n)$ is independent of $\boldsymbol{\beta} \sim \mathcal{N}(\mathbf{0}, \boldsymbol{\Gamma}^2)$, it follows that the joint distribution of $\boldsymbol{\varepsilon}$ and $\boldsymbol{\beta}$ is given by:

$$\begin{bmatrix} \boldsymbol{\varepsilon} \\ \boldsymbol{\beta} \end{bmatrix} \sim \mathcal{N}\left(\begin{bmatrix} \mathbf{0} \\ \mathbf{0} \end{bmatrix}, \sigma^2 \begin{bmatrix} \mathbf{I}_n & \mathbf{0} \\ \mathbf{0} & \boldsymbol{\Gamma}^2 \end{bmatrix} \right).$$

☞ 640 Use this and Theorem D.46 (for the transformation of normal random vectors) to derive the joint distribution of $\boldsymbol{Y} = \mathbf{1}\beta_0 + \mathbf{X}\boldsymbol{\beta} + \boldsymbol{\varepsilon}$ and $\boldsymbol{\beta}$. From this derive the (marginal) distribution of $\boldsymbol{Y}$ and the conditional distribution of $(\boldsymbol{\beta} \,|\, \boldsymbol{Y})$ by applying
☞ 641 Theorems D.50 and D.53, respectively and using part (a).

2. Prove the formula (6.11) for the covariance between features in the simulated data sets.

3. Prove that the KLC in (6.7) simplifies to (6.14), by using the Woodbury identity (A.60)
☞ 541 and the determinant identity (A.61).

4. In this exercise we establish that the James–Stein estimator $\overline{\boldsymbol{\beta}}$ in (6.16) has a smaller mean squared error than the ordinary least-squares estimator $\widehat{\boldsymbol{\beta}}$, when $p \geq 5$. In vector notation we have:

$$\overline{\boldsymbol{\beta}} = \left[1 - \frac{\varsigma^2 p/n}{\|\widehat{\boldsymbol{\beta}}\|^2}\right]_+ \widehat{\boldsymbol{\beta}},$$

where $\widehat{\boldsymbol{\beta}} = \mathbf{X}^\top \boldsymbol{Y}/n \sim \mathcal{N}(\boldsymbol{\beta}^*, \varsigma^2\mathbf{I}_p/n)$. To simplify the exposition, we rescale the parameters by a factor $\sqrt{n}/\varsigma$ and define $\boldsymbol{\alpha}^* := \boldsymbol{\beta}^*\sqrt{n}/\varsigma$ and $\boldsymbol{Z} := \widehat{\boldsymbol{\beta}}\sqrt{n}/\varsigma \sim \mathcal{N}(\boldsymbol{\alpha}^*, \mathbf{I}_p)$, so that it is enough to analyze estimators of $\boldsymbol{\alpha}^*$ of the form

$$\overline{\boldsymbol{\alpha}} = \left[1 - \frac{c}{\|\boldsymbol{Z}\|^2}\right]_+ \boldsymbol{Z} \tag{6.80}$$

for some $c \geq 0$. This includes the natural (maximum likelihood) estimator $\boldsymbol{Z}$.

(a) Show that $\mathbb{E}\|\boldsymbol{Z} - \boldsymbol{\alpha}^*\|^2 = p$; that is, the mean squared error of $\boldsymbol{Z}$ as an estimator of $\boldsymbol{\alpha}^*$ is p.

(b) Define the **Stein shrinkage estimator** of $\boldsymbol{\alpha}^*$:

$$\widetilde{\boldsymbol{\alpha}} := \left(1 - \frac{c}{\|\boldsymbol{Z}\|^2}\right)\boldsymbol{Z},$$

where $c > 0$ is a constant. Note that it is similar to (6.80), but without the positive part operation. Apply Theorem D.60 with the function $\boldsymbol{g}(\boldsymbol{z}) := \boldsymbol{z}/\|\boldsymbol{z}\|^2$ to show that ☞ 644

$$\mathbb{E}\|\widetilde{\boldsymbol{\alpha}} - \boldsymbol{\alpha}^*\|^2 = \mathbb{E}\|\boldsymbol{Z} - \boldsymbol{\alpha}^* - c\boldsymbol{g}(\boldsymbol{Z})\|^2 = p + (c^2 - 2c(p-2))\mathbb{E}\|\boldsymbol{Z}\|^{-2}.$$

Hence, conclude that for $p \geq 3$, we have

$$\mathbb{E}\|\widetilde{\boldsymbol{\alpha}} - \boldsymbol{\alpha}^*\|^2 < \mathbb{E}\|\boldsymbol{Z} - \boldsymbol{\alpha}^*\|^2, \quad \forall c \in (0, 2(p-2)).$$

(c) If $Z \sim \mathcal{N}(\alpha, 1)$, show that $\mathbb{E}[Z \mid Z^2 = z^2] = z\sinh(2z\alpha)/(1+\cosh(2z\alpha))$ for $z > 0$.

(d) Returning to the estimator $\overline{\boldsymbol{\alpha}}$ of $\boldsymbol{\alpha}^*$ in (6.80), we have $\overline{\boldsymbol{\alpha}} = \widetilde{\boldsymbol{\alpha}} \times \mathbb{1}_{\{\|\boldsymbol{Z}\|^2 > c\}}$ and $\overline{\boldsymbol{\alpha}} - \widetilde{\boldsymbol{\alpha}} = \frac{1}{2}\boldsymbol{Z}h(\|\boldsymbol{Z}\|)$, where $h(x) := |1 - c/x^2| - (1 - c/x^2) \geq 0$. Prove that the k-th components of $\overline{\boldsymbol{\alpha}}$ and $\widetilde{\boldsymbol{\alpha}}$ satisfy:

$$\mathbb{E}(\widetilde{\alpha}_k - \alpha_k^*)^2 - \mathbb{E}(\overline{\alpha}_k - \alpha_k^*)^2 \geq \alpha_k^*\mathbb{E}[h(\|\boldsymbol{Z}\|)Z_k].$$

(e) Next, show that in part (d) above $\alpha_k^*\mathbb{E}[h(\|\boldsymbol{Z}\|)Z_k] \geq 0$ for all $k = 1, \ldots, p$. Hence, deduce that $\mathbb{E}\|\widetilde{\boldsymbol{\alpha}} - \boldsymbol{\alpha}^*\|^2 \geq \mathbb{E}\|\overline{\boldsymbol{\alpha}} - \boldsymbol{\alpha}^*\|^2$. Hint: By conditioning on Z_k^2 and $\|\boldsymbol{Z}\|$ first show that $\mathbb{E}[h(\|\boldsymbol{Z}\|)Z_k] = \mathbb{E}[\, h(\|\boldsymbol{Z}\|)\,\mathbb{E}[Z_k \mid Z_k^2]\,]$ and then apply part (c).

(f) Using the above results, explain why the James–Stein estimator (6.16) has smaller mean squared error in comparison to the least-squares estimator $\widehat{\boldsymbol{\beta}} = \mathbf{X}^\top \boldsymbol{Y}/n$, provided that $p \geq 5$.

5. For the linear model $\boldsymbol{Y} = \mathbf{X}\boldsymbol{\beta}^* + \boldsymbol{\varepsilon}$, with $\boldsymbol{\varepsilon} \sim \mathcal{N}(\boldsymbol{0}, \varsigma^2\mathbf{I}_n)$, let $\widehat{\boldsymbol{\beta}}_\lambda = (\lambda\mathbf{I}_p + \mathbf{X}^\top\mathbf{X})^{-1}\mathbf{X}^\top\boldsymbol{y}$ be the ridge estimate of $\boldsymbol{\beta}^*$ in (6.17). In this exercise we prove Lemma 6.18, that is, the ridge learner $\mathbf{X}\widehat{\boldsymbol{\beta}}_{\lambda^*}$ at λ^* has a smaller expected in-sample risk than the learner $\mathbf{X}\widehat{\boldsymbol{\beta}}$ corresponding to the minimum-norm least-squares estimate $\widehat{\boldsymbol{\beta}} = \mathbf{X}^+\boldsymbol{y}$. ☞ 28

(a) If $\mathbf{X} = \mathbf{U}\mathbf{S}\mathbf{V}^\top = \sum_{i=1}^n s_i \boldsymbol{u}_i \boldsymbol{v}_i^\top$ is the full SVD of $\mathbf{X}$, show that: ☞ 532

$$\mathbf{U}^\top\boldsymbol{Y} \sim \mathcal{N}(\mathbf{S}\mathbf{V}^\top\boldsymbol{\beta}^*, \varsigma^2\mathbf{I}_n).$$

(b) Substitution of the SVD of $\mathbf{X}$ into (6.17) (with $\boldsymbol{Y}$ instead of $\boldsymbol{y}$) yields the (random) ridge estimator and learner:

$$\widehat{\boldsymbol{\beta}}_\lambda = \mathbf{V}(\lambda\mathbf{I}_p + \mathbf{S}^2)^{-1}\mathbf{S}\mathbf{U}^\top\boldsymbol{Y} = \sum_i \frac{s_i\langle \boldsymbol{u}_i, \boldsymbol{Y}\rangle}{s_i^2 + \lambda}\boldsymbol{v}_i$$

and

$$\mathbf{X}\widehat{\boldsymbol{\beta}}_\lambda = \mathbf{U}(\lambda\mathbf{I}_p + \mathbf{S}^2)^{-1}\mathbf{S}^2\mathbf{U}^\top\boldsymbol{Y} = \sum_i \frac{s_i^2\langle \boldsymbol{u}_i, \boldsymbol{Y}\rangle}{s_i^2 + \lambda}\boldsymbol{u}_i.$$

Using this, deduce the following squared bias and variance formulas:

$$\|\mathbf{X}\boldsymbol{\beta}^* - \mathbb{E}\mathbf{X}\widehat{\boldsymbol{\beta}}_\lambda\|^2 = \lambda^2 \sum_i \frac{s_i^2 \langle \boldsymbol{v}_i, \boldsymbol{\beta}^* \rangle^2}{(s_i^2 + \lambda)^2}$$

and

$$\mathbb{V}\text{ar}[\mathbf{X}\widehat{\boldsymbol{\beta}}_\lambda] = \varsigma^2 \sum_i \frac{s_i^4}{(s_i^2 + \lambda)^2} \boldsymbol{u}_i \boldsymbol{u}_i^\top .$$

(c) Show that $\lim_{\lambda \downarrow 0} \widehat{\boldsymbol{\beta}}_\lambda = \mathbf{X}^+ \boldsymbol{y}$; that is, in the limit the ridge estimate reduces to the minimum-norm least-squares estimate. Also, show that $\|\widehat{\boldsymbol{\beta}}_\lambda\|^2$ is a decreasing function of λ with $\lim_{\lambda \uparrow \infty} \|\widehat{\boldsymbol{\beta}}_\lambda\|^2 = 0$.

(d) Show that the ridge estimate (6.17) is the minimizer:

$$\widehat{\boldsymbol{\beta}}_\lambda = \underset{\boldsymbol{\beta}}{\operatorname{argmin}} \left\{ \|\boldsymbol{y} - \mathbf{X}\boldsymbol{\beta}\|^2 + \lambda \|\boldsymbol{\beta}\|^2 \right\}.$$

(e) Let $\boldsymbol{Y}'$ be a vector of test response data; that is, an independent realization of $\boldsymbol{Y}$. Show that the expected in-sample risk of the ridge learner, $\mathbb{E}\|\boldsymbol{Y}' - \mathbf{X}\widehat{\boldsymbol{\beta}}_\lambda\|^2/n$, can be written as

$$\frac{1}{n}\mathbb{E}\|\boldsymbol{Y}' - \mathbf{X}\widehat{\boldsymbol{\beta}}_\lambda\|^2 = \varsigma^2 + \frac{\lambda^2}{n} \sum_i \frac{s_i^2 \langle \boldsymbol{v}_i, \boldsymbol{\beta}^* \rangle^2}{(s_i^2 + \lambda)^2} + \frac{\varsigma^2}{n} \sum_i \frac{s_i^4}{(s_i^2 + \lambda)^2}.$$

(f) Prove Lemma 6.18 by using the inequalities for $x \geq 0$:

$$\sum_{k=0}^{2n+1} (-1)^k (k+1) x^k \leq (1+x)^{-2} \leq \sum_{k=0}^{2n} (-1)^k (k+1) x^k, \quad n = 0, 1, 2, \ldots.$$

6. For a given $\boldsymbol{\gamma}$, define the set $\mathcal{A} := \{i : \gamma_i > 0\}$. By using the full SVD of $\mathbf{X}_\mathcal{A} \boldsymbol{\Gamma}_\mathcal{A}$, ☞ 533 prove that $\mathbf{G} = (\mathbf{I}_n + \sum_{j \in \mathcal{A}} \gamma_j \boldsymbol{v}_j \boldsymbol{v}_j^\top)^{-1}$ can be written as:

$$\mathbf{G} = \mathbf{I}_n - \mathbf{X}_\mathcal{A} \mathbf{X}_\mathcal{A}^+ + (\mathbf{X}_\mathcal{A}^+)^\top ((\mathbf{X}_\mathcal{A}^\top \mathbf{X}_\mathcal{A})^+ + \boldsymbol{\Gamma}_\mathcal{A}^2)^{-1} \mathbf{X}_\mathcal{A}^+. \tag{6.81}$$

This identity is used in the derivation of the statistical properties of the klimax, James–Stein, and lasso estimators; see Exercises 23 to 28.

7. For the Bayesian model in (6.2), assume for simplicity that $\beta_0 = 0$, so that the true ☞ 67 model is $\boldsymbol{Y} \sim \mathcal{N}(\mathbf{X}\boldsymbol{\beta}^*, \varsigma^2 \mathbf{I}_n)$. Use the identity (2.101) to show that twice the Kullback–Leibler divergence between the predictive prior density $g(\boldsymbol{y})$ and the true model is:

$$2\mathcal{D}(\mathbf{X}\boldsymbol{\beta}^*, \varsigma^2 \mathbf{I}_n \,|\, \mathbf{0}, \sigma^2 \mathbf{G}^{-1}) = \frac{\varsigma^2}{\sigma^2} \operatorname{tr}(\mathbf{G}) - n - \ln|\mathbf{G}| - n \ln \frac{\varsigma^2}{\sigma^2} + \frac{\|\mathbf{X}\boldsymbol{\beta}^*\|_\mathbf{G}^2}{\sigma^2}. \tag{6.82}$$

Hence, deduce that minimizing the Kullback–Leibler divergence (6.82) for a fixed σ^2 is equivalent to minimizing $\mathbb{E}[\text{KLC}] = \mathbb{E}\|\boldsymbol{Y}\|_\mathbf{G}^2 - \sigma^2 \ln|\mathbf{G}|$:

$$2\mathcal{D}(\mathbf{X}\boldsymbol{\beta}^*, \varsigma^2 \mathbf{I}_n \,|\, \mathbf{0}, \sigma^2 \mathbf{G}^{-1}) = \frac{\mathbb{E}\|\boldsymbol{Y}\|_\mathbf{G}^2 - \sigma^2 \ln|\mathbf{G}|}{\sigma^2} - n\left(1 + \ln \frac{\varsigma^2}{\sigma^2}\right).$$

8. In this exercise we derive asymptotic formulas for the shrinkage and regularization parameters that minimize the Kullback–Leibler divergence (6.82). The assumed asymptotic regime is the classical setting where p is fixed and $n \uparrow \infty$. Since (6.82) also minimizes the expected KLC, these formulas provide insight into the best-case behavior of the klimax estimator in the classical setting where p is small and n is large. We assume that the limit of $\mathbf{\Sigma}_n := \mathbf{X}^\top \mathbf{X}/n \to \mathbf{\Sigma}^*$ is of full rank p, and that n is large enough so that $\mathbf{\Sigma}_n^{-1}$ exists.

(a) Let

$$(\boldsymbol{\gamma}^*, \sigma^*) := \underset{\boldsymbol{\gamma} \geq \mathbf{0}, \sigma > 0}{\operatorname{argmin}} \ \mathcal{D}(\mathbf{X}\boldsymbol{\beta}^*, \varsigma^2 \mathbf{I}_n \,|\, \mathbf{0}, \sigma^2 \mathbf{G}^{-1}(\boldsymbol{\gamma})),$$

where $\mathcal{D}$ is the Kullback–Leibler divergence (6.82). Assuming $\beta_i^* \neq 0$ for all i, show that as $n \uparrow \infty$:

$$\sigma^* = \varsigma, \quad \gamma_i^* = \left(\frac{\beta_i^*}{\sigma^*}\right)^2 + O\left(\frac{1}{n}\right).$$

In other words, when p is fixed and n is large, the optimal shrinkage parameter for the klimax estimate is (approximately) $\boldsymbol{\beta}^* \odot \boldsymbol{\beta}^*/\varsigma^2$ and a suitable value for the regularization parameter is $\sigma = \varsigma$.

Hint: If $\mathbf{C}_n := (\mathbf{\Gamma}\mathbf{\Sigma}_n\mathbf{\Gamma})^{-1}$ and $\mathbf{I}_p - \mathbf{C}_n/n \succ \mathbf{0}$, use the series expansions:

$$(\mathbf{I}_p + \mathbf{C}_n/n)^{-1} = \mathbf{I}_p - \frac{\mathbf{C}_n}{n} + \frac{\mathbf{C}_n^2}{n^2} + O(1/n^3)$$

$$\ln|\mathbf{I}_p + \mathbf{C}_n/n| = \frac{\operatorname{tr}(\mathbf{C}_n)}{n} - \frac{\operatorname{tr}(\mathbf{C}_n^2)}{2n^2} + O(1/n^3).$$

(b) Derive the following asymptotic behavior of the optimal shrinkage parameter λ and regularization parameter σ for the ridge regression estimator in Section 6.1.2.2:

$$\underset{(\lambda, \sigma) > \mathbf{0}}{\operatorname{argmin}} \ \mathcal{D}(\mathbf{X}\boldsymbol{\beta}^*, \varsigma^2 \mathbf{I}_n \,|\, \mathbf{0}, \sigma^2 \mathbf{G}^{-1}(\mathbf{1}/\lambda)) = \left(\frac{p\varsigma^2}{\|\boldsymbol{\beta}^*\|^2} + O(1/n), \ \varsigma\right).$$

(c) Show that the klimax estimator can be written as $\overline{\boldsymbol{\beta}} = \mathbf{\Gamma}^2(\mathbf{\Gamma}^2 + \frac{1}{n}\mathbf{\Sigma}_n^{-1})^{-1}(\boldsymbol{\beta}^* + \mathbf{X}^+\boldsymbol{\varepsilon})$. Hence, deduce the asymptotic expansions of the bias and variance of the klimax estimator:

$$\mathbf{\Sigma}_n(\mathbb{E}\overline{\boldsymbol{\beta}} - \boldsymbol{\beta}^*)(\mathbb{E}\overline{\boldsymbol{\beta}} - \boldsymbol{\beta}^*)^\top \mathbf{\Sigma}_n = \frac{1}{n^2}\mathbf{\Gamma}^{-2}\boldsymbol{\beta}^*(\boldsymbol{\beta}^*)^\top \mathbf{\Gamma}^{-2} + O(1/n^3),$$

$$\mathbb{V}\mathrm{ar}[\mathbf{\Sigma}_n\overline{\boldsymbol{\beta}}] = \frac{\varsigma^2}{n}\mathbf{\Sigma}_n - \frac{2\varsigma^2}{n^2}\mathbf{\Gamma}^{-2} + O(1/n^3).$$

These asymptotic formulas can be used to derive conditions under which the klimax estimator outperforms the more commonly used least-squares estimator (when p is fixed and n is large).

(d) If $\boldsymbol{\gamma} = \boldsymbol{\beta}^* \odot \boldsymbol{\beta}^*/\varsigma^2 + O(1/n)$ is set to its asymptotically optimal value in part (a), show that:

$$\mathbb{E}\|\overline{\boldsymbol{\beta}} - \boldsymbol{\beta}^*\|_{\mathbf{\Sigma}_n^2}^2 = \mathbb{E}\|\widehat{\boldsymbol{\beta}} - \boldsymbol{\beta}^*\|_{\mathbf{\Sigma}_n^2}^2 - \frac{\varsigma^2\operatorname{tr}(\mathbf{\Gamma}^{-2})}{n^2} + O(1/n^3).$$

In other words, the klimax estimator $\overline{\boldsymbol{\beta}}$ is asymptotically more efficient than the minimum-norm least-squares estimator $\widehat{\boldsymbol{\beta}}$, but only when the mean squared error criterion uses the weighted norm $\|\overline{\boldsymbol{\beta}} - \boldsymbol{\beta}^*\|_{\Sigma_n^2} = \|\Sigma_n(\overline{\boldsymbol{\beta}} - \boldsymbol{\beta}^*)\|^2$, rather than the usual Euclidean norm $\|\overline{\boldsymbol{\beta}} - \boldsymbol{\beta}^*\|$. This phenomenon is similar to the one observed for the simpler Stein estimator in Exercise 4.

(e) If $\lambda = p\varsigma^2/\|\boldsymbol{\beta}^*\|^2 + O(1/n)$, show that the ridge estimator is asymptotically more efficient than the minimum-norm least-squares one:

$$\mathbb{E}\|\widehat{\boldsymbol{\beta}}_\lambda - \boldsymbol{\beta}^*\|^2_{\Sigma_n^2} = \mathbb{E}\|\widehat{\boldsymbol{\beta}} - \boldsymbol{\beta}^*\|^2_{\Sigma_n^2} - \frac{\varsigma^4}{n^2}\frac{p^2}{\|\boldsymbol{\beta}^*\|^2} + O(1/n^3).$$

9. In this exercise we explore methods for the efficient computation of the ridge regression estimate (6.17) in Section 6.1.2.2. To this end, define the **ridge shrinkage operator**:

$$S_\lambda^{\text{RIDGE}}(x) := \frac{x}{1+\lambda} = \underset{\alpha}{\operatorname{argmin}}(\alpha - x)^2 + \lambda\alpha^2.$$

(a) If $f(\boldsymbol{\beta}) := \|\boldsymbol{y} - \mathbf{X}\boldsymbol{\beta}\|^2 + \lambda\|\boldsymbol{\beta}\|^2$ and $\boldsymbol{e}_k$ denotes the k-th unit vector, prove that

$$\underset{x}{\operatorname{argmin}} f(\boldsymbol{\beta} + \boldsymbol{e}_k(x - \beta_k)) = S^{\text{RIDGE}}_{\lambda/\|\boldsymbol{v}_k\|^2}\left(\beta_k + \frac{\boldsymbol{r}^\top \boldsymbol{v}_k}{\|\boldsymbol{v}_k\|^2}\right),$$

where $\boldsymbol{r} := \boldsymbol{y} - \mathbf{X}\boldsymbol{\beta}$ is the residual vector.

(b) Computing the ridge regression estimate $\widehat{\boldsymbol{\beta}}_\lambda = \operatorname{argmin}_{\boldsymbol{\beta}} f(\boldsymbol{\beta})$ via the coordinate-descent method involves iterating the coordinate-descent cycle:

$$\beta_k \leftarrow S^{\text{RIDGE}}_{\lambda/\|\boldsymbol{v}_k\|^2}\left(\beta_k + \frac{\boldsymbol{r}^\top \boldsymbol{v}_k}{\|\boldsymbol{v}_k\|^2}\right), \qquad k = 1, \ldots, p.$$

Implement the coordinate-descent method in Python and compare its output with the formula $\widehat{\boldsymbol{\beta}}_\lambda = (\mathbf{X}^\top\mathbf{X} + \lambda\mathbf{I}_n)^{-1}\mathbf{X}^\top\boldsymbol{y}$. The following pseudo-code might be helpful.

Algorithm 6.83: Simple CD to Minimize $\|y - \mathbf{X}\beta\|^2 + \lambda\|\beta\|^2$

Input: initial guess $\boldsymbol{b}$, residual $\boldsymbol{r} := \boldsymbol{y} - \mathbf{X}\boldsymbol{\beta}$, and $\mathbf{X}, \varepsilon, \lambda$
Output: (global) minimizer $\boldsymbol{\beta}$ and corresponding residual $\boldsymbol{r}$

repeat // iterate over CD cycles
  $\boldsymbol{\beta}_{\text{old}} \leftarrow \boldsymbol{\beta}$
  for $k = 1, \ldots, p$ **do** // coordinate-descent cycle
    $b_{\text{new}} \leftarrow S^{\text{RIDGE}}_{\lambda/\|\boldsymbol{v}_k\|^2}(b_k + \boldsymbol{v}_k^\top \boldsymbol{r}/\|\boldsymbol{v}_k\|^2)$ // this step costs $O(n)$
    $\boldsymbol{r} \leftarrow \boldsymbol{r} + (b_k - b_{\text{new}})\boldsymbol{v}_k$ // this update costs $O(n)$
    $b_k \leftarrow b_{\text{new}}$
until $\|\boldsymbol{\beta} - \boldsymbol{\beta}_{\text{old}}\| < \varepsilon$
return $\boldsymbol{\beta}$ & $\boldsymbol{r}$

(c) Let $\mathbf{U}_{k+1}$ and $\mathbf{T}_k$ for some $k < \operatorname{rank}(\mathbf{X})$ be the output of the Lanczos Decomposition Algorithm A.97 with inputs $\mathbf{A} = \mathbf{X}\mathbf{X}^\top$ and $\boldsymbol{b} = \mathbf{X}^\top\boldsymbol{y}$. Find the best approximation of ☞ 556 the ridge estimate (6.17) that lies in the Span $\{\mathbf{U}_k\}$ and minimizes the Mahalanobis distance $\|\cdot\|_{\mathbf{M}_\lambda}$ to $\widehat{\boldsymbol{\beta}}_\lambda$, where $\mathbf{M}_\lambda := \mathbf{X}^\top\mathbf{X} + \lambda\mathbf{I}_p$. In other words, find a formula for

$$\underset{\alpha}{\operatorname{argmin}} \|\widehat{\boldsymbol{\beta}}_\lambda - \mathbf{U}_k\boldsymbol{\alpha}\|_{\mathbf{M}_\lambda}.$$

Hint: Use the fact that $\mathbf{U}_k^\top(\mathbf{X}^\top\mathbf{X} + \lambda\mathbf{I}_p)\mathbf{U}_k = \mathbf{T}_k + \lambda\mathbf{I}_k$ and then apply formula (A.98).

10. Let $\boldsymbol{\beta} = \boldsymbol{\Gamma}^2\mathbf{X}^\top\mathbf{G}\boldsymbol{y}$, where $\boldsymbol{\Gamma}^2 = \operatorname{diag}(\boldsymbol{\gamma})$ and $\mathbf{G} = (\mathbf{I}_n + \sum_j \gamma_j \boldsymbol{v}_j\boldsymbol{v}_j^\top)^{-1}$. Recall the notation $\mathbf{G}_{\neg i} = (\mathbf{I}_n + \sum_{j\neq i} \gamma_j \boldsymbol{v}_j\boldsymbol{v}_j^\top)^{-1}$ from Table 6.21.

(a) Using the formula (6.27), prove the following identities for all $i = 1, \dots, p$:

$$\begin{aligned}
\|\boldsymbol{v}_i\|_{\mathbf{G}}^2 &= \frac{\boldsymbol{v}_i^\top\mathbf{G}_{\neg i}\boldsymbol{v}_i}{1 + \gamma_i\boldsymbol{v}_i^\top\mathbf{G}_{\neg i}\boldsymbol{v}_i},\\
\langle \boldsymbol{v}_i, \boldsymbol{y}\rangle_{\mathbf{G}} &= \frac{\boldsymbol{v}_i^\top\mathbf{G}_{\neg i}\boldsymbol{y}}{1 + \gamma_i\boldsymbol{v}_i^\top\mathbf{G}_{\neg i}\boldsymbol{v}_i},\\
\frac{\langle \boldsymbol{v}_i, \boldsymbol{y}\rangle_{\mathbf{G}}}{\|\boldsymbol{v}_i\|_{\mathbf{G}}^2} &= \frac{\boldsymbol{v}_i^\top\mathbf{G}_{\neg i}\boldsymbol{y}}{\boldsymbol{v}_i^\top\mathbf{G}_{\neg i}\boldsymbol{v}_i},\\
\beta_i &= \frac{\langle \boldsymbol{v}_i, \boldsymbol{y}\rangle_{\mathbf{G}}}{\|\boldsymbol{v}_i\|_{\mathbf{G}}^2}\left[1 - \frac{1}{1 + \gamma_i\boldsymbol{v}_i^\top\mathbf{G}_{\neg i}\boldsymbol{v}_i}\right].
\end{aligned}$$

(b) If $\lambda_i^2 := \sigma^2/\boldsymbol{v}_i^\top\mathbf{G}_{\neg i}\boldsymbol{v}_i$, show that the klimax-KKT conditions (6.22) imply that

$$\gamma_i = \frac{\beta_i^2 + \sqrt{\beta_i^4 + 4\lambda_i^2\beta_i^2}}{2\sigma^2}, \quad i = 1, \dots, p. \tag{6.84}$$

11. This exercise, along with Exercise 12, details the efficient computation of the klimax estimate, which is summarized in the coordinate-descent Algorithm 6.86. Let $\mathcal{A}$ with $|\mathcal{A}| = a$ be the set of activated features in klimax-KKT conditions (6.22), so that $\boldsymbol{\gamma}_{\mathcal{A}} > \mathbf{0}$ and $\boldsymbol{\gamma}_{\mathcal{A}^c} = \mathbf{0}$. Suppose that $\mathbf{X}_{\mathcal{A}} = \mathbf{Q}_{\mathcal{A}}\mathbf{R}_{\mathcal{A}}$ is the QR decomposition of the model matrix $\mathbf{X}_{\mathcal{A}}$. Define the $a \times p$ matrix $\mathbf{R} := \mathbf{Q}_{\mathcal{A}}^\top\mathbf{X} =: [\boldsymbol{r}_1, \dots, \boldsymbol{r}_p]$ and the $a \times a$ matrices ☞ 524
$\boldsymbol{\Gamma}_{\mathcal{A}} := \operatorname{diag}(\boldsymbol{\gamma}_{\mathcal{A}})^{1/2}$ and

$$\mathbf{G}_{\mathcal{A}} := (\mathbf{I}_a + \mathbf{R}_{\mathcal{A}}\boldsymbol{\Gamma}_{\mathcal{A}}^2\mathbf{R}_{\mathcal{A}}^\top)^{-1},$$

as well as the corresponding Mahalanobis inner product and norm for any $\boldsymbol{v}, \boldsymbol{u} \in \mathbb{R}^a$:

$$\begin{aligned}
\langle \boldsymbol{v}, \boldsymbol{u}\rangle_{\mathbf{G}_{\mathcal{A}}} &:= \boldsymbol{v}^\top\mathbf{G}_{\mathcal{A}}\boldsymbol{u},\\
\|\boldsymbol{v}\|_{\mathbf{G}_{\mathcal{A}}} &:= \sqrt{\boldsymbol{v}^\top\mathbf{G}_{\mathcal{A}}\boldsymbol{v}}.
\end{aligned}$$

(a) Using the Woodbury identity (A.60), show that ☞ 541

$$\mathbf{G} = \mathbf{I}_n - \mathbf{Q}_{\mathcal{A}}\mathbf{Q}_{\mathcal{A}}^\top + \mathbf{Q}_{\mathcal{A}}\mathbf{G}_{\mathcal{A}}\mathbf{Q}_{\mathcal{A}}^\top.$$

Hence, denoting $\boldsymbol{z} := \mathbf{Q}_{\mathcal{A}}^\top \boldsymbol{y} \in \mathbb{R}^a$, deduce that for all i:

$$\langle \boldsymbol{v}_i, \boldsymbol{y} \rangle_{\mathbf{G}} = (\langle \boldsymbol{v}_i, \boldsymbol{y} \rangle - \langle \boldsymbol{r}_i, \boldsymbol{z} \rangle)\mathbb{1}_{\{i \in \mathcal{A}^c\}} + \langle \boldsymbol{r}_i, \boldsymbol{z} \rangle_{\mathbf{G}_{\mathcal{A}}},$$

$$\|\boldsymbol{v}_i\|_{\mathbf{G}}^2 = (\|\boldsymbol{v}_i\|^2 - \|\boldsymbol{r}_i\|^2)\mathbb{1}_{\{i \in \mathcal{A}^c\}} + \|\boldsymbol{r}_i\|_{\mathbf{G}_{\mathcal{A}}}^2,$$

$$\|\boldsymbol{y}\|_{\mathbf{G}}^2 = \|\boldsymbol{y}\|^2 - \|\boldsymbol{z}\|^2 + \|\boldsymbol{z}\|_{\mathbf{G}_{\mathcal{A}}}^2,$$

$$\ln|\mathbf{G}| = \ln|\mathbf{G}_{\mathcal{A}}|.$$

(b) Suppose that we wish to add the feature k to the activated set $\mathcal{A}$ and correspondingly update the QR decomposition of $\mathbf{X}_{\mathcal{A}}$ to that of $[\mathbf{X}_{\mathcal{A}}, \boldsymbol{v}_k]$, as well as update $\boldsymbol{z} = \mathbf{Q}_{\mathcal{A}}^\top \boldsymbol{y}$. Using the Gram–Schmidt process in Algorithm A.22, verify that this update can be accomplished via the calculations:

$$\begin{aligned}
&\boldsymbol{r}_k \leftarrow \mathbf{Q}_{\mathcal{A}}^\top \boldsymbol{v}_k \\
&\boldsymbol{q}_k \leftarrow \boldsymbol{v}_k - \mathbf{Q}_{\mathcal{A}} \boldsymbol{r}_k \\
&\varrho_k \leftarrow \|\boldsymbol{q}_k\|, \quad (\varrho_k = \|\boldsymbol{\Pi}_{\mathcal{A}}^\perp \boldsymbol{v}_k\|) \\
&\boldsymbol{q}_k \leftarrow \boldsymbol{q}_k / \varrho_k, \quad (\mathbf{Q}_{\mathcal{A}}^\top \boldsymbol{q}_k = \mathbf{0}, \ \|\boldsymbol{q}_k\| = 1) \\
&\mathbf{Q}_{\mathcal{A}} \leftarrow [\mathbf{Q}_{\mathcal{A}}, \boldsymbol{q}_k] \\
&\mathbf{R} \leftarrow \begin{bmatrix} \mathbf{R} \\ \boldsymbol{q}_k^\top \mathbf{X} \end{bmatrix}, \quad (\text{upsize } \mathbf{Q}_{\mathcal{A}}) \\
&\boldsymbol{z} \leftarrow \begin{bmatrix} \boldsymbol{z} \\ \boldsymbol{q}_k^\top \boldsymbol{y} \end{bmatrix}, \quad (\text{upsize } \mathbf{Q}_{\mathcal{A}}^\top \boldsymbol{y}) \\
&\mathcal{A} \leftarrow \mathcal{A} \cup k, \quad (\text{add index to activated set}).
\end{aligned} \tag{6.85}$$

(c) Suppose that we have the Cholesky decomposition $\mathbf{L}_{\mathcal{A}} \mathbf{L}_{\mathcal{A}}^\top = \mathbf{G}_{\mathcal{A}}^{-1} \in \mathbb{R}^{a \times a}$. Verify that for all i:

$$\langle \boldsymbol{v}_i, \boldsymbol{y} \rangle_{\mathbf{G}} = (\langle \boldsymbol{v}_i, \boldsymbol{y} \rangle - \langle \boldsymbol{r}_i, \boldsymbol{z} \rangle)\mathbb{1}_{\{i \in \mathcal{A}^c\}} + \langle \mathbf{L}_{\mathcal{A}}^{-1} \boldsymbol{r}_i, \mathbf{L}_{\mathcal{A}}^{-1} \boldsymbol{z} \rangle,$$

$$\|\boldsymbol{v}_i\|_{\mathbf{G}}^2 = (\|\boldsymbol{v}_i\|^2 - \|\boldsymbol{r}_i\|^2)\mathbb{1}_{\{i \in \mathcal{A}^c\}} + \|\mathbf{L}_{\mathcal{A}}^{-1} \boldsymbol{r}_i\|^2,$$

$$\ln|\mathbf{G}| = -2 \sum_j \ln(\boldsymbol{e}_j^\top \mathbf{L}_{\mathcal{A}} \boldsymbol{e}_j).$$

(d) Suppose that $k \notin \mathcal{A}$ and we add the feature k, with corresponding shrinkage parameter $\gamma_k > 0$, to the activated set $\mathcal{A}$. It is desirable to compute the Cholesky factor of $\mathbf{G}_{\mathcal{A} \cup \{k\}}^{-1} \in \mathbb{R}^{(a+1) \times (a+1)}$ by updating and upsizing the Cholesky factor $\mathbf{L}_{\mathcal{A}} \in \mathbb{R}^{a \times a}$ from part (c). Using the notation from (6.85), verify that the Cholesky factor $\mathbf{L}_{\mathcal{A}}$ can be updated and upsized to $\mathbb{R}^{(a+1) \times (a+1)}$ at a cost of $O(a^2)$ via:

$$\begin{aligned}
\mathbf{L}_{\mathcal{A}} \mathbf{L}_{\mathcal{A}}^\top &\leftarrow \mathbf{L}_{\mathcal{A}} \mathbf{L}_{\mathcal{A}}^\top + \gamma_k \boldsymbol{r}_k \boldsymbol{r}_k^\top, \quad (\text{rank-1 Cholesky update via Algorithm A.76}) \\
\boldsymbol{l} &\leftarrow \mathbf{L}_{\mathcal{A}}^{-1} \boldsymbol{r}_k, \quad (\text{backward substitution}) \\
\mathbf{L}_{\mathcal{A}} &\leftarrow \begin{bmatrix} \mathbf{L}_{\mathcal{A}} & \mathbf{0} \\ \gamma_k \varrho_k \boldsymbol{l}^\top & \sqrt{1 + \gamma_k \varrho_k^2 (1 - \gamma_k \|\boldsymbol{l}\|^2)} \end{bmatrix}, \quad (\text{1 step of Algorithm A.69}) \\
\mathcal{A} &\leftarrow \mathcal{A} \cup \{k\}, \quad (\text{add feature to activated set}).
\end{aligned}$$

12. (Exercise 11 continued) Implement in Python the following algorithm:

Algorithm 6.86: Fast Computation of Klimax Estimate.

input: current $\boldsymbol{\beta}, \boldsymbol{\gamma}, \mathcal{E}, \mathbf{L}, \mathbf{Q}, \mathbf{R}, \boldsymbol{z}$, data $\{\mathbf{X}, \boldsymbol{y}\}$, tolerance ε, and regularization σ
output: optimal $\boldsymbol{\beta}$ and updated $\boldsymbol{\gamma}, \mathcal{E}, \mathbf{L}, \mathbf{Q}, \mathbf{R}, \boldsymbol{z}$

repeat // iterate over CD cycles
  $\boldsymbol{\beta}_{\text{old}} \leftarrow \boldsymbol{\beta}$ // store a copy before updating all coordinates
  for $i = 1, \ldots, p$ **do** // coordinate-descent cycle
    let $\boldsymbol{r}$ be the i-th column of $\mathbf{R} \in \mathbb{R}^{e \times p}$
    $\mathtt{Lr} \leftarrow \mathbf{L}^{-1}\boldsymbol{r}$, $\mathtt{vGv} \leftarrow \|\mathtt{Lr}\|^2$, $\mathtt{vGy} \leftarrow \langle \mathtt{Lr}, \mathbf{L}^{-1}\boldsymbol{z}\rangle$ // costs $O(e^2)$
    if $i \notin \mathcal{E}$ **then** // feature was never in working set
      $\mathtt{vGv} \leftarrow \mathtt{vGv} + (\|\boldsymbol{v}_i\|^2 - \|\boldsymbol{r}\|^2)_+$ // make sure norm always positive
      $\mathtt{vGy} \leftarrow \mathtt{vGy} + \langle \boldsymbol{v}_i, \boldsymbol{y}\rangle - \langle \boldsymbol{r}, \boldsymbol{z}\rangle$
    $\mathtt{vG_{\neg i}v} \leftarrow \mathtt{vGv}/(1 - \gamma_i \mathtt{vGv})$ // remove i-th feature via Sherman-Morrison
    $\lambda_i \leftarrow \frac{\sigma}{\sqrt{\mathtt{vG_{\neg i}v}}}$, $\beta_i \leftarrow S^{\text{JS}}_{\lambda_i}\left(\frac{\mathtt{vGy}}{\mathtt{vGv}}\right)$, $\gamma \leftarrow \frac{\beta_i}{\sigma^2}\frac{\mathtt{vGy}}{\mathtt{vGv}}$ // update via (6.26)
    if $\gamma \neq \gamma_i$ **then** // need for any updating?
      if $i \in \mathcal{E}$ **then** // feature in ever-activated set
        update $\mathbf{L}$ via Algorithm A.76, so that $\mathbf{L}\mathbf{L}^\top \leftarrow \mathbf{L}\mathbf{L}^\top + (\gamma - \gamma_i)\boldsymbol{r}\boldsymbol{r}^\top$
      else // $i \notin \mathcal{E}$ & incomplete basis $\Rightarrow$ upsize
        $\boldsymbol{q} \leftarrow \boldsymbol{v}_i - \mathbf{Q}\boldsymbol{r}$, $\varrho \leftarrow \|\boldsymbol{q}\|$, $\boldsymbol{q} \leftarrow \boldsymbol{q}/\varrho$ // Gram-Schmidt step
        if $\varrho > \varepsilon$ **then** // Is the new feature linearly independent?
          update $\mathbf{L}$ via Algorithm A.76, so that $\mathbf{L}\mathbf{L}^\top \leftarrow \mathbf{L}\mathbf{L}^\top + \gamma\boldsymbol{r}\boldsymbol{r}^\top$
          $\boldsymbol{l} \leftarrow \mathbf{L}^{-1}\boldsymbol{r}$ // costs $O(e^2)$
          $l \leftarrow 1 + \gamma\varrho^2(1 - \gamma\|\boldsymbol{l}\|^2)_+$ // ensure always positive
          $\mathbf{L} \leftarrow \begin{bmatrix} \mathbf{L} & \mathbf{0} \\ \gamma\varrho\boldsymbol{l}^\top & \sqrt{l} \end{bmatrix}$, $\mathbf{Q} \leftarrow [\mathbf{Q}, \boldsymbol{q}]$
          $\boldsymbol{z} \leftarrow \begin{bmatrix} \boldsymbol{z} \\ \boldsymbol{q}^\top\boldsymbol{y} \end{bmatrix}$, $\mathbf{R} \leftarrow \begin{bmatrix} \mathbf{R} \\ \boldsymbol{q}^\top\mathbf{X} \end{bmatrix}$ // $\boldsymbol{q}^\top\mathbf{X}$ costs $O(n \times p)$
        $\mathcal{E} \leftarrow \mathcal{E} \cup \{i\}$ // augment ever-activated set
      $\gamma_i \leftarrow \gamma$ // update shrinkage variable
until $\|\boldsymbol{\beta} - \boldsymbol{\beta}_{\text{old}}\| < \varepsilon$
return $\boldsymbol{\beta}, \boldsymbol{\gamma}, \mathcal{E}, \mathbf{L}, \mathbf{Q}, \mathbf{R}, \boldsymbol{z}$

This uses the coordinate-descent iteration (6.26), where during each coordinate update we store in memory the following:

- An **ever-activated set** of features $\mathcal{E}$. This set is the same as the activated set $\mathcal{A}$, except that once a feature (or more precisely, its index) enters the set $\mathcal{E}$, it never leaves it. In other words, as the iterations progress $e := |\mathcal{E}|$ is increasing.
- The QR decomposition $\mathbf{Q}_\mathcal{E}\mathbf{R}_\mathcal{E} = \mathbf{X}_\mathcal{E}$ and $\mathbf{R} = \mathbf{Q}_\mathcal{E}^\top\mathbf{X}$, as well as $\boldsymbol{z} = \mathbf{Q}_\mathcal{E}^\top\boldsymbol{y}$.
- The Cholesky factor $\mathbf{L}_\mathcal{E} \in \mathbb{R}^{e \times e}$ such that $\mathbf{L}_\mathcal{E}\mathbf{L}_\mathcal{E}^\top = \mathbf{I}_e + \mathbf{R}_\mathcal{E}\boldsymbol{\Gamma}_\mathcal{E}^2\mathbf{R}_\mathcal{E}^\top$.

When we update a given γ_i to γ_i^{new}, all of the three items above are updated accordingly, by noting that there are only two possibilities for the ever-activated set $\mathcal{E}$:

- $\mathcal{E}$ is unchanged, which happens when, either $k \in \mathcal{E}$, or $\gamma_i = \gamma_i^{\text{new}} = 0$.
- $\mathcal{E}$ is augmented to $\mathcal{E} \cup i$, which happens when $k \notin \mathcal{E}$ and $\gamma_i^{\text{new}} > 0$.

As an example, the Cholesky factor $\mathbf{L}_{\mathcal{E}}$ is updated as follows:

- If $\mathcal{E}$ is unchanged and $\gamma_i^{\text{new}} > 0$, then we use the rank-1 Cholesky update of $\mathbf{L}_{\mathcal{E}}$:

$$\mathbf{L}_{\mathcal{E}}\mathbf{L}_{\mathcal{E}}^\top \leftarrow \mathbf{L}_{\mathcal{E}}\mathbf{L}_{\mathcal{E}}^\top + (\gamma_i^{\text{new}} - \gamma_i)\boldsymbol{r}_i\boldsymbol{r}_i^\top.$$

- On the other hand, if $\mathcal{E}$ is augmented to $\mathcal{E} \cup i$, then the current Cholesky factor of size $e \times e$ is upsized to $(e+1) \times (e+1)$, as described in Exercise 11.

We never downsize the Cholesky factor, because we never remove elements from the ever-activated set $\mathcal{E}$.

The overall algorithm can thus be organized in such a way that each coordinate cycle costs $O(e^2 \times p)$ and adding a new feature to the ever-activated set costs $O(n \times p)$. Note that in Algorithm 6.86 the quantities $\|\boldsymbol{v}_i\|^2$ and $\langle \boldsymbol{v}_i, \boldsymbol{y}\rangle$ should ideally be precomputed and stored in memory, rather than being computed repeatedly from scratch within each coordinate-descent cycle.

13. (Exercise 12 continued) The coordinate-descent iteration (6.50) can be implemented efficiently by modifying Algorithm 6.86 to compute the lasso estimate. The modification replaces the update (6.26) in Line 10 of Algorithm 6.86 with the lasso update:

$$\lambda_i \leftarrow \frac{\sigma\|\boldsymbol{v}_i\|}{\mathsf{vG_{\neg i}v}}, \qquad \beta_i \leftarrow S_{\lambda_i}^{\text{LAS}}\left(\frac{\mathsf{vGy}}{\mathsf{vGv}}\right), \qquad \gamma_i \leftarrow \frac{|\beta_i|}{\sigma\|\boldsymbol{v}_i\|}.$$

Implement this lasso algorithm in Python and compute the regularization path for the artificial data set in Figure 6.52.

14. Use formula (6.25) to find a closed-from solution to the klimax-KKT conditions (6.22), provided that $\boldsymbol{v}_p^\top\boldsymbol{v}_i = 0$ for all $i < p$ and the regularization parameter σ satisfies:

$$\max_{i<p} \frac{|\langle \boldsymbol{v}_i, \boldsymbol{y}\rangle|}{\|\boldsymbol{v}_i\|} \leq \sigma < \frac{|\langle \boldsymbol{v}_p, \boldsymbol{y}\rangle|}{\|\boldsymbol{v}_p\|}, \quad \text{where} \quad p := \operatorname*{argmax}_i \frac{|\langle \boldsymbol{v}_i, \boldsymbol{y}\rangle|}{\|\boldsymbol{v}_i\|}.$$

15. This exercise investigates the behavior of the klimax method estimates as $\sigma \downarrow 0$. Let $\mathcal{A} := \{i : \gamma_i > 0\}$ be the set of activated features, so that $\mathbf{G}^{-1} = \mathbf{I}_n + \sum_{i\in\mathcal{A}} \gamma_i\boldsymbol{v}_i\boldsymbol{v}_i^\top$.

(a) Prove that for any $\boldsymbol{\beta} \in \mathbb{R}^p$, we have the inequality:

$$\|\boldsymbol{y}\|_{\mathbf{G}}^2 \leq \|\boldsymbol{y} - \mathbf{X}_{\mathcal{A}}\boldsymbol{\beta}_{\mathcal{A}}\|^2 + \|\boldsymbol{\Gamma}_{\mathcal{A}}^{-1}\boldsymbol{\beta}_{\mathcal{A}}\|^2,$$

with equality achieved at $\overline{\boldsymbol{\beta}} = \boldsymbol{\Gamma}^2\mathbf{X}^\top\mathbf{G}\boldsymbol{y}$, that is, $\overline{\boldsymbol{\beta}}_{\mathcal{A}^c} = \mathbf{0}$ and $\overline{\boldsymbol{\beta}}_{\mathcal{A}} = \boldsymbol{\Gamma}_{\mathcal{A}}^2\mathbf{X}_{\mathcal{A}}^\top\mathbf{G}\boldsymbol{y}$.

(b) Use the identity (6.81) to show that $\|\boldsymbol{y}\|_{\mathbf{G}}^2 \geq \|\boldsymbol{\Pi}_{\mathcal{A}}^{\perp}\boldsymbol{y}\|^2$ for all $\boldsymbol{\gamma} \geq \mathbf{0}$, with equality achieved in the limit as $\min_{i\in\mathcal{A}} \gamma_i \to \infty$ for any $\mathcal{A}$ such that $\text{rank}(\mathbf{X}_{\mathcal{A}}) = \text{rank}(\mathbf{X})$.

(c) Use the above results to explain why all klimax method (that is, klimax, James–Stein, and lasso) estimates converge to the minimum-norm least-squares estimate $\widehat{\boldsymbol{\beta}} = \mathbf{X}^{+}\boldsymbol{y}$ as the regularization parameter $\sigma \downarrow 0$.

16. Using Exercise 15, prove that the global minimizer of the penalized least-squares objective (6.35) satisfies the klimax-KKT conditions (6.22). Also, show that $\rho_{\mathbf{X}}$ satisfies for all $\beta_i \neq 0$:

$$\frac{\partial \rho_{\mathbf{X}}}{\partial \beta_i} = \frac{2\frac{\mathrm{d}}{\mathrm{d}\beta_i}|\beta_i|}{|\beta_i| + \sqrt{\beta_i^2 + 4\|\boldsymbol{v}_i\|^2/\boldsymbol{v}_i^\top \mathbf{G}_{\neg i}\boldsymbol{v}_i}}.$$

17. The **bridge estimate** is defined as the global minimizer of the penalized least-squares objective (6.35), where the multivariate penalty $\rho_{\mathbf{X}}$ is replaced by $\rho(\boldsymbol{\beta}) = \frac{1}{\alpha}\sum_{i=1}^p |\beta_i|^\alpha$ for $\alpha \in (0,1)$. Show that the bridge estimate can be written as $\overline{\boldsymbol{\beta}} = \mathrm{diag}(\widehat{\boldsymbol{\gamma}})\mathbf{X}^\top \mathbf{G}(\widehat{\boldsymbol{\gamma}})\boldsymbol{y}$, where $\widehat{\boldsymbol{\gamma}}$ is the global minimizer of $\ell(\boldsymbol{\gamma}) := \|\boldsymbol{y}\|_{\mathbf{G}}^2 + \sigma^2 \frac{2-\alpha}{\alpha}\sum_{i=1}^p \left(\|\boldsymbol{v}_i\|^2\gamma_i\right)^{\frac{\alpha}{2-\alpha}}$.

18. Recall the definition of the James–Stein penalty function (6.42):

$$\rho(\beta) := \frac{\sqrt{\beta^4 + 4\beta^2} - \beta^2}{4} + \frac{1}{2}\ln\left(1 + \frac{\sqrt{\beta^4 + 4\beta^2} + \beta^2}{2}\right).$$

(a) For $\beta \neq 0$, show that $\frac{\mathrm{d}\rho}{\mathrm{d}\beta} = \frac{2\frac{\mathrm{d}}{\mathrm{d}\beta}|\beta|}{|\beta| + \sqrt{\beta^2+4}}$ and $\frac{\mathrm{d}^2\rho}{\mathrm{d}\beta^2} = \frac{|\beta|}{2\sqrt{\beta^2+4}} - \frac{1}{2}$.

(b) Prove that the penalty satisfies: $|\beta| - \frac{\beta^2}{4} \leq \rho(\beta) \leq |\beta|$ and $\lim_{|\beta|\uparrow\infty} \frac{\rho(\beta)}{\ln|\beta|} = 1$.

(c) Prove that $\mathrm{argmin}_x \left\{\frac{(x-\alpha)^2}{2} + \lambda^2\rho(x/\lambda)\right\} = S_\lambda^{\mathrm{JS}}(\alpha)$.

(d) For the James–Stein estimator in Section 6.2.2, show that for each i the minimizer of the least-squares objective

$$\frac{\|\boldsymbol{y} - \mathbf{X}\boldsymbol{\beta}\|^2}{2} + \sigma^2 \sum_{i=1}^p \rho(\|\beta_i \boldsymbol{v}_i\|/\sigma)$$

with respect to β_i is given by $S_{\lambda_i}^{\mathrm{JS}}\left(\langle \boldsymbol{v}_i, \boldsymbol{r}_{\neg i}\rangle/\|\boldsymbol{v}_i\|^2\right)$, with $\lambda_i = \sigma/\|\boldsymbol{v}_i\|$, $i = 1, \ldots, p$, in correspondence with (6.41). This then justifies the coordinate-descent procedure in Algorithm 6.43.

19. Given the user-specified weights $\{\omega_i\}$, consider the lasso penalized least-squares objective:

$$f(\boldsymbol{\beta}) := \frac{1}{2}\|\boldsymbol{y} - \mathbf{X}\boldsymbol{\beta}\|^2 + \sum_i \omega_i |\beta_i|.$$

(a) Verify that computing $\min_{\boldsymbol{\beta}} f(\boldsymbol{\beta})$ is equivalent to solving the constrained optimization:

$$\min_{\boldsymbol{\beta},\boldsymbol{\gamma}} \quad \frac{1}{2}\|\boldsymbol{y} - \mathbf{X}\boldsymbol{\beta}\|^2 + \boldsymbol{\omega}^\top\boldsymbol{\gamma}$$
$$\text{subject to:} \quad \begin{bmatrix} \mathbf{I}_p & -\mathbf{I}_p \\ -\mathbf{I}_p & -\mathbf{I}_p \end{bmatrix} \begin{bmatrix} \boldsymbol{\beta} \\ \boldsymbol{\gamma} \end{bmatrix} \leq \mathbf{0},$$

where $\boldsymbol{\gamma}$ is functioning as a slack variable. ☞ 622

(b) Derive the KKT conditions satisfied by the optimal values for $(\boldsymbol{\beta}, \boldsymbol{\gamma})$. ☞ 595

(c) Let $\boldsymbol{\beta}^*$ be a minimizer of $f(\boldsymbol{\beta})$ and define $\mathcal{A} := \{i : \beta_i^* \neq 0\}$ to be the set of activated features. Show that the KKT conditions for $\boldsymbol{\beta}^*$ can be simplified to:

$$\begin{aligned} \boldsymbol{v}_i^\top(\boldsymbol{y} - \mathbf{X}\boldsymbol{\beta}^*) &= \omega_i \operatorname{sign}(\beta_i), \quad \forall i \in \mathcal{A}, \\ |\boldsymbol{v}_i^\top(\boldsymbol{y} - \mathbf{X}\boldsymbol{\beta}^*)| &\leq \omega_i, \quad \forall i \in \mathcal{A}^c. \end{aligned}$$

(d) If $\mathbf{X}_{\mathcal{A}}$ is of full column rank and $w_i := \omega_i \operatorname{sign}(\beta_i^*)$ for all $i \in \mathcal{A}$, show that:

$$\boldsymbol{\beta}_{\mathcal{A}^c}^* = \mathbf{0}, \quad \boldsymbol{\beta}_{\mathcal{A}}^* = \mathbf{X}_{\mathcal{A}}^+(\boldsymbol{y} - (\mathbf{X}_{\mathcal{A}}^+)^\top \boldsymbol{w}_{\mathcal{A}}).$$

☞ 606 20. This exercise derives the coordinate-descent method for the lasso estimate.

(a) Show that the minimizer

$$x^* := \underset{x}{\operatorname{argmin}} \left\{ \frac{(x-\alpha)^2}{2} + \lambda |x| \right\} = \begin{cases} 0, & |\alpha| \leq \lambda, \\ \alpha\left(1 - \frac{\lambda}{|\alpha|}\right), & |\alpha| > \lambda, \end{cases} \tag{6.87}$$

which for $\alpha \neq 0$ can be written as

$$x^* = S_\lambda^{\text{LAS}}(\alpha),$$

where $S_\lambda^{\text{LAS}}(x) := x(1 - \lambda/|x|)_+$ is the lasso shrinkage operator.

(b) Given the weight $\boldsymbol{\omega} \geq \mathbf{0}$ attached to $\boldsymbol{\beta}$, define the lasso penalized least-squares objective:

$$f(\boldsymbol{\beta}) := \frac{1}{2n}\|\boldsymbol{y} - \mathbf{X}\boldsymbol{\beta}\|^2 + \lambda \sum_i \omega_i |\beta_i|.$$

If $\lambda_k := n\omega_k\lambda/\|\boldsymbol{v}_k\|^2$ for all k and $\boldsymbol{r} := \boldsymbol{y} - \mathbf{X}\boldsymbol{\beta}$, use (6.87) to prove that:

$$\underset{x \in \mathbb{R}}{\operatorname{argmin}} f(\boldsymbol{\beta} + (x - \beta_k)\boldsymbol{u}_k) = S_{\lambda_k}^{\text{LAS}}\left(\beta_k + \boldsymbol{v}_k^\top \boldsymbol{r}/\|\boldsymbol{v}_k\|^2\right).$$

(c) A coordinate-descent cycle for computing $\operatorname{argmin}_{\boldsymbol{\beta}} f(\boldsymbol{\beta})$ consists of:

$$\beta_k \leftarrow S_{\lambda_k}^{\text{LAS}}\left(\beta_k + \boldsymbol{r}^\top \boldsymbol{v}_k/\|\boldsymbol{v}_k\|^2\right), \quad k = 1, \ldots, p.$$

Implement the coordinate-descent method in Python by storing and updating the residual vector $\boldsymbol{r} := \boldsymbol{y} - \mathbf{X}\boldsymbol{\beta}$ after each coordinate change. You may use the following pseudo-code in your implementation.

Algorithm 6.88: Simple CD to Minimize $\frac{1}{2n}\|\boldsymbol{y} - \mathbf{X}\boldsymbol{\beta}\|^2 + \lambda \sum_i \omega_i |\beta_i|$

Input: initial guess $\boldsymbol{\beta}$, residual $\boldsymbol{r} := \boldsymbol{y} - \mathbf{X}\boldsymbol{\beta}$, and $\mathbf{X}, \varepsilon, \lambda, \boldsymbol{\omega}$
Output: (global) minimizer $\boldsymbol{\beta}$ and corresponding residual $\boldsymbol{r}$

$\lambda_k \leftarrow n\omega_k\lambda/\|\boldsymbol{v}_k\|^2, \ k = 1, \ldots, p$
repeat // iterate over CD cycles
  $\boldsymbol{\beta}_{\text{old}} \leftarrow \boldsymbol{\beta}$
  for $k = 1, \ldots, p$ **do** // coordinate-descent cycle
    $\beta_{\text{new}} \leftarrow S_{\lambda_k}^{\text{LAS}}\left(\beta_k + \boldsymbol{v}_k^\top \boldsymbol{r}/\|\boldsymbol{v}_k\|^2\right)$ // this step costs $O(n)$
    $\boldsymbol{r} \leftarrow \boldsymbol{r} + (\beta_k - \beta_{\text{new}})\boldsymbol{v}_k$ // this update costs $O(n)$
    $\beta_k \leftarrow \beta_{\text{new}}$
until $\|\boldsymbol{\beta} - \boldsymbol{\beta}_{\text{old}}\| < \varepsilon$
return $\boldsymbol{\beta}$ & $\boldsymbol{r}$

(d) Let $\boldsymbol{g} := \mathbf{X}^\top(\mathbf{X}\boldsymbol{\beta} - \boldsymbol{y})$ be the gradient of the sum of squares: $\|\boldsymbol{y} - \mathbf{X}\boldsymbol{\beta}\|^2/2$. Show that

$$\frac{\boldsymbol{r}_{\neg k}^\top \boldsymbol{v}_k}{\|\boldsymbol{v}_k\|^2} = \beta_k - g_k/\|\boldsymbol{v}_k\|^2.$$

If β_k is updated to $\breve{\beta}$, then find a formula that correspondingly updates the gradient vector $\boldsymbol{g}$. Use this updating formula in a Python implementation to compute the lasso estimate via the coordinate-descent cycle:

$$\beta_k \leftarrow S^{\text{LAS}}_{n\omega_k\lambda/\|\boldsymbol{v}_k\|^2}\left(\beta_k - g_k/\|\boldsymbol{v}_k\|^2\right), \quad k = 1, \ldots, p.$$

The following pseudo-code may help with the implementation of this version of coordinate descent.

Algorithm 6.89: Covariance CD to Minimize $\frac{1}{2n}\|\boldsymbol{y}-\mathbf{X}\boldsymbol{\beta}\|^2+\lambda\sum_i \omega_i\,|\beta_i|$

Input: initial guess $\boldsymbol{\beta}$, gradient $\boldsymbol{g} := \mathbf{X}^\top(\mathbf{X}\boldsymbol{\beta} - \boldsymbol{y})$, and $\mathbf{X}, \varepsilon, \lambda, \boldsymbol{\omega}$
Output: (global) minimizer $\boldsymbol{\beta}$ and corresponding gradient $\boldsymbol{g}$

$\mathcal{E} \leftarrow \emptyset$ // ever-active set of features
$\boldsymbol{\Sigma} \leftarrow \text{diag}(\|\boldsymbol{v}_1\|^2, \ldots, \|\boldsymbol{v}_p\|^2)$ // store a sparse covariance matrix
$\lambda_k \leftarrow n\omega_k\lambda/\|\boldsymbol{v}_k\|^2,\ k = 1, \ldots, p$
repeat // iterate over CD cycles
  $\boldsymbol{\beta}_{\text{old}} \leftarrow \boldsymbol{\beta}$
  for $k = 1, \ldots, p$ **do** // coordinate-descent cycle
    $\beta_{\text{new}} \leftarrow S^{\text{LAS}}_{\lambda_k}(\beta_k - g_k/\|\boldsymbol{v}_k\|^2)$
    if $\beta_k \neq \beta_{\text{new}}$ **then** // Is there a need for updating?
      if $k \notin \mathcal{E}$ **then** // Is feature active for the first time?
        add $\mathbf{X}^\top\boldsymbol{v}_k$ as the k-th column of $\boldsymbol{\Sigma}$ // costs $O(np)$
        $\mathcal{E} \leftarrow \mathcal{E} \cup k$ // update ever-active set
      let $\boldsymbol{c}$ be the k-th column of $\boldsymbol{\Sigma}$
      $\boldsymbol{g} \leftarrow \boldsymbol{g} + (\beta_{\text{new}} - \beta_k)\boldsymbol{c}$ // gradient update costs $O(p)$
      $\beta_k \leftarrow \beta_{\text{new}}$
until $\|\boldsymbol{\beta} - \boldsymbol{\beta}_{\text{old}}\| < \varepsilon$
return $\boldsymbol{\beta}$ & $\boldsymbol{g}$

21. Let $\lambda_i := \lambda\|\boldsymbol{v}_i\|/\sqrt{n}$ for all i and consider computing the lasso estimate:

$$\underset{\boldsymbol{\beta}}{\text{argmin}}\ \frac{\|\boldsymbol{y} - \mathbf{X}\boldsymbol{\beta}\|^2}{2n} + \sum_{i=1}^{p} \lambda_i\,|\beta_i|. \tag{6.90}$$

(a) Show that the ADMM minimization method for computing (6.90) is equivalent to the iteration of the following steps: ☞ 618

$$\begin{aligned}
\boldsymbol{\beta} &\leftarrow \left(\mathbf{X}^\top\mathbf{X}/n + \varrho\mathbf{I}_p\right)^{-1}\left(\varrho\boldsymbol{z} + \mathbf{X}^\top\boldsymbol{y}/n - \boldsymbol{v}\right)\\
z_k &\leftarrow S^{\text{LAS}}_{\lambda_k/\varrho}(\beta_k + v_k/\varrho), \quad k = 1, \ldots, p\\
\boldsymbol{v} &\leftarrow \boldsymbol{v} + \varrho(\boldsymbol{\beta} - \boldsymbol{z}).
\end{aligned}$$

(b) Using the following pseudo-code, implement in Python the ADMM algorithm for the computation of (6.90).

Algorithm 6.91: ADMM Method for $\operatorname{argmin}_{\beta} \frac{\|y-X\beta\|^2}{2n} + \sum_{i=1}^{p} \lambda_i |\beta_i|$

Input: $\mathbf{X} \in \mathbb{R}^{n\times p}$, $\boldsymbol{y} \in \mathbb{R}^n$; parameter $\lambda > 0$ and $\varepsilon > 0$.

Output: $\widehat{\boldsymbol{\beta}}_\lambda$

$\varrho \leftarrow 0.01,\ \boldsymbol{v} \leftarrow \mathbf{0},\ \boldsymbol{z} \leftarrow \mathbf{0},\ \boldsymbol{\beta} \leftarrow \mathbf{0}$ // initial parameters

$\mathtt{Xy} \leftarrow \mathbf{X}^\top \boldsymbol{y}/n,\ \boldsymbol{\Sigma} \leftarrow \mathbf{X}^\top\mathbf{X}/n$ // precompute and store in memory

$\mathbf{L}\mathbf{L}^\top \leftarrow \boldsymbol{\Sigma} + \varrho \mathbf{I}_n$ // store Cholesky decomposition in memory

$\lambda_k \leftarrow \lambda\sqrt{\Sigma_{k,k}},\ k = 1, \ldots, p$

repeat // repeat until convergence

  $\boldsymbol{\beta}_{\text{old}} \leftarrow \boldsymbol{\beta}$

  $\boldsymbol{\beta} \leftarrow \varrho \boldsymbol{z} + \mathtt{Xy} - \boldsymbol{v}$

  $\boldsymbol{\beta} \leftarrow \mathbf{L}^{-1}\boldsymbol{\beta},\ \boldsymbol{\beta} \leftarrow \mathbf{L}^{-\top}\boldsymbol{\beta}$ // forward and backward substitution in $O(p^2)$

  $z_k \leftarrow S^{\text{LAS}}_{\lambda_k/\varrho}(\beta_k + v_k/\varrho), \quad k = 1, \ldots, p$

  $\boldsymbol{v} \leftarrow \boldsymbol{v} + \varrho(\boldsymbol{\beta} - \boldsymbol{z})$ // update dual/Lagrange variables

until $\|\boldsymbol{\beta} - \boldsymbol{\beta}_{\text{old}}\| < \varepsilon$ // one choice for the tolerance is $\varepsilon = 10^{-6}$

return $\boldsymbol{\beta}$

22. Let $\omega_i^2 := \|v_i\|^2/n$ for all i and define

$$f(\boldsymbol{\beta}, \mathbf{X}) := \frac{\|\boldsymbol{y} - \mathbf{X}\boldsymbol{\beta}\|^2}{2n} + \lambda \sum_i \omega_i |\beta_i|.$$

(a) Show that if $\lambda > \max_i \frac{|v_i^\top \boldsymbol{y}|}{n\omega_i}$, then $\operatorname{argmin}_{\boldsymbol{\beta}} f(\boldsymbol{\beta}, \mathbf{X}) = \mathbf{0}$.

(b) Assuming that the coordinate-descent algorithm in Exercise 20 converges to the unique lasso estimate and that $\min_k |v_k^\top \boldsymbol{y}| > 0$, prove that $f(\boldsymbol{b}, \mathbf{X}\boldsymbol{\Omega}) > f(\mathbf{0}, \mathbf{X}\boldsymbol{\Omega}),\ \forall \boldsymbol{b}$ implies that $\lambda \geq \max_i \frac{|v_i^\top \boldsymbol{y}|}{n\omega_i}$.

23. Show that if the model matrix is centered, that is, $\mathbf{X} = \boldsymbol{\Pi}_0^\perp \mathbf{X}$ (see Section 6.1.3.1), then

$$\overline{y} = \operatorname*{argmin}_{\beta_0} \|\boldsymbol{y} - \mathbf{1}\beta_0\|_{\mathbf{G}}^2.$$

24. Define $\mathbf{A}_{\mathcal{A}} := ((\mathbf{X}_{\mathcal{A}}^\top \mathbf{X}_{\mathcal{A}})^+ + \boldsymbol{\Gamma}_{\mathcal{A}}^2)^{-1}$ and $\boldsymbol{\Pi}_{\mathcal{A}}^\perp := \mathbf{I}_n - \mathbf{X}_{\mathcal{A}}\mathbf{X}_{\mathcal{A}}^+$, where $\mathcal{A}$ is a set of activated features.

(a) Use (6.81) to show that if $\mathcal{A} \equiv \mathcal{S} = \{i \geq 1 : \beta_i^* \neq 0\}$ and $\boldsymbol{y}$ is replaced with $\boldsymbol{Y}$ $(= \mathbf{X}\boldsymbol{\beta}^* + \boldsymbol{\varepsilon})$, then the klimax-KKT conditions (6.22) can be written as:

(6.92)

$$|\boldsymbol{e}_i^\top \mathbf{A}_{\mathcal{S}} \mathring{\boldsymbol{\beta}}_{\mathcal{S}}| = \sigma\sqrt{\boldsymbol{e}_i^\top \mathbf{A}_{\mathcal{S}} \boldsymbol{e}_i}, \quad \forall i \in \mathcal{S},$$

$$|v_i^\top \boldsymbol{\Pi}_{\mathcal{S}}^\perp \boldsymbol{\varepsilon} + (\mathbf{X}_{\mathcal{S}}^+ v_i)^\top \mathbf{A}_{\mathcal{S}} \mathring{\boldsymbol{\beta}}_{\mathcal{S}}| \leq \sigma\sqrt{\|\boldsymbol{\Pi}_{\mathcal{S}}^\perp v_i\|^2 + (\mathbf{X}_{\mathcal{S}}^+ v_i)^\top \mathbf{A}_{\mathcal{S}} \mathbf{X}_{\mathcal{S}}^+ v_i}, \quad \forall i \in \mathcal{S}^c.$$

(b) Show that if $\mathcal{A} \equiv \mathcal{S}$ and $\boldsymbol{y}$ is replaced with $\boldsymbol{Y}$, then the Hessian (6.23) simplifies to:

$$\mathbf{H}_{i,j}(\boldsymbol{\gamma}) = 2(\boldsymbol{e}_i^\top \mathbf{A}_\mathcal{S}\mathring{\boldsymbol{\beta}}_\mathcal{S})(\boldsymbol{e}_i^\top \mathbf{A}_\mathcal{S}\boldsymbol{e}_j)(\boldsymbol{e}_j^\top \mathbf{A}_\mathcal{S}\mathring{\boldsymbol{\beta}}_\mathcal{S}) - \sigma^2(\boldsymbol{e}_i^\top \mathbf{A}_\mathcal{S}\boldsymbol{e}_j)^2, \quad \forall i, j \in \mathcal{S}.$$

Hence, deduce that if $\gamma_i > 0$ for all $i \in \mathcal{S}$:

$$\boldsymbol{\Gamma}_\mathcal{S}^2 \mathbf{H}(\boldsymbol{\gamma})\boldsymbol{\Gamma}_\mathcal{S}^2 = 2\mathbf{P}_\mathcal{S} \odot \left(\boldsymbol{u}_\mathcal{S}\boldsymbol{u}_\mathcal{S}^\top\right) - \sigma^2 \mathbf{P}_\mathcal{S} \odot \mathbf{P}_\mathcal{S}$$

where $\mathbf{P}_\mathcal{S} := (\mathbf{I}_\mathcal{S} + (\boldsymbol{\Gamma}_\mathcal{S}\mathbf{X}_\mathcal{S}^\top\mathbf{X}_\mathcal{S}\boldsymbol{\Gamma}_\mathcal{S})^{-1})^{-1}$ and $u_i := \overline{\beta}_i/\sqrt{\gamma_i}$ for $i \in \mathcal{S}$.

(c) Suppose that $\widehat{\boldsymbol{\gamma}}$ satisfies the klimax-KKT conditions (6.92) and Assumptions 6.57 hold with *minimal signal strength* $1/\alpha_n = o_\mathbb{P}(1)$. Show that with probability approaching unity $\mathbf{H}(\widehat{\boldsymbol{\gamma}}) \succ \mathbf{0}$.

(d) If $\overline{\boldsymbol{\beta}}_\mathcal{S} := \boldsymbol{\Gamma}_\mathcal{S}^2\mathbf{X}_\mathcal{S}^\top\mathbf{G}\boldsymbol{Y}$, prove the identities:

$$\mathbf{A}_\mathcal{S}\mathring{\boldsymbol{\beta}}_\mathcal{S} = \boldsymbol{\Gamma}_\mathcal{S}^{-2}\overline{\boldsymbol{\beta}}_\mathcal{S} \quad \text{and} \quad \mathring{\boldsymbol{\beta}}_\mathcal{S} - \overline{\boldsymbol{\beta}}_\mathcal{S} = \frac{1}{n}\boldsymbol{\Sigma}_n^{-1}\mathbf{A}_\mathcal{S}\mathring{\boldsymbol{\beta}}_\mathcal{S}. \tag{6.93}$$

25. Let $\mathbf{B}_n^{-1} := (\mathring{\boldsymbol{\beta}}_\mathcal{S}\mathring{\boldsymbol{\beta}}_\mathcal{S}^\top) \odot (\mathbf{X}_\mathcal{S}^\top\mathbf{X}_\mathcal{S}/n)$. Using Assumptions 6.57, show that the solution to (6.92), that is: $|\boldsymbol{e}_i^\top\mathbf{A}_\mathcal{S}\mathring{\boldsymbol{\beta}}_\mathcal{S}|^2 = \sigma^2\boldsymbol{e}_i^\top\mathbf{A}_\mathcal{S}\boldsymbol{e}_i, \ \forall i \in \mathcal{S}$ satisfies the asymptotic relationship:

$$\frac{\sigma^2\gamma_i}{\mathring{\beta}_i^2} = 1 + \sigma^2\frac{\boldsymbol{e}_i^\top\mathbf{B}_n\boldsymbol{e}_i - 2\boldsymbol{e}_i^\top\mathbf{B}_n\mathbf{1}}{n} + O_\mathbb{P}(\alpha_n^{-4}), \quad \forall i \in \mathcal{S}.$$

26. Let $\widehat{\boldsymbol{\gamma}}$ be the shrinkage estimator from which the klimax oracle estimator is computed. In this exercise we work through the proof of Theorem 6.60 using Assumptions 6.57.

(a) Use Exercise 25 to prove that the probability of *false exclusions* of features in $\mathcal{S}$ vanishes, that is, $\mathbb{P}[\exists k \in \mathcal{S} : \widehat{\gamma}_k = 0] \downarrow 0$ as $n \uparrow \infty$.

(b) Use Exercise 24 to prove that (with probability approaching unity as $n \uparrow \infty$) $\widehat{\boldsymbol{\gamma}}$ satisfies the klimax-KKT conditions (6.92). Hence, deduce that $\widehat{\boldsymbol{\gamma}}$ is a strict local minimizer of $\text{KLC}(\boldsymbol{\gamma}, 0)$ in (6.7).

(c) Show that $\|\mathring{\boldsymbol{\beta}} - \overline{\boldsymbol{\beta}}\|^2 = O_\mathbb{P}(\alpha_n^{-2}\ln(p)/n)$, thus completing the proof of Theorem 6.60.

27. The purpose of this exercise is to prove Theorem 6.62 using all Assumptions 6.57, except the *klimax irrepresentability*, which is not needed.

(a) Show that if $\mathcal{A} \equiv \mathcal{S} = \{i \geq 1 : \beta_i^* \neq 0\}$ and $\boldsymbol{y}$ is replaced with $\boldsymbol{Y}$ $(= \mathbf{X}\boldsymbol{\beta}^* + \boldsymbol{\varepsilon})$, then the James–Stein KKT conditions (6.37) simplify to:

$$\begin{aligned} |\boldsymbol{e}_i^\top\mathbf{A}_\mathcal{S}\mathring{\boldsymbol{\beta}}_\mathcal{S}| &= \frac{\sigma\|\boldsymbol{v}_i\|}{\sqrt{1 + \|\boldsymbol{v}_i\|^2\gamma_i}}, \quad \forall i \in \mathcal{S}, \\ |\boldsymbol{v}_i^\top\boldsymbol{\Pi}_\mathcal{S}^\perp\boldsymbol{\varepsilon} + (\mathbf{X}_\mathcal{S}^+\boldsymbol{v}_i)^\top\mathbf{A}_\mathcal{S}\mathring{\boldsymbol{\beta}}_\mathcal{S}| &\leq \sigma\|\boldsymbol{v}_i\|, \quad \forall i \in \mathcal{S}^c. \end{aligned} \tag{6.94}$$

(b) If $\mathbf{B}_n^{-1} := (\mathring{\boldsymbol{\beta}}_\mathcal{S}\mathring{\boldsymbol{\beta}}_\mathcal{S}^\top) \odot (\mathbf{X}_\mathcal{S}^\top\mathbf{X}_\mathcal{S}/n)$, show that the solution to (6.94) satisfies:

$$\frac{\sigma^2\gamma_i}{\mathring{\beta}_i^2} = 1 + \frac{\sigma^2}{\|\boldsymbol{v}_i\|^2\mathring{\beta}_i^2} - 2\frac{\sigma^2}{n}\boldsymbol{e}_i^\top\mathbf{B}_n\mathbf{1} + O_\mathbb{P}(\alpha_n^{-4}), \quad \forall i \in \mathcal{S}.$$

Hence, conclude that $\mathbb{P}[\prod_{i\in\mathcal{S}}\gamma_i > 0] \to 1$ as $n \uparrow \infty$.

(c) Prove that a sufficient condition for the KKT inequality in (6.94) to hold is:

$$\sigma \geq \max_i \frac{|\varepsilon^\top \Pi_S^\perp v_i|}{\|v_i\|}(1 + O(\alpha_n^{-1})),$$

which is implied by the *klimax regularization* in Assumptions 6.57.

(d) Let $\widehat{\gamma}$ be the shrinkage parameter of the James–Stein oracle estimator. Mirroring the arguments in Exercise 26, prove that (with probability approaching unity as $n \uparrow \infty$) $\widehat{\gamma}$ satisfies the James–Stein KKT conditions (6.94). Hence, deduce that $\widehat{\gamma}$ is a strict local minimizer of $\text{JSC}(\gamma)$ in (6.36).

(e) Show that $\|\mathring{\beta}-\overline{\beta}\|^2 = O_{\mathbb{P}}(\alpha_n^{-2}\ln(p)/n)$, thus completing the proof of Theorem 6.62.

28. The purpose of this exercise is to work through the proof of Theorem 6.65, jointly using Assumptions 6.57 and 6.63.

(a) Show that if $\mathcal{A} \equiv \mathcal{S} = \{i \geq 1 : \beta_i^* \neq 0\}$ and y is replaced with Y $(= \mathbf{X}\beta^* + \varepsilon)$, then the lasso-klimax KKT conditions (6.46) simplify to:

$$\text{(6.95)}\qquad \begin{aligned} |e_i^\top \mathbf{A}_S \mathring{\beta}_S| &= \sigma\|v_i\|, \quad \forall i \in \mathcal{S},\\ |v_i^\top \Pi_S^\perp \varepsilon + (\mathbf{X}_S^+ v_i)^\top \mathbf{A}_S \mathring{\beta}_S| &\leq \sigma\|v_i\|, \quad \forall i \in \mathcal{S}^c. \end{aligned}$$

(b) Let $u_i := \|v_i\|\text{sign}(\mathring{\beta}_i)/\sqrt{n}$ for all i. Prove that the solution to (6.95) satisfies the asymptotic equation:

$$\frac{\sigma\|v_i\|\gamma_i}{|\mathring{\beta}_i|} = 1 - \frac{\sigma}{\sqrt{n}\mathring{\beta}_i} e_i^\top \Sigma_n^{-1} u_S + O_{\mathbb{P}}(\alpha_n^{-2}), \quad \forall i \in \mathcal{S}.$$

Hence, conclude that $\mathbb{P}[\prod_{i\in\mathcal{S}} \gamma_i > 0] \to 1$ as $n \uparrow \infty$.

(c) Show that the klimax and lasso regularization conditions are asymptotically equivalent in the sense that

$$\frac{\|Y^\top \Pi_S^\perp v_i\|}{\|v_i\|} = O_{\mathbb{P}}\left(\frac{\|Y^\top \Pi_S^\perp v_i\|}{\|\Pi_S^\perp v_i\|}\right) \quad \text{and} \quad \frac{\|Y^\top \Pi_S^\perp v_i\|}{\|\Pi_S^\perp v_i\|} = O_{\mathbb{P}}\left(\frac{\|Y^\top \Pi_S^\perp v_i\|}{\|v_i\|}\right).$$

Hint: It is sufficient to prove that $1 \geq \|\Pi_S^\perp v_i\|^2/\|v_i\|^2 \geq 1-(1-\delta)^2$ for all $i \in \mathcal{S}^c$.

(d) Prove that a sufficient condition for the KKT inequality in (6.95) to hold is the *lasso regularization* condition in Assumptions 6.63.

(e) Let $\widehat{\gamma}$ be the shrinkage parameter of the lasso oracle estimator. Mirroring the arguments in Exercise 26, prove that (with probability approaching unity as $n \uparrow \infty$) $\widehat{\gamma}$ satisfies the lasso-klimax KKT conditions (6.95). Hence, deduce that $\widehat{\gamma}$ is the global minimizer of $\text{LAS}(\gamma)$.

(f) Show that $\|\mathring{\beta} - \overline{\beta}\|^2 = O_{\mathbb{P}}(\ln(p)/n)$, thus completing the proof of Theorem 6.65.

29. Consider the following two conditions:

Gram–Schmidt Bound. As $n \uparrow \infty$ the following condition holds:

$$\max_{k\in\mathcal{S}^c} \frac{\|v_k\|}{\|\Pi_S^\perp v_k\|} = o_{\mathbb{P}}(\alpha_n).$$

Restricted Eigenvalue Bound. There are constants $\xi \geq 1$ and r_ξ such that:

$$\min_{\boldsymbol{v}\in\mathbb{R}^p} \left\{ \frac{\|\mathbf{X}\boldsymbol{v}\|^2}{n\|\boldsymbol{v}\|^2} : \|\boldsymbol{v}_{\mathcal{S}^c}\|_1 \leq \xi \|\boldsymbol{v}_{\mathcal{S}}\|_1 \right\} = r_\xi > 0. \tag{6.96}$$

Using Assumptions 6.57, prove the following.

(a) The *Gram–Schmidt bound* implies the *klimax irrepresentability*.

(b) The *klimax irrepresentability* and the *sparse eigenvalue upper bound* (6.64) jointly imply the *Gram–Schmidt bound*.

(c) The *restricted eigenvalue bound* implies the *sparse eigenvalue lower bound*: $\underline{\lambda} \geq r_\xi$.

(d) The **complete feature normalization**, that is $\|\boldsymbol{v}_i\|^2 = n$ for all i, and the *restricted eigenvalue bound*, used with $\xi \geq \sqrt{(1+\epsilon)|\mathcal{S}|}$ for some $\epsilon > 0$, jointly imply the *Gram–Schmidt bound*.

30. Suppose that $Z \geq 0$ is a positive random variable and $X \sim \chi^2_\nu \equiv \mathsf{Gamma}(\nu/2, 1/2)$.

(a) Prove the **Markov inequality**: $\mathbb{P}[Z \geq z] \leq \frac{\mathbb{E}Z}{z}$, $z > 0$.

(b) By applying the Markov inequality to

$$\mathbb{P}[\exp(t(X-\nu)) \geq \exp(tz)] \quad \text{and} \quad \mathbb{P}[\exp(t(\nu - X)) \geq \exp(tz)]$$

for suitable values of t and z, prove that

$$\mathbb{P}[X - \nu \geq 2\sqrt{\nu x}] \leq \exp(-x) \quad \text{and} \quad \mathbb{P}[\nu - X \geq 2\sqrt{\nu x}] \leq \exp(-x).$$

Hence, deduce the probability bound:

$$\mathbb{P}[|X-\nu| \geq 2\sqrt{\nu x}] \leq 2\exp(-x).$$

(c) If $\boldsymbol{Z}_1, \dots, \boldsymbol{Z}_p$ are iid draws from $\mathcal{N}(\mathbf{0}, \mathbf{I}_n)$, show that

$$\mathbb{P}\left[\min_{i\leq p} \|\boldsymbol{Z}_i\|^2/n \leq 1 + 2\sqrt{x/n}\right] \geq 1 - \exp(-px).$$

31. The purpose of this exercise is to prove Lemma 6.61 using Assumptions 6.57.

(a) Use Lemma 6.38 and the identities (6.81) and (6.93) to show that the klimax oracle estimator $\overline{\boldsymbol{\beta}}$ satisfies:

$$\frac{\|\boldsymbol{Y} - \mathbf{X}\overline{\boldsymbol{\beta}}\|^2 - \|\boldsymbol{\Pi}_{\mathcal{S}}^{\perp}\boldsymbol{\varepsilon}\|^2}{n - |\mathcal{S}|} = \mathcal{O}_{\mathbb{P}}(\alpha_n^{-2}\sigma^2/n).$$

(b) Use the results of Exercise 30 to show that

$$\frac{\|\boldsymbol{\Pi}_{\mathcal{S}}^{\perp}\boldsymbol{\varepsilon}\|^2}{n - |\mathcal{S}|} - \varsigma^2 = \mathcal{O}_{\mathbb{P}}(1/\sqrt{n}).$$

Hence, deduce the claim of Lemma 6.61.

(c) If additionally we use Assumptions 6.63, find a consistent estimator of ς^2 that uses the lasso oracle estimator.

32. Using Assumptions 6.57, show that the *minimal signal strength* can be deduced from the condition:

$$\frac{\sigma/\sqrt{n}}{\min_{i\in\mathcal{S}}|\beta_i^*|} = o_{\mathbb{P}}(1), \quad n \uparrow \infty.$$

33. Suppose that for each $i = 1, \ldots, k$ the function $h_i : \mathbb{R}^{k+1} \to \mathbb{R}$ is 1-th order continuously differentiable in the neighborhood of some (z, λ), where $z \in \mathbb{R}^k$. The **implicit function theorem** says that, if $\boldsymbol{h}(z, \lambda) = \mathbf{0}$, where $\boldsymbol{h} = [h_1, \ldots, h_k]^\top$, then

$$\frac{\partial \boldsymbol{h}}{\partial z^\top}\frac{\partial z}{\partial \lambda} + \frac{\partial \boldsymbol{h}}{\partial \lambda} = \mathbf{0}.$$

In the following exercises, we use the implicit function theorem to show that the klimax and lasso penalties are a decreasing function of the regularization parameter σ.

(a) If $\boldsymbol{\gamma}$ is a (local) minimizer of $\text{KLC}(\boldsymbol{\gamma}) = \text{KLC}(\boldsymbol{\gamma}, 0)$ in (6.7), prove that

$$\frac{\mathrm{d}}{\mathrm{d}\sigma^2} \ln|\mathbf{I}_n + \mathbf{X}\boldsymbol{\Gamma}^2\mathbf{X}^\top| < 0.$$

In other words, the log-determinant penalty in (6.7) is a decreasing function of the regularization parameter σ.

(b) If $\boldsymbol{\gamma}$ is the unique solution to the klimax-lasso conditions (6.46), show that the corresponding lasso estimate $\overline{\boldsymbol{\beta}} = \boldsymbol{\Gamma}^2\mathbf{X}^\top\mathbf{G}\boldsymbol{y}$ satisfies:

$$\frac{\mathrm{d}}{\mathrm{d}\sigma}\left[\sum_{i=1}^{p} \|\boldsymbol{v}_i\|\,|\overline{\beta}_i|/\sigma\right] < 0, \qquad \sigma > 0.$$

In other words, the lasso penalty $\rho^{\text{LAS}}(\mathbf{D}\overline{\boldsymbol{\beta}}/\sigma) = \|\mathbf{D}\overline{\boldsymbol{\beta}}/\sigma\|_1$ is a decreasing function of the regularization parameter σ.

34. Given a regularization parameter λ and weights $\omega_1, \ldots \omega_p$, the **elastic net** estimate $\widehat{\boldsymbol{\beta}}_{\text{NET}}$ is defined as the minimizer of the penalized least-squares objective (6.55).

(a) Define the diagonal matrix:

$$\mathbf{D} := \sigma(1-\alpha)\text{diag}([\omega_1\|\boldsymbol{v}_1\|, \ldots, \omega_p\|\boldsymbol{v}_p\|]),$$

the weights $\widetilde{\omega}_i := \alpha\omega_i\|\boldsymbol{v}_i\|$, and the augmented training set:

$$\widetilde{\boldsymbol{y}} := \begin{bmatrix}\boldsymbol{y}\\ \mathbf{0}\end{bmatrix}, \quad \widetilde{\mathbf{X}} := \begin{bmatrix}\mathbf{X}\\ \mathbf{D}^{1/2}\end{bmatrix}.$$

Show that $\widehat{\boldsymbol{\beta}}_{\text{NET}}$ is also the minimizer of the lasso penalized least-squares objective:

$$\frac{\|\widetilde{\boldsymbol{y}} - \widetilde{\mathbf{X}}\boldsymbol{\beta}\|^2}{2} + \sigma\sum_i \widetilde{\omega}_i|\beta_i|.$$

(b) Let $\boldsymbol{r} := \boldsymbol{y} - \mathbf{X}\boldsymbol{\beta}$ and $\xi_i := \sigma\omega_i/\|\boldsymbol{v}_i\|$ for all i. Use the results in Exercise 20 to show that $\widehat{\boldsymbol{\beta}}_{\text{NET}}$ satisfies the **net-KKT condition**:

$$\beta_i = \frac{\mathcal{S}^{\text{LAS}}_{\alpha\xi_i}\left(\beta_i + \langle \boldsymbol{r}, \boldsymbol{v}_i\rangle/\|\boldsymbol{v}_i\|^2\right)}{1 + (1-\alpha)\xi_i}, \qquad i = 1, \ldots, p.$$

(c) Let $\widetilde{\mathbf{G}} := \left(\mathbf{I}_{n+p} + \sum_{i=1}^{p} \gamma_i \widetilde{\boldsymbol{v}}_i \widetilde{\boldsymbol{v}}_i^\top\right)^{-1}$ and $\mathbf{\Gamma}^2 := \widetilde{\mathbf{\Gamma}}(\mathbf{I}_p + \widetilde{\mathbf{\Gamma}}\mathbf{D}\widetilde{\mathbf{\Gamma}})^{-1}\widetilde{\mathbf{\Gamma}}$. Show that

$$\begin{aligned}\widetilde{\boldsymbol{v}}_i^\top \widetilde{\mathbf{G}}_{\neg i}\widetilde{\boldsymbol{v}}_i &= \mathbf{D}_{i,i} + \boldsymbol{v}_i^\top \mathbf{G}_{\neg i}\boldsymbol{v}_i\\ \widetilde{\boldsymbol{v}}_i^\top \widetilde{\mathbf{G}}_{\neg i}\widetilde{\boldsymbol{y}} &= \boldsymbol{v}_i^\top \mathbf{G}_{\neg i}\boldsymbol{y}.\end{aligned}$$

(d) Using (6.50), show that $\widehat{\boldsymbol{\beta}}_{\text{NET}} = \mathbf{\Gamma}^2\mathbf{X}^\top\mathbf{G}\boldsymbol{y}$ satisfies the **net-klimax condition**:

$$\beta_i = \frac{\mathcal{S}^{\text{LAS}}_{\frac{\alpha\xi_i\|\boldsymbol{v}_i\|^2}{\boldsymbol{v}_i^\top\mathbf{G}_{\neg i}\boldsymbol{v}_i}}\left(\frac{\langle \boldsymbol{v}_i, \boldsymbol{y}\rangle_{\mathbf{G}}}{\|\boldsymbol{v}_i\|^2_{\mathbf{G}}}\right)}{1 + (1-\alpha)\frac{\xi_i\|\boldsymbol{v}_i\|^2}{\boldsymbol{v}_i^\top\mathbf{G}_{\neg i}\boldsymbol{v}_i}}, \qquad \gamma_i = \frac{|\beta_i|/(\xi_i\|\boldsymbol{v}_i\|^2)}{\alpha + (1-\alpha)|\beta_i|}, \qquad i = 1, \ldots, p.$$

35. Let $\mathbf{G} \succ \mathbf{0}$ be an $n \times n$ covariance matrix. For $\delta > 0$ and any $\boldsymbol{v}, \boldsymbol{y} \in \mathbb{R}^n$, define:

$$\begin{aligned}\mathbf{G}_\delta &:= \left(\mathbf{G}^{-1} + \delta \boldsymbol{v}\boldsymbol{v}^\top\right)^{-1},\\ \alpha &:= \frac{\left(\langle \boldsymbol{v}, \boldsymbol{y}\rangle_{\mathbf{G}} - \langle \boldsymbol{v}, \boldsymbol{y}\rangle_{\mathbf{G}_\delta}\right)^2}{\delta^2 \langle \boldsymbol{v}, \boldsymbol{y}\rangle_{\mathbf{G}}\langle \boldsymbol{v}, \boldsymbol{y}\rangle_{\mathbf{G}_\delta}}.\end{aligned}$$

Prove the following two identities:

$$\begin{aligned}\|\boldsymbol{v}\|^2_{\mathbf{G}} &= \frac{\alpha\delta + \sqrt{(\alpha\delta)^2 + 4\alpha}}{2},\\ \frac{\boldsymbol{y}^\top(\mathbf{G} - \mathbf{G}_\delta)\boldsymbol{y}}{\delta} &= \langle \boldsymbol{v}, \boldsymbol{y}\rangle_{\mathbf{G}}\langle \boldsymbol{v}, \boldsymbol{y}\rangle_{\mathbf{G}_\delta}.\end{aligned}$$

Since in the klimax Algorithm 6.28, each coordinate-descent update i requires computation of the two quantities $\langle \boldsymbol{v}_i, \boldsymbol{y}\rangle_{\mathbf{G}}$ and $\|\boldsymbol{v}_i\|^2_{\mathbf{G}}$, the above formulas show how these two quantities can be computed from two linear solves involving $\mathbf{G}$ and its perturbation along the directions of the model vector $\boldsymbol{v}_i$.

36. (Exercise 35 continued) The pseudo-code in Algorithm 6.97 below describes a coordinate-descent method for computing the klimax estimate using two applications of the conjugate gradient Algorithm A.103 within each coordinate update. Explain how Line 12 of Algorithm 6.97 must be modified in order to repurpose it for the computation of the lasso estimate. ☞ 560

Note that the multiplication of matrix $\mathbf{G}^{-1}$ with a vector $\boldsymbol{z} \in \mathbb{R}^n$ is accomplished via the formula

$$\mathbf{G}^{-1}\boldsymbol{z} = \boldsymbol{z} + \sum_{j \in \mathcal{A}} \gamma_j \langle \boldsymbol{v}_j, \boldsymbol{z}\rangle\, \boldsymbol{v}_j,$$

where $\mathcal{A} = \{i : \gamma_i > 0\}$ is the activated set. As a result, the cost of matrix vector multiplication in the conjugate gradient Algorithm A.103 is at most $O(n|\mathcal{A}|^2)$.

Algorithm 6.97: Klimax Estimate for a Given σ via Algorithm A.103

input: Data $\{\mathbf{X}, \boldsymbol{y}\}$, regularization parameter σ, and error tolerance ε.
output: Minimizer $\boldsymbol{\beta}$ and updated $\boldsymbol{\gamma}, \mathcal{A}, \mathsf{Gy}$.

$\boldsymbol{\beta} \leftarrow \mathbf{0},\ \boldsymbol{\gamma} \leftarrow \mathbf{0},\ \mathcal{A} \leftarrow \emptyset,\ \mathsf{Gy} \leftarrow \boldsymbol{y},\ \mathbf{X}_{\mathcal{A}} \leftarrow \emptyset$ // initialize with null solution
repeat // iterate over coordinate-descent cycles
  $\boldsymbol{\beta}_{\text{old}} \leftarrow \boldsymbol{\beta}$
  for $k = 1, \ldots, p$ **do** // coordinate-descent cycle
    $\delta \leftarrow 1/\|\boldsymbol{v}_k\|$ // perturbation for computing $\mathbf{G}_\delta \boldsymbol{y} = (\mathbf{G}^{-1} + \delta \boldsymbol{v}_k \boldsymbol{v}_k^\top)^{-1}\boldsymbol{y}$
    by initializing Algorithm A.103 from Gy, compute:

$$\mathsf{Gy}_\delta \leftarrow \left(\mathbf{I}_n + \mathbf{X}_{\mathcal{A}}\boldsymbol{\Gamma}_{\mathcal{A}}\mathbf{X}_{\mathcal{A}}^\top + \delta \boldsymbol{v}_k \boldsymbol{v}_k^\top\right)^{-1}\boldsymbol{y}$$

    $\mathsf{vGy} \leftarrow \boldsymbol{v}_k^\top \mathsf{Gy}$ // compute $\boldsymbol{v}_k^\top \mathbf{G}\boldsymbol{y}$
    $\mathsf{vGy}_\delta \leftarrow \boldsymbol{v}_k^\top \mathsf{Gy}_\delta$ // compute perturbation $\boldsymbol{v}_k^\top \mathbf{G}_\delta \boldsymbol{y}$
    $\alpha \leftarrow (\mathsf{vGy}_\delta - \mathsf{vGy})^2/(\delta^2\, \mathsf{vGy}_\delta\, \mathsf{vGy})$
    $\mathsf{vGv} \leftarrow \left(\alpha\delta + \sqrt{(\alpha\delta)^2 + 4\alpha}\right)/2$ // symmetric formula for $\boldsymbol{v}_k^\top \mathbf{G}\boldsymbol{v}_k$
    $\mathsf{vGy} \leftarrow \mathsf{vGy}/(1 - \gamma_k \mathsf{vGv})$ // compute $\boldsymbol{v}_k^\top \mathbf{G}_{/k}\boldsymbol{y}$
    $\mathsf{vGv} \leftarrow \mathsf{vGv}/(1 - \gamma_k \mathsf{vGv})$ // compute $\boldsymbol{v}_k^\top \mathbf{G}_{/k}\boldsymbol{v}_k$
    $\beta_k \leftarrow S^{\text{JS}}_{\frac{\sigma}{\sqrt{\mathsf{vGv}}}}\left(\frac{\mathsf{vGy}}{\mathsf{vGv}}\right),\ \widetilde{\gamma} \leftarrow \frac{\beta_k}{\sigma^2}\frac{\mathsf{vGy}}{\mathsf{vGv}}$ // update via (6.26)
    if $\widetilde{\gamma} \neq \gamma_k$ **then** // change in $\gamma_k \Rightarrow$ update $\mathcal{A}, \mathsf{Gy}$
      if $\gamma_k = 0$ **then** // upsize $\mathcal{A}, \mathbf{X}_{\mathcal{A}}$
        $\mathcal{A} \leftarrow \mathcal{A} \cup k,\ \mathbf{X}_{\mathcal{A}} \leftarrow [\mathbf{X}_{\mathcal{A}}, \boldsymbol{v}_k]$
      else if $\widetilde{\gamma} = 0$ **then** // downsize $\mathcal{A}, \mathbf{X}_{\mathcal{A}}$
        remove k and $\boldsymbol{v}_k$ from $\mathcal{A}$ and $\mathbf{X}_{\mathcal{A}}$, respectively
      $\gamma_k \leftarrow \widetilde{\gamma}$ // update k-th coordinate of γ
      by initializing Algorithm A.103 from Gy, compute:

$$\mathsf{Gy} \leftarrow (\mathbf{I}_n + \mathbf{X}_{\mathcal{A}}\boldsymbol{\Gamma}_{\mathcal{A}}\mathbf{X}_{\mathcal{A}}^\top)^{-1}\boldsymbol{y}$$

until $\|\boldsymbol{\beta} - \boldsymbol{\beta}_{\text{old}}\| < \varepsilon$
return $\boldsymbol{\beta}, \boldsymbol{\gamma}, \mathcal{A}, \mathsf{Gy}$

CHAPTER 7

Reproducing Kernel Methods

The purpose of this chapter is to familiarize the reader with reproducing kernel methods — a framework that offers a broad generalization of linear models. We start by explaining the essential concepts through the special example of ridge regression. We then introduce the more general setting of reproducing kernel Hilbert spaces and show that selecting the best prediction function in such spaces is in fact a finite-dimensional optimization problem. Applications to spline fitting, Gaussian process regression, and kernel PCA are given.

7.1 Introduction

In this chapter we return to the supervised learning setting in Chapter 5 (see also Chapters 2 and 6) and expand its scope. Given training data $\tau = \{(\boldsymbol{x}_1, y_1), \ldots, (\boldsymbol{x}_n, y_n)\}$, we wish to find a prediction function (the learner) g_τ that minimizes the (squared-error) training loss

$$\ell_\tau(g) = \frac{1}{n}\sum_{i=1}^{n}(y_i - g(\boldsymbol{x}_i))^2$$

within a class of functions $\mathcal{G}$. As noted in Chapter 2, if $\mathcal{G}$ is the set of all possible functions, then choosing *any* function g with the property that $g(\boldsymbol{x}_i) = y_i$ for all i will give zero training loss, but will likely have poor generalization performance (that is, suffer from overfitting).

Recall from Theorem 2.2 that the best possible prediction function (over all g) for the squared-error *risk* $\mathbb{E}(Y - g(\boldsymbol{X}))^2$ is given by $g^*(\boldsymbol{x}) = \mathbb{E}[Y \mid \boldsymbol{X} = \boldsymbol{x}]$. The class $\mathcal{G}$ should be simple enough to permit theoretical understanding and analysis but, at the same time, rich enough to contain the optimal function g^* (or a function close to g^*). This ideal can be realized by taking $\mathcal{G}$ to be a **Hilbert space** (i.e., a complete inner product space) of functions; see Appendix B.1. ☞ 21 ☞ 563

Many of the classes of functions that we have encountered so far are in fact Hilbert spaces. In particular, the set $\mathcal{G}$ of *linear* functions on $\mathbb{R}^p$ is a Hilbert space. To see this, identify with each element $\boldsymbol{\beta} \in \mathbb{R}^p$ the linear function $g_{\boldsymbol{\beta}} : \boldsymbol{x} \mapsto \boldsymbol{x}^\top\boldsymbol{\beta}$ and define the inner product on $\mathcal{G}$ as $\langle g_{\boldsymbol{\beta}}, g_{\boldsymbol{\gamma}} \rangle := \boldsymbol{\beta}^\top\boldsymbol{\gamma}$. In this way, $\mathcal{G}$ behaves in exactly the same way as (is isomorphic to) the space $\mathbb{R}^p$ equipped with the Euclidean inner product. The ☞ 517

DOI: 10.1201/9781003391173-7

latter is a Hilbert space, because it is *complete* with respect to the Euclidean norm. See Exercise 12 for a further discussion.

☞ 26 Let us now turn to our "running" polynomial regression Example 2.13, where the feature vector $\boldsymbol{x} = [1, u, u^2, \ldots, u^{p-1}]^\top =: \boldsymbol{\phi}(u)$ is itself a vector-valued function of another feature u. Then, the space of functions $h_{\boldsymbol{\beta}} : u \mapsto \boldsymbol{\phi}(u)^\top \boldsymbol{\beta}$ is a Hilbert space, through the identification $h_{\boldsymbol{\beta}} \equiv \boldsymbol{\beta}$. In fact, this is true for *any* feature mapping $\boldsymbol{\phi} : u \mapsto [\phi_1(u), \ldots, \phi_p(u)]^\top$.

This can be further generalized by considering **feature maps** $u \mapsto \kappa_u$, where each κ_u is a real-valued *function* $v \mapsto \kappa_u(v)$ on the feature space. As we shall soon see (in Section 7.3), functions of the form $u \mapsto \sum_{i=1}^{\infty} \beta_i \kappa_{v_i}(u)$ live in a Hilbert space of functions called a *reproducing kernel Hilbert space* (RKHS). We formally introduce the notion of an RKHS in Section 7.3, and in the subsequent section provide examples of the construction of an RKHS from linear and Gaussian kernels. The workhorse of any RKHS is the Representer Theorem 7.35, which is derived in Section 7.5. Applications of
☞ 337 the Representer Theorem include the *smoothing splines* (Section 7.6), Gaussian process regression (Section 7.7), kernel PCA (Section 7.8), and *support vector machines* for
☞ 373 classification (Section 8.7).

The RKHS formalism is intimately connected with the concept of **regularization**. This refers to a broad class of statistical techniques that aim to prevent overfitting when building predictive models. Regularization is often implemented by adding a tunable penalty term to the training loss, but there are various other ways to regularize the learning process. For example, the *early stopping* heuristic in gradient descent may be viewed as a regularization technique; see Exercise 18. One may even claim that Bayesian statistics
☞ 51 has an in-built regularization; see Remark 2.71. In the klimax method in Chapter 6 the strength of the regularization is controlled through the parameter σ, whose selection generates a spectrum of potential models. In the next section we introduce the main ideas behind L^2-regularization, which then segues into a discussion of kernel methods in the subsequent sections.

7.2 L^2-Regularization

Let $\mathcal{G}$ be the Hilbert space of functions over which we search for the minimizer, g_τ, of the training loss $\ell_\tau(g)$. Often, the Hilbert space $\mathcal{G}$ is rich enough so that we can find a learner g_τ within $\mathcal{G}$ such that the training loss is zero or close to zero. Consequently, if the space of functions $\mathcal{G}$ is sufficiently rich, we run the risk of overfitting. One way to avoid overfitting is to restrict attention to a subset of the space $\mathcal{G}$ by introducing a non-negative functional $J : \mathcal{G} \to \mathbb{R}_+$ which penalizes complex models (functions). In particular, we want to find functions $g \in \mathcal{G}$ such that $J(g) < c$ for some "regularization" constant $c > 0$. Thus, we can formulate the quintessential supervised learning problem as:

$$\min \{\ell_\tau(g) \; : \; g \in \mathcal{G}, J(g) < c\}, \tag{7.1}$$

the solution (argmin) of which is our learner. When this optimization problem is convex, it can be solved by first obtaining the Lagrangian dual function

$$\mathcal{L}^*(\alpha) := \min_{g \in \mathcal{G}} \{\ell_\tau(g) + \alpha(J(g) - c)\},$$

and then maximizing $\mathcal{L}^*(\alpha)$ with respect to $\alpha \geq 0$; see Section C.2.3. ☞ 597

In order to introduce the overall ideas of kernel methods, we will proceed by exploring (7.1) in the special case of **ridge regression**. We already encountered in (6.17) the ridge regression estimator

(7.2)
$$\widehat{\boldsymbol{\beta}}_\lambda = (\lambda \mathbf{I}_p + \mathbf{X}^\top \mathbf{X})^{-1} \mathbf{X}^\top \boldsymbol{y}$$

as an alternative to the least-squares estimator $\widehat{\boldsymbol{\beta}} = \lim_{\lambda \downarrow 0} \widehat{\boldsymbol{\beta}}_\lambda = \mathbf{X}^+ \boldsymbol{y}$; see identity (2.95). ☞ 62

Example 7.3 (Ridge Regression) Ridge regression is simply linear regression with an L^2 (or squared-norm) penalty functional. Suppose we have a training set $\tau = \{(\boldsymbol{x}_i, y_i), i = 1, \ldots, n\}$, with each $\boldsymbol{x}_i \in \mathbb{R}^p$ and we use a squared-norm penalty with regularization parameter $\lambda > 0$. Then, the problem is to solve

(7.4)
$$\min_{g \in \mathcal{G}} \sum_{i=1}^n (y_i - g(\boldsymbol{x}_i))^2 + \lambda \, \|g\|^2,$$

where $\mathcal{G}$ is the Hilbert space of linear functions on $\mathbb{R}^p$. As explained in Section 7.1, we can identify each $g \in \mathcal{G}$ with a vector $\boldsymbol{\beta} \in \mathbb{R}^p$ and, consequently, $\|g\|^2 = \langle \boldsymbol{\beta}, \boldsymbol{\beta} \rangle = \|\boldsymbol{\beta}\|^2$. The above *functional* optimization problem is thus equivalent to the *parametric* optimization problem

(7.5)
$$\min_{\boldsymbol{\beta} \in \mathbb{R}^p} \sum_{i=1}^n \left(y_i - \boldsymbol{x}_i^\top \boldsymbol{\beta}\right)^2 + \lambda \, \|\boldsymbol{\beta}\|^2,$$

which, in the notation of Chapter 5, further simplifies to

(7.6)
$$\min_{\boldsymbol{\beta} \in \mathbb{R}^p} \| \boldsymbol{y} - \mathbf{X}\boldsymbol{\beta} \|^2 + \lambda \, \|\boldsymbol{\beta}\|^2.$$

In other words, the solution to (7.4) is of the form $\boldsymbol{x} \mapsto \boldsymbol{x}^\top \boldsymbol{\beta}^*$, where $\boldsymbol{\beta}^*$ solves (7.5) (or equivalently (7.6)). Observe that as $\lambda \to \infty$, the regularization term becomes dominant and consequently the optimal g becomes identically zero.

The optimization problem in (7.6) is convex, and by setting the gradient equal to zero, we obtain

(7.7)
$$\mathbf{X}^\top (\mathbf{X}\boldsymbol{\beta} - \boldsymbol{y}) + \lambda \boldsymbol{\beta} = \mathbf{0}.$$

If $\lambda = 0$ these are simply the *normal equations*, albeit written in a slightly different form. ☞ 28
If the matrix $\mathbf{X}^\top \mathbf{X} + \lambda \mathbf{I}_p$ is invertible (which is the case for any $\lambda > 0$; see Exercise 13), then the solution to these modified normal equations is given by (7.2). ∎

When using regularization with respect to some Hilbert space $\mathcal{G}$, it is sometimes useful to decompose $\mathcal{G}$ into two orthogonal subspaces, $\mathcal{H}$ and C say, such that every $g \in \mathcal{G}$ can be uniquely written as $g = h + c$, with $h \in \mathcal{H}$, $c \in C$, and $\langle h, c \rangle = 0$. Such a $\mathcal{G}$ is said to be the **direct sum** of C and $\mathcal{H}$, and we write $\mathcal{G} = \mathcal{H} \oplus C$. Decompositions of this form become useful when functions in $\mathcal{H}$ are penalized but functions in C are not. We illustrate this decomposition with the ridge regression example where one of the features is a constant term, which we do not wish to penalize.

■ **Example 7.8 (Ridge Regression (cont.))** Suppose one of the features in Example 7.3 is the constant 1, which we do not wish to penalize. The reason for this is to ensure that when $\lambda \to \infty$, the optimal g becomes the "constant" model, $g(\boldsymbol{x}) = \beta_0$, rather than the "zero" model, $g(\boldsymbol{x}) = 0$. Let us alter the notation slightly by considering the feature vectors to be of the form $\widetilde{\boldsymbol{x}} = [1, \boldsymbol{x}^\top]^\top$, where $\boldsymbol{x} = [x_1, \dots, x_p]^\top$. We thus have $p + 1$ features, rather than p. Let $\mathcal{G}$ be the space of linear functions of $\widetilde{\boldsymbol{x}}$. Each linear function g of $\widetilde{\boldsymbol{x}}$ can be written as $g : \widetilde{\boldsymbol{x}} \mapsto \beta_0 + \boldsymbol{x}^\top \boldsymbol{\beta}$, which is the sum of the constant function $c : \widetilde{\boldsymbol{x}} \mapsto \beta_0$ and $h : \widetilde{\boldsymbol{x}} \mapsto \boldsymbol{x}^\top \boldsymbol{\beta}$. Moreover, the two functions are orthogonal with respect to the inner product on $\mathcal{G}$: $\langle c, h \rangle = [\beta_0, \mathbf{0}^\top][0, \boldsymbol{\beta}^\top]^\top = 0$, where $\mathbf{0}$ is a column vector of zeros.

As subspaces of $\mathcal{G}$, both C and $\mathcal{H}$ are again Hilbert spaces, and their inner products and norms follow directly from the inner product on $\mathcal{G}$. For example, each function $h : \widetilde{\boldsymbol{x}} \mapsto \boldsymbol{x}^\top \boldsymbol{\beta}$ in $\mathcal{H}$ has norm $\|h\|_{\mathcal{H}} = \|\boldsymbol{\beta}\|$, and the constant function $c : \widetilde{\boldsymbol{x}} \mapsto \beta_0$ in C has norm $|\beta_0|$.

The modification of the regularized optimization problem (7.4) where the constant term is not penalized can now be written as

$$\min_{g \in \mathcal{H} \oplus C} \sum_{i=1}^{n} (y_i - g(\widetilde{\boldsymbol{x}}_i))^2 + \lambda \, \|g\|_{\mathcal{H}}^2, \tag{7.9}$$

which further simplifies to

$$\min_{\beta_0, \boldsymbol{\beta}} \| \boldsymbol{y} - \beta_0 \mathbf{1} - \mathbf{X}\boldsymbol{\beta} \|^2 + \lambda \, \|\boldsymbol{\beta}\|^2,$$

where $\mathbf{1}$ is the $n \times 1$ vector of 1s. Observe that, in this case, as $\lambda \to \infty$ the optimal g tends to the sample mean $\overline{y}$ of the $\{y_i\}$; that is, we obtain the "default" regression model, without explanatory variables. Again, this is a convex optimization problem, and the solution follows from

$$\mathbf{X}^\top (\beta_0 \mathbf{1} + \mathbf{X}\boldsymbol{\beta} - \boldsymbol{y}) + \lambda \, \boldsymbol{\beta} = \mathbf{0}, \tag{7.10}$$

with

$$\beta_0 = \mathbf{1}^+ (\boldsymbol{y} - \mathbf{X}\boldsymbol{\beta}) = \mathbf{1}^\top (\boldsymbol{y} - \mathbf{X}\boldsymbol{\beta})/n. \tag{7.11}$$

This results in solving for $\boldsymbol{\beta}$ from

$$(\mathbf{X}^\top \mathbf{X} - n^{-1} \mathbf{X}^\top \mathbf{1}\mathbf{1}^\top \mathbf{X} + \lambda \, \mathbf{I}_p) \boldsymbol{\beta} = (\mathbf{X}^\top - n^{-1} \mathbf{X}^\top \mathbf{1}\mathbf{1}^\top) \boldsymbol{y}, \tag{7.12}$$

and determining β_0 from (7.11).

As a precursor to the kernel methods in the following sections, let us assume that $n \geq p$ and that $\mathbf{X}$ has full (column) rank p. Then any vector $\boldsymbol{\beta} \in \mathbb{R}^p$ can be written as a linear combination of the feature vectors $\{\boldsymbol{x}_i\}$; that is, as linear combinations of the columns of the matrix $\mathbf{X}^\top$. In particular, let $\boldsymbol{\beta} = \mathbf{X}^\top \boldsymbol{\alpha}$, where $\boldsymbol{\alpha} = [\alpha_1, \dots, \alpha_n]^\top \in \mathbb{R}^n$. In this case (7.12) reduces to

$$(\mathbf{X}\mathbf{X}^\top - n^{-1} \mathbf{1}\mathbf{1}^\top \mathbf{X}\mathbf{X}^\top + \lambda \, \mathbf{I}_n) \boldsymbol{\alpha} = (\mathbf{I}_n - n^{-1} \mathbf{1}\mathbf{1}^\top) \boldsymbol{y}.$$

Assuming invertibility of $(\mathbf{X}\mathbf{X}^\top - n^{-1}\mathbf{1}\mathbf{1}^\top\mathbf{X}\mathbf{X}^\top + \lambda\,\mathbf{I}_n)$, we have the solution

$$\widehat{\alpha} = (\mathbf{X}\mathbf{X}^\top - n^{-1}\mathbf{1}\mathbf{1}^\top\mathbf{X}\mathbf{X}^\top + \lambda\,\mathbf{I}_n)^{-1}(\mathbf{I}_n - n^{-1}\mathbf{1}\mathbf{1}^\top)\boldsymbol{y},$$

which depends on the training feature vectors $\{\boldsymbol{x}_i\}$ only through the $n \times n$ matrix of inner products: $\mathbf{X}\mathbf{X}^\top = [\langle \boldsymbol{x}_i, \boldsymbol{x}_j \rangle]$. This matrix is called the **Gram matrix** of the $\{\boldsymbol{x}_i\}$. From (7.11), the solution for the constant term is $\widehat{\beta}_0 = n^{-1}\mathbf{1}^\top(\boldsymbol{y} - \mathbf{X}\mathbf{X}^\top\widehat{\alpha})$. It follows that the learner is a linear combination of inner products $\{\langle \boldsymbol{x}_i, \boldsymbol{x}\rangle\}$ plus a constant:

$$g_\tau(\widetilde{\boldsymbol{x}}) = \widehat{\beta}_0 + \boldsymbol{x}^\top\mathbf{X}^\top\widehat{\alpha} = \widehat{\beta}_0 + \sum_{i=1}^n \widehat{\alpha}_i \,\langle \boldsymbol{x}_i, \boldsymbol{x}\rangle,$$

where the coefficients $\widehat{\beta}_0$ and $\widehat{\alpha}_i$ only depend on the inner products $\{\langle \boldsymbol{x}_i, \boldsymbol{x}_j\rangle\}$. We will see shortly that the representer Theorem 7.35 generalizes this result to a broad class of regularized optimization problems. ☞ 333 ■

7.3 Reproducing Kernel Hilbert Spaces

In this section, we formalize the idea outlined at the end of Section 7.1 of extending finite-dimensional feature maps to those that are *functions* by introducing a special type of Hilbert space of functions known as a *reproducing kernel Hilbert space* (RKHS). Although the theory extends naturally to Hilbert spaces of complex-valued functions, we restrict attention to Hilbert spaces of real-valued functions here.

To evaluate the loss of a learner g in some class of functions $\mathcal{G}$, we do not need to explicitly construct g — rather, it is only required that we can evaluate g at all the feature vectors $\boldsymbol{x}_1, \ldots, \boldsymbol{x}_n$ of the training set. A defining property of an RKHS is that function evaluation at a point $\boldsymbol{x}$ can be performed by simply taking the inner product of g with some feature function $\kappa_{\boldsymbol{x}}$ associated with $\boldsymbol{x}$. We will see that this property becomes particularly useful in light of the representer theorem (see Section 7.5), which states that the learner g itself can be represented as a linear combination of the set of feature functions $\{\kappa_{\boldsymbol{x}_i}, i = 1, \ldots, n\}$. Consequently, we can evaluate a learner g at the feature vectors $\{\boldsymbol{x}_i\}$ by taking linear combinations of terms of the form $\kappa(\boldsymbol{x}_i, \boldsymbol{x}_j) = \langle \kappa_{\boldsymbol{x}_i}, \kappa_{\boldsymbol{x}_j}\rangle_{\mathcal{G}}$. Collecting these inner products into a matrix $\mathbf{K} = [\kappa(\boldsymbol{x}_i, \boldsymbol{x}_j), i, j = 1, \ldots, n]$ (the Gram matrix of the $\{\kappa_{\boldsymbol{x}_i}\}$), we will see that the feature vectors $\{\boldsymbol{x}_i\}$ only enter the loss minimization problem through $\mathbf{K}$. ☞ 332

Definition 7.13 (Reproducing Kernel Hilbert Space): For a non-empty set $\mathcal{X}$, a Hilbert space $\mathcal{G}$ of functions $g : \mathcal{X} \to \mathbb{R}$ with inner product $\langle \cdot, \cdot \rangle_{\mathcal{G}}$ is called a **reproducing kernel Hilbert space** (RKHS) with **reproducing kernel** $\kappa : \mathcal{X} \times \mathcal{X} \to \mathbb{R}$ if:

1. for every $\boldsymbol{x} \in \mathcal{X}$, $\kappa_{\boldsymbol{x}} := \kappa(\boldsymbol{x}, \cdot)$ is in $\mathcal{G}$,
2. $\kappa(\boldsymbol{x}, \boldsymbol{x}) < \infty$ for all $\boldsymbol{x} \in \mathcal{X}$,
3. for every $\boldsymbol{x} \in \mathcal{X}$ and $g \in \mathcal{G}$, $g(\boldsymbol{x}) = \langle g, \kappa_{\boldsymbol{x}} \rangle_{\mathcal{G}}$.

The reproducing kernel of a Hilbert space of functions, if it exists, is unique; see Exercise 2. The main (third) condition in Definition 7.13 is known as the **reproducing property**. This property allows us to evaluate any function $g \in \mathcal{G}$ at a point $\boldsymbol{x} \in \mathcal{X}$ by taking the inner product of g and $\kappa_{\boldsymbol{x}}$; as such, $\kappa_{\boldsymbol{x}}$ is called the *representer of evaluation*. Further, by taking $g = \kappa_{\boldsymbol{x}'}$ and applying the reproducing property, we have $\langle \kappa_{\boldsymbol{x}'}, \kappa_{\boldsymbol{x}} \rangle_{\mathcal{G}} = \kappa(\boldsymbol{x}', \boldsymbol{x})$, and so by symmetry of the inner product it follows that $\kappa(\boldsymbol{x}, \boldsymbol{x}') = \kappa(\boldsymbol{x}', \boldsymbol{x})$. As a consequence, reproducing kernels are necessarily *symmetric* functions. Moreover, a reproducing kernel κ is a **positive semidefinite** function, meaning that for every $n \geq 1$ and every choice of $\alpha_1, \ldots, \alpha_n \in \mathbb{R}$ and $\boldsymbol{x}_1, \ldots, \boldsymbol{x}_n \in \mathcal{X}$, it holds that

$$\sum_{i=1}^{n} \sum_{j=1}^{n} \alpha_i \, \kappa(\boldsymbol{x}_i, \boldsymbol{x}_j) \, \alpha_j \geq 0. \tag{7.14}$$

In other words, *every* Gram matrix $\mathbf{K}$ associated with κ is a positive semidefinite matrix; that is $\boldsymbol{\alpha}^\top \mathbf{K} \boldsymbol{\alpha} \geq 0$ for all $\boldsymbol{\alpha}$. The proof is addressed in Exercise 1.

The following theorem gives an alternative characterization of an RKHS. The proof uses the Riesz representation Theorem B.20. Also note that in the theorem below we could have replaced the word "bounded" with "continuous", as the two are equivalent for linear functionals; see Theorem B.19.

☞ 569

Theorem 7.15 (Continuous Evaluation Functionals Characterize an RKHS): An RKHS $\mathcal{G}$ on a set $\mathcal{X}$ is a Hilbert space in which every **evaluation functional** $\delta_{\boldsymbol{x}} : g \mapsto g(\boldsymbol{x})$ is bounded. Conversely, a Hilbert space $\mathcal{G}$ of functions $\mathcal{X} \to \mathbb{R}$ for which every evaluation functional is bounded is an RKHS.

Proof. Note that, since evaluation functionals $\delta_{\boldsymbol{x}}$ are linear operators, showing boundedness is equivalent to showing continuity. Given an RKHS with reproducing kernel κ, suppose that we have a sequence $g_n \in \mathcal{G}$ converging to $g \in \mathcal{G}$, that is $\|g_n - g\|_{\mathcal{G}} \to 0$. We apply the Cauchy–Schwarz inequality (Theorem B.16) and the reproducing property of κ to find that for every $\boldsymbol{x} \in \mathcal{X}$ and any n:

☞ 568

$$\begin{aligned} |\delta_{\boldsymbol{x}} g_n - \delta_{\boldsymbol{x}} g| &= |g_n(\boldsymbol{x}) - g(\boldsymbol{x})| = |\langle g_n - g, \kappa_{\boldsymbol{x}} \rangle_{\mathcal{G}}| \leq \|g_n - g\|_{\mathcal{G}} \, \|\kappa_{\boldsymbol{x}}\|_{\mathcal{G}} \\ &= \|g_n - g\|_{\mathcal{G}} \sqrt{\langle \kappa_{\boldsymbol{x}}, \kappa_{\boldsymbol{x}} \rangle_{\mathcal{G}}} = \|g_n - g\|_{\mathcal{G}} \sqrt{\kappa(\boldsymbol{x}, \boldsymbol{x})}. \end{aligned}$$

Noting that $\sqrt{\kappa(\boldsymbol{x}, \boldsymbol{x})} < \infty$ by definition for every $\boldsymbol{x} \in \mathcal{X}$, and that $\|g_n - g\|_{\mathcal{G}} \to 0$ as $n \to \infty$, we have shown continuity of $\delta_{\boldsymbol{x}}$, that is $|\delta_{\boldsymbol{x}} g_n - \delta_{\boldsymbol{x}} g| \to 0$ as $n \to \infty$ for every $\boldsymbol{x} \in \mathcal{X}$.

Conversely, suppose that evaluation functionals are bounded. Then from the Riesz representation Theorem B.20, there exists some $g_{\delta_{\boldsymbol{x}}} \in \mathcal{G}$ such that $\delta_{\boldsymbol{x}} g = \langle g, g_{\delta_{\boldsymbol{x}}} \rangle_{\mathcal{G}}$ for all $g \in \mathcal{G}$ — the *representer* of evaluation. If we define $\kappa(\boldsymbol{x}, \boldsymbol{x}') = g_{\delta_{\boldsymbol{x}}}(\boldsymbol{x}')$ for all $\boldsymbol{x}, \boldsymbol{x}' \in \mathcal{X}$, then $\kappa_{\boldsymbol{x}} := \kappa(\boldsymbol{x}, \cdot) = g_{\delta_{\boldsymbol{x}}}$ is an element of $\mathcal{G}$ for every $\boldsymbol{x} \in \mathcal{X}$ and $\langle g, \kappa_{\boldsymbol{x}} \rangle_{\mathcal{G}} = \delta_{\boldsymbol{x}} g = g(\boldsymbol{x})$, so that the reproducing property in Definition 7.13 is verified. □

The fact that an RKHS has continuous evaluation functionals means that if two functions $g, h \in \mathcal{G}$ are "close" with respect to $\| \cdot \|_{\mathcal{G}}$, then their evaluations $g(\boldsymbol{x}), h(\boldsymbol{x})$

are close *for every* $\boldsymbol{x} \in \mathcal{X}$. Formally, convergence in $\|\cdot\|_{\mathcal{G}}$ norm implies pointwise convergence for all $\boldsymbol{x} \in \mathcal{X}$.

The following theorem shows that any finite function $\kappa : \mathcal{X} \times \mathcal{X} \to \mathbb{R}$ can serve as a reproducing kernel as long as it is finite, symmetric, and positive semidefinite. The corresponding (unique!) RKHS $\mathcal{G}$ is the completion of the set of all functions of the form $\sum_{i=1}^{n} \alpha_i \kappa_{\boldsymbol{x}_i}$ where $\alpha_i \in \mathbb{R}$ for all $i = 1, \ldots, n$.

Theorem 7.16 (Moore–Aronszajn): Given a non-empty set $\mathcal{X}$ and any finite symmetric positive semidefinite function $\kappa : \mathcal{X} \times \mathcal{X} \to \mathbb{R}$, there exists an RKHS $\mathcal{G}$ of functions $g : \mathcal{X} \to \mathbb{R}$ with reproducing kernel κ. Moreover, $\mathcal{G}$ is unique.

Proof. (Sketch) As the proof of uniqueness is treated in Exercise 2, the objective is to prove existence. The idea is to construct a pre-RKHS $\mathcal{G}_0$ from the given function κ that has the essential structure and then to extend $\mathcal{G}_0$ to an RKHS $\mathcal{G}$.

In particular, define $\mathcal{G}_0$ as the set of finite linear combinations of functions $\kappa_{\boldsymbol{x}}, \boldsymbol{x} \in \mathcal{X}$:

$$\mathcal{G}_0 := \left\{ g = \sum_{i=1}^{n} \alpha_i \kappa_{\boldsymbol{x}_i} \;\middle|\; \boldsymbol{x}_1, \ldots, \boldsymbol{x}_n \in \mathcal{X},\ \alpha_i \in \mathbb{R},\ n \in \mathbb{N} \right\}.$$

Define on $\mathcal{G}_0$ the following inner product:

$$\langle f, g \rangle_{\mathcal{G}_0} := \left\langle \sum_{i=1}^{n} \alpha_i \kappa_{\boldsymbol{x}_i}, \sum_{j=1}^{m} \beta_j \kappa_{\boldsymbol{x}'_j} \right\rangle_{\mathcal{G}_0} := \sum_{i=1}^{n} \sum_{j=1}^{m} \alpha_i \beta_j \kappa(\boldsymbol{x}_i, \boldsymbol{x}'_j).$$

Then $\mathcal{G}_0$ is an inner product space. In fact, $\mathcal{G}_0$ has the essential structure we require, namely that (i) evaluation functionals are bounded/continuous (Exercise 4) and (ii) Cauchy sequences in $\mathcal{G}_0$ that converge pointwise also converge in norm (see Exercise 5).

We then enlarge $\mathcal{G}_0$ to the set $\mathcal{G}$ of all functions $g : \mathcal{X} \to \mathbb{R}$ for which there exists a Cauchy sequence in $\mathcal{G}_0$ converging pointwise to g and define an inner product on $\mathcal{G}$ as the limit

$$\langle f, g \rangle_{\mathcal{G}} := \lim_{n \to \infty} \langle f_n, g_n \rangle_{\mathcal{G}_0}, \tag{7.17}$$

where $f_n \to f$ and $g_n \to g$. To show that $\mathcal{G}$ is an RKHS it remains to be shown that (1) this inner product is well defined; (2) evaluation functionals remain bounded; and (3) the space $\mathcal{G}$ is complete. A detailed proof is established in Exercises 6 and 7. □

7.4 Construction of Reproducing Kernels

In this section we describe various ways to construct a reproducing kernel $\kappa : \mathcal{X} \times \mathcal{X} \to \mathbb{R}$ for some feature space $\mathcal{X}$. Recall that κ needs to be a finite, symmetric, and positive semidefinite function (that is, it satisfies (7.14)). In view of Theorem 7.16, specifying the space $\mathcal{X}$ and a reproducing kernel $\kappa : \mathcal{X} \times \mathcal{X} \to \mathbb{R}$ corresponds to *uniquely* specifying an RKHS.

7.4.1 Reproducing Kernels via Feature Mapping

Perhaps the most fundamental way to construct a reproducing kernel κ is via a feature map $\boldsymbol{\phi} : \mathcal{X} \to \mathbb{R}^p$. We define $\kappa(\boldsymbol{x}, \boldsymbol{x}') := \langle \boldsymbol{\phi}(\boldsymbol{x}), \boldsymbol{\phi}(\boldsymbol{x}') \rangle$, where $\langle \; , \; \rangle$ denotes the Euclidean inner product. The function is clearly finite and symmetric. To verify that κ is positive semidefinite, let $\boldsymbol{\Phi}$ be the matrix with rows $\boldsymbol{\phi}(\boldsymbol{x}_1)^\top, \ldots, \boldsymbol{\phi}(\boldsymbol{x}_n)^\top$ and let $\boldsymbol{\alpha} = [\alpha_1, \ldots, \alpha_n]^\top \in \mathbb{R}^n$. Then,

$$\sum_{i=1}^{n} \sum_{j=1}^{n} \alpha_i \, \kappa(\boldsymbol{x}_i, \boldsymbol{x}_j) \, \alpha_j = \sum_{i=1}^{n} \sum_{j=1}^{n} \alpha_i \, \boldsymbol{\phi}^\top(\boldsymbol{x}_i) \, \boldsymbol{\phi}(\boldsymbol{x}_j) \, \alpha_j = \boldsymbol{\alpha}^\top \boldsymbol{\Phi} \boldsymbol{\Phi}^\top \boldsymbol{\alpha} = \|\boldsymbol{\Phi}^\top \boldsymbol{\alpha}\|^2 \geq 0.$$

■ **Example 7.18 (Linear Kernel)** Taking the identity feature map $\boldsymbol{\phi}(\boldsymbol{x}) = \boldsymbol{x}$ on $\mathcal{X} = \mathbb{R}^p$, gives the **linear kernel**

$$\kappa(\boldsymbol{x}, \boldsymbol{x}') = \langle \boldsymbol{x}, \boldsymbol{x}' \rangle = \boldsymbol{x}^\top \boldsymbol{x}'.$$

As can be seen from the proof of Theorem 7.16, the RKHS of functions corresponding to the linear kernel is the space of *linear* functions on $\mathbb{R}^p$. This space is isomorphic to $\mathbb{R}^p$ itself, as discussed in the introduction (see also Exercise 12). ■

■ **Example 7.19 (Theta Kernel)** If we use the feature map $\boldsymbol{\phi} : [0, 1] \to \mathbb{R}^p$, where

$$\boldsymbol{\phi}(x) := \left[1, \sqrt{2}\, \mathrm{e}^{-1^2\pi^2/4} \cos(\pi x), \cdots, \sqrt{2}\, \mathrm{e}^{-(p-1)^2\pi^2/4} \cos(\pi(p-1)x)\right]^\top,$$

we construct the kernel

$$\langle \boldsymbol{\phi}(x), \boldsymbol{\phi}(x') \rangle = 1 + 2 \sum_{k=1}^{p-1} \mathrm{e}^{-k^2\pi^2/2} \cos(\pi k x) \cos(\pi k x').$$

By taking the limit as $p \to \infty$, we then obtain the **theta kernel**:

$$\kappa(x, x') = \sum_{k=-\infty}^{\infty} \mathrm{e}^{-k^2\pi^2/2} \cos(\pi k x) \cos(\pi k x'),$$

☞ 153 which has important applications in kernel density estimation; see Section 4.5. ■

It is natural to wonder whether a given kernel function corresponds uniquely to a feature map. The answer is no, as we shall see by way of example.

■ **Example 7.20 (Feature Maps and Kernel Functions)** Let $\mathcal{X} = \mathbb{R}$ and consider feature maps $\phi_1 : \mathcal{X} \to \mathbb{R}$ and $\boldsymbol{\phi}_2 : \mathcal{X} \to \mathbb{R}^2$, with $\phi_1(x) := x$ and $\boldsymbol{\phi}_2(x) := [x, x]^\top/\sqrt{2}$. Then

$$\kappa_{\phi_1}(x, x') = \langle \phi_1(x), \phi_1(x') \rangle = x x',$$

but also

$$\kappa_{\boldsymbol{\phi}_2}(x, x') = \langle \boldsymbol{\phi}_2(x), \boldsymbol{\phi}_2(x') \rangle = x x'.$$

Thus, we arrive at the same kernel function defined for the same underlying set $\mathcal{X}$ via two different feature maps. ■

7.4.2 Kernels from Characteristic Functions

Another way to construct reproducing kernels on $\mathcal{X} = \mathbb{R}^p$ makes use of the properties of *characteristic functions*. In particular, we have the following result. We leave its proof as Exercise 10. ☞ 647

Theorem 7.21 (Reproducing Kernel from a Characteristic Function): Let $\boldsymbol{X} \sim \mu$ be an $\mathbb{R}^p$-valued random vector that is symmetric about the origin (that is, $\boldsymbol{X}$ and $-\boldsymbol{X}$ are identically distributed), and let ψ be its characteristic function: $\psi(\boldsymbol{t}) = \mathbb{E}\, \mathrm{e}^{\mathrm{i}\boldsymbol{t}^\top \boldsymbol{X}} = \int \mathrm{e}^{\mathrm{i}\boldsymbol{t}^\top \boldsymbol{x}}\, \mu(\mathrm{d}\boldsymbol{x})$ for $\boldsymbol{t} \in \mathbb{R}^p$. Then $\kappa(\boldsymbol{x}, \boldsymbol{x}') := \psi(\boldsymbol{x} - \boldsymbol{x}')$ is a valid reproducing kernel on $\mathbb{R}^p$.

■ **Example 7.22 (Gaussian Kernel)** The multivariate normal distribution with mean vector $\mathbf{0}$ and covariance matrix $b^2\, \mathbf{I}_p$ is clearly symmetric around the origin. Its characteristic function is

$$\psi(\boldsymbol{t}) = \exp\left(-\frac{1}{2} b^2\, \|\boldsymbol{t}\|^2\right), \quad \boldsymbol{t} \in \mathbb{R}^p.$$

Taking $b^2 = 1/\sigma^2$, this gives the popular **Gaussian kernel** on $\mathbb{R}^p$:

$$\kappa(\boldsymbol{x}, \boldsymbol{x}') = \exp\left(-\frac{1}{2} \frac{\|\boldsymbol{x} - \boldsymbol{x}'\|^2}{\sigma^2}\right). \tag{7.23}$$

The parameter σ is sometimes called the **bandwidth**. Note that in the machine learning literature, the Gaussian kernel is sometimes referred to as "the" **radial basis function (rbf) kernel**.[1]

From the proof of Theorem 7.16, we see that the RKHS $\mathcal{G}$ determined by the Gaussian kernel κ is the space of pointwise limits of functions of the form

$$g(\boldsymbol{x}) = \sum_{i=1}^{n} \alpha_i \exp\left(-\frac{1}{2} \frac{\|\boldsymbol{x} - \boldsymbol{x}_i\|^2}{\sigma^2}\right).$$

We can think of each point $\boldsymbol{x}_i$ having a feature $\kappa_{\boldsymbol{x}_i}$ that is a scaled multivariate Gaussian pdf centered at $\boldsymbol{x}_i$. ■

■ **Example 7.24 (Sinc Kernel)** The characteristic function of a $\mathcal{U}[-1, 1]$ random variable (which is symmetric around 0) is $\psi(t) = \mathrm{sinc}(t) := \sin(t)/t$, so

$$\kappa(x, x') = \mathrm{sinc}(x - x')$$

is a valid kernel. ■

Inspired by kernel density estimation (Section 4.4), we may be tempted to use the pdf of a random variable that is symmetric about the origin to construct a reproducing kernel. However, doing so will not work in general, as the next example illustrates. ☞ 144

[1]The term radial basis function is sometimes used more generally to mean kernels of the form $\kappa(\boldsymbol{x}, \boldsymbol{x}') = f(\|\boldsymbol{x} - \boldsymbol{x}'\|)$ for some function $f : \mathbb{R} \to \mathbb{R}$.

Example 7.25 (Uniform pdf Does not Construct a Valid Reproducing Kernel) Take the function $\psi(t) = \frac{1}{2}\mathbb{1}\{|t| \leq 1\}$, which is the pdf of $X \sim \mathcal{U}[-1, 1]$. Unfortunately, the function $\kappa(x, x') = \psi(x - x')$ is not positive semidefinite, as can be seen for example by constructing the matrix $\mathbf{A} = [\kappa(t_i, t_j), i, j = 1, 2, 3]$ for the points $t_1 = 0$, $t_2 = 0.75$, and $t_3 = 1.5$ as follows:

$$\mathbf{A} = \begin{pmatrix} \psi(0) & \psi(-0.75) & \psi(-1.5) \\ \psi(0.75) & \psi(0) & \psi(-0.75) \\ \psi(1.5) & \psi(0.75) & \psi(0) \end{pmatrix} = \begin{pmatrix} 0.5 & 0.5 & 0 \\ 0.5 & 0.5 & 0.5 \\ 0 & 0.5 & 0.5 \end{pmatrix}.$$

The eigenvalues of $\mathbf{A}$ are $\{1/2 - \sqrt{1/2}, 1/2, 1/2 + \sqrt{1/2}\} \approx \{-0.2071, 0.5, 1.2071\}$ and
☞ 531 so by Theorem A.38, $\mathbf{A}$ is not a positive semidefinite matrix, since it has a negative eigenvalue. Consequently, κ is not a valid reproducing kernel. ■

One of the reasons why the Gaussian kernel (7.23) is popular is that it enjoys the **universal approximation property** [126]: the space of functions spanned by the Gaussian kernel is dense in the space of continuous functions with support $\mathcal{Z} \subset \mathbb{R}^p$. Naturally, this is a desirable property especially if there is little prior knowledge about the properties of g^*. However, note that *every* function g in the RKHS $\mathcal{G}$ associated with a Gaussian kernel κ is infinitely differentiable. Moreover, a Gaussian RKHS does not contain non-zero constant functions. Indeed, if $A \subset \mathcal{Z}$ is non-empty and open, then the only function of the form $g(\boldsymbol{x}) = c\,\mathbb{1}\{\boldsymbol{x} \in A\}$ contained in $\mathcal{G}$ is the zero function ($c = 0$).

Consequently, if it is known that g is differentiable only to a certain order, one may prefer the **Matérn kernel** with parameters $\nu, \sigma > 0$:

$$\kappa_\nu(\boldsymbol{x}, \boldsymbol{x}') = \frac{2^{1-\nu}}{\Gamma(\nu)} \left(\sqrt{2\nu}\,\|\boldsymbol{x} - \boldsymbol{x}'\|/\sigma\right)^\nu K_\nu\left(\sqrt{2\nu}\,\|\boldsymbol{x} - \boldsymbol{x}'\|/\sigma\right), \tag{7.26}$$

which gives functions that are (weakly) differentiable to order $\lfloor \nu \rfloor$ (but not necessarily to order $\lceil \nu \rceil$). Here, K_ν denotes the modified Bessel function of the second kind; see
☞ 194 (4.102). The particular form of the Matérn kernel appearing in (7.26) ensures that $\lim_{\nu \to \infty} \kappa_\nu(\boldsymbol{x}, \boldsymbol{x}') = \kappa(\boldsymbol{x}, \boldsymbol{x}')$, where κ is the Gaussian kernel appearing in (7.23).

We remark that Sobolev spaces are closely related to the Matérn kernel. Up to constants (which scale the unit ball in the space), in dimension p and for a parameter $s > p/2$, these spaces can be identified with $\psi(\boldsymbol{t}) = \frac{2^{1-s}}{\Gamma(s)} \|\boldsymbol{t}\|^{s-p/2} K_{p/2-s}(\|\boldsymbol{t}\|)$, which in turn can be viewed as the characteristic function corresponding to the (radially symmetric)
☞ 193 multivariate Student's tdistribution with s degrees of freedom: that is, with pdf $f(\boldsymbol{x}) \propto (1 + \|\boldsymbol{x}\|^2)^{-s}$.

7.4.3 Reproducing Kernels Using Orthonormal Features

We have seen in Sections 7.4.1 and 7.4.2 how to construct reproducing kernels from feature maps and characteristic functions. Another way to construct kernels on a space $\mathcal{X}$ is to work directly from the function class $L^2(\mathcal{X}; \mu)$; that is, the set of square-integrable[2]
☞ 564 functions on $\mathcal{X}$ with respect to μ; see also Definition B.4. For simplicity, in what follows,

[2]A function $f : \mathcal{X} \to \mathbb{R}$ is said to be square-integrable if $\int f^2(\boldsymbol{x})\,\mu(\mathrm{d}\boldsymbol{x}) < \infty$, where μ is a measure on $\mathcal{X}$.

we will consider μ to be the Lebesgue measure, and will simply write $L^2(\mathcal{X})$ rather than $L^2(\mathcal{X};\mu)$. We will also assume that $\mathcal{X} \subseteq \mathbb{R}^p$.

Let $\{\xi_1, \xi_2, \ldots\}$ be an orthonormal basis of $L^2(\mathcal{X})$ and let $c_1, c_2, \ldots$ be a sequence of nonzero numbers. As discussed in Section 7.4.1, the kernel corresponding to a feature map $\boldsymbol{\phi} : \mathcal{X} \to \mathbb{R}^p$ is $\kappa(\boldsymbol{x}, \boldsymbol{x}') = \boldsymbol{\phi}(\boldsymbol{x})^\top \boldsymbol{\phi}(\boldsymbol{x}') = \sum_{i=1}^p \phi_i(\boldsymbol{x})\, \phi_i(\boldsymbol{x}')$. Now consider a (possibly infinite) sequence of feature functions $\phi_i = c_i\, \xi_i, i = 1, 2, \ldots$ and define

$$\kappa(\boldsymbol{x}, \boldsymbol{x}') := \sum_{i \geq 1} \phi_i(\boldsymbol{x})\, \phi_i(\boldsymbol{x}') = \sum_{i \geq 1} \lambda_i\, \xi_i(\boldsymbol{x})\, \xi_i(\boldsymbol{x}'), \tag{7.27}$$

where $\lambda_i = c_i^2, i = 1, 2, \ldots$. This is well-defined as long as $\sum_{i \geq 1} \lambda_i < \infty$, which we assume from now on. Let $\mathcal{H}$ be the linear space of functions of the form $f = \sum_{i \geq 1} \alpha_i \xi_i$, where

$$\sum_{i \geq 1} \frac{\alpha_i^2}{\lambda_i} < \infty.$$

As every $f \in L^2(\mathcal{X})$ can be represented as $f = \sum_{i \geq 1} \langle f, \xi_i \rangle \xi_i$, we see that $\mathcal{H}$ is a linear subspace of $L^2(\mathcal{X})$. On $\mathcal{H}$ define the inner product

$$\langle f, g \rangle_{\mathcal{H}} := \sum_{i \geq 1} \frac{\langle f, \xi_i \rangle \langle g, \xi_i \rangle}{\lambda_i}.$$

With this inner product, the squared norm of $f = \sum_{i \geq 1} \alpha_i\, \xi_i$ is $\|f\|_{\mathcal{H}}^2 = \sum_{i \geq 1} \alpha_i^2 / \lambda_i < \infty$. We show that $\mathcal{H}$ is actually an RKHS with kernel κ by verifying the conditions of Definition 7.13. First,

$$\kappa_{\boldsymbol{x}} = \sum_{i \geq 1} \lambda_i\, \xi_i(\boldsymbol{x})\, \xi_i \in \mathcal{H},$$

as $\sum_i \lambda_i < \infty$ by assumption, and so κ is finite. Second, the reproducing property holds. Namely, let $f = \sum_{i \geq 1} \alpha_i\, \xi_i$. Then,

$$\langle \kappa_{\boldsymbol{x}}, f \rangle_{\mathcal{H}} = \sum_{i \geq 1} \frac{\langle \kappa_{\boldsymbol{x}}, \xi_i \rangle \langle f, \xi_i \rangle}{\lambda_i} = \sum_{i \geq 1} \frac{\lambda_i\, \xi_i(\boldsymbol{x})\, \alpha_i}{\lambda_i} = \sum_{i \geq 1} \alpha_i \xi_i(\boldsymbol{x}) = f(\boldsymbol{x}).$$

The discussion above demonstrates that kernels can be constructed via (7.27). In fact, (under mild conditions) any given reproducing kernel κ can be written in the form (7.27), where this series representation enjoys desirable convergence properties. This result is known as Mercer's theorem, and is given below. We leave the full proof including the precise conditions to, e.g., [61], but the main idea is that a reproducing kernel κ can be thought of as a generalization of a positive semidefinite matrix $\mathbf{K}$, and can also be written in spectral form (see also Section A.7.3). In particular, by Theorem A.38, we can write ☞ 531 $\mathbf{K} = \mathbf{V}\mathbf{D}\mathbf{V}^\top$, where $\mathbf{V}$ is a matrix of orthonormal eigenvectors $[\boldsymbol{v}_\ell]$ and $\mathbf{D}$ the diagonal matrix of the (positive) eigenvalues $[\lambda_\ell]$; that is,

$$\mathbf{K}(i, j) = \sum_{\ell \geq 1} \lambda_\ell\, v_\ell(i)\, v_\ell(j).$$

In (7.29) below, $\boldsymbol{x}, \boldsymbol{x}'$ play the role of i, j, and ξ_ℓ plays the role of $\boldsymbol{v}_\ell$.

Theorem 7.28 (Mercer): Let $\kappa : \mathcal{X} \times \mathcal{X} \to \mathbb{R}$ be a reproducing kernel for a compact set $\mathcal{X} \subset \mathbb{R}^p$. Then (under mild conditions) there exists a countable sequence of non-negative numbers $\{\lambda_\ell\}$ decreasing to zero and functions $\{\xi_\ell\}$ orthonormal in $L^2(\mathcal{X})$ such that

$$\kappa(\boldsymbol{x}, \boldsymbol{x}') = \sum_{\ell \geq 1} \lambda_\ell \, \xi_\ell(\boldsymbol{x}) \, \xi_\ell(\boldsymbol{x}'), \qquad \text{for all } \boldsymbol{x}, \boldsymbol{x}' \in \mathcal{X}, \tag{7.29}$$

where (7.29) converges absolutely and uniformly on $\mathcal{X} \times \mathcal{X}$.

Further, if $\lambda_\ell > 0$, then (λ_ℓ, ξ_ℓ) is an (eigenvalue, eigenfunction) pair for the integral operator $K : L^2(\mathcal{X}) \to L^2(\mathcal{X})$ defined by $[Kf](\boldsymbol{x}) := \int_{\mathcal{X}} \kappa(\boldsymbol{x}, \boldsymbol{y}) f(\boldsymbol{y}) \, \mathrm{d}\boldsymbol{y}$ for $\boldsymbol{x} \in \mathcal{X}$.

Theorem 7.28 holds if (i) the kernel κ is continuous on $\mathcal{X} \times \mathcal{X}$, (ii) the function $\widetilde{\kappa}(\boldsymbol{x}) := \kappa(\boldsymbol{x}, \boldsymbol{x})$ defined for $\boldsymbol{x} \in \mathcal{X}$ is integrable. Extensions of Theorem 7.28 to more general spaces $\mathcal{X}$ and measures μ hold; see, e.g., [161] or [61].

The key importance of Theorem 7.28 lies in the fact that the series representation (7.29) converges absolutely and uniformly on $\mathcal{X} \times \mathcal{X}$. The uniform convergence is a much stronger condition than pointwise convergence, and means for instance that properties of the sequence of partial sums, such as continuity and integrability, are transferred to the limit.

Example 7.30 (Mercer) Suppose $\mathcal{X} = [-1, 1]$ and the kernel is $\kappa(x, x') = 1 + xx'$ which corresponds to the RKHS $\mathcal{G}$ of affine functions from $\mathcal{X} \to \mathbb{R}$. To find the (eigenvalue, eigenfunction) pairs for the integral operator appearing in Theorem 7.28, we need to find numbers $\{\lambda_\ell\}$ and orthonormal functions $\{\xi_\ell(x)\}$ that solve

$$\int_{-1}^{1} (1 + xx') \, \xi_\ell(x') \, \mathrm{d}x' = \lambda_\ell \, \xi_\ell(x), \quad \text{for all } x \in [-1, 1].$$

Consider first a constant function $\xi_1(x) = c$. Then, for all $x \in [-1, 1]$, we have that $2c = \lambda_1 c$, and the normalization condition requires that $\int_{-1}^{1} c^2 \, \mathrm{d}x = 1$. Together, these give $\lambda_1 = 2$ and $c = \pm 1/\sqrt{2}$. Next, consider an affine function $\xi_2(x) = a + bx$. Orthogonality requires that

$$\int_{-1}^{1} c(a + bx) \, \mathrm{d}x = 0,$$

which implies $a = 0$ (since $c \neq 0$). Moreover, the normalization condition then requires

$$\int_{-1}^{1} b^2 x^2 \, \mathrm{d}x = 1,$$

or, equivalently, $2b^2/3 = 1$, implying $b = \pm\sqrt{3/2}$. Finally, the integral equation reads

$$\int_{-1}^{1} (1 + xx') \, bx' \, \mathrm{d}x' = \lambda_2 \, bx \iff \frac{2bx}{3} = \lambda_2 bx,$$

implying that $\lambda_2 = 2/3$. We take the positive solutions (i.e., $c > 0$ and $b > 0$), and note that

$$\lambda_1\,\xi_1(x)\,\xi_1(x') + \lambda_2\,\xi_2(x)\,\xi_2(x') = 2\frac{1}{\sqrt{2}}\frac{1}{\sqrt{2}} + \frac{2}{3}\frac{\sqrt{3}}{\sqrt{2}}x\frac{\sqrt{3}}{\sqrt{2}}x' = 1 + xx' = \kappa(x, x'),$$

and so we have found the decomposition appearing in (7.29). As an aside, observe that ξ_1 and ξ_2 are orthonormal versions of the first two Legendre polynomials. The ☞ 567 corresponding feature map can be explicitly identified as $\phi_1(x) = \sqrt{\lambda_1}\,\xi_1(x) = 1$ and $\phi_2(x) = \sqrt{\lambda_2}\,\xi_2(x) = x$. ■

7.4.4 Kernels from Kernels

The following theorem lists some useful properties for constructing reproducing kernels from existing reproducing kernels.

Theorem 7.31 (Rules for Constructing Kernels from Other Kernels):

1. If $\kappa : \mathbb{R}^p \times \mathbb{R}^p \to \mathbb{R}$ is a reproducing kernel and $\boldsymbol{\phi} : \mathcal{X} \to \mathbb{R}^p$ is a function, then $\kappa(\boldsymbol{\phi}(\boldsymbol{x}), \boldsymbol{\phi}(\boldsymbol{x}'))$ is a reproducing kernel from $\mathcal{X} \times \mathcal{X} \to \mathbb{R}$.

2. If $\kappa : \mathcal{X} \times \mathcal{X} \to \mathbb{R}$ is a reproducing kernel and $f : \mathcal{X} \to \mathbb{R}_+$ is a function, then $f(\boldsymbol{x})\kappa(\boldsymbol{x}, \boldsymbol{x}')f(\boldsymbol{x}')$ is also a reproducing kernel from $\mathcal{X} \times \mathcal{X} \to \mathbb{R}$.

3. If κ_1 and κ_2 are reproducing kernels from $\mathcal{X} \times \mathcal{X} \to \mathbb{R}$, then so is their sum $\kappa_1 + \kappa_2$.

4. If κ_1 and κ_2 are reproducing kernels from $\mathcal{X} \times \mathcal{X} \to \mathbb{R}$, then so is their product $\kappa_1\kappa_2$.

5. If κ_1 and κ_2 are reproducing kernels from $\mathcal{X} \times \mathcal{X} \to \mathbb{R}$ and $\mathcal{Y} \times \mathcal{Y} \to \mathbb{R}$ respectively, then $\kappa_+((\boldsymbol{x}, \boldsymbol{y}), (\boldsymbol{x}', \boldsymbol{y}')) := \kappa_1(\boldsymbol{x}, \boldsymbol{x}') + \kappa_2(\boldsymbol{y}, \boldsymbol{y}')$ and $\kappa_\times((\boldsymbol{x}, \boldsymbol{y}), (\boldsymbol{x}', \boldsymbol{y}')) := \kappa_1(\boldsymbol{x}, \boldsymbol{x}')\kappa_2(\boldsymbol{y}, \boldsymbol{y}')$ are reproducing kernels from $(\mathcal{X} \times \mathcal{Y}) \times (\mathcal{X} \times \mathcal{Y}) \to \mathbb{R}$.

Proof. For Rules 1, 2, and 3 it is easy to verify that the resulting function is finite, symmetric, and positive semidefinite, and so is a valid reproducing kernel by Theorem 7.16. For example, for Rule 1 we have $\sum_{i=1}^n \sum_{j=1}^n \alpha_i\,\kappa(\boldsymbol{y}_i, \boldsymbol{y}_j)\alpha_j \geq 0$ for every choice of $\{\alpha_i\}_{i=1}^n$ and $\{\boldsymbol{y}_i\}_{i=1}^n \in \mathbb{R}^p$, since κ is a reproducing kernel. In particular, it holds true for $\boldsymbol{y}_i = \boldsymbol{\phi}(\boldsymbol{x}_i)$, $i = 1, \ldots, n$. Rule 4 is easy to show for kernels κ_1, κ_2 that admit a representation of the form (7.27), since

$$\begin{aligned}
\kappa_1(\boldsymbol{x}, \boldsymbol{x}')\,\kappa_2(\boldsymbol{x}, \boldsymbol{x}') &= \left(\sum_{i\geq 1} \phi_i^{(1)}(\boldsymbol{x})\,\phi_i^{(1)}(\boldsymbol{x}')\right)\left(\sum_{j\geq 1} \phi_j^{(2)}(\boldsymbol{x})\,\phi_j^{(2)}(\boldsymbol{x}')\right) \\
&= \sum_{i,j\geq 1} \phi_i^{(1)}(\boldsymbol{x})\,\phi_j^{(2)}(\boldsymbol{x})\,\phi_i^{(1)}(\boldsymbol{x}')\,\phi_j^{(2)}(\boldsymbol{x}') \\
&= \sum_{k\geq 1} \phi_k(\boldsymbol{x})\,\phi_k(\boldsymbol{x}') =: \kappa(\boldsymbol{x}, \boldsymbol{x}'),
\end{aligned}$$

showing that $\kappa = \kappa_1 \kappa_2$ also admits a representation of the form (7.27), where the new (possibly infinite) sequence of features (ϕ_k) is identified in a one-to-one way with the sequence $(\phi_i^{(1)} \phi_j^{(2)})$. We leave the proof of rule 5 as an exercise (Exercise 8). □

■ **Example 7.32 (Polynomial Kernel)** Consider $\boldsymbol{x}, \boldsymbol{x}' \in \mathbb{R}^2$ with

$$\kappa(\boldsymbol{x}, \boldsymbol{x}') = (1 + \langle \boldsymbol{x}, \boldsymbol{x}' \rangle)^2,$$

where $\langle \boldsymbol{x}, \boldsymbol{x}' \rangle = \boldsymbol{x}^\top \boldsymbol{x}'$. This is an example of a **polynomial kernel**. Combining the fact that sums and products of kernels are again kernels (rules 3 and 4 of Theorem 7.31), we find that, since $\langle \boldsymbol{x}, \boldsymbol{x}' \rangle$ and the constant function 1 are kernels, so are $1 + \langle \boldsymbol{x}, \boldsymbol{x}' \rangle$ and $(1 + \langle \boldsymbol{x}, \boldsymbol{x}' \rangle)^2$. By writing

$$\begin{aligned} \kappa(\boldsymbol{x}, \boldsymbol{x}') &= (1 + x_1 x_1' + x_2 x_2')^2 \\ &= 1 + 2x_1 x_1' + 2x_2 x_2' + 2x_1 x_2 x_1' x_2' + (x_1 x_1')^2 + (x_2 x_2')^2, \end{aligned}$$

we see that $\kappa(\boldsymbol{x}, \boldsymbol{x}')$ can be written as the inner product in $\mathbb{R}^6$ of the two feature vectors $\boldsymbol{\phi}(\boldsymbol{x})$ and $\boldsymbol{\phi}(\boldsymbol{x}')$, where the feature map $\boldsymbol{\phi} : \mathbb{R}^2 \to \mathbb{R}^6$ can be explicitly identified as

$$\boldsymbol{\phi}(\boldsymbol{x}) = [1, \sqrt{2}x_1, \sqrt{2}x_2, \sqrt{2}x_1 x_2, x_1^2, x_2^2]^\top.$$

Thus, the RKHS determined by κ can be explicitly identified with the space of functions $\boldsymbol{x} \mapsto \boldsymbol{\phi}(\boldsymbol{x})^\top \boldsymbol{\beta}$ for some $\boldsymbol{\beta} \in \mathbb{R}^6$. ■

In the above example we could explicitly identify the feature map. However, in general a feature map need not be explicitly available. Using a particular reproducing kernel corresponds to using an *implicit* (possibly infinite-dimensional!) feature map that never needs to be explicitly computed.

7.5 Representer Theorem

Recall the setting discussed at the beginning of this chapter: we are given training data $\tau = \{(\boldsymbol{x}_i, y_i)\}_{i=1}^n$ and a loss function that measures the fit to the data, and we wish to find a function g that minimizes the training loss, with the addition of a regularization term, as described in Section 7.2. To do this, we assume first that the class $\mathcal{G}$ of prediction functions can be decomposed as the direct sum of an RKHS $\mathcal{H}$, defined by a kernel function $\kappa : \mathcal{X} \times \mathcal{X} \to \mathbb{R}$, and another linear space of real-valued functions $\mathcal{H}_0$ on $\mathcal{X}$; that is,

$$\mathcal{G} = \mathcal{H} \oplus \mathcal{H}_0,$$

meaning that any element $g \in \mathcal{G}$ can be written as $g = h + h_0$, with $h \in \mathcal{H}$ and $h_0 \in \mathcal{H}_0$. In minimizing the training loss we wish to penalize the h term of g but not the h_0 term. Specifically, the aim is to solve the functional optimization problem

$$\min_{g \in \mathcal{H} \oplus \mathcal{H}_0} \sum_{i=1}^n \text{Loss}(y_i, g(\boldsymbol{x}_i)) + \lambda \, \|g\|_{\mathcal{H}}^2. \tag{7.33}$$

Here, we use a slight abuse of notation: $\|g\|_{\mathcal{H}}$ means $\|h\|_{\mathcal{H}}$ if $g = h + h_0$, as above. In this way, we can view $\mathcal{H}_0$ as the null space of the functional $g \mapsto \|g\|_{\mathcal{H}}$. This null space may be empty, but typically has a small dimension m; for example it could be the one-dimensional space of constant functions, as in Example 7.8. ☞ 322

■ **Example 7.34 (Null Space)** Consider again the setting of Example 7.8, for which we have feature vectors $\widetilde{\boldsymbol{x}} = [1, \boldsymbol{x}^\top]^\top$ and $\mathcal{G}$ consists of functions of the form $g : \widetilde{\boldsymbol{x}} \mapsto \beta_0 + \boldsymbol{x}^\top \boldsymbol{\beta}$. Each function g can be decomposed as $g = h + h_0$, where $h : \widetilde{\boldsymbol{x}} \mapsto \boldsymbol{x}^\top \boldsymbol{\beta}$, and $h_0 : \widetilde{\boldsymbol{x}} \mapsto \beta_0$.

Given $g \in \mathcal{G}$, we have $\|g\|_{\mathcal{H}} = \|\boldsymbol{\beta}\|$, and so the null space $\mathcal{H}_0$ of the functional $g \mapsto \|g\|_{\mathcal{H}}$ (that is, the set of all functions $g \in \mathcal{G}$ for which $\|g\|_{\mathcal{H}} = 0$) is the set of constant functions here, which has dimension $m = 1$. ■

Regularization favors elements in $\mathcal{H}_0$ and penalizes large elements in $\mathcal{H}$. As the regularization parameter λ varies between zero and infinity, solutions to (7.33) vary from "complex" ($g \in \mathcal{H} \oplus \mathcal{H}_0$) to "simple" ($g \in \mathcal{H}_0$).

A key reason why RKHSs are so useful is the following. By choosing $\mathcal{H}$ to be an RKHS in (7.33) this *functional* optimization problem effectively becomes a *parametric* optimization problem. The reason is that any solution to (7.33) can be represented as a finite-dimensional linear combination of kernel functions, evaluated at the training sample. This is known as the **kernel trick**.

Theorem 7.35 (Representer Theorem): The solution to the penalized optimization problem (7.33) is of the form

$$g(\boldsymbol{x}) = \sum_{i=1}^{n} \alpha_i \, \kappa(\boldsymbol{x}_i, \boldsymbol{x}) + \sum_{j=1}^{m} \eta_j \, q_j(\boldsymbol{x}), \tag{7.36}$$

where $\{q_1, \ldots, q_m\}$ is a basis of $\mathcal{H}_0$.

Proof. Let $\mathcal{F} = \operatorname{Span}\{\kappa_{\boldsymbol{x}_i}, i = 1, \ldots, n\}$. Clearly, $\mathcal{F} \subseteq \mathcal{H}$. Then, the Hilbert space $\mathcal{H}$ can be represented as $\mathcal{H} = \mathcal{F} \oplus \mathcal{F}^\perp$, where $\mathcal{F}^\perp$ is the orthogonal complement of $\mathcal{F}$. In other words, $\mathcal{F}^\perp$ is the class of functions

$$\{f^\perp \in \mathcal{H} : \langle f^\perp, f \rangle_{\mathcal{H}} = 0,\ f \in \mathcal{F}\} \equiv \{f^\perp : \langle f^\perp, \kappa_{\boldsymbol{x}_i} \rangle_{\mathcal{H}} = 0,\ \forall i\}.$$

It follows, by the reproducing kernel property, that for all $f^\perp \in \mathcal{F}^\perp$:

$$f^\perp(\boldsymbol{x}_i) = \langle f^\perp, \kappa_{\boldsymbol{x}_i} \rangle_{\mathcal{H}} = 0, \quad i = 1, \ldots, n.$$

Now, take any $g \in \mathcal{H} \oplus \mathcal{H}_0$, and write it as $g = f + f^\perp + h_0$, with $f \in \mathcal{F}$, $f^\perp \in \mathcal{F}^\perp$, and $h_0 \in \mathcal{H}_0$. By the definition of the null space $\mathcal{H}_0$, we have $\|g\|^2_{\mathcal{H}} = \|f + f^\perp\|^2_{\mathcal{H}}$. Moreover, by Pythagoras' theorem, the latter is equal to $\|f\|^2_{\mathcal{H}} + \|f^\perp\|^2_{\mathcal{H}}$. It follows that

$$\begin{aligned}\sum_{i=1}^{n} \mathrm{Loss}(y_i, g(\boldsymbol{x}_i)) + \lambda \|g\|^2_{\mathcal{H}} &= \sum_{i=1}^{n} \mathrm{Loss}(y_i, f(\boldsymbol{x}_i) + h_0(\boldsymbol{x}_i)) + \lambda \left(\|f\|^2_{\mathcal{H}} + \|f^\perp\|^2_{\mathcal{H}}\right) \\ &\geq \sum_{i=1}^{n} \mathrm{Loss}(y_i, f(\boldsymbol{x}_i) + h_0(\boldsymbol{x}_i)) + \lambda \|f\|^2_{\mathcal{H}}.\end{aligned}$$

Since we can obtain equality by taking $f^\perp = 0$, this implies that the minimizer of the penalized optimization problem (7.33) lies in the subspace $\mathcal{F} \oplus \mathcal{H}_0$ of $\mathcal{G} = \mathcal{H} \oplus \mathcal{H}_0$, and hence is of the form (7.36). □

Substituting the representation (7.36) of g into (7.33) gives the finite-dimensional optimization problem:

$$\min_{\boldsymbol{\alpha}\in\mathbb{R}^n, \boldsymbol{\eta}\in\mathbb{R}^m} \sum_{i=1}^{n} \mathrm{Loss}(y_i, (\mathbf{K}\boldsymbol{\alpha} + \mathbf{Q}\boldsymbol{\eta})_i) + \lambda\, \boldsymbol{\alpha}^\top \mathbf{K} \boldsymbol{\alpha}, \tag{7.37}$$

where

- $\mathbf{K}$ is the $n \times n$ (Gram) matrix with entries $[\kappa(\boldsymbol{x}_i, \boldsymbol{x}_j), i = 1, \ldots, n,\ j = 1, \ldots, n]$.
- $\mathbf{Q}$ is the $n \times m$ matrix with entries $[q_j(\boldsymbol{x}_i), i = 1, \ldots, n,\ j = 1, \ldots, m]$.

In particular, for the squared-error loss we have

$$\min_{\boldsymbol{\alpha}\in\mathbb{R}^n, \boldsymbol{\eta}\in\mathbb{R}^m} \| \boldsymbol{y} - (\mathbf{K}\boldsymbol{\alpha} + \mathbf{Q}\boldsymbol{\eta}) \|^2 + \lambda\, \boldsymbol{\alpha}^\top \mathbf{K} \boldsymbol{\alpha}. \tag{7.38}$$

This is a convex optimization problem, and its solution is found by differentiating (7.38) with respect to $\boldsymbol{\alpha}$ and $\boldsymbol{\eta}$ and equating to zero, leading to the following system of $(n + m)$ linear equations:

$$\begin{bmatrix} \mathbf{K}\mathbf{K} + \lambda\mathbf{K} & \mathbf{K}\mathbf{Q} \\ \mathbf{Q}^\top\mathbf{K} & \mathbf{Q}^\top\mathbf{Q} \end{bmatrix} \begin{bmatrix} \boldsymbol{\alpha} \\ \boldsymbol{\eta} \end{bmatrix} = \begin{bmatrix} \mathbf{K} \\ \mathbf{Q}^\top \end{bmatrix} \boldsymbol{y}. \tag{7.39}$$

As long as $\mathbf{Q}$ is of full column rank, the minimizing function is unique.

■ **Example 7.40 (Ridge Regression (cont.))** We return to Example 7.8 and identify that $\mathcal{H}$ is the RKHS with linear kernel function $\kappa(\boldsymbol{x}, \boldsymbol{x}') = \boldsymbol{x}^\top \boldsymbol{x}'$ and $C = \mathcal{H}_0$ is the linear space of constant functions. In this case, $\mathcal{H}_0$ is spanned by the function $q_1 \equiv 1$. Moreover, $\mathbf{K} = \mathbf{X}\mathbf{X}^\top$ and $\mathbf{Q} = \mathbf{1}$.

If we appeal to the representer theorem directly, then the problem in (7.9) becomes, as a result of (7.37):

$$\min_{\boldsymbol{\alpha}, \eta_0} \left\| \boldsymbol{y} - \eta_0 \mathbf{1} - \mathbf{X}\mathbf{X}^\top \boldsymbol{\alpha} \right\|^2 + \lambda \|\mathbf{X}^\top \boldsymbol{\alpha}\|^2.$$

This is a convex optimization problem, and so the solution follows by taking derivatives and setting them to zero. This gives the equations

$$\mathbf{X}\mathbf{X}^\top \left((\mathbf{X}\mathbf{X}^\top + \lambda \mathbf{I}_n)\, \boldsymbol{\alpha} + \eta_0 \mathbf{1} - \boldsymbol{y} \right) = \mathbf{0},$$

and

$$n\,\eta_0 = \mathbf{1}^\top(\boldsymbol{y} - \mathbf{X}\mathbf{X}^\top\boldsymbol{\alpha}).$$

Note that these are equivalent to (7.10) and (7.11) (once again assuming that $n \geq p$ and $\mathbf{X}$ has full rank p). Equivalently, the solution is found by solving (7.39):

$$\begin{bmatrix} \mathbf{X}\mathbf{X}^\top\mathbf{X}\mathbf{X}^\top + \lambda\,\mathbf{X}\mathbf{X}^\top & \mathbf{X}\mathbf{X}^\top\mathbf{1} \\ \mathbf{1}^\top\mathbf{X}\mathbf{X}^\top & n \end{bmatrix} \begin{bmatrix} \boldsymbol{\alpha} \\ \eta_0 \end{bmatrix} = \begin{bmatrix} \mathbf{X}\mathbf{X}^\top \\ \mathbf{1}^\top \end{bmatrix} \boldsymbol{y}.$$

This is a system of $(n+1)$ linear equations, and is typically of much larger dimension than the $(p+1)$ linear equations given by (7.10) and (7.11). As such, one may question the practicality of reformulating the problem in this way. However, the benefit of this formulation is that the problem can be expressed entirely through the Gram matrix $\mathbf{K}$, without having to explicitly compute the feature vectors — in turn permitting the (implicit) use of infinite-dimensional feature spaces. ■

■ **Example 7.41 (Estimating the Peaks Function)** Figure 7.43 shows the surface plot of the *peaks* function:

$$(7.42)\quad f(x_1,x_2) = 3(1-x_1)^2\mathrm{e}^{-x_1^2-(x_2+1)^2} - 10\left(\frac{x_1}{5} - x_1^3 - x_2^5\right)\mathrm{e}^{-x_1^2-x_2^2} - \frac{1}{3}\mathrm{e}^{-(x_1+1)^2-x_2^2}.$$

The goal is to learn the function $y = f(\boldsymbol{x})$ based on a small set of training data (pairs of $(\boldsymbol{x}, y)$ values). The red dots in the figure represent data $\tau = \{(\boldsymbol{x}_i, y_i)\}_{i=1}^{20}$, where $y_i = f(\boldsymbol{x}_i)$ and the $\{\boldsymbol{x}_i\}$ have been chosen in a **quasi-random** way, using *Hammersley points* (with bases 2 and 3) on the square $[-3, 3]^2$. Quasi-random point sets have better space-filling properties than either a regular grid of points or a set of pseudo-random points. We refer to [98] for details. Note that there is no observation noise in this particular problem.

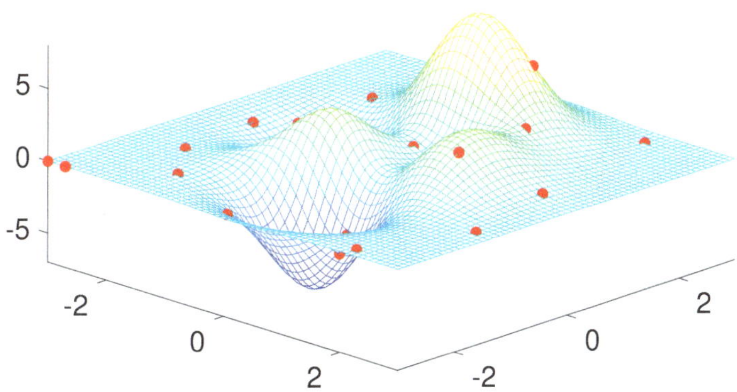

Figure 7.43: Peaks function sampled at 20 Hammersley points.

The purpose of this example is to illustrate how, using the small data set of size $n = 20$, the entire *peaks* function can be approximated well using kernel methods. In particular, we use the Gaussian kernel (7.23) on $\mathbb{R}^2$, and denote by $\mathcal{H}$ the unique RKHS corresponding to this kernel. We omit the regularization term in (7.33), and thus our objective is to find the solution to

$$\min_{g \in \mathcal{H}} \sum_{i=1}^{n} (y_i - g(\boldsymbol{x}_i))^2.$$

By the representer theorem, the optimal function is of the form

$$g(\boldsymbol{x}) = \sum_{i=1}^{n} \alpha_i \exp\left(-\frac{1}{2}\frac{\|\boldsymbol{x} - \boldsymbol{x}_i\|^2}{\sigma^2}\right),$$

where $\boldsymbol{\alpha} := [\alpha_1, \dots, \alpha_n]^\top$ is, by (7.39), the solution to the set of linear equations:

$$\mathbf{K}\mathbf{K}\boldsymbol{\alpha} = \mathbf{K}\boldsymbol{y}.$$

Note that we are performing regression over the class of functions $\mathcal{H}$ with an implicit feature space. Due to the representer theorem, the solution to this problem coincides with the solution to the linear regression problem for which the i-th feature (for $i = 1, \dots, n$) is chosen to be the vector $[\kappa(\boldsymbol{x}_1, \boldsymbol{x}_i), \dots, \kappa(\boldsymbol{x}_n, \boldsymbol{x}_i)]^\top$.

The following code performs these calculations and gives the contour plots of g and the peaks functions, shown in Figure 7.44. We see that the two are quite close. Code for the generation of Hammersley points is available from the book's GitHub site as genham.py.

peakskernel.py

```
from genham import hammersley
import numpy as np
import matplotlib.pyplot as plt
from mpl_toolkits.mplot3d import Axes3D
from matplotlib import cm
from numpy.linalg import norm

import numpy as np
def peaks(x,y):
    z =  (3*(1-x)**2 * np.exp(-(x**2) - (y+1)**2)
           - 10*(x/5 - x**3 - y**5) * np.exp(-x**2 - y**2)
           - 1/3 * np.exp(-(x+1)**2 - y**2))
    return(z)

n = 20
x = -3 + 6*hammersley([2,3],n)
z = peaks(x[:,0],x[:,1])
xx, yy = np.mgrid[-3:3:150j,-3:3:150j]
zz = peaks(xx,yy)
plt.contour(xx,yy,zz,levels=50)

fig=plt.figure()
ax = fig.add_subplot(111,projection='3d')
ax.plot_surface(xx,yy,zz,rstride=1,cstride=1,color='c',alpha=0.3,
    linewidth=0)
ax.scatter(x[:,0],x[:,1],z,color='k',s=20)
```

```
plt.show()

sig2 = 0.3 # kernel parameter
def k(x,u):
    return(np.exp(-0.5*norm(x- u)**2/sig2))
K = np.zeros((n,n))
for i in range(n):
    for j in range(n):
        K[i,j] = k(x[i,:],x[j])
alpha = np.linalg.solve(K@K.T, K@z)

N, = xx.flatten().shape
Kx = np.zeros((n,N))
for i in range(n):
    for j in range(N):
        Kx[i,j] = k(x[i,:],np.array([xx.flatten()[j],yy.flatten()[j]]))

g = Kx.T @ alpha
dim = np.sqrt(N).astype(int)
yhat = g.reshape(dim,dim)
plt.contour(xx,yy,yhat,levels=50)
```

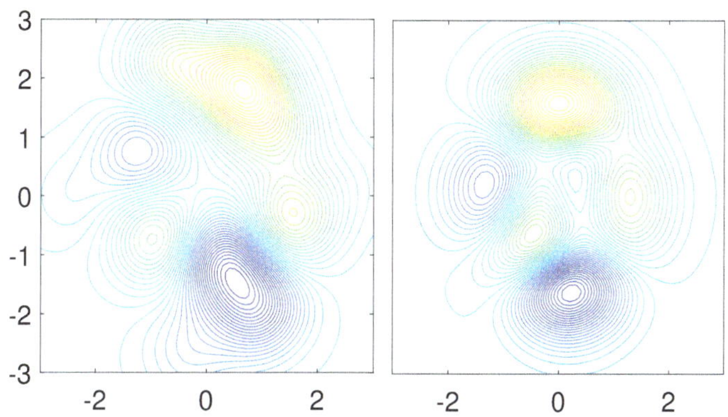

Figure 7.44: Contour plots for the prediction function g (left) and the *peaks* function given in (7.42) (right).

■

7.6 Smoothing Cubic Splines

A striking application of kernel methods is to fitting "well-behaved" functions to data. Key examples of "well-behaved" functions are those that do not have large second-order derivatives. Consider functions $g : [0, 1] \to \mathbb{R}$ that are twice differentiable and define

$$\|g''\|^2 := \int_0^1 (g''(x))^2 \, dx$$

as a measure of the size of the second derivative.

Example 7.45 (Behavior of $\|g''\|^2$) Intuitively, the larger $\|g''\|^2$ is, the more "wiggly" the function g will be. As an explicit example, consider $g(x) = \sin(\omega x)$ for $x \in [0, 1]$, where ω is a free parameter. We can explicitly compute $g''(x) = -\omega^2 \sin(\omega x)$, and consequently

$$\|g''\|^2 = \int_0^1 \omega^4 \sin^2(\omega x)\,\mathrm{d}x = \frac{\omega^4}{2}\,(1 - \mathrm{sinc}(2\omega))\,.$$

As $|\omega| \to \infty$, the frequency of g increases and we have $\|g''\|^2 \to \infty$. ■

Now, in the context of data fitting, consider the following penalized least-squares optimization problem on $[0, 1]$:

$$\min_{g\in\mathcal{G}} \sum_{i=1}^{n} (y_i - g(x_i))^2 + \lambda\,\|g''\|^2, \tag{7.46}$$

where we will specify $\mathcal{G}$ in what follows. In order to apply the kernel machinery, we want to write this in the form (7.33), for some RKHS $\mathcal{H}$ and null space $\mathcal{H}_0$. Clearly, the norm on $\mathcal{H}$ should be of the form $\|g\|_{\mathcal{H}} = \|g''\|$ and should be well-defined (i.e., finite and ensuring g and g' are absolutely continuous). This suggests that we take

$$\mathcal{H} = \{g \in L^2[0,1] : \|g''\| < \infty, g, g' \text{ absolutely continuous},\ g(0) = g'(0) = 0\},$$

with inner product

$$\langle f, g\rangle_{\mathcal{H}} := \int_0^1 f''(x)\,g''(x)\,\mathrm{d}x.$$

One rationale for imposing the boundary conditions $g(0) = g'(0) = 0$ is as follows: when expanding g about the point $x = 0$, Taylor's theorem (with integral remainder term) states that

$$g(x) = g(0) + g'(0)\,x + \int_0^x g''(s)\,(x - s)\,\mathrm{d}s.$$

Imposing the condition that $g(0) = g'(0) = 0$ for functions in $\mathcal{H}$ will ensure that $\mathcal{G} = \mathcal{H} \oplus \mathcal{H}_0$ where the null space $\mathcal{H}_0$ contains only linear functions, as we will see.

To see that this $\mathcal{H}$ is in fact an RKHS, we derive its reproducing kernel. Using integration by parts (or directly from the Taylor expansion above), write

$$g(x) = \int_0^x g'(s)\,\mathrm{d}s = \int_0^x g''(s)\,(x - s)\,\mathrm{d}s = \int_0^1 g''(s)\,(x - s)_+\,\mathrm{d}s,$$

where $y_+ := \max\{y, 0\}$. If κ is a kernel, then by the reproducing property it must hold that

$$g(x) = \langle g, \kappa_x\rangle_{\mathcal{H}} = \int_0^1 g''(s)\,\kappa_x''(s)\,\mathrm{d}s,$$

so that κ must satisfy $\frac{\partial^2}{\partial s^2}\kappa(x, s) = (x - s)_+$. Therefore, noting that $\kappa(x, u) = \langle \kappa_x, \kappa_u\rangle_{\mathcal{H}}$, we have (see Exercise 15)

$$\kappa(x, u) = \int_0^1 \frac{\partial^2\kappa(x, s)}{\partial s^2}\,\frac{\partial^2\kappa(u, s)}{\partial s^2}\,\mathrm{d}s = \frac{\max\{x, u\}\min\{x, u\}^2}{2} - \frac{\min\{x, u\}^3}{6}.$$

The last expression is a cubic function with quadratic and cubic terms that misses the constant and linear monomials. This is not surprising considering the Taylor's theorem interpretation of a function $g \in \mathcal{H}$. If we now take $\mathcal{H}_0$ as the space of functions of the following form (having zero second derivative):

$$h_0 = \eta_1 + \eta_2 x, \quad x \in [0, 1],$$

then (7.46) is exactly of the form (7.33).

As a consequence of the representer Theorem 7.35, the optimal solution to (7.46) is a linear combination of piecewise cubic functions:

$$g(x) = \eta_1 + \eta_2 x + \sum_{i=1}^{n} \alpha_i \, \kappa(x_i, x). \tag{7.47}$$

Such a function is called a **cubic spline** with n *knots* (with one knot at each data point x_i) — so called, because the piecewise cubic function between knots is required to be "tied together" at the knots. The parameters $\boldsymbol{\alpha}, \boldsymbol{\eta}$ are determined from (7.37) for instance by solving (7.39) with matrices $\mathbf{K} = [\kappa(x_i, x_j)]_{i,j=1}^n$ and $\mathbf{Q}$ with i-th row of the form $[1, x_i]$ for $i = 1, \ldots, n$.

Example 7.48 (Smoothing Spline) Figure 7.49 shows various cubic smoothing splines for the data $(0.05, 0.4), (0.2, 0.2), (0.5, 0.6), (0.75, 0.7), (1, 1)$. In the figure, we use the reparameterization $r = 1/(1 + \lambda)$ for the smoothing parameter. Thus $r \in [0, 1]$, where $r = 0$ means an infinite penalty for curvature (leading to the ordinary linear regression solution) and $r = 1$ does not penalize curvature at all and leads to a perfect fit via the so-called *natural spline*. Of course the latter will generally lead to overfitting. For r from 0 up to 0.8 the solutions will be close to the simple linear regression line, while only for r very close to 1, the shape of the curve changes significantly.

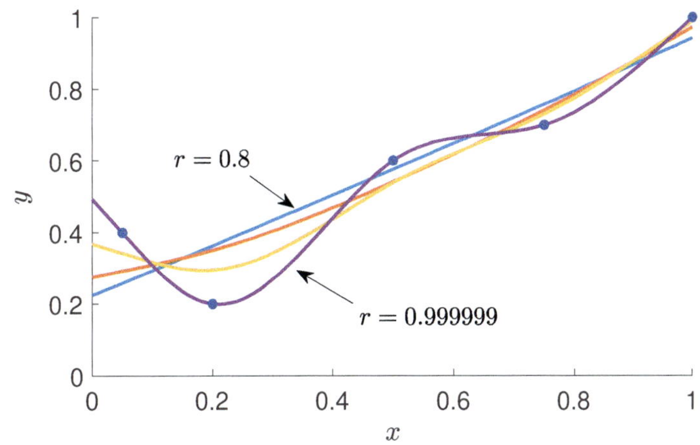

Figure 7.49: Various cubic smoothing splines for smoothing parameter $r = 1/(1 + \lambda) \in \{0.8, 0.99, 0.999, 0.999999\}$. For $r = 1$, the natural spline through the data points is obtained; for $r = 0$, the simple linear regression line is found.

The following code first computes the matrices $\mathbf{K}$ and $\mathbf{Q}$, and then solves the linear system (7.39). Finally, the smoothing curve is determined via (7.47), for selected points, and then plotted. Note that the code plots only a single curve corresponding to the specified value of p.

smoothspline.py

```
import matplotlib.pyplot as plt
import numpy as np

x = np.array([[0.05, 0.2, 0.5, 0.75, 1.]]).T
y = np.array([[0.4, 0.2, 0.6, 0.7, 1.]]).T

n = x.shape[0]
r = 0.999
ngamma = (1-r)/r

k = lambda x1, x2 : (1/2)* np.max((x1,x2)) * np.min((x1,x2)) ** 2 \
                          - ((1/6)* np.min((x1,x2))**3)
K = np.zeros((n,n))
for i in range(n):
    for j in range(n):
        K[i,j] = k(x[i], x[j])

Q = np.hstack((np.ones((n,1)), x))

m1 = np.hstack((K @ K.T + (ngamma * K), K @ Q))
m2 = np.hstack((Q.T @ K.T, Q.T @ Q))
M = np.vstack((m1,m2))

c = np.vstack((K, Q.T)) @ y

ad = np.linalg.solve(M,c)

# plot the curve
xx = np.arange(0,1+0.01,0.01).reshape(-1,1)

g = np.zeros_like(xx)
Qx = np.hstack((np.ones_like(xx), xx))
g = np.zeros_like(xx)
N = np.shape(xx)[0]

Kx = np.zeros((n,N))
for i in range(n):
    for j in range(N):
        Kx[i,j] = k(x[i], xx[j])

g = g + np.hstack((Kx.T, Qx)) @ ad

plt.ylim((0,1.15))
plt.plot(xx, g, label = 'r = {}'.format(r), linewidth = 2)
plt.plot(x,y, 'b.', markersize=15)
plt.xlabel('$x$')
plt.ylabel('$y$')
plt.legend()
```

■

7.7 Gaussian Process Regression

Another application of the kernel machinery is to Gaussian process regression. A **Gaussian process** (GP) on a space $\mathcal{X}$ is a stochastic process $\{Z_{\boldsymbol{x}}, \boldsymbol{x} \in \mathcal{X}\}$ where, for any choice of indices $\boldsymbol{x}_1, \ldots, \boldsymbol{x}_n$, the vector $[Z_{\boldsymbol{x}_1}, \ldots, Z_{\boldsymbol{x}_n}]^\top$ has a multivariate Gaussian distribution. As such, the distribution of a GP is completely specified by its mean and covariance functions $\mu : \mathcal{X} \to \mathbb{R}$ and $\kappa : \mathcal{X} \times \mathcal{X} \to \mathbb{R}$, respectively. The covariance function is a finite positive semidefinite function, and hence, in view of Theorem 7.16, can be viewed as a reproducing kernel on $\mathcal{X}$.

As for ordinary regression, the objective of GP regression is to learn a regression function g that predicts a response $y = g(\boldsymbol{x})$ for each feature vector $\boldsymbol{x}$. This is done in a Bayesian fashion, by establishing (1) a prior pdf for g and (2) the likelihood of the data, for a given g. From these two we then derive, via Bayes' formula, the posterior distribution of g given the data. We refer to Section 2.9 for the general Bayesian framework. ☞ 198 ☞ 49

A simple Bayesian model for GP regression is as follows: First, the prior distribution of g is taken to be the distribution of a GP with some known mean function μ and covariance function (that is, kernel) κ. Most often μ is taken to be a constant, and for simplicity of exposition, we take it to be 0. The Gaussian kernel (7.23) is often used for the covariance function. For radial basis function kernels (including the Gaussian kernel), points that are closer will be more highly correlated or "similar" [138], independent of translations in space.

Second, similar to standard regression, we view the observed feature vectors $\boldsymbol{x}_1, \ldots, \boldsymbol{x}_n$ as fixed and the responses $y_1, \ldots, y_n$ as outcomes of random variables $Y_1, \ldots, Y_n$. Specifically, given g, we model the $\{Y_i\}$ as

$$Y_i = g(\boldsymbol{x}_i) + \varepsilon_i, \quad i = 1, \ldots, n, \tag{7.50}$$

where $\{\varepsilon_i\} \overset{\text{iid}}{\sim} \mathcal{N}(0, \varsigma^2)$. To simplify the analysis, let us assume that ς^2 is known, so no prior needs to be specified for ς^2. Let $\boldsymbol{g} = [g(\boldsymbol{x}_1), \ldots, g(\boldsymbol{x}_n)]^\top$ be the (unknown) vector of regression values. Placing a GP prior on the function g is equivalent to placing a multivariate Gaussian prior on the vector $\boldsymbol{g}$:

$$\boldsymbol{g} \sim \mathcal{N}(\boldsymbol{0}, \mathbf{K}), \tag{7.51}$$

where the covariance matrix $\mathbf{K}$ of $\boldsymbol{g}$ is a Gram matrix (implicitly associated with a feature map through the kernel κ), given by:

$$\mathbf{K} = \begin{bmatrix} \kappa(\boldsymbol{x}_1, \boldsymbol{x}_1) & \kappa(\boldsymbol{x}_1, \boldsymbol{x}_2) & \ldots & \kappa(\boldsymbol{x}_1, \boldsymbol{x}_n) \\ \kappa(\boldsymbol{x}_2, \boldsymbol{x}_1) & \kappa(\boldsymbol{x}_2, \boldsymbol{x}_2) & \ldots & \kappa(\boldsymbol{x}_2, \boldsymbol{x}_n) \\ \vdots & \vdots & \ddots & \vdots \\ \kappa(\boldsymbol{x}_n, \boldsymbol{x}_1) & \kappa(\boldsymbol{x}_n, \boldsymbol{x}_2) & \ldots & \kappa(\boldsymbol{x}_n, \boldsymbol{x}_n) \end{bmatrix}. \tag{7.52}$$

The likelihood of our data given $\boldsymbol{g}$, denoted $p(\boldsymbol{y} \mid \boldsymbol{g})$, is obtained directly from the model (7.50):

$$(\boldsymbol{Y} \mid \boldsymbol{g}) \sim \mathcal{N}(\boldsymbol{g}, \varsigma^2 \mathbf{I}_n). \tag{7.53}$$

Solving this Bayesian problem involves deriving the posterior distribution of $(\boldsymbol{g} \mid \boldsymbol{Y})$. To do so, we first note that since $\boldsymbol{Y}$ has covariance matrix $\mathbf{K} + \varsigma^2 \mathbf{I}_n$ (which can be seen from

(7.50)), the joint distribution of $\boldsymbol{Y}$ and $\boldsymbol{g}$ is again normal, with mean $\mathbf{0}$ and covariance matrix:

$$\mathbf{K}_{y,g} = \begin{bmatrix} \mathbf{K} + \varsigma^2 \mathbf{I}_n & \mathbf{K} \\ \mathbf{K} & \mathbf{K} \end{bmatrix}. \tag{7.54}$$

☞ 641 The posterior can then be found by conditioning on $\boldsymbol{Y} = \boldsymbol{y}$, via Theorem D.53, giving

$$(\boldsymbol{g} \mid \boldsymbol{y}) \sim \mathcal{N}\left(\mathbf{K}^\top(\mathbf{K} + \varsigma^2\mathbf{I}_n)^{-1}\boldsymbol{y},\ \mathbf{K} - \mathbf{K}^\top(\mathbf{K} + \varsigma^2\mathbf{I}_n)^{-1}\mathbf{K}\right).$$

This only gives information about g at the observed points $\boldsymbol{x}_1, \ldots, \boldsymbol{x}_n$. It is more interesting to consider the posterior predictive distribution of $\widetilde{g} := g(\widetilde{\boldsymbol{x}})$ for a new input $\widetilde{\boldsymbol{x}}$. We can find the corresponding posterior predictive pdf $p(\widetilde{g} \mid \boldsymbol{y})$ by integrating out the joint posterior pdf $p(\widetilde{g}, \boldsymbol{g} \mid \boldsymbol{y})$, which is equivalent to taking the expectation of $p(\widetilde{g} \mid \boldsymbol{g})$ when $\boldsymbol{g}$ is distributed according to the posterior pdf $p(\boldsymbol{g} \mid \boldsymbol{y})$; that is,

$$p(\widetilde{g} \mid \boldsymbol{y}) = \int p(\widetilde{g} \mid \boldsymbol{g})\, p(\boldsymbol{g} \mid \boldsymbol{y})\, \mathrm{d}\boldsymbol{g}.$$

To do so more easily than direct evaluation via the above integral representation of $p(\widetilde{g} \mid \boldsymbol{y})$, we can begin with the joint distribution of $[\boldsymbol{y}^\top, \widetilde{g}]^\top$, which is multivariate normal with mean $\mathbf{0}$ and covariance matrix

$$\widetilde{\mathbf{K}} = \begin{bmatrix} \mathbf{K} + \varsigma^2 \mathbf{I}_n & \boldsymbol{\kappa} \\ \boldsymbol{\kappa}^\top & \kappa(\widetilde{\boldsymbol{x}}, \widetilde{\boldsymbol{x}}) \end{bmatrix}, \tag{7.55}$$

where $\boldsymbol{\kappa} = [\kappa(\widetilde{\boldsymbol{x}}, \boldsymbol{x}_1), \ldots, \kappa(\widetilde{\boldsymbol{x}}, \boldsymbol{x}_n)]^\top$. It now follows, again by using Theorem D.53, that $(\widetilde{g} \mid \boldsymbol{y})$ has a normal distribution with mean and variance given respectively by

$$\mu(\widetilde{\boldsymbol{x}}) = \boldsymbol{\kappa}^\top(\mathbf{K} + \varsigma^2\mathbf{I}_n)^{-1}\boldsymbol{y} \tag{7.56}$$

and

$$\sigma^2(\widetilde{\boldsymbol{x}}) = \kappa(\widetilde{\boldsymbol{x}}, \widetilde{\boldsymbol{x}}) - \boldsymbol{\kappa}^\top(\mathbf{K} + \varsigma^2\mathbf{I}_n)^{-1}\boldsymbol{\kappa}. \tag{7.57}$$

These are the **predictive** mean and variance parameters. It is important to note that we are predicting the *expected* response $\mathbb{E}\widetilde{Y} = g(\widetilde{\boldsymbol{x}})$ here, and not the actual response $\widetilde{Y}$.

■ **Example 7.58 (GP Regression)** Suppose the regression function is

$$g(x) = 2\sin(2\pi x), \quad x \in [0, 1].$$

We use GP regression to estimate g, using a Gaussian kernel of the form (7.23) with bandwidth parameter $\sigma = 0.2$. The explanatory variables $x_1, \ldots, x_{30}$ were drawn uniformly on the interval $[0, 1]$, and the responses were obtained from (7.50), with noise level $\varsigma = 0.5$. Figure 7.59 shows 10 samples from the prior distribution for g as well as the data points and the true sinusoidal regression function g.

Again assuming that the variance ς^2 is known, the predictive distribution as determined by (7.56) and (7.57) is shown in Figure 7.60 for bandwidth $\sigma = 0.2$ (left) and $\sigma = 0.02$ (right).

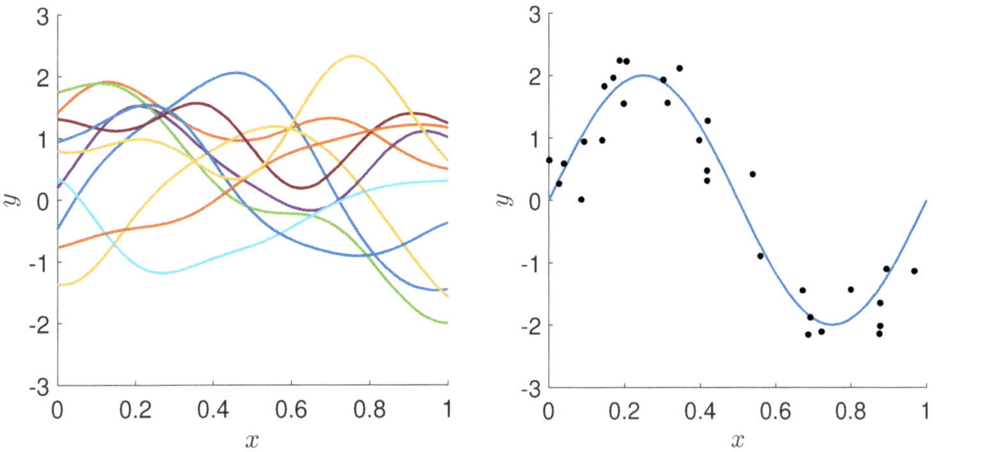

Figure 7.59: Left: samples drawn from the GP prior distribution. Right: the true regression function with the data points.

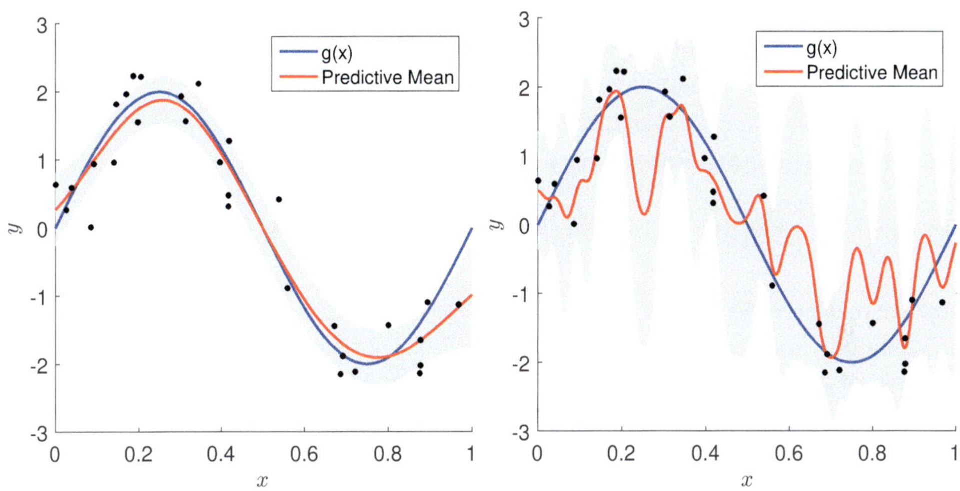

Figure 7.60: GP regression of synthetic data set with bandwidth $\sigma = 0.2$ (left) and $\sigma = 0.02$ (right). The black dots represent the data and the blue curve is the latent function $g(x) = 2\sin(2\pi x)$. The red curve is the mean of the GP predictive distribution given by (7.56), and the shaded region is the 95% confidence band, corresponding to the predictive variance given in (7.57).

Clearly, decreasing the bandwidth leads to the covariance between points x and x' decreasing at a faster rate with respect to the squared distance $\|x - x'\|^2$, leading to a predictive mean that is less smooth. ■

In the above exposition, we have taken the mean function for the prior distribution of g to be identically zero. If instead we have a general mean function m and write

$$\boldsymbol{m} = [m(\boldsymbol{x}_1), \ldots, m(\boldsymbol{x}_n)]^\top,$$

then the predictive variance (7.57) remains unchanged, and the predictive mean (7.56) is modified to read

$$\mu(\widetilde{x}) = m(\widetilde{x}) + \boldsymbol{\kappa}^\top (\mathbf{K} + \varsigma^2 \mathbf{I}_n)^{-1} (\boldsymbol{y} - \boldsymbol{m}) . \tag{7.61}$$

Typically, the variance ς^2 appearing in (7.50) is not known, and the kernel κ itself depends on several parameters — for instance a Gaussian kernel (7.23) with an unknown bandwidth parameter σ. In the Bayesian framework, one typically specifies a hierarchical model by introducing a prior $p(\boldsymbol{\theta})$ for the vector $\boldsymbol{\theta}$ of such **hyperparameters**. Now, the GP prior $(g \mid \boldsymbol{\theta})$ (equivalently, specifying $p(\boldsymbol{g} \mid \boldsymbol{\theta})$) and the model for the likelihood of the data given $\boldsymbol{Y}|\boldsymbol{g}, \boldsymbol{\theta}$, namely $p(\boldsymbol{y} \mid \boldsymbol{g}, \boldsymbol{\theta})$, are both dependent on $\boldsymbol{\theta}$. The posterior distribution of $(\boldsymbol{g} \mid \boldsymbol{y}, \boldsymbol{\theta})$ remains unchanged.

One approach to setting the hyperparameter $\boldsymbol{\theta}$ is to determine its posterior $p(\boldsymbol{\theta} \mid \boldsymbol{y})$ and obtain a point estimate, for instance via its maximum a posteriori estimate — a computationally demanding task. A practical alternative is to consider the *marginal likelihood* $p(\boldsymbol{y} \mid \boldsymbol{\theta})$ and maximize it with respect to $\boldsymbol{\theta}$. This procedure is called **empirical Bayes**.

Considering again the mean function m to be identically zero, from (7.54), we have that $(\boldsymbol{Y} \mid \boldsymbol{\theta})$ is multivariate normal with mean $\mathbf{0}$ and covariance matrix:

$$\mathbf{K}_y = \mathbf{K} + \varsigma^2 \mathbf{I}_n,$$

immediately giving an expression for the marginal log-likelihood:

$$\ln p(\boldsymbol{y} \mid \boldsymbol{\theta}) = -\frac{n}{2} \ln(2\pi) - \frac{1}{2} \ln |\mathbf{K}_y| - \frac{1}{2} \boldsymbol{y}^\top \mathbf{K}_y^{-1} \boldsymbol{y}, \tag{7.62}$$

where we recall that $|\mathbf{A}|$ stands for the absolute value of the determinant of matrix $\mathbf{A}$. We notice that only the second and third terms in (7.62) depend on $\boldsymbol{\theta}$. Considering a partial derivative of (7.62) with respect to a single element θ of the hyperparameter vector $\boldsymbol{\theta}$ yields

$$\frac{\partial}{\partial \theta} \ln p(\boldsymbol{y} \mid \boldsymbol{\theta}) = -\frac{1}{2} \mathrm{tr} \left(\mathbf{K}_y^{-1} \left[\frac{\partial}{\partial \theta} \mathbf{K}_y \right] \right) + \frac{1}{2} \boldsymbol{y}^\top \mathbf{K}_y^{-1} \left[\frac{\partial}{\partial \theta} \mathbf{K}_y \right] \mathbf{K}_y^{-1} \boldsymbol{y}, \tag{7.63}$$

where $\left[\frac{\partial}{\partial \theta} \mathbf{K}_y \right]$ is the element-wise derivative of matrix $\mathbf{K}_y$ with respect to θ. If these partial derivatives can be computed for each hyperparameter θ, gradient information could be used when maximizing (7.62).

Finally, we mention that when $\mathbf{K}_y$ is very large the computation of the trace in (7.63) can be very costly and one may have to use specialized approximation methods; see, for example, [119].

■ **Example 7.64 (GP Regression (cont.))** Continuing Example 7.58, we plot in Figure 7.65 the marginal log-likelihood as a function of the noise level ς and bandwidth parameter σ.

The maximum is attained for a bandwidth parameter σ around 0.20 and $\varsigma \approx 0.44$, which is very close to the left panel of Figure 7.60 for the case where ς was assumed to be known (and equal to 0.5). We note here that the marginal log-likelihood is extremely flat, perhaps owing to the small number of points.

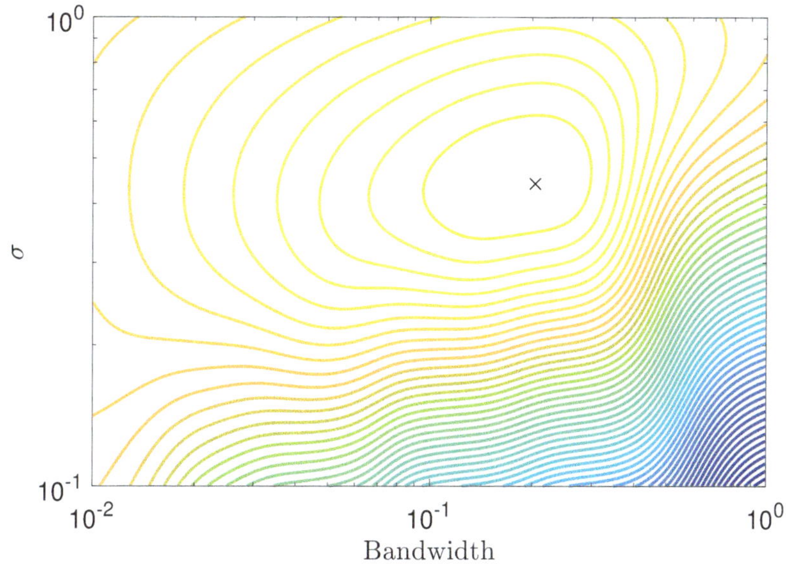

Figure 7.65: Contours of the marginal log-likelihood for the GP regression example. The maximum is denoted by a cross.

■

7.8 Kernel PCA

In its basic form, kernel PCA (principal component analysis) can be thought of as PCA in feature space. The main motivation for PCA introduced in Section 4.9 was ☞ 179 as a dimensionality reduction technique. There, the analysis rested on an SVD of the matrix $\widehat{\mathbf{\Sigma}} = \frac{1}{n}\mathbf{X}^\top\mathbf{X}$, where the data in $\mathbf{X}$ was first centered via $x'_{i,j} = x_{i,j} - \overline{x}_j$ where $\overline{x}_i = \frac{1}{n}\sum_{i=1}^n x_{i,j}$.

What we shall do is to first re-cast the problem in terms of the Gram matrix $\mathbf{K} = \mathbf{X}\mathbf{X}^\top = [\langle \boldsymbol{x}_i, \boldsymbol{x}_j\rangle]$ (note the different order of $\mathbf{X}$ and $\mathbf{X}^\top$), and subsequently replace the inner product $\langle \boldsymbol{x}, \boldsymbol{x}'\rangle$ with $\kappa(\boldsymbol{x}, \boldsymbol{x}')$ for a general reproducing kernel κ. To make the link, let us start with an SVD of $\mathbf{X}^\top$:

$$\mathbf{X}^\top = \mathbf{U}\mathbf{D}\mathbf{V}^\top. \tag{7.66}$$

The dimensions of $\mathbf{X}^\top$, $\mathbf{U}$, $\mathbf{D}$, and $\mathbf{V}$ are $d \times n$, $d \times d$, $d \times n$, and $n \times n$, respectively. Then an SVD of $\mathbf{X}^\top\mathbf{X}$ is

$$\mathbf{X}^\top\mathbf{X} = (\mathbf{U}\mathbf{D}\mathbf{V}^\top)(\mathbf{U}\mathbf{D}\mathbf{V}^\top)^\top = \mathbf{U}(\mathbf{D}\mathbf{D}^\top)\mathbf{U}^\top$$

and an SVD of $\mathbf{K}$ is

$$\mathbf{K} = (\mathbf{U}\mathbf{D}\mathbf{V}^\top)^\top(\mathbf{U}\mathbf{D}\mathbf{V}^\top) = \mathbf{V}(\mathbf{D}^\top\mathbf{D})\mathbf{V}^\top.$$

Let $\lambda_1 \geq \cdots \geq \lambda_r > 0$ denote the non-zero eigenvalues of $\mathbf{X}^\top\mathbf{X}$ (or, equivalently, of $\mathbf{K}$) and denote the corresponding $r \times r$ diagonal matrix by $\mathbf{\Lambda}$. Without loss of generality we can assume that the eigenvector of $\mathbf{X}^\top\mathbf{X}$ corresponding to λ_k is the k-th column of $\mathbf{U}$

☞ 179 and that the k-th column of $\mathbf{V}$ is an eigenvector of $\mathbf{K}$. Similar to Section 4.9, let $\mathbf{U}_k$ and $\mathbf{V}_k$ contain the first k columns of $\mathbf{U}$ and $\mathbf{V}$, respectively, and let $\mathbf{\Lambda}_k$ be the corresponding $k \times k$ submatrix of $\mathbf{\Lambda}$, $k = 1, \ldots, r$.

By the SVD (7.66), we have

$$\mathbf{X}^\top \mathbf{V}_k = \mathbf{U}\mathbf{D}\mathbf{V}^\top \mathbf{V}_k = \mathbf{U}_k \mathbf{\Lambda}_k^{1/2}.$$

Next, consider the projection of a point $\boldsymbol{x}$ onto the k-dimensional linear space spanned by the columns of $\mathbf{U}_k$ — the first k principal components. We saw in Section 4.9 that this projection simply is the linear mapping $\boldsymbol{x} \mapsto \mathbf{U}_k^\top \boldsymbol{x}$. Using the fact that $\mathbf{U}_k = \mathbf{X}^\top \mathbf{V}_k \mathbf{\Lambda}^{-1/2}$, we find that $\boldsymbol{x}$ is projected to a point $\boldsymbol{z}$ given by

$$\boldsymbol{z} = \mathbf{\Lambda}_k^{-1/2} \mathbf{V}_k^\top \mathbf{X} \boldsymbol{x} = \mathbf{\Lambda}_k^{-1/2} \mathbf{V}_k^\top \boldsymbol{\kappa}_{\boldsymbol{x}},$$

where we have (suggestively) defined

$$\boldsymbol{\kappa}_{\boldsymbol{x}} := [\langle \boldsymbol{x}_1, \boldsymbol{x} \rangle, \ldots, \langle \boldsymbol{x}_n, \boldsymbol{x} \rangle]^\top.$$

The important point is that $\boldsymbol{z}$ is completely determined by the vector of inner products $\boldsymbol{\kappa}_{\boldsymbol{x}}$ and the k principal eigenvalues and (right) eigenvectors of the Gram matrix $\mathbf{K}$. Note that each component z_m of $\boldsymbol{z}$ is of the form

$$z_m = \sum_{i=1}^{n} \alpha_{m,i}\, \kappa(\boldsymbol{x}_i, \boldsymbol{x}), \quad m = 1, \ldots, k. \tag{7.67}$$

The preceding discussion assumed centering of the columns of $\mathbf{X}$. Consider now an uncentered data matrix $\widetilde{\mathbf{X}}$. Then the centered data can be written as

$$\mathbf{X} = \widetilde{\mathbf{X}} - \frac{1}{n}\mathbf{E}_n \widetilde{\mathbf{X}},$$

where $\mathbf{E}_n$ is the $n \times n$ matrix of ones. Consequently,

$$\mathbf{X}\mathbf{X}^\top = \widetilde{\mathbf{X}}\widetilde{\mathbf{X}}^\top - \frac{1}{n}\mathbf{E}_n \widetilde{\mathbf{X}}\widetilde{\mathbf{X}}^\top - \frac{1}{n}\widetilde{\mathbf{X}}\widetilde{\mathbf{X}}^\top \mathbf{E}_n + \frac{1}{n^2}\mathbf{E}_n \widetilde{\mathbf{X}}\widetilde{\mathbf{X}}^\top \mathbf{E}_n,$$

or, more compactly, $\mathbf{X}\mathbf{X}^\top = \mathbf{H}\widetilde{\mathbf{X}}\widetilde{\mathbf{X}}^\top \mathbf{H}$, where

$$\mathbf{H} := \mathbf{I}_n - \frac{1}{n}\mathbf{1}_n \mathbf{1}_n^\top,$$

$\mathbf{I}_n$ is the $n \times n$ identity matrix, and $\mathbf{1}_n$ is the $n \times 1$ vector of ones.

To generalize to the kernel setting, we replace $\widetilde{\mathbf{X}}\widetilde{\mathbf{X}}^\top$ by

$$\mathbf{K} = [\kappa(\boldsymbol{x}_i, \boldsymbol{x}_j), i, j = 1, \ldots, n],$$

and set

$$\boldsymbol{\kappa}_{\boldsymbol{x}} = [\kappa(\boldsymbol{x}_1, \boldsymbol{x}), \ldots, \kappa(\boldsymbol{x}_n, \boldsymbol{x})]^\top,$$

so that $\mathbf{\Lambda}_k$ is the diagonal matrix of the k largest eigenvalues of $\mathbf{HKH}$ and $\mathbf{V}_k$ is the corresponding matrix of eigenvectors. Note that the "usual" PCA is recovered when we use the linear kernel $\kappa(\boldsymbol{x}, \boldsymbol{y}) = \boldsymbol{x}^\top \boldsymbol{y}$. However, instead of having only kernels that are explicitly inner products of feature vectors, we are now permitted to implicitly use *infinite* feature maps (functions) by using kernels.

■ **Example 7.68 (Kernel PCA)** We simulated 200 points, $\boldsymbol{x}_1, \ldots, \boldsymbol{x}_{200}$, from the uniform distribution on the set $B_1 \cup (B_4 \cap B_3^c)$, where $B_r := \{(x, y) \in \mathbb{R}^2 : x^2 + y^2 \leq r^2\}$ (disk with radius r). We apply kernel PCA with Gaussian kernel $\kappa(\boldsymbol{x}, \boldsymbol{x}') = \exp\left(-\|\boldsymbol{x} - \boldsymbol{x}'\|^2\right)$ and compute the functions $z_m(\boldsymbol{x}), m = 1, \ldots, 9$ in (7.67). Their density plots are shown in Figure 7.69. The data points are superimposed in each plot. From this we see that the principal components identify the radial structure present in the data.

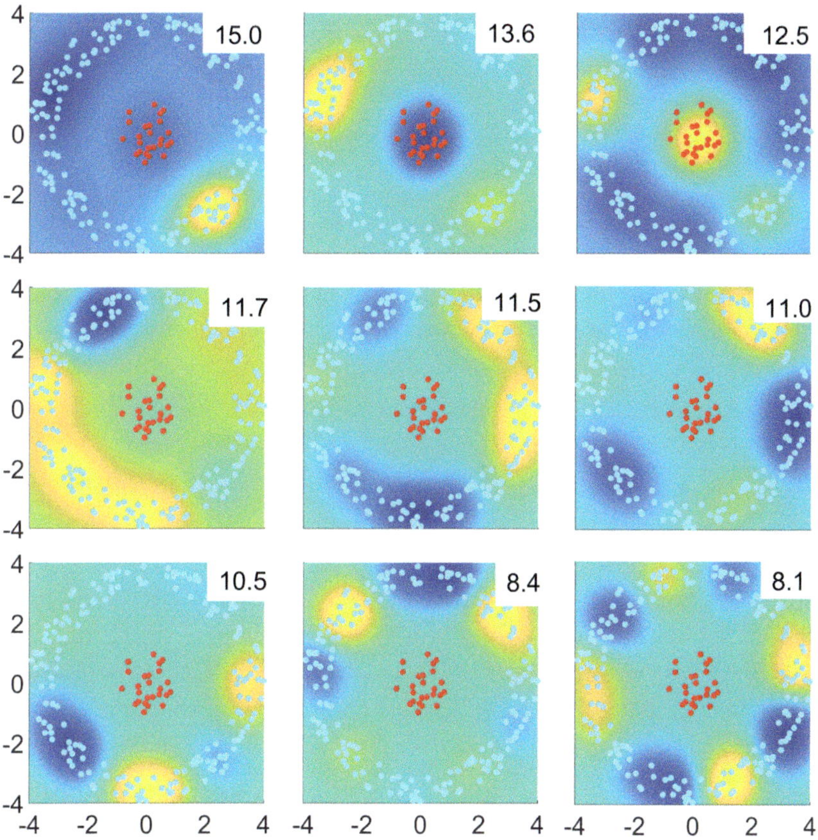

Figure 7.69: First nine eigenfunctions using a Gaussian kernel for the two-dimensional data set formed by the red and cyan points.

Finally, Figure 7.70 shows the projections $[z_1(\boldsymbol{x}_i), z_2(\boldsymbol{x}_i)]^\top, i = 1, \ldots, 200$ of the original data points onto the first two principal components. We see that the projected points can be separated by a straight line, whereas this is not possible for the original data; see also, Example 8.44 for a related problem. ☞ 375

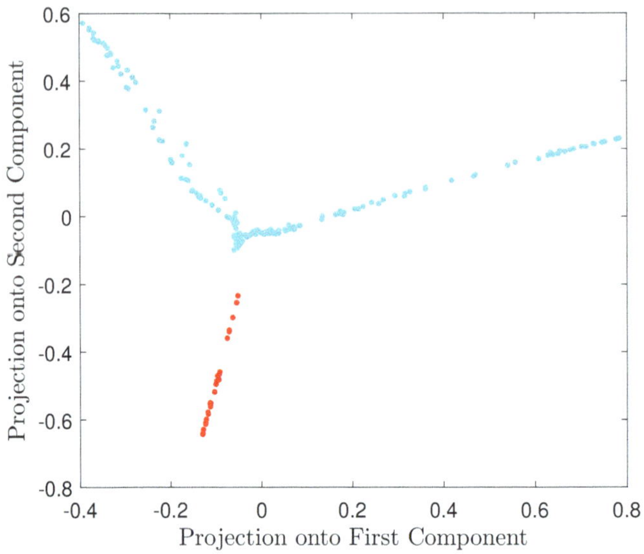

Figure 7.70: Projection of the data onto the first two principal components. Observe that already the projections of the inner and outer points are well separated.

■

Further Reading

For a good overview of the ridge regression and the lasso, we refer the reader to [54, 82]. For overviews of the theory of RKHS we refer to [4, 161, 181], and for in-depth background on splines and their connection to RKHSs we refer to [176]. For further details on GP regression we refer to [138] and for kernel PCA in particular we refer to [17, 130]. Finally, many facts about kernels and their corresponding RKHSs can be found in [161].

Exercises

1. Let $\mathcal{G}$ be an RKHS with reproducing kernel κ. Show that κ is a positive semidefinite function.

2. Show that a reproducing kernel, if it exists, is unique.

3. Let $\mathcal{G}$ be a Hilbert space of functions $g : \mathcal{X} \to \mathbb{R}$. Recall that the *evaluation functional* is the map $\delta_{\boldsymbol{x}} : g \mapsto g(\boldsymbol{x})$ for a given $\boldsymbol{x} \in \mathcal{X}$. Show that evaluation functionals are linear operators.

4. Let $\mathcal{G}_0$ be the pre-RKHS $\mathcal{G}_0$ constructed in the proof of Theorem 7.16. Thus, $g \in \mathcal{G}_0$ is of the form $g = \sum_{i=1}^{n} \alpha_i \, \kappa_{\boldsymbol{x}_i}$ and

$$\langle g, \kappa_{\boldsymbol{x}} \rangle_{\mathcal{G}_0} = \sum_{i=1}^{n} \alpha_i \, \langle \kappa_{\boldsymbol{x}_i}, \kappa_{\boldsymbol{x}} \rangle_{\mathcal{G}_0} = \sum_{i=1}^{n} \alpha_i \, \kappa(\boldsymbol{x}_i, \boldsymbol{x}) = g(\boldsymbol{x}).$$

Therefore, we may write the evaluation functional of $g \in \mathcal{G}_0$ at $\boldsymbol{x}$ as $\delta_{\boldsymbol{x}} g := \langle g, \kappa_{\boldsymbol{x}} \rangle_{\mathcal{G}_0}$. Show that $\delta_{\boldsymbol{x}}$ is bounded on $\mathcal{G}_0$ for every $\boldsymbol{x}$; that is, $|\delta_{\boldsymbol{x}} f| < \gamma \, \|f\|_{\mathcal{G}_0}$, for some $\gamma < \infty$.

5. Continuing Exercise 4, let (f_n) be a Cauchy sequence in $\mathcal{G}_0$ such that $|f_n(\boldsymbol{x})| \to 0$ for all $\boldsymbol{x}$. Show that $\|f_n\|_{\mathcal{G}_0} \to 0$.

6. Continuing Exercises 5 and 4, to show that the inner product (7.17) is well defined, a number of facts have to be checked.

(a) Verify that the limit converges.

(b) Verify that the limit is independent of the Cauchy sequences used.

(c) Verify that the properties of an inner product are satisfied. The only non-trivial property to verify is that $\langle f, f \rangle_{\mathcal{G}} = 0$ if and only if $f = 0$.

7. Exercises 4–6 show that $\mathcal{G}$ defined in the proof of Theorem 7.16 is an inner product space. It remains to prove that $\mathcal{G}$ is an RKHS. This requires us to prove that the inner product space $\mathcal{G}$ is complete (and thus Hilbert), and that its evaluation functionals are bounded and hence continuous (see Theorem B.19). This is done in a number of steps. ☞ 569

(a) Show that $\mathcal{G}_0$ is dense in $\mathcal{G}$ in the sense that every $f \in \mathcal{G}$ is a limit point (with respect to the norm on $\mathcal{G}$) of a Cauchy sequence (f_n) in $\mathcal{G}_0$.

(b) Show that every evaluation functional $\delta_{\boldsymbol{x}}$ on $\mathcal{G}$ is continuous at the 0 function. That is,

$$\forall \varepsilon > 0 : \exists \delta > 0 : \forall f \in \mathcal{G} : \|f\|_{\mathcal{G}} < \delta \Rightarrow |f(\boldsymbol{x})| < \varepsilon. \tag{7.71}$$

Continuity of $\delta_{\boldsymbol{x}}$ at all functions $g \in \mathcal{G}$ then follows automatically from linearity.

(c) Show that $\mathcal{G}$ is complete; that is, every Cauchy sequence $(f_n) \in \mathcal{G}$ converges in the norm $\|\cdot\|_{\mathcal{G}}$.

8. If κ_1 and κ_2 are kernels on $\mathcal{X}$ and $\mathcal{Y}$, then $\kappa_+((\boldsymbol{x}, \boldsymbol{y}), (\boldsymbol{x}', \boldsymbol{y}')) := \kappa_1(\boldsymbol{x}, \boldsymbol{x}') + \kappa_2(\boldsymbol{y}, \boldsymbol{y}')$ and $\kappa_\times((\boldsymbol{x}, \boldsymbol{y}), (\boldsymbol{x}', \boldsymbol{y}') := \kappa_1(\boldsymbol{x}, \boldsymbol{x}')\kappa_2(\boldsymbol{y}, \boldsymbol{y}')$ are kernels on the Cartesian product $\mathcal{X} \times \mathcal{Y}$. Prove this.

9. An RKHS enjoys the following desirable smoothness property: if (g_n) is a sequence belonging to RKHS $\mathcal{G}$ on $\mathcal{X}$, and $\|g_n - g\|_{\mathcal{G}} \to 0$, then $g(\boldsymbol{x}) = \lim_n g_n(\boldsymbol{x})$ for all $\boldsymbol{x} \in \mathcal{X}$. Prove this, using Cauchy–Schwarz.

10. Let $\boldsymbol{X}$ be an $\mathbb{R}^d$-valued random variable that is symmetric about the origin (that is, $\boldsymbol{X}$ and $(-\boldsymbol{X})$ are identically distributed). Denote by μ is its distribution and $\psi(\boldsymbol{t}) = \mathbb{E}\, \mathrm{e}^{\mathrm{i}\boldsymbol{t}^\top \boldsymbol{X}} = \int \mathrm{e}^{\mathrm{i}\boldsymbol{t}^\top \boldsymbol{x}} \, \mu(\mathrm{d}\boldsymbol{x})$ for $\boldsymbol{t} \in \mathbb{R}^d$ is its characteristic function. Verify that $\kappa(\boldsymbol{x}, \boldsymbol{x}') = \psi(\boldsymbol{x} - \boldsymbol{x}')$ is a real-valued positive semidefinite function.

11. Suppose an RKHS $\mathcal{G}$ of functions from $\mathcal{X} \to \mathbb{R}$ (with kernel κ) is invariant under a group $\mathcal{T}$ of transformations $T : \mathcal{X} \to \mathcal{X}$; that is, for all $f, g \in \mathcal{G}$ and $T \in \mathcal{T}$, we have (i) $f \circ T \in \mathcal{G}$ and (ii) $\langle f \circ T, g \circ T \rangle_{\mathcal{G}} = \langle f, g \rangle_{\mathcal{G}}$. Show that $\kappa(T\boldsymbol{x}, T\boldsymbol{x}') = \kappa(\boldsymbol{x}, \boldsymbol{x}')$ for all $\boldsymbol{x}, \boldsymbol{x}' \in \mathcal{X}$ and $T \in \mathcal{T}$.

12. Given two Hilbert spaces $\mathcal{H}$ and $\mathcal{G}$, we call a mapping $A : \mathcal{H} \to \mathcal{G}$ a **Hilbert space isomorphism** if it is

(i) a linear map; that is, $A(af + bg) = aA(f) + bA(g)$ for any $f, g \in \mathcal{H}$ and $a, b \in \mathbb{R}$.
(ii) a surjective map; and
(iii) an isometry; that is, for all $f, g \in \mathcal{H}$, it holds that $\langle f, g\rangle_{\mathcal{H}} = \langle Af, Ag\rangle_{\mathcal{G}}$.

Let $\mathcal{H} = \mathbb{R}^p$ (equipped with the usual Euclidean inner product) and construct its (continuous) *dual space* $\mathcal{G}$, consisting of all continuous linear functions from $\mathbb{R}^p$ to $\mathbb{R}$, as follows: (a) For each $\boldsymbol{\beta} \in \mathbb{R}^p$, define $g_{\boldsymbol{\beta}} : \mathbb{R}^p \to \mathbb{R}$ via $g_{\boldsymbol{\beta}}(\boldsymbol{x}) = \langle \boldsymbol{\beta}, \boldsymbol{x}\rangle = \boldsymbol{\beta}^\top \boldsymbol{x}$, for all $\boldsymbol{x} \in \mathbb{R}^p$. (b) Equip $\mathcal{G}$ with the inner product $\langle g_{\boldsymbol{\beta}}, g_{\boldsymbol{\gamma}}\rangle_{\mathcal{G}} := \boldsymbol{\beta}^\top \boldsymbol{\gamma}$.

Show that $A : \mathcal{H} \to \mathcal{G}$ defined by $A(\boldsymbol{\beta}) = g_{\boldsymbol{\beta}}$ for $\boldsymbol{\beta} \in \mathbb{R}^p$ is a Hilbert space isomorphism.

13. Let $\mathbf{X}$ be an $n \times p$ model matrix. Show that $\mathbf{X}^\top \mathbf{X} + \lambda \mathbf{I}_p$ for $\lambda > 0$ is invertible.

14. As Example 7.25 clearly illustrates, the pdf of a random variable that is symmetric about the origin is not in general a valid reproducing kernel. Take two such iid random variables X and X' with common pdf f, and define $Z = X + X'$. Denote by ψ_Z and f_Z the characteristic function and pdf of Z, respectively.

Show that if ψ_Z is in $L^1(\mathbb{R})$, f_Z is a positive semidefinite function. Use this to show that $\kappa(x, x') = f_Z(x - x') = \mathbb{1}\{|x - x'| \leq 2\}(1 - |x - x'|/2)$ is a valid reproducing kernel.

15. For the smoothing cubic spline of Section 7.6, show that

$$\kappa(x, u) = \frac{\max\{x, u\} \min\{x, u\}^2}{2} - \frac{\min\{x, u\}^3}{6}.$$

16. Let $\mathbf{X}$ be an $n \times p$ model matrix and let $\boldsymbol{u} \in \mathbb{R}^p$ be the unit-length vector with k-th entry equal to one ($u_k = \|\boldsymbol{u}\| = 1$). Suppose that the k-th column of $\mathbf{X}$ is $\boldsymbol{v}_k =: \boldsymbol{v}$ and that it is replaced with a new predictor $\boldsymbol{w}$, so that we obtain the new model matrix:

$$\widetilde{\mathbf{X}} = \mathbf{X} + (\boldsymbol{w} - \boldsymbol{v})\boldsymbol{u}^\top.$$

(a) Denoting

$$\boldsymbol{\delta} := \mathbf{X}^\top(\boldsymbol{w} - \boldsymbol{v}) + \frac{\|\boldsymbol{w} - \boldsymbol{v}\|^2}{2}\boldsymbol{u},$$

show that

$$\widetilde{\mathbf{X}}^\top \widetilde{\mathbf{X}} = \mathbf{X}^\top \mathbf{X} + \boldsymbol{u}\boldsymbol{\delta}^\top + \boldsymbol{\delta}\boldsymbol{u}^\top = \mathbf{X}^\top \mathbf{X} + \frac{(\boldsymbol{u} + \boldsymbol{\delta})(\boldsymbol{u} + \boldsymbol{\delta})^\top}{2} - \frac{(\boldsymbol{u} - \boldsymbol{\delta})(\boldsymbol{u} - \boldsymbol{\delta})^\top}{2}.$$

In other words, $\widetilde{\mathbf{X}}^\top \widetilde{\mathbf{X}}$ differs from $\mathbf{X}^\top \mathbf{X}$ by a symmetric matrix of rank two.

(b) Suppose that $\mathbf{B} := (\mathbf{X}^\top \mathbf{X} + \lambda \mathbf{I}_p)^{-1}$ is already computed. Explain how the Sherman–Morrison formulas in Theorem A.62 can be applied twice to compute the inverse and log-determinant of the matrix $\widetilde{\mathbf{X}}^\top \widetilde{\mathbf{X}} + \lambda \mathbf{I}_p$ in $\mathcal{O}((n + p)p)$ computing time, rather than the usual $\mathcal{O}((n + p^2)p)$ computing time.[3]

☞ 541

[3]This Sherman–Morrison updating is not always numerically stable. A more numerically stable method will perform two consecutive rank-one updates of the Cholesky decomposition of $\mathbf{X}^\top \mathbf{X} + \lambda \mathbf{I}_p$ via Algorithm A.76 in Appendix A.8.3.

(c) Write a Python program for updating a matrix $\mathbf{B} = (\mathbf{X}^\top \mathbf{X} + \lambda \mathbf{I}_p)^{-1}$ when we change the k-th column of $\mathbf{X}$, as shown in the following pseudo-code.

Algorithm 7.72: Updating via Sherman–Morrison Formula

Input: Matrices $\mathbf{X}$ and $\mathbf{B}$, index k, and replacement w for the k-th column of $\mathbf{X}$.
Output: Updated matrices $\mathbf{X}$ and $\mathbf{B}$.

Set $v \in \mathbb{R}^n$ to be the k-th column of $\mathbf{X}$.
Set $\boldsymbol{u} \in \mathbb{R}^p$ to be the unit-length vector such that $u_k = \|\boldsymbol{u}\| = 1$.
$\boldsymbol{\delta} \leftarrow \mathbf{X}^\top (w - v) + \frac{1}{2}\|w - v\|^2 \boldsymbol{u}$
$\mathbf{B} \leftarrow \mathbf{B} - \dfrac{\mathbf{B}\boldsymbol{u}\boldsymbol{\delta}^\top \mathbf{B}}{1 + \boldsymbol{\delta}^\top \mathbf{B}\boldsymbol{u}}$
$\mathbf{B} \leftarrow \mathbf{B} - \dfrac{\mathbf{B}\boldsymbol{\delta}\boldsymbol{u}^\top \mathbf{B}}{1 + \boldsymbol{u}^\top \mathbf{B}\boldsymbol{\delta}}$
Update the k-th column of $\mathbf{X}$ with w.
return $\mathbf{X}, \mathbf{B}$

17. Use Algorithm 7.72 from Exercise 16 to write Python code that computes the ridge regression coefficient $\boldsymbol{\beta}$ in (7.7). The following pseudo-code (with running cost of $O((n+p)p^2)$) may help with the writing of the Python code. ☞ 321

Algorithm 7.73: Ridge Regression via Sherman–Morrison Formula

Input: Training set $\{\mathbf{X}, \boldsymbol{y}\}$ and regularization parameter $\lambda > 0$.
Output: Solution $\widehat{\boldsymbol{\beta}} = (\mathbf{X}^\top \mathbf{X} + \lambda \mathbf{I}_p)^{-1} \mathbf{X}^\top \boldsymbol{y}$.

Set $\mathbf{A}$ to be an $n \times p$ matrix of zeros and $\mathbf{B} \leftarrow \mathbf{I}_p / \lambda$.
for $j = 1, \ldots, p$ **do**
  Set w to be the j-th column of $\mathbf{X}$.
  Update $\{\mathbf{A}, \mathbf{B}\}$ via Algorithm 7.72 with inputs $\{\mathbf{A}, \mathbf{B}, j, w\}$.
$\widehat{\boldsymbol{\beta}} \leftarrow \mathbf{B}(\mathbf{X}^\top \boldsymbol{y})$
return $\widehat{\boldsymbol{\beta}}$

18. In this exercise we explore how the *early stopping* of the *gradient descent* iterations (see Example C.61), ☞ 603

$$\boldsymbol{x}_{t+1} = \boldsymbol{x}_t - \alpha \nabla f(\boldsymbol{x}_t), \quad t = 0, 1, \ldots,$$

is (approximately) equivalent to the global minimization of $f(\boldsymbol{x}) + \frac{1}{2}\lambda\|\boldsymbol{x}\|^2$ for certain values of the *ridge regularization* parameter $\lambda > 0$ (see Example 7.3). We illustrate the **early stopping** idea on the quadratic function $f(\boldsymbol{x}) = \frac{1}{2}(\boldsymbol{x} - \boldsymbol{\mu})^\top \mathbf{H}(\boldsymbol{x} - \boldsymbol{\mu})$, where $\mathbf{H} \in \mathbb{R}^{n \times n}$ is a symmetric positive-definite (Hessian) matrix with eigenvalues $\{\lambda_k\}_{k=1}^n$.

(a) Verify that for a symmetric matrix $\mathbf{A} \in \mathbb{R}^n$ such that $\mathbf{I} - \mathbf{A}$ is invertible, we have

$$\mathbf{I} + \mathbf{A} + \cdots + \mathbf{A}^{t-1} = (\mathbf{I} - \mathbf{A}^t)(\mathbf{I} - \mathbf{A})^{-1}.$$

☞ 529

(b) Let $\mathbf{H} = \mathbf{Q}\mathbf{\Lambda}\mathbf{Q}^\top$ be the diagonalization of $\mathbf{H}$ as per Theorem A.34. If $\boldsymbol{x}_0 = \mathbf{0}$, show that the formula for $\boldsymbol{x}_t$ is

$$\boldsymbol{x}_t = \boldsymbol{\mu} - \mathbf{Q}(\mathbf{I} - \alpha\mathbf{\Lambda})^t\mathbf{Q}^\top\boldsymbol{\mu}.$$

Hence, deduce that a necessary condition for $\boldsymbol{x}_t$ to converge is $\alpha < 2/\max_k \lambda_k$.

(c) Show that the minimizer of $f(\boldsymbol{x}) + \frac{1}{2}\lambda\|\boldsymbol{x}\|^2$ can be written as

$$\boldsymbol{x}^* = \boldsymbol{\mu} - \mathbf{Q}(\mathbf{I} + \lambda^{-1}\mathbf{\Lambda})^{-1}\mathbf{Q}^\top\boldsymbol{\mu}.$$

(d) For a fixed value of t, let the learning rate $\alpha \downarrow 0$. Using part (b) and (c), show that if $\lambda = (1 + o(1))/(t\,\alpha)$ as $\alpha \downarrow 0$, then $\boldsymbol{x}_t = \boldsymbol{x}^*(1 + o(1))$. In other words, $\boldsymbol{x}_t$ is approximately equal to $\boldsymbol{x}^*$ for small α, provided that λ is inversely proportional to $t\,\alpha$.

CHAPTER 8

Classification

The purpose of this chapter is to explain the mathematical ideas behind well-known classification techniques such as the naïve Bayes method, linear and quadratic discriminant analysis, logistic/softmax classification, the K-nearest neighbors method, and support vector machines.

8.1 Introduction

Classification methods are supervised learning methods in which a categorical *response* variable Y takes one of c possible values (for example whether a person is sick or healthy), which is to be predicted from a vector $\boldsymbol{X}$ of *explanatory* variables (for example, the blood pressure, age, and smoking status of the person), using a *prediction function* g. In this sense, g classifies the input $\boldsymbol{X}$ into one of the classes, say in the set $\{0, \ldots, c-1\}$. For this reason, we will call g a **classification function** or simply **classifier**. As with any supervised learning technique (see Section 2.3), the goal is to minimize the expected loss or *risk*

$$\ell(g) = \mathbb{E}\,\text{Loss}(Y, g(\boldsymbol{X})) \tag{8.1}$$

for some loss function, $\text{Loss}(y, \widehat{y})$, that quantifies the impact of classifying a response y via $\widehat{y} = g(\boldsymbol{x})$. The natural loss function is the **zero–one** (also written 0–1) or **indicator loss**: $\text{Loss}(y, \widehat{y}) := \mathbb{1}\{y \neq \widehat{y}\}$; that is, there is no loss for a correct classification ($y = \widehat{y}$) and a unit loss for a misclassification ($y \neq \widehat{y}$). In this case the optimal classifier g^* is given in the following theorem.

Theorem 8.2 (Optimal Classifier): For the loss function $\text{Loss}(y, \widehat{y}) = \mathbb{1}\{y \neq \widehat{y}\}$, an optimal classification function is

$$g^*(\boldsymbol{x}) = \underset{y \in \{0,\ldots,c-1\}}{\text{argmax}}\ \mathbb{P}[Y = y \mid \boldsymbol{X} = \boldsymbol{x}]. \tag{8.3}$$

DOI: 10.1201/9781003391173-8

Proof. The goal is to minimize $\ell(g) = \mathbb{E}\,\mathbb{1}\{Y \neq g(\boldsymbol{X})\}$ over all functions g taking values in $\{0, \ldots, c-1\}$. Conditioning on $\boldsymbol{X}$ gives $\ell(g) = \mathbb{E}\,(\mathbb{P}[Y \neq g(\boldsymbol{X}) \,|\, \boldsymbol{X}]\,)$, by the tower property, and so minimizing $\ell(g)$ with respect to g can be accomplished by *maximizing* $\mathbb{P}[Y = g(\boldsymbol{x}) \,|\, \boldsymbol{X} = \boldsymbol{x}]$ with respect to $g(\boldsymbol{x})$, for every fixed $\boldsymbol{x}$. In other words, take $g(\boldsymbol{x})$ to be equal to the class label y for which $\mathbb{P}[Y = y \,|\, \boldsymbol{X} = \boldsymbol{x}]$ is maximal. □

☞ 635

The formulation (8.3) allows for "ties", when there is an equal probability between optimal classes for a feature vector $\boldsymbol{x}$. Assigning one of these tied classes arbitrarily (or randomly) to $\boldsymbol{x}$ does not affect the loss function and so we assume for simplicity that $g^*(\boldsymbol{x})$ is always a scalar value.

☞ 21

Note that, as was the case for the regression (see, e.g., Theorem 2.2), the optimal prediction function depends on the conditional pdf $f(y \,|\, \boldsymbol{x}) = \mathbb{P}[Y = y \,|\, \boldsymbol{X} = \boldsymbol{x}]$. However, since we assign $\boldsymbol{x}$ to class y if $f(y \,|\, \boldsymbol{x}) \geq f(z \,|\, \boldsymbol{x})$ for all z, we do not need to learn the entire surface of the function $f(y \,|\, \boldsymbol{x})$; we only need to estimate it well enough near the decision boundary $\{\boldsymbol{x} : f(y \,|\, \boldsymbol{x}) = f(z \,|\, \boldsymbol{x})\}$ for any choice of classes y and z. This is because the assignment (8.3) divides the feature space into c regions, $\mathcal{R}_y = \{\boldsymbol{x} : f(y \,|\, \boldsymbol{x}) = \max_z f(z \,|\, \boldsymbol{x})\}$, $y = 0, \ldots, c-1$.

Recall that for any supervised learning problem the smallest possible expected loss (that is, the irreducible risk) is given by $\ell^* = \ell(g^*)$. For the indicator loss, the irreducible risk is equal to $\mathbb{P}[Y \neq g^*(\boldsymbol{X})]$. This smallest possible probability of misclassification is often called the **Bayes error rate**.

> For a given training set τ, a classifier is often derived from a **pre-classifier** g_τ, which is a prediction function (learner) that can take any real value, rather than only values in the set of class labels. A typical situation is the case of binary classification with labels -1 and 1, where the prediction function g_τ is a function taking values in the interval $[-1, 1]$ and the actual classifier is given by $\text{sign}(g_\tau)$. It will be clear from the context whether a prediction function g_τ should be interpreted as a classifier or pre-classifier.

The indicator loss function may not always be the most appropriate choice of loss function for a given classification problem. For example, when diagnosing an illness, the mistake in misclassifying a person as being sick when in fact the person is healthy may be less serious than classifying the person as healthy when in fact the person is sick. In Section 8.2 we consider various classification metrics.

There are many ways to fit a classifier to a training set $\tau = \{(\boldsymbol{x}_1, y_1), \ldots, (\boldsymbol{x}_n, y_n)\}$. The approach taken in Section 8.3 is to use a Bayesian framework for classification. Here the conditional pdf $f(y \,|\, \boldsymbol{x})$ is viewed as a posterior pdf $f(y \,|\, \boldsymbol{x}) \propto f(\boldsymbol{x} \,|\, y) f(y)$ for a given class prior $f(y)$ and likelihood $f(\boldsymbol{x} \,|\, y)$. Section 8.4 discusses linear and quadratic discriminant analysis for classification, which assumes that the class of approximating functions for the conditional pdf $f(\boldsymbol{x} \,|\, y)$ is a parametric class $\mathcal{G}$ of Gaussian densities. As a result of this choice of $\mathcal{G}$, the marginal $f(\boldsymbol{x})$ is approximated via a Gaussian mixture density.

In contrast, in the logistic or soft-max classification in Section 8.5, the conditional pdf $f(y \,|\, \boldsymbol{x})$ is approximated using a more flexible class of approximating functions. As a result of this, the approximation to the marginal density $f(\boldsymbol{x})$ does not belong to a simple parametric class (such as a Gaussian mixture). As in unsupervised learning, the cross-entropy loss is the most common choice for training the learner.

The K-nearest neighbors method, discussed in Section 8.6, is yet another approach to classification that makes minimal assumptions on the class $\mathcal{G}$. Here the aim is to directly estimate the conditional pdf $f(y \,|\, \boldsymbol{x})$ from the training data, using only feature vectors in the neighborhood of $\boldsymbol{x}$. In Section 8.7 we explain the support vector methodology for classification; this is based on the same Reproducing Kernel Hilbert Space ideas that proved successful for regression analysis in Section 7.3. Finally, a versatile way to do both classification and regression is to use classification and regression trees. This is the topic of Chapter 9. Neural networks (Chapter 10) provide yet another way to perform classification.

☞ 323
☞ 391
☞ 429

8.2 Classification Metrics

The effectiveness of a classifier g is, theoretically, measured in terms of the risk (8.1), which depends on the loss function used. Fitting a classifier to iid training data $\tau = \{(\boldsymbol{x}_i, y_i)\}_{i=1}^n$ is established by minimizing the *training loss*

$$\ell_\tau(g) = \frac{1}{n}\sum_{i=1}^{n} \text{Loss}(y_i, g(\boldsymbol{x}_i)) \tag{8.4}$$

over some class of functions $\mathcal{G}$. As the training loss is often a poor estimator of the risk, the risk is usually estimated as in (8.4), using instead a test set $\tau' = \{(\boldsymbol{x}_i', y_i')\}_{i=1}^{n'}\}$ that is independent of the training set, as explained in Section 2.3. To measure the performance of a classifier on a training or test set, it is convenient to introduce the notion of a **loss matrix**. Consider a classification problem with classifier g, loss function Loss, and classes $0, \ldots, c-1$. If an input feature vector $\boldsymbol{x}$ is classified as $\widehat{y} = g(\boldsymbol{x})$ when the observed class is y, the loss incurred is, by definition, $\text{Loss}(y, \widehat{y})$. Consequently, we may identify the loss function with a matrix $\mathbf{L} = [\text{Loss}(j,k), j,k \in \{0,\ldots,c-1\}]$. For the indicator loss function, the matrix $\mathbf{L}$ has 0s on the diagonal and 1s everywhere else. Another useful matrix is the **confusion matrix**, denoted by $\mathbf{M}$, where the (j,k)-th element of $\mathbf{M}$ counts the number of times that, for the training or test data, the actual (observed) class is j whereas the predicted class is k. Table 8.5 shows the confusion matrix of some Dog/Cat/Possum classifier.

☞ 23

Table 8.5: Confusion matrix for three classes.

	Predicted		
Actual	Dog	Cat	Possum
Dog	30	2	6
Cat	8	22	15
Possum	7	4	41

We can now express the classifier performance (8.4) in terms of **L** and **M** as

$$\frac{1}{n}\sum_{j,k}[\mathbf{L}\odot\mathbf{M}]_{jk}, \tag{8.6}$$

where $\mathbf{L}\odot\mathbf{M}$ is the elementwise product of **L** and **M**. Note that for the indicator loss, (8.6) is simply $1-\text{tr}(\mathbf{M})/n$, and is called the **misclassification error**. The expression (8.6) makes it clear that both the counts and the loss are important in determining the performance of a classifier.

☞ 668 In the spirit of Table D.122 for hypothesis testing, it is sometimes useful to divide the elements of a confusion matrix into four groups. The diagonal elements are the **true positive** counts; that is, the numbers of correct classifications for each class. The true positive counts for the Dog, Cat, and Possum classes in Table 8.5 are $30, 22$, and 41, respectively. Similarly, the **true negative** count for a class is the sum of all matrix elements that do not belong to the row or the column of this particular class. For the Dog class it is $22+15+4+41=82$. The **false positive** count for a class is the sum of the corresponding column elements without the diagonal element. For the Dog class it is $8+7=15$. Finally, the **false negative** count for a specific class, can be calculated by summing over the corresponding row elements (again, without counting the diagonal element). For the Dog class it is $2+6=8$.

In terms of the elements of the confusion matrix, we have the following counts for class $j=0,\ldots,c-1$:

True positive	$\text{tp}_j=\mathbf{M}_{jj},$	
False positive	$\text{fp}_j=\sum_{k\neq j}\mathbf{M}_{kj},$	(column sum)
False negative	$\text{fn}_j=\sum_{k\neq j}\mathbf{M}_{jk},$	(row sum)
True negative	$\text{tn}_j=n-\text{fn}_j-\text{fp}_j-\text{tp}_j.$	

Note that in the binary classification case ($c=2$), and using the indicator loss function, the misclassification error (8.6) can be written as

$$\text{error}_j=\frac{\text{fp}_j+\text{fn}_j}{n}. \tag{8.7}$$

This does not depend on which of the two classes is considered, as $\text{fp}_0+\text{fn}_0=\text{fp}_1+\text{fn}_1$. Similarly, the **accuracy** measures the fraction of correctly classified objects:

$$\text{accuracy}_j=1-\text{error}_j=\frac{\text{tp}_j+\text{tn}_j}{n}. \tag{8.8}$$

In some cases, classification error (or accuracy) alone is not sufficient to adequately describe the effectiveness of a classifier.

As an example, consider the following two classification problems based on a fingerprint detection system:

1. Identification of authorized personnel in a top-secret military facility.
2. Identification to get an online discount for some retail chain.

Both problems are binary classification problems. However, a false positive in the first problem is extremely dangerous, while a false positive in the second problem will make a customer happy. Let us examine a classifier in the top-secret facility. The corresponding confusion matrix is given in Table 8.9.

Table 8.9: Confusion matrix for authorized personnel classification.

	Predicted	
Actual	authorized	non-authorized
authorized	100	400
non-authorized	50	100,000

From (8.8), we conclude that the accuracy of classification is equal to

$$\text{accuracy} = \frac{\text{tp} + \text{tn}}{\text{tp} + \text{tn} + \text{fp} + \text{fn}} = \frac{100 + 100,000}{100 + 100,000 + 50 + 400} \approx 99.55\%.$$

However, we can see that in this particular case, accuracy is a problematic metric, since the algorithm allowed 50 non-authorized personnel to enter the facility. One way to deal with this issue is to modify the loss function to give a much higher loss to non-authorized access. Thus, instead of an (indicator) loss matrix, we could for example take the loss matrix

$$\mathbf{L} = \begin{pmatrix} 0 & 1 \\ 1000 & 0 \end{pmatrix}.$$

An alternative approach is to keep the indicator loss function and consider additional classification metrics. Below we give a list of commonly used metrics. For simplicity we call an object whose actual class is j a "j-object".

- The **precision** (also called *positive predictive value*) is the fraction of all objects classified as j that are actually j-objects. Specifically,

$$\text{precision}_j = \frac{\text{tp}_j}{\text{tp}_j + \text{fp}_j}.$$

- The **recall** (also called *sensitivity*) is the fraction of all j-objects that are correctly classified as such. That is,

$$\text{recall}_j = \frac{\text{tp}_j}{\text{tp}_j + \text{fn}_j}.$$

- The **specificity** measures the fraction of all non-j-objects that are correctly classified as such. Specifically,

$$\text{specificity}_j = \frac{\text{tn}_j}{\text{fp}_j + \text{tn}_j}.$$

- The F_β **score** is a combination of the precision and the recall and is used as a single measurement for a classifier's performance. The F_β score is given by

$$F_{\beta,j} = \frac{(\beta^2 + 1)\,\text{tp}_j}{(\beta^2 + 1)\,\text{tp}_j + \beta^2\,\text{fn}_j + \text{fp}_j}.$$

For $\beta = 0$ we obtain the precision and for $\beta \to \infty$ we obtain the recall.

The particular choice of metric is clearly application dependent. For example, in the classification of authorized personnel in a top-secret military facility, suppose we have two classifiers. The first (Classifier 1) has a confusion matrix given in Table 8.9, and the second (Classifier 2) has a confusion matrix given in Table 8.10. Various metrics for these two classifiers are show in Table 8.11. In this case we prefer Classifier 1, which has a much higher precision.

Table 8.10: Confusion matrix for authorized personnel classification, using a different classifier (Classifier 2).

	Predicted	
Actual	Authorized	Non-Authorized
authorized	50	10
non-authorized	450	100,040

Table 8.11: Comparing the metrics for the confusion matrices in Tables 8.9 and 8.10.

Metric	Classifier 1	Classifier 2
accuracy	$9.955\,000 \times 10^{-1}$	$9.954\,000 \times 10^{-1}$
precision	$\mathbf{6.667 \times 10^{-1}}$	$\mathbf{1.000 \times 10^{-1}}$
recall	$2.000\,000 \times 10^{-1}$	$8.333\,000 \times 10^{-1}$
specificity	$9.995\,000 \times 10^{-1}$	$9.955\,000 \times 10^{-1}$
F_1	$3.077\,000 \times 10^{-1}$	$1.786\,000 \times 10^{-1}$

■ **Remark 8.12 (Multi-label and Hierarchical Classification)** In standard classification the classes are assumed to be mutually exclusive. For example a satellite image could be classified as "cloudy", "clear", or "foggy". In **multi-label classification** the classes (often called labels) do not have to be mutually exclusive. In this case the response is a subset $\mathcal{Y}$ of some collection of labels $\{0, \dots, c-1\}$. Equivalently, the response can be viewed as a binary vector of length c, where the y-th element is 1 if the response belongs to label y and 0 otherwise. Again, consider the satellite image example and add two labels, such as "road" and "river" to the previous three labels. Clearly, an image can contain both a road and a river. In addition, the image can be clear, cloudy, or foggy.

In **hierarchical classification** a hierarchical relation between classes/labels is taken into account during the classification process. Usually, the relations are modeled via a tree or a directed acyclic graph. A visual comparison between the hierarchical and non-hierarchical (flat) classification tasks for satellite image data is presented in Figure 8.13.

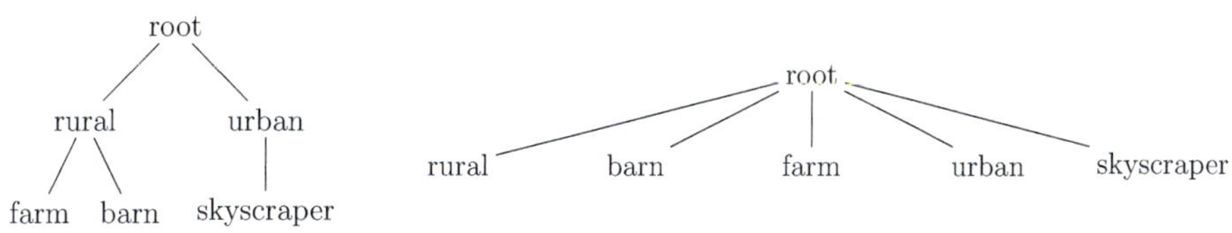

Figure 8.13: Hierarchical (left) and non-hierarchical (right) classification schemes. Barns and farms are common in rural areas, while skyscrapers are generally located in cities. While this relation can be clearly observed in the hierarchical model scheme, the connection is missing in the non-hierarchical design.

In multi-label classification, both the prediction $\widehat{\mathcal{Y}} := g(\boldsymbol{x})$ and the true response $\mathcal{Y}$ are *subsets* of the label set $\{0, \ldots, c-1\}$. A reasonable metric is the so-called **exact match ratio**, defined as

$$\text{exact match ratio} = \frac{\sum_{i=1}^{n} \mathbb{1}\{\widehat{\mathcal{Y}}_i = \mathcal{Y}_i\}}{n}.$$

The exact match ratio is rather stringent, as it requires a full match. In order to consider partial correctness, the following metrics could be used instead.

- The **accuracy** is defined as the ratio of correctly predicted labels and the total number of predicted and actual labels. The formula is given by

$$\text{accuracy} = \frac{\sum_{i=1}^{n} |\mathcal{Y}_i \cap \widehat{\mathcal{Y}}_i|}{\sum_{i=1}^{n} |\mathcal{Y}_i \cup \widehat{\mathcal{Y}}_i|}.$$

- The **precision** is defined as the ratio of correctly predicted labels and the total number of predicted labels. Specifically,

$$\text{precision} = \frac{\sum_{i=1}^{n} |\mathcal{Y}_i \cap \widehat{\mathcal{Y}}_i|}{\sum_{i=1}^{n} |\widehat{\mathcal{Y}}_i|}. \tag{8.14}$$

- The **recall** is defined as the ratio of correctly predicted labels and the total number of actual labels. Specifically,

$$\text{recall} = \frac{\sum_{i=1}^{n} |\mathcal{Y}_i \cap \widehat{\mathcal{Y}}_i|}{\sum_{i=1}^{n} |\mathcal{Y}_i|}. \tag{8.15}$$

- The **Hamming loss** counts the average number of incorrect predictions for all classes, calculated as

$$\text{Hamming} = \frac{1}{n\,c}\sum_{i=1}^{n}\sum_{y=0}^{c-1} \mathbb{1}\{y \in \widehat{\mathcal{Y}}_i\}\,\mathbb{1}\{y \notin \mathcal{Y}_i\} + \mathbb{1}\{y \notin \widehat{\mathcal{Y}}_i\}\,\mathbb{1}\{y \in \mathcal{Y}_i\}.$$

8.3 Classification via Bayes' Rule

We saw from Theorem 8.2 that the optimal classifier for classes $0,\dots,c-1$ divides the feature space into c regions, depending on $f(y \mid \boldsymbol{x})$: the conditional pdf of the response Y given the feature vector $\boldsymbol{X} = \boldsymbol{x}$. In particular, if $f(y \mid \boldsymbol{x}) > f(z \mid \boldsymbol{x})$ for all $z \neq y$, the feature vector $\boldsymbol{x}$ is classified as y. Classifying feature vectors on the basis of their conditional class probabilities is a natural thing to do, especially in a Bayesian learning context; see Section 2.9 for an overview of Bayesian terminology and usage. Specifically, the conditional probability $f(y \mid \boldsymbol{x})$ is interpreted as a *posterior* probability, of the form ☞ 49

$$f(y \mid \boldsymbol{x}) \propto f(\boldsymbol{x} \mid y) f(y), \tag{8.16}$$

where $f(\boldsymbol{x} \mid y)$ is the *likelihood* of obtaining feature vector $\boldsymbol{x}$ from class y and $f(y)$ is the *prior* probability[1] of class y. By making various modeling assumptions about the prior (e.g., all classes are *a priori* equally likely) and the likelihood function, one obtains the posterior pdf via Bayes' formula (8.16). A class $\widehat{y}$ is then assigned to a feature vector $\boldsymbol{x}$ according to the highest posterior probability; that is, we classify according to the **Bayes optimal decision rule**:

$$\widehat{y} = \underset{y}{\operatorname{argmax}}\, f(y \mid \boldsymbol{x}), \tag{8.17}$$

which is exactly (8.3). Since the discrete density $f(y \mid \boldsymbol{x})$, $y = 0,\dots,c-1$ is usually not known, the aim is to approximate it well with a function $g(y \mid \boldsymbol{x})$ from some class of functions $\mathcal{G}$. Note that in this context, $g(\cdot \mid \boldsymbol{x})$ refers to a discrete density (a probability mass function) for a given $\boldsymbol{x}$.

Suppose a feature vector $\boldsymbol{x} = [x_1,\dots,x_p]^\top$ of p features has to be classified into one of the classes $0,\dots,c-1$. For example, the classes could be different people and the features could be various facial measurements, such as the width of the eyes divided by the distance between the eyes, or the ratio of the nose height and mouth width. In the **naïve Bayes** method, the class of approximating functions $\mathcal{G}$ is chosen such that $g(\boldsymbol{x} \mid y) = g(x_1 \mid y) \cdots g(x_p \mid y)$, that is, conditional on the label, all features are independent. Assuming a uniform prior for y, the posterior pdf can thus be written as

$$g(y \mid \boldsymbol{x}) \propto \prod_{j=1}^{p} g(x_j \mid y),$$

where the marginal pdfs $g(x_j \mid y)$, $j = 1,\dots,p$ belong to a given class of approximating functions $\mathcal{G}$. To classify $\boldsymbol{x}$, simply take the y that maximizes the unnormalized posterior pdf.

[1]Here we have used the Bayesian notation convention of "overloading" the notation f.

For instance, suppose that the approximating class $\mathcal{G}$ is such that $(X_j \mid y) \sim \mathcal{N}(\mu_{yj}, \sigma^2)$, $y = 0, \ldots, c-1$, $j = 1, \ldots, p$. The corresponding posterior pdf is then

$$g(y \mid \boldsymbol{\theta}, \boldsymbol{x}) \propto \exp\left(-\frac{1}{2}\sum_{j=1}^{p}\frac{(x_j - \mu_{yj})^2}{\sigma^2}\right) = \exp\left(-\frac{1}{2}\frac{\|\boldsymbol{x} - \boldsymbol{\mu}_y\|^2}{\sigma^2}\right),$$

where $\boldsymbol{\mu}_y := [\mu_{y1}, \ldots, \mu_{yp}]^\top$ and $\boldsymbol{\theta} := \{\boldsymbol{\mu}_0, \ldots, \boldsymbol{\mu}_{c-1}, \sigma^2\}$ collects all model parameters. The probability $g(y \mid \boldsymbol{\theta}, \boldsymbol{x})$ is maximal when $\|\boldsymbol{x} - \boldsymbol{\mu}_y\|$ is minimal. Thus $\widehat{y} = \operatorname{argmin}_y \|\boldsymbol{x} - \boldsymbol{\mu}_y\|$ is the classifier that maximizes the posterior probability. That is, classify $\boldsymbol{x}$ as y when $\boldsymbol{\mu}_y$ is closest to $\boldsymbol{x}$ in Euclidean distance. Of course, the parameters (here, the $\{\boldsymbol{\mu}_y\}$ and σ^2) are unknown and have to be estimated from the training data.

We can extend the above idea to the case where also the variance σ^2 depends on the class y and feature j, as in the next example.

Example 8.18 (Naïve Bayes Classification) Table 8.19 lists the means μ and standard deviations σ of $p = 3$ normally distributed features, for $c = 4$ different classes. How should a feature vector $\boldsymbol{x} = [1.67, 2.00, 4.23]^\top$ be classified? The posterior pdf is

Table 8.19: Feature parameters.

	Feature 1		Feature 2		Feature 3	
Class	μ	σ	μ	σ	μ	σ
0	1.6	0.1	2.4	0.5	4.3	0.2
1	1.5	0.2	2.9	0.6	6.1	0.9
2	1.8	0.3	2.5	0.3	4.2	0.3
3	1.1	0.2	3.1	0.7	5.6	0.3

$$g(y \mid \boldsymbol{\theta}, \boldsymbol{x}) \propto (\sigma_{y1}\sigma_{y2}\sigma_{y3})^{-1} \exp\left(-\frac{1}{2}\sum_{j=1}^{3}\frac{(x_j - \mu_{yj})^2}{\sigma_{yj}^2}\right),$$

where $\boldsymbol{\theta} := \{\boldsymbol{\sigma}_j, \boldsymbol{\mu}_j\}_{j=0}^{c-1}$ again collects all model parameters. The (unscaled) values for $g(y \mid \boldsymbol{\theta}, \boldsymbol{x})$, $y = 0, 1, 2, 3$ are 53.5, 0.24, 8.37, and 3.5×10^{-6}, respectively. Hence, the feature vector should be classified as 0. The code follows.

naiveBayes.py

```
import numpy as np
x = np.array([1.67,2,4.23]).reshape(1,3)
mu = np.array([1.6, 2.4, 4.3,
               1.5, 2.9, 6.1,
               1.8, 2.5, 4.2,
               1.1, 3.1, 5.6]).reshape(4,3)
sig = np.array([0.1, 0.5, 0.2,
                0.2, 0.6, 0.9,
                0.3, 0.3, 0.3,
```

```
                0.2, 0.7, 0.3]).reshape(4,3)
g = lambda y: 1/np.prod(sig[y,:]) * np.exp(
      -0.5*np.sum((x-mu[y,:])**2/sig[y,:]**2));
for y in range(0,4):
    print('{:3.2e}'.format(g(y)))
```

```
5.35e+01
2.42e-01
8.37e+00
3.53e-06
```

8.4 Linear and Quadratic Discriminant Analysis

The Bayesian viewpoint for classification of the previous section (not limited to naïve Bayes) leads in a natural way to the well-established technique of **discriminant analysis**. We discuss the binary classification case first, with classes 0 and 1.

We consider a class of approximating functions $\mathcal{G}$ such that, conditional on the class $y \in \{0, 1\}$, the feature vector $\boldsymbol{X} = [X_1, \ldots, X_p]^\top$ has a $\mathcal{N}(\boldsymbol{\mu}_y, \boldsymbol{\Sigma}_y)$ distribution (see ☞ 47 (2.58)):

$$g(\boldsymbol{x} \mid \boldsymbol{\theta}, y) = \frac{1}{\sqrt{(2\pi)^p \, |\boldsymbol{\Sigma}_y|}} \, \mathrm{e}^{-\frac{1}{2} (\boldsymbol{x}-\boldsymbol{\mu}_y)^\top \boldsymbol{\Sigma}_y^{-1} (\boldsymbol{x}-\boldsymbol{\mu}_y)}, \quad \boldsymbol{x} \in \mathbb{R}^p, \quad y \in \{0, 1\}, \tag{8.20}$$

where $\boldsymbol{\theta} = \{\alpha_j, \boldsymbol{\mu}_j, \boldsymbol{\Sigma}_j\}_{j=0}^{c-1}$ collects all model parameters, including the probability vector $\boldsymbol{\alpha}$ (that is, $\sum_i \alpha_i = 1$ and $\alpha_i \geq 0$) which helps define the prior density: $g(y \mid \boldsymbol{\theta}) = \alpha_y, \; y \in \{0, 1\}$. Then, the posterior density is

$$g(y \mid \boldsymbol{\theta}, \boldsymbol{x}) \propto \alpha_y \times g(\boldsymbol{x} \mid \boldsymbol{\theta}, y),$$

and, according to the Bayes optimal decision rule (8.17), we classify $\boldsymbol{x}$ to come from class 0 if $\alpha_0 g(\boldsymbol{x} \mid \boldsymbol{\theta}, 0) > \alpha_1 g(\boldsymbol{x} \mid \boldsymbol{\theta}, 1)$ or, equivalently (by taking logarithms) if,

$$\ln \alpha_0 - \frac{1}{2} \ln |\boldsymbol{\Sigma}_0| - \frac{1}{2}(\boldsymbol{x} - \boldsymbol{\mu}_0)^\top \boldsymbol{\Sigma}_0^{-1} (\boldsymbol{x} - \boldsymbol{\mu}_0) > \ln \alpha_1 - \frac{1}{2} \ln |\boldsymbol{\Sigma}_1| - \frac{1}{2}(\boldsymbol{x} - \boldsymbol{\mu}_1)^\top \boldsymbol{\Sigma}_1^{-1} (\boldsymbol{x} - \boldsymbol{\mu}_1).$$

The function

$$\delta_y(\boldsymbol{x}) = \ln \alpha_y - \frac{1}{2} \ln |\boldsymbol{\Sigma}_y| - \frac{1}{2}(\boldsymbol{x} - \boldsymbol{\mu}_y)^\top \boldsymbol{\Sigma}_y^{-1} (\boldsymbol{x} - \boldsymbol{\mu}_y), \quad \boldsymbol{x} \in \mathbb{R}^p \tag{8.21}$$

is called the **quadratic discriminant function** for class $y = 0, 1$. A point $\boldsymbol{x}$ is classified to class y for which $\delta_y(\boldsymbol{x})$ is largest. The function is quadratic in $\boldsymbol{x}$ and so the decision boundary $\{\boldsymbol{x} \in \mathbb{R}^p : \delta_0(\boldsymbol{x}) = \delta_1(\boldsymbol{x})\}$ is quadratic as well. An important simplification arises for the case where the assumption is made that $\boldsymbol{\Sigma}_0 = \boldsymbol{\Sigma}_1 = \boldsymbol{\Sigma}$. Now, the decision boundary is the set of $\boldsymbol{x}$ for which

$$\ln \alpha_0 - \frac{1}{2}(\boldsymbol{x} - \boldsymbol{\mu}_0)^\top \boldsymbol{\Sigma}^{-1} (\boldsymbol{x} - \boldsymbol{\mu}_0) = \ln \alpha_1 - \frac{1}{2}(\boldsymbol{x} - \boldsymbol{\mu}_1)^\top \boldsymbol{\Sigma}^{-1} (\boldsymbol{x} - \boldsymbol{\mu}_1).$$

Expanding the above expression shows that the quadratic term in $\boldsymbol{x}$ is eliminated, giving a *linear* decision boundary in $\boldsymbol{x}$:

$$\ln \alpha_0 - \frac{1}{2}\boldsymbol{\mu}_0^\top \boldsymbol{\Sigma}^{-1}\boldsymbol{\mu}_0 + \boldsymbol{x}^\top \boldsymbol{\Sigma}^{-1}\boldsymbol{\mu}_0 = \ln \alpha_1 - \frac{1}{2}\boldsymbol{\mu}_1^\top \boldsymbol{\Sigma}^{-1}\boldsymbol{\mu}_1 + \boldsymbol{x}^\top \boldsymbol{\Sigma}^{-1}\boldsymbol{\mu}_1.$$

The corresponding **linear discriminant function** for class y is

$$\delta_y(\boldsymbol{x}) = \ln \alpha_y - \frac{1}{2}\boldsymbol{\mu}_y^\top \boldsymbol{\Sigma}^{-1}\boldsymbol{\mu}_y + \boldsymbol{x}^\top \boldsymbol{\Sigma}^{-1}\boldsymbol{\mu}_y, \quad \boldsymbol{x} \in \mathbb{R}^p. \tag{8.22}$$

Example 8.23 (Linear Discriminant Analysis) Consider the case where $\alpha_0 = \alpha_1 = 1/2$ and

$$\boldsymbol{\Sigma} = \begin{bmatrix} 2 & 0.7 \\ 0.7 & 2 \end{bmatrix}, \quad \boldsymbol{\mu}_0 = \begin{bmatrix} 0 \\ 0 \end{bmatrix}, \quad \boldsymbol{\mu}_1 = \begin{bmatrix} 2 \\ 4 \end{bmatrix}.$$

The distribution of $\boldsymbol{X}$ is a mixture of two bivariate normal distributions. Its pdf, ☞ 161

$$\frac{1}{2} g(\boldsymbol{x} \,|\, \boldsymbol{\theta}, y = 0) + \frac{1}{2} g(\boldsymbol{x} \,|\, \boldsymbol{\theta}, y = 1),$$

is depicted in Figure 8.24.

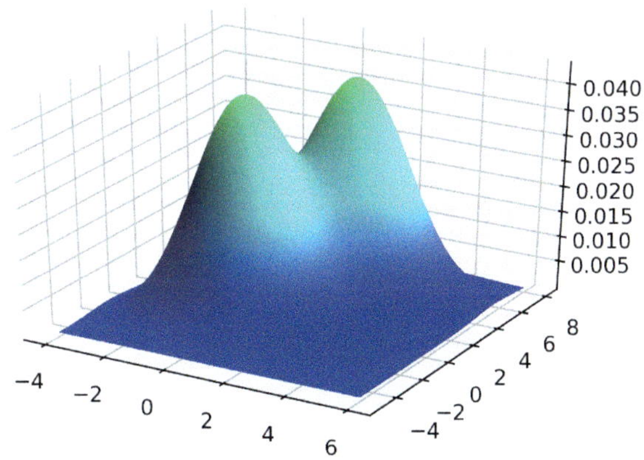

Figure 8.24: A Gaussian mixture density where the two mixture components have the same covariance matrix.

We used the following Python code to make this figure.

LDAmixture.py

```
import numpy as np
import matplotlib.pyplot as plt
from scipy.stats import multivariate_normal
from mpl_toolkits.mplot3d import Axes3D
from matplotlib.colors import LightSource

mu1 = np.array([0, 0])
mu2 = np.array([2, 4])
```

```
Sigma = np.array([[2, 0.7], [0.7, 2]])
x, y = np.mgrid[-4:6:150j, -5:8:150j]
mvn1 = multivariate_normal(mu1, Sigma)
mvn2 = multivariate_normal(mu2, Sigma)

xy = np.hstack((x.reshape(-1, 1), y.reshape(-1, 1)))
z = 0.5 * mvn1.pdf(xy).reshape(x.shape) + 0.5 * mvn2.pdf(xy).reshape(x.
    shape)

fig = plt.figure()
ax = fig.add_subplot(111, projection='3d')
ls = LightSource(azdeg=180, altdeg=65)
cols = ls.shade(z, plt.cm.winter)
surf = ax.plot_surface(x, y, z, rstride=1, cstride=1, linewidth=0,
                       antialiased=False, facecolors=cols)
plt.show()
```

The following Python code, which imports the previous code, draws a contour plot of the mixture density, simulates 1000 data points from the mixture density, and draws the decision boundary. To compute and display the linear decision boundary, let $[a_1, a_2]^\top = 2\Sigma^{-1}(\mu_1 - \mu_0)$ and $b = \mu_0^\top\Sigma^{-1}\mu_0 - \mu_1^\top\Sigma^{-1}\mu_1$. Then, the decision boundary can be written as $a_1x_1 + a_2x_2 + b = 0$ or, equivalently, $x_2 = -(a_1x_1 + b)/a_2$. We see in Figure 8.25 that the decision boundary nicely separates the two modes of the mixture density.

LDA.py

```
from LDAmixture import *
from numpy.random import rand
from numpy.linalg import inv

fig = plt.figure()
plt.contourf(x, y,z, cmap=plt.cm.Blues, alpha= 0.9,extend='both')
plt.ylim(-5.0,8.0)
plt.xlim(-4.0,6.0)
M = 1000
r = (rand(M,1) < 0.5)
for i in range(0,M):
    if r[i]:
        u = np.random.multivariate_normal(mu0,Sigma,1)
        plt.plot(u[0][0],u[0][1],'.r',alpha = 0.4)
    else:
        u = np.random.multivariate_normal(mu1,Sigma,1)
        plt.plot(u[0][0],u[0][1],'+k',alpha = 0.6)

a = 2*inv(Sigma) @ (mu1-mu0);
b = ( mu0.reshape(1,2) @ inv(Sigma) @ mu0.reshape(2,1)
    - mu1.reshape(1,2) @ inv(Sigma) @mu1.reshape(2,1) )
xx = np.linspace(-4,6,100)
yy = (-(a[0]*xx +b)/a[1])[0]
plt.plot(xx,yy,'m')
plt.show()
```

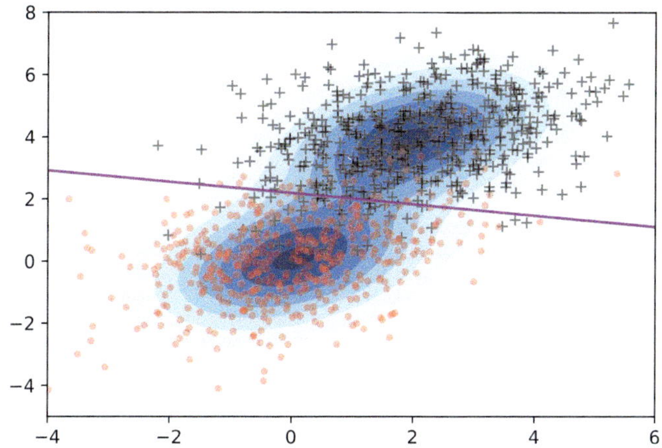

Figure 8.25: The linear discriminant boundary lies between the two modes of the mixture density and is linear.

■

To illustrate the difference between the linear and quadratic case, we specify different covariance matrices for the mixture components in the next example.

■ **Example 8.26 (Quadratic Discriminant Analysis)** As in Example 8.23 we consider a mixture of two Gaussians, but now with different covariance matrices. Figure 8.27 shows the quadratic decision boundary (a hyperbola). The Python code follows.

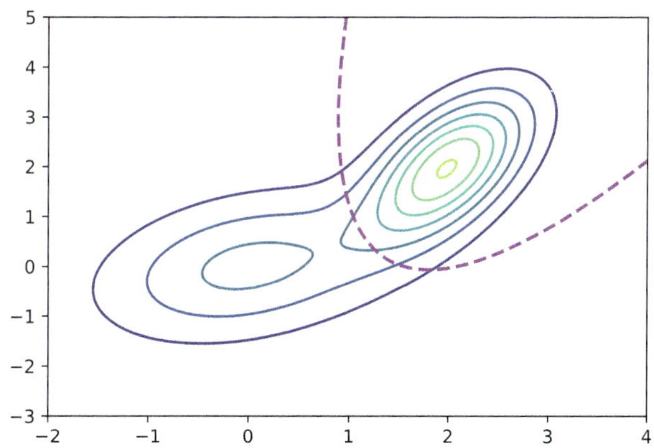

Figure 8.27: A quadratic decision boundary.

QDA.py

```
import numpy as np
import matplotlib.pyplot as plt
from scipy.stats import multivariate_normal

mu1 = np.array([0,0])
mu2 = np.array([2,2])
Sigma1 = np.array([[1,0.3],[0.3, 1]])
```

```
Sigma2 = np.array([[0.3,0.3],[0.3, 1]])
x, y = np.mgrid[-2:4:150j,-3:5:150j]
mvn1 = multivariate_normal( mu1, Sigma1 )
mvn2 = multivariate_normal( mu2, Sigma2 )

xy = np.hstack((x.reshape(-1,1),y.reshape(-1,1)))
z = ( 0.5*mvn1.pdf(xy).reshape(x.shape) +
      0.5*mvn2.pdf(xy).reshape(x.shape) )
plt.contour(x,y,z)

z1 = ( 0.5*mvn1.pdf(xy).reshape(x.shape) -
       0.5*mvn2.pdf(xy).reshape(x.shape))
plt.contour(x,y,z1, levels=[0],linestyles ='dashed',
            linewidths = 2, colors = 'm')
plt.show()
```

Of course, in practice the true parameter $\boldsymbol{\theta} = \{\alpha_j, \boldsymbol{\Sigma}_j, \boldsymbol{\mu}_j\}_{j=1}^c$ is not known and must be estimated from the training data — for example, by minimizing the *cross-entropy training loss* (4.4) with respect to $\boldsymbol{\theta}$:

☞ 135

$$\frac{1}{n}\sum_{i=1}^n \text{Loss}(f(\boldsymbol{x}_i, y_i), g(\boldsymbol{x}_i, y_i \mid \boldsymbol{\theta})) = -\frac{1}{n}\sum_{i=1}^n \ln g(\boldsymbol{x}_i, y_i \mid \boldsymbol{\theta}),$$

where

$$\ln g(\boldsymbol{x}, y \mid \boldsymbol{\theta}) = \ln \alpha_y - \frac{1}{2}\ln|\boldsymbol{\Sigma}_y| - \frac{1}{2}(\boldsymbol{x} - \boldsymbol{\mu}_y)^\top \boldsymbol{\Sigma}_y^{-1}(\boldsymbol{x} - \boldsymbol{\mu}_y) - \frac{p}{2}\ln(2\pi).$$

The corresponding estimates of the model parameters (see Exercise 2) are:

$$\begin{aligned} \widehat{\alpha}_y &= \frac{n_y}{n} \\ \widehat{\boldsymbol{\mu}}_y &= \frac{1}{n_y}\sum_{i:y_i=y} \boldsymbol{x}_i \\ \widehat{\boldsymbol{\Sigma}}_y &= \frac{1}{n_y}\sum_{i:y_i=y} (\boldsymbol{x}_i - \widehat{\boldsymbol{\mu}}_y)(\boldsymbol{x}_i - \widehat{\boldsymbol{\mu}}_y)^\top \end{aligned} \tag{8.28}$$

for $y = 0, \ldots, c-1$, where $n_y := \sum_{i=1}^n \mathbb{1}\{y_i = y\}$. For the case where $\boldsymbol{\Sigma}_y = \boldsymbol{\Sigma}$ for all y, we have $\widehat{\boldsymbol{\Sigma}} = \sum_y \widehat{\alpha}_y \widehat{\boldsymbol{\Sigma}}_y$.

When $c > 2$ classes are involved, the classification procedure carries through in exactly the same way, leading to quadratic and linear discriminant functions (8.21) and (8.22) for each class. The space $\mathbb{R}^p$ now is partitioned into c regions, determined by the linear or quadratic boundaries determined by each pair of Gaussians.

For the linear discriminant case (that is, when $\boldsymbol{\Sigma}_y = \boldsymbol{\Sigma}$ for all y), it is convenient to first "whiten" or **sphere the data** as follows: Let $\mathbf{B}$ be an invertible matrix such that $\boldsymbol{\Sigma} = \mathbf{B}\mathbf{B}^\top$, obtained, for example, via the Cholesky method. We linearly transform each data point $\boldsymbol{x}$ to $\boldsymbol{x}' := \mathbf{B}^{-1}\boldsymbol{x}$ and each mean $\boldsymbol{\mu}_y$ to $\boldsymbol{\mu}'_y := \mathbf{B}^{-1}\boldsymbol{\mu}_y$, $y = 0, \ldots, c-1$. Let the

☞ 543

random vector $\boldsymbol{X}$ be distributed according to the mixture pdf

$$g_{\boldsymbol{X}}(\boldsymbol{x} \mid \boldsymbol{\theta}) := \sum_y \alpha_y \frac{1}{\sqrt{(2\pi)^p \, |\boldsymbol{\Sigma}|}} \, \mathrm{e}^{-\frac{1}{2} (\boldsymbol{x}-\boldsymbol{\mu}_y)^\top \boldsymbol{\Sigma}^{-1} (\boldsymbol{x}-\boldsymbol{\mu}_y)}.$$

Then, by the transformation Theorem D.36, the vector $\boldsymbol{X}' = \mathbf{B}^{-1}\boldsymbol{X}$ has density ☞ 636

$$g_{\boldsymbol{X}'}(\boldsymbol{x}' \mid \boldsymbol{\theta}) = \frac{g_{\boldsymbol{X}}(\boldsymbol{x} \mid \boldsymbol{\theta})}{|\mathbf{B}^{-1}|} = \sum_{y=0}^{c-1} \frac{\alpha_y}{\sqrt{(2\pi)^p}} \, \mathrm{e}^{-\frac{1}{2} (\boldsymbol{x}-\boldsymbol{\mu}_y)^\top (\mathbf{B}\mathbf{B}^\top)^{-1} (\boldsymbol{x}-\boldsymbol{\mu}_y)}$$

$$= \sum_{y=0}^{c-1} \frac{\alpha_y}{\sqrt{(2\pi)^p}} \, \mathrm{e}^{-\frac{1}{2} (\boldsymbol{x}'-\boldsymbol{\mu}'_y)^\top (\boldsymbol{x}'-\boldsymbol{\mu}'_y)} = \sum_{y=0}^{c-1} \frac{\alpha_y}{\sqrt{(2\pi)^p}} \, \mathrm{e}^{-\frac{1}{2} \|\boldsymbol{x}'-\boldsymbol{\mu}'_y\|^2}.$$

This is the pdf of a mixture of standard p-dimensional normal distributions. The name "sphering" derives from the fact that the contours of each mixture component are perfect spheres. Classification of the transformed data is now particularly easy: classify $\boldsymbol{x}$ as $\widehat{y} := \mathrm{argmin}_y\{ \|\boldsymbol{x}' - \boldsymbol{\mu}'_y\|^2 - 2\ln \alpha_y\}$. Note that this rule only depends on the prior probabilities and the distance from $\boldsymbol{x}'$ to the transformed means $\{\boldsymbol{\mu}'_y\}$. This procedure can lead to a significant dimensionality reduction of the data. Namely, the data can be projected onto the space spanned by the differences between the mean vectors $\{\boldsymbol{\mu}'_y\}$. When there are c classes, this is a $(c-1)$-dimensional space, as opposed to the p-dimensional space of the original data. We explain the precise ideas via an example.

■ **Example 8.29 (Classification after Data Reduction)** Consider an equal mixture of three 3-dimensional Gaussian distributions with identical covariance matrices. After sphering the data, the covariance matrices are all equal to the identity matrix. Suppose the mean vectors of the sphered data are $\boldsymbol{\mu}_1 = [2, 1, -3]^\top$, $\boldsymbol{\mu}_2 = [1, -4, 0]^\top$, and $\boldsymbol{\mu}_3 = [2, 4, 6]^\top$. The left panel of Figure 8.30 shows the 3-dimensional (sphered) data from each of the three classes.

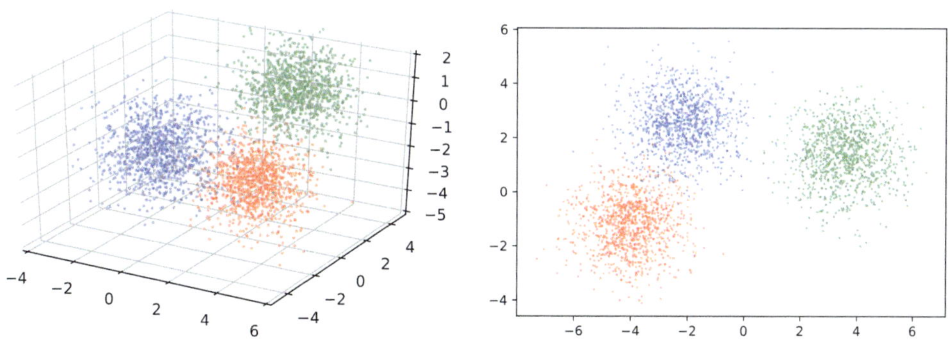

Figure 8.30: Left: original data. Right: projected data.

The data are stored in three 1000×3 matrices $\mathbf{X}_1$, $\mathbf{X}_2$, and $\mathbf{X}_3$. Here is how the data was generated and plotted.

datared.py

```
import numpy as np
from numpy.random import randn
import matplotlib.pyplot as plt
from mpl_toolkits.mplot3d import Axes3D

n = 1000
mu1 = np.array([2, 1, -3])
mu2 = np.array([1, -4, 0])
mu3 = np.array([2, 4, 0])
X1 = randn(n, 3) + mu1
X2 = randn(n, 3) + mu2
X3 = randn(n, 3) + mu3

fig = plt.figure()

ax = fig.add_subplot(111, projection='3d')
ax.plot(X1[:, 0], X1[:, 1], X1[:, 2], 'r.', alpha=0.5, markersize=2)
ax.plot(X2[:, 0], X2[:, 1], X2[:, 2], 'b.', alpha=0.5, markersize=2)
ax.plot(X3[:, 0], X3[:, 1], X3[:, 2], 'g.', alpha=0.5, markersize=2)
# ax.set_aspect("equal")

ax.set_xlim3d(-4, 6)
ax.set_ylim3d(-5, 5)
ax.set_zlim3d(-5, 2)
plt.show()
```

Since we have equal mixtures, we classify each data point $\boldsymbol{x}$ according to the closest distance to $\boldsymbol{\mu}_1$, $\boldsymbol{\mu}_2$, or $\boldsymbol{\mu}_3$. We can achieve a reduction in the dimensionality of the data by *projecting* the data onto the two-dimensional affine space spanned by the $\{\boldsymbol{\mu}_i\}$; that is, all vectors are of the form

$$\boldsymbol{\mu}_1 + \beta_1(\boldsymbol{\mu}_2 - \boldsymbol{\mu}_1) + \beta_2(\boldsymbol{\mu}_3 - \boldsymbol{\mu}_1), \quad \beta_1, \beta_2 \in \mathbb{R}.$$

In fact, one may just as well project the data onto the subspace spanned by the vectors $\boldsymbol{\mu}_{21} = \boldsymbol{\mu}_2 - \boldsymbol{\mu}_1$ and $\boldsymbol{\mu}_{31} = \boldsymbol{\mu}_3 - \boldsymbol{\mu}_1$. Let $\mathbf{W} = [\boldsymbol{\mu}_{21}, \boldsymbol{\mu}_{31}]$ be the 3×2 matrix whose columns are $\boldsymbol{\mu}_{21}$ and $\boldsymbol{\mu}_{31}$. The orthogonal projection matrix onto the subspace $\mathscr{W}$ spanned by the
☞ 521 columns of $\mathbf{W}$ is (see Theorem A.17):

$$\mathbf{P} = \mathbf{W}\mathbf{W}^+ = \mathbf{W}(\mathbf{W}^\top\mathbf{W})^{-1}\mathbf{W}^\top.$$

Let $\mathbf{U}\mathbf{D}\mathbf{V}^\top$ be the singular value decomposition of $\mathbf{W}$. Then $\mathbf{P}$ can also be written as

$$\mathbf{P} = \mathbf{U}\mathbf{D}(\mathbf{D}^\top\mathbf{D})^{-1}\mathbf{D}^\top\mathbf{U}^\top.$$

Note that $\mathbf{D}$ has dimension 3×2, so is not square. The first two columns of $\mathbf{U}$, say $\boldsymbol{u}_1$ and $\boldsymbol{u}_2$, form an orthonormal basis of the subspace $\mathscr{W}$. What we want to do is rotate this subspace to the $x - y$ plane, mapping $\boldsymbol{u}_1$ and $\boldsymbol{u}_2$ to $[1, 0, 0]^\top$ and $[0, 1, 0]^\top$, respectively. This is achieved via the rotation matrix $\mathbf{U}^{-1} = \mathbf{U}^\top$, giving the skewed projection matrix

$$\mathbf{R} = \mathbf{U}^\top\mathbf{P} = \mathbf{D}(\mathbf{D}^\top\mathbf{D})^{-1}\mathbf{D}^\top\mathbf{U}^\top,$$

whose 3rd row only contains zeros. Applying **R** to all the data points, and ignoring the 3rd component of the projected points (which is 0), gives the right panel of Figure 8.30. We see that the projected points are much better separated than the original ones. We have achieved dimensionality reduction of the data while retaining all the necessary information required for classification. Here is the rest of the Python code.

dataproj.py

```
from datared import *
from numpy.linalg import svd, pinv
mu21 = (mu2 - mu1).reshape(3,1)
mu31 = (mu3 - mu1).reshape(3,1)
W = np.hstack((mu21, mu31))
U,_,_ = svd(W)  # we only need U
P = W @ pinv(W)
R = U.T @ P

RX1 = (R @ X1.T).T
RX2 = (R @ X2.T).T
RX3 = (R @ X3.T).T
plt.plot(RX1[:,0],RX1[:,1],'b.',alpha=0.5,markersize=2)
plt.plot(RX2[:,0],RX2[:,1],'g.',alpha=0.5,markersize=2)
plt.plot(RX3[:,0],RX3[:,1],'r.',alpha=0.5,markersize=2)
plt.show()
```

8.5 Logistic Regression and Softmax Classification

In Example 5.77 we introduced the logistic (logit) regression model as a generalized linear model where, conditional on a p-dimensonal feature vector $\boldsymbol{x}$, the random response Y has a $\mathsf{Ber}(h(\boldsymbol{x}^\top\boldsymbol{\beta}))$ distribution with $h(u) = 1/(1+\mathrm{e}^{-u})$. The parameter $\boldsymbol{\beta}$ was then learned from the training data by maximizing the likelihood of the training responses or, equivalently, by minimizing the supervised version of the *cross-entropy training loss* (4.4): ☞ 243 ☞ 135

$$-\frac{1}{n}\sum_{i=1}^{n} \ln g(y_i \,|\, \boldsymbol{\beta}, \boldsymbol{x}_i),$$

where $g(y=1\,|\,\boldsymbol{\beta},\boldsymbol{x}) = 1/(1+\mathrm{e}^{-\boldsymbol{x}^\top\boldsymbol{\beta}})$ and $g(y=0\,|\,\boldsymbol{\beta},\boldsymbol{x}) = \mathrm{e}^{-\boldsymbol{x}^\top\boldsymbol{\beta}}/(1+\mathrm{e}^{-\boldsymbol{x}^\top\boldsymbol{\beta}})$. In particular, we have

$$\ln\frac{g(y=1\,|\,\boldsymbol{\beta},\boldsymbol{x})}{g(y=0\,|\,\boldsymbol{\beta},\boldsymbol{x})} = \boldsymbol{x}^\top\boldsymbol{\beta}. \tag{8.31}$$

In other words, the **log-odds ratio** is a linear function of the feature vector. As a consequence, the decision boundary

$$\{\boldsymbol{x} : g(y=0\,|\,\boldsymbol{\beta},\boldsymbol{x}) = g(y=1\,|\,\boldsymbol{\beta},\boldsymbol{x})\}$$

is the hyperplane $\boldsymbol{x}^\top\boldsymbol{\beta} = 0$. Note that $\boldsymbol{x}$ typically includes the constant feature. If the constant feature is considered separately, that is $\boldsymbol{x} = [1, \widetilde{\boldsymbol{x}}^\top]^\top$, then the boundary is an affine hyperplane in $\widetilde{\boldsymbol{x}}$.

Suppose that training on $\tau = \{(\boldsymbol{x}_i, y_i)\}$ yields the estimate $\widehat{\boldsymbol{\beta}}$ with the corresponding learner:

$$g_\tau(y = 1 \mid \boldsymbol{x}) = 1/(1 + \mathrm{e}^{-\boldsymbol{x}^\top \widehat{\boldsymbol{\beta}}}).$$

The learner can be used as a pre-classifier from which we obtain the classifier $\mathbb{1}\{g_\tau(y = 1 \mid \boldsymbol{x}) > 1/2\}$ or, equivalently,

$$\widehat{y} := \underset{j \in \{0,1\}}{\operatorname{argmax}}\, g_\tau(y = j \mid \boldsymbol{x}),$$

in accordance with the fundamental classification rule (8.3).

The above classification methodology for the logit model can be generalized to the **multi-logit** model where the response takes values in the set $\{0, \ldots, c-1\}$. The key idea is to replace (8.31) with

$$\ln \frac{g(y = j \mid \mathbf{W}, \boldsymbol{b}, \boldsymbol{x})}{g(y = 0 \mid \mathbf{W}, \boldsymbol{b}, \boldsymbol{x})} = \boldsymbol{x}^\top \boldsymbol{\beta}_j, \quad j = 1, \ldots, c-1, \tag{8.32}$$

where the matrix $\mathbf{W} \in \mathbb{R}^{(c-1)\times(p-1)}$ and vector $\boldsymbol{b} \in \mathbb{R}^{c-1}$ reparameterize all $\boldsymbol{\beta}_j \in \mathbb{R}^p$ such that (recall $\boldsymbol{x} = [1, \widetilde{\boldsymbol{x}}^\top]^\top$):

$$\mathbf{W}\widetilde{\boldsymbol{x}} + \boldsymbol{b} = [\boldsymbol{\beta}_1, \ldots, \boldsymbol{\beta}_{c-1}]^\top \boldsymbol{x}.$$

Observe that the random response Y is assumed to have a conditional probability distribution for which the log-odds ratio with respect to class j and a "reference" class (in this case 0) is *linear*. The separating boundaries between two pairs of classes are again affine hyperplanes.

The model (8.32) completely specifies the distribution of Y, namely:

$$g(y \mid \mathbf{W}, \boldsymbol{b}, \boldsymbol{x}) = \frac{\exp(z_{y+1})}{\sum_{k=1}^{c} \exp(z_k)}, \quad y = 0, \ldots, c-1,$$

where z_1 is an arbitrary constant, say 0, corresponding to the "reference" class $y = 0$, and

$$[z_2, \ldots, z_c]^\top := \mathbf{W}\widetilde{\boldsymbol{x}} + \boldsymbol{b}.$$

Note that $g(y \mid \mathbf{W}, \boldsymbol{b}, \boldsymbol{x})$ is the $(y+1)$-st component of $\boldsymbol{a} = \mathrm{softmax}(\boldsymbol{z})$, where

$$\mathrm{softmax} : \boldsymbol{z} \mapsto \frac{\exp(\boldsymbol{z})}{\sum_k \exp(z_k)}$$

is the **softmax** function and $\boldsymbol{z} = [z_1, \ldots, z_c]^\top$. Finally, we can write the classifier as

$$\widehat{y} = \underset{j \in \{0,\ldots,c-1\}}{\operatorname{argmax}}\; a_{j+1}.$$

In summary, we have the sequence of mappings transforming the input $\boldsymbol{x}$ into the output $\widehat{y}$:

$$\boldsymbol{x} \to \mathbf{W}\widetilde{\boldsymbol{x}} + \boldsymbol{b} \to \mathrm{softmax}(\boldsymbol{z}) \to \underset{j \in \{0,\ldots,c-1\}}{\operatorname{argmax}}\; a_{j+1} \to \widehat{y}.$$

In Example 10.18 we will revisit the multi-logit model and reinterpret this sequence of mappings as a *neural network*. In the context of neural networks, $\mathbf{W}$ is called a *weight* matrix and $\boldsymbol{b}$ is called a *bias* vector. ☞ 439

The parameters $\mathbf{W}$ and $\boldsymbol{b}$ have to be learned from the training data, which involves minimization of the supervised version of the *cross-entropy training loss* (4.4): ☞ 135

$$\frac{1}{n}\sum_{i=1}^{n} \text{Loss}(f(y_i \mid \boldsymbol{x}_i), g(y_i \mid \mathbf{W}, \boldsymbol{b}, \boldsymbol{x}_i)) = -\frac{1}{n}\sum_{i=1}^{n} \ln g(y_i \mid \mathbf{W}, \boldsymbol{b}, \boldsymbol{x}_i).$$

Using the softmax function, the *cross-entropy* loss can be simplified to:

$$\text{Loss}(f(y \mid \boldsymbol{x}), g(y \mid \mathbf{W}, \boldsymbol{b}, \boldsymbol{x})) = -z_{y+1} + \ln \sum_{k=1}^{c} \exp(z_k). \tag{8.33}$$

The discussion on training is postponed until Chapter 10, where we reinterpret the multi-logit model as a neural net, which can be trained using the *limited-memory BFGS* method (Exercise 10). Note that in the binary case ($c = 2$), where there is only one vector $\boldsymbol{\beta}$ to be estimated, Example 5.77 already established that minimization of the cross-entropy training loss is equivalent to likelihood maximization. ☞ 458

8.6 *K*-Nearest Neighbors Classification

Let $\tau = \{(\boldsymbol{x}_i, y_i)\}_{i=1}^n$ be the training set, with $y_i \in \{0, \dots, c-1\}$, and let $\boldsymbol{x}$ be a new feature vector. Define $\boldsymbol{x}_{(1)}, \boldsymbol{x}_{(2)}, \dots, \boldsymbol{x}_{(n)}$ as the feature vectors ordered by closeness to $\boldsymbol{x}$ in some distance $\text{dist}(\boldsymbol{x}, \boldsymbol{x}_i)$, e.g., the Euclidean distance $\|\boldsymbol{x} - \boldsymbol{x}'\|$. Let

$$\tau(\boldsymbol{x}) := \{(\boldsymbol{x}_{(1)}, y_{(1)}), \dots, (\boldsymbol{x}_{(K)}, y_{(K)})\}$$

be the subset of τ that contains K feature vectors $\boldsymbol{x}_i$ that are closest to $\boldsymbol{x}$. Then the **K-nearest neighbors** classification rule classifies $\boldsymbol{x}$ according to the most frequently occurring class labels in $\tau(\boldsymbol{x})$. If two or more labels receive the same number of votes, the feature vector is classified by selecting one of these labels randomly with equal probability. For the case $K = 1$ the set $\tau(\boldsymbol{x})$ contains only one element, say $(\boldsymbol{x}', y')$, and $\boldsymbol{x}$ is classified as y'. This divides the space into n regions

$$\mathcal{R}_i = \{\boldsymbol{x} : \text{dist}(\boldsymbol{x}, \boldsymbol{x}_i) \leq \text{dist}(\boldsymbol{x}, \boldsymbol{x}_j), j \neq i\}, \quad i = 1, \dots, n.$$

For a feature space $\mathbb{R}^p$ with the Euclidean distance, this gives a Voronoi tessellation of the feature space, similar to what was done for vector quantization in Section 4.7. ☞ 167

■ **Example 8.34 (Nearest Neighbor Classification)** The Python program below simulates 80 random points above and below the line $x_2 = x_1$. Points above the line $x_2 = x_1$ have label 0 and points below this line have label 1. Figure 8.35 shows the Voronoi tessellation obtained from the 1-nearest neighbor classification.

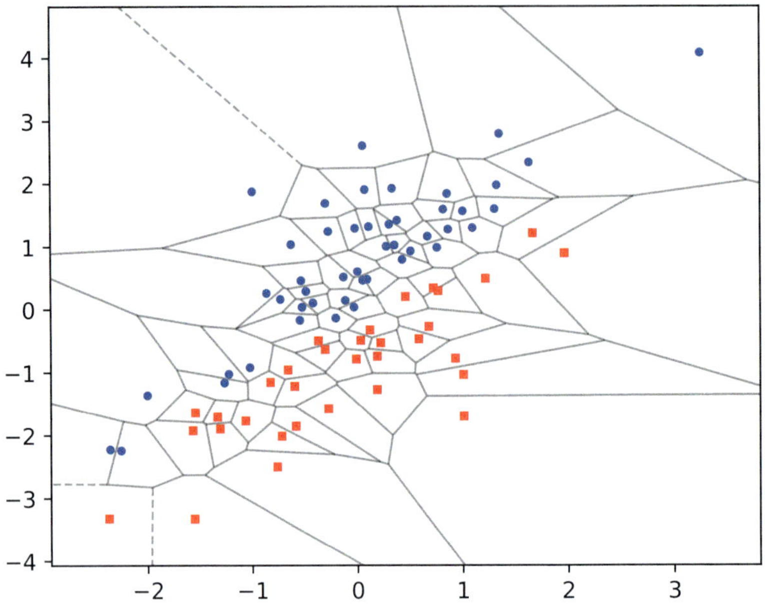

Figure 8.35: The 1-nearest neighbor algorithm divides up the space into Voronoi cells.

nearestnb.py

```
import numpy as np
from numpy.random import rand,randn
import matplotlib.pyplot as plt
from scipy.spatial import Voronoi, voronoi_plot_2d

np.random.seed(12345)
M = 80
x = randn(M,2)
y = np.zeros(M) # pre-allocate list

for i in range(M):
    if rand()<0.5:
        x[i,1], y[i] = x[i,0] + np.abs(randn()), 0
    else:
        x[i,1], y[i] = x[i,0] - np.abs(randn()), 1

vor = Voronoi(x)
plt_options = {'show_vertices':False, 'show_points':False,
               'line_alpha':0.5}
fig = voronoi_plot_2d(vor, **plt_options)
plt.plot(x[y==0,0], x[y==0,1],'bo',
         x[y==1,0], x[y==1,1],'rs', markersize=3)
```

■

8.7 Support Vector Machine

Suppose we are given the training set $\tau = \{(\boldsymbol{x}_i, y_i)\}_{i=1}^n$, where each response[2] y_i takes either the value -1 or 1, and we wish to construct a classifier taking values in $\{-1, 1\}$. As this merely involves a relabeling of the 0–1 classification problem in Section 8.1, by Theorem 8.2 the optimal classification function for the indicator loss, $\mathbb{1}\{y \neq \widehat{y}\}$, is:

$$g^*(\boldsymbol{x}) = \begin{cases} 1 & \text{if} \quad \mathbb{P}[Y = 1 \mid \boldsymbol{X} = \boldsymbol{x}] \geq 1/2, \\ -1 & \text{if} \quad \mathbb{P}[Y = 1 \mid \boldsymbol{X} = \boldsymbol{x}] < 1/2. \end{cases}$$

It is not difficult to show, see Exercise 5, that the function g^* can be viewed as the minimizer of the risk for the **hinge loss** function, $\text{Loss}(y, \widehat{y}) = (1 - y\,\widehat{y})_+$, over all prediction functions g (not necessarily taking values only in the set $\{-1, 1\}$). That is,

$$g^* = \underset{g}{\operatorname{argmin}}\, \mathbb{E}\,(1 - Y\,g(\boldsymbol{X}))_+. \tag{8.36}$$

Given the training set τ, we can approximate the risk $\ell(g) = \mathbb{E}\,(1 - Y\,g(\boldsymbol{X}))_+$ with the training loss

$$\ell_\tau(g) = \frac{1}{n}\sum_{i=1}^n (1 - y_i\, g(\boldsymbol{x}_i))_+,$$

and minimize this over a (smaller) class of functions to obtain the optimal prediction function g_τ. Finally, as the prediction function g_τ generally is not a classifier by itself (it usually does not only take values -1 or 1), we take the classifier

$$\operatorname{sign} g_\tau(\boldsymbol{x}).$$

Therefore, a feature vector $\boldsymbol{x}$ is classified according to 1 or -1 depending on whether $g_\tau(\boldsymbol{x}) \geq 0$ or < 0, respectively. The **optimal decision boundary** is given by the set of $\boldsymbol{x}$ for which $g_\tau(\boldsymbol{x}) = 0$.

Similar to the cubic smoothing spline or RKHS setting in (7.33), we can consider finding the best classifier, given the training data, via the penalized goodness-of-fit optimization:

$$\min_{g \in \mathcal{H} \oplus \mathcal{H}_0} \frac{1}{n}\sum_{i=1}^n [1 - y_i\, g(\boldsymbol{x}_i)]_+ + \widetilde{\gamma}\, \|g\|_{\mathcal{H}}^2,$$

for some regularization parameter $\widetilde{\gamma}$. It will be convenient to define $\gamma := 2n\widetilde{\gamma}$ and to solve the equivalent problem

$$\min_{g \in \mathcal{H} \oplus \mathcal{H}_0} \sum_{i=1}^n [1 - y_i\, g(\boldsymbol{x}_i)]_+ + \frac{\gamma}{2}\, \|g\|_{\mathcal{H}}^2.$$

We know from the Representer Theorem 7.35 that if κ is the reproducing kernel corresponding to $\mathcal{H}$, then the solution is of the form (assuming that the null space $\mathcal{H}_0$ has a constant term only): ☞ 333

$$g(\boldsymbol{x}) = \alpha_0 + \sum_{i=1}^n \alpha_i\, \kappa(\boldsymbol{x}_i, \boldsymbol{x}). \tag{8.37}$$

[2]The reason why we use responses -1 and 1 here, instead of 0 and 1, is that the notation becomes easier.

☞ 334 Substituting into the minimization expression yields the analogue of (7.37):

$$\min_{\alpha, \alpha_0} \sum_{i=1}^{n} [1 - y_i(\alpha_0 + \{\mathbf{K}\alpha\}_i)]_+ + \frac{\gamma}{2} \alpha^\top \mathbf{K} \alpha, \tag{8.38}$$

where $\mathbf{K}$ is the Gram matrix. This is a *convex* optimization problem, as it is the sum of a convex quadratic and piecewise linear term in $\boldsymbol{\alpha}$. Defining $\lambda_i := \gamma \alpha_i / y_i$, $i = 1, \ldots, n$ and $\boldsymbol{\lambda} := [\lambda_1, \ldots, \lambda_n]^\top$, we show in Exercise 10 that the optimal $\boldsymbol{\alpha}$ and α_0 in (8.38) can be obtained by solving the "dual" convex optimization problem

$$\begin{aligned} \max_{\lambda} \quad & \sum_{i=1}^{n} \lambda_i - \frac{1}{2\gamma} \sum_{i=1}^{n} \sum_{j=1}^{n} \lambda_i \lambda_j y_i y_j \, \kappa(\boldsymbol{x}_i, \boldsymbol{x}_j) \\ \text{subject to:} \quad & \boldsymbol{\lambda}^\top \boldsymbol{y} = 0, \ \mathbf{0} \le \boldsymbol{\lambda} \le \mathbf{1}, \end{aligned} \tag{8.39}$$

and $\alpha_0 = y_j - \sum_{i=1} \alpha_i \, \kappa(\boldsymbol{x}_i, \boldsymbol{x}_j)$ for any j for which $\lambda_j \in (0, 1)$. In view of (8.37), the optimal prediction function (pre-classifier) g_τ is then given by

$$g_\tau(\boldsymbol{x}) = \alpha_0 + \sum_{i=1}^{n} \alpha_i \, \kappa(\boldsymbol{x}_i, \boldsymbol{x}) = \alpha_0 + \frac{1}{\gamma} \sum_{i=1}^{n} y_i \lambda_i \, \kappa(\boldsymbol{x}_i, \boldsymbol{x}). \tag{8.40}$$

To mitigate possible numerical problems in the calculation of α_0 it is customary to take an overall average:

$$\alpha_0 = \frac{1}{|\mathcal{J}|} \sum_{j \in \mathcal{J}} \left\{ y_j - \sum_{i=1}^{n} \alpha_i \, \kappa(\boldsymbol{x}_i, \boldsymbol{x}_j) \right\},$$

where $\mathcal{J} := \{j : \lambda_j \in (0, 1)\}$.

Note that, from (8.40), the optimal pre-classifier $g(\boldsymbol{x})$ and the classifier sign $g(\boldsymbol{x})$ only depend on vectors $\boldsymbol{x}_i$ for which $\lambda_i \neq 0$. These vectors are called the **support vectors** of the support vector machine. It is also important to note that the quadratic function in (8.39) depends on the regularization parameter γ. By defining $\nu_i := \lambda_i / \gamma$, $i = 1, \ldots, n$, we can rewrite (8.39) as

$$\begin{aligned} \min_{\nu} \quad & \frac{1}{2} \sum_{i,j} \nu_i \nu_j y_i y_j \, \kappa(\boldsymbol{x}_i, \boldsymbol{x}_j) - \sum_{i=1}^{n} \nu_i \\ \text{subject to:} \quad & \sum_{i=1}^{n} \nu_i y_i = 0, \ \ 0 \le \nu_i \le 1/\gamma =: C, \quad i = 1, \ldots, n. \end{aligned} \tag{8.41}$$

For perfectly separable data, that is, data for which an affine plane can be drawn to perfectly separate the two classes, we may take $C = \infty$, as explained below. Otherwise, C needs to be chosen via cross-validation or a test data set, for example.

Geometric interpretation

For the linear kernel function $\kappa(\boldsymbol{x}, \boldsymbol{x}') = \boldsymbol{x}^\top \boldsymbol{x}'$, we have $g_\tau(\boldsymbol{x}) = \beta_0 + \boldsymbol{\beta}^\top \boldsymbol{x}$, with $\beta_0 = \alpha_0$ and $\boldsymbol{\beta} = \gamma^{-1} \sum_{i=1}^{n} \lambda_i y_i \boldsymbol{x}_i = \sum_{i=1}^{n} \alpha_i \boldsymbol{x}_i$, and so the decision boundary is an affine plane. The situation is illustrated in Figure 8.42.

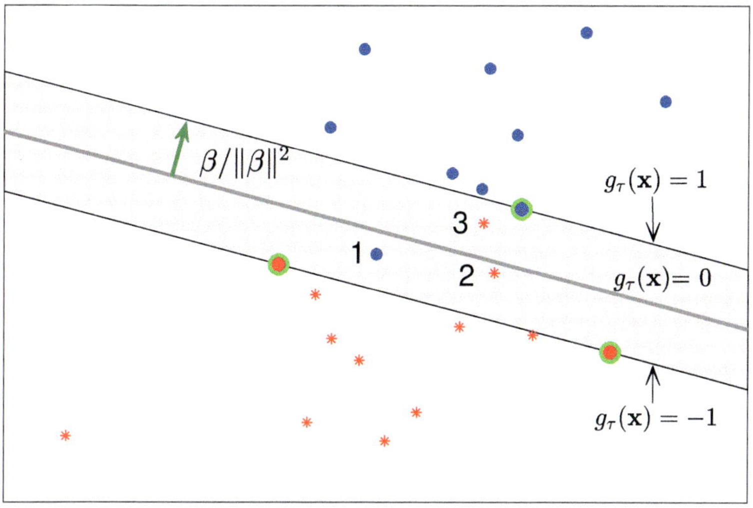

Figure 8.42: Classifying two classes (red and blue) using SVM.

The decision boundary is formed by the points $\boldsymbol{x}$ such that $g_\tau(\boldsymbol{x}) = 0$. The two sets $\{\boldsymbol{x} : g_\tau(\boldsymbol{x}) = -1\}$ and $\{\boldsymbol{x} : g_\tau(\boldsymbol{x}) = 1\}$ are called the *margins*. The distance from the points on a margin to the decision boundary is $1/\|\boldsymbol{\beta}\|$.

Based on the "multipliers" $\{\lambda_i\}$, we can divide the training samples $\{(\boldsymbol{x}_i, y_i)\}$ into three categories (see Exercise 11):

- Points for which $\lambda_i \in (0, 1)$. These are the support vectors on the margins (green encircled in the figure) and are correctly classified.
- Points for which $\lambda_i = 1$. These points, which are also support vectors, lie strictly inside the margins (points 1, 2, and 3 in the figure). Such points may or may not be correctly classified.
- Points for which $\lambda_i = 0$. These are the non-support vectors, which all lie outside the margins. Every such point is correctly classified.

If the classes of points $\{\boldsymbol{x}_i : y_i = 1\}$ and $\{\boldsymbol{x}_i : y_i = -1\}$ are perfectly separable by some affine plane, then there will be no points strictly inside the margins, so all support vectors will lie exactly on the margins. In this case (8.38) reduces to

$$\begin{aligned} &\min_{\boldsymbol{\beta},\beta_0} \ \|\boldsymbol{\beta}\|^2 \\ &\text{subject to:} \quad y_i(\beta_0 + \boldsymbol{x}_i^\top \boldsymbol{\beta}) \geq 1, \ i = 1, \ldots, n, \end{aligned} \tag{8.43}$$

using the fact that $\alpha_0 = \beta_0$ and $\mathbf{K}\boldsymbol{\alpha} = \mathbf{X}\mathbf{X}^\top\boldsymbol{\alpha} = \mathbf{X}\boldsymbol{\beta}$. We may replace $\min \|\boldsymbol{\beta}\|^2$ in (8.43) with $\max 1/\|\boldsymbol{\beta}\|$, as this gives the same optimal solution. As $1/\|\boldsymbol{\beta}\|$ is equal to half the margin width, the latter optimization problem has a simple interpretation: separate the points via an affine hyperplane such that the margin width is maximized.

■ **Example 8.44 (Support Vector Machine)** The data in Figure 8.45 was uniformly generated on the unit disc. Class-1 points (blue dots) have a radius less than 1/2 (y-values 1) and class-2 points (red crosses) have a radius greater than 1/2 (y-values -1).

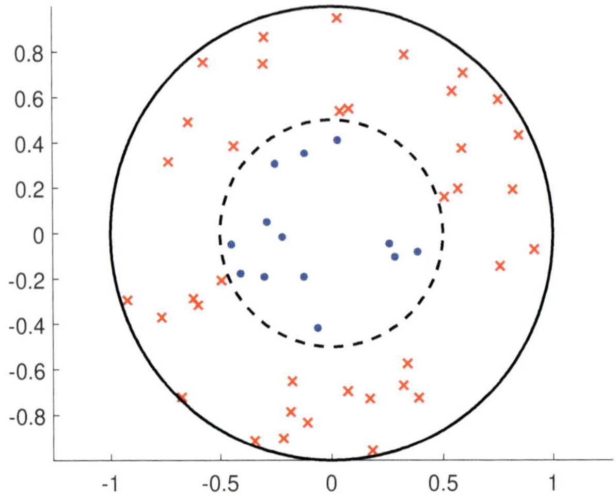

Figure 8.45: Separate the two classes.

Of course it is not possible to separate the two groups of points via a straight line in $\mathbb{R}^2$. However, it is possible to separate them in $\mathbb{R}^3$ by considering three-dimensional feature vectors

$$z = \begin{bmatrix} z_1 \\ z_2 \\ z_3 \end{bmatrix} = \begin{bmatrix} x_1 \\ x_2 \\ x_1^2 + x_2^2 \end{bmatrix}.$$

For any $\boldsymbol{x} = [x_1, x_2]^\top \in \mathbb{R}^2$, the corresponding feature vector $\boldsymbol{z}$ lies on a quadratic surface. In this space it is possible to separate the $\{z_i\}$ points into two groups by means of a planar surface, as illustrated in Figure 8.46.

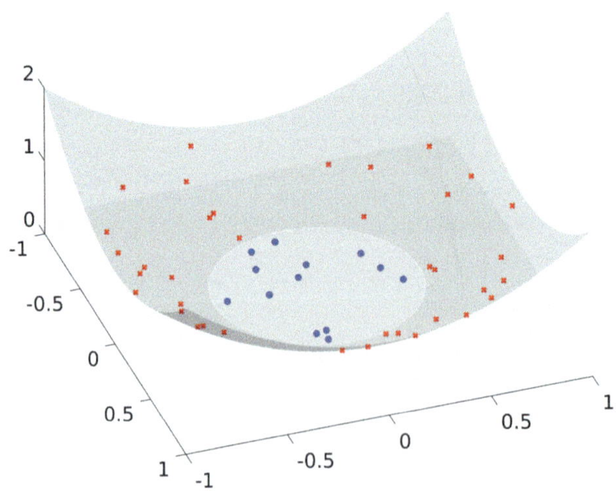

Figure 8.46: In feature space $\mathbb{R}^3$ the points can be separated by a plane.

We wish to find a separating plane in $\mathbb{R}^3$ using the transformed features. The following Python code uses the **SVC** function of the **sklearn** module to solve the quadratic optimization problem (8.41) (with $C = \infty$). The results are summarized in Table 8.47. The data is available from the book's GitHub site as svmcirc.csv.

svmquad.py

```
import numpy as np
from numpy import genfromtxt
from sklearn.svm import SVC

data = genfromtxt('svmcirc.csv', delimiter=',')
x = data[:,[0,1]] #vectors are rows
y = data[:,[2]].reshape(len(x),) #labels

tmp = np.sum(np.power(x,2),axis=1).reshape(len(x),1)
z = np.hstack((x,tmp))

clf = SVC(C = 10e20, kernel='linear')
clf.fit(z,y)

print("Support Vectors \n", clf.support_vectors_)
print("Support Vector Labels ",y[clf.support_])
print("Nu",clf.dual_coef_)
print("Bias",clf.intercept_)
```

```
 Support Vectors
 [[ 0.038758    0.53796     0.29090314]
 [-0.49116    -0.20563     0.28352184]
 [-0.45068    -0.04797     0.20541358]
 [-0.061107   -0.41651     0.17721465]]
Support Vector Labels  [-1. -1.  1.  1.]
Nu [[ -46.49249413 -249.01807328  265.31805855   30.19250886]]
Bias  [5.617891]
```

Table 8.47: Optimal support vector machine parameters for the $\mathbb{R}^3$ data.

$z^\top$			y	$\alpha = \nu y$
0.0388	0.5380	0.2909	−1	−46.4925
−0.4912	−0.2056	0.2835	−1	−249.0181
−0.4507	−0.0480	0.2054	1	265.3181
−0.0611	−0.4165	0.1772	1	30.1925

It follows that the normal vector of the plane is

$$\boldsymbol{\beta} = \sum_{i \in \mathcal{S}} \alpha_i \boldsymbol{z}_i = [-0.9128, 0.8917, -24.2764]^\top,$$

where $\mathcal{S}$ is the set of indices of the support vectors. We see that the plane is almost perpendicular to the z_1, z_2 plane. The bias term β_0 can also be found from the table above. In particular, for any $\boldsymbol{x}^\top$ and y in Table 8.47, we have $y - \boldsymbol{\beta}^\top \boldsymbol{z} = \beta_0 = 5.6179$.

To draw the separating boundary in $\mathbb{R}^2$ we need to project the intersection of the separating plane with the quadratic surface onto the z_1, z_2 plane. That is, we need to find all points (z_1, z_2) such that

$$5.6179 - 0.9128 z_1 + 0.8917 z_2 = 24.2764\,(z_1^2 + z_2^2). \tag{8.48}$$

This is the equation of a circle with (approximate) center $(0.019, -0.018)$ and radius 0.48, which is very close to the true circular boundary between the two groups, with center $(0, 0)$ and radius 0.5. This circle is drawn in Figure 8.49.

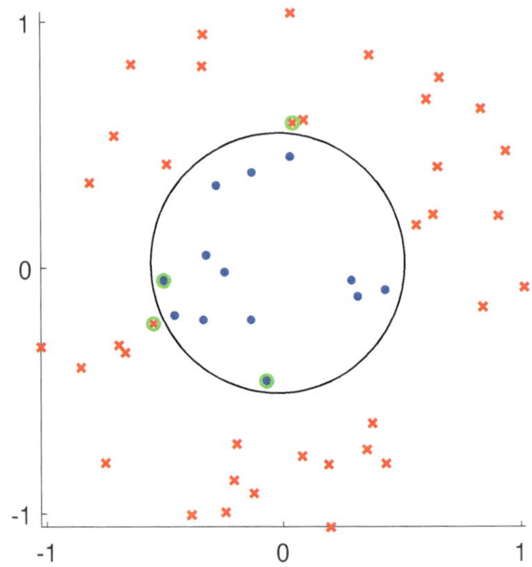

Figure 8.49: The circular decision boundary can be viewed equivalently as (a) the projection onto the x_1, x_2 plane of the intersection of the separating plane with the quadratic surface (both in $\mathbb{R}^3$), or (b) the set of points $\boldsymbol{x} = (x_1, x_2)$ for which $g_\tau(\boldsymbol{x}) = \beta_0 + \boldsymbol{\beta}^\top \boldsymbol{\phi}(\boldsymbol{x}) = 0$.

An equivalent way to derive this circular separating boundary is to consider the feature map $\boldsymbol{\phi}(\boldsymbol{x}) = [x_1, x_2, x_1^2 + x_2^2]^\top$ on $\mathbb{R}^2$, which defines a reproducing kernel

$$\kappa(\boldsymbol{x}, \boldsymbol{x}') = \boldsymbol{\phi}(\boldsymbol{x})^\top \boldsymbol{\phi}(\boldsymbol{x}'),$$

on $\mathbb{R}^2$, which in turn gives rise to a (unique) RKHS $\mathcal{H}$. The optimal prediction function (8.37) is now of the form

$$g_\tau(\boldsymbol{x}) = \alpha_0 + \frac{1}{\gamma} \sum_{i=1}^{n} y_i \, \lambda_i \, \boldsymbol{\phi}(\boldsymbol{x}_i)^\top \boldsymbol{\phi}(\boldsymbol{x}) = \beta_0 + \boldsymbol{\beta}^\top \boldsymbol{\phi}(\boldsymbol{x}), \tag{8.50}$$

where $\alpha_0 = \beta_0$ and

$$\boldsymbol{\beta} = \frac{1}{\gamma} \sum_{i=1}^{n} y_i \, \lambda_i \, \boldsymbol{\phi}(\boldsymbol{x}_i).$$

The decision boundary, $\{\boldsymbol{x} : g_\tau(\boldsymbol{x}) = 0\}$, is again a circle in $\mathbb{R}^2$. The following code determines the fitted model parameters and the decision boundary. Figure 8.49 shows the optimal decision boundary, which is identical to (8.48). The function `mykernel` specifies the custom kernel above.

svmkern.py

```
import numpy as np, matplotlib.pyplot as plt
from numpy import genfromtxt
from sklearn.svm import SVC

def mykernel(U,V):
    tmpU = np.sum(np.power(U,2),axis=1).reshape(len(U),1)
    U = np.hstack((U,tmpU))
    tmpV = np.sum(np.power(V,2),axis=1).reshape(len(V),1)
    V = np.hstack((V,tmpV))
    K = U @ V.T
    print(K.shape)
    return K

# read in the data
inp = genfromtxt('svmcirc.csv', delimiter=',')
data = inp[:,[0,1]] #vectors are rows
y = inp[:,[2]].reshape(len(data),) #labels

clf = SVC(C = 10e20, kernel=mykernel, gamma='auto') # custom kernel
# clf = SVC(C = 10e20, kernel="rbf", gamma='scale') # inbuilt

clf.fit(data,y)

print("Support Vectors \n", clf.support_vectors_)
print("Support Vector Labels ",y[clf.support_])
print("Nu ",clf.dual_coef_)
print("Bias ",clf.intercept_)

# plot
d = 0.001
x_min, x_max = -1,1
y_min, y_max = -1,1
xx, yy = np.meshgrid(np.arange(x_min, x_max, d), np.arange(y_min, y_max,
     d))
plt.plot(data[clf.support_,0],data[clf.support_,1],'go')
plt.plot(data[y==1,0],data[y==1,1],'b.')
plt.plot(data[y==-1,0],data[y==-1,1],'rx')
Z = clf.predict(np.c_[xx.ravel(), yy.ravel()])
Z = Z.reshape(xx.shape)
plt.contour(xx, yy, Z,colors ="k")
plt.show()
```

Finally, we illustrate the use of the Gaussian kernel

$$\kappa(\boldsymbol{x}, \boldsymbol{x}') = \mathrm{e}^{-c\,\|\boldsymbol{x}-\boldsymbol{x}'\|^2}, \tag{8.51}$$

where $c > 0$ is some tuning constant. This is an example of a *radial basis function kernel*, which are reproducing kernels of the form $\kappa(\boldsymbol{x}, \boldsymbol{x}') = f(\|\boldsymbol{x} - \boldsymbol{x}'\|)$, for some positive real-valued function f. Each feature vector $\boldsymbol{x}$ is now transformed to a *function* $\kappa_{\boldsymbol{x}} = \kappa(\boldsymbol{x}, \cdot)$. We can think of it as the (unnormalized) pdf of a Gaussian distribution centered around $\boldsymbol{x}$, and g_τ is a (signed) *mixture* of these pdfs, plus a constant; that is,

$$g_\tau(\boldsymbol{x}) = \alpha_0 + \sum_{i=1}^{n} \alpha_i\,\mathrm{e}^{-c\,\|\boldsymbol{x}_i-\boldsymbol{x}\|^2}.$$

Replacing in Line 2 of the previous code `mykernel` with `'rbf'` produces the SVM parameters given in Table 8.52. Figure 8.53 shows the decision boundary, which is not exactly circular, but is close to the true (circular) boundary $\{\boldsymbol{x} : \|\boldsymbol{x}\| = 1/2\}$. There are now seven support vectors, rather than the four in Figure 8.49.

Table 8.52: Optimal support vector machine parameters for the Gaussian kernel case.

$\boldsymbol{x}^\top$		y	α ($\times 10^9$)	$\boldsymbol{x}^\top$		y	α ($\times 10^9$)
0.0388	0.5380	−1	−0.0635	−0.4374	0.3854	−1	−1.4399
−0.4912	−0.2056	−1	−9.4793	0.3402	−0.5740	−1	−0.1000
0.5086	0.1576	−1	−0.5240	−0.4098	−0.1763	1	6.0662
−0.4507	−0.0480	1	5.5405				

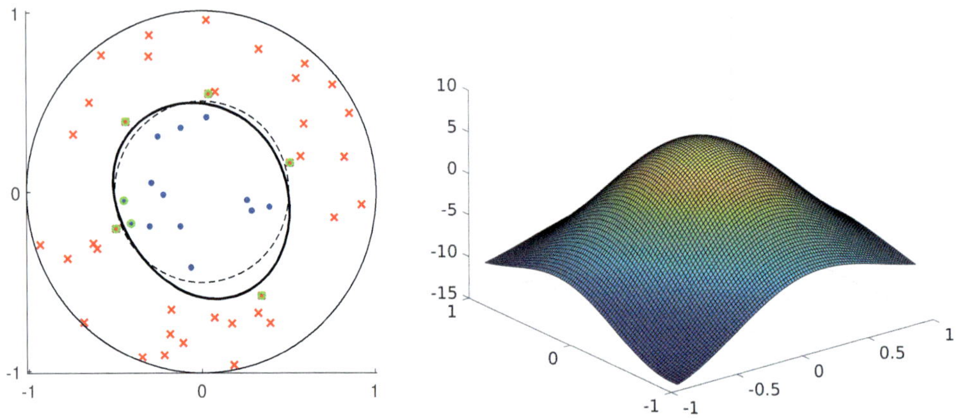

Figure 8.53: Left: The decision boundary $\{\boldsymbol{x} : g_\tau(\boldsymbol{x}) = 0\}$ is roughly circular, and separates the two classes well. There are seven support vectors, indicated by green circles. Right: The graph of g_τ is a scaled mixture of Gaussian pdfs plus a constant.

■

■ **Remark 8.54 (Scaling and Penalty Parameters)** When using a radial basis function in **SVC** in `sklearn`, the scaling c (8.51) can be set via the parameter `gamma`. Large values of `gamma` lead to highly peaked predicted functions, and small values lead to highly smoothed predicted functions. The parameter `C` in **SVC** refers $C = 1/\gamma$ in (8.41). ■

8.8 Classification with Scikit-Learn

In this section we apply several classification methods to a real-world data set, using the Python module `sklearn` (the package name is Scikit-Learn). Specifically, the data is obtained from UCI's **Breast Cancer Wisconsin** data set. This data set, first published and analyzed in [165], contains the measurements related to 569 images of 357 benign and 212 malignant breast masses. The goal is to classify a breast mass as benign or malignant based on 10 features: Radius, Texture, Perimeter, Area, Smoothness, Compactness,

Concavity, Concave Points, Symmetry, and Fractal Dimension of each mass. The mean, standard error, and "worst" of these attributes were computed for each image, resulting in 30 features. For instance, feature 1 is Mean Radius, feature 11 is Radius SE, feature 21 is Worst Radius.

The following Python code reads the data, extracts the response vector and model (feature) matrix and divides the data into a training and test set.

skclass1.py

```
from numpy import genfromtxt
from sklearn.model_selection import train_test_split
data = genfromtxt('breast_cancer_data.txt',delimiter=',', dtype=str)
y = data[:,1] #responses
X = data[:,2:].astype('float')#features as an ndarray matrix

X_train , X_test , y_train , y_test = train_test_split(
        X, y, test_size = 0.4, random_state = 1234)
```

To visualize the data we create, with the following code, a 3D scatterplot for the features mean *radius*, mean *texture*, and mean *concavity*, which correspond to the columns 0, 1, and 6 of the model matrix X. Figure 8.55 suggests that the malignant and benign breast masses could be well separated using these three features.

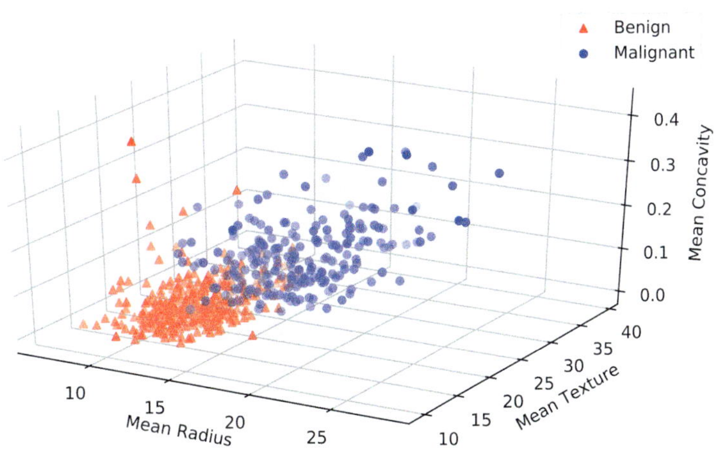

Figure 8.55: Scatterplot of three features of the benign and malignant breast masses.

skclass2.py

```
from skclass1 import X, y
import matplotlib.pyplot as plt
from mpl_toolkits.mplot3d import Axes3D
import numpy as np

Bidx = np.where(y == 'B')
```

```
Midx = np.where(y == 'M')

# plot features Radius (column 0), Texture (1), Concavity (6)
fig = plt.figure()
ax = fig.add_subplot(111, projection='3d')
ax.scatter(X[Bidx, 0], X[Bidx, 1], X[Bidx, 6],
           c='r', marker='^', label='Benign')
ax.scatter(X[Midx, 0], X[Midx, 1], X[Midx, 6],
           c='b', marker='o', label='Malignant')
ax.legend()
ax.set_xlabel('Mean Radius')
ax.set_ylabel('Mean Texture')
ax.set_zlabel('Mean Concavity')
plt.show()
```

The next code uses various classifiers to predict the category of breast masses (benign or malignant). In this case the training set has 341 elements and the test set has 228 elements. For each classifier the percentage of correct predictions (that is, the accuracy) in the test set is reported. We see that in this case quadratic discriminant analysis gives the highest accuracy (0.956). Exercise 18 explores the question whether this metric is the most appropriate for these data.

skclass3.py

```
from skclass1 import X_train, y_train, X_test, y_test
from sklearn.metrics import accuracy_score
import sklearn.discriminant_analysis as DA
from sklearn.naive_bayes import GaussianNB
from sklearn.neighbors import KNeighborsClassifier
#from sklearn.neighbors import DistanceMetric
from sklearn.linear_model import LogisticRegression
from sklearn.svm import SVC
#import numpy as np
#from sklearn.preprocessing import MinMaxScaler

names = ["Logit", "NBayes", "LDA", "QDA", "KNN", "SVM"]

classifiers = [LogisticRegression(C=1e5, max_iter=10000),  # Increased
    max_iter
               GaussianNB(),
               DA.LinearDiscriminantAnalysis(),
               DA.QuadraticDiscriminantAnalysis(),
               KNeighborsClassifier(n_neighbors=5),
               SVC(kernel='rbf', gamma=1e-4, max_iter=1000)]  #
                   Increased max_iter

print('Name  Accuracy\n' + 14 * '-')
for name, clf in zip(names, classifiers):
    clf.fit(X_train, y_train)
    y_pred = clf.predict(X_test)
    print('{:6}  {:3.3f}'.format(name, accuracy_score(y_test, y_pred)))
```

```
Name  Accuracy
--------------
```

```
Logit   0.947
NBayes  0.908
LDA     0.943
QDA     0.956
KNN     0.925
SVM     0.939
```

Further Reading

An excellent source for understanding various pattern recognition techniques is the book [53] by Duda et al. Theoretical foundations of classification, including the Vapnik–Chernovenkis dimension and the fundamental theorem of learning, are discussed in [153, 173, 174].

A popular measure for characterizing the performance of a binary classifier is the *receiver operating characteristic* (ROC) curve [58].

The naïve Bayes classification paradigm can be extended to handle explanatory variable dependency via graphical models such as Bayesian networks and Markov random fields [69, 93, 41]. For a detailed discussion on Bayesian decision theory, see [12].

Exercises

1. Let $0 \leq w \leq 1$. Show that the solution to the convex optimization problem

$$\begin{aligned} &\min_{p_1,\dots,p_n} \sum_{i=1}^{n} p_i^2 \\ &\text{subject to: } \sum_{i-1}^{n-1} p_i = w \text{ and } \sum_{i=1}^{n} p_i = 1, \end{aligned} \tag{8.56}$$

is given by:

$$p_i = w/(n-1), \quad i = 1, \dots, n-1$$

and $p_n = 1 - w$.

2. Derive the formulas (8.28) by minimizing the cross-entropy training loss:

$$-\frac{1}{n}\sum_{i=1}^{n} \ln g(\boldsymbol{x}_i, y_i \,|\, \boldsymbol{\theta}),$$

where $g(\boldsymbol{x}, y \,|\, \boldsymbol{\theta})$ is such that:

$$\ln g(\boldsymbol{x}, y \,|\, \boldsymbol{\theta}) = \ln \alpha_y - \frac{1}{2}\ln|\boldsymbol{\Sigma}_y| - \frac{1}{2}(\boldsymbol{x} - \boldsymbol{\mu}_y)^\top \boldsymbol{\Sigma}_y^{-1}(\boldsymbol{x} - \boldsymbol{\mu}_y) - \frac{p}{2}\ln(2\pi).$$

3. Adapt the code in Example 8.23 to plot the estimated decision boundary instead of the true one in Figure 8.25. Compare the true and estimated decision boundaries.

4. Recall from equation (8.32) that the decision boundaries of the multi-logit classifier are linear, and that the pre-classifier can be written as a conditional pdf of the form:

$$g(y \mid \mathbf{W}, \boldsymbol{b}, \boldsymbol{x}) = \frac{\exp(z_{y+1})}{\sum_{i=1}^{c} \exp(z_i)}, \quad y \in \{0, \dots, c-1\},$$

where $\boldsymbol{x}^\top = [1, \widetilde{\boldsymbol{x}}^\top]$ and $\boldsymbol{z} = \mathbf{W}\widetilde{\boldsymbol{x}} + \boldsymbol{b}$.

(a) Show that the linear discriminant pre-classifier in Section 8.4 can also be written as a conditional pdf of the form ($\boldsymbol{\theta} = \{\alpha_y, \boldsymbol{\Sigma}_y, \boldsymbol{\mu}_y\}_{y=0}^{c-1}$):

$$g(y \mid \boldsymbol{\theta}, \boldsymbol{x}) = \frac{\exp(z_{y+1})}{\sum_{i=1}^{c} \exp(z_i)}, \quad y \in \{0, \dots, c-1\},$$

where $\boldsymbol{x}^\top = [1, \widetilde{\boldsymbol{x}}^\top]$ and $\boldsymbol{z} = \mathbf{W}\widetilde{\boldsymbol{x}} + \boldsymbol{b}$. Find formulas for the corresponding $\boldsymbol{b}$ and $\mathbf{W}$ in terms of the linear discriminant parameters $\{\alpha_y, \boldsymbol{\mu}_y, \boldsymbol{\Sigma}_y\}_{y=0}^{c-1}$, where $\boldsymbol{\Sigma}_y = \boldsymbol{\Sigma}$ for all y.

(b) Explain which pre-classifier has smaller approximation error: the linear discriminant or multi-logit one? Justify your answer by proving an inequality between the two approximation errors.

5. Consider a binary classification problem where the response Y takes values in $\{-1, 1\}$. Show that optimal prediction function for the hinge loss

$$\text{Loss}(y, \widehat{y}) = (1 - y\widehat{y})_+ = \max\{0,\ 1 - y\widehat{y}\}$$

is the same as the optimal prediction function g^* for the indicator loss:

$$g^*(\boldsymbol{x}) = \begin{cases} 1 & \text{if} \quad \mathbb{P}[Y = 1 \mid \boldsymbol{X} = \boldsymbol{x}] > 1/2, \\ -1 & \text{if} \quad \mathbb{P}[Y = 1 \mid \boldsymbol{X} = \boldsymbol{x}] < 1/2. \end{cases}$$

That is, show that

$$\mathbb{E}\,(1 - Y\,h(\boldsymbol{X}))_+ \geq \mathbb{E}\,(1 - Y\,g^*(\boldsymbol{X}))_+ \qquad (8.57)$$

for all functions h.

☞ 184 6. In Example 4.92, we applied a principal component analysis (PCA) to the `iris` data, but refrained from classifying the flowers based on their feature vectors $\boldsymbol{x}$. Implement a 1-nearest neighbor algorithm, using a training set of 50 randomly chosen data pairs $(\boldsymbol{x}, y)$ from the `iris` data set. How many of the remaining 100 flowers are correctly classified? Now classify these entries with an off-the-shelf multi-logit classifier, e.g., such as can be found in the `sklearn` and `statsmodels` packages.

7. Figure 8.58 displays two groups of data points, given in Table 8.59. The convex hulls have also been plotted. It is possible to separate the two classes of points via a straight line. In fact, many such lines are possible. SVM gives the best separation, in the sense that the gap (margin) between the points is maximal.

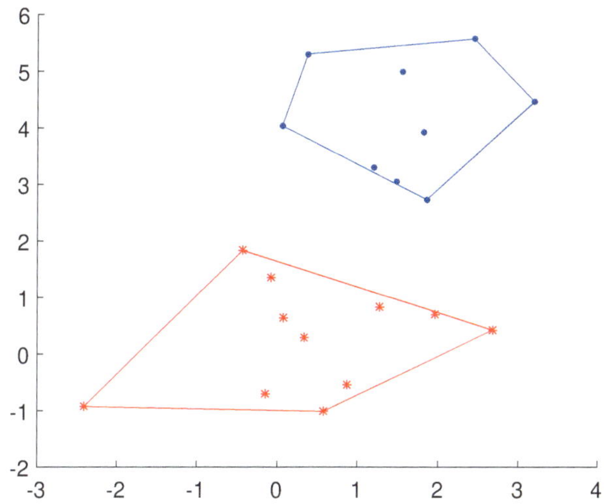

Figure 8.58: Separate the points by a straight line so that the separation between the two groups is maximal.

Table 8.59: Data for Figure 8.58.

x_1	x_2	y	x_1	x_2	y
2.4524	5.5673	−1	0.5819	−1.0156	1
1.2743	0.8265	1	1.2065	3.2984	−1
0.8773	−0.5478	1	2.6830	0.4216	1
1.4837	3.0464	−1	−0.0734	1.3457	1
0.0628	4.0415	−1	0.0787	0.6363	1
−2.4151	−0.9309	1	0.3816	5.2976	−1
1.8152	3.9202	−1	0.3386	0.2882	1
1.8557	2.7262	−1	−0.1493	−0.7095	1
−0.4239	1.8349	1	1.5554	4.9880	−1
1.9630	0.6942	1	3.2031	4.4614	−1

(a) Identify from the figure the three support vectors.

(b) For a separating boundary (line) given by $\beta_0 + \boldsymbol{\beta}^\top \boldsymbol{x} = 0$, show that the margin width is $2/\|\boldsymbol{\beta}\|$.

(c) Show that the parameters β_0 and $\boldsymbol{\beta}$ that solve the convex optimization problem (8.43) provide the maximal width between the margins.

(d) Solve (8.43) using a penalty approach; see Section C.4. In particular, minimize the penalty function ☞ 608

$$S(\boldsymbol{\beta}, \beta_0) = \|\boldsymbol{\beta}\|^2 - C \sum_{i=1}^{n} \min\left\{(\beta_0 + \boldsymbol{\beta}^\top \boldsymbol{x}_i)\, y_i - 1,\ 0\right\}$$

for some positive penalty constant C.

(e) Find the solution to the dual optimization problem (8.39) by using `sklearn`'s `SCV` method. Note that, as the two point sets are separable, the constraint $\lambda \leq 1$ may be removed, and the value of γ can be set to 1.

8. In Example 8.44 we used the feature map $\boldsymbol{\phi}(\boldsymbol{x}) = [x_1, x_2, x_1^2 + x_2^2]^\top$ to classify the points. An easier way is to map the points into $\mathbb{R}^1$ via the feature map $\boldsymbol{\phi}(\boldsymbol{x}) = \|\boldsymbol{x}\|$ or any monotone function thereof. Translated back into $\mathbb{R}^2$ this yields a circular separating boundary. Find the radius and center of this circle, using the fact that here the sorted norms for the two groups are $\ldots, 0.4889, 0.5528, \ldots$.

9. Let $Y \in \{0, 1\}$ be a response variable and let $h(\boldsymbol{x})$ be the regression function

$$h(\boldsymbol{x}) := \mathbb{E}[Y \mid \boldsymbol{X} = \boldsymbol{x}] = \mathbb{P}[Y = 1 \mid \boldsymbol{X} = \boldsymbol{x}].$$

Recall that the Bayes classifier is $g^*(\boldsymbol{x}) = \mathbb{1}\{h(\boldsymbol{x}) > 1/2\}$. Let $g : \mathbb{R} \to \{0, 1\}$ be any other classifier function. Below, we denote all probabilities and expectations conditional on $\boldsymbol{X} = \boldsymbol{x}$ as $\mathbb{P}_{\boldsymbol{x}}[\cdot]$ and $\mathbb{E}_{\boldsymbol{x}}[\cdot]$.

(a) Show that

$$\mathbb{P}_{\boldsymbol{x}}[g(\boldsymbol{x}) \neq Y] = \overbrace{\mathbb{P}_{\boldsymbol{x}}[g^*(\boldsymbol{x}) \neq Y]}^{\text{irreducible error}} + |2h(\boldsymbol{x}) - 1|\, \mathbb{1}\{g(\boldsymbol{x}) \neq g^*(\boldsymbol{x})\}.$$

Hence, deduce that for a learner $g_{\mathcal{T}}$ constructed from a training set $\mathcal{T}$, we have

$$\mathbb{E}[\mathbb{P}_{\boldsymbol{x}}[g_{\mathcal{T}}(\boldsymbol{x}) \neq Y \mid \mathcal{T}]] = \mathbb{P}_{\boldsymbol{x}}[g^*(\boldsymbol{x}) \neq Y] + |2h(\boldsymbol{x}) - 1|\, \mathbb{P}[g_{\mathcal{T}}(\boldsymbol{x}) \neq g^*(\boldsymbol{x})],$$

where the first expectation and last probability operations are with respect to $\mathcal{T}$.

(b) Using the previous result, deduce that for the unconditional error (that is, we no longer condition on $\boldsymbol{X} = \boldsymbol{x}$), we have

$$\mathbb{P}[g^*(\boldsymbol{X}) \neq Y] \leq \mathbb{P}[g_{\mathcal{T}}(\boldsymbol{X}) \neq Y].$$

(c) Show that, if $g_{\mathcal{T}} := \mathbb{1}\{h_{\mathcal{T}}(\boldsymbol{x}) > 1/2\}$ is a classifier function such that as $n \to \infty$

$$h_{\mathcal{T}}(\boldsymbol{x}) \xrightarrow{\text{d}} Z \sim \mathcal{N}(\mu(\boldsymbol{x}), \sigma^2(\boldsymbol{x}))$$

for some mean and variance functions $\mu(\boldsymbol{x})$ and $\sigma^2(\boldsymbol{x})$, respectively, then

$$\mathbb{P}_{\boldsymbol{x}}[g_{\mathcal{T}}(\boldsymbol{x}) \neq g^*(\boldsymbol{x})] \longrightarrow \Phi\left(\frac{\text{sign}(1 - 2h(\boldsymbol{x}))(2\mu(\boldsymbol{x}) - 1)}{2\sigma(\boldsymbol{x})}\right),$$

where Φ is the cdf of a standard normal random variable.

10. The purpose of this exercise is to derive the dual program (8.39) from the primal program (8.38). The starting point is to introduce a vector of auxiliary variables $\boldsymbol{\xi} := [\xi_1, \ldots, \xi_n]^\top$ and write the primal program as

$$\begin{aligned} &\min_{\boldsymbol{\alpha}, \alpha_0, \boldsymbol{\xi}} \sum_{i=1}^{n} \xi_i + \frac{\gamma}{2} \boldsymbol{\alpha}^\top \mathbf{K} \boldsymbol{\alpha} \\ &\text{subject to:} \quad \boldsymbol{\xi} \geq \mathbf{0}, \\ &\qquad y_i(\alpha_0 + \{\mathbf{K}\boldsymbol{\alpha}\}_i) \geq 1 - \xi_i, \ i = 1, \ldots, n. \end{aligned} \tag{8.60}$$

(a) Apply the Lagrangian optimization theory from Section C.2.2 to obtain the Lagrangian function $\mathcal{L}(\{\alpha_0, \boldsymbol{\alpha}, \boldsymbol{\xi}\}, \{\boldsymbol{\lambda}, \boldsymbol{\mu}\})$, where $\boldsymbol{\mu}$ and $\boldsymbol{\lambda}$ are the Lagrange multipliers corresponding to the first and second inequality constraints, respectively. ☞ 594

(b) Show that the Karush–Kuhn–Tucker (see Theorem C.42) conditions for optimizing $\mathcal{L}$ are: ☞ 595

$$\begin{aligned} \boldsymbol{\lambda}^\top \boldsymbol{y} &= 0 \\ \boldsymbol{\alpha} &= \boldsymbol{y} \odot \boldsymbol{\lambda}/\gamma \\ \mathbf{0} \leq \boldsymbol{\lambda} &\leq \mathbf{1} \\ (\mathbf{1} - \boldsymbol{\lambda}) \odot \boldsymbol{\xi} = \mathbf{0}, &\quad \lambda_i \,(y_i g(\boldsymbol{x}_i) - 1 + \xi_i) = 0, \; i = 1, \ldots, n \\ \boldsymbol{\xi} \geq \mathbf{0}, &\quad y_i g(\boldsymbol{x}_i) - 1 + \xi_i \geq 0, \; i = 1, \ldots, n. \end{aligned} \tag{8.61}$$

Here $\odot$ stands for componentwise multiplication; e.g., $\boldsymbol{y} \odot \boldsymbol{\lambda} = [y_1\lambda_1, \ldots, y_n\lambda_n]^\top$, and we have abbreviated $\alpha_0 + \{\mathbf{K}\boldsymbol{\alpha}\}_i$ to $g(\boldsymbol{x}_i)$, in view of (8.37). Hint: One of the KKT conditions is $\boldsymbol{\lambda} = \mathbf{1} - \boldsymbol{\mu}$; thus we can eliminate $\boldsymbol{\mu}$.

(c) Using the KKT conditions (8.61), reduce the Lagrange dual function $\mathcal{L}^*(\boldsymbol{\lambda}) := \min_{\alpha_0, \boldsymbol{\alpha}, \boldsymbol{\xi}} \mathcal{L}(\{\alpha_0, \boldsymbol{\alpha}, \boldsymbol{\xi}\}, \{\boldsymbol{\lambda}, \mathbf{1} - \boldsymbol{\lambda}\})$ to

$$\mathcal{L}^*(\boldsymbol{\lambda}) = \sum_{i=1}^{n} \lambda_i - \frac{1}{2\gamma} \sum_{i=1}^{n} \sum_{j=1}^{n} \lambda_i \lambda_j y_i y_j \, \kappa(\boldsymbol{x}_i, \boldsymbol{x}_j). \tag{8.62}$$

(d) As a consequence of (8.37) and (a)–(c), show that the optimal prediction function $g_\mathcal{T}$ is given by

$$g_\mathcal{T}(\boldsymbol{x}) = \alpha_0 + \frac{1}{\gamma} \sum_{i=1}^{n} y_i \lambda_i \, \kappa(\boldsymbol{x}_i, \boldsymbol{x}), \tag{8.63}$$

where $\boldsymbol{\lambda}$ is the solution to

$$\begin{aligned} &\max_{\boldsymbol{\lambda}} \quad \mathcal{L}^*(\boldsymbol{\lambda}) \\ &\text{subject to:} \quad \boldsymbol{\lambda}^\top \boldsymbol{y} = 0, \; \mathbf{0} \leq \boldsymbol{\lambda} \leq \mathbf{1}, \end{aligned} \tag{8.64}$$

and $\alpha_0 = y_j - \frac{1}{\gamma} \sum_{i=1}^{n} y_i \lambda_i \, \kappa(\boldsymbol{x}_i, \boldsymbol{x}_j)$ for any j such that $\lambda_j \in (0, 1)$.

11. Consider SVM classification as illustrated in Figure 8.42. The goal of this exercise is to classify the training points $\{(\boldsymbol{x}_i, y_i)\}$ based on the value of the multipliers $\{\lambda_i\}$ in Exercise 10. Let ξ_i be the auxiliary variable in Exercise 10, $i = 1, \ldots, n$.

(a) For $\lambda_i \in (0, 1)$ show that $(\boldsymbol{x}_i, y_i)$ lies exactly on the decision border.

(b) For $\lambda_i = 1$, show that $(\boldsymbol{x}_i, y_i)$ lies strictly inside the margins.

(c) Show that for $\lambda_i = 0$ the point $(\boldsymbol{x}_i, y_i)$ lies outside the margins and is correctly classified.

12. A well-known data set is the MNIST handwritten digit database, containing many thousands of digitalized numbers (from 0 to 9), each described by a 28×28 matrix of gray scales. A similar but much smaller data set is described in [90]. Here, each handwritten digit is summarized by a 8 × 8 matrix with integer entries from 0 (white) to 15 (black). Figure 8.65 shows the first 50 digitized images. The data set can be accessed with Python using the `sklearn` package as follows:

```
from sklearn import datasets
digits = datasets.load_digits()
x_digits = digits.data    # explanatory variables
y_digits = digits.target  # responses
```

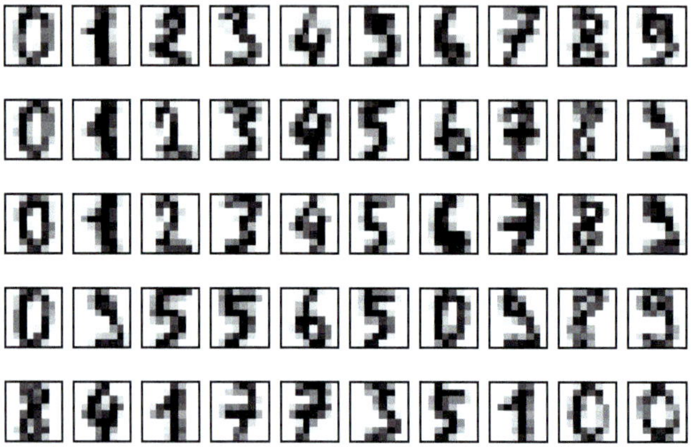

Figure 8.65: Classify the digitized images.

(a) Divide the data into a 75% training set and 25% test set.

(b) Compare the effectiveness of the K-nearest neighbors and naïve Bayes method to classify the data.

(c) Assess which K to use in the K-nearest neighbors classification.

13. Download the winequality-red.csv data set from UCI's **wine-quality** website. The response here is the wine quality (from 0 to 10) as specified by a wine "expert" and the explanatory variables are various characteristics such as acidity and sugar content. Use the SVC classifier of sklearn.svm with a linear kernel and penalty parameter $\texttt{C} = 1$ (see Remark 8.54) to fit the data. Use the method cross_val_score from sklearn.model_selection to obtain a five-fold cross-validation score as an estimate of the probability that the predicted class matches the expert's class.

14. Consider the credit approval data set crx.data from UCI's **credit approval** website. The data set is concerned with credit card applications. The last column in the data set indicates whether the application is approved (+) or not (−). With the view of preserving data privacy, all 15 explanatory variables were anonymized. Note that some explanatory variables are continuous and some are categorical.

(a) Load and prepare the data for analysis with sklearn. As a first step, eliminate data rows with missing values. Next, encode categorical explanatory variables using a OneHotEncoder object from sklearn.preprocessing to create a model matrix X with indicator variables for the categorical variables, as described in Section 5.3.5.

☞ 208

(b) The model matrix should contain 653 rows and 46 columns. The response variable should be a 0/1 variable (reject/approve). We will consider several classification algorithms and test their performance (using a zero-one loss) via ten-fold cross validation.

i. Write a function which takes 3 parameters: $\mathbf{X}$, $\mathbf{y}$, and a model, and returns the ten-fold cross-validation estimate of the expected generalization risk.

ii. Consider the following `sklearn` classifiers: `KNeighborsClassifier` ($k = 5$), `LogisticRegression`, and `MPLClassifier` (multi-layer perceptron). Use the function from (i) to identify the best performing classifier.

15. Consider a synthetic data set that was generated in the following fashion. The explanatory variable follows a standard normal distribution. The response label is 0 if the explanatory variable is between the 0.95 and 0.05 quantiles of the standard normal distribution, and 1, otherwise. The data set was generated using the following code.

```
import numpy as np
import scipy.stats
# generate data

np.random.seed(12345)
N = 100
X = np.random.randn(N)
q = scipy.stats.norm.ppf(0.95)
y = np.zeros(N)
y[X>=q] = 1
y[X<=-q] = 1
X = X.reshape(-1,1)
```

Compare the K-nearest neighbors classifier with $K = 5$ and logistic regression classifier. Without computation, which classifier is likely to be better for these data? Verify your answer by coding both classifiers and printing the corresponding training 0–1 loss.

16. Consider the `digits` data set from Exercise 12. In this exercise, we would like to train a binary classifier for the identification of digit 8.

(a) Divide the data such that the first 1000 rows are used as the training set and the rest are used as the test set.

(b) Train the `LogisticRegression` classifier from the `sklearn.linear_model` package.

(c) "Train" a naïve classifier that always returns 0. That is, the naïve classifier identifies each instance as being not 8.

(d) Compare the zero-one test losses of the logistic regression and the naïve classifiers.

(e) Find the confusion matrix, the precision, and the recall of the logistic regression classifier.

(f) Find the fraction of eights that are correctly detected by the logistic regression classifier.

17. Repeat Exercise 16 with the original MNIST data set. Use the first 60,000 rows as the train set and the remaining 10,000 rows as the test set. The original data set can be obtained using the following code.

```
from sklearn.datasets import fetch_openml

X, y = fetch_openml('mnist_784', version=1, return_X_y=True)
```

☞ 380 18. For the breast cancer data in Section 8.8, investigate and discuss whether *accuracy* is
☞ 355 the relevant metric to use or if other metrics discussed in Section 8.2 are more appropriate.

CHAPTER 9

Decision Trees and Ensemble Methods

Statistical learning methods based on decision trees have gained tremendous popularity due to their simplicity, intuitive representation, and predictive accuracy. This chapter gives an introduction to the construction and use of such trees. We also discuss two key ensemble methods, namely bootstrap aggregation and boosting, which can further improve the efficiency of decision trees and other learning methods.

9.1 Introduction

Tree-based methods provide a simple, intuitive, and powerful mechanism for both regression and classification. The main idea is to divide a (potentially complicated) feature space $\mathcal{X}$ into smaller regions and fit a simple prediction function to each region. For example, in a regression setting, one could take the mean of the training responses associated with the training features that fall in that specific region. In the classification setting, a commonly used prediction function takes the majority vote among the corresponding response variables. We start with a simple classification example.

■ **Example 9.1 (Decision Tree for Classification)** The left panel of Figure 9.2 shows a training set of 15 two-dimensional points (features) falling into two classes (red and blue). How should the new feature vector (black point) be classified?

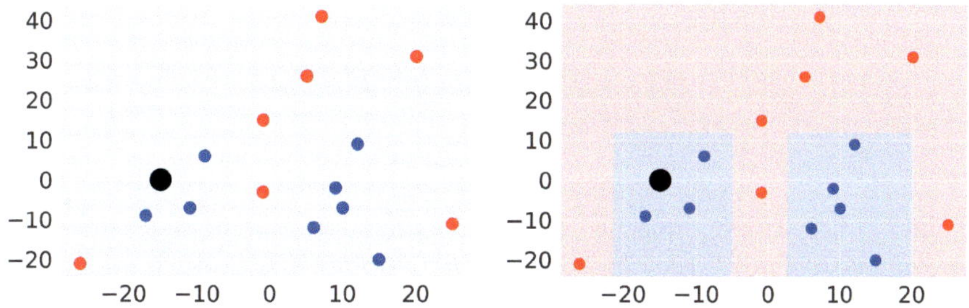

Figure 9.2: Left: training data and a new feature. Right: a partition of the feature space.

It is not possible to linearly separate the training set, but we can partition the feature

DOI: 10.1201/9781003391173-9

space $\mathcal{X} = \mathbb{R}^2$ into rectangular regions and assign a class (color) to each region, as shown in the right panel of Figure 9.2. Points in these regions are classified accordingly as blue or red. The partition thus defines a classifier (prediction function) g that assigns to each feature vector $\boldsymbol{x}$ a class "red" or "blue". For example, for $\boldsymbol{x} = [-15, 0]^\top$ (solid black point), $g(\boldsymbol{x}) =$ "blue", since it belongs to a blue region of the feature space.

Both the classification procedure and the partitioning of the feature space can be conveniently represented by a binary **decision tree**. This is a tree where each node v corresponds to a region (subset) $\mathcal{R}_v$ of the feature space $\mathcal{X}$ — the root node corresponding to the feature space itself.

Each internal node v contains a logical condition that divides $\mathcal{R}_v$ into two disjoint subregions. The leaf nodes (the terminal nodes of the tree) are not subdivided, and their corresponding regions form a partition of $\mathcal{X}$, as they are disjoint and their union is $\mathcal{X}$. Associated with each leaf node w is also a regional prediction function g^w on $\mathcal{R}_w$.

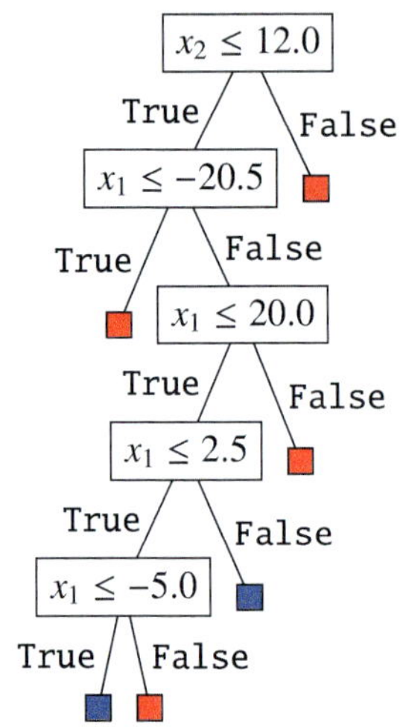

Figure 9.3: The decision-tree that corresponds to the partition in Figure 9.2.

The partitioning of Figure 9.2 was obtained from the decision tree shown in Figure 9.3. As an illustration of the decision procedure, consider again the input $\boldsymbol{x} = [x_1, x_2]^\top = [-15, 0]^\top$. The classification process starts from the tree root, which contains the condition $x_2 \leq 12.0$. As the second component of $\boldsymbol{x}$ is 0, the root condition is satisfied and we proceed to the left child, which contains the condition $x_1 \leq -20.5$. The next step is similar. As $-15 > -20.5$, the condition is not satisfied and we proceed to the right child. Such an evaluation of logical conditions along the tree path will eventually bring us to a leaf node and its associated region. In this case the process terminates in a leaf that corresponds to the left blue region in the right-hand panel of Figure 9.2. ■

More generally, a binary tree $\mathbb{T}$ will partition the feature space $\mathcal{X}$ into as many regions as there are leaf nodes. Denote the set of leaf nodes by $\mathcal{W}$. The overall prediction function g that corresponds to the tree can then be written as

$$g(\boldsymbol{x}) = \sum_{w \in \mathcal{W}} g^w(\boldsymbol{x})\, \mathbb{1}\{\boldsymbol{x} \in \mathcal{R}_w\}, \tag{9.4}$$

where $\mathbb{1}$ denotes the indicator function. The representation (9.4) is very general and depends on (1) how the regions $\{\mathcal{R}_w\}$ are constructed via the logical conditions in the decision tree, as well as (2) how the **regional prediction functions** of the leaf nodes are defined. Simple logical conditions of the form $x_j \leq \xi$ split a Euclidean feature space into rectangles aligned with the axes. For example, Figure 9.3 partitions the feature space into six rectangles: two blue and four red rectangles.

In a classification setting, the regional prediction function g^w corresponding to a leaf node w takes values in the set of possible class labels. In most cases, as in Example 9.2, it is taken to be constant on the corresponding region $\mathcal{R}_w$. In a regression setting, g^w is real-valued and also usually takes only one value. That is, every feature vector in $\mathcal{R}_w$ leads to the same predicted value. Of course, different regions will usually have different predicted values.

Constructing a tree with a training set $\tau = \{(\boldsymbol{x}_i, y_i)\}\}_{i=1}^n$ amounts to minimizing the training loss

$$\ell_\tau(g) = \frac{1}{n}\sum_{i=1}^n \text{Loss}(y_i, g(\boldsymbol{x}_i)) \tag{9.5}$$

for some loss function; see Chapter 2. With g of the form (9.4), we can write ☞ 19

$$\ell_\tau(g) = \frac{1}{n}\sum_{i=1}^n \text{Loss}(y_i, g(\boldsymbol{x}_i)) = \frac{1}{n}\sum_{i=1}^n \sum_{w\in\mathcal{W}} \mathbb{1}\{\boldsymbol{x}_i \in \mathcal{R}_w\}\,\text{Loss}(y_i, g(\boldsymbol{x}_i)) \tag{9.6}$$

$$= \sum_{w\in\mathcal{W}} \underbrace{\frac{1}{n}\sum_{i=1}^n \mathbb{1}\{\boldsymbol{x}_i \in \mathcal{R}_w\}\,\text{Loss}(y_i, g^w(\boldsymbol{x}_i))}_{(*)}, \tag{9.7}$$

where $(*)$ is the contribution by the regional prediction function g^w to the overall training loss. In the case where all $\{\boldsymbol{x}_i\}$ are different, finding a decision tree $\mathbb{T}$ that gives a zero squared-error or zero–one training loss is easy, see Exercise 1, but such an "overfitted" tree will have poor predictive behavior, expressed in terms of the generalization risk. Instead we consider a restricted class of decision trees and aim to minimize the training loss within that class. It is common to use a top-down greedy approach, which can only achieve an approximate minimization of the training loss.

9.2 Top-Down Construction of Decision Trees

Let $\tau = \{(\boldsymbol{x}_i, y_i)\}_{i=1}^n$ be the training set. The key to constructing a binary decision tree $\mathbb{T}$ is to specify a **splitting rule** for each node v, which can be defined as a logical function $s : \mathcal{X} \to \{\texttt{False}, \texttt{True}\}$ or, equivalently, a binary function $s : \mathcal{X} \to \{0, 1\}$. For example, in the decision tree of Figure 9.3 the root node has splitting rule $\boldsymbol{x} \mapsto \mathbb{1}\{x_2 \leq 12.0\}$, in correspondence with the logical condition $\{x_2 \leq 12.0\}$. During the construction of the tree, each node v is associated with a specific region $\mathcal{R}_v \subseteq \mathcal{X}$ and therefore also the training subset:

$$\{(\boldsymbol{x}, y) \in \tau : \boldsymbol{x} \in \mathcal{R}_v\} \subseteq \tau.$$

Using a splitting rule s, we can divide any subset σ of the training set τ into two sets:

$$\sigma_{\mathrm{T}} := \{(\boldsymbol{x}, y) \in \sigma \;:\; s(\boldsymbol{x}) = \texttt{True}\} \text{ and } \sigma_{\mathrm{F}} := \{(\boldsymbol{x}, y) \in \sigma \;:\; s(\boldsymbol{x}) = \texttt{False}\}. \tag{9.8}$$

Starting from an empty tree and the initial data set τ, a generic decision tree construction takes the form of the recursive Algorithm 9.9. Here we use the notation $\mathbb{T}_v$ for a subtree of $\mathbb{T}$ starting from node v. The final tree $\mathbb{T}$ is thus obtained via $\mathbb{T} =$ `Construct_Subtree`(v_0, τ), where v_0 is the root of the tree.

Algorithm 9.9: Construct_Subtree

Input: A node v and a subset of the training data: $\sigma \subseteq \tau$.
Output: A (sub) decision tree $\mathbb{T}_v$.

1 **if** termination criterion is met **then** // v is a leaf node
2 Train a regional prediction function g^v using the training data σ.
3 **else** // split the node
4 Find the best splitting rule s_v for node v.
5 Create successors v_{T} and v_{F} of v.
6 $\sigma_{\mathrm{T}} \leftarrow \{(\boldsymbol{x}, y) \in \sigma \ : \ s_v(\boldsymbol{x}) = \texttt{True}\}$
7 $\sigma_{\mathrm{F}} \leftarrow \{(\boldsymbol{x}, y) \in \sigma \ : \ s_v(\boldsymbol{x}) = \texttt{False}\}$
8 $\mathbb{T}_{v_{\mathrm{T}}} \leftarrow$ Construct_Subtree $(v_{\mathrm{T}}, \sigma_{\mathrm{T}})$ // left branch
9 $\mathbb{T}_{v_{\mathrm{F}}} \leftarrow$ Construct_Subtree $(v_{\mathrm{F}}, \sigma_{\mathrm{F}})$ // right branch
10 **return** $\mathbb{T}_v$

The splitting rule s_v divides the region $\mathcal{R}_v$ into two disjoint parts, say $\mathcal{R}_{v_{\mathrm{T}}}$ and $\mathcal{R}_{v_{\mathrm{F}}}$. The corresponding prediction functions, g^{T} and g^{F}, satisfy

$$g^v(\boldsymbol{x}) = g^{\mathrm{T}}(\boldsymbol{x})\, \mathbb{1}\{\boldsymbol{x} \in \mathcal{R}_{v_{\mathrm{T}}}\} + g^{\mathrm{F}}(\boldsymbol{x})\, \mathbb{1}\{\boldsymbol{x} \in \mathcal{R}_{v_{\mathrm{F}}}\}, \quad \boldsymbol{x} \in \mathcal{R}_v.$$

In order to implement the procedure described in Algorithm 9.9, we need to address the construction of the regional prediction functions g^v at the leaves (Line 2), the specification of the splitting rule (Line 4), and the termination criterion (Line 1). These important aspects are detailed in the following Sections 9.2.1, 9.2.2, and 9.2.3, respectively.

9.2.1 Regional Prediction Functions

In general, there is no restriction on how to choose the prediction function g^w for a leaf node $v = w$ in Line 2 of Algorithm 9.9. In principle we can train any model from the data; e.g., via linear regression. However, in practice very simple prediction functions are used. Below, we detail a popular choice for classification, as well as one for regression.

1. In the *classification* setting with class labels $0, \ldots, c-1$, the regional prediction function g^w for leaf node w is usually chosen to be *constant* and equal to the most common class label of the training data in the associated region $\mathcal{R}_w$ (ties can be broken randomly). More precisely, let n_w be the number of feature vectors in region $\mathcal{R}_w$ and let

$$p_z^w = \frac{1}{n_w} \sum_{\{(\boldsymbol{x},y)\in\tau \,:\, \boldsymbol{x}\in\mathcal{R}_w\}} \mathbb{1}_{\{y=z\}},$$

be the proportion of feature vectors in $\mathcal{R}_w$ that have class label $z = 0, \ldots, c-1$. The regional prediction function for node w is chosen to be the constant

$$g^w(\boldsymbol{x}) = \underset{z\in\{0,\ldots,c-1\}}{\operatorname{argmax}}\ p_z^w. \tag{9.10}$$

2. In the *regression* setting, g^w is usually chosen as the mean response in the region; that is,

$$g^w(\boldsymbol{x}) = \overline{y}_{\mathcal{R}_w} := \frac{1}{n_w} \sum_{\{(\boldsymbol{x},y)\in\tau \,:\, \boldsymbol{x}\in\mathcal{R}_w\}} y, \tag{9.11}$$

where n_w is again the number of feature vectors in $\mathcal{R}_w$. It is not difficult to show that $g^w(\boldsymbol{x}) = \overline{y}_{\mathcal{R}_w}$ minimizes the squared-error loss with respect to all constant functions, in the region $\mathcal{R}_w$; see Exercise 2.

9.2.2 Splitting Rules

In Line 4 in Algorithm 9.9, we divide region $\mathcal{R}_v$ into two sets, using a splitting rule (function) s_v. Consequently, the data set σ associated with node v (that is, the subset of the original data set τ whose feature vectors lie in $\mathcal{R}_v$), is also split — into σ_{T} and σ_{F}. What is the benefit of such a split in terms of a reduction in the training loss? If v were set to a leaf node, its contribution to the training loss would be (see (9.7)):

$$\frac{1}{n}\sum_{i=1}^{n} \mathbb{1}_{\{(\boldsymbol{x}_i,y_i)\in\sigma\}} \operatorname{Loss}(y_i, g^v(\boldsymbol{x}_i)). \tag{9.12}$$

If v were to be split instead, its contribution to the overall training loss would be:

$$\frac{1}{n}\sum_{i=1}^{n} \mathbb{1}_{\{(\boldsymbol{x}_i,y_i)\in\sigma_{\mathrm{T}}\}} \operatorname{Loss}(y_i, g^{\mathrm{T}}(\boldsymbol{x}_i)) + \frac{1}{n}\sum_{i=1}^{n} \mathbb{1}_{\{(\boldsymbol{x}_i,y_i)\in\sigma_{\mathrm{F}}\}} \operatorname{Loss}(y_i, g^{\mathrm{F}}(\boldsymbol{x}_i)), \tag{9.13}$$

where g^{T} and g^{F} are the prediction functions belonging to the child nodes v_{T} and v_{F}. A *greedy* heuristic is to pretend that the tree construction algorithm immediately terminates after the split, in which case v_{T} and v_{F} are leaf nodes, and g^{T} and g^{F} are readily evaluated — e.g., as in Section 9.2.1. Note that for any splitting rule the contribution (9.12) is always greater than or equal to (9.13). It therefore makes sense to choose the splitting rule such that (9.13) is minimized. Moreover, the termination criterion may involve comparing (9.13) with (9.12). If their difference is too small it may not be worth further splitting the feature space.

As an example, suppose the feature space is $\mathcal{X} = \mathbb{R}^p$ and we consider splitting rules of the form

$$s(\boldsymbol{x}) = \mathbb{1}\{x_j \leq \xi\}, \tag{9.14}$$

for some $1 \leq j \leq p$ and $\xi \in \mathbb{R}$, where we identify 0 with `False` and 1 with `True`. Due to the computational and interpretative simplicity, such binary splitting rules are implemented in many software packages and are considered to be the *de facto* standard. As we have seen, these rules divide up the feature space into rectangles, as in Figure 9.2. It is natural to ask how j and ξ should be chosen so as to minimize (9.13). For a regression problem, using a squared-error loss and a constant regional prediction function as in (9.11), the sum (9.13) is given by

$$\frac{1}{n}\sum_{(\boldsymbol{x},y)\in\sigma_{\mathrm{T}}:x_j\leq\xi} (y - \overline{y}_{\mathrm{T}})^2 + \frac{1}{n}\sum_{(\boldsymbol{x},y)\in\sigma_{\mathrm{F}}:x_j>\xi} (y - \overline{y}_{\mathrm{F}})^2, \tag{9.15}$$

where $\overline{y}_{\mathrm{T}}$ and $\overline{y}_{\mathrm{F}}$ are the average responses for the σ_{T} and σ_{F} data, respectively. Let $\{x_{j,k}\}_{k=1}^{m}$ denote the possible values of $x_j, j = 1, \ldots, p$ within the training subset σ (with $m \leq n$ elements). Note that, for a fixed j, (9.15) is a piecewise constant function of ξ, and that its minimal value is attained at some value $x_{j,k}$. As a consequence, to minimize (9.15) over all j and ξ, it suffices to evaluate (9.15) for each of the $m \times p$ values $x_{j,k}$ and then take the minimizing pair $(j, x_{j,k})$.

For a classification problem, using the indicator loss and a constant regional prediction function as in (9.10), the aim is to choose a splitting rule that minimizes

$$\frac{1}{n}\sum_{(\boldsymbol{x},y)\in\sigma_{\mathrm{T}}} \mathbb{1}\{y \neq y_{\mathrm{T}}^*\} + \frac{1}{n}\sum_{(\boldsymbol{x},y)\in\sigma_{\mathrm{F}}} \mathbb{1}\{y \neq y_{\mathrm{F}}^*\}, \tag{9.16}$$

where $y_{\mathrm{T}}^* = g^{\mathrm{T}}(\boldsymbol{x})$ is the most prevalent class (majority vote) in the data set σ_{T} and y_{F}^* is the most prevalent class in σ_{F}. If the feature space is $\mathcal{X} = \mathbb{R}^p$ and the splitting rules are of the form (9.14), then the optimal splitting rule can be obtained in the same way as described above for the regression case; the only difference is that (9.15) is replaced with (9.16).

We can view the minimization of (9.16) as minimizing a weighted average of "impurities" of nodes σ_{T} and σ_{F}. Namely, for an arbitrary training subset $\sigma \subseteq \tau$, if y^* is the most prevalent label, then

$$\frac{1}{|\sigma|}\sum_{(\boldsymbol{x},y)\in\sigma} \mathbb{1}\{y \neq y^*\} = 1 - \frac{1}{|\sigma|}\sum_{(\boldsymbol{x},y)\in\sigma} \mathbb{1}\{y = y^*\} = 1 - p_{y^*} = 1 - \max_{z\in\{0,\dots,c-1\}} p_z,$$

where p_z is the proportion of data points in σ that have class label z, $z = 0, \dots, c-1$. The quantity

$$1 - \max_{z\in\{0,\dots,c-1\}} p_z$$

measures the diversity of the labels in σ and is called the **misclassification impurity**. Consequently, (9.16) is the weighted sum of the misclassification impurities of σ_{T} and σ_{F}, with weights by $|\sigma_{\mathrm{T}}|/n$ and $|\sigma_{\mathrm{F}}|/n$, respectively. Note that the misclassification impurity only depends on the label proportions rather than on the individual responses. Instead of using the misclassification impurity to decide if and how to split a data set σ, we can use other impurity measures that only depend on the label proportions. Two popular choices are the **entropy impurity**:

$$-\sum_{z=0}^{c-1} p_z \log_2(p_z)$$

and the **Gini impurity**:

$$\frac{1}{2}\left(1 - \sum_{z=0}^{c-1} p_z^2\right).$$

All of these impurities are maximal when the label proportions are equal to $1/c$. Typical shapes of the above impurity measures are illustrated in Figure 9.17 for the two-label case, with class probabilities p and $1-p$. We see here the similarity of the different impurity measures. Note that impurities can be arbitrarily scaled, and so using $\ln(p_z) = \log_2(p_z)\ln(2)$ instead of $\log_2(p_z)$ above gives an equivalent entropy impurity.

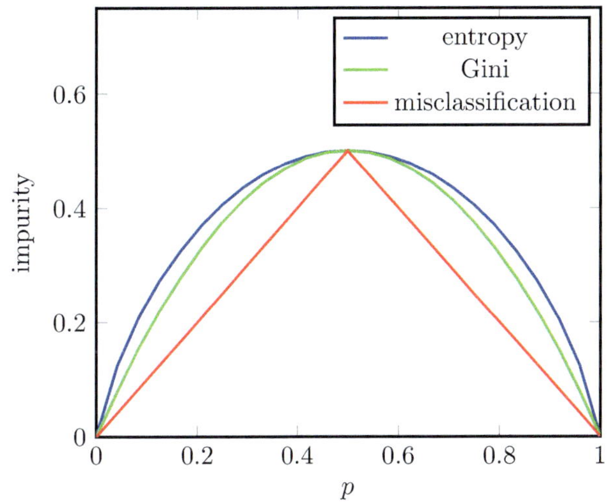

Figure 9.17: Entropy, Gini, and misclassification impurities for binary classification, with class frequencies $p_1 = p$ and $p_2 = 1 - p$. The entropy impurity was normalized (divided by 2), to ensure that all impurity measures attain the same maximum value of 1/2 at $p = 1/2$.

9.2.3 Termination Criterion

When building a tree, one can define various types of termination conditions. For example, we might stop when the number of data points in the tree node (the size of the input σ set in Algorithm 9.9) is less than or equal to some predefined number. Or we might choose the maximal depth of the tree in advance. Another possibility is to stop when there is no significant advantage, in terms of training loss, to split regions. Ultimately, the quality of a tree is determined by its predictive performance (generalization risk) and the termination condition should aim to strike a balance between minimizing the approximation error and minimizing the statistical error, as discussed in Section 2.4. ☞ 32

■ **Example 9.18 (Fixed Tree Depth)** To illustrate how the tree depth impacts on the generalization risk, consider Figure 9.19, which shows the typical behavior of the cross-validation loss as a function of the tree depth. Recall that the cross-validation loss is an estimate of the expected generalization risk. Complicated (deep) trees tend to overfit the training data by producing many divisions of the feature space. As we have seen, this overfitting problem is typical of all learning methods; see Chapter 2 and in particular Example 2.13. To conclude, increasing the maximal depth does not necessarily result in better performance. ☞ 26

To create Figure 9.19 we used[1] the Python method `make_blobs` from the `sklearn` module to produce a training set of size $n = 5000$ with ten-dimensional feature vectors (thus, $p = 10$ and $\mathcal{X} = \mathbb{R}^{10}$), each of which is classified into one of $c = 3$ classes. The full code is given below. ☞ 699

[1]The data used for Figure 9.2 was produced in a similar way.

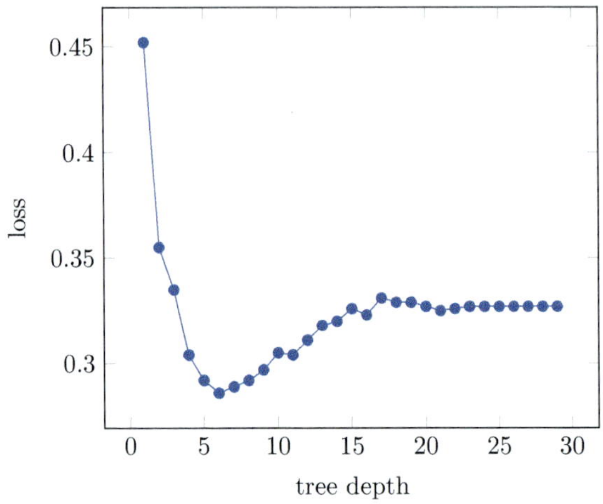

Figure 9.19: The ten-fold cross-validation loss as a function of the maximal tree depth for a classification problem. The optimal maximal tree depth is here 6.

TreeDepthCV.py

```
import numpy as np
from sklearn.datasets import make_blobs
from sklearn.model_selection import cross_val_score
from sklearn.tree import DecisionTreeClassifier
from sklearn.metrics import zero_one_loss
import matplotlib.pyplot as plt

def ZeroOneScore(clf, X, y):
    y_pred = clf.predict(X)
    return zero_one_loss(y, y_pred)

# Construct the training set
X, y =  make_blobs(n_samples=5000, n_features=10, centers=3,
                          random_state=10, cluster_std=10)

# construct a decision tree classifier
clf = DecisionTreeClassifier(random_state=0)

# Cross-validation loss as a function of tree depth (1 to 30)
xdepthlist = []
cvlist = []
tree_depth = range(1,30)
for d in tree_depth:
    xdepthlist.append(d)
    clf.max_depth=d
    cv = np.mean(cross_val_score(clf, X, y, cv=10, scoring=ZeroOneScore)
        )
    cvlist.append(cv)

plt.xlabel('tree depth', fontsize=18, color='black')
plt.ylabel('loss', fontsize=18, color='black')
plt.plot(xdepthlist, cvlist,'-*' , linewidth=0.5)
```

The code above relies heavily on sklearn and hides the implementation details. To show how decision trees are actually constructed using the previous theory, we proceed with a very basic implementation.

9.2.4 Basic Implementation

In this section we implement a regression tree, step by step. To run the program, amalgamate the code snippets below into one file, in the order presented. First, we import various packages and define a function to generate the training and test data.

BasicTree.py

```
import numpy as np
from sklearn.datasets import make_friedman1
from sklearn.model_selection import train_test_split

def makedata():
   n_points = 500 # number of samples

   X, y =  make_friedman1(n_samples=n_points, n_features=5,
                          noise=1.0, random_state=100)
   return train_test_split(X, y, test_size=0.5, random_state=3)
```

The "main" method calls the makedata method, uses the training data to build a regression tree, and then predicts the responses of the test set and reports the mean squared-error loss.

```
def main():
  X_train, X_test, y_train, y_test = makedata()
  maxdepth = 10 # maximum tree depth
  # Create tree root at depth 0
  treeRoot = TNode(0, X_train,y_train)

  # Build the regression tree with maximal depth equal to max_depth
  Construct_Subtree(treeRoot, maxdepth)

  # Predict
  y_hat = np.zeros(len(X_test))
  for i in range(len(X_test)):
     y_hat[i] = Predict(X_test[i],treeRoot)

  MSE = np.mean(np.power(y_hat - y_test,2))
  print("Basic tree: tree loss = ",  MSE)
```

The next step is to specify a tree node as a Python class. Each node has a number of attributes, including the features and the response data (X and y) and the depth at which the node is placed in the tree. The root node has depth 0. Each node w can calculate its contribution to the squared-error training loss $\sum_{i=1}^{n} \mathbb{1}\{\boldsymbol{x}_i \in \mathcal{R}^w\}(y_i - g^w(\boldsymbol{x}_i))^2$. Note that we have omitted the constant $1/n$ term when training the tree, which simply scales the loss (9.5).

```
class TNode:
   def __init__(self, depth, X, y):
      self.depth = depth
      self.X = X   # matrix of features
      self.y = y   # vector of response variables
      # initialize optimal split parameters
      self.j = None
      self.xi = None
      # initialize children to be None
      self.left = None
      self.right = None
      # initialize the regional predictor
      self.g = None

   def CalculateLoss(self):
      if(len(self.y)==0):
         return 0

      return np.sum(np.power(self.y - self.y.mean(),2))
```

The function below implements the training (tree-building) Algorithm 9.9.

```
def Construct_Subtree(node, max_depth):
    if(node.depth == max_depth or len(node.y) == 1):
        node.g  = node.y.mean()
    else:
        j, xi = CalculateOptimalSplit(node)
        node.j = j
        node.xi = xi
        Xt, yt, Xf, yf = DataSplit(node.X, node.y, j, xi)

        if(len(yt)>0):
            node.left = TNode(node.depth+1,Xt,yt)
            Construct_Subtree(node.left, max_depth)

        if(len(yf)>0):
            node.right = TNode(node.depth+1, Xf,yf)
            Construct_Subtree(node.right, max_depth)

    return node
```

This requires an implementation of the **CalculateOptimalSplit** function. To start, we implement a function **DataSplit** that splits the data according to $s(\boldsymbol{x}) = \mathbb{1}\{x_j \leq \xi\}$.

```
def DataSplit(X,y,j,xi):
    ids = X[:,j]<=xi
    Xt  = X[ids == True,:]
    Xf  = X[ids == False,:]
    yt  = y[ids == True]
    yf  = y[ids == False]
    return Xt, yt, Xf, yf
```

The **CalculateOptimalSplit** method runs through the possible splitting thresholds ξ from the set $\{x_{j,k}\}$ and finds the optimal split.

```
def CalculateOptimalSplit(node):
    X = node.X
    y = node.y
    best_var = 0
    best_xi = X[0,best_var]
    best_split_val = node.CalculateLoss()

    m, n  = X.shape

    for j in range(0,n):
        for i in range(0,m):
            xi = X[i,j]
            Xt, yt, Xf, yf = DataSplit(X,y,j,xi)
            tmpt = TNode(0, Xt, yt)
            tmpf = TNode(0, Xf, yf)
            loss_t = tmpt.CalculateLoss()
            loss_f = tmpf.CalculateLoss()
            curr_val =  loss_t + loss_f
            if (curr_val < best_split_val):
                best_split_val = curr_val
                best_var = j
                best_xi = xi
    return best_var,  best_xi
```

Finally, we implement the recursive method for prediction.

```
def Predict(X,node):
    if(node.right == None and node.left != None):
        return Predict(X,node.left)

    if(node.right != None and node.left == None):
        return Predict(X,node.right)

    if(node.right == None and node.left == None):
        return node.g
    else:
        if(X[node.j] <= node.xi):
            return Predict(X,node.left)
        else:
            return Predict(X,node.right)
```

Running the **main** function defined above gives a similar[2] result to what one would achieve with the **sklearn** package, using the **DecisionTreeRegressor** method.

```
main()  # run the main program

# compare with sklearn
```

[2]After establishing a best split $\xi = x_{j,k}$, sklearn assigns the corresponding feature vector randomly to one of the two child nodes, rather than to the True child.

```
from sklearn.tree import DecisionTreeRegressor

X_train, X_test, y_train, y_test = makedata() # use the same data
regTree = DecisionTreeRegressor(max_depth = 10, random_state=0)
regTree.fit(X_train,y_train)
y_hat = regTree.predict(X_test)
MSE2 = np.mean(np.power(y_hat - y_test,2))
print("DecisionTreeRegressor: tree loss = ", MSE2)
```

```
Basic tree: tree loss =  9.067077996170276
DecisionTreeRegressor: tree loss =  10.197991295531748
```

9.3 Additional Considerations

9.3.1 Binary Versus Non-Binary Trees

While it is possible to split a tree node into more than two groups (multiway splits), it generally produces inferior results compared to the simple binary split. The major reason is that multiway splits can lead to too many nodes near the tree root that have only a few data points, thus leaving insufficient data for later splits. As multiway splits can be represented as several binary splits, the latter is preferred [81].

9.3.2 Data Preprocessing

Sometimes, it can be beneficial to preprocess the data prior to the tree construction. For ☞ 179 example, PCA can be used with a view to identify the most important dimensions, which in turn will lead to simpler and possibly more informative splitting rules in the internal nodes.

9.3.3 Alternative Splitting Rules

We restricted our attention to splitting rules of the type $s(\boldsymbol{x}) = \mathbb{1}\{x_j \leq \xi\}$, where $j \in \{1, \ldots, p\}$ and $\xi \in \mathbb{R}$. These types of rules may not always result in a simple partition of the feature space, as illustrated by the binary data in Figure 9.20. In this case, the feature space could have been partitioned into just two regions, separated by a straight line.

In this case many classification methods discussed in Chapter 8, such as linear ☞ 362 discriminant analysis (Section 8.4), will work very well, whereas the classification tree is rather elaborate, dividing the feature set into too many regions. An obvious remedy is to use splitting rules of the form

$$s(\boldsymbol{x}) = \mathbb{1}\{\boldsymbol{a}^\top \boldsymbol{x} \leq \xi\}.$$

In some cases, such as the one just discussed, it may be useful to use a splitting rule that involves several variables, as opposed to a single one. The decision regarding the split type clearly depends on the problem domain. For example, for logical (binary) variables our domain knowledge may indicate that a different behavior is expected when both x_i and x_j $(i \neq j)$ are `True`. In this case, we will naturally introduce a decision rule of the form:

$$s(\boldsymbol{x}) = \mathbb{1}\{x_i = \texttt{True} \text{ and } x_j = \texttt{True}\}.$$

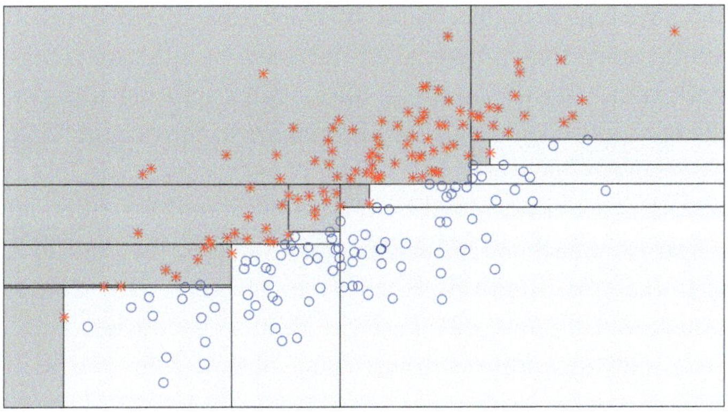

Figure 9.20: The two groups of points here can be separated by a straight line. Instead, the classification tree divides up the space into many rectangles, leading to an unnecessarily complicated classification procedure.

9.3.4 Categorical Variables

When an explanatory variable is categorical with labels (levels) say $\{1, \ldots, k\}$, the splitting rule is generally defined via a partition of the label set $\{1, \ldots, k\}$ into two subsets. Specifically, let L and R be a partition of $\{1, \ldots, k\}$. Then, the splitting rule is defined via

$$s(\boldsymbol{x}) = \mathbb{1}\{x_j \in L\}.$$

For the general supervised learning case, finding the optimal partition in the sense of minimal loss requires one to consider 2^k subsets of $\{1, \ldots, k\}$. Consequently, finding a good splitting rule for categorical variables can be challenging when the number of labels k is large.

9.3.5 Missing Values

Missing data is present in many real-life problems. Generally, when working with incomplete feature vectors, where one or more values are missing, it is typical to either completely delete the feature vector from the data (which may distort the data) or to impute (guess) its missing values from the available data; see e.g., [172]. Tree methods, however, allow an elegant approach for handling missing data. Specifically, in the general case, the missing data problem can be handled via *surrogate* splitting rules [32].

When dealing with categorical (factor) features, we can introduce an additional category "missing" for the absent data.

The main idea of surrogate rules is as follows: First, we construct a decision (regression or a classification) tree via Algorithm 9.9. During this construction process, the solution of the optimization problem (9.13) is calculated only over the observations that are not missing a particular variable. Suppose that a tree node v has a splitting rule $s^*(\boldsymbol{x}) = \mathbb{1}\{x_{j^*} \leq \xi^*\}$ for some $1 \leq j^* \leq k$ and threshold ξ^*.

For the node v we can introduce a set of alternative splitting rules that resemble the original splitting rule, sometimes called the *primary* splitting rule, using different variables and thresholds. Namely, we look for a binary splitting rule $s(\boldsymbol{x} \mid j, \xi)$, $j \neq j^*$ such that the data split introduced by s will be *similar* to the original data split from s^*. The similarity is generally measured via a binary misclassification loss, where the true classes of observations are determined by the primary splitting rule and the surrogate splitting rules serve as classifiers. Consider, for example, the data in Table 9.21 and suppose that the primary splitting rule at node v is $\mathbb{1}\{\text{Age} \leq 25\}$. That is, the five data points are split such that the left and the right child of v contains two and three data points, respectively. Next, the following surrogate splitting rules can be considered:

1. $\mathbb{1}\{\text{Salary} \leq 1500\}$, and
2. $\mathbb{1}\{\text{Height} \leq 173\}$.

Table 9.21: Example data with three variables (Age, Height, and Salary).

Id	Age	Height	Salary
1	20	173	1000
2	25	168	1500
3	38	191	1700
4	49	170	1900
5	62	182	2000

The $\mathbb{1}\{\text{Salary} \leq 1500\}$ surrogate rule completely mimics the primary rule, in the sense that the data splits induced by these rules are identical. Namely, both rules partition the data into two sets (by Id) $\{1, 2\}$ and $\{3, 4, 5\}$. On the other hand, the $\mathbb{1}\{\text{Height} \leq 173\}$ rule is less similar to the primary rule, since it causes the different partition $\{1, 2, 4\}$ and $\{3, 5\}$.

It is up to the user to define the number of surrogate rules for each tree node. As soon as these surrogate rules are available, we can use them to handle a new data point, even if the main rule cannot be applied due to a missing value of the primary variable x_{j^*}. Specifically, if the observation is missing the primary split variable, we apply the first (best) surrogate rule. If the first surrogate variable is also missing, we apply the second best surrogate rule, and so on.

9.4 Controlling the Tree Shape

Eventually, we are interested in getting the right-size tree. Namely, a tree that shows good generalization properties. It was already discussed in Section 9.2.3 (Figure 9.19) that shallow trees tend to underfit and deep trees tend to overfit the data. Basically, a shallow tree does not produce a sufficient number of splits and a deep tree will produce many partitions and thus many leaf nodes. If we grow the tree to a sufficient depth, each training sample will occupy a separate leaf and we will observe a zero loss with respect to the training data. The above phenomenon is illustrated in Figure 9.22, which presents the cross-validation loss and the training loss as a function of the tree depth.

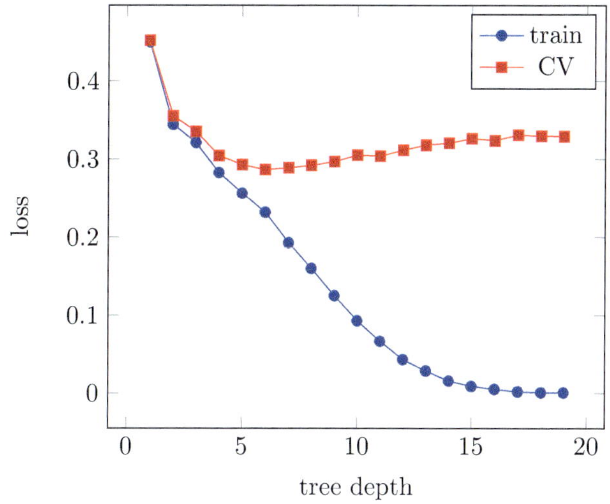

Figure 9.22: The cross-validation and the training loss as a function of the tree depth for a binary classification problem.

In order to overcome the under- and the overfitting problem, Breiman et al. [32] examined the possibility of stopping the tree from growing as soon as the decrease in loss due to a split of node v, as expressed in the difference of (9.12) and (9.13), is smaller than some predefined parameter $\delta \in \mathbb{R}$. Under this setting, the tree construction process will terminate when no leaf node can be split such that the contribution to the training loss after this split is greater than δ.

The authors found that this approach was unsatisfactory. Specifically, it was noted that a very small δ leads to an excessive amount of splitting and thus causes overfitting. Increasing δ did not work either. The problem is that the nature of the proposed rule is *one-step-look-ahead*. To see this, consider a tree node for which the best possible decrease in loss is smaller than δ. According to the proposed procedure, this node will not be split further. This may, however, be sub-optimal, because it could happen that one of the node's *descendants*, if split, could lead to a major decrease in loss.

To address these issues, a so-called *pruning* routine can be employed. The idea is as follows: We first grow a very deep tree and then prune (remove nodes) it upwards until we reach the root node. Consequently, the pruning process causes the number of tree nodes to decrease. While the tree is being pruned, the generalization risk gradually decreases up to the point where it starts increasing again, at which point the pruning is stopped. This decreasing/increasing behavior is due to the bias–variance tradeoff (2.35).

We next describe the details. To start with, let v and v' be tree nodes. We say that v' is a *descendant* of v if there is a path down the tree, which leads from v to v'. If such a path exists, we also say that v is an *ancestor* of v'. Consider the tree in Figure 9.23.

To formally define pruning, we will require the following Definition 9.24. An example of pruning is demonstrated in Figure 9.25.

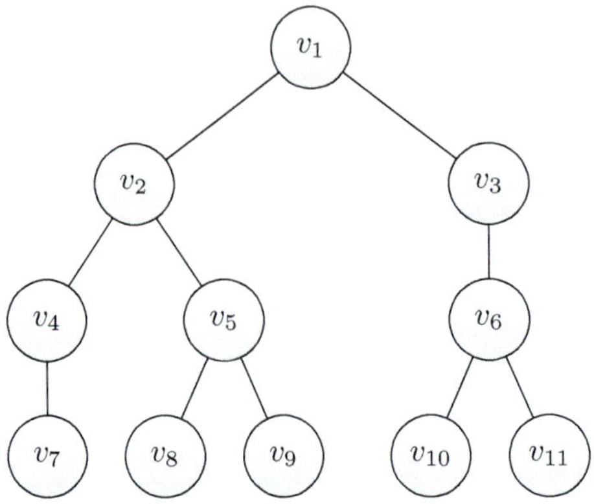

Figure 9.23: The node v_9 is a descendant of v_2, and v_2 is an ancestor of $\{v_4, v_5, v_7, v_8, v_9\}$, but v_6 is not a descendant of v_2.

Definition 9.24 (Branches and Pruning):

1. A **tree branch** $\mathbb{T}_v$ of the tree $\mathbb{T}$ is a sub-tree of $\mathbb{T}$ rooted at node $v \in \mathbb{T}$.
2. The pruning of branch $\mathbb{T}_v$ from a tree $\mathbb{T}$ is performed via deletion of the entire branch $\mathbb{T}_v$ from $\mathbb{T}$ except the branch's root node v. The resulting pruned tree is denoted by $\mathbb{T} - \mathbb{T}_v$.
3. A sub-tree $\mathbb{T} - \mathbb{T}_v$ is called a pruned sub-tree of $\mathbb{T}$. We indicate this with the notation $\mathbb{T} - \mathbb{T}_v < \mathbb{T}$ or $\mathbb{T} > \mathbb{T} - \mathbb{T}_v$.

(a) $\mathbb{T}$ (b) $\mathbb{T}_{v_2}$ (c) $\mathbb{T} - \mathbb{T}_{v_2}$

Figure 9.25: The pruned tree $\mathbb{T} - \mathbb{T}_{v_2}$ in (c) is the result of pruning the $\mathbb{T}_{v_2}$ branch in (b) from the original tree $\mathbb{T}$ in (a).

A basic decision tree pruning procedure is summarized in Algorithm 9.26.

Algorithm 9.26: Decision Tree Pruning

Input: Training set τ.

Output: Sequence of decision trees $\mathbb{T}^0 \succ \mathbb{T}^1 \succ \cdots$

1 Build a large decision tree $\mathbb{T}^0$ via Algorithm 9.9. [A possible termination criterion for that algorithm is to have some small predetermined number of data points at each terminal node of $\mathbb{T}^0$.]
2 $\mathbb{T}' \leftarrow \mathbb{T}^0$
3 $k \leftarrow 0$
4 **while** $\mathbb{T}'$ has more than one node **do**
5     $k \leftarrow k + 1$
6   Choose $v \in \mathbb{T}'$.
7   Prune the branch rooted at v from $\mathbb{T}'$.
8     $\mathbb{T}^k \leftarrow \mathbb{T}' - \mathbb{T}_v$ and $\mathbb{T}' \leftarrow \mathbb{T}^k$.
9 **return** $\mathbb{T}^0, \mathbb{T}^1, \ldots, \mathbb{T}^k$

Let $\mathbb{T}^0$ be the initial (deep) tree and let $\mathbb{T}^k$ be the tree obtained after the k-th pruning operation, for $k = 1, \ldots, K$. As soon as the sequence of trees $\mathbb{T}^0 \succ \mathbb{T}^1 \succ \cdots \succ \mathbb{T}^K$ is available, one can choose the best tree of $\{\mathbb{T}^k\}_{k=1}^K$ according to the smallest generalization risk. Specifically, we can split the data into training and validation sets. In this case, Algorithm 9.26 is executed using the training set and the generalization risks of $\{\mathbb{T}^k\}_{k=1}^K$ are estimated via the validation set.

While Algorithm 9.26 and the corresponding best tree selection process look appealing, there is still an important question to consider; namely, how to choose the node v and the corresponding branch $\mathbb{T}_v$ in Line 6 of the algorithm. In order to overcome this problem, Breiman proposed a method called *cost complexity pruning*, which we discuss next.

9.4.1 Cost-Complexity Pruning

Let $\mathbb{T} \prec \mathbb{T}^0$ be a tree obtained via pruning of a tree $\mathbb{T}^0$. Denote the set of leaf (terminal) nodes of $\mathbb{T}$ by $\mathcal{W}$. The number of leaves $|\mathcal{W}|$ is a measure for the complexity of the tree; recall that $|\mathcal{W}|$ is the number of regions $\{\mathcal{R}_w\}$ in the partition of $\mathcal{X}$. Corresponding to each tree $\mathbb{T}$ is a prediction function g, as in (9.4). In **cost-complexity pruning** the objective is to find a prediction function g (or, equivalently, tree $\mathbb{T}$) that minimizes the training loss $\ell_\tau(g)$ while taking into account the complexity of the tree. The idea is to regularize the training loss, similar to what was done in Chapter 7, by adding a penalty term for the complexity of the tree. This leads to the following definition.

Definition 9.27 (Cost-Complexity Measure): Let $\tau = \{(\boldsymbol{x}_i, y_i)\}_{i=1}^n$ be a data set and $\gamma \geq 0$ be a real number. For a given tree $\mathbb{T}$, the **cost-complexity measure** $C_\tau(\gamma, \mathbb{T})$ is defined as:

$$C_\tau(\gamma, \mathbb{T}) := \frac{1}{n} \sum_{w \in \mathcal{W}} \left(\sum_{i=1}^{n} \mathbb{1}\{\boldsymbol{x}_i \in \mathcal{R}_w\} \operatorname{Loss}(y_i, g^w(\boldsymbol{x}_i)) \right) + \gamma \, |\mathcal{W}| \tag{9.28}$$
$$= \ell_\tau(g) + \gamma \, |\mathcal{W}|,$$

where $\ell_\tau(g)$ is the training loss (9.5).

Small values of γ result in a small penalty for the tree complexity $|\mathcal{W}|$, and thus large trees (that fit the entire *training* data well) will minimize the measure $C_\tau(\gamma, \mathbb{T})$. In particular, for $\gamma = 0$, $\mathbb{T} = \mathbb{T}^0$ will be the minimizer of $C_\tau(\gamma, \mathbb{T})$. On the other hand, large values of γ will prefer smaller trees or, more precisely, trees with fewer leaves. For sufficiently large γ, the solution $\mathbb{T}$ will collapse to a single (root) node.

It can be shown that, for every value of γ, there exists a smallest minimizing sub-tree of $\mathbb{T}^0$ with respect to the cost-complexity measure. In practice, a suitable γ is selected via observing the performance of the learner on the validation set or by cross-validation.

These advantages and the corresponding limitations are detailed next.

9.4.2 Advantages and Limitations of Decision Trees

We list a number of advantages and disadvantages of decision trees, as compared with other supervised learning methods such as were discussed in Chapters 5, 7, and 8.

Advantages

1. The tree structure can handle both categorical and numerical features in a natural and straightforward way. Specifically, there is no need to pre-process categorical features, say via the introduction of dummy variables.

2. The final tree obtained after the training phase can be compactly stored for the purpose of making predictions for new feature vectors. The prediction process only involves a single tree traversal from the tree root to a leaf.

3. The hierarchical nature of decision trees allows for an efficient encoding of the feature's conditional information. Specifically, after an internal split of a feature x_j via the standard splitting rule (9.14), Algorithm 9.9 will only consider such subsets of data that were constructed based on this split, thus implicitly exploiting the corresponding conditional information from the initial split of x_j.

4. The tree structure can be easily understood and interpreted by domain experts with little statistical knowledge, since it is essentially a logical decision flow diagram.

5. The sequential decision tree growth procedure in Algorithm 9.9, and in particular the fact that the tree has been split using the most important features, provides an

implicit step-wise variable elimination procedure. In addition, the partition of the variable space into smaller regions results in simpler prediction problems in these regions.

6. Decision trees are invariant under monotone transformations of the data. To see this, consider the (optimal) splitting rule $s(\boldsymbol{x}) = \mathbb{1}\{x_3 \leq 2\}$, where x_3 is a positive feature. Suppose that x_3 is transformed to $x_3' = x_3^2$. Now, the optimal splitting rule will take the form $s(\boldsymbol{x}) = \mathbb{1}\{x_3' \leq 4\}$.

7. In the classification setting, it is common to report not only the predicted value of a feature vector, e.g., as in (9.10), but also the respective class probabilities. Decision trees handle this task without any additional effort. Specifically, consider a new feature vector. During the estimation process, we will perform a tree traversal and the point will end up in a certain leaf w. The probability of this feature vector lying in class z can be estimated as the proportion of training points in w that are in class z.

8. As each training point is treated equally in the construction of a tree, the structure of the tree will be relatively robust to outliers. In a way, trees exhibit a similar kind of robustness as the sample median does for real-valued data.

Limitations

Despite the fact that the decision trees are extremely interpretable, the predictive accuracy is generally inferior to other established statistical learning methods. In addition, decision trees, and in particular very deep trees that were not subject to pruning, are heavily reliant on their training set. A small change in the training set can result in a dramatic change of the resulting decision tree. Their inferior predictive accuracy, however, is a direct consequence of the bias–variance tradeoff. Specifically, a decision tree model generally exhibits a high variance. To overcome the above limitations, several promising approaches such as *bagging*, *random forest*, and *boosting* are introduced below.

> The bagging approach was initially introduced in the context of an ensemble of decision trees. However, both the bagging and the boosting methods can be applied to improve the accuracy of general prediction functions.

9.5 Bootstrap Aggregation

The major idea of the *bootstrap aggregation* or **bagging** method is to combine prediction functions learned from multiple data sets, with a view to improving overall prediction accuracy. Bagging is especially beneficial when dealing with predictors that tend to overfit the data, such as in decision trees, where the (unpruned) tree structure is very sensitive to small changes in the training set [55, 81].

To start with, consider an idealized setting for a regression tree, where we have access to B iid copies[3] $\mathcal{T}_1, \ldots, \mathcal{T}_B$ of a training set $\mathcal{T}$. Then, we can train B separate regression

[3]In this section $\mathcal{T}_k$ means the k-th training set, not a training set of size k.

models (B different decision trees) using these sets, giving learners $g_{\mathcal{T}_1}, \ldots, g_{\mathcal{T}_B}$, and take their average:

(9.29)
$$g_{\text{avg}}(\boldsymbol{x}) = \frac{1}{B} \sum_{b=1}^{B} g_{\mathcal{T}_b}(\boldsymbol{x}).$$

☞ 653 By the law of large numbers, as $B \to \infty$, the average prediction function converges to the expected prediction function $g^\dagger := \mathbb{E} g_{\mathcal{T}}$. The following result shows that using $g^\dagger$ as a prediction function (if it were known) would result in an expected squared-error ☞ 24 generalization risk that is less than or equal to the expected generalization risk for a general prediction function $g_{\mathcal{T}}$. It thus suggests that taking an average of prediction functions may lead to a better expected squared-error generalization risk.

Theorem 9.30 (Expected Squared-Error Generalization Risk): Let $\mathcal{T}$ be a random training set and let $\boldsymbol{X}, Y$ be a random feature vector and response that are independent of $\mathcal{T}$. Then,

$$\mathbb{E}\Big(Y - g_{\mathcal{T}}(\boldsymbol{X})\Big)^2 \geq \mathbb{E}\Big(Y - g^\dagger(\boldsymbol{X})\Big)^2.$$

Proof. We have

$$\mathbb{E}\Big[\Big(Y - g_{\mathcal{T}}(\boldsymbol{X})\Big)^2 \,\Big|\, \boldsymbol{X}, Y\Big] \geq \Big(\mathbb{E}[Y \mid \boldsymbol{X}, Y] - \mathbb{E}[g_{\mathcal{T}}(\boldsymbol{X}) \mid \boldsymbol{X}, Y]\Big)^2 = \Big(Y - g^\dagger(\boldsymbol{X})\Big)^2,$$

where the inequality follows from $\mathbb{E}U^2 \geq (\mathbb{E}U)^2$ for any (conditional) expectation. ☞ 635 Consequently, by the tower property,

$$\mathbb{E}\Big(Y - g_{\mathcal{T}}(\boldsymbol{X})\Big)^2 = \mathbb{E}\Big[\mathbb{E}[(Y - g_{\mathcal{T}}(\boldsymbol{X}))^2 \mid \boldsymbol{X}, Y]\Big] \geq \mathbb{E}\Big(Y - g^\dagger(\boldsymbol{X})\Big)^2.$$

□

Unfortunately, multiple independent data sets are rarely available. But we can substitute them by bootstrapped ones. Specifically, instead of the $\mathcal{T}_1, \ldots, \mathcal{T}_B$ sets, we can obtain ☞ 86 random training sets $\mathcal{T}_1^*, \ldots, \mathcal{T}_B^*$ by resampling them from a *single* (fixed) training set τ, similar to Algorithm 3.26, and use them to train B separate models. By model averaging as in (9.29) we obtain the bootstrapped aggregated estimator or **bagged estimator** of the form:

(9.31)
$$g_{\text{bag}}(\boldsymbol{x}) = \frac{1}{B} \sum_{b=1}^{B} g_{\mathcal{T}_b^*}(\boldsymbol{x}).$$

■ **Remark 9.33 (Bootstrap Aggregation for Classification Problems)** Note that (9.31) is suitable for handling regression problems. However, the bagging idea can be readily extended to handle classification settings as well. For example, g_{bag} can take the majority vote among $\{g_{\mathcal{T}_b^*}\}, b = 1, \ldots, B$; that is, to accept the most frequent class among B predictors. ■

Algorithm 9.32: Bootstrap Aggregation Sampling

Input: Training set $\tau = \{(\boldsymbol{x}_i, y_i)\}_{i=1}^n$ and resample size B.
Output: Bootstrapped data sets.

```
for b = 1 to B do
    T*_b ← ∅
    for i = 1 to n do
        Draw U ~ U(0, 1)
        I ← ⌈nU⌉                                  // select random index
        T*_b ← T*_b ∪ {(x_I, y_I)}.
return T*_b, b = 1, ..., B.
```

While bagging can be applied for any statistical model (such as decision trees, neural networks, linear regression, K-nearest neighbors, and so on), it is most effective for predictors that are sensitive to small changes in the training set. The reason becomes clear when we decompose the expected generalization risk as

$$\mathbb{E}\,\ell(g_{\mathcal{T}}) = \ell^* + \underbrace{\mathbb{E}\left(\mathbb{E}[g_{\mathcal{T}}(\boldsymbol{X}) \mid \boldsymbol{X}] - g^*(\boldsymbol{X})\right)^2}_{\text{expected squared bias}} + \underbrace{\mathbb{E}[\mathbb{V}\mathrm{ar}[g_{\mathcal{T}}(\boldsymbol{X}) \mid \boldsymbol{X}]]}_{\text{expected variance}}, \tag{9.34}$$

similar to (2.35). Compare this with the same decomposition for the average prediction ☞ 36
function g_{bag} in (9.29). As $\mathbb{E}g_{\mathrm{bag}}(\boldsymbol{x}) = \mathbb{E}g_{\mathcal{T}}(\boldsymbol{x})$, we see that any possible improvement in the generalization risk must be due to the expected variance term. Averaging and bagging are thus only useful for predictors with a large expected variance, relative to the other two terms. Examples of such "unstable" predictors include decision trees, neural networks, and subset selection in linear regression [34]. On the other hand, "stable" predictors are insensitive to small data changes, an example being the K-nearest neighbors method. ☞ 371
Note that for independent training sets $\mathcal{T}_1, \ldots, \mathcal{T}_B$ a reduction of the variance by a factor B is achieved: $\mathbb{V}\mathrm{ar}\, g_{\mathrm{bag}}(\boldsymbol{x}) = B^{-1}\mathbb{V}\mathrm{ar}\, g_{\mathcal{T}}(\boldsymbol{x})$. Again, it depends on the squared bias and irreducible loss how significant this reduction is for the generalization risk.

■ **Remark 9.35 (Limitations of Bagging)** It is important to remember that g_{bag} is not exactly equal to g_{avg}, which in turn is not exactly $g^\dagger$. Specifically, g_{bag} is constructed from the bootstrap approximation of the sampling pdf f. As a consequence, for stable predictors, it can happen that g_{bag} will perform worse than $g_{\mathcal{T}}$. In addition to the deterioration of the bagging performance for stable procedures, it can also happen that $g_{\mathcal{T}}$ has already achieved a near optimal predictive accuracy given the available training data. In this case, bagging will not introduce a significant improvement. ■

The bagging process provides an opportunity to estimate the generalization risk of the bagged model without an additional test set. Specifically, recall that we obtain the $\mathcal{T}_1^*, \ldots, \mathcal{T}_B^*$ sets from a single training set τ by sampling via Algorithm 9.32, and use them to train B separate models. It can be shown (see Exercise 8) that, for large sample sizes, on average about a third (more precisely, a fraction $\mathrm{e}^{-1} \approx 0.37$) of the original sample points are not included in bootstrapped set $\mathcal{T}_b^*$ for $1 \le b \le B$. Therefore, these samples can be used for the loss estimation. These samples are called **out-of-bag** (OOB) observations.

Specifically, for each sample from the original data set, we calculate the OOB loss using predictors that were trained without this particular sample. The estimation procedure is summarized in Algorithm 9.36. Hastie et al. [81] observe that, under certain conditions, the OOB loss is almost identical to the n-fold cross-validation loss. In addition, the OOB loss can be used to determine the number of trees required. Specifically, we can train predictors until the OOB loss stops changing. Namely, decision trees are added until the OOB loss stabilizes.

Algorithm 9.36: Out-of-Bag Loss Estimation

Input: The original data set $\tau = \{(\boldsymbol{x}_1, y_1), \ldots, (\boldsymbol{x}_n, y_n)\}$, the bootstrapped data sets $\{\mathcal{T}_1^*, \ldots, \mathcal{T}_B^*\}$, and the trained predictors $\left\{g_{\mathcal{T}_1^*}, \ldots, g_{\mathcal{T}_B^*}\right\}$.

Output: Out-of-bag loss for the averaged model.

for $i = 1$ **to** n **do**
- $C_i \leftarrow \emptyset$ // Indices of predictors not depending on $(\boldsymbol{x}_i, y_i)$
- **for** $b = 1$ **to** B **do**
  - **if** $(\boldsymbol{x}_i, y_i) \notin \mathcal{T}_b^*$ **then** $C_i \leftarrow C_i \cup \{b\}$
- $Y_i' \leftarrow |C_i|^{-1} \sum_{b \in C_i} g_{\mathcal{T}_b^*}(\boldsymbol{x}_i)$
- $L_i \leftarrow \text{Loss}\left(y_i, Y_i'\right)$

$L_{\text{OOB}} \leftarrow \frac{1}{n} \sum_{i=1}^{n} L_i$

return L_{OOB}.

■ **Example 9.37 (Bagging for a Regression Tree)** We next proceed with a basic bagging example for a regression tree, in which we compare the decision tree estimator with the corresponding bagged estimator. We use the R^2 metric (coefficient of determination) for comparison.

BaggingExample.py

```
import numpy as np
from sklearn.datasets import make_friedman1
from sklearn.tree import DecisionTreeRegressor
from sklearn.model_selection import train_test_split
from sklearn.metrics import r2_score

np.random.seed(100)

# Create regression problem
n_points = 1000
X, y = make_friedman1(n_samples=n_points, n_features=15, noise=1.0,
    random_state=100)

# Split to train/test set
X_train, X_test, y_train, y_test = train_test_split(X, y, test_size
    =0.33, random_state=100)
```

```
# Train single DecisionTreeRegressor
reg_tree = DecisionTreeRegressor(random_state=100)
reg_tree.fit(X_train, y_train)

# Test single DecisionTreeRegressor
y_pred_tree = reg_tree.predict(X_test)

# Bagging construction
n_estimators = 500
bag = [DecisionTreeRegressor(random_state=100) for _ in range(
    n_estimators)]
bootstrap_indices = [np.random.choice(len(X_train), size=len(X_train),
    replace=True) for _ in range(n_estimators)]

for i in range(n_estimators):
    X_bootstrap = X_train[bootstrap_indices[i]]
    y_bootstrap = y_train[bootstrap_indices[i]]
    bag[i].fit(X_bootstrap, y_bootstrap)

# Bagging prediction
y_pred_bagging = np.mean([tree.predict(X_test) for tree in bag], axis=0)

# Out-of-bag loss estimation
oob_predictions = np.zeros(len(X_train))
for i in range(len(X_train)):
    oob_trees = [bag[j] for j in range(n_estimators) if i not in
        bootstrap_indices[j]]
    if oob_trees:
        oob_predictions[i] = np.mean([tree.predict(X_train[i].reshape(1,
            -1)) for tree in oob_trees])

oob_score = r2_score(y_train, oob_predictions)

# Print results
print("DecisionTreeRegressor R^2 score =", r2_score(y_test, y_pred_tree)
    )
print("Bagging R^2 score =", r2_score(y_test, y_pred_bagging))
print("Bagging OOB R^2 score =", oob_score)
```

```
DecisionTreeRegressor R^2 score = 0.5652585102808715
Bagging R^2 score = 0.804368966748225
Bagging OOB R^2 score = 0.8230631980459328
```

The decision tree bagging improves the test-set R^2 score by about 42% (from 0.565 to 0.804). Moreover, the OOB score (0.82) is very close to the true generalization risk (0.80) of the bagged estimator. ■

The bagging procedure can be further enhanced by introducing random forests, which are discussed next.

9.6 Random Forests

In Section 9.5, we discussed the intuition behind the prediction averaging procedure. Specifically, for some feature vector $\boldsymbol{x}$ let $Z_b = g_{\mathcal{T}_b}(\boldsymbol{x}), b = 1, 2, \ldots, B$ be iid prediction values, obtained from independent training sets $\mathcal{T}_1, \ldots, \mathcal{T}_B$. Suppose that $\mathbb{V}\mathrm{ar}\, Z_b = \sigma^2$ for all $b = 1, \ldots, B$. Then the variance of the average prediction value $\overline{Z}_B$ is equal to σ^2/B. However, if bootstrapped data sets $\{\mathcal{T}_b^*\}$ are used instead, the corresponding random variables $\{Z_b\}$ will be *correlated*. In particular, $Z_b = g_{\mathcal{T}_b^*}(\boldsymbol{x})$ for $b = 1, \ldots, B$ are identically distributed (but not independent) with some positive pairwise correlation ϱ. It then holds that (see Exercise 9)

$$\mathbb{V}\mathrm{ar}\, \overline{Z}_B = \varrho\, \sigma^2 + \sigma^2 \frac{(1 - \varrho)}{B}. \tag{9.38}$$

While the second term of (9.38) goes to zero as the number of observation B increases, the first term remains constant.

This issue is particularly relevant for bagging with decision trees. For example, consider a situation in which there exists a feature that provides a very good split of the data. Such a feature will be selected and split for every $\{g_{\mathcal{T}_b^*}\}_{b=1}^B$ at the root level and we will consequently end up with highly correlated predictions. In such a situation, prediction averaging will not introduce the desired improvement in the performance of the bagged predictor.

The major idea of random forests is to perform bagging in combination with a "decorrelation" of the trees by including only a subset of features during the tree construction. For each bootstrapped training set $\mathcal{T}_b^*$ we build a decision tree using a randomly selected subset of $m \leq p$ features for the splitting rules. This simple but powerful idea will decorrelate the trees, since strong predictors will have a smaller chance to be considered at the root levels.

Consequentially, we can expect to improve the predictive performance of the bagged estimator. The resulting predictor (random forest) construction is summarized in Algorithm 9.39.

For regression problems, the output of Algorithm 9.39 is combined to yield the random forest prediction function:

$$g_{\mathrm{RF}}(\boldsymbol{x}) = \frac{1}{B} \sum_{b=1}^{B} g_{\mathcal{T}_b^*}(\boldsymbol{x}).$$

In the classification setting, similar to Remark 9.33, we take instead the majority vote from the $\{g_{\mathcal{T}_b^*}\}$.

Algorithm 9.39: Random Forest Construction

Input: Training set $\tau = \{(\boldsymbol{x}_i, y_i)\}_{i=1}^n$, the number of trees in the forest B, and the number $m \leq p$ of features to be included, where p is the total number of features in $\boldsymbol{x}$.

Output: Ensemble of trees.

Generate bootstrapped training sets $\{\mathcal{T}_1^*, \ldots, \mathcal{T}_B^*\}$ via Algorithm 9.32.
for $b = 1$ **to** B **do**
    Train a decision tree $g_{\mathcal{T}_b^*}$ via Algorithm 9.9, where each split is performed using m randomly selected features out of p.
return $\{g_{\mathcal{T}_b^*}\}_{b=1}^B$.

Example 9.40 (Random Forest for a Regression Tree) We continue with the basic bagging Example 9.37 for a regression tree, in which we compared the decision tree estimator with the corresponding bagged estimator. Here, however, we use the random forest with $B = 500$ trees and a subset size $m = 8$. It can be seen that the random forest's R^2 score outperforms that of the bagged estimator.

BaggingExampleRF.py

```
from sklearn.datasets import make_friedman1
from sklearn.model_selection import train_test_split
from sklearn.metrics import r2_score
from sklearn.ensemble import RandomForestRegressor

# create regression problem
n_points = 1000 # points
x, y =  make_friedman1(n_samples=n_points, n_features=15,
                       noise=1.0, random_state=100)
# split to train/test set
x_train, x_test, y_train, y_test = \
        train_test_split(x, y, test_size=0.33, random_state=100)
rf = RandomForestRegressor(n_estimators=500, oob_score = True,
    max_features=8,random_state=100)
rf.fit(x_train,y_train)
yhatrf = rf.predict(x_test)

print("RF R^2 score = ", r2_score(y_test, yhatrf),
      "\nRF OOB R^2 score = ", rf.oob_score_)
```

```
RF R^2 score =  0.8106589580845707
RF OOB R^2 score =  0.8260541058404149
```

Remark 9.41 (The Optimal Number of Subset Features m) The default values for m are $\lfloor p/3 \rfloor$ and $\lfloor \sqrt{p} \rfloor$ for regression and classification setting, respectively. However, the standard practice is to treat m as a hyperparameter that requires tuning, depending on the specific problem at hand [81].

Note that the procedure of bagging decision trees is a special case of a random forest construction (see Exercise 11). Consequently, the OOB loss is readily available for random forests.

While the advantage of bagging in the sense of enhanced accuracy is clear, we should also consider its negative aspects and, in particular, the loss of interpretability. Specifically a random forest consists of many trees, thus making the prediction process both hard to visualize and interpret. For example, given a random forest, it is not easy to determine a subset of features that are essential for accurate prediction.

The feature importance measure intends to address this issue. The idea is as follows: Each *internal* node of a decision tree induces a certain decrease in the training loss; see (9.13). Let us denote this decrease in the training loss by $\Delta_{\text{Loss}}(v)$, where v is not a leaf node of $\mathbb{T}$. In addition, recall that for splitting rules of the type $\mathbb{1}\{x_j \leq \xi\}$ $(1 \leq j \leq p)$, each node v is associated with a feature x_j that determines the split. Using the above definitions, we can define the **feature importance** of x_j as

$$\mathcal{I}_{\mathbb{T}}(x_j) = \sum_{v \text{ internal } \in \mathbb{T}} \Delta_{\text{Loss}}(v)\ \mathbb{1}\{x_j \text{ is associated with } v\}, \quad 1 \leq j \leq p. \tag{9.42}$$

While (9.42) is defined for a single tree, it can be readily extended to random forests. Specifically, the feature importance in that case will be averaged over all trees of the forest; that is, for a forest consisting of B trees $\{\mathbb{T}_1, \dots, \mathbb{T}_B\}$, the feature importance measure is:

$$\mathcal{I}_{\text{RF}}(x_j) = \frac{1}{B} \sum_{b=1}^{B} \mathcal{I}_{\mathbb{T}_b}(x_j), \quad 1 \leq j \leq p. \tag{9.43}$$

Example 9.44 (Feature Importance) We consider a classification problem with 15 features. The data is specifically designed to contain only 5 informative features out of 15. In the code below, we apply the random forest procedure and calculate the corresponding feature importance measures, which are summarized in Figure 9.45.

VarImportance.py

```
import numpy as np
from sklearn.datasets import make_classification
from sklearn.ensemble import RandomForestClassifier
import matplotlib.pyplot as plt, pylab

n_points = 1000 # create regression data with 1000 data points
n_feats = 15
x, y =  make_classification(n_samples=n_points, n_features=n_feats,
  n_informative=5, n_redundant=0, n_repeated=0, random_state=100,
      shuffle=False)
rf = RandomForestClassifier(n_estimators=200, max_features="log2")
rf.fit(x,y)
importances = rf.feature_importances_indices = np.argsort(importances)
    [::-1]

for f in range(15):
    print("Feature %d (%f)" % (indices[f]+1, importances[indices[f]]))

std = np.std([rf.feature_importances_ for tree in rf.estimators_],
             axis=0)
f = plt.figure()
plt.bar(range(n_feats), importances[indices],
       color="b", yerr=std[indices], align="center")
plt.xticks(range(n_feats), indices+1)
plt.xlim([-1, n_feats])
pylab.xlabel("feature index")
pylab.ylabel("importance")
plt.show()
```

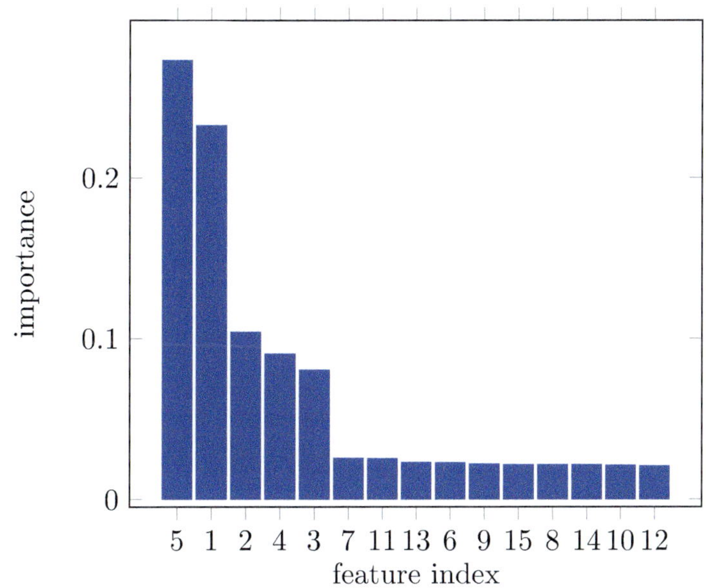

Figure 9.45: Importance measure for the 15-feature data set with only 5 informative features x_1, x_2, x_3, x_4, and x_5.

Clearly, it is hard to visualize and understand the prediction process based on 200 trees. However, Figure 9.45 shows that the features x_1, x_2, x_3, x_4, and x_5 were correctly identified as being important.

9.7 Boosting

Boosting is a powerful idea that aims to improve the accuracy of any learning algorithm, especially when involving **weak learners** — simple prediction functions that exhibit performance only slightly better than random guessing. Shallow decision trees typically yield weak learners.

Originally, boosting was developed for binary classification tasks, but it can be readily extended to handle general classification and regression problems. The boosting approach has some similarity with the bagging method in the sense that boosting uses an ensemble of prediction functions. Despite this similarity, there exists a fundamental difference between these methods. Specifically, while bagging involves the fitting of prediction functions to bootstrapped data, the predicting functions in boosting are learned *sequentially*. That is, each learner uses information from previous learners.

The idea is to start with a simple model (weak learner) g_0 for the data $\tau = \{(\boldsymbol{x}_i, y_i)\}_{i=1}^n$ and then to improve or "boost" this learner to a learner $g_1 := g_0 + h_1$. Here, the function h_1 is found by minimizing the training loss for $g_0 + h_1$ over all functions h in some class of functions $\mathcal{H}$. For example, $\mathcal{H}$ could be the set of prediction functions that can be obtained via a decision tree of maximal depth 2. Given a loss function Loss, the function

h_1 is thus obtained as the solution to the optimization problem

$$h_1 = \underset{h \in \mathcal{H}}{\operatorname{argmin}} \frac{1}{n} \sum_{i=1}^{n} \operatorname{Loss}\left(y_i, g_0(\boldsymbol{x}_i) + h\left(\boldsymbol{x}_i\right)\right). \tag{9.46}$$

This process can be repeated for g_1 to obtain $g_2 = g_1 + h_2$, and so on, yielding the boosted prediction function

$$g_B(\boldsymbol{x}) = g_0(\boldsymbol{x}) + \sum_{b=1}^{B} h_b(\boldsymbol{x}). \tag{9.47}$$

Instead of using the updating step $g_b = g_{b-1} + h_b$, one prefers to use the smooth updating step $g_b = g_{b-1} + \gamma\, h_b$, for some suitably chosen step-size parameter γ. As we shall see shortly, this helps reduce overfitting.

Boosting can be used for regression and classification problems. We start with a simple regression setting, using the squared-error loss; thus, $\operatorname{Loss}(y, \widehat{y}) = (y - \widehat{y})^2$. In this case, it is common to start with $g_0(\boldsymbol{x}) = n^{-1} \sum_{i=1}^{n} y_i$, and each h_b for $b = 1, \dots, B$ is chosen as a learner for the data set τ_b of residuals corresponding to g_{b-1}. That is,

$$\tau_b := \left\{\left(\boldsymbol{x}_i, e_i^{(b)}\right)\right\}_{i=1}^{n},$$

with

$$e_i^{(b)} := y_i - g_{b-1}(\boldsymbol{x}_i). \tag{9.48}$$

This leads to the following boosting procedure for regression with squared-error loss.

Algorithm 9.49: Regression Boosting with Squared-Error Loss

Input: Training set $\tau = \{(\boldsymbol{x}_i, y_i)\}_{i=1}^{n}$, the number of boosting rounds B, and a shrinkage step-size parameter γ.

Output: Boosted prediction function.

```
Set g_0(x) ← n^{-1} Σ_{i=1}^{n} y_i.
for b = 1 to B do
    Set e_i^{(b)} ← y_i − g_{b−1}(x_i) for i = 1,...,n, and let τ_b ← {(x_i, e_i^{(b)})}_{i=1}^{n}.
    Fit a prediction function h_b on the training data τ_b.
    Set g_b(x) ← g_{b−1}(x) + γ h_b(x).
return g_B.
```

The **step-size parameter** γ in Algorithm 9.49 controls the speed of the fitting process. Specifically, for small values of γ, boosting takes smaller steps towards the training loss minimization.

The step-size γ is of great practical importance, since it helps the boosting algorithm to avoid overfitting. This phenomenon is demonstrated in Figure 9.50.

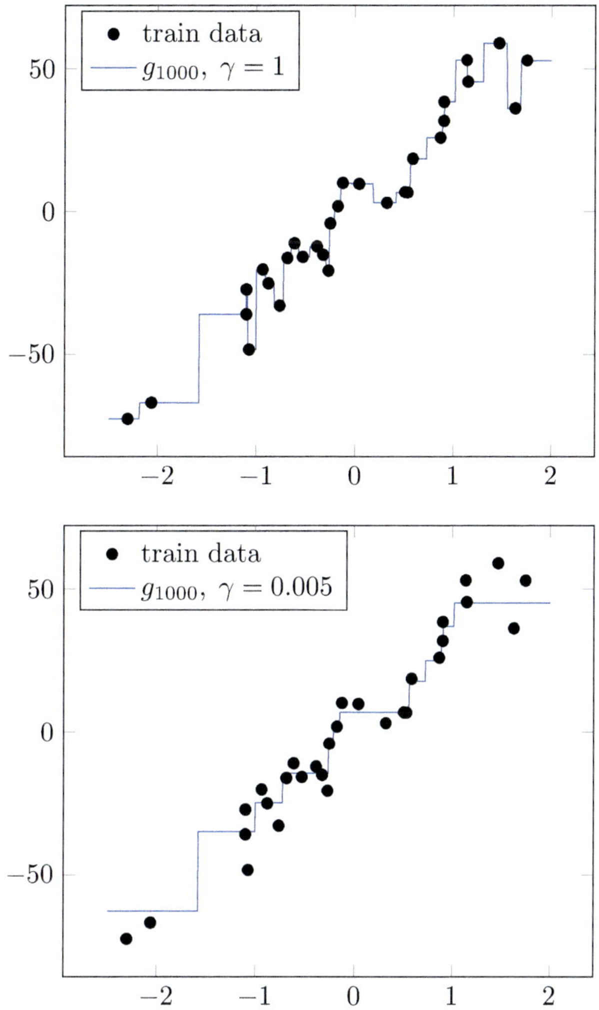

Figure 9.50: The upper and the lower panels show the fitted boosting regression model g_{1000} with $\gamma = 1.0$ and $\gamma = 0.005$, respectively. Note the overfitting on the upper panel.

A very basic implementation of Algorithm 9.49 which produces Figure 9.50 is provided below.

RegressionBoosting.py

```
import numpy as np
from sklearn.tree import DecisionTreeRegressor
from sklearn.model_selection import train_test_split
from sklearn.datasets import make_regression
import matplotlib.pyplot as plt
```

```
def TrainBoost(alpha,BoostingRounds,x,y):
    g_0 = np.mean(y)
    residuals  = y-alpha*g_0

    # list of basic regressor
    g_boost = []

    for i in range(BoostingRounds):
        h_i = DecisionTreeRegressor(max_depth=1)
        h_i.fit(x,residuals)
        residuals = residuals  - alpha*h_i.predict(x)
        g_boost.append(h_i)

    return g_0, g_boost

def Predict(g_0, g_boost,alpha, x):
    yhat = alpha*g_0*np.ones(len(x))
    for j in range(len(g_boost)):
        yhat = yhat+alpha*g_boost[j].predict(x)

    return yhat


np.random.seed(1)
sz = 30

# create data set
x,y = make_regression(n_samples=sz, n_features=1, n_informative=1,noise
    =10.0)

# boosting algorithm
BoostingRounds = 1000
alphas = [1, 0.005]

for alpha in alphas:
    g_0, g_boost = TrainBoost(alpha,BoostingRounds,x,y)
    yhat = Predict(g_0, g_boost, alpha, x)

    # plot
    tmpX =  np.reshape(np.linspace(-2.5,2,1000),(1000,1))
    yhatX = Predict(g_0, g_boost, alpha, tmpX)
    f = plt.figure()
    plt.plot(x,y,'*')
    plt.plot(tmpX,yhatX)
    plt.show()
```

The parameter γ can be viewed as a step size made in the direction of the negative gradient of the squared-error training loss. To see this, note that the negative gradient

$$-\left.\frac{\partial \operatorname{Loss}(y_i, z)}{\partial z}\right|_{z=g_{b-1}(\boldsymbol{x}_i)} = -\left.\frac{\partial (y_i - z)^2}{\partial z}\right|_{z=g_{b-1}(\boldsymbol{x}_i)} = 2(y_i - g_{b-1}(\boldsymbol{x}_i))$$

is two times the residual $e_i^{(b)}$ given in (9.48) that is used in Algorithm 9.49 to fit the prediction function h_b.

In fact, one of the major advances in the theory of boosting was the recognition that one can use a similar gradient descent method for any differentiable loss function. The resulting algorithm is called **gradient boosting**. The general gradient boosting algorithm is summarized in Algorithm 9.52.

The main idea is to mimic a gradient descent algorithm in the following sense. At ☞ 603 each stage of the boosting procedure, we calculate a negative gradient on n training points $\boldsymbol{x}_1, \ldots, \boldsymbol{x}_n$ (Lines 3–4). Then, we fit a simple model (such as a shallow decision tree) to approximate the gradient (Line 5) for any feature $\boldsymbol{x}$. Finally, similar to the gradient descent method, we make a γ-sized step in the direction of the negative gradient (Line 6).

Algorithm 9.52: Gradient Boosting

Input: Training set $\tau = \{(\boldsymbol{x}_i, y_i)\}_{i=1}^n$, the number of boosting rounds B, a differentiable loss function $\text{Loss}(y, \widehat{y})$, and a gradient step-size parameter γ.

Output: Gradient boosted prediction function.

1 Set $g_0(\boldsymbol{x}) \leftarrow 0$.

2 **for** $b = 1$ **to** B **do**

3   **for** $i = 1$ **to** n **do**

4   Evaluate the negative gradient of the loss at $(\boldsymbol{x}_i, y_i)$ via

$$r_i^{(b)} \leftarrow - \left. \frac{\partial \,\text{Loss}\,(y_i, z)}{\partial z} \right|_{z = g_{b-1}(\boldsymbol{x}_i)} \qquad i = 1, \ldots, n.$$

5   Approximate the negative gradient by solving

$$h_b = \underset{h \in \mathcal{H}}{\operatorname{argmin}} \frac{1}{n} \sum_{i=0}^{n} \left(r_i^{(b)} - h(\boldsymbol{x}_i) \right)^2. \tag{9.51}$$

6   Set $g_b(\boldsymbol{x}) \leftarrow g_{b-1}(\boldsymbol{x}) + \gamma\, h_b(\boldsymbol{x})$.

7 **return** g_B

■ **Example 9.53 (Gradient Boosting for a Regression Tree)** Let us continue with the basic bagging and random forest examples for a regression tree (Examples 9.37 and 9.40), where we compared the standard decision tree estimator with the corresponding bagging and random forest estimators. Now, we use the gradient boosting estimator from Algorithm 9.52, as implemented in `sklearn`. We use $\gamma = 0.1$ and perform $B = 100$ boosting rounds. As a prediction function h_b for $b = 1, \ldots, B$ we use small regression trees of depth at most 3. Note that such individual trees do not usually give good performance; that is, they are weak prediction functions. We can see that the resulting boosting prediction function gives the R^2 score equal to 0.899, which is better than R^2 scores of simple decision tree (0.5754), the bagged tree (0.761), and the random forest (0.8106).

GradientBoostingRegression.py

```
import numpy as np
from sklearn.datasets import make_friedman1
from sklearn.tree import DecisionTreeRegressor
from sklearn.model_selection import train_test_split
from sklearn.metrics import r2_score

# create regression problem
n_points = 1000 # points
x, y =  make_friedman1(n_samples=n_points, n_features=15,
                      noise=1.0, random_state=100)

# split to train/test set
x_train, x_test, y_train, y_test = \
        train_test_split(x, y, test_size=0.33, random_state=100)

# boosting sklearn
from sklearn.ensemble import GradientBoostingRegressor

breg = GradientBoostingRegressor(learning_rate=0.1,
            n_estimators=100, max_depth =3, random_state=100)
breg.fit(x_train,y_train)
yhat = breg.predict(x_test)
print("Gradient Boosting R^2 score = ",r2_score(y_test, yhat))
```

```
Gradient Boosting R^2 score =  0.8993055635639531
```

We proceed with the classification setting and consider the original boosting algorithm: **AdaBoost**. The inventors of the AdaBoost method considered a binary classification problem, where the response variable belongs to the $\{-1, 1\}$ set. The idea of AdaBoost is similar to the one presented in the regression setting, that is, AdaBoost fits a sequence of prediction functions:

$$g_0, g_1 = g_0 + h_1, g_2 = g_0 + h_1 + h_2, \ldots$$

with final prediction function

$$g_B(\boldsymbol{x}) = g_0(\boldsymbol{x}) + \sum_{b=1}^{B} h_b(\boldsymbol{x}), \tag{9.54}$$

where each function h_b is of the form $h_b(\boldsymbol{x}) = \alpha_b\, c_b(\boldsymbol{x})$, with $\alpha_b \in \mathbb{R}_+$ and where c_b is a proper (but weak) classifier in some class C. Thus, $c_b(\boldsymbol{x}) \in \{-1, 1\}$. Exactly as in (9.46), we solve at each boosting iteration the optimization problem

$$(\alpha_b, c_b) = \underset{\alpha \geq 0,\, c \in C}{\operatorname{argmin}} \frac{1}{n} \sum_{i=1}^{n} \operatorname{Loss}\left(y_i, g_{b-1}(\boldsymbol{x}_i) + \alpha\, c(\boldsymbol{x}_i)\right). \tag{9.55}$$

However, in this case the loss function is defined as $\operatorname{Loss}(y, \widehat{y}) = \mathrm{e}^{-y\,\widehat{y}}$. The algorithm starts with a simple model $g_0 := 0$ and for each successive iteration $b = 1, \ldots, B$ solves

(9.55). Thus,

$$(\alpha_b, c_b) = \underset{\alpha \geq 0,\, c \in C}{\operatorname{argmin}} \sum_{i=1}^{n} \underbrace{\mathrm{e}^{-y_i\, g_{b-1}(\boldsymbol{x}_i)}}_{w_i^{(b)}} \mathrm{e}^{-y_i\, \alpha\, c(\boldsymbol{x}_i)} = \underset{\alpha \geq 0,\, c \in C}{\operatorname{argmin}} \sum_{i=1}^{n} w_i^{(b)}\, \mathrm{e}^{-y_i\, \alpha\, c(\boldsymbol{x}_i)},$$

where $w_i^{(b)} := \exp\{-y_i\, g_{b-1}(\boldsymbol{x}_i)\}$ does not depend on α or c. It follows that

$$\begin{aligned}(\alpha_b, c_b) &= \underset{\alpha \geq 0,\, c \in C}{\operatorname{argmin}}\; \mathrm{e}^{-\alpha} \sum_{i=1}^{n} w_i^{(b)}\, \mathbb{1}\{c(\boldsymbol{x}_i) = y_i\} + \mathrm{e}^{\alpha} \sum_{i=1}^{n} w_i^{(b)}\, \mathbb{1}\{c(\boldsymbol{x}_i) \neq y_i\} \\ &= \underset{\alpha \geq 0,\, c \in C}{\operatorname{argmin}}\; (\mathrm{e}^{\alpha} - \mathrm{e}^{-\alpha})\, \ell_\tau^{(b)}(c) + \mathrm{e}^{-\alpha},\end{aligned} \tag{9.56}$$

where

$$\ell_\tau^{(b)}(c) := \frac{\sum_{i=1}^{n} w_i^{(b)}\, \mathbb{1}\{c(\boldsymbol{x}_i) \neq y_i\}}{\sum_{i=1}^{n} w_i^{(b)}}$$

can be interpreted as the weighted zero–one training loss at iteration b.

For any $\alpha \geq 0$, the program (9.56) is minimized by a classifier $c \in C$ that minimizes this weighted training loss; that is,

$$c_b(\boldsymbol{x}) = \underset{c \in C}{\operatorname{argmin}}\, \ell_\tau^{(b)}. \tag{9.57}$$

Substituting (9.57) into (9.56) and solving for the optimal α gives

$$\alpha_b = \frac{1}{2} \ln \left(\frac{1 - \ell_\tau^{(b)}(c_b)}{\ell_\tau^{(b)}(c_b)} \right). \tag{9.58}$$

This gives the AdaBoost algorithm, summarized below.

Algorithm 9.59: AdaBoost

Input: Training set $\tau = \{(\boldsymbol{x}_i, y_i)\}_{i=1}^{n}$, and the number of boosting rounds B.
Output: AdaBoost prediction function.

Set $g_0(\boldsymbol{x}) \leftarrow 0$.
for $i = 1$ **to** n **do**
  $w_i^{(1)} \leftarrow 1/n$
for $b = 1$ **to** B **do**
  Fit a classifier c_b on the training set τ by solving

$$c_b = \underset{c \in C}{\operatorname{argmin}}\, \ell_\tau^{(b)}(c) = \underset{c \in C}{\operatorname{argmin}} \frac{\sum_{i=1}^{n} w_i^{(b)}\, \mathbb{1}\{c(\boldsymbol{x}_i) \neq y_i\}}{\sum_{i=1}^{n} w_i^{(b)}}.$$

  Set $\alpha_b \leftarrow \frac{1}{2} \ln \left(\frac{1 - \ell_\tau^{(b)}(c_b)}{\ell_\tau^{(b)}(c_b)} \right)$. // Update weights
  for $i = 1$ **to** n **do**
    $w_i^{(b+1)} \leftarrow w_i^{(b)} \exp\{-y_i\, \alpha_b\, c_b(\boldsymbol{x}_i)\}$.
return $g_B(\boldsymbol{x}) := \sum_{b=1}^{B} \alpha_b\, c_b(\boldsymbol{x})$.

Algorithm 9.59 is quite intuitive. At the first step ($b = 1$), AdaBoost assigns an equal weight $w_i^{(1)} = 1/n$ to each training sample $(\boldsymbol{x}_i, y_i)$ in the set $\tau = \{(\boldsymbol{x}_i, y_i)\}_{i=1}^n$. Note that, in this case, the weighted zero–one training loss is equal to the regular zero–one training loss. At each successive step $b > 1$, the weights of observations that were incorrectly classified by the previous boosting prediction function g_b are increased, and the weights of correctly classified observations are decreased. Due to the use of the weighted zero–one loss, the set of incorrectly classified training samples will receive an extra weight and thus have a better chance of being classified correctly by the next classifier c_{b+1}. As soon as the AdaBoost algorithm finds the prediction function g_B, the final classification is delivered via

$$\operatorname{sign}\left(\sum_{b=1}^{B} \alpha_b \, c_b(\boldsymbol{x})\right).$$

The step-size parameter α_b found by the AdaBoost algorithm in Line 6 can be viewed as an optimal step-size in the sense of training loss minimization. However, similar to the regression setting, one can slow down the AdaBoost algorithm by setting α_b to be a fixed (small) value $\alpha_b = \gamma$. As usual, when the latter is done in practice, it is tackling the problem of overfitting.

We consider an implementation of Algorithm 9.59 for a binary classification problem. Specifically, during all boosting rounds, we use simple decision trees of depth 1 (also called decision tree **stumps**) as weak learners. The exponential and zero–one training losses as a function of the number of boosting rounds are presented in Figure 9.60.

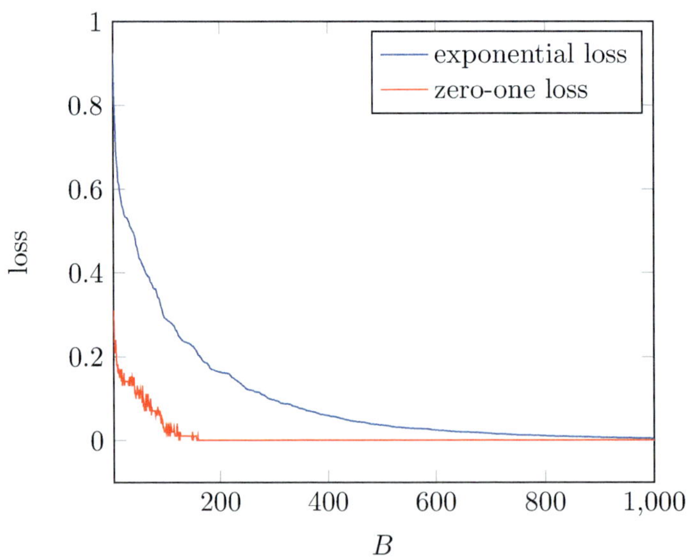

Figure 9.60: Exponential and zero–one training loss as a function of the number of boosting rounds B for a binary classification problem.

AdaBoost.py

```
from sklearn.datasets import make_blobs
from sklearn.tree import DecisionTreeClassifier
from sklearn.model_selection import train_test_split
from sklearn.metrics import zero_one_loss
import numpy as np


def ExponentialLoss(y,yhat):
    n = len(y)
    loss = 0
    for i in range(n):
        loss = loss+np.exp(-y[i]*yhat[i])
    loss = loss/n
    return loss

# create binary classification problem
np.random.seed(100)

n_points = 100 # points
x, y =  make_blobs(n_samples=n_points, n_features=5,  centers=2,
                      cluster_std=20.0, random_state=100)
y[y==0]=-1

# AdaBoost implementation
BoostingRounds = 1000
n = len(x)
W = 1/n*np.ones(n)

Learner = []
alpha_b_arr = []

for i in range(BoostingRounds):
    clf = DecisionTreeClassifier(max_depth=1)
    clf.fit(x,y, sample_weight=W)

    Learner.append(clf)

    train_pred = clf.predict(x)
    err_b = 0
    for i in range(n):
        if(train_pred[i]!=y[i]):
            err_b = err_b+W[i]

    err_b = err_b/np.sum(W)
    alpha_b = 0.5*np.log((1-err_b)/err_b)
    alpha_b_arr.append(alpha_b)

    for i in range(n):
        W[i] = W[i]*np.exp(-y[i]*alpha_b*train_pred[i])

yhat_boost = np.zeros(len(y))

for j in range(BoostingRounds):
```

```
    yhat_boost = yhat_boost+alpha_b_arr[j]*Learner[j].predict(x)


yhat = np.zeros(n)
yhat[yhat_boost>=0] = 1
yhat[yhat_boost<0] = -1
print("AdaBoost Classifier exponential loss = ", ExponentialLoss(y,
    yhat_boost))
print("AdaBoost Classifier zero--one loss = ",zero_one_loss(y,yhat))
```

```
AdaBoost Classifier exponential loss =  0.004224013663777142
AdaBoost Classifier zero--one loss =  0.0
```

Further Reading

Breiman's book on decision trees, [32], serves as a great starting point. Some additional advances can be found in [89, 137]. From the computational point of view, there exists an efficient recursive procedure for tree pruning; see Chapters 3 and 10 in [32]. Several advantages and disadvantages of using decision trees are debated in [55, 81]. A detailed discussion on bagging and random forests can be found in [33] and [35], respectively. Freund and Schapire [66] provide the first boosting algorithm, the AdaBoost. While AdaBoost was developed in the context of the computational complexity of learning, it was later discovered by Friedman [67] that AdaBoost is a special case of an additive model. In addition, it was shown that for any differentiable loss function, there exists an efficient boosting procedure which mimics the gradient descent algorithm. The foundation of the resulting gradient boosting method is detailed in [67]. Python packages that implement gradient boosting include `XGBoost` and `LightGBM`.

Exercises

1. Show that any training set $\tau = \{(\boldsymbol{x}, y_i), i = 1, \ldots, n\}$ can be fitted via a tree with zero training loss.

2. Suppose during the construction of a decision tree we wish to specify a constant regional prediction function g^w on the region $\mathcal{R}_w$, based on the training data in $\mathcal{R}_w$, say $\{(\boldsymbol{x}_1, y_1), \ldots, (\boldsymbol{x}_k, y_k)\}$. Show that $g^w(\boldsymbol{x}) := k^{-1} \sum_{i=1}^{k} y_i$ minimizes the squared-error loss.

3. Using the program from Section 9.2.4, write a basic implementation of a decision tree for a binary classification problem. Implement the misclassification, Gini index, and entropy impurity criteria to split nodes. Compare the results.

4. Suppose in the decision tree of Example 9.1, there are 3 blue and 2 red data points in a certain tree region. Calculate the misclassification impurity, the Gini impurity, and the entropy impurity. Repeat these calculations for 2 blue and 3 red data points.

5. Consider the procedure of finding the best splitting rule for a categorical variable with k labels from Section 9.3.4. Show that one needs to consider 2^k subsets of $\{1, \ldots, k\}$ to find the optimal partition of labels.

6. Reproduce Figure 9.22 using the following classification data.

```
from sklearn.datasets import make_blobs
X, y =  make_blobs(n_samples=5000, n_features=10, centers=3,
                           random_state=10, cluster_std=10)
```

7. Prove (9.28); that is, show that

$$\sum_{w \in \mathcal{W}} \left(\sum_{i=1}^{n} \mathbb{1}\{\boldsymbol{x}_i \in \mathcal{R}_w\} \operatorname{Loss}(y_i, g^w(\boldsymbol{x}_i)) \right) = n\, \ell_\tau(g)\,.$$

8. Suppose τ is a training set with n elements and τ^*, also of size n, is obtained from τ by bootstrapping; that is, resampling with replacement. Show that for large n, τ^* does not contain a fraction of about $\mathrm{e}^{-1} \approx 0.37$ of the points from τ.

9. Prove Equation (9.38).

10. Consider the following training/test split of the data. Construct a random forest regressor and identify the optimal subset size m in the sense of R^2 score (see Remark 9.41).

```
import numpy as np
from sklearn.datasets import make_friedman1
from sklearn.tree import DecisionTreeRegressor
from sklearn.model_selection import train_test_split
from sklearn.metrics import r2_score

# create regression problem
n_points = 1000 # points
x, y =  make_friedman1(n_samples=n_points, n_features=15,
                       noise=1.0, random_state=100)

# split to train/test set
x_train, x_test, y_train, y_test = \
        train_test_split(x, y, test_size=0.33, random_state=100)
```

11. Explain why bagging decision trees are a special case of random forests.

12. Show that (9.58) holds.

13. Consider the following classification data and module imports:

```
from sklearn.datasets import make_blobs
from sklearn.metrics import zero_one_loss
from sklearn.model_selection import train_test_split
import numpy as np
```

```
import matplotlib.pyplot as plt
from sklearn.ensemble import GradientBoostingClassifier

X_train, y_train =  make_blobs(n_samples=5000, n_features=10, centers=3,
     random_state=10, cluster_std=5)
```

Using the gradient boosting algorithm with $B = 100$ rounds, plot the training loss as a function of γ, for $\gamma = 0.1, 0.3, 0.5, 0.7, 1$. What is your conclusion regarding the relation between B and γ?

CHAPTER 10

Deep Learning

In this chapter, we show how one can construct a rich class of approximating functions called neural networks. The learners belonging to the neural-network class of functions have attractive properties that have made them ubiquitous in modern machine learning applications — their training is computationally feasible and their complexity is easy to control and fine-tune.

10.1 Introduction

In Chapter 2 we described the basic supervised learning task; namely, we wish to predict a random output Y from a random input $\boldsymbol{X}$, using a prediction function $g : \boldsymbol{x} \mapsto y$ that belongs to a suitably chosen class of approximating functions $\mathcal{G}$. More generally, we may wish to predict a vector-valued output $\boldsymbol{y}$ using a prediction function $\boldsymbol{g} : \boldsymbol{x} \mapsto \boldsymbol{y}$ from class $\mathcal{G}$.

In this chapter $\boldsymbol{y}$ denotes the vector-valued output for a given input $\boldsymbol{x}$. This differs from our previous use (e.g., in Table 2.11), where $\boldsymbol{y}$ denotes a vector of scalar outputs.

In the machine learning context, the class $\mathcal{G}$ is sometimes referred to as the *hypothesis space* or the *universe of possible models*, and the **representational capacity** of a hypothesis space $\mathcal{G}$ is simply its complexity.

Suppose that we have a class of functions $\mathcal{G}_L$, indexed by a parameter L that controls the complexity of the class, so that $\mathcal{G}_L \subset \mathcal{G}_{L+1} \subset \mathcal{G}_{L+2} \subset \cdots$. In selecting a suitable class of functions, we have to be mindful of the *approximation–estimation tradeoff*. On ☞ 32 the one hand, the class $\mathcal{G}_L$ must be complex (rich) enough to accurately represent the optimal unknown prediction function g^*, which may require a very large L. On the other hand, the learners in the class $\mathcal{G}_L$ must be simple enough to train with small estimation error and with minimal demands on computer memory, which may necessitate a small L.

In balancing these competing objectives, it helps if the more complex class $\mathcal{G}_{L+1}$ is easily constructed from an already existing and simpler $\mathcal{G}_L$. The simpler class of functions $\mathcal{G}_L$ may itself be constructed by modifying an even simpler class $\mathcal{G}_{L-1}$, and so on.

DOI: 10.1201/9781003391173-10

A class of functions that permits such a natural hierarchical construction is the class of **neural networks**. Conceptually, a neural network with L layers is a nonlinear parametric
☞ 219 regression model whose representational capacity can easily be controlled by L.

Alternatively, in (10.7) we will define the output of a neural network as the repeated composition of linear and (componentwise) nonlinear functions. As we shall see, this representation of the output will provide a flexible class of nonlinear functions that can be easily differentiated. As a result, the training of learners via gradient optimization
☞ 603 methods involves mostly standard matrix operations that can be performed very efficiently.

Historically, neural networks were originally intended to mimic the workings of the human brain, with the network nodes modeling neurons and the network links modeling the axons connecting neurons. For this reason, rather than using the terminology of the regression models in Chapter 5, we prefer to use a nomenclature inspired by the apparent resemblance of neural networks to structures in the human brain.

We note, however, that the attempts at building efficient machine learning algorithms by mimicking the functioning of the human brain have been as unsuccessful as the attempts at building flying aircraft by mimicking the flapping of birds' wings. Instead, many effective machine algorithms have been inspired by age-old mathematical ideas for function approximation. One such idea is the following fundamental result (see [171] for a proof).

Theorem 10.1 (Kolmogorov (1957)): Every continuous function $g^* : [0,1]^p \mapsto \mathbb{R}$ with $p \geq 2$ can be written as

$$g^*(\boldsymbol{x}) = \sum_{j=1}^{2p+1} h_j \left(\sum_{i=1}^{p} h_{ij}(x_i) \right),$$

where $\{h_j, h_{ij}\}$ is a set of univariate continuous functions that depend on g^*.

This result tells us that any continuous high-dimensional map can be represented as the function composition of much simpler (one-dimensional) maps. The composition of the maps needed to compute the output $g^*(\boldsymbol{x})$ for a given input $\boldsymbol{x} \in \mathbb{R}^p$ are depicted in Figure 10.2, showing a directed graph or *neural network* with three layers, denoted as $l = 0, 1, 2$.

In particular, each of the p components of the input $\boldsymbol{x}$ is represented as a node in the *input layer* ($l = 0$). In the **hidden layer** ($l = 1$) there are $q := 2p + 1$ nodes, each of which is associated with a pair of variables (z, a) with values

$$z_j := \sum_{i=1}^{p} h_{ij}(x_i) \quad \text{and} \quad a_j := h_j(z_j).$$

A link between nodes (z_j, a_j) and x_i with weight h_{ij} signifies that the value of z_j depends on the value of x_i via the function h_{ij}. Finally, the *output layer* ($l = 2$) represents the value $g^*(\boldsymbol{x}) = \sum_{j=1}^{q} a_j$. Note that the arrows on the graph remind us that the sequence of the computations is executed from left to right, or from the input layer $l = 0$ through to the output layer $l = 2$.

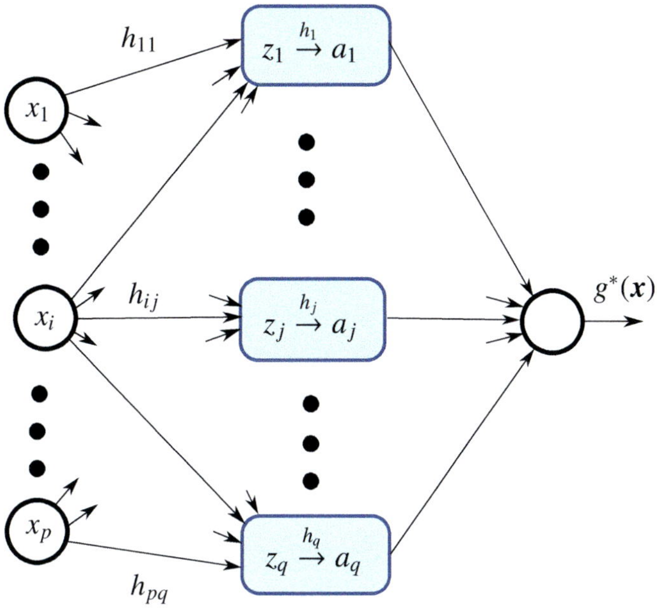

Figure 10.2: Every continuous function $g^* : [0,1]^p \mapsto \mathbb{R}$ can be represented by a neural network with one hidden layer ($l = 1$), an input layer ($l = 0$), and an output layer ($l = 2$).

In practice, we do not know the collection of functions $\{h_j, h_{ij}\}$, because they depend on the unknown g^*. In the unlikely event that g^* is linear, then all of the $(2p+1)(p+1)$ one-dimensional functions will be linear as well. However, in general, we should expect that each of the functions in $\{h_j, h_{ij}\}$ is nonlinear.

Unfortunately, Theorem 10.1 only asserts the existence of $\{h_j, h_{ij}\}$, and does not tell us how to construct these nonlinear functions. One way out of this predicament is to replace these $(2p+1)(p+1)$ unknown functions with a much larger number of known nonlinear functions called **activation functions.**[1] For example, a logistic activation function is

$$S(z) = (1 + \exp(-z))^{-1}.$$

We then hope that such a network, built from a sufficiently large number of activation functions, will have similar representational capacity as the neural network in Figure 10.2 with $(2p+1)(p+1)$ functions.

In general, we wish to use the simplest activation functions that will allow us to build a learner with large representational capacity and low training cost. The logistic function is merely one possible choice for an activation function from among infinite possibilities. Figure 10.3 shows a small selection of activation functions with different regularity or smoothness properties, including the popular **rectified linear unit** (ReLU).

In addition to choosing the type and number of activation functions in a neural network, we can improve its representational capacity in another important way: introduce more hidden layers. In the next section we explore this possibility in detail.

[1]Activation functions derive their name from models of a neuron's response when exposed to chemical or electric stimuli.

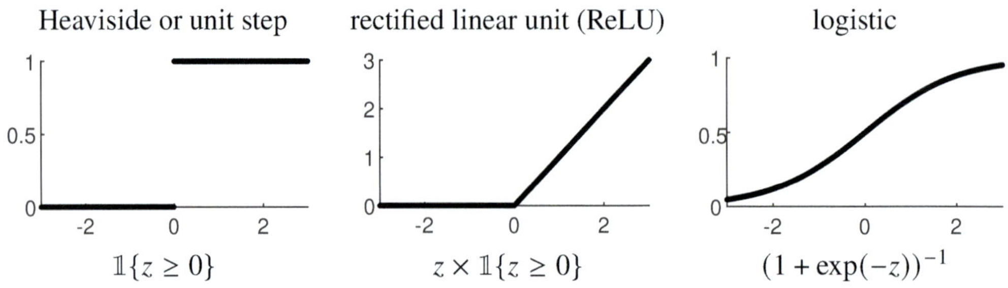

Figure 10.3: Some common activation functions $S(z)$ with their defining formulas and plots. The logistic function is an example of a *sigmoid* (that is, an S-shaped) function. Some books define the logistic function as $2S(z) - 1$ (in terms of our definition).

10.2 Feed-Forward Neural Networks

In a neural network with $L + 1$ layers, the zero or input layer ($l = 0$) encodes the input feature vector $\boldsymbol{x}$, and the last or output layer ($l = L$) encodes the (multivalued) output function $\boldsymbol{g}(\boldsymbol{x})$. The remaining layers are called *hidden layers*. Each layer has a number of nodes, say p_l nodes for layer $l = 0, \ldots, L$. In this notation, p_0 is the dimension of the input feature vector $\boldsymbol{x}$ and, for example, $p_L = 1$ signifies that $g(\boldsymbol{x})$ is a scalar output.

All nodes in the hidden layers ($l = 1, \ldots, L-1$) are associated with a pair of variables (z, a), which we gather into p_l-dimensional column vectors $\boldsymbol{z}_l$ and $\boldsymbol{a}_l$. In the so-called **feed-forward** networks, the variables in any layer l are simple functions of the variables in the preceding layer $l - 1$. In particular, $\boldsymbol{z}_l$ and $\boldsymbol{a}_{l-1}$ are related via the linear relation:

$$\boldsymbol{z}_l = \mathbf{W}_l \, \boldsymbol{a}_{l-1} + \boldsymbol{b}_l$$

for some **weight matrix** $\mathbf{W}_l$ and **bias vector** $\boldsymbol{b}_l$.

Within any hidden layer $l = 1, \ldots, L-1$, the components of the vectors $\boldsymbol{z}_l$ and $\boldsymbol{a}_l$ are related via $\boldsymbol{a}_l = \boldsymbol{S}_l(\boldsymbol{z}_l)$, where $\boldsymbol{S}_l : \mathbb{R}^{p_l} \mapsto \mathbb{R}^{p_l}$ is a nonlinear multivalued function. All of these multivalued functions are typically of the form

$$\boldsymbol{S}_l(\boldsymbol{z}) = [S(z_1), \ldots, S(z_{\dim(\boldsymbol{z})})]^\top, \quad l = 1, \ldots, L-1, \tag{10.4}$$

where S is an activation function common to all hidden layers.

The function $\boldsymbol{S}_L : \mathbb{R}^{p_{L-1}} \mapsto \mathbb{R}^{p_L}$ in the output layer is more general and its specification depends, for example, on whether the network is used for classification or for the prediction of a continuous output Y. A four-layer ($L = 3$) network is illustrated in Figure 10.5.

The output of this neural network is determined by the input vector $\boldsymbol{x}$, (nonlinear) functions $\{\boldsymbol{S}_l\}$, as well as weight matrices $\mathbf{W}_l = [w_{l,ij}]$ and bias vectors $\boldsymbol{b}_l = [b_{l,j}]$ for $l = 1, 2, 3$.

Here, the (i, j)-th element of the weight matrix $\mathbf{W}_l = [w_{l,ij}]$ is the weight that connects the j-th node in the $(l - 1)$-st layer with the i-th node in the l-th layer.

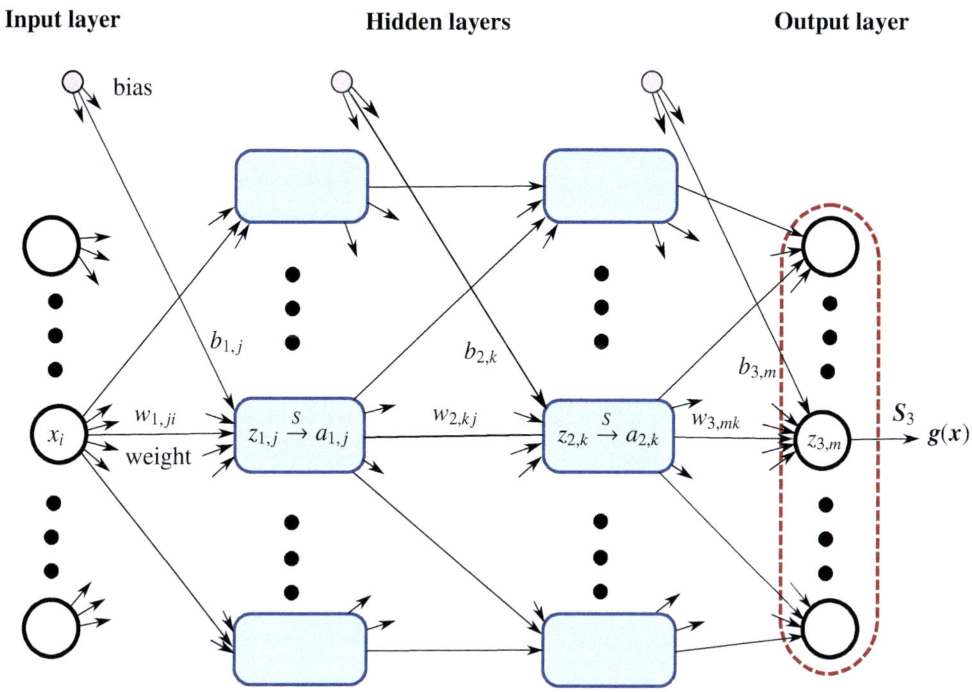

Figure 10.5: A neural network with $L = 3$: the $l = 0$ layer is the input layer, followed by two hidden layers, and the output layer. Hidden layers may have different numbers of nodes.

The name given to L (the number of layers without the input layer) is the **network depth** and $\max_l p_l$ is called the **network width**. While we mostly study networks that have an equal number of nodes in the hidden layers ($p_1 = \cdots = p_{L-1}$), in general there can be different numbers of nodes in each hidden layer.

The output $\boldsymbol{g}(\boldsymbol{x})$ of a multiple-layer neural network is obtained from the input $\boldsymbol{x}$ via the following sequence of computations:

$$\begin{aligned}(10.6)\qquad &\underbrace{\boldsymbol{x}}_{\boldsymbol{a}_0} \to \underbrace{\mathbf{W}_1\,\boldsymbol{a}_0 + \boldsymbol{b}_1}_{\boldsymbol{z}_1} \to \underbrace{\boldsymbol{S}_1(\boldsymbol{z}_1)}_{\boldsymbol{a}_1} \to \underbrace{\mathbf{W}_2\,\boldsymbol{a}_1 + \boldsymbol{b}_2}_{\boldsymbol{z}_2} \to \underbrace{\boldsymbol{S}_2(\boldsymbol{z}_2)}_{\boldsymbol{a}_2} \to \cdots \\ &\to \underbrace{\mathbf{W}_L\,\boldsymbol{a}_{L-1} + \boldsymbol{b}_L}_{\boldsymbol{z}_L} \to \underbrace{\boldsymbol{S}_L(\boldsymbol{z}_L)}_{\boldsymbol{a}_L} = \boldsymbol{g}(\boldsymbol{x}).\end{aligned}$$

Denoting the function $\boldsymbol{z} \mapsto \mathbf{W}_l\,\boldsymbol{z} + \boldsymbol{b}_l$ by $\boldsymbol{M}_l$, the output $\boldsymbol{g}(\boldsymbol{x})$ can thus be written as the function composition

$$(10.7)\qquad \boldsymbol{g}(\boldsymbol{x}) = \boldsymbol{S}_L \circ \boldsymbol{M}_L \circ \cdots \circ \boldsymbol{S}_2 \circ \boldsymbol{M}_2 \circ \boldsymbol{S}_1 \circ \boldsymbol{M}_1(\boldsymbol{x}).$$

The algorithm for computing the output $\boldsymbol{g}(\boldsymbol{x})$ for an input $\boldsymbol{x}$ is summarized next. Note that we leave open the possibility that the activation functions $\{\boldsymbol{S}_l\}$ have different definitions for each layer. In some cases, $\boldsymbol{S}_l$ may even depend on some or all of the already computed $\boldsymbol{z}_1, \boldsymbol{z}_2, \ldots$ and $\boldsymbol{a}_1, \boldsymbol{a}_2, \ldots$.

Algorithm 10.8: Feed-Forward Propagation for a Neural Network

Input: Feature vector $\boldsymbol{x}$; weights $\{w_{l,ij}\}$, biases $\{b_{l,i}\}$ for each layer $l = 1, \ldots, L$.

Output: The value of the prediction function $\boldsymbol{g}(\boldsymbol{x})$.

$\boldsymbol{a}_0 \leftarrow \boldsymbol{x}$ // the zero or input layer

for $l = 1$ **to** L **do**

  Compute the hidden variable $z_{l,i}$ for each node i in layer l:

$$\boldsymbol{z}_l \leftarrow \mathbf{W}_l\, \boldsymbol{a}_{l-1} + \boldsymbol{b}_l$$

  Compute the activation function $a_{l,i}$ for each node i in layer l:

$$\boldsymbol{a}_l \leftarrow \boldsymbol{S}_l(\boldsymbol{z}_l)$$

return $\boldsymbol{g}(\boldsymbol{x}) \leftarrow \boldsymbol{a}_L$ // the output layer

Example 10.9 (Nonlinear Multi-Output Regression) Given the input $\boldsymbol{x} \in \mathbb{R}^{p_0}$ and an activation function $S : \mathbb{R} \mapsto \mathbb{R}$, the output $\boldsymbol{g}(\boldsymbol{x}) := [g_1(\boldsymbol{x}), \ldots, g_{p_2}(\boldsymbol{x})]^\top$ of a *nonlinear* ☞ 255 *multi-output regression* model can be computed via a neural network with:

$$\begin{aligned}
\boldsymbol{z}_1 &= \mathbf{W}_1\, \boldsymbol{x} + \boldsymbol{b}_1, \quad \text{where } \mathbf{W}_1 \in \mathbb{R}^{p_1 \times p_0}, \boldsymbol{b}_1 \in \mathbb{R}^{p_1},\\
a_{1,k} &= S(z_{1,k}), \quad k = 1, \ldots, p_1,\\
\boldsymbol{g}(\boldsymbol{x}) &= \mathbf{W}_2\, \boldsymbol{a}_1 + \boldsymbol{b}_2, \quad \text{where } \mathbf{W}_2 \in \mathbb{R}^{p_2 \times p_1}, \boldsymbol{b}_2 \in \mathbb{R}^{p_2},
\end{aligned}$$

which is a neural network with one hidden layer and output function $\boldsymbol{S}_2(\boldsymbol{z}) = \boldsymbol{z}$. In the special case where $p_1 = p_2 = 1$, $\boldsymbol{b}_2 = 0$, $\mathbf{W}_2 = 1$, and we collect all parameters into the vector $\boldsymbol{\theta}^\top = [\boldsymbol{b}_1, \mathbf{W}_1] \in \mathbb{R}^{p_0+1}$, the neural network can be interpreted as a *generalized* ☞ 242 *linear model* with $\mathbb{E}[Y \mid \boldsymbol{X} = \boldsymbol{x}] = h([1, \boldsymbol{x}^\top]\, \boldsymbol{\theta})$ for some activation function h. ■

Example 10.10 (Multi-Logit Classification) Suppose that, for a classification problem, an input $\boldsymbol{x}$ has to be classified into one of c classes, labeled $0, \ldots, c-1$. We can perform the classification via a neural network with one hidden layer, with $p_1 = c$ nodes. In particular, we have

$$\boldsymbol{z}_1 = \mathbf{W}_1\, \boldsymbol{x} + \boldsymbol{b}_1, \quad \boldsymbol{a}_1 = \boldsymbol{S}_1(\boldsymbol{z}_1),$$

where $\boldsymbol{S}_1$ is the **softmax** function:

$$\text{softmax} : \boldsymbol{z} \mapsto \frac{\exp(\boldsymbol{z})}{\sum_k \exp(z_k)}.$$

For the output, we take $\boldsymbol{g}(\boldsymbol{x}) = [g_1(\boldsymbol{x}), \ldots, g_c(\boldsymbol{x})]^\top = \boldsymbol{a}_1$, which can then be used as a ☞ 354 *pre-classifier* of $\boldsymbol{x}$. The actual classifier of $\boldsymbol{x}$ into one of the categories $0, 1, \ldots, c-1$ is then

$$\underset{k \in \{0, \ldots, c-1\}}{\operatorname{argmax}} \; g_{k+1}(\boldsymbol{x}).$$

☞ 369 This is equivalent to the multi-logit classifier in Section 8.5. Note, however, that there we used a slightly different notation, with $\widetilde{\boldsymbol{x}}$ instead of $\boldsymbol{x}$ and we have a reference class; see Exercise 12. ■

In practical implementations, the softmax function can cause numerical over- and under-flow errors when either one of the $\exp(z_k)$ happens to be extremely large or $\sum_k \exp(z_k)$ happens to be very small. In such cases we can exploit the invariance property (Exercise 1):

$$\text{softmax}(\boldsymbol{z}) = \text{softmax}(\boldsymbol{z} + c \times \mathbf{1}) \quad \text{for any constant } c.$$

Using this property, we can compute $\text{softmax}(\boldsymbol{z})$ with greater numerical stability via $\text{softmax}(\boldsymbol{z} - \max_k\{z_k\} \times \mathbf{1})$.

When neural networks are used for classification into c classes and the number of output nodes is $c - 1$, then the $g_i(\boldsymbol{x})$ may be viewed as nonlinear *discriminant functions*. ☞ 363

Example 10.11 (Density Estimation) Estimating the density f of some random feature $X \in \mathbb{R}$ is the prototypical unsupervised learning task, which we tackled in Section 4.6.2 using Gaussian mixture models. We can view a Gaussian mixture model with ☞ 163 p_1 components and a common scale parameter $\sigma > 0$ as a neural network with two hidden layers, similar to the one on Figure 10.5. In particular, if the activation function in the first hidden layer, $\boldsymbol{S}_1$, is of the form (10.4) with $S(z) := \exp(-z^2/(2\sigma^2))/\sqrt{2\pi\sigma^2}$, then the density value $g(x)$ is computed via:

$$\begin{aligned}
\boldsymbol{z}_1 &= \mathbf{W}_1 x + \boldsymbol{b}_1, & \boldsymbol{a}_1 &= \boldsymbol{S}_1(\boldsymbol{z}_1),\\
\boldsymbol{z}_2 &= \mathbf{W}_2 \boldsymbol{a}_1 + \boldsymbol{b}_2, & \boldsymbol{a}_2 &= \boldsymbol{S}_2(\boldsymbol{z}_2),\\
g(x) &= \boldsymbol{a}_1^\top \boldsymbol{a}_2,
\end{aligned}$$

where $\mathbf{W}_1 = \mathbf{1}$ is a $p_1 \times 1$ column vector of ones, $\mathbf{W}_2 = \mathbf{O}$ is a $p_1 \times p_1$ matrix of zeros, and $\boldsymbol{S}_2$ is the softmax function. We identify the column vector $\boldsymbol{b}_1$ with the p_1 location parameters, $[\mu_1, \ldots, \mu_{p_1}]^\top$ of the Gaussian mixture and $\boldsymbol{b}_2 \in \mathbb{R}^{p_1}$ with the p_1 weights of the mixture. Note the unusual activation function of the output layer — it requires the value of $\boldsymbol{a}_1$ from the first hidden layer and $\boldsymbol{a}_2$ from the second hidden layer. ■

There are a number of key design characteristics of a feed-forward network. First, we need to choose the activation function(s). Second, we need to choose the loss function for the training of the network. As we shall explain in the next section, the most common choices are the ReLU activation function and the cross-entropy loss. Crucially, we need to carefully construct the **network architecture** — the number of connections among the nodes in different layers and the overall number of layers of the network.

For example, if the connections from one layer to the next are pruned (called *sparse connectivity*) and the links share the same weight values $\{w_{l,ij}\}$ (called *parameter sharing*) for all $\{(i, j) : |i - j| = 0, 1, \ldots\}$, then the weight matrices will be sparse and *Toeplitz*. ☞ 549

Intuitively, the parameter sharing and sparse connectivity can speed up the training of the network, because there are fewer parameters to learn, and the Toeplitz structure permits quick computation of the matrix-vector products in Algorithm 10.8. An important example of such a network is the **convolution neural network** (CNN), in which some or all of the network layers encode the linear operation of *convolution*:

$$\mathbf{W}_l \, \boldsymbol{a}_{l-1} = \boldsymbol{w}_l \star \boldsymbol{a}_{l-1},$$

☞ 550 where $[\boldsymbol{x} \star \boldsymbol{y}]_i := \sum_k x_k y_{i-k+1}$. As discussed in Example A.84, a convolution matrix is a special type of sparse Toeplitz matrix, and its action on a vector of learning parameters ☞ 576 can be evaluated quickly via the *fast Fourier transform*.

CNNs are particularly suited to image processing problems, because their *convolution layers* closely mimic the neurological properties of the visual cortex. In particular, the cortex partitions the visual field into many small regions and assigns a group of neurons to every such region. Moreover, some of these groups of neurons respond only to the presence of particular features (for example, edges).

This neurological property is naturally modeled via convolution layers in the neural network. Specifically, suppose that the input image is given by an $m_1 \times m_2$ matrix of pixels. Now, define a $k \times k$ matrix (sometimes called a *kernel*, where k is generally taken to be 3 or 5). Then, the convolution layer output can be calculated using the discrete convolution of all possible $k \times k$ input matrix regions and the kernel matrix; (see Example A.84). In particular, by noting that there are $(m_1 - k + 1) \times (m_2 - k + 1)$ possible regions in the original image, we conclude that the convolution layer output size is $(m_1 - k + 1) \times (m_2 - k + 1)$. In practice, we frequently define several kernel matrices, giving an output layer of size $(m_1 - k + 1) \times (m_2 - k + 1) \times$ (the number of kernels). Figure 10.12 shows a 5×5 input image and a 2×2 kernel with a 4×4 output matrix. An example of using a CNN for image classification is given in Section 10.5.2.

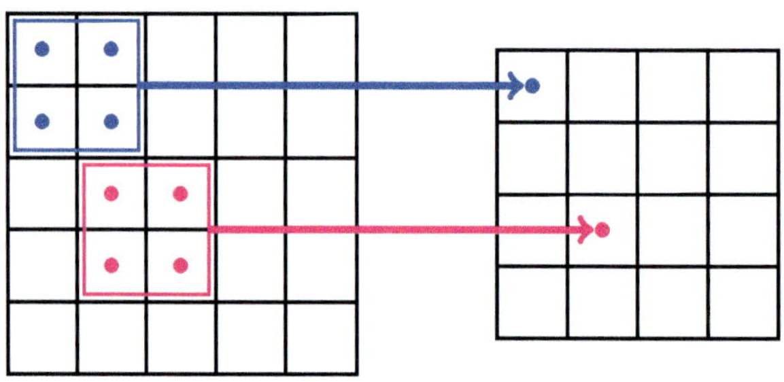

Figure 10.12: An example 5×5 input image and a 2×2 kernel. The kernel is applied to every 2×2 region of the original image.

10.3 Back-Propagation

The training of neural networks is a major challenge that requires both ingenuity and much experimentation. The algorithms for training neural networks with great depth are collectively referred to as **deep learning** methods. One of the simplest and most effective ☞ 603 methods for training is via *gradient descent* and its variations.

Gradient descent requires computation of the gradient with respect to all bias vectors and weight matrices. Given the potentially large number of parameters (weight and bias terms) in a neural network, we need to find an efficient method to calculate this gradient.

To illustrate the nature of the gradient computations, let $\boldsymbol{\theta} = \{\mathbf{W}_l, \boldsymbol{b}_l\}$ be a column vector of length $\dim(\boldsymbol{\theta}) = \sum_{l=1}^{L}(p_{l-1}p_l + p_l)$ that collects all the weight parameters (numbering $\sum_{l=1}^{L} p_{l-1}p_l$) and bias parameters (numbering $\sum_{l=1}^{L} p_l$) of a multiple-layer network with training loss:

$$\ell_\tau(\boldsymbol{g}(\cdot \mid \boldsymbol{\theta})) := \frac{1}{n}\sum_{i=1}^{n} \text{Loss}(\boldsymbol{y}_i, \boldsymbol{g}(\boldsymbol{x}_i \mid \boldsymbol{\theta})).$$

Writing $C_i(\boldsymbol{\theta}) := \text{Loss}(\boldsymbol{y}_i, \boldsymbol{g}(\boldsymbol{x}_i \mid \boldsymbol{\theta}))$ for short (using C for cost), we have

$$\ell_\tau(\boldsymbol{g}(\cdot \mid \boldsymbol{\theta})) = \frac{1}{n}\sum_{i=1}^{n} C_i(\boldsymbol{\theta}), \tag{10.13}$$

so that obtaining the gradient of ℓ_τ requires computation of $\partial C_i/\partial \boldsymbol{\theta}$ for every i. For activation functions of the form (10.4), define $\mathbf{D}_l$ as the diagonal matrix with the vector of derivatives

$$\boldsymbol{S}'(\boldsymbol{z}) := [S'(z_{l,1}), \ldots, S'(z_{l,p_l})]^\top$$

down its main diagonal; that is,

$$\mathbf{D}_l := \text{diag}(S'(z_{l,1}), \ldots, S'(z_{l,p_l})), \quad l = 1, \ldots, L-1.$$

The following theorem provides us with the formulas needed to compute the gradient of a typical $C_i(\boldsymbol{\theta})$.

Theorem 10.14 (Gradient of Training Loss): For a given (input, output) pair $(\boldsymbol{x}, \boldsymbol{y})$, let $\boldsymbol{g}(\boldsymbol{x} \mid \boldsymbol{\theta})$ be the output of Algorithm 10.8, and let $C(\boldsymbol{\theta}) = \text{Loss}(\boldsymbol{y}, \boldsymbol{g}(\boldsymbol{x} \mid \boldsymbol{\theta}))$ be an almost-everywhere differentiable loss function. Suppose $\{\boldsymbol{z}_l, \boldsymbol{a}_l\}_{l=1}^{L}$ are the vectors obtained during the feed-forward propagation ($\boldsymbol{a}_0 = \boldsymbol{x}, \boldsymbol{a}_L = \boldsymbol{g}(\boldsymbol{x} \mid \boldsymbol{\theta})$). Then, we have for $l = 1, \ldots, L$:

$$\frac{\partial C}{\partial \mathbf{W}_l} = \boldsymbol{\delta}_l\, \boldsymbol{a}_{l-1}^\top \quad \text{and} \quad \frac{\partial C}{\partial \boldsymbol{b}_l} = \boldsymbol{\delta}_l,$$

where $\boldsymbol{\delta}_l := \partial C/\partial \boldsymbol{z}_l$ is computed recursively for $l = L, \ldots, 2$:

$$\boldsymbol{\delta}_{l-1} = \mathbf{D}_{l-1}\mathbf{W}_l^\top\, \boldsymbol{\delta}_l \quad \text{with} \quad \boldsymbol{\delta}_L = \frac{\partial \boldsymbol{S}_L}{\partial \boldsymbol{z}_L}\frac{\partial C}{\partial \boldsymbol{g}}. \tag{10.15}$$

Proof. The scalar value C is obtained from the transitions (10.6), followed by the mapping $\boldsymbol{g}(\boldsymbol{x} \mid \boldsymbol{\theta}) \mapsto \text{Loss}(\boldsymbol{y}, \boldsymbol{g}(\boldsymbol{x} \mid \boldsymbol{\theta}))$. Using the chain rule (see Appendix C.1.1), we have ☞ 584

$$\boldsymbol{\delta}_L = \frac{\partial C}{\partial \boldsymbol{z}_L} = \frac{\partial \boldsymbol{g}(\boldsymbol{x})}{\partial \boldsymbol{z}_L}\frac{\partial C}{\partial \boldsymbol{g}(\boldsymbol{x})} = \frac{\partial \boldsymbol{S}_L}{\partial \boldsymbol{z}_L}\frac{\partial C}{\partial \boldsymbol{g}}.$$

Recall that the vector/vector derivative of a linear mapping $\boldsymbol{z} \mapsto \mathbf{W}\boldsymbol{z}$ is given by $\mathbf{W}^\top$; see (C.9). It follows that, since $\boldsymbol{z}_l = \mathbf{W}_l\, \boldsymbol{a}_{l-1} + \boldsymbol{b}_l$ and $\boldsymbol{a}_l = \boldsymbol{S}(\boldsymbol{z}_l)$, the chain rule gives ☞ 583

$$\frac{\partial \boldsymbol{z}_l}{\partial \boldsymbol{z}_{l-1}} = \frac{\partial \boldsymbol{a}_{l-1}}{\partial \boldsymbol{z}_{l-1}}\frac{\partial \boldsymbol{z}_l}{\partial \boldsymbol{a}_{l-1}} = \mathbf{D}_{l-1}\mathbf{W}_l^\top.$$

Hence, the recursive formula (10.15):

$$\delta_{l-1} = \frac{\partial C}{\partial z_{l-1}} = \frac{\partial z_l}{\partial z_{l-1}} \frac{\partial C}{\partial z_l} = \mathbf{D}_{l-1} \mathbf{W}_l^\top \delta_l, \quad l = L, \ldots, 3, 2.$$

Using the $\{\delta_l\}$, we can now compute the derivatives with respect to the weight matrices and the biases. In particular, applying the "scalar/matrix" differentiation rule (C.14) to $z_l = \mathbf{W}_l \, a_{l-1} + b_l$ gives:

$$\frac{\partial C}{\partial \mathbf{W}_l} = \frac{\partial C}{\partial z_l} \frac{\partial z_l}{\partial \mathbf{W}_l} = \delta_l \, a_{l-1}^\top, \quad l = 1, \ldots, L$$

and

$$\frac{\partial C}{\partial b_l} = \frac{\partial z_l}{\partial b_l} \frac{\partial C}{\partial z_l} = \delta_l, \quad l = 1, \ldots, L.$$

□

From the theorem we can see that for each pair (x, y) in the training set, we can compute the gradient $\partial C / \partial \theta$ in a sequential manner, by computing $\delta_L, \ldots, \delta_1$. This procedure is called **back-propagation**. Since back-propagation mostly involves simple matrix multiplication, it can be efficiently implemented using dedicated computing hardware such as graphical processor units (GPUs) and other parallel computing architecture. Note also that many matrix computations that run in quadratic time can be replaced with linear-time componentwise multiplication. Specifically, multiplication of a vector with a diagonal matrix is equivalent to componentwise multiplication:

$$\underbrace{\mathbf{A}}_{\text{diag}(a)} \; b = a \odot b.$$

Consequently, we can write $\delta_{l-1} = \mathbf{D}_{l-1} \mathbf{W}_l^\top \delta_l$ as: $\delta_{l-1} = S'(z_{l-1}) \odot \mathbf{W}_l^\top \delta_l$, $l = L, \ldots, 3, 2$.

We summarize the back-propagation algorithm for the computation of a typical $\partial C / \partial \theta$ in the Algorithm10.16. Lines 1 to 5 are the feed-forward part of the algorithm, and Lines 7 to 10 are the back-propagation part of the algorithm.

Note that for the gradient of $C(\theta)$ to exist at every point, we need the activation functions to be differentiable everywhere. This is the case, for example, for the logistic activation function in Figure 10.3. It is not the case for the ReLU function, which is differentiable everywhere, except at $z = 0$. However, in practice, the kink of the ReLU function at $z = 0$ is unlikely to trip the back-propagation algorithm, because rounding errors and the finite-precision computer arithmetic make it extremely unlikely that we will need to evaluate the ReLU at precisely $z = 0$. This is the reason why in Theorem 10.14 we merely required that $C(\theta)$ is almost-everywhere differentiable.

In spite of its kink at the origin, the ReLU has an important advantage over the logistic function. While the derivative of the logistic function decays exponentially fast to zero as we move away from the origin, a phenomenon referred to as **saturation**, the derivative of the ReLU function is always unity for positive z. Thus, for large positive z, the derivative

Algorithm 10.16: Computing the Gradient of a Typical $C(\boldsymbol{\theta})$

Input: Training example $(\boldsymbol{x}, \boldsymbol{y})$, weight matrices and bias vectors $\{\mathbf{W}_l, \boldsymbol{b}_l\}_{l=1}^L =: \boldsymbol{\theta}$, activation functions $\{\boldsymbol{S}_l\}_{l=1}^L$.

Output: The derivatives with respect to all weight matrices and bias vectors.

1 $\boldsymbol{a}_0 \leftarrow \boldsymbol{x}$

2 **for** $l = 1, \ldots, L$ **do** // feed-forward

3 $\quad \boldsymbol{z}_l \leftarrow \mathbf{W}_l\, \boldsymbol{a}_{l-1} + \boldsymbol{b}_l$

4 $\quad \boldsymbol{a}_l \leftarrow \boldsymbol{S}_l(\boldsymbol{z}_l)$

5 $\boldsymbol{\delta}_L \leftarrow \frac{\partial \boldsymbol{S}_L}{\partial \boldsymbol{z}_L} \frac{\partial C}{\partial \boldsymbol{g}}$

6 $\boldsymbol{z}_0 \leftarrow \boldsymbol{0}$ // arbitrary assignment needed to finish the loop

7 **for** $l = L, \ldots, 1$ **do** // back-propagation

8 $\quad \frac{\partial C}{\partial \boldsymbol{b}_l} \leftarrow \boldsymbol{\delta}_l$

9 $\quad \frac{\partial C}{\partial \mathbf{W}_l} \leftarrow \boldsymbol{\delta}_l\, \boldsymbol{a}_{l-1}^\top$

10 $\quad \boldsymbol{\delta}_{l-1} \leftarrow \boldsymbol{S}'(\boldsymbol{z}_{l-1}) \odot \mathbf{W}_l^\top \boldsymbol{\delta}_l$

11 **return** $\frac{\partial C}{\partial \mathbf{W}_l}$ and $\frac{\partial C}{\partial \boldsymbol{b}_l}$ for all $l = 1, \ldots, L$ and the value $\boldsymbol{g}(\boldsymbol{x}) \leftarrow \boldsymbol{a}_L$ (if needed)

of the logistic function does not carry any useful information, but the derivative of the ReLU can help guide a gradient optimization algorithm. The situation for the Heaviside function in Figure 10.3 is even worse, because its derivative is completely noninformative for any $z \neq 0$. In this respect, the lack of saturation of the ReLU function for $z > 0$ makes it a desirable activation function for training a network via back-propagation.

Finally, note that to obtain the gradient $\partial \ell_\tau / \partial \boldsymbol{\theta}$ of the training loss, we simply need to loop Algorithm 10.16 over all the n training examples, as follows:

Algorithm 10.17: Computing the Gradient of the Training Loss

Input: Training set $\tau = \{(\boldsymbol{x}_i, \boldsymbol{y}_i)\}_{i=1}^n$, weight matrices and bias vectors $\{\mathbf{W}_l, \boldsymbol{b}_l\}_{l=1}^L =: \boldsymbol{\theta}$, activation functions $\{\boldsymbol{S}_l\}_{l=1}^L$.

Output: The gradient of the training loss.

for $i = 1, \ldots, n$ **do** // loop over all training examples

$\quad$ Run Algorithm 10.16 with input $(\boldsymbol{x}_i, \boldsymbol{y}_i)$ to compute $\left\{\frac{\partial C_i}{\partial \mathbf{W}_l}, \frac{\partial C_i}{\partial \boldsymbol{b}_l}\right\}_{l=1}^L$

return $\frac{\partial C}{\partial \mathbf{W}_l} = \frac{1}{n}\sum_{i=1}^n \frac{\partial C_i}{\partial \mathbf{W}_l}$ and $\frac{\partial C}{\partial \boldsymbol{b}_l} = \frac{1}{n}\sum_{i=1}^n \frac{\partial C_i}{\partial \boldsymbol{b}_l}$ for all $l = 1, \ldots, L$

■ **Example 10.18 (Squared-Error and Cross-Entropy Loss)** The back-propagation Algorithm 10.16 requires a formula for $\boldsymbol{\delta}_L$ in line 5. In particular, to execute line 5 we need to specify both a loss function and an $\boldsymbol{S}_L$ that defines the output layer: $\boldsymbol{g}(\boldsymbol{x} \,|\, \boldsymbol{\theta}) = \boldsymbol{a}_L = \boldsymbol{S}_L(\boldsymbol{z}_L)$.

For instance, in the *multi-logit* classification of inputs $\boldsymbol{x}$ into p_L categories labeled $0, 1, \ldots, (p_L - 1)$, the output layer is defined via the softmax function: ☞ 369

$$\boldsymbol{S}_L : \boldsymbol{z}_L \mapsto \frac{\exp(\boldsymbol{z}_L)}{\sum_{k=1}^{p_L} \exp(z_{L,k})}.$$

In other words, $\boldsymbol{g}(\boldsymbol{x} \,|\, \boldsymbol{\theta})$ is a probability vector such that its $(y + 1)$-st component

$g_{y+1}(\boldsymbol{x} \mid \boldsymbol{\theta}) = g(y \mid \boldsymbol{\theta}, \boldsymbol{x})$ is the estimate or prediction of the true conditional probability $f(y \mid \boldsymbol{x})$. Combining the softmax output with the *cross-entropy* loss, as was done in ☞ 371 (8.33), yields:

$$\begin{aligned}\mathrm{Loss}(f(y \mid \boldsymbol{x}), g(y \mid \boldsymbol{\theta}, \boldsymbol{x})) &= -\ln g(y \mid \boldsymbol{\theta}, \boldsymbol{x}) \\ &= -\ln g_{y+1}(\boldsymbol{x} \mid \boldsymbol{\theta}) \\ &= -z_{y+1} + \ln \textstyle\sum_{k=1}^{p_L} \exp(z_k).\end{aligned}$$

Hence, we obtain the vector $\boldsymbol{\delta}_L$ with components $(k = 1, \ldots, p_L)$

$$\delta_{L,k} = \frac{\partial}{\partial z_k}\left(-z_{y+1} + \ln \textstyle\sum_{k=1}^{p_L} \exp(z_k)\right) = g_k(\boldsymbol{x} \mid \boldsymbol{\theta}) - \mathbb{1}\{y = k-1\}.$$

Note that we can remove a node from the final layer of the multi-logit network, because $g_1(\boldsymbol{x} \mid \boldsymbol{\theta})$ (which corresponds to the $y = 0$ class) can be eliminated, using the fact that $g_1(\boldsymbol{x} \mid \boldsymbol{\theta}) = 1 - \sum_{k=2}^{p_L} g_k(\boldsymbol{x} \mid \boldsymbol{\theta})$. For a numerical comparison, see Exercise 12.

As another example, in nonlinear multi-output regression (see Example 10.9), the output function $\boldsymbol{S}_L$ is often of the form (10.4), so that $\partial \boldsymbol{S}_L / \partial \boldsymbol{z} = \mathrm{diag}(S_L'(z_1), \ldots, S_L'(z_{p_L}))$. Combining the output $\boldsymbol{g}(\boldsymbol{x} \mid \boldsymbol{\theta}) = \boldsymbol{S}_L(\boldsymbol{z}_L)$ with the squared-error loss yields:

$$\mathrm{Loss}(\boldsymbol{y}, \boldsymbol{g}(\boldsymbol{x} \mid \boldsymbol{\theta})) = \|\boldsymbol{y} - \boldsymbol{g}(\boldsymbol{x} \mid \boldsymbol{\theta})\|^2 = \sum_{j=1}^{p_L} (y_j - g_j(\boldsymbol{x} \mid \boldsymbol{\theta}))^2.$$

Hence, line 5 in Algorithm 10.16 simplifies to:

$$\boldsymbol{\delta}_L = \frac{\partial \boldsymbol{S}_L}{\partial \boldsymbol{z}} \frac{\partial C}{\partial \boldsymbol{g}} = \boldsymbol{S}_L'(\boldsymbol{z}_L) \odot 2(\boldsymbol{g}(\boldsymbol{x} \mid \boldsymbol{\theta}) - \boldsymbol{y}).$$

■

10.4 Methods for Training

Neural networks have been studied for a long time, yet it is only recently that there have been sufficient computational resources to train them effectively. The training of neural networks requires minimization of a training loss, $\ell_\tau(\boldsymbol{g}(\cdot \mid \boldsymbol{\theta})) = \frac{1}{n}\sum_{i=1}^{n} C_i(\boldsymbol{\theta})$, which is typically a difficult high-dimensional optimization problem with multiple local minima. We next consider a number of simple training methods.

> In this section, the vectors $\boldsymbol{\delta}_t$ and $\boldsymbol{g}_t$ use the notation of Section C.3.2 and should not be confused with the derivative $\boldsymbol{\delta}$ and the prediction function $\boldsymbol{g}$, respectively.

10.4.1 Gradient Descent

If we can compute the gradient of $\ell_\tau(\boldsymbol{g}(\cdot \mid \boldsymbol{\theta}))$ via back-propagation, then we can apply ☞ 603 the gradient descent algorithm, which reads as follows: Starting from a guess $\boldsymbol{\theta}_1$, we iterate the following step until convergence:

$$\boldsymbol{\theta}_{t+1} = \boldsymbol{\theta}_t - \alpha_t \, \boldsymbol{u}_t, \qquad t = 1, 2, \ldots, \tag{10.19}$$

where $\boldsymbol{u}_t := \frac{\partial \ell_\tau}{\partial \boldsymbol{\theta}}(\boldsymbol{\theta}_t)$ and α_t is the **learning rate**.

Observe that, rather than operating directly on the weights and biases, we operate instead on $\boldsymbol{\theta} := \{\mathbf{W}_l, \boldsymbol{b}_l\}_{l=1}^L$ — a column vector of length $\sum_{l=1}^L (p_{l-1}p_l + p_l)$ that stores all the weight and bias parameters. The advantage of organizing the computations in this way is that we can easily compute the learning rate α_t; for example, via the *Barzilai–Borwein* formula in (C.62). ☞ 603

Algorithm 10.20: Training via Gradient Descent

Input: Training set $\tau = \{(\boldsymbol{x}_i, \boldsymbol{y}_i)\}_{i=1}^n$, initial weight matrices and bias vectors $\{\mathbf{W}_l, \boldsymbol{b}_l\}_{l=1}^L =: \boldsymbol{\theta}_1$, activation functions $\{S_l\}_{l=1}^L$.

Output: The parameters of the trained learner.

```
t ← 1, δ ← 0.1 × 1, u_{t-1} ← 0, α ← 0.1                 // initialization
while stopping condition is not met do
    compute the gradient u_t = ∂ℓ_τ/∂θ (θ_t) using Algorithm 10.17
    g ← u_t − u_{t-1}
    if δ^T g > 0 then                        // check if Hessian is positive-definite
        α ← δ^T g / ||g||^2                                     // Barzilai-Borwein
    else
        α ← 2 × α             // failing positivity, do something heuristic
    δ ← −α u_t
    θ_{t+1} ← θ_t + δ
    t ← t + 1
return θ_t as the minimizer of the training loss
```

Typically, we initialize the algorithm with small random values for $\boldsymbol{\theta}_1$, while being careful to avoid saturating the activation function. For example, in the case of the ReLU activation function, we will use small positive values to ensure that its derivative is not zero. A zero derivative of the activation function prevents the propagation of information useful for computing a good search direction.

Recall that computation of the gradient of the training loss via Algorithm 10.17 requires averaging over all training examples. When the size n of the training set τ_n is too large, computation of the gradient $\partial \ell_{\tau_n} / \partial \boldsymbol{\theta}$ via Algorithm 10.17 may be too costly. In such cases, we may employ the **stochastic gradient descent** algorithm. In this algorithm, we view the training loss as an expectation that can be approximated via Monte Carlo sampling. In particular, if K is a random variable with distribution $\mathbb{P}[K = k] = 1/n$ for $k = 1, \ldots, n$, then we can write

$$\ell_\tau(\boldsymbol{g}(\cdot \,|\, \boldsymbol{\theta})) = \frac{1}{n} \sum_{k=1}^n \mathrm{Loss}(\boldsymbol{y}_k, \boldsymbol{g}(\boldsymbol{x}_k \,|\, \boldsymbol{\theta})) = \mathbb{E}\, \mathrm{Loss}(\boldsymbol{y}_K, \boldsymbol{g}(\boldsymbol{x}_K \,|\, \boldsymbol{\theta})).$$

We can thus approximate $\ell_\tau(\boldsymbol{g}(\cdot \,|\, \boldsymbol{\theta}))$ via a Monte Carlo estimator using N iid copies of K:

$$\widehat{\ell_\tau}(\boldsymbol{g}(\cdot \,|\, \boldsymbol{\theta})) := \frac{1}{N} \sum_{i=1}^N \mathrm{Loss}(\boldsymbol{y}_{K_i}, \boldsymbol{g}(\boldsymbol{x}_{K_i} \,|\, \boldsymbol{\theta})).$$

The iid Monte Carlo sample $K_1, \ldots, K_N$ is called a **minibatch** (see also Exercise 3).

Typically, $n \gg N$ so that the probability of observing ties in a minibatch of size N is negligible.

Finally, note that if the learning rate of the stochastic gradient descent algorithm satisfies the conditions in (3.86), then the stochastic gradient descent algorithm is simply a version of the *stochastic approximation* Algorithm 3.85. ☞ 118

10.4.2 Levenberg–Marquardt Method

Since a neural network with squared-error loss is a special type of nonlinear regression model, it is possible to train it using classical nonlinear least-squares minimization methods, such as the *Levenberg–Marquardt* algorithm. ☞ 605

For simplicity of notation, suppose that the output of the net for an input $\boldsymbol{x}$ is a *scalar* $g(\boldsymbol{x})$. For a given input parameter $\boldsymbol{\theta}$ of dimension $d = \dim(\boldsymbol{\theta})$, the Levenberg–Marquardt Algorithm C.65 requires computation of the following *vector* of outputs:

$$\boldsymbol{g}(\tau \,|\, \boldsymbol{\theta}) := [g(\boldsymbol{x}_1 \,|\, \boldsymbol{\theta}), \ldots, g(\boldsymbol{x}_n \,|\, \boldsymbol{\theta})]^\top,$$

as well as the $n \times d$ matrix of Jacobi, $\mathbf{G}$, of $\boldsymbol{g}$ at $\boldsymbol{\theta}$. To compute these quantities, we can again use the back-propagation Algorithm 10.16, as follows:

Algorithm 10.21: Output for Training via Levenberg–Marquardt

Input: Training set $\tau = \{(\boldsymbol{x}_i, y_i)\}_{i=1}^n$, parameter $\boldsymbol{\theta}$.
Output: Vector $\boldsymbol{g}(\tau \,|\, \boldsymbol{\theta})$ and matrix of Jacobi $\mathbf{G}$ for use in Algorithm C.65.

```
for i = 1, ..., n do                                   // loop over all training examples
    Run Algorithm 10.16 with input (x_i, y_i) (using ∂C/∂g = 1 in line 5) to compute
    g(x_i | θ) and ∂g(x_i | θ)/∂θ.
g(τ | θ) ← [g(x_1 | θ), ..., g(x_n | θ)]^T
G ← [∂g(x_1 | θ)/∂θ, ..., ∂g(x_n | θ)/∂θ]^T
return g(τ | θ) and G
```

The Levenberg–Marquardt algorithm is not suitable for networks with a large number of parameters, because the cost of the matrix computations becomes prohibitive. For instance, obtaining the Levenberg–Marquardt search direction in (C.64) usually incurs an $O(d^3)$ cost. In addition, the Levenberg–Marquardt algorithm is applicable only when we wish to train the network using the squared-error loss. Both of these shortcomings are mitigated to an extent with the quasi-Newton or adaptive gradient methods described next.

10.4.3 Limited-Memory BFGS Method

All the methods discussed so far have been *first-order* optimization methods, that is, methods that only use the gradient vector $\boldsymbol{u}_t := \frac{\partial \ell_\tau}{\partial \boldsymbol{\theta}}(\boldsymbol{\theta}_t)$ at the current (and/or immediate past) candidate solution $\boldsymbol{\theta}_t$. In trying to design a more efficient *second-order* optimization method, we may be tempted to use Newton's method with a search direction: ☞ 600

$$-\mathbf{H}_t^{-1} \boldsymbol{u}_t,$$

where $\mathbf{H}_t$ is the $d \times d$ matrix of *second-order* partial derivatives of $\ell_{\mathcal{T}}(\boldsymbol{g}(\cdot \mid \boldsymbol{\theta}))$ at $\boldsymbol{\theta}_t$.

There are two problems with this approach. First, while the computation of $\boldsymbol{u}_t$ via Algorithm 10.17 typically costs $O(d)$, the computation of $\mathbf{H}_t$ costs $O(d^2)$. Second, even if we have somehow computed $\mathbf{H}_t$ very fast, computing the search direction $\mathbf{H}_t^{-1}\,\boldsymbol{u}_t$ still incurs an $O(d^3)$ cost. Both of these considerations make Newton's method impractical for large d.

Instead, a practical alternative is to use a *quasi-Newton method*, in which we directly aim to approximate $\mathbf{H}_t^{-1}$ via a matrix $\mathbf{C}_t$ that satisfies the *secant condition*: ☞ 601

$$\mathbf{C}_t\, \boldsymbol{g}_t = \boldsymbol{\delta}_t,$$

where $\boldsymbol{\delta}_t := \boldsymbol{\theta}_t - \boldsymbol{\theta}_{t-1}$ and $\boldsymbol{g}_t := \boldsymbol{u}_t - \boldsymbol{u}_{t-1}$.

An ingenious formula that generates a suitable sequence of approximating matrices $\{\mathbf{C}_t\}$ (each satisfying the secant condition) is the BFGS updating formula (C.57), which can be written as the recursion (see Exercise 8):

$$\mathbf{C}_t = \left(\mathbf{I} - \upsilon_t\, \boldsymbol{g}_t \boldsymbol{\delta}_t^\top\right)^\top \mathbf{C}_{t-1} \left(\mathbf{I} - \upsilon_t\, \boldsymbol{g}_t \boldsymbol{\delta}_t^\top\right) + \upsilon_t\, \boldsymbol{\delta}_t \boldsymbol{\delta}_t^\top, \quad \upsilon_t := (\boldsymbol{g}_t^\top \boldsymbol{\delta}_t)^{-1}. \tag{10.22}$$

This formula allows us to update $\mathbf{C}_{t-1}$ to $\mathbf{C}_t$ and then compute $\mathbf{C}_t\, \boldsymbol{u}_t$ in $O(d^2)$ time. While this quasi-Newton approach is better than the $O(d^3)$ cost of Newton's method, it may be still too costly in large-scale applications.

Instead, an approximate or *limited memory BFGS* updating can be achieved in $O(d)$ time. The idea is to store a few of the most recent pairs $\{\boldsymbol{\delta}_t, \boldsymbol{g}_t\}$ in order to evaluate its action on a vector $\boldsymbol{u}_t$ without explicitly constructing and storing $\mathbf{C}_t$ in computer memory. This is possible, because updating $\mathbf{C}_0$ to $\mathbf{C}_1$ in (10.22) requires only the pair $\boldsymbol{\delta}_1, \boldsymbol{g}_1$, and similarly computing $\mathbf{C}_t$ from $\mathbf{C}_0$ only requires the history of the updates $\boldsymbol{\delta}_1, \boldsymbol{g}_1 \ldots, \boldsymbol{\delta}_t, \boldsymbol{g}_t$, which can be shown as follows:

Define the matrices $\mathbf{A}_t, \ldots, \mathbf{A}_0$ via the backward recursion ($j = 1, \ldots, t$):

$$\mathbf{A}_t := \mathbf{I}, \quad \mathbf{A}_{j-1} := \left(\mathbf{I} - \upsilon_j\, \boldsymbol{g}_j \boldsymbol{\delta}_j^\top\right) \mathbf{A}_j,$$

and observe that all matrix vector products: $\mathbf{A}_j\, \boldsymbol{u} =: \boldsymbol{q}_j$, for $j = 0, \ldots, t$ can be computed efficiently via the backward recursion starting with $\boldsymbol{q}_t = \boldsymbol{u}$:

$$\tau_j := \boldsymbol{\delta}_j^\top \boldsymbol{q}_j, \quad \boldsymbol{q}_{j-1} = \boldsymbol{q}_j - \upsilon_j \tau_j\, \boldsymbol{g}_j, \quad j = t, t-1, \ldots, 1. \tag{10.23}$$

In addition to $\{\boldsymbol{q}_j\}$, we will make use of the vectors $\{\boldsymbol{r}_j\}$ defined via the recursion:

$$\boldsymbol{r}_0 := \mathbf{C}_0\, \boldsymbol{q}_0, \quad \boldsymbol{r}_j = \boldsymbol{r}_{j-1} + \upsilon_j \left(\tau_j - \boldsymbol{g}_j^\top \boldsymbol{r}_{j-1}\right) \boldsymbol{\delta}_j, \quad j = 1, \ldots, t. \tag{10.24}$$

At the final iteration t, the BFGS updating formula (10.22) can be rewritten in the form:

$$\mathbf{C}_t = \mathbf{A}_{t-1}^\top \mathbf{C}_{t-1} \mathbf{A}_{t-1} + \upsilon_t\, \boldsymbol{\delta}_t \boldsymbol{\delta}_t^\top.$$

By iterating the recursion (10.22) backwards to $\mathbf{C}_0$, we can write:

$$\mathbf{C}_t = \mathbf{A}_0^\top \mathbf{C}_0 \mathbf{A}_0 + \sum_{j=1}^{t} \upsilon_j\, \mathbf{A}_j^\top \boldsymbol{\delta}_j \boldsymbol{\delta}_j^\top \mathbf{A}_j,$$

that is, we can express $\mathbf{C}_t$ in terms of the initial $\mathbf{C}_0$ and the entire history of all BFGS values $\{\boldsymbol{\delta}_j, \boldsymbol{g}_j\}$, as claimed. Further, with the $\{\boldsymbol{q}_j, \boldsymbol{r}_j\}$ computed via (10.23) and (10.24), we can write:

$$\begin{aligned}
\mathbf{C}_t\, \boldsymbol{u} &= \mathbf{A}_0^\top \mathbf{C}_0\, \boldsymbol{q}_0 + \sum_{j=1}^{t} \upsilon_j \left(\boldsymbol{\delta}_j^\top \boldsymbol{q}_j\right) \mathbf{A}_j^\top \boldsymbol{\delta}_j \\
&= \mathbf{A}_0^\top \boldsymbol{r}_0 + \upsilon_1 \tau_1 \mathbf{A}_1^\top \boldsymbol{\delta}_1 + \sum_{j=2}^{t} \upsilon_j \tau_j \mathbf{A}_j^\top \boldsymbol{\delta}_j \\
&= \mathbf{A}_1^\top \left[\left(\mathbf{I} - \upsilon_1 \boldsymbol{\delta}_1 \boldsymbol{g}_1^\top\right) \boldsymbol{r}_0 + \upsilon_1 \tau_1 \boldsymbol{\delta}_1\right] + \sum_{j=2}^{t} \upsilon_j \tau_j \mathbf{A}_j^\top \boldsymbol{\delta}_j .
\end{aligned}$$

Hence, from the definition of the $\{\boldsymbol{r}_j\}$ in (10.24), we obtain

$$\begin{aligned}
\mathbf{C}_t \boldsymbol{u} &= \mathbf{A}_1^\top \boldsymbol{r}_1 + \sum_{j=2}^{t} \upsilon_j \tau_j \mathbf{A}_j^\top \boldsymbol{\delta}_j \\
&= \mathbf{A}_2^\top \boldsymbol{r}_2 + \sum_{j=3}^{t} \upsilon_j \tau_j \mathbf{A}_j^\top \boldsymbol{\delta}_j \\
&= \cdots = \mathbf{A}_t^\top \boldsymbol{r}_t + 0 = \boldsymbol{r}_t .
\end{aligned}$$

Given $\mathbf{C}_0$ and the history of all recent BFGS values $\{\boldsymbol{\delta}_j, \boldsymbol{g}_j\}_{j=1}^{h}$, the computation of the quasi-Newton search direction $\boldsymbol{d} = -\mathbf{C}_h\, \boldsymbol{u}$ can be accomplished via the recursions (10.23) and (10.24) as summarized in Algorithm 10.25.

Note that if $\mathbf{C}_0$ is a diagonal matrix, say the identity matrix, then $\mathbf{C}_0\, \boldsymbol{q}$ is cheap to compute and the cost of running Algorithm 10.25 is $\mathcal{O}(h\, d)$. Thus, for a fixed length of the BFGS history, the cost of the limited-memory BFGS updating grows linearly in d, making it a viable optimization algorithm in large-scale applications.

Algorithm 10.25: Limited-Memory BFGS Update

Input: BFGS history list $\{\boldsymbol{\delta}_j, \boldsymbol{g}_j\}_{j=1}^{h}$, initial $\mathbf{C}_0$, and input $\boldsymbol{u}$.

Output: $\boldsymbol{d} = -\mathbf{C}_h\, \boldsymbol{u}$, where $\mathbf{C}_t = \left(\mathbf{I} - \upsilon_t\, \boldsymbol{\delta}_t \boldsymbol{g}_t^\top\right) \mathbf{C}_{t-1} \left(\mathbf{I} - \upsilon_t\, \boldsymbol{g}_t \boldsymbol{\delta}_t^\top\right) + \upsilon_t\, \boldsymbol{\delta}_t \boldsymbol{\delta}_t^\top$.

```
q ← u
for i = h, h − 1, ..., 1 do        // backward recursion to compute A₀ u
    υ_i ← (δ_iᵀ g_i)^(−1)
    τ_i ← δ_iᵀ q
    q ← q − υ_i τ_i g_i
q ← C₀ q                           // compute C₀(A₀ u)
for i = 1, ..., h do               // compute recursion (10.24)
    q ← q + υ_i (τ_i − g_iᵀ q) δ_i
return d ← −q, the value of −C_h u
```

In summary, a quasi-Newton algorithm with limited-memory BFGS updating reads as follows:

Algorithm 10.26: Quasi-Newton Minimization with Limited-Memory BFGS

Input: Training set $\tau = \{(\boldsymbol{x}_i, \boldsymbol{y}_i)\}_{i=1}^n$, initial weight matrices and bias vectors $\{\mathbf{W}_l, \boldsymbol{b}_l\}_{l=1}^L =: \boldsymbol{\theta}_1$, activation functions $\{S_l\}_{l=1}^L$, and history parameter h.
Output: The parameters of the trained learner.

$t \leftarrow 1$, $\boldsymbol{\delta} \leftarrow 0.1 \times \mathbf{1}$, $\boldsymbol{u}_{t-1} \leftarrow \mathbf{0}$ // initialization
while stopping condition is not met **do**
  Compute $\ell_{\text{value}} = \ell_\tau(g(\cdot \mid \boldsymbol{\theta}_t))$ and $\boldsymbol{u}_t = \frac{\partial \ell_\tau}{\partial \boldsymbol{\theta}}(\boldsymbol{\theta}_t)$ via Algorithm 10.17.
  $\boldsymbol{g} \leftarrow \boldsymbol{u}_t - \boldsymbol{u}_{t-1}$
  Add $(\boldsymbol{\delta}, \boldsymbol{g})$ to the BFGS history as the newest BFGS pair.
  if the number of pairs in the BFGS history is greater than h **then**
    remove the oldest pair from the BFGS history
  Compute $\boldsymbol{d}$ via Algorithm 10.25 using the BFGS history, $\mathbf{C}_0 = \mathbf{I}$, and $\boldsymbol{u}_t$.
  $\alpha \leftarrow 1$
  while $\ell_\tau(g(\cdot \mid \boldsymbol{\theta}_t + \alpha\, \boldsymbol{d})) \geq \ell_{\text{value}} + 10^{-4}\alpha\, \boldsymbol{d}^\top \boldsymbol{u}_t$ **do**
    $\alpha \leftarrow \alpha/1.5$ // line-search along quasi-Newton direction
  $\boldsymbol{\delta} \leftarrow \alpha\, \boldsymbol{d}$
  $\boldsymbol{\theta}_{t+1} \leftarrow \boldsymbol{\theta}_t + \boldsymbol{\delta}$
  $t \leftarrow t + 1$
return $\boldsymbol{\theta}_t$ as the minimizer of the training loss

10.4.4 Adaptive Gradient Methods

Recall that the limited-memory BFGS method in the previous section determines a search direction using the recent history of previously computed gradients $\{\boldsymbol{u}_t\}$ and input parameters $\{\boldsymbol{\theta}_t\}$. This is because the BFGS pairs $\{\boldsymbol{\delta}_t, \boldsymbol{g}_t\}$ can be easily constructed from the identities: $\boldsymbol{\delta}_t = \boldsymbol{\theta}_t - \boldsymbol{\theta}_{t-1}$ and $\boldsymbol{g}_t = \boldsymbol{u}_t - \boldsymbol{u}_{t-1}$. In other words, using only past gradient computations and with little extra computation, it is possible to infer some of the second-order information contained in the Hessian matrix of $\ell_\tau(\boldsymbol{\theta})$. In addition to the BFGS method, there are other ways in which we can exploit the history of past gradient computations.

One approach is to use the *normal approximation method*, in which the Hessian of ℓ_τ at $\boldsymbol{\theta}_t$ is approximated via ☞ 604

$$\widehat{\mathbf{H}}_t = \gamma\, \mathbf{I} + \frac{1}{h} \sum_{i=t-h+1}^{t} \boldsymbol{u}_i \boldsymbol{u}_i^\top, \tag{10.27}$$

where $\boldsymbol{u}_{t-h+1}, \ldots, \boldsymbol{u}_t$ are the h most recently computed gradients and γ is a tuning parameter (for example, $\gamma = 1/h$). The search direction is then given by

$$-\widehat{\mathbf{H}}_t^{-1} \boldsymbol{u}_t,$$

which can be computed quickly in $O(h^2\, d)$ time either using the QR decomposition (Exercises 4 and 5), or the Sherman–Morrison Algorithm A.65. This approach requires that we store the last h gradient vectors in memory. ☞ 543

Another approach that completely bypasses the need to invert a Hessian approximation is the *Adaptive Gradient* or **AdaGrad** method, in which we only store the diagonal of $\widehat{\mathbf{H}}_t$ and use the search direction:

$$-\mathrm{diag}(\widehat{\mathbf{H}}_t)^{-1/2}\boldsymbol{u}_t.$$

We can avoid storing any of the gradient history by instead using the slightly different search direction[2]

$$-\boldsymbol{u}_t \Big/ \sqrt{\boldsymbol{v}_t + \gamma \times \mathbf{1}},$$

where the vector $\boldsymbol{v}_t$ is updated recursively via

$$\boldsymbol{v}_t = \left(1 - \frac{1}{h}\right)\boldsymbol{v}_{t-1} + \frac{1}{h}\boldsymbol{u}_t \odot \boldsymbol{u}_t.$$

With this updating of $\boldsymbol{v}_t$, the difference between the vector $\boldsymbol{v}_t + \gamma \times \mathbf{1}$ and the diagonal of the Hessian $\widehat{\mathbf{H}}_t$ will be negligible.

A more sophisticated version of AdaGrad is the *adaptive moment estimation* or **Adam** method, in which we not only average the vectors $\{\boldsymbol{v}_t\}$, but also average the gradient vectors $\{\boldsymbol{u}_t\}$, as follows:

Algorithm 10.28: Updating of Search Direction at Iteration t via Adam

Input: $\boldsymbol{u}_t, \widehat{\boldsymbol{u}}_{t-1}, \boldsymbol{v}_{t-1}, \boldsymbol{\theta}_t$, and parameters (α, h_v, h_u), equal to, e.g., $(10^{-3}, 10^{3}, 10)$.

Output: $\widehat{\boldsymbol{u}}_t, \boldsymbol{v}_t, \boldsymbol{\theta}_{t+1}$.

$\widehat{\boldsymbol{u}}_t \leftarrow \left(1 - \frac{1}{h_u}\right)\widehat{\boldsymbol{u}}_{t-1} + \frac{1}{h_u}\boldsymbol{u}_t$

$\boldsymbol{v}_t \leftarrow \left(1 - \frac{1}{h_v}\right)\boldsymbol{v}_{t-1} + \frac{1}{h_v}\boldsymbol{u}_t \odot \boldsymbol{u}_t$

$\boldsymbol{u}_t^* \leftarrow \widehat{\boldsymbol{u}}_t \Big/ \left(1 - (1 - h_u^{-1})^t\right)$

$\boldsymbol{v}_t^* \leftarrow \boldsymbol{v}_t \Big/ \left(1 - (1 - h_v^{-1})^t\right)$

$\boldsymbol{\theta}_{t+1} \leftarrow \boldsymbol{\theta}_t - \alpha\, \boldsymbol{u}_t^* \Big/ \left(\sqrt{\boldsymbol{v}_t^*} + 10^{-8} \times \mathbf{1}\right)$

return $\widehat{\boldsymbol{u}}_t, \boldsymbol{v}_t, \boldsymbol{\theta}_{t+1}$

Yet another computationally cheap approach is the **momentum method**, in which the gradient descent iteration (10.19) is modified to

$$\boldsymbol{\theta}_{t+1} = \boldsymbol{\theta}_t - \alpha_t\, \boldsymbol{u}_t + \gamma\, \boldsymbol{\delta}_t,$$

where $\boldsymbol{\delta}_t = \boldsymbol{\theta}_t - \boldsymbol{\theta}_{t-1}$ and γ is a tuning parameter. This strategy frequently performs better than the "vanilla" gradient descent method, because the search direction is less likely to change abruptly.

Numerical experience suggests that the vanilla gradient-descent Algorithm 10.20 and the Levenberg–Marquardt Algorithm C.65 are effective for networks with shallow architectures, but not for networks with deep architectures. In comparison, the stochastic gradient descent method, the limited-memory BFGS Algorithm 10.26, or any of the adaptive gradient methods in this section, can frequently handle networks with many hidden layers (provided that any tuning parameters and initialization values are carefully chosen via experimentation).

[2]Here we divide two vectors componentwise.

10.5 Examples in Python

In this section we provide two numerical examples in Python. In the first example, we train a neural network with the *stochastic gradient descent* method using the polynomial regression data from Example 2.13, and without using any specialized Python packages. ☞ 26

In the second example, we consider a realistic application of a neural network to image recognition and classification. Here we use the specialized open-source Python package **Pytorch**.

10.5.1 Simple Polynomial Regression

Consider again the polynomial regression data set depicted in Figure 2.14. We use a network with architecture

$$[p_0, p_1, p_2, p_3] = [1, 20, 20, 1].$$

In other words, we have two hidden layers with 20 neurons, resulting in a learner with a total of $\dim(\boldsymbol{\theta}) = 481$ parameters. To implement such a neural network, we first import the **numpy** and the **matplotlib** packages, then read the regression problem data and define the feed-forward neural network layers.

NeuralNetPurePython.py

```
import numpy as np
import matplotlib.pyplot as plt

# import data
data = np.genfromtxt('polyreg.csv',delimiter=',')
X = data[:,0].reshape(-1,1)
y = data[:,1].reshape(-1,1)

# Network setup
p = [X.shape[1], 20, 20, 1] # size of layers
L = len(p) - 1              # number of layers
```

Next, the **initialize** method generates random initial weight matrices and bias vectors $\{\mathbf{W}_l, \boldsymbol{b}_l\}_{l=1}^{L}$. Specifically, all parameters are initialized with values distributed according to the standard normal distribution.

```
def initialize(p, w_sig=1):
    W = np.zeros(len(p), dtype=object)
    b = np.zeros(len(p), dtype=object)
    for l in range(1, len(p)):
        W[l] = w_sig * np.random.randn(p[l], p[l-1])
        b[l] = w_sig * np.random.randn(p[l], 1)
    return W, b

W, b = initialize(p) # initialize weight matrices and bias vectors
```

The following code implements the ReLU activation function from Figure 10.3 and the squared error loss. Note that these functions return both the function values and the corresponding gradients.

```
def RELU(z, l):  # RELU activation function: value and derivative
    if l == L:
        return z, np.ones_like(z)
    else:
        val = np.maximum(0, z)
        J = np.array(z > 0, dtype=float)
        return val, J

def loss_fn(y,g):
    return (g - y)**2, 2 * (g - y)

S = RELU
```

Next, we implement the feed-forward and backward-propagation Algorithm 10.16. Here, we have implemented Algorithm 10.17 inside the backward-propagation loop.

```
def feedforward(x,W,b):
    a, z, gr_S = [0]*(L+1), [0]*(L+1), [0]*(L+1)
    a[0] = x.reshape(-1,1)
    for l in range(1,L+1):
        z[l] = W[l] @ a[l-1] + b[l] # affine transformation
        a[l], gr_S[l] = S(z[l], l) # activation function
    return a, z, gr_S

def backward(W,b,X,y):
    n =len(y)
    delta = [0]*(L+1)
    dC_db, dC_dW = [0]*(L+1), [0]*(L+1)
    loss=0
    for i in range(n): # loop over training examples
        a, z, gr_S = feedforward(X[i,:].T, W, b)
        cost, gr_C = loss_fn(y[i], a[L]) # cost i and gradient wrt g
        loss += cost/n
        delta[L] = gr_S[L] @ gr_C
        for l in range(L,0,-1): # l = L,...,1
            dCi_dbl = delta[l]
            dCi_dWl = delta[l] @  a[l-1].T

            # ---- sum up over samples ----
            dC_db[l] = dC_db[l] + dCi_dbl/n
            dC_dW[l] = dC_dW[l] + dCi_dWl/n
            # -----------------------------

            delta[l-1] =  gr_S[l-1] * W[l].T @ delta[l]
    return dC_dW, dC_db, loss
```

As explained in Section 10.4, it is sometimes more convenient to collect all the weight matrices and bias vectors $\{\mathbf{W}_l, \boldsymbol{b}_l\}_{l=1}^{L}$ into a single vector $\boldsymbol{\theta}$. Consequently, we code two functions that map the weight matrices and the bias vectors into a single parameter vector, and vice versa.

```
def list2vec(W,b):
    # converts list of weight matrices and bias vectors into
    # one column vector
    b_stack = np.vstack([b[i] for i in range(1, len(b))])
    W_stack = np.vstack([W[i].flatten().reshape(-1, 1) for i in range(1,
         len(W))])
    vec = np.vstack([b_stack, W_stack])
    return vec


def vec2list(vec, p):
    # converts vector to weight matrices and bias vectors
    W, b = [[]]*len(p),[[]]*len(p)
    p_count = 0

    for l in range(1,len(p)): # construct bias vectors
        b[l] = vec[p_count:(p_count+p[l])].reshape(-1,1)
        p_count = p_count + p[l]

    for l in range(1,len(p)): # construct weight matrices
        W[l] = vec[p_count:(p_count + p[l]*p[l-1])].reshape(p[l], p[l
            -1])
        p_count = p_count + (p[l]*p[l-1])
    return W, b
```

Finally, we run the *stochastic gradient descent* for 10^4 iterations using a minibatch of size 20 and a constant learning rate of $\alpha_t = 0.005$.

```
batch_size = 20
lr = 0.005
beta = list2vec(W,b)
loss_arr = []
n = len(X)
num_epochs = 10000
print("epoch | batch loss")
print("----------------------------")
for epoch in range(1,num_epochs+1):
    batch_idx = np.random.choice(n,batch_size)
    batch_X = X[batch_idx].reshape(-1,1)
    batch_y=y[batch_idx].reshape(-1,1)
    dC_dW, dC_db, loss = backward(W,b,batch_X,batch_y)
    d_beta = list2vec(dC_dW,dC_db)
    loss_arr.append(loss.flatten()[0])
    if(epoch==1 or np.mod(epoch,1000)==0):
        print(epoch,": ",loss.flatten()[0])
    beta = beta - lr*d_beta
    W,b = vec2list(beta,p)

# calculate the loss of the entire training set
dC_dW, dC_db, loss = backward(W,b,X,y)
print("entire training set loss = ",loss.flatten()[0])
xx = np.arange(0,1,0.01)
y_preds = np.zeros_like(xx)
```

```
for i in range(len(xx)):
    a, _, _ = feedforward(xx[i],W,b)
    y_preds[i],  = a[L]

plt.plot(X,y, 'r.', markersize = 4,label = 'y')
plt.plot(np.array(xx), y_preds, 'b',label = 'fit')
plt.legend()
plt.xlabel('x')
plt.ylabel('y')
plt.show()
plt.plot(np.array(loss_arr), 'b')
plt.xlabel('iteration')
plt.ylabel('Training Loss')
plt.show()
```

```
epoch | batch loss
----------------------------
1 :  158.6779278688539
1000 :  54.52430507401445
2000 :  38.346572088604965
3000 :  31.02036319180713
4000 :  22.91114276931535
5000 :  27.75810262906341
6000 :  22.296907007032928
7000 :  17.337367420038046
8000 :  19.233689945334195
9000 :  39.54261478969857
10000 :  14.754724387604416
entire training set loss =  28.904957963612727
```

The left panel of Figure 10.29 shows a trained neural network with a training loss of approximately 28.9. As seen from the right panel of Figure 10.29, the algorithm initially makes rapid progress until it settles down into a stationary regime after 400 iterations.

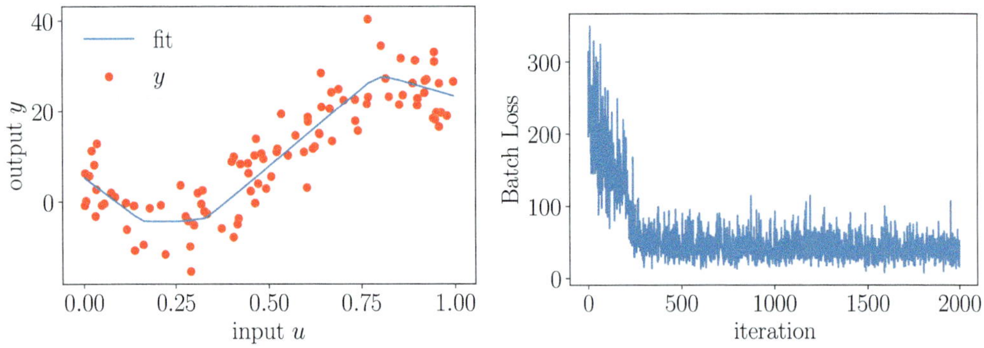

Figure 10.29: Left panel: The fitted neural network with training loss of $\ell_\tau(g_\tau) \approx 28.9$. Right panel: The evolution of the estimated loss, $\widehat{\ell_\tau}(g_\tau(\cdot \mid \boldsymbol{\theta}))$, over the gradient-descent iterations.

10.5.2 Image Classification

In this section, we will use the package `Pytorch`, which is an open-source machine learning library for Python. `Pytorch` can easily exploit any graphics processing unit

(GPU) for accelerated computation. As an example, we consider the Fashion-MNIST data set from https://www.kaggle.com/zalando-research/fashionmnist. The Fashion-MNIST data set contains 28×28 gray-scale images of clothing. Our task is to classify each image according to its label. Specifically, the labels are: T-Shirt, Trouser, Pullover, Dress, Coat, Sandal, Shirt, Sneaker, Bag, and Ankle Boot. Figure 10.30 depicts a typical ankle boot in the left panel and a typical dress in the right panel. To start with, we import the required libraries and load the Fashion-MNIST data set.

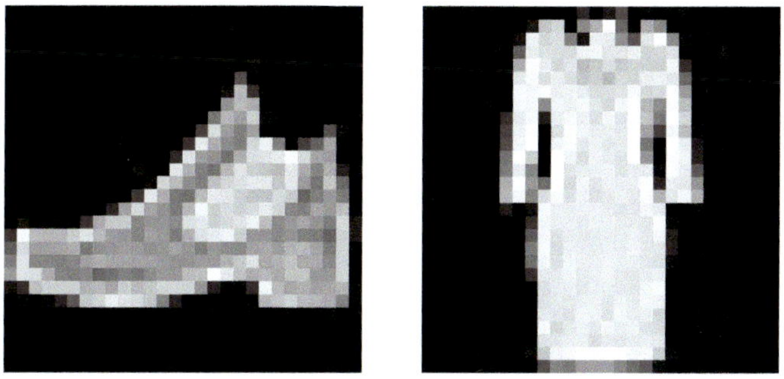

Figure 10.30: Left: an ankle boot. Right: a dress.

ImageClassificationPytorch.py

```
import torch
import torch.nn as nn
from torch.autograd import Variable
import pandas as pd
import numpy as np
import matplotlib.pyplot as plt
from torch.utils.data import Dataset, DataLoader
from PIL import Image
import zipfile

##########################################################
# data loader class
##########################################################
class LoadData(Dataset):
    def __init__(self, fName, transform=None):
        data = pd.read_csv(fName)
        self.X = np.array(data.iloc[:, 1:], dtype=np.uint8).reshape(-1,
            1, 28, 28)
        self.y = np.array(data.iloc[:, 0])

    def __len__(self):
        return len(self.X)

    def __getitem__(self, idx):
        img = self.X[idx]
        lbl = self.y[idx]
        return (img, lbl)
```

```
# unzip the zip file and load the image data
zip_file_path = 'fashionmnist.zip'
extract_to_path = '.'
with zipfile.ZipFile(zip_file_path, 'r') as zip_ref:
    zip_ref.extractall(extract_to_path)

train_ds = LoadData('fashion-mnist_train.csv')
test_ds  = LoadData('fashion-mnist_test.csv')

# set labels dictionary
labels = {0 : 'T-Shirt', 1 : 'Trouser', 2 : 'Pullover',
          3 : 'Dress', 4 : 'Coat', 5 : 'Sandal', 6 : 'Shirt',
          7 : 'Sneaker', 8 : 'Bag', 9 : 'Ankle Boot'}
```

Since an image input data is generally memory intensive, it is important to partition the data set into (mini-)batches. The code below defines a batch size of 100 images and initializes the Pytorch data loader objects. These objects will be used for efficient iteration over the data set.

```
# load the data in batches
batch_size = 100

train_loader = torch.utils.data.DataLoader(dataset=train_ds,
                                           batch_size=batch_size,
                                           shuffle=True)
test_loader = torch.utils.data.DataLoader(dataset=test_ds,
                                          batch_size=batch_size,
                                          shuffle=True)
```

Next, to define the network architecture in Pytorch all we need to do is define an instance of the `torch.nn.Module` class. Choosing a network architecture with good generalization properties can be a difficult task. Here, we use a network with two convolution layers (defined in the `cnn_layer` block), a 3×3 kernel, and three hidden layers (defined in the `flat_layer` block). Since there are ten possible output labels, the output layer has ten nodes. More specifically, the first and the second convolution layers have 16 and 32 output channels. Combining this with the definition of the 3×3 kernel, we conclude that the size of the first flat hidden layer should be:

$$\left(\overbrace{\underbrace{(28 - 3 + 1)}_{\text{first convolution layer}} \quad -3 + 1}^{\text{second convolution layer}} \right)^2 \times 32 = 18432,$$

where the multiplication by 32 follows from the fact that the second convolution layer has 32 output channels. Having said that, the `flat_fts` variable determines the number of output layers of the convolution block. This number is used to define the size of the first hidden layer of the `flat_layer` block. The rest of the hidden layers have 100 neurons and we use the ReLU activation function for all layers. Finally, note that the `forward` method in the `CNN` class implements the forward pass.

```
# define the network
class CNN(nn.Module):

    def __init__(self):
        super(CNN, self).__init__()

        self.cnn_layer = nn.Sequential(
                nn.Conv2d(1, 16, kernel_size=3, stride=(1,1)),
                nn.ReLU(),
                nn.Conv2d(16, 32, kernel_size=3, stride=(1,1)),
                nn.ReLU(),
        )
        self.flat_fts = (((28-3+1)-3+1)**2)*32


        self.flat_layer = nn.Sequential(
                nn.Linear(self.flat_fts, 100),
                nn.ReLU(),
                nn.Linear(100, 100),
                nn.ReLU(),
                nn.Linear(100, 100),
                nn.ReLU(),
                nn.Linear(100, 10))

    def forward(self, x):
        out = self.cnn_layer(x)
        out = out.view(-1, self.flat_fts)
        out = self.flat_layer(out)
        return out
```

Next, we specify how the network will be trained. We choose the device type, namely, the central processing unit (CPU) or the GPU (if available), the number of training iterations (epochs), and the learning rate. Then, we create an instance of the proposed convolution network and send it to the predefined device (CPU or GPU). Note how easily one can switch between the CPU or the GPU without major changes to the code.

In addition to the specifications above, we need to choose an appropriate loss function and training algorithm. Here, we use the *cross-entropy* loss and the *Adam* adaptive gradient Algorithm 10.28. Once these parameters are set, the learning proceeds to evaluate the gradient of the loss function via the back-propagation algorithm. ☞ 371

```
# learning parameters
num_epochs = 50
learning_rate = 0.001

#device = torch.device ('cpu') # use this to run on CPU
device = torch.device ('cuda') # use this to run on GPU

#instance of the Conv Net
cnn = CNN()
cnn.to(device=device)
```

```
#loss function and optimizer
criterion = nn.CrossEntropyLoss()
optimizer = torch.optim.Adam(cnn.parameters(), lr=learning_rate)

# the learning loop
losses = []
for epoch in range(1,num_epochs+1):
    for i, (images, labels) in enumerate(train_loader):
        images = Variable(images.float()).to(device=device)
        labels = Variable(labels).to(device=device)

        optimizer.zero_grad()
        outputs = cnn(images)
        loss = criterion(outputs, labels)
        loss.backward()
        optimizer.step()

        losses.append(loss.item())
    if(epoch==1 or epoch % 10 == 0):
        print ("Epoch : ", epoch, ", Training Loss: ",  loss.item())

# evaluate on the test set
cnn.eval()
correct = 0
total = 0
for images, labels in test_loader:
    images = Variable(images.float()).to(device=device)
    outputs = cnn(images)
    _, predicted = torch.max(outputs.data, 1)
    total += labels.size(0)
    correct += (predicted.cpu() == labels).sum()
print("Test Accuracy on the 10,000 training test images: ", (100 *
    correct.item() / total),"%")

# plot
plt.rc('text', usetex=True)
plt.rc('font', family='serif',size=20)
plt.tight_layout()

plt.plot(np.array(losses)[10:len(losses)])
plt.xlabel(r'{iteration}',fontsize=20)
plt.ylabel(r'{Batch Loss}',fontsize=20)
plt.subplots_adjust(top=0.8)
plt.show()
```

```
Epoch :  1 , Training Loss:  0.412550151348114
Epoch :  10 , Training Loss:  0.05452106520533562
Epoch :  20 , Training Loss:  0.07233225554227829
Epoch :  30 , Training Loss:  0.01696968264877796
Epoch :  40 , Training Loss:  0.0008199119474738836
Epoch :  50 , Training Loss:  0.006860652007162571
Test Accuracy on the 10,000 training test images: 91.02 %
```

Finally, we evaluate the network performance using the test data set. A typical minibatch loss as a function of iteration is shown in Figure 10.31 and the proposed neural network achieves about 91% accuracy on the test set.

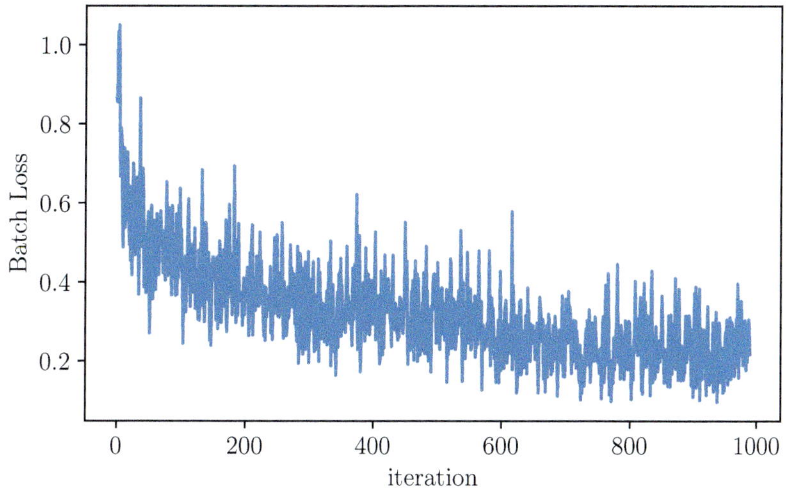

Figure 10.31: The batch loss history.

Further Reading

A popular book written by some of the pioneers of deep learning is [77]. For an excellent and gentle introduction to the intuition behind neural networks, we recommend [133] and [111]. A summary of many effective gradient descent methods for training of deep networks is given in [147]. An early resource on the limited-memory BFGS method is [113], and a more recent resource includes [18], which makes recommendations on the best choice for the length of the BFGS history (that is, the value of the parameter h).

Exercises

1. Show that the softmax function

$$\text{softmax} : \boldsymbol{z} \mapsto \frac{\exp(\boldsymbol{z})}{\sum_k \exp(z_k)}$$

satisfies the invariance property:

$$\text{softmax}(\boldsymbol{z}) = \text{softmax}(\boldsymbol{z} + c \times \mathbf{1}), \quad \text{for any constant } c.$$

2. *Projection pursuit* is a network with one hidden layer that can be written as:

$$g(\boldsymbol{x}) = S(\boldsymbol{\omega}^\top \boldsymbol{x}),$$

where S is a univariate *smoothing cubic spline*. If we use squared-error loss with $\tau_n = \{y_i, \boldsymbol{x}_i\}_{i=1}^n$, we need to minimize the training loss: ☞ 337

$$\frac{1}{n}\sum_{i=1}^{n}\left(y_i - S(\boldsymbol{\omega}^\top \boldsymbol{x}_i)\right)^2$$

with respect to ω and all cubic smoothing splines. This training of the network is ☞ 164 typically tackled iteratively in a manner similar to the *EM algorithm*. In particular, we iterate ($t = 1, 2, \ldots$) the following steps until convergence.

(a) Given the *missing data* ω_t, compute the spline S_t by training a cubic smoothing spline on $\{y_i, \omega_t^\top \boldsymbol{x}_i\}$. The smoothing coefficient of the spline may be determined as part of this step.

(b) Given the spline function S_t, compute the next projection vector ω_{t+1} via **iterative reweighted least squares**:

$$\omega_{t+1} = \underset{\boldsymbol{\beta}}{\operatorname{argmin}}\, (\boldsymbol{e}_t - \mathbf{X}\boldsymbol{\beta})^\top \boldsymbol{\Sigma}_t\, (\boldsymbol{e}_t - \mathbf{X}\boldsymbol{\beta}), \tag{10.32}$$

where

$$e_{t,i} := \omega_t^\top \boldsymbol{x}_i + \frac{y_i - S_t(\omega_t^\top \boldsymbol{x}_i)}{S_t'(\omega_t^\top \boldsymbol{x}_i)}, \quad i = 1, \ldots, n$$

is the adjusted response, and $\boldsymbol{\Sigma}_t^{1/2} = \operatorname{diag}(S_t'(\omega_t^\top \boldsymbol{x}_1), \ldots, S_t'(\omega_t^\top \boldsymbol{x}_n))$ is a diagonal matrix.

☞ 586 Apply Taylor's Theorem C.20 to the function S_t and derive the iterative reweighted least-squares optimization program (10.32).

☞ 441 3. Suppose that in the *stochastic gradient descent* method we wish to repeatedly draw minibatches of size N from τ_n, where we assume that $N \times m = n$ for some large integer m. Instead of repeatedly resampling from τ_n, an alternative is to reshuffle τ_n via a ☞ 127 random permutation $\boldsymbol{\Pi}$ and then advance sequentially through the reshuffled training set to construct m non-overlapping minibatches. A single traversal of such a reshuffled training set is called an **epoch**. The following pseudo-code describes the procedure.

Algorithm 10.33: Stochastic Gradient Descent with Reshuffling

Input: Training set $\tau_n = \{(\boldsymbol{x}_i, \boldsymbol{y}_i)\}_{i=1}^n$, initial weight matrices and bias vectors $\{\mathbf{W}_l, \boldsymbol{b}_l\}_{l=1}^L \to \boldsymbol{\theta}_1$, activation functions $\{S_l\}_{l=1}^L$, learning rates $\{\alpha_1, \alpha_2, \ldots\}$.

Output: The parameters of the trained learner.

```
t ← 1 and epoch ← 0
while stopping condition is not met do
    Draw U_1, ..., U_n ~iid U(0, 1).
    Let Π be the permutation of {1, ..., n} that satisfies U_Π1 < ··· < U_Πn.
    (x_i, y_i) ← (x_Πi, y_Πi) for i = 1, ..., n            // reshuffle τ_n
    for j = 1, ..., m do
        ℓ̂_τ ← (1/N) Σ_{i=(j-1)N+1}^{jN} Loss(y_i, g(x_i | θ))
        θ_{t+1} ← θ_t − α_t ∂ℓ̂_τ/∂θ (θ_t)
        t ← t + 1
    epoch ← epoch + 1                          // number of reshuffles or epochs
return θ_t as the minimizer of the training loss
```

Line 3: $U_1, \ldots, U_n \overset{\text{iid}}{\sim} \mathcal{U}(0,1)$; line 4: $U_{\Pi_1} < \cdots < U_{\Pi_n}$; line 7: $\widehat{\ell}_\tau \leftarrow \frac{1}{N}\sum_{i=(j-1)N+1}^{jN} \operatorname{Loss}(\boldsymbol{y}_i, \boldsymbol{g}(\boldsymbol{x}_i \mid \boldsymbol{\theta}))$; line 8: $\boldsymbol{\theta}_{t+1} \leftarrow \boldsymbol{\theta}_t - \alpha_t \frac{\partial \widehat{\ell}_\tau}{\partial \boldsymbol{\theta}}(\boldsymbol{\theta}_t)$.

Write Python code that implements the stochastic gradient descent with data reshuffling, and use it to train the neural net in Section 10.5.1. ☞ 447

4. Suppose that we wish to compute the inverse and log-determinant of the matrix

$$\mathbf{I}_n + \mathbf{U}\mathbf{U}^\top,$$

where $\mathbf{U}$ is an $n \times h$ matrix with $h \ll n$. Show that

$$(\mathbf{I}_n + \mathbf{U}\mathbf{U}^\top)^{-1} = \mathbf{I}_n - \mathbf{Q}_n\mathbf{Q}_n^\top,$$

where $\mathbf{Q}_n$ contains the first n rows of the $(n+h) \times h$ matrix $\mathbf{Q}$ in the QR factorization of the $(n+h) \times h$ matrix: ☞ 522

$$\begin{bmatrix} \mathbf{U} \\ \mathbf{I}_h \end{bmatrix} = \mathbf{Q}\mathbf{R}.$$

In addition, show that $\ln|\mathbf{I}_n + \mathbf{U}\mathbf{U}^\top| = \sum_{i=1}^h \ln r_{ii}^2$, where $\{r_{ii}\}$ are the diagonal elements of the $h \times h$ matrix $\mathbf{R}$.

5. Suppose that

$$\mathbf{U} = [\boldsymbol{u}_0, \boldsymbol{u}_1, \ldots, \boldsymbol{u}_{h-1}],$$

where all $\boldsymbol{u} \in \mathbb{R}^n$ are column vectors and we have computed $(\mathbf{I}_n + \mathbf{U}\mathbf{U}^\top)^{-1}$ via the QR factorization method in Exercise 4. If the columns of matrix $\mathbf{U}$ are updated to

$$[\boldsymbol{u}_1, \ldots, \boldsymbol{u}_{h-1}, \boldsymbol{u}_h],$$

show that the inverse $(\mathbf{I}_n + \mathbf{U}\mathbf{U}^\top)^{-1}$ can be updated in $O(h\,n)$ time (rather than computed from scratch in $O(h^2\,n)$ time). Deduce that the computing cost of updating the Hessian approximation (10.27) is the same as that for the *limited-memory BFGS* Algorithm 10.25.

In your solution you may use the following facts from [47]. Suppose we are given the $\mathbf{Q}$ and $\mathbf{R}$ factors in the QR factorization of a matrix $\mathbf{A} \in \mathbb{R}^{n\times h}$. If a row/column is added to matrix $\mathbf{A}$, then the $\mathbf{Q}$ and $\mathbf{R}$ factors need not be recomputed from scratch (in $O(h^2\,n)$ time), but can be updated efficiently in $O(h\,n)$ time. Similarly, if a row/column is removed from matrix $\mathbf{A}$, then the $\mathbf{Q}$ and $\mathbf{R}$ factors can be updated in $O(h^2)$ time.

6. Suppose that $\mathbf{U} \in \mathbb{R}^{n\times h}$ has its k-th column $\boldsymbol{v}$ replaced with $\boldsymbol{w}$, giving the updated $\widetilde{\mathbf{U}}$.

(a) If $\boldsymbol{e} \in \mathbb{R}^h$ denotes the unit-length vector such that $e_k = \|\boldsymbol{e}\| = 1$ and

$$\boldsymbol{r}_\pm := \frac{\sqrt{2}}{2}\mathbf{U}^\top(\boldsymbol{w} - \boldsymbol{v}) + \frac{\sqrt{2}\,\|\boldsymbol{w} - \boldsymbol{v}\|^2}{4}\boldsymbol{e} \pm \frac{\sqrt{2}}{2}\boldsymbol{e},$$

show that

$$\widetilde{\mathbf{U}}^\top\widetilde{\mathbf{U}} = \mathbf{U}^\top\mathbf{U} + \boldsymbol{r}_+\boldsymbol{r}_+^\top - \boldsymbol{r}_-\boldsymbol{r}_-^\top.$$

Hint: You may find Exercise 16 in Chapter 7 useful. ☞ 350

(b) Let $\mathbf{B} := (\mathbf{I}_h + \mathbf{U}^\top\mathbf{U})^{-1}$. Use the *Woodbury identity* (A.60) to show that ☞ 541

$$(\mathbf{I}_n + \widetilde{\mathbf{U}}\widetilde{\mathbf{U}}^\top)^{-1} = \mathbf{I}_n - \widetilde{\mathbf{U}}\left(\mathbf{B}^{-1} + \boldsymbol{r}_+\boldsymbol{r}_+^\top - \boldsymbol{r}_-\boldsymbol{r}_-^\top\right)^{-1}\widetilde{\mathbf{U}}^\top.$$

(c) Suppose that we have stored $\mathbf{B}$ in computer memory. Use Algorithm 7.72 and parts (a) and (b) to write pseudo-code that updates $(\mathbf{I}_n + \mathbf{U}\mathbf{U}^\top)^{-1}$ to $(\mathbf{I}_n + \widetilde{\mathbf{U}}\widetilde{\mathbf{U}}^\top)^{-1}$ in $O((n+h)h)$ computing time.

7. Equation (10.22) gives the rank-two BFGS update of the inverse Hessian $\mathbf{C}_{t-1}$ to $\mathbf{C}_t$. Instead of using a two-rank update, we can consider a one-rank update, in which $\mathbf{C}_{t-1}$ is updated to $\mathbf{C}_t$ by the general rank-one formula:

$$\mathbf{C}_t = \mathbf{C}_{t-1} + \upsilon_t \, \boldsymbol{r}_t \boldsymbol{r}_t^\top.$$

Find values for the scalar υ_t and vector $\boldsymbol{r}_t$, such that $\mathbf{C}_t$ satisfies the secant condition $\mathbf{C}_t \boldsymbol{g}_t = \boldsymbol{\delta}_t$.

8. Show that the *BFGS formula* (C.57) can be written as:

$$\mathbf{C} \leftarrow \left(\mathbf{I} - \upsilon \boldsymbol{g} \boldsymbol{\delta}^\top\right)^\top \mathbf{C} \left(\mathbf{I} - \upsilon \boldsymbol{g} \boldsymbol{\delta}^\top\right) + \upsilon \boldsymbol{\delta}\boldsymbol{\delta}^\top,$$

where $\upsilon := (\boldsymbol{g}^\top \boldsymbol{\delta})^{-1}$.

9. Show that the *BFGS formula* (C.57) is the solution to the constrained optimization problem:

$$\mathbf{C}_{\text{BFGS}} = \underset{\mathbf{A} \text{ subject to } \mathbf{A}\boldsymbol{g} = \boldsymbol{\delta},\ \mathbf{A} = \mathbf{A}^\top}{\operatorname{argmin}} \mathcal{D}(\mathbf{0}, \mathbf{C} \mid \mathbf{0}, \mathbf{A}),$$

where $\mathcal{D}$ is the Kullback–Leibler discrepancy defined in (C.56). On the other hand, show that the *DFP formula* (C.58) is the solution to the constrained optimization problem:

$$\mathbf{C}_{\text{DFP}} = \underset{\mathbf{A} \text{ subject to } \mathbf{A}\boldsymbol{g} = \boldsymbol{\delta},\ \mathbf{A} = \mathbf{A}^\top}{\operatorname{argmin}} \mathcal{D}(\mathbf{0}, \mathbf{A} \mid \mathbf{0}, \mathbf{C}).$$

☞ 254 10. Consider again the logistic regression model in Exercise 5.24, which used *iterative reweighted least squares* for training the learner. Repeat all the computations, but this time using the *limited-memory BFGS* Algorithm 10.26. Which training algorithm converges faster to the optimal solution?

11. Download the `seeds_dataset.txt` data set from the book's GitHub site, which contains 210 independent examples. The categorical output (response) here is the type of wheat grain: Kama, Rosa, and Canadian (encoded as 1, 2, and 3), so that $c = 3$. The seven continuous features (explanatory variables) are measurements of the geometrical properties of the grain (area, perimeter, compactness, length, width, asymmetry coefficient, and length of kernel groove). Thus, $\boldsymbol{x} \in \mathbb{R}^7$ (which does not include the
☞ 434 constant feature 1) and the multi-logit pre-classifier in Example 10.10 can be written as $\boldsymbol{g}(\boldsymbol{x}) = \text{softmax}(\mathbf{W}\boldsymbol{x} + \boldsymbol{b})$, where $\mathbf{W} \in \mathbb{R}^{3\times 7}$ and $\boldsymbol{b} \in \mathbb{R}^3$. Implement and train this pre-classifier on the first $n = 105$ examples of the seeds data set using, for example, Algorithm 10.20. Use the remaining $n' = 105$ examples in the data set to estimate the generalization risk of the learner using the cross-entropy loss. Hint: Use the cross-entropy loss formulas from Example 10.18.

12. In Exercise 11 above, we train the multi-logit classifier using a weight matrix $\mathbf{W} \in \mathbb{R}^{3\times 7}$ and bias vector $\boldsymbol{b} \in \mathbb{R}^3$. Repeat the training of the multi-logit model, but this time keeping z_1 as an arbitrary constant (say $z_1 = 0$), and thus setting $c = 0$ to be a "reference" class. This has the effect of removing a node from the output layer of the network, giving a weight matrix $\mathbf{W} \in \mathbb{R}^{2\times 7}$ and bias vector $\boldsymbol{b} \in \mathbb{R}^2$ of smaller dimensions than in (8.32). ☞ 370

13. Consider again Example 10.18, where we used a *softmax* output function $\boldsymbol{S}_L$ in ☞ 439 conjunction with the *cross-entropy* loss: $C(\boldsymbol{\theta}) = -\ln g_{y+1}(\boldsymbol{x} \mid \boldsymbol{\theta})$. Find formulas for $\frac{\partial C}{\partial \boldsymbol{g}}$ and $\frac{\partial \boldsymbol{S}_L}{\partial \boldsymbol{z}_L}$. Hence, verify that:

$$\frac{\partial \boldsymbol{S}_L}{\partial \boldsymbol{z}_L}\frac{\partial C}{\partial \boldsymbol{g}} = \boldsymbol{g}(\boldsymbol{x} \mid \boldsymbol{\theta}) - \boldsymbol{e}_{y+1},$$

where $\boldsymbol{e}_i$ is the unit length vector with an entry of 1 in the i-th position.

14. Derive the formula (C.60) for a diagonal Hessian update in a quasi-Newton method ☞ 603 for minimization. In other words, given a current minimizer $\boldsymbol{x}_t$ of $f(\boldsymbol{x})$, a diagonal matrix $\mathbf{C}$ of approximating the Hessian of f, and a gradient vector $\boldsymbol{u} = \nabla f(\boldsymbol{x}_t)$, find the solution to the constrained optimization program:

$$\begin{aligned} &\min_{\mathbf{A}} \mathcal{D}(\boldsymbol{x}_t, \mathbf{C} \,|\, \boldsymbol{x}_t - \mathbf{A}\boldsymbol{u}, \mathbf{A}) \\ &\text{subject to: } \mathbf{A}\boldsymbol{g} \geq \boldsymbol{\delta},\ \mathbf{A} \text{ is diagonal,} \end{aligned}$$

where $\mathcal{D}$ is the Kullback–Leibler distance defined in (C.56).

15. Consider again the Python implementation of the polynomial regression in Section 10.5.1, where the *stochastic gradient descent* was used for training.

Using the polynomial regression data set, implement and run the following four alternative training methods:

(a) the gradient-descent Algorithm 10.20;

(b) the Levenberg–Marquardt Algorithm C.65, in conjunction with Algorithm 10.21 ☞ 606 for computing the matrix of Jacobi;

(c) the *limited-memory BFGS* Algorithm 10.26;

(d) the *Adam* Algorithm 10.28, which uses past gradient values to determine the next search direction.

For each training algorithm, using trial and error, tune any algorithmic parameters so that the network training is as fast as possible. Comment on the relative advantages and disadvantages of each training/optimization method. For example, comment on which optimization method makes rapid initial progress, but gets trapped in a suboptimal solution, and which method is slower, but more consistent in finding good optima.

16. Consider again the `Pytorch` code in Section 10.5.2. Repeat all the computations, but this time using the *momentum* method for training of the network. Comment on which method is preferable: the *momentum* or the *Adam* method?

CHAPTER 11

REINFORCEMENT LEARNING

In this chapter, we provide an introduction to reinforcement learning. After a brief overview of Markov decision processes and Bellman's optimality principle, we discuss fundamental concepts such as value- and policy-iteration, dynamic programming, temporal difference learning, Monte Carlo tree search, and policy gradient methods. Throughout the chapter, we use a number of running examples to illustrate the wide variety of algorithms and their Python code.

11.1 Introduction

Reinforcement learning is based on the simple premise that knowledge is accumulated through repeated interactions between an actor or *agent* and its *environment*. For example, a child learns to speak by initially uttering random sounds and words, which provoke either encouraging or unfavorable responses from the parents, and eventually lead to meaningful sentences.

More generally, in a reinforcement learning setting, an agent interacts with the environment by performing particular *actions* and, in return, receives feedback in the form of information on the *state* of the environment as well as gaining an immediate *reward* for the action taken. By constantly interacting with the environment in this way, the ultimate aim of the agent is to optimize the rewards over the long term by iteratively fine-tuning its *policy*: the mechanism by which it chooses its actions.

A particular example is a robot operating in a two-dimensional maze where the ultimate goal could be to find the exit of the maze in as few steps as possible. At each step the robot is allowed to move up, down, left, or right, if this is possible (the robot cannot pass through a wall of the maze). After each action, the robot receives information of its position in the maze and the current allowable actions. Each move might carry a penalty (negative reward), and the game ends when the robot exits the maze. The objective would be to devise a strategy of moving within the maze that minimizes the total penalty.

DOI: 10.1201/9781003391173-11

The interaction of the agent with its environment forms a continuous loop (see Figure 11.1), where the agent iteratively

- follows its policy based on its current state and reward information,
- interacts with the environment through actions derived from this policy,
- collects experience in the form of state and reward information, and
- uses this new information to update its policy.

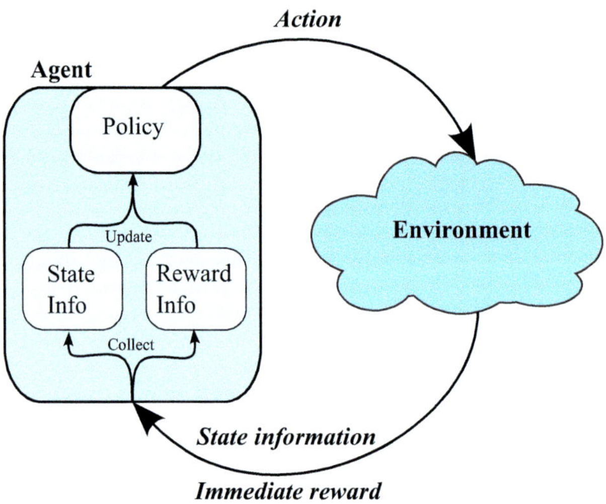

Figure 11.1: Schematic depiction of reinforcement learning.

When choosing its actions, the agent is constantly faced with the **exploration–exploitation dilemma**: the choice between trying out new actions with unknown rewards or focusing on actions with predictable rewards. Ideally, this process leads to an improvement in decision making over time, in which exploitation is gradually favored over exploration.

Since the action is taken based on the current state and reward, the mathematical description of the environment is typically provided by a *Markov decision process*, which we will discuss next. Throughout the rest of the chapter, we will utilize this model for the environment.

11.2 Markov Decision Processes

A mathematically principled framework for reinforcement learning is that of Markov decision processes (MDPs). We start with a simple running example to motivate its use.

■ **Example 11.2 (MDP)** Let $\mathbf{P}_0$ and $\mathbf{P}_1$ be two different transition matrix options for a
☞ 85 Markov chain with state space $\{0, 1, 2, 3\}$. Specifically, suppose that

$$\mathbf{P}_0 = \begin{bmatrix} 0 & 0.1 & 0.1 & 0.8 \\ 0.1 & 0.8 & 0.1 & 0 \\ 0.1 & 0.9 & 0 & 0 \\ 0.1 & 0 & 0 & 0.9 \end{bmatrix} \quad \text{and} \quad \mathbf{P}_1 = \begin{bmatrix} 0 & 0.4 & 0.2 & 0.4 \\ 0.1 & 0 & 0.9 & 0 \\ 0.1 & 0 & 0 & 0.9 \\ 0.9 & 0.1 & 0 & 0 \end{bmatrix}.$$

Note that under $\mathbf{P}_0$ sojourns (that is, stays) in state 3 are much more likely than under $\mathbf{P}_1$. Suppose that, while simulating the chain, we (the agent) can choose at any time step t whether we use $\mathbf{P}_0$ or $\mathbf{P}_1$ for our transitions. What *action* (0 or 1, the former selecting $\mathbf{P}_0$ and the latter selecting $\mathbf{P}_1$) should we take in each state in order to maximize the long-run fraction of sojourns in state 3? Note that there are only $2^4 = 16$ such possible action arrangements or *policies* that we can choose from, as there are two possible actions for each of the four states. Each policy $[a_0, a_1, a_2, a_3]$, with $a_i \in \{0, 1\}$, induces a different transition matrix $\widetilde{\mathbf{P}}$ for the Markov chain, the rows of which are equal to the rows of either $\mathbf{P}_0$ or $\mathbf{P}_1$. For example, the policy $[0, 0, 1, 1]$ would induce a transition matrix $\widetilde{\mathbf{P}}$ whose first two rows correspond to the first two rows of $\mathbf{P}_0$, and whose last two rows correspond to the last two rows of $\mathbf{P}_1$.

Given a policy and corresponding transition matrix $\widetilde{\mathbf{P}}$, the long-run fraction of sojourns in each state, $\boldsymbol{f} = [f_0, \ldots, f_3]$, can be obtained by solving the *global balance equations* $\boldsymbol{f} = \boldsymbol{f}\widetilde{\mathbf{P}}$, with $\sum_i f_i = 1$. For example, for the choice $\widetilde{\mathbf{P}} = \mathbf{P}_0$ we find $\boldsymbol{f} = [0.0909, 0.1570, 0.0248, 0.7273]$. The optimization of the long-run fraction of sojourns in state 3 can thus be accomplished by solving the global balance equations for each of the 16 policies and taking the largest value f_3. We find, see Exercise 1, the optimal policy $[0, 1, 1, 0]$ with maximum value $f_3 = 0.8827$. ☞ 661

Formally, a **Markov decision process** (MDP) is a Markov process $\{S_t, t = 0, 1, \ldots\}$ whose transition probabilities are governed by actions taken at each time step. Each action returns a reward, and the objective is to maximize some overall reward (to be detailed below). The main ingredients of an MDP are given by a 4-tuple $(\mathcal{S}, \mathcal{A}, p, r)$. Here, $\mathcal{S}$ is the **state space** of the Markov chain. Secondly, $\mathcal{A}$ is the **action space**; that is, the set of actions allowed for each state. Next, p is the **transition function**, which specifies the probability $p(s' \mid s, a)$ that the Markov chain transitions from state s to s' when action a is taken. Each such transition receives an immediate reward $r(s, a)$ where r is the **reward function** of the MDP.

In Example 11.2, the state space is $\{0, 1, 2, 3\}$, the action space is $\{0, 1\}$, and we have that $p(s' \mid s, a) = \mathbf{P}_a(s, s')$; i.e., the (s, s')-element of the one-step transition matrix $\mathbf{P}_a$, where $a = 0$ or 1. In this case the immediate reward is $r(s, a) = 1$ if $s = 3$ and 0 otherwise.

In the case where the immediate reward depends on the next state s' as well as on s and a, say it is of the form $\varrho(s, a, s')$, we can define $r(s, a) = \mathbb{E}\varrho(s, a, S')$ as the *expected* reward.

While $(\mathcal{S}, \mathcal{A}, p, r)$ describes the evolutionary mechanism of the MDP, it does not tell us how to choose the actions. In the terminology of Section 11.1: even if the agent has full knowledge of the environment process (the MDP), it is not directly clear what actions it should take, because the ultimate goal of the agent is not to select actions that maximize immediate reward, but rather to select actions that maximize a cumulative reward over the evolution of the process (the overall reward), which fundamentally depends on the time frame of the learning process. In **terminating** scenarios, the learning process ends after a finite number of steps T, which may be fixed or random. When T is fixed it is often called the **time horizon** of the MDP. A typical example of a random final time T

occurs when the Markov process enters a **terminating state** and then stops. We call such terminating scenarios **episodic**. A usual overall reward for a terminating scenario is the expected total reward $\mathbb{E}\sum_{t=0}^{T} R_t$, where R_t is the t-th reward.

In contrast, **continuing** scenarios do not have defined termination conditions and go on without end, effectively having $T = \infty$. In a continuing scenario, a common approach is to take the expectation of the *discounted total* of the individual rewards, $\mathbb{E}\sum_{t=0}^{\infty} \gamma^t R_t$, where $\gamma \in (0, 1)$ is a **discount rate**. In Example 11.2, the scenario is continuing, and an alternative approach is taken where the overall reward is the expected[1] long-run average reward $\mathbb{E}\lim_{T\to\infty} T^{-1} \sum_{t=0}^{T} \mathbb{1}_{\{S_t=3\}}$.

> The episodic and continuing schemes can be unified under the same framework by considering the terminal states of an episodic task as an absorbing state of a continuing MDP that generates a zero immediate reward.

Given an MDP $(\mathcal{S}, \mathcal{A}, p, r)$ and an overall reward function, the objective is to specify a rule or **policy** which prescribes what action to take when the MDP is in state s at time t. A policy is said to be **deterministic** if it always returns the same action for a given state and **stationary** if it does not depend on t. A stationary deterministic policy can thus be viewed as a function $\pi : \mathcal{S} \to \mathcal{A}$. In a **randomized** policy, actions are assigned randomly according to a probability distribution, in which case π maps $\mathcal{S}$ to the space of probability distributions on $\mathcal{A}$.

A major advantage of using discounted rewards in continuing and episodic scenarios is that one may assume that the optimal policy is *stationary*; see Section 11.3. For scenarios with a finite time horizon this is no longer the case; see Section 11.5. For simplicity, we consider for the remainder of this section only stationary policies (deterministic or randomized) and specify the overall reward function for a policy π via the **value function** (sometimes referred to as the **utility**):

$$V^\pi(s) = \mathbb{E}_\pi\left[\sum_{t=0}^{\infty} \gamma^t R_t \,\middle|\, S_0 = s\right], \qquad s \in \mathcal{S},$$

where $R_0, R_1, \ldots$ is the sequence of rewards and S_0 is the initial state of the MDP. The notation $\mathbb{E}_\pi$ is used to indicate that the MDP evolves under policy π. Thus, the value function, evaluated at state $s \in \mathcal{S}$, returns the expected discounted reward of the MDP when starting from state s and following policy π from there on. Note that for each t, we have $R_t = r(S_t, A_t)$, where $S_0, S_1, \ldots$ is the sequence of states and $A_0, A_1, \ldots$ the sequence of actions given by π.

The ultimate goal is to find a policy π^* that maximizes this value function (typically from the set of all deterministic policies Π); that is,

$$\pi^* \in \operatorname*{argmax}_{\pi \in \Pi} V^\pi(s).$$

Let $V^* = V^{\pi^*}$ denote the optimal value function. **Bellman's optimality principle** [11] states that an optimal policy has the property that, regardless of the initial state s and

[1]As the long-run average reward is a constant, we may remove the expectation sign.

corresponding action a (which leads to the the next state s'), the remaining actions must constitute an optimal policy with respect to the next state s'. For a stationary deterministic optimal policy, this principle translates into the following system of equations:

$$V^*(s) = \max_{a \in \mathcal{A}} \left\{ r(s,a) + \gamma \sum_{s' \in \mathcal{S}} p(s' \mid s,a)\, V^*(s') \right\}, \qquad s \in \mathcal{S}. \tag{11.3}$$

As we will see in Section 11.3, for MDPs with finite state and action spaces, there is a unique solution V^* to the Bellman optimality equation; however, the policy π that achieves this optimal value function may not be unique. The Bellman optimality equations say that the greedy policy of choosing actions that appear best after one step is actually optimal, if the optimal value function is already known. That is, the optimal (greedy) policy, derived from V^*, is given by

$$\pi^*(s) \in \underset{a \in \mathcal{A}}{\operatorname{argmax}} \left\{ r(s,a) + \gamma \sum_{s' \in \mathcal{S}} p(s' \mid s,a)\, V^*(s') \right\}, \qquad s \in \mathcal{S}. \tag{11.4}$$

In addition to the value function V it is useful to consider the **Q-function** for policy π, defined by

$$\begin{aligned} Q^\pi(s,a) &= \mathbb{E}_\pi \left[\sum_{t=0}^{\infty} \gamma^t R_t \,\middle|\, S_0 = s, A_0 = a \right] \\ &= r(s,a) + \gamma\, \mathbb{E}[V^\pi(S_1) \mid S_0 = s, A_0 = a], \qquad (s,a) \in \mathcal{S} \times \mathcal{A}. \end{aligned}$$

The Q-function is sometimes called the **action–value** function, as it assigns a value to different actions (for a fixed state s), as opposed to the state–value function V^π, which assigns a value to each state. Bellman's optimality principle, applied to $Q^* = Q^{\pi^*}$, leads to the equations:

$$Q^*(s,a) = r(s,a) + \gamma \sum_{s' \in \mathcal{S}} p(s' \mid s,a) \max_{a' \in \mathcal{A}} Q^*(s',a'). \tag{11.5}$$

If one has access to $Q^*(s,\cdot)$, the optimal policy for state s is simply the action a that maximizes $Q^*(s,a)$; that is, the optimal (greedy) policy is given by

$$\pi^*(s) = \underset{a \in \mathcal{A}}{\operatorname{argmax}}\, Q^*(s,a), \qquad s \in \mathcal{S}.$$

In the remainder of this chapter, we will mostly consider finite state and action spaces. Table 11.6 summarizes frequently used notation for MDPs. When one has continuous state and action spaces, a frequently employed approach is to discretize them (for example, via tile coding [167, Section 9.5.4]) and use the methods outlined here. Alternatively, parameterized policy methods can allow one to deal with continuous action spaces, by computing statistics associated with the probability distribution over actions; see Section 11.8.

Table 11.6: Summary of definitions for Markov decision processes.

$\mathcal{S}$	State space
$\mathcal{A}$	Action space
$\pi : \mathcal{S} \to \mathcal{A}$	Deterministic policy. $\pi(s)$ is the action in state s.
$p : \mathcal{S} \times \mathcal{S} \times \mathcal{A} \to [0, 1]$	Transition function. $p(s' \mid s, a)$ is the probability of transitioning from state s to s' when taking action a.
$r : \mathcal{S} \times \mathcal{A} \to \mathbb{R}$	Reward function. $r(s, a)$ is the reward received when using action a in state s.
$\gamma \in (0, 1)$	Discount rate.
$V^{\pi} : \mathcal{S} \to \mathbb{R}$	Value function. Expected discounted reward when starting in s and following policy π. The optimal value function V^* satisfies Bellman's equation (11.3).
$Q : \mathcal{S} \times \mathcal{A} \to \mathbb{R}$	Q-function, $Q^{\pi}(s, a)$ is the expected discounted reward, when starting in s, taking action a, and afterwards following policy π. The optimal Q-function Q^* satisfies Bellman's equation (11.5).

11.3 Value and Policy Iteration

The basic idea underpinning both value and policy iteration is to take the Bellman optimality equations (11.3) and to turn them into an iterative system of equations. Although in practice the optimal value function may not be readily available, it turns out that even if we simply have an *estimate* $\widehat{V}$ of V^* (via the iterative application of the Bellman equations), then a greedy approach will still result in a provably optimal policy.

11.3.1 Value Iteration

In **value iteration** we start from an initial guess V_0 of the optimal value function and iteratively evaluate

$$\begin{aligned} V_{n+1}(s) &= C[V_n](s) = \max_{a \in \mathcal{A}} Q_n(s, a) \\ &= \max_{a \in \mathcal{A}} \left\{ r(s, a) + \gamma \sum_{s' \in \mathcal{S}} p(s' \mid s, a)\, V_n(s') \right\}, \quad s \in \mathcal{S}, \end{aligned} \tag{11.7}$$

where $Q_n(s, a)$ is an estimate of the optimal Q-function at iteration n. The operator C which maps the function V_n to V_{n+1} is a **contraction** operator, and (11.3) shows that the optimal value function V^* is a fixed point of this operator; that is, $C[V^*] = V^*$. Exercise 2 demonstrates that, under the condition of boundedness of r, the sequence (V_n) of value functions in (11.7) converges to the optimal V^*; see also [141, 170, 167]. Once the optimal value function is determined, the optimal (greedy) policy follows from (11.4). The value iteration algorithm is summarized in the following pseudo-code.

Algorithm 11.8: Value Iteration

Input: Reward function r, transition function p, discount rate $\gamma \in (0, 1)$, and convergence threshold $\varepsilon > 0$.
Output: Optimal value function V^* and optimal policy π^*.

```
1 Arbitrarily initialize value function v. Initialize Δ > ε.
2 while Δ > ε do
3     Δ ← 0
4     for s ∈ S do
5         u ← v(s)
6         v(s) ← max_{a∈A} {r(s,a) + γ Σ_{s'∈S} p(s' | s,a) v(s')}
7         π(s) ← argmax_{a∈A} {r(s,a) + γ Σ_{s'∈S} p(s' | s,a) v(s')}
8         Δ ← max{Δ, |u − v(s)|}
9 return v, π
```

Instead of waiting until the end of each iteration to update the entire value function, Algorithm 11.8 improves efficiency by immediately using the updated values of states within the same iteration. Note also that in Step 7 the policy is updated "on the fly". It could also have been evaluated separately, after the optimal value function has been determined for all states $s \in \mathcal{S}$. Further, in Step 8, the convergence condition is updated — when $v(s)$ no longer changes by more than ε for all states $s \in \mathcal{S}$ then the algorithm with terminate.

> The convergence threshold Δ often has to be chosen very small (say 10^{-15}) to guarantee correct convergence to the value function.

Example 11.9 (MDP: Value Iteration) The Markov decision process in Example 11.2 has state space $\mathcal{S} = \{0, 1, 2, 3\}$ and action space $\mathcal{A} = \{0, 1\}$. The options for the transition function can be collated into a $2 \times 4 \times 4$ array (i.e., tensor) $\mathbf{P} = [\mathbf{P}_0; \mathbf{P}_1]$, where $\mathbf{P}_0$ and $\mathbf{P}_1$ are given in Example 11.2. For example, the policy $[0, 0, 1, 1]$ would then induce a transition matrix $\widetilde{\mathbf{P}}$ with rows

$$\mathbf{P}[0, 0, :],\ \mathbf{P}[0, 1, :],\ \mathbf{P}[1, 0, :],\ \mathbf{P}[1, 1, :],$$

where $\mathbf{P}[i, j, :]$ is the jth row of $\mathbf{P}_i$, $i = 0, 1$. The immediate reward function is given by $r(s) = \mathbb{1}_{\{s=3\}}$. Suppose we use an expected discounted total reward $\mathbb{E} \sum_{t=0}^{\infty} \gamma^t r(S_t)$ as our overall objective, rather than the long-run average reward. We can then use value iteration to determine the optimal value function and corresponding policy. In the Python code below, which will be reused several times, we define the state and action spaces, as well as the $2 \times 4 \times 4$ tensor $\mathbf{P}$.

MDP.py

```
import numpy as np

P = np.array([
  [0, 0.1, 0.1, 0.8],
  [0.1, 0.8, 0.1, 0],
  [0.1, 0.9, 0, 0],
  [0.1, 0, 0, 0.9],
  [0, 0.4, 0.2, 0.4],
  [0.1, 0, 0.9, 0],
  [0.1, 0, 0, 0.9],
  [0.9, 0.1, 0, 0]
])
P = P.reshape(2, 4, 4)

def r(s):  # Immediate reward function
    return float(s == 3)

States = np.array([0, 1, 2, 3], dtype=int)
Actions = np.array([0, 1], dtype=int)
```

Importing the definitions above and using a discount rate $\gamma = 0.9$, we find exactly the same optimal policy $\pi^* = [0, 1, 1, 0]$ as before.

MDP_Val_Iter.py

```
from MDP import *

gamma = 0.9 # Discount rate
epsilon = 1e-15  # Convergence threshold
v = np.zeros(len(States))  # Initialize value function
q = np.zeros(len(Actions))  # Initialize Q-value
pi = np.zeros(len(States), dtype=int)  # Initialize policy
Delta = np.infty  # Initialize Delta = infinity

while Delta > epsilon:
    Delta = 0
    for s in States:
        u = v[s]
        for a in Actions:
            q[a] = r(s)
            for s_next in States:
                q[a] += gamma * P[a, s, s_next] * v[s_next]

        v[s] = max(q)
        pi[s] = np.argmax(q)
        Delta = max(Delta, abs(u - v[s]))

print("Policy pi\n", pi)
print("\nValue function\n", v)
```

```
Policy pi
[0 1 1 0]
Value function
[7.81604587 7.15549541 7.96549541 8.96549541]
```

■ **Remark 11.10 (Unit Discount Rate)** If $\sup_{s,a} |r(s,a)| \leq M$ for some M, then $|V^*(s)| \leq \frac{M}{1-\gamma}$. A value of $\gamma = 1$ in the continuing case is not desirable as typically the sequence of rewards will be $\pm\infty$. However, for the episodic case (i.e., with terminal states) a value of $\gamma = 1$ is possible and corresponds to no discounting of future rewards; that is, each reward current or future is treated as equally important. In the continuing case there is a connection beteen the long-run average-reward value function (also called the *gain*) $\overline{V}^{\pi} = \lim_{T\to\infty} T^{-1} \sum_{t=0}^{T} R_t$ and the expected discounted reward value function $V^{\pi}(s;\gamma) = \mathbb{E} \sum_{t=0}^{\infty} \gamma^t R_t$, in that

$$\overline{V}^{\pi}(s) = \lim_{\gamma \uparrow 1^-} (1-\gamma) V^{\pi}(s;\gamma).$$

■

■ **Example 11.11 (Torus MDP: Value Iteration)** As a more elaborate example, consider a robot moving in a random terrain, as illustrated in Figure 11.13. The terrain has troughs and peaks and at each time step the robot receives a reward equal to the height of the terrain at the robot's position. The state space (possible positions) of the robot is the grid set $\mathcal{S} = \{0, \ldots, 19\}^2$ and the action space of the robot is $\mathcal{A} = \{0, 1, 2, 3\}$ corresponding to the directions "up", "right", "down", and "left", respectively. The borders "wrap around" like a torus. For example, if the robot is in state $(0, 10)$ and makes a left move (action 3), then it lands in state $(19, 10)$. The reward function only depends on the current state and not on the action and is based on the following modification of Matlab's *peaks* function:

$$\begin{aligned} f(x_1, x_2) = {} & 3(1-x_1)^2 \exp(-x_1^2 - (x_2+1)^2) \\ & - 10(x_1/5 + x_1^3 - x_2^5) \exp(-x_1^2 - x_2^2) \\ & - (1/3) \exp(-(x_1+1)^2 - x_2^2). \end{aligned} \tag{11.12}$$

Specifically, the reward function is given by

$$r(s) = f(-3 + 0.3i, \; -3 + 0.3j), \quad s = (i, j) \in \mathcal{S}.$$

Our goal is to find a stationary deterministic policy $\pi : \mathcal{S} \to \mathcal{A}$ that maximizes $\sum_{t=0}^{\infty} \gamma^t r(s_t)$. Note that in this case there is no randomness in the state transitions and in the total discounted reward. Starting from some state and following the optimal policy, the robot will trace out a specific path through the terrain that maximizes its total discounted height.

In the following Python code each state s is implemented as a tuple, $s = (i, j)$. The `numpy` method `ndindex` returns an iterator over all 20×20 possible states. Value iteration is performed to determine the optimal value function. This function is used to compute a path (i.e., sequence of states) of the MDP starting from state $(12, 5)$ and following the optimal policy for 50 time steps. Notice how the path travels first to the vicinity of the local maximizer $(7, 8)$ and then to the vicinity of the global maximizer $(10, 16)$ of the reward function. After 18 steps, the states in the path alternate between $(10, 15)$ and $(10, 16)$.

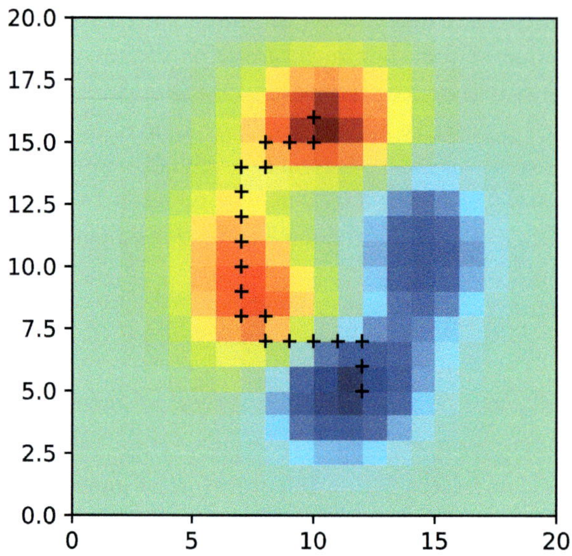

Figure 11.13: The reward function and optimal path (starting from $(12, 5)$) selected according to the optimal policy. The colors represent the height of the value function, with warmer colors being higher, and cooler colors being lower.

Torus_Val_Iter.py

```
import numpy as np
import matplotlib.pyplot as plt
from matplotlib import cm

def mypeaks(X, Y):
    return (3 * (1 - X)**2 * np.exp(-X**2 - (Y + 1)**2) -
            10 * (X / 5 + X**3 - Y**5) * np.exp(-X**2 - Y**2) -
            (1 / 3) * np.exp(-(X + 1)**2 - Y**2))

X, Y = np.meshgrid(np.arange(-3, 3, 0.3), np.arange(-3, 3, 0.3))
R = mypeaks(Y.T, X.T).T
dims = R.shape
dim = dims[0]

def r(s):
    return R[s]

def wrap(i, n):
    return i % n

z = np.array([[1, 0], [0, 1], [-1, 0], [0, -1]])

def next_state(s, a):
    s1 = s + z[a]
    return tuple(np.array([wrap(s1[0], dim), wrap(s1[1], dim)]))

gamma = 0.95
epsilon = 1e-15
```

```
v = np.zeros(dims)
q = np.zeros(4)
pi = np.zeros(dims, dtype=int)
Delta = np.inf

while Delta > epsilon:
    Delta = 0
    for s in np.ndindex(dims):
        u = v[s]
        for a in range(4):
            q[a] = r(s) + gamma * v[next_state(s, a)]
        v[s] = np.max(q)
        pi[s] = np.argmax(q)
        Delta = max(Delta, abs(u - v[s]))

print("Policy Pi\n", pi)
print("Value function\n", v)

plt.imshow(R.T, origin="lower", extent=[0, dim, 0, dim], aspect=1, cmap=
    cm.jet)

s = (12, 5)
path_x, path_y = [s[0]], [s[1]]
for _ in range(51):
    s = next_state(s, pi[s])
    path_x.append(s[0])
    path_y.append(s[1])

plt.plot(path_x, path_y, 'k+')
```

■

11.3.2 Policy Iteration

The key idea of **policy iteration** is to successively take a policy, evaluate its value function and then use this to improve the Q-function and matching policy. The value function of a policy π can be computed as the limit of the iterative procedure:

$$V_{n+1}(s) = r(s, \pi(s)) + \gamma \sum_{s' \in \mathcal{S}} p(s' \mid s, \pi(s)) V_n(s'). \tag{11.14}$$

Exercise 3 shows that V^π is a fixed point of (11.14); that is, V^π satisfies, for all $s \in \mathcal{S}$,

$$V^\pi(s) = r(s, \pi(s)) + \gamma \sum_{s' \in \mathcal{S}} p(s' \mid s, \pi(s)) V^\pi(s').$$

Once V^π has been determined, we also know the corresponding Q-function

$$Q^\pi(s, a) = r(s, a) + \gamma \sum_{s' \in \mathcal{S}} p(s' \mid s, a) V^\pi(s')$$

We can improve on the policy π (or at least do no worse) by choosing a new policy π_{new} via

$$\pi_{\text{new}}(s) = \underset{a \in \mathcal{A}}{\operatorname{argmax}}\, Q^\pi(s, a), \quad s \in \mathcal{S}.$$

The following policy iteration algorithm repeats this procedure until the policy no longer changes. Similar to value iteration, policy iteration is guaranteed to converge to the optimal policy. However, the number of policy improvements required may increase exponentially with the number of states.

Algorithm 11.15: Policy Iteration

Input: Immediate reward function r, state transition function p, discount rate $\gamma \in (0, 1)$, and convergence threshold $\varepsilon > 0$.
Output: Optimal deterministic policy π.

Arbitrarily initialize v, π, and $\pi_{\text{old}} \neq \pi$. Initialize $\Delta > \varepsilon$.
while $\max_{s \in \mathcal{S}} |\pi(s) - \pi_{\text{old}}(s)| > 0$ **do**
  $\pi_{\text{old}} \leftarrow \pi$
  while $\Delta > \varepsilon$ **do** `// Compute the value function of π`
    $\Delta \leftarrow 0$
    for $s \in \mathcal{S}$ **do**
      $u \leftarrow v(s)$
      $v(s) \leftarrow r(s, \pi(s)) + \gamma \sum_{s' \in \mathcal{S}} p(s' \mid s, \pi(s))\, v(s')$
      $\Delta \leftarrow \max\{\Delta, \max\{|u - v(s)|\}\}$
  for $s \in \mathcal{S}$ **do** `// Update the policy`
    $\pi(s) \leftarrow \text{argmax}_{a \in \mathcal{A}} \{r(s, a) + \gamma \sum_{s' \in \mathcal{S}} p(s' \mid s, a)\, v(s')\}$
return π

■ **Example 11.16 (MDP: Policy Iteration)** As in Example 11.9, we use a discounted total reward as our objective for the 4-state Markov chain. The following Python code uses policy iteration to find the solution of the MDP. Various objects were reused from the `MDP` module in Example 11.9.

MDP_Pol_Iter.py

```
from MDP import *

gamma = 0.9  # Discount rate


def calcv(pi):
    epsilon = 1e-15  # Convergence threshold
    v = np.zeros(len(States))  # Initialize value function with zeros
    Delta = np.inf
    while Delta > epsilon:
        Delta = 0
        for s in States:
            old_value = v[s]
            new_value = r(s)
            for s_next in States:
                new_value += gamma * P[pi[s], s, s_next] * v[s_next]
            v[s] = new_value
            Delta = max(Delta, abs(old_value - v[s]))
```

```
    return v


q = np.zeros(len(Actions))  # Initialize action values
pi = np.ones(len(States), dtype=int)  # Intialize policy
piold = np.zeros(len(States), dtype=int)

while max(abs(pi-piold)) > 0:
    pi, piold = piold, pi  # Swap objects
    v = calcv(pi)  # Calculate value function for pi

    for s in States:
        for a in Actions:  # compute Q function for state s
            q[a] = r(s)
            for s_next in States:
                q[a] += gamma * P[a, s, s_next] * v[s_next]

        pi[s] = np.argmax(q)  # Update policy

print("Policy pi\n", pi)
print("Value v\n", calcv(pi))
```

```
Policy pi
[0 1 1 0]
Value v
[7.81604587 7.15549541 7.96549541 8.96549541]
```

As expected, the resulting optimal policy and value function are the same as in Example 11.9. ■

11.4 Reinforcement Learning Concepts

In reinforcement learning, there are many possible ways to specify the agent's learning process. These can be classified in various ways, as described below.

In **model-based** scenarios, the agent tries to learn a model of the environment dynamics and uses this to inform its optimal policy (such as in Section 11.3). For instance, the agent may notice that, starting from some state, a particular action may lead to a transition to other states in a certain proportion. A model for these transitions can be built over time. In contrast, in **model-free** scenarios, the agent simply takes actions and experiences rewards, and must build its optimal policy from this information alone. Using our previous example, the agent takes actions at states according to its current policy, then collects rewards, but does not attempt to build a model of the transitions taking place. Model-free algorithms tend to be relatively simple, but they often require a large number of interactions with the environment to learn a policy, whereas model-based algorithm, while more complex, are more explainable, use information from the environment more efficiently, and enjoy greater stability. Model-based methods are discussed further in Sections 11.5 and 11.7, and we turn to model-free methods in Sections 11.6 and 11.8.

Reinforcement learning methods can also be divided into **off-policy** and **on-policy** methods. Off-policy methods allow the agent to learn from experience generated by a policy different to the one being optimized. For example, the agent policy could be a completely random exploration policy over available actions, while the target policy may be the one that maximizes the overall reward. A key example here is the epsilon-greedy policy which balances exploration and exploitation (see Algorithm 11.23). On-policy methods by contrast learn directly from the experience generated using the policy that is being optimized. An example here is the SARSA method discussed in Section 11.6.

Depending on how the policy is determined, reinforcement learning methods can be classified as **value-based**, **policy-based**, or **actor–critic**. Value-based methods learn the optimal values (i.e., utilities) of the states or (state, action)-pairs and derive the optimal policy from these (such as in Section 11.3, which contains the basic algorithms for value and policy iteration). They tend to be simple, but suffer in the high-dimensional or continuous settings. Policy-based algorithms learn the agent's policy in a more direct way. Such algorithms often involve parameterized policies. These are more effective for handling continuous action spaces, but can suffer from slow convergence and high variance in their estimates. Actor–critic algorithms are a combination of the previous two, where the "actor" represents the policy and the "critic" provides feedback on the quality of actions taken by the actor by estimating the value of the states or (state, action)-pairs. This leads to an accelerated learning process, good convergence properties, and more efficient use of sampled information. Section 11.8 describes various methods for policy improvement.

Finally, one can divide reinforcement learning methods into **online** or **offline**. Online is the setting where the agent is able to continuously interact with its environment and learn its policy based on immediate feedback. It is the natural framework where such real-time interaction is possible. The offline setting applies when the agent is able to pre-compute the optimal policy without interacting with the environment; for example, when the environment dynamics are completely known. Offline methods also occur in situations where interaction with the environment is dangerous, expensive, or impractical. In this setting, the optimal policy is derived only from historical interactions or makes use of known environments and rewards (see Section 11.3). We deal further with the offline setting in 11.5 and with the online setting in Sections 11.6, 11.7, and 11.8.

11.5 Dynamic Programming

We have so far focused on MDPs in continuing and episodic scenarios. Recall that episodic scenarios have one or more terminating states; and once the MDP enters such a state it will remain there forever, with subsequent zero immediate rewards. Thus, in both cases the cumulative discounted reward is of the form

$$\mathbb{E} \sum_{t=0}^{\infty} \gamma^t R_t .$$

The advantage of this infinite horizon formulation is that optimal stationary policies exist and may be computed via value iteration. For finite-horizon scenarios (that is, fixed $T < \infty$), the optimal policy is usually *non-stationary*. However, Bellman's optimality principle still applies in these finite-horizon scenarios, allowing the optimization problem to be broken up into a simpler subproblems in a recursive manner. Specifically, defining S_t, A_t, and R_t again as the state, action, and immediate reward at time t, suppose that the objective is to maximize the total expected reward $\mathbb{E}\sum_{t=0}^{T} R_t$, where $R_t = r(S_t, A_t), t = 0, \ldots, T-1$ and $R_T = F(S_T)$ for some known functions r and F. Let

$$V_t(s) = \mathbb{E}\Big[\sum_{k=t}^{T} R_k \,\Big|\, S_t = s\Big], \quad t = 0, \ldots, T,$$

be the optimal expected reward that can be obtained in the remaining time steps from t to T. Bellman's optimality principle yields the recursion

$$V_t(s) = \underset{a \in \mathcal{A}}{\operatorname{argmax}}\{r(s,a) + \mathbb{E}[V_{t+1}(S_{t+1}) \,|\, S_t = s, A_t = a]\}, \quad t = 0, \ldots, T-1,$$

with $V_T(s) = F(s)$. This recursion can, in principle, be solved *in reverse time order*; that is, starting from $t = T-1$, until at $t = 0$ the optimal value function V_0 is obtained. The optimal action at time t and state s then follows from

$$\pi_t(s) = \underset{a \in \mathcal{A}}{\operatorname{argmax}}\, V_t(s,a).$$

The resulting optimization procedure is called **dynamic programming**.

Example 11.17 (Robot Mining) Suppose that we have a gold mining facility with n robots. Each robot can either mine gold or make an identical copy of itself. Each robot that mines gold produces 1 unit of gold per time unit. To manufacture a new robot requires an amount c of gold. It takes 1 unit of time to manufacture a new robot, and during this time period the manufacturing robot cannot mine gold. The overall goal is to maximize the amount of gold mined over a predefined time horizon T.

This is a perfect example a sequential decision making problem where, at each time stage, we have to decide how many robots to allocate to mining gold, and how many robots to allocate to manufacturing more robots.

Let the system state at time t be (n_t, g_t), where n_t and g_t are respectively the number of robots and the amount of gold available at time t, where $t = 0, 1, \ldots, T$. Let $a_t \in \{0, 1, \ldots, \lfloor g_t/c \rfloor\}$ be the number of robots that will be manufacturing during the time period $[t, t+1)$. The expression in the floor function is the maximum number of robots that could be manufactured at that time stage with the gold available then.

Starting from some state (n_0, g_0) the question is how to choose a policy $a_0, a_1, \ldots, a_{T-1}$ that maximizes the total amount of gold g_T.

To solve this problem, define $V(n, g, t)$ as the maximum total amount of gold that can be obtained in the interval $[t, T]$, starting from state (n, g) at time t and following an optimal policy. For $t = T-1$, it is obvious what to do: only mine gold. Consequently, $V(n, g, T-1) = g + n$ and $a_{T-1} = 0$.

For $t = T-2, \ldots, 0$, Bellman's optimality principle gives the recursion:

$$V(n, g, t) = \max_{a \in \{0, \ldots, \lfloor g_t/c \rfloor\}} V(n+a, g+n-a-ca, t+1).$$

The recursion comes from considering the one-step behavior: if we manufacture a robots then at the next stage we will have $n+a$ robots available at time $t+1$, but the gold mined will be $g+n-a$ (as a fewer than n will be mining in the previous period) and gold will further be reduced by an amount ca. This recursion completely determines the value function V. However, the number of steps in the recursion grows exponentially in T, as the amount of gold available grows.

As a concrete example, suppose we start with one robot and no gold at time 0. Let the cost of a robot be $c = 1$ and the time horizon be $T = 10$. Then $V(1, 0, 1) = 32$.

Below is a Python implementation of the value function together with a function call to display the result for the situation just described. It uses the LRU cache feature which saves the called function results, so that repeated calls with the same function arguments do not have to be recomputed. The result of this is a speedup in the program.

robotmining1.py

```
from functools import lru_cache
import numpy as np

T = 10  # time horizon
cost_rob = 1  # cost of making a robot. Can be 0 or noninteger

@lru_cache(maxsize=None)
def valfun(n, g, t, cost_rob, T):
    if t >= T - 1:
        return g + n
    if cost_rob == 0:
        max_rob = n
    else:
        # determine how many robots can be manufactured at the most
        max_rob = min([n, int(np.floor(g / cost_rob))])
    rem_reward = np.zeros(max_rob + 1)
    for a in range(max_rob + 1):  # include max_rob in the range
        newg = g + n - a - cost_rob * a
        # remaining reward for action a
        rem_reward[a] = valfun(n + a, newg, t + 1, cost_rob, T)

    return max(rem_reward)


print(valfun(1, 0, 0, cost_rob, T))
```

```
(32.0, 0)
```

The policy is computed recursively. With an input as above, the function repeatedly calls itself until it hits the boundary condition: if we are at the next to last time stage $T - 1$, then the optimal action is for all robots to mine, and so the value is just $g + n$. Otherwise it determines the maximum number of robots that could be manufactured with the gold available (considering the cost of each robot to manufacture) and iterates over every possible choice, with the optimal action maximizing the reward at that stage.

robotmining2.py

```
from robotmining1 import *
n = 1
g = 0
t = 0  # starting values
```

```
actions = []  # store the actions
state = [n, g]  # n, gold
states = [state]  # store the states
for t in range(T):
    if cost_rob == 0:
        max_rob = n
    else:
        max_rob = min([n, int(np.floor(g / cost_rob))])
    rem_reward = np.zeros(max_rob + 1)
    for a in range(max_rob + 1):  # include max_rob in the range
        newg = g + n - a - cost_rob * a
        # remaining reward for action a
        rem_reward[a] = valfun(n + a, newg, t + 1, cost_rob, T)

    if t == T-2:
        a = 0
    else:
        a = np.argmax(rem_reward)

    actions.append(a)
    g = g + n - a - cost_rob*a
    n = n + a
    states.append([n, g])


print(f"{'Time':<5} {'Robots':<5} {'Gold':<5} {'Manuf.':<5} {'Mine':<5}"
    )

for i, s, a in zip(range(len(states)), states, actions):
    print(f"{i:<5} {s[0]:<6} {s[1]:<5} {a:<6} {s[0] - a}")
print(f"{T:<5} {states[-1][0]:<6} {states[-1][1]:<5} ")
```

An optimal policy for a time horizon of $T = 10$ starting with one robot and no gold at time 0, with the cost of a robot at $c = 1$, is given in Table 11.18.

Table 11.18: Optimal policy for the robot mining scenario.

time	robots	gold	manufacture	mine
0	1	0	0	1
1	1	1	1	0
2	2	0	0	2
3	2	2	1	1
4	3	2	2	1
5	5	1	1	4
6	6	4	4	2
7	10	2	0	10
8	10	12	0	10
9	10	22	0	10
10	10	32		

Note that the policy at each stage is a mix of robots mining for gold, and robots manufacturing more robots. Also note that this is just one of the optimal policies. For instance the last few lines of this table could have been:

Table 11.19: Optimal policy for the robot mining scenario (alternative).

time	robots	gold	manufacture	mine
7	10	2	2	8
8	12	8	0	12
9	12	20	0	12
10	12	32		

This approach will only work for relatively small T, as was mentioned earlier, as the number of states in the recursive dynamic programming expression explodes as T (and consequently the amount of gold available to manufacture robots) increases exponentially. For larger T a Monte Carlo based approach would scale well but lose the guarantee that an optimal policy is found. ■

11.6 Temporal Difference Learning

A **temporal–difference** method is a "trial-and-error" learning approach in which the value function (or, alternatively, the Q-function) is estimated in such a way that the "error" (that is, the difference between subsequent estimates) is used to correct the estimate in an online fashion. These methods have the advantage that they require no model (that is, specification of the transition function p), but only a mechanism to generate states, actions, and rewards. We will discuss three methods in this class: **TD**, **SARSA**, and **Q-learning**. All of them are temporal–difference algorithms, and TD directly updates the value estimates using the difference in value function estimates from successive states, while SARSA and Q-learning do the same with the Q-function, though their method for calculating differences is distinct.

We now describe the simplest (one-step) temporal difference (TD) method for evaluating a policy π. The starting point is Exercise 3, which shows that the value function V^π of policy π satisfies the equation

$$V^\pi(s) = r(s, \pi(s)) + \gamma \mathbb{E}_\pi\big[V^\pi(S_1) \,\big|\, S_0 = s\big], \quad s \in \mathcal{S}. \tag{11.20}$$

In Section 11.3.2 this motivated the iterative procedure (11.14) in policy iteration. Alternatively, (11.20) suggests that if $V(S)$ is a current estimate of the value function at a (random) state S then, after taking action $A = \pi(S)$ and observing the next state S' and reward $R = r(S, A)$, this estimate should be updated via $V(S) \leftarrow V(S) + D(S)$, where the difference $D(S)$ is given by

$$D(S) = R + \gamma V(S') - V(S).$$

However, it is customary to use instead a smoothed update

$$\begin{aligned} V(S) &= (1-\alpha)V(S) + \alpha D(S) \\ &= V(S) + \alpha(R + \gamma V(S') - V(S)), \end{aligned}$$

where $\alpha \in (0, 1]$ is the called the **learning rate**. This is the essence of the TD algorithm presented below. Although we have formulated it for episodic scenarios, the method can be adapted to continuing scenarios by taking the terminal condition to be $t \leq T$ for sufficiently large T.

Algorithm 11.21: TD Method for Estimating V^{π}

Input: Policy π, learning rate $\alpha \in (0, 1]$, discount rate $\gamma \in [0, 1]$, and terminal condition.
Output: Estimator V of V^{π}.

```
Arbitrarily initialize V except for terminal states which are initialized to zero.
for N episodes do
    Initialize S
    repeat
        A ← π(S)
        Take action A, observe R, S'
        V(S) ← V(S) + α (R + γV(S') − V(S))
        S ← S'
    until terminal condition is met
return V
```

> In practice, the learning rate α is often chosen to be a decaying function of the iteration; for example, $\alpha_t = 1/\sqrt{t+2}$ on iteration t.

Note that the form of the update appearing in Algorithm 11.21 is reminiscent of "smoothing" between the old estimate and the new estimate, as appears for example in the cross–entropy method; see Section 3.4.2. ☞ 112

Example 11.22 (MDP: Temporal Difference) We continue Examples 11.2, 11.9, and 11.16, using a discounted value function, with discount rate $\gamma = 0.9$. Although this is a continuing scenario, if we terminate each episode after $T = 100$ steps, then the difference between

$$V^{\pi}(s) = \mathbb{E}_{\pi}\left[\sum_{t=0}^{\infty} \gamma^t R_t \,\middle|\, S_0 = s\right]$$

and $\mathbb{E}_{\pi}[\sum_{t=0}^{T} \gamma^t R_t \mid S_0 = s]$ is less than $\gamma^{T+1}/(1-\gamma) \leq 3\times 10^{-4}$, irrespective of the policy π. In the Python code below, we estimate V^{π} for the optimal policy $\pi = [0, 1, 1, 0]$, using learning rate $\alpha = 0.1$. A uniform random policy, for example, will lead to significantly smaller values for the value function. The package `tqdm` provides a convenient progress bar for the computation.

MDP_Temp_Diff.py

```
from MDP import *
from tqdm import tqdm  # progress bar

gamma = 0.9   # Discount rate
```

```
v = np.zeros(len(States))  # Initialize value function
N_episodes = 1000


def policy(s):
    Pi = np.array([0, 1, 1, 0], dtype=int)  # the optimal policy
    out = Pi[s]
    # out = np.random.choice(Actions) # uniform random policy
    return out


T = 100
for _ in tqdm(range(N_episodes)):
    S = np.random.choice(States)
    t = 0
    for t in range(T):
        # alpha = 1/np.sqrt(t+1)
        alpha = 0.1
        A = policy(S)
        R = r(S)
        Sp = np.random.choice(States, p=P[A, S])
        v[S] += alpha * (R + gamma * v[Sp] - v[S])
        S = Sp

np.set_printoptions(precision=3)
print("\nEstimated value function:", v)
```

```
100%     1000/1000 [00:02<00:00, 425.35it/s]
Estimated value function: [7.9    7.099 8.028 8.902]
```

■

It is important to note that the TD algorithm above estimates the value function for a fixed policy π, but does not improve it. To find the optimal policy, one could iteratively update the current policy in a similar way as in policy iteration (Section 11.3.2), or ☞ 118 employ a noisy optimization scheme (Section 3.4.4).

Temporal-difference methods that do seek to improve the policy towards the optimal policy π^* usually update estimates of the Q-function rather than the value function. In SARSA one starts with an estimate Q of the optimal Q-function Q^* and uses Bellman's equations (11.5) to update this estimate via

$$\begin{aligned} Q(S, A) &= (1 - \alpha)Q(S, A) + \alpha(R + \gamma Q(S', A')) \\ &= Q(S, A) + \alpha(R + \gamma Q(S', A') - Q(S, A)), \end{aligned}$$

where S' and A' are the new state and action based on the current Q. In particular, when a *greedy* policy is used, the new action is given by $A' = \operatorname{argmax}_{a\in\mathcal{A}} Q(S, a)$. However, it may be more effective to replace the greedy with the following **ε-greedy** policy based on Q. Such a policy offers a balance between the *exploitation* of known actions and the *exploration* of new actions.

Algorithm 11.23: ε-Greedy Policy

Input: Q-function Q, parameter $\varepsilon \in [0, 1]$, action space $\mathcal{A}$, state $s \in \mathcal{S}$.
Output: Action according to the ε-greedy policy.

Draw $U \sim \mathcal{U}[0, 1]$
if $U \leq \varepsilon$ **then**
  Select action A uniformly from $\mathcal{A}$
else
  Select action A for state $s \in \mathcal{S}$ greedily, according to:
   $\pi(s) = \text{argmax}_{a\in\mathcal{A}} Q(s, a)$
return A

A common approach is to gradually decrease ε after each iteration, so as to increasingly favor exploitation over exploration. For example, an exponential decay is achieved with the scheme $\varepsilon_t = \beta^t$ for some decay rate $\beta \in (0, 1)$.

Algorithm 11.24 summarizes the SARSA method. Note that its name reflects the order in which states, actions, and reward are processed: S, A, R, S', A'.

Algorithm 11.24: SARSA for estimating Q^*

Input: Learning rate $\alpha \in (0, 1]$, small $\varepsilon > 0$, discount rate $\gamma \in (0, 1)$, and terminal condition.
Output: Estimator Q of Q^*.

Initialize Q (0 for terminal states)
for N episodes **do**
  Initialize state S
  Choose A from S using policy derived from Q (e.g., ε-greedy)
  repeat
   Take action A, observe R, S'
   Choose A' from S' using policy derived from Q (e.g., ε-greedy)
   $Q(S, A) \leftarrow Q(S, A) + \alpha(R + \gamma Q(S', A') - Q(S, A))$ // Update Q
   $S \leftarrow S'$, $A \leftarrow A'$
  until terminal condition is met
return Q

SARSA thus updates the Q-value $Q(S, A)$ based solely on the action A' taken by the agent, but not on other possible actions. As this action is derived from the current Q-value/policy, the method is said to be an **on-policy** temporal-difference learning algorithm. In contrast, Q-learning is an example of an **off-policy** temporal-difference learning algorithm, in which the agent can learn from actions that are not necessarily taken by the current policy. The Q-learning algorithm is very similar to the SARSA algorithm. The main change is that the difference term $R + \gamma Q(S', A') - Q(S, A)$ in SARSA is replaced with $R + \gamma \max_{a'} Q(S', a') - Q(S, A)$ in Q-learning. There is no need to sample a successive action A', as all possible actions are considered to update the Q-function. This results in the Q-learning method in Algorithm 11.25.

Algorithm 11.25: Q-learning for estimating Q^*

Input: Learning rate $\alpha \in (0, 1]$, small $\varepsilon > 0$, and discount rate $\gamma \in (0, 1)$, and terminal condition.

Output: Estimator Q of Q^*.

Initialize Q (0 for terminal states)
for N episodes **do**
  Initialize S
  repeat
    Choose A from S using policy derived from Q (e.g., ε-greedy)
    Take action A, observe R, S'
    $Q(S, A) \leftarrow Q(S, A) + \alpha(R + \gamma \max_{a'} Q(S', a') - Q(S, A))$ // Update Q
    $S \leftarrow S'$
  until terminal condition is met
return Q

The scheme converges to the optimal policy under the assumption that each state–action pair is visited an infinite number of times and the learning parameter α is appropriately decreased; see, for example, [179] and [13]. For both SARSA and Q-learning approaches, the final estimate Q of Q^* yields estimates of the value function V^* and optimal policy π^* simply by taking the maximum and argmax of Q.

■ **Example 11.26 (MDP: Q-Learning)** We continue with the running Examples 11.2, 11.9, 11.16, and 11.22, and now apply Q-learning to learn the optimal Q-function. This also gives an estimate of the optimal value function and policy, by taking the max and argmax with respect to actions, respectively.

The optimal value function found in Example 11.9 is given by $[V^*(0), \ldots, V^*(3)] = [7.82, 7.16, 7.97, 8.97]$, rounded to two decimal places. Combining this with (11.5), i.e.,

$$Q^*(s, a) = \mathbb{1}_{\{s=3\}} + \gamma\, \mathbb{E}[V^*(S_1) \mid S_0 = a, A_0 = a],$$

we find that the optimal Q-function is given by the matrix

$$\mathbf{Q}^* = \begin{bmatrix} 7.82 & 7.24 \\ 6.57 & 7.16 \\ 6.50 & 7.97 \\ 8.97 & 7.97 \end{bmatrix}. \tag{11.27}$$

In the Python code below, we adopt the ε-greedy strategy with a fixed $\varepsilon = 0.1$. To ensure proper convergence of the estimates, it is important to use a decaying schedule for the learning rate α. Here we apply an exponential decay with a decay parameter equal to $\beta = 0.9999$, similar to the *geometric cooling* scheme in simulated annealing;
☞ 107 see Section 3.4.1. The number of episodes is $N = 1000$, and each episode runs for $T = 100$ steps. For these parameter choices we observe a fairly accurate estimation of the optimal Q-function, optimal value function, and optimal policy. Figure 11.28 illustrates the convergence of the Q-function for state $s = 3$ and action $a = 0$.

MDP_Q-learn.py

```
from MDP import *
import matplotlib.pyplot as plt

np.random.seed(123)

gamma = 0.9  # Discount rate

q = np.random.rand(len(States),
                   len(Actions))  # Arbitrarily initialize estimate Q


def epsgreedy(Q, S, epsilon):
    if np.random.rand() < epsilon:
        A = np.random.choice(Actions)
    else:
        A = np.argmax(Q[S, :])
    return A


N = 1000  # Number of episodes
T = 100  # Max iterations
alpha = 1  # Initial alpha
eps = 0.1  # exploration probability
beta = 0.9999  # decay rate for alpha

q30_store = np.zeros(N)  # store for plotting
for n in range(N):
    S = np.random.choice(States)  # Initialize state
    for t in range(T):
        A = epsgreedy(q, S, eps)
        R = r(S)
        Sp = np.random.choice(States, p=P[A, S])
        q[S, A] += alpha * (R + gamma * np.max(q[Sp, :]) - q[S, A])
        S = Sp
        alpha *= beta  # reduce alpha by factor beta
    q30_store[n] = q[3,0]  # Store the Q-value for state len(States) and
         action 1

np.set_printoptions(precision=3)
print("Optimal Q-function estimate Q\n", q)
print("Optimal Value function estimate ", np.max(q, axis=1))
print("Learned policy pi\n", np.argmax(q, axis=1))
plt.plot(q30_store)
plt.axhline(8.97, color='r', linestyle='--')
```

```
Optimal Q-function estimate Q
 [[7.817 7.194]
  [6.621 7.146]
  [6.467 7.968]
  [8.982 7.98 ]]
Optimal Value function estimate  [7.817 7.146 7.968 8.982]
Learned policy pi
 [0 1 1 0]
```

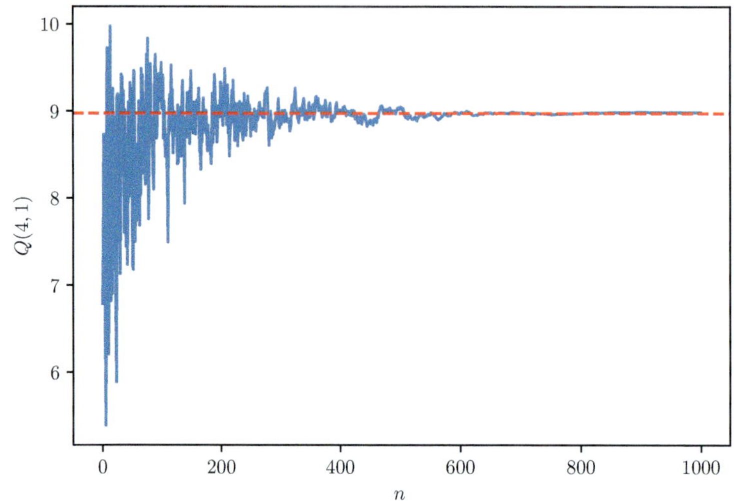

Figure 11.28: Evolution of the estimate of the optimal Q-function at state 3 and action 0. The true value is given by the red dashed line.

■

A key advantage of all three methods discussed here is that they do not require us to have a model for the transitions — we simply require a method of sampling states and actions based on the current state. This permits us to estimate the value function or Q-function directly, without estimating an MDP model first.

11.7 Monte Carlo Methods

Monte Carlo methods learn from simulated experience of sequences of states, actions, and rewards. These histories of agent interaction are used to form estimates of the value function or Q-function, which are then used for learning the agent's policy.

Given a policy, they can be used for estimating the associated value function by simulating sequences of states, actions, and rewards and then averaging discounted rewards, which is a direct estimate of the definition of the value function. This approach can be incorporated with, for example, policy iteration wherein the value function for the current policy is estimated over the episode.

As only a finite number of states can be simulated, Monte Carlo methods inherently apply to *terminating* scenarios, and in particular to *episodic* scenarios.

11.7.1 Estimating the Value Function via Monte Carlo

Recall that we have the value function for policy π, state s, and discount factor γ, defined as the following expectation:

$$V^{\pi}(s) = \mathbb{E}_{\pi}\left[\sum_{t=0}^{\infty} \gamma^t R_t \,\middle|\, S_0 = s\right].$$

In the episodic case, all rewards after reaching the terminal state are zero. Call the time of hitting the terminal state as T, after which no further rewards are issued. Then we may write

$$V^{\pi}(s) = \mathbb{E}_{\pi}\left[\sum_{t=0}^{T-1} \gamma^t R_t \,\middle|\, S_0 = s\right].$$

Appealing to the law of large numbers, as is usual in Monte Carlo methods, we may directly estimate this as

$$\widehat{V}^{\pi}(s) = \frac{1}{N}\sum_{k=1}^{N}\sum_{t=0}^{T_k-1} \gamma^t R_{k,t},$$

with $S_{k,0} = s$ being the state at time 0 on simulation k, T_k the time to hit the terminal state on simulation k, and $R_{k,t}$ is the immediate reward at time t on simulation k. The sequence

$$\left(S_0, A_0, R_0, \ldots, S_{T-1}, A_{T-1}, R_{T-1}, S_T\right)$$

is called a **trajectory** or **history** of the MDP. While each trajectory contributes to the estimate of $\widehat{V}^{\pi}(s)$ for a specific starting state s, it also contributes to estimates of every other state that is encountered during the trajectory. In particular, if τ is the first time that a state s' is encountered, then

$$\left(S_\tau, A_\tau, R_\tau \ldots, S_{T-1}, A_{T-1}, R_{T-1}, S_T\right)$$

is distributed as the trajectory of an MDP under policy π that starts at s' and terminates at time T. The corresponding discounted reward for this trajectory is $\sum_{t=\tau}^{T-1} \gamma^{t-\tau} R_t$. Algorithm 11.29 implements this idea by first generating the trajectory and then computing the discounted reward for any state that is encountered in the trajectory, by working backwards from the time T of termination.

Algorithm 11.29: Monte Carlo Estimation of V^{π}

Input: Policy π, discount factor $\gamma \in [0, 1]$.
Output: Estimate of V^{π}.

Arbitrarily initialize V, and empty list `Values`(s) for all $s \in \mathcal{S}$.
for each episode **do**
  Generate trajectory following π, namely $S_0, A_0, R_0, \ldots, A_{T-1}, R_{T-1}, S_T$
  $G \leftarrow 0$
  for $t = T-1, T-2, \ldots, 0$ **do**
    $G \leftarrow \gamma G + R_t$
    if S_t does not appear in $S_0, \ldots, S_{t-1}$ **then**
      Append G to `Values`(S_t)
      $V(S_t) \leftarrow$`Average(Values`(S_t)`)`
return V

■ **Example 11.30 (MDP: Monte Carlo Estimation of V^{π})** We continue with our running example of the 4-state MDP in Examples 11.2, 11.9, 11.16, 11.22, and 11.26. This time we estimate the value function for a given policy using Monte Carlo. Although Algorithm 11.29 requires an episodic scenario to obtain unbiased estimates for each trajectory, we may still accurately estimate the value function across the different states, by terminating each trajectory after a large enough number of steps. In Exercise 6 we consider a modified situation where a terminating state is added that can be reached from each of the 4 states with a small probability. This creates a proper episodic scenario.

The method **Vhat** in the following Python module implements Algorithm 11.29.

estimate_V.py

```
from MDP import *
np.set_printoptions(precision=3)


def Vhat(policy_func, n_episodes, t_horizon, gamma):
    V = np.zeros(len(States))
    Values = {s: [] for s in States}

    for _ in range(n_episodes):
        S = 1
        Ss, Rs = np.zeros(t_horizon+1, dtype=int), np.zeros(t_horizon+1)
        for t in range(t_horizon+1):
            Ss[t], Rs[t] = S, r(S)
            S = np.random.choice(States, p=P[policy_func(S), S])
        CR = 0  # Cumulated Discounted Remaining Reward at time T
        for t in range(t_horizon, -1, -1):
            CR = gamma * CR + Rs[t]  # CR at time t
            if Ss[t] not in Ss[:t]:  # If state not yet encountered
                Values[Ss[t]].append(CR)
                V[Ss[t]] = np.mean(Values[Ss[t]])

    return V, Values
```

Below, we use it to estimate the value function for the optimal policy $[0, 1, 1, 0]$. The number of times that state is encountered in each of the 1000 episodes is also reported.

MDP_MC.py

```
from estimate_V import *

gamma = 0.9
N_episodes, T = 1000, 100  # Parameters for MC estimation of V

def policy(s):
    return np.array([0, 1, 1, 0])[s]


Vest, Values = Vhat(policy, N_episodes, T, gamma)
print("Estimated value function:", Vest)
print("States encountered:", {s: len(vals) for s, vals in Values.items()
    })
```

```
Estimated value function: [7.835 7.165 7.943 8.978]
States encountered: {0: 1000, 1: 1000, 2: 987, 3: 1000}
```

Notice that, while noisy, the estimated value function for the optimal policy compares well with, for example, the values in Example 11.16. ■

This kind of direct estimate can be combined with a (noisy) optimization routine over policies to approach the optimal policy. In the next example, we employ the cross–entropy method to search for the best policy over the policy space using the above estimation function `Vhat`.

Example 11.31 (MDP: Monte Carlo Optimization of V^π) We continue with our running example of the 4-state MDP in Examples 11.2, 11.9, 11.16, 11.22, 11.26, 11.30. We use the Monte Carlo estimation of V^π from Example 11.30 and wrap a noisy optimization routine around it. In this case we use the cross–entropy method, as in Example 3.93. We parameterize the multivariate Bernoulli sampling distribution via a vector $\boldsymbol{p}$ of four probabilities, one for each state. Policies are then sampled and evaluated (noisily) by the Monte Carlo estimation of V^π routine. Note that we need to combine the values across all of the four states into an overall performance measure. This is done simply here as the average of all the four estimated values for the states. The best performers are then used to update the parameter vector $\boldsymbol{p}$. ☞ 122

MDP_MC_CE.py

```
from estimate_V import *

gamma = 0.9
p = np.array([0.5, 0.5, 0.5, 0.5])  # Initial CE success prob.

N_episodes, T = 20, 100  # Parameters for MC estimation of V
N_iter, n_samples, elite_frac = 5, 50, 0.1  # Parameters for CE

# Estimate the mean value function for a given policy
def Vmean(pol):
    return np.mean(Vhat(lambda s: pol[s], N_episodes, T, gamma)[0])

# Perform the CE optimization for the optimal policy
for _ in range(N_iter):
    samples = (np.random.rand(n_samples, 4) < p).astype(int)
    scores = np.array([Vmean(sample) for sample in samples])
    elite_threshold = np.percentile(scores, 100 * (1 - elite_frac))
    elite_samples = samples[scores >= elite_threshold]  # Maximization
        problem
    p = np.mean(elite_samples, axis=0)
    print(f"p = {p}, Best Score: {np.max(scores)}")
```

```
p =  [0.  1.  0.5 0.5] , Best Score: [5.04]
p =  [0. 1. 1. 0.] , Best Score: [8.294]
p =  [0. 1. 1. 0.] , Best Score: [8.211]
p =  [0. 1. 1. 0.] , Best Score: [8.312]
p =  [0. 1. 1. 0.] , Best Score: [8.213]
```

The algorithm finds the optimal policy quickly. The variation in the last four scores is simply due to the randomness inherent in the Monte Carlo estimation of V^π. ■

11.7.2 Estimating the Q-Function via Monte Carlo

Just as with direct estimation of the value function, one can directly estimate the Q-function via the law of large numbers. Recall that

$$Q^\pi(s,a) = \mathbb{E}_\pi\left[\sum_{t=0}^{\infty} \gamma^t R_t \,\middle|\, S_0 = s, A_0 = a\right].$$

Appealing to the law of large numbers, and considering episodes of length T, we may directly estimate this as

$$\widehat{Q}^\pi(s,a) = \frac{1}{N}\sum_{k=1}^{N}\sum_{t=0}^{T_k-1} \gamma^t R_{k,t},$$

where $S_{k,0} = s$, as before, and $A_{k,0} = a$, with the latter being the initial action of the k-th simulation. This is implemented in Algorithm 11.32 below in a recursive way.

Note that, in the non-episodic case, we can still obtain estimates $\widehat{V}^\pi$ and $\widehat{Q}^\pi$ by replacing T_k with some (large enough) fixed time horizon T. As noted in Example 11.22 the error in the expectations can be controlled via T, and the error due to Monte Carlo sampling via N.

Algorithm 11.32: Monte Carlo Exploring Starts (Episodic)

Input: Discount factor $\gamma \in [0, 1]$.
Output: Deterministic policy π.

```
Arbitrarily initialize policy π, Q, and empty list Updates(s, a) for all s ∈ S and
  a ∈ A.
for each episode do
    Choose (S_0, A_0) randomly (with non-zero probability for each possible pair)
    Complete trajectory S_0, A_0, R_0, ..., S_{T-1}, A_{T-1}, R_{T-1}, S_T following π
    G ← 0
    for t = T − 1, T − 2, ..., 0 do
        G ← γG + R_t
        if pair (S_t, A_t) does not appear in S_0, A_0, ..., S_{t-1}, A_{t-1} then
            Append G to Updates(S_t, A_t)
            Q(S_t, A_t) ← Average(Updates(S_t, A_t))
            π(S_t) ← argmax_a Q(S_t, a)
return π
```

■ **Example 11.33 (MDP: Monte Carlo Estimation of Q^*)** Here we implement Algorithm 11.32, which learns the optimal Q-function and policy via Monte Carlo simulation. To obtain a truly episodic MDP, we use the 5-state modification given in Exercise 6.

MDP_Epi_MC_Q.py

```python
import numpy as np
from MDP_Epi import *
import sys
```

```
np.set_printoptions(precision=3)

# Parameters
gamma = 0.9
N_episodes = 1000
Tmax = 2000

# Initialize policy and Q-values
Pi = np.random.randint(2, size=len(States))
Q = np.zeros((len(States), len(Actions)))
updates = { (s, a): [] for s in range(len(States)) for a in range(len(
    Actions)) }

# Initialize arrays for rewards, states, and actions
Rs = np.zeros(Tmax+1)
Ss = np.zeros(Tmax+1, dtype=int)
As = np.zeros(Tmax+1, dtype=int)


for n in range(N_episodes):
    S = np.random.choice([0,1,2,3])  # exclude absorbing state 4
    A = np.random.choice(Actions)
    t = 0
    while S != 4 and t < Tmax:
        Ss[t], As[t] = S, A
        Rs[t] = r(S)
        S = np.random.choice(States, p=P[A, S])
        A = Pi[S]
        t += 1
    G, T = 0, t
    for t in range(T, -1, -1):
        G = gamma * G + Rs[t]
        if (Ss[t], As[t]) not in [(Ss[tt], As[tt]) for tt in range(t)]:
            updates[(Ss[t], As[t])].append(G)
            Q[Ss[t], As[t]] = np.mean(updates[(Ss[t], As[t])])
            Pi[Ss[t]] = np.argmax(Q[Ss[t], :])

print("\nEstimated Policy:", Pi)
print("\nEstimated q-function:\n", Q)
print("\nEncountered:", {(s, a): len(updates[s, a]) for s in States for
    a in Actions})
```

```
Estimated Policy: [0 1 1 0 1]

Estimated q-function:
 [[7.009 6.389]
  [5.79  6.355]
  [5.779 7.164]
  [8.24  6.879]
  [0.    0.   ]]

Encountered: {(0, 0): 941, (0, 1): 141, (1, 0): 139, (1, 1): 697,
  (2, 0): 103, (2, 1): 839, (3, 0): 978, (3, 1): 135,
  (4, 0): 0, (4, 1): 0}
```

The algorithm consistently finds the optimal policy as in previous examples, and the estimates in the Q-table match those in Exercise 6. In this particular instance, it was not necessary to use an ε-greedy strategy. ■

11.7.3 Monte Carlo Tree Search

A Monte Carlo "rollout" refers to the process of simulating an MDP trajectory starting from a current state and following a specified policy until a terminal state is reached or a predefined number of steps have been taken. Rollout algorithms typically estimate the actions–values (Q-values) for the current state by using averages of returns (that is, total accumulated rewards) for many simulated trajectories which start with a given action and follow the specified policy. Based on these estimates, the action with the highest estimated action–value is taken. This leads to the next state, and the process is repeated. Unlike Q-learning, which learns the entire Q-function, these algorithms only learn estimates for action–values for the current state. The action–value estimates obtained are only used for the current action decision, and are then discarded.

Instead of discarding the action–value estimates after each trajectory, as in a standard rollout algorithm, **Monte Carlo tree search** (MCTS) accumulates these estimates to direct future simulations towards trajectories with higher reward. It does so by maintaining a dynamic tree structure, where the nodes of the tree are associated with states and the edges with actions. The tree root contains the state of interest. It is for this state only that we wish to select the best action. In addition to the state information, each node contains an estimate of the state *value* (e.g., the expected total discounted reward), and a list of child nodes, each of which is associated with a particular action. The values of the child nodes are thus estimates of the action–values (Q-values) of the parent node.

The trajectory generation and tree maintenance in MCTS is typically carried out in four phases, illustrated in Figure 11.34:

Selection. Starting from the root node, a trajectory is generated according to the current **tree policy**; that is, according to the action–values along the nodes of the tree. The trajectory terminates at a leaf node, corresponding to an untried action, which is then selected.

Expansion. The tree is expanded by adding a child node corresponding to the selected action.

Simulation. Starting from the newly added child node the trajectory is continued by following a **rollout policy**; for example by simulating a trajectory in which actions are randomly chosen. This is also called the *rollout* phase.

Backup. In this final phase, the return of the complete trajectory is backed up to the tree, updating the values of the nodes of the tree that were encountered before the rollout phase.

This sequence of operations is followed until some computational resource (time, number of simulations, tree depth, etc.) is used up. Then an action is selected, often simply the one that was most visited during the tree search.

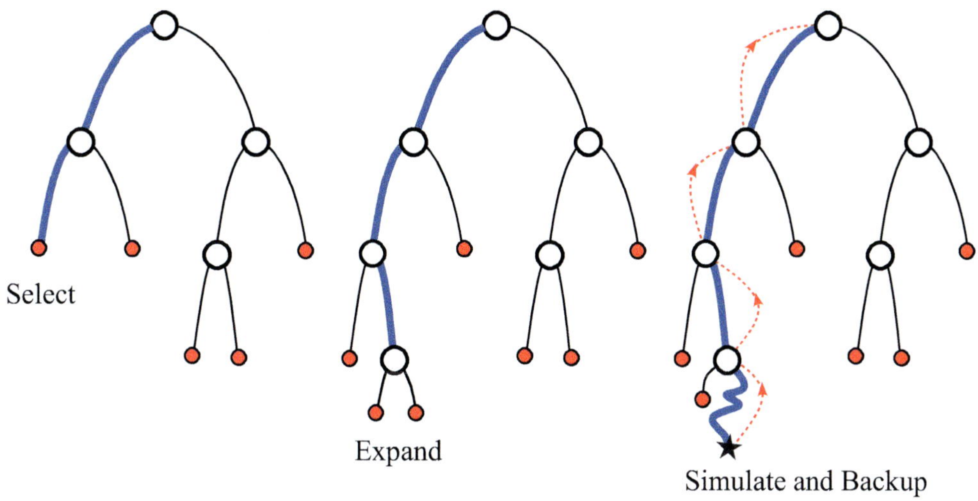

Figure 11.34: Illustration of the phases of Monte Carlo tree search.

After the action is selected and the new state observed, the process is repeated, either with a single root node being the new state, or sometimes with a tree constructed from the previous run.

For each parent node, the action selection in the tree policy involves more than simply choosing the action that gives the highest value of the child nodes. Instead, the selection is based on an **upper confidence bound** applied to trees (UCT) of the expected rewards, given by

$$\text{UCT}(v, n, t, c) = \frac{v}{n} + c\sqrt{\frac{2\ln t}{n}},$$

where v is the total value of the child node accrued over all visits, n is the total number of visits to the child node, t is the total number of times the parent node has been visited, and c is a parameter which controls the balance between exploration and exploitation. Higher values of c encourage more exploration. A commonly-used value for c is 1.5. The motivation behind UCT for action selection comes from the mathematical theory of **multi-armed bandits**; see Exercise 13.

■ **Example 11.35 (MDP: Monte Carlo Tree Search)** We continue our 4-state MDP running example, by implementing a Monte Carlo tree search for this problem. Our goal is to find the best action for each of the four states.

We break up the code into three parts. In part (a) we first import the usual Python libraries, as well as the state space, action space, transition matrices, and the immediate reward function of the MDP, the same as in Example 11.9. In addition, we define the UCT function and the Python class for a generic node in the MCTS tree. This node has a number of attributes, such as its state, value, and children. It also has a method for determining if all of its actions have been explored and a method that returns the best child in terms of its UCT value. This method will be used in the tree policy.

MDP_MCTS_a.py

```
from MDP import *
import numpy as np

rand_choice = np.random.choice

def uct(v: float, n: int, t: int, c_param: float = 1) -> float:
    """Upper Confidence Bound function for selection."""
    if t == 0 or n == 0:
        return np.inf
    return (v / n) + c_param * np.sqrt(2 * np.log(t) / n)


class MCTSNode:

    def __init__(self, state, parent=None):
        self.state = state
        self.parent = parent
        self.children = {}
        self.visits = 0
        self.value = 0.0  # value of the node
        self.depth = 0 if parent is None else parent.depth + 1
        self.reward = r(state)  # immediate reward

    def is_fully_expanded(self) -> bool:
        return len(self.children) == len(Actions)

    def best_child(self, c_param=1.5):
        choices_weights = [
            uct(child.value, child.visits, self.visits, c_param)
            for child in self.children.values()
        ]
        return list(self.children.values())[np.argmax(choices_weights)]
```

In part (b) we implement the principal ingredients of the MCTS algorithm. The main method, **mcts**, combines the selection, expansion, rollout and backup steps, and returns the best action for the root node. The depth of the MCTS tree is in this case limited to a maximal depth of 8. During the rollout phase, actions are chosen uniformly at random from the available actions at the node. The resulting total discounted reward for the rollout trajectory (which is only the last part of the entire trajectory) is returned. In the expansion phase a new node is created, including its associated attributes. In the backup or backpropagation phase the nodes in the entire trajectory are processed in reverse order, adding the cumulative reward to the value of each node. The value of a node is in this case the sum (rather than the average) of the cumulative rewards over all trajectories.

MDP_MCTS_b.py

```
from MDP_MCTS_a import *

gamma = 0.9
max_depth = 8
```

```
def rollout(state, depth: int) -> float:
    """Simulate a random rollout and return its total discounted reward
        """
    rollout_reward = 0.0
    for d in range(depth):
        action = rand_choice(Actions)
        rollout_reward += gamma**d * r(state)
        next_state = rand_choice(States, p=P[action, state])
        state = next_state
        if d == max_depth:
            break
    return rollout_reward


def expand(node: MCTSNode) -> MCTSNode:
    """Expand the node by trying an untried action and creating a new
        child."""
    untried_actions = [a for a in Actions if a not in node.children]
    action = rand_choice(untried_actions)
    next_state = rand_choice(States, p=P[action, node.state])
    child_node = MCTSNode(state=next_state, parent=node)
    child_node.value = child_node.reward + gamma * node.value
    node.children[action] = child_node
    return child_node


def backpropagate_trajectory(trajectory, rollout_reward: float):
    """Backpropagate the reward through the trajectory with correct
        discounting."""
    cumulative_reward = rollout_reward
    for node in reversed(trajectory):
        cumulative_reward = node.reward + gamma * cumulative_reward
        node.visits += 1
        node.value += cumulative_reward


def mcts(root: MCTSNode, iterations: int):
    """Perform MCTS and return the best action."""
    for _ in range(iterations):
        node = root
        trajectory = []

        while node.depth < max_depth:
            if node.is_fully_expanded():
                node = node.best_child()
            else:
                node = expand(node)
                trajectory.append(node)

        rollout_reward = rollout(node.state, max_depth - node.depth)
        backpropagate_trajectory(trajectory, rollout_reward)

    return max(root.children,
```

```
                key=lambda action: root.children[action].value / root.
                    children[
                    action].visits)
```

Finally, in part (c) we apply MCTS to our 4-state MDP problem. In particular, for each initial state we find the best action returned by `mcts`. This is repeated 20 times. The majority vote of the 20 decisions is reported, as well as the average number of times the correct (i.e., optimal) decision was made.

MDP_MCTS_c.py

```python
from MDP_MCTS_b import *

policy = [0, 1, 1, 0]
Ncounts = 20

print(f"{'State':<10} {'Vote':<10} {'Mean':<10}")
for initial_state in States:

    counts = 0
    acts = np.zeros(Ncounts, dtype=int)
    for i in range(Ncounts):
        root = MCTSNode(state=initial_state)
        acts[i] = mcts(root, iterations=1000)
        counts += (acts[i] == policy[initial_state])

    majority_vote = int(np.sum(acts) > Ncounts / 2)
    print(f"{initial_state:<10} {majority_vote:<10} {counts / Ncounts
        :<10.2f}")
```

```
State     Vote      Mean
0         0         0.60
1         1         0.75
2         1         1.00
3         0         0.90
```

Note that in this case the correct optimal policy was found, based on the majority vote. However, due to the stochastic nature of MCTS, it does *not* always find the optimal action for a given starting state. For example, selecting the correct action from state 0 only occurred 60% of the time over these 10 runs, and similarly only 75% of the time for state 1. This could be partially explained as the gap in values for actions in the Q-function (see Equation 11.27) for these states is the lowest (at 0.58 and −0.59) respectively, while state 2 has the highest gap at −1.47, and state 3 has the second highest at 1.00. ■

The previous example showcased how MCTS can be applied to the running example. It is a relatively simple application of the technique. However, MCTS, in combination with other heuristics, has been used for example in the well known AlphaGo program, which can compete with high level human players. The next example, while still relatively simple, highlights how MCTS can be used to make decisions in the game of Tic–Tac–Toe.

■ **Example 11.36 (MCTS for Tic–Tac–Toe)** In this example, we play a game of Tic–Tac–Toe (also called naughts–and–crosses) against a computer that uses MCTS to select actions. Starting with the human player, the two players take turns to place their symbols on a 3×3 board. The goal is to be the first player to get three of their symbols in a row, column, or diagonal.

In the implementation below, the board is represented as a list of 9 numbers, read row by row. Each number can be: 0 (empty), 1 (human), or 2 (computer). An action consists of choosing a number between 1 and 9 that has not been chosen yet and placing a symbol in the corresponding position on the board. For example, if the human in the first turn selects action 4, this places a 1 in the second row and first column of the board.

We divide the program into three parts. The first part mainly describes the system states. In contrast to the previous MDP example, the system state is here more elaborate and is described by a special Python class, which contains attributes such as the board and the current player. It also has functions for determining the available actions, applying an action and updating the state of the game, determining whether there is a winner given the current state, and printing the current state of the board to the console.

tic_tac_toe_mcts_a.py

```
import copy
import math
import random
import sys


class TicTacToeState():
    def __init__(self, current_player=None):
        self.board = [0 for i in range(9)]
        self.current_player = current_player
        self.next_player = 1 if current_player == 2 else 2
        self.is_terminal = False

    def available_actions(self):
        ''' Determine the actions available from this state (board cells
          that are still 0) '''
        actions = [i for i in range(9) if self.board[i] == 0]
        return actions

    def apply_action(self, action):
        ''' Get the next state by applying the action '''
        if action not in self.available_actions():
            print("Invalid action")
            sys.exit(0)

        next_state = TicTacToeState(current_player=self.next_player)
        next_state.board = self.board[:]
        if next_state.board[action] == 0:
            next_state.board[action] = self.current_player

        next_state.is_terminal = next_state.has_winner() or len(
            next_state.available_actions()) == 0
        return next_state
```

```
    def has_winner(self):
        ''' Determine if this state has a winner '''
        # Check rows
        for row in range(3):
            a = [self.board[row * 3 + i] == self.next_player for i in
                range(3)]
            if all(a) and len(set(a)) == 1:
                return True

        # Check columns
        for column in range(3):
            a = [self.board[column + i * 3] == self.next_player for i in
                range(3)]
            if all(a) and len(set(a)) == 1:
                return True

        # Check diagonals
        a = [self.board[i * 4] == self.next_player for i in range(3)]
        if all(a) and len(set(a)) == 1:
            return True

        a = [self.board[i * 2] == self.next_player for i in range(1, 4)]
        if all(a) and len(set(a)) == 1:
            return True

        return False

    def print(self):
        for row in range(3):
            a = [self.board[row * 3 + i] for i in range(3)]
            print(a)
        print("\n")
```

The code below implements the MCTS node class and the MCTS function itself. The node keeps track of its state, parent, available actions, children, number of visits to it, and cumulative reward. It has helper functions to select a random untried action, to calculate the UCT for the node, to determine the best action from average return, to return which child results from an action, and to add a child as needed.

The MCTS portion iterates through the stages of selection, expansion, simulation, and backpropagation.

tic_tac_toe_mcts_b.py

```
from tic_tac_toe_mcts_a import *

class MCTSNode():
    def __init__(self, state, parent=None):
        self.state = state
        self.parent = parent
        self.untried_actions = state.available_actions()
        self.children = {}
        self.visits = 0
```

```
        self.cumulative_reward = 0
        random.shuffle(self.untried_actions)

    def select_action(self):
        if len(self.untried_actions) != 0:
            return self.untried_actions.pop()

        log_parent_visits = math.log(self.visits)
        def uct(node):
            return node.cumulative_reward / node.visits + 2.0 * math.
                sqrt(2.0 * log_parent_visits / node.visits)

        action = max(self.children, key=lambda k: uct(self.children[k]))
        return action

    def best_action(self):
        def q(node):
            return node.cumulative_reward / node.visits

        # Get action with maximum average return
        action = max(self.children, key=lambda k: q(self.children[k]))
        return action

    def get_child(self, action):
        return self.children[action]

    def add_child(self, action):
        next_state = self.state.apply_action(action)
        self.children[action] = MCTSNode(next_state, parent=self)
        return self.get_child(action)

def mcts(state, n_iterations=1000, hard_difficulty=False):
    root = MCTSNode(state)
    for _ in range(n_iterations):
        node = root

        # Selection
        while node.state.is_terminal == False and len(node.
            untried_actions) == 0:
            action = node.select_action()
            node = node.get_child(action)

        # Expansion
        if node.state.is_terminal == False:
            action = node.select_action()
            node = node.add_child(action)

        # Simulation
        state = node.state
        while not state.is_terminal:
            action = random.choice(state.available_actions())
            state = state.apply_action(action)

        # Backpropagation
```

```
        player = state.next_player
        has_winner = state.has_winner()
        while node is not None:
            node.visits += 1
            if has_winner:
                if node.state.current_player != player:
                    node.cumulative_reward += 1.0
                elif hard_difficulty:
                    # If hard difficulty is True, the agent receives a
                        penalty for losing a game
                    node.cumulative_reward -= 1.0

            node = node.parent

    return root.best_action()
```

The final segment of the Tic–Tac–Toe code implements the play logic, including an option for difficulty setting and which player starts, as well as for printing the state of play and waiting for player input. Output of a game played with this code is given below.

tic_tac_toe_mcts_c.py

```
from tic_tac_toe_mcts_b import *

def play(first_player=1, hard_difficulty=False):
    '''
    first_player=1 => Player makes the first move
    first_player=2 => AI makes the first move

    hard_difficulty=False => The AI is trying to win
    hard_difficulty=True => The AI is trying to win and additionally
        trying not to lose
    '''
    state = TicTacToeState(current_player=first_player)
    while not state.is_terminal:
        if state.current_player == 1:
            state.print()
            action = int(input('select action [1-9]: '))-1
        else:
            action = mcts(state, hard_difficulty=hard_difficulty)
            print("AI chooses action " + str(action+1))

        state = state.apply_action(action)

    print("==================\n")
    state.print()
    if state.has_winner():
        if state.current_player == 1:
            print("AI wins")
        else:
            print("Player wins")
    else:
        print("draw")
```

```
play(first_player=1, hard_difficulty=False)
```

```
[0, 0, 0]
[0, 0, 0]
[0, 0, 0]

select action [1-9]: 1
AI chooses action 5
[1, 0, 0]
[0, 2, 0]
[0, 0, 0]

select action [1-9]: 3
AI chooses action 2
[1, 2, 1]
[0, 2, 0]
[0, 0, 0]

select action [1-9]: 7
AI chooses action 8
==================
[1, 2, 1]
[0, 2, 0]
[1, 2, 0]

AI wins
```

Note that with the code provided it is still possible for the human player to win, even though a perfect strategy would always result in a draw. ■

While MCTS tends to cope well when considering a deep sequence of moves, it tends to suffer when the number of choices per level of the tree is large, giving a wide search space. What happens is that there are simply large portions of the potential tree that are not discovered, and hence a high chance of not hitting upon the optimal sequence of actions. For example with AlphaGo, MCTS is integrated with parametric policy and value networks in the form of deep neural networks. This enables the scalability of MCTS to the massive search space inherent in the game of Go.

11.8 Parametric Policy Improvement

Policy iteration, SARSA, and Q-learning all aim to iteratively improve a given policy until no further improvement can be achieved. So far, we used the natural parameterization $\pi(s), s \in \mathcal{S}$ of a policy π, where for each state s, $\pi(s)$ represents an action. More generally, for a randomized policy π, the entity $\pi(s)$ represents a probability distribution on the action space, say with density[2] $p(a \mid s), a \in \mathcal{A}$. In a broader sense, we can consider parameterizations of the form $\pi_{\boldsymbol{\theta}}(s) = \{p_{\boldsymbol{\theta}}(a \mid s), a \in \mathcal{A}\}$, depending on a parameter vector $\boldsymbol{\theta} \in \mathbb{R}^k$. The task of learning an optimal parametric policy then

[2]We use here Bayesian notation, where we reuse the letter p for different (conditional) probability densities.

involves maximizing or minimizing some scalar performance measure $J(\boldsymbol{\theta})$, such as the expected total discounted reward. In **policy gradient** methods, this optimization is carried out via approximate gradient ascent (for maximization), leading to the iterative scheme

$$\boldsymbol{\theta}_{t+1} = \boldsymbol{\theta}_t + \alpha \widehat{\nabla J}(\boldsymbol{\theta}_t), \tag{11.37}$$

where $\widehat{\nabla J}(\boldsymbol{\theta}_t)$ is a stochastic gradient whose expectation approximates $\nabla J(\boldsymbol{\theta}_t)$, and α is
☞ 118 a learning rate, exactly as in the *stochastic approximation* Algorithm 3.85.

■ **Example 11.38 (MDP: Policy Gradient)** We return to Example 11.2 and consider policies that are parameterized by a vector $\boldsymbol{\theta} = [\theta_0, \dots, \theta_3]^\top \in \mathbb{R}^4$. In particular, let $\boldsymbol{p} = [p^{(0)}, \dots, p^{(3)}]^\top \in [0, 1]^4$, with

$$p^{(i)} = (1 - \mathrm{e}^{-\theta_i})^{-1}, \quad i = 0, \dots, 3$$

☞ 243 be the logistic transformation of $\boldsymbol{\theta}$. Each $\boldsymbol{p}$ corresponds to a random policy where, in state i, action 0 is taken with probability $1 - p^{(i)}$ and action 1 with probability $p^{(i)}$. In other words, policy $\boldsymbol{p}$ results in a transition matrix $\mathbf{P}$ for the Markov chain that is given by

$$\mathbf{P} = (\mathbf{1} - \boldsymbol{p}) \odot \mathbf{P}_0 + \boldsymbol{p} \odot \mathbf{P}_1,$$

where $\odot$ indicates the elementwise (vector–matrix) product. For $\boldsymbol{\theta} = \mathbf{0}$, we have $\boldsymbol{p} = [1/2, 1/2, 1/2, 1/2]^\top$, which corresponds to a uniformly random strategy.

Let $J(\boldsymbol{\theta})$ be the long-run fraction of sojourns in state 3, exactly as in Example 11.2. We wish to apply a gradient ascent procedure to maximize $J(\boldsymbol{\theta})$. In this case the gradient of $J(\boldsymbol{\theta})$ can be computed exactly, at least numerically. Using, a central difference approximation, we find, for example,

$$\nabla J(\mathbf{0}) = [-0.01929013, 0.0017486, 0.0649758, -0.02225997]^\top.$$

Taking an infinite step size in the direction of the gradient gives $\boldsymbol{\theta}_1 = [-\infty, \infty, \infty, -\infty]$, corresponding to the optimal policy $\pi = \boldsymbol{p}^\top = [0, 1, 1, 0]$, as found before. The following Python code details how the gradient and the policy are computed.

polygrad.py

```
from MDP import *
import numpy as np
from scipy.linalg import null_space


def logistic(x):
    return 1 / (1 + np.exp(-x))


def policy(theta):
    return logistic(theta)


def H(theta):
```

```
    p = policy(theta).reshape(-1, 1)
    Pnew = (1-p)*P[0] + p*P[1]
    ns = null_space(np.eye(4) - Pnew.T)
    f = ns / np.sum(ns)  # stationary distribution
    return f[3][0]


def gradient(f, x, h=1e-12):
    grad = np.zeros_like(x)
    for i in range(len(x)):
        x_forward = np.copy(x)
        x_backward = np.copy(x)
        x_forward[i] += h
        x_backward[i] -= h
        grad[i] = (f(x_forward) - f(x_backward)) / (2 * h)
    return grad


def gradient_ascent(f, x_init, step_size, num_iterations):
    x = np.copy(x_init)
    for _ in range(num_iterations):
        grad = gradient(f, x)
        x += step_size * grad
    return x


theta = np.zeros(4, dtype=np.float64)

theta_max = gradient_ascent(H, theta, np.inf, 1)

print("Policy:", policy(theta_max))
print("Value:", H(theta_max))

p = policy(theta)
Pnew = p*P[0] + (1-p)*P[1]
```

```
Policy: [0. 1. 1. 0.]
Value: 0.8827272727272729
```

■

A commonly-used type of parameterization is through the use of a real-valued **preference function** $h(s, a, \boldsymbol{\theta})$. These preferences are converted to probabilities in $\pi_{\boldsymbol{\theta}}(s) = \{p_{\boldsymbol{\theta}}(a \mid s), a \in \mathcal{A}\}$ via the softmax function:

$$p_{\boldsymbol{\theta}}(a \mid s) = \frac{\mathrm{e}^{h(s,a,\boldsymbol{\theta})}}{\sum_{a'} \mathrm{e}^{h(s,a',\boldsymbol{\theta})}}.$$

The preference function can for example be linear in $\boldsymbol{\theta}$ or the output of a neural network (see Chapter 10).

It is often the case that the performance function $J(\boldsymbol{\theta})$ can be written as an expectation $J(\boldsymbol{\theta}) = \mathbb{E}_{\boldsymbol{\theta}} H(\mathbf{Z})$ for some known function H and random vector $\mathbf{Z}$ whose pdf $p_{\boldsymbol{\theta}}$ can be evaluated. A typical case is where $\mathbf{Z}$ is the trajectory of an episodic MDP and $H(\mathbf{Z})$ is the discounted reward for this trajectory. That is,

$$\mathbf{Z} = (S_0, A_0, R_0, \ldots, S_{T-1}, A_{T-1}, R_{T-1}, S_T)$$

and $H(\mathbf{Z}) = \sum_{t=0}^{T-1} \gamma^t R_t$. If $p_{\boldsymbol{\theta}}(a \mid s)$, $s \in \mathcal{S}, a \in \mathcal{A}$ denotes the parameterized policy, then (again using Bayesian notation) we can write the probability density of the realization $z = (s_0, a_0, r_0, \ldots, s_{\tau-1}, a_{\tau-1}, r_{\tau-1}, s_\tau)$ as

$$p_{\boldsymbol{\theta}}(z) = p(s_0) \prod_{t=0}^{\tau} p_{\boldsymbol{\theta}}(a_t \mid s_t) p(s_{t+1} \mid s_t, a_t),$$

which involves the density of the initial state as well as the policy and the state transition densities. As the initial state density and the state transitions densities do not depend on

☞ 136 $\boldsymbol{\theta}$, we obtain the corresponding *score function*

$$\nabla_{\boldsymbol{\theta}} \ln p_{\boldsymbol{\theta}}(z) = \sum_{t=0}^{\tau-1} \nabla_{\boldsymbol{\theta}} \ln p_{\boldsymbol{\theta}}(a_t \mid s_t). \tag{11.39}$$

We can use (11.39) to estimate the gradient of $J(\boldsymbol{\theta})$ from iid copies $\mathbf{Z}^{(i)}, i = 1, \ldots, N$ of $\mathbf{Z}$. Namely, recalling that

$$\nabla_{\boldsymbol{\theta}} \ln p_{\boldsymbol{\theta}}(z) = \frac{\nabla_{\boldsymbol{\theta}} p_{\boldsymbol{\theta}}(z)}{p_{\boldsymbol{\theta}}(z)},$$

we have

$$\begin{aligned}
\nabla_{\boldsymbol{\theta}} J(\boldsymbol{\theta}) &= \nabla_{\boldsymbol{\theta}} \int H(z)\, p_{\boldsymbol{\theta}}(z)\, \mathrm{d}z = \int H(z)\, \nabla_{\boldsymbol{\theta}} p_{\boldsymbol{\theta}}(z)\, \mathrm{d}z \\
&= \int H(z) \frac{\nabla_{\boldsymbol{\theta}} p_{\boldsymbol{\theta}}(z)}{p_{\boldsymbol{\theta}}(z)} p_{\boldsymbol{\theta}}(z)\, \mathrm{d}z = \int H(z) \nabla_{\boldsymbol{\theta}} \ln p_{\boldsymbol{\theta}}(z)\, p_{\boldsymbol{\theta}}(z)\, \mathrm{d}z \\
&= \mathbb{E}_{\boldsymbol{\theta}}[H(\mathbf{Z})\, \nabla_{\boldsymbol{\theta}} \ln p_{\boldsymbol{\theta}}(\mathbf{Z})],
\end{aligned}$$

provided that the order of integration and differentiation may be interchanged in the second equation; see, for example, [105] for sufficient conditions. The gradient of $J(\boldsymbol{\theta}) = \mathbb{E} \sum_{t=0}^{T-1} \gamma^t R_t$ can now be estimated via the sample average

$$\begin{aligned}
\widehat{\nabla_{\boldsymbol{\theta}} J}(\boldsymbol{\theta}) &= \frac{1}{N} \sum_{i=1}^{N} H(\mathbf{Z}^{(i)})\, \nabla_{\boldsymbol{\theta}} \ln p_{\boldsymbol{\theta}}(\mathbf{Z}^{(i)}) \\
&= \frac{1}{N} \sum_{i=1}^{N} \sum_{t=0}^{T-1} \gamma^t R_t\, \nabla_{\boldsymbol{\theta}} \ln p_{\boldsymbol{\theta}}(A_t^{(i)} \mid S_t^{(i)}).
\end{aligned} \tag{11.40}$$

This technique for estimating the gradient of the expected reward with respect to the policy parameters is called the **score function method**. Higher-order derivatives (e.g., Hessians) can be estimated in a similar way. Moreover, the estimation procedure can be

☞ 104 readily modified to include *importance sampling*; see [146, Section 7.2] and [75].

Note that each subtrajectory $(S_k, A_k, R_k, \ldots, S_T)$ of a trajectory $(S_0, A_0, R_0, \ldots, S_T)$ is itself a trajectory of the episodic MDP, and hence can be used to estimate the gradient of expected discounted reward via (11.39), replacing $\sum_{t=0}^{T-1} \gamma^t$ with $\sum_{t=k}^{T-1} \gamma^{t-k}$. This is done in the following REINFORCE Algorithm 11.41.

Algorithm 11.41: REINFORCE (Episodic)

Input: Differentiable parameterized policy $\pi_\theta = \{p_\theta(a \mid s), s \in \mathcal{S}, a \in \mathcal{A}\}$, discount rate $\gamma \in (0,1)$, and learning rate $\alpha > 0$.
Output: Improved policy π_θ.

```
Initialize parameter vector θ.
for i = 1, ..., N do
    Generate trajectory S_0, A_0, R_0, ..., S_{T-1}, A_{T-1}, R_{T-1}, S_T using policy π_θ
    for k = 0, 1, ..., T − 1 do
        G ← Σ_{t=k}^{T-1} γ^{t-k} R_k
        θ ← θ + αG ∇_θ ln p_θ(A_t | S_t)
return π_θ
```

A major advantage of parameterized policies is their ability to deal with continuous state and action spaces, as illustrated in the following example.

Example 11.42 (Continuous Torus MDP: Policy Optimization) We consider a continuous version of the torus MDP in Example 11.11. In the current scenario, the robot is moving through a smooth landscape, again determined by the *modified peaks* function in (11.12). The state space is the square $\mathcal{S} = [-3, 3]^2$, with "wrapped-around" edges, to form a flat torus. The landscape is depicted in Figure 11.46. It has a global maximum of $r^* = 8.1062$ at state $\boldsymbol{x}^* = [-0.0093.1.5814]^\top$ and a local maximum 6.5581 at state $[-0.95206, -0.47165]^\top$.

Suppose our goal is to search, within in a class of parameterized policies $\{\pi_\theta\}$, for a policy that maximizes the performance measure

$$J(\boldsymbol{\theta}) = \int_0^1 r(\boldsymbol{x}(t))\,\mathrm{d}t, \tag{11.43}$$

where the path $(\boldsymbol{x}(t) := [x_1(t), x_2(t)]^\top, t \in [0,1])$ is completely determined by $\boldsymbol{\theta}$. Similar to Example 11.11, this performance evaluates the integrated amount of terrain height encountered during the robot's path. In order to obtain a meaningful optimization problem, we need to add some restrictions on the path. In particular, we assume that the robot moves at a fixed speed v; that is,

$$v = \|\dot{\boldsymbol{x}}(t)\| := \sqrt{\left(\tfrac{\mathrm{d}}{\mathrm{d}t}x_1(t)\right)^2 + \left(\tfrac{\mathrm{d}}{\mathrm{d}t}x_2(t)\right)^2}, \quad t \in [0,1].$$

The length of the curve is therefore v. Without this restriction on speed, the maximization of the integrated height is easy: take a curve $(\boldsymbol{x}(t))$ that travels at infinite speed to the global maximizer and then stays there for 1 unit of time. This gives a maximum integrated height of r^*.

We wish to use a class of parameterized policies that is rich enough to encompass a wide variety of curves with speed v. The approach that we will follow is to define each path as the gradient-trajectory of a topography function g_θ, which is different from the reward function r. In particular, we define g_θ as a mixture of Gaussian pdfs:

$$g_\theta(\boldsymbol{x}) = \sum_{i=1}^{m} w_i\, \phi(\boldsymbol{x} \mid \boldsymbol{\mu}_i, \boldsymbol{\Sigma}_i), \quad \boldsymbol{x} \in [-4.5, 4.5]^2, \tag{11.44}$$

where

$$\phi(\boldsymbol{x} \mid \boldsymbol{\mu}, \boldsymbol{\Sigma}) = \frac{1}{2\pi\sqrt{|\boldsymbol{\Sigma}|}} \mathrm{e}^{-\frac{1}{2}(\boldsymbol{x}-\boldsymbol{\mu})^\top \boldsymbol{\Sigma}^{-1}(\boldsymbol{x}-\boldsymbol{\mu})}$$

and $\boldsymbol{\theta}$ collects all the parameters. Note that the domain of g_θ is larger than the state space.

The gradient of ϕ with respect to $\boldsymbol{x}$ is easily found:

$$\nabla_{\boldsymbol{x}} \phi(\boldsymbol{x} \mid \boldsymbol{\mu}, \boldsymbol{\Sigma}) = -\phi(\boldsymbol{x}; \boldsymbol{\mu}, \boldsymbol{\Sigma}) \boldsymbol{\Sigma}^{-1}(\boldsymbol{x} - \boldsymbol{\mu})$$

Consequently, for a fixed $\boldsymbol{\theta}$, the gradient of $g_\theta(\boldsymbol{x})$ is given by

$$G_\theta(\boldsymbol{x}) := \nabla_{\boldsymbol{x}} g_\theta(\boldsymbol{x}) = -\sum_{i=1}^{m} w_i\, \phi(\boldsymbol{x} \mid \boldsymbol{\mu}_i, \boldsymbol{\Sigma}_i) \boldsymbol{\Sigma}_i^{-1}(\boldsymbol{x} - \boldsymbol{\mu}_i),$$

and the corresponding gradient-trajectory obeys the differential equation

$$\dot{\boldsymbol{x}}(t) = v \frac{G_\theta(\boldsymbol{x}(t))}{\|G_\theta(\boldsymbol{x}(t))\|}, \quad t \in [0, 1]. \tag{11.45}$$

The performance $J(\boldsymbol{\theta})$ of the gradient-trajectory is then computed via (11.43). Finally, we can use any optimization technique to attempt to maximize $J(\boldsymbol{\theta})$, such as the gradient ascent method (11.37). In the Python implementation below we use instead the CE
☞ 111 method to maximize $J(\boldsymbol{\theta})$, importing the `CEM` method from Section 3.4.2.

Specifically, we wish to maximize the performance (11.43) starting from state $\boldsymbol{x}_0 = [0, -1]^\top$. To reduce the number of parameters, while ensuring a rich enough collection of topographies in (11.44), we let $m = 20$, $\boldsymbol{\Sigma}_i = 5^2\, \mathbf{I}_2$ for all i, and distribute the kernel centers evenly over the square $[-4.5, 4.5]^2$ via a Hammersley point set, similar to what
☞ 335 was done in Example 7.41. The only free parameters to optimize are thus the weights $\{w_i\}$, which may be negative. The speed of the robot is chosen as $v = 20$, and the integral in (11.43) is approximated via the sum

$$\widehat{J}(\boldsymbol{\theta}) = \sum_{i=0}^{n-1} r(\boldsymbol{x}_i),$$

where

$$\boldsymbol{x}_{i+1} = \boldsymbol{x}_i + \delta v \frac{G_\theta(\boldsymbol{x}_i)}{\|G_\theta(\boldsymbol{x}_i)\|}, \quad i = 0, \ldots, n-1.$$

For a discretization of $n = 100$ steps we find a maximal performance of 7.39, with the optimal path given in Figure 11.46. We see that the robot travels towards the local maximizer, but then continues on to the global maximizer, where it remains from (roughly) time 0.2 until 1. The running time was about 90 seconds. For $n = 400$, the estimate of (11.43) increases to 7.50, with no noticeable change in the optimal path. The computation time increases by roughly a factor of 4.

torus_cont_kernel_CE.py

```
import numpy as np
import matplotlib.pyplot as plt
from matplotlib import cm
from ce_min import CEM


def radical_inverse_vdc(bits):
    bits = (bits << 16) | (bits >> 16)
    bits = ((bits & 0x55555555) << 1) | ((bits & 0xAAAAAAAA) >> 1)
    bits = ((bits & 0x33333333) << 2) | ((bits & 0xCCCCCCCC) >> 2)
    bits = ((bits & 0x0F0F0F0F) << 4) | ((bits & 0xF0F0F0F0) >> 4)
    bits = ((bits & 0x00FF00FF) << 8) | ((bits & 0xFF00FF00) >> 8)
    return float(bits) * 2.3283064365386963e-10


def hammersley2d(num_points):
    return np.array([[i / num_points, radical_inverse_vdc(i)]
                     for i in range(num_points)])


def mypeaks(x, y):
    return (3 * (1 - x)**2 * np.exp(-x**2 - (y + 1)**2) - 10 *
            (x / 5 + x**3 - y**5) * np.exp(-x**2 - y**2) -
            (1 / 3) * np.exp(-(x + 1)**2 - y**2))


def wrap(x, y):
    return np.mod(x + 3, 6) - 3, np.mod(y + 3, 6) - 3


def perf(weights, x_start=np.array([0, -1]),
         n_steps=100, b_ker=5.0, nu=20):
    delta = 1.0 / n_steps
    x_old = x_start
    J = 0.0
    path = [x_old]
    for _ in range(n_steps - 1):
        diff = x_old - c_ker
        grad_normal = np.exp(-0.5 * np.sum(diff**2, axis=1) /
                             b_ker).reshape(-1, 1) * diff / b_ker
        grad_gmm = (-weights.reshape(-1, 1) * grad_normal).sum(axis=0)
        grad = grad_gmm / np.linalg.norm(grad_gmm)
        x_new = x_old + delta * nu * grad
        x_new = np.mod(x_new + 3, 6) - 3
        J += mypeaks(x_new[0], x_new[1]) * delta
        path.append(x_old)
        x_old = x_new
    return J, path


def f(x, **kwargs):
    return -perf(x, **kwargs)[0]
```

```
num_ker = 20
c_ker = -4.5 + 9 * hammersley2d(num_ker)
mu = CEM(f, np.ones(num_ker), np.ones(num_ker),
         n_samples=500, n_el=50, n_iter=50)

X, Y = np.meshgrid(np.arange(-3, 3.01, 0.3), np.arange(-3, 3.01, 0.3))
Z = mypeaks(X.T, Y.T).T
plt.contourf(X, Y, Z, levels=40, cmap=cm.jet)
plt.gca().set_aspect('equal', adjustable='box')  # Set aspect ratio to 1
J, path = perf(mu)
print(J)
path = np.array(path)
plt.plot(path[:, 0], path[:, 1], 'k+')
plt.savefig('torus_cont.pdf', format='pdf')
plt.show()
```

```
7.386928306727683
```

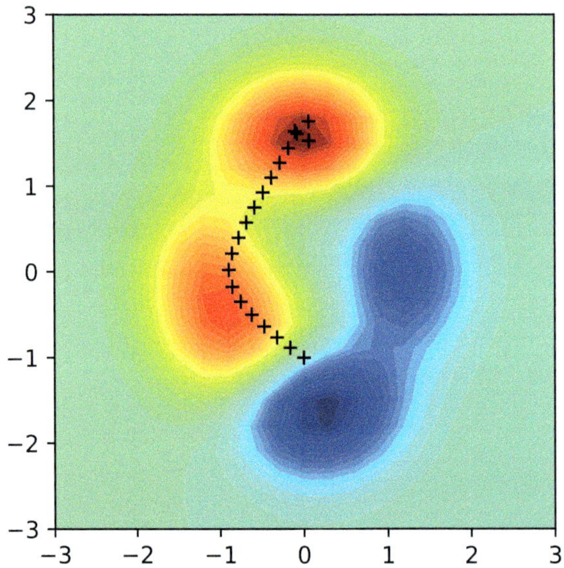

Figure 11.46: The reward function and optimal path, starting from $[0, -1]^\top$ and following the optimal policy.

Finally, **actor–critic** methods learn approximations to both the policy and the value function. Actor refers to the learned policy and critic refers to the learned value function (usually a parameterized approximation).

Here, the parameter vector for the policy π_θ is updated as

$$\boldsymbol{\theta}_{t+1} = \boldsymbol{\theta}_t + \alpha_\theta \underbrace{\left(R_{t+1} + \gamma \widehat{V}_{\boldsymbol{w}}(S_{t+1}) - \widehat{V}_{\boldsymbol{w}}(S_t)\right)}_{\delta_t} \nabla_\theta \ln p_{\boldsymbol{\theta}_t}(A_t \mid S_t).$$

The term $\widehat{V}_{\boldsymbol{w}}(s)$ is an approximate value for the state with parameter vector $\boldsymbol{w}$. It can be learned by a variety of methods.

Algorithm 11.47: Actor–Critic (Episodic)

Input: Differentiable parameterized policy $\pi_{\boldsymbol{\theta}} = \{p_{\boldsymbol{\theta}}(a \mid s), s \in \mathcal{S}, a \in \mathcal{A}\}$, differentiable parameterized value function $\widehat{V}_{\boldsymbol{w}}(s), s \in \mathcal{S}$, discount rate $\gamma \in (0, 1)$, and learning rates $\alpha_{\boldsymbol{\theta}} > 0$ and $\alpha_{\boldsymbol{w}} > 0$.

Output: Improved policy $\pi_{\boldsymbol{\theta}}$.

Initialize policy parameter vector $\boldsymbol{\theta} \in \mathbb{R}^k$ and state–value parameter vector $\boldsymbol{w} \in \mathbb{R}^d$.

for each episode **do**

  $G \leftarrow 1$

  repeat

    $A \sim p_{\boldsymbol{\theta}}(\cdot \mid S)$

    Take action A and observe S', R

    $\delta \leftarrow R + \gamma \widehat{V}_{\boldsymbol{w}}(S') - \widehat{V}_{\boldsymbol{w}}(S)$

    $\boldsymbol{w} \leftarrow \boldsymbol{w} + \alpha_{\boldsymbol{w}} \delta \nabla_{\boldsymbol{w}} \widehat{V}_{\boldsymbol{w}}(S)$

    $\boldsymbol{\theta} \leftarrow \boldsymbol{\theta} + \alpha_{\boldsymbol{\theta}} G \delta \nabla_{\boldsymbol{\theta}} \ln p_{\boldsymbol{\theta}}(A \mid S)$

    $G \leftarrow \gamma G$

    $S \leftarrow S'$

  until until S is terminal

return $\pi_{\boldsymbol{\theta}}$

Note that in Algorithm 11.47, if state S' is terminal, then $\widehat{V}_{\boldsymbol{w}}(S') = 0$. The non-episodic case can also be defined under the assumption that the system is in steady-state (an ergodic assumption).

Further Reading

An excellent overview of and introduction to reinforcement learning can be found in [167] and [149]. There are a number of classic works in dynamic programming and Markov decision processes. Chief among them are Bellman's 1957 book [11] which outlines his principle of optimality. Works focusing on stochastic models and the application of dynamic programming include Whittle's 1982 book [182], Ross's 1983 book [141], and Tijm's 1986 book [170]. Puterman's 2014 book [136] gives an in-depth account of Markov decision processes and dynamic programming in this context.

The temporal–difference methods discussed in Section 11.6 and Monte Carlo methods discussed in Section 11.7 both learn directly from experience. Both are linked through the TD(λ) algorithm (see e.g. [166, Chapter 12]) where the λ indicates the use of a so-called eligibility trace, and the '0' indicates no such use. The term $\lambda \in [0, 1]$ is a parameter which unifies Monte Carlo methods and one-step temporal–difference methods at each extreme. These bring temporal–difference methods and Monte Carlo methods under a single umbrella.

Although not covered in this chapter, *partially observable* Markov decision processes [160] are an important extension of the theory, frequently resulting in continuous state spaces. A convenient approach, which can also deal with continuous actions spaces, to solve such problems which employs the cross–entropy method can be found in [85]. Methods for MCTS applied to the partially observable setting can be found in [157, 185].

There are a number of other extensions — in particular offline reinforcement learning which relies on historical trajectories [62], deep reinforcement learning [6], Bayesian reinforcement learning [175], soft actor–critic methods (where the agent seeks to simultaneously maximize expected return and entropy) [80] multi-agent reinforcement learning [37], hierarchical reinforcement learning [9], and dealing with state and action representations [117] to name a few.

Exercises

1. Implement a Python program that enumerates all 16 policies in Example 11.2 and finds the optimal one.

2. Let $B(\mathcal{S})$ be the space of bounded functions on $\mathcal{S}$; that is, $B(\mathcal{S}) = \{v : \|v\| < \infty\}$, where $\|v\| = \sup_{s\in\mathcal{S}} |v(s)|$ denotes the supremum norm of v. Let C be the mapping in (11.7):

$$C[v](s) = \max_{a\in\mathcal{A}} \left\{ r(s,a) + \gamma \sum_{s'\in\mathcal{S}} p(s' \mid s,a)\, v(s') \right\}, \quad s \in \mathcal{S}.$$

We assume that $0 < \gamma < 1$ and that $\sup_{s,a} |r(s,a)| \leq M$ for some M.

(a) Show that if $\|v\| < \infty$ implies that $\|C[v]\| < \infty$. Thus, C maps $B(\mathcal{S})$ to $B(\mathcal{S})$.

(b) Show that $C : B(\mathcal{S}) \to B(\mathcal{S})$ is monotone; i.e., $u \leq v$ (elementwise) implies $C[u] \leq C[v]$.

(c) Show that C is a contraction operator:

$$\|C[u] - C[v]\| \leq \gamma \|u - v\|. \tag{11.48}$$

(d) Let $u_0 \in B(\mathcal{S})$ and define $u_{t+1} = C[u_t]$, $t = 0, 1, 2, \ldots$. Show that $\lim_{t\to\infty} \|u_t - u^*\| \to 0$, where u^* is the unique fixed point of C; i.e., it satisfies

$$C[u^*] = u^*.$$

3. Show that V^π is a fixed point of the iterative procedure (11.14). In particular, show that for all $s \in \mathcal{S}$:

$$V^\pi(s) = r(s, \pi(s)) + \gamma \mathbb{E}_\pi \left[V^\pi(S_1) \,\middle|\, S_0 = s \right]. \tag{11.49}$$

4. Implement a policy iteration algorithm for the Torus MDP in Example 11.11.

5. A robot moves in steps along a linear grid $\mathcal{S} = \{-2, -1, 0, 1, 2\}$ and is allowed one of three actions: move left (action -1), move right (action 1), or remain (action 0). In state

-2 only actions 0 and 1 are allowed, and in state 2 only actions -1 and 0 are allowed. The reward function is

$$r(s,a) = \begin{cases} -1\,, & s = -2 \text{ and } a = 0\,;\, s = -1 \text{ and } a = -1\,, \\ 1\,, & s = 2 \text{ and } a = 0\,;\, s = 1 \text{ and } a = 1\,, \\ 0\,, & \text{otherwise.} \end{cases}$$

That is, the robot receives a reward of 1 if it moves into state 2, a reward of -1 if it moves into state -2, and no reward otherwise.

(a) Give a table of the evolution of the value function, as in Algorithm 11.8, using $\varepsilon = 0.0001$ and $\gamma = 0.1$. Take the initial value function to be zero.

(b) Compute the optimal deterministic policy from the value function.

6. The following Python code provides an episodic modification of the 4-state MDP in Example 11.9, by introducing an absorbing 5th state, to which each of the other states can transition with a small probability $\delta = 0.01$, irrespective of the action. An episode terminates when the absorbing state is entered.

MDP_Epi.py

```
from MDP import * # Import P, r, and Actions
P_new = np.zeros((2, 5, 5)) # Add the fifth absorbing state
P_new[:, :4, :4] = P
P_new[:, 4, 4] = 1
delta = 0.01
P_new[:, :4, 4] = delta
P_new[:, :4, :4] *= (1 - delta)
for i in range(2):
    for j in range(4):
        row_sum = np.sum(P_new[i, j, :])
        P_new[i, j, :] /= row_sum

P = P_new
States = np.array([0, 1, 2, 3, 4], dtype=int)
```

(a) What is the probability distribution of the length T of an episode?

(b) Show that, using the same discount rate $\gamma = 0.9$ and reward function r as in Example 11.9, the optimal Q-table is given by

$$\mathbf{Q}^* = \begin{bmatrix} 7.0889648 & 6.5182169 \\ 5.8628851 & 6.4344794 \\ 5.7914358 & 7.2363794 \\ 8.2363794 & 7.2579530 \\ 0 & 0 \end{bmatrix}. \tag{11.50}$$

Compare with (11.27).

(c) What is the optimal policy?

7. Figure 11.51 shows the transition graphs for an MDP with two actions (0 and 1).

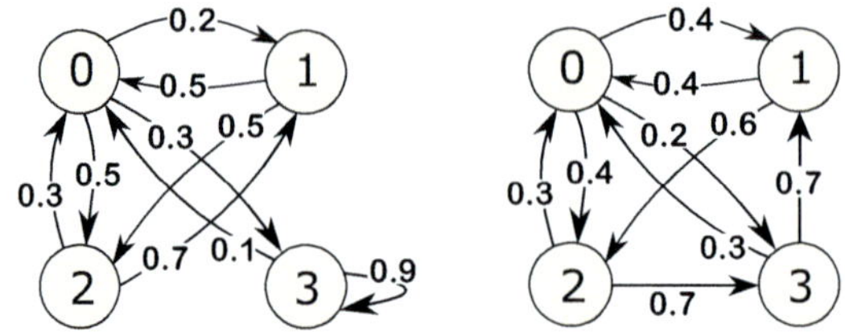

Figure 11.51: The transition graphs of the MDP for action 0 (left) and action 1 (right).

(a) From Example 11.9 modify the Python files `MDP.py` and `MDP_Val_Iter.py` and use these to find the optimal value function and Q-function for discount rate $\gamma = 0.9$ and reward function $r(s) = \mathbb{1}_{\{s=3\}}$.

(b) Modify the Python program `MDP_MC_Val_CE.py` in Example 11.31 to find the optimal policy using Monte Carlo estimation of the value function combined with the cross–entropy method for optimization. Compare with the true optimal policy found from the Q-function in (a).

☞ 128 8. Consider an MDP based on the 6-state transition graph given in Figure 3.96. Under action 0, the transition probabilities are given as specified in the figure. Under action 1, the same transition arcs apply, but the transition probabilities are equal, for example the probabilities to transition from state 1 to each of the states 2, 4, or 5 are 1/3. Under action 2, the arcs are reversed and the transition probabilities are again equal. A reward of 1 is awarded every time a transition takes place from either state 5 or 6. Take discount rate $\gamma = 0.99$.

9. Make a heatmap plot of the value function in Exercise 11.11.

10. Consider a Markov chain with the transition diagram given in Figure 11.52. At each step $t = 0, 1, \ldots$ we have to decide whether to stop the chain or to continue. If we stop the chain when it is in state s, a reward $r(s)$ is obtained. The question is to determine for each state whether it is better to stop at that state or to continue, in order to maximize the expected reward. Suppose the rewards are $6, 5, 3, 4, 0$ for states $1, 2, 3, 4, 5$. Determine the optimal stopping strategy using an MDP approach with two possible actions (stop and continue) in each state.

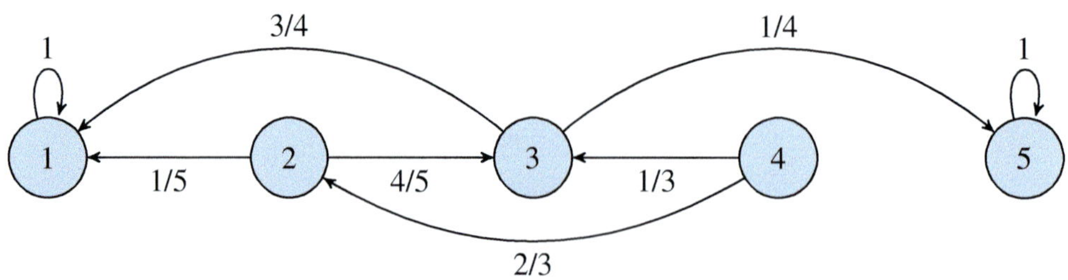

Figure 11.52: Determine at what states to stop in order to maximize the expected reward.

11. Modify Example 11.17 by changing the time to make a robot to 2. Hint: Consider the state of the system as $(n_t, g_t, n_{t-1}, g_{t-1})$.

12. Implement a Q-learning algorithm for the discrete torus MDP in Example 11.11. Discuss the advantages and disadvantages of this approach.

13. In the *multi-armed bandit* problem one must select between multiple actions (arms) which give a random reward. A key challenge is whether to *explore* (select untried arms) or *exploit* (select an arm with a known good reward). A simple mathematical model is that the rewards are independent random variables taking values in some interval $[a, b]$, with the expected reward for arm m being equal to μ_m, $m = 1, \dots, M$, where M is the number of arms. Let $R_1, R_2, \dots,$ be a sequence of rewards and let $S_n = \sum_{i=1}^n R_i$ be the total reward after n tries. Hoeffding's inequality states that for any $n \geq 1$ and $\varepsilon > 0$:

$$\mathbb{P}[S_n - \mathbb{E}S_n \geq \varepsilon] \leq \exp\left(-\frac{2\varepsilon^2}{n(b-a)^2}\right). \tag{11.53}$$

(a) Prove (11.53) via the following steps:

i. Using the (different) Hoeffding's inequality (2.105), show that for any $\lambda \in \mathbb{R}$: ☞ 68

$$\mathbb{E}\,\mathrm{e}^{\lambda(S_n - \mathbb{E}S_n)} \leq \exp\left(\frac{n\lambda^2(b-a)^2}{8}\right).$$

ii. Show that for $\lambda \geq 0$:

$$\mathbb{P}\left[S - \mathbb{E}S_n \geq \varepsilon\right] \leq \frac{\mathbb{E}\,\mathrm{e}^{\lambda(S_n - \mathbb{E}S_n)}}{\mathrm{e}^{\lambda\varepsilon}}.$$

iii. Combine i and ii to prove (11.53).

(b) Show that Hoeffding's inequality is symmetric, in the sense that also

$$\mathbb{P}[\mathbb{E}S_n - S_n \geq \varepsilon] \leq \exp\left(-\frac{2\varepsilon^2}{n(b-a)^2}\right). \tag{11.54}$$

(c) One heuristic to help the decision making in the multi-armed bandit problem is the following: Let $\widehat{\mu}_m$ be the average reward for arm m after n_m times that this arm was chosen, and let t be the total number of tries. For simplicity assume that all rewards take values in the interval $[0, 1]$. Use Hoeffding's inequality to show that

$$\mathbb{P}\left[\mu_m \geq \widehat{\mu}_m + \sqrt{2\ln(t)/n_m}\right] \leq t^{-4}.$$

The stochastic bound $\widehat{\mu}_m + \sqrt{2\ln(t)/n_m}$ is called the **upper confidence bound** for μ_m. The selection heuristic is to choose the arm with the highest upper confidence bound.

14. In the continuous torus MDP Example 11.42, modify the Python code given there to plot the height $r(\boldsymbol{x}(t))$ against time t for the optimal path. Use a step size of $1/n$, with $n = 400$.

15. In Example 11.42 we searched for an optimal policy within a given parametric family. If we were to optimize over *all* deterministic policies, subject to the corresponding path speed being v, the optimal policy would depend on the state as well as on the time. To obtain instead a *stationary* policy, we can alter the problem by using a discount factor ϱ, and maximize the performance function (i.e., value function):

$$V(\boldsymbol{x}(0)) = \int_0^\infty \mathrm{e}^{-\varrho t} r(\boldsymbol{x}(t))\,\mathrm{d}t.$$

We thus wish to find a deterministic stationary policy $\boldsymbol{v}(\boldsymbol{x}), \boldsymbol{x} \in [-3, 3]^\top$, in which $\boldsymbol{v}(\boldsymbol{x})$ is the velocity of the robot in state $\boldsymbol{x}$, under the constraint that $\|\boldsymbol{v}(\boldsymbol{x})\| = v$. Suppose that $\boldsymbol{v}^*(\boldsymbol{x})$ is the optimal policy in state $\boldsymbol{x}$ and $(\boldsymbol{z}(t), t \in [0, 1])$ is the optimal path, which satisfies the differential equation $\dot{\boldsymbol{z}}(t) = \boldsymbol{v}^*(\boldsymbol{z}(t))$ and $\boldsymbol{z}_0 = \boldsymbol{x}_0$.

Bellman's optimality principle stipulates that in a small time interval $[0, h]$ the optimal value function satisfies

$$V^*(\boldsymbol{x}) \approx \max_{\boldsymbol{v}, \|\boldsymbol{v}\|=v} \left\{ r(\boldsymbol{x})h + \mathrm{e}^{-\varrho h} V^*(\boldsymbol{x} + h\boldsymbol{v}) \right\}. \tag{11.55}$$

(a) Letting $h \downarrow 0$, show that (11.55) leads to the following partial differential equation for $V^*(\boldsymbol{x})$:

$$\varrho V^*(\boldsymbol{x}) = r(\boldsymbol{x}) + v\,\|\nabla V^*(\boldsymbol{x})\|. \tag{11.56}$$

(b) Show that the optimal policy (velocity) $\boldsymbol{v}(\boldsymbol{x})$ is given by

$$\boldsymbol{v}^*(\boldsymbol{x}) = v\,\frac{\nabla V^*(\boldsymbol{x})}{\|\nabla V^*(\boldsymbol{x})\|}.$$

That is, the optimal path is found via the steepest ascent of the value function V, at speed v.

APPENDIX A

LINEAR ALGEBRA

The purpose of this appendix is to review some important topics in linear algebra. We assume that the reader has some familiarity with matrix and vector operations, including matrix multiplication and the computation of determinants.

A.1 Vector Spaces, Bases, and Matrices

Linear algebra is the study of vector spaces and linear mappings. Vectors are, by definition, elements of some **vector space** $\mathcal{V}$ and satisfy the usual rules of addition and scalar multiplication, e.g.,

$$\text{if } \boldsymbol{x} \in \mathcal{V} \text{ and } \boldsymbol{y} \in \mathcal{V}, \text{ then } \alpha\boldsymbol{x} + \beta\boldsymbol{y} \in \mathcal{V} \text{ for all } \alpha, \beta \in \mathbb{R} \text{ (or } \mathbb{C}).$$

We will be dealing mostly with vectors in the Euclidean vector space $\mathbb{R}^n$ for some n. That is, we view the points of $\mathbb{R}^n$ as objects that can be added up and multiplied with a scalar, e.g., $(x_1, x_2) + (y_1, y_2) = (x_1 + y_1, x_2 + y_2)$ for points in $\mathbb{R}^2$. Sometimes it is convenient to work with the complex vector space $\mathbb{C}^n$ instead of $\mathbb{R}^n$; see also Section A.4.

Vectors $\boldsymbol{v}_1, \ldots, \boldsymbol{v}_k$ are called **linearly independent** if none of them can be expressed as a linear combination of the others; that is, if $\alpha_1\boldsymbol{v}_1 + \cdots + \alpha_n\boldsymbol{v}_n = \boldsymbol{0}$, then it must hold that $\alpha_i = 0$ for all $i = 1, \ldots, n$.

Definition A.1 (Basis of a Vector Space): A set of vectors $\mathcal{B} = \{\boldsymbol{v}_1, \ldots, \boldsymbol{v}_n\}$ is called a **basis** of the vector space $\mathcal{V}$ if every vector $\boldsymbol{x} \in \mathcal{V}$ can be written as a unique linear combination of the vectors in $\mathcal{B}$:

$$\boldsymbol{x} = \alpha_1\,\boldsymbol{v}_1 + \cdots + \alpha_n\,\boldsymbol{v}_n.$$

The (possibly infinite) number n is called the **dimension** of $\mathcal{V}$.

Using a basis $\mathcal{B}$ of $\mathcal{V}$, we can thus represent each vector $\boldsymbol{x} \in \mathcal{V}$ as a row or column of numbers

$$[\alpha_1, \ldots, \alpha_n] \quad \text{or} \quad \begin{bmatrix} \alpha_1 \\ \vdots \\ \alpha_n \end{bmatrix}. \tag{A.2}$$

Typically, vectors in $\mathbb{R}^n$ are represented via the **standard basis**, consisting of unit vectors (points) $\boldsymbol{e}_1 = (1, 0, \dots, 0), \dots, \boldsymbol{e}_n = (0, 0, \dots, 0, 1)$. As a consequence, any point $(x_1, \dots, x_n) \in \mathbb{R}^n$ can be represented, using the standard basis, as a row or column vector of the form (A.2) above, with $\alpha_i = x_i, i = 1, \dots, n$. We will also write $[x_1, x_2, \dots, x_n]^\top$, for the corresponding column vector, where $^\top$ denotes the **transpose**.

> To avoid confusion, we will use the convention from now on that a generic vector $\boldsymbol{x}$ is always represented via the standard basis as a *column* vector. The corresponding row vector is denoted by $\boldsymbol{x}^\top$.

A **matrix** can be viewed as an array of m rows and n columns that defines a **linear transformation** from $\mathbb{R}^n$ to $\mathbb{R}^m$ (or for complex matrices, from $\mathbb{C}^n$ to $\mathbb{C}^m$). The matrix is said to be *square* if $m = n$. If $\boldsymbol{a}_1, \boldsymbol{a}_2, \dots, \boldsymbol{a}_n$ are the columns of $\mathbf{A}$, that is, $\mathbf{A} = [\boldsymbol{a}_1, \boldsymbol{a}_2, \dots, \boldsymbol{a}_n]$, and if $\boldsymbol{x} = [x_1, \dots, x_n]^\top$, then $\mathbf{A}\boldsymbol{x} = x_1\, \boldsymbol{a}_1 + \cdots + x_n\, \boldsymbol{a}_n$. In particular, the standard basis vector $\boldsymbol{e}_k$ is mapped to the vector $\boldsymbol{a}_k$, $k = 1, \dots, n$. We sometimes use the notation $\mathbf{A} = [a_{ij}]$, to denote a matrix whose (i, j)-th element is a_{ij}. When we wish to emphasize that a matrix $\mathbf{A}$ is real-valued with m rows and n columns, we write $\mathbf{A} \in \mathbb{R}^{m \times n}$ (and $\mathbb{C}^{m \times n}$ for a complex $m \times n$ matrix). The **rank** of a matrix is the number of linearly independent rows or, equivalently, the number of linearly independent columns.

Example A.3 (Linear Transformation) Take the matrix

$$\mathbf{A} = \begin{bmatrix} 1 & 1 \\ -0.5 & -2 \end{bmatrix}.$$

It transforms the two basis vectors $[1, 0]^\top$ and $[0, 1]^\top$, shown in red and blue in the left panel of Figure A.4, to the vectors $[1, -0.5]^\top$ and $[1, -2]^\top$, shown on the right panel. Similarly, the points on the unit circle are transformed to an ellipse.

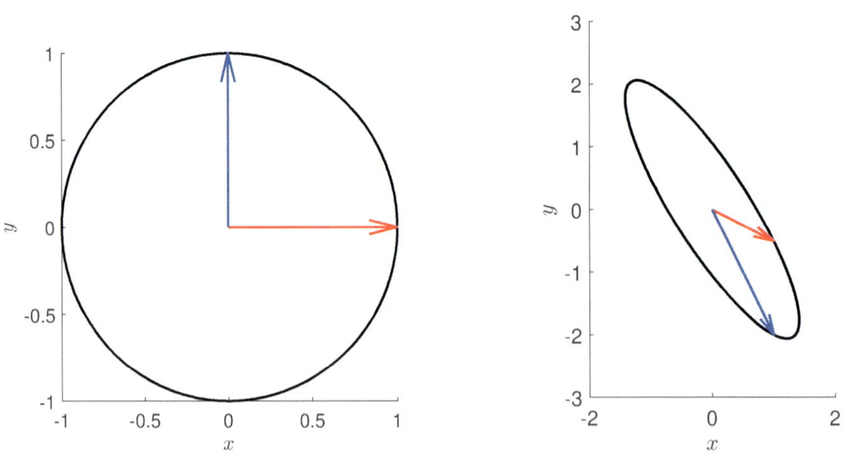

Figure A.4: A linear transformation of the unit circle.

Suppose $\mathbf{A} = [\boldsymbol{a}_1, \dots, \boldsymbol{a}_n]$, where the $\mathcal{A} = \{\boldsymbol{a}_i\}$ form a basis of $\mathbb{R}^n$. Take any vector $\boldsymbol{x} = [x_1, \dots, x_n]_{\mathcal{E}}^{\top}$ with respect to the standard basis $\mathcal{E}$ (we write subscript $\mathcal{E}$ to stress this). Then the representation of this vector with respect to $\mathcal{A}$ is simply

$$\boldsymbol{y} = \mathbf{A}^{-1}\boldsymbol{x},$$

where $\mathbf{A}^{-1}$ is the **inverse** of $\mathbf{A}$; that is, the matrix such that $\mathbf{A}\mathbf{A}^{-1} = \mathbf{A}^{-1}\mathbf{A} = \mathbf{I}_n$, where $\mathbf{I}_n$ is the n-dimensional identity matrix. To see this, note that $\mathbf{A}^{-1}\boldsymbol{a}_i$ gives the i-th unit vector representation, for $i = 1, \dots, n$, and recall that each vector in $\mathbb{R}^n$ is a unique linear combination of these basis vectors.

Example A.5 (Basis Representation) Consider the matrix

$$\mathbf{A} = \begin{bmatrix} 1 & 2 \\ 3 & 4 \end{bmatrix} \quad \text{with inverse} \quad \mathbf{A}^{-1} = \begin{bmatrix} -2 & 1 \\ 3/2 & -1/2 \end{bmatrix}. \tag{A.6}$$

The vector $\boldsymbol{x} = [1, 1]_{\mathcal{E}}^{\top}$ in the standard basis has representation $\boldsymbol{y} = \mathbf{A}^{-1}\boldsymbol{x} = [-1, 1]_{\mathcal{A}}^{\top}$ in the basis consisting of the columns of $\mathbf{A}$. Namely,

$$\mathbf{A}\boldsymbol{y} = -\begin{bmatrix} 1 \\ 3 \end{bmatrix} + \begin{bmatrix} 2 \\ 4 \end{bmatrix} = \begin{bmatrix} 1 \\ 1 \end{bmatrix}.$$

The **transpose** of a matrix $\mathbf{A} = [a_{ij}]$ is the matrix $\mathbf{A}^{\top} = [a_{ji}]$; that is, the (i, j)-th element of $\mathbf{A}^{\top}$ is the (j, i)-th element of $\mathbf{A}$. The **vectorization** of a matrix $\mathbf{A} \in \mathbb{R}^{m \times n}$ with n columns $\boldsymbol{a}_1, \dots, \boldsymbol{a}_n$ is the column vector of dimension mn:

$$\text{vec}(\mathbf{A}) := \begin{bmatrix} \boldsymbol{a}_1 \\ \vdots \\ \boldsymbol{a}_n \end{bmatrix}.$$

The **trace** of a square matrix $\mathbf{A}$, denoted $\text{tr}(\mathbf{A})$, is the sum of its diagonal elements. A useful result is the following cyclic property:

Theorem A.7 (Cyclic Property): The trace is invariant under cyclic permutations: $\text{tr}(\mathbf{ABC}) = \text{tr}(\mathbf{BCA}) = \text{tr}(\mathbf{CAB})$.

Proof. It suffices to show that $\text{tr}(\mathbf{DE})$ is equal to $\text{tr}(\mathbf{ED})$ for any $m \times n$ matrix $\mathbf{D} = [d_{ij}]$ and $n \times m$ matrix $\mathbf{E} = [e_{ij}]$. The diagonal elements of $\mathbf{DE}$ are $\sum_{j=1}^{n} d_{ij}\, e_{ji}, i = 1, \dots, m$ and the diagonal elements of $\mathbf{ED}$ are $\sum_{i=1}^{m} e_{ji}\, d_{ij}, j = 1, \dots, n$. They sum up to the same number $\sum_{i=1}^{m} \sum_{j=1}^{n} d_{ij}\, e_{ji}$. □

A square matrix has an inverse if and only if its columns (or rows) are linearly independent. This is the same as the matrix being of *full rank*; that is, its rank is equal to the number of columns. An equivalent statement is that its determinant is not zero. The **determinant** of an $n \times n$ matrix $\mathbf{A} = [a_{i,j}]$ is defined as

$$\det(\mathbf{A}) := \sum_{\pi} (-1)^{\zeta(\pi)} \prod_{i=1}^{n} a_{\pi_i, i}, \tag{A.8}$$

where the sum is over all permutations $\boldsymbol{\pi} = (\pi_1, \ldots, \pi_n)$ of $(1, \ldots, n)$, and $\zeta(\boldsymbol{\pi})$ is the number of pairs (i, j) for which $i < j$ and $\pi_i > \pi_j$. For example, $\zeta(2, 3, 4, 1) = 3$ for the pairs $(1, 4), (2, 4), (3, 4)$. The determinant of a **triangular matrix** — a matrix whose elements above (in a lower triangular matrix) or below (in an upper triangular matrix) the main diagonal are zeros — is simply the product of its diagonal elements. A **diagonal matrix** is a matrix with only zero elements off the diagonal. Geometrically, the determinant of a square matrix $\mathbf{A} = [\boldsymbol{a}_1, \ldots, \boldsymbol{a}_n]$ is the (signed) *volume* of the parallelepiped defined by the columns $\boldsymbol{a}_1, \ldots, \boldsymbol{a}_n$; that is, the set of points $\boldsymbol{x} = \sum_{i=1}^{n} \alpha_i \, \boldsymbol{a}_i$, where $0 \leq \alpha_i \leq 1, i = 1, \ldots, n$.

Theorem A.9 summarizes a number of useful matrix rules for the concepts that we have discussed so far. We leave the proofs, which typically involves "writing out" the equations, as an exercise for the reader; see also [163].

Theorem A.9 (Useful Matrix Rules):

1. $(\mathbf{AB})^\top = \mathbf{B}^\top \mathbf{A}^\top$
2. $(\mathbf{AB})^{-1} = \mathbf{B}^{-1} \mathbf{A}^{-1}$
3. $(\mathbf{A}^{-1})^\top = (\mathbf{A}^\top)^{-1} =: \mathbf{A}^{-\top}$
4. $\det(\mathbf{AB}) = \det(\mathbf{A}) \det(\mathbf{B})$ and $\det(\mathbf{A}^\top) = \det(\mathbf{A})$
5. $\boldsymbol{x}^\top \mathbf{A} \boldsymbol{x} = \operatorname{tr}(\mathbf{A} \boldsymbol{x} \boldsymbol{x}^\top)$
6. $\det(\mathbf{A}) = \prod_i a_{ii}$ if $\mathbf{A} = [a_{ij}]$ is triangular

Next, consider an $n \times p$ matrix $\mathbf{A}$ for which the matrix inverse fails to exist. That is, $\mathbf{A}$ is either non-square ($n \neq p$) or its determinant is 0. Instead of the inverse, we can use its so-called pseudo-inverse, which always exists.

Definition A.10 (Moore–Penrose Pseudo-Inverse): Let $\mathbf{A} \in \mathbb{R}^{n \times p}$. The **Moore-Penrose pseudo-inverse** of $\mathbf{A}$ is defined as the unique matrix $\mathbf{A}^+ \in \mathbb{R}^{p \times n}$ that satisfies the conditions:

1. $\mathbf{A}\mathbf{A}^+\mathbf{A} = \mathbf{A}$
2. $\mathbf{A}^+\mathbf{A}\mathbf{A}^+ = \mathbf{A}^+$
3. $(\mathbf{A}\mathbf{A}^+)^\top = \mathbf{A}\mathbf{A}^+$
4. $(\mathbf{A}^+\mathbf{A})^\top = \mathbf{A}^+\mathbf{A}$

We can write $\mathbf{A}^+$ explicitly in terms of $\mathbf{A}$ when $\mathbf{A}$ has a full column or row rank. For example, we always have

$$\mathbf{A}^\top\mathbf{A}\mathbf{A}^+ = \mathbf{A}^\top(\mathbf{A}\mathbf{A}^+)^\top = ((\mathbf{A}\mathbf{A}^+)\mathbf{A})^\top = (\mathbf{A})^\top = \mathbf{A}^\top. \tag{A.11}$$

If $\mathbf{A}$ has a full column rank p, then $(\mathbf{A}^\top\mathbf{A})^{-1}$ exists, so that from (A.11) it follows that

$$\mathbf{A}^+ = (\mathbf{A}^\top\mathbf{A})^{-1}\mathbf{A}^\top.$$

This is referred to as the **left pseudo-inverse**, as $\mathbf{A}^+\mathbf{A} = \mathbf{I}_p$. Similarly, if $\mathbf{A}$ has a full row rank n, that is, $(\mathbf{A}\mathbf{A}^\top)^{-1}$ exists, then it follows from

$$\mathbf{A}^+\mathbf{A}\mathbf{A}^\top = (\mathbf{A}^+\mathbf{A})^\top\mathbf{A}^\top = (\mathbf{A}(\mathbf{A}^+\mathbf{A}))^\top = \mathbf{A}^\top$$

that

$$\mathbf{A}^+ = \mathbf{A}^\top(\mathbf{A}\mathbf{A}^\top)^{-1}.$$

This is the **right pseudo-inverse**, as $\mathbf{A}\mathbf{A}^+ = \mathbf{I}_n$. Finally, if $\mathbf{A}$ is of full rank and square, then $\mathbf{A}^+ = \mathbf{A}^{-1}$.

A.2 Inner Product

The (Euclidean) **inner product** of two real vectors $\boldsymbol{x} = [x_1, \ldots, x_n]^\top$ and $\boldsymbol{y} = [y_1, \ldots, y_n]^\top$ is defined as the number

$$\langle \boldsymbol{x}, \boldsymbol{y} \rangle = \sum_{i=1}^{n} x_i\, y_i = \boldsymbol{x}^\top \boldsymbol{y}.$$

Here $\boldsymbol{x}^\top\boldsymbol{y}$ is the matrix multiplication of the $(1 \times n)$ matrix $\boldsymbol{x}^\top$ and the $(n \times 1)$ matrix $\boldsymbol{y}$. The inner product induces a geometry on the vector space $\mathbb{R}^n$, allowing for the definition of length, angle, and so on. The inner product satisfies the following properties:

1. $\langle \alpha\boldsymbol{x} + \beta\boldsymbol{y}, \boldsymbol{z} \rangle = \alpha \langle \boldsymbol{x}, \boldsymbol{z} \rangle + \beta \langle \boldsymbol{y}, \boldsymbol{z} \rangle$,
2. $\langle \boldsymbol{x}, \boldsymbol{y} \rangle = \langle \boldsymbol{y}, \boldsymbol{x} \rangle$,
3. $\langle \boldsymbol{x}, \boldsymbol{x} \rangle \geq 0$,
4. $\langle \boldsymbol{x}, \boldsymbol{x} \rangle = 0$ if and only if $\boldsymbol{x} = \mathbf{0}$.

Vectors $\boldsymbol{x}$ and $\boldsymbol{y}$ are called *perpendicular* (or **orthogonal**) if $\langle \boldsymbol{x}, \boldsymbol{y} \rangle = 0$. The **Euclidean norm** (or length) of a vector $\boldsymbol{x}$ is defined as

$$\|\boldsymbol{x}\| = \sqrt{x_1^2 + \cdots + x_n^2} = \sqrt{\langle \boldsymbol{x}, \boldsymbol{x} \rangle}.$$

If $\boldsymbol{x}$ and $\boldsymbol{y}$ are perpendicular, then **Pythagoras' theorem** holds:

$$\|\boldsymbol{x} + \boldsymbol{y}\|^2 = \langle \boldsymbol{x} + \boldsymbol{y},\, \boldsymbol{x} + \boldsymbol{y} \rangle = \langle \boldsymbol{x}, \boldsymbol{x} \rangle + 2\,\langle \boldsymbol{x}, \boldsymbol{y} \rangle + \langle \boldsymbol{y}, \boldsymbol{y} \rangle = \|\boldsymbol{x}\|^2 + \|\boldsymbol{y}\|^2. \tag{A.12}$$

A basis $\{\boldsymbol{v}_1, \ldots, \boldsymbol{v}_n\}$ of $\mathbb{R}^n$ in which all the vectors are pairwise perpendicular and have norm 1 is called an **orthonormal** (short for orthogonal and normalized) basis. For example, the standard basis is orthonormal.

Theorem A.13 (Orthonormal Basis Representation): If $\{\boldsymbol{v}_1, \ldots, \boldsymbol{v}_n\}$ is an orthonormal basis of $\mathbb{R}^n$, then any vector $\boldsymbol{x} \in \mathbb{R}^n$ can be expressed as

$$\boldsymbol{x} = \langle \boldsymbol{x}, \boldsymbol{v}_1 \rangle \, \boldsymbol{v}_1 + \cdots + \langle \boldsymbol{x}, \boldsymbol{v}_n \rangle \, \boldsymbol{v}_n. \tag{A.14}$$

Proof. Observe that, because the $\{\boldsymbol{v}_i\}$ form a basis, there exist unique $\alpha_1, \ldots, \alpha_n$ such that $\boldsymbol{x} = \alpha_1 \boldsymbol{v}_1 + \cdots + \alpha_n \boldsymbol{v}_n$. By the linearity of the inner product and the orthonormality of the $\{\boldsymbol{v}_i\}$ it follows that

$$\langle \boldsymbol{x}, \boldsymbol{v}_j \rangle = \Big\langle \sum_i \alpha_i \boldsymbol{v}_i, \boldsymbol{v}_j \Big\rangle = \alpha_j.$$

□

An $n \times n$ matrix $\mathbf{V}$ whose columns form an orthonormal basis is called an **orthogonal matrix**. Note that for an orthogonal matrix $\mathbf{V} = [\boldsymbol{v}_1, \ldots, \boldsymbol{v}_n]$, we have

$$\mathbf{V}^\top \mathbf{V} = \begin{bmatrix} \boldsymbol{v}_1^\top \\ \boldsymbol{v}_2^\top \\ \vdots \\ \boldsymbol{v}_n^\top \end{bmatrix} [\boldsymbol{v}_1, \boldsymbol{v}_2, \ldots, \boldsymbol{v}_n] = \begin{bmatrix} \boldsymbol{v}_1^\top \boldsymbol{v}_1 & \boldsymbol{v}_1^\top \boldsymbol{v}_2 & \ldots & \boldsymbol{v}_1^\top \boldsymbol{v}_n \\ \vdots & \vdots & \vdots & \vdots \\ \boldsymbol{v}_n^\top \boldsymbol{v}_1 & \boldsymbol{v}_n^\top \boldsymbol{v}_2 & \ldots & \boldsymbol{v}_n^\top \boldsymbol{v}_n \end{bmatrix} = \mathbf{I}_n.$$

Hence, $\mathbf{V}^{-1} = \mathbf{V}^\top$. Note also that an orthogonal transformation is **length preserving**; that is, $\mathbf{V}\boldsymbol{x}$ has the same length as $\boldsymbol{x}$. This follows from

$$\|\mathbf{V}\boldsymbol{x}\|^2 = \langle \mathbf{V}\boldsymbol{x}, \mathbf{V}\boldsymbol{x} \rangle = \boldsymbol{x}^\top \mathbf{V}^\top \mathbf{V} \boldsymbol{x} = \boldsymbol{x}^\top \boldsymbol{x} = \|\boldsymbol{x}\|^2.$$

If $\mathbf{V}$ is not a square matrix, say $\mathbf{V} \in \mathbb{R}^{m \times n}$ with $m > n$, but $\mathbf{V}^\top \mathbf{V} = \mathbf{I}_n$, then we call $\mathbf{V}$ an **orthonormal matrix**.

A.3 Kronecker Product

If $\mathbf{A} \in \mathbb{R}^{m \times n}$ and $\mathbf{B} \in \mathbb{R}^{p \times q}$, then the **Kronecker product** of $\mathbf{A}$ and $\mathbf{B}$, denoted as $\mathbf{A} \otimes \mathbf{B}$ is the $mp \times nq$ matrix

$$\mathbf{A} \otimes \mathbf{B} = \begin{bmatrix} a_{1,1}\mathbf{B} & \cdots & a_{1,n}\mathbf{B} \\ \vdots & \ddots & \vdots \\ a_{m,1}\mathbf{B} & \cdots & a_{m,n}\mathbf{B} \end{bmatrix}.$$

The Kronecker product behaves in the same way as multiplication of two scalars, except that it is not commutative, that is, in general $\mathbf{A} \otimes \mathbf{B} \neq \mathbf{B} \otimes \mathbf{A}$. Various basic properties of the Kronecker product and vectorization are summarized in the following theorem:

Theorem A.15 (Properties of Kronecker Product): For $\mathbf{A}, \mathbf{B}, \mathbf{C}, \mathbf{D}$ of commensurate dimension, the following properties hold.

1. *Transpose of Kronecker product:* $(\mathbf{A} \otimes \mathbf{B})^\top = \mathbf{A}^\top \otimes \mathbf{B}^\top$.
2. *Mixed-product property:* $(\mathbf{A} \otimes \mathbf{B})(\mathbf{C} \otimes \mathbf{D}) = (\mathbf{AC}) \otimes (\mathbf{BD})$.
3. *Vectorization in standard basis:* If $\boldsymbol{e}_1, \ldots, \boldsymbol{e}_n$ is the standard basis for $\mathbb{R}^n$ and $\boldsymbol{u}_1, \ldots, \boldsymbol{u}_m$ is the standard basis for $\mathbb{R}^m$, then ☞ 514

$$\text{vec}(\mathbf{A}) = \sum_{i=1}^{n} \boldsymbol{e}_i \otimes (\mathbf{A}\boldsymbol{e}_i) = \sum_{i=1}^{m} (\mathbf{A}^\top \boldsymbol{u}_i) \otimes \boldsymbol{u}_i.$$

4. *Trace via vectorization:* $\text{tr}(\mathbf{A}^\top \mathbf{B}) = \text{vec}(\mathbf{A})^\top \text{vec}(\mathbf{B})$.
5. *Vectorization and Kronecker Product:* $\text{vec}(\mathbf{ABC}) = (\mathbf{C}^\top \otimes \mathbf{A})\text{vec}(\mathbf{B})$.

Proof. The first three properties follow directly from writing out both sides of the identities using the definition of the Kronecker product. As for Property 4, using the standard basis representation of the vectorization, the inner product of $\text{vec}(\mathbf{A})$ and $\text{vec}(\mathbf{B})$ can be written as:

$$\begin{aligned}
\text{vec}(\mathbf{A})^\top \text{vec}(\mathbf{B}) &= \left[\sum_i \boldsymbol{e}_i \otimes (\mathbf{A}\boldsymbol{e}_i)\right]^\top \sum_j \boldsymbol{e}_j \otimes (\mathbf{B}\boldsymbol{e}_j) \\
&= \sum_{i,j} (\boldsymbol{e}_i^\top \otimes (\mathbf{A}\boldsymbol{e}_i)^\top)(\boldsymbol{e}_j \otimes (\mathbf{B}\boldsymbol{e}_j)) \\
&= \sum_{i,j} (\boldsymbol{e}_i^\top \boldsymbol{e}_j) \otimes (\boldsymbol{e}_i^\top \mathbf{A}^\top \mathbf{B}\boldsymbol{e}_j) = \sum_{i,j} \mathbb{1}_{\{i=j\}} \otimes ([\mathbf{A}^\top \mathbf{B}]_{i,j}) \\
&= \sum_i [\mathbf{A}^\top \mathbf{B}]_{i,i} = \text{tr}(\mathbf{A}^\top \mathbf{B}).
\end{aligned}$$

Finally, to prove Property 5, we first prove the special case:

$$\begin{aligned}
\text{vec}(\boldsymbol{u}\boldsymbol{v}^\top) &= \sum_i \boldsymbol{e}_i \otimes (\boldsymbol{u}\boldsymbol{v}^\top \boldsymbol{e}_i) = \sum_i \boldsymbol{e}_i \otimes (\boldsymbol{u} v_i) \\
&= \sum_i \boldsymbol{e}_i v_i \otimes \boldsymbol{u} = \boldsymbol{v} \otimes \boldsymbol{u}.
\end{aligned}$$

Hence, writing $\mathbf{B} = \sum_i \boldsymbol{p}_i \boldsymbol{q}_i^\top$ as a sum of rank-1 matrices and using $\text{vec}(\boldsymbol{u}\boldsymbol{v}^\top) = \boldsymbol{v} \otimes \boldsymbol{u}$, we obtain:

$$\begin{aligned}
\text{vec}(\mathbf{ABC}) &= \sum_i \text{vec}([\mathbf{A}\boldsymbol{p}_i][\boldsymbol{q}_i^\top \mathbf{C}]) = \sum_i (\mathbf{C}^\top \boldsymbol{q}_i) \otimes (\mathbf{A}\boldsymbol{p}_i) \\
&= \sum_i (\mathbf{C}^\top \otimes \mathbf{A})(\boldsymbol{q}_i \otimes \boldsymbol{p}_i) = (\mathbf{C}^\top \otimes \mathbf{A})\text{vec}(\mathbf{B}).
\end{aligned}$$

□

■ **Example A.16 (Pseudo-Inverse and Kronecker Product)** A number of other identities follow from the mixed-product identity in Theorem A.15. For example, if both $\mathbf{A} \in \mathbb{R}^{n\times n}$ and $\mathbf{B} \in \mathbb{R}^{m\times m}$ are square and invertible, then by the mixed-product identity:

$$(\mathbf{A} \otimes \mathbf{B})(\mathbf{A}^{-1} \otimes \mathbf{B}^{-1}) = \mathbf{I}_n \otimes \mathbf{I}_m = \mathbf{I}_{m\times n}.$$

Similarly, $(\mathbf{A}^{-1} \otimes \mathbf{B}^{-1})(\mathbf{A} \otimes \mathbf{B}) = \mathbf{I}_{m\times n}$ and hence $(\mathbf{A} \otimes \mathbf{B})^{-1} = (\mathbf{A}^{-1} \otimes \mathbf{B}^{-1})$. More generally, the mixed-product property implies that

$$(\mathbf{A} \otimes \mathbf{B})^{+} = \mathbf{A}^{+} \otimes \mathbf{B}^{+}.$$

■

A.4 Complex Vectors and Matrices

Instead of the vector space $\mathbb{R}^n$ of n-dimensional real vectors, it is useful to consider also the vector space $\mathbb{C}^n$ of n-dimensional complex vectors. In this case the **adjoint** or **conjugate transpose** operation ($*$) replaces the transpose operation ($\top$). This involves the usual transposition of the matrix or vector with the additional step that any complex number $z = x + \mathrm{i}\, y$ is replaced by its complex conjugate $\overline{z} = x - \mathrm{i}\, y$. For example, in $\mathbb{C}^2$, the conjugate transpose of

$$\boldsymbol{x} = \begin{bmatrix} a_1 + \mathrm{i}\, b_1 \\ a_2 + \mathrm{i}\, b_2 \end{bmatrix} \quad \text{is} \quad \boldsymbol{x}^* = [a_1 - \mathrm{i}\, b_1,\ a_2 - \mathrm{i}\, b_2].$$

The (Euclidean) inner product of $\boldsymbol{x}$ and $\boldsymbol{y}$ (viewed as column vectors in $\mathbb{C}^n$) is now defined as

$$\langle \boldsymbol{x}, \boldsymbol{y} \rangle = \boldsymbol{y}^* \boldsymbol{x} = \sum_{i=1}^{n} x_i\, \overline{y}_i,$$

which is no longer symmetric: $\langle \boldsymbol{x}, \boldsymbol{y} \rangle = \overline{\langle \boldsymbol{y}, \boldsymbol{x} \rangle}$. Note that this generalizes the real-valued inner product. Similarly, the **adjoint matrix** of a matrix $\mathbf{A} \in \mathbb{C}^{m\times n}$ is the matrix $\mathbf{A}^* \in \mathbb{C}^{n\times m}$ that satisfies

$$\langle \mathbf{A}\boldsymbol{x}, \boldsymbol{y} \rangle = \langle \boldsymbol{x}, \mathbf{A}^* \boldsymbol{y} \rangle$$

for all $\boldsymbol{x} \in \mathbb{C}^n$ and $\boldsymbol{y} \in \mathbb{C}^m$. For example, the adjoint matrix of

$$\mathbf{A} = \begin{bmatrix} a_{11} + \mathrm{i}\, b_{11} & a_{12} + \mathrm{i}\, b_{12} \\ a_{21} + \mathrm{i}\, b_{21} & a_{22} + \mathrm{i}\, b_{22} \end{bmatrix} \quad \text{is} \quad \mathbf{A}^* = \begin{bmatrix} a_{11} - \mathrm{i}\, b_{11} & a_{21} - \mathrm{i}\, b_{21} \\ a_{12} - \mathrm{i}\, b_{12} & a_{22} - \mathrm{i}\, b_{22} \end{bmatrix}.$$

The determinant of a complex matrix $\mathbf{A}$ is defined exactly as in (A.8). As a consequence of the results in Theorem A.9, we have that

$$\det(\mathbf{A}^*) = \overline{\det(\mathbf{A})}.$$

A complex matrix is said to be **Hermitian** or **self-adjoint** if $\mathbf{A}^* = \mathbf{A}$, and **unitary** if $\mathbf{A}^*\mathbf{A} = \mathbf{I}$ (that is, if $\mathbf{A}^* = \mathbf{A}^{-1}$). For real matrices "Hermitian" is the same as "symmetric", and "unitary" is the same as "orthogonal".

A.5 Orthogonal Projections

Let $\{\boldsymbol{u}_1, \ldots, \boldsymbol{u}_k\}$ be a set of linearly independent vectors in $\mathbb{R}^n$. The set

$$\mathcal{V} = \text{Span}\,\{\boldsymbol{u}_1, \ldots, \boldsymbol{u}_k\} = \{\alpha_1 \boldsymbol{u}_1 + \cdots + \alpha_k \boldsymbol{u}_k,\ \alpha_1, \ldots, \alpha_k \in \mathbb{R}\},$$

is called the **linear subspace** of $\mathbb{R}^n$ **spanned** by $\{\boldsymbol{u}_1, \ldots, \boldsymbol{u}_k\}$. The **orthogonal complement** of $\mathcal{V}$, denoted by $\mathcal{V}^\perp$, is the set of all vectors $\boldsymbol{w}$ that are orthogonal to $\mathcal{V}$, in the sense that $\langle \boldsymbol{w}, \boldsymbol{v} \rangle = 0$ for all $\boldsymbol{v} \in \mathcal{V}$. The matrix $\mathbf{P}$ such that $\mathbf{P}\boldsymbol{x} = \boldsymbol{x}$, for all $\boldsymbol{x} \in \mathcal{V}$, and $\mathbf{P}\boldsymbol{x} = \mathbf{0}$, for all $\boldsymbol{x} \in \mathcal{V}^\perp$ is called the **orthogonal projection matrix** onto $\mathcal{V}$. Suppose that $\mathbf{U} = [\boldsymbol{u}_1, \ldots, \boldsymbol{u}_k]$ has full rank, in which case $\mathbf{U}^\top\mathbf{U}$ is an invertible matrix. The orthogonal projection matrix $\mathbf{P}$ onto $\mathcal{V} = \text{Span}\,\{\boldsymbol{u}_1, \ldots, \boldsymbol{u}_k\}$ is then given by

$$\mathbf{P} = \mathbf{U}(\mathbf{U}^\top\mathbf{U})^{-1}\mathbf{U}^\top.$$

Namely, since $\mathbf{P}\mathbf{U} = \mathbf{U}$, the matrix $\mathbf{P}$ projects any vector in $\mathcal{V}$ onto itself. Moreover, $\mathbf{P}$ projects any vector in $\mathcal{V}^\perp$ onto the zero vector. Using the pseudo-inverse, it is possible to specify the projection matrix also for the case where $\mathbf{U}$ is not of full rank, leading to the following theorem.

Theorem A.17 (Orthogonal Projection): Let $\mathbf{U} = [\boldsymbol{u}_1, \ldots, \boldsymbol{u}_k]$. Then, the orthogonal projection matrix $\mathbf{P}$ onto $\mathcal{V} = \text{Span}\{\boldsymbol{u}_1, \ldots, \boldsymbol{u}_k\}$ is given by

$$\mathbf{P} = \mathbf{U}\,\mathbf{U}^+, \tag{A.18}$$

where $\mathbf{U}^+$ is the pseudo-inverse of $\mathbf{U}$.

Proof. By Property 1 of Definition A.10 we have

$$\mathbf{P}\mathbf{U} = \mathbf{U}\mathbf{U}^+\mathbf{U} = \mathbf{U},$$

so that $\mathbf{P}$ projects any vector in $\mathcal{V}$ onto itself. In addition, for any $\boldsymbol{w} \in \mathcal{V}^\perp$, we have that $\boldsymbol{w}^\top\mathbf{U}\boldsymbol{\alpha} = 0$ for all $\boldsymbol{\alpha} \in \mathbb{R}^k$, or equivalently, $\boldsymbol{w}^\top\mathbf{U} = \mathbf{0}^\top$. By the symmetry Property 3 of Definition A.10:

$$\mathbf{P}\boldsymbol{w} = (\boldsymbol{w}^\top\mathbf{P})^\top = (\boldsymbol{w}^\top\mathbf{U}\mathbf{U}^+)^\top = \mathbf{0}.$$

In other words, $\mathbf{P}$ projects any vector in $\mathcal{V}^\perp$ onto the zero vector. □

Note that in the special case where $\boldsymbol{u}_1, \ldots, \boldsymbol{u}_k$ above form an orthonormal basis of $\mathcal{V}$, then $\mathbf{U}^+ = \mathbf{U}^\top$ and the projection onto $\mathcal{V}$ is simple to describe, namely we have

$$\mathbf{P}\boldsymbol{x} = \mathbf{U}\mathbf{U}^\top\boldsymbol{x} = \sum_{i=1}^{k} \langle \boldsymbol{x}, \boldsymbol{u}_i \rangle\, \boldsymbol{u}_i. \tag{A.19}$$

For any point $\boldsymbol{x} \in \mathbb{R}^n$, the point in $\mathcal{V}$ that is closest to $\boldsymbol{x}$ is its orthogonal projection $\mathbf{P}\boldsymbol{x}$, as the following theorem shows.

Theorem A.20 (Orthogonal Projection and Minimal Distance): Let $\{\boldsymbol{u}_1, \ldots, \boldsymbol{u}_k\}$ be an orthonormal basis of $\mathcal{V} = \operatorname{Span}\{\boldsymbol{u}_1, \ldots, \boldsymbol{u}_k\}$ and let $\mathbf{P}$ be the orthogonal projection matrix onto $\mathcal{V}$. The solution to the minimization program

$$\min_{\boldsymbol{y} \in \mathcal{V}} \|\boldsymbol{x} - \boldsymbol{y}\|^2$$

is $\boldsymbol{y} = \mathbf{P}\boldsymbol{x}$. That is, $\mathbf{P}\boldsymbol{x} \in \mathcal{V}$ is closest to $\boldsymbol{x}$.

Proof. We can write each point $\boldsymbol{y} \in \mathcal{V}$ as $\boldsymbol{y} = \sum_{i=1}^k \alpha_i \boldsymbol{u}_i$. Consequently,

$$\|\boldsymbol{x} - \boldsymbol{y}\|^2 = \left\langle \boldsymbol{x} - \sum_{i=1}^k \alpha_i \boldsymbol{u}_i, \, \boldsymbol{x} - \sum_{i=1}^k \alpha_i \boldsymbol{u}_i \right\rangle = \|\boldsymbol{x}\|^2 - 2\sum_{i=1}^k \alpha_i \langle \boldsymbol{x}, \boldsymbol{u}_i \rangle + \sum_{i=1}^k \alpha_i^2.$$

Minimizing this with respect to the $\{\alpha_i\}$ gives $\alpha_i = \langle \boldsymbol{x}, \boldsymbol{u}_i \rangle, i = 1, \ldots, k$. In view of (A.19), the optimal $\boldsymbol{y}$ is thus $\mathbf{P}\boldsymbol{x}$. □

A.6 The Gram–Schmidt Procedure and QR Decomposition

Let $\mathbf{A} \in \mathbb{R}^{n \times p}$, where $p \leq n$. Then, there exists an orthonormal matrix $\mathbf{Q} \in \mathbb{R}^{n \times p}$ (hence satisfying $\mathbf{Q}^\top \mathbf{Q} = \mathbf{I}_p$) and an upper triangular matrix $\mathbf{R} \in \mathbb{R}^{p \times p}$, such that

$$\mathbf{A} = \mathbf{Q}\mathbf{R}.$$

This is the **QR decomposition** for real-valued matrices. When $\mathbf{A}$ has full column rank, such a decomposition can be obtained via the **Gram–Schmidt** procedure, which constructs an *orthonormal basis* $\{\boldsymbol{u}_1, \ldots, \boldsymbol{u}_p\}$ of the column space of $\mathbf{A}$ spanned by $\{\boldsymbol{a}_1, \ldots, \boldsymbol{a}_p\}$, using the following steps (see also Figure A.21):

1. Take $\boldsymbol{u}_1 = \boldsymbol{a}_1 / \|\boldsymbol{a}_1\|$.

2. Let

$$\boldsymbol{p}_1 = \langle \boldsymbol{u}_1, \boldsymbol{a}_2 \rangle \, \boldsymbol{u}_1 = \boldsymbol{u}_1 \boldsymbol{u}_1^\top \boldsymbol{a}_2$$

be the projection of $\boldsymbol{a}_2$ onto $\operatorname{Span}\{\boldsymbol{u}_1\}$. Now take

$$\boldsymbol{u}_2 = \frac{\boldsymbol{a}_2 - \boldsymbol{p}_1}{\|\boldsymbol{a}_2 - \boldsymbol{p}_1\|} = \frac{(\mathbf{I}_n - \boldsymbol{u}_1 \boldsymbol{u}_1^\top)\boldsymbol{a}_2}{\|(\mathbf{I}_n - \boldsymbol{u}_1 \boldsymbol{u}_1^\top)\boldsymbol{a}_2\|}.$$

This vector is perpendicular to $\boldsymbol{u}_1$ and has unit length.

3. Let $\boldsymbol{p}_2$ be the projection of $\boldsymbol{a}_3$ onto $\operatorname{Span}\{\boldsymbol{u}_1, \boldsymbol{u}_2\}$. That is,

$$\boldsymbol{p}_2 = \langle \boldsymbol{u}_1, \boldsymbol{a}_3 \rangle \, \boldsymbol{u}_1 + \langle \boldsymbol{u}_2, \boldsymbol{a}_3 \rangle \, \boldsymbol{u}_2 = \begin{bmatrix} \boldsymbol{u}_1 & \boldsymbol{u}_2 \end{bmatrix} \begin{bmatrix} \boldsymbol{u}_1^\top \\ \boldsymbol{u}_2^\top \end{bmatrix} \boldsymbol{a}_3.$$

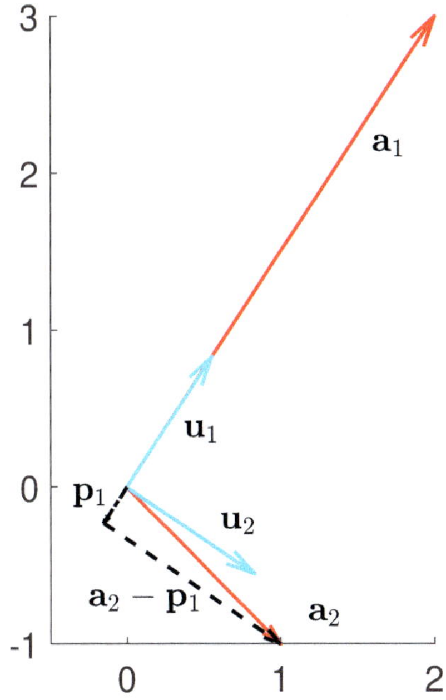

Figure A.21: Illustration of the Gram–Schmidt procedure.

Now take

$$\boldsymbol{u}_3 = \frac{\boldsymbol{a}_3 - \boldsymbol{p}_2}{\|\boldsymbol{a}_3 - \boldsymbol{p}_2\|} = \frac{(\mathbf{I}_n - \boldsymbol{u}_1\boldsymbol{u}_1^\top - \boldsymbol{u}_2\boldsymbol{u}_2^\top)\boldsymbol{a}_3}{\|(\mathbf{I}_n - \boldsymbol{u}_1\boldsymbol{u}_1^\top - \boldsymbol{u}_2\boldsymbol{u}_2^\top)\boldsymbol{a}_3\|}.$$

This vector is perpendicular to both $\boldsymbol{u}_1$ and $\boldsymbol{u}_2$ and has unit length.

4. Continue this process to obtain $\boldsymbol{u}_4, \ldots, \boldsymbol{u}_p$.

At the end of the procedure, a set $\{\boldsymbol{u}_1, \ldots, \boldsymbol{u}_p\}$ of p orthonormal vectors are obtained. Consequently, as a result of Theorem A.13,

$$\boldsymbol{a}_j = \sum_{i=1}^{j} \underbrace{\langle \boldsymbol{a}_j, \boldsymbol{u}_i \rangle}_{r_{ij}} \boldsymbol{u}_i, \quad j = 1, \ldots, p,$$

for some numbers $r_{ij}, j = 1, \ldots, i,\ i = 1, \ldots, p$. Denoting the corresponding upper triangular matrix $[r_{ij}]$ by $\mathbf{R}$, we have in matrix notation:

$$\mathbf{QR} = [\boldsymbol{u}_1, \ldots, \boldsymbol{u}_p] \begin{bmatrix} r_{11} & r_{12} & r_{13} & \ldots & r_{1p} \\ 0 & r_{22} & r_{23} & \ldots & r_{2p} \\ \vdots & 0 & \ddots & \ddots & \vdots \\ 0 & 0 & 0 & \ldots & r_{pp} \end{bmatrix} = [\boldsymbol{a}_1, \ldots, \boldsymbol{a}_p] = \mathbf{A},$$

which yields a QR decomposition. The following algorithm computes the QR decomposition via the Gram–Schmidt procedure.

Algorithm A.22: QR Decomposition via Gram–Schmidt: A = QR

Input: Matrix $\mathbf{A} = [\boldsymbol{a}_1, \dots, \boldsymbol{a}_p] \in \mathbb{R}^{n \times p}$ with $\text{rank}(\mathbf{A}) = p \leq n$.
Output: Orthonormal matrix $\mathbf{Q} \in \mathbb{R}^{n \times p}$ and upper triangular matrix $\mathbf{R} \in \mathbb{R}^{p \times p}$.

1 $\varrho \leftarrow \|\boldsymbol{a}_1\|, \mathbf{Q} \leftarrow \boldsymbol{a}_1/\varrho, \mathbf{R} \leftarrow \varrho$
2 **for** $k = 2, \dots, p$ **do**
3 $\quad \boldsymbol{r} \leftarrow \mathbf{Q}^\top \boldsymbol{a}_k$
4 $\quad \boldsymbol{u} \leftarrow \boldsymbol{a}_k - \mathbf{Q}\boldsymbol{r}$
5 $\quad \varrho \leftarrow \|\boldsymbol{u}\|$
6 $\quad \mathbf{Q} \leftarrow [\mathbf{Q} \;\; \boldsymbol{u}/\varrho]$
7 $\quad \mathbf{R} \leftarrow \begin{bmatrix} \mathbf{R} & \boldsymbol{r} \\ \mathbf{0}^\top & \varrho \end{bmatrix}$
8 **return** $\mathbf{Q}, \mathbf{R}$

Algorithm A.22 can easily be modified to compute the QR decomposition of a complex matrix $\mathbf{A} \in \mathbb{C}^{n \times p}$ by using the conjugate transpose $(*)$ in Section A.4.

When $\mathbf{A} \in \mathbb{C}^{n \times p}$ is a complex matrix, replace the transpose $(\top)$ in line 3 of the QR Algorithm A.22 with the conjugate transpose $(*)$ and compute the norms (ϱ) in lines 1 and 5 using the formula $\|\boldsymbol{u}\| = \sqrt{\boldsymbol{u}^*\boldsymbol{u}}$.

There exist various improvements of the Gram–Schmidt process (for example, the *Householder transformation* [76]) that not only improve the numerical stability of the QR decomposition, but also can be applied even when $\mathbf{A}$ is not full rank. A number of applications of the QR decomposition are given next.

■ **Example A.23 (Computation of Matrix Determinant)** The QR decomposition can be used to calculate the determinant of the matrix $\mathbf{A}$, whenever $\mathbf{A}$ is square. Namely,

$$\det(\mathbf{A}) = \det(\mathbf{Q})\det(\mathbf{R}) = \det(\mathbf{R}),$$

and since $\mathbf{R}$ is triangular, its determinant is the product of its diagonal elements. ■

■ **Example A.24 (Orthogonal Projection Matrix)** Recall from Theorem A.17 that the orthogonal projection matrix onto the space spanned by the columns of a model matrix $\mathbf{X} \in \mathbb{R}^{n \times p}$ is $\mathbf{P} = \mathbf{X}\mathbf{X}^+$. If $\mathbf{X}$ has full column rank with decomposition $\mathbf{X} = \mathbf{QR}$, then

$$\begin{aligned} \mathbf{P} &= \mathbf{QR}(\mathbf{QR})^+ = \mathbf{QRR}^+\mathbf{Q}^+ \\ &= \mathbf{QI}_p\mathbf{Q}^+ = \mathbf{QQ}^\top, \end{aligned}$$

where we used the identities from Exercise 2.7 that $(\mathbf{QR})^+ = \mathbf{R}^+\mathbf{Q}^+$ and $\mathbf{Q}^+ = (\mathbf{Q}^\top\mathbf{Q})^+\mathbf{Q}^\top = \mathbf{Q}^\top$. ☞ 63

■

■ **Example A.25 (QR Decomposition and Least Squares)** An important application of the QR decomposition is found in solving the least-squares problem in $O(p^2\, n)$ time:

$$\min_{\boldsymbol{\beta} \in \mathbb{R}^p} \|\mathbf{X}\boldsymbol{\beta} - \boldsymbol{y}\|^2$$

for some $\mathbf{X} \in \mathbb{R}^{n\times p}$ (model) matrix. Using the defining properties of the pseudo-inverse in Definition A.10, one can show that for any $\boldsymbol{\beta}$: ☞ 516

$$\|\mathbf{X}\mathbf{X}^{+}\mathbf{y} - \mathbf{y}\|^2 \leq \|\mathbf{X}\boldsymbol{\beta} - \mathbf{y}\|^2.$$

In other words, $\widehat{\boldsymbol{\beta}} := \mathbf{X}^{+}\mathbf{y}$ minimizes $\|\mathbf{X}\boldsymbol{\beta} - \mathbf{y}\|$. If we have the QR decomposition $\mathbf{X} = \mathbf{Q}\mathbf{R}$, then a numerically stable way to calculate $\widehat{\boldsymbol{\beta}}$ with an $\mathcal{O}(p^2\, n)$ cost is via

$$\widehat{\boldsymbol{\beta}} = (\mathbf{Q}\mathbf{R})^{+}\mathbf{y} = \mathbf{R}^{+}\mathbf{Q}^{+}\mathbf{y} = \mathbf{R}^{+}\mathbf{Q}^{\top}\mathbf{y}.$$

If $\mathbf{X}$ has full column rank, then $\mathbf{R}^{+} = \mathbf{R}^{-1}$.

While the QR decomposition is the method of choice for solving the ordinary least-squares regression problem, the *Sherman–Morrison recursion* is the method of choice for solving the regularized least-squares (or ridge) regression problem. ☞ 542 ☞ 321

■

■ **Example A.26 (Block Gram–Schmidt Process)** Suppose that the model matrix $\mathbf{X} \in \mathbb{R}^{n\times p}$ can be decomposed into d groups of model submatrices:

$$\mathbf{X} = [\mathbf{X}_1, \ldots, \mathbf{X}_d],$$

where $\mathbf{X}_i \in \mathbb{R}^{n\times p_i}$ for all i and $p = p_1 + \cdots + p_d$. This setting arises when features are selected in groups, as in Section 5.4.1.

When the first k blocks of features are of greater explanatory value than the rest, then we can frequently accelerate numerical computations by constructing the orthogonal projection $\mathbf{P}_k\mathbf{X}$ of $\mathbf{X}$ onto the $\mathbb{R}^{n\times(p_1+\cdots+p_k)}$-dimensional subspace $\mathrm{Span}\{\mathbf{X}_1, \ldots, \mathbf{X}_k\}$ of $\mathbb{R}^{n\times p}$.

In other words, we wish to compute the mutually orthonormal matrices $\mathbf{Q}_i \in \mathbb{R}^{n\times p_i}$, $i = 1, \ldots, k$ (that is, $\mathbf{Q}_i^{\top}\mathbf{Q}_j$ is an $p_i \times p_j$ matrix of zeros for $i \neq j$) with $\mathbf{Q}_i^{\top}\mathbf{Q}_i = \mathbf{I}_{p_i}$, as well as the matrices $\mathbf{R}_i = \mathbf{Q}_i^{\top}\mathbf{X} \in \mathbb{R}^{p_i\times p}$ for $i = 1, \ldots, k$, so that

$$\mathbf{P}_k\mathbf{X} = \sum_{i=1}^{k} \mathbf{Q}_i\mathbf{R}_i, \quad k = 1, \ldots, d.$$

This projection can be computed via the **block Gram–Schmidt process**, as follows.

Algorithm A.27: Block Orthogonalization via QR Decomposition

Input: $\mathbf{X} = [\mathbf{X}_1, \ldots, \mathbf{X}_d] \in \mathbb{R}^{n\times p}$ and $k \leq d$, where $\mathbf{X}_i \in \mathbb{R}^{n\times p_i}$ for all i
Output: $\mathbf{Q}_i \in \mathbb{R}^{n\times p_i}$ and $\mathbf{R}_i \in \mathbb{R}^{p_i\times p}$ for $i = 1, \ldots, k$

$\mathbf{Q}_1\mathbf{R}_1 \leftarrow \mathbf{X}_1$ // compute QR decomposition of $\mathbf{X}_1$
$\mathbf{R}_1 \leftarrow \mathbf{Q}_1^{\top}\mathbf{X}$, $\mathbf{Q} \leftarrow \mathbf{Q}_1$
for $i = 2, \ldots, k$ **do**
  $\mathbf{Q}_i\mathbf{R}_i \leftarrow \mathbf{X}_i - \mathbf{Q}\mathbf{Q}^{\top}\mathbf{X}_i$ // compute QR decomposition of $(\mathbf{I} - \mathbf{P}_i)\mathbf{X}_i$
  $\mathbf{R}_i \leftarrow \mathbf{Q}_i^{\top}\mathbf{X}$, $\mathbf{Q} \leftarrow [\mathbf{Q}, \mathbf{Q}_i]$
return $\mathbf{Q}, \mathbf{R}_1, \ldots, \mathbf{R}_k$

■

A.7 Eigenvalues and Eigenvectors

Let $\mathbf{A}$ be an $n \times n$ real or complex matrix. If $\mathbf{A}\boldsymbol{v} = \lambda \boldsymbol{v}$ for some number λ and non-zero vector $\boldsymbol{v}$, then λ is called an **eigenvalue** of $\mathbf{A}$ with **eigenvector** $\boldsymbol{v}$.

If $(\lambda, \boldsymbol{v})$ is an (eigenvalue, eigenvector) pair, the matrix $\lambda\mathbf{I} - \mathbf{A}$ maps any multiple of $\boldsymbol{v}$ to the zero vector. Consequently, the columns of $\lambda\mathbf{I} - \mathbf{A}$ are linearly *dependent*, and hence its determinant is 0. This provides a way to identify the eigenvalues, namely as the $r \leq n$ different roots $\lambda_1, \ldots, \lambda_r$ of the **characteristic polynomial**

$$\det(\lambda\mathbf{I} - \mathbf{A}) = (\lambda - \lambda_1)^{\alpha_1} \cdots (\lambda - \lambda_r)^{\alpha_r},$$

where $\alpha_1 + \cdots + \alpha_r = n$. The integer α_i is called the **algebraic multiplicity** of λ_i. The eigenvectors that correspond to an eigenvalue λ_i lie in the **kernel** or **null space** of the matrix $\lambda_i\mathbf{I} - \mathbf{A}$; that is, the linear space of vectors $\boldsymbol{v}$ such that $(\lambda_i\mathbf{I} - \mathbf{A})\boldsymbol{v} = \mathbf{0}$. This space is called the **eigenspace** of λ_i. Its dimension, $d_i \in \{1, \ldots, n\}$, is called the **geometric multiplicity** of λ_i. It always holds that $d_i \leq \alpha_i$. If $\sum_i d_i = n$, then we can construct a basis for $\mathbb{R}^n$ consisting of eigenvectors, as illustrated next.

Example A.28 (Linear Transformation (cont.)) We revisit the linear transformation in Figure A.4, where

$$\mathbf{A} = \begin{bmatrix} 1 & 1 \\ -1/2 & -2 \end{bmatrix}.$$

The characteristic polynomial is $(\lambda - 1)(\lambda + 2) + 1/2$, with roots $\lambda_1 = -1/2 - \sqrt{7}/2 \approx -1.8229$ and $\lambda_2 = -1/2 + \sqrt{7}/2 \approx 0.8229$. The corresponding unit eigenvectors are $\boldsymbol{v}_1 \approx [0.3339, -0.9426]^\top$ and $\boldsymbol{v}_2 \approx [0.9847, -0.1744]^\top$. The eigenspace corresponding to λ_1 is $\mathcal{V}_1 = \mathrm{Span}\{\boldsymbol{v}_1\} = \{\beta\boldsymbol{v}_1 : \beta \in \mathbb{R}\}$ and the eigenspace corresponding to λ_2 is $\mathcal{V}_2 = \mathrm{Span}\{\boldsymbol{v}_2\}$. The algebraic and geometric multiplicities are 1 in this case. Any pair of vectors taken from $\mathcal{V}_1$ and $\mathcal{V}_2$ forms a basis for $\mathbb{R}^2$. Figure A.29 shows how $\boldsymbol{v}_1$ and $\boldsymbol{v}_2$ are transformed to $\mathbf{A}\boldsymbol{v}_1 \in \mathcal{V}_1$ and $\mathbf{A}\boldsymbol{v}_2 \in \mathcal{V}_2$, respectively.

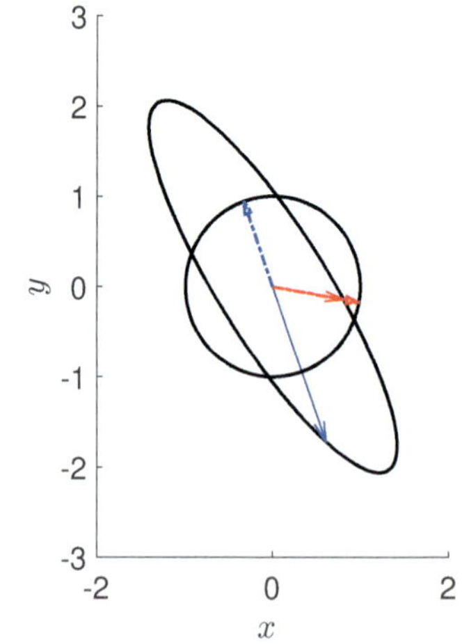

Figure A.29: The dashed arrows are the unit eigenvectors $\boldsymbol{v}_1$ (blue) and $\boldsymbol{v}_2$ (red) of matrix $\mathbf{A}$. Their transformed values $\mathbf{A}\boldsymbol{v}_1$ and $\mathbf{A}\boldsymbol{v}_2$ are indicated by solid arrows.

A matrix for which the algebraic and geometric multiplicities of all its eigenvalues are the same is called **semi-simple**. This is equivalent to the matrix being **diagonalizable**, meaning that there is a matrix $\mathbf{V}$ and a diagonal matrix $\mathbf{D}$ such that

$$\mathbf{A} = \mathbf{V}\mathbf{D}\mathbf{V}^{-1}.$$

To see that this so-called **eigen-decomposition** holds, suppose $\mathbf{A}$ is a semi-simple matrix with eigenvalues

$$\underbrace{\lambda_1, \ldots, \lambda_1}_{d_1}, \cdots, \underbrace{\lambda_r, \ldots, \lambda_r}_{d_r}.$$

Let $\mathbf{D}$ be the diagonal matrix whose diagonal elements are the eigenvalues of $\mathbf{A}$, and let $\mathbf{V}$ be a matrix whose columns are linearly independent eigenvectors corresponding to these eigenvalues. Then, for each (eigenvalue, eigenvector) pair $(\lambda, \boldsymbol{v})$, we have $\mathbf{A}\boldsymbol{v} = \lambda\boldsymbol{v}$. Hence, in matrix notation, we have $\mathbf{AV} = \mathbf{VD}$, and so $\mathbf{A} = \mathbf{VDV}^{-1}$.

■ **Example A.30 (Eigenvalues and Kronecker Product)** Suppose that we have the eigen-decompositions $\mathbf{A}_1 = \mathbf{V}_1\mathbf{D}_1\mathbf{V}_1^{-1}$ and $\mathbf{A}_2 = \mathbf{V}_2\mathbf{D}_2\mathbf{V}_2^{-1}$, where $\lambda_{1,j}, j = 1, \ldots, n_1$ and $\lambda_{2,j}, j = 1, \ldots, n_2$ are the eigenvalues of $\mathbf{A}_1$ and $\mathbf{A}_2$, respectively. Then, the $n_1 \times n_2$ eigenvalues of the Kronecker product $\mathbf{A}_1 \otimes \mathbf{A}_2$ are $\lambda_{1,i} \times \lambda_{2,j}$ for $i = 1, \ldots, n_1$ and $j = 1, \ldots, n_2$. To see this note that repeated application of the mixed-product identity in Theorem A.15 yields ☞ 518

$$\begin{aligned}(\mathbf{A}_1 \otimes \mathbf{A}_2)(\mathbf{V}_1 \otimes \mathbf{V}_2) &= (\mathbf{A}_1\mathbf{V}_1) \otimes (\mathbf{A}_2\mathbf{V}_2)\\ &= (\mathbf{V}_1\mathbf{D}_1) \otimes (\mathbf{V}_2\mathbf{D}_2)\\ &= (\mathbf{V}_1 \otimes \mathbf{V}_2)(\mathbf{D}_1 \otimes \mathbf{D}_2).\end{aligned}$$

Hence, the diagonal matrix $\mathbf{D}_1 \otimes \mathbf{D}_2$ of size $(n_1\, n_2) \times (n_1\, n_2)$ lists all the eigenvalues. ■

A.7.1 Left- and Right-Eigenvectors

The eigenvector as defined in the previous section is called a **right-eigenvector**, as it lies on the right of $\mathbf{A}$ in the equation $\mathbf{A}\boldsymbol{v} = \lambda\boldsymbol{v}$.

If $\mathbf{A}$ is a complex matrix with an eigenvalue λ, then the eigenvalue's complex conjugate $\overline{\lambda}$ is an eigenvalue of $\mathbf{A}^*$. To see this, define $\mathbf{B} := \lambda\mathbf{I} - \mathbf{A}$ and $\mathbf{B}^* := \overline{\lambda}\mathbf{I} - \mathbf{A}^*$. Since λ is an eigenvalue, we have $\det(\mathbf{B}) = 0$. Applying the identity

$$\det(\mathbf{B}) = \overline{\det(\mathbf{B}^*)},$$

we see that therefore $\det(\mathbf{B}^*) = 0$, and hence that $\overline{\lambda}$ is an eigenvalue of $\mathbf{A}^*$. Let $\boldsymbol{w}$ be an eigenvector corresponding to $\overline{\lambda}$. Then, $\mathbf{A}^*\boldsymbol{w} = \overline{\lambda}\boldsymbol{w}$ or, equivalently,

$$\boldsymbol{w}^*\mathbf{A} = \lambda\boldsymbol{w}^*.$$

For this reason, we call $\boldsymbol{w}^*$ the **left-eigenvector** of $\mathbf{A}$ for eigenvalue λ. If $\boldsymbol{v}$ is a (right-) eigenvector of $\mathbf{A}$, then its adjoint $\boldsymbol{v}^*$ is usually *not* a left-eigenvector, unless $\mathbf{A}^*\mathbf{A} = \mathbf{A}\mathbf{A}^*$ (such matrices are called **normal**; a real symmetric matrix is normal). However, the important property holds that left- and right-eigenvectors belonging to *different* eigenvalues are *orthogonal*. Namely, if $\boldsymbol{w}^*$ is a left-eigenvalue of λ_1 and $\boldsymbol{v}$ a right-eigenvalue of $\lambda_2 \neq \lambda_1$, then

$$\lambda_1 \boldsymbol{w}^* \boldsymbol{v} = \boldsymbol{w}^* \mathbf{A} \boldsymbol{v} = \lambda_2 \boldsymbol{w}^* \boldsymbol{v},$$

which can only be true if $\boldsymbol{w}^* \boldsymbol{v} = 0$.

Theorem A.31 (Schur Triangulation for Square Matrices): For any square matrix $\mathbf{A}$, there is a unitary matrix $\mathbf{U}$ such that $\mathbf{T} = \mathbf{U}^{-1}\mathbf{A}\mathbf{U}$ is upper triangular with the main diagonal containing the eigenvalues of $\mathbf{A}$.

Proof. The proof is by induction on the dimension n of the matrix. Clearly, the statement is true for $n = 1$, as $\mathbf{A}$ is simply a complex number and we can take $\mathbf{U}$ equal to 1. Suppose that the result is true for dimension n. We wish to show that it also holds for dimension $n + 1$. Any matrix $\mathbf{A}$ always has at least one eigenvalue λ with eigenvector $\boldsymbol{v}$, normalized to have length 1. Let $\mathbf{U}$ be any unitary matrix whose first column is $\boldsymbol{v}$. Such a matrix can always be constructed[1]. As $\mathbf{U}$ is unitary, the first row of $\mathbf{U}^{-1}$ is $\boldsymbol{v}^*$, and $\mathbf{U}^{-1}\mathbf{A}\mathbf{U}$ is of the form

$$\left[\begin{array}{c} \boldsymbol{v}^* \\ \hline * \end{array}\right] \mathbf{A} \underbrace{\left[\begin{array}{c|c} \boldsymbol{v} & * \end{array}\right]}_{\mathbf{U}} = \left[\begin{array}{c|c} \lambda & * \\ \hline \mathbf{0} & \mathbf{B} \end{array}\right],$$

for some matrix $\mathbf{B}$. By the induction hypothesis, there exists a unitary matrix $\mathbf{W}$ and an upper triangular matrix $\mathbf{T}$ such that $\mathbf{W}^{-1}\mathbf{B}\mathbf{W} = \mathbf{T}$. Now, define

$$\mathbf{V} := \left[\begin{array}{c|c} 1 & \mathbf{0}^\top \\ \hline \mathbf{0} & \mathbf{W} \end{array}\right].$$

Then,

$$\begin{aligned} \mathbf{V}^{-1}\left(\mathbf{U}^{-1}\mathbf{A}\mathbf{U}\right)\mathbf{V} &= \left[\begin{array}{c|c} 1 & \mathbf{0}^\top \\ \hline \mathbf{0} & \mathbf{W}^{-1} \end{array}\right] \left[\begin{array}{c|c} \lambda & * \\ \hline \mathbf{0} & \mathbf{B} \end{array}\right] \left[\begin{array}{c|c} 1 & \mathbf{0}^\top \\ \hline \mathbf{0} & \mathbf{W} \end{array}\right] \\ &= \left[\begin{array}{c|c} \lambda & * \\ \hline \mathbf{0} & \mathbf{W}^{-1}\mathbf{B}\mathbf{W} \end{array}\right] = \left[\begin{array}{c|c} \lambda & * \\ \hline \mathbf{0} & \mathbf{T} \end{array}\right], \end{aligned}$$

which is upper triangular of dimension $n + 1$. Since $\mathbf{UV}$ is unitary, this completes the induction, and hence the result is true for all n. □

■ **Example A.32 (Determinant and Matrix Eigenvalues)** Suppose that $\mathbf{A} \in \mathbb{R}^{n \times n}$ is a square matrix with Schur triangulation $\mathbf{A} = \mathbf{U}\mathbf{T}\mathbf{U}^{-1}$ (as per Theorem A.31) and eigenvalues $\lambda_1, \ldots, \lambda_r$ with corresponding algebraic multiplicities $\alpha_1, \ldots, \alpha_r$.

Using the properties of the matrix determinant in Theorem A.9, we can write

$$\begin{aligned} \det(\mathbf{A}) &= \det(\mathbf{U}\mathbf{T}\mathbf{U}^{-1}) = \det(\mathbf{U})\det(\mathbf{T})\det(\mathbf{U}^{-1}) \\ &= \det(\mathbf{U}\mathbf{U}^{-1})\det(\mathbf{T}) == \det(\mathbf{I}_n)\det(\mathbf{T}) = \det(\mathbf{T}). \end{aligned}$$

[1]After specifying $\boldsymbol{v}$ we can complete the rest of the unitary matrix via the Gram–Schmidt procedure, for example; see Section A.6.

Since **T** is upper triangular, its determinant is the product of its diagonal entries; see Theorem A.9. Moreover, since the main diagonal of **T** contains the $\alpha_1 + \cdots + \alpha_r = n$ (not necessarily distinct) eigenvalues of **A**, we conclude that the determinant of **A** is equal the product of the eigenvalues of **A**; that is,

$$\det(\mathbf{A}) = \prod_{i=1}^{r} \lambda_i^{\alpha_i}.$$

■

A.7.2 Eigen-Decomposition and Hermitian Matrices

Recall that a *Hermitian* matrix **A** is one for which $\mathbf{A}^* = \mathbf{A}$. Theorem A.31 can be used to prove the following important result.

Theorem A.33 (Eigenvalues of a Hermitian Matrix): Any $n \times n$ Hermitian matrix has real eigenvalues. The corresponding matrix of normalized eigenvectors is a unitary matrix.

Proof. Let **A** be a Hermitian matrix. By Theorem A.31 there exists a unitary matrix **U** such that $\mathbf{U}^{-1}\mathbf{A}\mathbf{U} = \mathbf{T}$, where **T** is upper triangular. It follows that the adjoint $(\mathbf{U}^{-1}\mathbf{A}\mathbf{U})^* = \mathbf{T}^*$ is lower triangular. However, $(\mathbf{U}^{-1}\mathbf{A}\mathbf{U})^* = \mathbf{U}^{-1}\mathbf{A}\mathbf{U}$, since $\mathbf{A}^* = \mathbf{A}$ and $\mathbf{U}^* = \mathbf{U}^{-1}$. Hence, **T** and $\mathbf{T}^*$ must be the same, which can only be the case if **T** is a *real diagonal* matrix **D**. Since $\mathbf{AU} = \mathbf{DU}$, the diagonal elements are exactly the eigenvalues and the corresponding eigenvectors are the columns of **U**. □

In particular, the eigenvalues of a real symmetric matrix are real. We can now repeat the proof of Theorem A.31 with real eigenvalues and eigenvectors, so that there exists an orthogonal matrix **Q** such that

$$\mathbf{Q}^{-1}\mathbf{A}\mathbf{Q} = \mathbf{Q}^\top\mathbf{A}\mathbf{Q} = \mathbf{D}.$$

The eigenvectors can be chosen as the columns of **Q**, which proves the eigen-decomposition:

Theorem A.34 (Real Symmetric Matrices are Orthogonally Diagonizable): Any real symmetric matrix **A** can be written as

$$\mathbf{A} = \mathbf{Q}\mathbf{D}\mathbf{Q}^\top,$$

where **D** is the diagonal matrix of (real) eigenvalues and **Q** is an orthogonal matrix whose columns are eigenvectors of **A**.

■ **Example A.35 (Real Symmetric Matrices and Ellipses)** As we have seen, linear transformations map circles into ellipses. We can use the above theory for real symmetric matrices to identify the principal axes. Consider, for example, the transformation with matrix

$$\mathbf{A} = \begin{bmatrix} 1 & 1 \\ -1/2 & -2 \end{bmatrix}$$

in (A.3). A point $\boldsymbol{x}$ on the unit circle is mapped to a point $\boldsymbol{y} = \mathbf{A}\boldsymbol{x}$. Since for such points $\|\boldsymbol{x}\|^2 = \boldsymbol{x}^\top \boldsymbol{x} = 1$, we have that $\boldsymbol{y}$ satisfies $\boldsymbol{y}^\top (\mathbf{A}^{-1})^\top \mathbf{A}^{-1} \boldsymbol{y} = 1$, which gives the equation for the ellipse

$$\frac{17\, y_1^2}{9} + \frac{20\, y_1 y_2}{9} + \frac{8\, y_2^2}{9} = 1.$$

Let $\mathbf{Q}$ be the orthogonal matrix of eigenvectors of the symmetric matrix

$$(\mathbf{A}^{-1})^\top \mathbf{A}^{-1} = (\mathbf{A}\mathbf{A}^\top)^{-1},$$

so

$$\mathbf{Q}^\top (\mathbf{A}\mathbf{A}^\top)^{-1} \mathbf{Q} = \mathbf{D}$$

for some diagonal matrix $\mathbf{D}$. Taking the inverse on both sides of the previous equation, we have

$$\mathbf{Q}^\top \mathbf{A}\mathbf{A}^\top \mathbf{Q} = \mathbf{D}^{-1},$$

which shows that $\mathbf{Q}$ is also the matrix of eigenvectors of $\mathbf{A}\mathbf{A}^\top$. These eigenvectors point precisely in the direction of the principal axes, as shown in Figure A.36. It turns out, see Section A.7.3, that the square roots of the eigenvalues of $\mathbf{A}\mathbf{A}^\top$, here approximately 2.4221 and 0.6193, correspond to the sizes of the principal axes of the ellipse, as illustrated in Figure A.36.

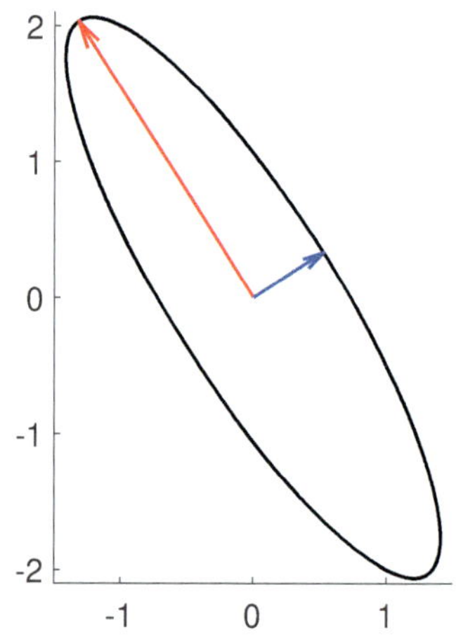

Figure A.36: The eigenvectors and eigenvalues of $\mathbf{A}\mathbf{A}^\top$ determine the principal axes of the ellipse.

■

The following definition generalizes the notion of positivity of a real variable to that of a Hermitian matrix, providing a crucial concept for multivariate differentiation and optimization; see Appendix C.

☞ 581

Definition A.37 (Positive (Semi)Definite Matrix): A matrix $\mathbf{A}$ is said to be **positive semidefinite** (we write $\mathbf{A} \succeq \mathbf{0}$) if $\langle \mathbf{A}\boldsymbol{x}, \boldsymbol{x} \rangle \geq 0$ for all $\boldsymbol{x}$. It is called **positive definite** (we write $\mathbf{A} \succ \mathbf{0}$) if $\langle \mathbf{A}\boldsymbol{x}, \boldsymbol{x} \rangle > 0$ for all $\boldsymbol{x} \neq \mathbf{0}$.

Any positive semidefinite matrix is necessarily Hermitian. Namely, $\boldsymbol{x}^*\mathbf{A}\boldsymbol{x} = \langle \mathbf{A}\boldsymbol{x}, \boldsymbol{x} \rangle$ is real-valued and is thus equal to its complex conjugate $\boldsymbol{x}^*\mathbf{A}^*\boldsymbol{x}$. As this holds for all $\boldsymbol{x}$, this can only be true if $\mathbf{A} = \mathbf{A}^*$.

The positive (semi)definiteness of a matrix can be directly related to the positivity of its eigenvalues, as follows:

Theorem A.38 (Eigenvalues of a Positive Semidefinite Matrix): All eigenvalues of a positive semidefinite matrix are non-negative and all eigenvalues of a positive definite matrix are strictly positive.

Proof. Let $\mathbf{A}$ be a positive semidefinite matrix. By Theorem A.33, the eigenvalues of $\mathbf{A}$ are all *real*. Suppose λ is an eigenvalue with eigenvector $\boldsymbol{v}$. As $\mathbf{A}$ is positive semidefinite, we have

$$0 \leq \langle \mathbf{A}\boldsymbol{v}, \boldsymbol{v} \rangle = \lambda \langle \boldsymbol{v}, \boldsymbol{v} \rangle = \lambda \|\boldsymbol{v}\|^2,$$

which can only be true if $\lambda \geq 0$. Similarly, for a positive definite matrix, λ must be strictly greater than 0. □

Theorem A.39 (Positive (Semi)definite Decomposition): Any real $\mathbf{A} \succeq \mathbf{0}$ can be written as

$$\mathbf{A} = \mathbf{B}\mathbf{B}^\top$$

for some real matrix $\mathbf{B}$. Conversely, for any real matrix $\mathbf{B}$, the matrix $\mathbf{B}\mathbf{B}^\top \succeq \mathbf{0}$. Additionally, $\mathbf{B}$ is full rank if and only if $\mathbf{A} \succ \mathbf{0}$.

Proof. The matrix $\mathbf{A}$ is both Hermitian (by definition) and real (by assumption) and hence it is symmetric. By Theorem A.34, we can write $\mathbf{A} = \mathbf{Q}\mathbf{D}\mathbf{Q}^\top$, where $\mathbf{D}$ is the diagonal matrix of (real) eigenvalues of $\mathbf{A}$. By Theorem A.38 all eigenvalues are non-negative when $\mathbf{A} \succeq \mathbf{0}$ or strictly positive when $\mathbf{A} \succ \mathbf{0}$, and thus their square root is real-valued. Now, define $\mathbf{B} = \mathbf{Q}\sqrt{\mathbf{D}}$, where $\sqrt{\mathbf{D}}$ is defined as the diagonal matrix whose diagonal elements are the *square roots* of the eigenvalues of $\mathbf{A}$. Then, $\mathbf{B}\mathbf{B}^\top = \mathbf{Q}\sqrt{\mathbf{D}}\,(\sqrt{\mathbf{D}})^\top\mathbf{Q}^\top = \mathbf{Q}\mathbf{D}\mathbf{Q}^\top = \mathbf{A}$ and $\mathbf{B}$ is full rank if and only if the eigenvalues of $\mathbf{A}$ are strictly positive. The converse statement follows from the fact that $\boldsymbol{x}^\top\mathbf{B}\mathbf{B}^\top\boldsymbol{x} = \|\mathbf{B}^\top\boldsymbol{x}\|^2 \geq 0$ for all $\boldsymbol{x}$. □

■ **Example A.40 (Bound on Log-Determinant)** From Theorem A.38 and the result of Example A.32, we can conclude that the logarithm of the determinant of any symmetric positive definite matrix $\mathbf{A} \in \mathbb{R}^{n \times n}$ can written as:

$$\ln |\mathbf{A}| = \ln \prod_{i=1}^{n} \lambda_i = \sum_{i=1}^{n} \ln \lambda_i,$$

where $\lambda_1, \ldots, \lambda_n$ are the (not necessarily distinct) positive eigenvalues of $\mathbf{A}$. As an application of this identity, we can prove the simple bound on the log-determinant:

$$\ln|\mathbf{A}| \leq \ln|\mathbf{A} \odot \mathbf{I}| = \sum_{i=1}^{n} \ln a_{ii}.$$

To see this, let

$$\widetilde{\mathbf{A}} := (\mathbf{A} \odot \mathbf{I}_n)^{-1/2}\mathbf{A}(\mathbf{A} \odot \mathbf{I}_n)^{-1/2}$$

be the matrix $\mathbf{A}$ scaled in such a way that its diagonal elements are unity. Then,

$$\ln|\mathbf{A}| - \ln|\mathbf{A} \odot \mathbf{I}| = \ln|\widetilde{\mathbf{A}}| = \sum_{i=1}^{n} \ln \widetilde{\lambda}_i,$$

where $\widetilde{\lambda}_1, \ldots, \widetilde{\lambda}_n$ are the eigenvalues of the positive definite matrix $\widetilde{\mathbf{A}}$. Since the sum $\widetilde{\lambda}_1 + \cdots + \widetilde{\lambda}_n = \mathrm{tr}(\widetilde{\mathbf{A}}) = n$, we can apply Jensen's inequality (see Exercise 19 in Chapter 2)
☞ 67 to obtain:

$$\frac{1}{n}\ln|\widetilde{\mathbf{A}}| = \frac{\ln\widetilde{\lambda}_1 + \cdots + \ln\widetilde{\lambda}_n}{n} \leq \ln\big((\widetilde{\lambda}_1 + \cdots + \widetilde{\lambda}_n)/n\big) = \ln 1 = 0,$$

and thus conclude that $\ln|\mathbf{A}| \leq \ln|\mathbf{A} \odot \mathbf{I}|$. ■

We finish with the following connection between positive definite matrices $\mathbf{A}, \mathbf{B}$, and their elementwise product — also called **Hadamard product** and denoted $\mathbf{A} \odot \mathbf{B}$.

Theorem A.41 (Positive Definiteness and Hadamard Product): If $\mathbf{A} \succ \mathbf{0}$ and $\mathbf{B} \succ \mathbf{0}$, then also $\mathbf{A} \odot \mathbf{B} \succ \mathbf{0}$.

Proof. Recall from the Theorem A.39 that we can write $\mathbf{A} = \mathbf{L}\mathbf{L}^\top$ and $\mathbf{B} = \mathbf{C}\mathbf{C}^\top$ for some full-rank matrices $\mathbf{L}$ and $\mathbf{C}$. We wish to show that $\boldsymbol{v}^\top[\mathbf{A} \odot \mathbf{B}]\boldsymbol{v} > 0$ for an arbitrary $\boldsymbol{v} \neq \mathbf{0}$. To this end, let $\mathbf{D} := \mathrm{diag}(\boldsymbol{v})$ and write

$$\begin{aligned}\boldsymbol{v}^\top[\mathbf{A} \odot \mathbf{B}]\boldsymbol{v} &= \mathrm{tr}\big(\mathbf{A}^\top\mathbf{D}\mathbf{B}\mathbf{D}\big) = \mathrm{tr}\big(\mathbf{L}\mathbf{L}^\top\mathbf{D}\mathbf{C}\mathbf{C}^\top\mathbf{D}\big) \\ &= \mathrm{tr}\big(\mathbf{L}^\top\mathbf{D}\mathbf{C}\mathbf{C}^\top\mathbf{D}\mathbf{L}\big) = \mathrm{tr}(\mathbf{R}\mathbf{R}^\top),\end{aligned}$$

where $\mathbf{R} := \mathbf{L}^\top\mathbf{D}\mathbf{C}$. Observe that $\mathbf{L}$ and $\mathbf{C}$ being full rank implies that $\mathrm{rank}(\mathbf{R}) = \mathrm{rank}(\mathbf{D}) \geq 1$. In other words, $\mathrm{tr}(\mathbf{R}\mathbf{R}^\top) = \|\mathrm{vec}(\mathbf{R})\|^2 > 0$, because only the zero matrix is of rank zero. □

A.7.3 Singular Value Decomposition

When a matrix is not square, then it is not diagonalizable and no eigen-decomposition exists. In such cases, we can instead use the more general **singular value decomposition** (SVD).

Theorem A.42 (Singular Value Decomposition): Any (complex) $m \times n$ matrix $\mathbf{A}$ admits a unique decomposition

$$\mathbf{A} = \mathbf{U}\boldsymbol{\Sigma}\mathbf{V}^*,$$

where $\mathbf{U}$ and $\mathbf{V}$ are unitary matrices of dimension m and n, respectively, and $\boldsymbol{\Sigma}$ is a real $m \times n$ diagonal matrix. If $\mathbf{A}$ is real, then $\mathbf{U}$ and $\mathbf{V}$ are both orthogonal matrices.

Proof. Without loss of generality we can assume that $m \geq n$ (otherwise, consider the transpose of $\mathbf{A}$). Then $\mathbf{A}^*\mathbf{A}$ is a positive semidefinite matrix, because $\langle \mathbf{A}^*\mathbf{A}\boldsymbol{v}, \boldsymbol{v}\rangle = \boldsymbol{v}^*\mathbf{A}^*\mathbf{A}\boldsymbol{v} = \|\mathbf{A}\boldsymbol{v}\|^2 \geq 0$ for all $\boldsymbol{v}$. Hence, $\mathbf{A}^*\mathbf{A}$ has non-negative real eigenvalues, $\lambda_1 \geq \lambda_2 \geq \cdots \geq \lambda_n \geq 0$. By Theorem A.33 the matrix $\mathbf{V} = [\boldsymbol{v}_1, \ldots, \boldsymbol{v}_n]$ of right-eigenvectors is a unitary matrix. Define the i-th **singular value** as $\sigma_i = \sqrt{\lambda_i}$, $i = 1, \ldots, n$ and suppose $\lambda_1, \ldots, \lambda_r$ are all greater than 0, and $\lambda_{r+1}, \ldots, \lambda_n = 0$. In particular, $\mathbf{A}\boldsymbol{v}_i = \mathbf{0}$ for $i = r+1, \ldots, n$. Let $\boldsymbol{u}_i = \mathbf{A}\boldsymbol{v}_i/\sigma_i$, $i = 1, \ldots, r$. Then, for $i, j \leq r$,

$$\langle \boldsymbol{u}_i, \boldsymbol{u}_j\rangle = \boldsymbol{u}_j^*\boldsymbol{u}_i = \frac{\boldsymbol{v}_j^*\mathbf{A}^*\mathbf{A}\boldsymbol{v}_i}{\sigma_i\,\sigma_j} = \frac{\lambda_i\,\mathbb{1}_{\{i=j\}}}{\sigma_i\sigma_j} = \mathbb{1}_{\{i=j\}}.$$

We can extend $\boldsymbol{u}_1, \ldots, \boldsymbol{u}_r$ to an orthonormal basis $\{\boldsymbol{u}_1, \ldots, \boldsymbol{u}_m\}$ of $\mathbb{C}^m$ (e.g., using the Gram–Schmidt procedure in Section A.6). Let $\mathbf{U} = [\boldsymbol{u}_1, \ldots, \boldsymbol{u}_n]$ be the corresponding unitary matrix. Defining $\boldsymbol{\Sigma}$ to be the $m \times n$ diagonal matrix with diagonal $(\sigma_1, \ldots, \sigma_r, 0, \ldots, 0)$, we have, ☞ 522

$$\mathbf{U}\boldsymbol{\Sigma} = [\mathbf{A}\boldsymbol{v}_1, \ldots, \mathbf{A}\boldsymbol{v}_r, \mathbf{0}, \ldots, \mathbf{0}] = \mathbf{A}\mathbf{V},$$

and hence $\mathbf{A} = \mathbf{U}\boldsymbol{\Sigma}\mathbf{V}^*$. □

Note that

$$\mathbf{A}\mathbf{A}^* = \mathbf{U}\boldsymbol{\Sigma}\mathbf{V}^*\mathbf{V}\boldsymbol{\Sigma}^*\mathbf{U}^* = \mathbf{U}\boldsymbol{\Sigma}\boldsymbol{\Sigma}^\top\mathbf{U}^* \quad \text{and} \quad \mathbf{A}^*\mathbf{A} = \mathbf{V}\boldsymbol{\Sigma}^*\mathbf{U}^*\mathbf{U}\boldsymbol{\Sigma}\mathbf{V}^* = \mathbf{V}\boldsymbol{\Sigma}^\top\boldsymbol{\Sigma}\mathbf{V}^*.$$

So, $\mathbf{U}$ is a unitary matrix whose columns are eigenvectors of $\mathbf{A}\mathbf{A}^*$ and $\mathbf{V}$ is a unitary matrix whose columns are eigenvectors of $\mathbf{A}^*\mathbf{A}$.

The SVD makes it possible to write the matrix $\mathbf{A}$ as a sum of rank-1 matrices, weighted by the singular values $\{\sigma_i\}$:

$$\text{(A.43)} \qquad \mathbf{A} = \begin{bmatrix}\boldsymbol{u}_1, \boldsymbol{u}_2, \ldots, \boldsymbol{u}_m\end{bmatrix} \begin{bmatrix} \sigma_1 & 0 & \ldots & \ldots & 0 \\ 0 & \ddots & 0 & \ldots & 0 \\ 0 & \ldots & \sigma_r & \ldots & 0 \\ 0 & \ldots & \ldots & 0\ \ldots & 0 \\ 0 & \ldots & \ldots & \ldots\ \ddots & 0 \end{bmatrix} \begin{bmatrix} \boldsymbol{v}_1^* \\ \boldsymbol{v}_2^* \\ \vdots \\ \boldsymbol{v}_n^* \end{bmatrix} = \sum_{i=1}^{r} \sigma_i\,\boldsymbol{u}_i\,\boldsymbol{v}_i^*,$$

which is called the **dyade** or **spectral representation** of $\mathbf{A}$.

Since $\text{rank}(\mathbf{A}) = r \leq \min\{m, n\}$, the representation (A.43) makes it possible to formally distinguish three different forms of the SVD.

Full SVD. This is simply the representation (A.43) where $\mathbf{U}$ and $\mathbf{V}$ are both square matrices, and $\mathbf{\Sigma} \in \mathbb{R}^{m\times n}$.

Thin SVD. This is the case when $m > n = r$ so that $\mathbf{A}$ can be written as

$$\mathbf{A} = \sum_{i=1}^{r} \sigma_i \, \boldsymbol{u}_i \, \boldsymbol{v}_i^* = \begin{bmatrix} \boldsymbol{u}_1, \dots, \boldsymbol{u}_n \end{bmatrix} \operatorname{diag}(\sigma_1, \dots, \sigma_n) \begin{bmatrix} \boldsymbol{v}_1^* \\ \boldsymbol{v}_2^* \\ \vdots \\ \boldsymbol{v}_n^* \end{bmatrix} = \mathbf{U\Sigma V}^*,$$

where both $\mathbf{V} \in \mathbb{R}^{n\times n}$ and $\mathbf{\Sigma} \in \mathbb{R}^{n\times n}$ are square, but $\mathbf{U} \in \mathbb{R}^{m\times n}$ is rectangular.

Compact SVD. This is the case when $r < \min\{m, n\}$ so that $\mathbf{A}$ can be written as

$$\mathbf{A} = \sum_{i=1}^{r} \sigma_i \, \boldsymbol{u}_i \, \boldsymbol{v}_i^* = \mathbf{U\Sigma V}^*,$$

where $\mathbf{\Sigma} \in \mathbb{R}^{r\times r}$ is square, but $\mathbf{U} \in \mathbb{R}^{m\times r}$ and $\mathbf{V} \in \mathbb{R}^{n\times r}$ are rectangular.

For real-valued matrices, the SVD has a nice geometric interpretation, illustrated in Figure A.44. The linear mapping defined by matrix $\mathbf{A}$ can be thought of as a succession of three linear operations: (1) an orthogonal transformation (i.e., a rotation with a possible flipping of some axes), corresponding to matrix $\mathbf{V}^\top$, followed by (2) a simple scaling of the unit vectors, corresponding to $\mathbf{\Sigma}$, followed by (3) another orthogonal transformation, corresponding to $\mathbf{U}$.

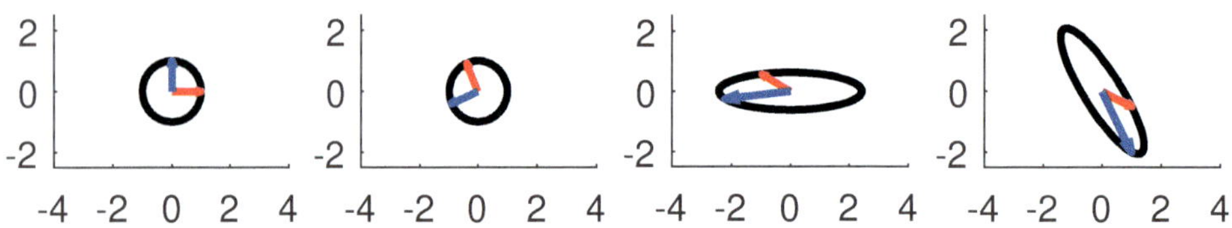

Figure A.44: The figure shows how the unit circle and unit vectors (first panel) are first rotated (second panel), then scaled (third panel), and finally rotated and flipped.

■ **Example A.45 (Ellipses)** We continue Example A.35. Using the **svd** method of the module `numpy.linalg`, we obtain the following SVD matrices for matrix $\mathbf{A}$:

$$\mathbf{U} = \begin{bmatrix} -0.5430 & 0.8398 \\ 0.8398 & 0.5430 \end{bmatrix}, \quad \mathbf{\Sigma} = \begin{bmatrix} 2.4221 & 0 \\ 0 & 0.6193 \end{bmatrix}, \quad \text{and} \quad \mathbf{V} = \begin{bmatrix} -0.3975 & 0.9176 \\ -0.9176 & -0.3975 \end{bmatrix}.$$

☞ 530 Figure A.36 shows the columns of the matrix $\mathbf{U\Sigma}$ as the two principal axes of the ellipse that is obtained by applying matrix $\mathbf{A}$ to the points of the unit circle. ■

■ **Example A.46 (Polar Decomposition)** Every rectangular matrix $\mathbf{A} \in \mathbb{R}^{m\times n}$ can be written in a **polar decomposition** form:

$$\mathbf{A} = \mathbf{QP},$$

where $\mathbf{P} \in \mathbb{R}^{n\times n}$ is a positive semi-definite matrix and $\mathbf{Q} \in \mathbb{R}^{m\times n}$ is an orthonormal matrix such that either $\mathbf{Q}^\top\mathbf{Q} = \mathbf{I}_n$ when $m \geq n$, or $\mathbf{Q}\mathbf{Q}^\top = \mathbf{I}_m$ when $n \geq m$. Using the full SVD of $\mathbf{A}$ in (A.43), we can explicitly construct such matrices via the formulas:

$$\mathbf{Q} := \sum_{i=1}^{\min\{n,m\}} \boldsymbol{u}_i \boldsymbol{v}_i^\top \quad \text{and} \quad \mathbf{P} := \sum_{j=1}^{r} \sigma_j \boldsymbol{v}_j \boldsymbol{v}_j^\top.$$

■

Theorem A.47 (Variational Characterization of Eigen-Values and -Vectors): Let $\mathbf{H}$ be a Hermitian matrix on an n-dimensional vector space $\mathcal{V}$. Denote the eigenvalues of $\mathbf{H}$ by $\lambda_1 \geq \lambda_2 \geq \cdots \geq \lambda_n \geq 0$, and let $\{\boldsymbol{v}_1, \ldots, \boldsymbol{v}_n\}$ be the corresponding orthonormal basis of eigenvectors. For each k, the solution to the optimization problem

$$\max_{\mathcal{U}_k \subseteq \mathcal{V}} \min_{\substack{\boldsymbol{x}\in\mathcal{U}_k \\ \|\boldsymbol{x}\|=1}} \langle \mathbf{H}\boldsymbol{x}, \boldsymbol{x}\rangle,$$

where the maximization is over all linear subspaces $\mathcal{U}_k$ of $\mathcal{V}$ with $k \leq n$, is given by $\mathcal{U}_k = \operatorname{Span}\{\boldsymbol{v}_1, \ldots, \boldsymbol{v}_k\}$ and $\boldsymbol{x} = \boldsymbol{v}_k$, giving the optimal value λ_k.

Proof. Pick any collection $\boldsymbol{v}_k, \ldots, \boldsymbol{v}_n$ of $n - k + 1$ unit-length eigenvectors. Since $n - k + 1 \geq 1$, there exists a vector $\boldsymbol{x} = \sum_{i=k}^n a_i \boldsymbol{v}_i \in \operatorname{Span}\{\boldsymbol{v}_k, \ldots, \boldsymbol{v}_n\} \cap \mathcal{U}_k$ with $\|\boldsymbol{x}\|^2 = \sum_{i=k}^n |a_i|^2 = 1$. For such a vector we have

$$\langle \mathbf{H}\boldsymbol{x}, \boldsymbol{x}\rangle = \sum_{i=k}^n \langle a_i\lambda_i\boldsymbol{v}_i,\, a_i\boldsymbol{v}_i\rangle = \sum_{i=k}^n |a_i|^2\lambda_i \leq \sum_{i=k}^n |a_i|^2\lambda_k = \lambda_k.$$

Thus, for any k-dimensional subspace $\mathcal{U}_k$ the minimum of $\langle \mathbf{H}\boldsymbol{x}, \boldsymbol{x}\rangle$ over all unit-length vectors in $\mathcal{U}_k$ is less than λ_k. In other words,

$$\max_{\mathcal{U}_k \subseteq \mathcal{V}} \min_{\substack{\boldsymbol{x}\in\mathcal{U}_k \\ \|\boldsymbol{x}\|=1}} \langle \mathbf{H}\boldsymbol{x}, \boldsymbol{x}\rangle \leq \lambda_k.$$

Taking $\mathcal{U}_k = \operatorname{Span}\{\boldsymbol{v}_1, \ldots, \boldsymbol{v}_k\}$ and $\boldsymbol{x} = \boldsymbol{v}_k$, we obtain $\langle \mathbf{H}\boldsymbol{x}, \boldsymbol{x}\rangle = \lambda_k$. □

■ **Example A.48 (Matrix Condition Number)** For the special case of Theorem A.47 where $\mathbf{H} = \mathbf{A}^*\mathbf{A}$, we have $\langle \mathbf{H}\boldsymbol{x}, \boldsymbol{x}\rangle = \boldsymbol{x}^*\mathbf{A}^*\mathbf{A}\boldsymbol{x} = \|\mathbf{A}\boldsymbol{x}\|^2$, so that the k-th singular value of $\mathbf{A}$ satisfies

$$\sigma_k = \max_{\mathcal{U}_k \subseteq \mathcal{V}} \min_{\substack{\boldsymbol{x}\in\mathcal{U}_k \\ \|\boldsymbol{x}\|=1}} \|\mathbf{A}\boldsymbol{x}\|.$$

Of particular importance are the largest and smallest singular values:

$$\sigma_1 = \max_{\|\boldsymbol{x}\|=1} \|\mathbf{A}\boldsymbol{x}\| \quad \text{and} \quad \sigma_n = \min_{\|\boldsymbol{x}\|=1} \|\mathbf{A}\boldsymbol{x}\|.$$

The quotient $\kappa(\mathbf{A}) = \sigma_1/\sigma_n$ is called the (Euclidean) **condition number** of the matrix $\mathbf{A}$. The condition number arises frequently in the analysis of the speed of convergence and accuracy of numerical methods that involve matrix computations; see Section A.8.5.1.

■

A practical method to compute the pseudo-inverse of a real-valued matrix $\mathbf{A}$ is via the singular value decomposition $\mathbf{A} = \mathbf{U\Sigma V}^\top$, where $\mathbf{\Sigma}$ is the diagonal matrix collecting all the positive singular values, say $\sigma_1, \ldots, \sigma_r$, as in Theorem A.42. In this case, $\mathbf{A}^+ = \mathbf{V\Sigma^+U}^\top$, where $\mathbf{\Sigma}^+$ is the $n \times m$ diagonal (pseudo-inverse) matrix:

$$\mathbf{\Sigma}^+ = \begin{bmatrix} \sigma_1^{-1} & 0 & \ldots & \ldots & 0 \\ 0 & \ddots & 0 & \ldots & 0 \\ 0 & \ldots & \sigma_r^{-1} & \ldots & 0 \\ 0 & \ldots & \ldots & 0\ \ldots & 0 \\ 0 & \ldots & \ldots & \ldots \ddots & 0 \end{bmatrix}.$$

We conclude with a typical application of the pseudo-inverse for a least-squares optimization problem from data science.

Example A.49 (Rank-Deficient Least Squares) Suppose in a linear regression problem we have a model matrix $\mathbf{X} \in \mathbb{R}^{n \times p}$ of full row rank (all rows of $\mathbf{X}$ are linearly independent) and with more features than observations: $p > n$. Under this setting, any solution to the equation $\mathbf{X}\boldsymbol{\beta} = \boldsymbol{y}$ provides a perfect fit to the data and minimizes (to 0) the least-squares problem $\widehat{\boldsymbol{\beta}} \in \operatorname{argmin}_{\boldsymbol{\beta} \in \mathbb{R}^p} \|\mathbf{X}\boldsymbol{\beta} - \boldsymbol{y}\|^2$.

☞ 45

In particular, if $\widehat{\boldsymbol{\beta}}$ minimizes $\|\mathbf{X}\boldsymbol{\beta} - \boldsymbol{y}\|^2$ then so does $\widehat{\boldsymbol{\beta}} + \boldsymbol{u}$ for all $\boldsymbol{u}$ in the null space $\mathcal{N}_{\mathbf{X}} := \{\boldsymbol{u} : \mathbf{X}\boldsymbol{u} = \mathbf{0}\}$, which has dimension $p - n$. To cope with the non-uniqueness of solutions, a possible approach is to solve instead the following optimization problem:

$$\begin{aligned} &\underset{\boldsymbol{\beta} \in \mathbb{R}^p}{\text{minimize}} && \frac{1}{2}\|\boldsymbol{\beta}\|^2 \\ &\text{subject to} && \boldsymbol{y} - \mathbf{X}\boldsymbol{\beta} = \mathbf{0}. \end{aligned}$$

That is, we are interested in a solution $\boldsymbol{\beta}$ with the smallest squared norm (or, equivalently, the smallest norm). The solution can be obtained via Lagrange's method (see Section C.2.2). Specifically, set $\mathcal{L}(\boldsymbol{\beta}, \boldsymbol{\lambda}) = \frac{1}{2}\|\boldsymbol{\beta}\|^2 + \boldsymbol{\lambda}^\top(\boldsymbol{y} - \mathbf{X}\boldsymbol{\beta})$, and solve

☞ 594

$$\begin{aligned} \nabla_{\boldsymbol{\beta}}\mathcal{L}(\boldsymbol{\beta}, \boldsymbol{\lambda}) &= \boldsymbol{\beta} - \mathbf{X}^\top\boldsymbol{\lambda} = \mathbf{0} \\ \nabla_{\boldsymbol{\lambda}}\mathcal{L}(\boldsymbol{\beta}, \boldsymbol{\lambda}) &= \boldsymbol{y} - \mathbf{X}\boldsymbol{\beta} = \mathbf{0}. \end{aligned}$$

The first equation yields $\boldsymbol{\beta} = \mathbf{X}^\top\boldsymbol{\lambda}$, and by substituting it in the second, we arrive at $\boldsymbol{\lambda} = (\mathbf{X}\mathbf{X}^\top)^{-1}\boldsymbol{y}$. Hence, $\boldsymbol{\beta} = \mathbf{X}^\top\boldsymbol{\lambda} = \mathbf{X}^\top(\mathbf{X}\mathbf{X}^\top)^{-1}\boldsymbol{y} = \mathbf{X}^+\boldsymbol{y}$. An example Python code is given below.

svdexample.py

```
from numpy import diag, zeros,vstack
from numpy.random import rand, seed
from numpy.linalg import svd, pinv
seed(12345)
n = 5
p = 8
X = rand(n,p)
y = rand(n,1)
```

```
U,S,VT = svd(X)
SI  = diag(1/S)
# compute pseudo inverse
pseudo_inv = VT.T @ vstack((SI, zeros((p-n,n)))) @ U.T
b = pseudo_inv @ y
#b = pinv(X) @ y #remove comment for built-in pseudo inverse
print(X @ b - y)
```

```
[[5.55111512e-16]
[1.11022302e-16]
[5.55111512e-16]
[8.60422844e-16]
[2.22044605e-16]]
```

A.8 Solving Linear Equations

In this section we describe a number of matrix decompositions that can assist in solving systems of linear equations. In particular, given the invertible matrix $\mathbf{A} \in \mathbb{R}^{n\times n}$ and vector $\boldsymbol{b} \in \mathbb{R}^n$, we wish to find the vector $\boldsymbol{x}$ that solves $\mathbf{A}\boldsymbol{x} = \boldsymbol{b}$. The choice of decomposition is typically guided by the structure of $\mathbf{A}$, as follows:

1. If $\mathbf{A}$ has no special structure, other than being invertible, then we can use the generally applicable LU decomposition in Section A.8.1.

2. If $\mathbf{A} = \mathbf{A}_0 + \sum_{k=1}^{p} \boldsymbol{a}_k \boldsymbol{a}_k^\top$ is a sum of a simple (for example, diagonal) matrix and a low-rank matrix, then we can use solution methods derived from the block LU decompositions in Section A.8.2. Matrices with this structure arise in, for example, the computation of regression estimates in Chapter 6. ☞ 261

3. If $\mathbf{A}$ is a positive definite matrix, then the Cholesky and LDL decompositions in Section A.8.3 are appropriate. If, additionally, $\mathbf{A}$ is a Toeplitz matrix, then the Levinson–Durbin decomposition in Section A.8.4 is preferable to the Cholesky one. Matrices with this structure arise when simulating *weakly stationary* Gaussian processes; see Example 3.7. ☞ 77

4. If $\mathbf{A}$ is amenable to fast matrix-vector multiplication, then we may use the *conjugate gradient* iterative solver, which is derived from the Hessenberg–Arnoldi and Lanczos decompositions in Section A.8.5. Circulant matrices and those that have a significant sparsity structure typically yield fast matrix-vector multiplication.

Finally, we note that matrix decompositions are not only used for solving linear equations, but are also used to simplify proofs in linear algebra, enhance numerical stability, and speed up computations.

A.8.1 (P)LU Decompositions

Every invertible matrix $\mathbf{A}$ can be written as the product of three matrices:

$$\mathbf{A} = \mathbf{PLU}, \tag{A.50}$$

where $\mathbf{L}$ is a lower triangular matrix, $\mathbf{U}$ an upper triangular matrix, and $\mathbf{P}$ a **permutation matrix**. A permutation matrix is a square matrix with a single 1 in each row and column, and zeros otherwise. The matrix product $\mathbf{PB}$ simply permutes the rows of a matrix $\mathbf{B}$ and, likewise, $\mathbf{BP}$ permutes its columns. A decomposition of the form (A.50) is called a **PLU decomposition**. As a permutation matrix is orthogonal, its transpose is equal to its inverse, and so we can write (A.50) as

$$\mathbf{P}^\top \mathbf{A} = \mathbf{LU}.$$

The decomposition is not unique, and in many cases $\mathbf{P}$ can be taken to be the identity matrix, in which case we speak of the **LU decomposition** of $\mathbf{A}$, also called the LR for left–right (triangular) decomposition.

A PLU decomposition of an invertible $n \times n$ matrix $\mathbf{A}_0$ can be obtained recursively as follows: The first step is to swap the rows of $\mathbf{A}_0$ such that the element in the first column and first row of the pivoted matrix is as large as possible in absolute value. Write the resulting matrix as

$$\widetilde{\mathbf{P}}_0\mathbf{A}_0 = \begin{bmatrix} a_1 & \boldsymbol{b}_1^\top \\ \boldsymbol{c}_1 & \mathbf{D}_1 \end{bmatrix},$$

where $\widetilde{\mathbf{P}}_0$ is the permutation matrix that swaps the first and k-th row, where k is the row that contains the largest element in the first column. Next, add the matrix $-\boldsymbol{c}_1[1, \boldsymbol{b}_1^\top/a_1]$ to the last $n-1$ rows of $\widetilde{\mathbf{P}}_0\mathbf{A}_0$, to obtain the matrix

$$\begin{bmatrix} a_1 & \boldsymbol{b}_1^\top \\ \mathbf{0} & \mathbf{D}_1 - \boldsymbol{c}_1\boldsymbol{b}_1^\top/a_1 \end{bmatrix} =: \begin{bmatrix} a_1 & \boldsymbol{b}_1^\top \\ \mathbf{0} & \mathbf{A}_1 \end{bmatrix}.$$

In effect, we add some multiple of the first row to each of the remaining rows in order to obtain zeros in the first column, except for the first element.

We now apply the same procedure to $\mathbf{A}_1$ as we did to $\mathbf{A}_0$ and then to subsequent smaller matrices $\mathbf{A}_2, \ldots, \mathbf{A}_{n-1}$:

1. Swap the first row with the row having the maximal absolute value element in the first column.

2. Make every other element in the first column equal to 0 by adding appropriate multiples of the first row to the other rows.

Suppose that $\mathbf{A}_t$ has a PLU decomposition $\mathbf{P}_t\mathbf{L}_t\mathbf{U}_t$. Then it is easy to check that

$$\underbrace{\widetilde{\mathbf{P}}_{t-1}^\top \begin{bmatrix} 1 & \mathbf{0}^\top \\ \mathbf{0} & \mathbf{P}_t \end{bmatrix}}_{\mathbf{P}_{t-1}} \underbrace{\begin{bmatrix} 1 & \mathbf{0}^\top \\ \mathbf{P}_t^\top \boldsymbol{c}_t/a_t & \mathbf{L}_t \end{bmatrix}}_{\mathbf{L}_{t-1}} \underbrace{\begin{bmatrix} a_t & \boldsymbol{b}_t^\top \\ \mathbf{0} & \mathbf{U}_t \end{bmatrix}}_{\mathbf{U}_{t-1}} \tag{A.51}$$

is a PLU decomposition of $\mathbf{A}_{t-1}$. Since the PLU decomposition for the scalar $\mathbf{A}_{n-1}$ is trivial, by working backwards we obtain a PLU decomposition $\mathbf{P}_0\mathbf{L}_0\mathbf{U}_0$ of $\mathbf{A}$.

Example A.52 (PLU Decomposition) Take

$$\mathbf{A} = \begin{bmatrix} 0 & -1 & 7 \\ 3 & 2 & 0 \\ 1 & 1 & 1 \end{bmatrix}.$$

Our goal is to modify **A** via Steps 1 and 2 above so as to obtain an upper triangular matrix with maximal elements on the diagonal. We first swap the first and second row. Next, we add $-1/3$ times the first row to the third row and $1/3$ times the second row to the third row:

$$\begin{bmatrix} 0 & -1 & 7 \\ 3 & 2 & 0 \\ 1 & 1 & 1 \end{bmatrix} \longrightarrow \begin{bmatrix} 3 & 2 & 0 \\ 0 & -1 & 7 \\ 1 & 1 & 1 \end{bmatrix} \longrightarrow \begin{bmatrix} 3 & 2 & 0 \\ 0 & -1 & 7 \\ 0 & 1/3 & 1 \end{bmatrix} \longrightarrow \begin{bmatrix} 3 & 2 & 0 \\ 0 & -1 & 7 \\ 0 & 0 & 10/3 \end{bmatrix}.$$

The final matrix is $\mathbf{U}_0$, and in the process we have applied the permutation matrices

$$\widetilde{\mathbf{P}}_0 = \begin{bmatrix} 0 & 1 & 0 \\ 1 & 0 & 0 \\ 0 & 0 & 1 \end{bmatrix}, \quad \widetilde{\mathbf{P}}_1 = \begin{bmatrix} 1 & 0 \\ 0 & 1 \end{bmatrix}.$$

Using the recursion (A.51) we can now recover $\mathbf{P}_0$ and $\mathbf{L}_0$. Namely, at the final iteration we have $\mathbf{P}_2 = 1, \mathbf{L}_2 = 1$, and $\mathbf{U}_2 = 10/3$. And subsequently,

$$\mathbf{P}_1 = \begin{bmatrix} 1 & 0 \\ 0 & 1 \end{bmatrix}, \quad \mathbf{L}_1 = \begin{bmatrix} 1 & 0 \\ -1/3 & 1 \end{bmatrix}, \quad \mathbf{P}_0 = \begin{bmatrix} 0 & 1 & 0 \\ 1 & 0 & 0 \\ 0 & 0 & 1 \end{bmatrix}, \quad \mathbf{L}_0 = \begin{bmatrix} 1 & 0 & 0 \\ 0 & 1 & 0 \\ 1/3 & -1/3 & 1 \end{bmatrix},$$

observing that $a_1 = 3, \boldsymbol{c}_1 = [0, 1]^\top, a_2 = -1$, and $\boldsymbol{c}_2 = 1/3$. ■

PLU decompositions can be used to solve large systems of linear equations of the form $\mathbf{A}\boldsymbol{x} = \boldsymbol{b}$ efficiently, especially when such an equation has to be solved for many different $\boldsymbol{b}$. This is done by first decomposing **A** into **PLU**, and then solving two triangular systems:

1. $\mathbf{L}\boldsymbol{y} = \mathbf{P}^\top\boldsymbol{b}$.
2. $\mathbf{U}\boldsymbol{x} = \boldsymbol{y}$.

The first equation can be solved efficiently via **forward substitution**, and the second via **backward substitution**, as illustrated in the following example.

■ **Example A.53 (Solving Linear Equations with an LU Decomposition)** Let $\mathbf{A} = \mathbf{PLU}$ be the same as in Example A.52. We wish to solve $\mathbf{A}\boldsymbol{x} = [1, 2, 3]^\top$. First, solving

$$\begin{bmatrix} 1 & 0 & 0 \\ 0 & 1 & 0 \\ 1/3 & -1/3 & 1 \end{bmatrix} \begin{bmatrix} y_1 \\ y_2 \\ y_3 \end{bmatrix} = \begin{bmatrix} 2 \\ 1 \\ 3 \end{bmatrix}$$

gives, $y_1 = 2$, $y_2 = 1$, and $y_3 = 3 - 2/3 + 1/3 = 8/3$, by forward substitution. Next,

$$\begin{bmatrix} 3 & 2 & 0 \\ 0 & -1 & 7 \\ 0 & 0 & 10/3 \end{bmatrix} \begin{bmatrix} x_1 \\ x_2 \\ x_3 \end{bmatrix} = \begin{bmatrix} 2 \\ 1 \\ 8/3 \end{bmatrix}$$

gives $x_3 = 4/5, x_2 = -1 + 28/5 = 23/5$, and $x_1 = 2(1 - 23/5)/3 = -12/5$, by backward substitution, so that $\boldsymbol{x} = [-12, 23, 4]^\top/5$. ■

A.8.2 Block LU Decompositions

LU (or more generally PLU) decompositions can also be applied to block matrices. A starting point is the following LU decomposition for a general 2×2 matrix:

$$\begin{bmatrix} a & b \\ c & d \end{bmatrix} = \begin{bmatrix} a & 0 \\ c & d - bc/a \end{bmatrix} \begin{bmatrix} 1 & b/a \\ 0 & 1 \end{bmatrix},$$

which holds as long as $a \neq 0$; this can be seen by simply writing out the matrix product. The block matrix generalization for matrices $\mathbf{A} \in \mathbb{R}^{n\times n}, \mathbf{B} \in \mathbb{R}^{n\times k}, \mathbf{C} \in \mathbb{R}^{k\times n}, \mathbf{D} \in \mathbb{R}^{k\times k}$ is

(A.54) $$\mathbf{\Sigma} := \begin{bmatrix} \mathbf{A} & \mathbf{B} \\ \mathbf{C} & \mathbf{D} \end{bmatrix} = \begin{bmatrix} \mathbf{A} & \mathbf{O}_{n\times k} \\ \mathbf{C} & \mathbf{D} - \mathbf{C}\mathbf{A}^{-1}\mathbf{B} \end{bmatrix} \begin{bmatrix} \mathbf{I}_n & \mathbf{A}^{-1}\mathbf{B} \\ \mathbf{O}_{k\times n} & \mathbf{I}_k \end{bmatrix},$$

provided that $\mathbf{A}$ is invertible (again, write out the block matrix product). Here, we use the notation $\mathbf{O}_{p\times q}$ to denote the $p \times q$ matrix of zeros. We can further rewrite this as:

(A.55) $$\mathbf{\Sigma} = \begin{bmatrix} \mathbf{I}_n & \mathbf{O}_{n\times k} \\ \mathbf{C}\mathbf{A}^{-1} & \mathbf{I}_k \end{bmatrix} \begin{bmatrix} \mathbf{A} & \mathbf{O}_{n\times k} \\ \mathbf{O}_{k\times n} & \mathbf{D} - \mathbf{C}\mathbf{A}^{-1}\mathbf{B} \end{bmatrix} \begin{bmatrix} \mathbf{I}_n & \mathbf{A}^{-1}\mathbf{B} \\ \mathbf{O}_{k\times n} & \mathbf{I}_k \end{bmatrix}.$$

Thus, inverting both sides, we obtain

$$\mathbf{\Sigma}^{-1} = \begin{bmatrix} \mathbf{I}_n & \mathbf{A}^{-1}\mathbf{B} \\ \mathbf{O}_{k\times n} & \mathbf{I}_k \end{bmatrix}^{-1} \begin{bmatrix} \mathbf{A} & \mathbf{O}_{n\times k} \\ \mathbf{O}_{k\times n} & \mathbf{D} - \mathbf{C}\mathbf{A}^{-1}\mathbf{B} \end{bmatrix}^{-1} \begin{bmatrix} \mathbf{I}_n & \mathbf{O}_{n\times k} \\ \mathbf{C}\mathbf{A}^{-1} & \mathbf{I}_k \end{bmatrix}^{-1}.$$

Inversion of the above block matrices gives (again write out)

(A.56) $$\mathbf{\Sigma}^{-1} = \begin{bmatrix} \mathbf{I}_n & -\mathbf{A}^{-1}\mathbf{B} \\ \mathbf{O}_{k\times n} & \mathbf{I}_k \end{bmatrix} \begin{bmatrix} \mathbf{A}^{-1} & \mathbf{O}_{n\times k} \\ \mathbf{O}_{k\times n} & (\mathbf{D} - \mathbf{C}\mathbf{A}^{-1}\mathbf{B})^{-1} \end{bmatrix} \begin{bmatrix} \mathbf{I}_n & \mathbf{O}_{n\times k} \\ -\mathbf{C}\mathbf{A}^{-1} & \mathbf{I}_k \end{bmatrix}.$$

■ **Example A.57 (Positive Definiteness and Block LU Decomposition)** Suppose that $\mathbf{B} = \mathbf{C}^\top$ in (A.55) and for each $\boldsymbol{x} \in \mathbb{R}^{n+k}$ we define

$$\boldsymbol{y} := \begin{bmatrix} \mathbf{I}_n & \mathbf{A}^{-1}\mathbf{C}^\top \\ \mathbf{O}_{k\times n} & \mathbf{I}_k \end{bmatrix} \boldsymbol{x}.$$

Since $\boldsymbol{x} = \mathbf{0}$ if and only if $\boldsymbol{y} = \mathbf{0}$, and

$$\boldsymbol{x}^\top \mathbf{\Sigma} \boldsymbol{x} = \boldsymbol{y}^\top \begin{bmatrix} \mathbf{A} & \mathbf{O}_{n\times k} \\ \mathbf{O}_{k\times n} & \mathbf{D} - \mathbf{C}\mathbf{A}^{-1}\mathbf{C}^\top \end{bmatrix} \boldsymbol{y},$$

we conclude that $\mathbf{\Sigma}$ is a (symmetric) positive definite matrix if and only if both $\mathbf{A}$ and $\mathbf{D} - \mathbf{C}\mathbf{A}^{-1}\mathbf{C}^\top$ are positive definite matrices. ■

Assuming that $\mathbf{D}$ is invertible, we could also perform a block UL (as opposed to LU) decomposition:

(A.58) $$\mathbf{\Sigma} = \begin{bmatrix} \mathbf{A} - \mathbf{B}\mathbf{D}^{-1}\mathbf{C} & \mathbf{B} \\ \mathbf{O}_{k\times n} & \mathbf{D} \end{bmatrix} \begin{bmatrix} \mathbf{I}_n & \mathbf{O}_{n\times k} \\ \mathbf{D}^{-1}\mathbf{C} & \mathbf{I}_k \end{bmatrix},$$

which, after a similar calculation as the one above, yields

$$\boldsymbol{\Sigma}^{-1} = \begin{bmatrix} \mathbf{I}_n & \mathbf{O}_{n\times k} \\ -\mathbf{D}^{-1}\mathbf{C} & \mathbf{I}_k \end{bmatrix} \begin{bmatrix} (\mathbf{A}-\mathbf{B}\mathbf{D}^{-1}\mathbf{C})^{-1} & \mathbf{O}_{n\times k} \\ \mathbf{O}_{k\times n} & \mathbf{D}^{-1} \end{bmatrix} \begin{bmatrix} \mathbf{I}_n & -\mathbf{B}\mathbf{D}^{-1} \\ \mathbf{O}_{k\times n} & \mathbf{I}_k \end{bmatrix}. \tag{A.59}$$

The upper-left block of $\boldsymbol{\Sigma}^{-1}$ from (A.59) must be the same as the upper-left block of $\boldsymbol{\Sigma}^{-1}$ from (A.56), leading to the **Woodbury identity**:

$$(\mathbf{A}-\mathbf{B}\mathbf{D}^{-1}\mathbf{C})^{-1} = \mathbf{A}^{-1} + \mathbf{A}^{-1}\mathbf{B}(\mathbf{D}-\mathbf{C}\mathbf{A}^{-1}\mathbf{B})^{-1}\mathbf{C}\mathbf{A}^{-1}. \tag{A.60}$$

From (A.54) and the fact that the determinant of a product is the product of the determinants, we see that $\det(\boldsymbol{\Sigma}) = \det(\mathbf{A})\det(\mathbf{D}-\mathbf{C}\mathbf{A}^{-1}\mathbf{B})$. Similarly, from (A.58) we have $\det(\boldsymbol{\Sigma}) = \det(\mathbf{A}-\mathbf{B}\mathbf{D}^{-1}\mathbf{C})\det(\mathbf{D})$, leading to the identity

$$\det(\mathbf{A}-\mathbf{B}\mathbf{D}^{-1}\mathbf{C})\det(\mathbf{D}) = \det(\mathbf{A})\det(\mathbf{D}-\mathbf{C}\mathbf{A}^{-1}\mathbf{B}). \tag{A.61}$$

The following special cases of (A.61) and (A.60) are of particular importance.

Theorem A.62 (Sherman–Morrison Formula): Suppose that $\mathbf{A} \in \mathbb{R}^{n\times n}$ is invertible and $\boldsymbol{x}, \boldsymbol{y} \in \mathbb{R}^n$. Then,

$$\det(\mathbf{A} + \boldsymbol{x}\boldsymbol{y}^\top) = \det(\mathbf{A})(1 + \boldsymbol{y}^\top\mathbf{A}^{-1}\boldsymbol{x}).$$

If in addition $\boldsymbol{y}^\top\mathbf{A}^{-1}\boldsymbol{x} \neq -1$, then the **Sherman–Morrison formula** holds:

$$(\mathbf{A} + \boldsymbol{x}\boldsymbol{y}^\top)^{-1} = \mathbf{A}^{-1} - \frac{\mathbf{A}^{-1}\boldsymbol{x}\boldsymbol{y}^\top\mathbf{A}^{-1}}{1 + \boldsymbol{y}^\top\mathbf{A}^{-1}\boldsymbol{x}}.$$

Proof. Take $\mathbf{B} = \boldsymbol{x}$, $\mathbf{C} = -\boldsymbol{y}^\top$, and $\mathbf{D} = 1$ in (A.61) and (A.60). □

One important application of the Sherman–Morrison formula is in the efficient solution of the linear system $\mathbf{A}\boldsymbol{x} = \boldsymbol{b}$, where $\mathbf{A}$ is an $n \times n$ matrix of the form:

$$\mathbf{A} = \mathbf{A}_0 + \sum_{j=1}^{p} \boldsymbol{a}_j\boldsymbol{a}_j^\top$$

for some column vectors $\boldsymbol{a}_1, \ldots, \boldsymbol{a}_p \in \mathbb{R}^n$ and $n \times n$ diagonal (or otherwise easily invertible) matrix $\mathbf{A}_0$. Such linear systems arise, for example, in the context of *ridge regression* and optimization. ☞ 321 ☞ 604

To see how the Sherman–Morrison formula can be exploited, define the matrices $\mathbf{A}_0, \ldots, \mathbf{A}_p$ via the recursion:

$$\mathbf{A}_k = \mathbf{A}_{k-1} + \boldsymbol{a}_k\boldsymbol{a}_k^\top, \quad k = 1, \ldots, p.$$

Application of Theorem A.62 for $k = 1, \ldots, p$ yields the identities:

$$\mathbf{A}_k^{-1} = \mathbf{A}_{k-1}^{-1} - \frac{\mathbf{A}_{k-1}^{-1}\boldsymbol{a}_k\boldsymbol{a}_k^\top\mathbf{A}_{k-1}^{-1}}{1 + \boldsymbol{a}_k^\top\mathbf{A}_{k-1}^{-1}\boldsymbol{a}_k}$$

$$\det(\mathbf{A}_k) = \det(\mathbf{A}_{k-1}) \times \left(1 + \boldsymbol{a}_k^\top\mathbf{A}_{k-1}^{-1}\boldsymbol{a}_k\right).$$

Therefore, by evolving the recursive relationships up until $k = p$, we obtain:

$$\mathbf{A}_p^{-1} = \mathbf{A}_0^{-1} - \sum_{j=1}^{p} \frac{\mathbf{A}_{j-1}^{-1} \boldsymbol{a}_j \boldsymbol{a}_j^\top \mathbf{A}_{j-1}^{-1}}{1 + \boldsymbol{a}_j^\top \mathbf{A}_{j-1}^{-1} \boldsymbol{a}_j}$$

$$\det(\mathbf{A}_p) = \det(\mathbf{A}_0) \times \prod_{j=1}^{p} \left(1 + \boldsymbol{a}_j^\top \mathbf{A}_{j-1}^{-1} \boldsymbol{a}_j\right).$$

These expressions will allow us to easily compute $\mathbf{A}^{-1} = \mathbf{A}_p^{-1}$ and $\det(\mathbf{A}) = \det(\mathbf{A}_p)$, provided the following quantities are available:

$$\boldsymbol{c}_{k,j} := \mathbf{A}_{k-1}^{-1} \boldsymbol{a}_j, \quad k = 1, \ldots, p-1, \quad j = k+1, \ldots, p.$$

Since, by Theorem A.62, we can write:

$$\mathbf{A}_{k-1}^{-1} \boldsymbol{a}_j = \mathbf{A}_{k-2}^{-1} \boldsymbol{a}_j - \frac{\mathbf{A}_{k-2}^{-1} \boldsymbol{a}_{k-1} \boldsymbol{a}_{k-1}^\top \mathbf{A}_{k-2}^{-1}}{1 + \boldsymbol{a}_{k-1}^\top \mathbf{A}_{k-2}^{-1} \boldsymbol{a}_{k-1}} \boldsymbol{a}_j,$$

the quantities $\{\boldsymbol{c}_{k,j}\}$ can be computed from the recursion:

$$\begin{aligned} \boldsymbol{c}_{1,j} &= \mathbf{A}_0^{-1} \boldsymbol{a}_j, \quad j = 1, \ldots, p \\ \boldsymbol{c}_{k,j} &= \boldsymbol{c}_{k-1,j} - \frac{\boldsymbol{a}_{k-1}^\top \boldsymbol{c}_{k-1,j}}{1 + \boldsymbol{a}_{k-1}^\top \boldsymbol{c}_{k-1,k-1}} \boldsymbol{c}_{k-1,k-1}, \quad k = 2, \ldots, p, \quad j = k, \ldots, p. \end{aligned} \tag{A.63}$$

Observe that this recursive computation takes $O(p^2 n)$ time and that once $\{\boldsymbol{c}_{k,j}\}$ are available, we can express $\mathbf{A}^{-1}$ and $\det(\mathbf{A})$ as:

$$\mathbf{A}^{-1} = \mathbf{A}_0^{-1} - \sum_{j=1}^{p} \frac{\boldsymbol{c}_{j,j}\, \boldsymbol{c}_{j,j}^\top}{1 + \boldsymbol{a}_j^\top \boldsymbol{c}_{j,j}}$$

$$\det(\mathbf{A}) = \det(\mathbf{A}_0) \times \prod_{j=1}^{p} \left(1 + \boldsymbol{a}_j^\top \boldsymbol{c}_{j,j}\right).$$

In summary, we have proved the following.

Theorem A.64 (Sherman–Morrison Recursion): The inverse and determinant of the $n \times n$ matrix $\mathbf{A} = \mathbf{A}_0 + \sum_{k=1}^{p} \boldsymbol{a}_k \boldsymbol{a}_k^\top$ are given respectively by:

$$\mathbf{A}^{-1} = \mathbf{A}_0^{-1} - \mathbf{C}\mathbf{D}^{-1}\mathbf{C}^\top \quad \text{and} \quad \det(\mathbf{A}) = \det(\mathbf{A}_0)\det(\mathbf{D}),$$

where $\mathbf{C} \in \mathbb{R}^{n \times p}$ and $\mathbf{D} \in \mathbb{R}^{p \times p}$ are the matrices

$$\mathbf{C} := \left[\boldsymbol{c}_{1,1}, \ldots, \boldsymbol{c}_{p,p}\right] \quad \text{and} \quad \mathbf{D} := \operatorname{diag}\left(1 + \boldsymbol{a}_1^\top \boldsymbol{c}_{1,1}, \ldots, 1 + \boldsymbol{a}_p^\top \boldsymbol{c}_{p,p}\right),$$

and all the $\{\boldsymbol{c}_{j,k}\}$ are computed from the recursion (A.63) in $O(p^2 n)$ time.

As a consequence of Theorem A.64, the solution to the linear system $\mathbf{A}\boldsymbol{x} = \boldsymbol{b}$ can be computed in $O(p^2 n)$ time via:

$$\boldsymbol{x} = \mathbf{A}_0^{-1}\boldsymbol{b} - \mathbf{C}\mathbf{D}^{-1}[\mathbf{C}^\top \boldsymbol{b}].$$

> If $n > p$, the Sherman–Morrison recursion can frequently be much faster than the $O(n^3)$ direct solution via the LU decomposition method in Section A.8.1.

In summary, the following algorithm computes the matrices $\mathbf{C}$ and $\mathbf{D}$ in Theorem A.64 via the recursion (A.63).

Algorithm A.65: Sherman–Morrison Recursion

Input: Easily invertible matrix $\mathbf{A}_0$ and column vectors $\boldsymbol{a}_1, \ldots, \boldsymbol{a}_p$.

Output: Matrices $\mathbf{C}$ and $\mathbf{D}$ such that $\mathbf{C}\mathbf{D}^{-1}\mathbf{C}^\top = \mathbf{A}_0^{-1} - \left(\mathbf{A}_0 + \sum_j \boldsymbol{a}_j \boldsymbol{a}_j^\top\right)^{-1}$.

$\boldsymbol{c}_k \leftarrow \mathbf{A}_0^{-1}\boldsymbol{a}_k$ for $k = 1, \ldots, p$ // assuming $\mathbf{A}_0$ is easily invertible matrix
for $k = 1, \ldots, p-1$ **do**
  $d_k \leftarrow 1 + \boldsymbol{a}_k^\top \boldsymbol{c}_k$
  for $j = k+1, \ldots, p$ **do**
    $\boldsymbol{c}_j \leftarrow \boldsymbol{c}_j - \dfrac{\boldsymbol{a}_k^\top \boldsymbol{c}_j}{d_k}\boldsymbol{c}_k$
$d_p \leftarrow 1 + \boldsymbol{a}_p^\top \boldsymbol{c}_p$
$\mathbf{C} \leftarrow [\boldsymbol{c}_1, \ldots, \boldsymbol{c}_p]$
$\mathbf{D} \leftarrow \mathrm{diag}(d_1, \ldots, d_p)$
return $\mathbf{C}$ and $\mathbf{D}$

Finally, note that if $\mathbf{A}_0$ is a diagonal matrix and we only store the diagonal elements of $\mathbf{D}$ and $\mathbf{A}_0$ (as opposed to storing the full matrices $\mathbf{D}$ and $\mathbf{A}_0$), then the storage or memory requirements of Algorithm A.65 are only $O(p\,n)$.

A.8.3 Cholesky and LDL Decompositions

If $\mathbf{A}$ is a real-valued positive definite matrix (and therefore symmetric), e.g., a covariance matrix, then an LU decomposition can be achieved with matrices $\mathbf{L}$ and $\mathbf{U} = \mathbf{L}^\top$.

Theorem A.66 (Cholesky Decomposition): A real-valued positive definite matrix $\mathbf{A} = [a_{ij}] \in \mathbb{R}^{n \times n}$ can be decomposed as

$$\mathbf{A} = \mathbf{L}\mathbf{L}^\top,$$

where the real $n \times n$ lower triangular matrix $\mathbf{L} = [l_{kj}]$ satisfies the recursive formula

$$l_{kj} = \frac{a_{kj} - \sum_{i=1}^{j-1} l_{ji}\, l_{ki}}{\sqrt{a_{jj} - \sum_{i=1}^{j-1} l_{ji}^2}}, \quad \text{where} \quad \sum_{i=1}^{0} l_{ji}\, l_{ki} := 0 \tag{A.67}$$

for $k = 1, \ldots, n$ and $j = 1, \ldots, k$.

Proof. The proof is by inductive construction. For $k = 1, \ldots, n$, let $\mathbf{A}_k$ be the left-upper $k \times k$ submatrix of $\mathbf{A} = \mathbf{A}_n$. With $\boldsymbol{e}_1 := [1, 0, \ldots, 0]^\top$, we have $\mathbf{A}_1 = a_{11} = \boldsymbol{e}_1^\top \mathbf{A} \boldsymbol{e}_1 > 0$ by the positive-definiteness of $\mathbf{A}$. It follows that $l_{11} = \sqrt{a_{11}}$. Suppose that $\mathbf{A}_{k-1}$ has a Cholesky factorization $\mathbf{L}_{k-1}\mathbf{L}_{k-1}^\top$ with $\mathbf{L}_{k-1}$ having strictly positive diagonal elements, we can construct a Cholesky factorization of $\mathbf{A}_k$ as follows: First write

$$\mathbf{A}_k = \begin{bmatrix} \mathbf{L}_{k-1}\mathbf{L}_{k-1}^\top & \boldsymbol{a}_{k-1} \\ \boldsymbol{a}_{k-1}^\top & a_{kk} \end{bmatrix}$$

and propose $\mathbf{L}_k$ to be of the form

$$\mathbf{L}_k = \begin{bmatrix} \mathbf{L}_{k-1} & \mathbf{0} \\ \boldsymbol{l}_{k-1}^\top & l_{kk} \end{bmatrix}$$

for some vector $\boldsymbol{l}_{k-1} \in \mathbb{R}^{k-1}$ and scalar l_{kk}, for which it must hold that

$$\begin{bmatrix} \mathbf{L}_{k-1}\mathbf{L}_{k-1}^\top & \boldsymbol{a}_{k-1} \\ \boldsymbol{a}_{k-1}^\top & a_{kk} \end{bmatrix} = \begin{bmatrix} \mathbf{L}_{k-1} & \mathbf{0} \\ \boldsymbol{l}_{k-1}^\top & l_{kk} \end{bmatrix} \begin{bmatrix} \mathbf{L}_{k-1}^\top & \boldsymbol{l}_{k-1} \\ \mathbf{0}^\top & l_{kk} \end{bmatrix}.$$

To establish that such an $\boldsymbol{l}_{k-1}$ and l_{kk} exist, we must verify that the set of equations

$$\begin{aligned} \mathbf{L}_{k-1}\boldsymbol{l}_{k-1} &= \boldsymbol{a}_{k-1} \\ \boldsymbol{l}_{k-1}^\top \boldsymbol{l}_{k-1} + l_{kk}^2 &= a_{kk} \end{aligned} \tag{A.68}$$

has a solution. The system $\mathbf{L}_{k-1}\boldsymbol{l}_{k-1} = \boldsymbol{a}_{k-1}$ has a unique solution, because (by assumption) $\mathbf{L}_{k-1}$ is lower diagonal with strictly positive entries down the main diagonal and we can solve for $\boldsymbol{l}_{k-1}$ using forward substitution: $\boldsymbol{l}_{k-1} = \mathbf{L}_{k-1}^{-1}\boldsymbol{a}_{k-1}$. We can solve the second equation as $l_{kk} = \sqrt{a_{kk} - \|\boldsymbol{l}_{k-1}\|^2}$, provided that the term within the square root is positive. We demonstrate this using the fact that $\mathbf{A}$ is a positive definite matrix. In particular, for $\boldsymbol{x} \in \mathbb{R}^n$ of the form $[\boldsymbol{x}_1^\top, x_2, \mathbf{0}^\top]^\top$, where $\boldsymbol{x}_1$ is a non-zero $(k-1)$-dimensional vector and x_2 a non-zero number, we have

$$0 < \boldsymbol{x}^\top \mathbf{A} \boldsymbol{x} = \left[\boldsymbol{x}_1^\top, x_2\right] \begin{bmatrix} \mathbf{L}_{k-1}\mathbf{L}_{k-1}^\top & \boldsymbol{a}_{k-1} \\ \boldsymbol{a}_{k-1}^\top & a_{kk} \end{bmatrix} \begin{bmatrix} \boldsymbol{x}_1 \\ x_2 \end{bmatrix} = \|\mathbf{L}_{k-1}^\top \boldsymbol{x}_1\|^2 + 2\boldsymbol{x}_1^\top \boldsymbol{a}_{k-1} x_2 + a_{kk} x_2^2.$$

Now take $\boldsymbol{x}_1 = -x_2 \mathbf{L}_{k-1}^{-\top} \boldsymbol{l}_{k-1}$ to obtain $0 < \boldsymbol{x}^\top \mathbf{A} \boldsymbol{x} = x_2^2 \left(a_{kk} - \|\boldsymbol{l}_{k-1}\|^2\right)$. Therefore, (A.68) can be uniquely solved. As we have already solved it for $k = 1$, we can solve it for any $k = 1, \ldots, n$, leading to the recursive formula (A.67) and Algorithm A.69 below. □

An implementation of Cholesky's decomposition that uses the notation in the proof of Theorem A.8.3 is the following algorithm, whose running cost is $O(n^3)$:

Algorithm A.69: Cholesky Decomposition

Input: Positive-definite $n \times n$ matrix $\mathbf{A}_n$ with entries $\{a_{ij}\}$.
Output: Lower triangular $\mathbf{L}_n$ such that $\mathbf{L}_n\mathbf{L}_n^\top = \mathbf{A}_n$.

$\mathbf{L}_1 \leftarrow \sqrt{a_{11}}$
for $k = 2, \ldots, n$ **do**
  $\boldsymbol{a}_{k-1} \leftarrow [a_{1k}, \ldots, a_{k-1,k}]^\top$
  $\boldsymbol{l}_{k-1} \leftarrow \mathbf{L}_{k-1}^{-1}\boldsymbol{a}_{k-1}$ // computed in $O(k^2)$ time via forward substitution
  $l_{kk} \leftarrow \sqrt{a_{kk} - \boldsymbol{l}_{k-1}^\top \boldsymbol{l}_{k-1}}$
  $\mathbf{L}_k \leftarrow \begin{bmatrix} \mathbf{L}_{k-1} & \mathbf{0} \\ \boldsymbol{l}_{k-1}^\top & l_{kk} \end{bmatrix}$
return $\mathbf{L}_n$

■ **Example A.70 (Block Gram–Schmidt Process (Continued))** Recall that in Example A.26 we computed the orthogonal projection $\mathbf{P}_k\mathbf{X}$ of $\mathbf{X}$ onto the Span $\{\mathbf{X}_1, \ldots, \mathbf{X}_k\}$ using Algorithm A.27. Instead of computing a QR decomposition for each $i = 1, \ldots, k$ in Algorithm A.27, one can instead compute a Cholesky factor $\mathbf{L}$ for each $i = 1, \ldots, k$, as described in the following Algorithm A.71.

Algorithm A.71: Block Gram–Schmidt Process: $\mathbf{Q}\mathbf{Q}^\top\mathbf{X} = \mathbf{Q}\mathbf{R}$

Input: $\mathbf{X} = [\mathbf{X}_1, \ldots, \mathbf{X}_d] \in \mathbb{R}^{n\times p}$ and $k \leq d$, where $\mathbf{X}_i \in \mathbb{R}^{n\times p_i}$ for all i
Output: $\mathbf{Q}_i \in \mathbb{R}^{n\times p_i}$ and $\mathbf{R}_i \in \mathbb{R}^{p_i\times p}$ for $i = 1, \ldots, k$

$\mathbf{L}\mathbf{L}^\top \leftarrow \mathbf{X}_1^\top\mathbf{X}_1$ // compute Cholesky factor $\mathbf{L}$ of $\mathbf{X}_1^\top\mathbf{X}_1$
$\mathbf{Q}_1 \leftarrow \mathbf{X}_1\mathbf{L}^{-\top}$ // compute orthonormal $\mathbf{Q}_1$ using forward substitution
$\mathbf{Q} \leftarrow \mathbf{Q}_1$, $\mathbf{R} \leftarrow \mathbf{Q}_1^\top\mathbf{X}$
for $i = 2, \ldots, k$ **do**
  $\mathbf{X}_i \leftarrow \mathbf{X}_i - \mathbf{Q}[\mathbf{Q}^\top\mathbf{X}_i]$ // compute $(\mathbf{I} - \mathbf{P}_i)\mathbf{X}_i$
  $\mathbf{L}\mathbf{L}^\top \leftarrow \mathbf{X}_i^\top\mathbf{X}_i$ // compute Cholesky factor $\mathbf{L}$ of $\mathbf{X}_i^\top(\mathbf{I} - \mathbf{P}_i)\mathbf{X}_i$
  $\mathbf{Q}_i \leftarrow \mathbf{X}_i\mathbf{L}^{-\top}$ // compute orthonormal $\mathbf{Q}_i$ using forward substitution
  $\mathbf{Q} \leftarrow [\mathbf{Q}, \mathbf{Q}_i]$ // append $\mathbf{Q}_i$ as a column-block of $\mathbf{Q}$
  $\mathbf{R} \leftarrow \begin{bmatrix} \mathbf{R} \\ \mathbf{Q}_i^\top\mathbf{X} \end{bmatrix}$ // append $\mathbf{Q}_i^\top\mathbf{X}$ as a row-block of $\mathbf{R}$
return $\mathbf{Q}, \mathbf{R}$

■

A.8.3.1 Cholesky Rank-1 Update

Suppose that we have the Cholesky factor $\mathbf{C}_n$ of $\mathbf{A}_n$ and we wish to compute the Cholesky factor $\mathbf{L}_n$ of $\mathbf{A}_n + \varrho\boldsymbol{x}_n\boldsymbol{x}_n^\top$, where $\varrho\boldsymbol{x}_n\boldsymbol{x}_n^\top$ is a rank-1 perturbation of $\mathbf{A}_n$ that preserves its positive definiteness. Rather than computing $\mathbf{L}_n$ via Algorithm A.69 at a cost of $O(n^3)$, we can compute $\mathbf{L}_n$ at a cost of $O(n^2)$ by taking advantage of the rank-1 structure of

$\varrho \boldsymbol{x}_n \boldsymbol{x}_n^\top$ and the availability of $\mathbf{C}_n$. This faster procedure relies on the following recursive result.

Theorem A.72 (Recursive Cholesky Factorization): Suppose that $\mathbf{L}_{k+1}$ is the $(k+1) \times (k+1)$ Cholesky factor satisfying

$$\mathbf{L}_{k+1}\mathbf{L}_{k+1}^\top = \mathbf{C}_{k+1}\mathbf{C}_{k+1}^\top + \varrho \boldsymbol{x}_{k+1}\boldsymbol{x}_{k+1}^\top, \tag{A.73}$$

or equivalently, in terms of smaller $k \times k$ matrix blocks:

$$\begin{bmatrix} l & \mathbf{0}^\top \\ \boldsymbol{l} & \mathbf{L}_k \end{bmatrix} \begin{bmatrix} l & \boldsymbol{l}^\top \\ \mathbf{0} & \mathbf{L}_k^\top \end{bmatrix} = \begin{bmatrix} c & \mathbf{0}^\top \\ \boldsymbol{c} & \mathbf{C}_k \end{bmatrix} \begin{bmatrix} c & \boldsymbol{c}^\top \\ \mathbf{0} & \mathbf{C}_k^\top \end{bmatrix} + \varrho \begin{bmatrix} x^2 & x\boldsymbol{x}_k^\top \\ x\boldsymbol{x}_k & \boldsymbol{x}_k\boldsymbol{x}_k^\top \end{bmatrix}. \tag{A.74}$$

Then, a formula of the same form as (A.73) holds for matrices of size $k \times k$:

$$\mathbf{L}_k\mathbf{L}_k^\top = \mathbf{C}_k\mathbf{C}_k^\top + \varrho\frac{c^2}{l^2}(\boldsymbol{x}_k - (x/c)\boldsymbol{c})(\boldsymbol{x}_k - (x/c)\boldsymbol{c})^\top. \tag{A.75}$$

Proof. Matching all commensurate blocks in (A.74) we derive the identities:

$$\begin{aligned}
l &= \sqrt{c^2 + \varrho x^2}, \\
\boldsymbol{l} &= \frac{c}{l}\boldsymbol{c} + \frac{\varrho x}{l}\boldsymbol{x}_k, \\
\mathbf{L}_k\mathbf{L}_k^\top &= \mathbf{C}_k\mathbf{C}_k^\top + \varrho\boldsymbol{x}_k\boldsymbol{x}_k^\top + \boldsymbol{c}\boldsymbol{c}^\top - \boldsymbol{l}\boldsymbol{l}^\top \\
&= \mathbf{C}_k\mathbf{C}_k^\top + \left(\varrho - \frac{\varrho^2 x^2}{l^2}\right)\boldsymbol{x}_k\boldsymbol{x}_k^\top + \left(1 - \frac{c^2}{l^2}\right)\boldsymbol{c}\boldsymbol{c}^\top - \frac{c\varrho x}{l^2}(\boldsymbol{c}\boldsymbol{x}_k^\top + \boldsymbol{x}_k\boldsymbol{c}^\top) \\
&= \mathbf{C}_k\mathbf{C}_k^\top + \frac{\varrho c^2}{l^2}\boldsymbol{x}_k\boldsymbol{x}_k^\top + \frac{\varrho x^2}{l^2}\boldsymbol{c}\boldsymbol{c}^\top - \varrho\frac{cx}{l^2}(\boldsymbol{c}\boldsymbol{x}_k^\top + \boldsymbol{x}_k\boldsymbol{c}^\top).
\end{aligned}$$

The latter implies (A.75), which is of the same form as (A.73). □

The theorem justifies the following recursive algorithm:

Algorithm A.76: Cholesky Rank-1 Update in $O(n^2)$ Time.

Input: Cholesky factor $\mathbf{L} \in \mathbb{R}^{n \times n}$ (so that $\mathbf{L}\mathbf{L}^\top = \mathbf{A}$), scalar ϱ, and vector $\boldsymbol{x}$.
Output: Lower triangular $\mathbf{L}$ such that $\mathbf{L}\mathbf{L}^\top = \mathbf{A} + \varrho\boldsymbol{x}\boldsymbol{x}^\top$.

for $k = 1, \ldots, n-1$ **do**
- Let $\boldsymbol{l}$ be the k-th column of $\mathbf{L}$
- $d_k \leftarrow \sqrt{l_k^2 + \varrho x_k^2}$
- Store $(l_k/d_k)\,\boldsymbol{l} + (\varrho x_k/d_k)\,\boldsymbol{x}$ as the k-th column of $\mathbf{L}$
- $\boldsymbol{x} \leftarrow \boldsymbol{x} - (x_k/l_k)\,\boldsymbol{l}$
- $\varrho \leftarrow \varrho\,(l_k/d_k)^2$

$\mathbf{L}_{n,n} \leftarrow \sqrt{\mathbf{L}_{n,n}^2 + \varrho x_n^2}$
return $\mathbf{L}$

A.8.3.2 LDL Decomposition

We can rewrite the Cholesky factorization in Algorithm A.69 to avoid taking square roots. To achieve this, we ensure that the diagonal elements of the lower triangular matrix $\mathbf{L}$ are always equal to one. This restriction is compensated for by introducing a diagonal matrix $\mathbf{D}^2$. As a result, the matrix $\mathbf{A}$ can be decomposed as: $\mathbf{A} = \mathbf{L}\mathbf{D}^2\mathbf{L}^\top$.

Algorithm A.77: LDL Decomposition

Input: Positive-definite $n \times n$ matrix $\mathbf{A}_n$ with entries $\{a_{ij}\}$.
Output: $\mathbf{L}_n$ and diagonal $\mathbf{D}_n^2$ such that $\mathbf{L}_n\mathbf{D}_n^2\mathbf{L}_n^\top = \mathbf{A}_n$.

```
1 L_1 ← 1, D_1^2 ← a_11
2 for k = 2, ..., n do
3     a_{k-1} ← [a_{1,k}, ..., a_{k-1,k}]^T
4     l_{k-1} ← D_{k-1}^{-2} L_{k-1}^{-1} a_{k-1}        // D^{-2} computed via elementwise division
5     d_k^2 ← a_{k,k} - l_{k-1}^T D_{k-1}^2 l_{k-1}
6     L_k ← [ L_{k-1}    0 ]
            [ l_{k-1}^T  1 ]
7     D_k^2 ← [ D_{k-1}^2  0     ]
              [ 0^T        d_k^2 ]
8 return L_n, D_n^2
```

In the time-series literature the LDL algorithm is sometimes known as the **innovations algorithm**.

Example A.78 (LDL Decomposition of a Dirichlet Matrix) Take the vectors $\boldsymbol{p}^\top = [p_1, \dots, p_n]^\top$ and $\boldsymbol{q}$, where $q_k := 1 - (p_1 + \cdots + p_k) > 0$ and $p_k > 0$ for $k = 1, \dots, n$. Then, the **Dirichlet covariance** matrix

$$\mathbf{A} := \mathrm{diag}(\boldsymbol{p}) - \boldsymbol{p}\boldsymbol{p}^\top$$

has the decomposition $\mathbf{A} = \mathbf{L}\mathbf{D}^2\mathbf{L}^\top$ with

$$\mathbf{D}^2 := \mathrm{diag}(p_1 q_1, p_2 q_2/q_1, \dots, p_n q_n/q_{n-1})$$

and:

$$\mathbf{L} := \begin{bmatrix} 1 & 0 & \cdots & & \cdots & 0 \\ -p_2/q_1 & 1 & 0 & & \cdots & 0 \\ \vdots & \ddots & \ddots & & \ddots & \vdots \\ \vdots & & \ddots & \ddots & \ddots & 0 \\ -p_n/q_1 & \cdots & \cdots & & -p_n/q_{n-1} & 1 \end{bmatrix}.$$

Incidentally, if $\boldsymbol{p} = \frac{\varrho}{1+(n-1)\varrho}\mathbf{1}$ for some correlation coefficient $\varrho \in (0, 1)$, then

$$\mathbf{A}^{-1} = (1 - \varrho)\mathbf{I}_n + \varrho\mathbf{1}\mathbf{1}^\top.$$

■

■ **Example A.79 (Tridiagonal Matrix)** Suppose that $\mathbf{A}_n$ is a *tridiagonal* positive semidefinite matrix. For example, for $n = 6$, we have:

$$\mathbf{A}_6 = \begin{bmatrix} a_{11} & a_{12} & 0 & 0 & 0 & 0 \\ a_{12} & a_{22} & a_{23} & 0 & 0 & 0 \\ 0 & a_{23} & a_{33} & a_{34} & 0 & 0 \\ 0 & 0 & a_{34} & a_{44} & a_{45} & 0 \\ 0 & 0 & 0 & a_{45} & a_{55} & a_{56} \\ 0 & 0 & 0 & 0 & a_{56} & a_{66} \end{bmatrix}.$$

In this case, the cost of the LDL Algorithm A.77 can be further reduced by observing that the lower triangular factor, say $\mathbf{L}_6$, must be of *bidiagonal* form:

$$\mathbf{L}_6 = \begin{bmatrix} 1 & 0 & 0 & 0 & 0 & 0 \\ l_{21} & 1 & 0 & 0 & 0 & 0 \\ 0 & l_{32} & 1 & 0 & 0 & 0 \\ 0 & 0 & l_{43} & 1 & 0 & 0 \\ 0 & 0 & 0 & l_{54} & 1 & 0 \\ 0 & 0 & 0 & 0 & l_{65} & 1 \end{bmatrix},$$

and both $\boldsymbol{a}_{k-1}$ and $\boldsymbol{l}_{k-1}$ in Algorithm A.77 are vectors of the form

$$\boldsymbol{a}_{k-1} = [0, \ldots, 0, a_{k-1,k}]^\top,$$
$$\boldsymbol{l}_{k-1} = [0, \ldots, 0, l_{k,k-1}]^\top.$$

Hence, $\boldsymbol{l}_{k-1} = \mathbf{D}_{k-1}^{-2}\mathbf{L}_{k-1}^{-1}\boldsymbol{a}_{k-1}$ in Line 4 of Algorithm A.77 can be replaced by $l_{k,k-1} = a_{k-1,k}/d_{k-1}^2$ and $d_k^2 = a_{k,k} - \boldsymbol{l}_{k-1}^\top \mathbf{D}_{k-1}^2 \boldsymbol{l}_{k-1} = a_{k,k} - (l_{k,k-1}d_{k-1})^2$, which are $\mathcal{O}(1)$ operations (the cost of computing them is independent of n). In other words, a properly modified version of Algorithm A.77 will run in $\mathcal{O}(n)$ time for a tridiagonal covariance, instead of the usual $\mathcal{O}(n^3)$ time. ■

■ **Example A.80 (Tridiagonal Linear System)** Suppose we wish to solve the linear system $\mathbf{T}\boldsymbol{x} = \boldsymbol{b}$, where $\mathbf{T} \in \mathbb{R}^{n\times n}$ is a tridiagonal covariance matrix. We can do this in $\mathcal{O}(n)$ time using the LDL decomposition $\mathbf{T} = \mathbf{L}\mathbf{D}^2\mathbf{L}^\top$. Note that we should not store the full matrices $\mathbf{L}$ and $\mathbf{D}$, as these are mostly filled with zeros (failing to account for this will cost $\mathcal{O}(n^2)$ in memory processing). We should instead simply compute and store the vectors $\widetilde{\boldsymbol{d}} = [d_1^2, \ldots, d_n^2]^\top$ and $\widetilde{\boldsymbol{l}} = [l_{2,1}, l_{3,2}, \ldots, l_{n,n-1}]^\top$. Formally, $\boldsymbol{x} = \mathbf{L}^{-\top}\mathbf{D}^{-2}\mathbf{L}^{-1}\boldsymbol{b}$, but we should also implement this without storing $\mathbf{L}^{-1}$ and $\mathbf{D}^{-2}$ explicitly, and only using the vectors $\widetilde{\boldsymbol{l}}$ and $\widetilde{\boldsymbol{d}}$. For example, we solve $\mathbf{L}\boldsymbol{y} = \boldsymbol{b}$ via the recursion $y_1 = b_1$, $y_k = b_k - l_{k,k-1}y_{k-1}$, $k = 2, 3, \ldots, n$. This strategy gives the following pseudo-code:

Algorithm A.81: LDL Decomposition for Tridiagonal Symmetric T

Input: diagonal $[t_{1,1}, \ldots, t_{n,n}]^\top$ and sub-diagonal $[t_{2,1}, \ldots, t_{n,n-1}]^\top$ of $\mathbf{T}$.
Output: $\widetilde{\boldsymbol{l}}$ and $\widetilde{\boldsymbol{d}}$ as compact representations of $\mathbf{L}$ and $\mathbf{D}^2$ in $\mathbf{T} = \mathbf{L}\mathbf{D}^2\mathbf{L}^\top$.

$\widetilde{d}_1 \leftarrow t_{1,1}$
for $k = 2, \ldots, n$ **do**
  $\widetilde{l}_{k-1} \leftarrow t_{k,k-1}/\widetilde{d}_{k-1}$
  $\widetilde{d}_k \leftarrow t_{k,k} - \widetilde{l}_{k-1}^2\widetilde{d}_{k-1}$
return $\widetilde{\boldsymbol{l}}$, $\widetilde{\boldsymbol{d}}$

Once we have computed the LDL decomposition of $\mathbf{T}$ in $O(n)$ time, then we can solve the linear system $\mathbf{T}\boldsymbol{x} = \boldsymbol{b}$ as follows:

Algorithm A.82: Solving $\mathbf{T}\boldsymbol{x} = \boldsymbol{b}$ in $O(n)$ Time

Input: $\boldsymbol{b}$, vectors $\widetilde{\boldsymbol{d}} \in \mathbb{R}^n$ and $\widetilde{\boldsymbol{l}} \in \mathbb{R}^{n-1}$ from LDL decomposition of $\mathbf{T}$.
Output: $\mathbf{T}^{-1}\boldsymbol{b}$.

```
x_1 ← b_1
for k = 2, ..., n do
    x_k ← b_k − l̃_{k−1} x_{k−1}          // forward pass
x ← x ÷ d̃                              // elementwise division
for k = 2, ..., n do
    x_k ← x_k − l̃_k x_{k+1}              // backward pass
return x
```

An important application of the solution of tridiagonal linear systems is given in Section A.8.5.1, where we describe Lanczos algorithm for solving large systems of linear equations. ■

A.8.4 Levinson–Durbin Decomposition

A special type of a structured matrix $\mathbf{A}$ is the one that satisfies the **Sylvester equation**:

$$\mathbf{M}_1\mathbf{A} - \mathbf{A}\mathbf{M}_2^* = \mathbf{G}_1\mathbf{G}_2^*, \tag{A.83}$$

where $\mathbf{M}_i \in \mathbb{C}^{n\times n}, i = 1,2$ are sparse matrices and $\mathbf{G}_i \in \mathbb{C}^{n\times r}$, $i = 1,2$ are matrices of rank $r \ll n$. The elements of $\mathbf{A}$ must be easy to recover from these matrices, e.g., with $O(1)$ operations. A typical example is a (square) **Toeplitz matrix**, which has the following structure:

$$\mathbf{A} = \begin{bmatrix} a_0 & a_{-1} & \cdots & a_{-(n-2)} & a_{-(n-1)} \\ a_1 & a_0 & a_{-1} & & a_{-(n-2)} \\ \vdots & a_1 & a_0 & \ddots & \vdots \\ a_{n-2} & & \ddots & \ddots & a_{-1} \\ a_{n-1} & a_{n-2} & \cdots & a_1 & a_0 \end{bmatrix}.$$

A general square Toeplitz matrix $\mathbf{A}$ is completely determined by the $2n - 1$ elements along its first row and column. If $\mathbf{A}$ is also Hermitian (i.e., $\mathbf{A}^* = \mathbf{A}$), then clearly it is determined by only n elements. If we define the matrices:

$$\mathbf{M}_1 = \begin{bmatrix} 0 & 0 & \cdots & 0 & 1 \\ 1 & 0 & 0 & & 0 \\ \vdots & 1 & 0 & \ddots & \vdots \\ 0 & & \ddots & \ddots & 0 \\ 0 & 0 & \cdots & 1 & 0 \end{bmatrix} \quad \text{and} \quad \mathbf{M}_2 = \begin{bmatrix} 0 & 1 & \cdots & 0 & 0 \\ 0 & 0 & 1 & & 0 \\ \vdots & 0 & 0 & \ddots & \vdots \\ 0 & & \ddots & \ddots & 1 \\ -1 & 0 & \cdots & 0 & 0 \end{bmatrix},$$

then (A.83) is satisfied with

$$\mathbf{G}_1\mathbf{G}_2^* := \begin{bmatrix} 1 & 0 \\ 0 & a_1 + a_{-(n-1)} \\ 0 & a_2 + a_{-(n-2)} \\ \vdots & \vdots \\ 0 & a_{n-1} + a_{-1} \end{bmatrix} \begin{bmatrix} a_{n-1} - a_{-1} & a_{n-2} - a_{-2} & \dots & a_1 - a_{-(n-1)} & 2a_0 \\ 0 & 0 & \dots & 0 & 1 \end{bmatrix}$$

$$= \begin{bmatrix} a_{n-1} - a_{-1} & a_{n-2} - a_{-2} & \dots & a_1 - a_{-(n-1)} & 2a_0 \\ 0 & 0 & \dots & 0 & a_1 + a_{-(n-1)} \\ \vdots & \vdots & \dots & \vdots & a_2 + a_{-(n-2)} \\ \vdots & \vdots & \dots & \vdots & \vdots \\ 0 & 0 & \dots & 0 & a_{n-1} + a_{-1} \end{bmatrix},$$

which has rank $r \leq 2$.

■ **Example A.84 (Discrete Convolution of Vectors)** The convolution of two vectors can be represented as multiplication of one of the vectors by a Toeplitz matrix. Suppose $\boldsymbol{x} = [x_1, \dots, x_m]^\top$ and $\boldsymbol{y} = [y_1, \dots, y_n]^\top$ are two complex-valued vectors. Then, their **convolution** is defined as the vector $\boldsymbol{z} = \boldsymbol{x} \star \boldsymbol{y}$ with i-th element

(A.85) $$z_i = [\boldsymbol{x} \star \boldsymbol{y}]_i = \sum_{k=1}^{m} x_k \, y_{i-k+1}, \quad i = 1, \dots, m + n - 1,$$

where $y_j := 0$ for $j \leq 0$. It is easy to verify that the convolution can be written as

$$\boldsymbol{x} \star \boldsymbol{y} = \mathbf{A}\boldsymbol{y},$$

where, denoting the d-dimensional column vector of zeros by $\mathbf{0}_d$, we have that

$$\mathbf{A} = \begin{bmatrix} \boldsymbol{x} & \mathbf{0} & & & \\ \mathbf{0}_{m-1} & \boldsymbol{x} & \ddots & & \\ & \mathbf{0}_{m-2} & \ddots & \mathbf{0}_{m-2} & \\ & & \ddots & \boldsymbol{x} & \mathbf{0}_{m-1} \\ & & & 0 & \boldsymbol{x} \end{bmatrix}.$$

Clearly, the matrix $\mathbf{A}$ is a (sparse) Toeplitz matrix. Convolutions can be computed
☞ 577 efficiently using fast Fourier transforms; see Example B.36. ■

A **circulant matrix** is a special Toeplitz matrix which is obtained from a vector $\boldsymbol{c}$ by circularly permuting its indices as follows:

(A.86) $$\mathbf{C} = \begin{bmatrix} c_0 & c_{n-1} & \dots & c_2 & c_1 \\ c_1 & c_0 & c_{n-1} & & c_2 \\ \vdots & c_1 & c_0 & \ddots & \vdots \\ c_{n-2} & & \ddots & \ddots & c_{n-1} \\ c_{n-1} & c_{n-2} & \dots & c_1 & c_0 \end{bmatrix}.$$

Note that $\mathbf{C}$ is completely determined by the n elements of its first column, $\boldsymbol{c}$.

To illustrate how structured matrices allow for faster matrix computations, consider solving the $n \times n$ linear system:

$$\mathbf{A}_n \boldsymbol{x}_n = \boldsymbol{a}_n$$

for $\boldsymbol{x}_n = [x_1, \ldots, x_n]^\top$, where $\boldsymbol{a}_n = [a_1, \ldots, a_n]^\top$, and

(A.87) $$\mathbf{A}_n := \begin{bmatrix} 1 & a_1 & \cdots & a_{n-2} & a_{n-1} \\ a_1 & 1 & \ddots & & a_{n-2} \\ \vdots & \ddots & \ddots & \ddots & \vdots \\ a_{n-2} & & \ddots & \ddots & a_1 \\ a_{n-1} & a_{n-2} & \cdots & a_1 & 1 \end{bmatrix}$$

is a real-valued symmetric positive-definite Toeplitz matrix (so that it is invertible). Note that the entries of $\mathbf{A}_n$ are completely determined by the right-hand side of the linear equation: vector $\boldsymbol{a}_n$. As we shall see shortly in Example A.91, the solution to the more general linear equation $\mathbf{A}_n \boldsymbol{x}_n = \boldsymbol{b}_n$, where $\boldsymbol{b}_n$ is arbitrary, can be efficiently computed using the solution to this specific system $\mathbf{A}_n \boldsymbol{x}_n = \boldsymbol{a}_n$, obtained via a special recursive algorithm (Algorithm A.88 below).

For every $k = 1, \ldots, n$ the $k \times k$ Toeplitz matrix $\mathbf{A}_k$ satisfies

$$\mathbf{A}_k = \mathbf{P}_k \, \mathbf{A}_k \, \mathbf{P}_k,$$

where $\mathbf{P}_k$ is a permutation matrix that "flips" the order of elements — rows when pre-multiplying and columns when post-multiplying. For example,

$$\begin{bmatrix} 1 & 2 & 3 & 4 & 5 \\ 6 & 7 & 8 & 9 & 10 \end{bmatrix} \mathbf{P}_5 = \begin{bmatrix} 5 & 4 & 3 & 2 & 1 \\ 10 & 9 & 8 & 7 & 6 \end{bmatrix}, \quad \text{where} \quad \mathbf{P}_5 = \begin{bmatrix} 0 & 0 & 0 & 0 & 1 \\ 0 & 0 & 0 & 1 & 0 \\ 0 & 0 & 1 & 0 & 0 \\ 0 & 1 & 0 & 0 & 0 \\ 1 & 0 & 0 & 0 & 0 \end{bmatrix}.$$

Clearly, $\mathbf{P}_k = \mathbf{P}_k^\top$ and $\mathbf{P}_k \mathbf{P}_k = \mathbf{I}_k$ hold, so that in fact $\mathbf{P}_k$ is an orthogonal matrix.

We can solve the $n \times n$ linear system $\mathbf{A}_n \boldsymbol{x}_n = \boldsymbol{a}_n$ in $O(n^2)$ time recursively, as follows: Assume that we have somehow solved for the upper $k \times k$ block $\mathbf{A}_k \boldsymbol{x}_k = \boldsymbol{a}_k$ and now we wish to solve for the $(k+1) \times (k+1)$ block:

$$\mathbf{A}_{k+1} \boldsymbol{x}_{k+1} = \boldsymbol{a}_{k+1} \quad \Longleftrightarrow \quad \begin{bmatrix} \mathbf{A}_k & \mathbf{P}_k \boldsymbol{a}_k \\ \boldsymbol{a}_k^\top \mathbf{P}_k & 1 \end{bmatrix} \begin{bmatrix} \boldsymbol{z} \\ \alpha \end{bmatrix} = \begin{bmatrix} \boldsymbol{a}_k \\ a_{k+1} \end{bmatrix}.$$

Therefore,

$$\begin{aligned} \alpha &= a_{k+1} - \boldsymbol{a}_k^\top \mathbf{P}_k \boldsymbol{z} \\ \mathbf{A}_k \boldsymbol{z} &= \boldsymbol{a}_k - \alpha \, \mathbf{P}_k \, \boldsymbol{a}_k. \end{aligned}$$

Since $\mathbf{A}_k^{-1} \mathbf{P}_k = \mathbf{P}_k \mathbf{A}_k^{-1}$, the second equation above simplifies to

$$\begin{aligned} \boldsymbol{z} &= \mathbf{A}_k^{-1} \boldsymbol{a}_k - \alpha \, \mathbf{A}_k^{-1} \mathbf{P}_k \, \boldsymbol{a}_k \\ &= \boldsymbol{x}_k - \alpha \, \mathbf{P}_k \, \boldsymbol{x}_k. \end{aligned}$$

Substituting $z = \boldsymbol{x}_k - \alpha\, \mathbf{P}_k\, \boldsymbol{x}_k$ into $\alpha = a_{k+1} - \boldsymbol{a}_k^\top\, \mathbf{P}_k\, z$ and solving for α yields:

$$\alpha = \frac{a_{k+1} - \boldsymbol{a}_k^\top\, \mathbf{P}_k\, \boldsymbol{x}_k}{1 - \boldsymbol{a}_k^\top\, \boldsymbol{x}_k}.$$

Finally, with the value of α computed above, we have

$$\boldsymbol{x}_{k+1} = \begin{bmatrix} \boldsymbol{x}_k - \alpha\, \mathbf{P}_k\, \boldsymbol{x}_k \\ \alpha \end{bmatrix}.$$

This gives the following **Levinson–Durbin** recursive algorithm for solving $\mathbf{A}_n\, \boldsymbol{x}_n = \boldsymbol{a}_n$.

Algorithm A.88: Levinson–Durbin Recursion for Solving $\mathbf{A}_n\, \boldsymbol{x}_n = \boldsymbol{a}_n$

Input: First row $[1, a_1, \ldots, a_{n-1}] = [1, \boldsymbol{a}_{n-1}^\top]$ of matrix $\mathbf{A}_n$.
Output: Solution $\boldsymbol{x}_n = \mathbf{A}_n^{-1}\, \boldsymbol{a}_n$.

$\boldsymbol{x}_1 \leftarrow a_1$
for $k = 1, \ldots, n-1$ **do**
  $\beta_k \leftarrow 1 - \boldsymbol{a}_k^\top\, \boldsymbol{x}_k$
  $\breve{\boldsymbol{x}} \leftarrow [x_{k,k}, x_{k,k-1}, \ldots, x_{k,1}]^\top$
  $\alpha \leftarrow (a_{k+1} - \boldsymbol{a}_k^\top\, \breve{\boldsymbol{x}})/\beta_k$
  $\boldsymbol{x}_{k+1} \leftarrow \begin{bmatrix} \boldsymbol{x}_k - \alpha\, \breve{\boldsymbol{x}} \\ \alpha \end{bmatrix}$
return $\boldsymbol{x}_n$

In the algorithm above, we have identified $\boldsymbol{x}_k = [x_{k,1}, x_{k,2}, \ldots, x_{k,k}]^\top$. The advantage of the Levinson–Durbin algorithm is that its running cost is $O(n^2)$, instead of the usual $O(n^3)$.

Using the $\{\boldsymbol{x}_k, \beta_k\}$ computed in Algorithm A.88, we construct the following lower triangular matrix recursively, setting $\mathbf{L}_1 = 1$ and

$$\mathbf{L}_{k+1} = \begin{bmatrix} \mathbf{L}_k & \mathbf{0}_k \\ -(\mathbf{P}_k \boldsymbol{x}_k)^\top & 1 \end{bmatrix}, \qquad k = 1, \ldots, n-1. \tag{A.89}$$

Then, we have the following factorization of $\mathbf{A}_n$.

Theorem A.90 (Levinson–Durbin Decomposition): For a real-valued symmetric positive-definite Toeplitz matrix $\mathbf{A}_n$ of the form (A.87), we have

$$\mathbf{L}_n\, \mathbf{A}_n\, \mathbf{L}_n^\top = \mathbf{D}_n,$$

where $\mathbf{L}_n$ is a the lower diagonal matrix (A.89) and $\mathbf{D}_n := \mathrm{diag}(1, \beta_1, \ldots, \beta_{n-1})$.

Proof. We give a proof by induction. Obviously, $\mathbf{L}_1 \mathbf{A}_1 \mathbf{L}_1^\top = 1 \cdot 1 \cdot 1 = 1 = \mathbf{D}_1$ is true.

Next, assume that the factorization $\mathbf{L}_k \mathbf{A}_k \mathbf{L}_k^\top = \mathbf{D}_k$ holds for a given k. Observe that

$$\mathbf{L}_{k+1}\mathbf{A}_{k+1} = \begin{bmatrix} \mathbf{L}_k & \mathbf{0}_k \\ -(\mathbf{P}_k \boldsymbol{x}_k)^\top & 1 \end{bmatrix} \begin{bmatrix} \mathbf{A}_k & \mathbf{P}_k\, \boldsymbol{a}_k \\ \boldsymbol{a}_k^\top \mathbf{P}_k & 1 \end{bmatrix}$$

$$= \begin{bmatrix} \mathbf{L}_k \mathbf{A}_k\,, & \mathbf{L}_k \mathbf{P}_k\, \boldsymbol{a}_k \\ -(\mathbf{P}_k \boldsymbol{x}_k)^\top \mathbf{A}_k + \boldsymbol{a}_k^\top \mathbf{P}_k\,, & -(\mathbf{P}_k \boldsymbol{x}_k)^\top \mathbf{P}_k\, \boldsymbol{a}_k + 1 \end{bmatrix}.$$

It is straightforward to verify that $[-(\mathbf{P}_k \boldsymbol{x}_k)^\top \mathbf{A}_k + \boldsymbol{a}_k^\top \mathbf{P}_k, -(\mathbf{P}_k \boldsymbol{x}_k)^\top \mathbf{P}_k\, \boldsymbol{a}_k + 1] = [\mathbf{0}_k^\top, \beta_k]$, yielding the recursion

$$\mathbf{L}_{k+1}\mathbf{A}_{k+1} = \begin{bmatrix} \mathbf{L}_k \mathbf{A}_k & \mathbf{L}_k \mathbf{P}_k\, \boldsymbol{a}_k \\ \mathbf{0}_k^\top & \beta_k \end{bmatrix}.$$

Secondly, observe that

$$\mathbf{L}_{k+1}\mathbf{A}_{k+1}\mathbf{L}_{k+1}^\top = \begin{bmatrix} \mathbf{L}_k \mathbf{A}_k & \mathbf{L}_k \mathbf{P}_k\, \boldsymbol{a}_k \\ \mathbf{0}_k^\top & \beta_k \end{bmatrix} \begin{bmatrix} \mathbf{L}_k^\top & -\mathbf{P}_k \boldsymbol{x}_k \\ \mathbf{0}_k^\top & 1 \end{bmatrix}$$

$$= \begin{bmatrix} \mathbf{L}_k \mathbf{A}_k \mathbf{L}_k^\top\,, & -\mathbf{L}_k \mathbf{A}_k \mathbf{P}_k \boldsymbol{x}_k + \mathbf{L}_k \mathbf{P}_k \boldsymbol{a}_k \\ \mathbf{0}_k^\top\,, & \beta_k \end{bmatrix}.$$

By noting that $\mathbf{A}_k \mathbf{P}_k \boldsymbol{x}_k = \mathbf{P}_k \mathbf{P}_k \mathbf{A}_k \mathbf{P}_k \boldsymbol{x}_k = \mathbf{P}_k \mathbf{A}_k \boldsymbol{x}_k = \mathbf{P}_k \boldsymbol{a}_k$, we obtain:

$$\mathbf{L}_{k+1}\mathbf{A}_{k+1}\mathbf{L}_{k+1}^\top = \begin{bmatrix} \mathbf{L}_k \mathbf{A}_k \mathbf{L}_k^\top & \mathbf{0}_k \\ \mathbf{0}_k^\top & \beta_k \end{bmatrix}.$$

Hence, the result follows by induction. □

Example A.91 (Solving $\mathbf{A}_n \boldsymbol{x}_n = \boldsymbol{b}_n$ in $O(n^2)$ Time) One application of the factorization in Theorem A.90 is in the fast solution of a linear system $\mathbf{A}_n \boldsymbol{x}_n = \boldsymbol{b}_n$, where the right-hand side is an arbitrary vector $\boldsymbol{b}_n$. Since the solution $\boldsymbol{x}_n$ can be written as

$$\boldsymbol{x}_n = \mathbf{A}_n^{-1} \boldsymbol{b}_n = \mathbf{L}_n^\top \mathbf{D}_n^{-1} \mathbf{L}_n \boldsymbol{b}_n,$$

we can compute $\boldsymbol{x}_n$ in $O(n^2)$ time, as follows:

Algorithm A.92: Solving $\mathbf{A}_n\, \boldsymbol{x}_n = \boldsymbol{b}_n$ for a General Right-Hand Side

Input: First row $[1, \boldsymbol{a}_{n-1}^\top]$ of matrix $\mathbf{A}_n$ and right-hand side $\boldsymbol{b}_n$.
Output: Solution $\boldsymbol{x}_n = \mathbf{A}_n^{-1}\, \boldsymbol{b}_n$.

Compute $\mathbf{L}_n$ in (A.89) and the numbers $\beta_1, \ldots, \beta_{n-1}$ via Algorithm A.88.
$[x_1, \ldots, x_n]^\top \leftarrow \mathbf{L}_n \boldsymbol{b}_n$ // computed in $O(n^2)$ time
$x_i \leftarrow x_i / \beta_{i-1}$ for $i = 2, \ldots, n$ // computed in $O(n)$ time
$[x_1, \ldots, x_n] \leftarrow [x_1, \ldots, x_n]\, \mathbf{L}_n$ // computed in $O(n^2)$ time
return $\boldsymbol{x}_n \leftarrow [x_1, \ldots, x_n]^\top$

Note that it is possible to avoid the explicit construction of the lower triangular matrix in (A.89) via the following modification of Algorithm A.88, which only stores an extra vector $\boldsymbol{y}$ at each recursive step of the Levinson–Durbin algorithm.

Algorithm A.93: Solving $\mathbf{A}_n \boldsymbol{x}_n = \boldsymbol{b}_n$ with $O(n)$ Memory Cost

Input: First row $[1, \boldsymbol{a}_{n-1}^\top]$ of matrix $\mathbf{A}_n$ and right-hand side $\boldsymbol{b}_n$.
Output: Solution $\boldsymbol{x}_n = \mathbf{A}_n^{-1}\, \boldsymbol{b}_n$.

$\boldsymbol{x} \leftarrow b_1$
$\boldsymbol{y} \leftarrow a_1$
for $k = 1, \dots, n-1$ **do**
  $\breve{\boldsymbol{x}} \leftarrow [x_k, x_{k-1}, \dots, x_1]$
  $\breve{\boldsymbol{y}} \leftarrow [y_k, y_{k-1}, \dots, y_1]$
  $\beta \leftarrow 1 - \boldsymbol{a}_k^\top \boldsymbol{y}$
  $\alpha_x \leftarrow (b_{k+1} - \boldsymbol{b}_k^\top \breve{\boldsymbol{x}})/\beta$
  $\alpha_y \leftarrow (a_{k+1} - \boldsymbol{a}_k^\top \breve{\boldsymbol{y}})/\beta$
  $\boldsymbol{x} \leftarrow [\boldsymbol{x} - \alpha_x \breve{\boldsymbol{x}}, \alpha_x]^\top$
  $\boldsymbol{y} \leftarrow [\boldsymbol{y} - \alpha_y \breve{\boldsymbol{y}}, \alpha_y]^\top$
return $\boldsymbol{x}$

A.8.5 Hessenberg–Arnoldi and Lanczos Decompositions

For a general positive definite matrix $\mathbf{A} \in \mathbf{R}^{n\times n}$, performing matrix-vector multiplications costs $O(n^2)$ operations; and solving linear equations via the Cholesky decomposition cost $O(n^3)$ operations. However, when $\mathbf{A}$ is *sparse* (that is, has relatively few non-zero elements) or is circulant, then the computational cost of matrix-vector multiplications can be significantly lower. The Lancsoz and conjugate gradient algorithms derived in this section are especially suitable for matrices with such structures. To derive these algorithms, we begin with a description of the more fundamental Hessenberg–Arnoldi decomposition.

A matrix $\mathbf{H}_n = [h_{ij}] \in \mathbb{R}^{n\times n}$ is called a **Hessenberg matrix** if it has zero entries below the first subdiagonal, that is, $h_{ij} = 0$ for all $i > j+1$. For example, we have

$$\mathbf{H}_4 = \begin{bmatrix} h_{11} & h_{12} & h_{13} & h_{14} \\ h_{21} & h_{22} & h_{23} & h_{24} \\ 0 & h_{32} & h_{33} & h_{34} \\ 0 & 0 & h_{43} & h_{14} \end{bmatrix}.$$

Suppose that $\mathbf{A} \in \mathbb{R}^{n\times n}$ is a square matrix and $\boldsymbol{b} \in \mathbb{R}^n$ is an arbitrary vector. The k-dimensional **Krylov subspace** of $\mathbf{A}$ that is generated by $\boldsymbol{b}$ is defined as:

$$\mathcal{K}_k(\mathbf{A}, \boldsymbol{b}) := \operatorname{Span}\left\{\boldsymbol{b}, \mathbf{A}\boldsymbol{b}, \mathbf{A}^2\boldsymbol{b}, \dots, \mathbf{A}^{k-1}\boldsymbol{b}\right\}.$$

Define $\boldsymbol{e}_k$ to be the unit vector of appropriate size with the k-th entry equal to one. Then, we have the following result.

Theorem A.94 (Hessenberg–Arnoldi Decomposition): Suppose that the Krylov subspace $\mathcal{K}_n(\mathbf{A}, \boldsymbol{b})$ spans $\mathbb{R}^n$ and that the columns of $\mathbf{U}_k := [\boldsymbol{u}_1, \dots, \boldsymbol{u}_k]$ form an orthonormal basis for $\mathcal{K}_k(\mathbf{A}, \boldsymbol{b})$ for each $k \le n$. Then, for each $k \le n$ there is a

Hessenberg matrix $\mathbf{H}_k$ such that ($h_{n+1,n} := 0$):

$$\text{(A.95)} \qquad \begin{aligned} \mathbf{A}\mathbf{U}_k &= \mathbf{U}_k\mathbf{H}_k + h_{k+1,k}\,\boldsymbol{u}_{k+1}\boldsymbol{e}_k^\top \\ \boldsymbol{u}_{k+1} &\propto (\mathbf{I}_n - \mathbf{U}_k\mathbf{U}_k^\top)\mathbf{A}\boldsymbol{u}_k. \end{aligned}$$

Proof. We proceed by induction. For $k = 1$ the identity (A.95) is equivalent to $\mathbf{A}\boldsymbol{u}_1 = \boldsymbol{u}_1\,h_{1,1} + h_{2,1}\boldsymbol{u}_2$ and $\boldsymbol{u}_2 \propto (\mathbf{I} - \boldsymbol{u}_1\boldsymbol{u}_1^\top)\mathbf{A}\boldsymbol{u}_1$. Direct substitution shows that the last identity is satisfied for $h_{1,1} := \boldsymbol{u}_1^\top\mathbf{A}\boldsymbol{u}_1$, which is a 1×1 Hessenberg matrix, $h_{2,1} := \|(\mathbf{I} - \boldsymbol{u}_1\boldsymbol{u}_1^\top)\mathbf{A}\boldsymbol{u}_1\|$, and the pair of orthonormal vectors $\boldsymbol{u}_1 := \boldsymbol{b}/\|\boldsymbol{b}\|$ and $\boldsymbol{u}_2 := (\mathbf{I} - \boldsymbol{u}_1\boldsymbol{u}_1^\top)\mathbf{A}\boldsymbol{u}_1/h_{2,1}$. The set $\{\boldsymbol{u}_1, \boldsymbol{u}_2\}$ is an orthonormal basis for the column space of $[\boldsymbol{b}, \mathbf{A}\boldsymbol{b}]$. Thus, (A.95) is verified for $k = 1$.

Now, assume that $\{\boldsymbol{u}_1, \ldots, \boldsymbol{u}_i\}$ is an orthonormal basis for $\mathcal{K}_i(\mathbf{A}, \boldsymbol{b})$ and the identity (A.95) is valid for $k = i - 1$. For $k = i$ we have to verify that there exists a Hessenberg matrix $\mathbf{H}_i$ and a $\boldsymbol{u}_{i+1} \propto (\mathbf{I}_n - \mathbf{U}_i\mathbf{U}_i^\top)\mathbf{A}\boldsymbol{u}_i$ such that

$$\begin{aligned} \mathbf{A}\mathbf{U}_i = \begin{bmatrix}\mathbf{A}\mathbf{U}_{i-1} & \mathbf{A}\boldsymbol{u}_i\end{bmatrix} &= \begin{bmatrix}\mathbf{U}_{i-1}\mathbf{H}_{i-1} + h_{i,i-1}\,\boldsymbol{u}_i\boldsymbol{e}_{i-1}^\top & \mathbf{A}\boldsymbol{u}_i\end{bmatrix} \\ &= \mathbf{U}_i \underbrace{\left[\begin{array}{c|c} \mathbf{H}_{i-1} & \boldsymbol{h}_i \\ h_{i,i-1}\,\boldsymbol{e}_{i-1}^\top & \end{array}\right]}_{=\,\mathbf{H}_i} + h_{i+1,i}\,\boldsymbol{u}_{i+1}\,\boldsymbol{e}_i^\top. \end{aligned}$$

In other words, we need to verify the existence of $\boldsymbol{h}_i \in \mathbb{R}^i$ and $h_{i+1,i}$ such that

$$\mathbf{A}\boldsymbol{u}_i = \mathbf{U}_i\boldsymbol{h}_i + h_{i+1,i}\,\boldsymbol{u}_{i+1}.$$

Direct substitution confirms that the last identity is satisfied for $\boldsymbol{h}_i := \mathbf{U}_i^\top\mathbf{A}\boldsymbol{u}_i$, $h_{i+1,i} := \|(\mathbf{I}_n - \mathbf{U}_i\mathbf{U}_i^\top)\mathbf{A}\boldsymbol{u}_i\|$, and $\boldsymbol{u}_{i+1} := (\mathbf{I}_n - \mathbf{U}_i\mathbf{U}_i^\top)\mathbf{A}\boldsymbol{u}_i/h_{i+1,i}$. It remains to verify that the orthonormal basis $\{\boldsymbol{u}_1, \ldots, \boldsymbol{u}_{i+1}\}$ spans $\mathcal{K}_{i+1}(\mathbf{A}, \boldsymbol{b})$, or equivalently that $\boldsymbol{u}_{i+1} \propto (\mathbf{I}_n - \mathbf{U}_i\mathbf{U}_i^\top)\mathbf{A}^i\boldsymbol{b}$. By assumption $\boldsymbol{u}_i \propto (\mathbf{I}_n - \mathbf{U}_{i-1}\mathbf{U}_{i-1}^\top)\mathbf{A}\boldsymbol{u}_{i-1}$, which implies

$$\begin{aligned} \boldsymbol{u}_{i+1} &\propto (\mathbf{I}_n - \mathbf{U}_i\mathbf{U}_i^\top)\mathbf{A}(\mathbf{I}_n - \mathbf{U}_{i-1}\mathbf{U}_{i-1}^\top)\mathbf{A}\boldsymbol{u}_{i-1} \\ &\propto (\mathbf{I}_n - \mathbf{U}_i\mathbf{U}_i^\top)\mathbf{A}^2\boldsymbol{u}_{i-1} - \underbrace{(\mathbf{I}_n - \mathbf{U}_i\mathbf{U}_i^\top)\mathbf{U}_{i-1}\boldsymbol{h}_{i-1}}_{=\,\mathbf{0}}. \end{aligned}$$

By repeating the last computation $j \leq i$ times we deduce that

$$\boldsymbol{u}_{i+1} \propto (\mathbf{I}_n - \mathbf{U}_i\mathbf{U}_i^\top)\mathbf{A}^j\boldsymbol{u}_{i-j+1}, \quad \forall j \leq i.$$

Thus, (A.95) is valid for $k = i$, which completes the inductive proof. □

A direct consequence of (A.95) is that $\mathbf{U}_k^\top\mathbf{A}\mathbf{U}_k = \mathbf{H}_k$ for each $k \leq n$. Further, if $\mathbf{A}$ is symmetric, then $\mathbf{H}_k = \mathbf{H}_k^\top$ implies that all Hessenberg matrices are tridiagonal. In other words, for any symmetric matrix $\mathbf{A} \in \mathbb{R}^{n\times n}$ with $\mathcal{K}_n(\mathbf{A}, \boldsymbol{b})$ spanning $\mathbb{R}^n$, we obtain the **Lanczos decomposition**:

$$\text{(A.96)} \qquad \mathbf{A}\mathbf{U}_k = \mathbf{U}_k\mathbf{T}_k + t_{k+1,k}\,\boldsymbol{u}_{k+1}\boldsymbol{e}_k^\top, \quad k \leq n,$$

where $\mathbf{T}_k$ is a symmetric invertible tridiagonal matrix and $t_{n+1,n} := 0$. The computation of the Lanczos decomposition with $\{\boldsymbol{u}_1, \ldots, \boldsymbol{u}_k\}$ spanning the Krylov subspace $\mathcal{K}_k(\mathbf{A}, \boldsymbol{b})$ is summarized in the following algorithm.

Algorithm A.97: Lanczos Decomposition: $\mathbf{A}\mathbf{U}_k = \mathbf{U}_k\mathbf{T}_k + t_{k+1,k}\,\boldsymbol{u}_{k+1}\boldsymbol{e}_k^\top$

Input: A method to multiply the symmetric $\mathbf{A} \in \mathbb{R}^{n\times n}$ with a vector, $\boldsymbol{b}$, $k \leq n$.
Output: Matrix $\mathbf{U}_{k+1}$, the diagonal of $\mathbf{T}_k$ and subdiagonal of $\mathbf{T}_{k+1}$

$\boldsymbol{u}_1 \leftarrow \boldsymbol{b}/\|\boldsymbol{b}\|$ // initial basis vector in Gram-Schmidt process
for $i = 1, \ldots, k$ **do**
  $\boldsymbol{u} \leftarrow \mathbf{A}\boldsymbol{u}_i$ // matrix-vector computation costs at most $O(n^2)$
  $\boldsymbol{h} \leftarrow \mathbf{U}_i^\top \boldsymbol{u}$ // Gram-Schmidt process costs $O(k^2 n)$
  $\boldsymbol{u} \leftarrow \boldsymbol{u} - \mathbf{U}_i\boldsymbol{h}$ // make $\boldsymbol{u}_{i+1}$ orthogonal to past basis vectors via Gram-Schmidt
  $t_{i,i} \leftarrow h_i$
  $t_{i+1,i} \leftarrow \|\boldsymbol{u}\|$
  $\boldsymbol{u} \leftarrow \boldsymbol{u}/t_{i+1,i}$
  $\mathbf{U}_{i+1} \leftarrow [\mathbf{U}_i, \boldsymbol{u}]$ // store $\boldsymbol{u}_{i+1}$ as the last column of $\mathbf{U}_{i+1}$
return $\mathbf{U}_{k+1}$, $[t_{1,1}, \ldots, t_{k,k}]^\top$, $[t_{2,1}, \ldots, t_{k+1,k}]^\top$

In the next section we provide an application of this decomposition to the solution of linear equations.

A.8.5.1 The Lanczos Algorithm for Linear Equations

Henceforth, we assume that $\mathbf{A} \in \mathbb{R}^{n\times n}$ is a positive definite matrix, which allows us to define the Mahalanobis inner product $\langle \boldsymbol{x}, \boldsymbol{y}\rangle_{\mathbf{A}} := (\mathbf{A}^{1/2}\boldsymbol{x})^\top(\mathbf{A}^{1/2}\boldsymbol{y})$ with corresponding Mahalanobis norm $\|\boldsymbol{x}\|_{\mathbf{A}}^2 := \boldsymbol{x}^\top\mathbf{A}\boldsymbol{x}$. In addition, define the quadratic function

$$f(\boldsymbol{x}) := \|\mathbf{A}^{-1}\boldsymbol{b} - \boldsymbol{x}\|_{\mathbf{A}}^2 = \boldsymbol{x}^\top\mathbf{A}\boldsymbol{x} - 2\boldsymbol{x}^\top\boldsymbol{b} + \boldsymbol{b}^\top\mathbf{A}^{-1}\boldsymbol{b},$$

and note that its unique global minimizer is $\boldsymbol{x}^* = \mathbf{A}^{-1}\boldsymbol{b}$ with $f(\boldsymbol{x}^*) = 0$.

The properties of f suggest that one way to compute the solution $\boldsymbol{x}^*$ to the linear system $\mathbf{A}\boldsymbol{x} = \boldsymbol{b}$ is via numerical minimization of f. If we have a good initial guess for the solution $\boldsymbol{x}^*$, say $\boldsymbol{x}_0$ with corresponding residual $\boldsymbol{r}_0 := \boldsymbol{b} - \mathbf{A}\boldsymbol{x}_0$, then we can improve on $\boldsymbol{x}_0$ by minimizing f over the linear subspace $\{\boldsymbol{x}_0 + \mathbf{U}_k\boldsymbol{\alpha},\ \boldsymbol{\alpha} \in \mathbb{R}^k\}$, where $\mathbf{U}_k$ is a judiciously chosen orthonormal matrix $\mathbf{U}_k \in \mathbb{R}^k$, $k \leq n$. Since

$$\operatorname*{argmin}_{\boldsymbol{\alpha}} f(\boldsymbol{x}_0 + \mathbf{U}_k\boldsymbol{\alpha}) = \operatorname*{argmin}_{\boldsymbol{\alpha}} \|\mathbf{A}^{-1/2}\boldsymbol{r}_0 - \mathbf{A}^{1/2}\mathbf{U}_k\boldsymbol{\alpha}\|^2 = (\mathbf{A}^{1/2}\mathbf{U}_k)^+\mathbf{A}^{-1/2}\boldsymbol{r}_0,$$

the minimizer $\boldsymbol{x}_k$ of f over the subspace $\{\boldsymbol{x}_0 + \mathbf{U}_k\boldsymbol{\alpha},\ \boldsymbol{\alpha} \in \mathbb{R}^k\}$ of $\mathbb{R}^n$ is:

$$\boldsymbol{x}_k = \boldsymbol{x}_0 + \mathbf{U}_k[\mathbf{U}_k^\top\mathbf{A}\mathbf{U}_k]^{-1}\mathbf{U}_k^\top\boldsymbol{r}_0.$$

The **Lanczos algorithm** chooses $\{\boldsymbol{u}_1, \ldots, \boldsymbol{u}_k\}$ to be the orthonormal basis that spans the Krylov subspace $\mathcal{K}_k(\mathbf{A}, \boldsymbol{r}_0)$. In view of the symmetry of $\mathbf{A}$ and Theorem A.94, this choice for $\mathbf{U}_k$ yields

$$\boldsymbol{x}_k = \boldsymbol{x}_0 + \mathbf{U}_k\mathbf{T}_k^{-1}\mathbf{U}_k^\top\boldsymbol{r}_0, \quad k = 1, 2, \ldots, \tag{A.98}$$

where by construction $\mathbf{U}_k^\top \boldsymbol{r}_0 = \boldsymbol{e}_1\|\boldsymbol{r}_0\|$ and $\mathbf{T}_k^{-1}$ is computed in $O(k)$ time via Algorithm A.82. How close $\boldsymbol{x}_k$ is to solving the linear system $\mathbf{A}\boldsymbol{x} = \boldsymbol{b}$ can be assessed via the norm of the k-th residual vector:

$$\begin{aligned} \boldsymbol{r}_k := \boldsymbol{b} - \mathbf{A}\boldsymbol{x}_k &= (\mathbf{I}_n - \mathbf{A}\mathbf{U}_k\mathbf{T}_k^{-1}\mathbf{U}_k^\top)\boldsymbol{r}_0 \\ &= (\mathbf{I}_n - (\mathbf{U}_k\mathbf{T}_k + t_{k+1,k}\boldsymbol{u}_{k+1}\boldsymbol{e}_k^\top)\mathbf{T}_k^{-1}\mathbf{U}_k^\top)\boldsymbol{r}_0 \\ &= \underbrace{(\mathbf{I}_n - \mathbf{U}_k\mathbf{U}_k^\top)\boldsymbol{r}_0}_{=\mathbf{0}} - (t_{k+1,k}\,\|\boldsymbol{r}_0\|\,\boldsymbol{e}_k^\top\mathbf{T}_k^{-1}\boldsymbol{e}_1)\,\boldsymbol{u}_{k+1}. \end{aligned} \tag{A.99}$$

Typically, the term $\boldsymbol{e}_k^\top\mathbf{T}_k^{-1}\boldsymbol{e}_1$ rapidly decays to zero as a function of $k < n$, making the norms of the residuals $\|\boldsymbol{r}_k\|^2 = \|\boldsymbol{r}_0\|^2\, t_{k+1,k}^2(\boldsymbol{e}_k^\top\mathbf{T}_k^{-1}\boldsymbol{e}_1)^2$ and

$$\|\boldsymbol{x}_k - \boldsymbol{x}^*\|_{\mathbf{A}} = \boldsymbol{r}_k^\top\mathbf{A}^{-1}\boldsymbol{r}_k = \boldsymbol{r}_k^\top\mathbf{U}_n\mathbf{T}_n^{-1}\mathbf{U}_n^\top\boldsymbol{r}_k = \|\boldsymbol{r}_k\|^2\,\boldsymbol{e}_{k+1}^\top\mathbf{T}_n^{-1}\boldsymbol{e}_{k+1}$$

so small that the approximation $\boldsymbol{x}_k$ may be acceptable as a numerical solution to the linear system $\mathbf{A}\boldsymbol{x} = \boldsymbol{b}$ even when $k \ll n$. More precisely, it can be proved [150] that the following error bound holds:

$$\frac{\|\boldsymbol{x}_k - \boldsymbol{x}^*\|_{\mathbf{A}}}{\|\boldsymbol{x}_0 - \boldsymbol{x}^*\|_{\mathbf{A}}} \leq 2\left[\frac{\kappa(\mathbf{A}) - 1}{\kappa(\mathbf{A}) + 1}\right]^k, \tag{A.100}$$

where $\kappa(\mathbf{A}) \geq 1$ is the (Euclidean) condition number of $\mathbf{A}$; see Example A.48. Since $\mathbf{A}$ is a symmetric positive definite matrix, it has n real positive eigenvalues $\lambda_1 \geq \cdots \geq \lambda_n > 0$, and therefore the condition number $\kappa(\mathbf{A}) = \sqrt{\lambda_1/\lambda_n}$.

Lastly, observe that $(x-1)/(x+1)$ is an increasing function for $x \geq 1$, and therefore (A.100) implies that the smaller the condition number $\kappa(\mathbf{A})$, the more rapid is the convergence of the approximation (A.98) to $\boldsymbol{x}^*$.

■ **Example A.101 (Regularization and the Lanczos Algorithm)** Suppose that we wish to compute the family of solutions $\{\boldsymbol{x}^*(\alpha) = (\mathbf{A} + \alpha\mathbf{I}_n)^{-1}\boldsymbol{b},\ \alpha \geq 0\}$. Given the Lanczos decomposition (A.96) of $\mathbf{A}$, the Lanczos decomposition of $\mathbf{A} + \alpha\mathbf{I}_n$ for any valid α is:

$$(\mathbf{A} + \alpha\mathbf{I}_n)\mathbf{U}_k = \mathbf{U}_k(\mathbf{T}_k + \alpha\mathbf{I}_k) + t_{k+1,k}\,\boldsymbol{u}_{k+1}\,\boldsymbol{e}_k^\top, \quad k \leq n.$$

Hence, from (A.98) the approximate solution to $\boldsymbol{x}^*(\alpha)$ with $\boldsymbol{x}_0 = \mathbf{0}$ is

$$\begin{aligned} \boldsymbol{x}_k(\alpha) &= \boldsymbol{x}_0 + \mathbf{U}_k(\mathbf{T}_k + \alpha\mathbf{I}_k)^{-1}\mathbf{U}_k^\top\boldsymbol{r}_0 \\ &= \mathbf{U}_k(\mathbf{T}_k + \alpha\mathbf{I}_k)^{-1}\mathbf{U}_k^\top\boldsymbol{b}. \end{aligned}$$

Assuming that

$$k > \frac{\ln(\varepsilon/2)}{\ln\frac{\kappa(\mathbf{A})-1}{\kappa(\mathbf{A})+1}},$$

then by (A.100) the relative error

$$\frac{\|\boldsymbol{x}_k(0) - \boldsymbol{x}^*(0)\|_{\mathbf{A}}}{\|\boldsymbol{x}^*(0)\|_{\mathbf{A}}} < \varepsilon.$$

From the fact that

$$\kappa(\mathbf{A} + \alpha \mathbf{I}_n) \le \kappa(\mathbf{A}), \quad \forall \alpha \ge 0,$$

we thus deduce that with the same number of steps k the worst-case relative error is

$$\sup_{\alpha \ge 0} \frac{\|\boldsymbol{x}_k(\alpha) - \boldsymbol{x}^*(\alpha)\|_{\mathbf{A}+\alpha\mathbf{I}_n}}{\|\boldsymbol{x}^*(\alpha)\|_{\mathbf{A}+\alpha\mathbf{I}_n}} < \varepsilon.$$

In other words, the same k-step Lanczos decomposition of $\mathbf{A}$ can be used to compute the entire family of solutions $\{\boldsymbol{x}^*(\alpha), \alpha \ge 0\}$ with the same relative error accuracy.

■

As seen from line 3 of Algorithm A.97, the Lanczos algorithm only requires the multiplication of $\mathbf{A}$ with a vector k times and the solution of the tridiagonal linear system $\mathbf{T}_k^{-1}\boldsymbol{e}_1$.

Whenever the approximation (A.98) is accurate for $k \ll n$, one can numerically solve the linear system $\mathbf{A}\boldsymbol{x} = \boldsymbol{b}$ for a cost of at most $O(k\,n^2)$. This cost can be significantly better than the usual cost of $O(n^3)$ using the LU decomposition in Section A.8.1.

The advantage of the Lanczos algorithm becomes even greater when multiplications of $\mathbf{A}$ with a vector are cheaper than the usual $O(n^2)$, as is the case when $\mathbf{A}$ is a sparse matrix, or has some special Toeplitz structure, as in Example B.33.

A.8.5.2 Conjugate Gradient Algorithm

Recall from (A.98) and (A.99) that $\boldsymbol{x}_k = \boldsymbol{x}_0 + \mathbf{U}_k\mathbf{T}_k^{-1}\mathbf{U}_k^\top\boldsymbol{r}_0$ and $\boldsymbol{r}_k = \boldsymbol{b} - \mathbf{A}\boldsymbol{x}_k$. Let $\mathbf{L}_k\mathbf{D}_k^2\mathbf{L}_k^\top$ be the LDL decomposition of the tridiagonal $\mathbf{T}_k$ (computed, for example, by Algorithm A.81), and define the matrix:

$$\mathbf{P}_k = [\boldsymbol{p}_1, \ldots, \boldsymbol{p}_k] := \mathbf{U}_k\mathbf{L}_k^{-\top}\mathbf{D}_k^{-1}.$$

Theorem A.102 (Conjugacy and Orthogonal Bases): The three sets of vectors $\{\boldsymbol{p}_1, \ldots, \boldsymbol{p}_k\}$, $\{\boldsymbol{r}_0, \ldots, \boldsymbol{r}_k\}$, and $\{\boldsymbol{x}_0, \ldots, \boldsymbol{x}_k\}$ satisfy the following for $k \ge 1$:

1. $\{\boldsymbol{r}_0, \ldots, \boldsymbol{r}_k\}$ is an orthogonal basis for $\mathcal{K}_{k+1}(\mathbf{A}, \boldsymbol{r}_0)$ with $\langle \boldsymbol{r}_i, \boldsymbol{r}_j \rangle = \mathbb{1}_{\{i=j\}}$;
2. $\{\boldsymbol{p}_1, \ldots, \boldsymbol{p}_k\}$ is an orthonormal basis for $\mathcal{K}_k(\mathbf{A}, \boldsymbol{r}_0)$ with $\langle \boldsymbol{p}_i, \boldsymbol{p}_j \rangle_{\mathbf{A}} = \mathbb{1}_{\{i=j\}}$;
3. $\boldsymbol{x}_k = \boldsymbol{x}_{k-1} + \alpha_k\boldsymbol{p}_k$ and $\boldsymbol{r}_k = \boldsymbol{r}_{k-1} - \alpha_k\mathbf{A}\boldsymbol{p}_k$ with $\alpha_k := \langle \boldsymbol{r}_0, \boldsymbol{p}_k \rangle$;
4. for some scalars β_k, $\gamma_{k+1} := \|\boldsymbol{r}_k + \beta_k\,\boldsymbol{p}_k\|_{\mathbf{A}}$, and $\gamma_1 := \|\boldsymbol{r}_0\|_{\mathbf{A}}$, we have

$$\boldsymbol{p}_1 = \frac{\boldsymbol{r}_0}{\gamma_1}, \quad \boldsymbol{p}_{k+1} = \frac{\boldsymbol{r}_k + \beta_k\,\boldsymbol{p}_k}{\gamma_{k+1}}, \quad k \ge 1.$$

Proof. We prove each property in turn.

1. The identity (A.99) implies that $\boldsymbol{r}_j \propto \boldsymbol{u}_{j+1}$ for all $j \leq k$, and hence the set of residual vectors $\{\boldsymbol{r}_0, \ldots, \boldsymbol{r}_k\}$ span $\mathcal{K}_{k+1}(\mathbf{A}, \boldsymbol{r}_0)$ and are orthogonal to each other with respect to the usual inner product: $\langle \boldsymbol{r}_i, \boldsymbol{r}_j \rangle = \boldsymbol{r}_i^\top \boldsymbol{r}_j$.

2. Since the columns of $\mathbf{P}_k$ are computed from an invertible linear transformation of $\mathbf{U}_k$, they span the Krylov subspace $\mathcal{K}_k(\mathbf{A}, \boldsymbol{r}_0)$. Moreover, since

$$\mathbf{P}_k^\top \mathbf{A} \mathbf{P}_k = \mathbf{D}_k^{-1} \mathbf{L}_k^{-1} \overbrace{\mathbf{U}_k^\top \mathbf{A} \mathbf{U}_k}^{\mathbf{T}_k} \mathbf{L}_k^{-\top} \mathbf{D}_k^{-1} = \mathbf{I}_k,$$

the $\{\boldsymbol{p}_k\}$ are orthonormal with respect to the inner product: $\langle \boldsymbol{p}_i, \boldsymbol{p}_j \rangle_{\mathbf{A}} = \mathbb{1}_{\{i=j\}}$.

3. From (A.98) and $\mathbf{U}_k = \mathbf{L}_k^\top \mathbf{D}_k \mathbf{P}_k$ we deduce that for $k = 1, 2, \ldots$

$$\begin{aligned} \boldsymbol{x}_k &= \boldsymbol{x}_0 + \mathbf{U}_k \mathbf{T}_k^{-1} \mathbf{U}_k^\top \boldsymbol{r}_0 \\ &= \boldsymbol{x}_0 + \mathbf{P}_k \mathbf{P}_k^\top \boldsymbol{r}_0 \\ &= \boldsymbol{x}_0 + (\mathbf{P}_{k-1} \mathbf{P}_{k-1}^\top + \boldsymbol{p}_k \boldsymbol{p}_k^\top) \boldsymbol{r}_0 \\ &= \boldsymbol{x}_{k-1} + \alpha_k \, \boldsymbol{p}_k. \end{aligned}$$

In other words, the $\boldsymbol{x}_k$ can be written solely as a linear combination of $\boldsymbol{x}_{k-1}$ and $\boldsymbol{p}_k$. A consequence of this is that $\boldsymbol{r}_k = \boldsymbol{r}_{k-1} - \alpha_k \mathbf{A} \boldsymbol{p}_k$ for $k \geq 1$.

4. First, the identity $\mathbf{U}_k = \mathbf{L}_k^\top \mathbf{D}_k \mathbf{P}_k$ is equivalent to writing $\boldsymbol{u}_1 = d_{1,1} \, \boldsymbol{p}_1$ and $\boldsymbol{u}_{i+1} = l_{i+1,i} \, d_{i,i} \, \boldsymbol{p}_i + d_{i+1,i+1} \, \boldsymbol{p}_{i+1}$ for all $i = 1, \ldots, k-1$. Second, from (A.99) we conclude that $\boldsymbol{p}_1 = \boldsymbol{r}_0 / \gamma_1$ and $\boldsymbol{p}_{i+1} = (\boldsymbol{r}_i + \beta_i \, \boldsymbol{p}_i) / \gamma_{i+1}$ for $i \geq 1$ and some scalars $\{\beta_i\}$ and $\{\gamma_i\}$. Since each $\boldsymbol{p}_i$ is normalized such that $\|\boldsymbol{p}_i\|_{\mathbf{A}} = 1$, we must have that $\gamma_1 = \|\boldsymbol{r}_0\|_{\mathbf{A}}$ and $\gamma_{i+1} = \|\boldsymbol{r}_i + \beta_i \, \boldsymbol{p}_i\|_{\mathbf{A}}$ for $i \geq 1$.

□

There are two consequences of Theorem A.102. First, since $\boldsymbol{p}_k \in \mathcal{K}_k(\mathbf{A}, \boldsymbol{r}_0)$ and each $\boldsymbol{r}_k \propto \boldsymbol{u}_{k+1}$ lies in the space orthogonal to $\mathcal{K}_k(\mathbf{A}, \boldsymbol{r}_0)$, we have that $\langle \boldsymbol{r}_j, \boldsymbol{p}_i \rangle = 0$ for $j \geq i$. Second, the four properties in Theorem A.102 together imply the following formulas:

$$\begin{aligned} \alpha_k &= \langle \boldsymbol{r}_{k-1}, \boldsymbol{p}_k \rangle = \frac{\|\boldsymbol{r}_{k-1}\|^2}{\gamma_k} = -\frac{\|\boldsymbol{r}_k\|^2}{\langle \boldsymbol{r}_k, \boldsymbol{p}_k \rangle_{\mathbf{A}}} \\ \beta_k &= -\langle \boldsymbol{r}_k, \boldsymbol{p}_k \rangle_{\mathbf{A}} = \frac{\gamma_k \, \|\boldsymbol{r}_k\|^2}{\|\boldsymbol{r}_{k-1}\|^2}. \end{aligned}$$

The **conjugate gradient algorithm** exploits these formulas to compute the approximation (A.98) via the recursive equations:

$$\begin{aligned} \boldsymbol{x}_k &= \boldsymbol{x}_{k-1} + \langle \boldsymbol{r}_{k-1}, \boldsymbol{p}_k \rangle \, \boldsymbol{p}_k \\ \boldsymbol{r}_k &= \boldsymbol{r}_{k-1} - \langle \boldsymbol{r}_{k-1}, \boldsymbol{p}_k \rangle \, \mathbf{A} \boldsymbol{p}_k \\ \boldsymbol{p}_{k+1} &\propto \boldsymbol{r}_k - \langle \boldsymbol{r}_k, \boldsymbol{p}_k \rangle_{\mathbf{A}} \, \boldsymbol{p}_k, \end{aligned}$$

where all the $\{\boldsymbol{p}_k\}$ satisfy the normalization condition $\|\boldsymbol{p}_k\|_{\mathbf{A}} = 1$. The following pseudo-code summarizes the conjugate gradient algorithm using a maximum number of iterations M_{iter}.

Algorithm A.103: Conjugate Gradient for Solving $\mathbf{A}x = b$

Input: A routine for evaluating $\boldsymbol{x} \mapsto \mathbf{A}\boldsymbol{x}$; $\boldsymbol{b}$; initial $\boldsymbol{x}$; tolerance ε and M_{iter}.
Output: The approximate solution to $\mathbf{A}\boldsymbol{x} = \boldsymbol{b}$.

1 $k \leftarrow 0,\ \boldsymbol{r} \leftarrow \boldsymbol{b} - \mathbf{A}\boldsymbol{x},\ \tau \leftarrow \|\boldsymbol{r}\|^2,\ \boldsymbol{p} \leftarrow \boldsymbol{r}$ // by construction the initial $\boldsymbol{p}_1 \propto \boldsymbol{r}_0$
2 **while** $\tau > \varepsilon$ and $k < M_{\text{iter}}$ **do**
3 $\quad k \leftarrow k + 1$ // iteration counter
4 $\quad \mathtt{Ap} \leftarrow \mathbf{A}\boldsymbol{p}$ // evaluate using the map $\boldsymbol{x} \mapsto \mathbf{A}\boldsymbol{x}$
5 $\quad \gamma \leftarrow \sqrt{\langle \boldsymbol{p}, \mathtt{Ap} \rangle}$ // normalizing constant
6 $\quad \boldsymbol{p} \leftarrow \boldsymbol{p}/\gamma$ // ensure that $\|\boldsymbol{p}_k\|_{\mathbf{A}} = 1$
7 $\quad \alpha \leftarrow \tau/\gamma^2$
8 $\quad \boldsymbol{x} \leftarrow \boldsymbol{x} + (\alpha\,\gamma)\,\boldsymbol{p}$ // update approximate solution to $\mathbf{A}\boldsymbol{x} = \boldsymbol{b}$
9 $\quad \boldsymbol{r} \leftarrow \boldsymbol{r} - \alpha\,\mathtt{Ap}$ // update residual
10 $\quad \beta \leftarrow \tau,\ \tau \leftarrow \|\boldsymbol{r}\|^2$
11 $\quad \beta \leftarrow \gamma\,\tau/\beta$
12 $\quad \boldsymbol{p} \leftarrow \boldsymbol{r} + \beta\,\boldsymbol{p}$ // update search direction
13 **return** $\boldsymbol{x}$

The normalization of the vector $\boldsymbol{p}$ in line 6 of Algorithm A.103 requires $\mathcal{O}(n)$ number of computations which are not necessary if one is only interested in the approximation to $\mathbf{A}^{-1}\boldsymbol{b}$. Thus, the computing cost can be further reduced by deleting line 6 and replacing lines 8 and 11 with $\boldsymbol{x} \leftarrow \boldsymbol{x} + \alpha\,\boldsymbol{p}$ and $\beta \leftarrow \tau/\beta$, respectively.

■ **Example A.104 (Least-Squares Estimate via Conjugate Gradient)** The conjugate gradient Algorithm A.103 can be used to find a solution to the normal equations (2.22) by setting $\mathbf{A} := \mathbf{X}^\top\mathbf{X}$ and $\boldsymbol{b} := \mathbf{X}^\top\boldsymbol{y}$. Note that the explicit construction of $\mathbf{A}$ is not required, as long as one can evaluate the map $\boldsymbol{x} \mapsto \mathbf{A}\boldsymbol{x}$ via, for example, the two matrix-vector multiplications: $\boldsymbol{x} \leftarrow \mathbf{X}\boldsymbol{x}$, followed by $\boldsymbol{x} \leftarrow \mathbf{X}^\top\boldsymbol{x}$.

If $\mathbf{X}$ is full rank, then Algorithm A.103 will converge to the unique least-squares solution $\mathbf{A}^{-1}\boldsymbol{b} = (\mathbf{X}^\top\mathbf{X})^{-1}\mathbf{X}^\top\boldsymbol{y}$. If, however, $\mathbf{X}$ fails to be full rank, then Algorithm A.103 will converge to the minimum-norm least-squares solution $\mathbf{X}^{+}\boldsymbol{y}$, provided that the algorithm is initialized from the zero vector. In other words, $\boldsymbol{x}_0 = \mathbf{0}$ in (A.98) implies that $\boldsymbol{x}_k$ belongs to both the column space of $\mathbf{A}$ and also the row space of $\mathbf{X}$. Exercise 2.8 shows that the least-squares solution is the minimum-norm one if and only if it lies in the row space of $\mathbf{X}$. ■

If we make all computations with infinite precision, then the k-th iteration of the conjugate gradient Algorithm A.103 will exactly equal the approximation (A.98) in the Lanczos algorithm. However, with finite precision computations in practice, the approximation of the conjugate gradient algorithm differs numerically to (A.98) due to the accumulation of round-off errors in the computation of the search directions $\boldsymbol{p}_1, \boldsymbol{p}_2, \ldots$, resulting in their eventual loss of orthogonality (that is, $\mathbf{P}_k^\top\mathbf{A}\mathbf{P}_k$ does not equal the identity matrix for large k). For this subtle and important issue, we refer the reader to [150].

As with the Lanczos algorithm, the greatest utility of the conjugate gradient Algorithm A.103 is in settings where multiplying $\mathbf{A}$ with a vector is significantly cheaper than the usual $O(n^2)$ cost. For example, when $\mathbf{A}$ is a Toeplitz matrix, each iteration of the conjugate gradient Algorithm A.103 only costs $O(n \ln n)$.

Further Reading

A good reference book on matrix computations is Golub and Van Loan [76]. A useful list of many common vector and matrix calculus identities can be found in [135]. Strang's introduction to linear algebra [163] is a classic textbook, and his recent book [164] combines linear algebra with the foundations of deep learning. Fast reliable algorithms for matrices with structure can be found in [91].

APPENDIX B

Functional Analysis

This appendix reviews a selection of topics from functional analysis, including Hilbert spaces and Fourier transforms. We also provide algorithms and Python code for the fast computation of discrete convolutions and discrete cosine transforms.

B.1 Introduction

Much of the theory on Euclidean vector spaces in Appendix A can be generalized to vector spaces of *functions*. Every element of a (real-valued) **function space** $\mathcal{H}$ is a function from some set $\mathcal{X}$ to $\mathbb{R}$, and elements can be added and scalar multiplied as if they were vectors. In other words, if $f \in \mathcal{H}$ and $g \in \mathcal{H}$, then $\alpha f + \beta g \in \mathcal{H}$ for all $\alpha, \beta \in \mathbb{R}$. On $\mathcal{H}$ we can impose an inner product as a mapping $\langle \cdot, \cdot \rangle$ from $\mathcal{H} \times \mathcal{H}$ to $\mathbb{R}$ that satisfies

1. $\langle \alpha f_1 + \beta f_2, g \rangle = \alpha \langle f_1, g \rangle + \beta \langle f_2, g \rangle$;
2. $\langle f, g \rangle = \langle g, f \rangle$;
3. $\langle f, f \rangle \geq 0$;
4. $\langle f, f \rangle = 0$ if and only if $f = 0$ (the zero function).

We focus on real-valued function spaces, although the theory for complex-valued function spaces is similar (and sometimes easier), under suitable modifications (e.g., $\langle f, g \rangle = \overline{\langle g, f \rangle}$).

Similar to the linear algebra setting in Section A.2, we say that two elements f and g in $\mathcal{H}$ are *orthogonal* to each other with respect to this inner product if $\langle f, g \rangle = 0$. Given an inner product, we can measure distances between elements of the function space $\mathcal{H}$ using the **norm**

$$\|f\| := \sqrt{\langle f, f \rangle}.$$

For example, the distance between two functions f_m and f_n is given by $\|f_m - f_n\|$. The space $\mathcal{H}$ is said to be **complete** if every sequence of functions $f_1, f_2, \ldots \in \mathcal{H}$ for which

$$\|f_m - f_n\| \to 0 \text{ as } m, n \to \infty, \qquad \text{(B.1)}$$

converges to some $f \in \mathcal{H}$; that is, $\|f - f_n\| \to 0$ as $n \to \infty$. A sequence that satisfies (B.1) is called a **Cauchy sequence**.

A complete inner product space is called a **Hilbert space**. The most fundamental Hilbert space of functions is the space L^2. An in-depth introduction to L^2 requires some measure theory [8]. For our purposes, it suffices to assume that $\mathcal{X} \subseteq \mathbb{R}^d$ and that on $\mathcal{X}$ a **measure** μ is defined which assigns to each suitable[1] set A a positive number $\mu(A) \geq 0$ (e.g., its volume). In many cases of interest μ is of the form

$$\mu(A) = \int_A w(\boldsymbol{x})\,\mathrm{d}\boldsymbol{x} \tag{B.2}$$

where $w \geq 0$ is a positive function on $\mathcal{X}$ which is called the **density** of μ with respect to the Lebesgue measure (the natural volume measure on $\mathbb{R}^d$). We write $\mu(\mathrm{d}\boldsymbol{x}) = w(\boldsymbol{x})\,\mathrm{d}\boldsymbol{x}$ to indicate that μ has density w. Another important case is where

$$\mu(A) = \sum_{\boldsymbol{x} \in A \cap \mathbb{Z}^d} w(\boldsymbol{x}), \tag{B.3}$$

where $w \geq 0$ is again called the density of μ, but now with respect to the counting measure on $\mathbb{Z}^d$ (which counts the points of $\mathbb{Z}^d$). Integrals with respect to measures μ in (B.2) and (B.3) can now be defined as

$$\int f(\boldsymbol{x})\,\mu(\mathrm{d}\boldsymbol{x}) = \int f(\boldsymbol{x})\,w(\boldsymbol{x})\,\mathrm{d}\boldsymbol{x},$$

and

$$\int f(\boldsymbol{x})\,\mu(\mathrm{d}\boldsymbol{x}) = \sum_{\boldsymbol{x}} f(\boldsymbol{x})\,w(\boldsymbol{x}),$$

respectively. We assume for simplicity that μ has the form (B.2). For measures of the form (B.3) (so-called discrete measures), replace integrals by sums in what follows.

Definition B.4 (L^2 Space): Let $\mathcal{X}$ be a subset of $\mathbb{R}^d$ with measure $\mu(\mathrm{d}\boldsymbol{x}) = w(\boldsymbol{x})\,\mathrm{d}\boldsymbol{x}$. The Hilbert space $L^2(\mathcal{X}, \mu)$ is the linear space of functions from $\mathcal{X}$ to $\mathbb{R}$ that satisfy

$$\int_{\mathcal{X}} f(\boldsymbol{x})^2\,w(\boldsymbol{x})\,\mathrm{d}\boldsymbol{x} < \infty, \tag{B.5}$$

and with inner product

$$\langle f, g \rangle = \int_{\mathcal{X}} f(\boldsymbol{x})\,g(\boldsymbol{x})\,w(\boldsymbol{x})\,\mathrm{d}\boldsymbol{x}. \tag{B.6}$$

Let $\mathcal{H}$ be a Hilbert space. A set of functions $\{f_i, i \in \mathcal{I}\}$ is called an **orthonormal system** if:

[1]Not all sets have a measure. Suitable sets are Borel sets, which can be thought of as countable unions of rectangles.

1. the norm of every f_i is 1; that is, $\langle f_i, f_i \rangle = 1$ for all $i \in \mathcal{I}$,

2. the $\{f_i\}$ are orthogonal; that is, $\langle f_i, f_j \rangle = 0$ for $i \neq j$.

It follows then that the $\{f_i\}$ are linearly independent; that is, the only linear combination $\sum_j \alpha_j f_j(\boldsymbol{x})$ that is equal to $f_i(\boldsymbol{x})$ for all $\boldsymbol{x}$ is the one where $\alpha_i = 1$ and $\alpha_j = 0$ for $j \neq i$. An orthonormal system $\{f_i\}$ is called an **orthonormal basis** if there is no other $f \in \mathcal{H}$ that is orthogonal to all the $\{f_i, i \in \mathcal{I}\}$ (other than the zero function). Although the general theory allows for uncountable bases, in practice[2] the set $\mathcal{I}$ is taken to be countable.

A Hilbert space $\mathcal{H}$ with an orthonormal basis $\{f_1, f_2, \ldots\}$ behaves very similarly to the familiar Euclidean vector space. In particular, every element (i.e., function) $f \in \mathcal{H}$ can be written as a unique linear combination of the basis vectors:

$$f = \sum_i \langle f, f_i \rangle \, f_i, \tag{B.7}$$

exactly as in Theorem A.13. The right-hand side of (B.7) is called a (generalized) **Fourier expansion** of f. Note that such a Fourier expansion holds for any orthonormal basis.

Example B.8 (Cosine Orthonormal Basis) The functions

$$u_0(x) = 1, \quad u_k(x) = \sqrt{2}\cos(k\pi x), \quad k = 1, 2, \ldots$$

form an orthonormal basis of the space $L^2[0,1]$ comprised of square integrable functions on the interval $[0,1]$. Hence, any function $f \in L^2[0,1]$ can be represented as a cosine series:

$$f(x) = c_0 + \sum_{k=1}^{\infty} c_k \cos(k\pi x),$$

where $c_0 = \int_0^1 f(x)\,\mathrm{d}x$ and $c_k = 2\int_0^1 f(x)\cos(k\pi x)\,\mathrm{d}x$, $k = 1, 2, \ldots$.

Alternatively, if we define $\beta_0 = 1$ and $\beta_k = \sqrt{2}$ for $k \neq 0$, then the collection

$$\{\beta_k \cos(k\pi x)\}_{k \in \mathbb{Z}}$$

forms a *redundant orthonormal set*. Despite the redundancy (since $\cos(k\pi x) = \cos(-k\pi x)$), every function $f \in L^2([0,1])$ can still be represented as

$$f(x) = \sum_{k \in \mathbb{Z}} c_k \beta_k \cos(k\pi x),$$

where the coefficients are given by

$$c_k = \int_0^1 f(x)\,\beta_k \cos(k\pi x)\,\mathrm{d}x.$$

[2]The function spaces typically encountered in machine learning and data science are usually *separable spaces*, which allows for the set $\mathcal{I}$ to be considered countable; see, e.g., [148].

In general, if $\{u_k\}_{k \in \mathbb{N}}$ is an orthonormal basis for $L^2[0, 1]$, then one can construct an orthonormal basis for the space $L^2([0, 1]^d)$ of square-integrable functions on the d-dimensional unit cube, simply by taking the *tensor product* of the $\{u_k\}_{k \in \mathbb{N}}$. Specifically, for each **multi-index** $\boldsymbol{k} = (k_1, \ldots, k_d) \in \mathbb{N}^d$, define the function

$$u_{\mathbf{k}}(x_1, \ldots, x_d) := u_{k_1}(x_1)u_{k_2}(x_2) \cdots u_{k_d}(x_d).$$

The collection $\{u_{\mathbf{k}}, \boldsymbol{k} \in \mathbb{N}^d\}$ forms an orthonormal basis for $L^2([0, 1]^d)$. Thus, each function $f \in L^2([0, 1]^d)$ can be represented as a linear combination of these basis functions. When using the cosine orthonormal basis, one can alternatively represent f using the redundant orthonormal set $\{\beta_{\boldsymbol{k}} \prod_{j=1}^d \cos(\pi k_j x_j),\ \boldsymbol{k} \in \mathbb{Z}^d\}$, where the $\{\beta_{\boldsymbol{k}}\}$ are appropriately chosen normalization constants. ■

■ **Example B.9 (Trigonometric Orthonormal Basis)** Let $\mathcal{H}$ be the Hilbert space $L^2((0, 2\pi), \mu)$, where $\mu(\mathrm{d}x) = w(x)\,\mathrm{d}x$ and w is the constant function $w(x) = 1, 0 < x < 2\pi$. Alternatively, take $\mathcal{X} = \mathbb{R}$ and w the indicator function on $(0, 2\pi)$. The trigonometric functions

$$g_0(x) = \frac{1}{\sqrt{2\pi}}, \quad g_k(x) = \frac{1}{\sqrt{\pi}} \cos(kx), \quad h_k(x) = \frac{1}{\sqrt{\pi}} \sin(kx), \quad k = 1, 2, \ldots$$

form an orthonormal basis of $\mathcal{H}$.

Consider the indicator function $f(x) = \mathbb{1}\{0 < x < \pi\}$. As the trigonometric functions $\{g_k\}$ and $\{h_k\}$ form a basis for $L^2((0, 2\pi), 1\mathrm{d}x)$, we can write

$$f(x) = a_0 \frac{1}{\sqrt{2\pi}} + \sum_{k=1}^{\infty} a_k \frac{1}{\sqrt{\pi}} \cos(kx) + \sum_{k=1}^{\infty} b_k \frac{1}{\sqrt{\pi}} \sin(kx), \tag{B.10}$$

where the parameters are $a_0 = \int_0^\pi 1/\sqrt{2\pi}\,\mathrm{d}x = \sqrt{\pi/2}$, $a_k = \int_0^\pi \cos(kx)/\sqrt{\pi}\,\mathrm{d}x$ and $b_k = \int_0^\pi \sin(kx)/\sqrt{\pi}\,\mathrm{d}x$, $k = 1, 2, \ldots$. This means that $a_k = 0$ for all k, $b_k = 0$ for even k, and $b_k = 2/(k\sqrt{\pi})$ for odd k. Consequently,

$$f(x) = \frac{1}{2} + \frac{2}{\pi} \sum_{k=1}^{\infty} \frac{\sin(kx)}{k}. \tag{B.11}$$

Figure B.12 shows several Fourier *approximations* obtained by truncating the sum (B.11).

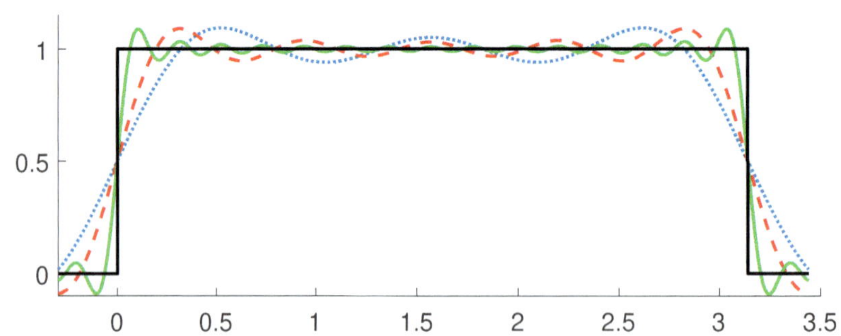

Figure B.12: Fourier approximations of the unit step function f on the interval $(0, \pi)$, truncating the infinite sum in (B.11) to $i = 2, 4$, and 14 terms, giving the dotted blue, dashed red, and solid green curves, respectively.

■

Starting from any countable basis, we can use the Gram–Schmidt procedure to obtain an orthonormal basis, as illustrated in the following example. ☞ 522

Example B.13 (Legendre Polynomials) Consider the space $L^2(\mathbb{R}, w(x)\,\mathrm{d}x)$, where $w(x) = \mathbb{1}\{-1 < x < 1\}$. We wish to construct an orthonormal basis of polynomial functions $g_0, g_1, g_2, \ldots$, starting from the collection of monomials: $\iota_0, \iota_1, \iota_2, \ldots$, where $\iota_k : x \mapsto x^k$. Using Gram–Schmidt, the first normalized zero-degree polynomial is $g_0 = \iota_0/\|\iota_0\| = \sqrt{1/2}$. To find g_1 (a polynomial of degree 1), project ι_1 (the identity function) onto the space spanned by g_0. The resulting projection is $p_1 := \langle g_0, \iota_1 \rangle g_0$, written out as

$$p_1(x) = \left(\int_{-1}^{1} x\, g_0(x)\,\mathrm{d}x\right) g_0(x) = \frac{1}{2}\int_{-1}^{1} x\,\mathrm{d}x = 0.$$

Hence, $g_1 = (\iota_1 - p_1)/\|\iota_1 - p_1\|$ is a linear function; that is, of the form $g_1(x) = ax$. The constant a is found by normalization:

$$1 = \|g_1\|^2 = \int_{-1}^{1} g_1^2(x)\,\mathrm{d}x = a^2 \int_{-1}^{1} x^2\,\mathrm{d}x = a^2\frac{2}{3},$$

so that $g_1(x) = \sqrt{3/2}x$. Continuing the Gram–Schmidt procedure, we find $g_2(x) = \sqrt{5/8}(3x^2 - 1), g_3(x) = \sqrt{7/8}(5x^3 - 3x)$ and, in general,

$$g_k(x) = \frac{\sqrt{2k+1}}{2^{k+\frac{1}{2}}k!}\frac{\mathrm{d}^k}{\mathrm{d}x^k}(x^2-1)^k, \quad k = 0, 1, 2, \ldots.$$

These are the (normalized) **Legendre polynomials**. The graphs of g_0, g_1, g_2, and g_3 are given in Figure B.14.

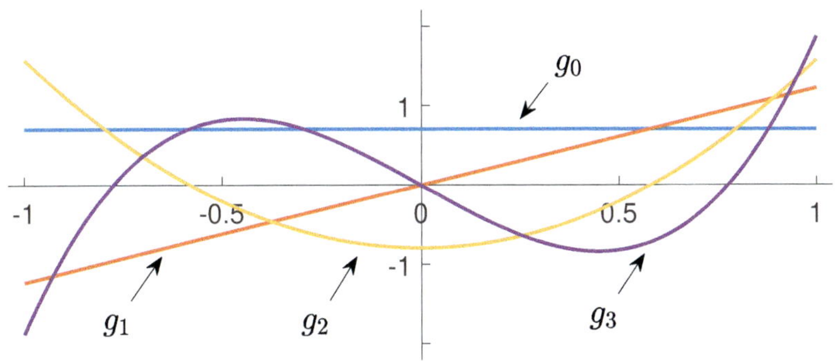

Figure B.14: The first 4 normalized Legendre polynomials.

As the Legendre polynomials form an orthonormal basis of $L^2(\mathbb{R}, \mathbb{1}\{-1 < x < 1\}\,\mathrm{d}x)$, they can be used to approximate arbitrary functions in this space. For example, Figure B.15 shows an approximation using the first 51 Legendre polynomials ($k = 0, 1, \ldots, 50$) of the Fourier expansion of the indicator function on the interval $(-1/2, 1/2)$. These Legendre polynomials form the basis of a 51-dimensional linear subspace onto which the indicator function is orthogonally projected.

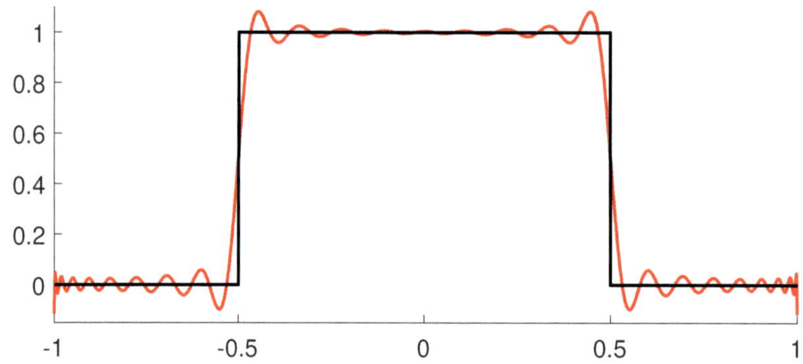

Figure B.15: Approximation of the indicator function on the interval $(-1/2, 1/2)$, using the Legendre polynomials $g_0, g_1, \ldots, g_{50}$.

The Legendre polynomials were produced in the following way: We started with
☞ 628 an unnormalized probability density on $\mathbb{R}$ — in this case the probability density of the uniform distribution on $(-1, 1)$. We then constructed a sequence of polynomials by applying the Gram–Schmidt procedure to the monomials $1, x, x^2, \ldots$.

By using exactly the same procedure, but with a different probability density, we can produce other such **orthogonal polynomials**. For example, the density of the standard exponential[3] distribution, $w(x) = \mathrm{e}^{-x}, x \geq 0$, gives the **Laguerre polynomials**, which are defined by the recurrence

$$(n+1)g_{n+1}(x) = (2n+1-x)g_n(x) - ng_{n-1}(x), \quad n = 1, 2, \ldots,$$

with $g_0(x) = 1$ and $g_1(x) = 1-x$, for $x \geq 0$. The **Hermite polynomials** are obtained when using instead the density of the standard normal distribution: $w(x) = \mathrm{e}^{-x^2/2}/\sqrt{2\pi}, x \in \mathbb{R}$. These polynomials satisfy the recursion

$$g_{n+1}(x) = xg_n(x) - \frac{\mathrm{d}g_n(x)}{\mathrm{d}x}, \quad n = 0, 1, \ldots,$$

with $g_0(x) = 1$, $x \in \mathbb{R}$. Note that the Hermite polynomials as defined above have not been normalized to have norm 1. To normalize, use the fact that $\|g_n\|^2 = n!$.

We conclude with a number of key results in functional analysis. The first one is the celebrated **Cauchy–Schwarz** inequality.

Theorem B.16 (Cauchy–Schwarz): Let $\mathcal{H}$ be a Hilbert space. For every $f, g \in \mathcal{H}$ it holds that

$$|\langle f, g\rangle| \leq \|f\| \, \|g\|.$$

Proof. The inequality is trivially true for $g = 0$ (zero function). For $g \neq 0$, we can write $f = \alpha g + h$, where $h \perp g$ and $\alpha = \langle f, g\rangle/\|g\|^2$. Consequently, $\|f\|^2 = |\alpha|^2 \|g\|^2 + \|h\|^2 \geq |\alpha|^2 \|g\|^2$. The result follows after rearranging this last inequality. □

[3]This can be further generalized to the density of a gamma distribution.

Let $\mathcal{V}$ and $\mathcal{W}$ be two linear vector spaces (for example, Hilbert spaces) on which norms $\|\cdot\|_{\mathcal{V}}$ and $\|\cdot\|_{\mathcal{W}}$ are defined. Suppose $A : \mathcal{V} \to \mathcal{W}$ is a mapping from $\mathcal{V}$ to $\mathcal{W}$. When $\mathcal{W} = \mathcal{V}$, such a mapping is often called an **operator**; when $\mathcal{W} = \mathbb{R}$ it is called a **functional**. Mapping A is said to be *linear* if $A(\alpha f + \beta g) = \alpha A(f) + \beta A(g)$. In this case we write Af instead of $A(f)$. If there exists $\gamma < \infty$ such that

$$\|Af\|_{\mathcal{W}} \leq \gamma \, \|f\|_{\mathcal{V}}, \quad f \in \mathcal{V}, \tag{B.17}$$

then A is said to be a **bounded mapping**. The smallest γ for which (B.17) holds is called the **norm** of A; denoted by $\|A\|$. A (not necessarily linear) mapping $A : \mathcal{V} \to \mathcal{W}$ is said to be *continuous* at f if for any sequence $f_1, f_2, \ldots$ converging to f the sequence $A(f_1), A(f_2), \ldots$ converges to $A(f)$. That is, if

$$\forall \varepsilon > 0, \exists \delta > 0 : \forall g \in \mathcal{V}, \|f - g\|_{\mathcal{V}} < \delta \Rightarrow \|A(f) - A(g)\|_{\mathcal{W}} < \varepsilon. \tag{B.18}$$

If the above property holds for every $f \in \mathcal{V}$, then the mapping A itself is called *continuous*.

Theorem B.19 (Continuity and Boundedness for Linear Mappings): For a linear mapping, continuity and boundedness are equivalent.

Proof. Let A be linear and bounded. We may assume that A is non-zero (otherwise the statement holds trivially), and that therefore $0 < \|A\| < \infty$. Taking $\delta < \varepsilon / \|A\|$ in (B.18) now ensures that $\|Af - Ag\|_{\mathcal{W}} \leq \|A\| \, \|f - g\|_{\mathcal{V}} < \|A\| \, \delta < \varepsilon$. This shows that A is continuous.

Conversely, suppose A is continuous. In particular, it is continuous at $f = 0$ (the zero-element of $\mathcal{V}$). Thus, take $f = 0$ and let ε and δ be as in (B.18). For any $g \neq 0$, let $h = \delta / (2\|g\|_{\mathcal{V}}) \, g$. As $\|h\|_{\mathcal{V}} = \delta/2 < \delta$, it follows from (B.18) that

$$\|Ah\|_{\mathcal{W}} = \frac{\delta}{2\|g\|_{\mathcal{V}}} \|Ag\|_{\mathcal{W}} < \varepsilon.$$

Rearranging the last inequality gives $\|Ag\|_{\mathcal{W}} < 2\varepsilon/\delta \|g\|_{\mathcal{V}}$, showing that A is bounded. □

Theorem B.20 (Riesz Representation Theorem): Any bounded linear functional ϕ on a Hilbert space $\mathcal{H}$ can be represented as $\phi(h) = \langle h, g \rangle$, for some $g \in \mathcal{H}$ (depending on ϕ).

Proof. Let P be the projection of $\mathcal{H}$ onto the nullspace $\mathcal{N}$ of ϕ; that is, $\mathcal{N} = \{g \in \mathcal{H} : \phi(g) = 0\}$. If ϕ is not the 0-functional, then there exists a $g_0 \neq 0$ with $\phi(g_0) \neq 0$. Let $g_1 = g_0 - Pg_0$. Then $g_1 \perp \mathcal{N}$ and $\phi(g_1) = \phi(g_0)$. Take $g_2 = g_1/\phi(g_1)$. For any $h \in \mathcal{H}$, $f := h - \phi(h) g_2$ lies in $\mathcal{N}$. As $g_2 \perp \mathcal{N}$ it holds that $\langle f, g_2 \rangle = 0$, which is equivalent to $\langle h, g_2 \rangle = \phi(h) \, \|g_2\|^2$. By defining $g = g_2 / \|g_2\|^2$ we have found our representation. □

B.2 Fourier Transforms

We will now briefly introduce the Fourier transform. Before doing so, we will extend the
☞ 564 concept of L^2 space of real-valued functions as follows:

Definition B.21 (L^p Space): Let $\mathcal{X}$ be a subset of $\mathbb{R}^d$ with measure $\mu(\mathrm{d}\boldsymbol{x}) = w(\boldsymbol{x})\,\mathrm{d}\boldsymbol{x}$ and $p \in [1, \infty)$. Then $L^p(\mathcal{X}, \mu)$ is the linear space of functions from $\mathcal{X}$ to $\mathbb{C}$ that satisfy

$$\int_{\mathcal{X}} |f(\boldsymbol{x})|^p \, w(\boldsymbol{x})\,\mathrm{d}\boldsymbol{x} < \infty. \tag{B.22}$$

When $p = 2$, $L^2(\mathcal{X}, \mu)$ is in fact a Hilbert space equipped with inner product

$$\langle f, g \rangle = \int_{\mathcal{X}} f(\boldsymbol{x})\,\overline{g(\boldsymbol{x})}\, w(\boldsymbol{x})\,\mathrm{d}\boldsymbol{x}. \tag{B.23}$$

We are now in a position to define the Fourier transform (with respect to the Lebesgue measure). Note that in the following Definitions B.24 and B.25 we have chosen a particular convention. Equivalent (but not identical) definitions exist that include scaling constants $(2\pi)^d$ or $(2\pi)^{-d}$ and where $-2\pi\boldsymbol{t}$ is replaced with $2\pi\boldsymbol{t}$, $\boldsymbol{t}$, or $-\boldsymbol{t}$.

Definition B.24 ((Multivariate) Fourier Transform): The **Fourier transform** $\mathcal{F}[f]$ of a (real- or complex-valued) function $f \in L^1(\mathbb{R}^d)$ is the function $\widetilde{f}$ defined as

$$\widetilde{f}(\boldsymbol{t}) := \int_{\mathbb{R}^d} \mathrm{e}^{-\mathrm{i}\,2\pi\,\boldsymbol{t}^\top \boldsymbol{x}} f(\boldsymbol{x})\,\mathrm{d}\boldsymbol{x}\,, \quad \boldsymbol{t} \in \mathbb{R}^d.$$

The Fourier transform $\widetilde{f}$ is continuous, uniformly bounded (since $f \in L^1(\mathbb{R}^d)$ implies that $|\widetilde{f}(\boldsymbol{t})| \leq \int_{\mathbb{R}^d} |f(\boldsymbol{x})|\,\mathrm{d}\boldsymbol{x} < \infty$), and satisfies $\lim_{\|\boldsymbol{t}\|\to\infty} \widetilde{f}(\boldsymbol{t}) = 0$ (a result known as the **Riemann–Lebesgue lemma**). However, $|\widetilde{f}|$ does not necessarily have a finite integral. A simple example in $\mathbb{R}^1$ is the Fourier transform of $f(x) = \mathbb{1}\{-1/2 < x < 1/2\}$. Then $\widetilde{f}(t) = \sin(\pi t)/(\pi t) = \operatorname{sinc}(\pi t)$, which is not absolutely integrable.

Definition B.25 ((Multivariate) Inverse Fourier Transform): The **inverse Fourier transform** $\mathcal{F}^{-1}[\widetilde{f}]$ of a (real- or complex-valued) function $\widetilde{f} \in L^1(\mathbb{R}^d)$ is the function $\check{f}$ defined as

$$\check{f}(\boldsymbol{x}) := \int_{\mathbb{R}^d} \mathrm{e}^{\mathrm{i}\,2\pi\,\boldsymbol{t}^\top \boldsymbol{x}}\, \widetilde{f}(\boldsymbol{t})\,\mathrm{d}\boldsymbol{t}, \quad \boldsymbol{x} \in \mathbb{R}^d.$$

As one would hope, it holds that if f and $\mathcal{F}[f]$ are both in $L^1(\mathbb{R}^d)$, then $f = \mathcal{F}^{-1}[\mathcal{F}[f]]$ almost everywhere.

The Fourier transform enjoys many interesting and useful properties, some of which we list below.

1. *Linearity*: For $f, g \in L^1(\mathbb{R}^d)$ and constants $a, b \in \mathbb{R}$,

$$\mathcal{F}[af + bg] = a\,\mathcal{F}[f] + b\,\mathcal{F}[g].$$

2. *Space Shifting and Scaling*: Let $\mathbf{A} \in \mathbb{R}^{d\times d}$ be an invertible matrix and $\boldsymbol{b} \in \mathbb{R}^d$ a constant vector. Let $f \in L^1(\mathbb{R}^d)$ and define $h(\boldsymbol{x}) := f(\mathbf{A}\boldsymbol{x} + \boldsymbol{b})$. Then

$$\mathcal{F}[h](\boldsymbol{t}) = \mathrm{e}^{\mathrm{i}\,2\pi\,(\mathbf{A}^{-\top}\boldsymbol{t})^\top \boldsymbol{b}}\,\widetilde{f}(\mathbf{A}^{-\top}\boldsymbol{t})/|\det(\mathbf{A})|,$$

where $\mathbf{A}^{-\top} := (\mathbf{A}^\top)^{-1} = (\mathbf{A}^{-1})^\top$.

3. *Frequency Shifting and Scaling*: Let $\mathbf{A} \in \mathbb{R}^{d\times d}$ be an invertible matrix and $\boldsymbol{b} \in \mathbb{R}^d$ a constant vector. Let $f \in L^1(\mathbb{R}^d)$ and define

$$h(\boldsymbol{x}) := \mathrm{e}^{-\mathrm{i}\,2\pi\,\boldsymbol{b}^\top\mathbf{A}^{-\top}\boldsymbol{x}}\,f(\mathbf{A}^{-\top}\boldsymbol{x})/|\det(\mathbf{A})|.$$

Then $\mathcal{F}[h](\boldsymbol{t}) = \widetilde{f}(\mathbf{A}\boldsymbol{t} + \boldsymbol{b})$.

4. *Differentiation*: Let $f \in L^1(\mathbb{R}^d) \cap C^1(\mathbb{R}^d)$ and let $f_k := \partial f/\partial x_k$ be the partial derivative of f with respect to x_k. If $f_k \in L^1(\mathbb{R}^d)$ for $k = 1, \dots, d$, then

$$\mathcal{F}[f_k](\boldsymbol{t}) = (\mathrm{i}\,2\pi\,t_k)\,\widetilde{f}(\boldsymbol{t}).$$

5. *Convolution*: Let $f, g \in L^1(\mathbb{R}^d)$ be real or complex valued functions. Their convolution, $f \star g$, is defined as

$$(f \star g)(\boldsymbol{x}) = \int_{\mathbb{R}^d} f(\boldsymbol{y})\,g(\boldsymbol{x} - \boldsymbol{y})\,\mathrm{d}\boldsymbol{y},$$

and is also in $L^1(\mathbb{R}^d)$. Moreover, the Fourier transform satisfies

$$\mathcal{F}[f \star g] = \mathcal{F}[f]\,\mathcal{F}[g].$$

6. *Duality*: Let f and $\mathcal{F}[f]$ both be in $L^1(\mathbb{R}^d)$. Then $\mathcal{F}[\mathcal{F}[f]](\boldsymbol{t}) = f(-\boldsymbol{t})$.

7. *Product Formula*: Let $f, g \in L^1(\mathbb{R}^d)$ and denote by $\widetilde{f}$, $\widetilde{g}$ their respective Fourier transforms. Then $\widetilde{f}\,g, f\,\widetilde{g} \in L^1(\mathbb{R}^d)$, and

$$\int_{\mathbb{R}^d} \widetilde{f}(\boldsymbol{z})\,g(\boldsymbol{z})\,\mathrm{d}\boldsymbol{z} = \int_{\mathbb{R}^d} f(\boldsymbol{z})\,\widetilde{g}(\boldsymbol{z})\,\mathrm{d}\boldsymbol{z}.$$

There are many additional properties which hold if $f \in L^1(\mathbb{R}^d) \cap L^2(\mathbb{R}^d)$. In particular, if $f, g \in L^1(\mathbb{R}^d) \cap L^2(\mathbb{R}^d)$, then $\widetilde{f}, \widetilde{g} \in L^2(\mathbb{R}^d)$ and $\langle \widetilde{f}, \widetilde{g}\rangle = \langle f, g\rangle$, a result often known as **Parseval's formula**. Putting $g = f$ gives the result often referred to as **Plancherel's theorem**.

The next result has important applications in kernel density estimation. ☞ 153

Theorem B.26 (Poisson Summation Formula): Let $f \in L^1(\mathbb{R})$ be a continuous function with Fourier transform $\widetilde{f}$. Then,

$$\sum_{k\in\mathbb{Z}} f(k+x)\,\mathrm{e}^{-\mathrm{i}\pi s(2k+x)} = \sum_{k\in\mathbb{Z}} \widetilde{f}(k+s)\,\mathrm{e}^{\mathrm{i}\pi x(2k+s)}, \quad x, s \in [0,1],$$

provided that $\sum_{k\in\mathbb{Z}} f(k+x)$ converges uniformly and $\sum_{k\in\mathbb{Z}} |\widetilde{f}(k)| < \infty$.

Proof. Consider the so-called **Dirichlet kernel** on $[0, 1]$:

$$D_n(x) := \sum_{k=-n}^{n} \mathrm{e}^{\mathrm{i}2\pi kx} = 1 + 2\sum_{k=1}^{n} \cos(2\pi kx), \quad x \in [0,1].$$

Define the Fourier coefficients: $a_0 := \int_0^1 f(u)\,\mathrm{d}u$, $b_0 := 0$, $a_k := 2\int_0^1 f(u)\cos(2\pi ku)\,\mathrm{d}u$, $b_k := 2\int_0^1 f(u)\sin(2\pi ku)\,\mathrm{d}u$ for $k \geq 1$. Then, under the conditions of the theorem, the Dirichlet kernel satisfies:

$$\int_0^1 f(u)D_n(u-x)\,\mathrm{d}u = \sum_{k=0}^{n} a_k\cos(2\pi kx) + b_k\sin(2\pi kx) \to f(x), \quad n\uparrow\infty.$$

Using the periodicity $\mathrm{e}^{\mathrm{i}2\pi k} = 1$ for $k \in \mathbb{Z}$, we then calculate:

$$\begin{aligned}
\mathrm{e}^{-\mathrm{i}\pi xs}\sum_{k\in\mathbb{Z}} \widetilde{f}(k+s)\,\mathrm{e}^{\mathrm{i}\pi x(2k+s)} &= \sum_{k\in\mathbb{Z}}\int_{\mathbb{R}} f(y)\,\mathrm{e}^{\mathrm{i}2\pi kx-\mathrm{i}2\pi(k+s)y}\,\mathrm{d}y\\
&= \sum_{k\in\mathbb{Z}}\sum_{m\in\mathbb{Z}}\int_m^{m+1} f(y)\,\mathrm{e}^{-\mathrm{i}2\pi sy}\mathrm{e}^{\mathrm{i}2\pi k(x-y)}\,\mathrm{d}y\\
(u \leftarrow y-m) \quad &= \lim_{n\uparrow\infty}\sum_{k=-n}^{n}\sum_{m\in\mathbb{Z}}\int_0^1 f(m+u)\,\mathrm{e}^{-\mathrm{i}2\pi s(u+m)}\mathrm{e}^{\mathrm{i}2\pi k(x-u)}\,\mathrm{d}u\\
(\text{change summation order}) \quad &= \lim_{n\uparrow\infty}\int_0^1\sum_{m\in\mathbb{Z}} f(m+u)\,\mathrm{e}^{-\mathrm{i}2\pi s(u+m)}D_n(x-u)\,\mathrm{d}u\\
&= \sum_{m\in\mathbb{Z}} f(m+x)\,\mathrm{e}^{-\mathrm{i}2\pi s(x+m)},
\end{aligned}$$

which completes the proof. □

In the next example, we make use of the Poisson summation formula and the following cosine-product formula:

$$\prod_{k=1}^{d}\cos(x_k) = \mathbb{E}\cos\left(\sum_{k=1}^{d} S_k\,x_k\right), \tag{B.27}$$

where $S_1, \ldots, S_d$ are iid random variables distributed according to the **Rademacher distribution**: $\mathbb{P}[S=-1] = \mathbb{P}[S=1] = 1/2$.

Example B.28 (Multivariate Theta Kernel Function) If f is the density of the $\mathcal{N}(0, \sigma^2/4)$ distribution, then $\mathcal{F}[f](m) = \mathrm{e}^{-\pi^2 m^2 \sigma^2/2}$ and Theorem B.26 yields (with $s = 0$):

$$\frac{1}{\sqrt{\pi\sigma^2/2}} \sum_{m\in\mathbb{Z}} \mathrm{e}^{-\frac{(x+m)^2}{\sigma^2/2}} = \sum_{m\in\mathbb{Z}} \mathrm{e}^{-\pi^2 m^2 \sigma^2/2} \mathrm{e}^{\mathrm{i}\pi 2mx} = \sum_{m\in\mathbb{Z}} \mathrm{e}^{-\pi^2 m^2 \sigma^2/2} \cos(2\pi m x).$$

Recall the theta kernel from Example 7.19. Using the identity (B.27) for $d = 2$, we obtain the **theta kernel identity** for $x, x' \in [0, 1]$: ☞ 326

$$\begin{aligned}\kappa_{\sigma^2}(x, x') &:= \sum_{m\in\mathbb{Z}} \mathrm{e}^{-\pi^2 m^2 \sigma^2/2} \cos(\pi m x)\cos(\pi m x') \\ &= \frac{1}{\sqrt{2\pi\sigma^2}} \sum_{m\in\mathbb{Z}} \left(\mathrm{e}^{-\frac{(x+x'+2m)^2}{2\sigma^2}} + \mathrm{e}^{-\frac{(x-x'+2m)^2}{2\sigma^2}} \right).\end{aligned}$$

This identity shows that $\kappa_{\sigma^2}(x, x') > 0$ for $\sigma > 0$ and, since $\int_0^1 \cos(\pi m u)\,\mathrm{d}u = 0$ for $m \neq 0$, $m \in \mathbb{Z}$, we conclude that $\kappa_{\sigma^2}(\cdot, x')$ is a probability density function on $[0, 1]$ for each $x' \in [0, 1]$. By taking the product of d theta kernels, we obtain the **multivariate theta kernel** on $[0, 1]^d$:

$$\begin{aligned}\kappa_{\sigma^2}(\boldsymbol{x}, \boldsymbol{x}') &:= \prod_{k=1}^{d} \sum_{m_k\in\mathbb{Z}} \mathrm{e}^{-\pi^2 m_k^2 \sigma^2/2} \cos(\pi m_k x_k)\cos(\pi m_k x_k') \\ &= \sum_{\boldsymbol{m}\in\mathbb{Z}^d} \prod_{k=1}^{d} \mathrm{e}^{-\pi^2 m_k^2 \sigma^2/2} \cos(\pi m_k x_k)\cos(\pi m_k x_k') \\ &= \sum_{\boldsymbol{m}\in\mathbb{Z}^d} \mathrm{e}^{-\pi^2 \|\boldsymbol{m}\|^2 \sigma^2/2} \cos(\pi \boldsymbol{m}^\top \boldsymbol{x}) \prod_{k=1}^{d} \cos(\pi m_k x_k'),\end{aligned}$$

where the last line follows from (B.27). The product construction ensures that $\kappa_{\sigma^2}(\cdot, \boldsymbol{x}')$ is a probability density function on $[0, 1]^d$ for each $\boldsymbol{x}' \in [0, 1]^d$.

The Fourier transform can be extended in several ways, in the first instance to functions in $L^2(\mathbb{R}^d)$ by continuity. A substantial extension of the theory is realized by replacing integration with respect to the Lebesgue measure (i.e., $\int_{\mathbb{R}^d} \cdots \mathrm{d}\boldsymbol{x}$) with integration with respect to a (finite Borel) measure μ (i.e., $\int_{\mathbb{R}^d} \cdots \mu(\mathrm{d}\boldsymbol{x})$). Moreover, there is a close connection between the Fourier transform and characteristic functions arising in probability theory. Indeed, if $\boldsymbol{X}$ is a random vector with pdf f, then its characteristic function ψ satisfies ☞ 647

$$\psi(\boldsymbol{t}) := \mathbb{E}\, \mathrm{e}^{\mathrm{i}\boldsymbol{t}^\top \boldsymbol{X}} = \mathcal{F}[f](-\boldsymbol{t}/(2\pi)).$$

B.2.1 Discrete Fourier Transform

Here, we introduce the (univariate) discrete Fourier transform, which can be viewed as a special case of the Fourier transform introduced in Definition B.24, where $d = 1$, integration is with respect to the counting measure, and $f(x) = 0$ for $x < 0$ and $x > (n - 1)$.

Definition B.29 (Discrete Fourier Transform): The **discrete Fourier transform** (DFT) of $\boldsymbol{x} = [x_0, \dots, x_{n-1}]^\top \in \mathbb{C}^n$ is the vector $\widetilde{\boldsymbol{x}} = [\widetilde{x}_0, \dots, \widetilde{x}_{n-1}]^\top$ whose elements are given by

$$\widetilde{x}_t = \sum_{s=0}^{n-1} \omega^{st}\, x_s, \quad t = 0, \dots, n-1, \tag{B.30}$$

where $\omega = \exp(-\mathrm{i}\, 2\pi/n)$.

In other words, $\widetilde{\boldsymbol{x}}$ is obtained from $\boldsymbol{x}$ via the linear transformation

$$\widetilde{\boldsymbol{x}} = \mathbf{F}\boldsymbol{x},$$

where

$$\mathbf{F} = \begin{bmatrix} 1 & 1 & 1 & \dots & 1 \\ 1 & \omega & \omega^2 & \dots & \omega^{n-1} \\ 1 & \omega^2 & \omega^4 & \dots & \omega^{2(n-1)} \\ \vdots & \vdots & \vdots & \ddots & \vdots \\ 1 & \omega^{n-1} & \omega^{2(n-1)} & \dots & \omega^{(n-1)^2} \end{bmatrix}.$$

The matrix $\mathbf{F}$ is a so-called **Vandermonde matrix**, and is clearly symmetric (i.e., $\mathbf{F} = \mathbf{F}^\top$). Moreover, $\mathbf{F}/\sqrt{n}$ is in fact a unitary matrix and hence its inverse is simply its complex conjugate $\overline{\mathbf{F}}/\sqrt{n}$. Thus, $\mathbf{F}^{-1} = \overline{\mathbf{F}}/n$ and we have that the **inverse discrete Fourier transform** (IDFT) is given by

$$x_t = \frac{1}{n}\sum_{s=0}^{n-1} \omega^{-st}\, \widetilde{x}_s, \quad t = 0, \dots, n-1, \tag{B.31}$$

or in terms of matrices and vectors,

$$\boldsymbol{x} = \overline{\mathbf{F}}\,\widetilde{\boldsymbol{x}}/n.$$

Observe that the IDFT of a vector $\boldsymbol{y}$ is related to the DFT of its complex conjugate $\overline{\boldsymbol{y}}$, since

$$\overline{\mathbf{F}}\,\boldsymbol{y}/n = \overline{\mathbf{F}\,\overline{\boldsymbol{y}}}/n.$$

Consequently, an IDFT can be computed via a DFT.

■ **Example B.32 (Fast Vector Multiplication with a Circulant Matrix)** There is a close connection between circulant matrices $\mathbf{C}$ and the DFT. To make this connection concrete, let $\mathbf{C}$ be the circulant matrix corresponding to the vector $\boldsymbol{c} \in \mathbb{C}^n$ and denote by $\boldsymbol{f}_t$ the t-th column of the discrete Fourier matrix $\mathbf{F}$, $t = 0, 1, \dots, n-1$. Then, the s-th element of $\mathbf{C}\boldsymbol{f}_t$ is

$$\sum_{k=0}^{n-1} c_{(s-k) \bmod n}\, \omega^{tk} = \sum_{y=0}^{n-1} c_y\, \omega^{t(s-y)} = \underbrace{\omega^{ts}}_{s\text{-th element of } \boldsymbol{f}_t} \underbrace{\sum_{y=0}^{n-1} c_y\, \omega^{-ty}}_{\lambda_s}.$$

Hence, the eigenvalues of $\mathbf{C}$ are

$$\lambda_t = \boldsymbol{c}^\top \overline{\boldsymbol{f}}_t, \quad t = 0, 1, \ldots, n-1,$$

with corresponding eigenvectors $\boldsymbol{f}_t$. Collecting the eigenvalues into the vector $\boldsymbol{\lambda} = [\lambda_0, \ldots, \lambda_{n-1}]^\top = \overline{\mathbf{F}}\boldsymbol{c}$, we therefore have the eigen-decomposition

$$\mathbf{C} = \mathbf{F}\,\mathrm{diag}(\boldsymbol{\lambda})\,\overline{\mathbf{F}}/n.$$

Consequently, one can compute the **circular convolution** of a vector $\boldsymbol{a} = [a_1, \ldots, a_n]^\top$ and $\boldsymbol{c} = [c_0, \ldots, c_{n-1}]^\top$ by a series of DFTs as follows: Construct the circulant matrix $\mathbf{C}$ corresponding to $\boldsymbol{c}$. Then, the circular convolution of $\boldsymbol{a}$ and $\boldsymbol{c}$ is given by $\boldsymbol{y} = \mathbf{C}\boldsymbol{a}$. Proceed in four steps:

1. Compute $\boldsymbol{z} = \overline{\mathbf{F}}\boldsymbol{a}/n$.
2. Compute $\boldsymbol{\lambda} = \overline{\mathbf{F}}\boldsymbol{c}$.
3. Compute $\boldsymbol{p} = \boldsymbol{z} \odot \boldsymbol{\lambda} = [z_1\,\lambda_0, \ldots, z_n\,\lambda_{n-1}]^\top$.
4. Compute $\boldsymbol{y} = \mathbf{F}\boldsymbol{p}$.

Steps 1 and 2 are (up to constants) in the form of an IDFT, and step 4 is in the form of a DFT. These are computable via the FFT (Section B.2.2) in $\mathcal{O}(n \ln n)$ time. Step 3 is a dot product computable in $\mathcal{O}(n)$ time. Thus, the circular convolution can be computed with the aid of the FFT in $\mathcal{O}(n \ln n)$ time. ☞ 576

■

■ **Example B.33 (Fast Vector Multiplication with a Toeplitz matrix)** One can also efficiently compute the product of an $n \times n$ Toeplitz matrix $\mathbf{A}$ and an $n \times 1$ vector $\boldsymbol{x}$ in $\mathcal{O}(n \ln n)$ time by embedding $\mathbf{A}$ into a circulant matrix $\mathbf{C}$ of size $2n \times 2n$. Namely, define

$$\mathbf{C} = \begin{bmatrix} \mathbf{A} & \mathbf{B} \\ \mathbf{B} & \mathbf{A} \end{bmatrix},$$

where

$$\mathbf{B} = \begin{bmatrix} 0 & a_{n-1} & \cdots & a_2 & a_1 \\ a_{-(n-1)} & 0 & a_{n-1} & & a_2 \\ \vdots & a_{-(n-1)} & 0 & \ddots & \vdots \\ a_{-2} & & \ddots & \ddots & a_{n-1} \\ a_{-1} & a_{-2} & \cdots & a_{-(n-1)} & 0 \end{bmatrix}.$$

Then a product of the form $\mathbf{A}\boldsymbol{x}$ can be computed in $\mathcal{O}(n \ln n)$ time, since we may write

$$\mathbf{C}\begin{bmatrix} \boldsymbol{x} \\ \mathbf{0} \end{bmatrix} = \begin{bmatrix} \mathbf{A} & \mathbf{B} \\ \mathbf{B} & \mathbf{A} \end{bmatrix}\begin{bmatrix} \boldsymbol{x} \\ \mathbf{0} \end{bmatrix} = \begin{bmatrix} \mathbf{A}\boldsymbol{x} \\ \mathbf{B}\boldsymbol{x} \end{bmatrix}.$$

The left-hand side is a product of a $2n \times 2n$ circulant matrix with vector of length $2n$, and so can be computed in $\mathcal{O}(n \ln n)$ time via the FFT, as previously discussed.

The following algorithm provides the pseudo-code for fast computation of $\mathbf{A}\boldsymbol{x}$ in the case when $\mathbf{A}$ is a symmetric Toeplitz matrix.

Algorithm B.34: $x \mapsto \mathbf{A}x$ for Symmetric Toeplitz A in $\mathcal{O}(n \ln n)$ Time

Input: $\boldsymbol{x}$ and first column $\boldsymbol{a}$ of symmetric Toeplitz $\mathbf{A} \in \mathbb{R}^{n \times n}$
Output: $\mathbf{A}\boldsymbol{x}$

$\boldsymbol{a} \leftarrow [\boldsymbol{a}^\top, 0, a_{n-1}, \ldots, a_2, a_1]^\top$
$\boldsymbol{x} \leftarrow [\boldsymbol{x}^\top, 0, \ldots, 0]^\top \in \mathbb{R}^{2n}$
$\boldsymbol{x} \leftarrow \mathbf{F}\overline{\boldsymbol{x}}/(2n)$ // in $\mathcal{O}(n \ln n)$ time using FFT algorithm
$\boldsymbol{\lambda} \leftarrow \mathbf{F}\overline{\boldsymbol{a}}$ // in $\mathcal{O}(n \ln n)$ time using FFT algorithm
$\boldsymbol{x} \leftarrow \boldsymbol{x} \odot \boldsymbol{\lambda}$ // in $\mathcal{O}(n)$ time
$\boldsymbol{x} \leftarrow \mathbf{F}\overline{\boldsymbol{x}}$ // in $\mathcal{O}(n \ln n)$ time using FFT algorithm
return $[x_1, \ldots, x_n]^\top$

Conceptually, one can also solve equations of the form $\mathbf{C}\boldsymbol{x} = \boldsymbol{b}$ for a given vector $\boldsymbol{b} \in \mathbb{C}^n$ and circulant matrix $\mathbf{C}$ (corresponding to $\boldsymbol{c} \in \mathbb{C}^n$, assuming all its eigenvalues are non-zero) via the following four steps:

1. Compute $\boldsymbol{z} = \overline{\mathbf{F}}\boldsymbol{b}/n$.
2. Compute $\boldsymbol{\lambda} = \overline{\mathbf{F}}\boldsymbol{c}$.
3. Compute $\boldsymbol{p} = \boldsymbol{z}/\boldsymbol{\lambda} = [z_1/\lambda_0, \ldots, z_n/\lambda_{n-1}]^\top$.
4. Compute $\boldsymbol{x} = \mathbf{F}\boldsymbol{p}$.

Once again, Steps 1 and 2 are (up to constants) in the form of an IDFT, and Step 4 is in the form of a DFT, all of which are computable via the FFT in $\mathcal{O}(n \ln n)$ time, and Step 3 is computable in $\mathcal{O}(n)$ time, meaning the solution $\boldsymbol{x}$ can be computed using the FFT in $\mathcal{O}(n \ln n)$ time.

B.2.2 Fast Fourier Transform

The **fast Fourier transform** (FFT) is a numerical algorithm for the fast evaluation of (B.30) and (B.31). By using a divide-and-conquer strategy, the algorithm reduces the computational complexity from $\mathcal{O}(n^2)$ (for the naïve evaluation of the linear transformation) to $\mathcal{O}(n \ln n)$ [87].

The essence of the algorithm lies in the following observation. Suppose $n = r_1 r_2$. Then one can express any index t appearing in (B.30) via a pair (t_0, t_1), with $t = t_1 r_1 + t_0$, where $t_0 \in \{0, 1, \ldots, r_1 - 1\}$ and $t_1 \in \{0, 1, \ldots, r_2 - 1\}$. Similarly, one can express any index s appearing in (B.30) via a pair (s_0, s_1), with $s = s_1 r_2 + s_0$, where $s_0 \in \{0, 1, \ldots, r_2 - 1\}$ and $s_1 \in \{0, 1, \ldots, r_1 - 1\}$.

Identifying $\widetilde{x}_t \equiv \widetilde{x}_{t_1, t_0}$ and $x_s \equiv x_{s_1, s_0}$, we may re-express (B.30) as

$$\text{(B.35)} \quad \widetilde{x}_{t_1,t_0} = \sum_{s_0=0}^{r_2-1} \omega^{s_0 t} \sum_{s_1=0}^{r_1-1} \omega^{s_1 r_2 t}\, x_{s_1,s_0}\,, \qquad t_0 = 0, 1, \ldots, r_1 - 1\,, t_1 = 0, 1, \ldots, r_2 - 1.$$

Observe that $\omega^{s_1 r_2 t} = \omega^{s_1 r_2 t_0}$ (because $\omega^{r_1 r_2} = 1$), so that the inner sum over s_1 depends only on s_0 and t_0. Define

$$y_{t_0,s_0} := \sum_{s_1=0}^{r_1-1} \omega^{s_1 r_2 t_0}\, x_{s_1,s_0}\,, \quad t_0 = 0, 1, \ldots, r_1 - 1\,, s_0 = 0, 1, \ldots, r_2 - 1.$$

Computing each y_{t_0,s_0} requires $O(n\, r_1)$ operations. In terms of the $\{y_{t_0,s_0}\}$, (B.35) can be written as

$$\widetilde{x}_{t_1,t_0} = \sum_{s_0=0}^{r_2-1} \omega^{s_0 t}\, y_{t_0,s_0}\,, \quad t_1 = 0, 1, \ldots, r_2 - 1\,, t_0 = 0, 1, \ldots, r_1 - 1,$$

requiring $O(n\, r_2)$ operations to compute. Thus, calculating the DFT using this two-step procedure requires $O(n\,(r_1 + r_2))$ operations, rather than $O(n^2)$.

Now supposing $n = r_1 r_2 \cdots r_m$, repeated application the above divide-and-conquer idea yields an m-step procedure requiring $O(n\,(r_1 + r_2 + \cdots + r_m))$ operations. In particular, if $r_k = r$ for all $k = 1, 2, \ldots, m$, we have that $n = r^m$ and $m = \log_r n$, so that the total number of operations is $O(r\, n\, m) \equiv O(r\, n\, \log_r(n))$. Typically, the *radix* r is a small (not necessarily prime) number, for instance $r = 2$.

Example B.36 (Discrete Convolution of Vectors) The FFT can be used to efficiently calculate the discrete convolution of two vectors $\boldsymbol{x}$ and $\boldsymbol{y}$, of lengths m and n, defined in (A.85). The key fact is that the DFT of the convolution is equal to the product of the DFTs ☞ 550 of $\boldsymbol{x}$ and $\boldsymbol{y}$, where both vectors are "padded" with zeros at the end. For some applications, such as kernel density estimation, it is required to truncate the *full* discrete convolution ☞ 144 $[z_1, \ldots, z_{m+n+1}]^\top$ in (A.85). In typical situations the length m of the "kernel" $\boldsymbol{x}$ is less than the length n of the "signal" $\boldsymbol{y}$, and the full convolution vector is truncated to a vector with the *same* length as $\boldsymbol{y}$, given by $[z_{\text{start}}, \ldots, z_{\text{start}+n}]^\top$, where the start index is $\lfloor (m+1)/2 \rfloor$. Here is a Python implementation for the "full" and "same" discrete convolutions.

Convolution.py

```python
import numpy as np
from scipy.fft import fft, ifft


def conv_full(x, y):
    dim = len(x) + len(y) - 1
    X = fft(x, dim)
    Y = fft(y, dim)
    return np.real(ifft(X * Y))


def conv_same(x, y):
    full_conv = conv_full(x, y)
    start_idx = (len(y) + 1) // 2
    return full_conv[start_idx : start_idx + len(x)]
```

■

B.2.3 Discrete Cosine Transform

The discrete cosine transform is important in signal processing due to its property of compressing highly correlated data very well [87, Pages 151–153]. It is closely related to the cosine series representation of a function. In particular, in Example B.8 we saw that the coefficients $\{c_t\}$ of the cosine series of a square-integrable function f on $[0, 1]$ are given by

$$c_0 = \int_0^1 f(x)\,\mathrm{d}x \quad \text{and} \quad c_t = 2\int_0^1 f(x)\cos(t\pi x)\,\mathrm{d}x, \quad t = 1, 2, \ldots.$$

In essence, the discrete cosine transform is simply a discretization of the above transformation (from f to $\{c_t\}$), whereby the interval $[0, 1]$ is divided into n subintervals:

$$[0, 1/n], \ldots [1 - 1/n, 1],$$

the integrals are replaced with sums, and the function f is only evaluated at the midpoints of the intervals.

Specifically, let $x_0, \ldots, x_{n-1}$ be a sequence of real numbers. The one-dimensional **discrete cosine transform** is defined as the sequence

$$\widetilde{x}_t = \alpha_t \sum_{k=0}^{n-1} x_k \cos\left(\frac{\pi(2k+1)t}{2n}\right), \quad 0 \leq t \leq n-1,$$

where $\alpha_0 = 1$ and $\alpha_t = 2,\ t \geq 1$. In vector notation:

$$\widetilde{\boldsymbol{x}} = \mathbf{B}\boldsymbol{x},$$

where the $n \times n$ matrix $\mathbf{B} = [b_{t,k}]$ is defined by

$$b_{t,k} = \alpha_t \cos\left(\pi(2k+1)t/(2n)\right), \quad 0 \leq t \leq n-1, \quad 0 \leq k \leq n-1.$$

The columns of $\mathbf{B}$ are perpendicular to each other and the norm of the tth column is $n\alpha_t$. Hence, if we define $\mathbf{A}$ as the diagonal matrix with diagonal elements $1/\sqrt{n\alpha_t}, t = 0, \ldots, n-1$, then the matrix $\mathbf{AB}$ is orthogonal. Consequently,

$$\boldsymbol{x} = (\mathbf{B})^{-1}\widetilde{\boldsymbol{x}} = \mathbf{B}^\top\mathbf{A}^\top\mathbf{A}\widetilde{\boldsymbol{x}},$$

which recovers the original sequence via the **inverse discrete cosine transform**:

$$x_k = \frac{1}{n}\sum_{t=0}^{n-1} \widetilde{x}_t \cos\left(\frac{\pi(2k+1)t}{2n}\right), \quad 0 \leq k \leq n-1.$$

Note that the discrete cosine transform is not the real part of the discrete Fourier transform. Nevertheless, the FFT can be used to compute the discrete cosine transform and its inverse in $O(n \ln n)$ operations, via Algorithms B.37 and B.38. For fastest performance one has to use $n = 2^m$ for some integer m. In the algorithms, we use the notation $a\!:\!b\!:\!\delta$ to denote a row vector of indices from a up to b in increments of δ. The reverse-order vector is denoted by $\overleftarrow{a\!:\!b\!:\!\delta}$. For example, $0\!:\!5\!:\!2 = [0, 2, 4]$ and $\overleftarrow{1\!:\!5\!:\!2} = [5, 3, 1]$. Both these constructs are readily translated in the Python code that follows the algorithms.

Algorithm B.37: Fast Discrete Cosine Transform

Input: Data vector $\boldsymbol{x} = [x_0, \dots, x_{n-1}]^\top$.
Output: Discrete cosine transform $\widetilde{\boldsymbol{x}} = [\widetilde{x}_0, \dots, \widetilde{x}_{n-1}]^\top$

$[i_0, \dots, i_{n-1}] \leftarrow [0:n-1:2, \overleftarrow{1:n-1:2}]$ // Reordering indices
for $k = 0, \dots, n-1$ **do**
  $r_k \leftarrow x_{i_k}$
Let $\widetilde{\boldsymbol{r}}$ be the DFT (B.30) of $\boldsymbol{r} = [r_0, \dots, r_{n-1}]^\top$, computed via the FFT.
Set $\alpha_0 \leftarrow 1$ and $\alpha_t \leftarrow 2$ for $1 \le t \le n-1$.
for $t = 0, \dots, n-1$ **do**
  $\widetilde{x}_t \leftarrow \Re\left[\alpha_t \, \widetilde{r}_t \, \mathrm{e}^{-\mathrm{i}\pi t/(2n)}\right]$ // Real part

Algorithm B.38: Fast Inverse Discrete Cosine Transform

Input: Vector $\widetilde{\boldsymbol{x}} = [\widetilde{x}_0, \dots, \widetilde{x}_{n-1}]^\top$.
Output: Inverse discrete cosine transform $\boldsymbol{x} = [x_0, \dots, x_{n-1}]^\top$

for $t = 0, \dots, n-1$ **do**
  $\widetilde{r}_t \leftarrow \widetilde{x}_t \, \mathrm{e}^{\mathrm{i}\pi t/(2n)}$
Let $\boldsymbol{z} = [z_0, \dots, z_{n-1}]^\top$ be the inverse DFT (B.31) of $\widetilde{\boldsymbol{r}}$, computed via the FFT.
$[i_0, \dots, i_{n-1}] \leftarrow [0:n-1:2, \overleftarrow{1:n-1:2}]$
Let $[j_0, \dots, j_{n-1}]$ be the vector of indices that sort $[i_0, \dots, i_{n-1}]$ in ascending order; that is, $i_{j_k} = k$, $k = 0, \dots, n-1$.
for $k = 0, \dots, n-1$ **do** // Reorder the output
  $x_k \leftarrow \Re[z_{j_k}]$

The following Python code implements both algorithms.

DCT.py

```
import numpy as np
from scipy.fft import fft, ifft

def dct(x):
    n = len(x)
    ind = np.hstack((np.arange(0, n, 2), np.flip(np.arange(1, n, 2))))
    r_tilde = fft(x[ind])
    alpha = 2 * np.ones(n)
    alpha[0] = 1
    w = np.exp(-1j * np.pi / (2 * n))
    return np.real(alpha * w ** np.arange(n) * r_tilde)


def idct(xtilde):
    n = len(xtilde)
    w = np.exp(1j * np.pi / (2 * n))
    x_scrambled = np.real(ifft(w ** np.arange(n) * xtilde))
    ind = np.hstack((np.arange(0, n, 2), np.flip(np.arange(1, n, 2))))
    jind = np.argsort(ind)
    return x_scrambled[jind]
```

Note that both algorithms take the real part of a sequence of complex variables whose imaginary part is in fact zero. However, from a numerical perspective it is advisable to keep this real part operation as there may be small numerical errors which result in a computational complex transform.

The connection between the DCT and the DFT can be seen as follows: Using the notation from Algorithm B.37, and assuming n is even, we have, for $0 \leq t \leq n-1$, the DCT as:

$$\widetilde{x}_t = \alpha_t \sum_{k=0}^{n-1} x_k \cos\left(\frac{\pi(2k+1)t}{2n}\right).$$

Splitting into even and odd terms, and using the sequence $\boldsymbol{r}$, we have:

$$\begin{aligned}\widetilde{x}_t &= \alpha_t \left[\sum_{k=0}^{n/2-1} x_{2k} \cos\left(\frac{\pi(4k+1)t}{2n}\right) + \sum_{k=0}^{n/2-1} x_{2k+1} \cos\left(\frac{\pi(4k+3)t}{2n}\right)\right] \\ &= \alpha_t \left[\sum_{k=0}^{n/2-1} r_k \cos\left(\frac{\pi(4k+1)t}{2n}\right) + \sum_{k=0}^{n/2-1} r_{n-k-1} \cos\left(\frac{\pi(4k+3)t}{2n}\right)\right].\end{aligned}$$

Reindexing the second sum via $k' = n - k - 1$ and combining the two sums we have

$$\widetilde{x}_t = \alpha_t \sum_{k=0}^{n-1} r_k \cos\left(\frac{\pi(4k+1)t}{2n}\right) = \Re\left[\alpha_t \mathrm{e}^{-\mathrm{i}\pi t/(2n)} \sum_{k=0}^{n-1} r_k \mathrm{e}^{-\mathrm{i}2\pi k t/n}\right],$$

where the sum on the right-hand-side is simply the t-th element of the DFT of the sequence $\boldsymbol{r}$, and these elements can be computed efficiently via the FFT.

Further Reading

Kolmogorov and Fomin's masterpiece on the theory of functions and functional analysis [94] still provides one of the best introductions to the topic. A popular choice for an advanced course in functional analysis is Rudin [148]. Jain's book on digital image processing [87] contains an in-depth treatment of fast Fourier and cosine transforms.

APPENDIX C

Multivariate Differentiation and Optimization

The purpose of this appendix is to review various aspects of multivariate differentiation and optimization. We assume the reader is familiar with differentiating a real-valued function.

C.1 Multivariate Differentiation

For a multivariate function f that maps a vector $\boldsymbol{x} = [x_1, \ldots, x_n]^\top$ to a real number $f(\boldsymbol{x})$, the **partial derivative** with respect to x_i, denoted $\frac{\partial f}{\partial x_i}$, is the derivative taken with respect to x_i while all other variables are held constant. We can write all the n partial derivatives neatly using the "scalar/vector" derivative notation:

(C.1) $$\text{scalar/vector:} \qquad \frac{\partial f}{\partial \boldsymbol{x}} := \begin{bmatrix} \frac{\partial f}{\partial x_1} \\ \vdots \\ \frac{\partial f}{\partial x_n} \end{bmatrix}.$$

This vector of partial derivatives is known as the **gradient** of f at $\boldsymbol{x}$ and is sometimes written as $\nabla f(\boldsymbol{x})$. Next, suppose that $\boldsymbol{f}$ is a multivalued (vector-valued) function taking values in $\mathbb{R}^m$, defined by

$$\boldsymbol{x} = \begin{bmatrix} x_1 \\ \vdots \\ x_n \end{bmatrix} \mapsto \begin{bmatrix} f_1(\boldsymbol{x}) \\ \vdots \\ f_m(\boldsymbol{x}) \end{bmatrix} =: \boldsymbol{f}(\boldsymbol{x}).$$

We can compute each of the partial derivatives $\partial f_i / \partial x_j$ and organize them neatly in a "vector/vector" derivative notation:

(C.2) $$\text{vector/vector:} \qquad \frac{\partial \boldsymbol{f}}{\partial \boldsymbol{x}} := \begin{bmatrix} \frac{\partial f_1}{\partial x_1} & \frac{\partial f_2}{\partial x_1} & \cdots & \frac{\partial f_m}{\partial x_1} \\ \frac{\partial f_1}{\partial x_2} & \frac{\partial f_2}{\partial x_2} & \cdots & \frac{\partial f_m}{\partial x_2} \\ \vdots & \vdots & \cdots & \vdots \\ \frac{\partial f_1}{\partial x_n} & \frac{\partial f_2}{\partial x_n} & \cdots & \frac{\partial f_m}{\partial x_n} \end{bmatrix}.$$

The transpose of this matrix is known as the **matrix of Jacobi** of $\boldsymbol{f}$ at $\boldsymbol{x}$ (sometimes called the *Fréchet derivative* of $\boldsymbol{f}$ at $\boldsymbol{x}$); that is,

$$\text{(C.3)} \qquad \mathbf{J}_{\boldsymbol{f}}(\boldsymbol{x}) := \left[\frac{\partial \boldsymbol{f}}{\partial \boldsymbol{x}}\right]^\top = \begin{bmatrix} \frac{\partial f_1}{\partial x_1} & \frac{\partial f_1}{\partial x_2} & \cdots & \frac{\partial f_1}{\partial x_n} \\ \frac{\partial f_2}{\partial x_1} & \frac{\partial f_2}{\partial x_2} & \cdots & \frac{\partial f_2}{\partial x_n} \\ \vdots & \vdots & \cdots & \vdots \\ \frac{\partial f_m}{\partial x_1} & \frac{\partial f_m}{\partial x_2} & \cdots & \frac{\partial f_m}{\partial x_n} \end{bmatrix}.$$

If we define $\boldsymbol{g}(\boldsymbol{x}) := \nabla f(\boldsymbol{x})$ and take the "vector/vector" derivative of $\boldsymbol{g}$ with respect to $\boldsymbol{x}$, we obtain the matrix of second-order partial derivatives of f:

$$\text{(C.4)} \qquad \mathbf{H}_f(\boldsymbol{x}) := \frac{\partial \boldsymbol{g}}{\partial \boldsymbol{x}} = \begin{bmatrix} \frac{\partial^2 f}{\partial^2 x_1} & \frac{\partial^2 f}{\partial x_1 \partial x_2} & \cdots & \frac{\partial^2 f}{\partial x_1 \partial x_m} \\ \frac{\partial^2 f}{\partial x_2 \partial x_1} & \frac{\partial^2 f}{\partial^2 x_2} & \cdots & \frac{\partial^2 f}{\partial x_2 \partial x_m} \\ \vdots & \vdots & \cdots & \vdots \\ \frac{\partial^2 f}{\partial x_m \partial x_1} & \frac{\partial^2 f}{\partial x_m \partial x_2} & \cdots & \frac{\partial^2 f}{\partial^2 x_m} \end{bmatrix},$$

which is known as the **Hessian matrix** of f at $\boldsymbol{x}$, also denoted as $\nabla^2 f(\boldsymbol{x})$. If these second-order partial derivatives are *continuous* in a region around $\boldsymbol{x}$, then $\frac{\partial f}{\partial x_i \partial x_j} = \frac{\partial f}{\partial x_j \partial x_i}$ and, hence, the Hessian matrix $\mathbf{H}_f(\boldsymbol{x})$ is *symmetric*. The trace of the Hessian matrix is known as the **Laplacian** of f and is denoted as:

$$\Delta f(\boldsymbol{x}) = \operatorname{tr}(\nabla^2 f(\boldsymbol{x})) = \sum_{i=1}^{m} \frac{\partial^2 f}{\partial x_i^2}.$$

We also mention that the Laplacian of the Laplacian is given by:

$$\Delta^2 f(\boldsymbol{x}) = \Delta(\Delta f(\boldsymbol{x})) = \sum_{i=1}^{m}\sum_{j=1}^{m} \frac{\partial^4 f}{\partial x_i^2 \partial x_j^2}.$$

Finally, note that we can also define a "scalar/matrix" derivative of y with respect to $\mathbf{X} \in \mathbb{R}^{m\times n}$ with (i,j)-th entry x_{ij}:

$$\frac{\partial y}{\partial \mathbf{X}} := \begin{bmatrix} \frac{\partial y}{\partial x_{11}} & \frac{\partial y}{\partial x_{12}} & \cdots & \frac{\partial y}{\partial x_{1n}} \\ \frac{\partial y}{\partial x_{21}} & \frac{\partial y}{\partial x_{22}} & \cdots & \frac{\partial y}{\partial x_{2n}} \\ \vdots & \vdots & \cdots & \vdots \\ \frac{\partial y}{\partial x_{m1}} & \frac{\partial y}{\partial x_{m2}} & \cdots & \frac{\partial y}{\partial x_{mn}} \end{bmatrix}$$

and a "matrix/scalar" derivative:

$$\frac{\partial \mathbf{X}}{\partial y} := \begin{bmatrix} \frac{\partial x_{11}}{\partial y} & \frac{\partial x_{12}}{\partial y} & \cdots & \frac{\partial x_{1n}}{\partial y} \\ \frac{\partial x_{21}}{\partial y} & \frac{\partial x_{22}}{\partial y} & \cdots & \frac{\partial x_{2n}}{\partial y} \\ \vdots & \vdots & \cdots & \vdots \\ \frac{\partial x_{m1}}{\partial y} & \frac{\partial x_{m2}}{\partial y} & \cdots & \frac{\partial x_{mn}}{\partial y} \end{bmatrix}.$$

■ **Example C.5 (Scalar/Matrix Derivative)** Let $y = \boldsymbol{a}^\top \mathbf{X} \boldsymbol{b}$, where $\mathbf{X} \in \mathbb{R}^{m \times n}$, $\boldsymbol{a} \in \mathbb{R}^m$, and $\boldsymbol{b} \in \mathbb{R}^n$. Since y is a scalar, we can write $y = \operatorname{tr}(y) = \operatorname{tr}(\mathbf{X}\boldsymbol{b}\boldsymbol{a}^\top)$, using the cyclic property of the trace (see Theorem A.7). Defining $\mathbf{C} := \boldsymbol{b}\boldsymbol{a}^\top$, we have ☞ 515

$$y = \sum_{i=1}^{m} [\mathbf{X}\mathbf{C}]_{ii} = \sum_{i=1}^{m} \sum_{j=1}^{n} x_{ij} c_{ji},$$

so that $\partial y / \partial x_{ij} = c_{ji}$ or, in matrix form,

$$\frac{\partial y}{\partial \mathbf{X}} = \mathbf{C}^\top = \boldsymbol{a}\boldsymbol{b}^\top.$$

■

■ **Example C.6 (Scalar/Matrix Derivative via the Woodbury Identity)** Let $y = \operatorname{tr}\left(\mathbf{X}^{-1}\mathbf{A}\right)$, where $\mathbf{X}, \mathbf{A} \in \mathbb{R}^{n \times n}$. We now prove that

$$\frac{\partial y}{\partial \mathbf{X}} = -\mathbf{X}^{-\top} \mathbf{A}^\top \mathbf{X}^{-\top}. \tag{C.7}$$

To show this, apply the Woodbury matrix identity to an infinitesimal perturbation, $\mathbf{X}+\varepsilon\mathbf{U}$, of $\mathbf{X}$, and take $\varepsilon \downarrow 0$ to obtain the following: ☞ 541

$$\frac{(\mathbf{X} + \varepsilon\, \mathbf{U})^{-1} - \mathbf{X}^{-1}}{\varepsilon} = -\mathbf{X}^{-1}\, \mathbf{U}\, (\mathbf{I} + \varepsilon\, \mathbf{X}^{-1}\, \mathbf{U})^{-1}\, \mathbf{X}^{-1} \longrightarrow -\mathbf{X}^{-1}\, \mathbf{U}\, \mathbf{X}^{-1}.$$

Therefore, as $\varepsilon \downarrow 0$

$$\frac{\operatorname{tr}\left((\mathbf{X} + \varepsilon\, \mathbf{U})^{-1}\mathbf{A}\right) - \operatorname{tr}\left(\mathbf{X}^{-1}\mathbf{A}\right)}{\varepsilon} \longrightarrow -\operatorname{tr}\left(\mathbf{X}^{-1}\mathbf{U}\,\mathbf{X}^{-1}\mathbf{A}\right) = -\operatorname{tr}\left(\mathbf{U}\,\mathbf{X}^{-1}\mathbf{A}\mathbf{X}^{-1}\right).$$

Now, suppose that $\mathbf{U}$ is an all zero matrix with a one in the (i, j)-th position. We can write,

$$\frac{\partial y}{\partial x_{ij}} = \lim_{\varepsilon \downarrow 0} \frac{\operatorname{tr}\left((\mathbf{X} + \varepsilon\, \mathbf{U})^{-1}\mathbf{A}\right) - \operatorname{tr}\left(\mathbf{X}^{-1}\mathbf{A}\right)}{\varepsilon} = -\operatorname{tr}\left(\mathbf{U}\mathbf{X}^{-1}\mathbf{A}\mathbf{X}^{-1}\right) = -\left[\mathbf{X}^{-1}\mathbf{A}\mathbf{X}^{-1}\right]_{ji}.$$

Therefore, $\frac{\partial y}{\partial \mathbf{X}} = -\left(\mathbf{X}^{-1}\mathbf{A}\mathbf{X}^{-1}\right)^\top$. ■

The following two examples specify multivariate derivatives for the important special cases of linear and quadratic functions.

■ **Example C.8 (Gradient of a Linear Function)** Let $\boldsymbol{f}(\boldsymbol{x}) = \mathbf{A}\boldsymbol{x}$ for some $m \times n$ constant matrix $\mathbf{A}$. Then, its vector/vector derivative (C.2) is the matrix

$$\frac{\partial \boldsymbol{f}}{\partial \boldsymbol{x}} = \mathbf{A}^\top. \tag{C.9}$$

To see this, let a_{ij} denote the (i, j)-th element of $\mathbf{A}$, so that

$$\boldsymbol{f}(\boldsymbol{x}) = \mathbf{A}\boldsymbol{x} = \begin{bmatrix} \sum_{k=1}^{n} a_{1k} x_k \\ \vdots \\ \sum_{k=1}^{n} a_{mk} x_k \end{bmatrix}.$$

To find the (j,i)-th element of $\frac{\partial \boldsymbol{f}}{\partial \boldsymbol{x}}$, we differentiate the i-th element of $\boldsymbol{f}$ with respect to x_j:

$$\frac{\partial f_i}{\partial x_j} = \frac{\partial}{\partial x_j} \sum_{k=1}^{n} a_{ik} x_k = a_{ij}.$$

In other words, the (i,j)-th element of $\frac{\partial \boldsymbol{f}}{\partial \boldsymbol{x}}$ is a_{ji}, the (i,j)-th element of $\mathbf{A}^\top$. ■

■ **Example C.10 (Gradient of a Quadratic Function)** Let $f(\boldsymbol{x}) = \boldsymbol{x}^\top \mathbf{A} \boldsymbol{x}$ for some $n \times n$ constant matrix $\mathbf{A}$. Then,

$$\nabla f(\boldsymbol{x}) = (\mathbf{A} + \mathbf{A}^\top)\boldsymbol{x}. \tag{C.11}$$

It follows immediately that if $\mathbf{A}$ is *symmetric*, that is, $\mathbf{A} = \mathbf{A}^\top$, then $\nabla(\boldsymbol{x}^\top \mathbf{A}\boldsymbol{x}) = 2\mathbf{A}\boldsymbol{x}$ and $\nabla^2(\boldsymbol{x}^\top \mathbf{A}\boldsymbol{x}) = 2\mathbf{A}$.

To prove (C.11), first observe that $f(\boldsymbol{x}) = \boldsymbol{x}^\top \mathbf{A}\boldsymbol{x} = \sum_{i=1}^n \sum_{j=1}^n a_{ij}\, x_i x_j$, which is a quadratic form in $\boldsymbol{x}$, is real-valued, with

$$\frac{\partial f}{\partial x_k} = \frac{\partial}{\partial x_k} \sum_{i=1}^{n}\sum_{j=1}^{n} a_{ij} x_i x_j = \sum_{j=1}^{n} a_{kj} x_j + \sum_{i=1}^{n} a_{ik} x_i.$$

The first term on the right-hand side is equal to the k-th element of $\mathbf{A}\boldsymbol{x}$, whereas the second term equals the k-th element of $\boldsymbol{x}^\top \mathbf{A}$, or equivalently the k-th element of $\mathbf{A}^\top \boldsymbol{x}$. ■

C.1.1 Chain Rule

Consider the functions $\boldsymbol{f} : \mathbb{R}^k \to \mathbb{R}^m$ and $\boldsymbol{g} : \mathbb{R}^m \to \mathbb{R}^n$. The function $\boldsymbol{x} \mapsto \boldsymbol{g}(\boldsymbol{f}(\boldsymbol{x}))$ is called the **composition** of $\boldsymbol{g}$ and $\boldsymbol{f}$, written as $\boldsymbol{g} \circ \boldsymbol{f}$, and is a function from $\mathbb{R}^k$ to $\mathbb{R}^n$. Suppose $\boldsymbol{y} = \boldsymbol{f}(\boldsymbol{x})$ and $\boldsymbol{z} = \boldsymbol{g}(\boldsymbol{y})$, as in Figure C.12. Let $\mathbf{J}_{\boldsymbol{f}}(\boldsymbol{x})$ and $\mathbf{J}_{\boldsymbol{g}}(\boldsymbol{y})$ be the (Fréchet) derivatives of $\boldsymbol{f}$ (at $\boldsymbol{x}$) and $\boldsymbol{g}$ (at $\boldsymbol{y}$), respectively. We may think of $\mathbf{J}_{\boldsymbol{f}}(\boldsymbol{x})$ as the matrix that describes how, in a neighborhood of $\boldsymbol{x}$, the function $\boldsymbol{f}$ can be approximated by a linear function: $\boldsymbol{f}(\boldsymbol{x} + \boldsymbol{h}) \approx \boldsymbol{f}(\boldsymbol{x}) + \mathbf{J}_{\boldsymbol{f}}(\boldsymbol{x})\boldsymbol{h}$, and similarly for $\mathbf{J}_{\boldsymbol{g}}(\boldsymbol{y})$. The well-known **chain rule** of calculus simply states that the derivative of the composition $\boldsymbol{g} \circ \boldsymbol{f}$ is the matrix product of the derivatives of $\boldsymbol{g}$ and $\boldsymbol{f}$; that is,

$$\mathbf{J}_{\boldsymbol{g}\circ\boldsymbol{f}}(\boldsymbol{x}) = \mathbf{J}_{\boldsymbol{g}}(\boldsymbol{y})\,\mathbf{J}_{\boldsymbol{f}}(\boldsymbol{x}).$$

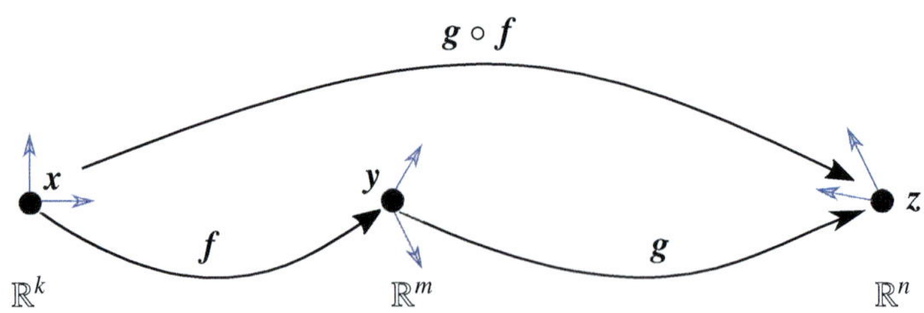

Figure C.12: Function composition. The blue arrows symbolize the linear mappings.

In terms of our vector/vector derivative notation, we have

$$\left[\frac{\partial z}{\partial \boldsymbol{x}}\right]^\top = \left[\frac{\partial z}{\partial \boldsymbol{y}}\right]^\top \left[\frac{\partial \boldsymbol{y}}{\partial \boldsymbol{x}}\right]^\top$$

or, more simply,

$$\frac{\partial z}{\partial \boldsymbol{x}} = \frac{\partial \boldsymbol{y}}{\partial \boldsymbol{x}} \frac{\partial z}{\partial \boldsymbol{y}}. \tag{C.13}$$

In a similar way we can establish a scalar/matrix chain rule. In particular, suppose $\mathbf{X}$ is an $n \times p$ matrix, which is mapped to $\boldsymbol{y} := \mathbf{X}\boldsymbol{\alpha}$ for a fixed p-dimensional vector $\boldsymbol{\alpha}$. In turn, $\boldsymbol{y}$ is mapped to a scalar $z := g(\boldsymbol{y})$ for some function g. Denote the columns of $\mathbf{X}$ by $\boldsymbol{x}_1, \ldots, \boldsymbol{x}_p$. Then,

$$\boldsymbol{y} = \mathbf{X}\boldsymbol{\alpha} = \sum_{j=1}^{p} \alpha_j \boldsymbol{x}_j,$$

and, therefore, $\partial \boldsymbol{y}/\partial \boldsymbol{x}_j = \alpha_j \mathbf{I}_n$. It follows by the chain rule (C.13) that

$$\frac{\partial z}{\partial \boldsymbol{x}_i} = \frac{\partial \boldsymbol{y}}{\partial \boldsymbol{x}_i} \frac{\partial z}{\partial \boldsymbol{y}} = \alpha_i \mathbf{I}_n \frac{\partial z}{\partial \boldsymbol{y}} = \alpha_i \frac{\partial z}{\partial \boldsymbol{y}}.$$

Therefore,

$$\frac{\partial z}{\partial \mathbf{X}} = \left[\frac{\partial z}{\partial \boldsymbol{x}_1}, \ldots, \frac{\partial z}{\partial \boldsymbol{x}_p}\right] = \left[\alpha_1 \frac{\partial z}{\partial \boldsymbol{y}}, \ldots, \alpha_p \frac{\partial z}{\partial \boldsymbol{y}}\right] = \frac{\partial z}{\partial \boldsymbol{y}} \boldsymbol{\alpha}^\top. \tag{C.14}$$

■ **Example C.15 (Derivative of the Log-Determinant)** Suppose we are given a positive definite matrix $\mathbf{X} \in \mathbb{R}^{n \times n}$ and wish to compute the scalar/matrix derivative $\partial \ln |\mathbf{X}|/\partial \mathbf{X}$. The result is

$$\frac{\partial \ln |\mathbf{X}|}{\partial \mathbf{X}} = \mathbf{X}^{-1}. \tag{C.16}$$

To see this, we can reason as follows: By Theorem A.34, we can write $\mathbf{X} = \mathbf{Q}\mathbf{D}\mathbf{Q}^\top$, ☞ 529 where $\mathbf{Q}$ is an orthogonal matrix and $\mathbf{D} = \mathrm{diag}(\lambda_1, \ldots, \lambda_n)$ is the diagonal matrix of eigenvalues of $\mathbf{X}$. The eigenvalues are strictly positive, since $\mathbf{X}$ is positive definite. Denoting the columns of $\mathbf{Q}$ by $\boldsymbol{q}_1, \ldots, \boldsymbol{q}_n$, we have

$$\lambda_i = \boldsymbol{q}_i^\top \mathbf{X} \boldsymbol{q}_i = \mathrm{tr}\left(\boldsymbol{q}_i \mathbf{X} \boldsymbol{q}_i^\top\right), \quad i = 1, \ldots, n. \tag{C.17}$$

From the properties of determinants, we have that the logarithm of the determinant of $\mathbf{X}$ is given by $y := \ln |\mathbf{X}| = \ln |\mathbf{Q}\mathbf{D}\mathbf{Q}^\top| = \ln(|\mathbf{Q}|\,|\mathbf{D}|\,|\mathbf{Q}^\top|) = \ln |\mathbf{D}| = \sum_{i=1}^n \ln \lambda_i$. We can thus write

$$\frac{\partial \ln |\mathbf{X}|}{\partial \mathbf{X}} = \sum_{i=1}^{n} \frac{\partial \ln \lambda_i}{\partial \mathbf{X}} = \sum_{i=1}^{n} \frac{\partial \lambda_i}{\partial \mathbf{X}} \frac{\partial \ln \lambda_i}{\partial \lambda_i} = \sum_{i=1}^{n} \frac{\partial \lambda_i}{\partial \mathbf{X}} \frac{1}{\lambda_i},$$

where the second equation follows from the chain rule applied to the function composition $\mathbf{X} \mapsto \lambda_i \mapsto y$. From (C.17) and Example C.5 we have $\partial \lambda_i/\partial \mathbf{X} = \boldsymbol{q}_i \boldsymbol{q}_i^\top$. It follows that

$$\frac{\partial y}{\partial \mathbf{X}} = \sum_{i=1}^{n} \boldsymbol{q}_i \boldsymbol{q}_i^\top \frac{1}{\lambda_i} = \mathbf{Q}\mathbf{D}^{-1}\mathbf{Q}^\top = \mathbf{X}^{-1}.$$

■

Example C.18 (Matrix/Scalar Derivative and the Chain Rule) A special instance of the identities (C.7) and (C.16) is when $\mathbf{A} := \boldsymbol{y}\boldsymbol{y}^\top$ and

$$\mathbf{X} := \mathbf{I}_n + \sum_{k=1}^{p} \gamma_k \boldsymbol{v}_k \boldsymbol{v}_k^\top$$

for some vectors $\boldsymbol{y}, \boldsymbol{v}_1, \ldots, \boldsymbol{v}_p$ in $\mathbb{R}^n$ and positive parameters $\gamma_1, \ldots, \gamma_p$. Using the identity:

$$\boldsymbol{y}^\top \mathbf{X}^{-1} \boldsymbol{y} = \mathrm{tr}(\mathbf{X}^{-1}\mathbf{A})$$

and applying the the chain rule in combination with the matrix/scalar derivative $\partial \mathbf{X}/\partial \gamma_i = \boldsymbol{v}_i \boldsymbol{v}_i^\top$ yields:

$$\begin{aligned} \frac{\partial(\boldsymbol{y}^\top \mathbf{X}^{-1}\boldsymbol{y})}{\partial \gamma_i} &= \mathrm{tr}\left(\frac{\partial \mathrm{tr}(\mathbf{X}^{-1}\mathbf{A})}{\partial \mathbf{X}} \frac{\partial \mathbf{X}}{\partial \gamma_i}\right) = -(\boldsymbol{v}_i^\top \mathbf{X}^{-1}\boldsymbol{y})^2 \\ \frac{\partial \ln |\mathbf{X}|}{\partial \gamma_i} &= \mathrm{tr}\left(\frac{\partial \ln |\mathbf{X}|}{\partial \mathbf{X}} \frac{\partial \mathbf{X}}{\partial \gamma_i}\right) = \boldsymbol{v}_i^\top \mathbf{X}^{-1} \boldsymbol{v}_i. \end{aligned} \tag{C.19}$$

■

C.1.2 Taylor Expansion

One of the most important results in approximation theory and statistics is Taylor's theorem. To set the stage, let $\boldsymbol{m} \in \mathbb{N}^n$ be a vector with positive integers and introduce the **multi-index notation**:

$$\boldsymbol{m}! := \prod_i (m_i!), \quad \boldsymbol{x}^{\boldsymbol{m}} := \prod_i x_i^{m_i}, \quad (\partial \boldsymbol{x})^{\boldsymbol{m}} := \prod_i (\partial x_i)^{m_i}.$$

We say that a function $f : \mathbb{R}^n \to \mathbb{R}$ is **k-th order continuously differentiable** at $\boldsymbol{x}$ if all the partial derivatives of the form $\partial^k f/(\partial \boldsymbol{x})^{\boldsymbol{m}}$ exist and are continuous at $\boldsymbol{x}$ for all $\boldsymbol{m} \in \mathbb{N}^n$ such that $|\boldsymbol{m}| := m_1 + \cdots + m_n = k$.

Theorem C.20 (Multi-dimensional Taylor Expansion): Let $\mathcal{X}$ be an open subset of $\mathbb{R}^n$ and let $\boldsymbol{a} \in \mathcal{X}$. If $f : \mathcal{X} \to \mathbb{R}$ is a k-th order continuously differentiable function, then for every $\boldsymbol{x} \in \mathcal{X}$ we have

$$f(\boldsymbol{x}) = f(\boldsymbol{a}) + \sum_{|\boldsymbol{m}| \le k} \frac{(\boldsymbol{x} - \boldsymbol{a})^{\boldsymbol{m}}}{\boldsymbol{m}!} \frac{\partial^{|\boldsymbol{m}|} f}{(\partial \boldsymbol{x})^{\boldsymbol{m}}}(\boldsymbol{a}) + o(\|\boldsymbol{x} - \boldsymbol{a}\|^k).$$

The matrix of Jacobi and the Hessian matrix feature prominently in multi-dimensional Taylor expansions, as shown in the following example.

Example C.21 (Linear and Quadratic Taylor Expansion) If f is a continuously twice differentiable function with Jacobian matrix $\mathbf{J}_f(\boldsymbol{x})$ and Hessian matrix $\mathbf{H}_f(\boldsymbol{x})$, then for every $\boldsymbol{x} \in \mathcal{X}$ Theorem C.20 yields the following first- and second-order Taylor expansions as $\|\boldsymbol{x} - \boldsymbol{a}\| \to 0$:

$$f(\boldsymbol{x}) = f(\boldsymbol{a}) + \mathbf{J}_f(\boldsymbol{a})\,(\boldsymbol{x} - \boldsymbol{a}) + o(\|\boldsymbol{x} - \boldsymbol{a}\|) \tag{C.22}$$

and

$$f(\boldsymbol{x}) = f(\boldsymbol{a}) + \mathbf{J}_f(\boldsymbol{a})\,(\boldsymbol{x}-\boldsymbol{a}) + \frac{1}{2}(\boldsymbol{x}-\boldsymbol{a})^\top\,\mathbf{H}_f(\boldsymbol{a})\,(\boldsymbol{x}-\boldsymbol{a}) + o(\|\boldsymbol{x}-\boldsymbol{a}\|^2). \tag{C.23}$$

This result is essentially saying that a smooth enough function behaves locally (in the neighborhood of a point $\boldsymbol{x}$) like a linear and quadratic function. Thus, the gradient or Hessian of an approximating linear or quadratic function is a basic building block of many approximation and optimization algorithms.

Note that there exists an $\boldsymbol{a}'$ that lies on the line segment between $\boldsymbol{x}$ and $\boldsymbol{a}$ such that (C.22) and (C.23) hold without remainder terms, with $\mathbf{J}_f(\boldsymbol{a})$ in (C.22) replaced by $\mathbf{J}_f(\boldsymbol{a}')$ and $\mathbf{H}_f(\boldsymbol{a})$ in (C.23) replaced by $\mathbf{H}_f(\boldsymbol{a}')$. ■

■ **Example C.24 (Error Estimation via Richardson Extrapolation)** Suppose that we have a three times continuously differentiable function f and we wish to approximate its derivative at x using the **central difference** approximation with step size $\varepsilon > 0$:

$$D_\varepsilon(x) := \frac{f(x+\varepsilon) - f(x-\varepsilon)}{2\varepsilon}.$$

An application of Theorem C.20 reveals the error in this approximation:

$$D_\varepsilon(x) - f'(x) = \frac{\varepsilon^2}{6} f^{(3)}(x) + o(\varepsilon^2).$$

Whilst the above expression tells us that the error is of order $O(\varepsilon^2)$, the exact value of the error term, $\frac{\varepsilon^2}{6} f^{(3)}(x)$, is typically unknown. Note that if we have a simple formula for the third derivative of f, then it is unlikely we need a central difference approximation of its first derivative.

Nevertheless, there is a way to accurately extrapolate the value of the unknown error from the observed behavior of the approximation itself. Consider computing $D_{c\varepsilon}(x)$, where $c \in (0, 1)$ is a constant, say $c = 1/2$. Then, an application of Taylor's Theorem C.20 to $D_{c\varepsilon}(x)$ reveals that

$$D_\varepsilon(x) - D_{c\varepsilon}(x) = (1 - c^2)\frac{\varepsilon^2}{6} f^{(3)}(x) + o(\varepsilon^2).$$

In other words, the error term is given by:

$$\frac{\varepsilon^2}{6} f^{(3)}(x) = \underbrace{\frac{D_\varepsilon(x) - D_{c\varepsilon}(x)}{1 - c^2}}_{\text{extrapolation}} + o(\varepsilon^2).$$

The right-hand-side expression in the equation above is called the **Richardson extrapolation** estimate of the error $\frac{\varepsilon^2}{6} f^{(3)}(x)$. Notice how the error is estimated or extrapolated from the two derivative approximations $D_\varepsilon(x)$ and $D_{c\varepsilon}(x)$, which use two step sizes that differ only via the multiplicative constant $c > 0$. ■

■ **Example C.25 (Density Estimation)** Suppose that $\mathcal{T}_n = \{\boldsymbol{X}_1, \ldots, \boldsymbol{X}_n\}$ is an iid sample from a smooth probability density f on $\mathbb{R}^d$ and $g_{\mathcal{T}_n}(\cdot \mid \sigma)$ is the kernel density

☞ 147 estimate in (4.35). Denoting $\mathbf{Z} \sim \mathcal{N}(\mathbf{0}, \mathbf{I}_d)$ and $\varphi_{\sigma^2}(\boldsymbol{z}) := \frac{1}{(2\pi\sigma^2)^{d/2}} \mathrm{e}^{-\frac{\|\boldsymbol{z}\|^2}{2\sigma^2}}$ to be the pdf of $\sigma \mathbf{Z}$, the expected value of the learner is:

$$\begin{aligned}\mathbb{E}\, g_{\mathcal{T}_n}(\boldsymbol{x} \mid \sigma) &= \int_{\mathbb{R}^d} f(\boldsymbol{y})\, \varphi_{\sigma^2}(\boldsymbol{y} - \boldsymbol{x})\, \mathrm{d}\boldsymbol{y} \\ &= \frac{1}{\sigma^d} \int_{\mathbb{R}^d} f(\boldsymbol{y})\, \varphi_1((\boldsymbol{y} - \boldsymbol{x})/\sigma)\, \mathrm{d}\boldsymbol{y} = \mathbb{E} f(\boldsymbol{x} + \sigma \mathbf{Z}),\end{aligned}$$

where we used the change of variable $\boldsymbol{z} = (\boldsymbol{y} - \boldsymbol{x})/\sigma$ in the integral. Applying Taylor's Theorem C.20 to $f(\boldsymbol{x} + \sigma \boldsymbol{z})$ under the assumption that $\sigma = \sigma_n \to 0$ as $n \uparrow \infty$ yields:

$$f(\boldsymbol{x} + \sigma \boldsymbol{z}) = \sum_{|\boldsymbol{m}| \leq 5} \frac{\sigma^{|\boldsymbol{m}|} \boldsymbol{z}^{\boldsymbol{m}}}{\boldsymbol{m}!} \frac{\partial^{|\boldsymbol{m}|} f}{(\partial \boldsymbol{x})^{\boldsymbol{m}}}(\boldsymbol{x}) + O(\sigma^6).$$

Since $\mathbb{E}\mathbf{Z}^{\boldsymbol{m}} = \mathbb{E} \prod_i Z_i^{m_i} = \prod_i \mathbb{E} Z_i^{m_i} = 0$ for all $\boldsymbol{m}$ such that $|\boldsymbol{m}|$ is an odd number and $\mathbb{E} Z_1^4 = 3$, we obtain:

$$\mathbb{E} f(\boldsymbol{x} + \sigma \mathbf{Z}) - f(\boldsymbol{x}) = \frac{\sigma^2}{2} \Delta f(\boldsymbol{x}) + \frac{\sigma^4}{8} \Delta^2 f(\boldsymbol{x}) + O(\sigma^6).$$

If f and its derivatives of all orders decay to zero as $\|\boldsymbol{x}\| \to \infty$, then integration by parts yields

$$-\int_{\mathbb{R}^d} \Delta f(\boldsymbol{x})\, \Delta^2 f(\boldsymbol{x})\, \mathrm{d}\boldsymbol{x} = \sum_{|\boldsymbol{m}|=3} \left\| \frac{\partial^3 f}{(\partial \boldsymbol{x})^{\boldsymbol{m}}} \right\|^2 =: \|\Delta^{3/2} f\|^2.$$

Therefore, the integrated squared bias is

$$\int_{\mathbb{R}^d} \left(\mathbb{E} f(\boldsymbol{x} + \sigma \mathbf{Z}) - f(\boldsymbol{x})\right)^2 \mathrm{d}\boldsymbol{x} = \frac{\sigma^4}{4} \|\Delta f\|^2 - \frac{\sigma^6}{8} \|\Delta^{3/2} f\|^2 + O(\sigma^8).$$

Similarly, using the identities:

$$\begin{aligned}\mathbb{V}\mathrm{ar}\, g_{\mathcal{T}_n}(\boldsymbol{x} \mid \sigma) &= \frac{1}{n} \mathbb{V}\mathrm{ar}\big(\varphi_{\sigma^2}(\boldsymbol{X} - \boldsymbol{x})\big) \\ &= \frac{\mathbb{E}\, \varphi_1^2((\boldsymbol{X} - \boldsymbol{x})/\sigma)}{n\sigma^{2d}} - \frac{\left(\mathbb{E}\, \varphi_{\sigma^2}(\boldsymbol{X} - \boldsymbol{x})\right)^2}{n} \\ &= \frac{\mathbb{E} f\left(\boldsymbol{x} + \sigma \mathbf{Z}/\sqrt{2}\right)}{n(4\pi\sigma^2)^{d/2}} - \frac{\left(\mathbb{E} f(\boldsymbol{x} + \sigma \mathbf{Z})\right)^2}{n},\end{aligned}$$

and interchanging the order of integration and expectation:

$$\int_{\mathbb{R}^d} \mathbb{E} f\left(\boldsymbol{x} + \sigma\, \mathbf{Z}/\sqrt{2}\right) \mathrm{d}\boldsymbol{x} = \mathbb{E} \int_{\mathbb{R}^d} f\left(\boldsymbol{x} + \sigma\, \mathbf{Z}/\sqrt{2}\right) \mathrm{d}\boldsymbol{x} = 1,$$

we obtain:

$$\int_{\mathbb{R}^d} \mathbb{V}\mathrm{ar}\, g_{\mathcal{T}_n}(\boldsymbol{x} \mid \sigma)\, \mathrm{d}\boldsymbol{x} = \frac{1}{(4\pi)^{d/2} n \sigma^d} - \frac{\int \left(\mathbb{E} f(\boldsymbol{x} + \sigma \mathbf{Z})\right)^2 \mathrm{d}\boldsymbol{x}}{n}.$$

■

C.2 Optimization Theory

Optimization is concerned with finding minimal or maximal solutions of a real-valued **objective function** f in some set $\mathcal{X}$:

$$\min_{x \in \mathcal{X}} f(\boldsymbol{x}) \quad \text{or} \quad \max_{x \in \mathcal{X}} f(\boldsymbol{x}). \tag{C.26}$$

Since any maximization problem can easily be converted into a minimization problem via the equivalence:

$$\max_{x} f(\boldsymbol{x}) \equiv -\min_{x} -f(\boldsymbol{x}),$$

we focus only on minimization problems. We use the following terminology. A **local minimizer** of $f(\boldsymbol{x})$ is an element $\boldsymbol{x}^* \in \mathcal{X}$ such that $f(\boldsymbol{x}^*) \leq f(\boldsymbol{x})$ for all $\boldsymbol{x}$ in some neighborhood of $\boldsymbol{x}^*$. If $f(\boldsymbol{x}^*) \leq f(\boldsymbol{x})$ for all $\boldsymbol{x} \in \mathcal{X}$, then $\boldsymbol{x}^*$ is called a **global minimizer** or **global solution**. The set of global minimizers is denoted by

$$\operatorname*{argmin}_{x \in \mathcal{X}} f(\boldsymbol{x}).$$

The function value $f(\boldsymbol{x}^*)$ corresponding to a local/global minimizer $\boldsymbol{x}^*$ is referred to as the **local/global minimum** of $f(\boldsymbol{x})$.

Optimization problems may be classified by the set $\mathcal{X}$ and the objective function f. If $\mathcal{X}$ is countable, the optimization problem is called **discrete** or **combinatorial**. If instead $\mathcal{X}$ is a nondenumerable set such as $\mathbb{R}^n$ and f takes values in a nondenumerable set, then the problem is said to be **continuous**. Optimization problems that are neither discrete nor continuous are said to be **mixed**.

The search set $\mathcal{X}$ is often defined by means of *constraints*. A standard setting for constrained optimization (minimization) is the following:

$$\begin{aligned} &\min_{x \in \mathcal{Y}} f(\boldsymbol{x}) \\ \text{subject to: } & h_i(\boldsymbol{x}) = 0, \quad i = 1, \ldots, m, \\ & g_i(\boldsymbol{x}) \leq 0, \quad i = 1, \ldots, k. \end{aligned} \tag{C.27}$$

Here, f is the objective function, and $\{g_i\}$ and $\{h_i\}$ are given functions so that $h_i(\boldsymbol{x}) = 0$ and $g_i(\boldsymbol{x}) \leq 0$ represent the *equality* and *inequality* constraints, respectively. The region $\mathcal{X} \subseteq \mathcal{Y}$ where the objective function is defined and where all the constraints are satisfied is called the **feasible region**. An optimization problem without constraints is said to be an **unconstrained** problem.

For an unconstrained continuous optimization problem, the search space $\mathcal{X}$ is often taken to be (a subset of) $\mathbb{R}^n$, and f is assumed to be a C^k function for sufficiently high k (typically $k = 2$ or 3 suffices); that is, its k-th order derivative is continuous. For a C^1 function the standard approach to minimizing $f(\boldsymbol{x})$ is to solve the equation

$$\nabla f(\boldsymbol{x}) = \boldsymbol{0}, \tag{C.28}$$

where $\nabla f(\boldsymbol{x})$ is the *gradient* of f at $\boldsymbol{x}$. The solutions $\boldsymbol{x}^*$ to (C.28) are called **stationary points**. Stationary points can be local/global minimizers, local/global maximizers, or **saddle points** (which are neither). If, in addition, the function is C^2, the condition ☞ 581

$$\boldsymbol{y}^\top (\nabla^2 f(\boldsymbol{x}^*))\, \boldsymbol{y} > 0 \quad \text{for all } \boldsymbol{y} \neq \boldsymbol{0} \tag{C.29}$$

ensures that the stationary point $\boldsymbol{x}^*$ is a local minimizer of f. The condition (C.29) states
☞ 582 that the *Hessian* matrix of f at $\boldsymbol{x}^*$ is *positive definite*. Recall that we write $\mathbf{H} \succ \mathbf{0}$ to indicate that a matrix $\mathbf{H}$ is positive definite.

In Figure C.30 we have a multiextremal objective function on $\mathcal{X} = \mathbb{R}$. There are four stationary points: two are local minimizers, one is a local maximizer, and one is neither a minimizer nor a maximizer, but a saddle point.

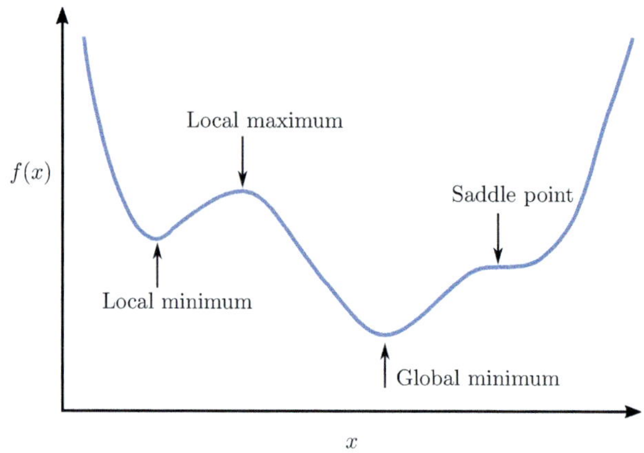

Figure C.30: A multiextremal objective function in one dimension.

C.2.1 Convexity and Optimization

An important class of optimization problems is related to the notion of *convexity*. A set $\mathcal{X}$ is said to be **convex** if for all $\boldsymbol{x}_1, \boldsymbol{x}_2 \in \mathcal{X}$ it holds that

$$\alpha\,\boldsymbol{x}_1 + (1-\alpha)\,\boldsymbol{x}_2 \in \mathcal{X} \quad \text{for all} \quad \alpha \in [0,1].$$

In addition, the objective function f is a **convex function** provided that for each $\boldsymbol{x}$ in the interior of $\mathcal{X}$ there exists a vector $\boldsymbol{v}$ such that

$$f(\boldsymbol{y}) \geq f(\boldsymbol{x}) + (\boldsymbol{y}-\boldsymbol{x})^\top \boldsymbol{v}, \quad \boldsymbol{y} \in \mathcal{X}. \tag{C.31}$$

The vector $\boldsymbol{v}$ in (C.31) may not be unique and is referred to as a **subgradient** of f. In fact, the set of all subgradients for which (C.31) holds is called the **subdifferential** of f at $\boldsymbol{x}$ and is denoted as $\partial f(\boldsymbol{x})$.

One crucial property of a convex function f is that *Jensen's inequality* holds (see
☞ 67 Exercise 19 in Chapter 2) for any random vector $\boldsymbol{X}$:

$$\mathbb{E} f(\boldsymbol{X}) \geq f(\mathbb{E}\boldsymbol{X}).$$

Another important property of a convex function f is that $f(\boldsymbol{x})$ is the global minimum if and only if $\mathbf{0} \in \partial f(\boldsymbol{x})$, that is, $\boldsymbol{v} = \mathbf{0}$ is a subgradient in (C.31).

■ **Example C.32 (Convexity and Directional Derivative)** The **directional derivative** of a multivariate function f at $\boldsymbol{x}$ in the unit direction $\boldsymbol{d}$ with $\|\boldsymbol{d}\| = 1$ is defined as the right derivative of $g(t) := f(\boldsymbol{x} + t\,\boldsymbol{d})$ at $t = 0$:

$$\lim_{t \downarrow 0} \frac{f(\boldsymbol{x} + t\,\boldsymbol{d}) - f(\boldsymbol{x})}{t} = \lim_{t \uparrow \infty} t\,(f(\boldsymbol{x} + \boldsymbol{d}/t) - f(\boldsymbol{x})).$$

This right derivative may not always exist. However, if f is a convex function, then the directional derivative of f at $\boldsymbol{x}$ in the interior of its domain always exists (in any direction $\boldsymbol{d}$).

To see this, let $t_1 \geq t_2 > 0$. By Jensen's inequality we have for any $\boldsymbol{x}$ and $\boldsymbol{y}$ in the interior of the domain:

$$\frac{t_2}{t_1} f(\boldsymbol{y}) + \left(1 - \frac{t_2}{t_1}\right) f(\boldsymbol{x}) \geq f\left(\frac{t_2}{t_1}\boldsymbol{y} + \left(1 - \frac{t_2}{t_1}\right)\boldsymbol{x}\right).$$

Making the substitution $\boldsymbol{y} = \boldsymbol{x} + t_1 \boldsymbol{d}$ and rearranging the last equation yields:

$$\frac{f(\boldsymbol{x} + t_1\,\boldsymbol{d}) - f(\boldsymbol{x})}{t_1} \geq \frac{f(\boldsymbol{x} + t_2\,\boldsymbol{d}) - f(\boldsymbol{x})}{t_2}.$$

In other words, the function:

$$t \mapsto \frac{f(\boldsymbol{x} + t\,\boldsymbol{d}) - f(\boldsymbol{x})}{t}$$

is increasing for $t > 0$ and therefore the directional derivative satisfies:

$$\lim_{t \downarrow 0} \frac{f(\boldsymbol{x} + t\,\boldsymbol{d}) - f(\boldsymbol{x})}{t} = \inf_{t>0} \frac{f(\boldsymbol{x} + t\,\boldsymbol{d}) - f(\boldsymbol{x})}{t}.$$

Hence, to show existence it is enough to show that $(f(\boldsymbol{x} + t\,\boldsymbol{d}) - f(\boldsymbol{x}))/t$ is bounded from below.

Since $\boldsymbol{x}$ lies in the interior of the domain of f, we can choose t small enough so that $\boldsymbol{x} + t\,\boldsymbol{d}$ also lies in the interior. Therefore, the convexity of f implies that there exists a subgradient vector $\boldsymbol{v}$ such that

$$f(\boldsymbol{x} + t\,\boldsymbol{d}) \geq f(\boldsymbol{x}) + \boldsymbol{v}^\top (t\,\boldsymbol{d}).$$

In other words,

$$\frac{f(\boldsymbol{x} + t\,\boldsymbol{d}) - f(\boldsymbol{x})}{t} \geq \boldsymbol{v}^\top \boldsymbol{d}$$

provides a lower bound for all $t > 0$, and the directional derivative of f at an interior $\boldsymbol{x}$ always exists (in any direction). Finally, note that if

$$\lim_{t \downarrow 0} \frac{f(\boldsymbol{x} + t\,\boldsymbol{d}) - f(\boldsymbol{x})}{t} = \boldsymbol{v}^\top \boldsymbol{d}$$

for all unit directions $\boldsymbol{d}$ and a unique subgradient $\boldsymbol{v}$, then f is in fact differentiable at $\boldsymbol{x}$. ■

A function f is said to be **δ-Lipschitz continuous**, with constant $\delta > 0$, if for all $\boldsymbol{x}, \boldsymbol{y} \in \mathbb{R}^d$,

$$|f(\boldsymbol{x}) - f(\boldsymbol{y})| \leq \delta\,\|\boldsymbol{x} - \boldsymbol{y}\|.$$

If f is a convex function, then the Lipschitz continuity of f is equivalent to boundedness of the subgradients of f.

Lemma C.33 (Bounded Subgradients and Lipschitz Continuity): Suppose that f is convex. Then, f is δ-Lipschitz continuous if and only if the norm of every subgradient $\boldsymbol{v}$ of f is bounded by δ: $\|\boldsymbol{v}\| \leq \delta$.

Proof. First, suppose that $\|\boldsymbol{v}\| \leq \delta$ for any subgradient $\boldsymbol{v}$ of f. If $\boldsymbol{v}_{\boldsymbol{x}}$ denotes the subgradient at an arbitrary $\boldsymbol{x}$, then (C.31) implies that

$$f(\boldsymbol{y}) - f(\boldsymbol{x}) \geq \langle \boldsymbol{y} - \boldsymbol{x}, \boldsymbol{v}_{\boldsymbol{x}} \rangle \geq -\|\boldsymbol{y} - \boldsymbol{x}\| \, \|\boldsymbol{v}_{\boldsymbol{x}}\| \geq -\delta \|\boldsymbol{y} - \boldsymbol{x}\|, \tag{C.34}$$

☞ 568 where we used the Cauchy–Schwarz inequality in Theorem B.16. Similarly, for the subgradient $\boldsymbol{v}_{\boldsymbol{y}}$ at $\boldsymbol{y}$, the inequality (C.31) yields:

$$f(\boldsymbol{x}) - f(\boldsymbol{y}) \geq \langle \boldsymbol{x} - \boldsymbol{y}, \boldsymbol{v}_{\boldsymbol{y}} \rangle \geq -\|\boldsymbol{y} - \boldsymbol{x}\| \, \|\boldsymbol{v}_{\boldsymbol{y}}\| \geq -\delta \|\boldsymbol{y} - \boldsymbol{x}\|. \tag{C.35}$$

Then, (C.34) and (C.35) together imply that for all $\boldsymbol{x}$ and $\boldsymbol{y}$ we have

$$\delta \|\boldsymbol{y} - \boldsymbol{x}\| \geq f(\boldsymbol{x}) - f(\boldsymbol{y}) \geq -\delta \|\boldsymbol{y} - \boldsymbol{x}\|, \tag{C.36}$$

which implies the δ-Lipschitz continuity of f. Second, assume that f is δ-Lipschitz continuous, so that (C.36) holds for all $\boldsymbol{x}$ and $\boldsymbol{y}$. Then, (C.36) and $f(\boldsymbol{x}) - f(\boldsymbol{y}) \geq \langle \boldsymbol{x} - \boldsymbol{y}, \boldsymbol{v}_{\boldsymbol{y}} \rangle$ imply that $\delta \|\boldsymbol{y} - \boldsymbol{x}\| \geq \langle \boldsymbol{x} - \boldsymbol{y}, \boldsymbol{v}_{\boldsymbol{y}} \rangle$ for all $\boldsymbol{x}, \boldsymbol{y}$. Since $\boldsymbol{x}$ is an arbitrary vector, we can select it as $\boldsymbol{x} = \boldsymbol{y} + \boldsymbol{v}_{\boldsymbol{y}}$ to deduce that the subgradient at $\boldsymbol{y}$ satisfies:

$$\|\boldsymbol{v}_{\boldsymbol{y}}\|^2 = \langle \boldsymbol{x} - \boldsymbol{y}, \boldsymbol{v}_{\boldsymbol{y}} \rangle \leq \delta \|\boldsymbol{x} - \boldsymbol{y}\| = \delta \|\boldsymbol{v}_{\boldsymbol{y}}\|.$$

Finally, since $\boldsymbol{y}$ is arbitrary, we deduce that the Euclidean norm of all subgradients must be bounded by δ. □

A function f satisfying (C.31) with strict inequality is said to be *strictly convex*. It is said to be a (strictly) **concave function** if $-f$ is (strictly) convex.

A property more stringent than strict convexity is **A-strong convexity**. A function f is said to be **A**-strongly convex if there exists a positive definite matrix **A** such that for all $\boldsymbol{x}$ and $\boldsymbol{v} \in \partial f(\boldsymbol{x})$:

$$f(\boldsymbol{y}) \geq f(\boldsymbol{x}) + (\boldsymbol{y} - \boldsymbol{x})^\top \boldsymbol{v} + \frac{1}{2} \|\boldsymbol{y} - \boldsymbol{x}\|_{\mathbf{A}}^2, \quad \boldsymbol{y} \in \mathcal{X}, \tag{C.37}$$

where $\|\boldsymbol{x}\|_{\mathbf{A}} := \sqrt{\boldsymbol{x}^\top \mathbf{A} \boldsymbol{x}}$ defines a Mahalanobis norm. A function f is said to be **A-strongly concave** if $-f$ is **A**-strongly convex.

Assuming that $\mathcal{X}$ is an open set, convexity for $f \in C^1$ is equivalent to

$$f(\boldsymbol{y}) \geq f(\boldsymbol{x}) + (\boldsymbol{y} - \boldsymbol{x})^\top \nabla f(\boldsymbol{x}) \quad \text{for all } \boldsymbol{x}, \boldsymbol{y} \in \mathcal{X}.$$

As matter of fact, f is differentiable at $\boldsymbol{x}$ if and only if the subdifferential $\partial f(\boldsymbol{x})$ contains only one element, that is, $\partial f(\boldsymbol{x}) = \{\nabla f(\boldsymbol{x})\}$.

Moreover, for $f \in C^2$ strict convexity is equivalent to the Hessian matrix being positive definite for all $\boldsymbol{x} \in \mathcal{X}$, and convexity is equivalent to the Hessian matrix being *positive semidefinite* for all $\boldsymbol{x}$; that is, $\boldsymbol{y}^\top \left(\nabla^2 f(\boldsymbol{x})\right) \boldsymbol{y} \geq 0$ for all $\boldsymbol{y}$ and $\boldsymbol{x}$. Recall that we
☞ 531 write $\mathbf{H} \succeq \mathbf{0}$ to indicate that a matrix $\mathbf{H}$ is positive semidefinite.

■ **Example C.38 (Convexity and Differentiability)** If f is a continuously differentiable multivariate function, then f is convex if and only if the univariate function

$$g(t) := f(\boldsymbol{x} + t\,\boldsymbol{d}), \quad t \in [0,1]$$

is a convex function for any $\boldsymbol{x}$ and $\boldsymbol{x} + \boldsymbol{d}$ in the interior of the domain of f. This property provides an alternative definition for convexity of a multivariate and differentiable function.

To see why it is true, first assume that f is convex and $t_1, t_2 \in [0,1]$. Then, using the subgradient definition of convexity in (C.31), we have

$$f(\boldsymbol{a}) \geq f(\boldsymbol{b}) + (\boldsymbol{a} - \boldsymbol{b})^\top \boldsymbol{v}$$

for some subgradient $\boldsymbol{v}$. Substituting with $\boldsymbol{a} = \boldsymbol{x} + t_1\boldsymbol{d}$ and $\boldsymbol{b} = \boldsymbol{x} + t_2\boldsymbol{d}$, we obtain

$$g(t_1) \geq g(t_2) + (t_1 - t_2)\,\boldsymbol{v}^\top \boldsymbol{d}$$

for any two points $t_1, t_2 \in [0,1]$. Therefore, g is convex, because we have identified the existence of a subgradient $\boldsymbol{v}^\top \boldsymbol{d}$ for each t_2.

Conversely, assume that g is convex for $t \in [0,1]$. Since f is differentiable, then so is g. Then, the convexity of g implies that there is a subgradient v at 0 such that: $g(t) \geq g(0) + t\,v$ for all $t \in [0,1]$. Rearranging,

$$v \geq \frac{g(t) - g(0)}{t},$$

and taking the right limit as $t \downarrow 0$ we obtain $v \geq g'(0) = \boldsymbol{d}^\top \nabla f(\boldsymbol{x})$. Therefore,

$$g(t) \geq g(0) + t\,v \geq g(0) + t\,\boldsymbol{d}^\top \nabla f(\boldsymbol{x})$$

and substituting $t = 1$ yields:

$$f(\boldsymbol{x} + \boldsymbol{d}) \geq f(\boldsymbol{x}) + \boldsymbol{d}^\top \nabla f(\boldsymbol{x}),$$

so that there exists a subgradient vector, namely $\nabla f(\boldsymbol{x})$, for each $\boldsymbol{x}$. Hence, f is convex by the definition in (C.31). ■

An optimization program of the form (C.27) is said to be a **convex programming problem** if:

1. The objective f is a *convex function*.
2. The inequality constraint functions $\{g_i\}$ are convex.
3. The equality constraint functions $\{h_i\}$ are *affine*, that is, of the form $\boldsymbol{e}_i^\top(\mathbf{A}\boldsymbol{x} - \boldsymbol{b})$. This is equivalent to both h_i and $-h_i$ being convex for all i.

Table C.39 summarizes some commonly encountered problems, all of which are convex, with the exception of the quadratic programs with $\mathbf{H} \not\succeq 0$.

Table C.39: Some common classes of optimization problems.

Name	$f(\boldsymbol{x})$	Constraints
Linear Program (LP)	$\boldsymbol{q}^\top \boldsymbol{x}$	$\mathbf{A}\boldsymbol{x} \leq \boldsymbol{b}$, or $\mathbf{A}\boldsymbol{x} \leq \boldsymbol{b}$ and $\boldsymbol{x} \geq \mathbf{0}$
Quadratic Program (QP)	$\frac{1}{2}\boldsymbol{x}^\top \mathbf{H}\boldsymbol{x} + \boldsymbol{q}^\top \boldsymbol{x}$	$\mathbf{A}\boldsymbol{x} \leq \boldsymbol{b}$ and $\mathbf{A}_{\text{eq}}\boldsymbol{x} = \boldsymbol{b}_{\text{eq}}$
Convex QP	$\frac{1}{2}\boldsymbol{x}^\top \mathbf{H}\boldsymbol{x} + \boldsymbol{q}^\top \boldsymbol{x}$	$\mathbf{A}\boldsymbol{x} \leq \boldsymbol{b}$ and $\mathbf{A}_{\text{eq}}\boldsymbol{x} = \boldsymbol{b}_{\text{eq}}$ $\quad (\mathbf{H} \succeq 0)$
Convex Program	$f(\boldsymbol{x})$ convex	$\{g_i(\boldsymbol{x})\}$ convex, $\{h_i(\boldsymbol{x})\}$ of the form $\boldsymbol{e}_i^\top(\mathbf{A}\boldsymbol{x} - \boldsymbol{b})$

Recognizing convex optimization problems or those that can be transformed to convex optimization problems can be challenging. However, once formulated as convex optimization problems, these can be efficiently solved using subgradient [156], bundle [83], and cutting-plane methods [86].

C.2.2 Lagrangian Method

The main components of the Lagrangian method are the Lagrange multipliers and the Lagrange function. The method was developed by Lagrange in 1797 for the optimization problem (C.27) with only equality constraints. In 1951 Kuhn and Tucker extended Lagrange's method to inequality constraints. Given an optimization problem (C.27) containing only equality constraints $h_i(\boldsymbol{x}) = 0,\ i = 1, \ldots, m$, the **Lagrange function** is defined as

$$\mathcal{L}(\boldsymbol{x}, \boldsymbol{\beta}) = f(\boldsymbol{x}) + \sum_{i=1}^{m} \beta_i\, h_i(\boldsymbol{x}),$$

where the coefficients $\{\beta_i\}$ are called the **Lagrange multipliers**. A necessary condition for a point $\boldsymbol{x}^*$ to be a local minimizer of $f(\boldsymbol{x})$ subject to the equality constraints $h_i(\boldsymbol{x}) = 0,\ i = 1, \ldots, m$, is

$$\nabla_{\boldsymbol{x}}\, \mathcal{L}(\boldsymbol{x}^*, \boldsymbol{\beta}^*) = \mathbf{0},$$
$$\nabla_{\boldsymbol{\beta}}\, \mathcal{L}(\boldsymbol{x}^*, \boldsymbol{\beta}^*) = \mathbf{0},$$

for some value $\boldsymbol{\beta}^*$. The above conditions are also sufficient if $\mathcal{L}(\boldsymbol{x}, \boldsymbol{\beta}^*)$ is a convex function of $\boldsymbol{x}$.

■ **Example C.40 (Quadratic Programming with Equality Constraints)** Consider minimizing the quadratic function:

$$f(\boldsymbol{x}) := \frac{1}{2}\boldsymbol{x}^\top \mathbf{H}\boldsymbol{x} + \boldsymbol{q}^\top \boldsymbol{x}$$

subject to the equality constraints $\mathbf{A}\boldsymbol{x} = \boldsymbol{b}$, where $\mathbf{A} \in \mathbb{R}^{m \times n}$ with $\text{rank}(\mathbf{A}) = m < n$. Then, the Lagrange function is

$$\mathcal{L}(\boldsymbol{x}, \boldsymbol{\beta}) = \frac{1}{2}\boldsymbol{x}^\top \mathbf{H}\boldsymbol{x} + \boldsymbol{q}^\top \boldsymbol{x} + \boldsymbol{\beta}^\top(\mathbf{A}\boldsymbol{x} - \boldsymbol{b}),$$

and a necessary condition for $\boldsymbol{x}^*$ to be a local minimizer of f is that it satisfies the linear system:

$$\begin{bmatrix} \mathbf{H} & \mathbf{A}^\top \\ \mathbf{A} & \mathbf{0} \end{bmatrix} \begin{bmatrix} \boldsymbol{x}^* \\ \boldsymbol{\beta}^* \end{bmatrix} = \begin{bmatrix} -\boldsymbol{q} \\ \boldsymbol{b} \end{bmatrix}. \tag{C.41}$$

The matrix arising from this linear system is called the **KKT matrix**, in reference to the title of Theorem C.42 below. If $\mathbf{H}$ is positive definite, then the blockwise inversion formula (A.56) implies that the KKT matrix is invertible. ■

Given the original optimization problem (C.27), containing both the equality and inequality constraints, the **generalized Lagrange function**, or **Lagrangian**, is defined as

$$\mathcal{L}(\boldsymbol{x}, \boldsymbol{\alpha}, \boldsymbol{\beta}) = f(\boldsymbol{x}) + \sum_{i=1}^{k} \alpha_i\, g_i(\boldsymbol{x}) + \sum_{i=1}^{m} \beta_i\, h_i(\boldsymbol{x}).$$

Theorem C.42 (Karush–Kuhn–Tucker (KKT) Conditions): A necessary condition for a point $\boldsymbol{x}^*$ to be a local minimizer of $f(\boldsymbol{x})$ in the optimization problem (C.27) is the existence of an $\boldsymbol{\alpha}^*$ and $\boldsymbol{\beta}^*$ satisfying the following conditions:

1. *Stationarity:* $\nabla_{\boldsymbol{x}} \mathcal{L}(\boldsymbol{x}^*, \boldsymbol{\alpha}^*, \boldsymbol{\beta}^*) = \mathbf{0}$.
2. *Primal equality feasibility:* $\nabla_{\boldsymbol{\beta}} \mathcal{L}(\boldsymbol{x}^*, \boldsymbol{\alpha}^*, \boldsymbol{\beta}^*) = \mathbf{0}$.
3. *Primal inequality feasibility:* $g_i(\boldsymbol{x}^*) \leq 0$ for $i = 1, \ldots, k$.
4. *Dual feasibility:* $\alpha_i^* \geq 0$ for $i = 1, \ldots, k$.
5. *Complementary slackness:* $\alpha_i^*\, g_i(\boldsymbol{x}^*) = 0$ for $i = 1, \ldots, k$.

For *convex* programs we have the following important results [30, 65]:

1. Every local solution $\boldsymbol{x}^*$ to a convex programming problem is a global solution and the set of global solutions is convex. If, in addition, the objective function is strictly convex, then any global solution is unique.

2. For a strictly convex programming problem with C^1 objective and constraint functions, the KKT conditions are necessary and sufficient for a unique global solution.

Finally, observe that the condition $\nabla_{\boldsymbol{\beta}} \mathcal{L}(\boldsymbol{x}^*, \boldsymbol{\alpha}^*, \boldsymbol{\beta}^*) = \mathbf{0}$ in Theorem C.42 is equivalent to requiring that $h_i(\boldsymbol{x}^*) = 0$ for $i = 1, \ldots, m$. Hence, the label *primal feasibility*.

■ **Example C.43 (Nonnegative Least Squares)** Consider finding the minimizer of the sum of squares $\|\mathbf{X}\boldsymbol{b} - \boldsymbol{y}\|^2$, subject to the constraint that $\boldsymbol{b} \geq \mathbf{0}$:

$$\boldsymbol{b}^* = \underset{\boldsymbol{b} \geq \mathbf{0}}{\operatorname{argmin}}\, \|\mathbf{X}\boldsymbol{b} - \boldsymbol{y}\|^2,$$

where $\mathbf{X} \in \mathbb{R}^{n \times p}$ with $\operatorname{rank}(\mathbf{X}) = p < n$. This is equivalent to a convex quadratic programming problem with $\mathbf{H} := \mathbf{X}^\top \mathbf{X}$ and $\boldsymbol{q} := -\mathbf{X}^\top \boldsymbol{y}$. According to Theorem C.42, the minimizer $\boldsymbol{b}^*$ must then satisfy the KKT conditions:

$$\begin{aligned}\mathbf{H}\boldsymbol{b}^* + \boldsymbol{q} - \boldsymbol{\alpha}^* &= \mathbf{0}\\ \boldsymbol{\alpha}^* \geq \mathbf{0},\ \boldsymbol{b}^* &\geq \mathbf{0}\\ \boldsymbol{\alpha}^* \odot \boldsymbol{b}^* &= \mathbf{0}.\end{aligned}$$

Eliminating $\boldsymbol{\alpha}^*$ yields the simplest form of the KKT equations:

$$\begin{aligned}\mathbf{H}\boldsymbol{b}^* + \boldsymbol{q} &\geq \mathbf{0}\\ \boldsymbol{b}^* \odot (\mathbf{H}\boldsymbol{b}^* + \boldsymbol{q}) &= \mathbf{0}\\ \boldsymbol{b}^* &\geq \mathbf{0}.\end{aligned}$$

In particular, this means that for each k, either $b_k^* > 0$ and $\boldsymbol{e}_k^\top \mathbf{H}\boldsymbol{b}^* + q_k = 0$, or $b_k^* = 0$ and $\boldsymbol{e}_k^\top \mathbf{H}\boldsymbol{b}^* + q_k \geq 0$. ■

When the optimization program is not convex, then the KKT conditions are merely a necessary condition for a *local minimizer*. However, when the Lagrangian is second order continuously differentiable at the KKT point, that is, the Hessian matrix

$$\mathbf{H} := \frac{\partial^2}{\partial \boldsymbol{x} \partial \boldsymbol{x}^\top} \mathcal{L}(\boldsymbol{x}^*, \boldsymbol{\alpha}^*, \boldsymbol{\beta}^*)$$

is well defined, then a sufficient condition for a strict local minimizer is the following.

Theorem C.44 (Second-Order Sufficient Condition): Let $(\boldsymbol{x}^*, \boldsymbol{\alpha}^*, \boldsymbol{\beta}^*)$ satisfy the KKT conditions in Theorem C.42. The KKT point $\boldsymbol{x}^*$ is a local minimizer of (C.27), provided that $\boldsymbol{v}^\top \mathbf{H} \boldsymbol{v} > 0$ for all $\boldsymbol{v} \neq \mathbf{0}$ that satisfy the linear equations:

$$\begin{aligned}\boldsymbol{v}^\top \nabla g_i(\boldsymbol{x}^*) &= 0, \quad \forall i \in \{j : \alpha_j^* > 0\}\\ \boldsymbol{v}^\top \nabla h_i(\boldsymbol{x}^*) &= 0, \quad i = 1, \ldots, m.\end{aligned}$$

■ **Example C.45 (Minimization with Nonnegativity Constraints)** Suppose that we wish to minimize the second-order continuously differentiable function f subject to the constraint that $\boldsymbol{x} \geq \mathbf{0}$. Similarly to Example C.43, the KKT conditions in Theorem C.42 can be simplified to:

$$\begin{aligned}\nabla f(\boldsymbol{x}^*) &\geq \mathbf{0}\\ \boldsymbol{x}^* \odot \nabla f(\boldsymbol{x}^*) &= \mathbf{0}\\ \boldsymbol{x}^* &\geq \mathbf{0}.\end{aligned}$$

In other words, if $\boldsymbol{g} := \nabla f(\boldsymbol{x}^*)$ and $\mathcal{A} := \{j : x_j^* > 0\}$ is the set of nonzero variables, then the KKT point $\boldsymbol{x}^*$ satisfies:

$$\begin{aligned}g_i &\geq 0, \quad \forall i \in \mathcal{A}^c\\ g_i &= 0, \quad \forall i \in \mathcal{A}.\end{aligned}$$

If $\mathbf{H} := \nabla^2 f(\boldsymbol{x}^*)$ is the Hessian of f at $\boldsymbol{x}^*$, then Theorem C.44 asserts that a sufficient condition for $\boldsymbol{x}^*$ to be a local minimizer is that $\boldsymbol{v}^\top \mathbf{H} \boldsymbol{v} > 0$ for all nonzero vectors $\boldsymbol{v}$ such that $\boldsymbol{v}_{\mathcal{A}^c} = \mathbf{0}$. In other words, the Hessian matrix needs to be positive definite only within the $|\mathcal{A}|$-dimensional vector subspace spanned by $\{\boldsymbol{e}_i, i \in \mathcal{A}\}$, where $\boldsymbol{e}_1, \ldots, \boldsymbol{e}_n$ are the unit-length *standard basis* vectors. ■

C.2.3 Duality

The aim of duality is to provide an alternative formulation of an optimization problem which is often more computationally efficient or has some theoretical significance (see [65, Page 219]). The original problem (C.27) is referred to as the **primal** problem whereas the reformulated problem, based on Lagrange multipliers, is called the **dual** problem. Duality theory is most relevant to convex optimization problems. It is well known that if the primal optimization problem is (strictly) convex then the dual problem is (strictly) concave and has a (unique) solution from which the (unique) optimal primal solution can be deduced.

The **Lagrange dual program** (also called the **Wolfe dual**) of the primal program (C.27), is:

$$\begin{aligned} \max_{\boldsymbol{\alpha}, \boldsymbol{\beta}} \quad & \mathcal{L}^*(\boldsymbol{\alpha}, \boldsymbol{\beta}) \\ \text{subject to:} \quad & \boldsymbol{\alpha} \geq \mathbf{0}, \end{aligned}$$

where $\mathcal{L}^*$ is the *Lagrange dual function*:

$$\mathcal{L}^*(\boldsymbol{\alpha}, \boldsymbol{\beta}) = \inf_{\boldsymbol{x} \in \mathcal{X}} \mathcal{L}(\boldsymbol{x}, \boldsymbol{\alpha}, \boldsymbol{\beta}), \tag{C.46}$$

giving the greatest lower bound (infimum) of $\mathcal{L}(\boldsymbol{x}, \boldsymbol{\alpha}, \boldsymbol{\beta})$ over all possible $\boldsymbol{x} \in \mathcal{X}$.

It is not difficult to see that if f^* is the minimal value of the primal problem, then $\mathcal{L}^*(\boldsymbol{\alpha}, \boldsymbol{\beta}) \leq f^*$ for any $\boldsymbol{\alpha} \geq \mathbf{0}$ and any $\boldsymbol{\beta}$. This property is called **weak duality**. The Lagrangian dual program thus determines the best lower bound on f^*. If d^* is the optimal value for the dual problem then $d^* \leq f^*$. The difference $f^* - d^*$ is called the **duality gap**.

The duality gap is extremely useful for providing lower bounds for the solutions of primal problems that may be impossible to solve directly. It is important to note that for linearly constrained problems, if the primal is infeasible (does not have a solution satisfying the constraints), then the dual is either infeasible or unbounded. Conversely, if the dual is infeasible then the primal has no solution. Of crucial importance is the **strong duality** theorem, which states that for convex programs (C.27) with linear constrained functions h_i and g_i the duality gap is zero, and any $\boldsymbol{x}^*$ and $(\boldsymbol{\alpha}^*, \boldsymbol{\beta}^*)$ satisfying the KKT conditions are (global) solutions to the primal and dual programs, respectively. In particular, this holds for linear and convex quadratic programs (note that not all quadratic programs are convex).

■ **Example C.47 (Linear Programming)** Consider the **linear programming** optimization $\min_{\boldsymbol{x}:\mathbf{A}\boldsymbol{x} \leq \boldsymbol{b}} \boldsymbol{q}^\top \boldsymbol{x}$, using the notation in Table C.39. The Lagrangian is:

$$\mathcal{L}(\boldsymbol{x}, \boldsymbol{\alpha}) = \boldsymbol{q}^\top \boldsymbol{x} + \boldsymbol{\alpha}^\top (\mathbf{A}\boldsymbol{x} - \boldsymbol{b})$$

and its dual function is

$$\mathcal{L}^*(\alpha) = \inf_{x} \mathcal{L}(x, \alpha) = \begin{cases} -\alpha^\top b, & \text{if } \mathbf{A}^\top \alpha + q = \mathbf{0} \\ -\infty, & \text{otherwise.} \end{cases}$$

Therefore, the Lagrange dual program is:

$$\begin{aligned} \min_{\alpha} \quad & \alpha^\top b \\ \text{subject to:} \quad & \mathbf{A}^\top \alpha + q = \mathbf{0} \\ & \alpha \geq \mathbf{0}. \end{aligned}$$

■

■ **Example C.48 (Convex Quadratic Programming)** Consider the convex **quadratic programming** optimization from Table C.39:

$$\min_{x : \mathbf{A}x \leq b} \frac{1}{2} x^\top \mathbf{H} x + q^\top x.$$

The Lagrange dual function is:

$$\mathcal{L}^*(\alpha) = -\frac{1}{2}(\mathbf{A}^\top \alpha + q)^\top \mathbf{H}^{-1}(\mathbf{A}^\top \alpha + q) - \alpha^\top b.$$

Hence, the Wolfe dual is

$$\min_{\alpha \geq \mathbf{0}} \frac{1}{2} \alpha^\top \mathbf{A}\mathbf{H}^{-1}\mathbf{A}^\top \alpha + \alpha^\top (b + \mathbf{A}\mathbf{H}^{-1} q) + \text{const.},$$

where the constant absorbs all terms not involving α. ■

For a convex primal program with C^1 objective and constraint functions, the Lagrangian dual function (C.46) can be obtained by simply setting the gradient (with respect to x) of the Lagrangian $\mathcal{L}(x, \alpha, \beta)$ to zero. One can further simplify the dual program by substituting into the Lagrangian the relations between the variables thus obtained.

Further, for a convex primal problem, if there is a *strictly feasible* point $\widetilde{x}$ (that is, a feasible point satisfying all of the inequality constraints with strict inequality), then the duality gap is zero, and strong duality holds. This is known as **Slater's condition** [30, Page 226].

The Lagrange dual problem is an important example of a *saddle-point problem* or *minimax* problem. In such problems the aim is to locate a point $(x^*, y^*) \in \mathcal{X} \times \mathcal{Y}$ that satisfies

$$\sup_{y \in \mathcal{Y}} \inf_{x \in \mathcal{X}} f(x, y) = \inf_{x \in \mathcal{X}} f(x, y^*) = f(x^*, y^*) = \sup_{y \in \mathcal{Y}} f(x^*, y) = \inf_{x \in \mathcal{X}} \sup_{y \in \mathcal{Y}} f(x, y).$$

The equation

$$\sup_{y \in \mathcal{Y}} \inf_{x \in \mathcal{X}} f(x, y) = \inf_{x \in \mathcal{X}} \sup_{y \in \mathcal{Y}} f(x, y)$$

is known as the **minimax** equality. Other problems that fall into this framework are zero-sum games in game theory; see also [38] for a number of combinatorial optimization problems that can be viewed as minimax problems.

C.3 Numerical Root-Finding and Minimization

In order to minimize a C^1 function $f : \mathbb{R}^n \to \mathbb{R}$ one may solve

$$\nabla f(\boldsymbol{x}) = \boldsymbol{0},$$

which gives a stationary point of f. As a consequence, any technique for root-finding can be transformed into an unconstrained optimization method by attempting to locate roots of the gradient. However, as noted in Section C.2, not all stationary points are minima, and so additional information (such as is contained in the Hessian, if f is C^2) needs to be considered in order to establish the type of stationary point.

Alternatively, a root of a continuous function $\boldsymbol{g} : \mathbb{R}^n \to \mathbb{R}^n$ may be found by minimizing the norm of $\boldsymbol{g}(\boldsymbol{x})$ over all $\boldsymbol{x}$; that is, by solving $\min_{\boldsymbol{x}} f(\boldsymbol{x})$, with $f(\boldsymbol{x}) := \|\boldsymbol{g}(\boldsymbol{x})\|_p$, where for $p \geq 1$ the **p-norm** of $\boldsymbol{y} = [y_1, \ldots, y_n]^\top$ is defined as

$$\|\boldsymbol{y}\|_p := \left(\sum_{i=1}^{n} |y_i|^p \right)^{1/p}.$$

Hence, any (un)constrained optimization method can be transformed into a technique for locating the roots of a function.

Starting with an initial guess $\boldsymbol{x}_0$, most minimization and root-finding algorithms create a sequence $\boldsymbol{x}_0, \boldsymbol{x}_1, \ldots$ using the iterative updating rule:

$$\boldsymbol{x}_{t+1} = \boldsymbol{x}_t + \alpha_t \, \boldsymbol{d}_t, \quad t = 0, 1, 2, \ldots, \tag{C.49}$$

where $\alpha_t > 0$ is a (typically small) step size, called the **learning rate**, and the vector $\boldsymbol{d}_t$ is the search direction at step t. The iteration (C.49) continues until the sequence $\{\boldsymbol{x}_t\}$ is deemed to have converged to a solution, or a computational budget has been exhausted. The performance of all such iterative methods depends crucially on the quality of the initial guess $\boldsymbol{x}_0$.

There are three broad categories of iterative algorithms of the form (C.49):

- Those of **line search** type, where at iteration t we first compute a direction $\boldsymbol{d}_t$ and then determine a reasonable step size α_t along this direction. For example, in the case of minimization, $\alpha_t > 0$ may be chosen to approximately minimize $f(\boldsymbol{x}_t + \alpha \, \boldsymbol{d}_t)$ for fixed $\boldsymbol{x}_t$ and $\boldsymbol{d}_t$.

- Those of **trust region** type, where at each iteration t we first determine a suitable step size α_t and then compute an approximately optimal direction $\boldsymbol{d}_t$.

- Those of **coordinate descent** type, where at iteration t we pick up the k-th coordinate of $\boldsymbol{x}_t$ according to some formula, say $k = 1 + (t \mod n)$, and determine the optimal search direction α_t with the k-th coordinate search direction: $\boldsymbol{d}_t = \boldsymbol{e}_k$.

In the following sections, we review several widely-used root-finding and optimization algorithms of the line search type, and finish with brief introduction to the coordinate descent method.

C.3.1 Newton-Like Methods

Suppose we wish to find roots of a function $\boldsymbol{f} : \mathbb{R}^n \to \mathbb{R}^n$. If $\boldsymbol{f}$ is in C^1, we can approximate $\boldsymbol{f}$ around a point $\boldsymbol{x}_t$ as

$$\boldsymbol{f}(\boldsymbol{x}) \approx \boldsymbol{f}(\boldsymbol{x}_t) + \mathbf{J}_{\boldsymbol{f}}(\boldsymbol{x}_t)(\boldsymbol{x} - \boldsymbol{x}_t),$$

where $\mathbf{J}_{\boldsymbol{f}}$ is the *matrix of Jacobi* — the matrix of partial derivatives of $\boldsymbol{f}$; see (C.3). ☞ 582 When $\mathbf{J}_{\boldsymbol{f}}(\boldsymbol{x}_t)$ is invertible, this linear approximation has root $\boldsymbol{x}_t - \mathbf{J}_{\boldsymbol{f}}^{-1}(\boldsymbol{x}_t)\,\boldsymbol{f}(\boldsymbol{x}_t)$. This gives the iterative updating formula (C.49) for finding roots of $\boldsymbol{f}$ with direction $\boldsymbol{d}_t = -\mathbf{J}_{\boldsymbol{f}}^{-1}(\boldsymbol{x}_t)\,\boldsymbol{f}(\boldsymbol{x}_t)$ and learning rate $\alpha_t = 1$. This is known as **Newton's method** (or the *Newton–Raphson method*) for root-finding.

Instead of a unit learning rate, sometimes it is more effective to use an α_t that satisfies the **Armijo inexact line search** condition:

$$\|\boldsymbol{f}(\boldsymbol{x}_t + \alpha_t\,\boldsymbol{d}_t)\| < (1 - \varepsilon_1\,\alpha_t)\,\|\boldsymbol{f}(\boldsymbol{x}_t)\|,$$

where ε_1 is a small heuristically chosen constant, say $\varepsilon_1 = 10^{-4}$. For C^1 functions, such an α_t always exists by continuity and can be computed as in the following algorithm.

Algorithm C.50: Newton–Raphson for Finding Roots of $\boldsymbol{f}(\boldsymbol{x}) = \mathbf{0}$

Input: An initial guess $\boldsymbol{x}$ and stopping error $\varepsilon > 0$.
Output: The approximate root of $\boldsymbol{f}(\boldsymbol{x}) = \mathbf{0}$.

while $\|\boldsymbol{f}(\boldsymbol{x})\| > \varepsilon$ and budget is not exhausted **do**
  Solve the linear system $\mathbf{J}_{\boldsymbol{f}}(\boldsymbol{x})\,\boldsymbol{d} = -\boldsymbol{f}(\boldsymbol{x})$.
  $\alpha \leftarrow 1$
  while $\|\boldsymbol{f}(\boldsymbol{x} + \alpha\,\boldsymbol{d})\| > (1 - 10^{-4}\alpha)\,\|\boldsymbol{f}(\boldsymbol{x})\|$ **do**
    $\alpha \leftarrow \alpha/2$
  $\boldsymbol{x} \leftarrow \boldsymbol{x} + \alpha\,\boldsymbol{d}$
return $\boldsymbol{x}$

We can adapt a root-finding Newton-like method in order to minimize a differentiable function $f : \mathbb{R}^n \to \mathbb{R}$. We simply try to locate a zero of the gradient of f. When f is a C^2 function, the function $\nabla f : \mathbb{R}^n \to \mathbb{R}^n$ is continuous, and so the root of ∇f leads to the search direction

$$\boldsymbol{d}_t = -\mathbf{H}_t^{-1}\,\nabla f(\boldsymbol{x}_t), \tag{C.51}$$

where $\mathbf{H}_t$ is the Hessian matrix at $\boldsymbol{x}_t$ (the matrix of Jacobi of the gradient is the Hessian). When the learning rate α_t is equal to 1, the update $\boldsymbol{x}_t - \mathbf{H}_t^{-1}\,\nabla f(\boldsymbol{x}_t)$ can alternatively be derived by assuming that $f(\boldsymbol{x})$ is approximately quadratic and convex in the neighborhood of $\boldsymbol{x}_t$, that is,

$$f(\boldsymbol{x}) \approx f(\boldsymbol{x}_t) + (\boldsymbol{x} - \boldsymbol{x}_t)^\top \nabla f(\boldsymbol{x}_t) + \frac{1}{2}(\boldsymbol{x} - \boldsymbol{x}_t)^\top \mathbf{H}_t(\boldsymbol{x} - \boldsymbol{x}_t), \tag{C.52}$$

and then minimizing the right-hand side of (C.52) with respect to $\boldsymbol{x}$.

The following algorithm uses an *Armijo inexact line search* for minimization and guards against the possibility that the Hessian may not be positive definite (that is, its ☞ 543 Cholesky decomposition does not exist).

Algorithm C.53: Newton–Raphson for Minimizing $f(x)$

Input: initial $\boldsymbol{x}$; tolerance $\varepsilon > 0$; line search parameter $\xi \in (0,1)$.
Output: An approximate minimizer of $f(\boldsymbol{x})$.

```
L ← I_n                                        // the identity matrix
while ||∇f(x)|| > ε and budget is not exhausted do
    Compute the Hessian H at x.
    if H ≻ 0 then                              // Cholesky is successful
        LL^T ← H                               // update/recompute Cholesky factor of Hessian
    d ← −L^{-1}∇f(x)                           // computed by forward substitution
    d ← L^{-T}d                                // computed by backward substitution
    α ← 1
    while f(x + αd) > f(x) + α 10^{-4}∇f(x)^T d do
        α ← α × ξ
    x ← x + α d
return x
```

A downside with all Newton-like methods is that at each step they require the calculation and inversion of an $n \times n$ Hessian matrix, which has computing time of $O(n^3)$, and is thus infeasible for large n. One way to avoid this cost is to use quasi-Newton methods, described next.

C.3.2 Quasi-Newton Methods

The idea behind quasi-Newton methods is to replace the inverse Hessian in (C.51) at iteration t by an $n \times n$ matrix $\mathbf{C}$ satisfying the **secant condition**:

$$\mathbf{C}\,\boldsymbol{g} = \boldsymbol{\delta}, \tag{C.54}$$

where $\boldsymbol{\delta} \leftarrow \boldsymbol{x}_t - \boldsymbol{x}_{t-1}$ and $\boldsymbol{g} \leftarrow \nabla f(\boldsymbol{x}_t) - \nabla f(\boldsymbol{x}_{t-1})$ are vectors stored in memory at each iteration t. The secant condition is satisfied, for example, by the *Broyden's family* of matrices:

$$\mathbf{A} + \frac{1}{\boldsymbol{u}^\top \boldsymbol{g}}(\boldsymbol{\delta} - \mathbf{A}\,\boldsymbol{g})\,\boldsymbol{u}^\top$$

for some $\boldsymbol{u} \neq \mathbf{0}$ and $\mathbf{A}$. Since there is an infinite number of matrices that satisfy the condition (C.54), we need a way to determine a unique $\mathbf{C}$ at each iteration t such that computing and storing $\mathbf{C}$ from one step to the next is fast and avoids any costly matrix inversion. The following examples illustrate how, starting with an initial guess $\mathbf{C} = \mathbf{I}$ at $t = 0$, such a matrix $\mathbf{C}$ can be efficiently updated from one iteration to the next.

■ **Example C.55 (Low-Rank Hessian Update)** The quadratic model (C.52) can be strengthened by further assuming that $\exp(-f(\boldsymbol{x}))$ is proportional to a probability density that can be approximated in the neighborhood of $\boldsymbol{x}_t$ by the pdf of the $\mathcal{N}(\boldsymbol{x}_{t+1}, \mathbf{H}_t^{-1})$ distribution. This normal approximation allows us to measure the *discrepancy* between two pairs $(\boldsymbol{x}_1, \mathbf{H}_0)$ and $(\boldsymbol{x}_2, \mathbf{H}_1)$ using the Kullback–Leibler divergence between the pdfs ☞ 43

of the $\mathcal{N}(\boldsymbol{x}_1, \mathbf{H}_0^{-1})$ and $\mathcal{N}(\boldsymbol{x}_2, \mathbf{H}_1^{-1})$ distributions (see formula (2.101) on page 67):
(C.56)
$$\mathcal{D}(\boldsymbol{x}_1, \mathbf{H}_0^{-1} \,|\, \boldsymbol{x}_2, \mathbf{H}_1^{-1}) := \frac{1}{2}\Big(\mathrm{tr}(\mathbf{H}_1\mathbf{H}_0^{-1}) - \ln|\mathbf{H}_1\mathbf{H}_0^{-1}| + (\boldsymbol{x}_2 - \boldsymbol{x}_1)^\top \mathbf{H}_1 (\boldsymbol{x}_2 - \boldsymbol{x}_1) - n\Big).$$

Suppose that the latest approximation to the inverse Hessian is $\mathbf{C}$ and we wish to compute an updated approximation for step t. One approach is to find the symmetric matrix that minimizes its Kullback–Leibler discrepancy from $\mathbf{C}$, as defined above, subject to the constraint (C.54). In other words,

$$\min_{\mathbf{A}} \mathcal{D}(\mathbf{0}, \mathbf{C} \,|\, \mathbf{0}, \mathbf{A})$$
$$\text{subject to: } \mathbf{A}\boldsymbol{g} = \boldsymbol{\delta}, \ \mathbf{A} = \mathbf{A}^\top.$$

The solution to this constrained optimization (see Exercise 9 on page 458) yields the Broyden–Fletcher–Goldfarb–Shanno or **BFGS formula** for updating the matrix $\mathbf{C}$ from one iteration to the next:

$$\mathbf{C}_{\mathrm{BFGS}} = \mathbf{C} + \underbrace{\frac{\boldsymbol{g}^\top\boldsymbol{\delta} + \boldsymbol{g}^\top\mathbf{C}\boldsymbol{g}}{(\boldsymbol{g}^\top\boldsymbol{\delta})^2}\boldsymbol{\delta}\boldsymbol{\delta}^\top - \frac{1}{\boldsymbol{g}^\top\boldsymbol{\delta}}\left(\boldsymbol{\delta}\boldsymbol{g}^\top\mathbf{C} + (\boldsymbol{\delta}\boldsymbol{g}^\top\mathbf{C})^\top\right)}_{\text{BFGS update}}. \tag{C.57}$$

In a practical implementation, we keep a single copy of $\mathbf{C}$ in memory and apply the BFGS update to it at every iteration. Note that if the current $\mathbf{C}$ is symmetric, then so is the updated matrix. Moreover, the BFGS update is a matrix of rank two.

Since the Kullback–Leibler divergence is not symmetric, it is possible to flip the roles of $\mathbf{H}_0$ and $\mathbf{H}_1$ in (C.56) and instead solve

$$\min_{\mathbf{A}} \mathcal{D}(\mathbf{0}, \mathbf{A} \,|\, \mathbf{0}, \mathbf{C})$$
$$\text{subject to: } \mathbf{A}\boldsymbol{g} = \boldsymbol{\delta}, \ \mathbf{A} = \mathbf{A}^\top.$$

The solution (see Exercise 9 on page 458) gives the Davidon–Fletcher–Powell or *DFP formula* for updating the matrix $\mathbf{C}$ from one iteration to the next:

$$\mathbf{C}_{\mathrm{DFP}} = \mathbf{C} + \underbrace{\frac{\boldsymbol{\delta}\boldsymbol{\delta}^\top}{\boldsymbol{g}^\top\boldsymbol{\delta}} - \frac{\mathbf{C}\boldsymbol{g}\boldsymbol{g}^\top\mathbf{C}}{\boldsymbol{g}^\top\mathbf{C}\boldsymbol{g}}}_{\text{DFP update}}. \tag{C.58}$$

Note that if the curvature condition $\boldsymbol{g}^\top\boldsymbol{\delta} > 0$ holds and the current $\mathbf{C}$ is positive definite, then so is its update. ■

■ **Example C.59 (Diagonal Hessian Update)** The original BFGS formula requires $\mathcal{O}(n^2)$ storage and computation, which may be unmanageable for large n. One way to circumvent the prohibitive quadratic cost is to only store and update a diagonal Hessian matrix from one iteration to the next. If $\mathbf{C}$ is diagonal, then we may not be able to satisfy the secant condition (C.54) and maintain positive definiteness. Instead the secant condition (C.54) can be relaxed to the set of inequalities $\boldsymbol{g} \geq \mathbf{C}^{-1}\boldsymbol{\delta}$, which are related to
☞ 590 the definition of a *subgradient* for convex functions. We can then find a unique diagonal

matrix by minimizing $\mathcal{D}(\boldsymbol{x}_t, \mathbf{C} \,|\, \boldsymbol{x}_{t+1}, \mathbf{A})$ with respect to $\mathbf{A}$ and subject to the constraints that $\mathbf{A}\boldsymbol{g} \geq \boldsymbol{\delta}$ and $\mathbf{A}$ is diagonal. The solution (Exercise 14 on page 459) yields the updating formula for a diagonal element c_i of $\mathbf{C}$:

$$c_i \leftarrow \begin{cases} \dfrac{2c_i}{1+\sqrt{1+4c_i u_i^2}}, & \text{if } \dfrac{2c_i}{1+\sqrt{1+4c_i u_i^2}} \geq \delta_i/g_i \\ \delta_i/g_i, & \text{otherwise,} \end{cases} \tag{C.60}$$

where $\boldsymbol{u} := \nabla f(\boldsymbol{x}_t)$ and we assume a unit learning rate: $\boldsymbol{x}_{t+1} = \boldsymbol{x}_t - \mathbf{A}\boldsymbol{u}$. ■

■ **Example C.61 (Scalar Hessian Update)** If the identity matrix is used in place of the Hessian in (C.51), one obtains **gradient descent** methods, in which the iteration (C.49) reduces to $\boldsymbol{x}_{t+1} = \boldsymbol{x}_t - \alpha_t \nabla f(\boldsymbol{x}_t)$.

The rationale for the name *gradient descent* is as follows: If we start from any point $\boldsymbol{x}$ and make an infinitesimal move in some direction, then the function value is reduced by the largest magnitude in the (unit norm) direction: $\boldsymbol{u}^* := -\nabla f(\boldsymbol{x})/\|\nabla f(\boldsymbol{x})\|$. This is seen from the following inequality for all unit vectors $\boldsymbol{u}$ (that is, $\|\boldsymbol{u}\| = 1$):

$$\frac{\mathrm{d}}{\mathrm{d}t} f(\boldsymbol{x} + t\,\boldsymbol{u}^*)\Big|_{t=0} \leq \frac{\mathrm{d}}{\mathrm{d}t} f(\boldsymbol{x} + t\,\boldsymbol{u})\Big|_{t=0}.$$

Observe that equality is achieved if and only if $\boldsymbol{u} = \boldsymbol{u}^*$. This inequality is an easy consequence of the Cauchy–Schwarz inequality: ☞ 568

$$-\nabla f^\top \boldsymbol{u} \leq |\nabla f^\top \boldsymbol{u}| \underbrace{\leq}_{\text{Cauchy–Schwartz}} \|\boldsymbol{u}\|\,\|\nabla f\| = \|\nabla f\| = -\nabla f^\top \boldsymbol{u}^*.$$

The gradient descent iteration, $\boldsymbol{x}_{t+1} = \boldsymbol{x}_t - \alpha_t \nabla f(\boldsymbol{x}_t)$, still requires a suitable choice of the learning rate α_t. An alternative way to think about the iteration is to assume that the learning rate is always unity, and that at each iteration we use an inverse Hessian matrix of the form $\alpha_t \mathbf{I}$ for some positive constant α_t. Satisfying the secant condition (C.54) with a matrix of the form $\mathbf{C} = \alpha\mathbf{I}$ is not possible. However, it is possible to choose α so that the secant condition (C.54) is satisfied in the direction of $\boldsymbol{g}$ (or alternatively $\boldsymbol{\delta}$). This gives the **Barzilai–Borwein formulas** for the learning rate at iteration t:

$$\alpha_t = \frac{\boldsymbol{g}^\top \boldsymbol{\delta}}{\|\boldsymbol{g}\|^2} \quad \left(\text{or alternatively} \quad \alpha_t = \frac{\|\boldsymbol{\delta}\|^2}{\boldsymbol{\delta}^\top \boldsymbol{g}}\right). \tag{C.62}$$

■

C.3.3 Normal Approximation Method

Let $\varphi_{\mathbf{H}_t^{-1}}(\boldsymbol{x} - \boldsymbol{x}_{t+1})$ denote the pdf of the $\mathcal{N}(\boldsymbol{x}_{t+1}, \mathbf{H}_t^{-1})$ distribution. As we already saw in Example C.55, the quadratic approximation (C.52) of f in the neighborhood of $\boldsymbol{x}_t$ is equivalent (up to a constant) to the minus of the logarithm of the pdf $\varphi_{\mathbf{H}_t^{-1}}(\boldsymbol{x} - \boldsymbol{x}_{t+1})$. In other words, we use $\varphi_{\mathbf{H}_t^{-1}}(\boldsymbol{x} - \boldsymbol{x}_{t+1})$ as a simple model for the density

$$\exp(-f(\boldsymbol{x})) \Big/ \int \exp(-f(\boldsymbol{y}))\,\mathrm{d}\boldsymbol{y}.$$

One consequence of the normal approximation is that for $\boldsymbol{x}$ in the neighborhood of $\boldsymbol{x}_{t+1}$, we can write:

$$-\nabla f(\boldsymbol{x}) \approx \frac{\partial}{\partial \boldsymbol{x}} \ln \varphi_{\mathbf{H}_t^{-1}}(\boldsymbol{x} - \boldsymbol{x}_{t+1}) = -\mathbf{H}_t(\boldsymbol{x} - \boldsymbol{x}_{t+1}).$$

In other words, using the fact that $\mathbf{H}_t^\top = \mathbf{H}_t$,

$$\nabla f(\boldsymbol{x})[\nabla f(\boldsymbol{x})]^\top \approx \mathbf{H}_t(\boldsymbol{x} - \boldsymbol{x}_{t+1})(\boldsymbol{x} - \boldsymbol{x}_{t+1})^\top \mathbf{H}_t,$$

and taking expectations on both sides with respect to $\boldsymbol{x} \sim \mathcal{N}(\boldsymbol{x}_{t+1}, \mathbf{H}_t^{-1})$ gives:

$$\mathbb{E}\, \nabla f(\boldsymbol{x})\, [\nabla f(\boldsymbol{x})]^\top \approx \mathbf{H}_t.$$

This suggests that, given the gradient vectors computed in the past h (where h stands for **h**istory) of Newton iterations:

$$\boldsymbol{u}_i := \nabla f(\boldsymbol{x}_i), \quad i = t-(h-1), \ldots, t,$$

the Hessian matrix $\mathbf{H}_t$ can be approximated via the average

$$\frac{1}{h} \sum_{i=t-h+1}^{t} \boldsymbol{u}_i \boldsymbol{u}_i^\top.$$

A shortcoming of this approximation is that, unless h is large enough, the Hessian approximation:

$$\sum_{i=t-h+1}^{t} \boldsymbol{u}_i \boldsymbol{u}_i^\top$$

may not be full rank and hence not invertible. To ensure that the Hessian approximation
☞ 321 is invertible, we add a suitable diagonal matrix $\mathbf{A}_0$ to obtain the *regularized* version of the approximation:

$$\mathbf{H}_t \approx \mathbf{A}_0 + \frac{1}{h} \sum_{i=t-h+1}^{t} \boldsymbol{u}_i \boldsymbol{u}_i^\top.$$

With this full-rank approximation of the Hessian, the Newton search direction in (C.51) becomes:

$$\boldsymbol{d}_t = -\left(\mathbf{A}_0 + \frac{1}{h} \sum_{i=t-h+1}^{t} \boldsymbol{u}_i \boldsymbol{u}_i^\top\right)^{-1} \boldsymbol{u}_t. \tag{C.63}$$

☞ 543 Thus, $\boldsymbol{d}_t$ can be computed in $\mathcal{O}(h^2 n)$ time via the Sherman–Morrison Algorithm A.65. Further to this, the search direction (C.63) can be efficiently updated to the next one:

$$\boldsymbol{d}_{t+1} = -\left(\mathbf{A}_0 + \frac{1}{h} \sum_{i=t-h+2}^{t+1} \boldsymbol{u}_i \boldsymbol{u}_i^\top\right)^{-1} \boldsymbol{u}_{t+1}$$

in $\mathcal{O}(h\, n)$ time, thus avoiding the usual $\mathcal{O}(h^2 n)$ cost (see Exercise 5 on page 457).

C.3.4 Nonlinear Least Squares

Consider the squared-error training loss in nonlinear regression: ☞ 219

$$\ell_\tau(g(\cdot \mid \boldsymbol{\beta})) = \frac{1}{n}\sum_{i=1}^{n}(g(\boldsymbol{x}_i \mid \boldsymbol{\beta}) - y_i)^2,$$

where $g(\cdot \mid \boldsymbol{\beta})$ is a nonlinear prediction function that depends on the parameter $\boldsymbol{\beta}$ (for example, (5.47) shows the nonlinear logistic prediction function). The training loss can be written as $\frac{1}{n}\|\boldsymbol{g}(\tau \mid \boldsymbol{\beta}) - \boldsymbol{y}\|^2$, where

$$\boldsymbol{g}(\tau \mid \boldsymbol{\beta}) := [g(\boldsymbol{x}_1 \mid \boldsymbol{\beta}), \dots, g(\boldsymbol{x}_n \mid \boldsymbol{\beta})]^\top$$

is the vector of outputs.

We wish to minimize the training loss in terms of $\boldsymbol{\beta}$. In the Newton-like methods in Section C.3.1, one derives an iterative minimization algorithm that is inspired by a Taylor expansion of $\ell_\tau(g(\cdot \mid \boldsymbol{\beta}))$. Instead, given a current guess $\boldsymbol{\beta}_t$, we can consider the Taylor expansion of the nonlinear prediction function $\boldsymbol{g}$:

$$\boldsymbol{g}(\tau \mid \boldsymbol{\beta}) \approx \boldsymbol{g}(\tau \mid \boldsymbol{\beta}_t) + \mathbf{G}_t(\boldsymbol{\beta} - \boldsymbol{\beta}_t),$$

where $\mathbf{G}_t := \mathbf{J}_{\boldsymbol{g}}(\boldsymbol{\beta}_t)$ is the matrix of Jacobi of $\boldsymbol{g}(\tau \mid \boldsymbol{\beta})$ at $\boldsymbol{\beta}_t$. Denoting the residual $\boldsymbol{e}_t := \boldsymbol{g}(\tau \mid \boldsymbol{\beta}_t) - \boldsymbol{y}$ and replacing $\boldsymbol{g}(\tau \mid \boldsymbol{\beta})$ with its Taylor approximation in $\ell_\tau(g(\cdot \mid \boldsymbol{\beta}))$, we obtain the approximation to the training loss in the neighborhood of $\boldsymbol{\beta}_t$: ☞ 582

$$\ell_\tau(g(\cdot \mid \boldsymbol{\beta})) \approx \frac{1}{n}\left\|\mathbf{G}_t(\boldsymbol{\beta} - \boldsymbol{\beta}_t) + \boldsymbol{e}_t\right\|^2.$$

The minimization of the right-hand side is a linear least-squares problem and therefore $\boldsymbol{d}_t := \boldsymbol{\beta} - \boldsymbol{\beta}_t$ satisfies the normal equations: ☞ 28

$$\mathbf{G}_t^\top \mathbf{G}_t \boldsymbol{d}_t = \mathbf{G}_t^\top(-\boldsymbol{e}_t).$$

Assuming that $\mathbf{G}_t^\top \mathbf{G}_t$ is invertible, the normal equations yield the **Gauss–Newton** search direction:

$$\boldsymbol{d}_t = -(\mathbf{G}_t^\top \mathbf{G}_t)^{-1}\mathbf{G}_t^\top \boldsymbol{e}_t.$$

Unlike the search direction (C.51) for Newton-like algorithms, the search direction of a Gauss–Newton algorithm does not require the computation of a Hessian matrix.

Observe that in the Gauss–Newton approach we determine $\boldsymbol{d}_t$ by viewing the search direction as coefficients in a linear regression with model matrix $\mathbf{G}_t$ and response $-\boldsymbol{e}_t$. This suggests that instead of using a linear regression, we can compute $\boldsymbol{d}_t$ via a *ridge regression* with a suitable choice for the regularization parameter γ: ☞ 321

$$\boldsymbol{d}_t = -(\mathbf{G}_t^\top \mathbf{G}_t + n\gamma \mathbf{I}_p)^{-1}\mathbf{G}_t^\top \boldsymbol{e}_t.$$

If we replace $n\mathbf{I}_p$ with the diagonal matrix $\mathrm{diag}(\mathbf{G}_t^\top \mathbf{G}_t)$, we then obtain the **Levenberg–Marquardt** search direction:

$$\boldsymbol{d}_t = -(\mathbf{G}_t^\top \mathbf{G}_t + \gamma\, \mathrm{diag}(\mathbf{G}_t^\top \mathbf{G}_t))^{-1}\mathbf{G}_t^\top \boldsymbol{e}_t. \tag{C.64}$$

Recall that the ridge regularization parameter γ has the following effect on the least-squares solution: When it is zero, then the solution $\boldsymbol{d}_t$ coincides with the search direction of the Gauss–Newton method, and when γ tends to infinity, then $\|\boldsymbol{d}_t\|$ tends to zero. Thus, γ controls both the magnitude and direction of vector $\boldsymbol{d}_t$. A simple version of the Levenberg–Marquardt algorithm is the following.

Algorithm C.65: Levenberg–Marquardt for Minimizing $\frac{1}{n}\|g(\tau \mid \beta) - y\|^2$

Input: An initial guess $\boldsymbol{\beta}_0$; stopping error $\varepsilon > 0$; training set τ.
Output: An approximate minimizer of $\frac{1}{n}\|\boldsymbol{g}(\tau \mid \boldsymbol{\beta}) - \boldsymbol{y}\|^2$.

$t \leftarrow 0$ and $\gamma \leftarrow 0.01$ (or another default value)
while stopping condition is not met **do**
  Compute the search direction $\boldsymbol{d}_t$ via (C.64).
  $\boldsymbol{e}_{t+1} \leftarrow \boldsymbol{g}(\tau \mid \boldsymbol{\beta}_t + \boldsymbol{d}_t) - \boldsymbol{y}$
  if $\|\boldsymbol{e}_{t+1}\| < \|\boldsymbol{e}_t\|$ **then**
    $\gamma \leftarrow \gamma/10, \quad \boldsymbol{e}_{t+1} \leftarrow \boldsymbol{e}_t, \quad \boldsymbol{\beta}_{t+1} \leftarrow \boldsymbol{\beta}_t + \boldsymbol{d}_t$
  else
    $\gamma \leftarrow \gamma \times 10$
  $t \leftarrow t + 1$
return $\boldsymbol{\beta}_t$

C.3.5 Coordinate Descent Method

The **coordinate descent** method of minimizing $f : \mathbb{R}^n \to \mathbb{R}$ consists in iterating until convergence the **coordinate cycle**:

$$x_k \leftarrow x_k + \operatorname*{argmin}_{\alpha \in \mathbb{R}} f(\boldsymbol{x} + \alpha\, \boldsymbol{e}_k), \quad k = 1, \ldots, n, \tag{C.66}$$

where $\boldsymbol{e}_k \in \mathbb{R}^n$ is the unit vector with 1 in the k-th coordinate. Note that (C.66) can also be written as

$$x_k \leftarrow \operatorname*{argmin}_{\alpha \in \mathbb{R}} f(\boldsymbol{x} + (\alpha - x_k)\boldsymbol{e}_k), \quad k = 1, \ldots, n.$$

In the cyclic coordinate descent (C.66), a single cycle consists of going through the sequence of indexes $1, \ldots, n$. More generally, in the **essentially cyclic** coordinate descent, a single cycle consists in going through any finite sequence $j_1, j_2, \ldots, j_m$, which contains all the integers from 1 to n at least once (that is, for any $i \in \{1, \ldots, n\}$ there exists an k such that $j_k = i$).

Iterating the coordinate cycle (C.66) produces a sequence of vectors $(\boldsymbol{x}_t, t \geq 0)$. Whether the sequence $(\boldsymbol{x}_t)$ converges, and whether the desired minimizer is located, cannot be guaranteed, except in a few special cases.

For example, the function:

$$f(x_1, x_2) = x_1^2 + x_2^2 - 5x_1x_2$$

does not have a local minimum, because $f(x, x) = -3x^2$ is not bounded from below. Yet, the saddle point $(x_1, x_2) = (0, 0)$ is also a coordinate-wise minimizer; that is, if the coordinate descent cycles (C.66) are initialized at $\boldsymbol{x}_0 = \boldsymbol{0}$, then $\boldsymbol{x}_t = \boldsymbol{0}$ for all t.

■ **Example C.67 (Gauss–Seidel For $\mathbf{A}x = b$)** Consider minimizing the quadratic function

$$f(x) := x^\top \mathbf{A} x - 2x^\top b,$$

where

$$\mathbf{A} = [a_1, \dots, a_n] \in \mathbb{R}^{n \times n}$$

is a positive semi-definite matrix with strictly positive diagonal elements (that is, $a_{k,k} > 0$ for all k). This setting is similar to that of the Lanczos algorithm in Section A.8.5.1, with the main difference being that $\mathbf{A}$ was assumed to be positive definite in Section A.8.5.1, ensuring that f was strictly convex and $\operatorname{argmin}_x f(x)$ was a singleton equal to $\mathbf{A}^{-1}b$.

In the current example, $\operatorname{argmin}_x f(x)$ is not necessarily a singleton. The convexity of f implies that $\operatorname{argmin}_x f(x)$ is the set of stationary points of f. This set of stationary points is the same as the set of solutions of the linear equation

$$\nabla f(x) = 2(\mathbf{A}x - b) = \mathbf{0}.$$

In fact, the coordinate descent iteration (C.66) for minimizing the quadratic function f is equivalent to the classical **Gauss–Seidel** method for solving the linear system $\mathbf{A}x = b$.

If $r := b - \mathbf{A}x = -\frac{1}{2}\nabla f(x)$ denotes the residual vector at x, then we have:

$$\operatorname*{argmin}_{\alpha} f(x + \alpha\, e_k) = \operatorname*{argmin}_{\alpha} \alpha^2 e_k^\top \mathbf{A} e_k - 2\alpha\, e_k^\top r = \frac{r_k}{a_{k,k}},$$

so that the coordinate descent cycle (C.66) simplifies to $x_k \leftarrow x_k + r_k/a_{k,k}$ for $k = 1, \dots, n$. The residual vector r can be updated after each coordinate modification at the cost of $O(n)$ via $r \leftarrow r - (r_k/a_{k,k})\, a_k$. A simple implementation of this coordinate descent algorithm is given as follows:

Algorithm C.68: Coordinate Descent for Computing $\min_x x^\top \mathbf{A} x - 2x^\top b$

Input: $\mathbf{A} = [a_{i,j}] \in \mathbb{R}^{n \times n}$; vector b; initial guess x_0; tolerance $\varepsilon > 0$.
Output: An element in $\operatorname{argmin}_x x^\top \mathbf{A} x - 2x^\top b$.

$r \leftarrow b - \mathbf{A}x$ // residual vector
repeat // each cycle costs $O(n^2)$
  for $k = 1, \dots, n$ **do**
    $\alpha \leftarrow r_k/a_{k,k}$ // optimal step size
    $x_k \leftarrow x_k + \alpha$ // update k-th coordinate
    $r \leftarrow r - \alpha a_k$ // update residual using k-th column of $\mathbf{A}$
until $\|r\| < \varepsilon$
return x

Algorithm C.68 is guaranteed to converge to an element of $\operatorname{argmin}_x f(x)$, provided that $\mathbf{A}$ is a positive semi-definite matrix with strictly positive diagonal elements [115]. ■

■ **Example C.69 (Coordinate Descent for Nonnegative Least Squares)** As in Example C.67, consider minimizing the quadratic function $x^\top \mathbf{A} x - 2x^\top b$, but this time with the constraint that $x \geq \mathbf{0}$. in other words, we wish to compute

$$\min_{x \geq \mathbf{0}} x^\top \mathbf{A} x - 2x^\top b.$$

This optimization problem arises, for example, when we wish to compute the least-squares estimate $\|\boldsymbol{y} - \mathbf{X}\boldsymbol{b}\|^2$ subject to the constraint that the regression coefficients $\boldsymbol{b}$ are positive.

Recall that without the constraint $x_k \geq 0$ for each k, the coordinate descent cycle (C.66) simplifies to $x_k \leftarrow x_k + r_k/a_{k,k}$ for $k = 1, \ldots, n$. With the addition of the positivity constraint the coordinate update becomes:

$$x_k \leftarrow (x_k + r_k/a_{k,k})_+, \quad k = 1, \ldots, n,$$

where $x_+ := \max\{x, 0\}$. The method of Lagrange multipliers in Example C.43 suggest one possible stopping criterion, namely,

$$\|\boldsymbol{x} \odot \boldsymbol{r}\|^2 = \sum_{k:x_k>0} (x_k\, r_k)^2 < \varepsilon^2$$

for some small $\varepsilon > 0$. The pseudo-code in Algorithm C.70 summarizes the resulting coordinate descent algorithm.

Similar to Example C.67, Algorithm C.70 is guaranteed to converge to an element of the set $\mathrm{argmin}_{\boldsymbol{x}\geq\boldsymbol{0}}\, \boldsymbol{x}^\top\mathbf{A}\boldsymbol{x} - 2\boldsymbol{x}^\top\boldsymbol{b}$, provided that $\mathbf{A}$ is a positive semi-definite matrix with positive diagonal elements [114]. If additionally, $\mathbf{A}$ is a positive definite matrix, then Algorithm C.70 will converge to the unique minimizer.

Algorithm C.70: Coordinate Descent for Computing $\min_{\boldsymbol{x}\geq\boldsymbol{0}} \boldsymbol{x}^\top\mathbf{A}\boldsymbol{x} - 2\boldsymbol{x}^\top\boldsymbol{b}$

Input: $\mathbf{A} = [a_{i,j}] \in \mathbb{R}^{n\times n}$; vector $\boldsymbol{b}$; initial guess $\boldsymbol{x}_0$; tolerance $\varepsilon > 0$.
Output: An element in $\mathrm{argmin}_{\boldsymbol{x}\geq\boldsymbol{0}}\, \boldsymbol{x}^\top\mathbf{A}\boldsymbol{x} - 2\boldsymbol{x}^\top\boldsymbol{b}$.

```
r ← b − Ax                                   // residual vector
repeat                                       // each cycle costs O(n^2)
    for k = 1, ..., n do
        x_new ← (x_k + r_k/a_{k,k})_+        // new k-th coordinate value
        if x_new ≠ x_k then                  // updates only if necessary
            r ← r + (x_k − x_new) a_k        // update residual, if needed
            x_k ← x_new
until ||x ⊙ r|| < ε                          // one possible stopping criterion
return x
```

■

C.4 Constrained Numerical Minimization

A constrained optimization problem of the form (C.27) can sometimes be reformulated as a simpler unconstrained problem — for example, the unconstrained set $\mathcal{Y}$ can be transformed to the feasible region $\mathcal{X}$ of the constrained problem via a function $\boldsymbol{\phi} : \mathbb{R}^n \to \mathbb{R}^n$ such that $\mathcal{X} = \boldsymbol{\phi}(\mathcal{Y})$. Then, (C.27) is equivalent to the minimization problem

$$\min_{\boldsymbol{y}\in\mathcal{Y}} f(\boldsymbol{\phi}(\boldsymbol{y})),$$

in the sense that a solution $\boldsymbol{x}^*$ of the original problem is obtained from a transformed solution $\boldsymbol{y}^*$ via $\boldsymbol{x}^* = \boldsymbol{\phi}(\boldsymbol{y}^*)$. Table C.71 lists some examples of possible transformations.

Table C.71: Some transformations to eliminate constraints.

Constrained	Unconstrained
$x > 0$	$\exp(y)$
$x \geq 0$	y^2
$l \leq x \leq u$	$l + (u - l)\sin^2(y)$

Unfortunately, an unconstrained minimization method used in combination with these transformations is rarely effective. Instead, it is more common to use penalty functions.

C.4.1 The Majorize-Minimize Method

The overarching idea of **penalty functions** is to transform a constrained problem into an unconstrained problem by adding weighted constraint-violation terms to the original objective function, with the premise that the new problem has a solution that is identical or close to the original one. For example, if there are only equality constraints, then

$$\widetilde{f}(\boldsymbol{x}) := f(\boldsymbol{x}) + \sum_{i=1}^{m} \tau_i \, |h_i(\boldsymbol{x})|$$

for any $\tau_1, \ldots, \tau_m > 0$, gives an *exact penalty function*, in the sense that the minimizer of the penalized function $\widetilde{f}$ is equal to the minimizer of f subject to the m equality constraints $h_1, \ldots, h_m$. Similarly, when the problem involves only inequality constraints, the penalized function

$$\widetilde{f}(\boldsymbol{x}) := f(\boldsymbol{x}) + 2\sum_{j=1}^{k} \upsilon_j [g_j(\boldsymbol{x})]_+,$$

where $z_+ := \max(z, 0)$ denotes the **rectified linear unit (ReLU)**, yields the same minimizer as the original problem of minimizing $f(\boldsymbol{x})$ subject to the inequality constraints $g_1, \ldots, g_k$, for any choice of penalty parameters $\upsilon_1, \ldots, \upsilon_k > 0$.

In both constrained scenarios the non-differentiability of the penalty terms poses a challenge to the direct optimization of $\widetilde{f}$. In the majorization–minimization approach below, the absolute value and ReLU terms are majorized using the following simple upper bounds.

Lemma C.72 (Penalty Function Bounds): For all $z \in \mathbb{R}$ and $\gamma > 0$,

$$2|z| \leq \frac{z^2 + \gamma^2}{\gamma} \quad \text{and} \quad 2z_+ \leq \frac{(z+\gamma)^2}{2\gamma}.$$

Proof. The first inequality follows by expanding $(|z| - \gamma)^2/\gamma \geq 0$, $\gamma > 0$ and the second follows from the first inequality and the fact that $2z_+ = z + |z|$. □

By replacing the absolute value and ReLU terms in the penalized objective with their respective bounds in Lemma C.72, we arrive at the following majorized penalty function, which incorporates both equality and inequality constraints:

$$\widetilde{f}(\boldsymbol{x},\boldsymbol{\gamma},\check{\boldsymbol{\gamma}}) := f(\boldsymbol{x}) + \frac{1}{2}\sum_{i=1}^{m} \tau_i \frac{h_i^2(\boldsymbol{x}) + \gamma_i^2}{\gamma_i} + \frac{1}{2}\sum_{j=1}^{k} \upsilon_j \frac{(g_j(\boldsymbol{x}) + \check{\gamma}_j)^2}{\check{\gamma}_j}.$$

The **Majorization–Minimization (MM)** method for constrained optimization iterates between the unconstrained minimization over $\boldsymbol{x}$ and the simple minimization over $(\boldsymbol{\gamma},\check{\boldsymbol{\gamma}})$, whereby the penalty parameters $\{\tau_i\}$ and $\{\upsilon_j\}$ may be strengthened or weakened after each iteration.

Algorithm C.73: MM Method for Constrained Optimization

Input: Initial guess $\boldsymbol{x}$ and positive penalty parameters $\boldsymbol{\tau}$, $\boldsymbol{\upsilon}$.
Output: Minimizer of $f(\boldsymbol{x})$ subject to equality and inequality constraints.

repeat
- $(\boldsymbol{\gamma},\check{\boldsymbol{\gamma}}) \leftarrow \operatorname{argmin}_{(\boldsymbol{\gamma},\check{\boldsymbol{\gamma}})\geq \mathbf{0}} \widetilde{f}(\boldsymbol{x},\boldsymbol{\gamma},\check{\boldsymbol{\gamma}})$ `// simple minimization`
- $\boldsymbol{x} \leftarrow \operatorname{argmin}_{\boldsymbol{x}} \widetilde{f}(\boldsymbol{x},\boldsymbol{\gamma},\check{\boldsymbol{\gamma}})$ `// unconstrained optimization`
- Strengthen penalties for violated constraints and weaken penalties for satisfied constraints.

until a convergence criterion is met
return $\boldsymbol{x}$

We illustrate this methodology with a common optimization setting.

Example C.74 (Example C.40 Continued) The minimization of the quadratic function in Example C.40 with m equality constraints $\mathbf{A}\boldsymbol{x} = \boldsymbol{b}$ can be accomplished via the unconstrained minimization of

$$\widetilde{f}(\boldsymbol{x},\boldsymbol{\gamma}) := \frac{1}{2}\boldsymbol{x}^\top\mathbf{H}\boldsymbol{x} + \boldsymbol{q}^\top\boldsymbol{x} + \frac{1}{2}(\mathbf{A}\boldsymbol{x}-\boldsymbol{b})^\top\mathbf{D}(\mathbf{A}\boldsymbol{x}-\boldsymbol{b}) + \boldsymbol{\tau}^\top\boldsymbol{\gamma},$$

where $\mathbf{D} \in \mathbb{R}^{m\times m}$ is a diagonal matrix with i-th diagonal element equal to τ_i/γ_i. Alternating between computing $\operatorname{argmin}_{\boldsymbol{\gamma}\geq\mathbf{0}} \widetilde{f}(\boldsymbol{x},\boldsymbol{\gamma})$ and $\operatorname{argmin}_{\boldsymbol{x}} \widetilde{f}(\boldsymbol{x},\boldsymbol{\gamma})$ yields the updating formulas for the iterative unconstrained minimization of $\widetilde{f}$:

$$\begin{aligned}
\boldsymbol{\gamma} &\leftarrow |\mathbf{A}\boldsymbol{x}-\boldsymbol{b}| = \operatorname*{argmin}_{\boldsymbol{\gamma}\geq\mathbf{0}} \widetilde{f}(\boldsymbol{x},\boldsymbol{\gamma})\\
\boldsymbol{x} &\leftarrow (\mathbf{H}+\mathbf{A}^\top\mathbf{D}\mathbf{A})^{-1}(\mathbf{A}^\top\mathbf{D}\boldsymbol{b}-\boldsymbol{q}) = \operatorname*{argmin}_{\boldsymbol{x}} \widetilde{f}(\boldsymbol{x},\boldsymbol{\gamma}).
\end{aligned}$$

Since the diagonal matrix $\mathbf{D}$ can cause numerical instability when the γ_i's are very small and converging to zero, we can instead compute the step

$$\boldsymbol{x} \leftarrow \operatorname*{argmin}_{\boldsymbol{x}} \widetilde{f}(\boldsymbol{x},\boldsymbol{\gamma})$$

from the KKT system of equations:

$$\begin{bmatrix}\mathbf{H} & \mathbf{A}^\top\\ \mathbf{A} & -\mathbf{D}^{-1}\end{bmatrix}\begin{bmatrix}\boldsymbol{x}\\ \boldsymbol{\beta}\end{bmatrix} = \begin{bmatrix}-\boldsymbol{q}\\ \boldsymbol{b}\end{bmatrix}, \tag{C.75}$$

because the identity (A.59) with $\mathbf{H}_\gamma := \mathbf{H} + \mathbf{A}^\top \mathbf{D}\mathbf{A}$ implies that

$$\begin{bmatrix} \mathbf{H}_\gamma^{-1} & \mathbf{H}_\gamma^{-1}\mathbf{A}^\top\mathbf{D} \\ \mathbf{D}\mathbf{A}\mathbf{H}_\gamma^{-1} & \mathbf{D}\mathbf{A}\mathbf{H}_\gamma^{-1}\mathbf{A}^\top\mathbf{D} - \mathbf{D} \end{bmatrix} = \begin{bmatrix} \mathbf{H} & \mathbf{A}^\top \\ \mathbf{A} & -\mathbf{D}^{-1} \end{bmatrix}^{-1}.$$

In particular, when $\mathbf{H} \succ \mathbf{0}$ the $\boldsymbol{x}$ that solves (C.75) can be written as

$$\boldsymbol{x} = (\mathbf{H}^{-1} - \mathbf{H}^{-1}\mathbf{A}^\top\mathbf{S}\mathbf{A}\mathbf{H}^{-1})(-\boldsymbol{q}) + \mathbf{H}^{-1}\mathbf{A}^\top\mathbf{S}\boldsymbol{b},$$

where

$$\mathbf{S} := (\mathbf{D}^{-1} + \mathbf{A}\mathbf{H}^{-1}\mathbf{A}^\top)^{-1}.$$

Note that the additional variable $\boldsymbol{\beta}$ in (C.75) plays a role similar to that of the Lagrange multipliers in (C.41). The following pseudo-code summarizes the procedure.

Algorithm C.76: MM Method for Computing $\text{argmin}_{\boldsymbol{x}:\mathbf{A}\boldsymbol{x}=\boldsymbol{b}}\, \boldsymbol{x}^\top\mathbf{H}\boldsymbol{x} + 2\boldsymbol{q}^\top\boldsymbol{x}$

Input: $\mathbf{A} \in \mathbb{R}^{m\times n}$, $\boldsymbol{b} \in \mathbb{R}^m$, $\mathbf{H} \in \mathbb{R}^{n\times n}$, $\boldsymbol{q} \in \mathbb{R}^n$; initial guess $\boldsymbol{x}$; $\varepsilon > 0$.
Output: The unique vector $\boldsymbol{x} = \text{argmin}_{\boldsymbol{x}:\mathbf{A}\boldsymbol{x}=\boldsymbol{b}}\, \boldsymbol{x}^\top\mathbf{H}\boldsymbol{x} + 2\boldsymbol{q}^\top\boldsymbol{x}$.

1 $\boldsymbol{\tau} \leftarrow \mathbf{1}$ // initial penalty parameter
2 **repeat** // repeat until convergence
3 $\quad\boldsymbol{x}_{\text{old}} \leftarrow \boldsymbol{x}$
4 $\quad\boldsymbol{\gamma} \leftarrow |\mathbf{A}\boldsymbol{x} - \boldsymbol{b}|$ // compute $\text{argmin}_{\boldsymbol{\gamma}\geq\mathbf{0}} \widetilde{f}(\boldsymbol{x},\boldsymbol{\gamma})$
5 $\quad\mathcal{A} \leftarrow \{j : \gamma_j < \varepsilon\}$ // find the almost active constraints
6 $\quad\tau_j \leftarrow 2\tau_j,\ \forall j \notin \mathcal{A}$ // strengthen penalties for violated constraints
7 $\quad\tau_j \leftarrow \tau_j/2,\ \forall j \in \mathcal{A}$ // weaken penalties for satisfied constraints
8 $\quad\mathbf{D}^{-1} \leftarrow \text{diag}(\gamma_1/\tau_1, \ldots, \gamma_m/\tau_m)$
9 $\quad$let $\boldsymbol{x}$ be the solution to the KKT system (C.75)
10 **until** $\|\boldsymbol{x} - \boldsymbol{x}_{\text{old}}\| < \varepsilon$ // one choice for the tolerance is $\varepsilon = 10^{-6}$
11 **return** $\boldsymbol{x}$

Lines 6 and 7 in Algorithm C.76 heuristically adjust the magnitude of the penalty parameter $\boldsymbol{\tau}$. For instance, if the j-th constraint is almost satisfied (that is, $|\boldsymbol{e}_j^\top\mathbf{A}\boldsymbol{x} - b_j| < \varepsilon$), then its penalty τ_j is halved. ■

Having illustrated the usage of the absolute value bound in Lemma C.72, in the following example we illustrate the application of the ReLU bound for tacking inequality constraints.

■ **Example C.77 (Quadratic Programming with Inequality Constraints)** Consider minimizing the quadratic function

$$f(\boldsymbol{x}) := \frac{1}{2}\boldsymbol{x}^\top\mathbf{H}\boldsymbol{x} + \boldsymbol{q}^\top\boldsymbol{x}$$

subject to the inequality constraints $\mathbf{A}\boldsymbol{x} \leq \boldsymbol{b}$, where $\mathbf{A} \in \mathbb{R}^{k\times n}$. Following the MM method for constrained optimization, we aim to minimize the following function jointly over $\boldsymbol{x}$ and $\boldsymbol{\gamma} \geq \mathbf{0}$:

$$\widetilde{f}(\boldsymbol{x},\boldsymbol{\gamma}) := \frac{1}{2}\boldsymbol{x}^\top\mathbf{H}\boldsymbol{x} + \boldsymbol{q}^\top\boldsymbol{x} + \frac{1}{2}(\mathbf{A}\boldsymbol{x} - \boldsymbol{b} + \boldsymbol{\gamma})^\top\mathbf{D}\,(\mathbf{A}\boldsymbol{x} - \boldsymbol{b} + \boldsymbol{\gamma}),$$

where $\mathbf{D} := \mathrm{diag}(\upsilon_1/\gamma_1, \ldots, \upsilon_k/\gamma_k)$ is a diagonal matrix. Minimization over $\boldsymbol{\gamma} \geq \mathbf{0}$ for a fixed $\boldsymbol{x}$ yields

$$|\mathbf{A}\boldsymbol{x} - \boldsymbol{b}| = \underset{\boldsymbol{\gamma} \geq \mathbf{0}}{\mathrm{argmin}}\, \widetilde{f}(\boldsymbol{x}, \boldsymbol{\gamma}).$$

On the other hand, unconstrained minimization over $\boldsymbol{x}$ for a fixed $\boldsymbol{\gamma}$ yields:

$$\boldsymbol{x}^* = \left(\mathbf{H} + \mathbf{A}^\top \mathbf{D}\mathbf{A}\right)^{-1} \left(\mathbf{A}^\top \mathbf{D}(\boldsymbol{b} - \boldsymbol{\gamma}) - \boldsymbol{q}\right). \tag{C.78}$$

Define

$$\mathcal{A} := \{j : \gamma_j < \varepsilon\} = \{j : |\boldsymbol{e}_j^\top \mathbf{A}\boldsymbol{x} - b_j| < \varepsilon\}$$

to be the set of constraints that is almost active, as well as matrices $\overline{\mathbf{A}}$ and $\underline{\mathbf{A}}$ that collect the active and inactive set of columns of $\mathbf{A}$, respectively, so that:

$$\overline{\mathbf{A}}\boldsymbol{x} = \sum_{j \in \mathcal{A}} x_j \mathbf{A}\boldsymbol{e}_j \quad \text{and} \quad \underline{\mathbf{A}}\boldsymbol{x} = \sum_{j \notin \mathcal{A}} x_j \mathbf{A}\boldsymbol{e}_j.$$

Similarly, we let $\overline{\mathbf{D}} \in \mathbb{R}^{|\mathcal{A}| \times |\mathcal{A}|}$ and $\underline{\mathbf{D}} \in \mathbb{R}^{|\overline{\mathcal{A}}| \times |\overline{\mathcal{A}}|}$ be the diagonal matrices that collect the active and inactive set of diagonal elements of $\mathbf{D}$, respectively. If the set $\mathcal{A} \neq \emptyset$, then we can split $(\mathbf{H} + \mathbf{A}^\top \mathbf{D}\mathbf{A})$ into two matrices:

$$\mathbf{H} + \mathbf{A}^\top \mathbf{D}\mathbf{A} = \underline{\mathbf{H}} + \overline{\mathbf{A}}^\top \overline{\mathbf{D}}\, \overline{\mathbf{A}},$$

where

$$\underline{\mathbf{H}} := \mathbf{H} + \sum_{j \notin \mathcal{A}} \frac{\upsilon_j}{\gamma_j} (\boldsymbol{e}_j^\top \mathbf{A})^\top \boldsymbol{e}_j^\top \mathbf{A} \quad \text{and} \quad \overline{\mathbf{A}}^\top \overline{\mathbf{D}}\, \overline{\mathbf{A}} = \sum_{j \in \mathcal{A}} \frac{\upsilon_j}{\gamma_j} (\boldsymbol{e}_j^\top \mathbf{A})^\top \boldsymbol{e}_j^\top \mathbf{A}.$$

☞ 541 Applying Woodbury's identity (A.60) to $(\underline{\mathbf{H}} + \overline{\mathbf{A}}^\top \overline{\mathbf{D}}\, \overline{\mathbf{A}})^{-1}$ and simplifying (C.78) yields

$$\boldsymbol{x}^* = \left(\underline{\mathbf{H}}^{-1} - \underline{\mathbf{H}}^{-1} \overline{\mathbf{A}}^\top \mathbf{S}\, \overline{\mathbf{A}}\, \underline{\mathbf{H}}^{-1}\right) \left(\underline{\mathbf{A}}^\top \underline{\mathbf{D}}\, (\underline{\boldsymbol{b}} - \underline{\boldsymbol{\gamma}}) - \boldsymbol{q}\right) + \underline{\mathbf{H}}^{-1} \overline{\mathbf{A}}^\top \mathbf{S}\, (\overline{\boldsymbol{b}} - \overline{\boldsymbol{\gamma}}), \tag{C.79}$$

where

$$\mathbf{S} := \left((\overline{\mathbf{D}})^{-1} + \overline{\mathbf{A}}\, \underline{\mathbf{H}}^{-1} \overline{\mathbf{A}}^\top\right)^{-1}.$$

Note that while formula (C.78) does not permit computations with $\gamma_j = 0$ for any $j \in \mathcal{A}$, due to the quantities υ_j/γ_j, $j \in \mathcal{A}$ in (C.78), there are no such difficulties when using the alternative formula (C.79), because (C.79) calls for the computation of the reciprocal quantities γ_j/υ_j, $j \in \mathcal{A}$.

In summary, the MM procedure for computing $\mathrm{argmin}_{\boldsymbol{x}:\mathbf{A}\boldsymbol{x} \leq \boldsymbol{b}}\, \boldsymbol{x}^\top \mathbf{H}\boldsymbol{x} + 2\boldsymbol{q}^\top \boldsymbol{x}$ using the Cholesky Algorithm A.69 is given in Algorithm C.80.

In lines 5 and 6 of Algorithm C.80 we use a heuristic in selecting a suitable value of the penalization parameter $\boldsymbol{\upsilon}$. Specifically, if a particular constraint is violated, that is, it lies in the set C, then its penalization level is doubled. Contrariwise, if a constraint is satisfied, then its penalization level is halved.

This heuristic adjustment of the values of $\boldsymbol{\upsilon}$ has the same goal as the adjustment of the ridge regularization parameter in the Levenberg–Marquardt Algorithm C.65, namely, to

Algorithm C.80: MM Method for Computing $\mathrm{argmin}_{x:\mathbf{A}x\le b}\, x^\top \mathbf{H} x + 2q^\top x$

Input: $\mathbf{A}\in\mathbb{R}^{k\times n}$, $b\in\mathbb{R}^k$, $\mathbf{H}\in\mathbb{R}^{n\times n}$, $q\in\mathbb{R}^n$; initial guess x; $\varepsilon>0$.

Output: The vector $x=\mathrm{argmin}_{x:\mathbf{A}x\le b}\, x^\top\mathbf{H}x+2q^\top x$.

```
υ ← 1                                   // initial penalty parameter
repeat                                  // repeat until convergence
    x_old ← x
    C ← {j : e_j^T A x > b_j}           // find the set of violated constraints
    υ_j ← 2υ_j, ∀j ∈ C                  // strengthen penalties for violated constraints
    υ_j ← υ_j/2, ∀j ∉ C                 // weaken penalties for satisfied constraints
    γ ← |Ax − b|                        // compute argmin_{γ≥0} f̃(x, γ)
    𝒜 ← {j : γ_j < ε}                   // find the almost active constraints
    use 𝒜 to construct D̄, Ā, b̄, γ̄
    𝒜̲ ← {j : γ_j > ε} ∩ {j : υ_j > ε²}   // numerically inactive constraints
    use 𝒜̲ to construct D̲, A̲, b̲, γ̲
    H̲ ← H + Σ_{j∉𝒜̲} (υ_j/γ_j) (e_j^T A)^T e_j^T A
    LL^T ← H̲                            // Cholesky of H̲ in O(n³) time
    B ← L^{-1} Ā^T                       // forward substitution in O(|𝒜| n²) time
    CC^T ← (D̄)^{-1} + B^T B              // Cholesky of (D̄)^{-1} + Ā H̲^{-1} Ā^T in O(|𝒜|³) time
    x ← A̲^T D̲ (b̲ − γ̲) − q
    x ← L^{-1} x
    x ← x + B C^{-T} C^{-1} (b̄ − γ̄ − B^T x)   // forward and backward substitution
    x ← L^{-T} x                        // compute argmin_x f̃(x, γ)
until ||x − x_old|| < ε                 // one choice for the tolerance is ε = 10^{-6}
return x
```

accelerate the convergence rate of the algorithm. This is because penalizing constraints that are already satisfied is not only unnecessary, but slows down the overall convergence of Algorithm C.80.

Algorithm C.80 implements the updating formula (C.79) via the computation of two Cholesky decompositions at each iteration for a cost of $O(n^3 \wedge |\mathcal{A}|^3)$. Instead of using the Cholesky Algorithm A.69, formula (C.79) can be implemented via large-scale iterative methods such as the conjugate gradient Algorithm A.103. In particular, from (A.56) we have that ☞ 540

$$\begin{bmatrix} \underline{\mathbf{H}} & \overline{\mathbf{A}}^\top \\ \overline{\mathbf{A}} & -(\overline{\mathbf{D}})^{-1} \end{bmatrix}^{-1} = \begin{bmatrix} \underline{\mathbf{H}}^{-1} - \underline{\mathbf{H}}^{-1}\overline{\mathbf{A}}^\top \mathbf{S}\overline{\mathbf{A}}\underline{\mathbf{H}}^{-1} & \underline{\mathbf{H}}^{-1}\overline{\mathbf{A}}^\top\mathbf{S} \\ \mathbf{S}\overline{\mathbf{A}}\underline{\mathbf{H}}^{-1} & -\mathbf{S} \end{bmatrix},$$

which implies that (C.79) can be computed from the solution of the KKT system of equations:

$$\begin{bmatrix} \underline{\mathbf{H}} & \overline{\mathbf{A}}^\top \\ \overline{\mathbf{A}} & -(\overline{\mathbf{D}})^{-1} \end{bmatrix} \begin{bmatrix} x \\ \beta \end{bmatrix} = \begin{bmatrix} \underline{\mathbf{A}}^\top\underline{\mathbf{D}}(\underline{b}-\underline{\gamma}) - q \\ \overline{b}-\overline{\gamma} \end{bmatrix}. \tag{C.81}$$

The KKT system (C.81) can either be solved via the (P)LU method in Section A.8.1 at a

☞ 537 cost of $\mathcal{O}((n+|\mathcal{A}|)^3)$, or via an iterative method such as the conjugate gradient approach
☞ 560 in Example A.104. Either way, we simply replace lines 13 to 19 in Algorithm C.80 with the single line: "set $\boldsymbol{x}$ to be the solution to (C.81)".

If $\mathbf{H}$ is invertible, then the minimizer computed by the MM Algorithm C.80 corresponds to the following (local) minimum of $\boldsymbol{x}^\top \mathbf{H}\boldsymbol{x} + \boldsymbol{q}^\top \boldsymbol{x}$:

$$\left(\overline{\boldsymbol{b}} + \overline{\mathbf{A}}\mathbf{H}^{-1}\boldsymbol{q}\right)^\top \left(\overline{\mathbf{A}}\mathbf{H}^{-1}\overline{\mathbf{A}}^\top\right)^{-1} \left(\overline{\boldsymbol{b}} + \overline{\mathbf{A}}\mathbf{H}^{-1}\boldsymbol{q}\right) - \boldsymbol{q}^\top\mathbf{H}^{-1}\boldsymbol{q}.$$

■

■ **Example C.82 (Nonnegative Least Squares)** Consider again Example C.77 in the special case where $\mathbf{A} := -\mathbf{I}_n$ and $\boldsymbol{b} = \mathbf{0}$. In this case, we have

$$\underset{\boldsymbol{\gamma}\geq\mathbf{0}}{\operatorname{argmin}}\, \widetilde{f}(\boldsymbol{x},\boldsymbol{\gamma}) = |\boldsymbol{x}|$$

and (C.78) can be written as:

$$\boldsymbol{x}^* = \boldsymbol{\Gamma}(\boldsymbol{\Gamma}\mathbf{H}\boldsymbol{\Gamma} + \operatorname{diag}(\boldsymbol{\upsilon}))^{-1}\boldsymbol{\Gamma}(\boldsymbol{\upsilon} - \boldsymbol{q}),$$

where $\boldsymbol{\Gamma}$ is a diagonal matrix with i-th diagonal element equal to $\sqrt{\gamma_i}$. An adaptation of Algorithm C.80 to this nonnegative quadratic minimization is the following.

Algorithm C.83: MM Method for Computing $\operatorname{argmin}_{\boldsymbol{x}\geq\mathbf{0}} \boldsymbol{x}^\top\mathbf{H}\boldsymbol{x} + 2\boldsymbol{q}^\top\boldsymbol{x}$

Input: $\mathbf{H} \in \mathbb{R}^{n\times n}$, $\boldsymbol{q} \in \mathbb{R}^n$; initial guess $\boldsymbol{x} > \mathbf{0}$; $\varepsilon > 0$.
Output: The vector $\boldsymbol{x} = \operatorname{argmin}_{\boldsymbol{x}\geq\mathbf{0}} \boldsymbol{x}^\top\mathbf{H}\boldsymbol{x} + 2\boldsymbol{q}^\top\boldsymbol{x}$.

```
1  υ ← 1                                    // initial penalty parameter
2  repeat                                   // repeat until convergence
3      x_old ← x
4      C ← {j : x_j < 0}                    // find the set of violated constraints
5      υ_j ← 10(υ_j + ε), ∀j ∈ C            // strengthen penalty for violated constraints
6      υ_j ← υ_j/2, ∀j ∉ C                  // weaken penalty for satisfied constraints
7      γ ← |x|, A ← {j : γ_j > ε}           // inactive constraints
8      x_{A^c} ← 0                          // set x_j ← 0, ∀j ∈ A^c
9      use A to construct H̲, Γ̲, υ̲, q̲         // inactive constraint vectors/matrices
10     LL^T ← Γ̲H̲Γ̲ + diag(υ̲)                 // Cholesky decomposition
11     x̲ ← Γ̲(υ̲ − q̲), x̲ ← L^{-T}(L^{-1}x̲)
12     x̲ ← Γ̲x̲, x_A ← x̲                      // update x_j, ∀j ∈ A
13 until ||x − x_old|| < ε
14 return x
```

When $\mathbf{H}$ has some special structure like sparsity, then it may be more efficient to replace the Cholesky factorization in steps 10 and 11 of Algorithm C.83 with the
☞ 560 conjugate gradient Algorithm A.103. In this case, we never explicitly form the matrix:

$$\underline{\boldsymbol{\Gamma}}\,\underline{\mathbf{H}}\,\underline{\boldsymbol{\Gamma}} + \operatorname{diag}(\underline{\boldsymbol{\upsilon}}),$$

but simply use the matrix-vector multiplication mapping:

$$x \mapsto \underline{v} \odot x + \underline{\gamma} \odot \left[\underline{\mathbf{H}} \left(\underline{\gamma} \odot x \right) \right].$$

Table C.84 compares the performance of the MM Algorithm C.83 with the TNT-NN algorithm in [131], which is a faster version of the popular Lawson and Hanson [104] active-set nonnegative least-squares algorithm for computing $\mathrm{argmin}_{\beta \geq \mathbf{0}} \|y - \mathbf{X}\beta\|^2$. The experiments were performed on a Dell Inspiron desktop computer with 16 GB RAM and `Intel(R) Core(TM) i7-9700 CPU @ 3.00GHz` CPU.

Given an n in Table C.84, we simulate the regression model $Y = \mathbf{X}\beta + \varepsilon$ with random model matrix $\mathbf{X} \in \mathbb{R}^{n \times n}$, all elements of which are iid $\mathcal{N}(0, 1/\sqrt{n})$, $\varepsilon \sim \mathcal{N}(\mathbf{0}, \mathbf{I}_n)$, and

$$\beta_1, \ldots, \beta_{10} \overset{\text{iid}}{\sim} \mathcal{U}(0, \sqrt{p}),$$

with $\beta_j = 0$ for $j \geq 11$. Algorithm C.83 is initialized with an x whose components are iid $\mathcal{U}(0, 1)$ distributed. In addition, $\varepsilon = 10^{-7}$ and we employed the Cholesky factorization in Algorithm C.83.

Table C.84: A comparison of the run time between the TNT-NN and MM algorithms for computing $\mathrm{argmin}_{\beta \geq \mathbf{0}} \|Y - \mathbf{X}\beta\|^2$.

CPU time (seconds)			CPU time (seconds)		
n	MM	TNT-NN	n	MM	TNT-NN
500	0.039	0.11	5500	6.89	8.42
1000	0.11	0.17	6000	8.87	11.04
1500	0.25	0.43	6500	10.71	13.85
2000	0.48	0.85	7000	12.83	16.14
2500	0.79	1.50	7500	15.69	19.59
3000	1.26	2.12	8000	19.45	22.15
3500	2.00	3.03	8500	21.95	26.30
4000	3.07	4.54	9000	24.86	28.95
4500	4.18	5.33	9500	30.55	36.10
5000	5.39	6.89	10000	34.82	38.81

Next, we repeat the same experimental setup, except that $\mathbf{X} \in \mathbb{R}^{n \times n}$ is a sparse matrix with $n^2/100$ nonzero entries, all of which are iid $\mathcal{N}(0, 1/\sqrt{n})$ distributed. To take advantage of the sparsity of $\mathbf{X}$, we never explicitly construct $\mathbf{H} = \mathbf{X}^\top \mathbf{X}$ and, instead of the Cholesky factorization, use the conjugate gradient Algorithm A.103 with maximum number of iterations equal to 40.

Table C.85 reports the results of a comparison with the TNT-NN algorithm [131]. Note that the TNT-NN algorithm runs out of memory for $n \geq 3.5 \times 10^4$, suggesting that it is not able to take advantage of the sparsity of the model matrix $\mathbf{X}$. In contrast, the MM Algorithm C.84 exhibits a running cost of $\mathcal{O}((n/100)^3)$.

Table C.85: A comparison of the run time between the TNT-NN and MM algorithms on a sparse nonnegative least-squares minimization.

CPU time (seconds)			CPU time (seconds)		
n	MM	TNT-NN	n	MM	TNT-NN
0.5×10^4	0.60851	3.6837	4×10^4	85.518	–
1×10^4	2.4546	21.355	5×10^4	140.71	–
1.5×10^4	7.682	55.829	6×10^4	201.27	–
2×10^4	16.396	144.64	7×10^4	408.72	–
2.5×10^4	30.626	303.26	8×10^4	467.78	–
3×10^4	37.434	557.87	9×10^4	584.7	–
3.5×10^4	56.094	–	10^5	843.23	–

■

■ **Example C.86 (Linear Programming)** Take the linear program in Table C.39, where we seek to minimize the function $f(\boldsymbol{x}) := \boldsymbol{q}^\top \boldsymbol{x}$ subject to the inequality constraints $\boldsymbol{x} \geq \boldsymbol{0}$ and $\mathbf{A}\boldsymbol{x} \leq \boldsymbol{b}$, where $\mathbf{A} \in \mathbb{R}^{k \times n}$. The MM method then requires the minimization of the following function jointly over $\boldsymbol{x}$ and $\boldsymbol{\gamma}, \check{\boldsymbol{\gamma}} \geq \boldsymbol{0}$:

$$\widetilde{f}(\boldsymbol{x}, \boldsymbol{\gamma}, \check{\boldsymbol{\gamma}}) := \boldsymbol{q}^\top \boldsymbol{x} + \frac{1}{2}(\mathbf{A}\boldsymbol{x} - \boldsymbol{b} + \boldsymbol{\gamma})^\top \mathbf{D}(\mathbf{A}\boldsymbol{x} - \boldsymbol{b} + \boldsymbol{\gamma}) + \sum_{j=1}^{n} \frac{\check{\upsilon}_j(\check{\gamma}_j - x_j)^2}{2\check{\gamma}_j},$$

where $\mathbf{D} := \mathrm{diag}(\upsilon_1/\gamma_1, \ldots, \upsilon_k/\gamma_k)$. Minimization over $\boldsymbol{\gamma}$ for a fixed $\boldsymbol{x}, \check{\boldsymbol{\gamma}}$ yields

$$|\mathbf{A}\boldsymbol{x} - \boldsymbol{b}| = \underset{\boldsymbol{\gamma} \geq \boldsymbol{0}}{\mathrm{argmin}}\, \widetilde{f}(\boldsymbol{x}, \boldsymbol{\gamma}, \check{\boldsymbol{\gamma}}),$$

and minimization over $\check{\boldsymbol{\gamma}}$ for fixed $\boldsymbol{x}, \boldsymbol{\gamma}$ yields

$$|\boldsymbol{x}| = \underset{\check{\boldsymbol{\gamma}} \geq \boldsymbol{0}}{\mathrm{argmin}}\, \widetilde{f}(\boldsymbol{x}, \boldsymbol{\gamma}, \check{\boldsymbol{\gamma}}).$$

Further, with $\check{\boldsymbol{\Gamma}} := \mathrm{diag}(\sqrt{\check{\gamma}_1}, \ldots, \sqrt{\check{\gamma}_n})$ the unconstrained minimization of $\widetilde{f}$ over $\boldsymbol{x}$ for fixed $\boldsymbol{\gamma}, \check{\boldsymbol{\gamma}}$ yields:

$$\boldsymbol{x}^* = \check{\boldsymbol{\Gamma}}(\mathrm{diag}(\check{\boldsymbol{\upsilon}}) + \check{\boldsymbol{\Gamma}}\mathbf{A}^\top \mathbf{D}\mathbf{A}\check{\boldsymbol{\Gamma}})^{-1}(\check{\boldsymbol{\Gamma}}\mathbf{A}^\top \mathbf{D}(\boldsymbol{b} - \boldsymbol{\gamma}) - \check{\boldsymbol{\Gamma}}(\boldsymbol{q} - \check{\boldsymbol{\upsilon}})). \tag{C.87}$$

As in Example C.77, define the set $\mathcal{A} := \{j : |\boldsymbol{e}_j^\top(\mathbf{A}\boldsymbol{x} - \boldsymbol{b})| < \varepsilon\}$ of almost active constraints and split the matrix $\mathrm{diag}(\check{\boldsymbol{\upsilon}}) + \check{\boldsymbol{\Gamma}}\mathbf{A}^\top \mathbf{D}\mathbf{A}\check{\boldsymbol{\Gamma}}$ into two parts:

$$\mathrm{diag}(\check{\boldsymbol{\upsilon}}) + \check{\boldsymbol{\Gamma}}\mathbf{A}^\top \mathbf{D}\mathbf{A}\check{\boldsymbol{\Gamma}} = \underline{\mathbf{H}} + \check{\boldsymbol{\Gamma}}\,\overline{\mathbf{A}}^\top\, \overline{\mathbf{D}}\,\overline{\mathbf{A}}\,\check{\boldsymbol{\Gamma}},$$

where $\underline{\mathbf{H}} := \mathrm{diag}(\check{\boldsymbol{\upsilon}}) + \sum_{j \notin \mathcal{A}} \frac{\upsilon_j}{\gamma_j} (\boldsymbol{e}_j^\top \mathbf{A}\check{\boldsymbol{\Gamma}})^\top \boldsymbol{e}_j^\top \mathbf{A}\check{\boldsymbol{\Gamma}}$. Similar to the derivation of (C.81), the updating formula (C.87) suggests that we can alternatively update via $\boldsymbol{x}^* = \check{\boldsymbol{\Gamma}}\check{\boldsymbol{x}}$, where $\check{\boldsymbol{x}}$ is the solution of the KKT system of equations:

$$\begin{bmatrix} \underline{\mathbf{H}} & \check{\boldsymbol{\Gamma}}\,\overline{\mathbf{A}}^\top \\ \overline{\mathbf{A}}\,\check{\boldsymbol{\Gamma}} & -(\overline{\mathbf{D}})^{-1} \end{bmatrix} \begin{bmatrix} \check{\boldsymbol{x}} \\ \boldsymbol{\beta} \end{bmatrix} = \begin{bmatrix} \check{\boldsymbol{\Gamma}}\underline{\mathbf{A}}^\top \underline{\mathbf{D}}\,(\underline{\boldsymbol{b}} - \underline{\boldsymbol{\gamma}}) - \check{\boldsymbol{\Gamma}}(\boldsymbol{q} - \check{\boldsymbol{\upsilon}}) \\ \overline{\boldsymbol{b}} - \overline{\boldsymbol{\gamma}} \end{bmatrix}.$$

The following pseudo-code summarizes the algorithm using the analogue of the formula (C.79) for computing $\operatorname{argmin}_{\boldsymbol{x}} \widetilde{f}(\boldsymbol{x}, \boldsymbol{\gamma}, \check{\boldsymbol{\gamma}})$.

Algorithm C.88: MM Method for Computing $\operatorname{argmin}_{\boldsymbol{x}:\boldsymbol{x}\geq\boldsymbol{0},\mathbf{A}\boldsymbol{x}\leq\boldsymbol{b}} \boldsymbol{q}^\top\boldsymbol{x}$

Input: $\mathbf{A} \in \mathbb{R}^{k\times n}$, $\boldsymbol{b} \in \mathbb{R}^k$, $\boldsymbol{q} \in \mathbb{R}^n$; initial guess $\boldsymbol{x}$; $\varepsilon = 10^{-5}$.
Output: The vector $\boldsymbol{x} = \operatorname{argmin}_{\boldsymbol{x}:\boldsymbol{x}\geq\boldsymbol{0},\mathbf{A}\boldsymbol{x}\leq\boldsymbol{b}} \boldsymbol{q}^\top\boldsymbol{x}$.

$\boldsymbol{\upsilon} \leftarrow \varepsilon\mathbf{1}_k,\ \check{\boldsymbol{\upsilon}} \leftarrow \varepsilon\mathbf{1}_n$ // initial penalty parameters
repeat // repeat until convergence
  $\boldsymbol{x}_{\text{old}} \leftarrow \boldsymbol{x},\ \check{\boldsymbol{\gamma}} \leftarrow |\boldsymbol{x}|$ // compute $\operatorname{argmin}_{\check{\boldsymbol{\gamma}}\geq\boldsymbol{0}} \widetilde{f}(\boldsymbol{x},\boldsymbol{\gamma},\check{\boldsymbol{\gamma}})$
  $\check{C} \leftarrow \{j : x_j < 0\}$ // find the set of violated constraints
  $\check{\upsilon}_j \leftarrow 10\,(\check{\upsilon}_j + \varepsilon),\ \forall j \in \check{C}$ // strengthen penalty for violated constraint
  $\check{\upsilon}_j \leftarrow \varepsilon \vee \check{\upsilon}_j/2,\ \forall j \notin \check{C}$ // weaken penalties for satisfied constraints
  $\boldsymbol{\gamma} \leftarrow \mathbf{A}\boldsymbol{x} - \boldsymbol{b},\ C \leftarrow \{j : \gamma_j > 0\},\ \boldsymbol{\gamma} \leftarrow |\boldsymbol{\gamma}|$ // compute $\operatorname{argmin}_{\boldsymbol{\gamma}\geq\boldsymbol{0}} \widetilde{f}(\boldsymbol{x},\boldsymbol{\gamma},\check{\boldsymbol{\gamma}})$
  $\upsilon_j \leftarrow 10\,(\upsilon_j + \varepsilon),\ \forall j \in C$
  $\upsilon_j \leftarrow \upsilon_j/2,\ \forall j \notin C$
  $\mathcal{A} \leftarrow \{j : \gamma_j < \varepsilon\}$ // almost active constraints in $\mathbf{A}\boldsymbol{x} \leq \boldsymbol{b}$
  use $\mathcal{A}$ to construct $\overline{\mathbf{D}}, \overline{\mathbf{A}}, \overline{\boldsymbol{b}}, \overline{\boldsymbol{\gamma}}$
  $\underline{\mathcal{A}} \leftarrow \{j : \gamma_j > \varepsilon\} \cap \{j : \upsilon_j > \varepsilon^2\}$ // numerically inactive constraints
  use $\underline{\mathcal{A}}$ to construct $\underline{\mathbf{D}}, \underline{\mathbf{A}}, \underline{\boldsymbol{b}}, \underline{\boldsymbol{\gamma}}$
  $\mathbf{L}\mathbf{L}^\top \leftarrow \operatorname{diag}(\check{\boldsymbol{\upsilon}}) + \sum_{j\in\underline{\mathcal{A}}} \frac{\underline{\upsilon_j}}{\underline{\gamma_j}} (\boldsymbol{e}_j^\top \mathbf{A}\check{\boldsymbol{\Gamma}})\, \boldsymbol{e}_j^\top \mathbf{A}\check{\boldsymbol{\Gamma}}$ // Cholesky decomposition of $\underline{\mathbf{H}}$
  $\mathbf{B} \leftarrow \mathbf{L}^{-1}\check{\boldsymbol{\Gamma}}\overline{\mathbf{A}}^\top$ // forward substitution in $O(n^2|\mathcal{A}|)$ time
  $\mathbf{C}\mathbf{C}^\top \leftarrow \left(\overline{\mathbf{D}}\right)^{-1} + \mathbf{B}^\top\mathbf{B}$ // Cholesky decomposition in $O(|\mathcal{A}|^3)$ time
  $\boldsymbol{x} \leftarrow \check{\boldsymbol{\Gamma}}\underline{\mathbf{A}}^\top\underline{\mathbf{D}}\,(\underline{\boldsymbol{b}} - \underline{\boldsymbol{\gamma}}) - \check{\boldsymbol{\Gamma}}(\boldsymbol{q} - \check{\boldsymbol{\upsilon}})$
  $\boldsymbol{x} \leftarrow \mathbf{L}^{-1}\boldsymbol{x}$
  $\boldsymbol{x} \leftarrow \boldsymbol{x} + \mathbf{B}\mathbf{C}^{-\top}\mathbf{C}^{-1}\left(\overline{\boldsymbol{b}} - \overline{\boldsymbol{\gamma}} - \mathbf{B}^\top\boldsymbol{x}\right)$ // forward and backward substitution
  $\boldsymbol{x} \leftarrow \check{\boldsymbol{\Gamma}}\,\mathbf{L}^{-\top}\boldsymbol{x}$ // compute $\operatorname{argmin}_{\boldsymbol{x}} \widetilde{f}(\boldsymbol{x},\boldsymbol{\gamma},\check{\boldsymbol{\gamma}})$
until $\|\boldsymbol{x} - \boldsymbol{x}_{\text{old}}\| < \varepsilon$
return $\boldsymbol{x}$

Table C.89 compares the run time of the MM Algorithm C.88 with three alternative methods implemented in the GUROBI commercial optimization software [79]. The three GUROBI methods are the primal simplex, dual simplex, and interior-point methods.

Given an n, the **packing linear program** in Table C.89 is defined as:

$$k = 10\,n, \quad \boldsymbol{q} = -\mathbf{1}_n, \quad \boldsymbol{b} = n\mathbf{1}_k$$

and matrix $\mathbf{A} \in \mathbb{R}^{k\times n}$ with iid random entries drawn from the $\mathsf{Exp}(1)$ distribution. If $\boldsymbol{x} \geq \boldsymbol{0}$ represents a resource, then the packing linear program can be interpreted as maximizing the overall resource transport $\sum_j x_j$ subject to the cost constraint: $\max_j \boldsymbol{e}_j^\top\mathbf{A}\boldsymbol{x} \leq 1$.

Note that the interior-point method failed to produce a solution for $n \geq 3500$ due to excessive memory demands on the Dell Inspiron desktop computer with 16 GB RAM and `Intel(R) Core(TM) i7-9700 CPU @ 3.00GHz` CPU.

The MM Algorithm C.88 was initialized from an $\boldsymbol{x}$ with iid random entries drawn from the $\mathcal{U}(0, 1)$ distribution.

Table C.89: A comparison of the run time for the GUROBI (primal simplex, dual simplex, interior-point) algorithms and the MM Algorithm C.88 on the packing linear program $\max_{\boldsymbol{x} \geq \mathbf{0}, \mathbf{A}\boldsymbol{x} \leq n \mathbf{1}_k} \sum_j x_j$ with $k = 10\,n$.

	CPU time (seconds)			
n	MM	primal simplex	dual simplex	interior-point
500	1.28	15.40	2.20	3.51
1000	5.26	61.22	10.52	16.11
1500	12.50	210.05	23.39	35.26
2000	32.11	551.97	46.16	70.76
2500	55.26	1202.2	80.41	142.27
3000	88.22	2216.9	129.52	239.96
3500	137.65	3976.1	206.66	–
4000	196.43	6262.5	346.48	–
4500	305.64	9455	477.19	–

■

C.4.2 Alternating Direction Method of Multipliers

The **alternating direction method of multipliers** (ADMM) [29, 162] is a method that combines the ideas of penalty functions and the Lagrange multiplier method. The ADMM is designed to solve problems of the form:

☞ 594

$$\text{(C.90)} \qquad \begin{aligned} \min_{\boldsymbol{x} \in \mathbb{R}^n, \boldsymbol{z} \in \mathbb{R}^m} \quad & f(\boldsymbol{x}) + g(\boldsymbol{z}) \\ \text{subject to:} \quad & \mathbf{A}\boldsymbol{x} + \mathbf{B}\boldsymbol{z} = \boldsymbol{b}, \end{aligned}$$

where $\mathbf{A} \in \mathbb{R}^{k \times n}$, $\mathbf{B} \in \mathbb{R}^{k \times m}$, and $\boldsymbol{b} \in \mathbb{R}^k$, and $f : \mathbb{R}^n \to \mathbb{R}$ and $g : \mathbb{R}^m \to \mathbb{R}$ are convex functions. The approach is to form an **augmented Lagrangian**

$$\mathcal{L}_\varrho(\boldsymbol{x}, \boldsymbol{z}, \boldsymbol{v}) := f(\boldsymbol{x}) + g(\boldsymbol{z}) + \boldsymbol{v}^\top (\mathbf{A}\boldsymbol{x} + \mathbf{B}\boldsymbol{z} - \boldsymbol{b}) + \frac{\varrho}{2} \|\mathbf{A}\boldsymbol{x} + \mathbf{B}\boldsymbol{z} - \boldsymbol{b}\|^2,$$

where $\varrho > 0$ is a penalty parameter, and $\boldsymbol{v} \in \mathbb{R}^k$ are Lagrange multiplier variables. The ADMM then iterates through updates of the following form:

$$\text{(C.91)} \qquad \boldsymbol{x}^{(t+1)} = \underset{\boldsymbol{x} \in \mathbb{R}^n}{\operatorname{argmin}} \, \mathcal{L}_\varrho(\boldsymbol{x}, \boldsymbol{z}^{(t)}, \boldsymbol{v}^{(t)})$$

$$\text{(C.92)} \qquad \boldsymbol{z}^{(t+1)} = \underset{\boldsymbol{z} \in \mathbb{R}^m}{\operatorname{argmin}} \, \mathcal{L}_\varrho(\boldsymbol{x}^{(t+1)}, \boldsymbol{z}, \boldsymbol{v}^{(t)})$$

$$\text{(C.93)} \qquad \boldsymbol{v}^{(t+1)} = \boldsymbol{v}^{(t)} + \varrho \left(\mathbf{A}\boldsymbol{x}^{(t+1)} + \mathbf{B}\boldsymbol{z}^{(t+1)} - \boldsymbol{b} \right).$$

One justification of the updating equation (C.93) can be provided when g is a smooth differentiable function. In that case, the convexity assumption implies that the minimizer $\boldsymbol{z}^{(t+1)}$ in (C.92) satisfies the equation:

$$\nabla_{\boldsymbol{z}} \mathcal{L}_\varrho(\boldsymbol{x}^{(t+1)}, \boldsymbol{z}^{(t+1)}, \boldsymbol{v}^{(t)}) = \nabla_{\boldsymbol{z}} g(\boldsymbol{z}^{(t+1)}) + \mathbf{B}^\top \boldsymbol{v}^{(t)} + \varrho \mathbf{B}^\top (\mathbf{A}\boldsymbol{x}^{(t+1)} + \mathbf{B}\boldsymbol{z}^{(t+1)} - \boldsymbol{b}) = \mathbf{0}.$$

As a consequence, the updating equation (C.93) ensures that the latest $\boldsymbol{v}^{(t+1)}$ satisfies the Lagrange multiplier condition: ☞ 595

$$\nabla_z \mathcal{L}_0(\boldsymbol{x}^{(t+1)}, \boldsymbol{z}^{(t+1)}, \boldsymbol{v}^{(t+1)}) = \nabla_z g(\boldsymbol{z}^{(t+1)}) + \mathbf{B}^\top \boldsymbol{v}^{(t+1)} = \mathbf{0}.$$

Example C.94 (Quadratic Programming with Inequality Constraints) Consider again Example C.77. To apply the ADMM algorithm we define $\mathbf{B} := -\mathbf{I}_k$ with $m = k$ and

$$g(\boldsymbol{z}) := \begin{cases} 0, & \max_j z_j \leq 0 \\ \infty, & \max_j z_j > 0. \end{cases}$$

Since

$$\underset{\boldsymbol{z} \in \mathbb{R}^k}{\text{argmin}} \, \|\boldsymbol{z} - \boldsymbol{c}\|^2 + g(\boldsymbol{z}) = \underset{\boldsymbol{z} \leq \mathbf{0}}{\text{argmin}} \, \|\boldsymbol{z} - \boldsymbol{c}\|^2 = [0 \wedge c_1, \ldots, 0 \wedge c_k]^\top,$$

the ADMM iteration over $\boldsymbol{x}, \boldsymbol{z}, \boldsymbol{v}$ can be written as:

$$\begin{aligned} \boldsymbol{x} &\leftarrow (\mathbf{H} + \varrho \mathbf{A}^\top \mathbf{A})^{-1} (\varrho \mathbf{A}^\top \boldsymbol{z} + \varrho \mathbf{A}^\top \boldsymbol{b} - \boldsymbol{q} - \mathbf{A}^\top \boldsymbol{v}) \\ \boldsymbol{z} &\leftarrow \mathbf{A}\boldsymbol{x} - \boldsymbol{b} + \varrho^{-1} \boldsymbol{v}, \quad z_j \leftarrow 0 \wedge z_j, \; j = 1, \ldots, k \\ \boldsymbol{v} &\leftarrow \boldsymbol{v} + \varrho \, (\mathbf{A}\boldsymbol{x} - \boldsymbol{b} - \boldsymbol{z}) \, . \end{aligned}$$

This ADMM recipe is summarized in the following pseudo-code.

Algorithm C.95: ADMM Method for $\text{argmin}_{\boldsymbol{x}:\mathbf{A}\boldsymbol{x} \leq \boldsymbol{b}} \, \boldsymbol{x}^\top \mathbf{H} \boldsymbol{x} + 2\boldsymbol{q}^\top \boldsymbol{x}$

Input: $\mathbf{A} \in \mathbb{R}^{k \times n}$, $\boldsymbol{b} \in \mathbb{R}^k$, $\mathbf{H} \in \mathbb{R}^{n \times n}$, $\boldsymbol{q} \in \mathbb{R}^n$; initial guess $\boldsymbol{x}$; $\varepsilon > 0$.
Output: The vector $\boldsymbol{x} = \text{argmin}_{\boldsymbol{x}:\mathbf{A}\boldsymbol{x} \leq \boldsymbol{b}} \, \boldsymbol{x}^\top \mathbf{H} \boldsymbol{x} + 2\boldsymbol{q}^\top \boldsymbol{x}$.

$\varrho \leftarrow 0.01$, $\boldsymbol{v} \leftarrow \mathbf{0}$, $\boldsymbol{z} \leftarrow \mathbf{0}$ // initial parameters
$\mathbf{L}\mathbf{L}^\top \leftarrow \mathbf{H} + \varrho \mathbf{A}^\top \mathbf{A}$ // store Cholesky decomposition in memory
repeat // repeat until convergence
  $\boldsymbol{x}_{\text{old}} \leftarrow \boldsymbol{x}$
  $\boldsymbol{x} \leftarrow \mathbf{A}^\top (\varrho \boldsymbol{z} + \varrho \boldsymbol{b} - \boldsymbol{v}) - \boldsymbol{q}$
  $\boldsymbol{x} \leftarrow \mathbf{L}^{-1} \boldsymbol{x}$, $\boldsymbol{x} \leftarrow \mathbf{L}^{-\top} \boldsymbol{x}$ // forward and backward substitution costs $O(n^2)$
  $\widetilde{\boldsymbol{z}} \leftarrow \mathbf{A}\boldsymbol{x} - \boldsymbol{b} + \varrho^{-1} \boldsymbol{v}$
  $z_j \leftarrow 0 \wedge \widetilde{z}_j$ for all j
  $\boldsymbol{v} \leftarrow \varrho(\widetilde{\boldsymbol{z}} - \boldsymbol{z})$ // update dual/Lagrange variables
until $\|\boldsymbol{x} - \boldsymbol{x}_{\text{old}}\| < \varepsilon$ // one choice for the tolerance is $\varepsilon = 10^{-6}$
return $\boldsymbol{x}$

On the one hand, each iteration in the ADMM Algorithm C.95 costs $O(n^2)$, and the corresponding iteration in the MM Algorithm C.80 costs $O(n^3)$. On the other hand, the MM Algorithm C.80 typically requires fewer iterations to converge; see Table C.96. Thus, it is not apriori obvious whether the ADMM is always preferable to the MM method.

In Algorithm C.95 above, we set the parameter ϱ to 0.01. In practice, this parameter has to be chosen very carefully as it affects the overall convergence rate of the ADMM algorithm. A more sophisticated version of Algorithm C.95 that updates ϱ dynamically is described in [162] with accompanying state-of-the-art software implementation.

Table C.96 shows a time comparison of the MM Algorithm C.80 (with $\varepsilon = 10^{-5}$) and the software provided in [162] on a number of randomly generated numerical test cases with increasing number of variables n; see Algorithm C.97. Overall, for this particular example, the MM algorithm is about ten times faster. In addition, Table C.96 also provides a comparison with the GUROBI commercial optimization software [79], which implements an interior-point method similar to Algorithm C.102 in the next section.

Table C.96: A comparison of the run time and total number of iterations for the ADMM, GUROBI, and MM algorithms using the quadratic optimization setup in Algorithm C.97. The MM algorithm is roughly ten times faster than both the ADMM and GUROBI algorithms.

	CPU time (seconds)			number of iterations		
n	MM	ADMM	GUROBI	MM	ADMM	GUROBI
1000	0.45868	4.2662	5.9841	24	200	14
1500	1.1403	15.479	18.602	24	425	15
2000	2.5236	32.089	39.696	24	325	15
2500	4.0432	69.148	72.063	24	375	15
3000	6.925	103.47	121.88	25	475	15
3500	12.397	180.4	192.98	29	925	15
4000	15.559	233.87	277.57	25	550	14
4500	21.15	376.35	409.68	25	1175	14
5000	29.732	452.56	591.01	26	700	15

The inputs $\mathbf{A}, \boldsymbol{b}, \mathbf{H}, \boldsymbol{q}$ and initial guess $\boldsymbol{x}$ were generated according to the following recipe.

Algorithm C.97: Generation of Random Test Cases

Input: Number of variables n.

Output: $\mathbf{A}, \boldsymbol{b}, \mathbf{H}, \boldsymbol{q}$ and initial guess $\boldsymbol{x}$ for $\operatorname{argmin}_{\boldsymbol{x}:\mathbf{A}\boldsymbol{x}\leq\boldsymbol{b}} \boldsymbol{x}^\top\mathbf{H}\boldsymbol{x} + 2\boldsymbol{q}^\top\boldsymbol{x}$

$\mathbf{Z} \in \mathbb{R}^{n\times n}$ such that $\mathrm{vec}(\mathbf{A}) \sim \mathcal{N}(\mathbf{0}, \mathbf{I}_n \otimes \mathbf{I}_n)$ // matrix normal distribution

$\mathbf{QR} \leftarrow \mathbf{Z}$ // QR decomposition via Algorithm A.22

$d_1 \leftarrow 0, d_2 \leftarrow 0$

$d_3, \ldots, d_n \overset{\text{iid}}{\sim} \mathsf{Exp}(1)$

$\mathbf{D} \leftarrow \mathrm{diag}(d_1, \ldots, d_n)$

$\mathbf{H} \leftarrow \mathbf{QDQ}^\top$

$k \leftarrow n + 100$ // more constraints than variables

$\mathbf{A} \in \mathbb{R}^{k\times n}$ such that $\mathrm{vec}(\mathbf{A}) \sim \mathcal{N}(\mathbf{0}, \mathbf{I}_n \otimes \mathbf{I}_k)$

$\boldsymbol{b} \leftarrow \mathcal{N}(\mathbf{0}, \mathbf{I}_k)$

$\boldsymbol{q} \leftarrow \mathcal{N}(\mathbf{0}, \mathbf{I}_n)$

$\boldsymbol{x} \leftarrow \mathcal{N}(\mathbf{0}, \mathbf{I}_n)$

return $\mathbf{A}, \boldsymbol{b}, \mathbf{H}, \boldsymbol{q}, \boldsymbol{x}$

■

Example C.98 (ADMM for Linear Programming) Consider the linear program of minimizing $\boldsymbol{q}^\top \boldsymbol{x}$ with respect to $\boldsymbol{x}$ and subject to the constraint

$$\begin{bmatrix} \mathbf{A} \\ -\mathbf{I}_n \end{bmatrix} \boldsymbol{x} \leq \begin{bmatrix} \boldsymbol{b} \\ \mathbf{0} \end{bmatrix}$$

for some $\mathbf{A} \in \mathbb{R}^{k \times n}$ and $\boldsymbol{b} \in \mathbb{R}^k$. This form coincides with the linear program in Table C.39 and is a special case of the quadratic program in Example C.94. We simply need to replace the matrix $\mathbf{A}^\top$ in Example C.94 with $[\mathbf{A}^\top, -\mathbf{I}_n]$ and $\mathbf{H}$ with a matrix of zeros.

The resulting ADMM iteration over $\boldsymbol{x} \in \mathbb{R}^n, \boldsymbol{z} = [\boldsymbol{z}_1^\top, \boldsymbol{z}_2^\top]^\top \in \mathbb{R}^{k+n}$, and $\boldsymbol{v} = [\boldsymbol{v}_1^\top, \boldsymbol{v}_2^\top]^\top \in \mathbb{R}^{k+n}$ then simplifies to:

$$\begin{aligned}
&\boldsymbol{x} \leftarrow (\mathbf{I}_n + \mathbf{A}^\top \mathbf{A})^{-1} (\mathbf{A}^\top (z_1 + \boldsymbol{b} - \varrho^{-1} \boldsymbol{v}_1) - z_2 - \varrho^{-1} \boldsymbol{q} + \varrho^{-1} \boldsymbol{v}_2) \\
&z_1 \leftarrow \mathbf{A}\boldsymbol{x} - \boldsymbol{b} + \varrho^{-1} \boldsymbol{v}_1, \quad z_2 \leftarrow \varrho^{-1} \boldsymbol{v}_2 - \boldsymbol{x}, \\
&z_j \leftarrow 0 \wedge z_j, \quad j = 1, \ldots, k+n \\
&\boldsymbol{v}_1 \leftarrow \boldsymbol{v}_1 + \varrho \left(\mathbf{A}\boldsymbol{x} - \boldsymbol{b} - z_1 \right), \quad \boldsymbol{v}_2 \leftarrow \boldsymbol{v}_2 - \varrho (\boldsymbol{x} + z_2).
\end{aligned}$$

The resulting pseudo-code is given below.

Algorithm C.99: ADMM Method for $\operatorname{argmin}_{\boldsymbol{x}: \boldsymbol{x} \geq \mathbf{0}, \mathbf{A}\boldsymbol{x} \leq \boldsymbol{b}} \boldsymbol{q}^\top \boldsymbol{x}$

Input: $\mathbf{A} \in \mathbb{R}^{k \times n}$, $\boldsymbol{b} \in \mathbb{R}^k$, $\boldsymbol{q} \in \mathbb{R}^n$; initial guess $\boldsymbol{x}$; $\varepsilon > 0$.
Output: The vector $\boldsymbol{x} = \operatorname{argmin}_{\boldsymbol{x}: \boldsymbol{x} \geq \mathbf{0}, \mathbf{A}\boldsymbol{x} \leq \boldsymbol{b}} \boldsymbol{q}^\top \boldsymbol{x}$.

$\varrho \leftarrow 0.01$, $\boldsymbol{v} \leftarrow \mathbf{0}$, $z \leftarrow \mathbf{0}$ // initial parameters
$\mathbf{L}\mathbf{L}^\top \leftarrow \mathbf{I}_n + \mathbf{A}^\top \mathbf{A}$ // store Cholesky decomposition in memory
repeat // repeat until convergence
  $\boldsymbol{x}_{\text{old}} \leftarrow \boldsymbol{x}$
  $\boldsymbol{x} \leftarrow \mathbf{A}^\top (z_1 + \boldsymbol{b} - \varrho^{-1} \boldsymbol{v}_1) - z_2 - \varrho^{-1} \boldsymbol{q} + \varrho^{-1} \boldsymbol{v}_2$
  $\boldsymbol{x} \leftarrow \mathbf{L}^{-1} \boldsymbol{x}$, $\boldsymbol{x} \leftarrow \mathbf{L}^{-\top} \boldsymbol{x}$ // forward and backward substitution costs $O(n^2)$
  $\widetilde{z}_1 \leftarrow \mathbf{A}\boldsymbol{x} - \boldsymbol{b} + \varrho^{-1} \boldsymbol{v}_1$, $z_1 \leftarrow \mathbf{0} \wedge \widetilde{z}_1$
  $z_2 \leftarrow \varrho^{-1} \boldsymbol{v}_2 - \boldsymbol{x}$, $z_2 \leftarrow \mathbf{0} \wedge z_2$
  $\boldsymbol{v}_1 \leftarrow \varrho (\widetilde{z}_1 - z_1)$, $\boldsymbol{v}_2 \leftarrow \boldsymbol{v}_2 - \varrho (\boldsymbol{x} + z_2)$ // update dual variables
until $\|\boldsymbol{x} - \boldsymbol{x}_{\text{old}}\| < \varepsilon$ // one choice for the tolerance is $\varepsilon = 10^{-6}$
return $\boldsymbol{x}$

C.4.3 Barrier Function Method

Suppose that (C.27) has inequality constraints only. **Barrier functions** are an important example of penalty functions that can handle inequality constraints. The prototypical example is a *logarithmic barrier function* which gives the unconstrained optimization:

$$\widetilde{f}(\boldsymbol{x}) = f(\boldsymbol{x}) - \nu \sum_{j=1}^{k} \ln(-g_j(\boldsymbol{x})), \quad \nu > 0,$$

such that the minimizer of $\widetilde{f}$ tends to the minimizer of f as $\nu \to 0$. Direct minimization of $\widetilde{f}$ via an unconstrained minimization algorithm is frequently too difficult. Instead, it is common to combine the logarithmic barrier function with the Lagrangian method as follows:

The idea is to introduce k nonnegative auxiliary or **slack variables** $s_1, \ldots, s_k$ that satisfy the equalities $g_j(\boldsymbol{x}) + s_j = 0$ for all j. These equalities ensure that the inequality constraints are maintained: $g_j(\boldsymbol{x}) = -s_j \leq 0$ for all j. Then, instead of the unconstrained optimization of $\widetilde{f}$, we consider the unconstrained optimization of the Lagrangian:

$$\mathcal{L}(\boldsymbol{x}, \boldsymbol{s}, \boldsymbol{\beta}) = f(\boldsymbol{x}) - \nu \sum_{j=1}^{k} \ln s_j + \sum_{j=1}^{k} \beta_j \left(g_j(\boldsymbol{x}) + s_j\right), \tag{C.100}$$

where $\nu > 0$ and $\boldsymbol{\beta}$ are the Lagrange multipliers for the equalities $g_j(\boldsymbol{x}) + s_j = 0, j = 1, \ldots, k$.

Observe how the logarithmic barrier function keeps the slack variables positive. In addition, while the optimization of $\widetilde{f}$ is over n dimensions (recall that $\boldsymbol{x} \in \mathbb{R}^n$), the optimization of the Lagrangian function $\mathcal{L}$ is over $n + 2k$ dimensions. Despite this enlargement of the search space with the variables $\boldsymbol{s}$ and $\boldsymbol{\beta}$, the optimization of the Lagrangian $\mathcal{L}$ is easier in practice than the direct optimization of $\widetilde{f}$.

Example C.101 (Interior-Point Method for Nonnegativity) One of the simplest and most common constrained optimization problems can be formulated as the minimization of $f(\boldsymbol{x})$ subject to nonnegative $\boldsymbol{x}$, that is: $\min_{\boldsymbol{x} \geq \boldsymbol{0}} f(\boldsymbol{x})$. In this case, the Lagrangian with logarithmic barrier (C.100) is:

$$\mathcal{L}(\boldsymbol{x}, \boldsymbol{s}, \boldsymbol{\beta}) = f(\boldsymbol{x}) - \nu \sum_{k} \ln s_k + \boldsymbol{\beta}^\top (\boldsymbol{s} - \boldsymbol{x}).$$

The KKT conditions in Theorem C.42 are a necessary condition for a minimizer, and yield the nonlinear system for $[\boldsymbol{x}^\top, \boldsymbol{s}^\top, \boldsymbol{\beta}^\top]^\top \in \mathbb{R}^{3n}$:

$$\begin{bmatrix} \nabla f(\boldsymbol{x}) - \boldsymbol{\beta} \\ -\nu/\boldsymbol{s} + \boldsymbol{\beta} \\ \boldsymbol{s} - \boldsymbol{x} \end{bmatrix} = \boldsymbol{0},$$

where $\nu/\boldsymbol{s}$ is a shorthand notation for a column vector with components $\{\nu/s_j\}$. To solve this system, we can use Newton's method for root finding (see, for example, Algorithm C.50), which requires a formula for the matrix of Jacobi of $\mathcal{L}$. Here, this $(3n) \times (3n)$ matrix is:

$$\mathbf{J}_{\mathcal{L}}(\boldsymbol{x}, \boldsymbol{s}, \boldsymbol{\beta}) = \begin{bmatrix} \mathbf{H} & \mathbf{O} & -\mathbf{I} \\ \mathbf{O} & \mathbf{D} & \mathbf{I} \\ -\mathbf{I} & \mathbf{I} & \mathbf{O} \end{bmatrix} = \begin{bmatrix} \mathbf{H} & \mathbf{B} \\ \mathbf{B}^\top & \mathbf{E} \end{bmatrix},$$

where $\mathbf{H}$ is the $n \times n$ Hessian of f at $\boldsymbol{x}$; $\mathbf{D} := \operatorname{diag}(\nu/(\boldsymbol{s} \odot \boldsymbol{s}))$ is an $n \times n$ diagonal matrix; $\mathbf{B} := [\mathbf{O}, -\mathbf{I}]$ is an $n \times (2n)$ matrix, and[1]

$$\mathbf{E} := \begin{bmatrix} \mathbf{D} & \mathbf{I} \\ \mathbf{I} & \mathbf{O} \end{bmatrix} = \begin{bmatrix} \mathbf{O} & \mathbf{I} \\ \mathbf{I} & -\mathbf{D} \end{bmatrix}^{-1}.$$

[1]Here $\mathbf{O}$ is an $n \times n$ matrix of zeros and $\mathbf{I}$ is the $n \times n$ identity matrix.

Further, we define

$$\mathbf{H}_\nu := (\mathbf{H} - \mathbf{B}\mathbf{E}^{-1}\mathbf{B}^\top)^{-1} = (\mathbf{H}+\mathbf{D})^{-1}.$$

Using this notation and applying the matrix blockwise inversion formula (A.59), we obtain the inverse of the matrix of Jacobi: ☞ 541

$$\begin{bmatrix} \mathbf{H} & \mathbf{B} \\ \mathbf{B}^\top & \mathbf{E} \end{bmatrix}^{-1} = \begin{bmatrix} \mathbf{H}_\nu & -\mathbf{H}_\nu\mathbf{B}\mathbf{E}^{-1} \\ -\mathbf{E}^{-1}\mathbf{B}^\top\mathbf{H}_\nu & \mathbf{E}^{-1}+\mathbf{E}^{-1}\mathbf{B}^\top\mathbf{H}_\nu\mathbf{B}\mathbf{E}^{-1} \end{bmatrix}$$

$$= \begin{bmatrix} \mathbf{H}_\nu & \mathbf{H}_\nu & -\mathbf{H}_\nu\mathbf{D} \\ \mathbf{H}_\nu & \mathbf{H}_\nu & \mathbf{I}-\mathbf{H}_\nu\mathbf{D} \\ -\mathbf{D}\mathbf{H}_\nu & \mathbf{I}-\mathbf{D}\mathbf{H}_\nu & \mathbf{D}\mathbf{H}_\nu\mathbf{D}-\mathbf{D} \end{bmatrix}.$$

Therefore, the search direction in Newton's root-finding method is given by:

$$-\mathbf{J}_{\mathcal{L}}^{-1}\begin{bmatrix} \nabla f(\boldsymbol{x}) - \boldsymbol{\beta} \\ -\nu/\boldsymbol{s} + \boldsymbol{\beta} \\ \boldsymbol{s} - \boldsymbol{x} \end{bmatrix} = \begin{bmatrix} \mathrm{d}\boldsymbol{x} \\ \mathrm{d}\boldsymbol{x} + \boldsymbol{x} - \boldsymbol{s} \\ \nu/\boldsymbol{s} - \boldsymbol{\beta} - \mathbf{D}(\mathrm{d}\boldsymbol{x} + \boldsymbol{x} - \boldsymbol{s}) \end{bmatrix},$$

where

$$\mathrm{d}\boldsymbol{x} := -(\mathbf{H}+\mathbf{D})^{-1}\left[\nabla f(\boldsymbol{x}) - 2\nu/\boldsymbol{s} + \mathbf{D}\boldsymbol{x}\right],$$

and we have assumed that $\mathbf{H}+\mathbf{D}$ is a positive-definite matrix. If at any step of the iteration the matrix $\mathbf{H}+\mathbf{D}$ fails to be positive-definite, then Newton's root-finding algorithm may fail to converge. Thus, any practical implementation will have to include a fail-safe feature to guard against this possibility.

In summary, for a given penalty parameter $\nu > 0$, we can locate the approximate nonnegative minimizer of f using, for example, the version of the Newton–Raphson root-finding method given in Algorithm C.102.

Algorithm C.102: Logarithmic Barrier Method for $\mathrm{argmin}_{\boldsymbol{x}\geq 0} f(\boldsymbol{x})$

Input: An initial guess $\boldsymbol{x}$ and stopping error $\varepsilon > 0$.
Output: The approximate nonnegative minimizer $\boldsymbol{x}_\nu$ of f.

$\boldsymbol{s} \leftarrow \boldsymbol{x}$, $\boldsymbol{\beta} \leftarrow \nu/\boldsymbol{s}$, $\mathrm{d}\boldsymbol{x} \leftarrow \boldsymbol{\beta}$
while $\|\mathrm{d}\boldsymbol{x}\| > \varepsilon$ and budget is not exhausted **do**
  Compute the gradient $\boldsymbol{u}$ and the Hessian $\mathbf{H}$ of f at $\boldsymbol{x}$.
  $\boldsymbol{s}_1 \leftarrow \nu/\boldsymbol{s}$, $\boldsymbol{s}_2 \leftarrow \boldsymbol{s}_1/\boldsymbol{s}$, $\boldsymbol{w} \leftarrow 2\boldsymbol{s}_1 - \boldsymbol{u} - \boldsymbol{s}_2 \odot \boldsymbol{x}$
  if $(\mathbf{H} + \mathrm{diag}(\boldsymbol{s}_2)) \succ \mathbf{0}$ **then** // if Cholesky successful
    Compute the Cholesky factor $\mathbf{L}$ satisfying $\mathbf{L}\mathbf{L}^\top = \mathbf{H} + \mathrm{diag}(\boldsymbol{s}_2)$.
    $\mathrm{d}\boldsymbol{x} \leftarrow \mathbf{L}^{-1}\boldsymbol{w}$ // computed by forward substitution
    $\mathrm{d}\boldsymbol{x} \leftarrow \mathbf{L}^{-\top}\mathrm{d}\boldsymbol{x}$ // computed by backward substitution
  else
    $\mathrm{d}\boldsymbol{x} \leftarrow \boldsymbol{w}/\boldsymbol{s}_2$ // if Cholesky fails, do gradient descent
  $\mathrm{d}\boldsymbol{s} \leftarrow \mathrm{d}\boldsymbol{x} + \boldsymbol{x} - \boldsymbol{s}$, $\mathrm{d}\boldsymbol{\beta} \leftarrow \boldsymbol{s}_1 - \boldsymbol{\beta} - \boldsymbol{s}_2 \odot \mathrm{d}\boldsymbol{s}$, $\alpha \leftarrow 1$
  while $\min_j\{s_j + \alpha\, \mathrm{d}s_j\} < 0$ **do**
    $\alpha \leftarrow \alpha/2$ // ensure nonnegative slack variables
  $\boldsymbol{x} \leftarrow \boldsymbol{x} + \alpha\, \mathrm{d}\boldsymbol{x}$, $\boldsymbol{s} \leftarrow \boldsymbol{s} + \alpha\, \mathrm{d}\boldsymbol{s}$, $\boldsymbol{\beta} \leftarrow \boldsymbol{\beta} + \alpha\, \mathrm{d}\boldsymbol{\beta}$
return $\boldsymbol{x}_\nu \leftarrow \boldsymbol{x}$

In practice, one needs to choose a sufficiently small value for ν, so that the output $\boldsymbol{x}_\nu$ of Algorithm C.102 is a good approximation to $\boldsymbol{x}^* = \mathrm{argmin}_{\boldsymbol{x}\geq\boldsymbol{0}} f(\boldsymbol{x})$. Alternatively, one can create a decreasing sequence of penalty parameters $\nu_1 > \nu_2 > \cdots$ and compute the corresponding solutions $\boldsymbol{x}_{\nu_1}, \boldsymbol{x}_{\nu_2}, \ldots$ of the penalized problems. In the so-called **interior-point method**, a given $\boldsymbol{x}_{\nu_i}$ is used as an initial guess for computing $\boldsymbol{x}_{\nu_{i+1}}$ and so on until the approximation to the minimizer $\boldsymbol{x}^* = \mathrm{argmin}_{\boldsymbol{x}\geq\boldsymbol{0}} f(\boldsymbol{x})$ is deemed accurate. ■

Further Reading

For an excellent introduction to convex optimization and Lagrangian duality see [30]. A classical text on optimization algorithms and, in particular, on quasi-Newton methods is [65]. For more details on the *alternating direction method of multipliers* see [29].

APPENDIX D

Probability and Statistics

The purpose of this chapter is to establish the baseline probability and statistics background for this book. We review basic concepts such as the sum and product rules of probability, random variables and their probability distributions, expectations, independence, conditional probability, transformation rules, limit theorems, and Markov chains. The properties of the multivariate normal distribution are discussed in more detail. The main ideas from statistics are also reviewed, including estimation techniques (such as maximum likelihood estimation), confidence intervals, and hypothesis testing.

D.1 Random Experiments and Probability Spaces

The basic notion in probability theory is that of a **random experiment**: an experiment whose outcome cannot be determined in advance. Mathematically, a random experiment is modeled via a triplet $(\Omega, \mathcal{H}, \mathbb{P})$, where:

- Ω is the set of all possible outcomes of the experiment, called the **sample space**.
- $\mathcal{H}$ is the collection of all subsets of Ω to which a probability can be assigned; such subsets are called **events**.
- $\mathbb{P}$ is a **probability measure**, which assigns to each event A a number $\mathbb{P}[A]$ between 0 and 1, indicating the likelihood that the outcome of the random experiment lies in A.

Any probability measure $\mathbb{P}$ must satisfy the following **Kolmogorov axioms**:

1. $\mathbb{P}[A] \geq 0$ for every event A.
2. $\mathbb{P}[\Omega] = 1$.
3. For any sequence $A_1, A_2, \ldots$ of events,

   (D.1) $$\mathbb{P}\Big[\bigcup_i A_i\Big] \leq \sum_i \mathbb{P}[A_i],$$

   with strict *equality* whenever the events are *disjoint* (that is, non-overlapping).

When (D.1) holds as an equality, it is often referred to as the **sum rule** of probability. It simply states that if an event can happen in a number of different but not simultaneous ways, the probability of that event is the sum of the probabilities of the comprising events. If the events are allowed to overlap, then the inequality (D.1) is called the **union bound**.

In many applications the sample space is *countable*; that is, $\Omega = \{a_1, a_2, \ldots\}$. In this case the easiest way to specify a probability measure $\mathbb{P}$ is to first assign a number p_i to each **elementary event** $\{a_i\}$, with $\sum_i p_i = 1$, and then to define

$$\mathbb{P}[A] = \sum_{i:a_i \in A} p_i \quad \text{for all } A \subseteq \Omega.$$

Here the collection of events $\mathcal{H}$ can be taken to be equal to the collection of *all* subsets of Ω. The triple $(\Omega, \mathcal{H}, \mathbb{P})$ is called a **discrete probability space**. This idea is graphically represented in Figure D.2. Each element a_i, represented by a dot, is assigned a weight (that is, probability) p_i, indicated by the size of the dot. The probability of the event A is simply the sum of the weights of all the outcomes in A.

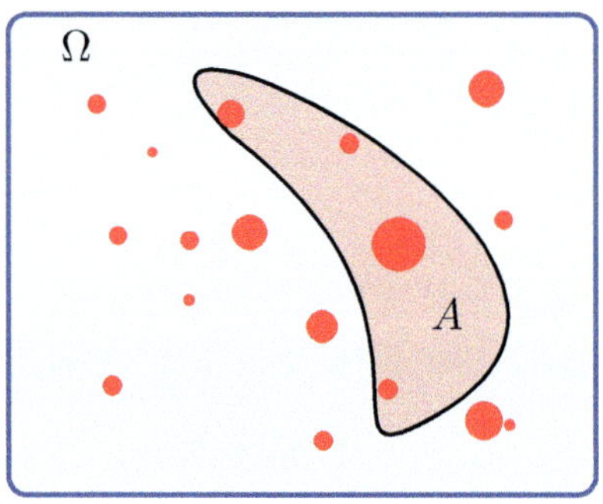

Figure D.2: A discrete probability space.

■ **Remark D.3 (Equilikely Principle)** A special case of a discrete probability space occurs when a random experiment has finitely many outcomes that are all *equally likely*. In this case the probability measure is given by

$$\mathbb{P}[A] = \frac{|A|}{|\Omega|}, \tag{D.4}$$

where $|A|$ denotes the number of outcomes in A and $|\Omega|$ is the total number of outcomes. Thus, the calculation of probabilities reduces to counting numbers of outcomes in events. This is called the **equilikely principle**. ■

D.2 Random Variables and Probability Distributions

It is often convenient to describe a random experiment via "random variables", representing numerical measurements of the experiment. Random variables are usually denoted by capital letters from the last part of the alphabet. From a mathematical point of view, a **random variable** X is a function from Ω to $\mathbb{R}$ such that sets of the form $\{a < X \leq b\} := \{\omega \in \Omega : a < X(\omega) \leq b\}$ are events (and so can be assigned a probability).

All probabilities involving a random variable X can be computed, in principle, from its **cumulative distribution function** (cdf), defined by

$$F(x) = \mathbb{P}[X \le x], \quad x \in \mathbb{R}.$$

For example $\mathbb{P}[a < X \le b] = \mathbb{P}[X \le b] - \mathbb{P}[X \le a] = F(b) - F(a)$. Figure D.5 shows a generic cdf. Note that any cdf is right-continuous, increasing, and lies between 0 and 1.

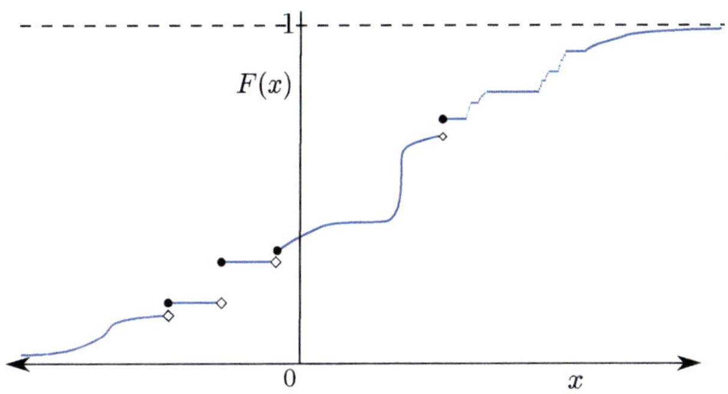

Figure D.5: A cumulative distribution function (cdf).

A cdf F_d is called *discrete* if there exist numbers $x_1, x_2, \ldots$ and probabilities $0 < f(x_i) \le 1$ summing up to 1, such that for all x

$$F_d(x) = \sum_{x_i \le x} f(x_i). \tag{D.6}$$

Such a cdf is piecewise constant and has jumps of sizes $f(x_1), f(x_2), \ldots$ at points $x_1, x_2, \ldots$, respectively. The function $f(x)$ is called a **probability mass function** or **discrete probability density function** (pdf). It is often easier to use the pdf rather than the cdf, since probabilities can simply be calculated from it via summation:

$$\mathbb{P}[X \in B] = \sum_{x \in B} f(x),$$

as illustrated in Figure D.7.

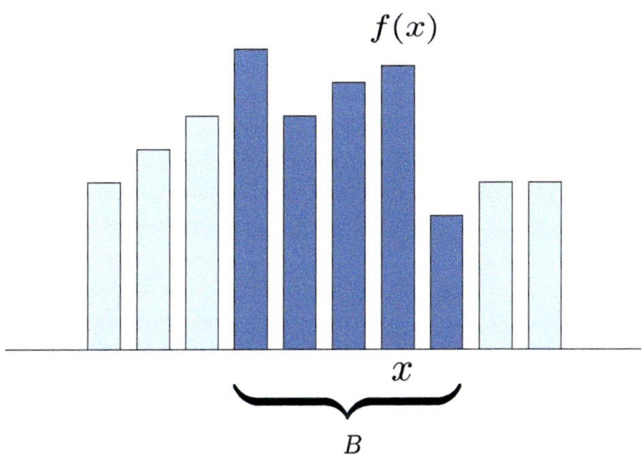

Figure D.7: Discrete probability density function (pdf). The darker area corresponds to the probability $\mathbb{P}[X \in B]$.

A cdf F_c is called *continuous*[1], if there exists a positive function f such that for all x

$$F_c(x) = \int_{-\infty}^{x} f(u)\,\mathrm{d}u. \tag{D.8}$$

Note that such an F_c is differentiable (and hence continuous) with derivative f. The function f is called the *probability density function (continuous pdf)*. By the fundamental theorem of integration, we have

$$\mathbb{P}[a < X \leq b] = F(b) - F(a) = \int_a^b f(x)\,\mathrm{d}x.$$

Thus, calculating probabilities reduces to integration, as illustrated in Figure D.9.

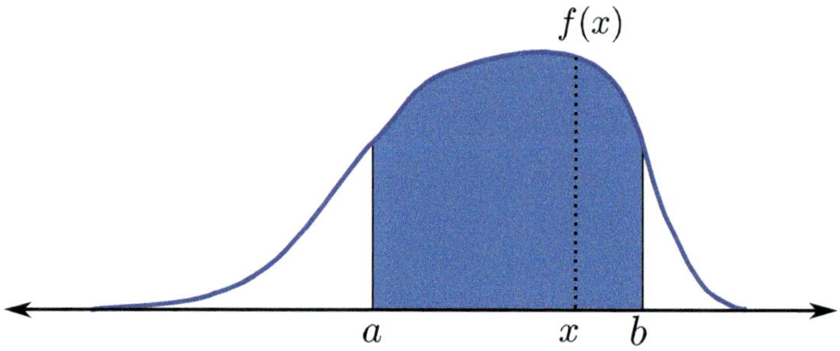

Figure D.9: Continuous probability density function (pdf). The shaded area corresponds to the probability $\mathbb{P}[X \in B]$, with B being here the interval $(a, b]$.

Remark D.10 (Probability Density and Probability Mass) It is important to note that we deliberately use the *same* name, "pdf", and symbol, f, in both the discrete and the continuous case, rather than distinguish between a probability mass function (pmf) and probability density function (pdf). From a theoretical point of view the pdf plays exactly the same role in the discrete and continuous cases. We use the notation $X \sim$ Dist, $X \sim f$, and $X \sim F$ to indicate that X has distribution Dist, pdf f, and cdf F. ■

Tables D.11 and D.12 list a number of important continuous and discrete distributions. Note that in Table D.11, Γ is the gamma function:

$$\Gamma(\alpha) = \int_0^\infty \mathrm{e}^{-x} x^{\alpha-1}\,\mathrm{d}x, \quad \alpha > 0.$$

The Gamma$(n/2, 1/2)$ distribution is called the *chi-squared distribution* with n degrees of freedom, denoted χ^2_n. The t_1 distribution is also called the *Cauchy* distribution.

[1]In advanced probability, we would say "*absolutely continuous with respect to the Lebesgue measure*".

Table D.11: Commonly used continuous distributions.

Name	Notation	$f(x)$	$x \in$	Parameters
Uniform	$\mathcal{U}[\alpha,\beta]$	$\dfrac{1}{\beta-\alpha}$	$[\alpha,\beta]$	$\alpha<\beta$
Normal	$\mathcal{N}(\mu,\sigma^2)$	$\dfrac{1}{\sigma\sqrt{2\pi}}\mathrm{e}^{-\frac{1}{2}\left(\frac{x-\mu}{\sigma}\right)^2}$	$\mathbb{R}$	$\sigma>0,\ \mu\in\mathbb{R}$
Gamma	$\mathsf{Gamma}(\alpha,\lambda)$	$\dfrac{\lambda^\alpha x^{\alpha-1}\mathrm{e}^{-\lambda x}}{\Gamma(\alpha)}$	$\mathbb{R}_+$	$\alpha,\lambda>0$
Inverse Gamma	$\mathsf{InvGamma}(\alpha,\lambda)$	$\dfrac{\lambda^\alpha x^{-\alpha-1}\mathrm{e}^{-\lambda x^{-1}}}{\Gamma(\alpha)}$	$\mathbb{R}_+$	$\alpha,\lambda>0$
Exponential	$\mathsf{Exp}(\lambda)$	$\lambda\,\mathrm{e}^{-\lambda x}$	$\mathbb{R}_+$	$\lambda>0$
Beta	$\mathsf{Beta}(\alpha,\beta)$	$\dfrac{\Gamma(\alpha+\beta)}{\Gamma(\alpha)\Gamma(\beta)}x^{\alpha-1}(1-x)^{\beta-1}$	$[0,1]$	$\alpha,\beta>0$
Weibull	$\mathsf{Weib}(\alpha,\lambda)$	$\alpha\lambda\,(\lambda x)^{\alpha-1}\mathrm{e}^{-(\lambda x)^\alpha}$	$\mathbb{R}_+$	$\alpha,\lambda>0$
Pareto	$\mathsf{Pareto}(\alpha,\lambda)$	$\alpha\lambda\,(1+\lambda x)^{-(\alpha+1)}$	$\mathbb{R}_+$	$\alpha,\lambda>0$
Student	t_ν	$\dfrac{\Gamma(\frac{\nu+1}{2})}{\sqrt{\nu\pi}\,\Gamma(\frac{\nu}{2})}\left(1+\dfrac{x^2}{\nu}\right)^{-(\nu+1)/2}$	$\mathbb{R}$	$\nu>0$
F	$\mathsf{F}(m,n)$	$\dfrac{\Gamma(\frac{m+n}{2})\,(m/n)^{m/2}x^{(m-2)/2}}{\Gamma(\frac{m}{2})\,\Gamma(\frac{n}{2})\,[1+(m/n)x]^{(m+n)/2}}$	$\mathbb{R}_+$	$m,n\in\mathbb{N}_+$

Table D.12: Commonly used discrete distributions.

Name	Notation	$f(x)$	$x \in$	Parameters
Bernoulli	$\mathsf{Ber}(p)$	$p^x(1-p)^{1-x}$	$\{0,1\}$	$0\le p\le 1$
Binomial	$\mathsf{Bin}(n,p)$	$\binom{n}{x}p^x(1-p)^{n-x}$	$\{0,1,\ldots,n\}$	$0\le p\le 1$, $n\in\mathbb{N}$
Discrete uniform	$\mathcal{U}\{1,\ldots,n\}$	$\dfrac{1}{n}$	$\{1,\ldots,n\}$	$n\in\{1,2,\ldots\}$
Geometric	$\mathsf{Geom}(p)$	$p(1-p)^{x-1}$	$\{1,2,\ldots\}$	$0\le p\le 1$
Poisson	$\mathsf{Poi}(\lambda)$	$\mathrm{e}^{-\lambda}\dfrac{\lambda^x}{x!}$	$\mathbb{N}$	$\lambda>0$

D.3 Expectation

It is often useful to consider different kinds of numerical characteristics of a random variable. One such quantity is the expectation, which measures the "average" value of the distribution.

The **expectation** (or expected value or mean) of a random variable X with pdf f, denoted by $\mathbb{E}X$ or[2] $\mathbb{E}[X]$ (and sometimes μ), is defined by

$$\mathbb{E}X = \begin{cases} \sum_x x\, f(x) & \text{discrete case,} \\ \int_{-\infty}^{\infty} x\, f(x)\,\mathrm{d}x & \text{continuous case.} \end{cases}$$

If X is a random variable, then a function of X, such as X^2 or $\sin(X)$, is again a random variable. Moreover, the expected value of a function of X is simply a weighted average of the possible values that this function can take. That is, for any real function h

$$\mathbb{E}h(X) = \begin{cases} \sum_x h(x)\, f(x) & \text{discrete case,} \\ \int_{-\infty}^{\infty} h(x)\, f(x)\,\mathrm{d}x & \text{continuous case,} \end{cases}$$

provided that the sum or integral are well-defined.

The **variance** of a random variable X, denoted by $\mathbb{V}\mathrm{ar}\, X$ (and sometimes σ^2), is defined by

$$\mathbb{V}\mathrm{ar}\, X = \mathbb{E}(X - \mathbb{E}[X])^2 = \mathbb{E}X^2 - (\mathbb{E}X)^2.$$

The square root of the variance is called the **standard deviation**. Table D.13 lists the expectations and variances for some well-known distributions. Both variance and standard deviation measure the spread or dispersion of the distribution. Note, however, that the standard deviation measures the dispersion in the same units as the random variable, unlike the variance, which uses squared units.

Table D.13: Expectations and variances for some well-known distributions.

Dist.	$\mathbb{E}X$	$\mathbb{V}\mathrm{ar}\, X$	Dist.	$\mathbb{E}X$	$\mathbb{V}\mathrm{ar}\, X$
$\mathsf{Bin}(n,p)$	np	$np(1-p)$	$\mathsf{Gamma}(\alpha,\lambda)$	$\frac{\alpha}{\lambda}$	$\frac{\alpha}{\lambda^2}$
$\mathsf{Geom}(p)$	$\frac{1}{p}$	$\frac{1-p}{p^2}$	$\mathcal{N}(\mu,\sigma^2)$	μ	σ^2
$\mathsf{Poi}(\lambda)$	λ	λ	$\mathsf{Beta}(\alpha,\beta)$	$\frac{\alpha}{\alpha+\beta}$	$\frac{\alpha\beta}{(\alpha+\beta)^2(1+\alpha+\beta)}$
$\mathcal{U}[\alpha,\beta]$	$\frac{\alpha+\beta}{2}$	$\frac{(\beta-\alpha)^2}{12}$	$\mathsf{Weib}(\alpha,\lambda)$	$\frac{\Gamma(1/\alpha)}{\alpha\,\lambda}$	$\frac{2\Gamma(2/\alpha)}{\alpha} - \left(\frac{\Gamma(1/\alpha)}{\alpha\,\lambda}\right)^2$
$\mathsf{Exp}(\lambda)$	$\frac{1}{\lambda}$	$\frac{1}{\lambda^2}$	$\mathsf{F}(m,n)$	$\frac{n}{n-2}\ (n>2)$	$\frac{2n^2(m+n-2)}{m(n-2)^2(n-4)}\ (n>4)$
t_ν	$0\ (\nu>1)$	$\frac{\nu}{\nu-2}\ (\nu>2)$			

It is sometimes useful to consider the **moment generating function** of a random variable X. This is the function M defined by

$$M(s) = \mathbb{E}\,\mathrm{e}^{sX}, \quad s \in \mathbb{R}. \tag{D.14}$$

The moment generating functions of two random variables coincide if and only if the random variables have the same distribution; see also Theorem D.69.

[2]We only use brackets in an expectation if it is unclear with respect to which random variable the expectation is taken.

Example D.15 (MGF of the Gamma(α, λ) Distribution) Let $X \sim \mathsf{Gamma}(\alpha, \lambda)$. For $s < \lambda$, the moment generating function of X at s is given by

$$\begin{aligned} M(s) = \mathbb{E}e^{sX} &= \int_0^\infty e^{sx} \frac{e^{-\lambda x} \lambda^\alpha x^{\alpha-1}}{\Gamma(\alpha)} \, dx \\ &= \left(\frac{\lambda}{\lambda - s}\right)^\alpha \int_0^\infty \underbrace{\frac{e^{-(\lambda-s)x} (\lambda - s)^\alpha x^{\alpha-1}}{\Gamma(\alpha)}}_{\text{pdf of Gamma}(\alpha,\lambda-s)} \, dx = \left(\frac{\lambda}{\lambda - s}\right)^\alpha, \end{aligned}$$

where in the last line we used the fact that the integral of a pdf equals unity. For $s \geq \lambda$, $M(s) = \infty$. Interestingly, the moment generating function has a much simpler formula than the pdf. ■

D.4 Joint Distributions

Distributions for random vectors and stochastic processes can be specified in much the same way as for random variables. In particular, the distribution of a random vector $\boldsymbol{X} = [X_1, \ldots, X_n]^\top$ is completely determined by specifying the **joint cdf** F, defined by

$$F(x_1, \ldots, x_n) = \mathbb{P}[X_1 \leq x_1, \ldots, X_n \leq x_n], \quad x_i \in \mathbb{R}, \ i = 1, \ldots, n.$$

Similarly, the distribution of a **stochastic process**, that is, a collection of random variables $\{X_t, t \in \mathscr{T}\}$, for some index set $\mathscr{T}$, is completely determined by its finite-dimensional distributions; specifically, the distributions of the random vectors $[X_{t_1}, \ldots, X_{t_n}]^\top$ for every choice of n and $t_1, \ldots, t_n$.

By analogy to the one-dimensional case, a random vector $\boldsymbol{X} = [X_1, \ldots, X_n]^\top$ taking values in $\mathbb{R}^n$ is said to have a pdf f if, in the continuous case,

$$\mathbb{P}[\boldsymbol{X} \in B] = \int_B f(\boldsymbol{x}) \, d\boldsymbol{x}, \tag{D.16}$$

for all n-dimensional rectangles B. Replace the integral with a sum for the discrete case. The pdf is also called the **joint pdf** of $X_1, \ldots, X_n$. The pdfs of the individual components — called *marginal pdfs* — can be recovered from the joint pdf by "integrating out the other variables". For example, for a continuous random vector $[X, Y]^\top$ with pdf f, the pdf f_X of X is given by

$$f_X(x) = \int f(x, y) \, dy.$$

D.5 Conditioning and Independence

Conditional probabilities and conditional distributions are used to model additional information on a random experiment. Independence is used to model *lack* of such information.

D.5.1 Conditional Probability

Suppose some event $B \subseteq \Omega$ occurs. Given this fact, event A will occur if and only if $A \cap B$ occurs, and the relative chance of A occurring is therefore $\mathbb{P}[A \cap B]/\mathbb{P}[B]$, provided $\mathbb{P}[B] > 0$. This leads to the definition of the **conditional probability** of A given B:

$$\mathbb{P}[A \mid B] = \frac{\mathbb{P}[A \cap B]}{\mathbb{P}[B]}, \quad \text{if } \mathbb{P}[B] > 0. \tag{D.17}$$

The above definition breaks down if $\mathbb{P}[B] = 0$. Such conditional probabilities must be treated with more care [16].

Three important consequences of the definition of conditional probability are:

1. **Product rule**: For any sequence of events $A_1, A_2, \ldots, A_n$,

$$\mathbb{P}[A_1 \cdots A_n] = \mathbb{P}[A_1]\,\mathbb{P}[A_2 \mid A_1]\,\mathbb{P}[A_3 \mid A_1 A_2] \cdots \mathbb{P}[A_n \mid A_1 \cdots A_{n-1}], \tag{D.18}$$

   using the abbreviation $A_1 A_2 \cdots A_k := A_1 \cap A_2 \cap \cdots \cap A_k$.

2. **Law of total probability**: If $\{B_i\}$ forms a *partition* of Ω (that is, $B_i \cap B_j = \emptyset, i \neq j$ and $\cup_i B_i = \Omega$), then for any event A

$$\mathbb{P}[A] = \sum_i \mathbb{P}[A \mid B_i]\,\mathbb{P}[B_i]. \tag{D.19}$$

3. **Bayes' rule**: Let $\{B_i\}$ form a partition of Ω. Then, for any event A with $\mathbb{P}[A] > 0$,

$$\mathbb{P}[B_j \mid A] = \frac{\mathbb{P}[A \mid B_j]\,\mathbb{P}[B_j]}{\sum_i \mathbb{P}[A \mid B_i]\,\mathbb{P}[B_i]}. \tag{D.20}$$

D.5.2 Independence

Two events A and B are said to be **independent** if the knowledge that B has occurred does not change the probability that A occurs. That is, A, B independent $\Leftrightarrow \mathbb{P}[A \mid B] = \mathbb{P}[A]$. Since $\mathbb{P}[A \mid B]\,\mathbb{P}[B] = \mathbb{P}[A \cap B]$, an alternative definition of independence is

$$A, B \text{ independent} \Leftrightarrow \mathbb{P}[A \cap B] = \mathbb{P}[A]\,\mathbb{P}[B].$$

This definition covers the case where $\mathbb{P}[B] = 0$ and can be extended to arbitrarily many events: events $A_1, A_2, \ldots$ are said to be (mutually) independent if for any k and any choice of distinct indices $i_1, \ldots, i_k$,

$$\mathbb{P}[A_{i_1} \cap A_{i_2} \cap \cdots \cap A_{i_k}] = \mathbb{P}[A_{i_1}]\,\mathbb{P}[A_{i_2}] \cdots \mathbb{P}[A_{i_k}].$$

The concept of independence can also be formulated for random variables. Random variables $X_1, X_2, \ldots$ are said to be **independent** if the events $\{X_{i_1} \leq x_{i_1}\}, \ldots, \{X_{i_n} \leq x_{i_n}\}$ are independent for all finite choices of n distinct indices $i_1, \ldots, i_n$ and values $x_{i_1}, \ldots, x_{i_n}$.

An important characterization of independent random variables is the following (for a proof, see [143], for example).

Theorem D.21 (Independence): Random variables $X_1, \ldots, X_n$ with marginal pdfs $f_{X_1}, \ldots, f_{X_n}$ and joint pdf f are independent if and only if

$$f(x_1, \ldots, x_n) = f_{X_1}(x_1) \cdots f_{X_n}(x_n) \quad \text{for all } x_1, \ldots, x_n. \tag{D.22}$$

Many probabilistic models involve random variables $X_1, X_2, \ldots$ that are **independent and identically distributed**, abbreviated as **iid**. We use this abbreviation throughout this book.

D.5.3 Expectation and Covariance

Similar to the univariate case, the expected value of a real-valued function h of a random vector $\boldsymbol{X} \sim f$ is a weighted average of all values that $h(\boldsymbol{X})$ can take. Specifically, in the continuous case, $\mathbb{E}h(\boldsymbol{X}) = \int h(\boldsymbol{x}) f(\boldsymbol{x}) \, \mathrm{d}\boldsymbol{x}$. In the discrete case replace this multi-dimensional integral with a sum. Using this result, it is not difficult to show that for any collection of dependent or independent random variables $X_1, \ldots, X_n$,

$$\mathbb{E}[a + b_1 X_1 + b_2 X_2 + \cdots + b_n X_n] = a + b_1 \mathbb{E}X_1 + \cdots + b_n \mathbb{E}X_n \tag{D.23}$$

for all constants $a, b_1, \ldots, b_n$. Moreover, for *independent* random variables,

$$\mathbb{E}[X_1 X_2 \cdots X_n] = \mathbb{E}X_1 \, \mathbb{E}X_2 \cdots \mathbb{E}X_n. \tag{D.24}$$

We leave the proofs as an exercise.

The **covariance** of two random variables X and Y with expectations μ_X and μ_Y, respectively, is defined as

$$\mathbb{C}\text{ov}(X, Y) = \mathbb{E}[(X - \mu_X)(Y - \mu_Y)].$$

This is a measure of the amount of linear dependency between the variables. Let $\sigma_X^2 = \mathbb{V}\text{ar}\, X$ and $\sigma_Y^2 = \mathbb{V}\text{ar}\, Y$. A scaled version of the covariance is given by the **correlation coefficient**,

$$\varrho(X, Y) = \frac{\mathbb{C}\text{ov}(X, Y)}{\sigma_X \, \sigma_Y}.$$

The following properties follow directly from the definitions of variance and covariance.

1. $\mathbb{V}\text{ar}\, X = \mathbb{E}X^2 - \mu_X^2$.
2. $\mathbb{V}\text{ar}[aX + b] = a^2 \sigma_X^2$.
3. $\mathbb{C}\text{ov}(X, Y) = \mathbb{E}[XY] - \mu_X \mu_Y$.
4. $\mathbb{C}\text{ov}(X, Y) = \mathbb{C}\text{ov}(Y, X)$.
5. $-\sigma_X \sigma_Y \leq \mathbb{C}\text{ov}(X, Y) \leq \sigma_X \sigma_Y$.
6. $\mathbb{C}\text{ov}(aX + bY, Z) = a \, \mathbb{C}\text{ov}(X, Z) + b \, \mathbb{C}\text{ov}(Y, Z)$.
7. $\mathbb{C}\text{ov}(X, X) = \sigma_X^2$.

8. $\mathbb{V}\mathrm{ar}[X+Y] = \sigma_X^2 + \sigma_Y^2 + 2\,\mathbb{C}\mathrm{ov}(X,Y)$.

9. If X and Y are independent, then $\mathbb{C}\mathrm{ov}(X,Y) = 0$.

As a consequence of Properties 2 and 8 we have that for any sequence of *independent* random variables $X_1, \ldots, X_n$ with variances $\sigma_1^2, \ldots, \sigma_n^2$,

$$\mathbb{V}\mathrm{ar}[a_1 X_1 + a_2 X_2 + \cdots + a_n X_n] = a_1^2\,\sigma_1^2 + a_2^2\,\sigma_2^2 + \cdots + a_n^2\,\sigma_n^2, \tag{D.25}$$

for any choice of constants $a_1, \ldots, a_n$.

For random column vectors, such as $\boldsymbol{X} = [X_1, \ldots, X_n]^\top$, it is convenient to write the expectations and covariances in vector and matrix notation. For a random vector $\boldsymbol{X}$ we define its **expectation vector** as the vector of expectations

$$\boldsymbol{\mu} = [\mu_1, \ldots, \mu_n]^\top = [\mathbb{E}X_1, \ldots, \mathbb{E}X_n]^\top.$$

Similarly, if the expectation of a matrix is the matrix of expectations, then given two random vectors $\boldsymbol{X} \in \mathbb{R}^n$ and $\boldsymbol{Y} \in \mathbb{R}^m$, the $n \times m$ matrix

$$\mathbb{C}\mathrm{ov}(\boldsymbol{X}, \boldsymbol{Y}) = \mathbb{E}[(\boldsymbol{X} - \mathbb{E}\boldsymbol{X})(\boldsymbol{Y} - \mathbb{E}\boldsymbol{Y})^\top] \tag{D.26}$$

has (i,j)-th element $\mathbb{C}\mathrm{ov}(X_i, Y_j) = \mathbb{E}[(X_i - \mathbb{E}X_i)(Y_j - \mathbb{E}Y_j)]$. A consequence of this definition is that

$$\mathbb{C}\mathrm{ov}(\mathbf{A}\boldsymbol{X}, \mathbf{B}\boldsymbol{Y}) = \mathbf{A}\,\mathbb{C}\mathrm{ov}(\boldsymbol{X}, \boldsymbol{Y})\mathbf{B}^\top,$$

where $\mathbf{A}$ and $\mathbf{B}$ are two matrices with n and m columns, respectively.

The **covariance matrix** of the vector $\boldsymbol{X}$ is defined as the $n \times n$ matrix $\mathbb{C}\mathrm{ov}(\boldsymbol{X}, \boldsymbol{X})$. The covariance matrix is also denoted as $\mathbb{V}\mathrm{ar}(\boldsymbol{X}) = \mathbb{C}\mathrm{ov}(\boldsymbol{X}, \boldsymbol{X})$, in analogy with the scalar identity $\mathbb{V}\mathrm{ar}(X) = \mathbb{C}\mathrm{ov}(X, X)$.

A useful application of the cyclic property of the trace of a matrix (see Theorem A.7)
☞ 515 is the following.

Theorem D.27 (Expectation of a Quadratic Form): Let $\mathbf{A}$ be an $n \times n$ matrix and $\boldsymbol{X}$ an n-dimensional random vector with expectation vector $\boldsymbol{\mu}$ and covariance matrix $\boldsymbol{\Sigma}$. The random variable $Y := \boldsymbol{X}^\top \mathbf{A} \boldsymbol{X}$ has expectation $\mathrm{tr}(\mathbf{A}\boldsymbol{\Sigma}) + \boldsymbol{\mu}^\top \mathbf{A} \boldsymbol{\mu}$.

Proof. Since Y is a scalar, it is equal to its trace. Now, using the cyclic property: $\mathbb{E}Y = \mathbb{E}\,\mathrm{tr}(Y) = \mathbb{E}\,\mathrm{tr}(\boldsymbol{X}^\top \mathbf{A} \boldsymbol{X}) = \mathbb{E}\,\mathrm{tr}(\mathbf{A}\boldsymbol{X}\boldsymbol{X}^\top) = \mathrm{tr}(\mathbf{A}\,\mathbb{E}[\boldsymbol{X}\boldsymbol{X}^\top]) = \mathrm{tr}(\mathbf{A}(\boldsymbol{\Sigma} + \boldsymbol{\mu}\boldsymbol{\mu}^\top)) = \mathrm{tr}(\mathbf{A}\boldsymbol{\Sigma}) + \mathrm{tr}(\mathbf{A}\boldsymbol{\mu}\boldsymbol{\mu}^\top) = \mathrm{tr}(\mathbf{A}\boldsymbol{\Sigma}) + \boldsymbol{\mu}^\top \mathbf{A} \boldsymbol{\mu}$. □

D.5.4 Conditional Density and Conditional Expectation

Suppose X and Y are both discrete or both continuous, with joint pdf f, and suppose $f_X(x) > 0$. Then, the **conditional pdf** of Y given $X = x$ is given by

$$f_{Y|X}(y\,|\,x) = \frac{f(x,y)}{f_X(x)} \quad \text{for all } y. \tag{D.28}$$

In the discrete case, the formula is a direct translation of (D.17), with $f_{Y|X}(y \mid x) = \mathbb{P}[Y = y \mid X = x]$. In the continuous case, a similar interpretation in terms of densities can be used; see, for example, [143, Page 221]. The corresponding distribution is called the **conditional distribution** of Y given $X = x$. Note that (D.28) implies that

$$f(x, y) = f_X(x)\, f_{Y \mid X}(y \mid x).$$

This is useful when the marginal and conditional pdfs are given, rather than the joint one. More generally, for the n-dimensional case we have

$$f(x_1, \ldots, x_n) = f_{X_1}(x_1)\, f_{X_2 \mid X_1}(x_2 \mid x_1) \cdots f_{X_n \mid X_1, \ldots, X_{n-1}}(x_n \mid x_1, \ldots, x_{n-1}), \tag{D.29}$$

which is in essence a rephrasing of the product rule (D.18) in terms of probability densities. ☞ 632

As a conditional pdf has all the properties of an ordinary pdf, we may define expectations with respect to it. The **conditional expectation** of a random variable Y given $X = x$ is defined as

$$\mathbb{E}[Y \mid X = x] = \begin{cases} \sum_y y\, f_{Y|X}(y \mid x) & \text{discrete case,} \\ \int y\, f_{Y|X}(y \mid x)\, \mathrm{d}y & \text{continuous case.} \end{cases} \tag{D.30}$$

Note that $\mathbb{E}[Y \mid X = x]$ is a function of x. The corresponding random variable is written as $\mathbb{E}[Y \mid X]$. A similar formalism can be used when conditioning on a sequence of random variables $X_1, \ldots, X_n$. The conditional expectation has similar properties to the ordinary expectation. Other useful properties (see, for example, [183]) are:

1. *Tower property*: If $\mathbb{E}Y$ exists, then

$$\mathbb{E}\,\mathbb{E}[Y \mid X] = \mathbb{E}Y. \tag{D.31}$$

2. *Taking out what is known*: If $\mathbb{E}Y$ exists, then

$$\mathbb{E}\,[XY \mid X] = X\mathbb{E}[Y \mid X].$$

D.6 Functions of Random Variables

Let $\boldsymbol{x} = [x_1, \ldots, x_n]^\top$ be a column vector in $\mathbb{R}^n$ and $\mathbf{A}$ an $m \times n$ matrix. The mapping $\boldsymbol{x} \mapsto \boldsymbol{z}$, with $\boldsymbol{z} = \mathbf{A}\boldsymbol{x}$, is a linear transformation, as discussed in Section A.1. Now consider a random vector $\boldsymbol{X} = [X_1, \ldots, X_n]^\top$ and let $\boldsymbol{Z} := \mathbf{A}\boldsymbol{X}$. Then $\boldsymbol{Z}$ is a random vector in $\mathbb{R}^m$. The following theorem details how the distribution of $\boldsymbol{Z}$ is related to that of $\boldsymbol{X}$. ☞ 513

Theorem D.32 (Linear Transformation): If $\boldsymbol{X}$ has an expectation vector $\boldsymbol{\mu}_{\boldsymbol{X}}$ and covariance matrix $\boldsymbol{\Sigma}_{\boldsymbol{X}}$, then the expectation vector of $\boldsymbol{Z}$ is

$$\boldsymbol{\mu}_{\boldsymbol{Z}} = \mathbf{A}\, \boldsymbol{\mu}_{\boldsymbol{X}} \tag{D.33}$$

and the covariance matrix of $\mathbf{Z}$ is

(D.34) $$\boldsymbol{\Sigma}_{\boldsymbol{Z}} = \mathbf{A}\,\boldsymbol{\Sigma}_{\boldsymbol{X}}\,\mathbf{A}^\top.$$

If, in addition, $\mathbf{A}$ is an invertible $n \times n$ matrix and $\boldsymbol{X}$ is a continuous random vector with pdf $f_{\boldsymbol{X}}$, then the pdf of the continuous random vector $\boldsymbol{Z} = \mathbf{A}\boldsymbol{X}$ is given by

(D.35) $$f_{\boldsymbol{Z}}(\boldsymbol{z}) = \frac{f_{\boldsymbol{X}}(\mathbf{A}^{-1}\boldsymbol{z})}{|\det(\mathbf{A})|}, \quad \boldsymbol{z} \in \mathbb{R}^n,$$

where $|\det(\mathbf{A})|$ denotes the absolute value of the determinant of $\mathbf{A}$.

Proof. We have $\boldsymbol{\mu}_{\boldsymbol{Z}} = \mathbb{E}\boldsymbol{Z} = \mathbb{E}[\mathbf{A}\boldsymbol{X}] = \mathbf{A}\,\mathbb{E}\boldsymbol{X} = \mathbf{A}\boldsymbol{\mu}_{\boldsymbol{X}}$ and

$$\begin{aligned}\boldsymbol{\Sigma}_{\boldsymbol{Z}} &= \mathbb{E}[(\boldsymbol{Z} - \boldsymbol{\mu}_{\boldsymbol{Z}})(\boldsymbol{Z} - \boldsymbol{\mu}_{\boldsymbol{Z}})^\top] = \mathbb{E}[\mathbf{A}(\boldsymbol{X} - \boldsymbol{\mu}_{\boldsymbol{X}})(\mathbf{A}(\boldsymbol{X} - \boldsymbol{\mu}_{\boldsymbol{X}}))^\top] \\ &= \mathbf{A}\,\mathbb{E}[(\boldsymbol{X} - \boldsymbol{\mu}_{\boldsymbol{X}})(\boldsymbol{X} - \boldsymbol{\mu}_{\boldsymbol{X}})^\top]\mathbf{A}^\top \\ &= \mathbf{A}\,\boldsymbol{\Sigma}_{\boldsymbol{X}}\,\mathbf{A}^\top.\end{aligned}$$

For $\mathbf{A}$ invertible and $\boldsymbol{X}$ continuous (as opposed to discrete), let $\boldsymbol{z} = \mathbf{A}\boldsymbol{x}$ and $\boldsymbol{x} = \mathbf{A}^{-1}\boldsymbol{z}$. Consider the n-dimensional cube $C = [z_1, z_1 + h] \times \cdots \times [z_n, z_n + h]$. Then,

$$\mathbb{P}[\boldsymbol{Z} \in C] \approx h^n\, f_{\boldsymbol{Z}}(\boldsymbol{z}),$$

by definition of the joint density of $\boldsymbol{Z}$. Let D be the image of C under $\mathbf{A}^{-1}$ — that is, all
☞ 513 points $\boldsymbol{x}$ such that $\mathbf{A}\boldsymbol{x} \in C$. Recall from Section A.1 that any matrix B linearly transforms an n-dimensional rectangle with volume V into an n-dimensional parallelepiped with volume $V\,|\det(B)|$. Thus, in addition to the above expression for $\mathbb{P}[\boldsymbol{Z} \in C]$, we also have

$$\mathbb{P}[\boldsymbol{Z} \in C] = \mathbb{P}[\boldsymbol{X} \in D] \approx h^n|\det(\mathbf{A}^{-1})|\, f_{\boldsymbol{X}}(\boldsymbol{x}) = h^n|\det(\mathbf{A})|^{-1}\, f_{\boldsymbol{X}}(\boldsymbol{x}).$$

Equating these two expressions for $\mathbb{P}[\boldsymbol{Z} \in C]$, dividing both sides by h^n, and letting h go to 0, we obtain (D.35). □

For a generalization of the linear transformation rule (D.35), consider an arbitrary mapping $\boldsymbol{x} \mapsto \boldsymbol{g}(\boldsymbol{x})$, written out:

$$\begin{bmatrix} x_1 \\ x_2 \\ \vdots \\ x_n \end{bmatrix} \mapsto \begin{bmatrix} g_1(\boldsymbol{x}) \\ g_2(\boldsymbol{x}) \\ \vdots \\ g_n(\boldsymbol{x}) \end{bmatrix}.$$

Theorem D.36 (Transformation Rule): Let $\boldsymbol{X}$ be an n-dimensional vector of continuous random variables with pdf $f_{\boldsymbol{X}}$. Let $\boldsymbol{Z} = \boldsymbol{g}(\boldsymbol{X})$, where $\boldsymbol{g}$ is an invertible mapping with inverse $\boldsymbol{g}^{-1}$ and *matrix of Jacobi* $\mathbf{J}_{\boldsymbol{g}}$; that is, the matrix of partial derivatives of $\boldsymbol{g}$. Then, at $z = \boldsymbol{g}(\boldsymbol{x})$ the random vector $\boldsymbol{Z}$ has pdf

$$f_{\boldsymbol{Z}}(z) = \frac{f_{\boldsymbol{X}}(\boldsymbol{x})}{|\det(\mathbf{J}_{\boldsymbol{g}}(\boldsymbol{x}))|} = f_{\boldsymbol{X}}(\boldsymbol{g}^{-1}(z))\,|\det(\mathbf{J}_{\boldsymbol{g}^{-1}}(z))|, \quad z \in \mathbb{R}^n. \tag{D.37}$$

Proof. For a fixed $\boldsymbol{x}$, let $z = \boldsymbol{g}(\boldsymbol{x})$; and thus $\boldsymbol{x} = \boldsymbol{g}^{-1}(z)$. In the neighborhood of $\boldsymbol{x}$, the function $\boldsymbol{g}$ behaves like a linear function, in the sense that $\boldsymbol{g}(\boldsymbol{x} + \boldsymbol{\delta}) \approx \boldsymbol{g}(\boldsymbol{x}) + \mathbf{J}_{\boldsymbol{g}}(\boldsymbol{x})\,\boldsymbol{\delta}$ for small vectors $\boldsymbol{\delta}$; see also Section C.1. Consequently, an infinitesimally small n-dimensional rectangle at $\boldsymbol{x}$ with volume V is transformed into an infinitesimally small n-dimensional parallelepiped at z with volume $V\,|\det(\mathbf{J}_{\boldsymbol{g}}(\boldsymbol{x}))|$. Now, as in the proof of the linear case, let C be a small cube around $z = \boldsymbol{g}(\boldsymbol{x})$ with volume h^n. Let D be the image of C under $\boldsymbol{g}^{-1}$. Then, ☞ 581

$$h^n\, f_{\boldsymbol{Z}}(z) \approx \mathbb{P}[\boldsymbol{Z} \in C] \approx h^n |\det(\mathbf{J}_{\boldsymbol{g}^{-1}}(z))|\, f_{\boldsymbol{X}}(\boldsymbol{x}),$$

and since $|\det(\mathbf{J}_{\boldsymbol{g}^{-1}}(z))| = 1/|\det(\mathbf{J}_{\boldsymbol{g}}(\boldsymbol{x}))|$, (D.37) follows as h goes to 0. □

Typically, in coordinate transformations it is $\boldsymbol{g}^{-1}$ that is given — that is, an expression for $\boldsymbol{x}$ as a function of z.

Example D.38 (Polar Transform) Suppose X, Y are independent and have standard normal distribution. The joint pdf is

$$f_{X,Y}(x, y) = \frac{1}{2\pi}\mathrm{e}^{-\frac{1}{2}(x^2+y^2)}, \quad (x, y) \in \mathbb{R}^2.$$

In polar coordinates we have

$$X = R\cos\Theta \quad \text{and} \quad Y = R\sin\Theta, \tag{D.39}$$

where $R \geq 0$ is the radius and $\Theta \in [0, 2\pi)$ the angle of the point (X, Y). What is the joint pdf of R and Θ? By the radial symmetry of the bivariate normal distribution, we would expect Θ to be uniform on $(0, 2\pi)$. But what is the pdf of R? To work out the joint pdf, consider the inverse transformation $\mathbf{g}^{-1}$, defined by

$$\begin{bmatrix} r \\ \theta \end{bmatrix} \overset{\mathbf{g}^{-1}}{\longmapsto} \begin{bmatrix} r\cos\theta \\ r\sin\theta \end{bmatrix} = \begin{bmatrix} x \\ y \end{bmatrix}.$$

The corresponding matrix of Jacobi is

$$\mathbf{J}_{\mathbf{g}^{-1}}(r, \theta) = \begin{bmatrix} \cos\theta & -r\sin\theta \\ \sin\theta & r\cos\theta \end{bmatrix},$$

which has determinant r. Since $x^2 + y^2 = r^2(\cos^2\theta + \sin^2\theta) = r^2$, it follows by the transformation rule (D.37) that the joint pdf of R and Θ is given by

$$f_{R,\Theta}(r, \theta) = f_{X,Y}(x, y)\, r = \frac{1}{2\pi}\mathrm{e}^{-\frac{1}{2}r^2}\, r, \quad \theta \in (0, 2\pi), \quad r \geq 0.$$

By integrating out θ and r, respectively, we find $f_R(r) = r\,\mathrm{e}^{-r^2/2}$ and $f_\Theta(\theta) = 1/(2\pi)$. Since $f_{R,\Theta}$ is the product of f_R and f_Θ, the random variables R and Θ are independent. ■

D.7 Multivariate Normal Distribution

The normal (or Gaussian) distribution — especially its multi-dimensional version — plays a central role in data science and machine learning. Recall from Table D.11 that a random variable X is said to have a *normal* distribution with parameters μ and σ^2 if its pdf is given by

$$f(x) = \frac{1}{\sigma\sqrt{2\pi}}\, \mathrm{e}^{-\frac{1}{2}\left(\frac{x-\mu}{\sigma}\right)^2}, \quad x \in \mathbb{R}. \tag{D.40}$$

We write $X \sim \mathcal{N}(\mu, \sigma^2)$. The parameters μ and σ^2 are the expectation and variance of the distribution, respectively. If $\mu = 0$ and $\sigma = 1$ then

$$f(x) = \frac{1}{\sqrt{2\pi}}\, \mathrm{e}^{-x^2/2},$$

and the distribution is known as the **standard normal** distribution. The cdf of the standard normal distribution is often denoted by Φ and its pdf by φ. In Figure D.41 the pdf of the $\mathcal{N}(\mu, \sigma^2)$ distribution for various μ and σ^2 is plotted.

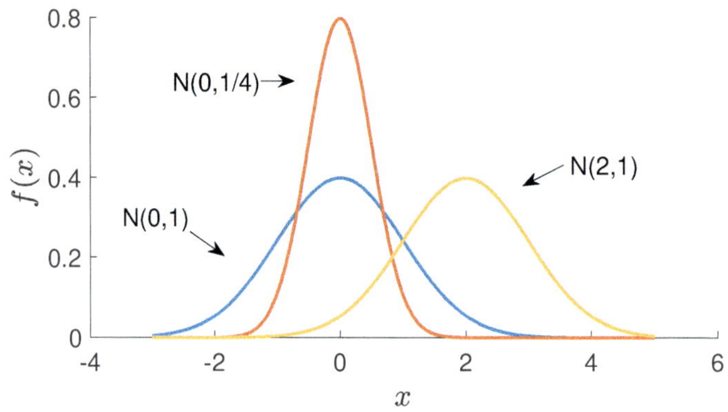

Figure D.41: The pdf of the $\mathcal{N}(\mu, \sigma^2)$ distribution for various μ and σ^2.

We next consider some important properties of the normal distribution.

Theorem D.42 (Standardization): Let $X \sim \mathcal{N}(\mu, \sigma^2)$ and define $Z = (X-\mu)/\sigma$. Then Z has a standard normal distribution.

Proof. The cdf of Z is given by

$$\begin{aligned}\mathbb{P}[Z \leq z] &= \mathbb{P}[(X-\mu)/\sigma \leq z] = \mathbb{P}[X \leq \mu + \sigma z] \\ &= \int_{-\infty}^{\mu+\sigma z} \frac{1}{\sigma\sqrt{2\pi}}\, \mathrm{e}^{-\frac{1}{2}\left(\frac{x-\mu}{\sigma}\right)^2} \mathrm{d}x = \int_{-\infty}^{z} \frac{1}{\sqrt{2\pi}}\, \mathrm{e}^{-y^2/2} \mathrm{d}y = \Phi(z),\end{aligned}$$

where we make a change of variable $y = (x-\mu)/\sigma$ in the fourth equation. Hence, $Z \sim \mathcal{N}(0, 1)$. □

The rescaling procedure in Theorem D.42 is called **standardization**. It follows from Theorem D.42 that any $X \sim \mathcal{N}(\mu, \sigma^2)$ can be written as

$$X = \mu + \sigma Z, \qquad \text{where } Z \sim \mathcal{N}(0, 1).$$

In other words, any normal random variable can be viewed as an **affine transformation** — that is, a linear transformation plus a constant — of a standard normal random variable.

The next result is about the function $\overline{\Phi} := 1 - \Phi$, called the **complementary cdf** of the standard normal distribution, which has applications in the theory of feature selection.

Theorem D.43 (Mill's Ratio): The **Mill's ratio** is defined as $\overline{\Phi}(z)/\varphi(z)$ and satisfies:

$$\frac{z}{z^2+1} \leq \frac{\overline{\Phi}(z)}{\varphi(z)} \leq \frac{1}{z}, \quad \forall z > 0.$$

Proof. Let $\overline{h}(z) := \varphi(z) - z\overline{\Phi}(z)$. Then, for the upper bound we need to show that $\overline{h}(z) \geq 0$ for $z > 0$. The latter follows from $\lim_{z\uparrow\infty} \overline{h}(z) = 0$ and the monotonicity of $\overline{h}$:

$$\frac{\mathrm{d}}{\mathrm{d}z}\overline{h}(z) = -z\varphi(z) - \overline{\Phi}(z) + z\varphi(z) = -\overline{\Phi}(z) \leq 0, \quad \forall z.$$

Similarly, we prove the lower bound by defining

$$\underline{h}(z) := z\varphi(z) - (z^2+1)\overline{\Phi}(z)$$

and noting that $\underline{h}(z) \leq 0$, because $\frac{\mathrm{d}}{\mathrm{d}z}\underline{h}(z) = 2\overline{h}(z) \geq 0$ and $\lim_{z\uparrow\infty} \underline{h}(z) = 0$. □

We now generalize the standard normal density to n dimensions. Let $Z_1, \ldots, Z_n$ be independent and standard normal random variables. The joint pdf of $\boldsymbol{Z} = [Z_1, \ldots, Z_n]^\top$ is given by

$$f_{\boldsymbol{Z}}(\boldsymbol{z}) = \prod_{i=1}^{n} \frac{1}{\sqrt{2\pi}} \mathrm{e}^{-\frac{1}{2}z_i^2} = (2\pi)^{-\frac{n}{2}} \mathrm{e}^{-\frac{1}{2}\boldsymbol{z}^\top \boldsymbol{z}}, \quad \boldsymbol{z} \in \mathbb{R}^n. \tag{D.44}$$

We write $\boldsymbol{Z} \sim \mathcal{N}(\boldsymbol{0}, \mathbf{I})$, where $\mathbf{I}$ is the identity matrix. Consider the affine transformation

$$\boldsymbol{X} = \boldsymbol{\mu} + \mathbf{B}\,\boldsymbol{Z} \tag{D.45}$$

for some $m \times n$ matrix $\mathbf{B}$ and m-dimensional vector $\boldsymbol{\mu}$. Note that, by (D.33) and (D.34), $\boldsymbol{X}$ has expectation vector $\boldsymbol{\mu}$ and covariance matrix $\boldsymbol{\Sigma} = \mathbf{B}\mathbf{B}^\top$. We say that $\boldsymbol{X}$ has a **multivariate normal** or *multivariate Gaussian* distribution with mean vector $\boldsymbol{\mu}$ and covariance matrix $\boldsymbol{\Sigma}$. We write $\boldsymbol{X} \sim \mathcal{N}(\boldsymbol{\mu}, \boldsymbol{\Sigma})$. ☞ 636

The following theorem states that any affine combination of independent multivariate normal random variables is again multivariate normal.

Theorem D.46 (Affine Transformation of Normal Random Vectors): Let $\boldsymbol{X}_1, \boldsymbol{X}_2, \ldots, \boldsymbol{X}_r$ be independent m_i-dimensional normal random vectors, with $\boldsymbol{X}_i \sim \mathcal{N}(\boldsymbol{\mu}_i, \boldsymbol{\Sigma}_i)$, $i = 1, \ldots, r$. Then, for any $n \times 1$ vector $\boldsymbol{a}$ and $n \times m_i$ matrices $\mathbf{B}_1, \ldots, \mathbf{B}_r$,

$$\boldsymbol{a} + \sum_{i=1}^{r} \mathbf{B}_i \boldsymbol{X}_i \sim \mathcal{N}\left(\boldsymbol{a} + \sum_{i=1}^{r} \mathbf{B}_i \boldsymbol{\mu}_i, \sum_{i=1}^{r} \mathbf{B}_i \boldsymbol{\Sigma}_i \mathbf{B}_i^\top\right). \tag{D.47}$$

Proof. Denote the n-dimensional random vector in the left-hand side of (D.47) by $\boldsymbol{Y}$. By definition, each $\boldsymbol{X}_i$ can be written as $\boldsymbol{\mu}_i + \mathbf{A}_i \boldsymbol{Z}_i$, where the $\{\boldsymbol{Z}_i\}$ are independent (because the $\{\boldsymbol{X}_i\}$ are independent), so that

$$\boldsymbol{Y} = \boldsymbol{a} + \sum_{i=1}^{r} \mathbf{B}_i (\boldsymbol{\mu}_i + \mathbf{A}_i \boldsymbol{Z}_i) = \boldsymbol{a} + \sum_{i=1}^{r} \mathbf{B}_i \boldsymbol{\mu}_i + \sum_{i=1}^{r} \mathbf{B}_i \mathbf{A}_i \boldsymbol{Z}_i,$$

which is an affine combination of independent standard normal random vectors. Hence, $\boldsymbol{Y}$ is multivariate normal. Its expectation vector and covariance matrix can be found
☞ 635 easily from Theorem D.32. □

■ **Example D.48 (Matrix Normal Distribution)** Let $\mathbf{B}_1 \in \mathbb{R}^{m \times m}$, $\mathbf{B}_2 \in \mathbb{R}^{n \times n}$, and $\mathbf{M} \in \mathbb{R}^{m \times n}$ be real matrices, and let $\mathbf{Z} \in \mathbb{R}^{m \times n}$ be a random matrix whose entries are iid $\mathcal{N}(0, 1)$. By Theorem D.46 the entries of the $m \times n$ matrix

$$\mathbf{X} = \mathbf{M} + \mathbf{B}_1 \mathbf{Z} \mathbf{B}_2^\top \tag{D.49}$$

☞ 518 are also normally distributed such that the vectorization $\boldsymbol{Y} = \text{vec}(\mathbf{X})$ has mean $\mathbb{E}\boldsymbol{Y} = \text{vec}(\mathbf{M})$ and covariance matrix:

$$\mathbb{V}\text{ar}(\boldsymbol{Y}) = \boldsymbol{\Sigma}_2 \otimes \boldsymbol{\Sigma}_1,$$

☞ 519 where $\boldsymbol{\Sigma}_1 = \mathbf{B}_1 \mathbf{B}_1^\top$ and $\boldsymbol{\Sigma}_2 = \mathbf{B}_2 \mathbf{B}_2^\top$. To see this, use the identity (5) to write

$$\boldsymbol{Y} - \mathbb{E}\boldsymbol{Y} = \text{vec}(\mathbf{B}_1 \mathbf{Z} \mathbf{B}_2^\top) = (\mathbf{B}_2 \otimes \mathbf{B}_1)\text{vec}(\mathbf{Z}).$$

Hence, by the mixed-product Kronecker identity,

$$\begin{aligned} \mathbb{V}\text{ar}(\boldsymbol{Y}) &= (\mathbf{B}_2 \otimes \mathbf{B}_1)\mathbb{V}\text{ar}(\text{vec}(\mathbf{Z}))(\mathbf{B}_2 \otimes \mathbf{B}_1)^\top \\ &= (\mathbf{B}_2 \otimes \mathbf{B}_1)(\mathbf{B}_2^\top \otimes \mathbf{B}_1^\top) \\ &= (\mathbf{B}_2 \mathbf{B}_2^\top) \otimes (\mathbf{B}_1 \mathbf{B}_1^\top). \end{aligned}$$

The matrix $\mathbf{X}$ in (D.49) is said to have **matrix normal** distribution with mean $\mathbf{M}$ and covariance $\boldsymbol{\Sigma}_2 \otimes \boldsymbol{\Sigma}_1$, or simply

$$\text{vec}(\mathbf{X}) \sim \mathcal{N}(\text{vec}(\mathbf{M}), \boldsymbol{\Sigma}_2 \otimes \boldsymbol{\Sigma}_1).$$

■

The next theorem shows that the distribution of a subvector of a multivariate normal random vector is again normal.

Theorem D.50 (Marginal Distributions of Normal Random Vectors): Let $\boldsymbol{X} \sim \mathcal{N}(\boldsymbol{\mu}, \boldsymbol{\Sigma})$ be an n-dimensional normal random vector. Decompose $\boldsymbol{X}$, $\boldsymbol{\mu}$, and $\boldsymbol{\Sigma}$ as

$$\boldsymbol{X} = \begin{bmatrix} \boldsymbol{X}_p \\ \boldsymbol{X}_q \end{bmatrix}, \quad \boldsymbol{\mu} = \begin{bmatrix} \boldsymbol{\mu}_p \\ \boldsymbol{\mu}_q \end{bmatrix}, \quad \boldsymbol{\Sigma} = \begin{bmatrix} \boldsymbol{\Sigma}_p & \boldsymbol{\Sigma}_r \\ \boldsymbol{\Sigma}_r^\top & \boldsymbol{\Sigma}_q \end{bmatrix}, \tag{D.51}$$

where $\boldsymbol{\Sigma}_p$ is the upper left $p \times p$ corner of $\boldsymbol{\Sigma}$ and $\boldsymbol{\Sigma}_q$ is the lower right $q \times q$ corner of $\boldsymbol{\Sigma}$. Then, $\boldsymbol{X}_p \sim \mathcal{N}(\boldsymbol{\mu}_p, \boldsymbol{\Sigma}_p)$.

Proof. We give a proof assuming that $\boldsymbol{\Sigma}$ is positive definite. Let $\mathbf{B}\mathbf{B}^\top$ be the (lower) Cholesky decomposition of $\boldsymbol{\Sigma}$. We can write ☞ 543

$$\begin{bmatrix} \boldsymbol{X}_p \\ \boldsymbol{X}_q \end{bmatrix} = \begin{bmatrix} \boldsymbol{\mu}_p \\ \boldsymbol{\mu}_q \end{bmatrix} + \underbrace{\begin{bmatrix} \mathbf{B}_p & \mathbf{O} \\ \mathbf{C}_r & \mathbf{C}_q \end{bmatrix}}_{\mathbf{B}} \begin{bmatrix} \boldsymbol{Z}_p \\ \boldsymbol{Z}_q \end{bmatrix}, \tag{D.52}$$

where $\boldsymbol{Z}_p$ and $\boldsymbol{Z}_q$ are independent p- and q-dimensional standard normal random vectors. In particular, $\boldsymbol{X}_p = \boldsymbol{\mu}_p + \mathbf{B}_p\boldsymbol{Z}_p$, which means that $\boldsymbol{X}_p \sim \mathcal{N}(\boldsymbol{\mu}_p, \boldsymbol{\Sigma}_p)$, since $\mathbf{B}_p\mathbf{B}_p^\top = \boldsymbol{\Sigma}_p$. □

By relabeling the elements of $\boldsymbol{X}$ we see that Theorem D.50 implies that *any* subvector of $\boldsymbol{X}$ has a multivariate normal distribution. For example, $\boldsymbol{X}_q \sim \mathcal{N}(\boldsymbol{\mu}_q, \boldsymbol{\Sigma}_q)$.

The following theorem shows that not only the marginal distributions of a normal random vector are normal, but also its *conditional distributions*.

Theorem D.53 (Conditional Distributions of Normal Random Vectors): Let $\boldsymbol{X} \sim \mathcal{N}(\boldsymbol{\mu}, \boldsymbol{\Sigma})$ be an n-dimensional normal random vector with $\det(\boldsymbol{\Sigma}) > 0$. If $\boldsymbol{X}$ is decomposed as in (D.51), then

$$(\boldsymbol{X}_q \,|\, \boldsymbol{X}_p = \boldsymbol{x}_p) \sim \mathcal{N}(\boldsymbol{\mu}_q + \boldsymbol{\Sigma}_r^\top\boldsymbol{\Sigma}_p^{-1}(\boldsymbol{x}_p - \boldsymbol{\mu}_p),\ \boldsymbol{\Sigma}_q - \boldsymbol{\Sigma}_r^\top\boldsymbol{\Sigma}_p^{-1}\boldsymbol{\Sigma}_r). \tag{D.54}$$

As a consequence, $\boldsymbol{X}_p$ and $\boldsymbol{X}_q$ are *independent* if and only if they are *uncorrelated*; that is, if $\boldsymbol{\Sigma}_r = \mathbf{O}$ (zero matrix).

Proof. From (D.52) we see that $\boldsymbol{X}_p = \boldsymbol{\mu}_p + \mathbf{B}_p\boldsymbol{Z}_p$ and $\boldsymbol{X}_q = \boldsymbol{\mu}_q + \mathbf{C}_r\boldsymbol{Z}_p + \mathbf{C}_q\boldsymbol{Z}_q$. Consequently,

$$(\boldsymbol{X}_q \,|\, \boldsymbol{X}_p = \boldsymbol{x}_p) = \boldsymbol{\mu}_q + \mathbf{C}_r\,\mathbf{B}_p^{-1}(\boldsymbol{x}_p - \boldsymbol{\mu}_p) + \mathbf{C}_q\boldsymbol{Z}_q,$$

where $\boldsymbol{Z}_q$ is a q-dimensional multivariate standard normal random vector. It follows that $\boldsymbol{X}_q$ conditional on $\boldsymbol{X}_p = \boldsymbol{x}_p$ has a $\mathcal{N}(\boldsymbol{\mu}_q + \mathbf{C}_r\,\mathbf{B}_p^{-1}(\boldsymbol{x}_p - \boldsymbol{\mu}_p), \mathbf{C}_q\mathbf{C}_q^\top)$ distribution. The proof of (D.54) is completed by observing that $\boldsymbol{\Sigma}_r^\top\boldsymbol{\Sigma}_p^{-1} = \mathbf{C}_r\mathbf{B}_p^\top(\mathbf{B}_p^\top)^{-1}\mathbf{B}_p^{-1} = \mathbf{C}_r\,\mathbf{B}_p^{-1}$, and

$$\boldsymbol{\Sigma}_q - \boldsymbol{\Sigma}_r^\top\boldsymbol{\Sigma}_p^{-1}\boldsymbol{\Sigma}_r = \mathbf{C}_r\mathbf{C}_r^\top + \mathbf{C}_q\mathbf{C}_q^\top - \mathbf{C}_r\mathbf{B}_p^{-1}\underbrace{\boldsymbol{\Sigma}_r}_{\mathbf{B}_p\mathbf{C}_r^\top} = \mathbf{C}_q\mathbf{C}_q^\top.$$

If $\boldsymbol{X}_p$ and $\boldsymbol{X}_q$ are independent, then they are obviously uncorrelated, as $\boldsymbol{\Sigma}_r = \mathbb{E}[(\boldsymbol{X}_p - \boldsymbol{\mu}_p)(\boldsymbol{X}_q - \boldsymbol{\mu}_q)^\top] = \mathbb{E}(\boldsymbol{X}_p - \boldsymbol{\mu}_p)\,\mathbb{E}(\boldsymbol{X}_q - \boldsymbol{\mu}_q)^\top = \mathbf{O}$. Conversely, if $\boldsymbol{\Sigma}_r = \mathbf{O}$, then by (D.54) the conditional distribution of $\boldsymbol{X}_q$ given $\boldsymbol{X}_p$ is the same as the unconditional distribution of $\boldsymbol{X}_q$; that is, $\mathcal{N}(\boldsymbol{\mu}_q, \boldsymbol{\Sigma}_q)$. In other words, $\boldsymbol{X}_q$ is independent of $\boldsymbol{X}_p$. □

The next few results are about the relationships between the normal, chi-squared, Student, and F distributions, defined in Table D.11. Recall that the chi-squared family of distributions, denoted by χ^2_n, are simply Gamma$(n/2, 1/2)$ distributions, where the parameter $n \in \{1, 2, 3, \ldots\}$ is called the **degrees of freedom**.

Theorem D.55 (Relationship Between Normal and χ^2 Distributions): If $\boldsymbol{X} \sim \mathcal{N}(\boldsymbol{\mu}, \boldsymbol{\Sigma})$ is an n-dimensional normal random vector with $\det(\boldsymbol{\Sigma}) > 0$, then

(D.56) $$(\boldsymbol{X} - \boldsymbol{\mu})^\top \boldsymbol{\Sigma}^{-1} (\boldsymbol{X} - \boldsymbol{\mu}) \sim \chi^2_n.$$

Proof. Let $\mathbf{B}\mathbf{B}^\top$ be the Cholesky decomposition of $\boldsymbol{\Sigma}$, where $\mathbf{B}$ is invertible. Since $\boldsymbol{X}$ can be written as $\boldsymbol{\mu} + \mathbf{B}\boldsymbol{Z}$, where $\boldsymbol{Z} = [Z_1, \ldots, Z_n]^\top$ is a vector of independent standard normal random variables, we have

$$(\boldsymbol{X} - \boldsymbol{\mu})^\top \boldsymbol{\Sigma}^{-1} (\boldsymbol{X} - \boldsymbol{\mu}) = (\boldsymbol{X} - \boldsymbol{\mu})^\top (\mathbf{B}\mathbf{B}^\top)^{-1} (\boldsymbol{X} - \boldsymbol{\mu}) = \boldsymbol{Z}^\top \boldsymbol{Z} = \sum_{i=1}^n Z_i^2.$$

☞ 630 Using the independence of $Z_1, \ldots, Z_n$, the moment generating function of $Y = \sum_{i=1}^n Z_i^2$ is given by

$$\mathbb{E}\,\mathrm{e}^{sY} = \mathbb{E}\,\mathrm{e}^{s(Z_1^2 + \cdots + Z_n^2)} = \mathbb{E}\,[\mathrm{e}^{sZ_1^2} \cdots \mathrm{e}^{sZ_n^2}] = \left(\mathbb{E}\,\mathrm{e}^{sZ^2}\right)^n,$$

where $Z \sim \mathcal{N}(0, 1)$. The moment generating function of Z^2 is

$$\mathbb{E}\,\mathrm{e}^{sZ^2} = \int_{-\infty}^{\infty} \mathrm{e}^{sz^2} \frac{1}{\sqrt{2\pi}} \mathrm{e}^{-z^2/2} \mathrm{d}z = \frac{1}{\sqrt{2\pi}} \int_{-\infty}^{\infty} \mathrm{e}^{-\frac{1}{2}(1-2s)z^2} \mathrm{d}z = \frac{1}{\sqrt{1 - 2s}},$$

so that $\mathbb{E}\,\mathrm{e}^{sY} = \left(\frac{1}{2}/(\frac{1}{2} - s)\right)^{\frac{n}{2}}$, $s < \frac{1}{2}$, which is the moment generating function of the
☞ 631 Gamma$(n/2, 1/2)$ distribution; that is, the χ^2_n distribution — see Example D.15. The result now follows from the uniqueness of the moment generating function. □

A consequence of Theorem D.55 is that if $\boldsymbol{X} = [X_1, \ldots, X_n]^\top$ is n-dimensional standard normal, then the squared length $\|\boldsymbol{X}\|^2 = X_1^2 + \cdots + X_n^2$ has a χ^2_n distribution. If instead $X_i \sim \mathcal{N}(\mu_i, 1), i = 1, \ldots$, then $\|\boldsymbol{X}\|^2$ is said to have a **noncentral χ^2_n distribution**. This distribution depends on the $\{\mu_i\}$ only through the norm $\|\boldsymbol{\mu}\|$. We write $\|\boldsymbol{X}\|^2 \sim \chi^2_n(\theta)$, where $\theta = \|\boldsymbol{\mu}\|$ is the **noncentrality parameter**.

Such distributions frequently occur when considering *projections* of multivariate normal random variables, as summarized in the following theorem.

Theorem D.57 (Relationship Between Normal and Noncentral χ^2 Distributions): Let $X \sim \mathcal{N}(\mu, \mathbf{I}_n)$ be an n-dimensional normal random vector and let $\mathcal{V}_k \subset \mathcal{V}_m$ be linear subspaces of dimensions k and m, respectively, with $k < m \le n$. Let X_k and X_m be orthogonal projections of X onto $\mathcal{V}_k$ and $\mathcal{V}_m$, and let μ_k and μ_m be the corresponding projections of μ. Then, the following holds.

1. The random vectors X_k, $X_m - X_k$, and $X - X_m$ are independent.
2. $\|X_k\|^2 \sim \chi^2_k(\|\mu_k\|)$, $\|X_m - X_k\|^2 \sim \chi^2_{m-k}(\|\mu_m - \mu_k\|)$, and $\|X - X_m\|^2 \sim \chi^2_{n-m}(\|\mu - \mu_m\|)$.

Proof. Let $v_1, \ldots, v_n$ be an orthonormal basis of $\mathbb{R}^n$ such that $v_1, \ldots, v_k$ spans $\mathcal{V}_k$ and $v_1, \ldots, v_m$ spans $\mathcal{V}_m$. By (A.19) we can write the orthogonal projection matrices onto $\mathcal{V}_j$, as $\mathbf{P}_j = \sum_{i=1}^{j} v_i v_i^\top$, $j = k, m, n$, where $\mathcal{V}_n$ is defined as $\mathbb{R}^n$. Note that $\mathbf{P}_n$ is simply the identity matrix. Let $\mathbf{V} := [v_1, \ldots, v_n]$ and define $Z := [Z_1, \ldots, Z_n]^\top = \mathbf{V}^\top X$. Recall from Section A.2 that any orthogonal transformation such as $z = \mathbf{V}^\top x$ is *length preserving*; that is, $\|z\| = \|x\|$. ☞ 521 ☞ 518

To prove the first statement of the theorem, note that $\mathbf{V}^\top X_j = \mathbf{V}^\top \mathbf{P}_j X = [Z_1, \ldots, Z_j, 0, \ldots, 0]^\top$, $j = k, m$. It follows that $\mathbf{V}^\top (X_m - X_k) = [0, \ldots, 0, Z_{k+1}, \ldots, Z_m, 0, \ldots, 0]^\top$ and $\mathbf{V}^\top (X - X_m) = [0, \ldots, 0, Z_{m+1}, \ldots, Z_n]^\top$. Moreover, being a linear transformation of a normal random vector, Z is also normal, with covariance matrix $\mathbf{V}^\top \mathbf{V} = \mathbf{I}_n$. In particular, the $\{Z_i\}$ are *independent*. This shows that X_k, $X_m - X_k$ and $X - X_m$ are independent as well.

Next, observe that $\|X_k\| = \|\mathbf{V}^\top X_k\| = \|Z_k\|$, where $Z_k := [Z_1, \ldots, Z_k]^\top$. The latter vector has independent components with variances 1, and its squared norm has therefore (by definition) a $\chi^2_k(\theta)$ distribution. The noncentrality parameter is $\theta = \|\mathbb{E} Z_k\| = \|\mathbb{E} X_k\| = \|\mu_k\|$, again by the length-preserving property of orthogonal transformations. This shows that $\|X_k\|^2 \sim \chi^2_k(\|\mu_k\|)$. The distributions of $\|X_m - X_k\|^2$ and $\|X - X_m\|^2$ follow by analogy. □

Theorem D.57 is frequently used in the statistical analysis of *normal linear models*; see Section 5.4. In typical situations μ lies in the subspace $\mathcal{V}_m$ or even $\mathcal{V}_k$ — in which case $\|X_m - X_k\|^2 \sim \chi^2_{m-k}$ and $\|X - X_m\|^2 \sim \chi^2_{n-m}$, independently. The (scaled) quotient then turns out to have an F distribution — a consequence of the following theorem. ☞ 213

Theorem D.58 (Relationship Between χ^2 and F Distributions): Let $U \sim \chi^2_m$ and $V \sim \chi^2_n$ be independent. Then,

$$\frac{U/m}{V/n} \sim \mathsf{F}(m, n).$$

Proof. For notational simplicity, let $c = m/2$ and $d = n/2$. The pdf of $W = U/V$ is given by $f_W(w) = \int_0^\infty f_U(wv)\, v\, f_V(v)\, dv$. Substituting the pdfs of the corresponding Gamma distributions, we have

$$\begin{aligned} f_W(w) &= \int_0^\infty \frac{(wv)^{c-1}\,\mathrm{e}^{-wv/2}}{\Gamma(c)\,2^c}\, v\, \frac{v^{d-1}\mathrm{e}^{-v/2}}{\Gamma(d)\,2^d}\,\mathrm{d}v \\ &= \frac{w^{c-1}}{\Gamma(c)\,\Gamma(d)\,2^{c+d}} \int_0^\infty v^{c+d-1}\,\mathrm{e}^{-(1+w)v/2}\,\mathrm{d}v \\ &= \frac{\Gamma(c+d)}{\Gamma(c)\,\Gamma(d)}\,\frac{w^{c-1}}{(1+w)^{c+d}}, \end{aligned}$$

where the last equality follows from the fact that the integrand is equal to $\Gamma(\alpha)\lambda^{-\alpha}$ times the density of the $\mathsf{Gamma}(\alpha, \lambda)$ distribution with $\alpha = c + d$ and $\lambda = (1 + w)/2$. The density of $Z = \frac{n}{m}\frac{U}{V}$ is given by

$$f_Z(z) = f_W(z\,m/n)\,m/n.$$

The proof is completed by comparing the resulting expression with the pdf of the F
☞ 629 distribution given in Table D.11. □

■ **Corollary D.59 (Relationship Between Normal, χ^2, and t Distributions)** Suppose that $Z \sim \mathcal{N}(0, 1)$ and $V \sim \chi^2_n$ are independent. Then,

$$\frac{Z}{\sqrt{V/n}} \sim \mathsf{t}_n.$$

■

Proof. Let $T = Z/\sqrt{V/n}$. Because $Z^2 \sim \chi^2_1$, we have by Theorem D.58 that $T^2 \sim \mathsf{F}(1, n)$. The result follows now from the symmetry around 0 of the pdf of T and the fact that the square of a t_n random variable has an $\mathsf{F}(1, n)$ distribution. □

Finally, the next result shows that computing expectations under the multivariate normal distribution may be significantly simplified.

Theorem D.60 (Covariance and Normality): Suppose that $\boldsymbol{Y} \sim \mathcal{N}(\boldsymbol{\mu}, \boldsymbol{\Sigma})$, where $\boldsymbol{\Sigma} \in \mathbb{R}^{n\times n}$ is an invertible covariance matrix, and $\boldsymbol{g} : \mathbb{R}^n \to \mathbb{R}^n$ is a continuously differentiable function with derivative matrix $\mathbf{G}(\boldsymbol{y}) := \frac{\partial \boldsymbol{g}}{\partial \boldsymbol{y}}(\boldsymbol{y})$. Then,

$$\operatorname{tr}\left[\mathbb{C}\mathrm{ov}(\boldsymbol{g}(\boldsymbol{Y}), \boldsymbol{Y})\right] = \mathbb{E}\operatorname{tr}\left[\mathbf{G}(\boldsymbol{Y})^\top \boldsymbol{\Sigma}\right].$$

Proof. Let $f(\boldsymbol{y})$ be the density of $\boldsymbol{Y}$. Computing the gradient

$$\frac{\partial}{\partial \boldsymbol{y}}(\boldsymbol{y}-\boldsymbol{\mu})^\top \boldsymbol{\Sigma}^{-1}(\boldsymbol{y}-\boldsymbol{\mu}) = 2\boldsymbol{\Sigma}^{-1}(\boldsymbol{y}-\boldsymbol{\mu}),$$

we deduce that $\frac{\nabla f(\boldsymbol{y})}{f(\boldsymbol{y})} = \nabla \ln f(\boldsymbol{y}) = -\boldsymbol{\Sigma}^{-1}(\boldsymbol{y} - \boldsymbol{\mu})$. Hence,

$$\begin{aligned}
\mathbb{E}[(\boldsymbol{Y} - \boldsymbol{\mu})^\top \boldsymbol{g}(\boldsymbol{Y})] &= \int_{\mathbb{R}^n} f(\boldsymbol{y})(\boldsymbol{y} - \boldsymbol{\mu})^\top \boldsymbol{g}(\boldsymbol{y})\, \mathrm{d}\boldsymbol{y} \\
&= \int_{\mathbb{R}^n} f(\boldsymbol{y})[\boldsymbol{\Sigma}^{-1}(\boldsymbol{y} - \boldsymbol{\mu})]^\top \boldsymbol{\Sigma} \boldsymbol{g}(\boldsymbol{y})\, \mathrm{d}\boldsymbol{y} \\
&= -\int_{\mathbb{R}^n} [\nabla f(\boldsymbol{y})]^\top \boldsymbol{\Sigma} \boldsymbol{g}(\boldsymbol{y})\, \mathrm{d}\boldsymbol{y} \\
&= -\sum_{i,j} \int_{\mathbb{R}^n} \frac{\partial f}{\partial y_i} \Sigma_{i,j} g_j(\boldsymbol{y})\, \mathrm{d}\boldsymbol{y}.
\end{aligned}$$

By integration by parts, $-\int_{\mathbb{R}^n} \frac{\partial f}{\partial y_i} \Sigma_{i,j} g_j(\boldsymbol{y})\, \mathrm{d}\boldsymbol{y} = \int_{\mathbb{R}^n} f(\boldsymbol{y}) \Sigma_{i,j} \frac{\partial g_j}{\partial y_i}\, \mathrm{d}\boldsymbol{y}$. Therefore,

$$\begin{aligned}
\operatorname{tr}\left[\mathbb{C}\mathrm{ov}(\boldsymbol{g}(\boldsymbol{Y}), \boldsymbol{Y})\right] &= \mathbb{E}[(\boldsymbol{Y} - \boldsymbol{\mu})^\top \boldsymbol{g}(\boldsymbol{Y})] \\
&= \mathbb{E} \sum_{i,j} \Sigma_{i,j} \frac{\partial g_j}{\partial y_i}(\boldsymbol{Y}) = \mathbb{E} \operatorname{tr}\left[\mathbf{G}(\boldsymbol{Y})^\top \boldsymbol{\Sigma}\right].
\end{aligned}$$

□

D.8 Convergence of Random Variables

Recall that a random variable X is a function from Ω to $\mathbb{R}$. If we have a sequence of random variables $X_1, X_2, \ldots$ (for instance, $X_n(\omega) = X(\omega) + \frac{1}{n}$ for each $\omega \in \Omega$), then one can consider the pointwise convergence:

$$\lim_{n \to \infty} X_n(\omega) = X(\omega), \quad \text{for all } \omega \in \Omega,$$

in which case we say that $X_1, X_2, \ldots$ *converges surely* to X. A more interesting type of convergence uses the probability measure $\mathbb{P}$ associated with X.

Definition D.61 (Convergence in Probability): The sequence of random variables $X_1, X_2, \ldots$ *converges in probability* to a random variable X if, for all $\varepsilon > 0$,

$$\lim_{n \to \infty} \mathbb{P}\left[|X_n - X| > \varepsilon\right] = 0.$$

We denote the **convergence in probability** as $X_n \xrightarrow{\mathbb{P}} X$.

Convergence in probability refers only to the distribution of X_n. Instead, if the sequence $X_1, X_2, \ldots$ is defined on a common probability space, then we can consider the following mode of convergence that uses the joint distribution of the sequence of random variables.

Definition D.62 (Almost Sure Convergence): The sequence of random variables $X_1, X_2, \ldots$ **converges almost surely** to a random variable X if for every $\varepsilon > 0$

$$\lim_{n\to\infty} \mathbb{P}\left[\sup_{k\geq n} |X_k - X| > \varepsilon\right] = 0.$$

We denote the **almost sure convergence** as $X_n \xrightarrow{\text{a.s.}} X$.

Note that in accordance with these definitions $X_n \xrightarrow{\text{a.s.}} 0$ is equivalent to $\sup_{k\geq n} |X_k| \xrightarrow{\mathbb{P}} 0$.

Example D.63 (Convergence in Probability and Almost Sure Convergence) Since the event $\{|X_n - X| > \varepsilon\}$ is contained in $\{\sup_{k\geq n} |X_k - X| > \varepsilon\}$, we can conclude that almost sure convergence implies convergence in probability. However, the converse is not true in general. For instance, consider the iid sequence $X_1, X_2, \ldots$ with marginal distribution

$$\mathbb{P}[X_n = 1] = 1 - \mathbb{P}[X_n = 0] = 1/n.$$

Clearly, $X_n \xrightarrow{\mathbb{P}} 0$. However, for $\varepsilon < 1$ and any $n = 1, 2, \ldots$ we have,

$$\begin{aligned}
\mathbb{P}\left[\sup_{k\geq n} |X_k| \leq \varepsilon\right] &= \mathbb{P}[X_n \leq \varepsilon, X_{n+1} \leq \varepsilon, \ldots] \\
&= \mathbb{P}[X_n \leq \varepsilon] \times \mathbb{P}[X_{n+1} \leq \varepsilon] \times \cdots \text{ (using independence)} \\
&= \lim_{m\to\infty} \prod_{k=n}^{m} \mathbb{P}[X_k \leq \varepsilon] = \lim_{m\to\infty} \prod_{k=n}^{m} \left(1 - \frac{1}{k}\right) \\
&= \lim_{m\to\infty} \frac{n-1}{n} \times \frac{n}{n+1} \times \cdots \times \frac{m-1}{m} = 0.
\end{aligned}$$

It follows that $\mathbb{P}[\sup_{k\geq n} |X_k - 0| > \varepsilon] = 1$ for any $0 < \varepsilon < 1$ and all $n \geq 1$. In other words, it is *not* true that $X_n \xrightarrow{\text{a.s.}} 0$.

Another important type of convergence is useful when we are interested in estimating
☞ 73 expectations or multi-dimensional integrals via Monte Carlo methodology.

Definition D.64 (Convergence in Distribution): The sequence of random variables $X_1, X_2, \ldots$ is said to *converge in distribution* to a random variable X with distribution function $F_X(x) = \mathbb{P}[X \leq x]$ provided that:

(D.65) $$\lim_{n\to\infty} \mathbb{P}[X_n \leq x] = F_X(x) \text{ for all } x \text{ such that } \lim_{a\to x} F_X(a) = F_X(x).$$

We denote the **convergence in distribution** as either $X_n \xrightarrow{\text{d}} X$, or $X_n \xrightarrow{\text{d}} F_X$.

The generalization to random vectors replaces (D.65) with

(D.66) $$\lim_{n\to\infty} \mathbb{P}[\boldsymbol{X}_n \in A] = \mathbb{P}[\boldsymbol{X} \in A] \text{ for all } A \subset \mathbb{R}^n \text{ such that } \mathbb{P}[\boldsymbol{X} \in \partial A] = 0,$$

where ∂A denotes the boundary of the set A.

A useful tool for demonstrating convergence in distribution is the **characteristic function** ψ_X of a random vector X, defined as the expectation: ☞ 327

(D.67) $$\psi_X(t) := \mathbb{E}\, e^{i t^\top X}, \quad t \in \mathbb{R}^n.$$

The moment generating function in (D.14) is a special case of the characteristic function evaluated at $t = -is$. Note that while the moment generating function of a random variable may not exist, its characteristic function always exists. The characteristic function of a random vector $X \sim f$ is closely related to the Fourier transform of its pdf f. ☞ 570

Example D.68 (Characteristic Function of a Gaussian Random Vector) The density of the multivariate standard normal distribution is given in (D.44) and thus the characteristic function of $\mathbf{Z} \sim \mathcal{N}(\mathbf{0}, \mathbf{I}_n)$ is

$$\begin{aligned}
\psi_{\mathbf{Z}}(t) = \mathbb{E}\, e^{i t^\top \mathbf{Z}} &= (2\pi)^{-n/2} \int_{\mathbb{R}^n} e^{i t^\top z - \frac{1}{2}\|z\|^2}\, dz \\
&= e^{-\|t\|^2/2} (2\pi)^{-n/2} \int_{\mathbb{R}^n} e^{-\frac{1}{2}\|z - i t^\top\|^2}\, dz \\
&= e^{-\|t\|^2/2}, \quad t \in \mathbb{R}^n.
\end{aligned}$$

Hence, the characteristic function of the random vector $X = \mu + \mathbf{B}\mathbf{Z}$ in (D.45) with multivariate normal distribution $\mathcal{N}(\mu, \Sigma)$ is given by ☞ 639

$$\begin{aligned}
\psi_X(t) &= \mathbb{E}\, e^{i t^\top X} \\
&= \mathbb{E}\, e^{i t^\top (\mu + \mathbf{B}\mathbf{Z})} \\
&= e^{i t^\top \mu}\, \mathbb{E}\, e^{i (\mathbf{B}^\top t)^\top \mathbf{Z}} \\
&= e^{i t^\top \mu}\, \psi_{\mathbf{Z}}(\mathbf{B}^\top t) \\
&= e^{i t^\top \mu - \|\mathbf{B}^\top t\|^2/2} = e^{i t^\top \mu - t^\top \Sigma t/2}.
\end{aligned}$$

■

The importance of the characteristic function is mainly derived from the following result, for which a proof can be found, for example, in [16].

Theorem D.69 (Characteristic Function): Suppose that $\psi_{X_1}(t), \psi_{X_2}(t), \ldots$ are the characteristic functions of the sequence of d-dimensional random vectors $X_1, X_2, \ldots$ and $\psi_X(t)$ is the characteristic function of X. Then, the following three statements are equivalent:

1. $\lim \psi_{X_n}(t) = \psi_X(t)$ for all $t \in \mathbb{R}^d$.
2. $X_n \xrightarrow{d} X$.
3. $\lim \mathbb{E} h(X_n) = \mathbb{E} h(X)$ for all bounded continuous functions $h : \mathbb{R}^d \to \mathbb{R}$.

Example D.70 (Convergence in Distribution) Define the random variables $Y_1, Y_2, \ldots$ as

$$Y_n := \sum_{k=1}^{n} X_k \left(\frac{1}{2}\right)^k, \qquad n = 1, 2, \ldots,$$

where $X_1, X_2, \ldots \stackrel{\text{iid}}{\sim} \mathsf{Ber}(1/2)$. We now show that $Y_n \xrightarrow{\text{d}} \mathcal{U}(0,1)$. First, note that

$$\mathbb{E}\exp(\mathrm{i}tY_n) = \prod_{k=1}^{n} \mathbb{E}\exp(\mathrm{i}tX_k/2^k) = 2^{-n}\prod_{k=1}^{n}(1+\exp(\mathrm{i}t/2^k)).$$

Second, from the collapsing product, $(1-\exp(\mathrm{i}t/2^n))\prod_{k=1}^n(1+\exp(\mathrm{i}t/2^k)) = 1-\exp(\mathrm{i}t)$, we have

$$\mathbb{E}\exp(\mathrm{i}tY_n) = (1-\exp(\mathrm{i}t))\frac{1/2^n}{1-\exp(\mathrm{i}t/2^n)}.$$

It follows that $\lim_{n\to\infty}\mathbb{E}\exp(\mathrm{i}tY_n) = (\exp(\mathrm{i}t)-1)/(\mathrm{i}t)$, which we recognize as the ☞ 647 characteristic function of the $\mathcal{U}(0,1)$ distribution. ■

Yet another mode of convergence is the following.

Definition D.71 (Convergence in L^p-norm): The sequence of random variables $X_1, X_2, \ldots$ *converges in L^p-norm* to a random variable X if

$$\lim_{n\to\infty}\mathbb{E}|X_n - X|^p = 0, \quad p \geq 1.$$

We denote the **convergence in L^p-norm** as $X_n \xrightarrow{L^p} X$.

The case for $p = 2$ corresponds to convergence in mean squared error. The following example illustrates that convergence in L^p-norm is qualitatively different from convergence in distribution.

Example D.72 (Comparison of Modes of Convergence) Define $X_n := 1 - X$, where X has a uniform distribution on the interval (0,1). Clearly, $X_n \xrightarrow{\text{d}} \mathcal{U}(0,1)$. However, $\mathbb{E}|X_n - X| \longrightarrow \mathbb{E}|1-2X| = 1/2$ and so the sequence does not converge in L^1-norm. In addition, $\mathbb{P}[|X_n - X| > \varepsilon] \longrightarrow 1 - \varepsilon \neq 0$ and so X_n does not converge in probability as well.

Thus, in general $X_n \xrightarrow{\text{d}} X$ implies neither $X_n \xrightarrow{\mathbb{P}} X$, nor $X_n \xrightarrow{L^1} X$. ■

Definition D.73 (Complete Convergence): The sequence of random variables $X_1, X_2, \ldots$ is said to *converge completely* to X if for all $\varepsilon > 0$

$$\sum_n \mathbb{P}[|X_n - X| > \varepsilon] < \infty.$$

We denote the **complete convergence** as $X_n \xrightarrow{\text{cpl.}} X$.

■ **Example D.74 (Complete and Almost Sure Convergence)** We show that complete convergence implies almost sure convergence. We can bound the criterion for almost sure convergence as follows:

$$\begin{aligned}
\mathbb{P}[\sup_{k\geq n}|X_k - X| > \varepsilon] &= \mathbb{P}[\cup_{k\geq n}\{|X_k - X| > \varepsilon\}] \\
&\leq \sum_{k\geq n}\mathbb{P}[|X_k - X| > \varepsilon] \qquad \text{by union bound in (D.1)} \\
&\leq \underbrace{\sum_{k=1}^{\infty}\mathbb{P}[|X_k - X| > \varepsilon]}_{=c<\infty \text{ from } X_n \xrightarrow{\text{cpl.}} X} - \sum_{k=1}^{n-1}\mathbb{P}[|X_k - X| > \varepsilon] \\
&\leq c - \sum_{k=1}^{n-1}\mathbb{P}[|X_k - X| > \varepsilon] \longrightarrow c - c = 0, \qquad n \to \infty.
\end{aligned}$$

Hence, by definition $X_n \xrightarrow{\text{a.s.}} X$. ■

The next theorem shows how the different types of convergence are related to each other. For example, in the diagram below, the notation $\overset{q\geq p}{\Longrightarrow}$ means that L^q-norm convergence implies L^p-norm convergence under the assumption that $q \geq p \geq 1$.

Theorem D.75 (Modes of Convergence): The most general relationships among the various modes of convergence for numerical random variables are shown on the following hierarchical diagram:

$$\begin{array}{ccccc}
\boxed{X_n \xrightarrow{\text{cpl.}} X} & \Rightarrow & \boxed{X_n \xrightarrow{\text{a.s.}} X} & & \\
 & & \Downarrow & & \\
 & & \boxed{X_n \xrightarrow{\mathbb{P}} X} & \Rightarrow & \boxed{X_n \xrightarrow{\text{d}} X} \\
 & & \Uparrow & & \\
\boxed{X_n \xrightarrow{L^q} X} & \overset{q\geq p}{\Longrightarrow} & \boxed{X_n \xrightarrow{L^p} X} & &
\end{array}.$$

Proof. **1.** First, we show that $X_n \xrightarrow{\mathbb{P}} X \Rightarrow X_n \xrightarrow{\text{d}} X$ using the inequality $\mathbb{P}[A\cap B] \leq \mathbb{P}[A]$ for any event B. To this end, consider the distribution function F_X of X:

$$\begin{aligned}
F_{X_n}(x) = \mathbb{P}[X_n \leq x] &= \mathbb{P}[X_n \leq x, |X_n - X| > \varepsilon] + \mathbb{P}[X_n \leq x, |X_n - X| \leq \varepsilon] \\
&\leq \mathbb{P}[|X_n - X| > \varepsilon] + \mathbb{P}[X_n \leq x, X \leq X_n + \varepsilon] \\
&\leq \mathbb{P}[|X_n - X| > \varepsilon] + \mathbb{P}[X \leq x + \varepsilon].
\end{aligned}$$

Now, in the arguments above we can switch the roles of X_n and X (there is a symmetry) to deduce the analogous result: $F_X(x) \leq \mathbb{P}[|X - X_n| > \varepsilon] + \mathbb{P}[X_n \leq x + \varepsilon]$. Therefore, making the switch $x \to x - \varepsilon$ gives $F_X(x - \varepsilon) \leq \mathbb{P}[|X - X_n| > \varepsilon] + F_{X_n}(x)$. Putting it all together gives:

$$F_X(x - \varepsilon) - \mathbb{P}[|X - X_n| > \varepsilon] \leq F_{X_n}(x) \leq \mathbb{P}[|X_n - X| > \varepsilon] + F_X(x + \varepsilon).$$

Taking $n \longrightarrow \infty$ on both sides yields for any $\varepsilon > 0$:

$$F_X(x-\varepsilon) \le \lim_{n\to\infty} F_{X_n}(x) \le F_X(x+\varepsilon).$$

Since F_X is continuous at x, by assumption, we can take $\varepsilon \downarrow 0$ to conclude that $F_{X_n}(x)$ converges to $F_X(x)$.

2. Second, we show that $X_n \xrightarrow{L^q} X \Rightarrow X_n \xrightarrow{L^p} X$ for $q \ge p \ge 1$. Since the function

☞ 67 $f(x) = x^{p/q}$ is concave for $p/q \le 1$, Jensen's inequality yields:

$$(\mathbb{E}|X|^q)^{p/q} = f(\mathbb{E}|X|^q) \ge \mathbb{E} f(|X|^q) = \mathbb{E}|X|^p.$$

In other words, $(\mathbb{E}|X_n - X|^p)^{1/p} \le (\mathbb{E}|X_n - X|^q)^{1/q} \longrightarrow 0$, proving the statement of the theorem.

3. Third, we show that $X_n \xrightarrow{L^1} X \Rightarrow X_n \xrightarrow{\mathbb{P}} X$. First note that for any random variable Y, we can write: $\mathbb{E}|Y| \ge \mathbb{E}[|Y|\,\mathbb{1}_{\{|Y|>\varepsilon\}}] \ge \mathbb{E}[|\varepsilon|\,\mathbb{1}_{\{|Y|>\varepsilon\}}] = \varepsilon\,\mathbb{P}[|Y| > \varepsilon]$. Therefore, we obtain **Chebyshev's inequality**:

$$\mathbb{P}[|Y| > \varepsilon] \le \frac{\mathbb{E}|Y|}{\varepsilon}. \tag{D.76}$$

Using Chebyshev's inequality and $X_n \xrightarrow{L^1} X$, we can write

$$\mathbb{P}[|X_n - X| > \varepsilon] \le \frac{\mathbb{E}|X_n - X|}{\varepsilon} \longrightarrow 0, \quad n \to \infty.$$

Hence, by definition $X_n \xrightarrow{\mathbb{P}} X$.

4. Finally, $X_n \xrightarrow{\text{cpl.}} X \Rightarrow X_n \xrightarrow{\text{a.s.}} X \Rightarrow X_n \xrightarrow{\mathbb{P}} X$ is proved in Examples D.74 and D.63. □

Under extra conditions, converse statements of Theorem D.75 can be established. The following are two typical examples.

■ **Example D.77 (Convergence in Distribution to a Constant)** If $X_n \xrightarrow{\text{d}} c$ for some constant c, then $X_n \xrightarrow{\mathbb{P}} c$ as well. To see this, note that $X_n \xrightarrow{\text{d}} c$ stands for

$$\lim_{n\to\infty} \mathbb{P}[X_n \le x] = \begin{cases} 1, & x > c, \\ 0, & x < c. \end{cases}$$

In other words, we can write:

$$\mathbb{P}[|X_n - c| > \varepsilon] \le 1 - \mathbb{P}[X_n \le c + \varepsilon] + \mathbb{P}[X_n \le c - \varepsilon] \longrightarrow 1 - 1 + 0 = 0, \quad n \to \infty,$$

which shows that $X_n \xrightarrow{\mathbb{P}} c$ by definition. ■

■ **Example D.78 (Convergence in Probability and Boundedness)** Suppose that $X_n \xrightarrow{\mathbb{P}} 0$ and also $|X_n| \le c$ for some constant c. The boundedness then helps us prove that $\mathbb{E}X_n \to 0$, as follows: For any arbitrary $\varepsilon > 0$, we have:

$$\begin{aligned}|\mathbb{E}X_n| &\le \mathbb{E}|X_n| = \mathbb{E}|X_n|\mathbb{1}_{\{|X_n|\le\varepsilon\}} + \mathbb{E}|X_n|\mathbb{1}_{\{|X_n|>\varepsilon\}} \\ &\le \mathbb{E}\,\varepsilon\,\mathbb{1}_{\{|X_n|\le\varepsilon\}} + \mathbb{E}\,c\,\mathbb{1}_{\{|X_n|>\varepsilon\}} \\ &\le \varepsilon + c\,\mathbb{P}[|X_n| > \varepsilon] \to \varepsilon + c \times 0.\end{aligned}$$

Since ε is arbitrary, we conclude that $\mathbb{E}X_n \to 0$ as $n \uparrow \infty$. ■

Finally, we will make use of the following theorem.

Theorem D.79 (Slutsky): Let $g(\boldsymbol{x}, \boldsymbol{y})$ be a continuous scalar function of vectors $\boldsymbol{x}$ and $\boldsymbol{y}$. Suppose that $\boldsymbol{X}_n \xrightarrow{\mathrm{d}} \boldsymbol{X}$ and $\boldsymbol{Y}_n \xrightarrow{\mathbb{P}} \boldsymbol{c}$ for some finite constant $\boldsymbol{c}$. Then,

$$g(\boldsymbol{X}_n, \boldsymbol{Y}_n) \xrightarrow{\mathrm{d}} g(\boldsymbol{X}, \boldsymbol{c}).$$

Proof. We prove the theorem for scalar X and Y. The proof for random vectors is analogous. First, we show that $\boldsymbol{Z}_n := \begin{bmatrix} X_n \\ Y_n \end{bmatrix} \xrightarrow{\mathrm{d}} \begin{bmatrix} X \\ c \end{bmatrix} =: \boldsymbol{Z}$ using, for example, Theorem D.69. In other words, we wish to show that the characteristic function of the joint distribution of X_n and Y_n converges pointwise as $n \to \infty$: ☞ 647

$$\psi_{X_n,Y_n}(\boldsymbol{t}) = \mathbb{E}\,\mathrm{e}^{\mathrm{i}(t_1X_n+t_2Y_n)} \longrightarrow \mathrm{e}^{\mathrm{i}t_2c}\mathbb{E}\,\mathrm{e}^{\mathrm{i}t_1X} = \psi_{X,c}(\boldsymbol{t}), \quad \forall \boldsymbol{t} \in \mathbb{R}^2.$$

To show the limit above, consider

$$\begin{aligned}
|\psi_{X_n,Y_n}(\boldsymbol{t}) - \psi_{X,c}(\boldsymbol{t})| &\leq |\psi_{X_n,c}(\boldsymbol{t}) - \psi_{X,c}(\boldsymbol{t})| + |\psi_{X_n,Y_n}(\boldsymbol{t}) - \psi_{X_n,c}(\boldsymbol{t})| \\
&= |\mathrm{e}^{\mathrm{i}t_2c}\,\mathbb{E}\,(\mathrm{e}^{\mathrm{i}t_1X_n} - \mathrm{e}^{\mathrm{i}t_1X})| + |\mathbb{E}\,\mathrm{e}^{\mathrm{i}(t_1X_n+t_2c)}(\mathrm{e}^{\mathrm{i}t_2(Y_n-c)} - 1)| \\
&\leq |\mathrm{e}^{\mathrm{i}t_2c}| \times |\mathbb{E}(\mathrm{e}^{\mathrm{i}t_1X_n} - \mathrm{e}^{\mathrm{i}t_1X})| + \mathbb{E}\,|\mathrm{e}^{\mathrm{i}(t_1X_n+t_2c)}| \times |\mathrm{e}^{\mathrm{i}t_2(Y_n-c)} - 1| \\
&\leq |\psi_{X_n}(t_1) - \psi_X(t_1)| + \mathbb{E}\,|\mathrm{e}^{\mathrm{i}t_2(Y_n-c)} - 1|.
\end{aligned}$$

Since $X_n \xrightarrow{\mathrm{d}} X$, Theorem D.69 implies that $\psi_{X_n}(t_1) \longrightarrow \psi_X(t_1)$, and the first term $|\psi_{X_n}(t_1) - \psi_X(t_1)|$ goes to zero. For the second term we use the fact that

$$|\mathrm{e}^{\mathrm{i}x} - 1| = \left|\int_0^x \mathrm{i}\,\mathrm{e}^{\mathrm{i}\theta}\,\mathrm{d}\theta\right| \leq \left|\int_0^x |\mathrm{i}\,\mathrm{e}^{\mathrm{i}\theta}|\,\mathrm{d}\theta\right| = |x|, \quad x \in \mathbb{R}$$

to obtain the bound:

$$\begin{aligned}
\mathbb{E}|\mathrm{e}^{\mathrm{i}t_2(Y_n-c)} - 1| &= \mathbb{E}|\mathrm{e}^{\mathrm{i}t_2(Y_n-c)} - 1|\mathbb{1}_{\{|Y_n-c|>\varepsilon\}} + \mathbb{E}|\mathrm{e}^{\mathrm{i}t_2(Y_n-c)} - 1|\mathbb{1}_{\{|Y_n-c|\leq\varepsilon\}} \\
&\leq 2\mathbb{E}\,\mathbb{1}_{\{|Y_n-c|>\varepsilon\}} + \mathbb{E}|t_2(Y_n - c)|\mathbb{1}_{\{|Y_n-c|\leq\varepsilon\}} \\
&\leq 2\mathbb{P}[|Y_n - c| > \varepsilon] + |t_2|\varepsilon \longrightarrow |t_2|\varepsilon, \quad n \to \infty.
\end{aligned}$$

Since ε is arbitrary, we can let $\varepsilon \downarrow 0$ to conclude that $\lim_{n\to\infty} |\psi_{X_n,Y_n}(\boldsymbol{t}) - \psi_{X,c}(\boldsymbol{t})| = 0$. In other words, $\boldsymbol{Z}_n \xrightarrow{\mathrm{d}} \boldsymbol{Z}$, and by the continuity of g, we have $g(\boldsymbol{Z}_n) \xrightarrow{\mathrm{d}} g(\boldsymbol{Z})$ or $g(\boldsymbol{X}_n, \boldsymbol{Y}_n) \xrightarrow{\mathrm{d}} g(\boldsymbol{X}, \boldsymbol{c})$. □

■ **Example D.80 (Necessity of Slutsky's Condition)** The condition that Y_n converges in probability to a constant cannot be relaxed. For example, suppose that $g(x, y) = x + y$, $X_n \xrightarrow{\mathrm{d}} X \sim \mathcal{N}(0, 1)$ and $Y_n \xrightarrow{\mathrm{d}} Y \sim \mathcal{N}(0, 1)$. Then, our intuition tempts us to *incorrectly* conclude that $X_n + Y_n \xrightarrow{\mathrm{d}} \mathcal{N}(0, 2)$. This intuition is false, because we can have $Y_n = -X_n$ for all n so that $X_n + Y_n = 0$, while both X and Y have the same marginal distribution (in this case standard normal). ■

Just like we use a big- and small-order notation for real sequences, we can also use such notation for random variables.

Definition D.81 (Small-Order Convergence in Probability): If for any arbitrarily small $\varepsilon > 0$ and $\delta > 0$ there exists an $n_{\varepsilon,\delta}$ (depending on ε, δ) such that:

$$\mathbb{P}[|X_n/\alpha_n| > \delta] \le \varepsilon, \qquad \forall n \ge n_{\varepsilon,\delta},$$

then the sequence (X_n) is said to be of **small order in probability** with respect to the real sequence (α_n). We write $X_n = o_{\mathbb{P}}(\alpha_n)$ or, equivalently, $|X_n/\alpha_n| \xrightarrow{\mathbb{P}} 0$.

■ **Example D.82 (Small-Order Convergence in Probability)** Suppose that we have $\alpha_n/X_n = o_{\mathbb{P}}(1)$. Then, for an arbitrary constant $\delta > 0$,

$$\lim_{n\uparrow\infty} \mathbb{P}[|X_n| \ge |\alpha_n|/\delta] = 1.$$

In other words, $|X_n|$ is larger than a multiple of $|\alpha_n|$ with probability approaching one. ■

Another useful notation for the convergence of sequences of random variables is the following.

Definition D.83 (Stochastic Boundedness): If for any arbitrary $\varepsilon > 0$ there exists an integer n_ε and $\delta_\varepsilon > 0$ (both depending on ε) such that:

$$\mathbb{P}[|X_n/\alpha_n| > \delta_\varepsilon] \le \varepsilon, \quad \forall n \ge n_\varepsilon,$$

then the sequence (X_n) is said to be of **big order in probability** with respect to the real sequence (α_n). We write $X_n = O_{\mathbb{P}}(\alpha_n)$.

☞ 48 ■ **Example D.84 (Least-Squares Estimator Error Analysis)** Consider the Gaussian linear regression $\boldsymbol{Y} = \mathbf{X}\boldsymbol{\beta}^* + \boldsymbol{\varepsilon}$ with $\boldsymbol{\varepsilon} \sim \mathcal{N}(\mathbf{0}, \varsigma^2\mathbf{I}_n)$ and under the assumption that $\|\mathbf{X}\boldsymbol{v}\|^2/\|\boldsymbol{v}\|^2 \ge n\underline{\lambda} > 0$ for all $\boldsymbol{v} \in \mathbb{R}^p$ and n. If $\widehat{\boldsymbol{\beta}} := \mathbf{X}^+\boldsymbol{Y}$ is the ordinary least-squares estimator, then

$$\|\widehat{\boldsymbol{\beta}} - \boldsymbol{\beta}^*\|_\infty = O_{\mathbb{P}}(1/\sqrt{n}).$$

To show this, note that $\widehat{\boldsymbol{\beta}} - \boldsymbol{\beta}^* = \mathbf{X}^+\boldsymbol{\varepsilon}$, and define $\boldsymbol{Z} := \sqrt{n\underline{\lambda}}\mathbf{X}^+\boldsymbol{\varepsilon}/\varsigma$, so that $\boldsymbol{Z} \sim \mathcal{N}(\mathbf{0}, n\underline{\lambda}(\mathbf{X}^\top\mathbf{X})^{-1})$ and $\max_i \mathbb{V}\mathrm{ar}(Z_i) \le 1$. The latter implies that

$$\max_i \mathbb{P}[|Z_i| \ge z] \le 1 - \Phi(z),$$

where Φ is the cdf of the standard normal distribution. Then, by the union bound $\mathbb{P}[\cup_i A_i] \le \sum_i \mathbb{P}[A_i]$, we have

$$\mathbb{P}[\max_{i\le p} |Z_i| > z] \le \sum_{i=1}^{p} \mathbb{P}[|Z_i| > z] \le p \max_i \mathbb{P}[|Z_i| > z] \le p(1 - \Phi(z)).$$

It follows that if $z := \Phi^{-1}(1 - \varepsilon/p)$ for a given $\varepsilon > 0$, then

$$\mathbb{P}[\|\mathbf{X}^+\boldsymbol{\varepsilon}\|_\infty > z\varsigma/\sqrt{n\underline{\lambda}}] = \mathbb{P}[\|\boldsymbol{Z}\|_\infty > z) \le p(1 - \Phi(z)] = \varepsilon.$$

In other words, $\|\mathbf{X}^+\boldsymbol{\varepsilon}\|_\infty = O_{\mathbb{P}}(1/\sqrt{n})$. ■

D.9 Law of Large Numbers and Central Limit Theorem

Two main results in probability are the *law of large numbers* and *the central limit theorem*. Both are limit theorems involving sums of independent random variables. In particular, consider a sequence $X_1, X_2, \ldots$ of iid random variables with finite expectation μ and finite variance σ^2. For each n define $\overline{X}_n := (X_1 + \cdots + X_n)/n$. What can we say about the (random) sequence of averages $\overline{X}_1, \overline{X}_2, \overline{X}_3, \ldots$? By (D.23) and (D.25) we have $\mathbb{E}\,\overline{X}_n = \mu$ ☞ 633 and $\mathbb{V}\mathrm{ar}\,\overline{X}_n = \sigma^2/n$. Hence, as n increases, the variance of the (random) average $\overline{X}_n$ goes to 0. This means that by Definition D.8, the average $\overline{X}_n$ converges to μ in L^2-norm as $n \to \infty$, that is, $\overline{X}_n \xrightarrow{L^2} \mu$.

In fact, to obtain *convergence in probability* the variance need not be finite — it is sufficient to assume that $\mu = \mathbb{E}X < \infty$.

Theorem D.85 (Weak Law of Large Numbers): If $X_1, \ldots, X_n$ are iid with finite expectation μ, then for all $\varepsilon > 0$

$$\lim_{n\to\infty} \mathbb{P}\left[|\overline{X}_n - \mu| > \varepsilon\right] = 0.$$

In other words, $\overline{X}_n \xrightarrow{\mathbb{P}} \mu$.

The theorem has a natural generalization for random vectors. Namely, if $\boldsymbol{\mu} = \mathbb{E}\boldsymbol{X} < \infty$, then $\mathbb{P}\left[\|\overline{\boldsymbol{X}}_n - \boldsymbol{\mu}\| > \varepsilon\right] \to 0$, where $\|\cdot\|$ is the Euclidean norm. We give a proof in ☞ 513 the scalar case.

Proof. Let $Z_k := X_k - \mu$ for all k, so that $\mathbb{E}Z = 0$. We thus need to show that $\overline{Z}_n \xrightarrow{\mathbb{P}} 0$. We use the properties of the characteristic function of Z denoted as ψ_Z. Due to the iid ☞ 647 assumption, we have

$$\text{(D.86)} \quad \psi_{\overline{Z}_n}(t) = \mathbb{E}\,\mathrm{e}^{\mathrm{i}t\overline{Z}_n} = \mathbb{E}\prod_{i=1}^{n} \mathrm{e}^{\mathrm{i}tZ_i/n} = \prod_{i=1}^{n} \mathbb{E}\,\mathrm{e}^{\mathrm{i}Z_i t/n} = \prod_{i=1}^{n} \psi_Z(t/n) = [\psi_Z(t/n)]^n.$$

An application of Taylor's Theorem C.20 in the neighborhood of $t = 0$ yields

$$\psi_Z(t/n) = \psi_Z(0) + o(t/n).$$

Since $\psi_Z(0) = 1$, we have:

$$\psi_{\overline{Z}_n}(t) = [\psi_Z(t/n)]^n = [1 + o(1/n)]^n \longrightarrow 1, \quad n \to \infty.$$

The characteristic function of a random variable that always equals zero is 1. Therefore, Theorem D.69 implies that $\overline{Z}_n \xrightarrow{\mathrm{d}} 0$. However, according to Example D.72, convergence in distribution to a constant implies convergence in probability. Hence, $\overline{Z}_n \xrightarrow{\mathbb{P}} 0$. □

There is also a stronger version of this theorem, as follows:

Theorem D.87 (Strong Law of Large Numbers): If $X_1, \ldots, X_n$ are iid with expectation μ and $\mathbb{E}X^2 < \infty$, then for all $\varepsilon > 0$

$$\lim_{n\to\infty} \mathbb{P}\left[\sup_{k\geq n} |\overline{X}_k - \mu| > \varepsilon\right] = 0.$$

In other words, $\overline{X}_n \xrightarrow{\text{a.s.}} \mu$.

Proof. First, note that any random variable X can be written as the difference of two positive random variables: $X = X_+ - X_-$, where $X_+ := \max\{X, 0\}$ and $X_- := -\min\{X, 0\}$. Thus, without loss of generality, we assume that the random variables in the theorem above are nonnegative.

Second, from the sequence $\{\overline{X}_1, \overline{X}_2, \overline{X}_3, \ldots\}$ we can select the subsequence $\{\overline{X}_1, \overline{X}_4, \overline{X}_9, \overline{X}_{16}, \ldots\} =: \{\overline{X}_{j^2}\}$. Then, from Chebyshev's inequality (D.76) and the iid condition, we have

$$\sum_{j=1}^{\infty} \mathbb{P}\left[\left|\overline{X}_{j^2} - \mu\right| > \varepsilon\right] \leq \frac{\mathbb{V}\text{ar}\, X}{\varepsilon^2} \sum_{j=1}^{\infty} \frac{1}{j^2} < \infty.$$

Therefore, by definition $\overline{X}_{n^2} \xrightarrow{\text{cpl.}} \mu$ and from Theorem D.75 we conclude that $\overline{X}_{n^2} \xrightarrow{\text{a.s.}} \mu$.

Third, for any arbitrary n, we can find a k, say $k = \lfloor\sqrt{n}\rfloor$, so that $k^2 \leq n \leq (k+1)^2$. For such a k and nonnegative $X_1, X_2, \ldots$, it holds that

$$\frac{k^2}{(k+1)^2}\overline{X}_{k^2} \leq \overline{X}_n \leq \overline{X}_{(k+1)^2}\frac{(k+1)^2}{k^2}.$$

Since $\overline{X}_{k^2}$ and $\overline{X}_{(k+1)^2}$ converge almost surely to μ as k (and hence n) goes to infinity, we conclude that $\overline{X}_n \xrightarrow{\text{a.s.}} \mu$. □

Note that the condition $\mathbb{E}X^2 < \infty$ in Theorem D.87 can be weakened to $\mathbb{E}\,|X| < \infty$ and the iid condition on the variables $X_1, \ldots, X_n$ can be relaxed to mere pairwise independence. The corresponding proof, however, is significantly more difficult.

The *Central Limit Theorem* describes the approximate distribution of $\overline{X}_n$, and it applies to both continuous and discrete random variables. Loosely, it states that

the average of a large number of iid random variables approximately has a normal distribution.

Specifically, the random variable $\overline{X}_n$ has a distribution that is approximately normal, with expectation μ and variance σ^2/n.

Theorem D.88 (Central Limit Theorem): If $X_1, \ldots, X_n$ are iid with finite expectation μ and finite variance σ^2, then for all $x \in \mathbb{R}$,

$$\lim_{n\to\infty} \mathbb{P}\left[\frac{\overline{X}_n - \mu}{\sigma/\sqrt{n}} \leq x\right] = \Phi(x),$$

where Φ is the cdf of the standard normal distribution.

Proof. Let $Z_k := (X_k - \mu)/\sigma$ for all k, so that $\mathbb{E}\, Z = 0$ and $\mathbb{E}\, Z^2 = 1$. We thus need to show that $\sqrt{n}\,\overline{Z}_n \xrightarrow{\mathrm{d}} \mathcal{N}(0,1)$. We again use the properties of the characteristic function. Let ψ_Z be the characteristic function of an iid copy of Z, then due to the iid assumption a similar calculation to the one in (D.86) yields: ☞ 647

$$\psi_{\sqrt{n}\,\overline{Z}_n}(t) = \mathbb{E}\, \mathrm{e}^{\mathrm{i}t\sqrt{n}\,\overline{Z}_n} = [\psi_Z(t/\sqrt{n})]^n.$$

An application of Taylor's Theorem C.20 in the neighborhood of $t = 0$ yields

$$\psi_Z(t/\sqrt{n}) = 1 + \frac{t}{\sqrt{n}}\psi'_Z(0) + \frac{t^2}{2n}\psi''_Z(0) + o(t^2/n).$$

Since $\psi'_Z(0) = \mathbb{E}\, \frac{\mathrm{d}}{\mathrm{d}t}\mathrm{e}^{\mathrm{i}tZ}\big|_{t=0} = \mathrm{i}\,\mathbb{E}Z = 0$ and $\psi''_Z(0) = \mathrm{i}^2\,\mathbb{E}Z^2 = -1$, we have:

$$\psi_{\sqrt{n}\,\overline{Z}_n}(t) = \left[\psi_Z(t/\sqrt{n})\right]^n = \left[1 - \frac{t^2}{2n} + o(1/n)\right]^n \longrightarrow \mathrm{e}^{-t^2/2}, \quad n \to \infty.$$

From Example D.68, we recognize $\mathrm{e}^{-t^2/2}$ as the characteristic function of the standard normal distribution. Thus, from Theorem D.69 we conclude that $\sqrt{n}\,\overline{Z}_n \xrightarrow{\mathrm{d}} \mathcal{N}(0,1)$. □

■ **Example D.89 (Exponential and Uniform Random Variables)** Figure D.90 shows the central limit theorem in action. The left part shows the pdfs of $\overline{X}_1, 2\overline{X}_2, \ldots, 4\overline{X}_4$ for the case where the $\{X_i\}$ have a $\mathcal{U}[0,1]$ distribution. The right part shows the same for the $\mathsf{Exp}(1)$ distribution. In both cases, we clearly see convergence to a bell-shaped curve, characteristic of the normal distribution.

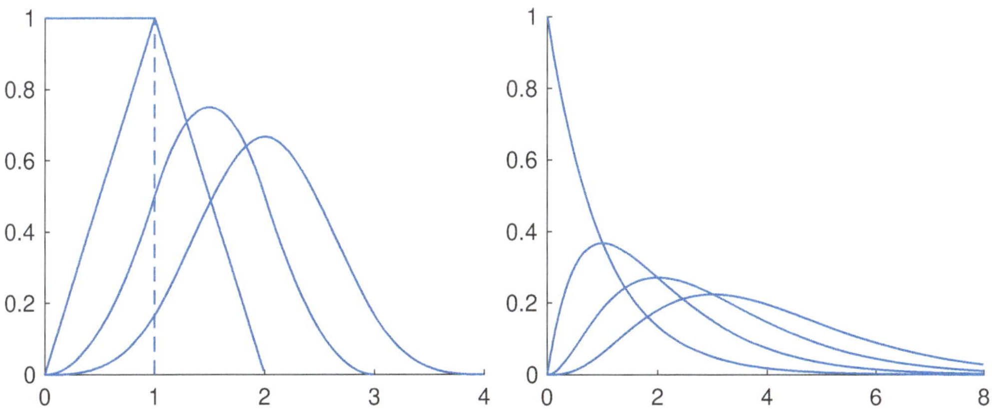

Figure D.90: Illustration of the central limit theorem for (left) the uniform distribution and (right) the exponential distribution.

■

The multivariate version of the central limit theorem is the basis for many asymptotic (in the size of the training set) results in machine learning and data science.

Theorem D.91 (Multivariate Central Limit Theorem): Let $X_1, \ldots, X_n$ be iid random vectors with expectation vector μ and finite covariance matrix Σ. Define $\overline{X}_n := (X_1 + \cdots + X_n)/n$. Then,

$$\sqrt{n}\,(\overline{X}_n - \mu) \xrightarrow{\text{d}} \mathcal{N}(\mathbf{0}, \Sigma) \quad \text{as } n \to \infty.$$

One application is as follows: Suppose that a parameter of interest, θ^*, is the unique solution of the system of equations $\mathbb{E}\,\psi(X \mid \theta^*) = \mathbf{0}$, where ψ is a continuously differentiable vector-valued (or multivalued) function and the distribution of X does not depend on θ. An **M-estimator** of θ^*, denoted $\widehat{\theta}_n$, is the solution to the system of equations that results from approximating the expectation with respect to X using an average of n iid copies of X:

$$\overline{\psi}_n(\theta) := \frac{1}{n}\sum_{i=1}^{n} \psi(X_i \mid \theta).$$

Thus, $\overline{\psi}_n(\widehat{\theta}_n) = \mathbf{0}$.

Theorem D.92 (M-estimator): The M-estimator is asymptotically normal as $n \to \infty$:

$$\sqrt{n}\,(\widehat{\theta}_n - \theta^*) \xrightarrow{\text{d}} \mathcal{N}(\mathbf{0}, \mathbf{A}^{-1}\mathbf{B}\mathbf{A}^{-\top}), \tag{D.93}$$

☞ 582 where $\mathbf{A} := -\mathbb{E}\frac{\partial \psi}{\partial \theta}(X \mid \theta^*)$ and $\mathbf{B} := \mathbb{E}\,[\psi(X \mid \theta^*)\,\psi(X \mid \theta^*)^\top]$ is the covariance matrix of $\psi(X \mid \theta^*)$.

Proof. We give a proof under the simplifying assumption[3] that $\widehat{\theta}_n$ is a unique root, that is, for any θ and ε, there exists a $\delta > 0$ such that $\|\widehat{\theta}_n - \theta\| > \varepsilon$ implies that $\|\overline{\psi}_n(\theta)\| > \delta$.

First, we argue that $\widehat{\theta}_n \xrightarrow{\mathbb{P}} \theta^*$; that is, $\mathbb{P}[\|\widehat{\theta}_n - \theta^*\| > \varepsilon] \to 0$. From the multivariate extension of Theorem D.85, we have that

$$\overline{\psi}_n(\theta^*) \xrightarrow{\mathbb{P}} \mathbb{E}\,\overline{\psi}_n(\theta^*) = \mathbb{E}\,\psi(X \mid \theta^*) = \mathbf{0}.$$

Therefore, using the uniqueness of $\widehat{\theta}_n$, we can show that $\widehat{\theta}_n \xrightarrow{\mathbb{P}} \theta^*$ via the bound:

$$\mathbb{P}\left[\|\widehat{\theta}_n - \theta^*\| > \varepsilon\right] \leq \mathbb{P}\left[\|\overline{\psi}_n(\theta^*)\| > \delta\right] = \mathbb{P}\left[\|\overline{\psi}_n(\theta^*) - \mathbb{E}\,\overline{\psi}_n(\theta^*)\| > \delta\right] \to 0,\ n \to \infty.$$

Second, we take a Taylor expansion of each component of the vector $\overline{\psi}_n(\widehat{\theta}_n)$ around θ^* to obtain:

$$\overline{\psi}_n(\widehat{\theta}_n) = \overline{\psi}_n(\theta^*) + \mathbf{J}_n(\theta')(\widehat{\theta}_n - \theta^*),$$

where $\mathbf{J}_n(\theta)$ is the Jacobian of $\overline{\psi}_n$ at θ, and θ' lies on the line segment joining $\widehat{\theta}_n$ and θ^*. Rearrange the last equation and multiply both sides by $\sqrt{n}\,\mathbf{A}^{-1}$ to obtain:

$$-\mathbf{A}^{-1}\mathbf{J}_n(\theta')\sqrt{n}\,(\widehat{\theta}_n - \theta^*) = \mathbf{A}^{-1}\sqrt{n}\,\overline{\psi}_n(\theta^*).$$

[3]The result holds under far less stringent assumptions.

By the central limit theorem, $\sqrt{n}\,\overline{\psi}(\boldsymbol{\theta}^*)$ converges in distribution to $\mathcal{N}(\mathbf{0}, \mathbf{B})$. Therefore,

$$-\mathbf{A}^{-1}\mathbf{J}_n(\boldsymbol{\theta}')\sqrt{n}\,(\widehat{\boldsymbol{\theta}}_n - \boldsymbol{\theta}^*) \xrightarrow{\mathrm{d}} \mathcal{N}(\mathbf{0}, \mathbf{A}^{-1}\mathbf{B}\mathbf{A}^{-\top}).$$

Theorem D.85 (the weak law of large numbers) applied to the iid random matrices $\{\frac{\partial}{\partial\boldsymbol{\theta}}\psi(\boldsymbol{X}_i \mid \boldsymbol{\theta})\}$ shows that

$$\mathbf{J}_n(\boldsymbol{\theta}) \xrightarrow{\mathbb{P}} \mathbb{E}\frac{\partial}{\partial\boldsymbol{\theta}}\psi(\boldsymbol{X} \mid \boldsymbol{\theta}).$$

Moreover, since $\widehat{\boldsymbol{\theta}}_n \xrightarrow{\mathbb{P}} \boldsymbol{\theta}^*$ and $\mathbf{J}_n$ is continuous in $\boldsymbol{\theta}$, we have that $\mathbf{J}_n(\boldsymbol{\theta}') \xrightarrow{\mathbb{P}} -\mathbf{A}$. Therefore, by Slutsky's theorem, $-\mathbf{A}^{-1}\mathbf{J}_n(\boldsymbol{\theta}')\sqrt{n}\,(\widehat{\boldsymbol{\theta}}_n - \boldsymbol{\theta}^*) - \sqrt{n}\,(\widehat{\boldsymbol{\theta}}_n - \boldsymbol{\theta}^*) \xrightarrow{\mathbb{P}} \mathbf{0}$. □ ☞ 651

Finally, we mention **Laplace's approximation** for large n to the following logarithm of the **Laplace integral**:

$$l_n := \ln \int_{\mathbb{R}^p} \mathrm{e}^{-n\,\ell_n(\boldsymbol{\theta})}\,\mathrm{d}\boldsymbol{\theta},$$

where $\ell_n : \mathbb{R}^p \to \mathbb{R}$. Typically, in Bayesian statistics $\ell_n(\boldsymbol{\theta}) = -\frac{1}{n}\ln[g(\tau_n \mid \boldsymbol{\theta})g(\boldsymbol{\theta})]$, where $g(\boldsymbol{\theta})$ is the prior density for some parameter $\boldsymbol{\theta}$ and $g(\tau_n \mid \boldsymbol{\theta})$ is the likelihood of the data τ_n. In that case, l_n is simply the logarithm of the marginal likelihood of the data: $l_n = \ln g(\tau_n)$. ☞ 49

The following theorem establishes lower and upper bounds for l_n under mild assumptions on ℓ_n. For this purpose, recall the definition of $\mathbf{A}$-strong convexity provided in (C.37). ☞ 592

Theorem D.94 (Laplace's Approximation): If $\boldsymbol{Z}_n \sim \mathcal{N}(\boldsymbol{\theta}_n, \boldsymbol{\Sigma}_n/n)$ for some positive definite matrices $\{\boldsymbol{\Sigma}_n\}$, then:

$$l_n \geq -n\,\mathbb{E}\,\ell_n(\boldsymbol{Z}_n) + \frac{p}{2}\ln(2\pi\mathrm{e}/n) + \frac{1}{2}\ln|\boldsymbol{\Sigma}_n|. \tag{D.95}$$

Further, if each ℓ_n is $\mathbf{A}$-strongly convex with global minimizer $\boldsymbol{\theta}_n$, then:

$$l_n \leq -n\,\ell_n(\boldsymbol{\theta}_n) + \frac{p}{2}\ln(2\pi/n) - \frac{1}{2}\ln|\mathbf{A}|. \tag{D.96}$$

Proof. Let $\varphi_{\boldsymbol{\Sigma}_n/n}(\boldsymbol{z}) := \frac{\exp(-n\,\boldsymbol{z}^\top\boldsymbol{\Sigma}_n^{-1}\boldsymbol{z}/2)}{(2\pi/n)^{p/2}|\boldsymbol{\Sigma}_n|^{1/2}}$, $\boldsymbol{z} \in \mathbb{R}^p$ denote the pdf of the $\mathcal{N}(\mathbf{0}, \boldsymbol{\Sigma}_n/n)$ distribution, where $\boldsymbol{\Sigma}_n$ is positive definite. Writing the integral in l_n as an expectation under $\boldsymbol{Z}_n$, we have

$$l_n = \ln \mathbb{E}\,\mathrm{e}^{-n\,\ell_n(\boldsymbol{Z}_n) - \ln\varphi_{\boldsymbol{\Sigma}_n/n}(\boldsymbol{Z}_n - \boldsymbol{\theta}_n)},$$

and applying Jensen's inequality (see Exercise 19 in Chapter 2) with the logarithmic function yields the lower bound: ☞ 67

$$l_n \geq -n\,\mathbb{E}\,\ell_n(\boldsymbol{Z}_n) - \mathbb{E}\ln\varphi_{\boldsymbol{\Sigma}_n/n}(\boldsymbol{Z}_n - \boldsymbol{\theta}_n).$$

The last term in the lower bound is the differential entropy of $\boldsymbol{Z}_n$; see (2.102): ☞ 67

$$\begin{aligned}\mathcal{H}(\boldsymbol{Z}_n) &= -\int_{\mathbb{R}^p} \varphi_{\boldsymbol{\Sigma}_n/n}(\boldsymbol{z})\ln\varphi_{\boldsymbol{\Sigma}_n/n}(\boldsymbol{z})\,\mathrm{d}\boldsymbol{z} \\ &= \frac{p}{2}\ln(2\pi\mathrm{e}) + \frac{1}{2}\ln|\boldsymbol{\Sigma}_n/n| = \frac{p}{2}\ln(2\pi\mathrm{e}/n) + \frac{1}{2}\ln|\boldsymbol{\Sigma}_n|.\end{aligned}$$

This proves the lower bound (D.95).

To prove the upper bound (D.96), define the squared Mahalanobis norm $\|z\|_{\mathbf{A}}^2 := z^\top \mathbf{A} z$. By the $\mathbf{A}$-strong convexity, for each $\boldsymbol{\theta}_n$ there exists a subgradient $\boldsymbol{v}_n$ such that:

$$\ell_n(z) \geq \ell_n(\boldsymbol{\theta}_n) + (z - \boldsymbol{\theta}_n)^\top \boldsymbol{v}_n + \frac{1}{2}\|z - \boldsymbol{\theta}_n\|_{\mathbf{A}}^2, \quad \forall z \in \mathbb{R}^p.$$

Since each $\boldsymbol{\theta}_n$ is the minimizer of ℓ_n, we can set $\boldsymbol{v}_n = \mathbf{0}$ and bound the Laplace integral:

$$\begin{aligned} \mathrm{e}^{l_n} &= \int_{\mathbb{R}^p} \mathrm{e}^{-n\ell_n(z)}\, \mathrm{d}z \\ &\leq \mathrm{e}^{-n\ell_n(\boldsymbol{\theta}_n)} \int_{\mathbb{R}^p} \mathrm{e}^{-\frac{n}{2}\|z - \boldsymbol{\theta}_n\|_{\mathbf{A}}^2}\, \mathrm{d}z \\ &\leq \mathrm{e}^{-n\ell_n(\boldsymbol{\theta}_n)} (2\pi)^{p/2} |n\mathbf{A}|^{-1/2}, \end{aligned}$$

which yields (D.96) after taking the logarithm on both sides. □

In many applications of Theorem D.94, the sequence $\{\boldsymbol{\theta}_n\}$ converges to the global minimizer of the limiting function $\ell := \lim \ell_n$. A corollary of the bounds in Theorem D.94 is the following lemma, which builds on the concept of Lipschitz continuity;
☞ 592 see Lemma C.33. In particular, we extend the notion of Lipschitz continuity to sequences of functions by defining a sequence $\{f_n\}$ to be **uniformly Lipschitz continuous** if there exists a constant $\delta > 0$ such that:

$$\sup_n \sup_{\boldsymbol{x} \neq \boldsymbol{y}} \frac{|f_n(\boldsymbol{x}) - f_n(\boldsymbol{y})|}{\|\boldsymbol{x} - \boldsymbol{y}\|} \leq \delta.$$

Lemma D.97 (Asymptotics of the Laplace Integral): Suppose that each ℓ_n is first-order differentiable. If the sequence of gradients $\{\nabla \ell_n\}$ is uniformly Lipschitz continuous and $\ln |\boldsymbol{\Sigma}_n| = \mathcal{O}(1)$ as $n \uparrow \infty$, then Theorem D.94 implies that:

$$l_n = -n\,\ell_n(\boldsymbol{\theta}_n) - \frac{p}{2}\ln n + \mathcal{O}(1). \tag{D.98}$$

Proof. First, note that $\mathbb{E}\,\ell_n(\boldsymbol{Z}_n) = \mathbb{E}\,\ell_n\big(\boldsymbol{\theta}_n + \boldsymbol{\Sigma}_n^{1/2}\boldsymbol{V}/\sqrt{n}\big)$ for some $\boldsymbol{V} \sim \mathcal{N}(\mathbf{0}, \mathbf{I}_p)$, so that without loss of generality we can set $\boldsymbol{Z}_n = \boldsymbol{\theta}_n + \boldsymbol{\Sigma}_n^{1/2}\boldsymbol{V}/\sqrt{n}$, where $\boldsymbol{\theta}_n$ is the minimizer of ℓ_n. Second, the difference between the upper (D.96) and lower (D.95) bounds is precisely

$$n\big(\mathbb{E}\,\ell_n(\boldsymbol{Z}_n) - \ell_n(\boldsymbol{\theta}_n)\big) - \underbrace{\frac{1}{2}\big(p + \ln|\mathbf{A}| + \ln|\boldsymbol{\Sigma}_n|\big)}_{=\,\mathcal{O}(1)},$$

and so (D.98) will be established if we show that $n\big(\mathbb{E}\,\ell_n(\boldsymbol{Z}_n) - \ell_n(\boldsymbol{\theta}_n)\big) = \mathcal{O}(1)$ as $n \uparrow \infty$.
☞ 586 To this end, apply Taylor's Theorem C.20 with $\boldsymbol{V} = \sqrt{n}\,\boldsymbol{\Sigma}_n^{-1/2}(\boldsymbol{Z}_n - \boldsymbol{\theta}_n)$:

$$\ell_n(\boldsymbol{Z}_n) = \ell_n(\boldsymbol{\theta}_n) + \frac{1}{\sqrt{n}}\boldsymbol{V}^\top \boldsymbol{\Sigma}_n^{1/2} \nabla \ell_n(\overline{\boldsymbol{\theta}}_n),$$

where $\overline{\boldsymbol{\theta}}_n$ is a point on the line segment connecting $\boldsymbol{\theta}_n$ and $\boldsymbol{Z}_n$. Rearranging, we obtain

$$\begin{aligned}
|\ell_n(\boldsymbol{Z}_n) - \ell_n(\boldsymbol{\theta}_n)| &= \ell_n(\boldsymbol{Z}_n) - \ell_n(\boldsymbol{\theta}_n)\\
&= \frac{1}{\sqrt{n}}(\boldsymbol{\Sigma}_n^{1/2}\boldsymbol{V})^\top \nabla\ell_n(\overline{\boldsymbol{\theta}}_n)\\
&\leq \frac{1}{\sqrt{n}}\|\boldsymbol{V}\|_{\boldsymbol{\Sigma}_n}\|\nabla\ell_n(\overline{\boldsymbol{\theta}}_n)\|, \quad \text{(Cauchy–Schwarz in Theorem B.16)}\\
&\leq \frac{1}{\sqrt{n}}\|\boldsymbol{V}\|_{\boldsymbol{\Sigma}_n}\|\nabla\ell_n(\overline{\boldsymbol{\theta}}_n) - \nabla\ell_n(\boldsymbol{\theta}_n)\|\\
&\leq \frac{\delta}{\sqrt{n}}\|\boldsymbol{V}\|_{\boldsymbol{\Sigma}_n}\|\overline{\boldsymbol{\theta}}_n - \boldsymbol{\theta}_n\|, \quad \text{(using the uniform Lipschitz continuity)}\\
&\leq \frac{\delta}{\sqrt{n}}\|\boldsymbol{V}\|_{\boldsymbol{\Sigma}_n}\|\boldsymbol{Z}_n - \boldsymbol{\theta}_n\| = \delta\,\|\boldsymbol{V}\|^2_{\boldsymbol{\Sigma}_n}/n.
\end{aligned}$$

The proof is then completed, because

$$\mathbb{E}|\ell_n(\boldsymbol{Z}_n) - \ell_n(\boldsymbol{\theta}_n)| \leq \delta\,\mathbb{E}\|\boldsymbol{V}\|^2_{\boldsymbol{\Sigma}_n}/n = \delta\,\mathrm{tr}(\boldsymbol{\Sigma}_n)/n = O(1/n).$$

□

We can further sharpen the asymptotic result in Lemma D.97 by invoking the concept of uniform continuity. A sequence of functions $\{f_n(\boldsymbol{x})\}$ is said to be **uniformly continuous in both n and the argument $\boldsymbol{x}$** if for any $\varepsilon > 0$, there exits a $\delta > 0$ such that:

$$\sup_n \sup_{(\boldsymbol{x},\boldsymbol{y}):\|\boldsymbol{x}-\boldsymbol{y}\|<\delta} |f_n(\boldsymbol{x}) - f_n(\boldsymbol{y})| < \varepsilon.$$

This condition ensures that the continuity of the functions is uniform not only over the domain but also across the entire sequence.

Theorem D.99 (Finer Asymptotics of the Laplace Integral): Suppose that each ℓ_n is $\mathbf{A}$-strongly convex and the second-order derivatives of $\{\ell_n(\boldsymbol{\theta})\}$ are uniformly continuous in both n and $\boldsymbol{\theta}$. If $\boldsymbol{\Sigma}_n := [\nabla^2\ell_n(\boldsymbol{\theta}_n)]^{-1}$ is the inverse of the Hessian of ℓ_n at the unique global minimizer $\boldsymbol{\theta}_n$, then as $n \uparrow \infty$:

(D.100) $$l_n = -n\,\ell_n(\boldsymbol{\theta}_n) + \frac{p}{2}\ln(2\pi/n) + \frac{1}{2}\ln|\boldsymbol{\Sigma}_n| + o(1)$$

(D.101) $$\mathbb{E}\,\ell_n(\boldsymbol{Z}_n) = \ell_n(\boldsymbol{\theta}_n) + \frac{p}{2n} + o(1/n).$$

Proof. Taylor's Theorem C.20 with $\boldsymbol{v}_n := \sqrt{n}\,\boldsymbol{\Sigma}_n^{-1/2}(\boldsymbol{\theta} - \boldsymbol{\theta}_n)$ yields: ☞ 586

$$\ell_n(\boldsymbol{\theta}) = \ell_n(\boldsymbol{\theta}_n) + \tfrac{1}{\sqrt{n}}\boldsymbol{v}_n^\top\boldsymbol{\Sigma}_n^{1/2}\underbrace{\nabla\ell_n(\boldsymbol{\theta}_n)}_{=\,\mathbf{0}} + \tfrac{1}{2n}\boldsymbol{v}_n^\top\underbrace{\boldsymbol{\Sigma}_n^{1/2}\nabla^2\ell_n(\boldsymbol{\theta}_n)\boldsymbol{\Sigma}_n^{1/2}}_{\mathbf{I}_p}\boldsymbol{v}_n + \tfrac{1}{2n}\boldsymbol{v}_n^\top\mathbf{R}_n(\boldsymbol{\theta}_n,\boldsymbol{v}_n)\boldsymbol{v}_n,$$

where, given some $\overline{\boldsymbol{\theta}}_n$ on the line segment between $\boldsymbol{\theta}_n$ and $\boldsymbol{\theta} = \boldsymbol{\theta}_n + \frac{1}{\sqrt{n}}\boldsymbol{\Sigma}_n^{1/2}\boldsymbol{v}_n$, we define:

$$\mathbf{R}_n(\boldsymbol{\theta}_n,\boldsymbol{v}_n) := \boldsymbol{\Sigma}_n^{1/2}\big[\nabla^2\ell_n(\overline{\boldsymbol{\theta}}_n) - \nabla^2\ell_n(\boldsymbol{\theta}_n)\big]\boldsymbol{\Sigma}_n^{1/2}.$$

Substituting the Taylor expansion for $\ell_n(\boldsymbol{\theta})$ in the Laplace integral and making the change of variable $\boldsymbol{\theta} = \boldsymbol{\theta}_n + \frac{1}{\sqrt{n}}\boldsymbol{\Sigma}_n^{1/2}\boldsymbol{v}_n$, so that $\mathrm{d}\boldsymbol{\theta} = |\boldsymbol{\Sigma}_n/n|^{1/2}\mathrm{d}\boldsymbol{v}_n$, we obtain:

$$\begin{aligned}
\mathrm{e}^{l_n} &= |\boldsymbol{\Sigma}_n/n|^{1/2} \int_{\mathbb{R}^p} \mathrm{e}^{-n\ell_n(\boldsymbol{\theta}_n) - \frac{1}{2}\|\boldsymbol{v}_n\|^2 - \frac{1}{2}\boldsymbol{v}_n^\top \mathbf{R}_n(\boldsymbol{\theta}_n, \boldsymbol{v}_n)\boldsymbol{v}_n}\,\mathrm{d}\boldsymbol{v}_n \\
&= |2\pi\boldsymbol{\Sigma}_n/n|^{1/2}\mathrm{e}^{-n\ell_n(\boldsymbol{\theta}_n)} \int_{\mathbb{R}^p} \varphi_{\mathbf{I}_p}(\boldsymbol{v}_n)\,\mathrm{e}^{-\frac{1}{2}\boldsymbol{v}_n^\top \mathbf{R}_n(\boldsymbol{\theta}_n, \boldsymbol{v}_n)\boldsymbol{v}_n}\,\mathrm{d}\boldsymbol{v}_n \\
&= |2\pi\boldsymbol{\Sigma}_n/n|^{1/2}\mathrm{e}^{-n\ell_n(\boldsymbol{\theta}_n)}\,\mathbb{E}\,\mathrm{e}^{-\frac{1}{2}\boldsymbol{V}^\top \mathbf{R}_n(\boldsymbol{\theta}_n, \boldsymbol{V})\boldsymbol{V}},
\end{aligned}$$

where $\boldsymbol{V} \sim \mathcal{N}(\mathbf{0}, \mathbf{I}_p)$. The proof of (D.100) will then be completed, provided that the term $\mathbb{E}\,\mathrm{e}^{-\frac{1}{2}\boldsymbol{V}^\top \mathbf{R}_n(\boldsymbol{\theta}_n, \boldsymbol{V})\boldsymbol{V}} = 1 + o(1)$. To show this, note that $\{\boldsymbol{\Sigma}_n\}$ is bounded ($\boldsymbol{\Sigma}_n \preceq \mathbf{A}^{-1}$ for all n), and hence the entries of the sequence of matrices $\{\mathbf{R}_n\}$ are uniformly continuous in both n and the arguments $\boldsymbol{\theta}_n$ and $\boldsymbol{v}_n$. As a result, $\|\overline{\boldsymbol{\theta}}_n - \boldsymbol{\theta}_n\| \leq \|\boldsymbol{V}\|_{\boldsymbol{\Sigma}_n}/n \stackrel{\text{a.s.}}{\to} 0$ implies that for any $\varepsilon > 0$ there is a large enough n_ε such that for all $\boldsymbol{\theta}_n$, $\boldsymbol{V}$, and $n > n_\varepsilon$:

$$\varepsilon\mathbf{I}_p \succ \mathbf{R}_n(\boldsymbol{\theta}_n, \boldsymbol{V}) \succ -\varepsilon\mathbf{I}_p.$$

Using $\|\boldsymbol{V}\|^2 \sim \chi_p^2$ (see Theorem D.55) and Example D.15 (with $\alpha = p/2$ and $\lambda = 1$), we thus obtain for all $n \geq n_\varepsilon$:

$$(1+\varepsilon)^{p/2} = \mathbb{E}\,\mathrm{e}^{-\frac{\varepsilon}{2}\|\boldsymbol{V}\|^2} \leq \mathbb{E}\,\mathrm{e}^{-\frac{1}{2}\boldsymbol{V}^\top \mathbf{R}_n(\boldsymbol{\theta}_n, \boldsymbol{V})\boldsymbol{V}} \leq \mathbb{E}\,\mathrm{e}^{\frac{\varepsilon}{2}\|\boldsymbol{V}\|^2} = (1-\varepsilon)^{p/2},$$

which completes the proof of (D.100) by taking $\varepsilon \downarrow 0$. As for (D.101), the fact that for $n > n_\varepsilon$

$$-\varepsilon\, p = -\varepsilon\,\mathbb{E}\|\boldsymbol{V}\|^2 \leq \mathbb{E}[\boldsymbol{V}^\top \mathbf{R}_n(\boldsymbol{\theta}_n, \boldsymbol{V})\boldsymbol{V}] \leq \varepsilon\,\mathbb{E}\|\boldsymbol{V}\|^2 = \varepsilon\, p,$$

and Taylor's expansion:

$$\begin{aligned}
\mathbb{E}\,\ell_n(\boldsymbol{Z}_n) &= \ell_n(\boldsymbol{\theta}_n) + \frac{1}{2n}\mathbb{E}\|\boldsymbol{V}\|^2 + \frac{1}{2n}\mathbb{E}[\boldsymbol{V}^\top \mathbf{R}_n(\boldsymbol{\theta}_n, \boldsymbol{V})\boldsymbol{V}] \\
&= \ell_n(\boldsymbol{\theta}_n) + \frac{p}{2n} + O(\varepsilon/n) = \ell_n(\boldsymbol{\theta}_n) + \frac{p}{2n} + o(1/n),
\end{aligned}$$

completes the proof. □

D.10 Markov Chains

Definition D.102 (Markov Chain): A **Markov chain** is a collection $\{X_t, t = 0, 1, 2, \ldots\}$ of random variables (or random vectors) whose futures are conditionally independent of their pasts given their present values. That is,

$$\text{(D.103)} \qquad \mathbb{P}[X_{t+1} \in A \,|\, X_s, s \leq t] = \mathbb{P}[X_{t+1} \in A \,|\, X_t] \quad \text{for all } t.$$

In other words, the conditional distribution of the future variable X_{t+1}, given the entire past $\{X_s, s \leq t\}$, is the same as the conditional distribution of X_{t+1} given only the present X_t. Property (D.103) is called the **Markov property**.

The index t in X_t is usually seen as a "time" or "step" parameter. The index set $\{0, 1, 2, \ldots\}$ in the definition above was chosen out of convenience. It can be replaced by any countable index set. We restrict ourselves to **time-homogeneous** Markov chains — Markov chains for which the conditional pdfs $f_{X_{t+1} \mid X_t}(y \mid x)$ do not depend on t; we abbreviate these as $q(y \mid x)$. The $\{q(y \mid x)\}$ are called the **(one-step) transition densities** of the Markov chain. Note that the random variables or vectors $\{X_t\}$ may be *discrete* (e.g., taking values in some set $\{1, \ldots, r\}$) or *continuous* (e.g., taking values in an interval $[0, 1]$ or $\mathbb{R}^d$). In particular, in the *discrete* case, each $q(y \mid x)$ is a probability: $q(y \mid x) = \mathbb{P}[X_{t+1} = y \mid X_t = x]$.

The distribution of X_0 is called the **initial distribution** of the Markov chain. The one-step transition densities and the initial distribution completely specify the distribution of the random vector $[X_0, X_1, \ldots, X_t]^\top$. Namely, we have by the product rule (D.29) and the Markov property that the joint pdf is given by ☞ 635

$$\begin{aligned} f_{X_0,\ldots,X_t}(x_0, \ldots, x_t) &= f_{X_0}(x_0)\, f_{X_1 \mid X_0}(x_1 \mid x_0) \cdots f_{X_t \mid X_{t-1},\ldots,X_0}(x_t \mid x_{t-1}, \ldots, x_0) \\ &= f_{X_0}(x_0)\, f_{X_1 \mid X_0}(x_1 \mid x_0) \cdots f_{X_t \mid X_{t-1}}(x_t \mid x_{t-1}) \\ &= f_{X_0}(x_0)\, q(x_1 \mid x_0)\, q(x_2 \mid x_1) \cdots q(x_t \mid x_{t-1}). \end{aligned}$$

A Markov chain is said to be **ergodic** if the probability distribution of X_t converges to a fixed distribution as $t \to \infty$. Ergodicity is a property of many Markov chains. Intuitively, the probability of encountering the Markov chain in a state x at a time t far into the future should not depend on the t, provided that the Markov chain can reach every state from any other state — such Markov chains are said to be **irreducible** — and does not "escape" to infinity. Thus, for an ergodic Markov chain the pdf $f_{X_t}(x)$ converges to a fixed **limiting pdf** $f(x)$ as $t \to \infty$, irrespective of the starting state. For the discrete case, $f(x)$ corresponds to the long-run fraction of times that the Markov process visits x.

Under mild conditions (such as irreducibility) the limiting pdf $f(x)$ can be found by solving the **global balance equations**:

$$f(x) = \begin{cases} \sum_y f(y)\, q(x \mid y) & \text{(discrete case)}, \\ \int f(y)\, q(x \mid y)\, \mathrm{d}y & \text{(continuous case)}. \end{cases} \tag{D.104}$$

For the discrete case the rationale behind this is as follows: Since $f(x)$ is the long-run proportion of time that the Markov chain spends in x, the proportion of transitions *out of* x is $f(x)$. This should be balanced with the proportion of transitions *into* state x, which is $\sum_y f(y)\, q(x \mid y)$.

One is often interested in a stronger type of balance equations. Imagine that we have taken a video of the evolution of the Markov chain, which we may run in forward and reverse time. If we cannot determine whether the video is running forward or backward (we cannot determine any systematic "looping", which would indicate in which direction time is flowing), the chain is said to be time-reversible or simply **reversible**.

Although not every Markov chain is reversible, each ergodic Markov chain, when run backwards, gives another Markov chain — the **reverse Markov chain** — with transition densities $\widetilde{q}(y \mid x) = f(y)\, q(x \mid y)/f(x)$. To see this, first observe that $f(x)$ is the long-run proportion of time spent in x for both the original and reverse Markov chain. Secondly, the "probability flux" from x to y in the reversed chain must be equal to the probability flux from y to x in the original chain, meaning $f(x)\, \widetilde{q}(y \mid x) = f(y)\, q(x \mid y)$, which yields the stated transition probabilities for the reversed chain. In particular, for a *reversible*

Markov chain we have

(D.105) $$f(x)\, q(y \mid x) = f(y)\, q(x \mid y) \quad \text{for all } x, y.$$

These are the **detailed (or local) balance equations**. Note that the detailed balance equations imply the global balance equations. Hence, if a Markov chain is irreducible and there exists a pdf such that (D.105) holds, then $f(x)$ must be the limiting pdf. In the discrete state space case an additional condition is that the chain must be **aperiodic**, meaning that the return times to the same state cannot always be a multiple of some integer ≥ 2.

Example D.106 (Random Walk on a Graph) Consider a Markov chain that performs a "random walk" on the graph in Figure D.107, at each step jumping from the current vertex (node) to one of the adjacent vertices, with equal probability. Clearly this Markov chain is reversible. It is also irreducible and aperiodic. Let $f(x)$ denote the limiting probability that the chain is in vertex x. By symmetry, $f(1) = f(2) = f(7) = f(8)$, $f(4) = f(5)$ and $f(3) = f(6)$. Moreover, by the detailed balance equations, $f(4)/5 = f(1)/3$, and $f(3)/4 = f(1)/3$. It follows that $f(1) + \cdots + f(8) = 4f(1) + 2 \times 5/3\, f(1) + 2 \times 4/3\, f(1) = 10\, f(1) = 1$, so that $f(1) = 1/10$, $f(3) = 2/15$, and $f(4) = 1/6$.

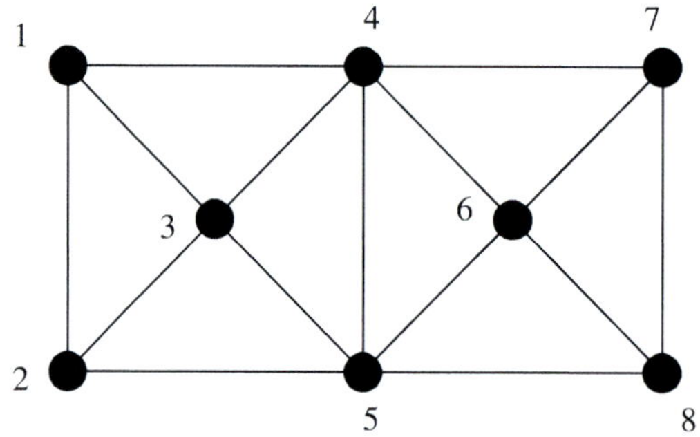

Figure D.107: The random walk on this graph is reversible.

D.11 Statistics

Statistics deals with the gathering, summarization, analysis, and interpretation of data. The two main branches of statistics are:

1. **Classical or frequentist statistics**: Here the observed data τ is viewed as the outcome of random data $\mathcal{T}$ described by a probabilistic model — usually the model is specified up to a (multi-dimensional) parameter; that is, $\mathcal{T} \sim g(\cdot \mid \boldsymbol{\theta})$ for some $\boldsymbol{\theta}$. The statistical inference is then purely concerned with the model and in particular with the parameter $\boldsymbol{\theta}$. For example, on the basis of the data one may wish to

   (a) estimate the parameter,

(b) perform statistical tests on the parameter, or

(c) validate the model.

2. **Bayesian statistics**: In this approach we average over all possible values of the parameter $\boldsymbol{\theta}$ using a user-specified weight function $g(\boldsymbol{\theta})$ and obtain the model $\mathcal{T} \sim \int g(\cdot \mid \boldsymbol{\theta})\, g(\boldsymbol{\theta})\, \mathrm{d}\boldsymbol{\theta}$. For practical computations, this means that we can treat $\boldsymbol{\theta}$ as a random variable with pdf $g(\boldsymbol{\theta})$. Bayes' formula $g(\boldsymbol{\theta} \mid \tau) \propto g(\tau \mid \boldsymbol{\theta})\, g(\boldsymbol{\theta})$ is used to learn $\boldsymbol{\theta}$ based on the observed data τ. ☞ 49

■ **Example D.108 (Iid Sample)** The most fundamental statistical model is where the data $\mathcal{T} = X_1, \ldots, X_n$ is such that the random variables $X_1, \ldots, X_n$ are assumed to be independent and identically distributed:

$$X_1, \ldots, X_n \overset{\text{iid}}{\sim} \mathsf{Dist},$$

according to some known or unknown distribution Dist. An iid sample is often called a **random sample** in the statistics literature. Note that the word "sample" can refer to both a collection of random variables and to a single random variable. It should be clear from the context which meaning is being used.

Often our guess or model for the true distribution is specified up to an unknown parameter $\boldsymbol{\theta}$, with $\boldsymbol{\theta} \in \Theta$. The most common model is:

$$X_1, \ldots, X_n \overset{\text{iid}}{\sim} \mathcal{N}(\mu, \sigma^2),$$

in which case $\boldsymbol{\theta} = (\mu, \sigma^2)$ and $\Theta = \mathbb{R} \times \mathbb{R}_+$. ■

D.12 Estimation

Suppose the model $g(\cdot \mid \boldsymbol{\theta})$ for the data $\mathcal{T}$ is completely specified up to an unknown parameter vector $\boldsymbol{\theta}$. The aim is to estimate $\boldsymbol{\theta}$ on the basis of the observed data τ only (an alternative goal could be to estimate $\boldsymbol{\eta} = \boldsymbol{\psi}(\boldsymbol{\theta})$ for some vector-valued function $\boldsymbol{\psi}$). Specifically, the goal is to find an **estimator** $\boldsymbol{T} = \boldsymbol{T}(\mathcal{T})$ that is close to the unknown $\boldsymbol{\theta}$. The corresponding outcome $\boldsymbol{t} = \boldsymbol{T}(\tau)$ is the **estimate** of $\boldsymbol{\theta}$. The **bias** of an estimator $\boldsymbol{T}$ of $\boldsymbol{\theta}$ is defined as $\mathbb{E}\boldsymbol{T} - \boldsymbol{\theta}$. An estimator $\boldsymbol{T}$ of $\boldsymbol{\theta}$ is said to be **unbiased** if $\mathbb{E}_{\boldsymbol{\theta}}\boldsymbol{T} = \boldsymbol{\theta}$. We often write $\widehat{\boldsymbol{\theta}}$ for both an estimator and estimate of $\boldsymbol{\theta}$. The **mean squared error** (MSE) of a real-valued estimator T is defined as

$$\text{MSE} = \mathbb{E}_\theta (T - \theta)^2.$$

An estimator T_1 is said to be more **efficient** than an estimator T_2 if the MSE of T_1 is smaller than the MSE of T_2. The MSE can be written as the sum

$$\text{MSE} = (\mathbb{E}_\theta T - \theta)^2 + \mathbb{V}\text{ar}_\theta T.$$

The first term measures the unbiasedness and the second is the variance of the estimator. In particular, for an unbiased estimator the MSE of an estimator is simply equal to its variance.

For simulation purposes it is often important to include the running time of the estimator in efficiency comparisons. One way to compare two unbiased estimators T_1 and T_2 is to compare their **relative time variance products**,

$$\frac{r_i \, \mathbb{V}\mathrm{ar}\, T_i}{(\mathbb{E}\, T_i)^2}, \quad i = 1, 2, \tag{D.109}$$

where r_1 and r_2 are the times required to calculate the estimators T_1 and T_2, respectively. In this scheme, T_1 is considered more efficient than T_2 if its relative time variance product is smaller. We discuss next two systematic approaches for constructing sound estimators.

D.12.1 Method of Moments

Suppose $x_1, \ldots, x_n$ are outcomes from an iid sample $X_1, \ldots, X_n \sim_{\text{iid}} g(x \,|\, \boldsymbol{\theta})$, where $\boldsymbol{\theta} = [\theta_1, \ldots, \theta_k]^\top$ is unknown. The moments of the sampling distribution can be easily estimated. Namely, if $X \sim g(x \,|\, \boldsymbol{\theta})$, then the r-th moment of X, that is $\mu_r(\boldsymbol{\theta}) = \mathbb{E}_{\boldsymbol{\theta}} X^r$ (assuming it exists), can be estimated through the **sample r-th moment**: $\frac{1}{n}\sum_{i=1}^n x_i^r$. The **method of moments** involves choosing the estimate $\widehat{\boldsymbol{\theta}}$ of $\boldsymbol{\theta}$ such that each of the first k sample and true moments are matched:

$$\frac{1}{n}\sum_{i=1}^n x_i^r = \mu_r(\widehat{\boldsymbol{\theta}}), \quad r = 1, 2, \ldots, k.$$

In general, this set of equations is nonlinear and so its solution often has to be found numerically.

Example D.110 (Sample Mean and Sample Variance) Suppose the data is given by $\mathcal{T} = \{X_1, \ldots, X_n\}$, where the $\{X_i\}$ form an iid sample from a general distribution with mean μ and variance $\sigma^2 < \infty$. Matching the first two moments gives the set of equations

$$\frac{1}{n}\sum_{i=1}^n x_i = \mu,$$
$$\frac{1}{n}\sum_{i=1}^n x_i^2 = \mu^2 + \sigma^2.$$

The method of moments estimates for μ and σ^2 are therefore the **sample mean**

$$\widehat{\mu} = \overline{x} = \frac{1}{n}\sum_{i=1}^n x_i, \tag{D.111}$$

and

$$\widehat{\sigma^2} = \frac{1}{n}\sum_{i=1}^n x_i^2 - (\overline{x})^2 = \frac{1}{n}\sum_{i=1}^n (x_i - \overline{x})^2. \tag{D.112}$$

The corresponding estimator for μ, $\overline{X}$, is unbiased. However, the estimator for σ^2 is biased: $\mathbb{E}\,\widehat{\sigma^2} = \sigma^2(n-1)/n$. An unbiased estimator is the **sample variance**

$$S^2 = \widehat{\sigma^2}\frac{n}{n-1} = \frac{1}{n-1}\sum_{i=1}^n (X_i - \overline{X})^2.$$

Its square root, $S = \sqrt{S^2}$, is called the **sample standard deviation**. ■

■ **Example D.113 (Sample Covariance Matrix)** The method of moments can also be used to estimate the covariance matrix of a random vector. In particular, let the $\boldsymbol{X}_1, \ldots, \boldsymbol{X}_n$ be iid copies of a d-dimensional random vector $\boldsymbol{X}$ with mean vector $\boldsymbol{\mu}$ and covariance matrix $\boldsymbol{\Sigma}$. We assume $n \geq d$. The moment estimator for $\boldsymbol{\mu}$ is, as in the $d = 1$ case, $\overline{\boldsymbol{X}} = (\boldsymbol{X}_1 + \cdots + \boldsymbol{X}_n)/n$. As the covariance matrix can be written (see (D.26)) as ☞ 634

$$\boldsymbol{\Sigma} = \mathbb{E}(\boldsymbol{X} - \boldsymbol{\mu})(\boldsymbol{X} - \boldsymbol{\mu})^\top,$$

the method of moments yields the estimator

$$\widehat{\boldsymbol{\Sigma}} = \frac{1}{n}\sum_{i=1}^{n}(\boldsymbol{X}_i - \overline{\boldsymbol{X}})(\boldsymbol{X}_i - \overline{\boldsymbol{X}})^\top. \tag{D.114}$$

Similar to the one-dimensional case ($d = 1$), replacing the factor $1/n$ with $1/(n-1)$ gives an unbiased estimator, called the **sample covariance matrix**. ■

D.12.2 Maximum Likelihood Method

The concept of *likelihood* is central in statistics. It describes in a precise way the information about model parameters that is contained in the observed data.

Let $\mathcal{T}$ be a (random) data object that is modeled as a draw from the pdf $g(\tau \,|\, \boldsymbol{\theta})$ (discrete or continuous) with parameter vector $\boldsymbol{\theta} \in \Theta$. Let τ be an outcome of $\mathcal{T}$. The function $L(\boldsymbol{\theta} \,|\, \tau) := g(\tau \,|\, \boldsymbol{\theta})$, $\boldsymbol{\theta} \in \Theta$, is called the **likelihood function** of $\boldsymbol{\theta}$, based on τ. The (natural) logarithm of the likelihood function is called the **log-likelihood function** and is often denoted by a lower case l.

> $L(\boldsymbol{\theta} \,|\, \tau)$ and $g(\tau \,|\, \boldsymbol{\theta})$ have the same formula, but the first is viewed as a function of $\boldsymbol{\theta}$ for fixed τ, where the second is viewed as a function of τ for fixed $\boldsymbol{\theta}$.

The concept of likelihood is particularly useful when $\mathcal{T}$ is modeled as an iid sample $\{X_1, \ldots, X_n\}$ from some pdf $\mathring{g}$. In that case, the likelihood of the data $\tau = \{x_1, \ldots, x_n\}$, as a function of $\boldsymbol{\theta}$, is given by the product

$$L(\boldsymbol{\theta} \,|\, \tau) = \prod_{i=1}^{n} \mathring{g}(x_i \,|\, \boldsymbol{\theta}). \tag{D.115}$$

Let τ be an observation from $\mathcal{T} \sim g(\tau \,|\, \boldsymbol{\theta})$, and suppose that $g(\tau \,|\, \boldsymbol{\theta})$ takes its largest value at $\boldsymbol{\theta} = \widehat{\boldsymbol{\theta}}$. In a way this $\widehat{\boldsymbol{\theta}}$ is our best estimate for $\boldsymbol{\theta}$, as it maximizes the probability (density) for the observation τ. It is called the **maximum likelihood estimate** (MLE) of $\boldsymbol{\theta}$. Note that $\widehat{\boldsymbol{\theta}} = \widehat{\boldsymbol{\theta}}(\tau)$ is a function of τ. The corresponding random variable, also denoted $\widehat{\boldsymbol{\theta}}$ is the **maximum likelihood estimator** (also abbreviated as MLE).

Maximization of $L(\boldsymbol{\theta} \,|\, \tau)$ as a function of $\boldsymbol{\theta}$ is equivalent (when searching for the maximizer) to maximizing the log-likelihood $l(\boldsymbol{\theta} \,|\, \tau)$, as the natural logarithm is an increasing function. This is often easier, especially when $\mathcal{T}$ is an iid sample from some sampling distribution. For example, for L of the form (D.115), we have

$$l(\boldsymbol{\theta} \mid \tau) = \sum_{i=1}^{n} \ln \mathring{g}(x_i \mid \boldsymbol{\theta}).$$

If $l(\boldsymbol{\theta} \mid \tau)$ is a differentiable function with respect to $\boldsymbol{\theta}$ and the maximum is attained in the *interior* of Θ, *and there exists a unique maximum point*, then we can find the MLE of $\boldsymbol{\theta}$ by solving the equations

$$\frac{\partial}{\partial \theta_i} l(\boldsymbol{\theta} \mid \tau) = 0, \quad i = 1, \ldots, d.$$

Example D.116 (Bernoulli Random Sample) Suppose we have data $\tau_n = \{x_1, \ldots, x_n\}$ and assume the model $X_1, \ldots, X_n \sim_{\text{iid}} \mathsf{Ber}(\theta)$. Then, the likelihood function is given by

$$L(\theta \mid \tau) = \prod_{i=1}^{n} \theta^{x_i}(1-\theta)^{1-x_i} = \theta^s (1-\theta)^{n-s}, \quad 0 < \theta < 1, \tag{D.117}$$

where $s := x_1 + \cdots + x_n =: n\overline{x}$. The log-likelihood is $l(\theta) = s \ln \theta + (n-s) \ln(1-\theta)$. Through differentiation with respect to θ, we find the derivative

$$\frac{s}{\theta} - \frac{n-s}{1-\theta} = \frac{s}{\theta(1-\theta)} - \frac{n}{1-\theta}. \tag{D.118}$$

Solving $l'(\theta) = 0$ gives the ML estimate $\widehat{\theta} = \overline{x}$ and ML estimator $\widehat{\theta} = \overline{X}$.

D.13 Confidence Intervals

An essential part in any estimation procedure is to provide an assessment of the *accuracy* of the estimate. Indeed, without information on its accuracy the estimate itself would be meaningless. Confidence intervals (also called **interval estimates**) provide a precise way of describing the uncertainty in the estimate.

Let $X_1, \ldots, X_n$ be random variables with a joint distribution depending on a parameter $\theta \in \Theta$. Let $T_1 < T_2$ be statistics; that is, $T_i = T_i(X_1, \ldots, X_n)$, $i = 1, 2$ are functions of the data, but not of θ.

1. The random interval (T_1, T_2) is called a **stochastic confidence interval** for θ with confidence $1 - \alpha$ if

$$\mathbb{P}_\theta[T_1 < \theta < T_2] \geq 1 - \alpha \quad \text{for all } \theta \in \Theta. \tag{D.119}$$

2. If t_1 and t_2 are the observed values of T_1 and T_2, then the interval (t_1, t_2) is called the **(numerical) confidence interval** for θ with confidence $1 - \alpha$ for every $\theta \in \Theta$.

3. If the right-hand side of (D.119) is merely a heuristic estimate or approximation of the true probability, then the resulting interval is called an **approximate confidence interval**.

4. The probability $\mathbb{P}_\theta[T_1 < \theta < T_2]$ is called the **coverage probability**. For a $1-\alpha$ confidence interval, it must be at least $1-\alpha$.

For multi-dimensional parameters $\boldsymbol{\theta} \in \mathbb{R}^d$ the stochastic confidence interval is replaced with a stochastic **confidence region** $C \subset \mathbb{R}^d$ such that $\mathbb{P}_{\boldsymbol{\theta}}[\boldsymbol{\theta} \in C] \geq 1-\alpha$ for all $\boldsymbol{\theta}$.

■ **Example D.120 (Approximate Confidence Interval for the Mean)** Take an iid sample $X_1, X_2, \ldots, X_n$ from a distribution with mean μ and variance $\sigma^2 < \infty$ (both assumed to be unknown). By the central limit theorem and the law of large numbers, ☞ 654

$$T = \frac{\overline{X} - \mu}{S/\sqrt{n}} \overset{\text{approx.}}{\sim} \mathcal{N}(0,1),$$

for large n, where S is the sample standard deviation. Rearranging the approximate equality $\mathbb{P}[|T| \leq z_{1-\alpha/2}] \approx 1-\alpha$, where $z_{1-\alpha/2}$ is the $1-\alpha/2$ quantile of the standard normal distribution, yields

$$\mathbb{P}\left[\overline{X} - z_{1-\alpha/2}\frac{S}{\sqrt{n}} \leq \mu \leq \overline{X} + z_{1-\alpha/2}\frac{S}{\sqrt{n}}\right] \approx 1-\alpha,$$

so that

$$\left(\overline{X} - z_{1-\alpha/2}\frac{S}{\sqrt{n}},\ \overline{X} + z_{1-\alpha/2}\frac{S}{\sqrt{n}}\right), \text{ abbreviated as } \overline{X} \pm z_{1-\alpha/2}\frac{S}{\sqrt{n}}, \tag{D.121}$$

is an approximate stochastic $(1-\alpha)$ confidence interval for μ. ■

Since (D.121) is an asymptotic result only, care should be taken when applying it to cases where the sample size is small or moderate and the sampling distribution is heavily skewed.

D.14 Hypothesis Testing

Suppose the model for the data $\mathcal{T}$ is described by a family of probability distributions that depend on a parameter $\boldsymbol{\theta} \in \Theta$. The aim of **hypothesis testing** is to decide, on the basis of the observed data τ, which of two competing hypotheses holds true; these being the **null hypothesis**, $H_0 : \boldsymbol{\theta} \in \Theta_0$, and the **alternative hypothesis**, $H_1 : \boldsymbol{\theta} \in \Theta_1$.

In classical statistics the null hypothesis and alternative hypothesis do not play equivalent roles. H_0 contains the "status quo" statement and is only rejected if the observed data are very unlikely to have happened under H_0.

The decision whether to accept or reject H_0 is dependent on the outcome of a **test statistic** $\boldsymbol{T} = \boldsymbol{T}(\mathcal{T})$. For simplicity, we discuss only the one-dimensional case $\boldsymbol{T} \equiv T$. Two (related) types of decision rules are generally used:

1. **Decision rule 1**: *Reject H_0 if T falls in the critical region.*

   Here the **critical region** is any appropriately chosen region in $\mathbb{R}$. In practice a critical region is one of the following:

- *left one-sided*: $(-\infty, c]$,
- *right one-sided*: $[c, \infty)$,
- *two-sided*: $(-\infty, c_1] \cup [c_2, \infty)$.

For example, for a right one-sided test, H_0 is rejected if the outcome of the test statistic is too large. The endpoints c, c_1, and c_2 of the critical regions are called **critical values**.

2. **Decision rule 2**: *Reject* H_0 *if the* P-value *is smaller than some* significance level α.

   The **P-value** is the probability that, under H_0, the (random) test statistic takes a value as extreme as or more extreme than the one observed. In particular, if t is the observed outcome of the test statistic T, then

   - *left one-sided test*: $P := \mathbb{P}_{H_0}[T \leq t]$,
   - *right one-sided*: $P := \mathbb{P}_{H_0}[T \geq t]$,
   - *two-sided*: $P := \min\{2\mathbb{P}_{H_0}[T \leq t],\ 2\mathbb{P}_{H_0}[T \geq t]\}$.

   The smaller the P-value, the greater the strength of the evidence against H_0 provided by the data. As a rule of thumb:

   $$\begin{array}{ll} P < 0.10 & \text{suggestive evidence,} \\ P < 0.05 & \text{reasonable evidence,} \\ P < 0.01 & \text{strong evidence.} \end{array}$$

Whether the first or the second decision rule is used, one can make two types of errors, as depicted in Table D.122.

Table D.122: Type I and II errors in hypothesis testing.

	True statement	
Decision	H_0 *is true*	H_1 *is true*
Accept H_0	Correct	Type II Error
Reject H_0	Type I Error	Correct

The choice of the test statistic and the corresponding critical region involves a multiobjective optimization criterion, whereby both the probabilities of a type I and type II error should, ideally, be chosen as small as possible. Unfortunately, these probabilities compete with each other. For example, if the critical region is made larger (smaller), the probability of a type II error is reduced (increased), but at the same time the probability of a type I error is increased (reduced).

Since the type I error is considered more serious, Neyman and Pearson [132] suggested the following approach: choose the critical region such that the probability of a type II error is as small as possible, while keeping the probability of a type I error below a predetermined small **significance level** α.

■ **Remark D.123 (Equivalence of Decision Rules)** Note that decision rule 1 and 2 are equivalent in the following sense:

$$\text{Reject } H_0 \text{ if } T \text{ falls in the critical region, at significance level } \alpha.$$
$$\Leftrightarrow$$
$$\text{Reject } H_0 \text{ if the P-value is} \leq \text{significance level } \alpha.$$

In other words, the P-value of the test is the smallest level of significance that would lead to the rejection of H_0. ■

In general, a statistical test involves the following steps:

1. Formulate an appropriate statistical model for the data.
2. Give the null (H_0) and alternative (H_1) hypotheses in terms of the parameters of the model.
3. Determine the test statistic (a function of the data only).
4. Determine the (approximate) distribution of the test statistic under H_0.
5. Calculate the outcome of the test statistic.
6. Calculate the P-value *or* the critical region, given a preselected significance level α.
7. Accept or reject H_0.

The actual choice of an appropriate test statistic is akin to selecting a good estimator for the unknown parameter $\boldsymbol{\theta}$. The test statistic should summarize the information about $\boldsymbol{\theta}$ and make it possible to distinguish between the alternative hypotheses.

■ **Example D.124 (Hypothesis Testing)** We are given outcomes $x_1, \ldots, x_m$ and $y_1, \ldots, y_n$ of two simulation studies obtained via independent runs, with $m = 100$ and $n = 50$. The sample means and standard deviations are $\overline{x} = 1.3$, $s_X = 0.1$ and $\overline{y} = 1.5$, $s_Y = 0.3$. Thus, the $\{x_i\}$ are outcomes of iid random variables $\{X_i\}$, the $\{y_i\}$ are outcomes of iid random variables $\{Y_i\}$, and the $\{X_i\}$ and $\{Y_i\}$ are independent. We wish to assess whether the expectations $\mu_X = \mathbb{E}X_i$ and $\mu_Y = \mathbb{E}Y_i$ are the same or not. Going through the 7 steps above, we have:

1. The model is already specified above.
2. $H_0 : \mu_X - \mu_Y = 0$ versus $H_1 : \mu_X - \mu_Y \neq 0$.

3. For similar reasons as in Example D.120, take

$$T = \frac{\overline{X} - \overline{Y}}{\sqrt{S_X^2/m + S_Y^2/n}}.$$

4. By the central limit theorem, the statistic T has, under H_0, approximately a standard normal distribution (assuming the variances are finite).

5. The outcome of T is $t = (\overline{x} - \overline{y})/\sqrt{s_X^2/m + s_Y^2/n} \approx -4.59$.

6. As this is a two-sided test, the P-value is $2\mathbb{P}_{H_0}[T \leq -4.59] \approx 4 \cdot 10^{-6}$.

7. Because the P-value is extremely small, there is overwhelming evidence that the two expectations are not the same.

■

Further Reading

Accessible treatises on probability and stochastic processes include [40, 43, 59, 78, 143]. Kallenberg's book [88] provides a complete graduate-level overview of the foundations of modern probability. An advanced course in probability theory and stochastic processes at an upper undergraduate and graduate level can be found in [95]. Details on the convergence of probability measures and limit theorems are given in [16]. For an accessible introduction to mathematical statistics with simple applications see, for example, [41, 101, 178]. For a more detailed overview of statistical inference, see [15, 39]. A standard reference for classical (frequentist) statistical inference is [108].

APPENDIX E

PYTHON PRIMER

Python has become the programming language of choice for many researchers and practitioners in data science and machine learning. This appendix gives a brief introduction to the language. As the language is under constant development and each year many new packages are being released, we do not pretend to be exhaustive in this introduction. Instead, we hope to provide enough information for novices to get started with this beautiful and carefully thought-out language.

E.1 Getting Started

The main website for Python is

https://www.python.org/,

where you will find documentation, a tutorial, beginners' guides, software examples, and so on. It is important to note that there are two incompatible "branches" of Python, called Python 3 and Python 2. Further development of the language will involve only Python 3, and in this appendix (and indeed the rest of the book) we only consider Python 3. As there are many interdependent packages that are frequently used with a Python installation, it is convenient to install a distribution — for instance, the **Anaconda** Python distribution, available from

https://www.anaconda.com/.

The Anaconda installer automatically installs the most important packages and also provides a convenient interactive development environment (IDE), called *Spyder*.

> Use the *Anaconda Navigator* to launch *Spyder*, *Jupyter notebook*, install and update packages, or open a command-line terminal.

To get started[1], try out the Python statements in the input boxes that follow. You can either type these statements at the IPython command prompt or run them as (very

[1]We assume that you have installed all the necessary files and have launched *Spyder*.

short) Python programs. The output for these two modes of input can differ slightly. For example, typing a variable name in the console causes its contents to be automatically printed, whereas in a Python program this must be done explicitly by calling the **print** function. Selecting (highlighting) several program lines in *Spyder* and then pressing function key[2] F9 is equivalent to executing these lines one by one in the console.

In Python, data is represented as an **object** or relation between objects (see also Section E.2). Basic data types are numeric types (including integers, booleans, and floats), sequence types (including strings, tuples, and lists), sets, and mappings (currently, dictionaries are the only built-in mapping type).

Strings are sequences of characters, enclosed by single or double quotes. We can print strings via the **print** function.

```
print("Hello World!")
```
```
Hello World!
```

For pretty-printing output, Python strings can be formatted using the **format** function. The bracket syntax {i} provides a placeholder for the i-th variable to be printed, with 0 being the first index. Individual variables can be formatted separately and as desired; ☞ 684 formatting syntax is discussed in more detail in Section E.9.

```
print("Name:{1} (height {2} m, age {0})".format(111,"Bilbo",0.84))
```
```
Name:Bilbo (height 0.84 m, age 111)
```

Lists can contain different types of objects, and are created using square brackets as in the following example:

```
x = [1,'string',"another string"]  # Quote type is not important
```
```
[1, 'string', 'another string']
```

Elements in lists are indexed starting from 0, and are **mutable** (can be changed):

```
x = [1,2]
x[0] = 2  # Note that the first index is 0
x
```
```
[2,2]
```

In contrast, tuples (with round brackets) are **immutable** (cannot be changed). Strings are immutable as well.

```
x = (1,2)
x[0] = 2
```
```
TypeError: 'tuple' object does not support item assignment
```

[2]This may depend on the keyboard and operating system.

Lists can be accessed via the **slice** notation `[start:end]`. It is important to note that `end` is the index of the first element that will *not* be selected, and that the first element has index 0. To gain familiarity with the slice notation, execute each of the following lines.

```
a = [2, 3, 5, 7, 11, 13, 17, 19, 23]
a[1:4]  # Elements with index from 1 to 3
a[:4]   # All elements with index less than 4
a[3:]   # All elements with index 3 or more
a[-2:]  # The last two elements
```

```
[3, 5, 7]
[2, 3, 5, 7]
[7, 11, 13, 17, 19, 23]
[19, 23]
```

An **operator** is a programming language construct that performs an action on one or more operands. The action of an operator in Python depends on the type of the operand(s). For example, operators such as +, $*$, $-$, and % that are arithmetic operators when the operands are of a numeric type, can have different meanings for objects of non-numeric type (such as strings).

```
'hello' + 'world'  # String concatenation
```

```
'helloworld'
```

```
'hello' * 2   # String repetition
```

```
'hellohello'
```

```
[1,2] * 2       # List repetition
```

```
[1, 2, 1, 2]
```

```
15 % 4   # Remainder of 15/4
```

```
3
```

Some common Python operators are given in Table E.1. ☞ 676

E.2 Python Objects

As mentioned in the previous section, data in Python is represented by objects or relations between objects. We recall that basic data types included strings and numeric types (such as integers, booleans, and floats).

As Python is an object-oriented programming language, functions are objects too (everything is an object!). Each object has an identity (unique to each object and immutable — that is, cannot be changed — once created), a type (which determines which operations can be applied to the object, and is considered immutable), and a value (which is either mutable or immutable). The unique identity assigned to an object obj can be found by calling **id**, as in id(obj).

Each object has a list of **attributes**, and each attribute is a reference to another object. The function **dir** applied to an object returns the list of attributes. For example, a string object has many useful attributes, as we shall shortly see. Functions are objects with the __call__ attribute.

A class (see Section E.8) can be thought of as a template for creating a custom type of object.

```
s = "hello"
d = dir(s)
print(d,flush=True)  # Print the list in "flushed" format
```

```
['__add__', '__class__', '__contains__', '__delattr__', '__dir__',
 ... (many left out) ... 'replace', 'rfind',
  'rindex', 'rjust', 'rpartition', 'rsplit', 'rstrip', 'split',
  'splitlines', 'startswith', 'strip', 'swapcase', 'title',
  'translate', 'upper', 'zfill']
```

Any attribute attr of an object obj can be accessed via the **dot notation**: obj.attr. To find more information about any object use the **help** function.

```
s = "hello"
help(s.replace)
```

```
replace(...) method of builtins.str instance
    S.replace(old, new[, count]) -> str

    Return a copy of S with all occurrences of substring
    old replaced by new.  If the optional argument count is
    given, only the first count occurrences are replaced.
```

This shows that the attribute **replace** is in fact a function. An attribute that is a function is called a **method**. We can use the **replace** method to create a new string from the old one by changing certain characters.

```
s = 'hello'
s1 = s.replace('e','a')
print(s1)
```

```
hallo
```

In many Python editors, pressing the TAB key, as in objectname.<TAB>, will bring up a list of possible attributes via the editor's autocompletion feature.

E.3 Types and Operators

Each object has a **type**. Three basic data types in Python are `str` (for string), `int` (for integers), and `float` (for floating point numbers). The function `type` returns the type of an object.

```
t1 = type([1,2,3])
t2 = type((1,2,3))
t3 = type({1,2,3})
print(t1,t2,t3)
```

```
<class 'list'> <class 'tuple'> <class 'set'>
```

The **assignment** operator, =, assigns an object to a variable; e.g., `x = 12`. An **expression** is a combination of values, operators, and variables that yields another value or variable.

> Variable names are case sensitive and can only contain letters, numbers, and underscores. They must start with either a letter or underscore. Note that reserved words such as `True` and `False` are case sensitive as well.

Python is a dynamically typed language, and the type of a variable at a particular point during program execution is determined by its most recent object assignment. That is, the type of a variable does not need to be explicitly declared from the outset (as is the case in C or Java), but instead the type of the variable is determined by the object that is currently assigned to it.

It is important to understand that a variable in Python is a **reference** to an object — think of it as a label on a shoe box. Even though the label is a simple entity, the *contents* of the shoe box (the object to which the variable refers) can be arbitrarily complex. Instead of moving the contents of one shoe box to another, it is much simpler to merely move the label.

```
x = [1,2]
y = x    # y refers to the same object as x
print(id(x) == id(y))  # check that the object id's are the same
y[0] = 100  # change the contents of the list that y refers to
print(x)
```

```
True
[100,2]
```

```
x = [1,2]
y = x    # y refers to the same object as x
y = [100,2]  # now y refers to a different object
print(id(x) == id(y))
print(x)
```

```
False
[1,2]
```

Table E.1 shows a selection of Python operators for numerical and logical variables.

Table E.1: Common numerical (left) and logical (right) operators.

+	addition	~	binary NOT
-	subtraction	&	binary AND
*	multiplication	^	binary XOR
**	power	|	binary OR
/	division	==	equal to
//	integer division	!=	not equal to
%	modulus		

Several of the numerical operators can be combined with an assignment operator, as in `x += 1` to mean `x = x + 1`. Operators such as + and * can be defined for other data types as well, where they take on a different meaning. This is called operator *overloading*, an example of which is the use of `<List> * <Integer>` for list repetition as we saw earlier.

E.4 Functions and Methods

Functions make it easier to divide a complex program into simpler parts. To create a **function**, use the following syntax:

```
def <function name>(<parameter_list>):
    <statements>
```

A function takes a list of input variables that are references to objects. Inside the function, a number of statements are executed which may modify the objects, but not the reference itself. In addition, the function may return an output object (or will return the value `None` if not explicitly instructed to return output). Think again of the shoe box analogy. The input variables of a function are labels of shoe boxes, and the objects to which they refer are the contents of the shoe boxes. The following program highlights some of the subtleties of variables and objects in Python.

> Note that the statements within a function must be indented. This is Python's way to define where a function begins and ends.

```python
x = [1,2,3]

def change_list(y):
    y.append(100) # Append an element to the list referenced by y
    y[0]=0        # Modify the first element of the same list
    y = [2,3,4]   # The local y now refers to a different list
            # The list to which y first referred does not change
    return sum(y)
```

```
print(change_list(x))
print(x)
```

```
9
[0, 2, 3, 100]
```

Variables that are defined inside a function only have *local scope*; that is, they are recognized only within that function. This allows the same variable name to be used in different functions without creating a conflict. If any variable is used within a function, Python first checks if the variable has local scope. If this is not the case (the variable has not been defined inside the function), then Python searches for that variable outside the function (the global scope). The following program illustrates several important points.

```
from numpy import array, square, sqrt

x = array([1.2,2.3,4.5])

def stat(x):
    n = len(x)          #the length of x
    meanx = sum(x)/n
    stdx = sqrt(sum(square(x - meanx))/n)
    return [meanx,stdx]

print(stat(x))
```

```
[2.6666666666666665, 1.3719410418171119]
```

1. Basic math functions such as **sqrt** are unknown to the standard Python interpreter and need to be imported. More on this in Section E.5 below.

2. As was already mentioned, indentation is crucial. It shows where the function begins and ends.

3. No semicolons[3] are needed to end lines, but the first line of the function definition (here line 5) must end with a colon (:).

4. Lists are not arrays (vectors of numbers), and vector operations cannot be performed on lists. However, the **numpy** module is designed specifically with efficient vector/matrix operations in mind. On the second code line, we define `x` as a vector (`ndarray`) object. Functions such as **square**, **sum**, and **sqrt** are then applied to such arrays. Note that we used the default Python functions **len** and **sum**. More on **numpy** in Section E.10.

5. Running the program with `stat(x)` instead of `print(stat(x))` in line 11 will not show any output in the console.

[3] Semicolons can be used to put multiple commands on a single line.

> To display the complete list of built-in functions, type (using double underscores) dir(__builtin__) .

E.5 Modules

A Python **module** is a programming construct that is useful for organizing code into manageable parts. To each module with name module_name is associated a Python file module_name.py containing any number of definitions, e.g., of functions, classes, and variables, as well as executable statements. Modules can be imported into other programs using the syntax: import <module_name> as <alias_name>, where <alias_name> is a shorthand name for the module.

When imported into another Python file, the module name is treated as a **namespace**, providing a naming system where each object has its unique name. For example, different modules mod1 and mod2 can have different sum functions, but they can be distinguished by prefixing the function name with the module name via the dot notation, as in mod1.sum and mod2.sum. For example, the following code uses the sqrt function of the numpy module.

```
import numpy as np
np.sqrt(2)
```

```
1.4142135623730951
```

A Python *package* is simply a directory of Python modules; that is, a collection of modules with additional startup information (some of which may be found in its __path__ attribute). Python's built-in module is called __builtins__. Of the great many useful Python modules, Table E.2 gives a few.

The numpy package contains various subpackages, such as random, linalg, and fft. More details are given in Section E.10.

> When using *Spyder*, press Ctrl+I in front of any object, to display its help file in a separate window.

As we have already seen, it is also possible to import only specific functions from a module using the syntax: from <module_name> import <fnc1, fnc2, ...>.

```
from numpy import sqrt, cos
sqrt(2)
cos(1)
```

```
1.4142135623730951
0.5403023058681398
```

This avoids the tedious prefixing of functions via the (alias) of the module name. However, for large programs it is good practice to always use the prefix/alias name construction, to be able to clearly ascertain precisely which module a function being used belongs to.

Table E.2: A few useful Python modules/packages.

datetime	Module for manipulating dates and times.
matplotlib	MATLAB-type plotting package.
numpy	Fundamental package for scientific computing, including random number generation and linear algebra tools. Defines the ubiquitous ndarray class.
os	Python interface to the operating system.
pandas	Fundamental module for data analysis. Defines the powerful DataFrame class.
pytorch	Machine learning library that supports GPU computation.
scipy	Ecosystem for mathematics, science, and engineering, containing many tools for numerical computing, including those for integration, solving differential equations, and optimization.
requests	Library for performing HTTP requests and interfacing with the web.
seaborn	Package for statistical data visualization.
sklearn	Easy to use machine learning library.
statsmodels	Package for the analysis of statistical models.

E.6 Flow Control

Flow control in Python is similar to that of many programming languages, with conditional statements as well as while and for loops. The syntax for if-then-else flow control is as follows:

```
if <condition1>:
   <statements>
elif <condition2>:
   <statements>
else:
   <statements>
```

Here, <condition1> and <condition2> are logical conditions that are either True or False; logical conditions often involve comparison operators (such as ==, >, <=, !=). In the example above, there is one elif part, which allows for an "else if" conditional statement. In general, there can be more than one elif part, or it can be omitted. The else part can also be omitted. The colons are essential, as are the indentations.

The while and for loops have the following syntax.

```
while <condition>:
   <statements>

for <variable> in <collection>:
   <statements>
```

Above, <collection> is an iterable object (see Section E.7 below). For further control in for and while loops, one can use a break statement to exit the current loop, and the continue statement to continue with the next iteration of the loop, while abandoning any remaining statements in the current iteration. Here is an example.

```
import numpy as np
ans = 'y'
while ans != 'n':
  outcome = np.random.randint(1,6+1)
  if outcome == 6:
      print("Hooray a 6!")
      break
  else:
      print("Bad luck, a", outcome)
  ans = input("Again? (y/n) ")
```

E.7 Iteration

Iterating over a sequence of objects, such as used in a for loop, is a common operation. To better understand how iteration works, we consider the following code.

```
s = "Hello"
for c in s:
    print(c,'*', end=' ')
```

```
H * e * l * l * o *
```

A string is an example of a Python object that can be iterated. One of the methods of a string object is __iter__. Any object that has such a method is called an **iterable**. Calling this method creates an **iterator** — an object that returns the next element in the sequence to be iterated. This is done via the method __next__.

```
s = "Hello"
t = s.__iter__()    # t is now an iterator. Same as iter(s)
print(t.__next__() ) # same as next(t)
print(t.__next__() )
print(t.__next__() )
```

```
H
e
l
```

The inbuilt functions **next** and **iter** simply call these corresponding double-underscore functions of an object. When executing a for loop, the sequence/collection over which to iterate must be an iterable. During the execution of the for loop, an iterator is created and the **next** function is executed until there is no next element. An iterator is also an iterable, so can be used in a for loop as well. Lists, tuples, and strings are so-called **sequence** objects and are iterables, where the elements are iterated by their index.

The most common iterator in Python is the **range** iterator, which allows iteration over a range of indices. Note that `range` returns a `range` object, not a list.

```
for i in range(4,20):
    print(i, end=' ')
print(range(4,20))
```

```
4 5 6 7 8 9 10 11 12 13 14 15 16 17 18 19
range(4,20)
```

Similar to Python's slice operator $[i : j]$, the iterator `range`(i, j) ranges from i to j, *not including* the index j.

Two other common iterables are sets and dictionaries. Python **sets** are, as in mathematics, unordered collections of unique objects. Sets are defined with curly brackets { }, as opposed to round brackets ( ) for tuples, and square brackets [ ] for lists. Unlike lists, sets do not have duplicate elements. Many of the usual set operations are implemented in Python, including the union `A | B` and intersection `A & B`.

```
A = {3, 2, 2, 4}
B = {4, 3, 1}
C = A & B
for i in A:
    print(i)
print(C)
```

```
2
3
4
{3, 4}
```

A useful way to construct lists is by **list comprehension**; that is, by expressions of the form

`<expression> for <element> in <list> if <condition>`

For sets a similar construction holds. In this way, lists and sets can be defined using very similar syntax as in mathematics. Compare, for example, the mathematical definition of the sets $A := \{3, 2, 4, 2\} = \{2, 3, 4\}$ (no order and no duplication of elements) and $B := \{x^2 : x \in A\}$ with the Python code below.

```
setA = {3, 2, 4, 2}
setB = {x**2 for x in setA}
print(setB)
listA = [3, 2, 4, 2]
listB = [x**2 for x in listA]
print(listB)
```

```
{16, 9, 4}
[9, 4, 16, 4]
```

A **dictionary** is a set-like data structure, containing one or more `key:value` pairs enclosed in curly brackets. The keys are often of the same type, but do not have to be; the same holds for the values. Here is a simple example, storing the ages of Lord of the Rings characters in a dictionary.

```
DICT = {'Gimly': 140, 'Frodo':51, 'Aragorn': 88}
for key in DICT:
     print(key, DICT[key])
```

```
Gimly 140
Frodo 51
Aragorn 88
```

E.8 Classes

Recall that objects are of fundamental importance in Python — indeed, data types and functions are all objects. A **class** is an object type, and writing a class definition can be thought of as creating a template for a new type of object. Each class contains a number of attributes, including a number of inbuilt methods. The basic syntax for the creation of a class is:

```
class <class_name>:
   def __init__(self):
       <statements>
   <statements>
```

The main inbuilt method is `__init__`, which creates an **instance** of a class object. For example, `str` is a class object (string class), but `s = str('Hello')` or simply `s = 'Hello'`, creates an instance, `s`, of the `str` class. Instance attributes are created during initialization and their values may be different for different instances. In contrast, the values of class attributes are the same for every instance. The variable `self` in the initialization method refers to the current instance that is being created. Here is a simple example, explaining how attributes are assigned.

```
class shire_person:
   def __init__(self,name): # initialization method
        self.name = name     # instance attribute
        self.age = 0             # instance attribute
   address = 'The Shire'         # class attribute

print(dir(shire_person)[1:5],'...',dir(shire_person)[-2:])
                                      # list of class attributes

p1 = shire_person('Sam')   # create an instance
p2 = shire_person('Frodo') # create another instance
print(p1.__dict__)    # list of instance attributes

p2.race = 'Hobbit'    # add another attribute to instance p2
p2.age = 33           # change instance attribute
```

```
print(p2.__dict__)

print(getattr(p1,'address'))   # content of p1's class attribute
```

```
['__delattr__', '__dict__', '__dir__', '__doc__'] ...
['__weakref__', 'address']
{'name': 'Sam', 'age': 0}
{'name': 'Frodo', 'age': 33, 'race': 'Hobbit'}
The Shire
```

It is good practice to create all the attributes of the class object in the `__init__` method, but, as seen in the example above, attributes can be created and assigned everywhere, even outside the class definition. More generally, attributes can be added to any object that has a `__dict__`.

An "empty" class can be created via

```
class <class_name>:
    pass
```

Python classes can be derived from a parent class by **inheritance**, via the following syntax.

```
class <class_name>(<parent_class_name>):
    <statements>
```

The derived class (initially) inherits all of the attributes of the parent class.

As an example, the class `shire_person` below inherits the attributes `name`, `age`, and `address` from its parent class `person`. This is done using the `super` function, used here to refer to the parent class `person` without naming it explicitly. When creating a new object of type `shire_person`, the `__init__` method of the parent class is invoked, and an additional instance attribute `Shire_address` is created. The `dir` function confirms that `Shire_address` is an attribute only of `shire_person` instances.

```
class person:
   def __init__(self,name):
       self.name = name
       self.age = 0
       self.address= ' '

class shire_person(person):
   def __init__(self,name):
       super().__init__(name)
       self.Shire_address = 'Bag End'

p1 = shire_person("Frodo")
p2 = person("Gandalf")
print(dir(p1)[:1],dir(p1)[-3:] )
print(dir(p2)[:1],dir(p2)[-3:] )
```

```
['Shire_address'] ['address', 'age', 'name']
['__class__'] ['address', 'age', 'name']
```

E.9 Files

To write to or read from a file, a file first needs to be opened. The **open** function in Python creates a file object that is iterable, and thus can be processed in a sequential manner in a for or while loop. Here is a simple example.

```
fout = open('output.txt','w')
for i in range(0,41):
    if i%10 == 0:
        fout.write('{:3d}\n'.format(i))
fout.close()
```

The first argument of **open** is the name of the file. The second argument specifies if the file is opened for reading ('r'), writing ('w'), appending ('a'), and so on. See help(open). Files are written in text mode by default, but it is also possible to write in binary mode. The above program creates a file output.txt with 5 lines, containing the strings 0, 10, ..., 40. Note that if we had written fout.write(i) in the fourth line of the code above, an error message would be produced, as the variable i is an integer, and not a string. Recall that the expression string.format() is Python's way to specify the format of the output string.

The formatting syntax {:3d} indicates that the output should be constrained to a specific width of three characters, each of which is a decimal value. As mentioned in the introduction, bracket syntax {i} provides a placeholder for the i-th variable to be printed, with 0 being the first index. The format for the output is further specified by {i:format}, where format is typically[4] of the form:

[width][.precision][type]
In this specification:

- width specifies the minimum width of output;
- precision specifies the number of digits to be displayed after the decimal point for a floating point values of type f, or the number of digits before *and* after the decimal point for a floating point values of type g;
- type specifies the type of output. The most common types are s for strings, d for integers, b for binary numbers, f for floating point numbers (floats) in fixed-point notation, g for floats in general notation, e for floats in scientific notation.

The following illustrates some behavior of formatting on numbers.

[4]More formatting options are possible.

```
'{:5d}'.format(123)
'{:.4e}'.format(1234567890)
'{:.2f}'.format(1234567890)
'{:.2f}'.format(2.718281828)
'{:.3f}'.format(2.718281828)
'{:.3g}'.format(2.718281828)
'{:.3e}'.format(2.718281828)
'{0:3.3f}; {2:.4e};'.format(123.456789,  0.00123456789)
```

```
'  123'
'1.2346e+09'
'1234567890.00'
'2.72'
'2.718'
'2.72'
'2.718e+00'
'123.457; 1.2346e-03;'
```

The following code reads the text file `output.txt` line by line, and prints the output on the screen. To remove the newline \n character, we have used the **strip** method for strings, which removes any whitespace from the start and end of a string.

```
fin = open('output.txt','r')
for line in fin:
    line = line.strip()   # strips a newline character
    print(line)
fin.close()
```

```
0
10
20
30
40
```

When dealing with file input and output it is important to always close files. Files that remain open, e.g., when a program finishes unexpectedly due to a programming error, can cause considerable system problems. For this reason it is recommended to open files via *context management*. The syntax is as follows:

```
with open('output.txt', 'w') as f:
      f.write('Hi there!')
```

Context management ensures that a file is correctly closed even when the program is terminated prematurely. An example is given in the next program, which outputs the most-frequent words in Dicken's *A Tale of Two Cities*, which can be downloaded from the book's GitHub site as `ataleof2cities.txt`.

Note that in the next program, the file `ataleof2cities.txt` must be placed in the current working directory. The current working directory can be determined via `import os` followed by `cwd = os.getcwd()`.

```
numline = 0
DICT = {}
with open('ataleof2cities.txt', encoding="utf8") as fin:
    for line in fin:
        words = line.split()
        for w in words:
           if w not in DICT:
               DICT[w] = 1
           else:
               DICT[w] +=1
        numline += 1

sd = sorted(DICT,key=DICT.get,reverse=True) #sort the dictionary

print("Number of unique words: {}\n".format(len(DICT)))
print("Ten most frequent words:\n")
print("{:8} {}".format("word", "count"))
print(15*'-')
for i in range(0,10):
    print("{:8} {}".format(sd[i], DICT[sd[i]]))
```

```
Number of unique words: 19091

Ten most frequent words:

word     count
---------------
the      7348
and      4679
of       3949
to       3387
a        2768
in       2390
his      1911
was      1672
that     1650
I        1444
```

E.10 NumPy

The package NumPy (module name `numpy`) provides the building blocks for scientific computing in Python. It contains all the standard mathematical functions, such as `sin`, `cos`, `tan`, etc., as well as efficient functions for random number generation, linear algebra, and statistical computation.

```
import numpy as np   #import the package
x = np.cos(1)
data = [1,2,3,4,5]
y = np.mean(data)
z = np.std(data)
print('cos(1) = {0:1.8f}  mean = {1}  std = {2}'.format(x,y,z))
```

```
cos(1) = 0.54030231  mean = 3.0  std = 1.4142135623730951
```

E.10.1 Creating and Shaping Arrays

The fundamental data type in numpy is the ndarray. This data type allows for fast matrix operations via highly optimized numerical libraries such as LAPACK and BLAS; this in contrast to (nested) lists. As such, numpy is often essential when dealing with large amounts of quantitative data.

ndarray objects can be created in various ways. The following code creates a 2×3×2 array of zeros. Think of it as a 3-dimensional matrix or two stacked 3 × 2 matrices.

```
A = np.zeros([2,3,2])  # 2 by 3 by 2 array of zeros
print(A)
print(A.shape)    # number of rows and columns
print(type(A))    # A is an ndarray
```

```
[[[ 0.   0.]
  [ 0.   0.]
  [ 0.   0.]]

 [[ 0.   0.]
  [ 0.   0.]
  [ 0.   0.]]]
(2, 3, 2)
<class 'numpy.ndarray'>
```

We will be mostly working with 2D arrays; that is, ndarrays that represent ordinary matrices. We can also use the range method and lists to create ndarrays via the array method. Note that arange is numpy's version of range, with the difference that arange returns an ndarray object.

```
a = np.array(range(4))     # equivalent to np.arange(4)
b = np.array([0,1,2,3])
C = np.array([[1,2,3],[3,2,1]])
print(a, '\n', b,'\n' , C)
```

```
[0 1 2 3]
[0 1 2 3]
 [[1 2 3]
 [3 2 1]]
```

The dimension of an ndarray can be obtained via its shape method, which returns a tuple. Arrays can be reshaped via the reshape method. This does not change the current ndarray object. To make the change permanent, a new instance needs to be created.

```
a = np.array(range(9)) #a is an ndarray of shape (9,)
print(a.shape)
A = a.reshape(3,3)   #A is an ndarray of shape (3,3)
print(a)
print(A)
```

```
[0 1 2 3 4 5 6 7 8]
(9,)
[[0, 1, 2]
 [3, 4, 5]
 [6, 7, 8]]
```

> One shape dimension for **reshape** can be specified as -1. The dimension is then inferred from the other dimension(s).

The 'T' attribute of an ndarray gives its transpose. Note that the transpose of a "vector" with shape $(n,)$ is the same vector. To distinguish between column and row vectors, reshape such a vector to an $n \times 1$ and $1 \times n$ array, respectively.

```
a = np.arange(3)   #1D array (vector) of shape (3,)
print(a)
print(a.shape)
b = a.reshape(-1,1) # 3x1 array (matrix) of shape (3,1)
print(b)
print(b.T)
A = np.arange(9).reshape(3,3)
print(A.T)
```

```
[0 1 2]
(3,)
[[0]
 [1]
 [2]]
[[0 1 2]]
[[0 3 6]
 [1 4 7]
 [2 5 8]]
```

Two useful methods of joining arrays are **hstack** and **vstack**, where the arrays are joined horizontally and vertically, respectively.

```
A = np.ones((3,3))
B = np.zeros((3,2))
C = np.hstack((A,B))
print(C)
```

```
[[ 1.  1.  1.  0.  0.]
 [ 1.  1.  1.  0.  0.]
 [ 1.  1.  1.  0.  0.]]
```

E.10.2 Slicing

Arrays can be sliced similarly to Python lists. If an array has several dimensions, a slice for each dimension needs to be specified. Recall that Python indexing starts at '0' and ends at 'len(obj)-1'. The following program illustrates various slicing operations.

```
A = np.array(range(9)).reshape(3,3)
print(A)
print(A[0])     # first row
print(A[:,1])   # second column
print(A[0,1])   # element in first row and second column
print(A[0:1,1:2])  # (1,1) ndarray containing A[0,1] = 1
print(A[1:,-1]) # elements in 2nd and 3rd rows, and last column
```

```
[[0 1 2]
 [3 4 5]
 [6 7 8]]
[0 1 2]
[1 4 7]
1
[[1]]
[5 8]
```

Note that ndarrays are mutable objects, so that elements can be modified directly, without having to create a new object.

```
A[1:,1] = [0,0] # change two elements in the matrix A above
print(A)
```

```
[[0, 1, 2]
 [3, 0, 5]
 [6, 0, 8]]
```

E.10.3 Array Operations

Basic mathematical operators and functions act *element-wise* on ndarray objects.

```
x = np.array([[2,4],[6,8]])
y = np.array([[1,1],[2,2]])
print(x+y)
```

```
[[ 3,  5]
 [ 8, 10]]
```

```
print(np.divide(x,y))   # same as x/y
```

```
[[ 2.  4.]
 [ 3.  4.]]
```

```
print(np.sqrt(x))
```

```
[[1.41421356  2.        ]
 [2.44948974  2.82842712]]
```

In order to compute matrix multiplications and compute inner products of vectors, numpy's **dot** function can be used, either as a method of an ndarray instance or as a method of np.

```
print(np.dot(x,y))
```

```
[[10, 10]
 [22, 22]]
```

```
print(x.dot(x))    # same as np.dot(x,x)
```

```
[[28, 40]
 [60, 88]]
```

Since version 3.5 of Python, it is possible to multiply two `ndarrays` using the **@ operator** (which implements the `np.matmul` method). For matrices, this is similar to using the **`dot`** method. For higher-dimensional arrays the two methods behave differently.

```
print(x @ y)
```

```
[[10 10]
 [22 22]]
```

NumPy allows arithmetic operations on arrays of different shapes (dimensions). Specifically, suppose two arrays have dimensions $(m_1, m_2, \ldots, m_p)$ and $(n_1, n_2, \ldots, n_p)$, respectively. The arrays or shapes are said to be **aligned** if for all $i = 1, \ldots, p$ it holds that

- $m_i = n_i$, or
- $\min\{m_i, n_i\} = 1$, or
- either m_i or n_i, or both are missing.

For example, shapes $(1, 2, 3)$ and $(4, 2, 1)$ are aligned, as are $(2, ,)$ and $(1, 2, 3)$. However, $(2, 2, 2)$ and $(1, 2, 3)$ are not aligned. NumPy "duplicates" the array elements across the smaller dimension to match the larger dimension. This process is called **broadcasting** and is carried out without actually making copies, thus providing efficient memory use. Below are some examples.

```
import numpy as np
A= np.arange(4).reshape(2,2) # (2,2) array

x1 = np.array([40,500])         # (2,) array
x2 = x1.reshape(2,1)        # (2,1) array

print(A + x1) # shapes (2,2) and (2,)
print(A * x2) # shapes (2,2) and (2,1)
```

```
[[ 40 501]
 [ 42 503]]
[[   0   40]
 [1000 1500]]
```

Note that above `x1` is duplicated row-wise and `x2` column-wise. Broadcasting also applies to the matrix-wise operator @, as illustrated below. Here, the matrix b is duplicated across the third dimension resulting in the two matrix multiplications

$$\begin{bmatrix} 0 & 1 \\ 2 & 3 \end{bmatrix} \begin{bmatrix} 0 & 1 \\ 2 & 3 \end{bmatrix} \quad \text{and} \quad \begin{bmatrix} 4 & 5 \\ 6 & 7 \end{bmatrix} \begin{bmatrix} 0 & 1 \\ 2 & 3 \end{bmatrix}.$$

```
B = np.arange(8).reshape(2,2,2)
b = np.arange(4).reshape(2,2)
print(B@b)
```

```
[[[ 2  3]
  [ 6 11]]

 [[10 19]
  [14 27]]]
```

Functions such as **sum**, **mean**, and **std** can also be executed as methods of an `ndarray` instance. The argument `axis` can be passed to specify along which dimension the function is applied. By default `axis=None`.

```
a = np.array(range(4)).reshape(2,2)
print(a.sum(axis=0)) #summing over rows gives column totals
```

```
[2, 4]
```

E.10.4 Random Numbers

One of the sub-modules in `numpy` is `random`. It contains many functions for random variable generation.

```
import numpy as np
np.random.seed(123)  # set the seed for the random number generator
x = np.random.random()      # uniform (0,1)
y = np.random.randint(5,9)  # discrete uniform 5,...,8
z = np.random.randn(4)      # array of four standard normals
print(x,y,'\n',z)
```

```
0.6964691855978616 7
 [ 1.77399501 -0.66475792 -0.07351368  1.81403277]
```

For more information on random variable generation in `numpy`, see

https://docs.scipy.org/doc/numpy/reference/random/index.html.

E.11 Matplotlib

The main Python graphics library for 2D and 3D plotting is `matplotlib`, and its sub-package `pyplot` contains a collection of functions that make plotting in Python similar to that in MATLAB.

E.11.1 Creating a Basic Plot

The code below illustrates various possibilities for creating plots. The style and color of lines and markers can be changed, as well as the font size of the labels. Figure E.3 shows the result.

sqrtplot.py

```
import matplotlib.pyplot as plt
import numpy as np
x = np.arange(0, 10, 0.1)
u = np.arange(0,10)
y = np.sqrt(x)
v = u/3
plt.figure(figsize = [4,2])  # size of plot in inches
plt.plot(x,y, 'g--')         # plot green dashed line
plt.plot(u,v,'r.')           # plot red dots
plt.xlabel('x')
plt.ylabel('y')
plt.tight_layout()
plt.savefig('sqrtplot.pdf',format='pdf')  # saving as pdf
plt.show()                   # both plots will now be drawn
```

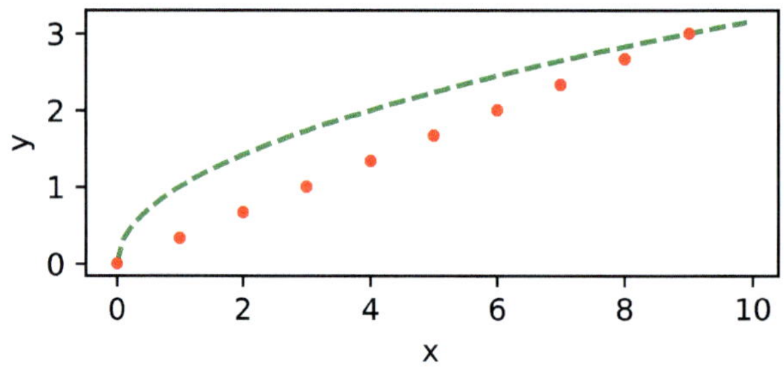

Figure E.3: A simple plot created using pyplot.

The library **matplotlib** also allows the creation of subplots. The scatterplot and histogram in Figure E.4 have been produced using the code below. When creating a histogram there are several optional arguments that affect the layout of the graph. The number of bins is determined by the parameter `bins` (the default is 10). Scatterplots also take a number of parameters, such as a string `c` which determines the color of the dots, and `alpha` which affects the transparency of the dots.

histscat.py

```
import matplotlib.pyplot as plt
import numpy as np
x = np.random.randn(1000)
u = np.random.randn(100)
v = np.random.randn(100)
plt.subplot(121)             # first subplot
plt.hist(x,bins=25, facecolor='b')
plt.xlabel('X Variable')
plt.ylabel('Counts')
plt.subplot(122)             # second subplot
plt.scatter(u,v,c='b', alpha=0.5)
plt.show()
```

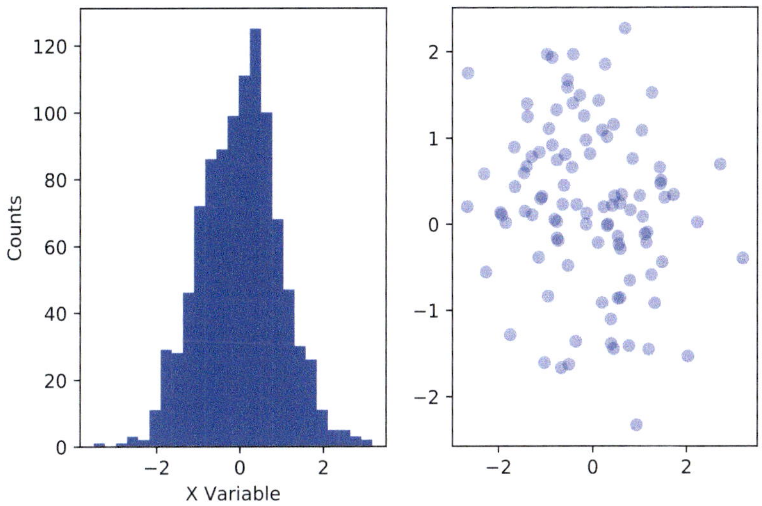

Figure E.4: A histogram and scatterplot.

One can also create three-dimensional plots as illustrated below.

surf3dscat.py

```
import matplotlib.pyplot as plt
import numpy as np
from mpl_toolkits.mplot3d import Axes3D

def npdf(x,y):
    return np.exp(-0.5*(pow(x,2)+pow(y,2)))/np.sqrt(2*np.pi)

x, y  = np.random.randn(100), np.random.randn(100)
z = npdf(x,y)

xgrid, ygrid = np.linspace(-3,3,100), np.linspace(-3,3,100)

Xarray, Yarray = np.meshgrid(xgrid,ygrid)
Zarray = npdf(Xarray,Yarray)

fig = plt.figure(figsize=plt.figaspect(0.4))
ax1 = fig.add_subplot(121, projection='3d')
ax1.scatter(x,y,z, c='g')
ax1.set_xlabel('$x$')
ax1.set_ylabel('$y$')
ax1.set_zlabel('$f(x,y)$')
ax2 = fig.add_subplot(122, projection='3d')
ax2.plot_surface(Xarray,Yarray,Zarray,cmap='viridis',edgecolor='none')
ax2.set_xlabel('$x$')
ax2.set_ylabel('$y$')
ax2.set_zlabel('$f(x,y)$')

plt.show()
```

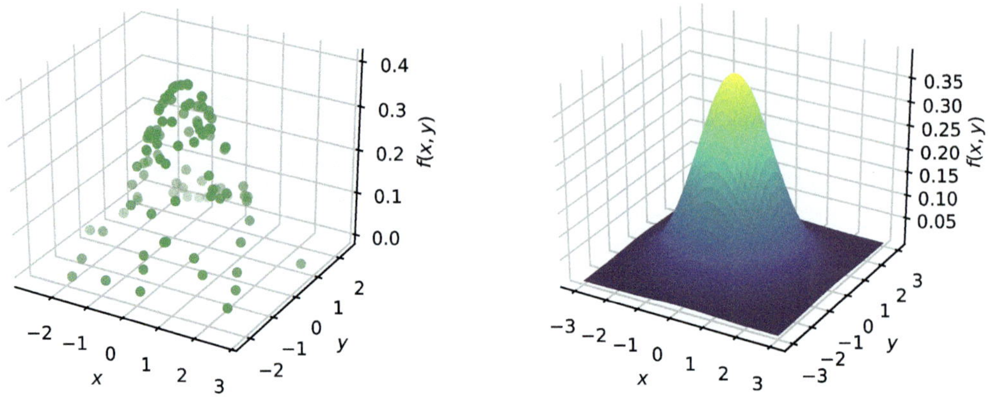

Figure E.5: Three-dimensional scatter- and surface plots.

E.12 Pandas

The Python package Pandas (module name `pandas`) provides various tools and data structures for data analytics, including the fundamental `DataFrame` class.

> For the code in this section we assume that `pandas` has been imported via
> `import pandas as pd`.

E.12.1 Series and DataFrame

The two main data structures in `pandas` are `Series` and `DataFrame`. A `Series` object can be thought of as a combination of a dictionary and a one-dimensional `ndarray`. The syntax for creating a `Series` object is

```
series = pd.Series(<data>, index=['index'])
```

Here, <data> is some one-dimensional data structure, such as a one-dimensional `ndarray`, a list, or a dictionary, and `index` is a list of names of the same length as <data>. When <data> is a dictionary, the `index` is created from the keys of the dictionary. When <data> is an `ndarray` and `index` is omitted, the default index will be `[0, ..., len(data)-1]`.

```
DICT = {'one':1, 'two':2, 'three':3, 'four':4}
print(pd.Series(DICT))
```

```
one      1
two      2
three    3
four     4
dtype: int64
```

```
years =  ['2000','2001','2002']
cost = [2.34, 2.89, 3.01]
print(pd.Series(cost,index = years, name = 'MySeries')) #name it
```

```
2000    2.34
2001    2.89
2002    3.01
Name: MySeries, dtype: float64
```

The most commonly-used data structure in pandas is the `DataFrame`, which can be thought of as pandas' implementation of a spreadsheet or as a dictionary in which each "key" of the dictionary corresponds to a column name and the dictionary "value" is the data in that column. To create a `DataFrame` one can use the pandas `DataFrame` method, which has three main arguments: data, index (row labels), and columns (column labels).

```
DataFrame(<data>, index=['<row_name>'], columns=['<column_name>'])
```

If the index is not specified, the default index is `[0, ..., len(data)-1]`. Data can also be read directly from a CSV or Excel file, as is done in Section 1.1. If a dictionary is used to create the data frame (as below), the dictionary keys are used as the column names. ☞ 1

```
DICT = {'numbers':[1,2,3,4], 'squared':[1,4,9,16] }
df = pd.DataFrame(DICT, index = list('abcd'))
print(df)
```

```
   numbers  squared
a        1        1
b        2        4
c        3        9
d        4       16
```

E.12.2 Manipulating Data Frames

Often data encoded in `DataFrame` or `Series` objects need to be extracted, altered, or combined. Getting, setting, and deleting columns works in a similar manner as for dictionaries. The following code illustrates various operations.

```
ages = [6,3,5,6,5,8,0,3]
d={'Gender':['M', 'F']*4, 'Age': ages}
df1 = pd.DataFrame(d)
df1.at[0,'Age']= 60  # change an element
df1.at[1,'Gender'] = 'Female' # change another element
df2 = df1.drop('Age', axis=1)  # drop a column
df3 = df2.copy();    # create a separate copy of df2
df3['Age'] = ages   # add the original column
dfcomb = pd.concat([df1,df2,df3],axis=1)  # combine the three dfs
print(dfcomb)
```

```
  Gender  Age  Gender  Gender  Age
0      M   60       M       M    6
1 Female    3  Female  Female    3
2      M    5       M       M    5
3      F    6       F       F    6
4      M    5       M       M    5
5      F    8       F       F    8
6      M    0       M       M    0
7      F    3       F       F    3
```

Note that the above `DataFrame` object has two `Age` columns. The expression `dfcomb['Age']` will return a DataFrame with both these columns.

Table E.6: Useful **pandas** methods for data manipulation.

`agg`	Aggregate the data using one or more functions.
`apply`	Apply a function to a column or row.
`astype`	Change the data type of a variable.
`concat`	Concatenate data objects.
`replace`	Find and replace values.
`read_csv`	Read a CSV file into a DataFrame.
`sort_values`	Sort by values along rows or columns.
`stack`	Stack a DataFrame.
`to_excel`	Write a DataFrame to an Excel file.

It is important to correctly specify the data type of a variable before embarking on data summarization and visualization tasks, as Python may treat different types of objects in dissimilar ways. Common data types for entries in a `DataFrame` object are `float`, `category`, `datetime`, `bool`, and `int`. A generic object type is `object`.

```
d={'Gender':['M', 'F', 'F']*4, 'Age': [6,3,5,6,5,8,0,3,6,6,7,7]}
df=pd.DataFrame(d)
print(df.dtypes)
df['Gender'] = df['Gender'].astype('category')  #change the type
print(df.dtypes)
```

```
Gender    object
Age        int64
dtype: object
Gender    category
Age          int64
dtype: object
```

E.12.3 Extracting Information

Extracting statistical information from a `DataFrame` object is facilitated by a large collection of methods (functions) in **pandas**. Table E.7 gives a selection of data inspection methods. See Chapter 1 for their practical use. The code below provides several examples of useful methods. The **apply** method allows one to apply general functions to

☞ 1

columns or rows of a DataFrame. These operations do not change the data. The **loc** method allows for accessing elements (or ranges) in a data frame and acts similar to the slicing operation for lists and arrays, *with the difference that the "stop" value is included*, as illustrated in the code below.

```
import numpy as np
import pandas as pd
ages = [6,3,5,6,5,8,0,3]
np.random.seed(123)
df = pd.DataFrame(np.random.randn(3,4), index = list('abc'),
                  columns = list('ABCD'))
print(df)
df1 = df.loc["b":"c","B":"C"]      # create a partial data frame
print(df1)
meanA = df['A'].mean()            # mean of 'A' column
print('mean of column A = {}'.format(meanA))
expA = df['A'].apply(np.exp)  # exp of all elements in 'A' column
print(expA)
```

```
          A         B         C         D
a -1.085631  0.997345  0.282978 -1.506295
b -0.578600  1.651437 -2.426679 -0.428913
c  1.265936 -0.866740 -0.678886 -0.094709
          B         C
b  1.651437 -2.426679
c -0.866740 -0.678886
mean of column A = -0.13276486552118785
a    0.337689
b    0.560683
c    3.546412
Name: A, dtype: float64
```

The **groupby** method of a DataFrame object is useful for summarizing and displaying the data in manipulated ways. It groups data according to one or more specified columns, such that methods such as **count** and **mean** can be applied to the grouped data.

```
df = pd.DataFrame({'W':['a','a','b','a','a','b'],
         'X':np.random.rand(6),
         'Y':['c','d','d','d','c','c'], 'Z':np.random.rand(6)})
print(df)
```

```
   W         X  Y         Z
0  a  0.993329  c  0.641084
1  a  0.925746  d  0.428412
2  b  0.266772  d  0.460665
3  a  0.201974  d  0.261879
4  a  0.529505  c  0.503112
5  b  0.006231  c  0.849683
```

```
print(df.groupby('W').mean(numeric_only=True))
```

```
          X         Z
```

Table E.7: Useful pandas methods for data inspection.

columns	Column names.
count	Counts number of non-NA cells.
crosstab	Cross-tabulate two or more categories.
describe	Summary statistics.
dtypes	Data types for each column.
head	Display the top rows of a DataFrame.
groupby	Group data by column(s).
info	Display information about the DataFrame.
loc	Access a group or rows or columns.
mean	Column/row mean.
plot	Plot of columns.
std	Column/row standard deviation.
sum	Returns column/row sum.
tail	Display the bottom rows of a DataFrame.
value_counts	Counts of different non-null values.
var	Variance.

```
W
a   0.662639   0.458622
b   0.136502   0.655174
```

```
print(df.groupby(['W', 'Y']).mean(numeric_only=True))
```

```
            X          Z
W Y
a c   0.761417   0.572098
  d   0.563860   0.345145
b c   0.006231   0.849683
  d   0.266772   0.460665
```

To allow for multiple functions to be calculated at once, the agg method can be used. It can take a list, dictionary, or string of functions.

```
numeric_cols = df.select_dtypes(include=np.number).columns
print(df.groupby('W')[numeric_cols].agg ([sum ,np.mean]))
```

```
           X                    Z
         sum      mean        sum      mean
W
a   2.650555  0.662639   1.834487  0.458622
b   0.273003  0.136502   1.310348  0.655174
```

E.12.4 Plotting

The plot method of a DataFrame makes plots of a DataFrame using Matplotlib. Different types of plot can be accessed via the kind = 'str' construction, where str is one of line (default), bar, hist, box, kde, and several more. Finer control, such as modifying the font, is obtained by using matplotlib directly. The following code produces the line and box plots in Figure E.8.

```
import numpy as np
import pandas as pd
import matplotlib
df = pd.DataFrame({'normal':np.random.randn(100),
         'Uniform':np.random.uniform(0,1,100)})
font = {'family' : 'serif', 'size'   : 14} #set font
matplotlib.rc('font', **font)   # change font
df.plot()   # line plot (default)
df.plot(kind = 'box')  # box plot
matplotlib.pyplot.show()  #render plots
```

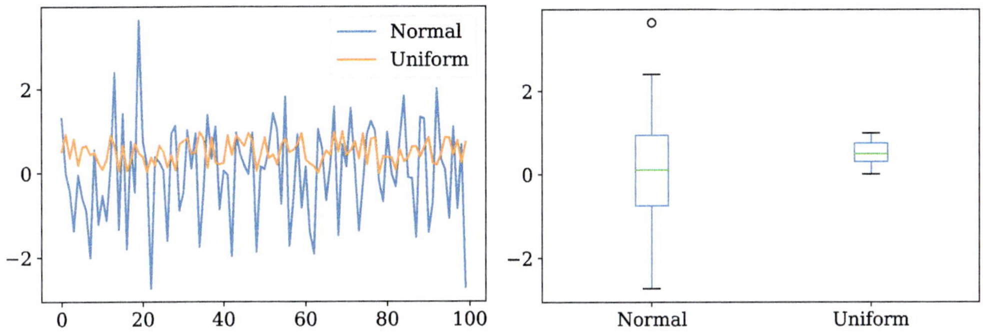

Figure E.8: A line and box plot using the plot method of DataFrame.

E.13 Scikit-learn

Scikit-learn is an open-source machine learning and data science library for Python. The library includes a range of algorithms relating to the chapters in this book. It is widely used due to its simplicity and its breadth. The module name is sklearn. Below is a brief introduction into modeling the data with sklearn. The full documentation can be found at

https://scikit-learn.org/.

E.13.1 Partitioning the Data

Randomly partitioning the data in order to test the model may be achieved easily with sklearn's function train_test_split. For example, suppose that the training data is described by the matrix X of explanatory variables and the vector y of responses. Then the following code splits the data set into training and testing sets, with the testing set being half of the total set.

```
from sklearn.model_selection import train_test_split
X_train, X_test, y_train, y_test = train_test_split(X, y,
                                                    test_size = 0.5)
```

As an example, the following code generates a synthetic data set and splits it into equally-sized training and test sets.

syndat.py

```
import numpy as np
import matplotlib.pyplot as plt
from sklearn.model_selection import train_test_split

np.random.seed(1234)

X=np.pi*(2*np.random.random(size=(400,2))-1)
y=(np.cos(X[:,0])*np.sin(X[:,1])>=0)

X_train , X_test , y_train , y_test = train_test_split(X, y, test_size
    =0.5)


fig = plt.figure()
ax = fig.add_subplot(111)
ax.scatter(X_train[y_train==0,0],X_train[y_train==0,1], c='g',
            marker='o',alpha=0.5)
ax.scatter(X_train[y_train==1,0],X_train[y_train==1,1], c='b',
            marker='o',alpha=0.5)
ax.scatter(X_test[y_test==0,0],X_test[y_test==0,1], c='g',
            marker='s',alpha=0.5)
ax.scatter(X_test[y_test==1,0],X_test[y_test==1,1], c='b',
            marker='s',alpha=0.5)

plt.savefig('sklearntraintest.pdf',format='pdf')
plt.show()
```

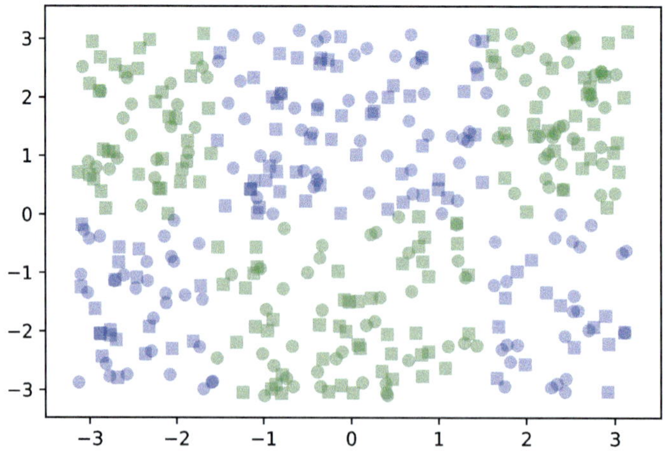

Figure E.9: Example training (circles) and test (squares) set for two class classification. Explanatory variables are the (x, y) coordinates, classes are zero (green) or one (blue).

E.13.2 Standardization

In some instances it may be necessary to standardize the data. This may be done in sklearn with scaling methods such as MinMaxScaler or StandardScaler. Scaling may improve the convergence of gradient-based estimators and is useful when visualizing data on vastly different scales. For example, suppose that X is our explanatory data (e.g., stored as a numpy array), and we wish to standardize such that each value lies between 0 and 1.

```
from sklearn import preprocessing
min_max_scaler = preprocessing.MinMaxScaler(feature_range=(0, 1))
x_scaled = min_max_scaler.fit_transform(X)
# equivalent to:
x_scaled = (X - X.min(axis=0)) / (X.max(axis=0) - X.min(axis=0))
```

E.13.3 Fitting and Prediction

Once the data has been partitioned and standardized if necessary, the data may be fitted to a statistical model, e.g., a classification or regression model. For example, continuing with our data from above, the following fits a model to the data and predicts the responses for the test set.

```
from sklearn.someSubpackage import someClassifier
clf = someClassifier()         # choose appropriate classifier
clf.fit(X_train, y_train)      # fit the data
y_prediction = clf.predict(X_test)  # predict
```

Specific classifiers for logistic regression, naïve Bayes, linear and quadratic discriminant analysis, K-nearest neighbors, and support vector machines are given in Section 8.8.

☞ 380

E.13.4 Testing the Model

Once the model has made its prediction we may test its effectiveness, using relevant metrics. For example, for classification we may wish to produce the confusion matrix for the test data. The following code does this for the data shown in Figure E.9, using a support vector machine classifier.

```
from sklearn import svm
clf = svm.SVC(kernel = 'rbf')
clf.fit(X_train , y_train)
y_prediction = clf.predict(X_test)

from sklearn.metrics import confusion_matrix
print(confusion_matrix(y_test , y_prediction))
```

```
[[102  12]
 [  1  85]]
```

E.14 System Calls, URL Access, and Speed-Up

Operating system commands (whether in Windows, MacOS, or Linux) for creating directories, copying or removing files, or executing programs from the system shell can be issued from within Python by using the package os. Another useful package is requests which enables direct downloads of files and webpages from URLs. The following Python script uses both. It also illustrates a simple example of exception handling in Python.

misc.py

```
import os
import requests
for c in "123456":
  try:                        # if it does not yet exist
     os.mkdir("MyDir"+ c)  # make a directory
  except:                     # otherwise
      pass                    # do nothing

uname = "https://github.com/DSML-book/Programs2e/tree/master/Appendices/
    Python Primer/"
fname = "ataleof2cities.txt"
r = requests.get(uname + fname)
print(r.text)
open('MyDir1/ato2c.txt', 'wb').write(r.content) #write to a file
                                               # bytes mode is important here
```

The package numba can significantly speed up calculations via smart compilation. First run the following code.

jitex.py

```
import timeit
import numpy as np
from numba import jit

n = 10**8

#@jit(nopython=True)
def myfun(s, n):
    for i in range(1, n):
        s = s + 1/i
    return s

start = timeit.default_timer()
print("Euler's constant is approximately {:9.8f}".format(
    myfun(0, n) - np.log(n)))
end = timeit.default_timer()
print("elapsed time: {:3.2f} seconds".format(end - start))
```

```
Euler's constant is approximately 0.57721566
elapsed time: 5.71 seconds
```

Now remove the # character before the @ character in the code above, in order to activate the "just in time" compiler. This gives a 38-fold speedup:

```
Euler's constant is approximately 0.57721566
elapsed time: 0.15 seconds
```

Further Reading

To learn Python, we recommend [116] and [154]. However, as Python is constantly evolving, the most up-to-date references will be available from the Internet.

BIBLIOGRAPHY

[1] S. C. Ahalt, A. K. Krishnamurthy, P. Chen, and D. E. Melton. Competitive learning algorithms for vector quantization. *Neural Networks*, 3:277–290, 1990.

[2] J. Aitchison and C. G. G. Aitken. Multivariate binary discrimination by the kernel method. *Biometrika*, 63(3):413–420, 1976.

[3] H. Akaike. A new look at the statistical model identification. *IEEE Transactions on Automatic Control*, 19(6):716–723, 1974.

[4] N. Aronszajn. Theory of reproducing kernels. *Transactions of the American Mathematical Society*, 68:337–404, 1950.

[5] D. Arthur and S. Vassilvitskii. K-means++: The advantages of careful seeding. In *Proceedings of the Eighteenth Annual ACM-SIAM Symposium on Discrete Algorithms*, pages 1027–1035, Philadelphia, 2007. Society for Industrial and Applied Mathematics.

[6] K. Arulkumaran, M. P. Deisenroth, M. Brundage, and A. A. Bharath. Deep reinforcement learning: A brief survey. *IEEE Signal Processing Magazine*, 34(6):26–38, 2017.

[7] S. Asmussen and P. W. Glynn. *Stochastic Simulation: Algorithms and Analysis*. Springer, New York, 2007.

[8] R. G. Bartle. *The Elements of Integration and Lebesgue Measure*. John Wiley & Sons, Hoboken, 1995.

[9] A. G. Barto and S. Mahadevan. Recent advances in hierarchical reinforcement learning. *Discrete Event Dynamic Systems*, 13:341–379, 2003.

[10] D. Bates and D. Watts. *Nonlinear Regression Analysis and Its Applications*. John Wiley & Sons, Hoboken, 1988.

[11] R. Bellman. *Dynamic Programming*. Princeton University Press, Princeton, NJ, 1957.

[12] J. O. Berger. *Statistical Decision Theory and Bayesian Analysis*. Springer, New York, second edition, 1985.

[13] D. P. Bertsekas and J. N. Tsitsiklis. *Neuro-Dynamic Programming*. Optimization and Neural Computation Series. Athena Scientific, 1996.

[14] J. Bezdek. *Pattern Recognition with Fuzzy Objective Function Algorithms*. Plenum Press, New York, 1981.

[15] P. J. Bickel and K. A. Doksum. *Mathematical Statistics*, volume I. Pearson Prentice Hall, Upper Saddle River, second edition, 2007.

[16] P. Billingsley. *Probability and Measure*. John Wiley & Sons, New York, third edition, 1995.

[17] C. M. Bishop. *Pattern Recognition and Machine Learning*. Springer, New York, 2006.

[18] P. T. Boggs and R. H. Byrd. Adaptive, limited-memory BFGS algorithms for unconstrained optimization. *SIAM Journal on Optimization*, 29(2):1282–1299, 2019.

[19] Z. I. Botev. Nonparametric density estimation via diffusion mixing. Technical report, Department of Mathematics, University of Queensland, 2007.

[20] Z. I. Botev, Y.-L. Chen, P. L'Ecuyer, S. MacNamara, and D. P. Kroese. Exact posterior simulation from the linear Lasso regression. In M. Rabe, A. A. Juan, N. Mustafee, A. Skoogh, S. Jain, and B. Johansson, editors, *2018 Winter Simulation Conference (WSC)*, pages 1706–1717, Gothenburg, Sweden, December 2018.

[21] Z. I. Botev, J. F. Grotowski, and D. P. Kroese. Kernel density estimation via diffusion. *Annals of Statistics*, 38(5):2916–2957, 2010.

[22] Z. I. Botev and D. P. Kroese. Global likelihood optimization via the cross-entropy method, with an application to mixture models. In R. G. Ingalls, M. D. Rossetti, J. S. Smith, and B. A. Peters, editors, *Proceedings of the 2004 Winter Simulation Conference*, pages 529–535, Washington, DC, December 2004.

[23] Z. I. Botev and D. P. Kroese. Non-asymptotic bandwidth selection for density estimation of discrete data. *Methodology and Computing in Applied Probability*, 10(3):435–451, 2008.

[24] Z. I. Botev, D. P. Kroese, R. Y. Rubinstein, and P. L'Ecuyer. The cross-entropy method for optimization. In V. Govindaraju and C.R. Rao, editors, *Machine Learning: Theory and Applications*, volume 31 of *Handbook of Statistics*, pages 35–59. Elsevier, 2013.

[25] Z. I. Botev and P. L'Ecuyer. Simulation from the normal distribution truncated to an interval in the tail. In A. Puliafito and K. S. Trivedi, editors, *Proceedings of the 10th EAI International Conference on Performance Evaluation Methodologies and Tools (VALUETOOLS 2016)*, pages 23–29. ACM, 2017.

[26] Z. I. Botev and P. L'Ecuyer. Simulation from the tail of the univariate and multivariate normal distribution. In A. Puliafito and K. S. Trivedi, editors, *Systems Modeling: Methodologies and Tools*, pages 115–132. Springer, 2019.

[27] Z. I. Botev and C. J. Lloyd. Importance accelerated Robbins–Monro recursion with applications to parametric confidence limits. *Electronic Journal of Statistics*, 9:2058–2075, 2015.

[28] Z. I. Botev and A. Ridder. Variance reduction. *Wiley StatsRef: Statistics Reference Online*, pages 1–6, 2017.

[29] S. Boyd, N. Parikh, E. Chu, B. Peleato, and J. Eckstein. Distributed optimization and statistical learning via the alternating direction method of multipliers. *Foundations and Trends in Machine Learning*, 3:1–122, 2010.

[30] S. Boyd and L. Vandenberghe. *Convex Optimization*. Cambridge University Press, Cambridge, 2004. Seventh printing with corrections, 2009.

[31] R. A. Boyles. On the convergence of the EM algorithm. *Journal of the Royal Statistical Society, Series B*, 45(1):47–50, 1983.

[32] L. Breiman. *Classification and Regression Trees*. CRC Press, Boca Raton, 1987.

[33] L. Breiman. Bagging predictors. *Machine Learning*, 24(2):123–140, 1996.

[34] L. Breiman. Heuristics of instability and stabilization in model selection. *Annals of Statistics*, 24(6):2350–2383, 12 1996.

[35] L. Breiman. Random forests. *Machine Learning*, 45(1):5–32, 2001.

[36] E. J. Candes, M. B. Wakin, and S. P. Boyd. Enhancing sparsity by reweighted l_1 minimization. *Journal of Fourier Analysis and Applications*, 14:877–905, 2008.

[37] L. Canese, G. C. Cardarilli, L. Di Nunzio, R. Fazzolari, D. Giardino, M. Re, and S. Spanó. Multi-agent reinforcement learning: A review of challenges and applications. *Applied Sciences*, 11(11), 2021.

[38] F. Cao, D.-Z. Du, B. Gao, P.-J. Wan, and P. M. Pardalos. Minimax problems in combinatorial optimization. In D.-Z. Du and P. M. Pardalos, editors, *Minimax and Applications*, pages 269–292. Kluwer, Dordrecht, 1995.

[39] G. Casella and R. L. Berger. *Statistical Inference*. Chapman and Hall/CRC, second edition, 2024.

[40] E. Çinlar. *Introduction to Stochastic Processes*. Prentice Hall, Englewood Cliffs, 1975.

[41] J. C. C. Chan and D. P. Kroese. *Statistical Modeling and Computation*. Springer, New York, second edition, 2025.

[42] N. Chopin. Fast simulation of truncated Gaussian distributions. *Statistics and Computing*, 21:275–288, 2011.

[43] K. L. Chung. *A Course in Probability Theory*. Academic Press, New York, second edition, 1974.

[44] M. J. Colbrook, Z. I. Botev, K. Kuritz, and S. MacNamara. Kernel density estimation with linked boundary conditions. *Studies in Applied Mathematics*, 145(3):357–396, 2020.

[45] T. M. Cover and J. A. Thomas. *Elements of Information Theory*. John Wiley & Sons, New York, 1991.

[46] L. Dai, K. Chen, Z. Sun, Z. Liu, and G. Li. Broken adaptive ridge regression and its asymptotic properties. *Journal of multivariate analysis*, 168:334–351, 2018.

[47] J. W. Daniel, W. B. Gragg, L. Kaufman, and G. W. Stewart. Reorthogonalization and stable algorithms for updating the Gram-Schmidt QR factorization. *Mathematics of Computation*, 30(136):772–795, 1976.

[48] P.-T. de Boer, D. P. Kroese, S. Mannor, and R. Y. Rubinstein. A tutorial on the cross-entropy method. *Annals of Operations Research*, 134(1):19–67, 2005.

[49] A. P. Dempster, N. M. Laird, and D. B. Rubin. Maximum likelihood from incomplete data via the EM algorithm. *Journal of the Royal Statistical Society: Series B (Methodological)*, 39(1):1 – 38, 1977.

[50] L. Devroye. *Non-Uniform Random Variate Generation*. Springer, New York, 1986.

[51] N. R. Draper and H. Smith. *Applied Regression Analysis*. John Wiley & Sons, New York, third edition, 1998.

[52] Q. Duan and D. P. Kroese. Splitting for optimization. *Computers & Operations Research*, 73:119–131, 2016.

[53] R. O. Duda, P. E. Hart, and D. G. Stork. *Pattern Classification*. John Wiley & Sons, New York, third edition, 2024.

[54] B. Efron and T. J. Hastie. *Computer Age Statistical Inference: Algorithms, Evidence, and Data Science*. Cambridge University Press, Cambridge, 2016.

[55] B. Efron and R. Tibshirani. *An Introduction to the Bootstrap*. Chapman & Hall, New York, 1994.

[56] J. Fan and I. Gijbels. *Local polynomial modelling and its applications: monographs on statistics and applied probability 66*. Routledge, 1996.

[57] J. Fan, R. Li, C.-H. Zhang, and H. Zou. *Statistical foundations of data science*. Chapman and Hall/CRC, 2020.

[58] T. Fawcett. An introduction to ROC analysis. *Pattern Recognition Letters*, 27(8):861–874, June 2006.

[59] W. Feller. *An Introduction to Probability Theory and Its Applications*, volume I. John Wiley & Sons, Hoboken, second edition, 1970.

[60] L. Feng and C.-H. Zhang. Sorted concave penalized regression. *Annals of Statistics*, 47(6):3069–3098, 2019.

[61] J. C. Ferreira and V. A. Menegatto. Eigenvalues of integral operators defined by smooth positive definite kernels. *Integral Equations and Operator Theory*, 64:61–81, 2009.

[62] R. Figueiredo Prudencio, M. R. O. A. Maximo, and E. L. Colombini. A survey on offline reinforcement learning: Taxonomy, review, and open problems. *IEEE Transactions on Neural Networks and Learning Systems*, 35(8):10237–10257, 2024.

[63] N. I. Fisher and P. K. Sen, editors. *The Collected Works of Wassily Hoeffding*. Springer, New York, 1994.

[64] G. S. Fishman. *Monte Carlo: Concepts, Algorithms and Applications*. Springer, New York, 1996.

[65] R. Fletcher. *Practical Methods of Optimization*. John Wiley & Sons, New York, 1987.

[66] Y. Freund and R. E. Schapire. A decision-theoretic generalization of on-line learning and an application to boosting. *J. Comput. Syst. Sci.*, 55(1):119–139, 1997.

[67] J. H. Friedman. Greedy function approximation: A gradient boosting machine. *Annals of Statistics*, 29(5):1189–1232, 2001.

[68] W. Fu and K. Knight. Asymptotics for lasso-type estimators. *The Annals of Statistics*, 28(5):1356 – 1378, 2000.

[69] A. Gelman. *Bayesian Data Analysis*. Chapman & Hall, New York, second edition, 2004.

[70] A. Gelman and J. Hall. *Data Analysis Using Regression and Multilevel/Hierarchical Models*. Cambridge University Press, Cambridge, 2006.

[71] S. Geman and D. Geman. Stochastic relaxation, Gibbs distribution and the Bayesian restoration of images. *IEEE Transactions on Pattern Analysis and Machine Intelligence*, 6(6):721–741, 1984.

[72] J. E. Gentle. *Random Number Generation and Monte Carlo Methods*. Springer, New York, second edition, 2003.

[73] W. R. Gilks, S. Richardson, and D. J. Spiegelhalter. *Markov Chain Monte Carlo in Practice*. Chapman & Hall, New York, 1996.

[74] P. Glasserman. *Monte Carlo Methods in Financial Engineering*. Springer, New York, 2003.

[75] P. Glasserman. Combined derivative estimators. In Z. I. Botev, A. Keller, C. Lemieux, and B. Tuffin, editors, *Advances in Modeling and Simulation: Festschrift for Pierre L'Ecuyer*, pages 193–210. Springer, 2022.

[76] G. H. Golub and C. F. Van Loan. *Matrix Computations*. Johns Hopkins University Press, Baltimore, fourth edition, 2013.

[77] I. Goodfellow, Y. Bengio, and A. Courville. *Deep Learning*. MIT Press, Cambridge, 2016.

[78] G. R. Grimmett and D. R. Stirzaker. *Probability and Random Processes*. Oxford University Press, third edition, 2001.

[79] Gurobi Optimization, LLC. Gurobi Optimizer Reference Manual, 2023.

[80] T. Haarnoja, A. Zhou, K. Hartikainen, G. Tucker, S. Ha, J. Tan, V. Kumar, H. Zhu, A. Gupta, P. Abbeel, and S. Levine. Soft actor-critic algorithms and applications. *CoRR*, abs/1812.05905, 2018.

[81] T. J. Hastie, R. J. Tibshirani, and J. H. Friedman. *The Elements of Statistical Learning: Data mining, Inference, and Prediction*. Springer, New York, 2009.

[82] T. J. Hastie, R. J. Tibshirani, and M. Wainwright. *Statistical Learning with Sparsity: The Lasso and Generalizations*. CRC Press, Boca Raton, 2015.

[83] J.-B. Hiriart-Urruty and C. Lemarèchal. *Fundamentals of Convex Analysis*. Springer, New York, 2001.

[84] W. Hock and K. Schittkowski. *Test Examples for Nonlinear Programming Codes*. Springer, New York, 1981.

[85] M. Hoerger, H. Kurniawati, D. P. Kroese, and N. Ye. A surprisingly simple continuous-action POMDP solver: Lazy cross-entropy search over policy trees. In *The Thirty-Eighth Conference on Artificial Intelligence (AAAI-24)*, volume 38, pages 20134–20142, Washington, DC, 2024. Association for the Advancement of Artificial Intelligence.

[86] J. E. Kelley, Jr. The cutting-plane method for solving convex programs. *Journal of the Society for Industrial and Applied Mathematics*, 8(4):703–712, 1960.

[87] A. K. Jain. *Fundamentals of Digital Image Processing*. Prentice Hall, Englewood Cliffs, 1989.

[88] O. Kallenberg. *Foundations of Modern Probability*. Springer, New York, second edition, 2002.

[89] A. Karalic. Linear regression in regression tree leaves. In *Proceedings of ECAI-92*, pages 440–441, Hoboken, 1992. John Wiley & Sons.

[90] C. Kaynak. Methods of combining multiple classifiers and their applications to handwritten digit recognition. Master's thesis, Institute of Graduate Studies in Science and Engineering, Bogazici University, Istanbul, Türkiye, 1995.

[91] T. Keilath and A. H. Sayed, editors. *Fast Reliable Algorithms for Matrices with Structure*. SIAM, Pennsylvania, 1999.

[92] C. Nussbaumer Knaflic. *Storytelling with Data: A Data Visualization Guide for Business Professionals*. John Wiley & Sons, Hoboken, 2015.

[93] D. Koller and N. Friedman. *Probabilistic Graphical Models: Principles and Techniques - Adaptive Computation and Machine Learning*. The MIT Press, Cambridge, 2009.

[94] A. N. Kolmogorov and S. V. Fomin. *Elements of the Theory of Functions and Functional Analysis*. Dover Publications, Mineola, 1999.

[95] D. P. Kroese and Z. I. Botev. *An Advanced Course in Probability and Stochastic Processes*. Chapman & Hall/CRC Press, Boca Raton, 2023.

[96] D. P. Kroese, T. Brereton, T. Taimre, and Z. I. Botev. Why the Monte Carlo method is so important today. *Wiley Interdisciplinary Reviews: Computational Statistics*, 6(6):386–392, 2014.

[97] D. P. Kroese, S. Porotsky, and R. Y. Rubinstein. The cross-entropy method for continuous multi-extremal optimization. *Methodology and Computing in Applied Probability*, 8(3):383–407, 2006.

[98] D. P. Kroese, T. Taimre, and Z. I. Botev. *Handbook of Monte Carlo Methods*. John Wiley & Sons, New York, 2011.

[99] H. J. Kushner and G. G. Yin. *Stochastic Approximation and Recursive Algorithms and Applications*. Springer, New York, second edition, 2003.

[100] P. Lafaye de Micheaux, R. Drouilhet, and B. Liquet. *The R Software: Fundamentals of Programming and Statistical Analysis*. Springer, New York, 2014.

[101] R. J. Larsen and M. L. Marx. *An Introduction to Mathematical Statistics and Its Applications*. Prentice Hall, New York, third edition, 2001.

[102] P. J. Laub, R. Salomone, and Z. I. Botev. Monte Carlo estimation of the density of the sum of dependent random variables. *Mathematics and Computers in Simulation*, 161:23–31, 2019.

[103] A. M. Law and W. D. Kelton. *Simulation Modeling and Analysis*. McGraw-Hill, New York, third edition, 2000.

[104] C. L. Lawson and R. J. Hanson. *Solving least squares problems*. SIAM, 1995.

[105] P. L'Ecuyer. A unified view of IPA, SF, and LR gradient estimation techniques. *Management Science*, 36(11):1364–1383, 1990.

[106] P. L'Ecuyer. Good parameters and implementations for combined multiple recursive random number generators. *Operations Research*, 47(1):159 – 164, 1999.

[107] P. L'Ecuyer, F. Puchhammer, and A. B. Abdellah. Monte Carlo and Quasi–Monte Carlo density estimation via conditioning. *INFORMS Journal on Computing*, 34(3):1729–1748, 2022.

[108] E. L. Lehmann and G. Casella. *Theory of Point Estimation.* Springer, New York, second edition, 1998.

[109] C. Lemieux. Control variates. *Wiley StatsRef: Statistics Reference Online*, pages 1–8, 2017.

[110] T. G. Lewis and W. H. Payne. Generalized feedback shift register pseudorandom number algorithm. *Journal of the ACM*, 20(3):456–468, 1973.

[111] B. Liquet, S. Moka, and Y. Nazarathy. *Mathematical engineering of deep learning*. CRC Press, 2024.

[112] R. J. A. Little and D. B. Rubin. *Statistical Analysis with Missing Data.* John Wiley & Sons, Hoboken, second edition, 2002.

[113] D. C. Liu and J. Nocedal. On the limited memory BFGS method for large scale optimization. *Mathematical Programming*, 45(1-3):503–528, 1989.

[114] Z-Q. Luo and P. Tseng. On the convergence of the coordinate descent method for convex differentiable minimization. *Journal of Optimization Theory and Applications*, 72(1):7–35, 1992.

[115] Z-Q. Luo and P. Tseng. Error bounds and convergence analysis of feasible descent methods: a general approach. *Annals of Operations Research*, 46(1):157–178, 1993.

[116] M. Lutz. *Learning Python.* O'Reilly, fifth edition, 2013.

[117] O.-A. Maillard, D. Ryabko, and R. Munos. Selecting the state-representation in reinforcement learning. In J. Shawe-Taylor, R. Zemel, P. Bartlett, F. Pereira, and K. Q. Weinberger, editors, *Advances in Neural Information Processing Systems*, volume 24, pages 1–9. Curran Associates, Inc., 2011.

[118] S. J. Marron and M. P. Wand. Exact mean integrated squared error. *The Annals of Statistics*, 20(2):712–736, 1992.

[119] A. Mathur, S. Moka, and Z. I. Botev. Variance reduction for matrix computations with applications to gaussian processes. In Q. Zhao and L. Xia, editors, *Proceedings of the 14th EAI International Conference on Performance Evaluation Methodologies and Tools (VALUETOOLS 2021)*, pages 243–261. Springer, 2021.

[120] M. Matsumoto and T. Nishimura. Mersenne twister: A 623-dimensionally equidistributed uniform pseudo-random number generator. *ACM Transactions on Modeling and Computer Simulation*, 8(1):3–30, 1998.

[121] W. McKinney. *Python for Data Analysis*. O'Reilly Media, Inc., second edition, 2017.

[122] G. J. McLachlan and T. Krishnan. *The EM Algorithm and Extensions*. John Wiley & Sons, Hoboken, second edition, 2008.

[123] G. J. McLachlan and D. Peel. *Finite Mixture Models*. John Wiley & Sons, New York, 2000.

[124] K. L. Mengersen and R. L. Tweedie. Rates of convergence of the Hastings and Metropolis algorithms. *The Annals of Statistics*, 24(1):101–121, 1996.

[125] N. Metropolis, A. W. Rosenbluth, M. N. Rosenbluth, A. H. Teller, and E. Teller. Equations of state calculations by fast computing machines. *Journal of Chemical Physics*, 21(6):1087–1092, 1953.

[126] C. A. Micchelli, Y. Xu, and H. Zhang. Universal kernels. *Journal of Machine Learning Research*, 7:2651–2667, 2006.

[127] Z. Michalewicz. *Genetic Algorithms + Data Structures = Evolution Programs*. Springer, New York, third edition, 1996.

[128] J. F. Monahan. *Numerical Methods of Statistics*. Cambridge University Press, London, 2010.

[129] T. A. Mroz. The sensitivity of an empirical model of married women's hours of work to economic and statistical assumptions. *Econometrica*, 55(4):765–799, 1987.

[130] K. P. Murphy. *Machine Learning: A Probabilistic Perspective*. The MIT Press, Cambridge, 2012.

[131] J. M. Myre, E. Frahm, D. J. Lilja, and M. O. Saar. Tnt-nn: a fast active set method for solving large non-negative least squares problems. *Procedia Computer Science*, 108:755–764, 2017.

[132] J. Neyman and E. Pearson. On the problem of the most efficient tests of statistical hypotheses. *Philosophical Transactions of the Royal Society of London, Series A*, 231:289–337, 1933.

[133] M. A. Nielsen. *Neural Networks and Deep Learning*, volume 25. Determination Press, 2015.

[134] T. Park and G. Casella. The Bayesian Lasso. *Journal of the American Statistical Association*, 103(482):681–686, 2008.

[135] K. B. Petersen and M. S. Pedersen. The Matrix Cookbook. *Technical University of Denmark*, 2008.

[136] M. L. Puterman. *Markov decision processes: discrete stochastic dynamic programming*. John Wiley & Sons, 2014.

[137] J. R. Quinlan. Learning with continuous classes. In A. Adams and L. Sterling, editors, *Proceedings AI'92*, pages 343–348, Singapore, 1992. World Scientific.

[138] C. E. Rasmussen and C. K. I. Williams. *Gaussian Processes for Machine Learning*. MIT Press, Cambridge, 2006.

[139] B. D. Ripley. *Stochastic Simulation*. John Wiley & Sons, New York, 1987.

[140] C. P. Robert and G. Casella. *Monte Carlo Statistical Methods*. Springer, New York, second edition, 2004.

[141] S. M. Ross. *Introduction to Stochastic Dynamic Programming*. Academic Press, New York, 1983.

[142] S. M. Ross. *Simulation*. Academic Press, New York, third edition, 2002.

[143] S. M. Ross. *A First Course in Probability*. Prentice Hall, Englewood Cliffs, seventh edition, 2005.

[144] R. Y. Rubinstein. The cross-entropy method for combinatorial and continuous optimization. *Methodology and Computing in Applied Probability*, 1:127–190, 1999.

[145] R. Y. Rubinstein and D. P. Kroese. *The Cross-Entropy Method: A Unified Approach to Combinatorial Optimization, Monte-Carlo Simulation and Machine Learning*. Springer, New York, 2004.

[146] R. Y. Rubinstein and D. P. Kroese. *Simulation and the Monte Carlo Method*. John Wiley & Sons, New York, third edition, 2017.

[147] S. Ruder. An overview of gradient descent optimization algorithms. *arXiv: 1609.04747*, 2016.

[148] W. Rudin. *Functional Analysis*. McGraw–Hill, Singapore, second edition, 1991.

[149] S. Russell and P. Norvig. *Artificial Intelligence: A Modern Approach*. Prentice Hall, Upper Saddle River, third edition, 2010.

[150] Y. Saad. *Iterative methods for sparse linear systems*. SIAM, 2003.

[151] D. Salomon. *Data Compression: The Complete Reference*. Springer, New York, 2000.

[152] G. A. F. Seber and A. J. Lee. *Linear Regression Analysis*. John Wiley & Sons, Hoboken, second edition, 2003.

[153] S. Shalev-Shwartz and S. Ben-David. *Understanding Machine Learning: From Theory to Algorithms*. Cambridge University Press, Cambridge, 2014.

[154] Z. A. Shaw. *Learning Python 3 the Hard Way*. Addison–Wesley, Boston, 2017.

[155] Y. Shen, S. Kiatsupaibul, Z. B. Zabinsky, and R. L. Smith. An analytically derived cooling schedule for simulated annealing. *Journal of Global Optimization*, 38(2):333–365, 2007.

[156] N. Z. Shor. *Minimization Methods for Non-differentiable Functions*. Springer, Berlin, 1985.

[157] D. Silver and J. Veness. Monte-Carlo planning in large POMDPs. In J. Lafferty, C. Williams, J. Shawe-Taylor, R. Zemel, and A. Culotta, editors, *Advances in Neural Information Processing Systems*, volume 23, pages 2164–2172. Curran Associates, Inc., 2010.

[158] B. W. Silverman. *Density Estimation for Statistics and Data Analysis*. Chapman & Hall, New York, 1986.

[159] J. S. Simonoff. *Smoothing Methods in Statistics*. Springer, New York, 1996.

[160] M. T. J. Spaan. Partially observable Markov decision processes. In *Reinforcement Learning. Adaptation, Learning, and Optimization*, volume 12. Springer, Berlin, 2012.

[161] I. Steinwart and A. Christmann. *Support Vector Machines*. Springer, New York, 2008.

[162] B. Stellato, G. Banjac, P. Goulart, A. Bemporad, and S. Boyd. Osqp: An operator splitting solver for quadratic programs. *Mathematical Programming Computation*, 12(4):637–672, 2020.

[163] G. Strang. *Introduction to Linear Algebra*. Wellesley–Cambridge Press, Cambridge, fifth edition, 2016.

[164] G. Strang. *Linear Algebra and Learning from Data*. Wellesley–Cambridge Press, Cambridge, 2019.

[165] W. N. Street, W. H. Wolberg, and O. L. Mangasarian. Nuclear feature extraction for breast tumor diagnosis. In *IS&T/SPIE 1993 International Symposium on Electronic Imaging: Science and Technology, San Jose, CA*, pages 861–870, 1993.

[166] R. S. Sutton and A. G. Barto. *Reinforcement Learning: An Introduction*. MIT Press, Cambridge, MA, 2010.

[167] R. S. Sutton and A. G. Barto. *Reinforcement Learning: An Introduction*. MIT Press, Cambridge, MA, second edition, 2020.

[168] T. Taimre, D. P. Kroese, and Z. I. Botev. Monte Carlo methods. *Wiley StatsRef: Statistics Reference Online*, pages 1–10, 2014.

[169] R. Tibshirani. Regression shrinkage and selection via the lasso. *Journal of the Royal Statistical Society Series B: Statistical Methodology*, 58(1):267–288, 1996.

[170] H. C. Tijms. *Stochastic Modeling and Analysis: A Computational Approach.* John Wiley & Sons, New York, 1986.

[171] V. M. Tikhomirov. On the representation of continuous functions of several variables as superpositions of continuous functions of one variable and addition. In *Selected Works of A. N. Kolmogorov*, pages 383–387. Springer, Berlin, 1991.

[172] S. van Buuren. *Flexible Imputation of Missing Data.* CRC Press, Boca Raton, second edition, 2018.

[173] V. N. Vapnik. *The Nature of Statistical Learning Theory.* Springer, New York, 1995.

[174] V. N. Vapnik and A. Ya. Chervonenkis. On the uniform convergence of relative frequencies of events to their probabilities. *Theory of Probability and Its Applications*, 16(2):264–280, 1971.

[175] N. Vlassis, M. Ghavamzadeh, S. Mannor, and P. Poupart. Bayesian reinforcement learning. In *Reinforcement Learning. Adaptation, Learning, and Optimization*, volume 12. Springer, Berlin, 2012.

[176] G. Wahba. *Spline Models for Observational Data.* SIAM, Philadelphia, 1990.

[177] M. P. Wand and M. C. Jones. *Kernel smoothing.* CRC press, 1994.

[178] L. Wasserman. *All of Statistics: A Concise Course in Statistical Inference.* Springer, 2010.

[179] J. C. H. Watkins and P. Dayan. Q-learning. *Machine Learning*, 8(3–4):279–292, 1992.

[180] A. Webb. *Statistical Pattern Recognition.* Arnold, London, 1999.

[181] H. Wendland. *Scattered Data Approximation.* Cambridge University Press, Cambridge, 2005.

[182] P. Whittle. *Optimization Over Time: Dynamic Programming and Stochastic Control.* John Wiley & Sons, Hoboken, 1982.

[183] D. Williams. *Probability with Martingales.* Cambridge University Press, Cambridge, 1991.

[184] C. F. J. Wu. On the convergence properties of the EM algorithm. *The Annals of Statistics*, 11(1):95–103, 1983.

[185] N. Ye, A. Somani, D. Hsu, and W. Lee. DESPOT: Online POMDP planning with regularization. *Journal of Artificial Intelligence Research*, 58:231–266, 2017.

[186] H. Zou and T. Hastie. Regularization and variable selection via the elastic net. *Journal of the Royal Statistical Society Series B: Statistical Methodology*, 67(2):301–320, 2005.

INDEX

D

E

F

M

N

O

P

Q

R

S

T

U

V

W